T0181816

Graduate Texts in Physics

Graduate Texts in Physics

Graduate Texts in Physics publishes core learning/teaching material for graduate- and advanced-level undergraduate courses on topics of current and emerging fields within physics, both pure and applied. These textbooks serve students at the MS- or PhD-level and their instructors as comprehensive sources of principles, definitions, derivations, experiments and applications (as relevant) for their mastery and teaching, respectively. International in scope and relevance, the textbooks correspond to course syllabi sufficiently to serve as required reading. Their didactic style, comprehensiveness and coverage of fundamental material also make them suitable as introductions or references for scientists entering, or requiring timely knowledge of, a research field.

More information about this series at http://www.springer.com/series/8431

Gianluca Calcagni

Classical and Quantum Cosmology

Gianluca Calcagni
Instituto de Estructura de la Materia
Consejo Superior de Investigaciones
Científicas (CSIC)
Madrid, Spain

ISSN 1868-4513 ISSN 1868-4521 (electronic)
Graduate Texts in Physics
ISBN 978-3-319-82273-0 ISBN 978-3-319-41127-9 (eBook)
DOI 10.1007/978-3-319-41127-9

Printed on acid-free paper

This Springer imprint is published by Springer Nature
The registered company is Springer International Publishing AG Switzerland

Acknowledgements

Several people helped me to improve this work through their careful reading of its parts and, last but not least, through their encouragement. I thank Emanuele Alesci, Raúl Carballo, Dario Francia, Steffen Gielen, Renate Loll, Johannes Mosig, Giuseppe Nardelli, Daniele Oriti, Thanu Padmanabhan, Johannes Thürigen, Aleksey Toporensky and, in particular, Claus Kiefer and Edward Wilson-Ewing for valuable feedback. Special thanks go to Sachiko Kuroyanagi (gravitational waves consulting), Claus Kiefer, my editor Angela Lahee and all those who believed in this project. Angela's patience with my delays in delivering the book has been, to put it in one word, exemplary.

Chapters 2, 3, 4, 5, 6, 9 and 10 are partly based on three series of lectures on quantum gravity and cosmology given at Penn State University during Spring Term 2009 and at the University of Potsdam in Summer Semester 2010 and Winter Semester 2011–2012.

Contents

Chapter 1
Introduction

Nel suo profondo vidi che s'interna
legato con amore in un volume,
ciò che per l'universo si squaderna;
sustanze e accidenti e lor costume,
quasi conflati insieme, per tal modo
che ciò ch'i' dico è un semplice lume.

— Dante, *Paradiso,* XXXIII, 85–90

In its depths enclosed I saw
bound with love in a volume,
that which enfolds in the universe;
substances and accidents and their bonds,
almost united together, so that
what I say is a simple glimpse.

Contents

1.1 Micro from Macro

According to modern data, the large-scale structure of the Universe and the anisotropies of the cosmic microwave background can be explained by an early stage of accelerated expansion, called *inflation*, driven by an effective cosmological constant. The latter is often identified with a scalar field (generically dubbed "inflaton") slowly rolling down its potential. The same dynamical mechanism may also provide an explanation for the present phase of acceleration.

G. Calcagni, *Classical and Quantum Cosmology*, Graduate Texts in Physics,
DOI 10.1007/978-3-319-41127-9_1

Inflation was originally devised for solving a number of problems afflicting the hot big bang model, in particular the flatness or entropy problem (Why is the Universe flat? Why does it have such a high entropy?), the horizon problem (Why are distant, causally disconnected regions in thermal equilibrium?) and the monopole problem (Where are the topological defects we would expect emerging from cosmological phase transitions?). However, the reasons of the success of inflation rely on an aspect which is far more than a side bonus. In fact, an immediate consequence of this scenario is that cosmological large-scale structures were originated by the exponential dilatation of quantum fluctuations of the inflaton up to macroscopic scales. In a *coniunctio oppositorum*, the study of the macrocosm allows us to investigate the microcosm. Thus, cosmological observations are complementary to those with ground-based particle accelerators and they clarify both the composition and geometric structure of the Universe and the primordial role of the known elementary interactions. Inflation is a promising playground whereon to understand the high-energy, small-scale behaviour of Nature and, in particular, gravity at quantum scales. And how does gravity behave at such scales? Is it a quantum force or not?

During the last years, quantum gravity has been receiving a great amount of attention from the community of theoreticians. The driving motivation, familiar to anyone who has tried their fortune in this broad subject, is to realize a consistent, predictive merging of general relativity with quantum mechanics. The programme can be carried out in various forms, from ambitious theories of everything (such as string theory) where all forces are unified, to more minimalistic approaches aiming to quantize gravity alone, not to mention proposals where gravity emerges as an effective phenomenon from more fundamental degrees of freedom.[1] A problem endemic in most of these scenarios is their difficulty in making contact with observations. This stems from the highly technical nature of the theoretical frameworks, where the notions of conventional geometry and matter, continuum spacetime, general covariance and physical observables are typically deformed, modified, or disappear altogether. The lack of experimental feedback makes it quite difficult to discriminate among different models and to characterize them as *falsifiable*. It is natural to turn to cosmology in an attempt to bridge this gap and advance our knowledge.

The early Universe is not the only context wherein to accept these challenges: also some unexplained facts concerning the composition of the late-time and present Universe might require new physics. The state of the art in cosmology has been experiencing an interesting phase. On one hand, there are a number of

[1]Often, but not always, we will reserve the umbrella name "quantum gravity" to the second type of approaches, where matter is introduced by hand. This is done only for convenience in certain chapters but it not a clear-cut terminology. In a sense, string theory is a theory of quantum gravity (and it was presented as such in many early papers), although it may be argued that it is also a realization of emergent gravity. Group field theory will be included among quantum-gravity theories (Chap. 11) but it is also a theory of everything, since it aims to derive both geometry and matter degrees of freedom from a unified structure. And so on.

phenomenological models which have been formulated in great detail and fit the available data. On the other hand, observational data obtained in ground-based observatories and space missions (satellites and balloon probes), have the task and possibility to discriminate among the many cosmic theories on the market. The predictions of what we will come to know as the *cold big bang model* are in excellent agreement with observations but, despite the progress in precision cosmology, we know less than 5 % of the content of the present universe. The rest is divided into two components which are dubbed "dark matter" and "dark energy" in homage to our ignorance. While dark matter may be understood within models of particle physics, the nature of dark energy is more elusive. Is it a cosmological constant? Why is its energy density so small but non-zero? Why did the universe start accelerating only in its recent history? Many believe that a true solution of the cosmological constant and dark energy problems is deeply tied with a consistent formulation of a quantum theory of gravity. In other words, the cosmological constant might be a yet unrecognized manifestation of physics beyond the Standard Model. In turn, this theory should also be able to address other questions: Can we avoid the big-bang singularity and, more generally, spacetime singularities? How did inflation happen? What is the inflaton?

It might as well be that most or perhaps all these questions are being formulated in a wrong way. Progress in the knowledge of a unified quantum theory of Nature, including both particle physics and general relativity in some effective limits, should help answering these questions or, at least, asking them correctly. We are of course working upon the hope, motivated by aesthetic arguments, that such a theory does exist.

This book discusses models stemming from classical and quantum theories at the frontier of modern cosmology. Applications of quantum mechanics and quantum field theory range from the theory of inflationary perturbations to mini-superspace quantizations, from the big-bang problem in loop quantum cosmology to inflation in string theory, down to the relation between a quantum theory of gravity and the cosmological constant problem. We will review how the early universe can be described within candidate theories of quantum gravity, in particular string theory and loop quantum gravity. However, we shall only touch the tip of the iceberg and will not embark on an in-depth study of these theories. We will consider only cosmological phenomenology and the main problems a complete model of the early universe should be able to face. We forewarn the reader that none of the proposals advanced to tackle issues such as the big-bang and the cosmological constant problem is, for several reasons, commonly regarded as satisfactory. Nonetheless, they constitute attempts of resolution of these problems and, as such, they deserve attention.

The text is meant for advanced undergraduate and graduate students planning to work on cosmology of the early universe, theories of quantum gravity or string cosmology, but also for specialists in quantum gravity or string theory who would not dislike a cosmological venture. On the other hand, the more traditional cosmologist can find an updated non-technical overview of modern approaches to inflation, the big-bang and the cosmological constant problems, theories of quantum gravity at large and their cosmology, string theory and string cosmology. Researchers looking for a reference book with extensive bibliographic resources will also find benefit in the reading. Most chapters are accompanied by partially or fully solved problems.

For these reasons, this is not a standard textbook on cosmology. Many important branches of cosmology will either be reviewed qualitatively (cosmic microwave background anisotropies) or just mentioned briefly here and there (dark matter, structure formation, observational and statistical techniques, astrophysical processes). The reader can find further material in the main bibliography, although not all themes are covered by extant books or review articles.

1.2 Outline of the Topics

1. **Hot big bang model.** Cosmological standard model; content of the universe; thermal history. Textbooks: [1–3] (general relativity), [4] (Chaps. 2–3 on general relativity), [5–8] (cosmology).
2. **Cosmological perturbations.** Linear and non-linear cosmological perturbations; separate universe approach; Gaussian random fields. Textbooks: [5] (linear perturbations, separate universe and $\delta\mathcal{N}$ formalism). Reviews: [9, 10] (linear perturbations).
3. **Cosmic microwave background.** CMB primer; Gaussian and non-Gaussian spectra; polarization. Textbooks: [5, 11] (CMB). Reviews: [12–15] (anisotropies), [16, 17] (polarization), [18–20] (non-Gaussianity).
4. **Inflation.** Problems of the cosmological standard model; standard scalar-field models; classical and quantum dynamics; model building; spectra and non-Gaussianity; open issues. Textbooks: [5, 6, 21] (basic and multi-field inflation, perturbations), [22] (quantum field theory in curved spacetime), [23] (eternal inflation). Reviews: [24] (particle-physics models of inflation), [25] (inflationary perturbations), [26] (more updated review on inflation).
5. **Big-bang problem.** Globally hyperbolic spacetimes; classification of singularities; singularity theorems; mixmaster universe and BKL conjecture. Textbooks: [27] (Chap. 8 on singularity theorems, Chap. 10 on big-bang singularity). Reviews: [28] (BKL singularity).
6. **Cosmological constant problem.** The Λ problem in quantum field theory; quintessence; acceleration from modified gravity and alternative models. Textbooks: [29] (dark energy). Reviews: [30–34] (Λ problem), [35–38] (dark energy), [39–42] (scalar-tensor, $f(R)$ and higher-order models).
7. **The problem of quantum gravity.** Do we need to quantize gravity?; perturbative gravity; approaches to quantum gravity.
8. **Canonical quantum gravity.** Hamiltonian formalism in Ashtekar–Barbero and ADM variables; Wheeler–DeWitt equation; cosmological constant problem in canonical quantum gravity. Textbooks: [4] (Chap. 4 on first-order formalism, Chaps. 6–9 on loop quantum gravity and spin foams), [43] (Chaps. 1 and

2 on constrained Hamiltonian systems), [44] (mathematical introduction to canonical gravity and loop quantum gravity), [45] (canonical quantum gravity). Reviews: [46, 47] (loop quantum gravity), [48] (Wheeler–DeWitt equation).

9. **Canonical quantum cosmology.** Mini-superspace; Wheeler–DeWitt quantum cosmology; loop quantum cosmology; big-bang and cosmological constant problems revisited; inhomogeneities and inflation. Textbooks: [49]. Reviews: [48, 50] (Wheeler–DeWitt cosmology), [51–53] (loop quantum cosmology).
10. **Cosmology of quantum gravities.** Asymptotic safety; causal dynamical triangulation; spin foams; group field theory; causal sets; non-commutative spacetimes; non-local gravity. Review references will be given at the beginning of Chap. 11.
11. **String theory.** Overview of string theory; bosonic string; superstring; low-energy limits; branes and fluxes; compactification; Calabi–Yau spaces and orbifolds; moduli stabilization; dualities; M-theory. Textbooks: [54–57] (string theory). Reviews: [58] (basics of string theory), [59] (open strings), [60, 61] (string field theory), [62] (M-theory), [63–67] (flux compactification).
12. **String cosmology.** String landscape; de Sitter vacua; cosmological constant problem; KKLT uplifting scenarios; large-volume uplifting scenarios; inflation in string theory; moduli inflation; axion inflation; slow-roll D-brane inflation; DBI inflation; braneworld cosmology; cosmological tachyon; higher-order string gravity; non-local string cosmology; pre-big-bang cosmology; string-gas cosmology; ekpyrotic universe; big-bang problem in string theory; cosmological billiards. Books: [68] (string inflation). Reviews: [63–67] (string landscape), [69–74] (moduli and D-brane inflation), [75] (cosmic strings), [76] (pre-landscape perspective on the cosmological constant), [71] (big-bang problem), [77] (cosmological billiards).

The list of cosmological models in Chaps. 7 and 11 is by no means complete. Notable absentees include, among others, the Dvali–Gabadadze–Porrati scenario [78–80], massive and bimetric gravity [81–89] and Hořava–Lifshitz gravity [90–95] (although the latter is mentioned in a couple of places in Chap. 11).

An elementary knowledge of quantum field theory is a helpful pre-requisite [96–99]. Important branches of theoretical physics such as supersymmetry and supergravity will be touched upon in the context of inflation (Sect. 5.12), the cosmological constant problem (Sect. 7.1.3), renormalization properties of perturbative gravity (Sect. 8.2) and string theory (Chaps. 12 and 13). To get a deeper insight into supersymmetry, the interested reader can consult specialized books [100, 101] and reviews [102–106]. Applications of supersymmetry to canonical quantum cosmology can be found in [107, 108].

The recent advances in the experimental determination of natural observables have been spectacular. It is impressive to think back at the number of breakthroughs we witnessed in physics during the years 2012–2016: the discovery of the Higgs boson at the LHC, culminating a never-ending chain of achievements for the Standard Model of particle interactions; the great improvement in our knowledge of the cosmic microwave background temperature spectrum first by WMAP and then by PLANCK, and the ensuing restriction of the parameter space of inflation; a false alarm about the detection of gravitational waves, followed some months later by the actual discovery of such waves by Advanced LIGO, from the first black-hole merger ever observed —not only a major success for Einstein's theory of general relativity, but also the first time we could literally *hear* the voice of gravity. Through this chain of events, I found myself in the embarrassing situation of rewriting some parts of the book time and again. All the above observations carry important consequences for the discrimination of astro-particle and cosmological models and, hence, for our investigation of quantum-gravity and string phenomenology. Most probably, and hopefully, more of this ongoing story will have to be revised in the next few years.

1.3 About Citations

Whenever practical and up to exceptions indicated in the text, an exhaustive list of references (updated November 2016) has been provided in order to offer ample documentation for those readers seriously interested in learning more about a subject to work on it actively. The simplified presentation of many models in the book, especially in string cosmology, does not do justice to the tremendous mole of work behind these models.

The main disadvantage of this choice is that the most important references marking historical landmarks may be lost, at a first reading, in a sea of data quite often not directly used in the book. The main tools to obviate to such a problem have been, on one hand, to cite foundational papers and stress their role separately from the others and, on the other hand, to order the references in each group by submission date. For a few topics, the group of references is disproportionately large with respect to the importance of such topics in the economy of the book. Perhaps one factor that influenced this outcome is the child's insatiable curiosity many of us share when starting to learn about the scope and story of some research theme we come across. A curiosity which makes us ask: How did it continue? How will it end?

This system, with all its flaws, omissions and imprecisions the author may have introduced, may be best suited for a reader who enjoys a critical exploration of the literature.

1.4 Conventions

$:=$ and $=:$ define the symbol or function on the side of the colon: e.g., $f(x) := \ln(1+x)^2 =: g(x)$. The equality $=$ is used in a standard way, when two things are the same (logically, this includes also objects already defined): $f(x) = \ln(1+x)^2 = 2\ln(1+x)$. Depending on the context, $\equiv$ is an identity ($1 \equiv 1, f(x) \equiv g(x)$) or an equivalence relation ($a \equiv b$). Except in Chap. 9, $\approx$ means approximation of a function or a number to a number (e.g., $\pi \approx 3.14$, $\mathcal{N} \approx 60$ e-folds, or (1.2), (1.3) and (1.4) below). $\propto$ and $\sim$ denote proportionality between functions, respectively exact and asymptotic, while $\simeq$ means "asymptotically equal to" and is employed between functions. For instance, when $x \ll 1$ we have $g(x) \equiv f(x) = \ln(1+x)^2 \simeq x(2-x) \simeq 2x \propto x \approx 0$. To make implicit use of the asymptotic expansion, we would write $f(x) \sim x$. Sometimes, $\sim$ also appears in order-of-magnitude estimates, e.g., if $\epsilon \approx 3.2 \times 10^{-2}$ we will say that $\epsilon = O(10^{-2})$ or $\epsilon \sim 10^{-2}$. Finally, $\cong$ means isomorphism between sets or groups. The mathematician will forgive me if I have not been completely consistent with the above usage of symbols.

Greek indices $\mu, \nu, \ldots = 0, 1, \ldots, D-1$ from the middle of the alphabet denote spacetime directions, while Greek indices $\alpha, \beta, \ldots$ from the beginning of the alphabet label spatial directions. Whenever the vielbein formalism is employed, Latin indices $a, b, \ldots$ from the beginning of the alphabet denote directions in the internal (tangent) space; we reserve the middle-alphabet letters $i, j, \ldots$ to the spatial directions. In Chaps. 12 and 13, $\mu, \nu, \ldots$ are labels of 26- or 10-dimensional target spacetime, reserving $M, N, \cdots = 0, 1, \ldots, 10$ for dimensions in M-theory. In the same chapters, $a, b, \ldots$ indicate world-sheet and world-volume coordinates or, in the case of compactifications, the coordinates in the non-compact four-dimensional manifold. Directions in Calabi–Yau spaces are labelled by $l, m, \ldots$. Exceptions to these rules may occur, in which case awareness of the context will avoid confusion.

Unless stated otherwise, the Einstein summation notation for repeated indices is employed, $a^\mu b_\mu = \sum_\mu a^\mu b_\mu$.

We choose the "mostly plus" spacetime signature $(-, +, \cdots, +)$.

1.5 Measure Units

A length unit commonly used in cosmology is the megaparsec (Mpc):

$$1\,\text{Mpc} = 3.26156 \times 10^6 \text{ light years} \approx 3.0857 \times 10^{22}\,\text{m}\,.$$

To describe phenomena at quantum scales, the Planck length is better suited. In units of mass m, length l and time t, the scaling of the speed of light, gravitational constant and Planck constant in D dimensions is, respectively, $[c] = lt^{-1}$, $[G] = l^{D-1}t^{-2}m^{-1}$ and $[\hbar] = l^2t^{-1}m$. The Planck length, time and mass are then

$$l_{\text{Pl}} := \left(\frac{\hbar G}{c^3}\right)^{\frac{1}{D-2}}, \qquad t_{\text{Pl}} := \left(\frac{\hbar G}{c^{D+1}}\right)^{\frac{1}{D-2}}, \qquad m_{\text{Pl}} := \left(\frac{\hbar^{D-3}c^{5-D}}{G}\right)^{\frac{1}{D-2}}. \tag{1.1}$$

In four dimensions,

$$l_{\rm Pl} = \sqrt{\frac{\hbar G}{c^3}} \approx 1.6163 \times 10^{-35}\,{\rm m}\,, \tag{1.2}$$

$$t_{\rm Pl} = \sqrt{\frac{\hbar G}{c^5}} \approx 5.3912 \times 10^{-44}\,{\rm s}\,, \tag{1.3}$$

$$m_{\rm Pl} = \sqrt{\frac{\hbar c}{G}} \approx 1.2209 \times 10^{19}\,{\rm GeV}c^{-2}\,, \tag{1.4}$$

to which we add the reduced Planck mass $M_{\rm Pl} := m_{\rm Pl}/\sqrt{8\pi} \approx 2.435 \times 10^{18}\,{\rm GeV}$ and the Planck temperature

$$T_{\rm Pl} \approx 1.4 \times 10^{32}\,{\rm K}\,.$$

Here we used the conversion formula

$$T_{\rm K} = \frac{T_{\rm eV}}{k_{\rm B}} \approx 1.1605 \times 10^4\,T_{\rm eV}\,,$$

where $k_{\rm B} \approx 8.6173 \times 10^{-5}\,{\rm eV\,K^{-1}}$ is Boltzmann constant.

We shall adopt natural units in which the speed of light and the Planck constant are dimensionless:

$$c = 1, \qquad \hbar = 1\,.$$

These units will be occasionally restored.

References

1. S. Weinberg, *Gravitation and Cosmology* (Wiley, New York, 1972)
2. C.W. Misner, K.S. Thorne, J.A. Wheeler, *Gravitation* (Freeman, New York, 1973)
3. L.D. Landau, E.M. Lifshitz, *The Classical Theory of Fields* (Butterworth–Heinemann, London, 1980)
4. C. Rovelli, *Quantum Gravity* (Cambridge University Press, Cambridge, 2007)
5. D.H. Lyth, A.R. Liddle, *The Primordial Density Perturbation* (Cambridge University Press, Cambridge, 2009)
6. S. Weinberg, *Cosmology* (Oxford University Press, Oxford, 2008)
7. V. Mukhanov, *Physical Foundations of Cosmology* (Cambridge University Press, Cambridge, 2005)
8. S. Dodelson, *Modern Cosmology* (Academic Press, San Diego, 2003)
9. H. Kodama, M. Sasaki, Cosmological perturbation theory. Prog. Theor. Phys. Suppl. **78**, 1 (1984)
10. V.F. Mukhanov, H.A. Feldman, R.H. Brandenberger, Theory of cosmological perturbations. Phys. Rep. **215**, 203 (1992)

11. R. Durrer, *The Cosmic Microwave Background* (Cambridge University Press, Cambridge, 2008)
12. M.J. White, D. Scott, J. Silk, Anisotropies in the cosmic microwave background. Ann. Rev. Astron. Astrophys. **32**, 319 (1994)
13. W.T. Hu, Wandering in the Background: A CMB Explorer. Ph.D. thesis, UC Berkeley, Berkeley (1995). [arXiv:astro-ph/9508126]
14. W. Hu, N. Sugiyama, J. Silk, The physics of microwave background anisotropies. Nature **386**, 37 (1997). [arXiv:astro-ph/9604166]
15. W. Hu, S. Dodelson, Cosmic microwave background anisotropies. Ann. Rev. Astron. Astrophys. **40**, 171 (2002). [arXiv:astro-ph/0110414]
16. M. Kamionkowski, A. Kosowsky, A. Stebbins, Statistics of cosmic microwave background polarization. Phys. Rev. D **55**, 7368 (1997). [arXiv:astro-ph/9611125]
17. W. Hu, M.J. White, A CMB polarization primer. New Astron. **2**, 323 (1997). [arXiv:astro-ph/9706147]
18. N. Bartolo, E. Komatsu, S. Matarrese, A. Riotto, Non-Gaussianity from inflation: theory and observations. Phys. Rep. **402**, 103 (2004). [arXiv:astro-ph/0406398]
19. E. Komatsu et al., Non-Gaussianity as a probe of the physics of the primordial universe and the astrophysics of the low redshift universe. arXiv:0902.4759
20. M. Liguori, E. Sefusatti, J.R. Fergusson, E.P.S. Shellard, Primordial non-Gaussianity and bispectrum measurements in the cosmic microwave background and large-scale structure. Adv. Astron. **2010**, 980523 (2010). [arXiv:1001.4707]
21. A.D. Linde, *Particle Physics and Inflationary Cosmology* (Harwood, Chur, 1990). [arXiv:hep-th/0503203]
22. N.D. Birrell, P.C.W. Davies, *Quantum Fields in Curved Space* (Cambridge University Press, Cambridge, 1982)
23. S. Winitzki, *Eternal Inflation* (World Scientific, Singapore, 2009)
24. D.H. Lyth, A. Riotto, Particle physics models of inflation and the cosmological density perturbation. Phys. Rep. **314**, 1 (1999). [arXiv:hep-ph/9807278]
25. A. Riotto, Inflation and the theory of cosmological perturbations. arXiv:hep-ph/0210162
26. A.D. Linde, Inflationary cosmology. Lect. Notes Phys. **738**, 1 (2008). [arXiv:0705.0164]
27. S.W. Hawking, G.F.R. Ellis, *The Large Scale Structure of Space-Time* (Cambridge University Press, Cambridge, 1973)
28. G. Montani, M.V. Battisti, R. Benini, G. Imponente, Classical and quantum features of the mixmaster singularity. Int. J. Mod. Phys. A **23**, 2353 (2008). [arXiv:0712.3008]
29. L. Amendola, S. Tsujikawa, *Dark Energy* (Cambridge University Press, Cambridge, 2010)
30. S. Weinberg, The cosmological constant problem. Rev. Mod. Phys. **61**, 1 (1989)
31. V. Sahni, A.A. Starobinsky, The case for a positive cosmological Λ-term. Int. J. Mod. Phys. D **9**, 373 (2000). [arXiv:astro-ph/9904398]
32. S.M. Carroll, The cosmological constant. Living Rev. Relat. **4**, 1 (2001)
33. S. Nobbenhuis, Categorizing different approaches to the cosmological constant problem. Found. Phys. **36**, 613 (2006). [arXiv:gr-qc/0411093]
34. J. Martin, Everything you always wanted to know about the cosmological constant problem (but were afraid to ask). C. R. Phys. **13**, 566 (2012). [arXiv:1205.3365]
35. T. Padmanabhan, Cosmological constant: the weight of the vacuum. Phys. Rep. **380**, 235 (2003). [arXiv:hep-th/0212290]
36. P.J.E. Peebles, B. Ratra, The cosmological constant and dark energy. Rev. Mod. Phys. **75**, 559 (2003). [arXiv:astro-ph/0207347]
37. E.J. Copeland, M. Sami, S. Tsujikawa, Dynamics of dark energy. Int. J. Mod. Phys. D **15**, 1753 (2006). [arXiv:hep-th/0603057]
38. S. Tsujikawa, Quintessence: a review. Class. Quantum Grav. **30**, 214003 (2013). [arXiv:1304.1961]
39. T.P. Sotiriou, V. Faraoni, $f(R)$ theories of gravity. Rev. Mod. Phys. **82**, 451 (2010). [arXiv:0805.1726]
40. A. De Felice, S. Tsujikawa, $f(R)$ theories. Living Rev. Relat. **13**, 3 (2010)

41. S. Tsujikawa, Modified gravity models of dark energy. Lect. Notes Phys. **800**, 99 (2010). [arXiv:1101.0191]
42. T. Clifton, P.G. Ferreira, A. Padilla, C. Skordis, Modified gravity and cosmology. Phys. Rep. **513**, 1 (2012). [arXiv:1106.2476]
43. M. Henneaux, C. Teitelboim, *Quantization of Gauge Systems* (Princeton University Press, Princeton, 1994)
44. T. Thiemann, *Modern Canonical Quantum General Relativity* (Cambridge University Press, Cambridge, 2007); Introduction to modern canonical quantum general relativity. arXiv:gr-qc/0110034
45. C. Kiefer, *Quantum Gravity* (Oxford University Press, Oxford, 2012)
46. T. Thiemann, Quantum gravity: from theory to experimental search. Lect. Notes Phys. **631**, 412003 (2003). [arXiv:gr-qc/0210094]
47. A. Ashtekar, J. Lewandowski, Background independent quantum gravity: a status report. Class. Quantum Grav. **21**, R53 (2004). [arXiv:gr-qc/0404018]
48. D.L. Wiltshire, An introduction to quantum cosmology, in *Cosmology: The Physics of the Universe*, ed. by B. Robson, N. Visvanathan, W.S. Woolcock (World Scientific, Singapore, 1996). [arXiv:gr-qc/0101003]
49. M. Bojowald, Quantum cosmology. Lect. Notes Phys. **835**, 1 (2011)
50. C. Kiefer, B. Sandhöfer, Quantum cosmology, in *Beyond the Big Bang*, ed. by R. Vaas (Springer, Berlin, 2008). [arXiv:0804.0672]
51. M. Bojowald, Loop quantum cosmology. Living Rev. Relat. **11**, 4 (2008)
52. A. Ashtekar, P. Singh, Loop quantum cosmology: a status report. Class. Quantum Grav. **28**, 213001 (2011). [arXiv:1108.0893]
53. K. Banerjee, G. Calcagni, M. Martín-Benito, Introduction to loop quantum cosmology. SIGMA **8**, 016 (2012). [arXiv:1109.6801]
54. M.B. Green, J.H. Schwarz, E. Witten, *Superstring Theory* (Cambridge University Press, Cambridge, 1987)
55. J. Polchinski, *String Theory* (Cambridge University Press, Cambridge, 1998)
56. K. Becker, M. Becker, J.H. Schwarz, *String Theory and M-Theory* (Cambridge University Press, Cambrdige, 2007)
57. B. Zwiebach, *A First Course in String Theory* (Cambridge University Press, Cambridge, 2009)
58. D. Tong, String theory. arXiv:0908.0333
59. C. Angelantonj, A. Sagnotti, Open strings. Phys. Rep. **371**, 1 (2002); Erratum-ibid. **376**, 339 (2003). [arXiv:hep-th/0204089]
60. K. Ohmori, A review on tachyon condensation in open string field theories. arXiv:hep-th/0102085
61. E. Fuchs, M. Kroyter, Analytical solutions of open string field theory. Phys. Rep. **502**, 89 (2011). [arXiv:0807.4722]
62. M.J. Duff, M theory (the theory formerly known as strings). Int. J. Mod. Phys. A **11**, 5623 (1996). [arXiv:hep-th/9608117]
63. M. Graña, Flux compactifications in string theory: a comprehensive review. Phys. Rep. **423**, 91 (2006). [arXiv:hep-th/0509003]
64. M.R. Douglas, S. Kachru, Flux compactification. Rev. Mod. Phys. **79**, 733 (2007). [arXiv:hep-th/0610102]
65. R. Blumenhagen, B. Körs, D. Lüst, S. Stieberger, Four-dimensional string compactifications with D-branes, orientifolds and fluxes. Phys. Rep. **445**, 1 (2007). [arXiv:hep-th/0610327]
66. F. Denef, M.R. Douglas, S. Kachru, Physics of string flux compactifications. Ann. Rev. Nucl. Part. Sci. **57**, 119 (2007). [arXiv:hep-th/0701050]
67. F. Denef, Course 12 – Lectures on constructing string vacua. Les Houches **87**, 483 (2008). [arXiv:0803.1194]
68. D. Baumann, L. McAllister, *Inflation and String Theory* (Cambridge University Press, Cambridge, 2015). [arXiv:1404.2601]

69. S.-H.H. Tye, Brane inflation: string theory viewed from the cosmos. Lect. Notes Phys. **737**, 949 (2008). [arXiv:hep-th/0610221]
70. R. Kallosh, On inflation in string theory. Lect. Notes Phys. **738**, 119 (2008). [arXiv:hep-th/0702059]
71. L. McAllister, E. Silverstein, String cosmology: a review. Gen. Relat. Grav. **40**, 565 (2008). [arXiv:0710.2951]
72. M. Cicoli, F. Quevedo, String moduli inflation: an overview. Class. Quantum Grav. **28**, 204001 (2011). [arXiv:1108.2659]
73. C.P. Burgess, L. McAllister, Challenges for string cosmology. Class. Quantum Grav. **28**, 204002 (2011). [arXiv:1108.2660]
74. C.P. Burgess, M. Cicoli, F. Quevedo, String inflation after Planck 2013. JCAP **1311**, 003 (2013). [arXiv:1306.3512]
75. E.J. Copeland, L. Pogosian, T. Vachaspati, Seeking string theory in the cosmos. Class. Quantum Grav. **28**, 204009 (2011). [arXiv:1105.0207]
76. E. Witten, The cosmological constant from the viewpoint of string theory. arXiv:hep-ph/0002297
77. M. Henneaux, D. Persson, P. Spindel, Spacelike singularities and hidden symmetries of gravity. Living Rev. Relat. **11**, 1 (2008)
78. G.R. Dvali, G. Gabadadze, M. Porrati, 4D gravity on a brane in 5D Minkowski space. Phys. Lett. B **485**, 208 (2000). [arXiv:hep-th/0005016]
79. D. Gorbunov, K. Koyama, S. Sibiryakov, More on ghosts in DGP model. Phys. Rev. D **73**, 044016 (2006). [arXiv:hep-th/0512097]
80. W. Fang, S. Wang, W. Hu, Z. Haiman, L. Hui, M. May, Challenges to the DGP model from horizon-scale growth and geometry. Phys. Rev. D **78**, 103509 (2008). [arXiv:0808.2208]
81. S. Deser, R. Jackiw, S. Templeton, Topologically massive gauge theories. Ann. Phys. (N.Y.) **140**, 372 (1982)
82. E.A. Bergshoeff, O. Hohm, P.K. Townsend, Massive gravity in three dimensions. Phys. Rev. Lett. **102**, 201301 (2009). [arXiv:0901.1766]
83. C. de Rham, G. Gabadadze, Generalization of the Fierz–Pauli action. Phys. Rev. D **82**, 044020 (2010). [arXiv:1007.0443]
84. C. de Rham, G. Gabadadze, A.J. Tolley, Resummation of massive gravity. Phys. Rev. Lett. **106**, 231101 (2011). [arXiv:1011.1232]
85. S.F. Hassan, R.A. Rosen, On non-linear actions for massive gravity. JHEP **1107**, 009 (2011). [arXiv:1103.6055]
86. S.F. Hassan, R.A. Rosen, Resolving the ghost problem in non-linear massive gravity. Phys. Rev. Lett. **108**, 041101 (2012). [arXiv:1106.3344]
87. G. D'Amico, C. de Rham, S. Dubovsky, G. Gabadadze, D. Pirtskhalava, A.J. Tolley, Massive cosmologies. Phys. Rev. D **84**, 124046 (2011). [arXiv:1108.5231]
88. S.F. Hassan, R.A. Rosen, Bimetric gravity from ghost-free massive gravity. JHEP **1202**, 126 (2012). [arXiv:1109.3515]
89. A. De Felice, A.E. Gümrükçüoğlu, C. Lin, S. Mukohyama, On the cosmology of massive gravity. Class. Quantum Grav. **30**, 184004 (2013). [arXiv:1304.0484]
90. P. Hořava, Quantum gravity at a Lifshitz point. Phys. Rev. D **79**, 084008 (2009). [arXiv:0901.3775]
91. G. Calcagni, Cosmology of the Lifshitz universe. JHEP **0909**, 112 (2009). [arXiv:0904.0829]
92. E. Kiritsis, G. Kofinas, Hořava–Lifshitz cosmology. Nucl. Phys. B **821**, 467 (2009). [arXiv:0904.1334]
93. R. Iengo, J.G. Russo, M. Serone, Renormalization group in Lifshitz-type theories. JHEP **0911**, 020 (2009). [arXiv:0906.3477]
94. D. Blas, O. Pujolas, S. Sibiryakov, Consistent extension of Hořava gravity. Phys. Rev. Lett. **104**, 181302 (2010). [arXiv:0909.3525]
95. P. Hořava, C.M. Melby-Thompson, General covariance in quantum gravity at a Lifshitz point. Phys. Rev. D **82**, 064027 (2010). [arXiv:1007.2410]
96. P. Ramond, *Field Theory: A Modern Primer* (Westview Press, Boulder, 1997)

97. S. Weinberg, *The Quantum Theory of Fields*, vol. I (Cambridge University Press, Cambridge, 1995)
98. S. Weinberg, *The Quantum Theory of Fields*, vol. II (Cambridge University Press, Cambridge, 1997)
99. M. Srednicki, *Quantum Field Theory* (Cambridge University Press, Cambridge, 2007)
100. P.C. West, *Introduction to Supersymmetry and Supergravity* (World Scientific, Singapore, 1990)
101. J. Wess, J. Bagger, *Supersymmetry and Supergravity* (Princeton University Press, Princeton, 1992)
102. P. Van Nieuwenhuizen, Supergravity. Phys. Rep. **68**, 189 (1981)
103. S.J. Gates, M.T. Grisaru, M. Roček, W. Siegel, Superspace, or one thousand and one lessons in supersymmetry. Front. Phys. **58**, 1 (1983). [arXiv:hep-th/0108200]
104. J.D. Lykken, Introduction to supersymmetry. arXiv:hep-th/9612114
105. S.P. Martin, A supersymmetry primer, in *Perspectives on Supersymmetry*, ed. by G.L. Kane (World Scientific, Singapore, 1998). [arXiv:hep-ph/9709356]
106. A. Van Proeyen, Tools for supersymmetry. arXiv:hep-th/9910030
107. P.D. D'Eath, *Supersymmetric Quantum Cosmology* (Cambridge University Press, Cambridge, 2005)
108. P. Vargas Moniz, *Quantum Cosmology – The Supersymmetric Perspective*. Lect. Notes Phys. **803**, 1 (2010); Lect. Notes Phys. **804**, 1 (2010)

Chapter 2
Hot Big Bang Model

This to attain, whether heav'n move or earth,
Imports not, if thou reckon right; the rest
From man or angel the great Architect
Did wisely to conceal, and not divulge
His secrets to be scanned by them who aught
Rather admire; or if they list to try
Conjecture, he his fabric of the heav'ns
Hath left to their disputes, perhaps to move
His laughter at their quaint opinions wide
Hereafter, when they come to model heav'n
And calculate the stars . . .

— Milton, *Paradise Lost,* VIII, 70–80

Contents

G. Calcagni, *Classical and Quantum Cosmology*, Graduate Texts in Physics,
DOI 10.1007/978-3-319-41127-9_2

2.1 Cosmic Expansion and Cosmological Principle

2.1.1 The Universe at Large Scales

Cosmology is the study of Nature at very large scales. Conventionally, "large scales" span a range of about 1–8000 Mpc, from our local group of galaxies to the most distant light we can possibly detect (Fig. 2.1). Phenomena occurring within a galaxy (ours or another) and associated with galactic media, stars, supernovæ, black holes, gamma-ray bursts and so on are also part of cosmology, in so far as they determine relative distances, ages, composition and gravitational properties of local patches of the universe, as well as important information on elementary particle physics and gravity.

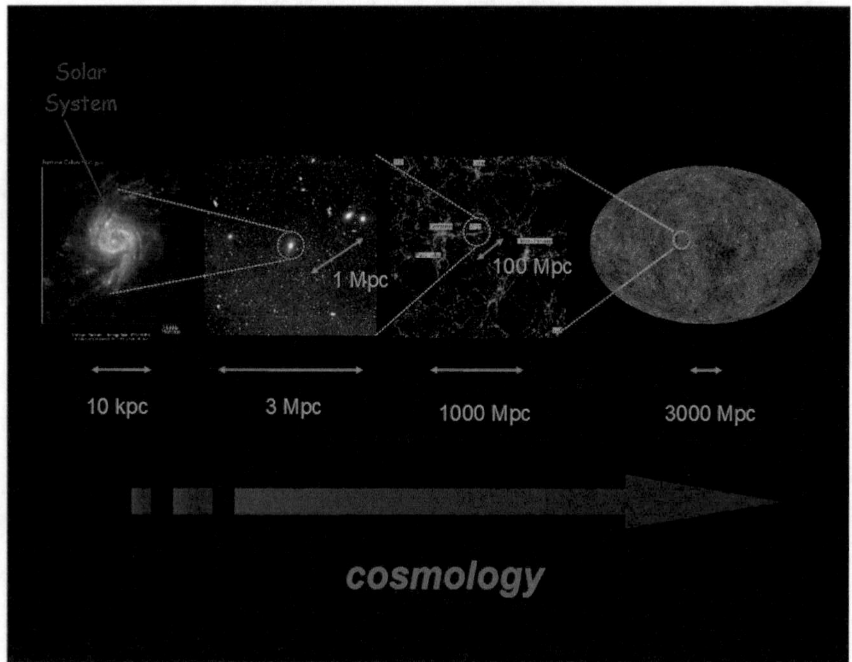

Fig. 2.1 Cosmological scales (Source: adaptation of a figure by C. Schimd)

The large-scale structure of the universe has been studied by several experiments, including the 2-degree Field Galaxy Redshift Survey (2dFGRS) [1], the 6-degree Field Galaxy Survey (6dFGS) [2, 3] and the Sloan Digital Sky Survey (SDSS) [4, 5], later combined with the Supernova Legacy Survey (SNLS) [6, 7]. The first mapped about 4 % of the sky, recording over 220,000 galaxies and 100 quasars as far away as 800 Mpc (see Problem 2.7). The 6dFGS covered a fraction of the sky ten times larger, sampling 125,000 galaxies. The database of the SDSS-III survey covers about one quarter of the sky, over 1,800,000 galaxies and over 300,000 quasars.

The galaxy distribution near the local group is rather irregular and characterized by several super-clusters of galaxies where visible matter is much more concentrated than in other almost-empty regions called giant vacua (Figs. 2.2 and 2.3). However, at larger scales the galaxy distribution becomes more uniform in all directions (Fig. 2.4). With good approximation, the large-scale matter distribution of the universe is *homogeneous* (independent of the point at which the observation takes place) and *isotropic* (independent of the direction of observation). This is all the more apparent at even larger scales. In fact, the sky is also uniformly filled with a radiation background of cosmic origin, called *cosmic microwave background* (CMB). Its distribution has been mapped with great accuracy by the Wilkinson Microwave Anisotropy Probe (WMAP) [10–12] and by the PLANCK satellite [13, 14], and it shows a remarkable degree of isotropy (Fig. 2.5).

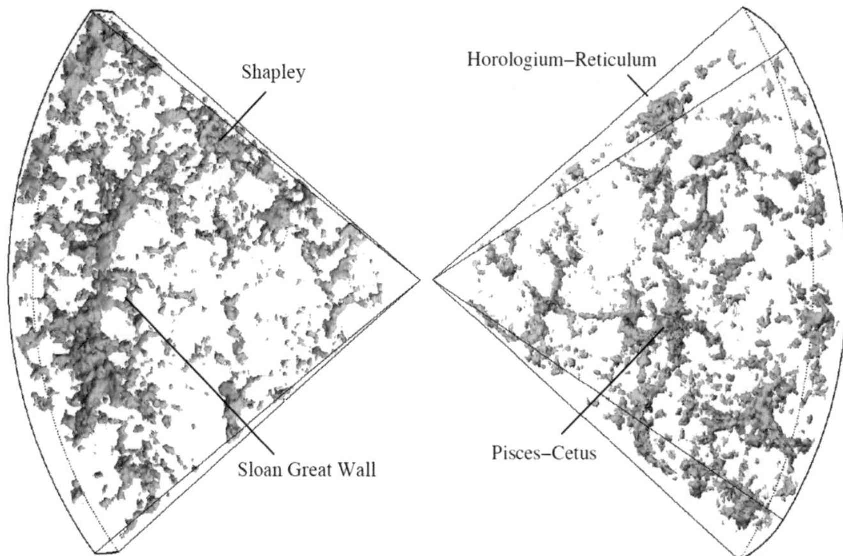

Fig. 2.2 The Sloan Great Wall in a DTFE reconstruction of the inner parts of the 2dF Galaxy Redshift Survey. The most distant galaxies in the figure are at about one billion light years (∼300 Mpc) from Earth (Source: [8])

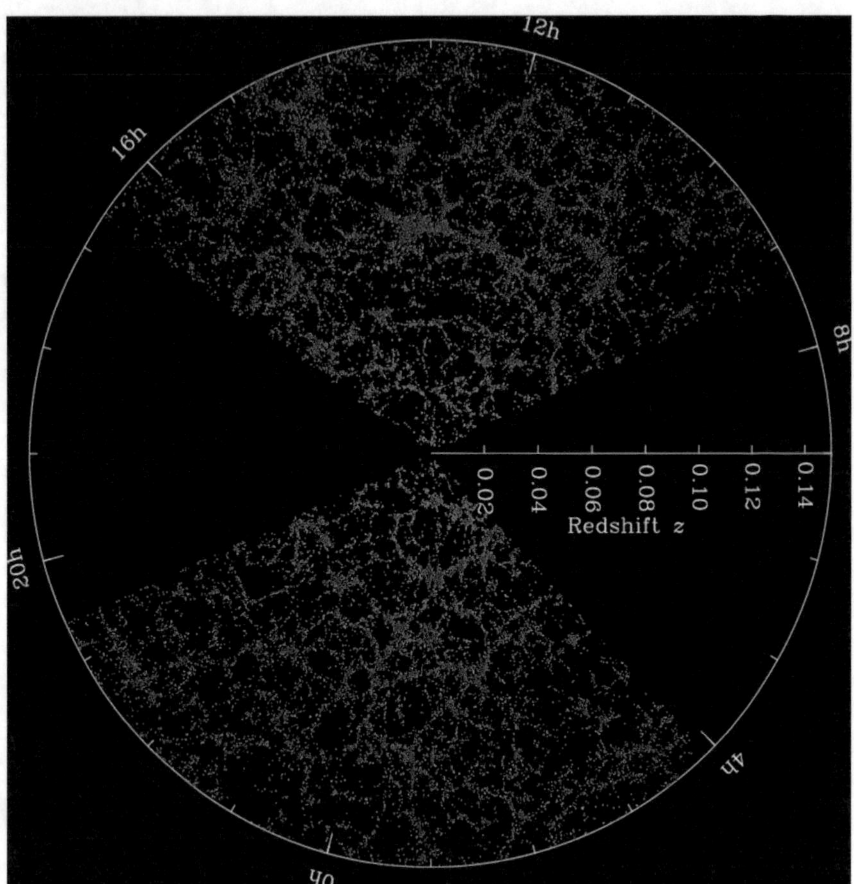

Fig. 2.3 Slices through the SDSS 3-dimensional map of the distribution of galaxies. Earth is at the centre and each point represents a galaxy, typically containing about 100 billion stars. Galaxies are colored according to the ages of their stars, with the redder, more strongly clustered points showing galaxies that are made of older stars. The outer circle is at a distance of two billion light years. The region between the wedges was not mapped by the SDSS because dust in our own Galaxy obscures the view of the distant universe in these directions (Credit: M. Blanton and the Sloan Digital Sky Survey [4])

The CMB is regarded as a snapshot of the universe dated back to more than 13 billion years ago, when matter, a hot homogeneous plasma of baryons and other particles, became transparent to radiation for the first time. Today, the CMB has cooled down to a mean temperature of [15, 16]

$$T_0 = (2.7255 \pm 0.0006)\,\mathrm{K}\,. \tag{2.1}$$

Looking in any direction, the observer measures the same value (isotropy). Since there is no reason why we should occupy a privileged place in the universe, the

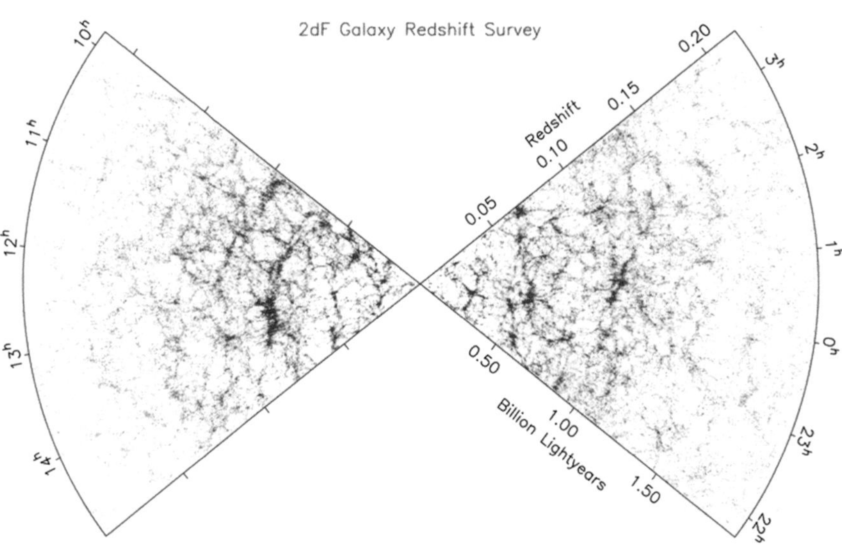

Fig. 2.4 2dFGRS Galaxy map (Credit: 2dF Galaxy Redshift Survey [9])

Fig. 2.5 Temperature anisotropy map of the cosmic microwave background radiation in the 2013 PLANCK data release. Colors indicate warmer (*red*) and cooler (*blue*) spots with respect to the mean temperature (©ESA and the Planck Collaboration [13])

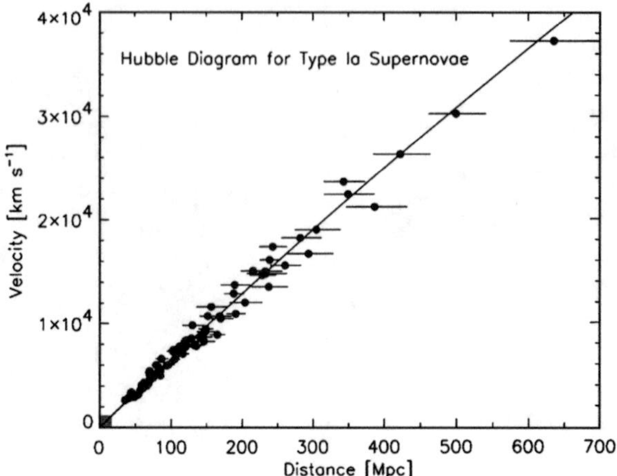

Fig. 2.6 Hubble diagram: recession velocity of extragalactic supernovæ as a function of distance. ©2004 National Academy of Sciences, USA (Source: [17])

same value should be measured also by observers positioned at any other point (homogeneity). Deviations from T_0 are of the order of micro-Kelvins, so that the very distant (i.e., early) universe is, for any observer, homogeneous and isotropic up to one part over 10^5.

We also see other large-scale phenomena. The rotation curves of galaxies and gravitational lensing effects strongly indicate the presence of more matter than what we observe. Baryons and leptons are typically interested by electromagnetic interactions, so that this extra component (called *dark matter*) is believed to be made of exotic particles not yet detected in the laboratory.

Finally, galaxies are seen to recede away from us according to Hubble's law. This law states that the farther the galaxy, the greater the recession speed (Fig. 2.6).[1] Observations of type Ia supernovæ (SNe Ia) show a deviation from the linear Hubble law [18–23], thus indicating that the universe not only expands, but it does so with increasing rate.

To summarize, observations establish the ingredients of the cosmological standard model known, for reasons which shall soon become clear, as *hot big bang* (henceforth "standard model;" we shall use capital initials for the Standard Model of particles):

(I) At sufficiently large scales, the universe is isotropic: its properties are independent of the direction of observation.

(II) Copernican principle: our location is not special. Consequently, if the universe is observed as isotropic from everywhere, it is also homogeneous: its thermal

[1] The distance of a galaxy is determined by astrophysics standard candles such as variable stars and supernovæ. For a historical introduction on Hubble law, see [17].

properties are the same at every point. By "point" we mean a sufficiently large local patch.

(III) The universe is composed of radiation and baryonic as well as non-baryonic matter. At early times, it was in thermal equilibrium.

(IV) The universe expands. At late times, the expansion is accelerated.

Points **(I)** and **(II)** go under the name of cosmological principle:

> ***Cosmological principle.*** *The universe does not possess a privileged point or direction; it is therefore homogeneous and isotropic (approximately).*

2.1.2 Friedmann–Lemaître–Robertson–Walker Background

The above ingredients can be formalized in Einstein's theory of general relativity. We assume spacetime is a "globally hyperbolic" Lorentzian manifold $(\mathcal{M}, g)$ endowed with a metric $g_{\mu\nu}$. Definitions of global hyperbolicity, space- and time-like curves and geodesics will be given later in the technical Sect. 6.1.1. Here, we only need to grasp the intuitive meaning of global hyperbolicity, namely, the possibility to foliate spacetime into spatial sections stacked along a natural time direction.

Homogeneity and isotropy are realized by the Friedmann–Lemaître–Robertson–Walker (FLRW) metric:

$$
\mathrm{d}s^2 = g_{\mu\nu}\mathrm{d}x^\mu \mathrm{d}x^\nu = -\mathrm{d}t^2 + a^2(t)\gamma_{\alpha\beta}\mathrm{d}x^\alpha \mathrm{d}x^\beta \,, \tag{2.2a}
$$

where $t = x^0$ is *synchronous* (or *proper*) *time*, $a(t)$ is the expansion parameter or *scale factor*, and

$$
\gamma_{\alpha\beta}\mathrm{d}x^\alpha \mathrm{d}x^\beta = \frac{\mathrm{d}r^2}{1 - \textsc{k}\, r^2} + r^2 \mathrm{d}\Omega_{D-2}^2 \tag{2.2b}
$$

is the line element, in hyperspherical coordinates, of the maximally symmetric $(D-1)$-dimensional space Σ of constant sectional curvature $\textsc{k}$ (intrinsic curvature). The latter is equal to -1 for an open universe, 0 for a flat universe and $+1$ for a closed universe with radius a (Fig. 2.7). In four dimensions (polar coordinates), $\mathrm{d}\Omega_2^2 = \mathrm{d}\theta^2 + \sin^2\theta \mathrm{d}\varphi^2$, where $0 \leqslant \theta \leqslant \pi$ and $0 \leqslant \varphi \leqslant 2\pi$.

The fact that the universe follows a homogeneous evolution allows us to use the same *universal clock* at each point, so that physical properties which are equal in different places imply synchronized clocks.[2] Sometimes, the choice of a coordinate system (equivalently, of a metric $g_{\mu\nu}$) is referred to as a *gauge* choice. The FLRW

[2]This is true, of course, for large-scale observations, while at very small scales, where non-FLRW metrics are better descriptions of the environment, the problem of synchronization of clocks persists.

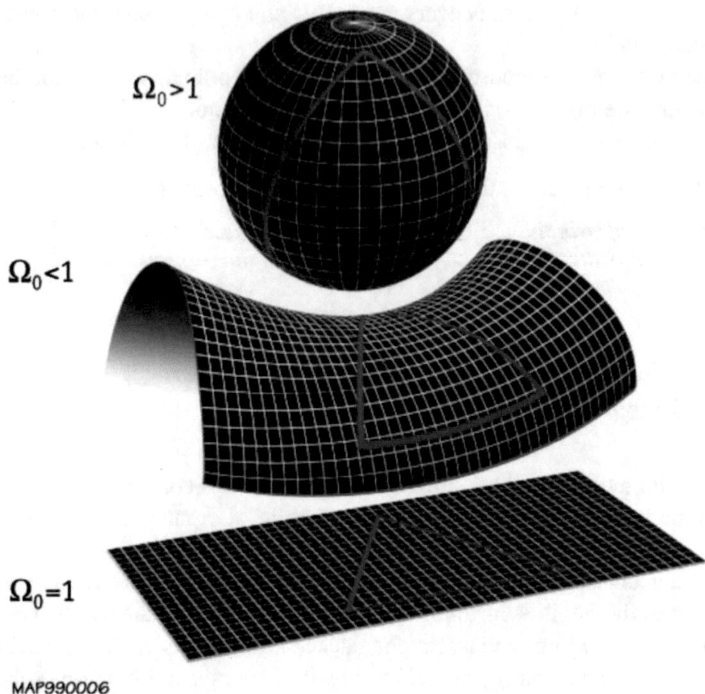

Fig. 2.7 Three possible geometries of the universe illustrated for *two-dimensional* spatial sections: closed, open and flat from top to bottom, corresponding to a total density parameter Ω_0 today respectively greater than, less than or equal to 1 (see (2.84)). In a closed or open $(2+1)$-dimensional universe, the sum of the internal angles of a triangle is, respectively, greater than or less than π. Only in a universe with flat spatial curvature the sum is π (Credit: NASA/WMAP Science Team [10])

metric corresponds to synchronous gauge, which we will meet again later. Another gauge makes use of *conformal time* τ, defined as

$$d\tau := \frac{dt}{a(t)}. \tag{2.3}$$

The line element (2.2) becomes

$$ds^2 = a^2(\tau)(-d\tau^2 + \gamma_{\alpha\beta}dx^\alpha dx^\beta).$$

For a free particle moving at the speed of light, the coordinate distance spanned in a conformal time interval $\Delta\tau$ is $c\Delta\tau$.

Since the time coordinate is not physical time, we can choose different universal clocks according to the problem at hand. Apart from t, τ, or the redshift z below, one can also use matter as relational time, provided it evolves monotonically.

For instance, in the presence of a classical scalar field one can make a time reparametrization whenever $\dot{\phi}$ (dots denote differentiation with respect to synchronous time t) does not change sign (non-singular Jacobian), so that the Hubble parameter and other geometrical quantities are thought of as depending on ϕ.

The spatial part of the metric is homogeneous and isotropic by construction. The coordinates x^α are "glued" to the continuous fluid elements representing the matter content. Therefore, a coordinate point x^α in space represents a fluid element passing at time t on the given point. This is why spatial coordinates are called *comoving*. It is easy to relate comoving distances with *proper* distances:

$$\boxed{(\text{proper distance}) = a(t) \times (\text{comoving distance})\,.} \tag{2.4}$$

Proper distances are also called dynamical or, more often, physical, but we will not use these alternative names here (strictly speaking, comoving distances are also physical and dynamical).

Observations tell us that the universe is expanding, so we assume that the scale factor $a(t) > 0$ increases, $\dot{a}(t) > 0$. Cosmological distances (see Problem 2.7) are better characterized by the *redshift*

$$1 + z := \frac{a_0}{a}\,, \tag{2.5}$$

where $a_0 := a(t_0)$ is the scale factor today. The scale factor is not a physical observable and its normalization is arbitrary. Typically, one chooses $a_0 = 1$. A local observer O is at $z = 0$ and distant objects are at $z > 0$. For large redshift, $1/z$ roughly indicates the size of a closed universe with respect to its radius today.

The name redshift comes from the fact that

$$z = \frac{\lambda_0 - \lambda}{\lambda}\,,$$

where λ is the wave-length of radiation emitted by a source and λ_0 is the observed wave-length. To show this, consider two nearby objects separated by a small proper distance $\mathrm{d}R$, where $\mathrm{d}R = a(1 - \textsc{K}r^2)^{-1/2}\mathrm{d}r$. The relative velocity of the two objects is given by the Hubble law, $\mathrm{d}v = H\mathrm{d}R$. By Doppler law, the infinitesimal change in wave-length is $\mathrm{d}\lambda/\lambda = \mathrm{d}v/c = H\mathrm{d}t = \mathrm{d}a/a$, where we used $c = \dot{R}$. Integrating, we get

$$\frac{\lambda}{a} = \frac{\lambda_0}{a_0}. \tag{2.6}$$

The momentum of a photon scales as a^{-1} or, in other words, the comoving wavelength of the signal is the same at the source and for the observer.

Comoving coordinates define a frame where the universe is isotropic and all observers move along the Hubble flow. On the other hand, a proper observer sees nearby galaxies receding but this does not mean that O is the "centre" of the universe, since there is no privileged point. Redshift measurements and distance statements such as (2.4) always refer to the relative position of the observed object with respect to O. To use a simple analogy, one can imagine the two-dimensional surface of a balloon (the universe) dotted with spots (galaxies). As long as the balloon inflates, the distance between the spots, measured on the balloon surface by a two-dimensional observer, increases.

The *Hubble parameter*

$$H(t) := \frac{\dot{a}(t)}{a(t)} \tag{2.7}$$

describes the expansion rate of the universe and defines the *Hubble distance* (or *radius* or, improperly, *horizon*; c units temporarily restored)

$$R_H(t) := \frac{c}{H(t)}, \tag{2.8}$$

which is the distance between the observer at time t and an object moving with the cosmological expansion at the speed of light. The Hubble horizon is a crucial quantity in the theory of structure formation and in the notion of quantum measurement in cosmology. It marks the boundary between the causal region centered at the observer O and the external region and roughly corresponds to the size of the observable universe (not to be confused with the physical dimensions of a closed universe parametrized by a). Actually, the true causal horizon is the *particle horizon*, defined as the sphere, centered at O, containing all the points which could have interacted with O through light signals since the "beginning" at $t = t_\mathrm{i}$ [24]. The radius of the sphere is

$$R_\mathrm{p} = a(t)\, r_\mathrm{p} := a(t) \int_{t_\mathrm{i}}^{t} \mathrm{d}t' \frac{c}{a(t')} = a(t) \int_{a_\mathrm{i}}^{a(t)} \mathrm{d}a \frac{c}{Ha^2}. \tag{2.9}$$

We will see that, in some cases, the particle and Hubble horizon are approximately the same. Also, when the comoving particle horizon is zero at t_i,

$$r_\mathrm{p} = c\tau. \tag{2.10}$$

We will almost always use natural units, so that from now on we trade the symbol $r_{\rm p}$ with τ. The context will make it clear whether τ is a distance (comoving particle horizon) or conformal time.[3]

The value of the Hubble parameter today is measured combining data on type Ia supernovæ, galaxy distributions and CMB observations:

$$\begin{aligned} H_0 &= 100\,h\,\mathrm{km\,s^{-1}\,Mpc^{-1}} \\ &\approx 3.336\,h \times 10^{-1}\,\mathrm{Gpc}^{-1} \\ &\approx 1.747\,h \times 10^{-61} m_{\rm Pl} \\ &\approx 2.133\,h \times 10^{-42}\mathrm{GeV}\,. \end{aligned} \tag{2.11}$$

The PLANCK 2015 value combining information on the temperature spectrum, on the polarization spectra at low multipoles and on lensing reconstruction ("PLANCK TT+lowP+lensing" likelihood) is [14]

$$h = 0.678 \pm 0.009 \quad (68\,\%\,\mathrm{CL})\,, \tag{2.12}$$

where the error is at the 68 % confidence level (CL) and includes statistical and systematic uncertainties.[4] Estimates change for different experiments and data sets in the same experiments, but they generally agree (see [12] for WMAP and [14] for other likelihood analyses within PLANCK).

Lower bounds on the age of the universe t_0 can be obtained from estimates of the age of astrophysical objects (e.g., globular clusters) or gamma-ray bursts, but a more precise value is found from the Hubble parameter:

$$t_0 \simeq H_0^{-1} = 9.786 h^{-1}\,\mathrm{Gyr}\,, \tag{2.13}$$

where Gyr is one billion years. With (2.12), one obtains $t_0 \sim 14.4\,\mathrm{Gyr}$. An $O(1)$ correction factor in the above formula, depending on the various energy components, can refine this estimate (see Problem 2.1). According to the PLANCK TT+lowP+lensing likelihood [14],

$$t_0 = 13.799 \pm 0.038\,\mathrm{Gyr} \quad (68\,\%\,\mathrm{CL})\,. \tag{2.14}$$

[3]Throughout this book, we shall call "universe" the spacetime region causally connected with the observer, while storing the name "Universe" with capital U for the whole spacetime, inclusive of the regions outside the horizon.

[4]Although errors are bound to change as soon as new measurements become available, we report them to give the reader an idea of the level of accuracy of modern observations. The student should take these numbers with a critical attitude. Where do they come from? How do they change if one varies the data samples and the prior constraints? Is a model still acceptable if it predicts numbers outside the 1σ experimental interval? (The answer to the last question is Yes. One should start worrying, or getting excited, only when the model offshoots beyond the 3σ level.)

2.2 Einstein and Continuity Equations

In this section, we introduce the classical total action and equations of motion of general relativity in D dimensions. Our conventions for the Levi-Civita connection, Riemann and Ricci tensors, and Ricci scalar are

$$\Gamma^{\rho}_{\mu\nu} := \tfrac{1}{2}g^{\rho\sigma}\left(\partial_\mu g_{\nu\sigma} + \partial_\nu g_{\mu\sigma} - \partial_\sigma g_{\mu\nu}\right), \tag{2.15}$$

$$R^{\rho}{}_{\mu\sigma\nu} := \partial_\sigma \Gamma^{\rho}_{\mu\nu} - \partial_\nu \Gamma^{\rho}_{\mu\sigma} + \Gamma^{\tau}_{\mu\nu}\Gamma^{\rho}_{\sigma\tau} - \Gamma^{\tau}_{\mu\sigma}\Gamma^{\rho}_{\nu\tau}, \tag{2.16}$$

$$R_{\mu\nu} := R^{\rho}{}_{\mu\rho\nu}, \qquad R := R_{\mu\nu}g^{\mu\nu}. \tag{2.17}$$

The *Ansatz* of general relativity is the Einstein–Hilbert gravitational action

$$\boxed{S_g = \frac{1}{2\kappa^2}\int d^D x\, \sqrt{-g}\,(R - 2\Lambda)\,,} \tag{2.18}$$

where $g := \det(g_{\mu\nu})$ is the determinant of the dimensionless metric $g_{\mu\nu}$, $\kappa^2 = 8\pi G$ is D-dimensional Newton's constant, and Λ is the *cosmological constant*. The couplings have dimension

$$[\kappa^2] = 2 - D\,, \qquad [\Lambda] = 2\,. \tag{2.19}$$

In a spacetime with $D = 2$ dimensions, the Newton constant is dimensionless, the Einstein–Hilbert action is a topological invariant and there are no dynamical gravitational degrees of freedom. The Newton constant has negative energy dimensionality for $D > 2$, a fact of utmost importance for the renormalization properties of a quantum theory of gravity (Sect. 8.2).

The gravitational action is second order in spacetime derivatives. It can be augmented by higher-order curvature terms (Sect. 7.5) but there is no immediate need to do so because standard general relativity already explains most of the large-scale observations.

The constant Λ in (2.18) was originally introduced by Einstein in order to have stationary-universe solutions. After the discovery of cosmic expansion by Hubble in 1929, stationary cosmology was abandoned in favour of the big bang model and Λ was dropped. In the last years, however, the cosmological constant has been regaining the interest of the community thanks also to the advancements in astronomical and cosmological observations.

Assuming that matter is *minimally coupled* with gravity, the total action is

$$S = S_g + S_\partial + S_m\,, \tag{2.20}$$

where S_g is (2.18) and $S_m = \int d^D x \sqrt{-g}\mathcal{L}_m$ is the matter action. The piece S_∂, which will be further discussed in Sect. 9.1.4, is the *York–Gibbons–Hawking boundary term* (first written by Einstein [25]) added for consistency with the variational principle [26, 27]. In fact, the latter only requires that variations of the metric be zero at the boundary (i.e., the geometry of the boundary is fixed) but, in general, the normal first derivatives may be non-vanishing. In order to take this issue into account and obtain the Einstein equations correctly, one must add a specific S_∂ to the total action. Unless stated otherwise, throughout the book we will ignore the boundary term either because we consider closed manifolds (which have no boundary) or just for simplicity of presentation.

To find the equations of motion (Einstein equations) we need the variations (see Sect. 3.1.1)

$$\delta\sqrt{-g} = -\tfrac{1}{2}\, g_{\mu\nu}\, \sqrt{-g}\, \delta g^{\mu\nu}\,, \tag{2.21}$$

$$\delta R = (R_{\mu\nu} + g_{\mu\nu}\,\Box - \nabla_\mu\nabla_\nu)\, \delta g^{\mu\nu}\,, \tag{2.22}$$

where $\nabla_\nu V_\mu := \partial_\nu V_\mu - \Gamma^\sigma_{\mu\nu} V_\sigma$ is the covariant derivative of a vector V_μ and $\Box = \nabla_\mu\nabla^\mu$ is the curved d'Alembertian or Laplace–Beltrami operator. From (2.20), the Einstein equations $\delta S/\delta g^{\mu\nu} = 0$ read

$$\boxed{G_{\mu\nu} + g_{\mu\nu}\Lambda = \kappa^2 T_{\mu\nu}\,,} \tag{2.23}$$

where

$$G_{\mu\nu} := R_{\mu\nu} - \frac{1}{2} g_{\mu\nu} R\,, \tag{2.24}$$

$$T_{\mu\nu} := -\frac{2}{\sqrt{-g}}\frac{\delta S_{\rm m}}{\delta g^{\mu\nu}} = -2\frac{\partial \mathcal{L}_{\rm m}}{\partial g^{\mu\nu}} + g_{\mu\nu}\mathcal{L}_{\rm m}\,. \tag{2.25}$$

Taking the trace of (2.23) gives

$$-\left(\frac{D}{2} - 1\right) R + D\Lambda = \kappa^2 T_\mu^{\ \mu}\,. \tag{2.26}$$

$T_{\mu\nu}$ is the *energy-momentum tensor* (or stress-energy tensor) of matter. Its definition determines the continuity equation. In fact, let

$$\delta S_{\rm m} = \frac{1}{2}\int {\rm d}^D x\, \sqrt{-g}\, T^{\mu\sigma}\, \delta g_{\mu\sigma} \tag{2.27}$$

be the infinitesimal variation of the matter action with respect to the external field $\delta g_{\mu\sigma}$. For a constant infinitesimal coordinate transformation

$$x^{\mu\prime} = x^\mu + \delta x^\mu\,, \qquad \delta x^\nu = -a^\nu\,, \tag{2.28}$$

one has

$$\delta g_{\mu\sigma} = g_{\nu\sigma}\partial_\mu a^\nu + g_{\mu\nu}\partial_\sigma a^\nu + a^\nu \partial_\nu g_{\mu\sigma} \,, \tag{2.29}$$

where we used the definition of the Lie derivative for rank-2 and rank-0 tensors (see Chap. 3). Plugging (2.29) into (2.27) and integrating by parts, we get

$$\delta S_{\rm m} = -\int \mathrm{d}^D x\, a^\nu \left[\partial_\mu(\sqrt{-g}T^\mu_{\ \nu}) - \frac{1}{2}\sqrt{-g}T^{\mu\sigma}\partial_\nu g_{\mu\sigma}\right]. \tag{2.30}$$

Invariance of the action under diffeomorphisms requires $\delta S_{\rm m}$ to vanish *on shell* (i.e., when the dynamical equations are satisfied). Using the properties of the Levi-Civita connection (2.15) and the definition of the covariant derivative of a rank-2 tensor,

$$\begin{aligned}\nabla_\mu T^\mu_{\ \nu} &= \partial_\mu T^\mu_{\ \nu} + \Gamma^\mu_{\mu\sigma}T^\sigma_{\ \nu} - \Gamma^\sigma_{\mu\nu}T^\mu_{\ \sigma}\\ &= \frac{1}{\sqrt{-g}}\partial_\mu\left(\sqrt{-g}T^\mu_{\ \nu}\right) - \frac{1}{2}(\partial_\nu g_{\mu\sigma})T^{\mu\sigma}\,,\end{aligned}$$

one finally obtains the continuity equation

$$\boxed{\nabla_\mu T^\mu_{\ \nu} = 0\,.} \tag{2.31}$$

The energy-momentum tensor carries the contribution of all forms of energy in the universe, except gravity.

The continuity and Einstein equations are not independent because of the contracted Bianchi identities

$$2\nabla^\mu R_{\mu\nu} = \nabla_\nu R\,. \tag{2.32}$$

The divergence of (2.23) correctly reproduces (2.31).

2.2.1 Energy Conditions

The matter sector can be classified according to a set of covariant conditions [28]. Consider an arbitrary future-directed time-like vector t^μ ($t_\mu t^\mu < 0$; see Sect. 6.1.1) and a null vector n^μ ($n_\mu n^\mu = 0$). The *null energy condition* (NEC) is

$$T_{\mu\nu}n^\mu n^\nu \geq 0 \qquad \forall\, n^\mu \text{ null}\,. \tag{2.33}$$

When the Einstein equations (2.23) hold, this is equivalent to the *null convergence condition*

$$R_{\mu\nu}n^\mu n^\nu \geqslant 0 \qquad \forall\, n^\mu \quad \text{null}\,. \tag{2.34}$$

We will use this form of the NEC in Chap. 6.

The *weak energy condition* (WEC) states that

$$T_{\mu\nu}t^\mu t^\nu \geqslant 0 \qquad \forall\, t^\mu \text{ time-like}\,. \tag{2.35}$$

In particular, given the unit time-like vector u^μ ($u_\mu u^\mu = -1$), the matter energy density

$$\rho := T_{\mu\nu}u^\mu u^\nu \tag{2.36}$$

is always non-negative. Most of the known matter fields obey the NEC and WEC. Exceptions are condensates, which admit negative-energy states. Violations of the WEC can also come from quantum effects (Casimir energy, squeezed states, and so on).

The *dominant energy condition* (DEC) requires that $-T_{\mu\nu}t^\mu$ is either future-directed or null,

$$-T_{\mu\nu}t^\mu = Au^\mu + Bn^\mu\,, \tag{2.37}$$

where $A, B > 0$. In other words, momentum cannot (be observed to) flow faster than light (causal flux). In particular, the DEC implies the WEC.

The *time-like convergence condition* requires that

$$R_{\mu\nu}t^\mu t^\nu \geqslant 0 \qquad \forall\, t^\mu \text{ time-like}\,. \tag{2.38}$$

If the Einstein equations hold, this is equivalent to the *strong energy condition* (SEC)

$$\left(T_{\mu\nu} - \frac{1}{D-2}g_{\mu\nu}T_\sigma^{\ \sigma} + \frac{2}{D-2}g_{\mu\nu}\frac{\Lambda}{\kappa^2}\right)t^\mu t^\nu \geqslant 0\,. \tag{2.39}$$

2.3 Perfect Fluid

The energy-momentum tensor assumes a simple form when the only matter content of the universe is a *perfect fluid* (zero heat flow and anisotropic stress). Its definition is

$$\boxed{T_{\mu\nu} = (\rho + P)\, u_\mu u_\nu + P\, g_{\mu\nu}\,,} \tag{2.40}$$

where $\rho = T_{00}$ and $P = T_\alpha^{\ \alpha}/(D-1)$ are the energy density and pressure of the fluid, $u^\mu = \mathrm{d}x^\mu/\mathrm{d}t$ is the comoving D-velocity (unit time-like vector, $u_\mu u^\mu = -1$) tangent to a fluid element's world-line, and t is proper time along the fluid world-line. Since we are in a globally hyperbolic spacetime, u^μ also corresponds to the unit vector normal to spatial Cauchy surfaces. The metric $g_{\mu\nu}$ induces a Riemannian metric $h_{\mu\nu}$ defined by the *first fundamental form*

$$h_{\mu\nu} = g_{\mu\nu} + u_\mu u_\nu\,, \qquad h_\mu^{\ \nu} = h_\mu^{\ \sigma} h_\sigma^{\ \nu}\,, \qquad u^\mu h_\mu^{\ \nu} = 0\,, \tag{2.41}$$

which is the metric spatial projection orthogonal to the comoving velocity.

At every point of spacetime, the local rest frame of a fluid is defined as the coordinate system where the off-diagonal components of the fluid vanish, $T^{0\alpha} = 0$, and

$$u^\mu = \delta_0^\mu = ((-g_{00})^{-1/2}, 0, \dots, 0)^\mu\,. \tag{2.42}$$

In covariant formalism [29–31], the gradient of the four-velocity field is decomposed as

$$\nabla_\nu u_\mu = \sigma_{\mu\nu} + \omega_{\mu\nu} + \frac{1}{D-1}\theta h_{\mu\nu} - \dot{u}_\mu u_\nu\,. \tag{2.43}$$

Here,

$$\sigma_{\mu\nu} := \nabla_{(\nu} u_{\mu)} - \frac{1}{D-1}\theta h_{\mu\nu} + \dot{u}_{(\mu} u_{\nu)} = h_{\sigma(\nu}\nabla^\sigma u_{\mu)} - \frac{1}{D-1}\theta h_{\mu\nu} \tag{2.44}$$

is the symmetric *shear* tensor,

$$\omega_{\mu\nu} := \nabla_{[\nu} u_{\mu]} + \dot{u}_{[\mu} u_{\nu]} = h_{\sigma[\nu}\nabla^\sigma u_{\mu]} \tag{2.45}$$

is the anti-symmetric *vorticity* tensor,

$$\theta := \nabla^\mu u_\mu \tag{2.46}$$

is the *volume expansion* (or expansion of the family of time-like geodesics u_μ) and a dot defines the proper-time derivative

$$\dot{} := u_\mu \nabla^\mu\,. \tag{2.47}$$

Integrating the volume expansion along a world-line with respect to t, one defines the *number of e-foldings*

$$\mathcal{N}_a := \frac{1}{D-1}\int_{t_\mathrm{i}}^t \mathrm{d}t'\theta\,, \tag{2.48}$$

where $t_{\rm i}$ is some initial time. This definition is unique up to an integration constant for each world-line, which can be fixed by choosing a reference hypersurface where $\mathcal{N}_a = 0$.

The continuity equation (2.31) contracted with $-u^\nu$ is

$$\dot{\rho} + (\theta - \sigma)(\rho + p) = 0\,, \tag{2.49}$$

where $\sigma = \sigma_\mu{}^\mu = u^\mu \dot{u}_\mu$. This equation is exact and valid at all scales. Contracting (2.31) with $h_\rho{}^\nu$, we get

$$D_\mu p + h_\mu{}^\nu \dot{u}_\nu (\rho + p) = 0\,, \tag{2.50}$$

where

$$D_\mu := h_\mu{}^\nu \nabla_\nu = \nabla_\mu + u_\mu u^\nu \nabla_\nu \tag{2.51}$$

is the spatial projection of the covariant derivative. Equation (2.50) can also be written as

$$u_\mu \dot{P} + (\dot{u}_\mu + \sigma u_\mu)(\rho + P) + \nabla_\mu P = 0\,. \tag{2.52}$$

Defining the effective *barotropic index*

$$w := \frac{P}{\rho}\,, \tag{2.53}$$

the energy conditions of the previous section are readily translated into conditions on (2.53). The NEC is

$$\text{NEC:} \qquad \rho(1 + w) \geqslant 0\,, \tag{2.54}$$

and is implied by all the other conditions. In fact, one can write any time-like vector of modulus $|t|$ as $t^\mu = |t| u^\mu + n^\mu$. Contracting t^μ with (2.40) according to the given condition (2.35), (2.37) or (2.39), and then sending $|t|$ to zero, one always reobtains (2.54). The WEC, DEC and SEC then are

$$\text{WEC:} \qquad \rho \geqslant 0\,, \qquad w \geqslant -1\,, \tag{2.55}$$

$$\text{DEC:} \qquad \rho \geqslant |P|\,, \qquad -1 \leqslant w \leqslant 1\,, \tag{2.56}$$

$$\text{SEC:} \qquad \rho[(D-3) + (D-1)w] - \frac{2\Lambda}{\kappa^2} \geqslant 0\,, \tag{2.57}$$

while

$$\text{DEC} + \text{SEC}(\Lambda = 0): \qquad -\frac{D-3}{D-1} \leqslant w \leqslant 1\,. \tag{2.58}$$

In $D = 4$, (2.58) reads $-1/3 \leqslant w \leqslant 1$.

Summarizing,

$$\text{DEC} \implies \text{WEC} \implies \text{NEC} \impliedby \text{SEC}(\Lambda = 0)\,.$$

2.3.1 Scalar Field

If matter is a scalar field with potential $V(\phi)$, its Lagrangian density is

$$\mathcal{L}_{\rm m} = \mathcal{L}_\phi = -\frac{1}{2}\partial_\mu\phi\partial^\mu\phi - V(\phi)\,. \tag{2.59}$$

Scaling dimension (or engineering, or canonical dimension; in momentum units) of ϕ is

$$[\phi] = \frac{D-2}{2}\,, \tag{2.60}$$

while for a polynomial potential

$$V(\phi) = \sum_{n=0}^{N} \sigma_n \phi^n\,, \tag{2.61}$$

the engineering dimension of the couplings is

$$[\sigma_n] = D - \frac{n(D-2)}{2}\,. \tag{2.62}$$

It is straightforward to see that the scalar equation of motion $\delta S_{\rm m}/\delta\phi = 0$ is in agreement with (2.31). Invariance of the total action under the infinitesimal shift

$$\phi \to \phi + \delta\phi \tag{2.63}$$

yields the equation of motion

$$0 = \frac{\partial \mathcal{L}_\phi}{\partial \phi} - \frac{\mathrm{d}}{\mathrm{d}x^\mu}\frac{\partial \mathcal{L}_\phi}{\partial(\partial_\mu\phi)}\,. \tag{2.64}$$

From (2.59), we get

$$T^\mu_{\ \nu} = \delta^\mu_\nu \mathcal{L}_\phi + \partial^\mu\phi\partial_\nu\phi \tag{2.65}$$

and

$$\boxed{\Box\phi - V_{,\phi} = 0\,,} \tag{2.66}$$

where $V_{,\phi} = \partial V/\partial\phi$ and

$$\Box\phi = \frac{1}{\sqrt{-g}}\partial_\mu(\sqrt{-g}\partial^\mu\phi)\,. \tag{2.67}$$

The scalar field is a particular case of perfect fluid, with world-lines orthogonal to $\phi = \text{const}$ hypersurfaces [32]:

$$u_\mu = -\frac{\partial_\mu\phi}{\dot{\phi}}\,. \tag{2.68}$$

When $V \propto \phi^2$, (2.66) is called Klein–Gordon equation.

2.4 Friedmann Equations

On an FLRW background, the only non-vanishing Levi-Civita and Ricci components are

$$\Gamma^0_{\alpha\beta} = Hg_{\alpha\beta}\,, \qquad \Gamma^\beta_{\alpha 0} = H\delta^\beta_\alpha\,, \qquad \Gamma^\lambda_{\alpha\beta} = \Gamma^\lambda_{\alpha\beta}(\gamma_{\alpha\beta})\,, \tag{2.69}$$

and

$$R_{00} = -(D-1)(H^2 + \dot{H}) = -(D-1)\frac{\ddot{a}}{a}\,, \tag{2.70}$$

$$R_{\alpha\beta} = \left[\frac{2\textsf{K}}{a^2} + (D-1)H^2 + \dot{H}\right]g_{\alpha\beta}\,, \tag{2.71}$$

$$\begin{aligned} R &= {}^{(D-1)}R + (D-1)\left(DH^2 + 2\dot{H}\right) \\ &= (D-1)\left(\frac{2\textsf{K}}{a^2} + DH^2 + 2\dot{H}\right), \end{aligned} \tag{2.72}$$

where we have exploited the symmetries of the space-like hypersurface $\tilde{\Sigma}$ [33] and ${}^{(D-1)}R$ is its Ricci scalar.

For simplicity we assume that the universe is filled by a perfect fluid, (2.40). The FLRW shear and viscosity vanish, while the volume expansion and the Hubble parameter coincide,

$$\theta = (D-1)H\,, \qquad \mathcal{N}_a = \ln\frac{a}{a_\mathrm{i}}\,. \tag{2.73}$$

The 00 component of the Einstein equations (2.23) is

$$\left(\frac{D}{2}-1\right)H^2 = \frac{\kappa^2}{D-1}\rho + \frac{\Lambda}{D-1} - \frac{\mathrm{K}}{a^2}\,, \tag{2.74}$$

while combining that with the trace equation (2.26) one obtains

$$-(D-2)\left(H^2+\dot{H}\right) + \frac{2\Lambda}{D-1} = \frac{\kappa^2}{D-1}\left[(D-3)\rho + (D-1)P\right]. \tag{2.75}$$

The continuity equation (2.49) has no shear and becomes

$$\dot{\rho} + (D-1)H(\rho+P) = 0\,, \tag{2.76}$$

while (2.52) is trivial in the fluid rest frame. For a barotropic fluid $p = w\rho$, w is constant and the continuity equation is solved by

$$\rho = \rho_0\, a^{-(D-1)(1+w)}\,, \tag{2.77}$$

where $a_0 = a(t_0) = 1$. From (2.74) with $\Lambda = 0 = \mathrm{K}$, it follows that

$$a = \left(\frac{t}{t_0}\right)^{\frac{2}{(D-1)(1+w)}}\,. \tag{2.78}$$

For a scalar field,

$$\rho_\phi = \frac{\dot{\phi}^2}{2} + V\,, \qquad P_\phi = \mathcal{L}_\phi = \frac{\dot{\phi}^2}{2} - V\,, \tag{2.79}$$

and (2.76) becomes (2.66),

$$\ddot{\phi} + (D-1)H\dot{\phi} + V_{,\phi} = 0\,. \tag{2.80}$$

From now on we specialize to four dimensions, $D = 4$. In this case, we obtain the first and second *Friedmann equations*

$$\boxed{H^2 = \frac{\kappa^2}{3}\rho + \frac{\Lambda}{3} - \frac{\mathrm{K}}{a^2}}\,, \tag{2.81}$$

$$\boxed{\frac{\ddot{a}}{a} = H^2 + \dot{H} = -\frac{\kappa^2}{6}(\rho + 3P) + \frac{\Lambda}{3}\,.} \tag{2.82}$$

Equation (2.82) is the FLRW version of the Raychaudhuri equation (Problem 6.1).

2.5 Content of the Universe

Equation (2.82) can be rewritten in terms of the parameter Ω, defined as the ratio between the total energy density ρ and the critical density $\rho_{\rm crit}$ sufficient to stop the expansion:

$$\Omega := \frac{\rho}{\rho_{\rm crit}}\,, \qquad \rho_{\rm crit} := \frac{3H^2}{\kappa^2}\,. \tag{2.83}$$

Equation (2.81) with $\Lambda = 0$ becomes

$$\Omega - 1 = \Omega_{\rm K}\,, \tag{2.84}$$

where

$$\Omega_{\rm K} := \frac{\rm K}{a^2H^2} \tag{2.85}$$

is the deviation from the critical density. If the universe is spatially flat, $\Omega = 1$.

Let us now take point **(III)** of the recipe of the universe into account (Sect. 2.1.1). Apart from the intrinsic curvature term, the contributions to the total energy density ρ is typically divided into radiation ($\Omega_{\rm r}$, which includes photons), baryonic matter and non-baryonic matter ($\Omega_{\rm m} = \Omega_{\rm b} + \Omega_{\rm nb} + \Omega_\nu$, where we include also massive neutrinos).[5] In the view of explaining point **(IV)**, we also add an extra component we call Ω_Λ (which might or might not correspond to a non-vanishing Λ in (2.81)):

$$\Omega = \Omega_{\rm r} + \Omega_{\rm m} + \Omega_\Lambda + \Omega_{\rm K}\,. \tag{2.86}$$

2.5.1 *Dust and Radiation*

Regarded as a perfect fluid, each contribution obeys an independent continuity equation (2.76). The case $w = 1/3$ is radiation which redshifts away

[5]In general relativity and quantum gravity, one calls "matter" everything which is not geometry. We did so in Sect. 2.2 but here we use cosmology jargon and separate radiation from the rest of the "matter."

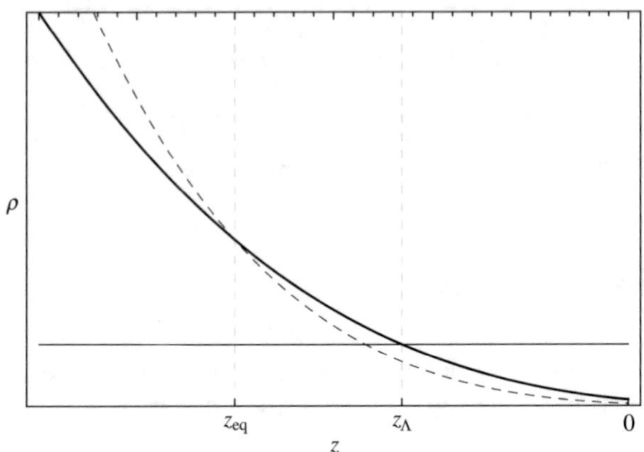

Fig. 2.8 Matter (*thick curve*), radiation (*dashed curve*), and cosmological constant (*thin line*) energy densities as functions of redshift. Radiation-matter and matter-dark energy equality times are indicated. The redshift scale is arbitrary

as (see (2.77))

$$\rho_r = \rho_{r0}(1+z)^4\,, \tag{2.87}$$

while $w = 0$ is pressureless (dust) matter,

$$\rho_m = \rho_{m0}(1+z)^3\,. \tag{2.88}$$

The radiation density decreases faster than matter density during the expansion ($z \to 0$). Therefore, the radiation component dominates over matter at early times, but matter eventually takes over (Fig. 2.8).

The moment[6] t_{eq} when the two densities coincide is called *equality*. When the universe is filled only with radiation and dust, the total energy density evolves as (2.87) when $z \gtrsim z_{eq}$, and as (2.88) otherwise. The redshift at equality is constrained by CMB observations (from the ratio of the first peak to the third peak of the power spectrum) combined with other observations. For the PLANCK TT+lowP+lensing likelihood [14],

$$z_{eq} = 3365 \pm 44 \quad (68\,\%\,\text{CL}). \tag{2.89}$$

Different data sets all converge to the rounded value

$$z_{eq} \approx 3400\,, \tag{2.90}$$

[6]In an approximate sense: the transition is smooth, not point-wise.

which we will use in Sect. 2.7. Once the redshift of an object is found, one can calculate the corresponding age. In particular, radiation-matter equality happened when the universe was less than 10,000 years old (see Problem 2.2).

From (2.78), one can see that

$$a \sim t^{\frac{2}{3}}\,, \qquad w = 0\,, \tag{2.91}$$

$$a \sim t^{\frac{1}{2}}\,, \qquad w = \frac{1}{3}\,. \tag{2.92}$$

In these scenarios where the universe expands as a *power law* (p is a constant, not to be confused with pressure),

$$\boxed{a \sim t^p\,, \qquad p = \frac{2}{3(1+w)}\,, \qquad H = \frac{p}{t}\,,} \tag{2.93}$$

and when

$$w > -\frac{1}{3} \qquad (0 < p < 1) \tag{2.94}$$

holds, the particle horizon (2.9) is about the same as the Hubble horizon (2.8):

$$\begin{aligned} R_\mathrm{p} &= t^p \int_{t_\mathrm{i}}^{t} \mathrm{d}t'\,t'^{-p} \overset{t \gg t_\mathrm{i}}{\simeq} \frac{1}{1-p}t = \frac{p}{1-p}R_H \\ &\simeq R_H\,. \end{aligned} \tag{2.95}$$

Several properties of the particle and Hubble horizons and the size at different epochs can be found in Problems 2.3–2.6 (see also Sect. 3.1.3). Exact power-law solutions are the subject of Problem 2.11.

Recent estimates of radiation and matter density today are (for the PLANCK TT+lowP+lensing likelihood at 68 % CL) [14]

$$\Omega_{\mathrm{r}0}h^2 = 2.469 \times 10^{-5}\,, \tag{2.96}$$

$$\Omega_{\mathrm{b}0}h^2 = 0.02226 \pm 0.00023\,, \tag{2.97}$$

$$\Omega_{\mathrm{c}0}h^2 = 0.1186 \pm 0.0020\,, \tag{2.98}$$

$$\Omega_{\mathrm{m}0}h^2 = 0.1415 \pm 0.0019\,. \tag{2.99}$$

The difference between the total contribution of matter (measured from the dynamics of galaxy clusters) and baryonic matter (from the visible galaxy distribution) is ascribed to the presence of another non-baryonic component, dark matter (see [34] for a review). Dark matter does not interact with photons and it can be observed only indirectly. In (2.98), we indicated it as $\Omega_\mathrm{c} = \Omega_\mathrm{nb} + \Omega_\nu$. Ω_m is a derived quantity obtained from the sum of baryonic and dark matter densities. Evidence is in favour of dark matter being made of particles moving at non-relativistic speed and lying outside the Standard Model of particle physics. Because of the weak or absent interaction with radiation, density fluctuations of non-baryonic dark matter can start growing much earlier than for ordinary matter, thus having time to tune their amplitudes at the level observed today in large-scale structures. Among the particle candidates from minimal extensions (supersymmetric and not) of the Standard Model are axions, sterile neutrinos and WIMPs (weakly interactive massive particles) such as neutralinos; experiments can place bounds on the abundances of these particles [35, 36]. There is also a possibility that gravity itself, via some modified action, may account for dark-matter effects. A small part of Ω_c can be due to massive neutrinos, $\Omega_{\nu 0}h^2 < 0.0025$ (95 % CL) [14].

2.5.2 *Hot Big Bang and the Big-Bang Problem*

The quasi isotropy of the CMB temperature indicates that the early universe was in thermal equilibrium. Therefore, we can express the energy density of radiation and matter in terms of thermodynamical quantities such as the temperature T.

The most important processes through which radiation interacts with matter are three: single and double Compton scattering and Bremsstrahlung. *Single Compton scattering* describes the collision of a photon with a free electron,

$$\gamma(\nu) + e^- \longrightarrow \gamma(\nu') + e^- .$$

When the photon energy is small with respect to the rest mass of the electron, the scattering process can be approximated by *Thomson scattering*.

When an electron at rest collides with a photon, a negative energy transfer occurs from the electron to the photon, whose frequency ν red-shifts ($\nu' < \nu$). The combination of this effect with the inverse scattering, where a photon gains energy from a relativistic electron ($\nu' > \nu$), leads to thermal equilibrium. These are elastic processes and the number of photons is conserved. In the *double* or *inelastic Compton scattering*, on the other hand, photons are emitted or absorbed,

$$\gamma + e^- \longleftrightarrow \gamma + \gamma + e^- .$$

In the Coulomb interaction between an electron and an atomic nucleus X, the former is accelerated in the rest frame of the ion X and emits radiation. In this case, one has *thermal Bremsstrahlung*,

$$e^- + X \longleftrightarrow e^- + X + \gamma .$$

The inverse reaction may also happen, since charged particles can absorb photons.

The energy (temperature) of these processes determines which is the dominant effect:

- Bremsstrahlung for $1\,\text{eV} < T < 90$ eV.
- Single Compton scattering for 90 eV $< T < 1$ keV.
- Double Compton scattering for $T \gtrsim 1$ keV,

where $T = k_\text{B} T_\text{K}$ is measured in eV. All these processes have a characteristic velocity greater than the cosmic expansion and hence they modify the photon distribution of the cosmological plasma.

At equilibrium, the photons involved in Compton scattering obey a Bose–Einstein distribution

$$f(\omega) = \frac{1}{e^{\frac{\hbar\omega}{T}-\mu} - 1}, \tag{2.100}$$

where $\omega = 2\pi\nu$ is the angular frequency and μ is the chemical potential, non-vanishing if the total number of photons is conserved. However, the number of low-frequency photons change due to inelastic Compton scattering and Bremsstrahlung, thus sending effectively μ to zero. In the meanwhile, elastic Compton scattering redistributes the photons over the spectrum. One then obtains a *black-body spectrum* of intensity (in $\hbar = 1$ units)

$$I(\omega, T) = \frac{\omega^3}{2\pi^2} f(\omega) = \frac{1}{2\pi^2} \frac{\omega^3}{e^{\frac{\omega}{T}} - 1}. \tag{2.101}$$

Once equilibrium and the black-body spectrum are achieved, they are *preserved* both by the above processes and by cosmic expansion, whose effect is to rescale the spectrum while maintaining its form. We have seen in (2.6) that $\lambda \sim a$, so that $\omega \sim 1 + z$ and the ratio ω/T in (2.101) is unchanged if $T \sim 1 + z$. We can find the proportionality constant as follows. The distribution (2.100) defines the number density of bosons of the i-th species per unit volume, $dn_i(\omega) = d^3\omega\, g_i (2\pi)^{-3} f(\omega)$, where g_i is the number of spin states ($g_\gamma = 2$ for photons). Therefore, the energy density of a species with black-body distribution is the integral of $dn(\omega)\,\omega$ over all frequencies ($d^3\omega = 4\pi d\omega\, \omega^2$), which is the integrated intensity

$$\begin{aligned} \rho_i &= \int_0^{+\infty} dn_i(\omega)\, \omega = g_i \int_0^{+\infty} d\omega\, I(\omega, T) \\ &\stackrel{x=\omega/T}{=} \frac{g_i(T)}{2\pi^2} T^4 \int_0^{+\infty} dx \frac{x^3}{e^x - 1} = \frac{\pi^2}{30} g_i T^4 . \end{aligned}$$

One should sum over all species i in thermal equilibrium ($\mu_i \ll T$), including fermions. The latter obey a Fermi–Dirac distribution $f(\omega) = (e^{\omega/T-\mu} + 1)^{-1}$, resulting in a contribution $(7/8)g_i$ for each species. The total energy density of radiation is the Stefan–Boltzmann law with g_i replaced by a total effective $g_*(T)$:

$$\rho_r = \frac{\pi^2}{30} g_*(T)\, T^4 . \tag{2.102}$$

The coefficient $g_*(T) = O(1) - O(10)$ is temperature dependent because a species with mass m_i no longer participates in (2.102) when the temperature falls below m_i. Away from these *decoupling* events, g_* can be regarded as constant. For a radiation-dominated universe, comparing (2.102) with (2.87) one gets the temperature as a function of redshift:

$$\begin{aligned} T &= \left(\frac{30}{\pi^2}\frac{\rho_{r0}}{g_*}\right)^{1/4}(1+z) \simeq 3.96\left(\frac{\Omega_{r0}h^2}{g_*}\right)^{1/4}(1+z)\times 10^{-3}\,\text{eV} \\ &\overset{(2.96)}{\approx} \frac{2.79}{g_*^{1/4}}(1+z)\times 10^{-4}\,\text{eV} \qquad (2.103) \\ &\approx \frac{3.24}{g_*^{1/4}}(1+z)\,\text{K}. \qquad (2.104) \end{aligned}$$

Using (2.81) and (2.92), one has

$$H = \frac{m_{\text{Pl}} t_{\text{Pl}}}{2t} , \tag{2.105}$$

so that, with (2.81),

$$t = \left(\frac{45}{16\pi^3 g_*}\right)^{1/2} \frac{m_{\text{Pl}}^2 t_{\text{Pl}}}{T^2} \approx \frac{2.42}{g_*^{1/2}}\left(\frac{10^6\text{eV}}{T}\right)^2 \text{s}. \tag{2.106}$$

With these formulæ, one can calculate the temperature at a given redshift and the age of the universe at a given temperature (see Problem 2.8).

Since $T \propto 1 + z$, one concludes that the universe was a high-temperature, radiation-dominated plasma in early epochs. This is why the standard model is called *hot*. As in the great majority of cosmological solutions, the power-law profile (2.93) tends to $a \to 0$ in the past. At the *big bang* $t = 0$ (the earliest instant ever), the metric is singular and general relativity breaks down. The Universe reduces to a point of infinite temperature and energy density. Generic classical cosmological solutions do possess a big-bang singularity and one cannot trust the theory at the very beginning. This is clearly a problem of self-consistency which is desirable to solve. We will talk about it much later (Chaps. 6, 10 and 13).

2.5.3 Dark Energy and the cosmological Constant Problem

The hot big bang model works very well from early ages until almost today. This is the main reason why we assumed that only dust and radiation are present. Other types of fluid, if added, should first prove themselves necessary to explain experimental data, and then find their place in a field-theory model. For example, an extra *stiff matter* component ($w = 1$) would scale as $\rho_{\rm stiff} = \rho_{\rm stiff0}(1+z)^6$ and it would dominate at early times. Let $\tilde{z}_{\rm eq}$ be the redshift when $\rho_{\rm stiff} = \rho_{\rm r}$. The hot big bang model is verified to a high degree of accuracy starting from the *big-bang nucleosynthesis* (BBN; see Problem 2.8) at $z_{\rm BBN} \sim 10^9$, so that $\tilde{z}_{\rm eq} > z_{\rm BBN}$. From (2.96), one must have $\Omega_{\rm stiff0}h^2 \lesssim 10^{-23}$, and any such component would be unobservable today. Therefore, there is no need to consider stiff matter.

On the other hand, from the positivity of ρ, (2.82) and the strong energy condition (2.57) with $\Lambda = 0$, it follows that a universe filled only with matter and radiation must decelerate, $\ddot{a} < 0$. This is in contrast with point **(IV)** so that, after all, we do have to add some other contribution to the total energy density. So far, we have ignored the cosmological constant. Let us rewrite the Friedmann equations (2.81)–(2.82):

$$\frac{\ddot{a}}{a} = -\frac{\kappa^2}{6}[(\rho+\rho_\Lambda)+3(P+P_\Lambda)]\,, \tag{2.107}$$

$$H^2 = \frac{\kappa^2}{3}(\rho+\rho_\Lambda) - \frac{\textsc{K}}{a^2}\,, \tag{2.108}$$

where

$$\rho_\Lambda = -P_\Lambda := \frac{\Lambda}{\kappa^2}\,. \tag{2.109}$$

A particular solution of these Friedmann equations is the *de Sitter universe* [37], a cosmological model without matter ($\rho = 0$) and flat spatial sections ($\textsc{K} = 0$):

$$\boxed{H = \sqrt{\frac{\Lambda}{3}} \qquad \Rightarrow \qquad a(t) = a(0)\,\mathrm{e}^{Ht}\,.} \tag{2.110}$$

The Hubble rate is *constant* and the scale factor expands exponentially. It is the prototypical inflationary background and the cosmology with the simplest analytic properties (see Problem 2.10). It has an analogue also for a closed universe ($\textsc{K} = 1$):

$$a(t) = H^{-1}\cosh(Ht)\,, \tag{2.111}$$

where $H = \sqrt{\Lambda/3}$.

In a quantum field theory context, ρ_Λ and p_Λ represent the energy density and pressure of quantum fluctuations of the vacuum. At the semi-classical level, gravity can be treated as classical, while Einstein equations take the form [38]

$$G_{\mu\nu} = \kappa^2 \langle T_{\mu\nu} \rangle \,, \tag{2.112}$$

which can be argued to be valid when dispersion in the phase of matter wave-functions is negligible [39]. In a local inertial frame, Lorentz invariance requires that the vacuum expectation value of the energy-momentum tensor be proportional to the Minkowski metric, $\langle T_{\mu\nu} \rangle \propto \eta_{\mu\nu}$. Therefore, in a general frame [40, 41]

$$\langle T_{\mu\nu} \rangle = -\rho_{\rm vac}\, g_{\mu\nu} \,. \tag{2.113}$$

This is a cosmological-constant contribution $\rho_{\rm vac} = \Lambda/\kappa^2 = \rho_\Lambda$ and the "vacuum equation of state" is (2.109), corresponding to the barotropic index

$$w_\Lambda = -1 \,. \tag{2.114}$$

The fact that the pressure P_Λ is negative should not worry the reader. In fact, in gravity the quantity P loses its usual meaning of thermodynamical pressure; here, it encodes a generic effect of "anti-gravitational repulsion."

A notable feature of the de Sitter universe is that it accelerates,

$$\frac{\ddot{a}}{a} = H^2 \Omega_\Lambda > 0 \,. \tag{2.115}$$

Measuring the redshift of type I supernovæ, in 1997 it was discovered that the universe is indeed accelerating [18, 42, 43]. First estimates gave $0.3 \leqslant \Omega_{\Lambda 0} \leqslant 0.9$. The advance in what is now called precision cosmology can be appreciated by comparing this early constraint with the most recent one to date from CMB observations [14] (PLANCK TT+lowP+lensing),

$$\Omega_{\Lambda 0} = \frac{\Lambda}{3H_0^2} = 0.692 \pm 0.012 \quad (68\,\%\ \text{CL}) \,, \tag{2.116}$$

or from the combination of data [14] about the CMB, supernovæ [6], baryon acoustic oscillations (BAO) [5, 44–48] and H_0 ("PLANCK TT+lowP+lensing+ext" likelihood):

$$\Omega_{\Lambda 0} = 0.6935 \pm 0.0072 \quad (68\,\%\ \text{CL}) \,. \tag{2.117}$$

Using (2.116),

$$\rho_\Lambda = \Omega_{\Lambda 0} \frac{3H_0^2}{\kappa^2} \approx 5.4 \times 10^{-123} m_{\rm Pl}^4 \approx (3.3 \times 10^{-3} \text{eV})^4 \,. \tag{2.118}$$

This is a very low density, about $10^{-29}\,\mathrm{g\,cm^{-3}}$, which would be extremely difficult to detect in the laboratory. Nevertheless, it constitutes about 70 % of the total density, since it uniformly fills the universe and all the regions otherwise empty of matter.

The experimental value w_0 of w_Λ today varies according to the prior constraints (zero or non-zero curvature, constant or varying w_Λ, and so on). For instance, for a constant w_Λ and a flat universe [14] (PLANCK TT+lowP+lensing+ext),

$$w_0 = -1.006^{+0.085}_{-0.091} \quad (95\,\%\ \mathrm{CL})\,. \tag{2.119}$$

Somewhat weaker bounds come from redshift-space distortions [49–51]. Letting $a(t_0) = a_0 = 1$ be the present value of the scale factor, a two-parameter model of a varying w_Λ is [52, 53]

$$w_\Lambda =: w_0 + w_a(a-1)\,, \tag{2.120}$$

so that today $w = w_0$. In this case, the PLANCK TT+lowP+lensing+ext posteriors are

$$-1.2 < w_0 < -0.6\,, \qquad -1.5 < w_a < 0.6 \qquad (95\,\%\ \mathrm{CL})\,. \tag{2.121}$$

A cosmological constant is the simplest explanation of late-time acceleration. However, Λ is a problematic object in many respects. Other "small" numbers do appear in cosmology but, usually, they are described by known physics and are not perceived as troublesome as (2.118) [54]. An instance is the energy density of matter or radiation at the time of equality,

$$\rho_{\mathrm{eq}} \approx 2.4\times 10^{-113} m_{\mathrm{Pl}}^4\,, \tag{2.122}$$

which depends on the number density of photons and matter particles and can be calculated within the framework of standard high-energy physics [55]. On the other hand, the cosmological constant problem consists in the "unnatural" smallness of the quantity (2.118). There is an issue of fine tuning both at the classical level (via the ratio $\rho_\Lambda/\rho_{\mathrm{eq}} \approx 10^{-10}$) and in an effective quantum field theory set-up. Consider microscopic quantum degrees of freedom described by states with proper momentum $p < p_{\max}$, and assume a uniform distribution of the number of states per unit volume in momentum space, $\mathrm{d}^3\mathbf{p}\, n(p) \sim \mathrm{d}^3\mathbf{p}$. Then, the vacuum energy density due to these quantum states is (see Sect. 7.1)

$$\rho_{\mathrm{vac}} \sim \rho_{\mathrm{vac},0} \sim \int_0^{p_{\max}} \mathrm{d}p\, p^3 \sim p_{\max}^4\,, \tag{2.123}$$

where $p = |\mathbf{p}|$ and $p_{\max}$ is a cut-off above which the effective theory breaks down. We would expect that

$$p_{\max} \sim m_{\mathrm{Pl}} \sim 10^{19}\,\mathrm{GeV} \tag{2.124}$$

at the Planck scale, or that

$$p_{\max} = M_{\rm SUSY} > 10^{-16}\, m_{\rm Pl} \sim 10^3\,{\rm GeV} \tag{2.125}$$

for supersymmetry. If, as suggested above, Λ is the energy contribution of vacuum, then observations impose that $\rho_{\rm vac}^{1/4} = \rho_\Lambda^{1/4} \sim 10^{-31} m_{\rm Pl} \sim 10^{-3}\,{\rm eV}$, much smaller than (2.124) or (2.125). Therefore, there must be a contribution V_0 to the potential of the effective theory cancelling vacuum effects almost exactly. A calculation in the Standard Model of particles predicts too small a V_0, insufficient to meet the experimental value (2.118). On the other hand, supersymmetry overshoots the target and vacuum effects are cancelled exactly ($\Lambda = 0$); however, supersymmetry is broken at low energy scales, thus reducing but not removing the fine tuning. This is one formulation of the cosmological constant problem, which we will study in greater detail in Chap. 7.

Also, the universe has been accelerating only since $\ddot{a} = 0$, at (varying-w_Λ parametrization [56])

$$z_{\rm acc} = 0.81 \pm 0.30 \quad (95\,\%\ {\rm CL})\,, \tag{2.126}$$

that is, 6.8 ± 1.4 billion years ago. For a constant $\Lambda = \Lambda_0$, that epoch would be marked as somewhat special in the history of the universe, because it would be triggered by a non-thermodynamical effect. In other words, the model would depend on the initial condition $\Lambda = \Lambda_0$ (*coincidence problem*).

Notice that $z_{\rm acc}$ is larger than the redshift z_Λ at which $\rho_{\rm m} = \rho_\Lambda$: the onset of acceleration is before the cosmological constant dominates over matter (see [56] for a discussion). Figure 2.8 shows the evolution of the matter, radiation, and Λ energy densities.

There may be other candidates, cloaked under the mysterious name of *dark energy*, driving acceleration and accounting for the contribution (2.116) (see Problem 2.9). In the vacuum interpretation, the cosmological "constant" naturally varies with time, since the vacuum energy is temperature dependent. A running $\Lambda(t)$ better accommodates the coincidence problem and can be realized by a number of mechanisms. One of the most popular is *quintessence*, a dynamical scalar field operating like the inflaton but at different energy scales. The equation of state of a single homogeneous scalar spans the whole range of the DEC and, in general, models dominated by an energy component with $w < -1/3$ always accelerate:

$$\frac{\ddot{a}}{a} = -\frac{1}{2}H^2(1+3w)\Omega\,. \tag{2.127}$$

However, due to the observational constraint $w \approx -1$ the quintessence field must be very similar to a cosmological constant and its dynamics is subject to a certain amount of fine tuning.

2.5.4 Spatial Curvature and Topology

The curvature contribution Ω_K can be constrained by combining CMB data (mainly the position of the temperature power spectrum peaks, plus polarization) with distance indicators. In the so-called ΛCDM *concordance model* (Cold Dark Matter with a pure cosmological constant, $w_\Lambda = -1$), one has (PLANCK TT+lowP+lensing+BAO [14])

$$\Omega_{K0} = 0.000 \pm 0.005 \quad (95\,\%\ \text{CL})\,. \tag{2.128}$$

The constraint changes according to data pools, but it is always compatible with flat geometry.

From observations, one can also constrain the "shape of the Universe." A non-trivial cosmic topology [57–62] arises if the Universe is not simply connected. In the latter case, it can have disconnected components or be multi-connected, i.e., some spacetime points are identified. The closed compact (hypersphere), flat and open non-compact FLRW topologies are only three special cases among many other possibilities, including some where spatial sections are flat and compact (3-torus $\mathbb{T}^3$; e.g., [63]), flat and non-compact (some Clifford–Klein spaces [57, 64]) or with positive curvature (some Clifford–Klein spaces or, e.g., Poincaré dodecahedral space [65]).

Detecting topology is difficult because the dynamics is, in general, the same. For example, the hypertorus $\mathbb{T}^3$ is locally isomorphic to three-dimensional Euclidean space, so that spatial flat sections and the 3-torus share the same metric; the only change is in the boundary conditions. If the characteristic curvature scale of the topology is larger than the observed universe, then there is little or no hope to see any effect. However, if the observed patch is larger than physical space, one could observe multiple images of luminous sources in large-scale structures [66] and the CMB [67].

The impact of topology on the cosmic microwave background and its interplay with the inflationary mechanism have been studied extensively [65, 68–99]. A total density parameter close to the critical value $\Omega \sim 1$ implies that many topologies are undetectable, while others are already excluded. For some compact and semi-compact topologies, one can place bounds on their characteristic scale [99]. Although, so far, there is no evidence for a non-trivial topology and the scale of a compact topology is tightly constrained [100, 101], it may be important to keep an eye on this direction. A Universe with non-trivial topology can arise, for instance, via quantum creation in Wheeler–DeWitt quantum cosmology [63, 90, 102–104], in string theory [105, 106] or in quantum gravity and quantum cosmology models of "third quantization" (Sect. 10.2.4). The discovery of an imprint of a specific cosmic topology could be relevant in the big quest of a quantum theory of Nature, since it could justify, constrain or refine some of the models mentioned in this book.

2.6 An Obscure Big Picture

To summarize, the total density parameter measured today is

$$\Omega_0 = \Omega_{K0} + \Omega_{r0} + (\Omega_{b0} + \Omega_{c0}) + \Omega_{\Lambda 0}\,. \tag{2.129}$$

Using the estimate of the Hubble parameter h in (2.12), and neglecting radiation and curvature contributions (which amount to less than 1 % of the total), one ends up with the following picture:

$$\Omega_{b0} \sim 4.8\,\%\,, \qquad \Omega_{c0} \sim 26\,\%\,, \qquad \Omega_{m0} \sim 31\,\%\,, \qquad \Omega_{\Lambda 0} \sim 69\,\%\,. \tag{2.130}$$

We do not have striking evidence in favour of any specific model of cold dark matter or dark energy over the others, so we could say that we know less than 5 % of our world! The first cosmological models did not have dark components at all, but observations forced themselves upon our conception, which was then changed to make room for new ingredients. As Figs. 2.9 and 2.10 illustrate, our picture of the universe has been evolving dramatically in the last fifty years. Science is a measure of our ignorance.

To the best of our knowledge, the concordance model still fits all data without the compelling need to invoke exotic physics such as supersymmetry, string theory or quantized gravity. In this sense, cosmology does not really need quantum gravity at large. However, certain ingredients create a sense of uneasiness in the researcher, mainly for theoretical reasons. The cosmological constant problem, both in its old version of Sect. 2.5.3 and in the new one (Why is $\Omega_{\Lambda 0}$ of the same order of magnitude as Ω_{m0}?) is one example and we will soon see others (the big-bang

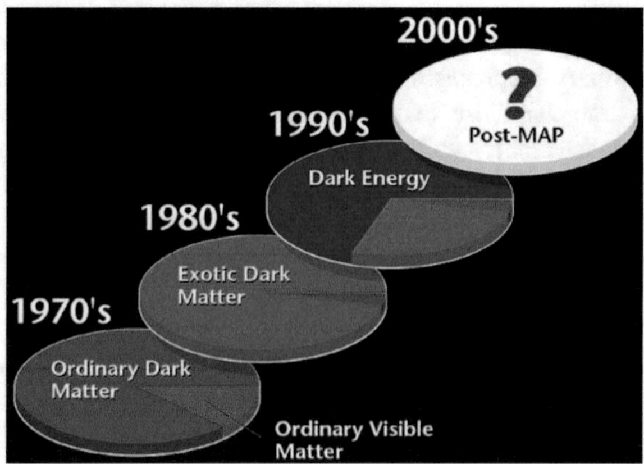

Fig. 2.9 Evolution of our knowledge of the content of the universe from the 1970s (Credit: NASA/WMAP Science Team)

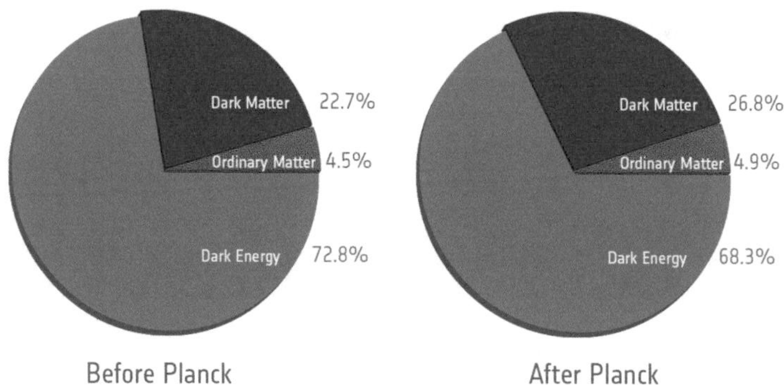

Fig. 2.10 Content of the universe as estimated by WMAP [12] (*left pie chart*) and PLANCK 2013 [107] (*right*) (©ESA and the Planck Collaboration [13])

problem and the inflation-related problems). So, in this other respect, quantum gravity might need cosmology. One duty of modern theoretical physics beyond the Standard Model is to be in agreement with observations, in particular cosmological. Suppose, for instance, one has a model of matter and gravity at hand which can explain a late stage of acceleration (this is a typical goal of phenomenology). One starts from the Friedmann equations or their analogue in, e.g., synchronous time, which is an unphysical parameter. Switching to expressions in redshift z such as (2.87) and (2.88), rescaling the energy densities $\rho_i \to \Omega_i$ and plugging the measured values Ω_{i0} for all the components inserted "by hand" (e.g., matter and radiation), one can follow the dynamical evolution from large redshift until today or to the future, and constrain the parameters of the model (in this example, those determining the dark energy density).

However, it is often difficult to falsify phenomenological models on the basis of constraints placed *a posteriori*. It would be greatly desirable to construct models with *predictive* power, i.e., with few or no free parameters, or such that its free parameters determine the cosmological observables only within certain intervals. Therefore, a second, tremendously challenging task of theoretical physics is to enhance its predictive power by asking the right questions on one hand (What are the physical observables? How can we formulate the cosmological constant and big-bang problems in a non-misleading, operational way?) and looking towards the right directions on the other hand: Does the model agree with observations? How are dark energy and inflation realized? Can we build an arsenal of smoking guns to favour one model against another?

2.7 Problems and Solutions

2.1 Age of the universe. Calculate the formula for the age of the universe in the presence of matter and a cosmological constant. Compare it with the estimate $t_0 \sim H_0^{-1}$ given in (2.13) (ignore error bars).

Solution From (2.81) and (2.88), it follows that

$$H^2 = \frac{\kappa^2}{3}\left(\rho_m + \frac{\Lambda}{\kappa^2}\right) = H^2\left(\Omega_m + \Omega_\Lambda\right) = H_0^2\left(1 - \Omega_{m0} + \Omega_{m0}a^{-3}\right).$$

In the last step, we have used the fact that $\Omega_{\Lambda 0} = 1 - \Omega_{m0}$. The above equation can be integrated from the beginning ($a(t_i) = a_i$) until today ($a_0 = 1$). Since $t_i \ll t_0$, one then has $a_i \approx 0$ and

$$\begin{aligned} t_0 &= \frac{1}{H_0}\int_0^1 \frac{da}{\sqrt{(1-\Omega_{m0})a^2 + \Omega_{m0}a^{-1}}} \\ &= \frac{1}{H_0}\frac{2}{3\sqrt{1-\Omega_{m0}}}\ln\left(\frac{1+\sqrt{1-\Omega_{m0}}}{\sqrt{\Omega_{m0}}}\right) \\ &\approx 0.955 \times \frac{1}{H_0} \\ &\approx 13.79\,\text{Gyr}, \end{aligned}$$

where we used (2.12), (2.13) and (2.99). This estimate is 96 % the crude one $t_0 \sim H_0^{-1}$ and is very close to (2.14). Adding also radiation does not change the numbers because the radiation-dominated era is only a small fraction of the present age.

2.2 Equality and decoupling. An important epoch in the history of the early universe is *decoupling*, when radiation last scattered with matter. Observations center this period at

$$z_{\rm dec} \approx 1091\,. \qquad (2.131)$$

How old was the universe at $z_{\rm dec}$ and $z_{\rm eq}$? Use the experimental estimate of the age of the universe today t_0 and ignore the errors.

Solution We must express the redshift z as a function of time t, and to do so we have to choose a profile $a(t)$. For redshifts $z \gtrsim z_{\text{eq}}$, we can use (2.93):

$$z(t) + 1 = \frac{a_0}{a(t)} = \left(\frac{t_0}{t}\right)^p \qquad \Rightarrow \qquad t = \frac{t_0}{[z(t) + 1]^{1/p}} \,. \tag{2.132}$$

The universe is dominated by dust and $p = 2/3$. For the age given in (2.14), $t_0 \approx$ 13.8 Gyr, one obtains

$$t_{\text{dec}} \approx 3.83 \times 10^5 \text{ yr} \,. \tag{2.133}$$

At decoupling time, the universe was about 380,000 years old. Equation (2.91) is still an acceptable approximation near matter-radiation equality,

$$t_{\text{eq}} \approx 7.0 \times 10^4 \text{ yr} \,. \tag{2.134}$$

2.3 Cosmological horizons 1. Describe the behaviour of the proper and comoving particle horizons for flat power-law cosmologies (2.93) with $0 < p < 1$ and a barotropic fluid (no cosmological constant).

Solution From the Friedmann equations, one can see that the Hubble parameter always decreases if $w > -1$:

$$\dot{H} = -\frac{3}{2}H^2(1 + w) \,. \tag{2.135}$$

Therefore, the Hubble horizon always increases. In particular, for $w > -1/3$ ($0 < p < 1$) the particle horizon is well defined and follows the same evolution as R_H. From (2.95) and $t_{\text{i}} = 0$,

$$R_{\text{p}} = \frac{t}{1-p} = \frac{p}{1-p}\frac{1}{H} \,. \tag{2.136}$$

Also the comoving particle horizon increases with time,

$$\tau = \frac{R_{\text{p}}}{a} = \frac{t^{1-p}}{1-p} = \frac{p}{1-p} r_H \,, \tag{2.137}$$

but at a lower rate.

2.4 Cosmological horizons 2. Determine the recession speed (in c units) of the particle horizon for power-law cosmologies (2.93) with $0 < p < 1$ and a barotropic fluid (no cosmological constant). Does the recession speed violate any principle of special or general relativity?

Solution From (2.136),

$$\dot{R}_\mathrm{p} = \frac{c}{1-p} > c\,. \tag{2.138}$$

This result does not violate special relativity, because the latter is valid only in local inertial frames and does not apply to relative speeds of distant objects. On the other hand, one requires that signals do not travel faster than light, but no signal is interchanged between us and the horizon.

2.5 Horizons and distances 1. Determine the size of the particle and Hubble horizons at z_eq, z_dec and today for a matter-dominated universe. How much did the observable universe increase from decoupling to equality and from equality until today?

Solution At these redshifts, (2.91) is a very good approximation of the cosmic expansion. From (2.95) and (2.132), one has

$$z+1 = \left(\frac{H}{H_0}\right)^p \qquad \Rightarrow \qquad \frac{1}{H} = \frac{1}{H_0(z+1)^{1/p}}\,. \tag{2.139}$$

We know that

$$\frac{1}{H_0} \approx 3h^{-1}\,\mathrm{Gpc} \approx 4.42\,\mathrm{Gpc}\,. \tag{2.140}$$

Therefore, for $p = 2/3$ the size of the particle horizon is $R_\mathrm{p} = 2/H$ and

$$R_\mathrm{p}(z_\mathrm{eq}) \approx 45\,\mathrm{kpc}\,, \qquad R_\mathrm{p}(z_\mathrm{dec}) \approx 245\,\mathrm{kpc}\,, \qquad R_\mathrm{p}(0) \approx 8.8\,\mathrm{Gpc}\,.$$

The Hubble horizon H^{-1} is simply half the particle horizon. The growth rate of the horizons from redshift z_1 to redshift z_2 is

$$\frac{R_\mathrm{p}(z_2)}{R_\mathrm{p}(z_1)} = \frac{R_H(z_2)}{R_H(z_1)} = \left(\frac{1+z_1}{1+z_2}\right)^{3/2}, \tag{2.141}$$

so that

$$\frac{R_{\rm p}(z_{\rm dec})}{R_{\rm p}(z_{\rm eq})} \approx 5.5\,, \qquad \frac{R_{\rm p}(0)}{R_{\rm p}(z_{\rm dec})} \approx 36 \times 10^3\,, \qquad \frac{R_{\rm p}(0)}{R_{\rm p}(z_{\rm eq})} \approx 2 \times 10^5\,.$$

The comoving particle horizon is

$$\tau(z) = (1+z)R_{\rm p}(z) = \frac{1}{H_0}(z+1)^{1-\frac{1}{p}}\,. \tag{2.142}$$

It follows that $\tau_0 = R_{\rm p}(0)$, while

$$\tau_{\rm eq} \approx 152\,{\rm Mpc}\,, \qquad \tau_{\rm dec} \approx 268\,{\rm Mpc}\,.$$

The growth ratios are

$$\frac{\tau_{\rm dec}}{\tau_{\rm eq}} \approx 1.8\,, \qquad \frac{\tau_0}{\tau_{\rm dec}} \approx 33\,, \qquad \frac{\tau_0}{\tau_{\rm eq}} \approx 58\,.$$

2.6 Horizons and distances 2. Repeat the previous exercise for the ΛCDM model with radiation included. Find analytic expressions for the correction factors.

Solution A better estimate of particle horizons should also take radiation and the cosmological constant into account:

$$\begin{aligned}\tau(z) &= \int_0^{a(z)} \frac{{\rm d}a}{Ha^2}\\ &= \frac{1}{H_0}\int_0^{(1+z)^{-1}} \frac{{\rm d}a}{\sqrt{(1-\Omega_{\rm m0}-\Omega_{\rm r0})a^4 + \Omega_{\rm m0}a + \Omega_{\rm r0}}}\,.\end{aligned} \tag{2.143}$$

The integral can be done exactly in certain regimes. At high redshifts $10 < z < 10^4$, one can ignore the cosmological constant and obtain

$$\begin{aligned}\tau(z) &\simeq \frac{1}{H_0}\int_0^{(1+z)^{-1}} \frac{{\rm d}a}{\sqrt{\Omega_{\rm m0}a + \Omega_{\rm r0}}}\\ &= \tau_{\rm CDM}(z)\,\frac{\sqrt{\Omega_{\rm r0}(1+z)+\Omega_{\rm m0}} - \sqrt{\Omega_{\rm r0}(1+z)}}{\Omega_{\rm m0}}\,,\end{aligned} \tag{2.144}$$

where

$$\tau_{\rm CDM}(z) = \frac{2}{H_0\sqrt{1+z}}\,. \tag{2.145}$$

In particular, the correction factors with respect to those found in the previous exercise are

$$\tau_{\rm eq} \approx 0.89\,\tau_{\rm CDM,eq}\,, \qquad \tau_{\rm dec} \approx 1.18\,\tau_{\rm CDM,dec}\,. \tag{2.146}$$

As expected, the correction factors are close to 1.

At late times, dark energy is causing the recent expansion of the universe to be greater than in the past. Therefore, we expect to obtain a higher value for τ_0. In fact,

$$\begin{aligned}\tau(z) &= \frac{1}{H_0}\int_0^{(1+z)^{-1}} \frac{da}{\sqrt{(1-\Omega_{\rm m0})a^4+\Omega_{\rm m0}a}}\\ &= \tau_{\rm CDM,0}\frac{1}{\sqrt{\Omega_{\rm m0}}}F\left[\frac{1}{6},\frac{1}{2};\frac{7}{6};-\frac{1-\Omega_{\rm m0}}{\Omega_{\rm m0}(1+z)^3}\right],\end{aligned} \tag{2.147}$$

where F is the *hypergeometric function*

$$F(a,b;c;x) := \sum_{n=0}^{\infty}\frac{\Gamma(a+n)\Gamma(b+n)}{\Gamma(a)\Gamma(b)}\frac{\Gamma(c)}{\Gamma(c+n)}\frac{x^n}{n!}\,. \tag{2.148}$$

Today, $\tau_0 \approx 1.63\,\tau_{\rm CDM,0}$. The effect, actually, turns out to be quite large because the universe has been accelerating for about half its age. Including all contributions (radiation, matter, Λ), (2.143) yields

$$\tau_0 \approx 1.61\,\tau_{\rm CDM,0}\,. \tag{2.149}$$

Using (2.143), we summarize the results in Table 2.1.

The growth ratios do not change much and are

$$\frac{\tau_{\rm dec}}{\tau_{\rm eq}} \approx 2.3\,, \qquad \frac{\tau_0}{\tau_{\rm dec}} \approx 45\,, \qquad \frac{\tau_0}{\tau_{\rm eq}} \approx 106\,, \tag{2.150}$$

Table 2.1 Comoving and proper particle horizons

z	τ	$R_{\rm p}$
$z_{\rm eq}$	134 Mpc	39 kpc
$z_{\rm dec}$	315 Mpc	289 kpc
0	14.2 Gpc	

and

$$\frac{R_\mathrm{p}(z_\mathrm{dec})}{R_\mathrm{p}(z_\mathrm{eq})} \approx 7.3\,, \qquad \frac{R_\mathrm{p}(0)}{R_\mathrm{p}(z_\mathrm{dec})} \approx 5\times 10^4\,, \qquad \frac{R_\mathrm{p}(0)}{R_\mathrm{p}(z_\mathrm{eq})} \approx 3.6\times 10^5\,. \tag{2.151}$$

From matter-radiation equality until today, the linear size of the causal patch (observable universe) has been increasing 360,000 times.

2.7 Distances. Determine the distance of luminous objects at redshift z. Write an approximate formula for $z \ll 1$. How far are sources at z_dec in a flat universe? And galaxies at $z = 0.2$?

Solution The comoving distance $\chi(t)$ of an object at redshift $z(t)$ from us is the distance light covered from time t until today. This is related to the particle horizon via the photon geodesic equation. In conformal time ($c = 1$),

$$0 = \mathrm{d}s^2 = a^2(\tau)\left(-\mathrm{d}\tau^2 + \frac{\mathrm{d}r^2}{1-\mathrm{K}r^2}\right)\,, \tag{2.152}$$

where r is the spatial coordinate interval. Taking the square root and integrating,

$$\int_\tau^{\tau_0} \mathrm{d}\tau' = \int_0^\chi \frac{\mathrm{d}r}{\sqrt{1-\mathrm{K}r^2}}\,. \tag{2.153}$$

Inverting with respect to χ,

$$\chi(\tau) = \begin{cases} |\mathrm{K}|^{-1/2}\sinh\left[|\mathrm{K}|^{1/2}(\tau_0-\tau)\right]\,, & \mathrm{K}<0 \\ \tau_0-\tau\,, & \mathrm{K}=0\;. \\ \mathrm{K}^{-1/2}\sin\left[\mathrm{K}^{1/2}(\tau_0-\tau)\right]\,, & \mathrm{K}>0 \end{cases} \tag{2.154}$$

In a flat or closed universe, signals emitted inside the particle horizon could not have travelled a distance greater than the radius of the horizon today.

For high redshifts, the comoving distance in a *flat* universe is well approximated by the particle horizon today. For instance, at $z = z_\mathrm{dec}$ the comoving horizon is one order of magnitude smaller that the horizon today τ_0 (see (2.150)) and the comoving distance of the $z = z_\mathrm{dec}$ surface (*last-scattering surface*) is

$$\chi(z_\mathrm{dec}) \simeq \tau_0 \approx 9.6h^{-1}\,\mathrm{Gpc} \approx 14.2\,\mathrm{Gpc}\,, \tag{2.155}$$

where we presented the result also in the standard form with the h factor reinstated.

The proper distance

$$d(z) := a(z)\chi(z) = (1+z)\,\chi(z) \tag{2.156}$$

$$= \frac{R_{\rm p}(0)}{1+z} - R_{\rm p}(z) \tag{2.157}$$

is always smaller than χ. We will see in Chap. 4 that $z_{\rm dec}$ corresponds to the time when the CMB was originated. Since the CMB is isotropic only in the comoving coordinate frame,[7] statements about the "distance" of the last scattering surface implicitly refer to the comoving distance $\chi(z)$.

At small redshift, curvature effects can be neglected. The proper and comoving distances are about the same and one can expand (2.157) to get

$$d(z) = a_0\chi(z) + O(z^2) = -\frac{c}{H_0 a_0}\partial_z a(z)\big|_{z=0} z + O(z^2)$$

$$= \frac{cz}{H_0} + O(z^2) \approx 3z\,h^{-1}\,\text{Gpc}\,. \tag{2.158}$$

Objects at $z = 0.2$ are $600\,h^{-1}$ Mpc away from us. The 2dFGRS survey covers redshifts $z \lesssim 0.3$, so that the largest redshift observed by the survey corresponds to a distance of about 1.3 Gpc. The SDSS main galaxy sample contains galaxies with $z \lesssim 0.4$ and quasars as far as $z \sim 5$.

Another distance of great interest in astrophysics is the *luminosity distance*, which can be expressed either via the absolute magnitude $M_{\rm abs}$ of an object and its apparent magnitude $M_{\rm app}$ or via the luminosity L (in Watts) and the energy flux $\mathcal{F}$ (Watts per area):

$$d_L := 10^{\frac{M_{\rm app}-M_{\rm abs}}{5}-5}\,\text{Mpc} = \sqrt{\frac{L}{4\pi\mathcal{F}}}\,. \tag{2.159}$$

The luminosity of certain objects such as type I supernovæ is known. For these "standard candles," the luminosity distance can be determined with a certain accuracy. In these cases, $d_L = \chi(\tau) \simeq \tau_0 - \tau$ and, via (2.156), one can extract valuable information on the expansion properties of the universe.

2.8 Thermal history of the universe. Determine the temperature of radiation at $z_{\rm eq}$, $z_{\rm dec}$ and today. Calculate the age and the order of magnitude of

(continued)

[7] Sometimes, this frame preference is perceived as a contradiction of general relativity. However, global coordinate frames may be defined once we fix our metric. The observed CMB frame is a solution of Einstein's equations.

redshift corresponding to $T \sim 1\,\text{MeV}$ and $T \sim 0.1\text{MeV}$. Assume $g_* = 1$ and round up the numbers from below, since $g_* > 1$.

Solution From (2.103) and (2.104),

$$T_{\text{eq}} \approx 1\,\text{eV} \approx 10^4\,\text{K}\,, \tag{2.160}$$

$$T_{\text{dec}} \approx 0.3\,\text{eV} \approx 3 \times 10^3\,\text{K}\,, \tag{2.161}$$

$$T_0 \approx 3 \times 10^{-4}\,\text{eV} \approx 3\,\text{K}\,. \tag{2.162}$$

More precise numbers can be obtained by inserting realistic values for g_* [108]. In particular, (2.162) agrees with the measured temperature (2.1) of the microwave background.

$T \sim 1\,\text{MeV} \sim 10^{10}\,\text{K}$ is the typical binding energy of nuclei. When the temperature of the universe is lower than that, atomic nuclei are being synthesized. From (2.103) and (2.106) at high redshift ($z \sim 10^4 T\,\text{eV}^{-1}$), we get

$$t \approx 2\,\text{s}\,, \qquad z \sim 10^{10}\,. \tag{2.163}$$

However, the inverse process also occurs and nuclei are destroyed until the universe cools down enough. One can show that ^{4}He nuclei begin to form at about $T \sim 0.1\,\text{MeV}$. This temperature roughly marks the onset of the epoch known as big-bang nucleosynthesis:

$$t_{\text{BBN}} \approx 200\,\text{s}\,, \qquad z_{\text{BBN}} \sim 10^9\,. \tag{2.164}$$

This era begins three minutes after the big bang and lasts about 17 minutes, after which the nuclear fusion reaction rate drops off ($T \sim 20$–$50\,\text{keV}$). Measurements of light elements have confirmed these calculations. Nucleosynthesis has been taking place again in the core of stars since their formation, 100 million years after the big bang.

We summarize the thermal history of the universe in Table 2.2. The particle horizon before equivalence is found from (2.139) and (2.105),

$$R_{\text{p.}} = \frac{1}{H} = \frac{1}{H_0 \Omega_{\text{r0}}^{1/2}(1+z)^2} \approx 6 \times 10^8 z^{-2}\,\text{kpc} \approx 2 \times 10^{25} z^{-2}\,\text{km}\,, \tag{2.165}$$

which also correctly reproduces the large-z asymptotic limit of (2.143).

The reader should not take these values too seriously. Depending on both the available experimental data and the details of the high-energy particle physics involved during the early stages, the big-bang timeline can change from time to time, and from book to book. However, it is still remarkable that we have been able

Table 2.2 Simplified thermal history of the universe from BBN until today. The first line corresponds to the highest energy probed in ground-based laboratories, at the Large Hadron Collider (LHC) [109]. The values of this table should be taken only as indicative

T (K)	T (eV)	t	z	R_p	Event
10^{16}	10^{12}	10^{-12} s	10^{16}	0.2 mm	Highest energy probed in laboratory (LHC).
10^{10}	10^5	2 s	10^{10}	10^5 km	Formation and destruction of nuclei begins.
10^9	10^4	200 s	10^9	10^7 km	**Nucleosynthesis** of light ions.
10^8	3×10^3	20 min	10^8	10^9 km	Big-bang nucleosynthesis ends.
10^4	1	7.0×10^4 yr	3400	39 kpc	Radiation-matter **equality**. Matter domination begins.
3000	0.3	3.8×10^5 yr	1090	289 kpc	**Decoupling** of matter and radiation. CMB forms. Atoms form.
100	10^{-2}	10^8 yr	25	110 Mpc	First stars.
3	3×10^{-4}	14×10^9 yr	0	14.2 Gpc	Today.

to reconstruct the history of the early universe in such a detail from a handful of formulæ. We collect them here again in an approximated fashion, for a radiation-dominated universe:

$$T_\mathrm{K} \sim 10^4 \left(\frac{T_\mathrm{eV}}{1\,\mathrm{eV}}\right) \mathrm{K}\,, \qquad t \sim \left(\frac{10^6\,\mathrm{eV}}{T_\mathrm{eV}}\right)^2 \mathrm{s}\,,$$

$$z \sim 10^4 \left(\frac{T_\mathrm{eV}}{1\,\mathrm{eV}}\right), \qquad R_\mathrm{p} \sim 10^{25} z^{-2}\,\mathrm{km}\,.$$

2.9 Accelerating universe. Consider a $D = 4$ flat universe filled only with a barotropic fluid. Discuss the meaning and behaviour of the parameter

$$\epsilon := -\frac{\mathrm{d}\ln H}{\mathrm{d}\ln a} = -\frac{\dot{H}}{H^2} = 1 - \frac{\ddot{a}}{aH^2}\,. \tag{2.166}$$

Can an accelerating universe be dominated by matter or radiation?

Solution Equation (2.166) is the definition of the so called *first slow-roll parameter*, and is nothing but the recession speed of the particle horizon:

$$\epsilon = \dot{R}_H\,. \tag{2.167}$$

From (2.135), one sees that $\epsilon = 0$ for $w = -1$ (cosmological constant, de Sitter expansion), $\epsilon > 0$ for $w > -1$, and $\epsilon < 0$ for $w < -1$. In the latter case the Hubble parameter (horizon) increases (respectively, decreases) with time, a situation called *super-acceleration*. Moreover, (2.127) and (2.166) yield the additional constraint $\epsilon > 1$ when $w > -1/3$. Summarizing,

$$\epsilon < 0 \qquad \text{if} \qquad w < -1\,, \tag{2.168a}$$

$$\epsilon = 0 \qquad \text{if} \qquad w = -1\,, \tag{2.168b}$$

$$0 < \epsilon < 1 \qquad \text{if} \qquad -1 < w < -\tfrac{1}{3}\,, \tag{2.168c}$$

$$\epsilon > 1 \qquad \text{if} \qquad w > -\tfrac{1}{3}\,. \tag{2.168d}$$

In particular, we have acceleration only if $\epsilon < 1$ and power-law cosmologies (2.93) with $0 < p = 1/\epsilon < 1$ cannot accelerate. Equations (2.168b) and (2.168c) define a period of *inflation* ($\ddot{a} > 0$), while (2.168a) corresponds to *super-inflation* ($\dot{H} > 0$).

2.10 Exact solutions: H = const. Consider a D-dimensional universe with $\Lambda = 0$ filled with a scalar field with potential V. Find the exact solutions of the equations of motion for a constant Hubble parameter, listing $a(t)$, $\epsilon(t)$, $\phi(t)$ and $V(\phi)$. Recast the solutions in conformal time.

Solution The Friedmann and continuity equations are

$$\left(\frac{D}{2}-1\right)H^2 = \frac{\kappa^2}{D-1}\left(\frac{\dot{\phi}^2}{2}+V\right)-\frac{\textsf{K}}{a^2}\,, \tag{2.169}$$

$$H^2+\dot{H} = \frac{\kappa^2}{D-1}\left[\frac{2}{D-2}V-\dot{\phi}^2\right], \tag{2.170}$$

$$0 = \ddot{\phi}+(D-1)H\dot{\phi}+V_{,\phi}\,. \tag{2.171}$$

For a constant Hubble parameter $H(t) = H$,

$$a(t) = \mathrm{e}^{Ht}\,, \qquad \epsilon(t) = 0\,. \tag{2.172}$$

In a flat universe, the exact solution of the Friedmann and scalar equations is just a cosmological constant (de Sitter spacetime),

$$\phi(t) = \phi_0\,, \qquad V(\phi) = \frac{(D-1)(D-2)H^2}{2\kappa^2}\,, \qquad \textsc{k} = 0\,. \tag{2.173}$$

The Friedmann equations show that there is no solution if $\textsc{k} = -1$, while there is one for a closed universe, but only in $D = 4$:

$$\phi_\pm(t) = \pm\sqrt{\frac{2}{\kappa^2 H^2}}\,\mathrm{e}^{-Ht}\,, \qquad \textsc{k} = 1\,,$$

while the continuity equation fixes the potential:

$$V(\phi) = \frac{3H^2}{\kappa^2} + H^2\phi^2\,. \tag{2.174}$$

This solution is *not* de Sitter because spatial sections are not flat. (We recall that the $H =$ const cosmology corresponds mathematically to de Sitter spacetime only if spatial sections are flat. In that case, the de Sitter hyperboloid is only half covered by FLRW coordinates.) The scalar field $\phi_\pm$ rolls down its potential from $t = -\infty$ and climbs it again after passing the global minimum. The solution is actually unique, since cosmological equations of motion are invariant under *time reversal*,

$$t \to -t\,, \tag{2.175}$$

and the direction of the rolling in a symmetric potential does not matter.

We can recast proper-time solutions into expressions in conformal time by inverting $\tau(t)$. For an $H =$ const background,

$$\tau = \int \mathrm{d}t\,\mathrm{e}^{-Ht} = -\frac{\mathrm{e}^{-Ht}}{H}\,, \tag{2.176}$$

so that

$$t = -\frac{\ln(-H\tau)}{H}\,. \tag{2.177}$$

Notice that τ runs from $-\infty$ to 0, so that the above expression is well defined. The geometric background in τ is

$$a(\tau) = \frac{1}{H|\tau|}\,, \qquad \mathcal{H}(\tau) := \frac{a'}{a} = aH = \frac{1}{|\tau|}\,, \tag{2.178}$$

where a prime denotes differentiation with respect to τ. The solution in the closed universe is linear in τ,

$$\phi(\tau) = \phi_0|\tau|\,. \tag{2.179}$$

2.11 Exact solutions: power-law expansion. Repeat the same calculations of the previous exercise for a power-law expansion, $a \propto t^p$, where $p > 0$.

Solution In this case,

$$a(t) = \left(\frac{t}{\bar{t}}\right)^p\,, \qquad H(t) = \frac{p}{t}\,, \qquad \epsilon(t) = \frac{1}{p}\,, \tag{2.180}$$

where $\bar{t}$ is some reference time. Trying the profile $\phi(t) = (\phi_0/q)t^q$ in the (sum of the) Friedmann equations, one finds that it must be $q = 0$. This suggests to consider the limit $q \to 0$, which is a logarithmic profile:

$$\phi(t) = \phi_0 \ln(t/\bar{t})\,. \tag{2.181}$$

From now on, $\bar{t} = 1$. This gives the exponential potential [110–113]

$$V(\phi) = \frac{(D-1)p-1}{2}\phi_0^2\, \mathrm{e}^{-2\phi/\phi_0}\,. \tag{2.182}$$

If the universe is flat, then

$$\phi_0 = \pm\sqrt{\frac{(D-2)p}{\kappa^2}}\,, \qquad \mathrm{K} = 0\,, \tag{2.183}$$

while for a curved universe only the case $p = 1$ is a solution:

$$\phi_0 = \pm\sqrt{\frac{D-2+2\mathrm{K}}{\kappa^2}}\,, \qquad p = 1\,. \tag{2.184}$$

This solution is real only if $D > 2(1-\mathrm{K})$. Therefore, it is always valid for a closed universe, while for an open universe it exists only in $D > 4$.

For a power-law expansion, conformal time is

$$\tau = \int \mathrm{d}t\, t^{-p} = \frac{t^{1-p}}{1-p}\,, \qquad p \neq 1\,, \tag{2.185}$$

which is positive if $p < 1$. Inverting this expression,

$$t = [(1-p)\tau]^{\frac{1}{1-p}} . \tag{2.186}$$

The solution with $K = 0$ is

$$a(\tau) = \left(\frac{\tau}{\tau_0}\right)^{\frac{p}{1-p}} , \qquad \mathcal{H}(\tau) = \frac{p}{1-p}\frac{1}{\tau} , \qquad \phi(\tau) = \phi_0 \ln \tau , \tag{2.187}$$

while for $p = 1$

$$\tau = \int \frac{\mathrm{d}t}{t} = \ln t , \tag{2.188}$$

and the solution is

$$a(\tau) = \mathrm{e}^{\tau} , \qquad \mathcal{H}(\tau) = 1 , \qquad \phi(\tau) = \phi_0 \tau . \tag{2.189}$$

References

1. S. Cole et al. [The 2dFGRS Collaboration], The 2dF Galaxy Redshift Survey: power-spectrum analysis of the final dataset and cosmological implications. Mon. Not. R. Astron. Soc. **362**, 505 (2005). [arXiv:astro-ph/0501174]
2. http://www.aao.gov.au/local/www/6df
3. D. Heath Jones et al. [The 6dFGS Collaboration], The 6dF Galaxy Survey: final redshift release (DR3) and southern large-scale structures. Mon. Not. R. Astron. Soc. **399**, 683 (2009). [arXiv:0903.5451]
4. http://www.sdss3.org, http://classic.sdss.org, http://www.youtube.com/watch?v=08LBltePDZw
5. C.P. Ahn et al. [SDSS Collaboration], The tenth data release of the Sloan Digital Sky Survey: first spectroscopic data from the SDSS-III Apache Point Observatory Galactic Evolution Experiment. Astrophy. J. Suppl. **211**, 17 (2014). [arXiv:1307.7735]
6. M. Betoule et al., Improved photometric calibration of the SNLS and the SDSS supernova surveys. Astron. Astrophys. **552**, A124 (2013). [arXiv:1212.4864]
7. M. Betoule et al. [SDSS Collaboration], Improved cosmological constraints from a joint analysis of the SDSS-II and SNLS supernova samples. Astron. Astrophys. **568**, A22 (2014). [arXiv:1401.4064]
8. "2dfdtfe" by Willem Schaap – http://en.wikipedia.org/wiki/Sloan_Great_Wall. Licensed under Creative Commons Attribution-Share Alike 3.0 via Wikimedia Commons – http://commons.wikimedia.org/wiki/File:2dfdtfe.gif#mediaviewer/File:2dfdtfe.gif
9. http://www2.aao.gov.au/~TDFgg/Public/Pics/2dFzcone.jpg
10. http://map.gsfc.nasa.gov
11. C.L. Bennett et al. [WMAP Collaboration], Nine-year Wilkinson Microwave Anisotropy Probe (WMAP) observations: final maps and results. Astrophys. J. Suppl. **208**, 20 (2013). [arXiv:1212.5225]

12. G. Hinshaw et al. [WMAP Collaboration], Nine-year Wilkinson Microwave Anisotropy Probe (WMAP) observations: cosmological parameter results. Astrophys. J. Suppl. **208**, 19 (2013). [arXiv:1212.5226]
13. http://www.esa.int/Our_Activities/Space_Science/Planck
14. P.A.R. Ade et al. [Planck Collaboration], Planck 2015 results. XIII. Cosmological parameters. Astron. Astrophys. **594**, A13 (2016). [arXiv:1502.01589]
15. J.C. Mather, D.J. Fixsen, R.A. Shafer, C. Mosier, D.T. Wilkinson, Calibrator design for the COBE Far Infrared Absolute Spectrophotometer (FIRAS). Astrophys. J. **512**, 511 (1999). [arXiv:astro-ph/9810373]
16. D.J. Fixsen, The temperature of the cosmic microwave background. Astrophys. J. **707**, 916 (2009). [arXiv:0911.1955]
17. R.P. Kirshner, Hubble's diagram and cosmic expansion. Proc. Natl. Acad. Sci. **101**, 8 (2004)
18. A.G. Riess et al. [Supernova Search Team Collaboration], Observational evidence from supernovae for an accelerating universe and a cosmological constant. Astron. J. **116**, 1009 (1998). [arXiv:astro-ph/9805201]
19. J.L. Tonry et al. [Supernova Search Team Collaboration], Cosmological results from high-z supernovæ. Astrophys. J. **594**, 1 (2003). [arXiv:astro-ph/0305008]
20. R.A. Knop et al. [Supernova Cosmology Project Collaboration], New constraints on Ω_m, Ω_λ, and w from an independent set of eleven high-redshift supernovae observed with HST. Astrophys. J. **598**, 102 (2003). [arXiv:astro-ph/0309368]
21. A.G. Riess et al. [Supernova Search Team Collaboration], Type Ia supernova discoveries at $z > 1$ from the Hubble Space Telescope: evidence for past deceleration and constraints on dark energy evolution. Astrophys. J. **607**, 665 (2004). [arXiv:astro-ph/0402512]
22. W.M. Wood-Vasey et al. [ESSENCE Collaboration], Observational constraints on the nature of the dark energy: first cosmological results from the ESSENCE supernova survey. Astrophys. J. **666**, 694 (2007). [arXiv:astro-ph/0701041]
23. T.M. Davis et al., Scrutinizing exotic cosmological models using ESSENCE supernova data combined with other cosmological probes. Astrophys. J. **666**, 716 (2007). [arXiv:astro-ph/0701510]
24. W. Rindler, Visual horizons in world-models. Mon. Not. R. Astron. Soc. **116**, 662 (1956)
25. A. Einstein, Hamiltonsches Prinzip und allgemeine Relativitätstheorie. Sitz.-ber. Kgl. Preuss. Akad. Wiss. **1916**, 1111 (1916)
26. J.W. York, Role of conformal three-geometry in the dynamics of gravitation. Phys. Rev. Lett. **28**, 1082 (1972)
27. G.W. Gibbons, S.W. Hawking, Action integrals and partition functions in quantum gravity. Phys. Rev. D **15**, 2752 (1977)
28. S.W. Hawking, G.F.R. Ellis, *The Large Scale Structure of Space-Time* (Cambridge University Press, Cambridge, 1973)
29. S.W. Hawking, Perturbations of an expanding universe. Astrophys. J. **145**, 544 (1966)
30. G.F.R. Ellis, Relativistic cosmology, in *General Relativity and Cosmology, Proceedings of the XLVII Enrico Fermi Summer School*, ed. by R.K. Sachs (Academic Press, New York, 1971)
31. G.F.R. Ellis, M. Bruni, Covariant and gauge-invariant approach to cosmological density fluctuations. Phys. Rev. D **40**, 1804 (1989)
32. M.S. Madsen, Scalar fields in curved spacetimes. Class. Quantum Grav. **5**, 627 (1988)
33. S. Weinberg, *Gravitation and Cosmology* (Wiley, New York, 1972)
34. S. Colafrancesco, Dark matter in modern cosmology. AIP Conf. Proc. **1206**, 5 (2010). [arXiv:1004.3869]
35. Z. Ahmed et al. [CDMS-II Collaboration], Results from a low-energy analysis of the CDMS II Germanium data. Phys. Rev. Lett. **106**, 131302 (2011). [arXiv:1011.2482]
36. S. Galli, F. Iocco, G. Bertone, A. Melchiorri, CMB constraints on dark matter models with large annihilation cross-section. Phys. Rev. D **80**, 023505 (2009). [arXiv:0905.0003]
37. W. de Sitter, Einstein's theory of gravitation and its astronomical consequences. Third paper. Mon. Not. R. Astron. Soc. **78**, 3 (1917)

38. R. Utiyama, B.S. DeWitt, Renormalization of a classical gravitational field interacting with quantized matter fields. J. Math. Phys. **3**, 608 (1962)
39. T.P. Singh, T. Padmanabhan, Notes on semiclassical gravity. Ann. Phys. (N.Y.) **196**, 296 (1989)
40. Ya.B. Zel'dovich, The cosmological constant and the theory of elementary particles. Sov. Phys. Usp. **11**, 381 (1968)
41. Ya.B. Zel'dovitch, I.D. Novikov, *Relativistic Astrophysics*, vol. 1 (University of Chicago Press, Chicago, 1971)
42. S. Perlmutter et al. [The Supernova Cosmology Project], Discovery of a supernova explosion at half the age of the universe and its cosmological implications. Nature **391**, 51 (1998). [arXiv:astro-ph/9712212]
43. S. Perlmutter et al. [The Supernova Cosmology Project], Measurements of Ω and Λ from 42 high-redshift supernovæ. Astrophys. J. **517**, 565 (1999). [arXiv:astro-ph/9812133]
44. D.J. Eisenstein et al. [SDSS Collaboration], Detection of the baryon acoustic peak in the large-scale correlation function of SDSS luminous red galaxies. Astrophys. J. **633**, 560 (2005). [arXiv:astro-ph/0501171]
45. W.J. Percival et al. [SDSS Collaboration], Baryon acoustic oscillations in the Sloan Digital Sky Survey Data Release 7 galaxy sample. Mon. Not. R. Astron. Soc. **401**, 2148 (2010). [arXiv:0907.1660]
46. N.G. Busca et al., Baryon acoustic oscillations in the Ly-α forest of BOSS quasars. Astron. Astrophys. **552**, A96 (2013). [arXiv:1211.2616]
47. L. Anderson et al. [BOSS Collaboration], The clustering of galaxies in the SDSS-III Baryon Oscillation Spectroscopic Survey: baryon acoustic oscillations in the Data Release 10 and 11 galaxy samples. Mon. Not. R. Astron. Soc. **441**, 24 (2014). [arXiv:1312.4877]
48. A.J. Ross, L. Samushia, C. Howlett, W.J. Percival, A. Burden M. Manera, The clustering of the SDSS DR7 main galaxy sample I: a 4 per cent distance measure at $z = 0.15$. Mon. Not. R. Astron. Soc. **449**, 835 (2015). [arXiv:1409.3242]
49. M. Tegmark et al. [SDSS Collaboration], Cosmological constraints from the SDSS luminous red galaxies. Phys. Rev. D **74**, 123507 (2006). [arXiv:astro-ph/0608632]
50. L.-M. Wang, P.J. Steinhardt, Cluster abundance constraints on quintessence models. Astrophys. J. **508**, 483 (1998). [arXiv:astro-ph/9804015]
51. S. Tsujikawa, A. De Felice, J. Alcaniz, Testing for dynamical dark energy models with redshift-space distortions. JCAP **1301**, 030 (2013). [arXiv:1210.4239]
52. M. Chevallier, D. Polarski, Accelerating universes with scaling dark matter. Int. J. Mod. Phys. D **10**, 213 (2001). [arXiv:gr-qc/0009008]
53. E.V. Linder, Exploring the expansion history of the universe. Phys. Rev. Lett. **90**, 091301 (2003). [arXiv:astro-ph/0208512]
54. T. Padmanabhan, H. Padmanabhan, Cosmological constant from the emergent gravity perspective. Int. J. Mod. Phys. D **23**, 1430011 (2014). [arXiv:1404.2284]
55. A. Mazumdar, The origin of dark matter, matter-anti-matter asymmetry, and inflation. arXiv:1106.5408
56. A. Melchiorri, L. Pagano, S. Pandolfi, When did cosmic acceleration start? Phys. Rev. D **76**, 041301 (2007). [arXiv:0706.1314]
57. G.F.R. Ellis, Topology and cosmology. Gen. Relat. Grav. **2**, 7 (1971)
58. M. Lachièze-Rey, J.-P. Luminet, Cosmic topology. Phys. Rep. **254**, 135 (1995). [arXiv:gr-qc/9605010]
59. G.D. Starkman, Topology and cosmology. Class. Quantum Grav. **15**, 2529 (1998)
60. J.-P. Luminet, B.F. Roukema, Topology of the universe: theory and observation, in *Proceedings of the NATO Advanced Study Institute on Theoretical and Observational Cosmology*, ed. by M. Lachièze-Rey (Kluwer, Dordrecht, 1999); NATO Sci. Ser. C **541**, 117 (1999). [arXiv:astro-ph/9901364]
61. J.J. Levin, Topology and the cosmic microwave background. Phys. Rep. **365**, 251 (2002). [arXiv:gr-qc/0108043]

62. J.-P. Luminet, Cosmic topology: twenty years after. Grav. Cosmol. **20**, 15 (2014). [arXiv:1310.1245]
63. Ya.B. Zeldovich, A.A. Starobinsky, Quantum creation of a universe in a nontrivial topology. Sov. Astron. Lett. **10**, 135 (1984)
64. B.S. DeWitt, C.F. Hart, C.J. Isham, Topology and quantum field theory. Physica A **96**, 197 (1979)
65. J.-P. Luminet, J. Weeks, A. Riazuelo, R. Lehoucq, J.-P. Uzan, Dodecahedral space topology as an explanation for weak wide-angle temperature correlations in the cosmic microwave background. Nature **425**, 593 (2003). [arXiv:astro-ph/0310253]
66. E. Gausmann, R. Lehoucq, J.-P. Luminet, J.-P. Uzan, J. Weeks, Topological lensing in spherical spaces. Class. Quantum Grav. **18**, 5155 (2001). [arXiv:gr-qc/0106033]
67. R. Lehoucq, J. Weeks, J.-P. Uzan, E. Gausmann, J.-P. Luminet, Eigenmodes of three-dimensional spherical spaces and their application to cosmology. Class. Quantum Grav. **19**, 4683 (2002). [arXiv:gr-qc/0205009]
68. D.D. Sokolov, V.F. Shvartsman, An estimate of the size of the universe from a topological point of view. Zh. Eksp. Teor. Fiz. **66**, 412 (1974) [Sov. Phys. JETP **39**, 196 (1975)]
69. D.D. Sokolov, A.A. Starobinsky, Globally inhomogeneous "spliced" universes. Sov. Astron. **19**, 629 (1976)
70. N.J. Cornish, D.N. Spergel, G.D. Starkman, Does chaotic mixing facilitate $\Omega < 1$ inflation? Phys. Rev. Lett. **77**, 215 (1996). [arXiv:astro-ph/9601034]
71. J.J. Levin, E. Scannapieco, J. Silk, The topology of the universe: the biggest manifold of them all. Class. Quantum Grav. **15**, 2689 (1998). [arXiv:gr-qc/9803026]
72. G.I. Gomero, A.F.F. Teixeira, M.J. Rebouças, A. Bernui, Spikes in cosmic crystallography. Int. J. Mod. Phys. D **11**, 869 (2002). [arXiv:gr-qc/9811038]
73. J.R. Bond, D. Pogosian, T. Souradeep, CMB anisotropy in compact hyperbolic universes. 1. Computing correlation functions. Phys. Rev. D **62**, 043005 (2000). [arXiv:astro-ph/9912124]
74. J.R. Bond, D. Pogosian, T. Souradeep, CMB anisotropy in compact hyperbolic universes. 2. COBE maps and limits. Phys. Rev. D **62**, 043006 (2000). [arXiv:astro-ph/9912144]
75. J. Barrow, H. Kodama, The isotropy of compact universes. Class. Quantum Grav. **18**, 1753 (2001). [arXiv:gr-qc/0012075]
76. G.I. Gomero, M.J. Rebouças, R.K. Tavakol, Detectability of cosmic topology in almost flat universes. Class. Quantum Grav. **18**, 4461 (2001). [arXiv:gr-qc/0105002]
77. J.D. Barrow, H. Kodama, All universes great and small. Int. J. Mod. Phys. D **10**, 785 (2001). [arXiv:gr-qc/0105049]
78. G.I. Gomero, M.J. Rebouças, R.K. Tavakol, Are small hyperbolic universes observationally detectable? Class. Quantum Grav. **18**, L145 (2001). [arXiv:gr-qc/0106044]
79. G.I. Gomero, M.J. Rebouças, Detectability of cosmic topology in flat universes. Phys. Lett. A **311**, 319 (2003). [arXiv:gr-qc/0202094]
80. J. Weeks, R. Lehoucq, J.-P. Uzan, Detecting topology in a nearly flat spherical universe. Class. Quantum Grav. **20**, 1529 (2003). [arXiv:astro-ph/0209389]
81. G.I. Gomero, M.J. Rebouças, R. Tavakol, Limits on the detectability of cosmic topology in hyperbolic universes. Int. J. Mod. Phys. A **17**, 4261 (2002). [arXiv:gr-qc/0210016]
82. J.R. Weeks, Detecting topology in a nearly flat hyperbolic universe. Mod. Phys. Lett. A **18**, 2099 (2003). [arXiv:astro-ph/0212006]
83. A. Riazuelo, J.-P. Uzan, R. Lehoucq, J. Weeks, Simulating cosmic microwave background maps in multi-connected spaces. Phys. Rev. D **69**, 103514 (2004). [arXiv:astro-ph/0212223]
84. A. de Oliveira-Costa, M. Tegmark, M. Zaldarriaga, A. Hamilton, The significance of the largest scale CMB fluctuations in WMAP. Phys. Rev. D **69**, 063516 (2004). [arXiv:astro-ph/0307282]
85. B. Mota, M.J. Rebouças, R. Tavakol, Constraints on the detectability of cosmic topology from observational uncertainties. Class. Quantum Grav. **20**, 4837 (2003). [arXiv:gr-qc/0308063]
86. B. Mota, G.I. Gomero, M.J. Rebouças, R. Tavakol, What do very nearly flat detectable cosmic topologies look like? Class. Quantum Grav. **21**, 3361 (2004). [arXiv:astro-ph/0309371]

87. N.J. Cornish, D.N. Spergel, G.D. Starkman, E. Komatsu, Constraining the topology of the universe. Phys. Rev. Lett. **92**, 201302 (2004). [arXiv:astro-ph/0310233]
88. A. Riazuelo, J. Weeks, J.-P. Uzan, R. Lehoucq, J.-P. Luminet, Cosmic microwave background anisotropies in multi-connected flat spaces. Phys. Rev. D **69**, 103518 (2004). [arXiv:astro-ph/0311314]
89. B.F. Roukema, B. Lew, M. Cechowska, A. Marecki, S. Bajtlik, A hint of Poincaré dodecahedral topology in the WMAP first year sky map. Astron. Astrophys. **423**, 821 (2004). [arXiv:astro-ph/0402608]
90. A.D. Linde, Creation of a compact topologically nontrivial inflationary universe. JCAP **0410**, 004 (2004). [arXiv:hep-th/0408164]
91. R. Aurich, S. Lustig, F. Steiner, CMB anisotropy of the Poincaré dodecahedron. Class. Quantum Grav. **22**, 2061 (2005). [arXiv:astro-ph/0412569]
92. B. Mota, M.J. Rebouças, R. Tavakol, The local shape of the universe in the inflationary limit. Int. J. Mod. Phys. A **20**, 2415 (2005). [arXiv:astro-ph/0503683]
93. R. Aurich, S. Lustig, F. Steiner, CMB anisotropy of spherical spaces. Class. Quantum Grav. **22**, 3443 (2005). [arXiv:astro-ph/0504656]
94. R. Aurich, S. Lustig, F. Steiner, The circles-in-the-sky signature for three spherical universes. Mon. Not. R. Astron. Soc. **369**, 240 (2006). [arXiv:astro-ph/0510847]
95. M. Kunz, N. Aghanim, L. Cayon, O. Forni, A. Riazuelo, J.-P. Uzan, Constraining topology in harmonic space. Phys. Rev. D **73**, 023511 (2006). [arXiv:astro-ph/0510164]
96. R. Aurich, H.S. Janzer, S. Lustig, F. Steiner, Do we live in a small universe? Class. Quantum Grav. **25**, 125006 (2008). [arXiv:0708.1420]
97. B. Mota, M.J. Rebouças, R. Tavakol, Circles-in-the-sky searches and observable cosmic topology in the inflationary limit. Phys. Rev. D **78**, 083521 (2008). [arXiv:0808.1572]
98. B. Mota, M.J. Rebouças, R. Tavakol, Circles-in-the-sky searches and observable cosmic topology in a flat universe. Phys. Rev. D **81**, 103516 (2010). [arXiv:1002.0834]
99. G. Aslanyan, A.V. Manohar, The topology and size of the universe from the cosmic microwave background. JCAP **1206**, 003 (2012). [arXiv:1104.0015]
100. P.A.R. Ade et al. [Planck Collaboration], Planck 2013 results. XXVI. Background geometry and topology of the Universe. Astron. Astrophys. **571**, A26 (2014). [arXiv:1303.5086]
101. P.A.R. Ade et al. [Planck Collaboration], Planck 2015 results. XVIII. Background geometry and topology of the Universe. Astron. Astrophys. **594**, A18 (2016). [arXiv:1502.01593]
102. Yu.P. Goncharov, A.A. Bytsenko, The supersymmetric Casimir effect and quantum creation of the universe with nontrivial topology. Phys. Lett. B **160**, 385 (1985)
103. Yu.P. Goncharov, A.A. Bytsenko, The supersymmetric Casimir effect and quantum creation of the universe with nontrivial topology (II). Phys. Lett. B **169**, 171 (1986)
104. Yu.P. Goncharov, A.A. Bytsenko, Casimir effect in supergravity theories and the quantum birth of the Universe with non-trivial topology. Class. Quantum Grav. **4**, 555 (1987)
105. R.H. Brandenberger, C. Vafa, Superstrings in the early universe. Nucl. Phys. B **316**, 391 (1989)
106. B. McInnes, Inflation, large branes, and the shape of space. Nucl. Phys. B **709**, 213 (2005). [arXiv:hep-th/0410115]
107. P.A.R. Ade et al. [Planck Collaboration], Planck 2013 results. XVI. Cosmological parameters. Astron. Astrophys. **571**, A16 (2014). [arXiv:1303.5076]
108. D.H. Lyth, A.R. Liddle, *The Primordial Density Perturbation* (Cambridge University Press, Cambridge, 2009)
109. http://lhc.web.cern.ch/lhc
110. F. Lucchin, S. Matarrese, Power-law inflation. Phys. Rev. D **32**, 1316 (1985)
111. J.J. Halliwell, Scalar fields in cosmology with an exponential potential. Phys. Lett. B **185**, 341 (1987)
112. J. Yokoyama, K.-i. Maeda, On the dynamics of the power law inflation due to an exponential potential. Phys. Lett. B **207**, 31 (1988)
113. Y. Kitada, K.-i. Maeda, Cosmic no-hair theorem in power-law inflation. Phys. Rev. D **45**, 1416 (1992)

Chapter 3
Cosmological Perturbations

Yet it is possible to see peril in the finding of ultimate perfection. It is clear that the ultimate pattern contains its own fixity. In such perfection, all things move toward death.

— Frank Herbert, *Dune*

Contents

G. Calcagni, *Classical and Quantum Cosmology*, Graduate Texts in Physics,
DOI 10.1007/978-3-319-41127-9_3

The cosmological principle is an idealization of what we observe in nature. At late times, matter does not occupy the whole space uniformly due to gravitational clustering. Near the big bang, a homogeneous and isotropic universe is not the most general initial condition. Even if one chose FLRW as the classical initial state at very early times, quantum fluctuations of particle fields would be of the same order of the particle horizon, and they would give rise to strong inhomogeneities in the metric and matter energy density. On the other hand, CMB anisotropies are rather small and a perfect FLRW background is still a good lowest-order approximation. A reasonable hypothesis, verified by experiments, is that CMB anisotropies come from these metric and density perturbations. Because the effect is so small, the formalism of *linear perturbations* is well suited for our purpose [1–4]. One should also explain why *non-linear* inhomogeneities [5] do not play a major role in the early universe. We shall postpone the issue to Chap. 5. In parallel, late-time non-linear large-scale structures and the possible detection of tiny primordial non-linear effects require to go beyond the first perturbative order. In preparation of this, we review some results on linear and non-linear perturbation theory.

3.1 Metric Perturbations

According to the background field method, we decompose the metric in two parts, an unperturbed background and a perturbation:

$$g_{\mu\nu} = \tilde{g}_{\mu\nu} + h_{\mu\nu} \,. \tag{3.1}$$

In this chapter, the background $\tilde{g}_{\mu\nu}$ is assumed to be a *flat* FLRW metric. The assumption $\textsc{k} = 0$ is made both for simplicity and because the universe can be argued to have been effectively almost flat during most part of its life.[1] The perturbation $h_{\mu\nu}$ should not be confused with the spatial projection tensor (2.41).

3.1.1 Linearized Einstein Equations

We now calculate the variation of the Riemann tensor, the Ricci tensor and the Ricci scalar with respect to the generic symmetric linear perturbation (3.1) of the D-dimensional background metric $\tilde{g}_{\mu\nu}$.[2] Preservation of the trace equation $\tilde{g}_{\mu\nu}\tilde{g}^{\mu\nu} =$

[1]This actually constitutes a major problem in the hot big bang model, which we shall meet soon. The problem will be solved eventually, hence the flatness assumption is justified.

[2]The calculus of metric variations runs along exactly the same steps when $h_{\mu\nu}$ is interpreted as an infinitesimal variation of the metric rather than a physical metric perturbation. In that case, $\delta g_{\mu\nu} \longleftrightarrow h_{\mu\nu}$, $\delta g^{\mu\nu} \longleftrightarrow -h^{\mu\nu}$ and (2.22) stems from (3.87).

$D = g_{\mu\nu}g^{\mu\nu}$ implies that the inverse metric is

$$g^{\mu\nu} = \tilde{g}^{\mu\nu} - h^{\mu\nu}, \tag{3.2}$$

where $h^{\mu\nu} = \tilde{g}^{\mu\rho}\tilde{g}^{\nu\sigma}h_{\rho\sigma}$. With this definition of the perturbed inverse metric, $h_{\mu\nu}$ is a genuine tensor and indices can be lowered and raised without changing the overall sign,

$$A^{\mu\nu\rho\cdots}h_{\mu\nu} = A_{\mu\nu}{}^{\rho\cdots}h^{\mu\nu} = A_{\mu}{}^{\nu\cdots}h^{\mu}{}_{\nu}.$$

In the following, a tilde will denote quantities in the background metric, while indices in brackets represent symmetrized and anti-symmetrized combinations:

$$A_{(\mu}B_{\nu)} := \frac{1}{2}\left(A_\mu B_\nu + A_\nu B_\mu\right), \qquad A_{[\mu}B_{\nu]} := \frac{1}{2}\left(A_\mu B_\nu - A_\nu B_\mu\right).$$

In Problem 3.1, we derive step by step the linearized Einstein equations from the Einstein–Hilbert action (2.18) plus matter:

$$\delta G_{\mu\nu} + h_{\mu\nu}\Lambda = \kappa^2\delta T_{\mu\nu}, \tag{3.3}$$

where

$$\delta G_{\mu\nu} = \delta R_{\mu\nu} - \tfrac{1}{2}h_{\mu\nu}\tilde{R} - \tfrac{1}{2}\tilde{g}_{\mu\nu}\delta R. \tag{3.4}$$

The space-space off-diagonal component $\Pi_{\alpha\beta} = \delta T_{\alpha\beta}$, $\alpha \neq \beta$, is the *anisotropic stress*.

The perturbation $h_{\mu\nu}$ can be split in three sectors:

- *Tensor* perturbations, described by the symmetric, transverse, traceless spatial tensor $h_{\alpha\beta}$.
- *Scalar* perturbations, described by four scalar fields Φ, Ψ, B and E.
- *Vector* perturbations, consisting in two divergence-free spatial vectors S_α and F_α.

In the linear approximation, the three types of perturbations evolve independently and they can be studied separately. Vector perturbations will play no role in the following, since they are typically damped away in the early universe. All these functions represent corrections to the FLRW line element, and the identification of $h_{\alpha\beta}$, Φ, Ψ, B and E with the metric components will be shown later, in (3.19) and (3.52).

3.1.2 Gauge Invariance and Gauge Fixing

Following [2, Sect. I.3], let $\mathcal{M}$ be the spacetime manifold in which a coordinate system $\{x^\mu\}$ has been fixed, and let A be a tensor. At any point of $\mathcal{M}$, we define a background function $\tilde{A}(x)$ whose form does not depend on the coordinate choice: given another system $\{\bar{x}^\mu\}$, one has $\bar{\tilde{A}}(\bar{x}) = \tilde{A}(\bar{x})$. We perturb A at linear order in both coordinate frames,

$$A(x) = \tilde{A}(x) + \delta A(x)\,, \qquad \bar{A}(\bar{x}) = \tilde{A}(\bar{x}) + \delta\bar{A}(\bar{x})\,.$$

Diffeomorphisms $x^\mu \mapsto \bar{x}^\mu$, which change coordinate systems but leave the background invariant, are also called gauge transformations. For an infinitesimal transformation $x^\mu \mapsto \bar{x}^\mu = x^\mu + v^\mu(x)$,

$$\delta\bar{A} - \delta A = \mathcal{L}_v \tilde{A}\,, \tag{3.5}$$

where $\mathcal{L}_v$ is the Lie derivative along the vector v^μ (see below). If $\delta\bar{A} - \delta A = 0$, the perturbation is the same in both frames and the linear perturbation theory is *gauge invariant*; in this case, one avoids the so-called gauge modes, unphysical perturbative terms due to a bad choice of coordinates.

Alternatively, one can consider two manifolds, one physical and the other the background with respect to which we perturb the former. A *gauge choice* is then a one-to-one correspondence between points of the background and points in the physical spacetime. A gauge transformation is a change in the mapping which keeps background points fixed. This is different from a coordinate transformation, which would change the point. A *perturbation* of A is the difference between the value $A(x)$ at the point x in spacetime and the value $\tilde{A}(x)$ in the gauge-equivalent point in the background.

3.1.3 Cosmological Horizons

Perturbations are quite often studied in momentum space. In conformal time, comoving coordinates and *flat* spatial slices, expand a $D = 4$ random scalar field $\varphi(\tau, \mathbf{x})$ in a Fourier integral,

$$\varphi(\tau, \mathbf{x}) = \int_{-\infty}^{+\infty} \frac{\mathrm{d}^3\mathbf{k}}{(2\pi)^3} \varphi_{\mathbf{k}}(\tau)\, \mathrm{e}^{\mathrm{i}\mathbf{k}\cdot\mathbf{x}}\,. \tag{3.6}$$

If the coefficients $\varphi_{\mathbf{k}}$ depend only on the *comoving wave-number*

$$k := |\mathbf{k}| = \frac{2\pi}{\lambda_{\mathrm{com}}}\,, \tag{3.7}$$

they describe an *isotropic* perturbation of comoving wave-length λ_{com}.

One can compare the wave-number k of a given perturbation mode with the cosmological horizon in any epoch. We want to define the cosmological horizon distinguishing between the causal region of the universe and the "outside." For power-law scenarios with $0 < p < 1$, the particle horizon is the natural choice. However, for a de Sitter background or power-law scenarios with $p > 1$, conformal time is negative ((2.176) and (2.185)) and the particle horizon, meant as the radius of a causally connected region, is ill defined. In these cases, it is convenient to choose the comoving Hubble horizon

$$r_H = \frac{1}{aH} = \frac{1}{\mathcal{H}} \tag{3.8}$$

as the boundary of the causal patch. Here $\mathcal{H} := a'/a$. From the definition of particle horizon, we have

$$\tau = \int_{r_{H\mathrm{i}}}^{r_H} \frac{\mathrm{d}r'_H}{\epsilon - 1}, \tag{3.9}$$

where $\epsilon = 1 - \mathcal{H}'/\mathcal{H}^2$ is defined in (2.166). Assuming that $r_{H\mathrm{i}} \approx 0$ and ϵ be constant, one has

$$\tau = \frac{r_H}{\epsilon - 1}.$$

From (2.172), the comoving Hubble horizon of the de Sitter universe is

$$\boxed{r_H = |\tau|,} \tag{3.10}$$

while the power-law horizon (2.137) is

$$r_H = \frac{1-p}{p}\tau. \tag{3.11}$$

We have already seen that for p in the radiation- or matter-domination eras, the coefficient $(1-p)/p$ is $O(1)$. But this is true also for $p \gg 1$ ($\epsilon \ll 1$) and, without loss of generality, we can set (3.10) as the *definition* of our horizon both for de Sitter and power-law expansion for *any* p. In particular, the power-law Hubble horizon is indeed (3.10) in the limit $p \to +\infty$.

Therefore, we identify *small scales* (modes well inside the Hubble horizon) as those modes k obeying the inequality

$$k|\tau| \gg 1, \tag{3.12}$$

and *large scales* (modes outside the horizon) as

$$k|\tau| \ll 1\,. \tag{3.13}$$

The moment when a perturbation enters or exits the horizon at a given time τ_* is called *horizon crossing* and is characterized by

$$\boxed{k|\tau_*| = 1\,.} \tag{3.14}$$

Sometimes, one modifies the definition of horizon crossing with $O(1)$ coefficients (see Sect. 3.2.3) but the physics is unchanged.

Once an horizon has been specified, we can split any perturbation field $\varphi(\tau, \mathbf{x})$ into a *coarse-grained* and a *fine-grained* part, integrating over wave-lengths outside and inside the horizon, respectively:

$$\begin{aligned}\varphi(\tau,\mathbf{x}) &= \int_{k<k_*} \frac{\mathrm{d}^3\mathbf{k}}{(2\pi)^3}\varphi_{\mathbf{k}}(\tau)\mathrm{e}^{\mathrm{i}\mathbf{k}\cdot\mathbf{x}} + \int_{k>k_*} \frac{\mathrm{d}^3\mathbf{k}}{(2\pi)^3}\varphi_{\mathbf{k}}(\tau)\mathrm{e}^{\mathrm{i}\mathbf{k}\cdot\mathbf{x}}\\ &=: \varphi_{\mathrm{c}}(\tau,\mathbf{x}) + \varphi_{\mathrm{q}}(\tau,\mathbf{x})\,.\end{aligned} \tag{3.15}$$

The subscripts c and q stem for "classical" and "quantum." We shall explain these names in Chap. 5.

3.1.4 Separate Universe Approach

An intuitive way of dealing with inhomogeneities at large scales is the *separate universe* approach [6] (Fig. 3.1). Consider a perturbation of proper wave-length λ greater than the Hubble horizon R_H at time $\tau = \tau_1$ centered at the (proper) point $\mathbf{x}$ (*a* or *b* in the figure). Define a patch $U_\mathbf{x}$ of the Universe as a region of proper size λ_s slightly larger than R_H. Inside the patch, the physics can be described by a set of homogeneous fields $f_\mathbf{x}(\tau)$, where the subscript denotes the center of the patch, not a spatial dependence. Thus, for each patch $U_\mathbf{x}$ we have a "local" scale factor $a_\mathbf{x}(\tau)$, a Hubble parameter $H_\mathbf{x}(\tau)$, and so on. The scale factor defines a local number of e-foldings $\mathcal{N}_\mathbf{x}(\tau) = \ln a_\mathbf{x}(\tau)$ (in this chapter, we shall omit the subscript "a" in $\mathcal{N}_a$).

Consider now a collection of patches of overall size λ_0, much larger than λ and within which spatial gradients are not negligible but still mild. At ultra-large scales $\lambda_{\mathrm{inhom}} > \lambda_0$, there may be strong inhomogeneities (according to the stochastic model of eternal inflation described later). Then, we have the hierarchy of scales

$$R_H \lesssim \lambda_s < \lambda \ll \lambda_0 \lesssim \lambda_{\mathrm{inhom}}\,. \tag{3.16}$$

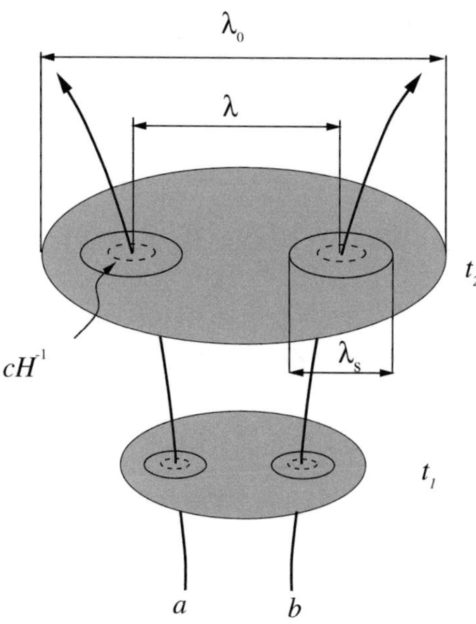

Fig. 3.1 The separate universe approach. A perturbation described by two locally FLRW patches is let evolve in time (Reprinted figure with permission from [6]. ©2000 by the American Physical Society)

The region within λ_0 is coarse-grained by patches of size λ_s which are locally FLRW and characterized by perturbations of wave-length λ. In this context, the definition of a perturbation is clear and, as a great added bonus with respect to the traditional definition (3.1), it goes *beyond linear order*. Let $\psi_{\mathbf{x}}(\tau)$ be some locally measured potential which enters the metric. A perturbation at time t along the spatial direction α with wave-length $\lambda^\alpha = |x_1^\alpha - x_2^\alpha|$ is then

$$\delta\psi_{(\alpha)}(\tau) := \psi_{x_1^\alpha}(\tau) - \psi_{x_2^\alpha}(\tau). \tag{3.17}$$

If the perturbation is non-linear, the coarse graining in terms of local patches is refined and the spatial index in local parameters is promoted to an actual spatial dependence,

$$\psi_{\mathbf{x}}(\tau) = \psi(\tau, \mathbf{x}) .$$

Perturbations can then be identified with gradients. One has

$$\frac{\psi(\tau, \mathbf{x}_1) - \psi(\tau, \mathbf{x}_2)}{x_1^\alpha - x_2^\alpha} \simeq \partial_\alpha \psi(\tau, \mathbf{x}) ,$$

so that, up to a numerical factor (no summation over α),

$$\delta\psi_{(\alpha)} \sim \lambda^\alpha \partial_\alpha \psi(\tau, \mathbf{x}) .$$

The time evolution of $\psi(\tau, \mathbf{x})$ between two slices $\tau = \tau_a$ and $\tau = \tau_b$ is determined by the cosmic expansion via the local number of e-foldings, as shown later. Thus, one has an intuitive setting where to study the dynamics of long-wave-length inhomogeneities.

3.2 Linear Tensor Perturbations

In this section, we give a simple example of calculation in the linear theory: tensor perturbations. The separate universe picture is not necessarily invoked at the linear level.

3.2.1 *Transverse-Traceless Gauge*

Since the background metric defines a global Lorentz frame, one can fix the gauge for the tensor perturbation to the *transverse-traceless* (or *traceless-harmonic*) *gauge*. The proof of this statement can be sketched as follows on a flat FLRW background. In conformal time the metric is conformally flat, so that we can do a conformal transformation (Sect. 7.4.2) and consider Minkowski spacetime, where the problem is simplified. There, one can always choose the traceless-harmonic gauge [7]

$$h_{\mu 0} = 0\,, \qquad \partial^\nu h_{\mu\nu} = 0\,, \qquad h = 0\,, \tag{3.18}$$

which is discussed in Problem 3.2. Then, the perturbed line element in conformal time is

$$\mathrm{d}s^2 = a^2(\tau)\left[-\mathrm{d}\tau^2 + \left(\delta_{\alpha\beta} + h_{\alpha\beta}\right)\mathrm{d}x^\alpha \mathrm{d}x^\beta\right]. \tag{3.19}$$

In four dimensions, a gravitational wave $h_{\alpha\beta}$ propagating along, e.g., the x^3 direction can be decomposed into two polarization scalar modes,

$$h_{\alpha\beta}(x) = h_+(x)e^+_{\alpha\beta} + h_\times(x)e^\times_{\alpha\beta}\,, \tag{3.20}$$

where

$$e^+ = \begin{pmatrix} 1 & 0 \\ 0 & -1 \end{pmatrix}, \qquad e^\times = \begin{pmatrix} 0 & 1 \\ 1 & 0 \end{pmatrix}. \tag{3.21}$$

Interpreted as a particle, the gravitational wave $h_{\alpha\beta}$ is called the *graviton*.

3.2.2 Equation of Motion

On a conformally flat background, the traceless-harmonic gauge greatly simplifies (3.86) and (3.87):

$$\delta R_{\mu\nu} = -\tfrac{1}{2}\tilde{\Box} h_{\mu\nu}\,, \qquad \delta R = 0\,. \tag{3.22}$$

Therefore, (3.4) becomes [8]

$$\delta G_{\alpha\beta} = -\tfrac{1}{2}\tilde{\Box} h_{\alpha\beta}\,. \tag{3.23}$$

The same result could have been obtained by perturbing the action at second order in linear perturbations (Problem 3.3). In four dimensions, we have two degrees of freedom:

$$\delta^{(2)} S_h = \frac{1}{4\kappa^2} \sum_{\lambda=+,\times} \int \mathrm{d}^4 x \sqrt{-\tilde{g}}\, \left(h_\lambda \tilde{\Box} h_\lambda - 2\Lambda h_\lambda^2\right)\,. \tag{3.24}$$

The effective action of the independent polarization modes is that of two massive scalars.

3.2.3 Mukhanov–Sasaki Equation and Solution

Ignoring the anisotropic stress $\Pi_{\alpha\beta}$, we can solve the perturbed Einstein equations in this approximation for a de Sitter and a power-law flat background. We can consider each polarization mode separately. Call $\varphi = h_\lambda$ and $m^2 = 2\Lambda$. The equation of motion for each mode is that for a massive scalar field. Since this equation appears also in the scalar sector, from now on we forget we have derived it for the tensor modes in four dimensions and leave the dimensionality D arbitrary. In linear perturbation theory on a flat FLRW background, we have to solve the generic equation

$$0 = -\tilde{\Box}\varphi + m^2\varphi = \ddot{\varphi} - \frac{1}{a^2}\nabla^2\varphi + (D-1)H\dot{\varphi} + m^2\varphi\,,$$

where $\nabla^2 = \sum_\alpha \partial_\alpha^2$ is the spatial Laplacian. It is convenient to go to momentum space, where $\nabla^2 \to -|\mathbf{k}|^2 = -k^2$. Switching to conformal time, we need the useful formulæ

$$\partial_t = \frac{1}{a}\partial_\tau\,, \qquad \partial_t^2 = \frac{1}{a^2}(\partial_\tau^2 - \mathcal{H}\partial_\tau)\,, \tag{3.25}$$

so that

$$\varphi_{\mathbf{k}}'' + (D-2)\mathcal{H}\varphi_{\mathbf{k}}' + (k^2 + m^2a^2)\varphi_{\mathbf{k}} = 0\,. \tag{3.26}$$

Before looking for analytic solutions, we can determine the qualitative asymptotic behaviour of the perturbation $\varphi_{\mathbf{k}}$ by appealing to the results of Sect. 3.1.3. Let us consider a massless scalar field, $m^2 = 0$. At very small scales ($k|\tau| \gg 1$), the curvature of the manifold is negligible and one can ignore the Hubble friction term. The perturbation $\varphi_{\mathbf{k}}$ is well inside the horizon and obeys the harmonic oscillator equation

$$\varphi_{\mathbf{k}}'' + k^2\varphi_{\mathbf{k}} \simeq 0\,, \qquad k|\tau| \gg 1\,, \tag{3.27}$$

whose solutions are incoming and outgoing plane waves,

$$\varphi_{\mathbf{k}} \simeq A_k\, \mathrm{e}^{\pm ik\tau}\,, \qquad k|\tau| \gg 1\,, \tag{3.28}$$

where A_k is some normalization constant. On the other hand, outside the horizon the effective mass term can be ignored and the perturbation is approximately constant,

$$\varphi_{\mathbf{k}} \simeq C_k\,, \qquad k|\tau| \ll 1\,. \tag{3.29}$$

The normalization C_k can be determined by a *junction condition* at horizon crossing:

$$\big|\varphi(k|\tau| \ll 1)\big| \simeq \big|\varphi(k|\tau| \gg 1)\big| \qquad \Rightarrow \qquad |C_k| = |A_k|\,. \tag{3.30}$$

Classically, the plane-wave normalization is $A_k = (2k)^{-1/2}$.

In order to solve (3.26) analytically, one remembers that the engineering dimension of a scalar field in D dimensions is $\alpha := (D-2)/2$. Since in natural units $[a] = -1$, the scaling dimension of the field

$$w_k := a^{\alpha}\varphi_{\mathbf{k}} \tag{3.31}$$

is zero (in the left-hand side, we anticipated that w will depend only on k by isotropy). Scalar fields have zero scaling dimension in an effective two-dimensional background, where there is no Hubble friction term (see (3.26)). Indeed, using

$$\varphi_{\mathbf{k}}' = a^{-\alpha}\left(w_k' - \alpha\mathcal{H}w_k\right)\,, \qquad \varphi_{\mathbf{k}}'' = a^{-\alpha}\left[w_k'' - 2\alpha\mathcal{H}w_k' + \alpha(\alpha-1+\epsilon)\mathcal{H}^2 w_k\right],$$

we obtain an equation without friction:

$$\boxed{w_k'' + (k^2 - M^2)w_k = 0\,,} \tag{3.32}$$

where

$$\boxed{M^2 := \frac{D-2}{2}\left(\frac{D}{2}-\epsilon\right)\mathcal{H}^2 - m^2a^2\,.} \tag{3.33}$$

The field w is often called *Mukhanov–Sasaki variable* and (3.32) *Mukhanov–Sasaki equation*.

On a de Sitter background (2.178), the effective mass reads

$$M^2 = \left[D(D-2) - 4\frac{m^2}{H^2}\right]\frac{1}{4\tau^2} = \frac{4\nu^2-1}{4\tau^2}\,, \tag{3.34}$$

where

$$\nu = \sqrt{\frac{D(D-2)+1}{4} - \frac{m^2}{H^2}}\,. \tag{3.35}$$

Thus, (3.32) has been rewritten as a Bessel-type equation. The general solution is a superposition of Bessel functions [9, formula 8.491.5]:

$$w_k = C_1\sqrt{k|\tau|}\,J_\nu(k|\tau|) + C_2\sqrt{k|\tau|}\,Y_\nu(k|\tau|)\,. \tag{3.36}$$

Asymptotically, at large scales

$$J_\nu(y) \overset{y\ll 1}{\simeq} \frac{1}{2^\nu\Gamma(\nu+1)}y^\nu\,,$$

and

$$Y_\nu(y) \overset{y\ll 1}{\simeq} \begin{cases} \frac{2}{\pi}\left[\ln\left(\frac{y}{2}\right) + \gamma_{\rm EM}\right] & \text{if} \quad \nu = 0 \\ -\frac{2^\nu\Gamma(\nu)}{\pi}y^{-\nu} & \text{if} \quad \nu > 0 \end{cases}\,,$$

where $\gamma_{\rm EM} \approx 0.577$ is the Euler–Mascheroni constant. At small scales,

$$J_\nu(y) \overset{y\gg 1}{\simeq} \sqrt{\frac{2}{\pi y}}\cos\left(y - \frac{\nu\pi}{2} - \frac{\pi}{4}\right)\,,$$

$$Y_\nu(y) \overset{y\gg 1}{\simeq} \sqrt{\frac{2}{\pi y}}\sin\left(y - \frac{\nu\pi}{2} - \frac{\pi}{4}\right)\,.$$

On a power-law background (2.187), the Mukhanov–Sasaki equation is no longer a Bessel equation and cannot be solved exactly, unless $m^2 = 0$. One has

$$M^2 = \frac{4\nu^2 - 1}{4\tau^2}, \qquad \nu = \sqrt{\frac{D-2}{2}\left(\frac{D}{2} - \frac{1}{p}\right)\left(\frac{p}{1-p}\right)^2 + \frac{1}{4}}. \tag{3.37}$$

In the limit $p \to +\infty$, one obtains (3.35). In this sense, the large-p limit of power-law cosmology is de Sitter.

3.2.4 *Discovery of Gravitational Waves*

One hundred years after Einstein's prediction that weak gravitational signals propagate as transverse waves at the speed of light [10, 11], gravitational waves have been detected for the first time in September 2015 by Advanced LIGO, the Laser Interferometer Gravitational-Wave Observatory [12–15]. The instrument measured, with a signal-to-noise ratio of about 24, the $O(10^{-21})$ spatial strain created by the waves coming from a binary black-hole system distant 1.3 billion light years from us. Two black holes of about 30–40 solar masses each coalesced into a larger black hole radiating a total energy in gravitational waves equivalent to 3 solar masses. The strain pattern recorded by Advanced LIGO was in complete agreement with general relativity (Fig. 3.2). Possible violations of Einstein's theory in the event GW150914 cannot exceed 4 % in the noise-weighted signal correlation [16]. In particular, the upper bound on a mass for the graviton results to be $m_g < 1.2 \times 10^{-22}$ eV at the 90 % confidence level. The first announcement of the discovery was soon followed by the study of another significant event, detected in December 2015 and generated by a merger with smaller masses [17, 18].

The discovery of gravitational waves marked one of the most momentous successes of general relativity as well as the first direct confirmation of the existence of binary black-hole systems. Yet, this is only the beginning of modern gravitational-wave astronomy. Primordial gravitational waves, i.e., those produced in the early universe during inflation, have not been detected yet and there are only (but quite important) upper bounds on their amplitude (Sect. 4.4.2).

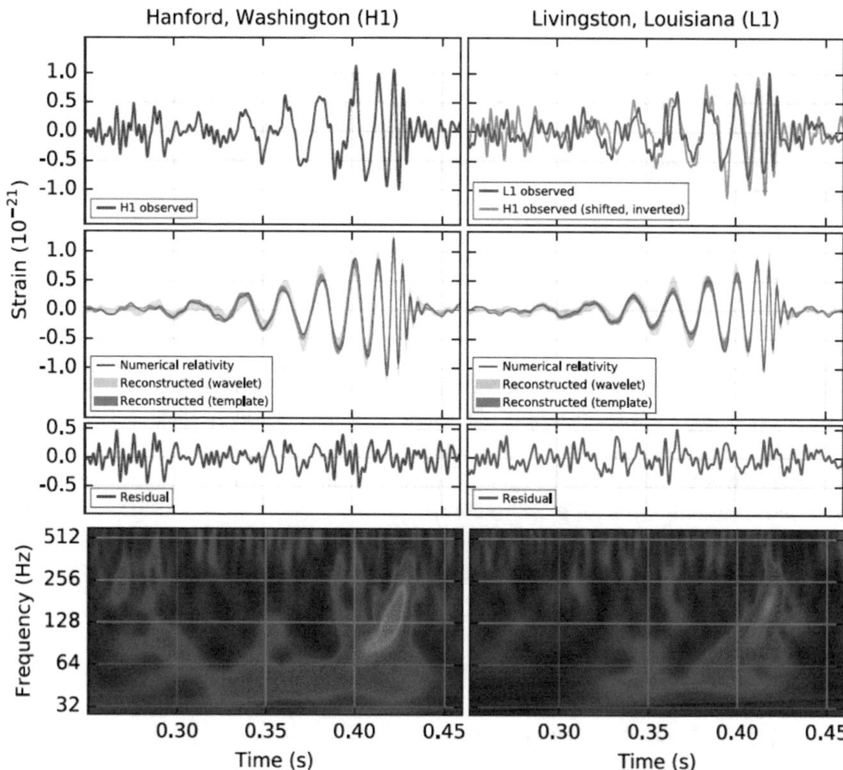

Fig. 3.2 The gravitational-wave event GW150914 observed by the LIGO Hanford (H1, *left column panels*) and Livingston (L1, *right column panels*) detectors. Times are shown relative to September 14, 2015 at 09:50:45 UTC. For visualization, all time series are filtered with a 35–350 Hz band-pass filter to suppress other signals. *Top row, left*: H1 strain. *Top row, right*: L1 strain. GW150914 arrived first at L1 and $6.9^{+0.5}_{-0.4}$ ms later at H1; for a visual comparison the H1 data are also shown, shifted in time by this amount and inverted (to account for the detectors' relative orientations). *Second row*: gravitational-wave strain projected onto each detector in the 35–350 Hz band. *Solid lines* show a numerical relativity wave-form for a system with parameters consistent with those recovered from GW150914 confirmed by an independent calculation. *Shaded areas* show 90 % credible regions for two wave-form reconstructions: one that models the signal as a set of sine-Gaussian wavelets and one that models the signal using binary-black-hole template wave-forms. These reconstructions have a 95 % overlap. *Third row*: residuals after subtracting the filtered numerical relativity wave-form from the filtered detector time series. *Bottom row*: a time-frequency decomposition of the signal power associated with GW150914. Both plots show a signal with frequency increasing with time (Source: [12, 15])

3.3 Scalar Perturbations

Scalar perturbations generate matter density fluctuations and it is important to identify the primordial fluctuation fields seeding cosmic structures. To this and other aims, we move to fully non-linear perturbations, later specializing to linear variables.

3.3.1 *Non-linear Perturbations*

The most general treatment of non-linear perturbations is the *covariant formalism* [19–28]. At large scales, it reduces to other approaches named *gradient expansion* [29–33] and $\delta\mathcal{N}$ *formalism* [34–37]. The separate universe approach in its original formulation is the lowest-order approximation of the gradient expansion [31], but in practice it encompasses all these methods.

Consider a non-FLRW universe filled with a perfect fluid defining the metric decomposition (2.41). For two scalars $f_{1,2}$ and from (2.47) and (2.51),

$$D_\mu f_1 - \frac{\dot{f}_1}{\dot{f}_2} D_\mu f_2 = \partial_\mu f_1 - \frac{\dot{f}_1}{\dot{f}_2} \partial_\mu f_2 \,. \tag{3.38}$$

We define the *Lie derivative* along the world-line u^μ (we recall that $u_\mu u^\mu = -1$) on a scalar f and on a covector v_μ as

$$\boxed{\mathcal{L}_u f := \dot{f}\,, \qquad \mathcal{L}_u v_\mu := \dot{v}_\mu + v_\nu \partial_\mu u^\nu \,.} \tag{3.39}$$

Assuming only the continuity equation (2.31), one can show that the non-linear *curvature perturbation on uniform density hypersurfaces*

$$\zeta_\mu := D_\mu \mathcal{N} - \frac{\dot{\mathcal{N}}}{\dot{\rho}} D_\mu \rho = \partial_\mu \mathcal{N} - \frac{\dot{\mathcal{N}}}{\dot{\rho}} \partial_\mu \rho = \partial_\mu \mathcal{N} + \frac{1}{D-1} \frac{1}{\rho + P} \partial_\mu \rho \tag{3.40}$$

obeys a conservation equation. In the second and third steps we have used, respectively, (3.38) and the projected continuity equation, which is (2.76) with H replaced by the local Hubble parameter $H(t, \mathbf{x}) = \dot{\mathcal{N}}$:

$$\dot{\rho} = -(D-1)\dot{\mathcal{N}}(\rho + P) \,. \tag{3.41}$$

Notice that ζ_μ is purely geometric and depends neither on the background nor on the form of the total action. On a FLRW background, it vanishes.

Taking the Lie derivative of ζ_μ, we find (Problem 3.4)

$$\mathcal{L}_u(\zeta_\mu) = -\frac{\dot{\mathcal{N}}}{\rho + P}\Gamma_\mu\,, \tag{3.42}$$

where in the right-hand side we defined the non-linear *non-adiabatic pressure*

$$\Gamma_\mu := D_\mu p - \frac{\dot{P}}{\dot{\rho}} D_\mu \rho = \partial_\mu p - \frac{\dot{P}}{\dot{\rho}}\partial_\mu \rho\,. \tag{3.43}$$

Another important quantity is the non-linear curvature perturbation on comoving hypersurfaces, or *comoving curvature perturbation*

$$\mathcal{R}_\mu := -D_\mu \mathcal{N}\,. \tag{3.44}$$

It obeys another continuity equation, which can be written in terms of matter variables after using the Einstein equations [25]. By definition,

$$\zeta_\mu + \mathcal{R}_\mu = \frac{D_\mu \rho}{(D-1)(\rho + P)}\,. \tag{3.45}$$

From this expression, the name of ζ_μ is clarified: this quantity coincides with (minus) the curvature perturbation on hypersurfaces where the energy density is unperturbed. These hypersurfaces are called uniform-density slices.

The comoving curvature perturbation is fundamental in the study of cosmologies dominated by a scalar field. From (2.68), one notices that the projected covariant derivative vanishes identically on a scalar, $D_\mu \phi = 0$. By virtue of (3.38), the definition (3.44) is equivalent to

$$\mathcal{R}_\mu = -\partial_\mu \mathcal{N} + \frac{\dot{\mathcal{N}}}{\dot{\phi}}\partial_\mu \phi\,. \tag{3.46}$$

3.3.2 Non-linear Perturbations at Large Scales

In the separate universe picture, we can find a rather useful formula for the curvature perturbation at large scales, where gradient terms can be neglected. By "large" we mean slightly larger than the Hubble horizon of the local patch, where long-wave-length inhomogeneities are still smoothed out.

Define the perturbed spatial metric as

$$g_{\alpha\beta} = \tilde{a}^2 \mathrm{e}^{-2\Psi_{\rm NL}}\gamma_{\alpha\beta}\,, \tag{3.47}$$

where $\Psi_{\rm NL} = \Psi_{\rm NL}(t,\mathbf{x})$, $\tilde{a}$ is the background scale factor and $\gamma_{\alpha\beta}$ is the 3-metric with tensor perturbations included. The "local" scale factor

$$a(t,\mathbf{x}) = \tilde{a}(t)\,\mathrm{e}^{-\Psi_{\rm NL}(t,\mathbf{x})} \tag{3.48}$$

encodes also the perturbation $\delta\kappa$ of the intrinsic curvature, which allows one to identify $\Psi_{\rm NL}$ with the non-linear curvature perturbation $\mathcal{R}_{\rm NL}$.

At super-horizon scales, we can ignore gradient terms. The expansion rate is ($D = 4$ from now on)

$$\frac{\theta}{3} \simeq \frac{\dot{a}}{a} = \tilde{H} - \dot{\Psi}_{\rm NL}\,. \tag{3.49}$$

Integrating this in time, one obtains the non-linear, large-scale perturbation of the number of e-foldings,

$$\delta\mathcal{N} := \mathcal{N}(t,\mathbf{x}) - \tilde{\mathcal{N}}(t) = \Psi_{\rm NL}(t_{\rm i},\mathbf{x}) - \Psi_{\rm NL}(t,\mathbf{x})\,. \tag{3.50}$$

This is called the $\delta\mathcal{N}$ *formula*. Notice that we can always choose the initial slicing to be flat, $\Psi_{\rm NL}(t_{\rm i},\mathbf{x}) = 0$. On uniform density slices, $-\Psi_{\rm NL} = -\mathcal{R}_{\rm NL} = \zeta_{\rm NL}$ is the Salopek–Bond non-linear curvature perturbation [5]. Thus, the $\delta\mathcal{N}$ formula can be recast as

$$\boxed{\delta\mathcal{N} = \zeta_{\rm NL}\,.} \tag{3.51}$$

The non-linear curvature perturbation on a uniform-density slice at time t is given by the perturbation of the number of e-foldings of the expansion of a local patch centered in $\mathbf{x}$, starting from any flat slice. Below, we shall see that $\zeta_\alpha \simeq \partial_\alpha\zeta_{\rm NL}$ at linear order.

3.3.3 *Linear Perturbations at Large Scales*

We now consider linear fluctuations. The line element in conformal time for linear scalar perturbations on a flat background is

$$\begin{aligned} \mathrm{d}s^2 = a^2(\tau)\,&\big\{-(1+2\Phi)\mathrm{d}\tau^2 + 2(\partial_\alpha B)\,\mathrm{d}x^\alpha\mathrm{d}\tau \\ &+ \left[(1-2\Psi)\delta_{\alpha\beta} + 2\partial_\alpha\partial_\beta E\right]\mathrm{d}x^\alpha\mathrm{d}x^\beta\big\}\,, \end{aligned} \tag{3.52}$$

while the fluid perturbations are

$$\rho = \tilde{\rho} + \delta\rho\,, \qquad P = \tilde{P} + \delta P\,. \tag{3.53}$$

Using suitable combinations of Φ (called *gravitational potential*), Ψ (called Bardeen potential), B and E, for a given perturbative order one can construct gauge-invariant quantities under infinitesimal coordinate transformations. These quantities determine which gauge choice we can (or cannot) make. Some of the most common gauge choices are:

- *Synchronous gauge*, where $g_{00} = -1$ and $g_{0\alpha} = 0$. Coordinate time coincides with proper time and the foliation is orthogonal (space-like hypersurfaces are orthogonal to time-like world-lines).
- *Newtonian* or *longitudinal gauge*, where $B = 0 = E$. It corresponds to a time-like hypersurface where expansion is isotropic. The fields Φ and Ψ are then automatically gauge invariant and, from the form of the line element, one identifies Φ with the gravitational potential associated with the perturbation. If the anisotropic stress of the matter energy-momentum tensor vanishes, then $\Phi = \Psi$.
- *Flat gauge*, where $\Psi = 0 = E$ and space-like hypersurfaces remain flat.
- *Total-matter gauge*, where $E = 0$ and space-like hypersurfaces (rest frame of free-falling particles) are orthogonal to matter velocity $v = B$.
- *Uniform density gauge*, where the perturbation of the total energy density vanishes, $\delta\rho = 0$.

The set of linearized dynamical equations is rather complicated and we shall not present it here. However, we are in a position to relate the non-linear variables of Sects. 3.3.1 and 3.3.2 with the scalar potentials Φ and Ψ in longitudinal gauge. The zero component ζ_0 vanishes at linear order, since

$$\zeta_0 \simeq \delta\dot{\mathcal{N}} - \frac{\dot{\bar{\mathcal{N}}}}{\dot{\bar{\rho}}}\delta\dot{\rho} - \delta\left(\frac{\dot{\mathcal{N}}}{\dot{\rho}}\right)\dot{\bar{\rho}} = 0\,.$$

On the other hand, the spatial components are just gradients of a scalar,

$$\zeta_\alpha \simeq \partial_\alpha\zeta\,, \qquad \zeta := \delta\mathcal{N} - \frac{\dot{\bar{\mathcal{N}}}}{\dot{\bar{\rho}}}\delta\rho\,. \tag{3.54}$$

To show that this is the perhaps more familiar linear curvature perturbation on uniform density slices, ignore tensor and vector perturbations (which decouple from the equations anyway) and linearize (3.47) to get

$$a = \tilde{a} + \delta a \simeq \tilde{a}(1 - \Psi)\,. \tag{3.55}$$

Thus, Ψ in the line element (3.52) is the same $\Psi_{\rm NL}$ in (3.47) at linear order.

From now, on we omit tildes since all expressions will be at linear order and there is no risk of confusion. The linear version of (3.50) is now $\delta\mathcal{N} = -\Psi$. Plugging that into (3.54), we finally obtain

$$\boxed{\zeta = -\Psi - H\frac{\delta\rho}{\dot{\rho}}}, \qquad (3.56)$$

which corresponds, as promised, to the definition of the linear ζ [4]. Simply going to the uniform density gauge, one immediately sees that ζ is the linear version of $\zeta_{\rm NL}$. Therefore, $\zeta_{\rm NL}$ is the large-scale approximation of ζ_μ. To recast the conservation equation (3.42) in linear form, consider (3.43). The left-hand side is $\mathcal{L}_u(\zeta_\alpha) \simeq \partial_\alpha\dot{\zeta}$. Again, the 0 component vanishes at linear order, while

$$\Gamma_\alpha \simeq \partial_\alpha \delta P_{\rm nad}\,, \qquad \delta P_{\rm nad} := \dot{P}\left(\frac{\delta P}{\dot{P}} - \frac{\delta\rho}{\dot{\rho}}\right). \qquad (3.57)$$

Then,

$$\dot{\zeta} \simeq -\frac{H}{\rho + P}\delta P_{\rm nad} \qquad \text{at large scales}\,. \qquad (3.58)$$

In the case of *adiabatic* (or *isentropic*) *perturbations*, different particle species (baryons, photons, neutrinos, dark matter) have the same spatial distribution and a density perturbation $\delta\rho$ depending on the total background density and on the equation of state. As a consequence,

$$\delta P_{\rm nad} = 0 \qquad (3.59)$$

and the linear curvature perturbation is constant at sufficiently large scales.

Furthermore, the curvature perturbation is the same for all species (in particular, for dust matter and radiation, $\zeta_{\rm m} = \zeta_{\rm r}$). In flat gauge and for barotropic fluids,

$$\zeta = -H\frac{\delta\rho}{\dot{\rho}} = \frac{1}{3(1+w)}\frac{\delta\rho}{\rho} = -\frac{\delta a}{a}\,. \qquad (3.60)$$

Also, one can show that, at large scales where one can neglect the anisotropic stress and in any era dominated by a barotropic fluid, the Bardeen potential Ψ is related to ζ as

$$\Phi \simeq \Psi \simeq -\frac{3+3w}{5+3w}\zeta\,. \qquad (3.61)$$

During matter domination, on large scales

$$\Psi = \Phi = -\frac{3}{5}\zeta\,. \tag{3.62}$$

In the presence of a scalar field, the scalar part of the linearized $\mathcal{R}_\alpha$ (3.46) is a gradient, $\mathcal{R}^{\rm s}_\alpha \simeq \partial_\alpha \mathcal{R}$, where

$$\boxed{\mathcal{R} := \Psi + \frac{H}{\dot\phi}\delta\phi} \tag{3.63}$$

is the linear comoving curvature perturbation. Since the gradient of the scalar field vanishes for a comoving observer, (3.63) relates two quantities defined on different hypersurfaces, the curvature perturbation on comoving slices $\mathcal{R}$ and the fluctuation $\delta\phi$ on other suitably defined slices, in particular flat.

Since at large scales the density perturbation $\delta\rho$ is negligible, from (3.45) we get

$$\mathcal{R} \simeq -\zeta\,. \tag{3.64}$$

Comoving and uniform density curvature perturbations coincide at the linear level and at large scales. The intuitive reason is that, for a canonical Klein–Gordon field, $\delta\rho - \delta P \simeq 2V_{,\phi}\delta\phi$ on large scales. The left-hand side vanishes on uniform density slices, implying that these slices are also comoving.

3.4 Gaussian Random Fields

Before applying perturbation theory to cosmological observations, in this section we describe some statistical properties of $D = 4$ *isotropic Gaussian distributions* on space-like manifolds [38, 39]. The time dependence of the generic random scalar field $\varphi(t, \mathbf{x})$ will be omitted.

1. If the field $\varphi(\mathbf{x})$ can be expressed as the Fourier superposition (3.6) of coefficients $\varphi_{\mathbf{k}}$ and if the real and imaginary part of the latter are statistically independent and with the same distribution for all $\mathbf{k}$ (isotropy condition), then the probability distribution of $\varphi(\mathbf{x})$ is Gaussian. In particular, if $\mathrm{Re}(\varphi_{\mathbf{k}})$ and $\mathrm{Im}(\varphi_{\mathbf{k}})$ obey an isotropic Gaussian distribution, so will $\varphi(\mathbf{x})$.
2. The statistical properties of isotropic causal Gaussian fields are completely described by the two-point correlation function.
3. A random field is *ergodic* (that is, its spatial averages in a given realization are equal to the expectation values on the entire ensemble) if, and only if, its spectrum is continuous.

The last property will be discussed in the next chapter. As far as the first is concerned, the Gaussian isotropic distribution with variance σ^2 is defined as

$$f[\varphi(\mathbf{x})] = \frac{1}{\sqrt{2\pi\sigma^2}}\,\mathrm{e}^{-\frac{\varphi^2(\mathbf{x})}{2\sigma^2}}\,, \tag{3.65}$$

where σ^2 does not depend on $\mathbf{x}$. The distribution f is normalized to 1,

$$\int_{-\infty}^{+\infty} \mathrm{d}\varphi\, f[\varphi] = 1\,. \tag{3.66}$$

We denote with angular brackets the average of an operator $\mathcal{O}(\varphi)$

$$\langle \mathcal{O}(\varphi)\rangle := \int_{-\infty}^{+\infty} \mathrm{d}\varphi\, f[\varphi]\mathcal{O}(\varphi)\,. \tag{3.67}$$

In particular, if φ is Gaussian, then φ_{c} and φ_{q} in (3.15) are separately Gaussian in large- and small-wave-length ensemble averages,

$$\langle\cdot\rangle = \langle\cdot\rangle_{\mathrm{c}} + \langle\cdot\rangle_{\mathrm{q}}\,. \tag{3.68}$$

Since f is even in φ, *correlation functions* of odd order vanish,

$$\xi_{2n+1}^{(\varphi)} := \left\langle \varphi^{2n+1}\right\rangle = 0\,, \tag{3.69}$$

while

$$\xi_{2n}^{(\varphi)} = \left\langle \varphi^{2n}\right\rangle = \frac{2^n\Gamma\left(\frac{1}{2}+n\right)}{\sqrt{\pi}}\sigma^{2n}\,. \tag{3.70}$$

In particular, $\xi_2^{(\varphi)} = \sigma^2$.

3.4.1 Power Spectrum

Random fields and their correlation functions can also be described in momentum space. If $\varphi(\mathbf{x})$ obeys a Gaussian isotropic statistics, then

$$\langle \varphi_{\mathbf{k}}\varphi_{\mathbf{k}'}\rangle =: (2\pi)^3\delta(\mathbf{k}+\mathbf{k}')P_\varphi(k)\,, \tag{3.71}$$

where P_φ is, by definition, the Fourier transform of the two-point correlation function (Problem 3.5). It is customary to rescale P_φ and define a quantity $\mathcal{P}_\varphi$, the *power spectrum*, which will play a fundamental role:

$$\boxed{\mathcal{P}_\varphi(k) := \frac{k^3}{2\pi^2} P_\varphi(k)\,.} \tag{3.72}$$

In Problem 3.5 we prove the *Wiener–Khintchine theorem*

$$\xi_2^{(\varphi)}(\varrho) := \langle \varphi(\mathbf{x}_1)\varphi(\mathbf{x}_2)\rangle = \int_0^{+\infty} \frac{\mathrm{d}k}{k}\mathcal{P}_\varphi(k)\frac{\sin(k\varrho)}{k\varrho}\,, \tag{3.73}$$

where $\varrho := |\mathbf{x}_1 - \mathbf{x}_2|$. For a *power-law spectrum*

$$\boxed{\mathcal{P}_\varphi(k) = A_\varphi k^{n_\varphi}\,,} \tag{3.74}$$

with n_φ constant, an evaluation of (3.71) yields [9, formula 3.761.4]

$$\xi_2^{(\varphi)}(\varrho) = \left[A_\varphi \Gamma(n_\varphi - 1)\sin\frac{(n_\varphi - 1)\pi}{2}\right]\frac{1}{\varrho^{n_\varphi}}\,. \tag{3.75}$$

When the spectrum (3.74) is *scale invariant*,

$$n_\varphi = 0\,, \tag{3.76}$$

an expansion of (3.75) around $n_\varphi \sim 0$ gives, up to an irrelevant additive constant,

$$\xi_2^{(\varphi)}(\varrho) \sim \ln\varrho\,. \tag{3.77}$$

If the spectral index is positive, $n_\varphi > 0$, the spectrum (3.74) will have an ultraviolet (UV) divergence and there will be more power at small scales (large k). Then the spectrum is said to be *blue-tilted*. On the other hand, if $n_\varphi < 0$ the spectrum will diverge in the infrared (IR) and will be said to be *red-tilted*.

3.4.2 Bispectrum and Trispectrum

In the next chapter, we shall be interested also in higher-order correlators. The *bispectrum* in momentum space is

$$\langle \varphi_{\mathbf{k}_1}\varphi_{\mathbf{k}_2}\varphi_{\mathbf{k}_3}\rangle =: (2\pi)^3\delta(\mathbf{k}_1 + \mathbf{k}_2 + \mathbf{k}_3)B_\varphi(k_1, k_2, k_3)\,, \tag{3.78}$$

where $k_i := |\mathbf{k}_i|$. One can show that B_φ is the Fourier transform of the three-point correlation function (Problem 3.5). For a Gaussian distribution, the bispectrum is identically zero. Experimentally, its estimate can help to constrain the statistical distribution of an almost Gaussian observable. Another estimator is the four-point correlation function. For a Gaussian field,

$$\langle\varphi(\mathbf{x}_1)\varphi(\mathbf{x}_2)\varphi(\mathbf{x}_3)\varphi(\mathbf{x}_4)\rangle_{\rm G} \overset{(3.73)}{=} \xi_2(\varrho_{12})\xi_2(\varrho_{34}) + \xi_2(\varrho_{13})\xi_2(\varrho_{24}) + \xi_2(\varrho_{14})\xi_2(\varrho_{23})\,. \tag{3.79}$$

In momentum space,

$$\langle\varphi_{\mathbf{k}_1}\varphi_{\mathbf{k}_2}\varphi_{\mathbf{k}_3}\varphi_{\mathbf{k}_4}\rangle_{\rm G} = (2\pi)^6\delta(\mathbf{k}_1+\mathbf{k}_2)\delta(\mathbf{k}_3+\mathbf{k}_4)P_\varphi(k_1)P_\varphi(k_3) + (\mathbf{k}_3 \leftrightarrow \mathbf{k}_2) + (\mathbf{k}_3 \leftrightarrow \mathbf{k}_1)\,. \tag{3.80}$$

In the case of coincident directions $\mathbf{x}_i = \mathbf{x}$, (3.79) becomes

$$\langle\varphi^4(\mathbf{x})\rangle = 3\langle\varphi^2(\mathbf{x})\rangle^2 \qquad \text{(Gaussian)}\,. \tag{3.81}$$

The four-point correlation function (3.79) is always non-zero irrespectively of the mutual angular separations. This is called the disconnected trispectrum. If the statistics deviates from Gaussianity, there will also be a connected contribution which vanishes at large angular separation. In momentum space, this non-Gaussian *trispectrum* is defined as the difference between the total trispectrum and the four-point auto-correlation function (3.81):

$$\langle\varphi_{\mathbf{k}_1}\varphi_{\mathbf{k}_2}\varphi_{\mathbf{k}_3}\varphi_{\mathbf{k}_4}\rangle_{\rm NG} =: (2\pi)^3\delta(\mathbf{k}_1+\mathbf{k}_2+\mathbf{k}_3+\mathbf{k}_4)T_\varphi(k_1,k_2,k_3,k_4)\,. \tag{3.82}$$

3.5 Problems and Solutions

3.1 Linearized Einstein equations. Expanding the Einstein–Hilbert action (2.18) with matter around a generic background as in (3.1), find the linearized Einstein equations (3.3).

Solution From the definitions of the Levi-Civita connection (2.15) and the Riemann tensor (2.16), one has

$$\delta\sqrt{-g} = \tfrac{1}{2}\,\tilde{g}^{\mu\nu}\,\sqrt{-\tilde{g}}\,h_{\mu\nu} = \tfrac{1}{2}\,\tilde{g}_{\mu\nu}\,\sqrt{-\tilde{g}}\,h^{\mu\nu}\,, \tag{3.83}$$

$$\delta\Gamma^\rho{}_{\sigma\gamma} = \tfrac{1}{2}\,\tilde{g}^{\rho\mu}\left(\tilde{\nabla}_\gamma h_{\mu\sigma} + \tilde{\nabla}_\sigma h_{\mu\gamma} - \tilde{\nabla}_\mu h_{\sigma\gamma}\right), \tag{3.84}$$

and

$$\begin{aligned}
\delta R^{\rho}{}_{\sigma\mu\nu} &= \tilde{\nabla}_{\mu}\delta\Gamma^{\rho}{}_{\sigma\nu} - \tilde{\nabla}_{\nu}\delta\Gamma^{\rho}{}_{\sigma\mu} \\
&= \tfrac{1}{2}\,\tilde{g}^{\rho\tau}\left(\tilde{\nabla}_{\mu}\tilde{\nabla}_{\nu}h_{\tau\sigma} + \tilde{\nabla}_{\mu}\tilde{\nabla}_{\sigma}h_{\tau\nu} - \tilde{\nabla}_{\mu}\tilde{\nabla}_{\tau}h_{\sigma\nu}\right) \\
&\quad -\tfrac{1}{2}\,\tilde{g}^{\rho\tau}\left(\tilde{\nabla}_{\nu}\tilde{\nabla}_{\mu}h_{\tau\sigma} + \tilde{\nabla}_{\nu}\tilde{\nabla}_{\sigma}h_{\tau\mu} - \tilde{\nabla}_{\nu}\tilde{\nabla}_{\tau}h_{\sigma\mu}\right) \\
&= \tilde{g}^{\rho\tau}\left(\tilde{\nabla}_{[\mu}\tilde{\nabla}_{\nu]}h_{\tau\sigma} + \tilde{\nabla}_{\mu}\tilde{\nabla}_{[\sigma}h_{\tau]\nu} + \tilde{\nabla}_{\nu}\tilde{\nabla}_{[\tau}h_{\sigma]\mu}\right), \qquad (3.85) \\
\delta R_{\sigma\nu} &= \tilde{\nabla}_{\mu}\delta\Gamma^{\mu}{}_{\sigma\nu} - \tilde{\nabla}_{\nu}\delta\Gamma^{\mu}{}_{\sigma\mu} \\
&= \tilde{g}^{\mu\tau}\left(\tilde{\nabla}_{[\mu}\tilde{\nabla}_{\nu]}h_{\tau\sigma} + \tilde{\nabla}_{\mu}\tilde{\nabla}_{[\sigma}h_{\tau]\nu} + \tilde{\nabla}_{\nu}\tilde{\nabla}_{[\tau}h_{\sigma]\mu}\right), \qquad (3.86) \\
\delta R &= -\tilde{R}_{\mu\nu}h^{\mu\nu} + \tilde{g}^{\sigma\nu}\delta R_{\sigma\nu} \\
&= -\tilde{R}_{\mu\nu}h^{\mu\nu} + \tilde{g}^{\sigma\nu}\tilde{g}^{\mu\tau}\left(\tilde{\nabla}_{[\mu}\tilde{\nabla}_{\nu]}h_{\tau\sigma} + 2\tilde{\nabla}_{\mu}\tilde{\nabla}_{[\sigma}h_{\tau]\nu}\right) \\
&= (-\tilde{R}^{\mu\nu} + \tilde{\nabla}^{\mu}\tilde{\nabla}^{\nu} - \tilde{g}^{\mu\nu}\,\tilde{\Box})\,h_{\mu\nu}\,. \qquad (3.87)
\end{aligned}$$

In the last line, we used the fact that the commutator of two vector operators contracted with a symmetric rank-2 tensor vanishes. This can be seen also by remembering the commutator of covariant derivatives on a tensor:

$$\begin{aligned}
2\nabla_{[\mu}\nabla_{\nu]}A^{\rho_1\cdots\rho_n}_{\sigma_1\cdots\sigma_m} &= \sum_{p=1}^{n} R^{\rho_p}{}_{\tau\mu\nu}A^{\rho_1\cdots\rho_{p-1}\tau\rho_{p+1}\cdots\rho_n}_{\sigma_1\cdots\sigma_m} \\
&\quad -\sum_{l=1}^{m} R^{\tau}{}_{\sigma_l\mu\nu}A^{\rho_1\cdots\rho_n}_{\sigma_1\cdots\sigma_{l-1}\tau\sigma_{l+1}\cdots\sigma_m}\,. \qquad (3.88)
\end{aligned}$$

In particular, if $A_{\rho\sigma}$ is symmetric, then

$$\nabla_{[\mu}\nabla_{\nu]}A_{\rho\sigma} = -A_{\tau(\sigma}R^{\tau}{}_{\rho)\mu\nu}\,. \qquad (3.89)$$

From (3.83), (3.86) and (3.87), we obtain (3.3).

3.2 Transverse-traceless gauge. Show that it is always possible to choose the transverse-traceless gauge 3.18 in *D* dimensions.

Solution Suppose to select a global D-dimensional Lorentz frame with D-velocity $u^\mu = \delta^\mu_0$. If the (linearized) gravity theory is covariant, it is by definition invariant under a general coordinate transformation. In particular, a gauge transformation of the gravitational wave $h_{\mu\nu}$ is

$$\tilde{h}_{\mu\nu} = h_{\mu\nu} + \partial_\mu \xi_\nu + \partial_\nu \xi_\mu. \tag{3.90}$$

Let $h := \eta^{\mu\nu} h_{\mu\nu}$. The choice of ξ_μ such that $\partial_\sigma \partial^\sigma \xi_\mu = -\partial^\nu h_{\mu\nu} + (1/2)\partial_\mu h$ is called harmonic:

$$\partial^\nu \tilde{h}_{\mu\nu} = \tfrac{1}{2}\, \partial_\mu \tilde{h}. \tag{3.91}$$

We can always add an arbitrary harmonic function to ξ^μ, which corresponds to a residual gauge freedom: $\bar{h}_{\mu\nu} = \tilde{h}_{\mu\nu} + \partial_\mu \zeta_\nu + \partial_\nu \zeta_\mu$, where $\partial_\nu \partial^\nu \zeta_\mu = 0$. We want to fix the gauge so that $\bar{h}_{\mu\nu}\, u^\nu = \bar{h}_{\mu 0} = 0$. Hence,

$$\tilde{h}_{\mu 0} + \partial_\mu\, \zeta_0 + \zeta'_\mu = 0\,. \tag{3.92}$$

Also, $\bar{h} = \tilde{h} + 2\partial_\mu \zeta^\mu$, so that the trace of $\bar{h}$ vanishes if

$$\partial^\mu \zeta_\mu = -\tfrac{1}{2}\, \tilde{h}\,. \tag{3.93}$$

Both conditions (3.92) and (3.93) are satisfied when $\partial^\mu\, \tilde{h}_{\mu 0} = (1/2)\tilde{h}'$, but this relation always holds because $\tilde{h}_{\mu\nu}$ satisfies the harmonic gauge (3.91). Therefore, in a global Lorentz frame it is always possible to find a gauge in which $h_{\mu\nu}$ satisfies the constraints (3.18).

3.3 Graviton action. Derive (3.24) by perturbing the Einstein–Hilbert action (2.18) in four dimensions at second order.

Solution We get

$$\begin{aligned}
\delta^{(2)} S_h &= \frac{1}{2\kappa^2} \int \mathrm{d}^4 x \left[(\tilde{R} - 2\Lambda)\, \delta^{(2)} \sqrt{-g} + \sqrt{-\tilde{g}}\delta^{(2)} R \right] \\
&= \frac{1}{2\kappa^2} \int \mathrm{d}^4 x \sqrt{-\tilde{g}} \left[(\tilde{R} - 2\Lambda)\, \frac{1}{2}\, h_{\mu\nu} - \delta R_{\mu\nu} \right] h^{\mu\nu} \\
&= \frac{1}{2\kappa^2} \int \mathrm{d}^4 x \sqrt{-\tilde{g}} \left(\frac{1}{2} h_{\alpha\beta} \tilde{\Box} h^{\alpha\beta} - \Lambda h_{\alpha\beta} h^{\alpha\beta} \right),
\end{aligned} \tag{3.94}$$

where in the first line we dropped a vanishing term proportional to the trace of $h_{\mu\nu}$. Using (3.20), we obtain (3.24).

3.4 Covariant formalism. Prove (3.42).

Solution From definition (3.39) of the Lie derivative, and given a scalar f, one finds the useful relation

$$\begin{aligned}\mathcal{L}_u(\partial_\mu f) &= u^\nu\partial_\nu\partial_\mu f + (\partial_\mu u^\nu)\partial_\nu f = u^\nu\partial_\mu\partial_\nu f + (\partial_\mu u^\nu)\partial_\nu f \\ &= \partial_\mu \dot{f}\,.\end{aligned} \tag{3.95}$$

Then, from (3.40) we get

$$\begin{aligned}\mathcal{L}_u(\zeta_\mu) &= \mathcal{L}_u(\partial_\mu\mathcal{N}) + \frac{1}{D-1}\frac{1}{\rho+P}\mathcal{L}_u(\partial_\mu\rho) + \frac{1}{D-1}\mathcal{L}_u\left(\frac{1}{\rho+P}\right)\partial_\mu\rho \\ &\overset{(3.95)}{=} \partial_\mu\dot{\mathcal{N}} + \frac{1}{D-1}\frac{1}{\rho+P}\partial_\mu\dot{\rho} - \frac{\dot{\rho}+\dot{P}}{(D-1)(\rho+P)^2}\partial_\mu\rho \\ &\overset{(2.76)}{=} -\frac{\dot{\mathcal{N}}}{\rho+P}\left(\partial_\mu P - \frac{\dot{P}}{\dot{\rho}}\partial_\mu\rho\right)\,.\end{aligned} \tag{3.96}$$

The bracket in the right-hand side coincides with (3.43).

3.5 Wiener–Khintchine theorem. Show that P_φ and B_φ, defined in (3.71) and (3.78), are the Fourier transform of the two-point and three-point correlation function, respectively. Prove the Wiener–Khintchine theorem (3.73).

Solution From definition (3.71),

$$\begin{aligned}\langle\varphi(\mathbf{x}_1)\varphi(\mathbf{x}_2)\rangle &= \int\frac{\mathrm{d}^3\mathbf{k}_1}{(2\pi)^3}\frac{\mathrm{d}^3\mathbf{k}_2}{(2\pi)^3}\langle\varphi_{\mathbf{k}_1}\varphi_{\mathbf{k}_2}\rangle\mathrm{e}^{\mathrm{i}(\mathbf{k}_1\cdot\mathbf{x}_1+\mathbf{k}_2\cdot\mathbf{x}_2)} \\ &\overset{(3.71)}{=} \int\frac{\mathrm{d}^3\mathbf{k}}{(2\pi)^3}P_\varphi(k)\mathrm{e}^{\mathrm{i}\mathbf{k}\cdot(\mathbf{x}_1-\mathbf{x}_2)}\,,\end{aligned} \tag{3.97}$$

while from definition (3.78)

$$\begin{aligned}\langle\varphi(\mathbf{x}_1)\varphi(\mathbf{x}_2)\varphi(\mathbf{x}_3)\rangle &= \int\frac{\mathrm{d}^3\mathbf{k}_1}{(2\pi)^3}\frac{\mathrm{d}^3\mathbf{k}_2}{(2\pi)^3}\frac{\mathrm{d}^3\mathbf{k}_3}{(2\pi)^3}\mathrm{e}^{\mathrm{i}(\mathbf{k}_1\cdot\mathbf{x}_1+\mathbf{k}_2\cdot\mathbf{x}_2+\mathbf{k}_3\cdot\mathbf{x}_3)} \\ &\qquad\times\langle\varphi_{\mathbf{k}_1}\varphi_{\mathbf{k}_2}\varphi_{\mathbf{k}_3}\rangle\end{aligned}$$

$$\stackrel{(3.78)}{=} \int \frac{d^3\mathbf{k}_1}{(2\pi)^3} \frac{d^3\mathbf{k}_2}{(2\pi)^3} e^{i[\mathbf{k}_1\cdot(\mathbf{x}_1-\mathbf{x}_3)+\mathbf{k}_2\cdot(\mathbf{x}_2-\mathbf{x}_3)]} \times B_\varphi(k_1,k_2,k_3)\,. \tag{3.98}$$

In particular,

$$\langle\varphi^3(\mathbf{x})\rangle = \int \frac{d^3\mathbf{k}_1}{(2\pi)^3} \frac{d^3\mathbf{k}_2}{(2\pi)^3} B_\varphi(k_1,k_2,k_3)\,. \tag{3.99}$$

Let β be the angle between $\mathbf{k}$ and $\mathbf{x}_1-\mathbf{x}_2$ and let $\varrho = |\mathbf{x}_1-\mathbf{x}_2|$. We recall that the infinitesimal "surface" ($(D-2)$-dimensional volume) element of the $(D-2)$-sphere (boundary of the unit $(D-1)$-ball) is

$$d\Omega_{D-2} = d\theta \sin\varphi_1 d\varphi_1 \sin^2\varphi_2 d\varphi_2 \cdots \sin^{D-3}\varphi_{D-3} d\varphi_{D-3}\,. \tag{3.100}$$

The surface of the $(D-2)$-sphere is

$$\Omega_{D-2} = \int d\Omega_{D-2} = \frac{2\pi^{(D-1)/2}}{\Gamma[(D-1)/2]}\,. \tag{3.101}$$

In particular, $\Omega_2 = 4\pi$. Since $d^3\mathbf{k} = d\Omega_2 dk k^2$, (3.97) gives

$$\begin{aligned}\xi_2^{(\varphi)}(\varrho) &:= \langle\varphi(\mathbf{x}_1)\varphi(\mathbf{x}_2)\rangle \\ &= \int_0^{+\infty} dk k^2 \frac{2\pi}{(2\pi)^3} P_\varphi(k) \int_{-1}^{1} d(\cos\beta) e^{ik\varrho\cos\beta} \\ &= \int_0^{+\infty} \frac{dk}{k} \left[\frac{k^3}{2\pi^2} P_\varphi(k)\right] \frac{\sin(k\varrho)}{k\varrho}\,,\end{aligned}$$

as anticipated.

References

1. H. Kodama, M. Sasaki, Cosmological perturbation theory. Prog. Theor. Phys. Suppl. **78**, 1 (1984)
2. V.F. Mukhanov, H.A. Feldman, R.H. Brandenberger, Theory of cosmological perturbations. Phys. Rep. **215**, 203 (1992)
3. J.M. Bardeen, Gauge-invariant cosmological perturbations. Phys. Rev. D **22**, 1882 (1980)
4. J.M. Bardeen, P.J. Steinhardt, M.S. Turner, Spontaneous creation of almost scale-free density perturbations in an inflationary universe. Phys. Rev. D **28**, 679 (1983)
5. D.S. Salopek, J.R. Bond, Nonlinear evolution of long wavelength metric fluctuations in inflationary models. Phys. Rev. D **42**, 3936 (1990)

6. D. Wands, K.A. Malik, D.H. Lyth, A.R. Liddle, A new approach to the evolution of cosmological perturbations on large scales. Phys. Rev. D **62**, 043527 (2000). [arXiv:astro-ph/0003278]
7. C.W. Misner, K.S. Thorne, J.A. Wheeler, *Gravitation* (Freeman, New York, 1973)
8. E. Lifshitz, On the gravitational stability of the expanding universe. Zh. Eksp. Teor. Fiz. **16**, 587 (1946) [J. Phys. JETP **10**, 116 (1946)]
9. I.S. Gradshteyn, I.M. Ryzhik, *Table of Integrals, Series, and Products* (Academic Press, London, 2007)
10. A. Einstein, Näherungsweise Integration der Feldgleichungen der Gravitation. Sitz.-ber. Kgl. Preuss. Akad. Wiss. **1916**, 688 (1916)
11. A. Einstein, Über Gravitationswellen. Sitz.-ber. Kgl. Preuss. Akad. Wiss. **1918**, 154 (1918)
12. B.P. Abbott et al. [LIGO Scientific and Virgo Collaborations], Observation of gravitational waves from a binary black hole merger. Phys. Rev. Lett. **116**, 061102 (2016). [arXiv:1602.03837]
13. B.P. Abbott et al. [LIGO Scientific and Virgo Collaborations], Properties of the binary black hole merger GW150914. Phys. Rev. Lett. **116**, 241102 (2016). [arXiv:1602.03840]
14. B.P. Abbott et al. [LIGO Scientific and Virgo Collaborations], Observing gravitational-wave transient GW150914 with minimal assumptions. Phys. Rev. D **93**, 122004 (2016). [arXiv:1602.03843]
15. https://losc.ligo.org/events/GW150914
16. B.P. Abbott et al. [LIGO Scientific and Virgo Collaborations], Tests of general relativity with GW150914. Phys. Rev. Lett. **116**, 221101 (2016). [arXiv:1602.03841]
17. B.P. Abbott et al. [LIGO Scientific and Virgo Collaborations], GW151226: observation of gravitational waves from a 22-solar-mass binary black hole coalescence. Phys. Rev. Lett. **116**, 241103 (2016). [arXiv:1606.04855]
18. B.P. Abbott et al. [LIGO Scientific and Virgo Collaborations], Binary black hole mergers in the first Advanced LIGO observing run. Phys. Rev. X **6**, 041015 (2016). [arXiv:1606.04856]
19. G.F.R. Ellis, M. Bruni, Covariant and gauge-invariant approach to cosmological density fluctuations. Phys. Rev. D **40**, 1804 (1989)
20. G.F.R. Ellis, J. Hwang, M. Bruni, Covariant and gauge-independent perfect-fluid Robertson–Walker perturbations. Phys. Rev. D **40**, 1819 (1989)
21. M. Bruni, G.F.R. Ellis, P.K.S. Dunsby, Gauge invariant perturbations in a scalar field dominated universe. Class. Quantum Grav. **9**, 921 (1992)
22. M. Bruni, P.K.S. Dunsby, G.F.R. Ellis, Cosmological perturbations and the physical meaning of gauge-invariant variables. Astrophys. J. **395**, 34 (1992)
23. P.K.S. Dunsby, M. Bruni, G.F.R. Ellis, Covariant perturbations in a multifluid cosmological medium. Astrophys. J. **395**, 54 (1992)
24. D. Langlois, F. Vernizzi, Evolution of non-linear cosmological perturbations. Phys. Rev. Lett. **95**, 091303 (2005). [arXiv:astro-ph/0503416]
25. D. Langlois, F. Vernizzi, Conserved non-linear quantities in cosmology. Phys. Rev. D **72**, 103501 (2005). [arXiv:astro-ph/0509078]
26. D. Langlois, F. Vernizzi, Nonlinear perturbations for dissipative and interacting relativistic fluids. JCAP **0602**, 014 (2006). [arXiv:astro-ph/0601271]
27. D. Langlois, F. Vernizzi, Nonlinear perturbations of cosmological scalar fields. JCAP **0702**, 017 (2007). [arXiv:astro-ph/0610064]
28. A. Naruko, A general proof of the equivalence between the δN and covariant formalisms. Europhys. Lett. **98**, 69001 (2012). [arXiv:1202.1516]
29. G.L. Comer, N. Deruelle, D. Langlois, J. Parry, Growth or decay of cosmological inhomogeneities as a function of their equation of state. Phys. Rev. D **49**, 2759 (1994)
30. J. Parry, D.S. Salopek, J.M. Stewart, Solving the Hamilton–Jacobi equation for general relativity. Phys. Rev. D **49**, 2872 (1994). [arXiv:gr-qc/9310020]
31. G.I. Rigopoulos, E.P.S. Shellard, The separate universe approach and the evolution of nonlinear superhorizon cosmological perturbations. Phys. Rev. D **68**, 123518 (2003). [arXiv:astro-ph/0306620]
32. G.I. Rigopoulos, E.P.S. Shellard, Non-linear inflationary perturbations. JCAP **0510**, 006 (2005). [arXiv:astro-ph/0405185]

33. G.I. Rigopoulos, E.P.S. Shellard, B.J.W. van Tent, Non-linear perturbations in multiple-field inflation. Phys. Rev. D **73**, 083521 (2006). [arXiv:astro-ph/0504508]
34. A.A. Starobinsky, Dynamics of phase transition in the new inflationary universe scenario and generation of perturbations. Phys. Lett. B **117**, 175 (1982)
35. M. Sasaki, E.D. Stewart, A general analytic formula for the spectral index of the density perturbations produced during inflation. Prog. Theor. Phys. **95**, 71 (1996). [arXiv:astro-ph/9507001]
36. M. Sasaki, T. Tanaka, Super-horizon scale dynamics of multi-scalar inflation. Prog. Theor. Phys. **99**, 763 (1998). [arXiv:gr-qc/9801017]
37. D.H. Lyth, K.A. Malik, M. Sasaki, A general proof of the conservation of the curvature perturbation. JCAP **0505**, 004 (2005). [arXiv:astro-ph/0411220]
38. R.J. Adler, *The Geometry of Random Fields* (Wiley, London, 1981)
39. R.J. Adler, J. Taylor, *Random Fields and Geometry* (Springer, New York, 2007)

Chapter 4
Cosmic Microwave Background

Philolaus puts fire in the middle, around the center, which he calls furnace of everything and abode of Zeus and mother of the gods and altar and junction and measure of nature. And then another fire at the top, surrounding the whole.

— Aëtius (ed. H. Diels), *Doxographi Graeci*, II 7,7

Contents

G. Calcagni, *Classical and Quantum Cosmology*, Graduate Texts in Physics,
DOI 10.1007/978-3-319-41127-9_4

The standard hot big bang model predicts that today the universe has a temperature of a few Kelvin [1]. In 1964, a background signal was discovered and found consistent with a black-body spectrum at the temperature of about 3 K [2], which was soon recognized as radiation from the primordial universe [3]. Later observations confirmed the black-body spectrum and defined the main characteristics of this radiation, such as the presence of tiny anisotropies in an otherwise extremely isotropic background (Fig. 4.1).

Until two decades ago, theoretical predictions of cosmological models were so precise with respect to the data then available, and the number of well-motivated scenarios so large, that estimates of cosmological parameters did not allow to make a decisive choice among the models. With the last generations of experiments inaugurated by COBE, observational cosmology has radically changed face by keeping up with, and in some cases going beyond, the theory. By now, the cosmic microwave background is recognized as one of the best bridges between models of structure formations, high-energy scenarios within and beyond the Standard Model, and their experimental verification. Its physics, therefore, is important not only for cosmology and astrophysics. In this chapter, we will give a qualitative description of CMB processes.

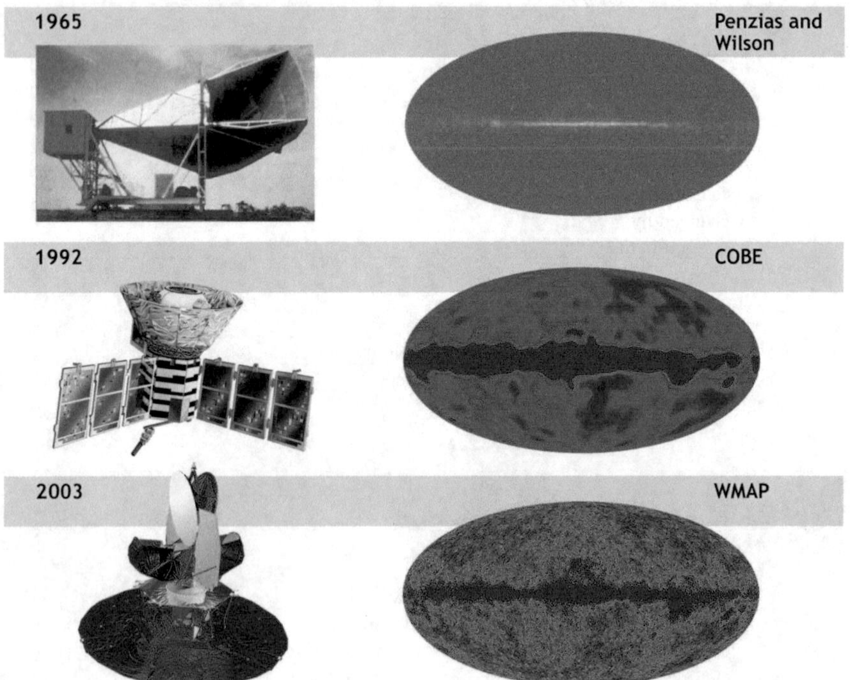

Fig. 4.1 The microwave sky from its discovery to WMAP. The *red band* in the COBE map is the microwave emissions from the Milky Way (Credit: NASA/WMAP Science Team [4])

4.1 Cosmic Background Radiation

After less than a second since the big bang, the universe presents itself as a hot plasma of matter and radiation in thermal equilibrium. Known matter is made of free baryons and electrons, not bound in atomic structures and which interact with themselves, dark matter and radiation. Nucleosynthesis of light elements (D, He, Li, ...) happens a few minutes after the birth of the universe. Baryons remain in the form of plasma until about 360,000 years, when the universe has cooled down to a temperature $T \sim 3000\,\mathrm{K} \approx 0.3\,\mathrm{eV}$. At this point, protons can capture electrons and form neutral hydrogen atoms; this process is called *recombination* and happens during decoupling at $z = z_{\mathrm{dec}}$, (2.131). The composition of the universe at decoupling is shown in Fig. 4.2. Recombination does not begin at the temperature corresponding to the ionization energy of hydrogen, 13.6 eV, mainly due to the strong compensation between recombination rate and photo-dissociation. Each electron captured in the fundamental state emits a photon which reionizes other atoms already formed. Up to recombination, photons and free electrons scatter and the universe is opaque to electromagnetic radiation. After that, the density of diffusion centers decreases, matter and radiation decouple, and the universe becomes transparent to light. Therefore, it is not possible to observe by light the cosmos beyond the *last-scattering surface* (Fig. 4.3).

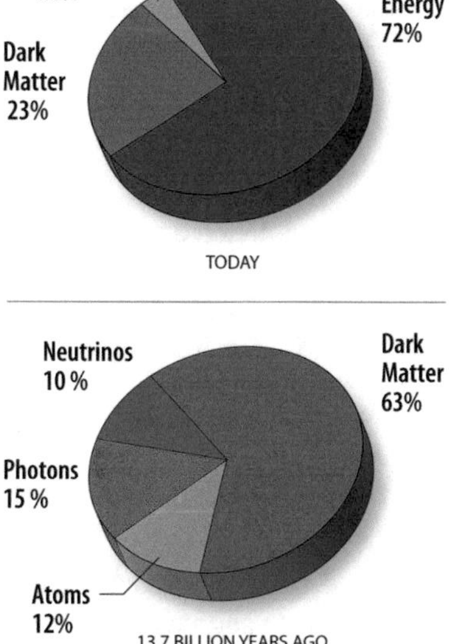

Fig. 4.2 Composition of the universe today and at decoupling. The percentages in the top pie chart are based on WMAP data and have been slightly changed by PLANCK, as reported in the text (Credit: NASA/WMAP Science Team [4])

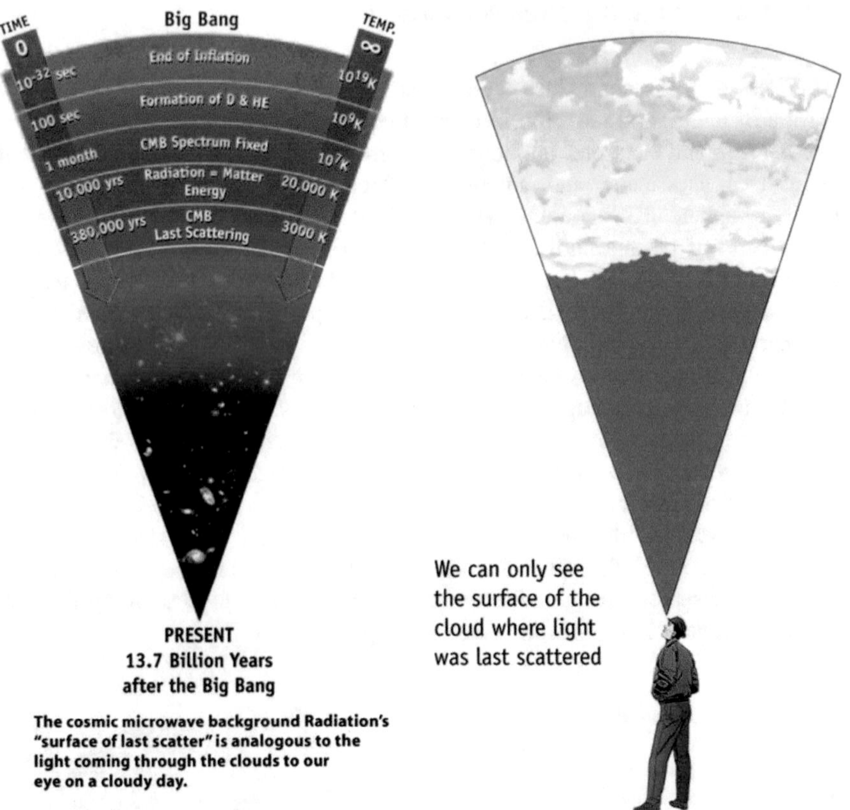

Fig. 4.3 Last-scattering surface (Credit: NASA/WMAP Science Team [4])

Cosmic radiation emitted at decoupling reaches us almost unperturbed from the last-scattering surface and its intensity peak lies in the microwave spectral region. The almost perfect CMB black-body spectrum [5, 6], which we have described in Sect. 2.5.2, is shown in Fig. 4.4.

4.1.1 Boltzmann Equation and Spectral Distortions

The physics of the interactions between matter and radiation during the history of the universe, in particular during recombination, is based upon the *radiative transfer* or *Boltzmann equation*,

$$\frac{\mathrm{d}f}{\mathrm{d}t} = C[f]\,, \tag{4.1}$$

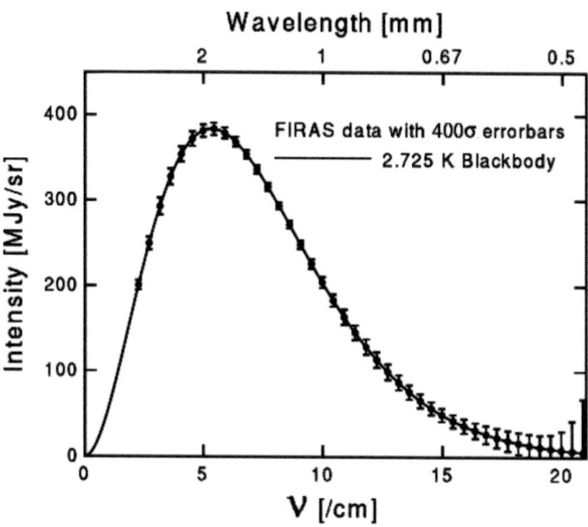

Fig. 4.4 Black-body CMB spectrum measured by the COBE FIRAS instrument [6]. The *horizontal-axis* variable is the frequency in cm^{-1}. The *vertical axis* is the power per unit area per unit frequency per unit solid angle in megajanskies per steradian ($1\,\mathrm{Jy} = 10^{-26}$ watts per square meter per hertz). The error bars are magnified 400 times! (Credit: E.L. Wright [11])

which governs the dynamics of the photon distribution $f(x^\mu, p^\mu)$. The left-hand side takes gravity into account:

$$\frac{\mathrm{d}f}{\mathrm{d}t} = \frac{\mathrm{d}x^\mu}{\mathrm{d}t}\frac{\partial f}{\partial x^\mu} + \frac{\mathrm{d}p^\mu}{\mathrm{d}t}\frac{\partial f}{\partial p^\mu} = p^\mu \frac{\partial f}{\partial x^\mu} - \Gamma^\sigma_{\mu\nu} p^\mu p^\nu \frac{\partial f}{\partial p^\sigma}\,,$$

where we used the geodesics equation. The right-hand side describes interactions with other particles. If the collision operator C vanishes, then the photon number density is conserved along geodesics in the state space.

The simple form of (4.1) is deceiving, since the mechanisms involved are extremely heterogeneous. For instance, in the collision term one must include the contributions of Compton scattering and thermal Bremsstrahlung. Each of these processes then branches into a number of secondary effects; for example, Compton scattering gives rise to Doppler shifts, distortion effects, and so on.

The construction of realistic models and the solution of radiative transfer equations, propagating through matter-radiation decoupling on an expanding cosmological background, demands a remarkable computing effort. In fact, one has to solve from a dozen coupled differential equations in analytic approaches [7–10] to several thousands of equations in numerical systems involving baryon distributions and dynamics, dark matter, photons and neutrinos. In this sense, the anthology of phenomena proposed below is only qualitative.

The thermal equilibrium established by all three processes discussed in Sect. 2.5.2 is preserved until $z \sim 10^7$ ($T \sim 1\,\mathrm{keV}$), after which any mechanism modifying radiation energy will distort the spectrum. The fact that it is difficult to thermalize the spectrum after this epoch, and that we observe an almost perfect black-body emission, constitutes solid evidence in favour of the primordial origin of the CMB. Among the distortion mechanisms, we mention three.

4.1.1.1 Sunyaev–Zel'dovich Effect

If double Compton scattering and Bremsstrahlung are dominated by inverse Compton scattering with high-energy electrons, low-frequency photons are blue-shifted towards Wien's region. When photons pass through a hot relativistic plasma (e.g., in dense clusters of galaxies) they are diffused at higher energies. Observing the plasma in the Rayleigh–Jeans region of the CMB spectrum, one sees less photons and a spot colder than the average [12, 13]. The effect is independent of the redshift of the plasma. Distortions of this kind measured in the CMB black-body spectrum allow to detect galaxy clusters.

4.1.1.2 Non-thermal Sources

Consider photons emitted by non-thermal processes such as the decay of massive particles [14] or electromagnetic phenomena. If these occur at $z \gtrsim 10^7$, they reach thermal equilibrium with CMB photons thanks to Compton scattering and Bremsstrahlung, and they leave no imprint. However, if this happens at $10^5 \lesssim z \lesssim 10^7$ ($t \sim 10^7 - 10^9$ s), non-thermal photons can reach kinetic but not chemical equilibrium, since interactions changing the number of photons are slower than the cosmic expansion rate. One will then obtain a Bose–Einstein spectrum (2.100) with non-zero chemical potential, deviating from the Planck distribution at low frequencies. Data from the COBE spectrophotometer FIRAS give an upper limit for the absolute value of the chemical potential, $|\mu| < 9 \times 10^{-5}$ at 95 % CL [6].

4.1.1.3 Compton Distortion (Comptonization)

Photons emitted before recombination at $10^3 \lesssim z \lesssim 10^5$ ($t \sim 10^9 - 10^{12}$ s) are unable to reach either kinetic or thermal equilibrium. They produce a spectral distortion due to the energy transfer from the baryonic fluid to radiation [15]:

$$I(\omega, x, y) = \frac{1}{2\pi^2} \frac{\omega^3}{e^x - 1} \left\{ 1 + y \frac{x}{1 - e^{-x}} \left[x \coth \frac{x}{2} - 4 \right] \right\},$$

where $x := \omega/T$ and y is a distortion parameter. If $y \sim 1$, the photon energy after last scattering is about twice the initial one. The same effect takes place when CMB photons are rediffused by intergalactic plasma after recombination. The FIRAS constraint is $|y| < 15 \times 10^{-6}$ at 95 % CL [6].

4.1.2 *Last-Scattering Surface*

Since recombination does not happen instantaneously, the last-scattering surface has a finite thickness Δz. To determine it, one defines the total ionization fraction $x_e(t) := n_e(t)/n_p(t)$ (n_e and n_p are the number density of free electrons and protons, respectively) and the *optical depth*

$$\tau(t) := \sigma_{\mathrm{T}} \int_t^{t_0} \mathrm{d}t' n_e(t') , \tag{4.2}$$

where σ_{T} is Thomson scattering cross-section. Since $\sigma_{\mathrm{T}} n_e$ is the probability per unit time that a photon scatters, $\exp(-\tau)$ is the photon flux attenuation factor, i.e., the probability that a photon last scattered at time t. Eventually, one finds [16–18] that the *visibility function*

$$g(t) := \sigma_{\mathrm{T}} n_e(t)\, \mathrm{e}^{-\tau(t)}$$

can be approximated, in redshift units, by a Gaussian centered at z_{dec} and of width $\Delta z \approx 80$. Therefore, the last scattering surface can be regarded as a very thin shell with radius (2.155) of about 30 billion light years.

4.2 Temperature Anisotropies: Formalism

The temperature anisotropies detected by COBE, WMAP, PLANCK and other experiments are the result of inhomogeneities of matter distribution at the time of last scattering. Compton scattering is an isotropic process in the electron rest frame, so that earlier anisotropies were cancelled before decoupling of matter and radiation. What remains are only *inhomogeneities*, perceived by the observer as temperature *anisotropies*. Later on, gravitational collapse of super-dense regions has given rise to the formation of large-scale structures such as galaxies and clusters of galaxies.

4.2.1 *Spherical Harmonics*

The microwave sky is conveniently described using a chart in polar coordinates (ϑ, φ) of the last-scattering sphere S^2, with metric $\mathfrak{g}_{ab}$

$$\mathfrak{g} = \begin{pmatrix} 1 & 0 \\ 0 & (\sin\vartheta)^2 \end{pmatrix} . \tag{4.3}$$

The *spherical harmonics* are defined as

$$Y_{\ell m}(\vartheta,\varphi)=(-1)^{\frac{m+|m|}{2}}\sqrt{\frac{2\ell+1}{4\pi}\frac{(\ell-|m|)!}{(\ell+|m|)!}}\,P_{\ell|m|}(\cos\vartheta)\,\mathrm{e}^{im\varphi}\,, \tag{4.4}$$

where $0\leqslant\vartheta\leqslant\pi$, $0\leqslant\varphi\leqslant 2\pi$, $\ell=0,1,2,\ldots$ are called *multipoles*, $m=-\ell,\ldots,\ell$ and $P_{\ell|m|}(x)$ are the associated Legendre functions:

$$P_{\ell m}(y):=\frac{(1-y^2)^{m/2}}{2^\ell \ell!}\frac{\mathrm{d}^{\ell+m}}{\mathrm{d}y^{\ell+m}}(y^2-1)^\ell\,,$$

$$P_{\ell,-m}(y)=(-1)^m\frac{(\ell-m)!}{(\ell+m)!}P_{\ell m}(y)\,.$$

These obey the orthogonality relation

$$\int_{-1}^{1}\mathrm{d}y\,P_{\ell' m}(y)P_{\ell m}(y)=\frac{2}{2\ell+1}\frac{(\ell+m)!}{(\ell-m)!}\delta_{\ell\ell'}\,. \tag{4.5}$$

The spherical harmonics have parity $(-1)^m$ under complex conjugation,

$$Y^*_{\ell m}(\vartheta,\varphi)=(-1)^m Y_{\ell,-m}(\vartheta,\varphi)\,,$$

and constitute a complete orthonormal set,

$$\int_{S^2}\mathrm{d}\Omega_2\,Y^*_{\ell m}(\vartheta,\varphi)Y_{\ell' m'}(\vartheta,\varphi)=\delta_{mm'}\delta_{\ell\ell'}\,,$$

where * denotes complex conjugation. The first spherical harmonics are

$$Y_{00}=\frac{1}{\sqrt{4\pi}}\,,\qquad Y_{10}=\sqrt{\frac{3}{4\pi}}\cos\vartheta\,,\qquad Y_{11}=-\sqrt{\frac{3}{8\pi}}\sin\vartheta\,\mathrm{e}^{i\varphi}\,.$$

If ϑ_{12} is the angle between the unit vectors $\mathbf{e}_1$ and $\mathbf{e}_2$ corresponding to directions with coordinates (ϑ_1,φ_1) and (ϑ_2,φ_2), then $\cos\vartheta_{12}=\mathbf{e}_1\cdot\mathbf{e}_2$ and one can show that

$$\sum_{m=-\ell}^{\ell}Y^*_{\ell m}(\vartheta_1,\varphi_1)Y_{\ell m}(\vartheta_2,\varphi_2)=\frac{1}{4\pi}(2\ell+1)P_\ell(\cos\vartheta_{12})\,, \tag{4.6}$$

where P_ℓ is the *Legendre polynomial* of degree ℓ and parity $(-1)^\ell$:

$$P_\ell(y):=P_{\ell 0}(y)=\frac{1}{2^\ell \ell!}\frac{\mathrm{d}^\ell}{\mathrm{d}y^\ell}(y^2-1)^\ell\,.$$

They are orthogonal according to (4.5),

$$\int_{-1}^{1} \mathrm{d}y\, P_{\ell'}(y) P_\ell(y) = \frac{2}{2\ell+1}\delta_{\ell\ell'}\,.$$

The first Legendre polynomials are

$$P_0(y) = 1\,, \quad P_1(y) = y\,, \quad P_2(y) = \tfrac{1}{2}(3y^2 - 1)\,, \quad P_3(y) = \tfrac{1}{2}(5y^3 - 3y)\,. \tag{4.7}$$

A formula we shall use once is the Legendre expansion

$$\frac{\sin(\varrho|\mathbf{e}_1 - \mathbf{e}_2|)}{\varrho|\mathbf{e}_1 - \mathbf{e}_2|} = \sum_{\ell=0}^{+\infty}(2\ell+1) j_\ell^2(\varrho) P_\ell(\mathbf{e}_1 \cdot \mathbf{e}_2)\,, \tag{4.8}$$

where ϱ is a variable and

$$j_\ell(\varrho) := (-\varrho)^\ell \left(\frac{1}{\varrho}\frac{\mathrm{d}}{\mathrm{d}\varrho}\right)^\ell \frac{\sin\varrho}{\varrho} \tag{4.9}$$

are the spherical Bessel functions j of integer order ℓ.

4.2.2 Gaussian Spectrum

Let us denote with

$$\Theta(\mathbf{e}\chi) := \frac{\delta T(\mathbf{e}\chi)}{T_0} \tag{4.10}$$

the temperature fluctuation (with respect to the mean value T_0) observed in the direction of the unit vector $\mathbf{e}$ whose tip lies on the sphere S^2 of comoving radius χ centered at the observer. The CMB temperature fluctuation has been produced at last scattering, so that $\chi \simeq \tau_0$ and the field

$$\Theta(\mathbf{e}) := \Theta(\mathbf{e}\tau_0) \tag{4.11}$$

can be parametrized by the inclination and azimuth angles (ϑ, φ). In terms of spherical harmonics,

$$\Theta(\vartheta, \varphi) = \sum_{\ell=0}^{+\infty}\sum_{m=-\ell}^{\ell} a_{\ell m} Y_{\ell m}(\vartheta, \varphi)\,, \tag{4.12}$$

where

$$a_{\ell m}=\int_{S^2}\mathrm{d}\Omega_2\,\Theta(\mathbf{e})Y^*_{\ell m}(\mathbf{e}) \tag{4.13}$$

are complex coefficients whose statistics is not known a priori. For (4.12) to be real, it must be

$$a^*_{\ell m}=(-1)^m a_{\ell,-m}\,. \tag{4.14}$$

In particular, $a_{\ell 0}$ is real.

At this point, we make an assumption which is justified experimentally, and that will be explained later by a theoretical model. Namely, we assume that temperature anisotropies come from a causal process with isotropic Gaussian distribution. From (4.13), this means that the real and imaginary parts of the coefficients $a_{\ell m}$ are statistically independent (for $m \geqslant 0$, by virtue of (4.14)) and have isotropic Gaussian distributions. For each ℓ, defining the $(2\ell+1)$-dimensional set

$$\left\{\alpha_{\ell,\bar m}\,\middle|\,\bar m=1,2,\dots,2\ell+1\right\}:=\left\{a_{\ell 0},\,\sqrt{2}\mathrm{Re}(a_{\ell m}),\,\sqrt{2}\mathrm{Im}(a_{\ell m})\,\middle|\,1\leqslant m\leqslant \ell\right\}, \tag{4.15}$$

one has [19–21]

$$f_\ell[\alpha_{\ell,\bar m}]=\frac{1}{\sqrt{2\pi\sigma_\ell^2}}\,\mathrm{e}^{-\frac{\alpha^2_{\ell,\bar m}}{2\sigma_\ell^2}}\,,\qquad \ell\in\mathbb{N}\,, \tag{4.16}$$

where σ_ℓ^2 is the variance of the distributions f_ℓ. Written as

$$C_\ell^{TT}=\sigma_\ell^2\,, \tag{4.17}$$

it is called *temperature angular spectrum.*

The isotropy assumption (statistical rotation invariance, Fig. 4.5) ensures that the variance does not depend on m, so that it is the same for all $\alpha_{\ell,\bar m}$ for a given multipole ℓ. As in Sect. 3.4, the distribution f_ℓ is normalized to 1 and correlation functions of odd order vanish. On the other hand, even-order correlation functions are completely determined by C_ℓ^{TT}.

We denote with angular brackets the average over an "ensemble of skies," i.e., a collection of skies observed at different points in the Universe for a given ℓ. For an observable $\mathcal{O}(\alpha)$ measured in the realization $\{\alpha\}\subseteq\{\alpha_{\ell,\bar m}\,|\,\ell\in\mathbb{N}\}$, the average is

$$\langle\mathcal{O}(\alpha)\rangle:=\int_{-\infty}^{+\infty}\left(\prod_{\bar m=1}^{2\ell+1}\mathrm{d}\alpha_{\ell,\bar m}f_\ell[\alpha_{\ell,\bar m}]\right)\mathcal{O}(\alpha)\,. \tag{4.18}$$

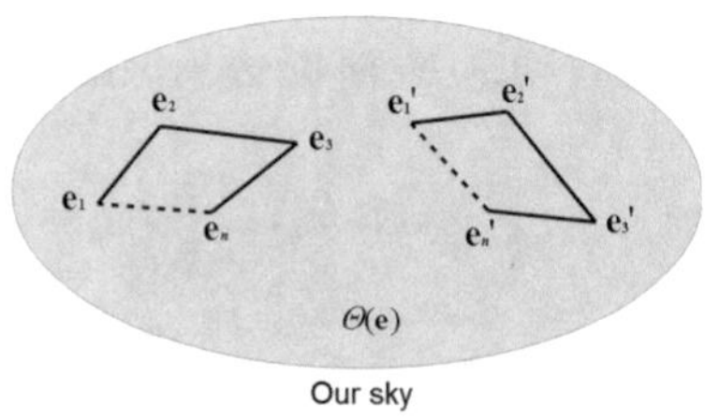

Fig. 4.5 Statistical isotropy of n-point angular correlation functions. As long as its configuration is preserved, we can average $\Theta(\mathrm{e}_1)\ldots\Theta(\mathrm{e}_n)$ over all possible orientations and positions on the sky (Credit: caption from [22])

The results of Sect. 3.4 hold with the random field φ replaced by $a_{\ell m}$: the average of the coefficients $a_{\ell m}$ over an ensemble is zero,

$$\langle a_{\ell m}\rangle = 0\,, \tag{4.19}$$

while (Problem 4.1)

$$\langle a^*_{\ell m} a_{\ell' m'}\rangle = C_\ell^{TT}\delta_{\ell'\ell}\delta_{m'm}. \tag{4.20}$$

From these formulæ, there descend the properties of the anisotropies: all the correlation functions of odd order vanish,

$$\langle\Theta(\mathbf{e}_1)\Theta(\mathbf{e}_2)\cdots\Theta(\mathbf{e}_{2n+1})\rangle = 0\,,$$

while the correlation functions of even order are completely described by the angular spectrum C_ℓ^{TT}. In particular, the *temperature-temperature spectrum* and the four-point correlation function are (Problem 4.1)

$$K(\vartheta_{12}) := \langle\Theta(\mathbf{e}_1)\Theta(\mathbf{e}_2)\rangle = \frac{1}{4\pi}\sum_{\ell=0}^{+\infty}(2\ell+1)C_\ell^{TT}P_\ell(\cos\vartheta_{12})\,, \tag{4.21}$$

$$\begin{aligned}\langle\Theta(\mathbf{e}_1)\Theta(\mathbf{e}_2)\Theta(\mathbf{e}_3)\Theta(\mathbf{e}_4)\rangle &= K(\vartheta_{12})K(\vartheta_{34}) + K(\vartheta_{13})K(\vartheta_{24})\\ &\quad + K(\vartheta_{14})K(\vartheta_{23})\,.\end{aligned} \tag{4.22}$$

Orthogonality of the P_ℓ ensures that the modes ℓ do not couple with each other and can be treated separately.

Since all non-vanishing correlation functions depend only on ℓ for a perfectly isotropic Gaussian distribution, it is convenient to define the quantity

$$a_\ell^2 := \sum_{m=-\ell}^{\ell}|a_{\ell m}|^2\,, \tag{4.23}$$

whose ensemble average is proportional to the angular spectrum:

$$\langle a_\ell^2\rangle = (2\ell+1)C_\ell^{TT}\,. \tag{4.24}$$

From (4.16), it can be shown that $a_\ell := \sqrt{a_\ell^2}$ obeys the normalized distribution

$$F_\ell[a_\ell] = \left\langle \delta\left(a_\ell - \sqrt{\sum_{\bar{m}} \alpha_{\ell,\bar{m}}^2}\right)\right\rangle$$
$$= \sqrt{\frac{2}{\pi}} \frac{1}{(2\ell-1)!!} \frac{a_\ell^{2\ell}}{\sigma_\ell^{2\ell+1}} \mathrm{e}^{-\frac{a_\ell^2}{2\sigma_\ell^2}}, \qquad \ell \in \mathbb{N}. \tag{4.25}$$

This can be used instead of (4.18) to calculate ensemble averages of observables which depend only on a_ℓ, such as temperature correlation functions:

$$\langle \mathcal{O}(a_\ell)\rangle = \int_0^{+\infty} \mathrm{d}a_\ell \, F_\ell[a_\ell]\mathcal{O}(a_\ell). \tag{4.26}$$

In comoving coordinates and on flat spatial slices, the Fourier expansion of Θ is

$$\Theta(\mathbf{e}\chi) = \int_{-\infty}^{+\infty} \frac{\mathrm{d}^3\mathbf{k}}{(2\pi)^3} \Theta_{\mathbf{k}} \, \mathrm{e}^{\mathrm{i}\mathbf{k}\cdot\mathbf{e}\chi}, \tag{4.27}$$

where the coefficients $\Theta_{\mathbf{k}}$ obey a statistics inherited from the temperature field Θ in comoving space. For the CMB fluctuation field at last scattering ($\chi = \tau_0$) and from (4.13),

$$a_{\ell m} = 4\pi(-\mathrm{i})^\ell \int_{-\infty}^{+\infty} \frac{\mathrm{d}^3\mathbf{k}}{(2\pi)^3} \Theta_{\mathbf{k}} Y_{\ell m}^*(\hat{\mathbf{k}}), \tag{4.28}$$

where $\hat{\mathbf{k}} = \mathbf{k}/|\mathbf{k}|$. The analogue of (4.20) is (3.71) with the replacements $\varphi_{\mathbf{k}} \to \Theta_{\mathbf{k}}$ and $\mathbf{x} \to \tau_0\mathbf{e}$. Let β be the angle between $\mathbf{k}$ and $\mathbf{e}_1 - \mathbf{e}_2$. From the Wiener–Khintchine theorem (3.73),

$$\langle \Theta(\mathbf{e}_1)\Theta(\mathbf{e}_2)\rangle = \int_0^{+\infty} \frac{\mathrm{d}k}{k} \mathcal{P}_\Theta(k) \frac{\sin(k\tau_0|\mathbf{e}_1 - \mathbf{e}_2|)}{k\tau_0|\mathbf{e}_1 - \mathbf{e}_2|}. \tag{4.29}$$

Using (4.8) and comparing (4.29) with (4.21), we obtain

$$C_\ell^{TT} = 4\pi \int_0^{+\infty} \frac{\mathrm{d}k}{k} \mathcal{P}_\Theta(k) j_\ell^2(k\tau_0). \tag{4.30}$$

4.2.3 Ergodic Hypothesis and Cosmic Variance

The averages considered above are to be taken over the ensemble of all possible skies. Since the available data belong to just one position, we have to extract all the information from just one element of the ensemble and to keep track of an unavoidable theoretical error. At first, one might hope to achieve a great simplification by making the following assumption:

Ergodic hypothesis. *Averages over the ensemble of all possible skies are equivalent to spatial averages over one sky:*

$$\langle \mathcal{O} \rangle \stackrel{?}{=} \langle \mathcal{O} \rangle_{\rm sky} := \frac{1}{4\pi} \int_{S^2} d\Omega_2\, \mathcal{O}\,. \tag{4.31}$$

For a causal process to be spatially ergodic, it is necessary (but not sufficient) that correlation functions decay in the large-distance limit. In this case, averages over the ensemble can be replaced by spatial integrals over one realization. However, a causal process on a two-dimensional sphere (as is the case for CMB anisotropies) *cannot* be ergodic because the angular spectrum C_ℓ^{TT} is labelled on a discrete set, while a random field is ergodic if, and only if, its spectrum is continuous (Sect. 3.4). We can also sketch a direct proof of the statement. A necessary (but not sufficient) condition for the fluctuation field $\delta T/T$ on the sphere to be ergodic is that it be homogeneous and isotropic. This is true by virtue of (4.21). Furthermore, the integral over S^2 of the two-point correlation function yields

$$\langle K(\vartheta_{12}) \rangle_{\rm sky} = \sigma_0^2\,, \tag{4.32}$$

which vanishes if, and only if, the monopole variance is identically zero, i.e., when $a_{00} = 0$. In this case, the average of the fluctuation field is equal to the average over the sphere, since

$$\langle \Theta(\mathbf{e}) \rangle_{\rm sky} = \frac{a_{00}}{\sqrt{4\pi}} = 0 = \langle \Theta(\mathbf{e}) \rangle\,. \tag{4.33}$$

However, because of (4.32) it is not true that $\langle \Theta(\mathbf{e}_1)\Theta(\mathbf{e}_2) \rangle_{\rm sky} = K(\vartheta_{12})$: integrating $\Theta(\mathbf{e}_1)\Theta(\mathbf{e}_2)$ by parts and using (4.33), one finds that the left-hand side of this equation vanishes, $\langle \Theta(\mathbf{e}_1)\Theta(\mathbf{e}_2) \rangle_{\rm sky} = 0$.

Despite this result, we are forced to perform spatial averages over the only available sky. In doing so, we make an error which can be estimated. This error will be smaller for small angular scales, where it is possible to average over a large number of pairs of independent directions separated by the same angle. This corresponds to have many modes m for a given $\ell \gg 1$. At large scales, there are fewer samples and estimates of averages are more difficult. This effect is called *cosmic variance* and stems from the present impossibility to perform complete measurements, no matter how precise and accurate, of theoretical quantities.

Let us quantify cosmic variance for the two-point correlation function. From (4.21) and (4.26), the observed temperature-temperature spectrum is determined by a single realization of coefficients $a_{\ell m}$,

$$K_{\rm obs}(\vartheta_{12}) = \frac{1}{4\pi}\sum_{\ell=0}^{+\infty}(2\ell+1)C_\ell^{\rm obs}P_\ell(\cos\vartheta_{12})\,,$$

where $C_\ell^{\rm obs} := a_\ell^2/(2\ell+1)$. Then,

$$\begin{aligned}\mathrm{Var}[K(\vartheta_{12})] &:= \left\langle [K_{\rm obs}(\vartheta_{12}) - K(\vartheta_{12})]^2\right\rangle \\ &= \left(\frac{1}{4\pi}\right)^2 \sum_{\ell=0}^{+\infty}(2\ell+1)^2\sigma_{C_\ell}^2 P_\ell^2(\cos\vartheta_{12})\,,\end{aligned}$$

where (superscript TT omitted) [23]

$$\sigma_{C_\ell}^2 := \left\langle (C_\ell^{\rm obs} - C_\ell)^2\right\rangle = \left\langle C_\ell^{\rm obs\,2}\right\rangle - C_\ell^2 = \frac{2}{2\ell+1}C_\ell^2\,. \tag{4.34}$$

The cosmic variance $\sigma_{C_\ell}^2$ is smaller at large ℓ and lowest-order multipoles are afflicted by a greater theoretical uncertainty.

4.3 Temperature Power Spectrum

4.3.1 *What we Observe*

The comparison between experimental data and theoretical predictions is rather complex and can be divided into several steps illustrated in Fig. 4.6 for the temperature-temperature spectrum [24, 25]:

1. Compression of the set of time-ordered data into sky maps at different frequencies in order to minimize the effect of various noise signals. The PLANCK satellite [26–29] is sensitive to $O(10^3)$ band-powers, with 9 main band channels, from which about $O(10^5)$ maps with $O(10^7)$ pixels each is generated.
2. Compression of multi-frequency maps into a single map to minimize the contribution of instrumental noise and foreground contamination.
3. Compression of the CMB map into the angular spectrum. In the PLANCK case, the spectrum has about $O(10^3)$ points.
4. Conversion of spectrum measurements into constraints of $O(10)$ cosmological parameters.

The quantities of Sects. 4.2.2 and 4.2.3 do not take into account instrumental effects and experimental conditions. Temperature fluctuations (and, as we shall see later, polarization) cannot be determined at an exact point in the sky. A receiver has a finite *angular resolution* limiting the determination of the spectrum at momenta

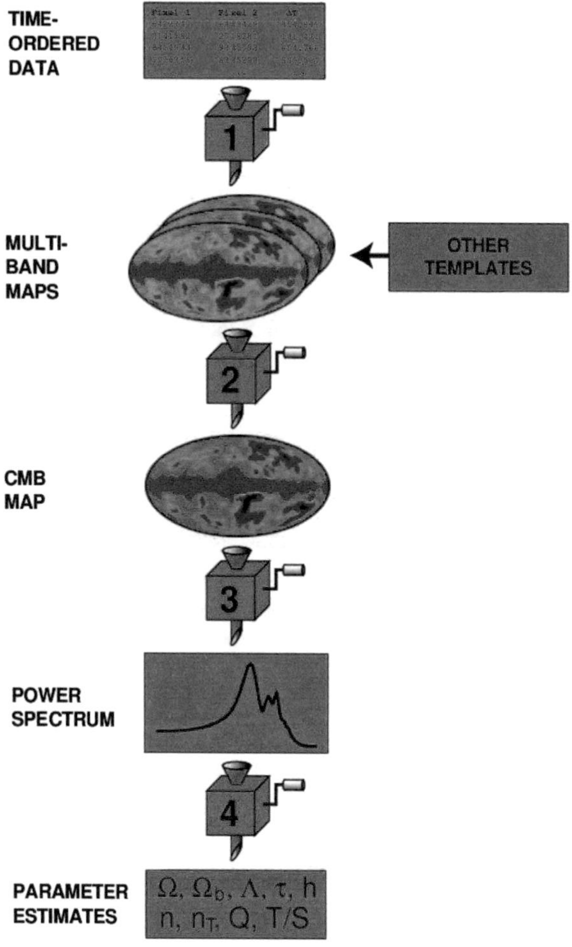

Fig. 4.6 The analysis of a large CMB data set is conveniently broken down into four steps: map-making, foreground removal, power spectrum extraction and parameter estimation (see also [24]) (Credit: [25])

smaller than a certain threshold. Conventionally, the sensitivity of the instrument is encoded in a window function W_ℓ. For instance, the temperature two-point correlation function would be smeared as

$$K(\vartheta_{12}) = \frac{1}{4\pi}\sum_{\ell=0}^{+\infty}(2\ell+1)C_\ell^{TT}W_\ell(\vartheta_{12})P_\ell(\cos\vartheta_{12}),$$

where the symmetry of the apparatus allows one to consider W_ℓ as a function of the separation angle only. In general, the window function depends on several factors, including the position of the sensor as a function of time and the observational strategy. If the sensor has a Gaussian response of width σ, the window function is $W_\ell = \exp[-\ell(\ell+1)\sigma^2]$ [30], which cuts high multipoles. A low-multipole cut for an experiment taking pairwise measurements separated by an angle θ is controlled by the window function [23]

$$W_\ell = 2[1 - P_\ell(\cos\theta)] \exp[-\ell(\ell+1)\sigma^2] .$$

There are several other sources of uncertainty which depend on the experiment. For instance, thermal maps covering only a limited portion of the sky carry an error $\sigma^2_{\rm sample}$, called sample variance, due to the fact that one is taking measurements for a multipole ℓ using only some of the momenta m. The error is determined by cosmic variance and the observed solid angle $\Omega_{\rm obs}$ [31]:

$$\sigma^2_{\rm sample} \simeq \frac{4\pi}{\Omega_{\rm obs}} \sigma^2_{C_\ell} .$$

Just to give the reader an idea of the progress of full-sky CMB experiments as far as temperature anisotropies are concerned, we can compare the angular resolution θ of COBE, WMAP and PLANCK. The DMR experiment (Differential Microwave Radiometer) on board of COBE, which produced the first thermal map of the sky, had $\theta = 7°$, corresponding to measurements of the spectrum at $\ell \lesssim 15$. The angular resolution of WMAP was about 33 times better, $\theta \gtrsim 0.2°$, $\ell \lesssim 1000$. The level of detail of PLANCK maps has been improved by a factor of 2, $\theta \approx 0.07°$ ($\ell < 2500$). We already have a satisfactory knowledge of the temperature spectrum at large and intermediate scales. At small scales, further observations will be necessary to place stronger constraints on some cosmological parameters [32].

To date, the most recent determination of the angular spectrum is based upon the 2015 PLANCK data release. The PLANCK experimental (binned) points are shown in Fig. 4.7 for the conventionally rescaled power spectrum

$$\frac{\ell(\ell+1)}{2\pi} C_\ell^{TT} , \tag{4.35}$$

together with the effect of cosmic variance. The solid line in the plots is the best-fit curve of the ΛCDM concordance model, which contains $O(10)$ free parameters. To understand why the spectrum has this particular shape, we have to give a qualitative explanation of the CMB physical mechanisms.

4.3.2 Angular Scales

Suppose to observe a small patch in the sky of angular scale ϑ (in radians; the angle in degrees will be denoted as $\theta = 360°\vartheta/(2\pi)$). One can show that, in the patch,

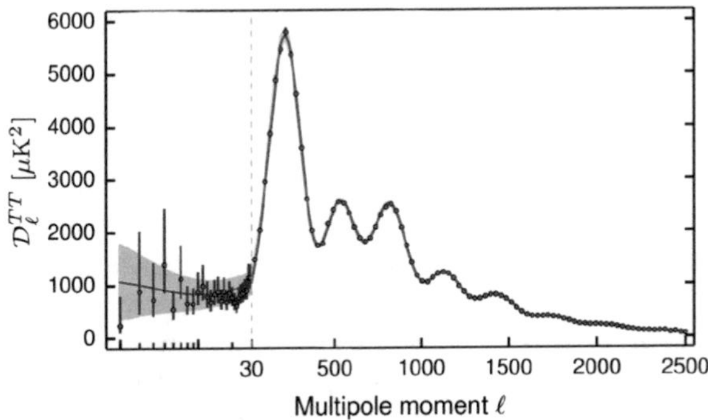

Fig. 4.7 The temperature (*TT*) power spectrum (4.35) detected by PLANCK. The curve is the ΛCDM model best fit. The *colored band* represents cosmic variance. The error bars show $\pm 1\,\sigma$ uncertainties and include cosmic variance (©ESA and the Planck Collaboration [33])

the approximate relation $\vartheta \simeq 2\pi/\ell$ holds [16],

$$\theta \simeq \frac{360^\circ}{\ell}. \tag{4.36}$$

The greater the multipole ℓ, the smaller is the angular scale.

We have seen in (2.155) that the comoving distance of the last-scattering surface is approximately τ_0. Then, the comoving distance subtending an angle θ in the microwave sky is

$$\chi \simeq \tau_0\vartheta \simeq 250\,\frac{\theta}{1^\circ}\,\text{Mpc}. \tag{4.37}$$

For large ℓ, there is a simple relation between comoving wave-number and multipole order,

$$k \simeq \frac{\ell}{\tau_0}. \tag{4.38}$$

The first term of the harmonic expansion (4.12) is the *monopole*, $\ell = 0$, a correction of the mean temperature measured in a single sky with respect to the average over the ensemble. This contribution is not measurable.

The *dipole* term, $\ell = 1$, is dominated by an anisotropy of order $\Theta \sim 10^{-3}$ due to the relative motion of our galaxy with respect to the CMB frame. This anisotropy varies as $1+\cos\theta$, where θ is the angle between the line of sight and the direction of motion. The velocity of the Local Group (the cluster of galaxies to which the Milky Way belongs) with respect to the CMB frame turns out to be $(627 \pm 22)\,\text{km}\,\text{s}^{-1}$ towards the Hydra–Centaurus super-cluster [34]. The velocity of the Sun, on the other hand, is $v \approx 371\,\text{km}\,\text{s}^{-1} \approx 1.2 \times 10^{-3}c$. This term is the largest source of anisotropies, $\delta T/T = O(\text{mK})$; it can be removed from CMB maps and will be ignored from now on.

The first contribution of intrinsic anisotropies is the *quadrupole*, $\ell = 2$. Below, we will see that the concordance-model best fit of the CMB spectrum predicts a quadrupole amplitude rather different from the observed value. Sometimes, this fact is invoked to justify, or constrain, theoretical models beyond the concordance scenario. However, we should stress that the cosmic variance $\sigma^2_{C_\ell}$ is still too high at these scales. This does not mean that we have to throw away data at low multipoles, because best-fit curves also interpolate points at intermediate scales, where error bars are small. A deviation of the concordance model from low-ℓ data can be ascribed to cosmic variance and other contaminations [35].

In Table 4.1 and Fig. 4.8, we make a rough distinction between large, intermediate and small scales, according to (4.36) and (4.37). Very-large-scale anisotropies are greater than the particle horizon at last scattering (Problem 4.2), while those on small scales are typically non-linear and associated with galactic formations. Consequently, fluctuations smaller than a few degrees strongly depend on the details of the cosmological model (matter content, reionization, and so on), while large-scale predictions are founded upon more general assumptions.

Due to the finite width of the last-scattering surface, anisotropies smaller than about 2.4′ are washed out (Problem 4.3).

4.3.2.1 Geometry

If the universe is not flat globally, projection effects modify temperature anisotropies. A comoving distance on the last-scattering surface is approximately given by (2.154) for $\tau = 0$. Equation (4.37) is therefore replaced by

$$\vartheta \simeq \frac{\chi}{\chi(0)} \simeq \vartheta_{\text{flat}} \frac{\tau_0}{\chi(0)} . \tag{4.39}$$

Figure 4.9 shows that $\tau_0/\chi(0) > 1$ (< 1) for a closed (respectively, open) universe. Intuitively, light rays are focussed (respectively, diverge) in closed (open) geometry. Therefore, in a closed (open) universe a given comoving scale subtends a larger (smaller) angle with respect to the flat case. The aspect of actual CMB maps is very close to that simulated for zero curvature (Fig. 4.10).

Table 4.1 Multipoles versus angular scales versus comoving scales

	ℓ	θ (°)	Scales (Mpc)
Large scales	2–60	>6	>1500
Intermediate scales	60–1000	0.4–6	100–1500
Small scales	>1000	<0.4	<100

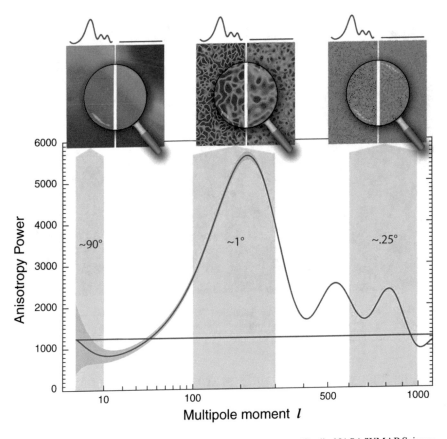

Fig. 4.8 Angular scales of the CMB temperature power spectrum (Credit: NASA/WMAP Science Team [4])

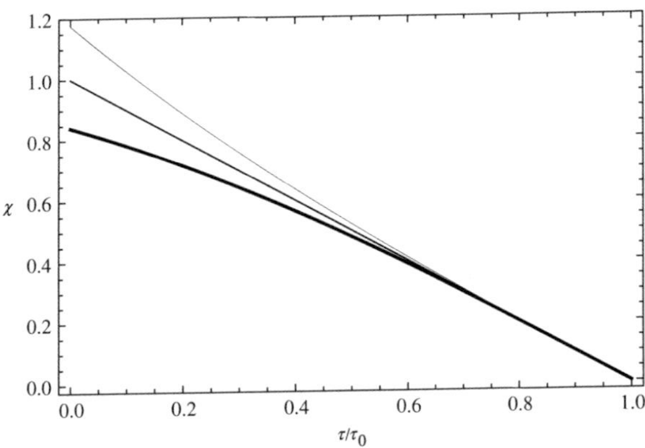

Fig. 4.9 Comoving distance $\chi(\tau)$ as a function of conformal time for an open, flat and closed universe (increasing thickness). At last scattering $\tau \approx 0$

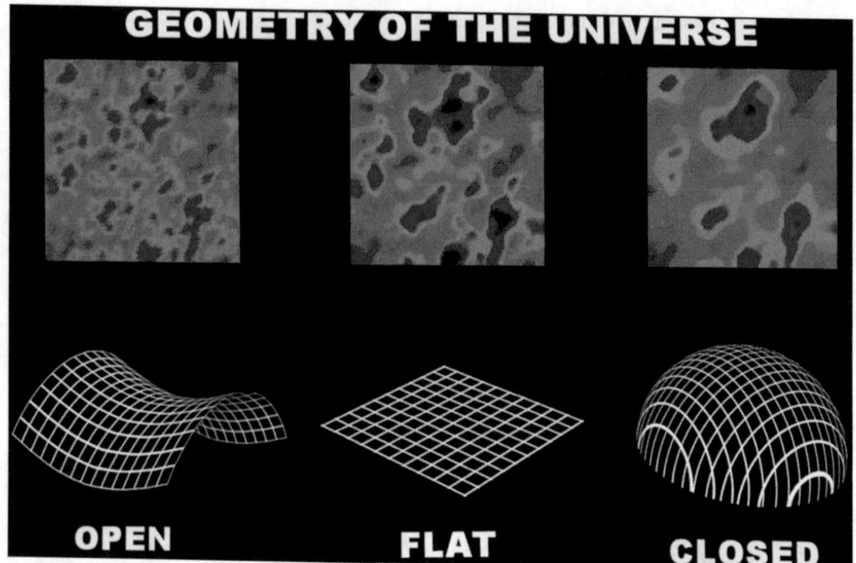

Fig. 4.10 Anisotropy distortion from geometry. Simulated CMB features are characterized by larger angular scales in a closed universe. Observations favour the flat universe (Credit: NASA/WMAP Science Team [4])

4.3.3 *Sachs–Wolfe Plateau* ($\ell \lesssim 60$)

Having described the statistics of temperature fluctuations, we review some of the most important physical mechanisms giving rise to anisotropies. The latter depend on the observed angular scale, both because they stem from different mechanisms and because the details of the same phenomenon can change with the scale. CMB anisotropies can be divided in two groups: *primary anisotropies*, originated by processes taking place before and during decoupling, and later-time *secondary anisotropies*.

4.3.3.1 Scalar Sachs–Wolfe Effect

At large scales, the dominant contribution to anisotropies is the *Sachs–Wolfe effect* [36, 37]: photons travelling from the last-scattering surface to the observer pass through metric perturbations of the background which lower their frequency. Overdense regions of the last-scattering surface correspond to deeper potential wells, so that the gravitational redshift of the photons leaving these regions is partly compensated by the intrinsic temperature excess of the gas compressed in the wells. This gives rise to cold spots in CMB maps.

Consider a perturbed metric $g_{\mu\nu} = \tilde{g}_{\mu\nu} + h_{\mu\nu}$, where $\tilde{g}_{\mu\nu}$ is the background metric and the perturbation $h_{\mu\nu}$ contains both scalar and tensor modes. Ignore for the moment the tensor contribution and consider the gravitational potential $\Phi = h_{00}/2$. A calculation of the photon geodesics in the perturbed background yields, up to a dipole term,

$$\Theta_{\rm SW} = \Theta|_{\rm e} + \Phi|_{\rm o}^{\rm e} + \frac{1}{2}\int_{\tau_0}^{\tau_{\rm e}} \mathrm{d}\tau h'_{\mu\nu}\frac{\mathrm{d}\tilde{x}^\mu}{\mathrm{d}\tau}\frac{\mathrm{d}\tilde{x}^\nu}{\mathrm{d}\tau}, \tag{4.40}$$

where e and o indicate values at the emission and observation points ($\tau_{\rm e} = \tau_{\rm dec}$) and $\tilde{x}^\mu(\tau)$ is the photon unperturbed geodesic path parametrized by conformal time.

The first term is the intrinsic fluctuation of the last-scattering surface due to the local photon density perturbations. For adiabatic perturbations, the ratio of the number densities of two species is everywhere the same and the specific entropy s is constant for every species i:

$$\delta s := \delta\left(\frac{n_\gamma}{n_i}\right) \propto \frac{\delta\rho_i}{\rho_i} - \frac{3}{4}\frac{\delta\rho_\gamma}{\rho_\gamma} = 0. \tag{4.41}$$

Then, one can show that the first term in (4.40) is

$$\Theta|_{\rm e} = \frac{1}{4}\frac{\delta\rho_\gamma}{\rho_\gamma} = \frac{1}{3}\frac{\delta\rho_{\rm m}}{\rho_{\rm m}} \simeq -\frac{2}{3}\Phi_{\rm e}. \tag{4.42}$$

The second term in (4.40) represents the gravitational redshift due to the net change of potential from the emission point to the observer. Since $\Phi_{\rm o}$ would only give an isotropic temperature shift, we can set it to zero. The last term in (4.40) is responsible for secondary anisotropies and will be discussed later. Thus, one obtains (subscript e omitted)

$$\Theta_{\rm SW} = \frac{1}{3}\Phi. \tag{4.43}$$

In the case of *isocurvature perturbations*, the total density remains unperturbed while pressure (or, equivalently, entropy) fluctuates. One can show that isocurvature fluctuations produce an anisotropy six times larger, $\delta T/T \simeq 2\Phi$.

Equation (4.43) means that the temperature fluctuation at large scales is determined by the scalar primordial perturbation via the Sachs–Wolfe effect. The observed angular spectrum in Fig. 4.7 tells us that $\ell(\ell+1)C_\ell^{TT} \approx$ const in this region, which is called *Sachs–Wolfe plateau*. How do we explain this result?

Recall that C_ℓ^{TT} can be written as an integral over wave-number modes at last scattering, (4.30). As we have just seen, the primordial spectrum $\mathcal{P}_\Theta(k)$ is actually proportional to the spectrum of primordial scalar perturbations, $\mathcal{P}_\Theta(k) \simeq \mathcal{P}_\Phi(k)/9$. At asymptotically large scales, we can approximate the spectrum to a power law,

$$\mathcal{P}_\Theta(k) \simeq A_\Theta(\tau_0 k)^{n_\Theta}, \tag{4.44}$$

where $A_\Theta = \text{const}$ is the dimensionless amplitude of the perturbation and n_Θ is the spectral index. Equation (4.30) can be integrated exactly to give [21, 23]

$$C_\ell^{TT} \simeq 4\pi A_\Theta \int_0^{+\infty} \mathrm{d}\varrho\, \varrho^{n_\Theta-1} j_\ell^2(\varrho) = \left[\frac{\sqrt{\pi}\,\Gamma(\frac{2-n_\Theta}{2})\Gamma(\ell+\frac{n_\Theta}{2})}{\Gamma(\frac{3-n_\Theta}{2})\Gamma(\ell+\frac{4-n_\Theta}{2})}\right]\pi A_\Theta\,.$$

If the primordial spectrum is exactly scale invariant, $n_\Theta = 0$, then the quantity in square brackets is equal to $2[\ell(\ell+1)]^{-1}$ and we obtain

$$\frac{\ell(\ell+1)}{2\pi} C_\ell^{TT} \simeq A_\Theta\,, \qquad \ell \ll 100\,. \tag{4.45}$$

Here is the reason why the left-hand side of this equation is preferred in spectral plots such as Fig. 4.7: at large scales, a scale-invariant (*Harrison–Zel'dovich* [38, 39]) scalar primordial spectrum generates a flat angular spectrum. This is only an approximation, since $n_\Theta \neq 0$ and the primordial spectrum is not perfectly power-law in general (the spectral index depends on the momentum) but it is a good qualitative description of what we observe at large scales.

Of course, we have not explained much in terms of physics. We have simply shifted the original question "Why is the Sachs–Wolfe region a plateau?" to a different level: Why is the primordial spectrum of scalar perturbations approximately scale invariant? Or, in other words and to yet another level: What are the primordial seeds of metric perturbations? We will answer that in the next chapter.

4.3.3.2 Gravitational Spectrum

Tensor modes $h_{+,\times}$ do not give rise to perturbations of scalar densities and their only contribution to the temperature spectrum is the last term of (4.40). Solving the perturbed Einstein equations shows that tensor modes decay with the comoving wave-number scale (Problem 4.4), so that the main effect is to increase the amplitude of the total spectrum in the region $\ell < 100$. Perturbations with wave-numbers corresponding to low ℓ were outside the horizon at last scattering. At small scales, the tensor spectrum rapidly decays by cosmic expansion.

Measurements of the CMB temperature-temperature spectrum do not distinguish between the contributions of the scalar (C_ℓ^{s}, described above) and the tensor (C_ℓ^{t}) primordial seeds. The observed angular spectrum is actually

$$C_\ell^{TT} = C_\ell^{TT,\mathrm{s}} + C_\ell^{TT,\mathrm{t}}\,. \tag{4.46}$$

As a consequence, in (4.45) one has

$$A_\Theta \simeq \beta_{\mathrm{s}} A_{\mathrm{s}} + \beta_{\mathrm{t}} A_{\mathrm{t}}\,, \qquad \ell \ll 100\,, \tag{4.47}$$

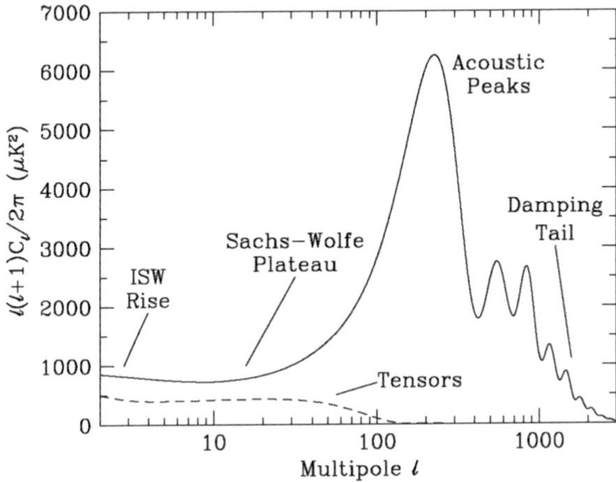

Fig. 4.11 Scalar and tensor temperature-temperature spectra for a realistic cosmological model. The main features of the spectrum at large, intermediate and small angular scales are, respectively, the Sachs–Wolfe plateau, acoustic oscillations and the damping tail (Credit: [40])

where A_s and A_t are the amplitudes of the primordial scalar and tensor perturbations, while β_s and β_t are approximately constant $O(1)$ coefficients. Therefore, scale invariance at low multipoles is guaranteed if both the scalar and the tensor primordial spectrum are scale independent. Actually, this requirement is too stringent because it turns out that realistic models predict a very low tensor-to-scalar amplitude ratio A_t/A_s (Fig. 4.11). To date, we have only an upper bound on this ratio ($A_t/A_s \ll 1$) which we will discuss in Sect. 4.4.2.

4.3.3.3 Late Integrated Sachs–Wolfe Effect

The last term in (4.40), called *integrated Sachs–Wolfe effect* (ISW), is due to the time variation of the metric. The tensor contribution to the angular spectrum is indeed part of the ISW effect. In the scalar ISW effect, the gravitational potential between source and observer decays along the signal path.

The *late ISW effect* modifies large angular scales and takes place in models where today's matter density is smaller than the critical density. In particular, $\Omega_{m0} < 1$ is a characteristic feature of open universes or flat universes with a cosmological constant. In these cases, matter-curvature (respectively, matter-Λ) equality is at

$$1 + z_K = \frac{\Omega_{K0}}{\Omega_{m0}} \qquad \text{or} \qquad 1 + z_\Lambda = \left(\frac{\Omega_{\Lambda 0}}{\Omega_{m0}}\right)^{\frac{1}{3}} . \tag{4.48}$$

Due to opposite contributions of under- and over-densities, the late ISW effect is cancelled for fluctuations of wave-length smaller than the horizon at this epoch. Since $z_K > z_\Lambda$ for a fixed density contribution, matter domination ends (and the potential decays to zero) at earlier times in open models. Thus, $\tau_{\mathrm{K}} < \tau_{\Lambda}$, $\ell_{\mathrm{K}} > \ell_{\Lambda}$ and, for a fixed Ω_{m0}, the late ISW effect in flat Λ models suffers cancellation at lower multipoles ℓ with respect to open models. In other words, at large scales the late ISW effect is greater for open universes than for flat universes with Λ. Also, in the first case there is a more pronounced shift in the position of the acoustic peaks [24].

In Fig. 4.7, one can see a feature in the Sachs–Wolfe plateau at very low multipoles called *ISW rise*. Observations ascribe the ISW rise to a cosmological constant.

4.3.4 Acoustic Peaks ($60 \lesssim \ell \lesssim 1000$)

4.3.4.1 Oscillations

Take a simplified model of photons interacting with baryonic matter. Before recombination, both components feel a gravitational force $\mathcal{F}$ in regions where the total density is perturbed. Density perturbations are directly associated with scalar perturbations of the metric, in such a way that in conformal time and momentum space

$$\frac{\mathcal{F}_{\mathbf{k}}}{1+\mathrm{R}} \simeq -\frac{k^2}{3}\Phi_{\mathbf{k}} + \Psi''_{\mathbf{k}}\,, \tag{4.49}$$

where

$$\mathrm{R} := \frac{3\rho_{\mathrm{b}}}{4\rho_{\gamma}} \simeq 3\times 10^4\,(1+z)^{-1}\,\Omega_{\mathrm{b0}}h^2\,. \tag{4.50}$$

In the longitudinal gauge, ignoring matter pressure one has $\Phi_{\mathbf{k}} \simeq \Psi_{\mathbf{k}}$. In the ensuing collapse, the cosmic fluid is compressed until photon pressure starts dominating. Then, one can show that the temperature fluctuation obeys the dynamics of a driven harmonic oscillator,

$$\Theta''_{\mathbf{k}} + k^2 c_{\mathrm{s}}^2 \Theta_{\mathbf{k}} \simeq \frac{\mathcal{F}_{\mathbf{k}}}{1+\mathrm{R}}\,, \tag{4.51}$$

where

$$c_{\mathrm{s}} := \frac{\dot{p}}{\dot{\rho}} = \frac{1}{\sqrt{3(1+\mathrm{R})}} \tag{4.52}$$

is the speed of sound of the total fluid. Equation (4.51) can be obtained from the Boltzmann and Euler equations of the fluid, in the approximation $\mathrm{R} \approx$ const on small and intermediate scales and during recombination.

For adiabatic perturbations, the initial condition is (4.42), $\Theta_{\mathbf{k}}(0) = -2\Phi_{\mathbf{k}}/3$ and $\Theta'_{\mathbf{k}}(0) = 0$. In a static potential ($\Phi'_{\mathbf{k}} = 0$) and if $R \ll 1$, the constant force shifts the zero point of the oscillation to $-\Phi_{\mathbf{k}}$: (4.51) becomes

$$(\Theta_{\mathbf{k}} + \Phi_{\mathbf{k}})'' + k^2 c_s^2 (\Theta_{\mathbf{k}} + \Phi_{\mathbf{k}}) \simeq 0\,. \tag{4.53}$$

Therefore, the initial amplitude is $\Theta_{\mathbf{k}}(0) + \Phi_{\mathbf{k}} = \Phi_{\mathbf{k}}/3$ and evolves as

$$\Theta_{\mathbf{k}}(\tau) + \Phi_{\mathbf{k}} = \frac{1}{3}\Phi_{\mathbf{k}} \cos(k c_s \tau) \simeq \frac{1}{3}\Phi_{\mathbf{k}} \cos(k r_s)\,, \tag{4.54}$$

where

$$r_s(\tau) := \int_0^\tau d\tau' c_s(\tau') \simeq c_s \tau = O(1)\,\tau \tag{4.55}$$

is the comoving *sound horizon*, which is of the same order of magnitude than the comoving particle horizon.

At last scattering, photons decouple from baryons, climb the potential wells, and suffer a redshift equal to Φ cancelling the zero-point offset. The behaviour of a perturbation of wave-number $k = \pi/(c_s\tau)$ is determined according its size with respect to r_s.

Fluctuations of comoving wave-length longer than the horizon at last scattering, $k r_s(\tau_0) \ll 1$, do not oscillate and evolve from the initial condition only by cosmological redshift: the amplitude is then fixed at $\Theta_{\mathbf{k}}(\tau) \simeq \Phi_{\mathbf{k}}/3$, which is the Sachs–Wolfe effect (4.43).

Modes within the horizon leave an oscillatory pattern in the cosmic radiation [41, 42]. From (4.54), the fluctuation peaks are located at $k r_s = n\pi$. Odd peaks $n = 1, 3, 5, \ldots$ correspond to compression phases (temperature maxima), while even peaks $n = 2, 4, 6, \ldots$ represent rarefaction phases (temperature minima).

When two causally disconnected regions merge, the relative pressure gradient forms matter structures. An isocurvature Fourier mode entering the horizon would begin to oscillate from its minimum rather than its maximum. For isocurvature initial conditions ($\Theta(0) = 0$, $\Theta'(0) = \text{const}$), one has $\Theta_{\mathbf{k}}(\tau) + \Phi_{\mathbf{k}} = \Theta'(0) \sin(k r_s)$. The peaks of the fluctuation correspond to $k r_s = (n-1/2)\pi$ and compression occurs at even n. With respect to the adiabatic scenario, acoustic peaks are out of phase by $\pi/2$.

Summarizing, the scale determining the behaviour of a mode is the comoving wave-number at horizon crossing (Sect. 3.1.3),

$$k_* = \frac{\pi}{c_s \tau_*}\,.$$

The amplitude of each mode is constant while outside the sound horizon. As soon as a mode enters the causal region, its amplitude starts oscillating like a stationary acoustic wave. Large-wave-length modes (small k, small ℓ) are the last to enter the horizon and they oscillate more slowly. Therefore, the first acoustic peak in the spectrum C_ℓ^{TT} corresponds to the *last* mode entering the horizon before decoupling.

4.3.4.2 Baryon Drag

The previous calculation did not take baryons into account, $\mathrm{R} \ll 1$. The effective mass $m_{\rm eff} \propto 1 + \mathrm{R}$ of the oscillator (4.51) increases with the baryonic matter density. When baryons collapse in the gravitational potential wells, they drag photons along, causing a displacement from the zero point, $\Theta + \Phi \rightarrow \Theta + (1 + \mathrm{R})\Phi$, and thus a greater compression. The effect survives after last scattering and heightens compression peaks with respect to under-density peaks (Fig. 4.12). For what said above, the order of the peaks allows us to discriminate between adiabatic and isocurvature perturbations. A visual inspection of Fig. 4.7 is not sufficient, but a cross-check of different features of the CMB spectrum shows that it is the adiabatic scenario to be realized in Nature.

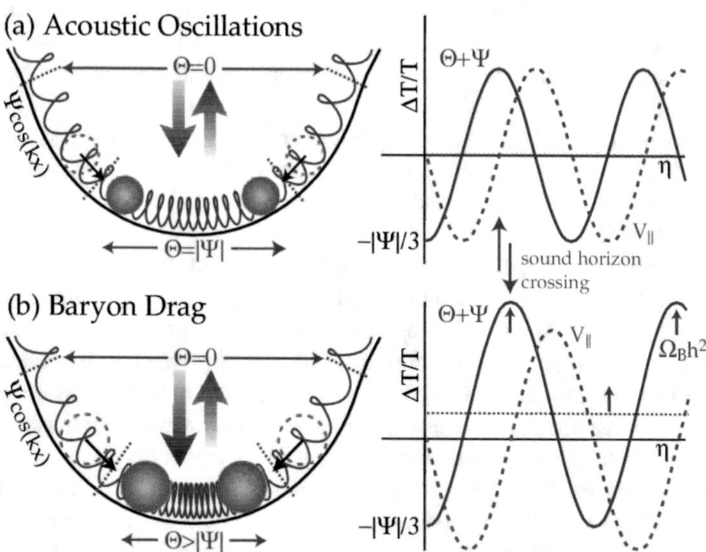

Fig. 4.12 (**a**) Acoustic oscillations (Ψ in the figure is the perturbation Φ in the text). Photon pressure resists gravitational compression of the fluid setting up acoustic oscillations (*left panel*, real space $-\pi/2 \lesssim kx \lesssim \pi/2$). Springs and balls schematically represent the fluid pressure and the effective mass, respectively. Gravity displaces the zero point such that $\Theta\cos(kx) = -\Psi\cos(kx)$ at equilibrium with oscillations in time of amplitude $\Psi/3$ (*right panel*). The displacement is cancelled by the redshift $\Psi\cos(kx)$ a photon experiences when climbing out of the well. Velocity oscillations lead to a Doppler effect $V_\parallel$ shifted by $\pi/2$ with respect to the phase of the temperature perturbation. (**b**) Baryon drag increases the gravitating mass, causing more infall and a net zero-point displacement, even after redshift. Temperature crests (compression) are enhanced over troughs (rarefaction) and Doppler contributions (Reprinted by permission from Macmillan Publishers Ltd: Nature [43], ©1997)

4.3.4.3 Doppler Effect

Each Fourier mode of the density field induces a mode in the field of peculiar velocities, whose oscillations are out of phase by $\pi/2$ with respect to the oscillations of the baryonic fluid. This Doppler effect fills the valleys of the angular spectrum, which would go to zero otherwise. While the velocity of the observer induces a dipole anisotropy, the velocity of the fluid is associated with smaller scales and with anisotropies dependent on the line-of-sight.

4.3.4.4 Early Integrated Sachs–Wolfe Effect

The *early ISW effect* happens just after recombination if matter and radiation densities are still comparable. In the adiabatic case, the potential Φ decays between $t_{\rm dec}$ and complete matter domination. For the particular imprint of this secondary anisotropy on C_ℓ^{TT}, the first acoustic peak is widened at scales larger than the particle horizon at recombination (Problem 4.2).

4.3.5 Damping Tail ($\ell \gtrsim 1000$)

4.3.5.1 Photon Diffusion and Reionization

At small scales, the details of the interaction between matter and radiation affect the CMB temperature anisotropies [8–10]. In Fig. 4.7, we see that small-scale fluctuations are progressively damped for $\ell \gtrsim \ell_{\rm dec}$, where $\ell_{\rm dec}$ is the multipole scale characterizing the horizon at recombination (Problem 4.2). This is because photons diffuse within the horizon at $t_{\rm dec}$, when their mean free path is of order of the particle horizon. The greater the thickness of the last-scattering surface, the greater the optical depth (4.2) and the damping of the spectrum. At scales smaller than $\tau_{\rm dec}$, Θ is suppressed by a factor $e^{-\tau}$. The details of the *Silk damping* [44, 45] strongly depend on the cosmological parameters and their study can break the degeneracy between models of structure formation.

At some point during its history, the neutral hydrogen of the Universe was ionized by highly energetic objects associated with stellar evolution. Photons interacted with free electrons and diffused via Thomson scattering. In the case of an early *reionization*, fluctuations generated at recombination would be deleted, while others would be created due to the motion of the new diffusion centers. In the simple model of "instantaneous reionization" at $z_{\rm rei}$ from the neutral state to the fully ionized state, one has [16]

$$\tau \simeq 2.5 \times 10^{-3}(1 + z_{\rm rei})^{3/2}.$$

Again, Θ is suppressed by a factor $e^{-\tau}$ at scales smaller than the horizon at this epoch (large multipoles ℓ). The farther reionization is in the past, the greater the damping of the spectrum. Also, a non-negligible ionization fraction during the history of the universe would lengthen the period of Compton drag [46–49] after recombination, thus slowing down the growth of density perturbations. Actually this does not happen and reionization has a negligible effect on the CMB, since the optical depth is very small [50] (PLANCK TT+lowP+lensing):

$$\tau = 0.066 \pm 0.016\,, \qquad z_{\rm rei} = 8.8^{+1.7}_{-1.4} \qquad (68\,\%\ \mathrm{CL})\,. \tag{4.56}$$

4.3.6 Secondary Anisotropies

Secondary anisotropies are due to processes happening after decoupling and their origin is typically rather different from that of intrinsic CMB fluctuations. We mention only a few:

- Sunyaev–Zel'dovich effect (Sect. 4.1.1) and kinematic (or velocity) Sunyaev–Zel'dovich effect [51]. See [52, 53] for first detections.
- Late and early ISW effects (Sects. 4.3.3 and 4.3.4).
- The *Rees–Sciama effect* [54], i.e., the late ISW effect for potentials associated with the gravitational collapse of non-linear structures such as clusters of galaxies. In this case, one has to go beyond linear perturbation theory. The blueshift experienced by photons when falling into the potential is not cancelled by a redshift at the exit. In general, the effect is smaller than intrinsic anisotropies and is localized in correspondence with galaxy structures or voids [55, 56]. The ΛCDM model does not produce large voids, so that the size of hot and cold spots in the CMB, together with the projected galaxy distribution therein, can constrain the dark-energy paradigm. At the time of writing, no evidence of super-voids has been found [35, 50].
- The *Ostriker–Vishniac effect* [57], a Doppler shift due to the peculiar velocity of the diffusing intergalactic medium.
- *Weak gravitational lensing* [58–60]. CMB photons can be deflected by potential gradients (non-linear structures, galaxy clusters) along our line of sight. Their frequency does not change and the spectrum remains black-body. However, a non-linear distortion (mode coupling) is produced which redistributes the temperature-temperature spectrum at small angular scales ($\sim 10'$). Gravitational lenses can produce and distort non-Gaussian and $O(\mu\mathrm{K})$ polarization signals.
- Contributions from several extra-galactic sources, detectable by multi-frequency observations.

4.4 Cosmological Parameters and Observational Constraints

It is time to define the standard cosmological parameters tabulated in experimental results and entering analytic and numerical manipulations. We begin with density and equation-of-state parameters.

4.4.1 Shape of the Angular Spectrum and Parameters

So far, we got acquainted with some parameters such as

$$h,\ \Omega_{\mathrm{m}0},\ \Omega_{\mathrm{b}0},\ \Omega_{\Lambda 0},\ \Omega_{\nu 0},\ \tau\ \text{(optical depth)},\ w_\Lambda,\ \dots\ .$$

The position $\ell_1 \simeq 200\Omega_0^{-1/2}$ [61] of the first acoustic peak gives information on the total density parameter and, in particular, on the geometry of the universe and on the cosmological constant (Fig. 4.13). The BOOMERanG [62, 63] and MAXIMA experiments [64–66] took the first precision measurement of ℓ_1 [67]. In the estimate of [68], $\ell_1 = 212 \pm 17$. This alone is not sufficient to conclude that $\Omega_0 \approx 1$. In

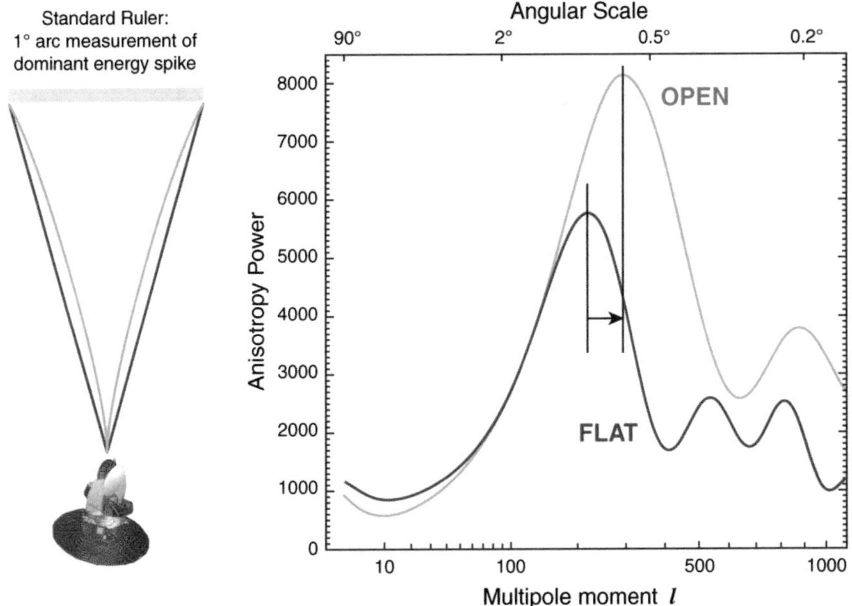

Fig. 4.13 Dependence of the CMB temperature spectrum on the geometry (Credit: NASA/WMAP Science Team [4])

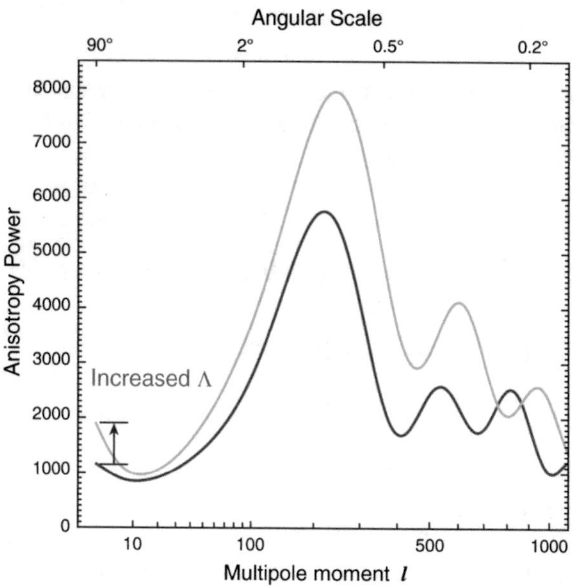

Fig. 4.14 Dependence of the CMB temperature spectrum on the cosmological constant (Credit: NASA/WMAP Science Team [4])

fact, only the joint determination of the amplitude and the position of the peaks can constrain the allowed region in the parameter space [24].[1]

Also a cosmological constant induces a shift in the position of the acoustic peaks because, for a given normalization, the anisotropies of models with $\Lambda \neq 0$ correspond to smaller angular scales. However, the effect is less pronounced than for curved models [24]. On the other hand, Λ changes the large-scale normalization via the late ISW effect (Fig. 4.14).

From (4.50) and the discussion on baryon drag, it follows that measurements of the relative height of the peaks determine $\Omega_{b0}h^2$. Changing the baryon density and h modifies the CMB spectrum accordingly, in particular the height of the first peak (Fig. 4.15).

4.4.2 Primordial Spectra

The set of standard cosmological parameters must be completed with information on the primordial spectra. For scalar perturbations, we elect the curvature perturbation

[1]For instance, a non-scale-invariant primordial spectrum shifts (but only mildly) the position of the peaks, $\Delta\ell_i/\ell_i \simeq n_i(n_s - 1)$, where $n_i \ll 1$ [67] and n_s will be defined in (4.58).

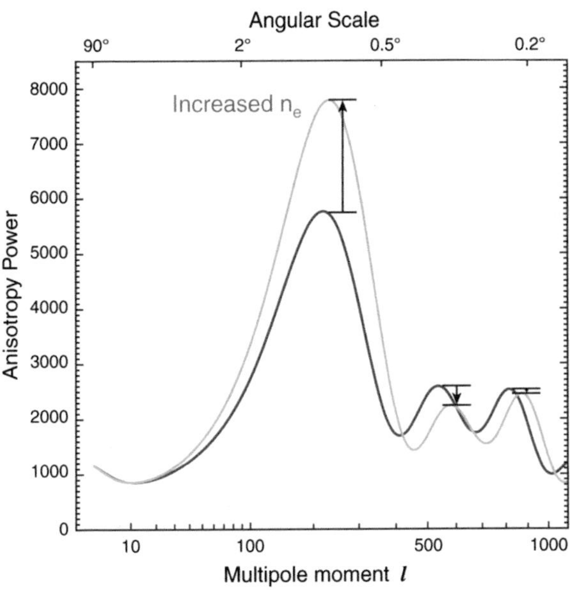

Fig. 4.15 Dependence of the peaks of the CMB temperature spectrum on baryon density (Credit: NASA/WMAP Science Team [4])

ζ, equation (3.54), as the primordial scalar degree of freedom, and its spectrum

$$\boxed{\mathcal{P}_\mathrm{s}(k) := \mathcal{P}_\zeta(k)} \tag{4.57}$$

as the primordial spectrum of scalar perturbations. Its scale dependence is governed by the *scalar spectral index*

$$\boxed{n_\mathrm{s} - 1 := \frac{\mathrm{d}\ln\mathcal{P}_\mathrm{s}}{\mathrm{d}\ln k}\,.} \tag{4.58}$$

The -1 in the left-hand side is by convention. For a power-law spectrum, n_s is constant but, in general, it will be k-dependent and the *running of the scalar index* will be non-vanishing:

$$\boxed{\alpha_\mathrm{s} := \frac{\mathrm{d}n_\mathrm{s}}{\mathrm{d}\ln k}\,.} \tag{4.59}$$

The power spectrum of ζ generates $\mathcal{P}_\Theta$ through a series of evolution processes encoded in a scale-dependent *transfer function*,

$$\Theta_{\mathbf{k}} = \mathcal{T}_\ell(k)\zeta_{\mathbf{k}} \,. \tag{4.60}$$

$\mathcal{T}_\ell(k)$ depends only on the modulus of the momentum because the dynamical equations are rotation invariant. From (3.62) and (4.43), $\Theta \simeq \Phi/3 \simeq -\zeta/5$ and one can see that the transfer function is constant at large scales,

$$\mathcal{T}_\ell(k) \approx -\frac{1}{5} \,. \tag{4.61}$$

To make contact with observations, we assume (4.57) to be the power law

$$\mathcal{P}_\mathrm{s}(k) = A_\mathrm{s}(k\tau_0)^{n_\mathrm{s}-1} \,, \tag{4.62}$$

where A_s is the large-scale *scalar amplitude* which appeared in (4.47). Comparing this expression with (4.44) and $n_\Theta = n_\mathrm{s} - 1$, and noting that $\mathcal{P}_\Theta(k) \simeq \mathcal{P}_\zeta(k)/25$, we find that the coefficient β_s in (4.47) is

$$\beta_\mathrm{s} = \frac{1}{25} \,.$$

The primordial spectrum of tensor perturbations is

$$\boxed{\mathcal{P}_\mathrm{t}(k) := \sum_{\lambda=+,\times} \mathcal{P}_{h_\lambda}(k) = 2\mathcal{P}_{h_\lambda}(k) \,,} \tag{4.63}$$

while the *tensor spectral index* is

$$\boxed{n_\mathrm{t} := \frac{\mathrm{d}\ln \mathcal{P}_\mathrm{t}}{\mathrm{d}\ln k} \,.} \tag{4.64}$$

The large-scale *tensor amplitude* A_t in (4.47) is defined via

$$\mathcal{P}_\mathrm{t}(k) = A_\mathrm{t}(k\tau_0)^{n_\mathrm{t}} \,, \tag{4.65}$$

while the coefficient β_t in (4.47) has a mild ℓ-dependence which was calculated in [69]:

$$\beta_\mathrm{t} = \left(1 + \frac{48\pi^2}{385}\right) \frac{\pi}{18} \beta_\ell \,,$$

where $\beta_2 \approx 1.118$, $\beta_3 \approx 0.878$, $\beta_4 \approx 0.819$ and $\beta_\ell \sim 1$ at large ℓ.

Another physical observable is the *tensor-to-scalar ratio*

$$\boxed{r := \frac{\mathcal{P}_\mathrm{t}}{\mathcal{P}_\mathrm{s}}\,,} \tag{4.66}$$

which replaces A_t as an observable upon using (4.62) and (4.65) in the scale-invariant limit. To summarize, there are five parameters associated with early-time physics:

$$A_\mathrm{s},\ n_\mathrm{s},\ \alpha_\mathrm{s},\ r,\ n_\mathrm{t}\,.$$

One could define other parameters such as the running of the tensor index α_t and higher-order momentum derivatives of the spectrum, but presently one can place significant constraints only on the above set of observables. In practice, observations can estimate the parameters in a finite range of momentum scales and, for a given set of data, one can choose a characteristic *pivot scale* k_0 within this range. Note that k_0 is not fixed observationally, except from the fact that we can choose any value among the scales relevant to the experiment. In general, the constraints on the parameter space, and in particular the likelihood contours, depend (even strongly) on the choice of the pivot scale [70].

The scalar spectrum with non-trivial running is parametrized as [71]

$$\mathcal{P}_\mathrm{s}(k) = \mathcal{P}_\mathrm{s}(k_0)\left(\frac{k}{k_0}\right)^{n_\mathrm{s}(k_0)-1+\frac{1}{2}\ln\left(\frac{k}{k_0}\right)\alpha_\mathrm{s}(k_0)}\,, \tag{4.67}$$

while the tensor spectrum is written as

$$\mathcal{P}_\mathrm{t}(k) = \mathcal{P}_\mathrm{t}(k_0)\left(\frac{k}{k_0}\right)^{n_\mathrm{t}(k_0)}\,, \tag{4.68}$$

where we have already set $\alpha_\mathrm{t} = 0$.

Together with the uncertainty due to cosmic variance, we should also consider that observations yield a finite discrete set of quantities C_ℓ^{TT}, $\ell = 2, 3, 4, \ldots$, which can be fitted by an infinite number of possible spectra. The ensuing degeneration is sometimes called "cosmic confusion" and only the combined experimental efforts on CMB, large-scale structure, matter distribution and astronomical explorations are able to reduce it.

In computer simulations, it is customary to restrict the parameter space either by discretizing the prior interval of some of the parameters or by fixing them altogether (*marginalization*). We can give an example of how parameter constraints change with the enlargement of the parameter space. The experimental values of some of the parameters have been quoted in Chap. 2. Here we report on the others.

The PLANCK mean for the scalar amplitude ($r = 0$) is [50] (PLANCK TT+lowP+lensing)

$$\mathcal{P}_s(k_0) = (2.139 \pm 0.063) \times 10^{-9} \qquad \text{at} \quad k_0 = 0.05\,\text{Mpc}^{-1} \quad (68\,\%\,\text{CL}). \tag{4.69}$$

The pivot scale k_0 corresponds to $\ell \approx 700$ via (4.38). Including tensor modes but excluding scalar running ($\alpha_s = 0$) and assuming a flat ΛCDM model, the PLANCK TT+lowP+lensing likelihood yields a scalar spectral index and a tensor-to-scalar ratio at the pivot scale $k_0 = 0.002\,\text{Mpc}^{-1}$ ($\ell \approx 30$) [33]

$$n_s = 0.9688 \pm 0.0061 \qquad (68\,\%\,\text{CL}), \tag{4.70}$$

$$r_{0.002} < 0.114 \qquad (95\,\%\,\text{CL}). \tag{4.71}$$

This result excludes the Harrison–Zel'dovich spectrum above 5σ.

These bounds can be slightly modified in at least two ways. One is to consider generalized reionization scenarios [72]. The other is to allow for a scalar running, as one can see in Fig. 4.16. For PLANCK TT+lowP+lensing, one gets [33]

$$\begin{aligned} n_s(k_0) &= 0.9690 \pm 0.0063 && (68\,\%\,\text{CL}), \\ \alpha_s &= -0.0076^{+0.0092}_{-0.0080} && (68\,\%\,\text{CL}), \\ r &< 0.176 && (95\,\%\,\text{CL}), \end{aligned} \tag{4.72}$$

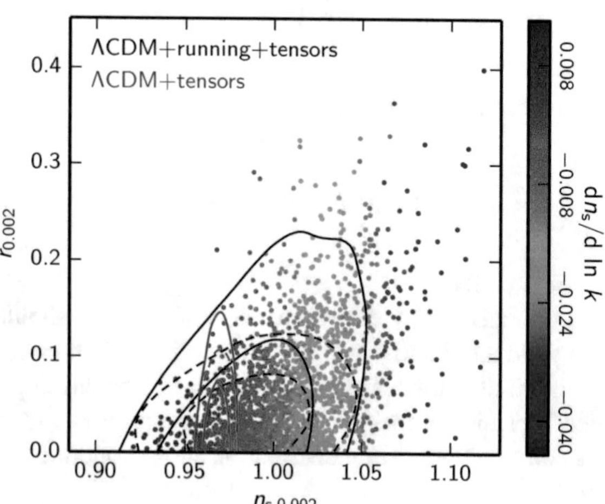

Fig. 4.16 Constraints on the tensor-to-scalar ratio r at $k_0 = 0.002\,\text{Mpc}^{-1}$ in the ΛCDM model with running, using PLANCK TT (samples, colored by the running parameter) and PLANCK TT+lensing+BAO (*black contours*). *Dashed contours* show the corresponding constraints also including the BKP B-mode likelihood. These are compared to the constraints when $\alpha_s = 0$ (*blue contours*) (Credit: [50], reproduced with permission ©ESO)

at $k_0 = 0.05\,\text{Mpc}^{-1}$, while $r_{0.002} < 0.186$ at 95 % CL. Constraints on the running depend on both prior assumptions and the data set. For instance, for $r = 0$ the PLANCK TT+lowP+lensing likelihood gives $\alpha_s = -0.0033 \pm 0.0074$ at 68 % CL, while for PLANCK TT+lowP+lensing+ext+BKP $\alpha_s(k_0) = -0.0065 \pm 0.0076$ [50]. This and other estimates all agree on the fact that a power-law spectrum with constant index is consistent with observations.

Information about the spectrum of gravitational waves may be gained both from polarization spectra (as we shall see in the next section) and from independent sources other than the CMB. Observations by the telescopes BICEP2 and *Keck Array* at the South Pole agree with PLANCK. A joint analysis (BKP) in early 2015 placed the upper bound

$$r < 0.12 \quad (95\,\%\ \text{CL}) \tag{4.73}$$

at $k_0 = 0.05\,\text{Mpc}^{-1}$ [73], while a PLANCK 2015 TT+lowP+lensing+ext+BKP likelihood analysis yields $r_{0.002} < 0.09$ at $k_0 = 0.002\,\text{Mpc}^{-1}$ [50], the same upper bound found for $k_0 = 0.05\,\text{Mpc}^{-1}$ by the more recent BICEP2+*Keck Array* data analysis [74]. Combining the latter data with PLANCK constraints gives the most stringent upper bound on r to date, at $k_0 = 0.05\,\text{Mpc}^{-1}$ [74]:

$$r < 0.07 \quad (95\,\%\ \text{CL})\,. \tag{4.74}$$

The tensor-to-scalar ratio r can be bound indirectly also through observations of large-scale structures (galaxy correlation, cluster abundance, Lyman-α forest statistics, and so on) [75]. Also, upper bounds on the tensor spectral index are obtained by combining the CMB bound on r with other experiments. The spectrum of primordial gravitational waves is not directly observed because, after inflation, it evolves into a stochastic background Ω_{gw} (gravitational-wave energy spectrum) through radiation and matter eras. This process can be encoded by a transfer function $\mathcal{T}(k)$ [76], so that [77, 78]

$$\Omega_{\text{gw}} := \frac{1}{\rho_{\text{crit},0}} \frac{d\rho_{\text{gw}}}{d\ln k} = \mathcal{T}^2(k)\mathcal{P}_{\text{t}}(k)\,, \tag{4.75}$$

where $\rho_{\text{crit},0} = 3H_0^2/\kappa^2$ is the critical energy density today and ρ_{gw} is the energy density of the gravitational waves. Then [79],

$$n_{\text{t}} \simeq \frac{1}{\ln f - \ln f_0} \ln\left(2.29 \times 10^{14} \frac{h^2 \Omega_{\text{gw}}}{r}\right), \tag{4.76}$$

where $f = k/(2\pi)$ is the frequency of the signal and $f_0 = k_0/(2\pi) = 3.10 \times 10^{-18}$ Hz corresponds to the pivot scale $k_0 = 0.002\,\text{Mpc}^{-1}$. Observations of pulsar timing (slightly changed by gravitational waves passing between the pulsar and the observer), interferometer experiments (upgrades of Advanced LIGO [80, 81], eLISA [82]) and the theory of big-bang nucleosynthesis can place constraints on the tensor index [79]. For instance, taking an upper bound $r < 0.30$, from pulsar timing $n_t \lesssim 0.79$, while from BBN $n_t \lesssim 0.15$. These bounds constrain models of primordial perturbations predicting blue-tilted spectra. Standard inflation typically does not fall into this category.

4.5 Polarization

The processes responsible for temperature anisotropies can also give rise to polarization of CMB photons [23, 83–97]. In particular, Thomson scattering always produces polarized radiation even if the incoming photons are not polarized.

In general, the Thomson differential cross-section depends on the polarization directions $\boldsymbol{\varepsilon}'$ and $\boldsymbol{\varepsilon}$ (directions of the electric field E) of the incident and scattered light, $d\sigma_T/d\Omega_2 \propto |\boldsymbol{\varepsilon}' \cdot \boldsymbol{\varepsilon}|^2$. Consider the simple case where the photon scatters at $\pi/2$ with respect to the incident direction. The outgoing polarization must be orthogonal to the outgoing direction, so that the polarization component of incoming radiation parallel to the outgoing direction does not scatter. Thus, only one polarization state is left.

Thomson scattering produces a net *linear* polarization only if the incident radiation has a quadrupole temperature anisotropy. In fact, if incoming light were isotropic, contributions from orthogonal directions would cancel each other and the scattered radiation would be unpolarized (Fig. 4.17). Therefore, one expects to find a polarization pattern in the cosmic microwave background (Fig. 4.18).

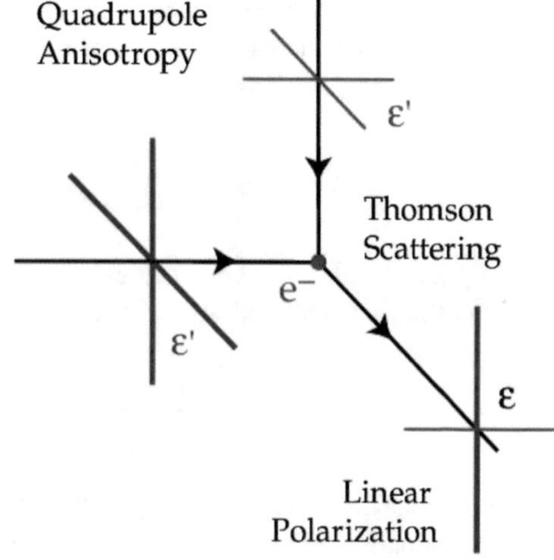

Fig. 4.17 Thomson scattering of radiation with a quadrupole anisotropy generates linear polarization. The bars orthogonal to scattering directions are the intensity of the electric field in the two polarization directions. *Blue colors* (*thick lines*) and *red colors* (*thin lines*) represent, respectively, hot and cold radiation (Credit: [97])

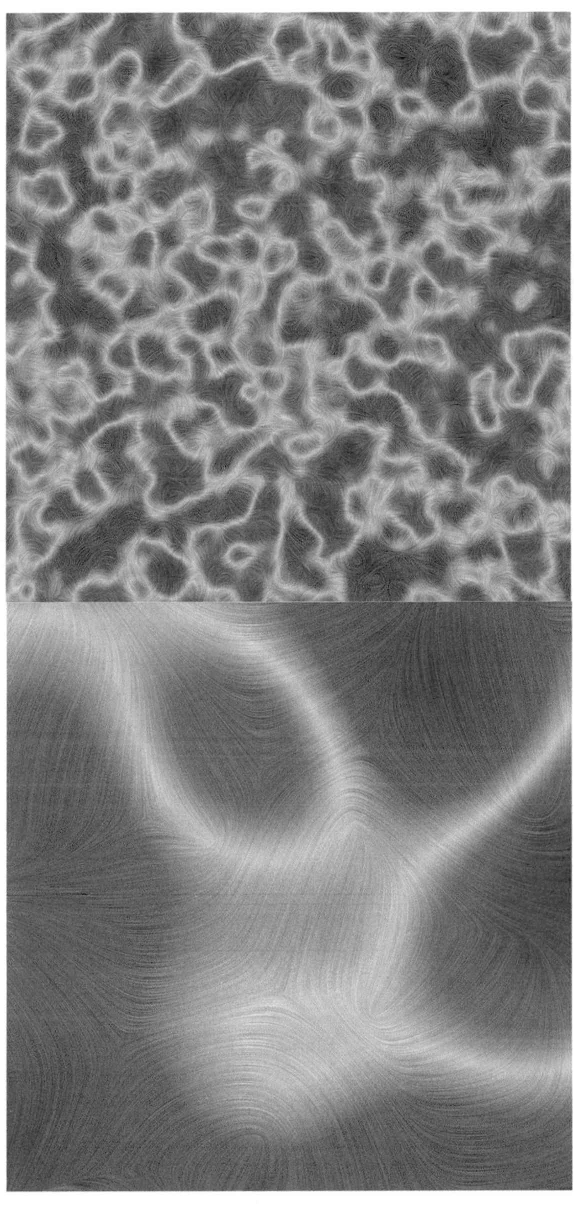

Fig. 4.18 The polarization of the CMB as detected by PLANCK, on a patch of the sky measuring 20 degrees across and a zoom. The colors represent the temperature variations above (*red*) and below (*blue*) the average temperature of the CMB, while the textures that cut across the colors show the direction and intensity of the polarized light. The curly textures are characteristic of E-mode polarization, which is the dominant type for the CMB (©ESA and the Planck Collaboration [26])

4.5.1 Formalism

Following a classical treatment, let

$$E_1 = E_{01}(t)\cos[\omega t - \theta_1(t)]\,, \qquad E_2 = E_{02}(t)\cos[\omega t - \theta_2(t)]$$

be the components of the electric field at a given point of space for a monochromatic electromagnetic wave of angular frequency ω, propagating along the x^3 direction. This wave is imagined to come from a series of Thomson scattering events.

Polarized radiation is described by the *Stokes parameters* I, Q, U, V. These are the temporal averages

$$I := \left\langle E_{01}^2\right\rangle_t + \left\langle E_{02}^2\right\rangle_t\,, \qquad Q := \left\langle E_{01}^2\right\rangle_t - \left\langle E_{02}^2\right\rangle_t\,,$$
$$U := \langle 2E_{01}E_{02}\cos(\theta_1-\theta_2)\rangle_t\,, \qquad V := \langle 2E_{01}E_{02}\sin(\theta_1-\theta_2)\rangle_t\,,$$

where I is the radiation intensity and V characterizes circular polarization. The Stokes parameters are additive in plane-wave superpositions. In unpolarized radiation, phases and amplitudes are not correlated and $Q = U = V = 0$.

For CMB radiation, the time averages do not depend on the choice of the time interval; also, there is no circular polarization and we set $V = 0$. While I and V are invariant under coordinate transformations, Q and U transform as the independent components of a rank-2 tensor P_{ab}:[2]

$$\mathrm{P}(\mathbf{e}) = \frac{1}{\sqrt{2}}\left[Q(\mathbf{e})g^+ - U(\mathbf{e})g^\times\right] = \frac{1}{\sqrt{2}}\begin{pmatrix} Q(\mathbf{e}) & -U(\mathbf{e})\sin\vartheta \\ -U(\mathbf{e})\sin\vartheta & -Q(\mathbf{e})\sin^2\vartheta \end{pmatrix}, \tag{4.77}$$

where

$$g^+ = \mathfrak{g}\,e^+ = \begin{pmatrix} 1 & 0 \\ 0 & -(\sin\vartheta)^2 \end{pmatrix}, \qquad g^\times = \sqrt{\det\mathfrak{g}}\,e^\times = \begin{pmatrix} 0 & \sin\vartheta \\ \sin\vartheta & 0 \end{pmatrix}, \tag{4.78}$$

are the polarization tensors $e_{ab}^{+,\times}$ (equation (3.21)) on the sphere. P_{ab} is traceless (with respect to the metric $\mathfrak{g}$ (4.3)) and symmetric, and as such it can be decomposed into a curl and a gradient part. These are conventionally denoted, respectively, with a superscript "*B*" and "*E*" (in some early literature, "*C*" and "*G*") in analogy with electric and magnetic fields (not to be confused with the radiation field itself). Therefore, the polarization tensor P can be expanded in spherical tensor harmonics,

$$\mathrm{P}_{ab}(\mathbf{e}) = T_0\sum_{\ell=2}^{+\infty}\sum_{m=-\ell}^{\ell}\left[a_{\ell m}^E Y_{\ell m(ab)}^E(\mathbf{e}) + a_{\ell m}^B Y_{\ell m(ab)}^B(\mathbf{e})\right], \tag{4.79}$$

[2] This definition differs by a factor $1/\sqrt{2}$ from that given in [95].

where we have omitted the monopole and dipole terms $\ell = 0, 1$ because they do not contribute [98]. $Y^E_{\ell m(ab)}$ and $Y^B_{\ell m(ab)}$ are constructed from (4.4),

$$Y^E_{\ell m(ab)} := \sqrt{\frac{2(\ell-2)!}{(\ell+2)!}} \left(Y_{\ell m;a;b} - \frac{1}{2} \mathfrak{g}_{ab} Y_{\ell m;c}{}^{;c} \right),$$

$$Y^B_{\ell m(ab)} := \sqrt{\frac{(\ell-2)!}{2(\ell+2)!}} \left(Y_{\ell m;a;c}\, \epsilon^c{}_b + Y_{\ell m;b;c}\, \epsilon^c{}_a \right),$$

where a semicolon denotes the covariant derivative with respect to $\mathfrak{g}_{ab}$ and $\epsilon = \sqrt{\det \mathfrak{g}}\, e^+ e^\times$ is the anti-symmetric tensor with components $\epsilon_{12} = -\epsilon_{21} = \sin\vartheta$. Notice that under a *parity transformation*

$$(\vartheta, \varphi) \to (\pi - \vartheta, -\varphi), \tag{4.80}$$

the Y's transform as

$$\begin{aligned} Y_{\ell m}(\pi - \vartheta, -\varphi) &= (-1)^\ell Y_{\ell,-m}(\vartheta, \varphi), \\ Y^E_{\ell m(ab)}(\pi - \vartheta, -\varphi) &= (-1)^\ell Y_{\ell,-m(ab)}(\vartheta, \varphi), \\ Y^B_{\ell m(ab)}(\pi - \vartheta, -\varphi) &= -(-1)^\ell Y_{\ell,-m(ab)}(\vartheta, \varphi). \end{aligned} \tag{4.81}$$

The tensor harmonics form a complete orthonormal basis for, respectively, the irrotational and curl components of the polarization field P:

$$\begin{aligned} &\int_{S^2} \mathrm{d}\Omega_2\, Y^{E*}_{\ell m(ab)}(\mathbf{e}) Y^{E\,(ab)}_{\ell' m'}(\mathbf{e}) = \delta_{mm'} \delta_{\ell\ell'}, \\ &\int_{S^2} \mathrm{d}\Omega_2\, Y^{B*}_{\ell m(ab)}(\mathbf{e}) Y^{B\,(ab)}_{\ell' m'}(\mathbf{e}) = \delta_{mm'} \delta_{\ell\ell'}, \\ &\int_{S^2} \mathrm{d}\Omega_2\, Y^{E*}_{\ell m(ab)}(\mathbf{e}) Y^{B\,(ab)}_{\ell' m'}(\mathbf{e}) = 0. \end{aligned} \tag{4.82}$$

Explicit expressions of the tensor harmonics in terms of the associated Legendre functions have been obtained in [95, 98, 99]:

$$Y^E_{\ell m(ab)}(\mathbf{e}) = \sqrt{\frac{(\ell-2)!}{2(\ell+2)!}} \left[W_{\ell m}(\mathbf{e}) g^+ + \mathrm{i} X_{\ell m}(\mathbf{e}) g^\times \right],$$

$$Y^B_{\ell m(ab)}(\mathbf{e}) = \sqrt{\frac{(\ell-2)!}{2(\ell+2)!}} \left[W_{\ell m}(\mathbf{e}) g^\times - \mathrm{i} X_{\ell m}(\mathbf{e}) g^+ \right],$$

where

$$W_{\ell m}(\mathbf{e}) = \left[\partial_\vartheta^2 - \cot\vartheta\,\partial_\vartheta + \frac{m^2}{(\sin\vartheta)^2}\right] Y_{\ell m}(\vartheta,\varphi) = 2\left[\partial_\vartheta^2 - \ell(\ell+1)\right] Y_{\ell m}(\vartheta,\varphi)$$

$$= -2(-1)^m \sqrt{\frac{2\ell+1}{4\pi}\frac{(\ell-m)!}{(\ell+m)!}}\, \mathrm{e}^{im\varphi} \left\{(\ell+m)\frac{\cos\vartheta}{(\sin\vartheta)^2} P_{\ell-1,m}(\cos\vartheta) - \left[\frac{\ell-m^2}{(\sin\vartheta)^2} + \frac{\ell(\ell-1)}{2}\right] P_{\ell m}(\cos\vartheta)\right\}$$

and

$$X_{\ell m}(\mathbf{e}) = 2\frac{m}{\sin\vartheta}\left(\partial_\vartheta - \cot\vartheta\right) Y_{\ell m}(\vartheta,\varphi)$$

$$= 2(-1)^m \sqrt{\frac{2\ell+1}{4\pi}\frac{(\ell-m)!}{(\ell+m)!}}\, \mathrm{e}^{im\varphi} \times \frac{m}{(\sin\vartheta)^2}\left[(\ell-1)\cos\vartheta\, P_{\ell m}(\cos\vartheta) - (\ell+m)P_{\ell-1,m}(\cos\vartheta)\right] .$$

Another formalism, based on spin-weighted spherical harmonics [100, 101], has also become a standard [93].

4.5.2 *Spectra*

The coefficients in (4.79) can be obtained by the orthonormality properties (4.82):

$$a^E_{\ell m} = \frac{1}{T_0}\int_{S^2} \mathrm{d}\Omega_2\, \mathrm{P}_{ab}(\mathbf{e})\, Y^{E\,(ab)\,*}_{\ell m}(\mathbf{e})\,, \tag{4.83}$$

$$a^B_{\ell m} = \frac{1}{T_0}\int_{S^2} \mathrm{d}\Omega_2\, \mathrm{P}_{ab}(\mathbf{e})\, Y^{B\,(ab)\,*}_{\ell m}(\mathbf{e})\,. \tag{4.84}$$

Given a polarization field $\mathrm{P}_{ab}(\mathbf{e})$, these two equations permit to separate the E and B components in the polarization signal and find the corresponding spectra.

Since temperature anisotropies and polarization are both generated by the same primordial density fluctuations, the statistical properties of the coefficients (4.83) and (4.84) will be the same as for the temperature coefficients (4.13), which we denote with a superscript T. In particular, for an isotropic Gaussian statistics one can define the mixed angular spectra (cross-correlated two-point functions) as

$$\left\langle a^{T\,*}_{\ell' m'} a^T_{\ell m}\right\rangle = C^{TT}_\ell \delta_{\ell\ell'}\delta_{mm'}\,, \qquad \left\langle a^{E\,*}_{\ell' m'} a^E_{\ell m}\right\rangle = C^{EE}_\ell \delta_{\ell\ell'}\delta_{mm'}\,, \tag{4.85}$$

$$\left\langle a^{T*}_{\ell'm'} a^{E}_{\ell m} \right\rangle = C^{TE}_{\ell} \delta_{\ell\ell'} \delta_{mm'} \,, \qquad \left\langle a^{B*}_{\ell'm'} a^{B}_{\ell m} \right\rangle = C^{BB}_{\ell} \delta_{\ell\ell'} \delta_{mm'} \,, \tag{4.86}$$

$$\left\langle a^{T*}_{\ell'm'} a^{B}_{\ell m} \right\rangle = C^{TB}_{\ell} \delta_{\ell\ell'} \delta_{mm'} \,, \qquad \left\langle a^{E*}_{\ell'm'} a^{B}_{\ell m} \right\rangle = C^{EB}_{\ell} \delta_{\ell\ell'} \delta_{mm'} \,. \tag{4.87}$$

From the parity properties (4.81), it follows that *if parity is preserved in the physics responsible for anisotropies and polarization*, then

$$C^{TB}_{\ell} = 0 = C^{EB}_{\ell} \,. \tag{4.88}$$

If this is the case, as we assume, then the statistics is completely specified by the spectra (4.85) and (4.86). In general, the polarization amplitude is smaller than the temperature amplitude because polarization has been formed after recombination, when the number of diffusors was already reduced. The thicker the last-scattering surface, the greater the signal.

Studying the Boltzmann equation for scalar, vector and tensor perturbations, one finds that also polarization spectra display a sequence of oscillations and a damping tail. Intuitively, this is because each perturbation mode still obeys a cosmological harmonic oscillator equation, but with a non-zero source coming from the anisotropic stress of radiation. Due to the velocity field at last scattering, the peaks of the polarization spectra are out of phase with respect to the thermal spectrum [91].

As in (4.30), polarization spectra are related to the primordial power spectrum $\mathcal{P}_\Theta(k)$ by transfer functions $\mathcal{T}_{i\ell}(k)$,

$$C^{ij}_{\ell} = 4\pi \int_0^{+\infty} \frac{\mathrm{d}k}{k} \mathcal{T}_{i\ell}(k) \mathcal{T}_{j\ell}(k) \mathcal{P}_\Theta(k) \,, \qquad i,j = T,E,B \,. \tag{4.89}$$

The polarization pattern changes according to the type of primordial perturbation seeding it. For instance, the curl component $a^{B}_{\ell m}$ is zero for scalar perturbations, which produce only *TE* and *EE* spectra [95]. Tensor modes generate a *BB* spectrum, which is somewhat smaller than the tensor-produced *EE* spectrum ($C^{BB}_{\ell}/C^{EE}_{\ell} \sim 8/13$ at large ℓ [16]).[3]

4.5.3 What we Observe

As in the case of the temperature spectrum, observations of polarization angular spectra suffer from a number of experimental uncertainties due to cosmic variance, instrumental noise, limited sky coverage and angular resolution [95]. From what

[3] Also vector modes can potentially give rise to a curl-curl signal (for them, $C^{BB}_{\ell}/C^{EE}_{\ell} \sim 6$ at large ℓ [16]).

Fig. 4.19 WMAP 3-year CMB map. The *white bars* show the polarization direction of the oldest light. Polarization direction is usually depicted as a discontinuous vector field, as explained in [102] or [95] (Credit: NASA/WMAP Science Team [4])

said above the observed spectra are, ignoring vector perturbations, (4.46) and

$$C_\ell^{EE} = C_\ell^{EE,\mathrm{s}} + C_\ell^{EE,\mathrm{t}}, \tag{4.90}$$

$$C_\ell^{TE} = C_\ell^{TE,\mathrm{s}} + C_\ell^{TE,\mathrm{t}}, \tag{4.91}$$

$$C_\ell^{BB} = C_\ell^{BB,\mathrm{t}}. \tag{4.92}$$

These spectra can be obtained from the two-point correlation functions of Q, U and Θ, as described in [95] (see also [52]).

The first experiment to positively detect a polarization pattern in the CMB was WMAP, first at large scales (Fig. 4.19) and then in correspondence with hot and cold spots of angular size $\theta \approx 1°$ (Fig. 4.20), as predicted by the theory [89, 90].

In models with a standard reionization scenario, the (re-normalized) *EE* spectrum is flat at low multipoles. Evaluating the posterior distribution of a band power with constant spectrum over the range $2 \leqslant \ell \leqslant 7$, the 7-year WMAP data yielded [103]

$$\frac{\ell(\ell+1)}{2\pi} C_\ell^{EE} = 0.074^{+0.034}_{-0.025}\,\mu\mathrm{K}^2\,, \qquad 2 \leqslant \ell \leqslant 7\,,$$

at 68 % CL and with cosmic variance taken into account. Comparing with the temperature quadrupole, one sees that

$$\frac{C_\ell^{EE}}{C_\ell^{TT}} \sim 10^{-3}$$

at large scales. Indeed, polarization is more difficult to detect.

The *TE* cross-spectrum observed by PLANCK is shown in Fig. 4.21 together with the ΛCDM best fit. The acoustic oscillations are clearly visible. The cosmological

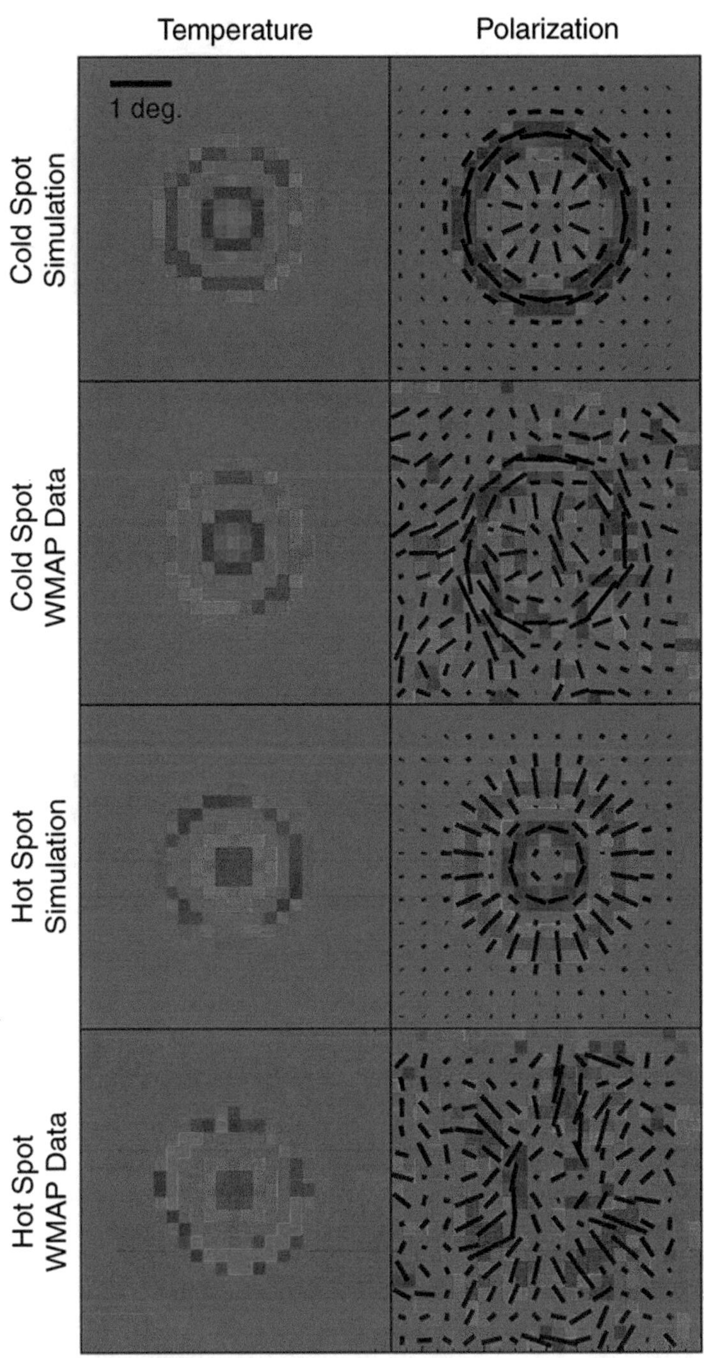

Fig. 4.20 With the 7-year results, WMAP has produced a visual demonstration that the polarization pattern around hot and cold spots (angular size $\theta \approx 1^\circ$) follows the pattern expected in the cosmic concordance model [89, 90] (Credit: NASA/WMAP Science Team [4])

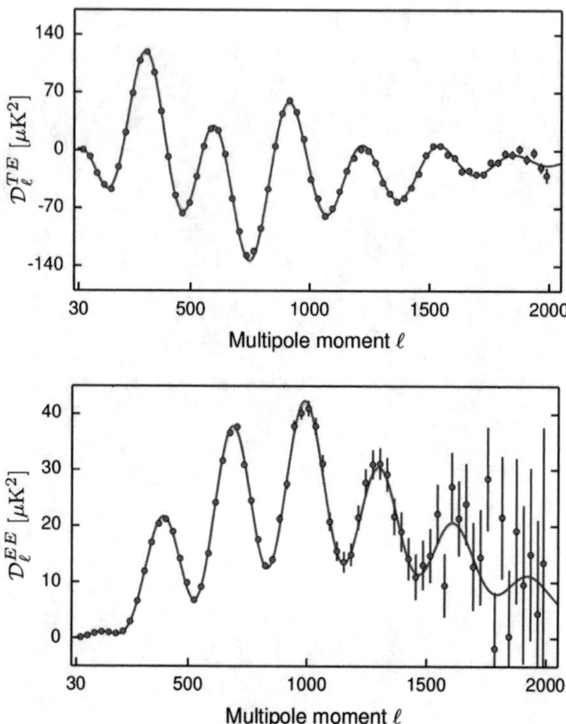

Fig. 4.21 The *TE* (*top figure*) and *EE* (*bottom figure*) polarization spectra $\ell(\ell+1)C_\ell^{TE,EE}/(2\pi)$ measured by PLANCK. *Solid lines* are the theoretical spectra computed from the ΛCDM best fit to the PLANCK data. The error bars show $\pm 1\,\sigma$ errors. A detailed discussion on the polarization signal in the PLANCK 2015 release can be found in [112] (©ESA and the Planck Collaboration [33])

primordial B-mode polarization has not been detected by WMAP or PLANCK 2015. In contrast, non-primordial B-modes sourced by gravitational lensing have been observed at high multipoles (intermediate and small scales) [104], also by SPTPOL (the polarization-sensitive receiver mounted on the South Pole Telescope) [105] and the POLARBEAR experiment in Chile [106].

The *TB* cross-spectrum in Fig. 4.22 is consistent with the null result (4.88). By late 2016, no *TB* nor *EB* spectra have been detected by the Antarctic telescopes BICEP2 and *Keck Array* either [73]. These results are in agreement both with parity conservation [52] and the notion that the primordial tensor amplitude A_t (the one capable of seeding a *BB* spectrum) is much smaller than the scalar amplitude A_s. Foreground signals and systematic effects can produce non-zero B signals, so that observations provide also a test for residual polarization contamination.

In general, given a set of values for the spectral observables, one can obtain the form of the C_ℓ^{ij}'s via dedicated numerical codes such as CAMB [107] (also available as an online tool [108]) implemented in the Fortran code COSMOMC [109], or the

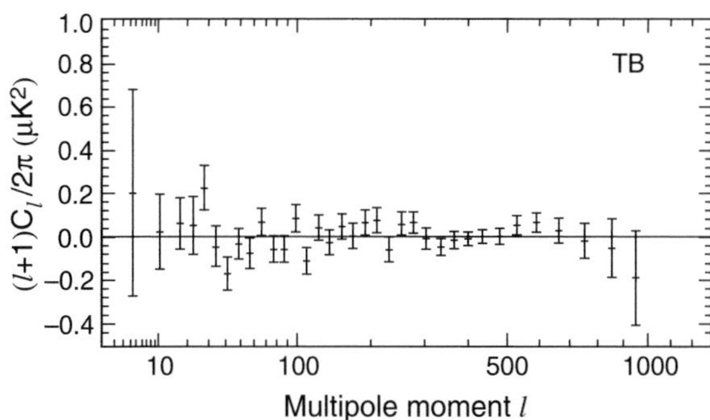

Fig. 4.22 The temperature-polarization (*TB*) cross-power spectrum measured by WMAP. This spectrum is predicted to be zero in the basic ΛCDM model and the measured spectrum is consistent with zero. Note that the plotted spectrum is $(\ell+1)C_\ell^{TB}/(2\pi)$, not $\ell(\ell+1)C_\ell^{TB}/(2\pi)$ (Credit: NASA/WMAP Science Team [113])

analogous CLASS package [110] for the Python code MONTE PYTHON [111]. The input values may either come as predictions of a model, in which case the output curve can be compared with data, or as best-fit values from data analysis.

4.6 Non-Gaussianity

So far, the results of this chapter have been obtained by assuming that the statistical distribution of the multipole coefficients $a^i_{\ell m}$, $i = T, E, B$, is isotropic and Gaussian. In this case, all odd-point correlation functions vanish, while all even-point correlation functions can be found through the two-point correlation function (Sect. 3.4).

The Gaussian hypothesis has been verified by observations, but experimental uncertainties do not exclude small deviations from it. These deviations in the temperature fluctuation Θ (and the multipole coefficients $a_{\ell m}$, via (4.28)) could be generated by exotic physics in the primordial perturbation spectrum. A non-Gaussian statistical distribution of the primordial field or fields can be quantified by the three-point and higher-order correlation functions. In parallel, an observational campaign on CMB, large-scale structure, high-z galaxies, Lyman-α forest, and gravitational lensing can detect non-Gaussian signals and clarify their physical origin [114, 115].

4.6.1 Bispectrum

The counterpart in momentum space of the three-point temperature correlation function (the bispectrum introduced in Sect. 3.4.2) is [116–118]

$$B^{m_1 m_2 m_3}_{\ell_1 \ell_2 \ell_3} := \langle a_{\ell_1 m_1} a_{\ell_2 m_2} a_{\ell_3 m_3} \rangle \ . \tag{4.93}$$

We can construct an *estimator* of the bispectrum observed in a single sky. Maintaining the isotropy hypothesis, one can write [118]

$$B^{m_1m_2m_3}_{\ell_1\ell_2\ell_3} = \mathcal{G}^{m_1m_2m_3}_{\ell_1\ell_2\ell_3}\, b_{\ell_1\ell_2\ell_3}\,, \tag{4.94}$$

where $b_{\ell_1\ell_2\ell_3}$ is a real and symmetric arbitrary function of the momenta ℓ_i called reduced bispectrum, while $\mathcal{G}^{m_1m_2m_3}_{\ell_1\ell_2\ell_3}$ is the Gaunt integral, real and symmetric under permutations of momenta ℓ_i and m_i:

$$\begin{aligned}\mathcal{G}^{m_1m_2m_3}_{\ell_1\ell_2\ell_3} &:= \int_{S^2} \mathrm{d}\Omega_2\, Y_{\ell_1m_1}(\mathbf{e}_1)Y_{\ell_2m_2}(\mathbf{e}_2)Y_{\ell_3m_3}(\mathbf{e}_3)\\ &= \sqrt{\frac{(2\ell_1+1)(2\ell_2+1)(2\ell_3+1)}{4\pi}}\begin{pmatrix}\ell_1 & \ell_2 & \ell_3\\ 0 & 0 & 0\end{pmatrix}\begin{pmatrix}\ell_1 & \ell_2 & \ell_3\\ m_1 & m_2 & m_3\end{pmatrix},\end{aligned}$$

where the two tables are the $3j$ Wigner symbol. Equation (4.94) is the generic rotation-invariant form of the bispectrum. From the properties of the Wigner symbol, it follows that the bispectrum $B^{m_1m_2m_3}_{\ell_1\ell_2\ell_3}$ must obey three conditions:

- $m_1 + m_2 + m_3 = 0$ (triangular closure condition);
- $\ell_1 + \ell_2 + \ell_3 = 2n$, where $n \in \mathbb{N}$;
- $|\ell_i - \ell_j| \leqslant \ell_k \leqslant \ell_i + \ell_j$ for every cyclic permutation.

If one of these properties is not satisfied, the Gaunt integral (and the bispectrum) vanishes identically. Using the identity

$$\sum_{m_1,m_2,m_3}\begin{pmatrix}\ell_1 & \ell_2 & \ell_3\\ m_1 & m_2 & m_3\end{pmatrix}\mathcal{G}^{m_1m_2m_3}_{\ell_1\ell_2\ell_3} = \sqrt{\frac{(2\ell_1+1)(2\ell_2+1)(2\ell_3+1)}{4\pi}}\begin{pmatrix}\ell_1 & \ell_2 & \ell_3\\ 0 & 0 & 0\end{pmatrix},$$

one can see that the estimator (bispectrum averaged over momenta m) is

$$\begin{aligned}B_{\ell_1\ell_2\ell_3} &:= \sum_{m_1,m_2,m_3}\begin{pmatrix}\ell_1 & \ell_2 & \ell_3\\ m_1 & m_2 & m_3\end{pmatrix}B^{m_1m_2m_3}_{\ell_1\ell_2\ell_3}\\ &= \sqrt{\frac{(2\ell_1+1)(2\ell_2+1)(2\ell_3+1)}{4\pi}}\begin{pmatrix}\ell_1 & \ell_2 & \ell_3\\ 0 & 0 & 0\end{pmatrix}b_{\ell_1\ell_2\ell_3}\,.\end{aligned}$$

It is affected by cosmic variance, as discussed in [16, 22].

A quantity which is sometimes used is the limit of zero angular separation of the bispectrum, called skewness: $\mathcal{S}_3 := \langle\Theta^3(\mathbf{e})\rangle/\langle\Theta^2(\mathbf{e})\rangle^{3/2}$. In alternative, one can take the equilateral three-point correlation function, where $\mathbf{e}_1\cdot\mathbf{e}_2 = \mathbf{e}_3\cdot\mathbf{e}_1 = \mathbf{e}_2\cdot\mathbf{e}_3$.

In momentum space, we parametrize the level of non-Gaussianity at leading order via the momentum-dependent *non-linear parameter* $f_{\rm NL}$:

$$B_\zeta(k_1,k_2,k_3) = \frac{6}{5} f_{\rm NL}(k_1,k_2,k_3) \sum_{\alpha<\beta} P_\zeta(k_\alpha) P_\zeta(k_\beta)\,, \tag{4.95}$$

where $\alpha,\beta = 1,2,3$ and we hint, through the subscripts ζ, that the origin of primordial non-Gaussianity can be traced back to the statistical properties of the curvature perturbation on uniform density slices. In spherical momentum space, the corresponding reduced bispectrum is [116, 118]

$$b_{\ell_1\ell_2\ell_3} = 2\int_0^{+\infty} d\varrho\,\varrho^2 \left[b_{\ell_1} b_{\ell_2} b^{\rm N}_{\ell_3} + (\ell_3 \leftrightarrow \ell_2) + (\ell_3 \leftrightarrow \ell_1)\right]\,, \tag{4.96}$$

where

$$b_\ell(\varrho) := \frac{2}{\pi}\int_0^{+\infty} dk\,k^2 \mathcal{P}_\zeta(k) \mathcal{T}_\ell(k) j_\ell(k\varrho)\,,$$

$$b^{\rm N}_\ell(\varrho) := \frac{2}{\pi}\int_0^{+\infty} dk\,k^2 f_{\rm NL}(k_1,k_2,k_3) \mathcal{T}_\ell(k) j_\ell(k\varrho)\,.$$

4.6.2 *Trispectrum*

The connected trispectrum in momentum space has been introduced in Sect. 3.4.2. In spherical momenta space, a rotationally invariant estimator $T^{\ell_3\ell_4}_{\ell_1\ell_2}$ of the connected trispectrum is [119, 120]

$$\langle a_{\ell_1 m_1} a_{\ell_2 m_2} a_{\ell_3 m_3} a_{\ell_4 m_4}\rangle_{\rm connected} = \sum_{L=0}^{+\infty} \sum_{M=-L}^{L} (-1)^M \begin{pmatrix} \ell_1 & \ell_2 & L \\ m_1 & m_2 & -M \end{pmatrix} \times \begin{pmatrix} \ell_3 & \ell_4 & L \\ m_3 & m_4 & M \end{pmatrix} T^{\ell_3\ell_4}_{\ell_1\ell_2}(L)\,, \tag{4.97}$$

where L is the third edge of the triangles formed with ℓ_1 and ℓ_2 and ℓ_3 and ℓ_4. Parity invariance requires that $\ell_1+\ell_2+L = 2n$, $\ell_1+\ell_2+L = 2n'$, $m_1+m_2-M = 0$ and $m_3+m_4+M = 0$.

In momentum space, we parametrize the next-to-leading order of non-Gaussianity via the non-linear parameters $\tau_{\rm NL}$ and $g_{\rm NL}$:

$$T_\zeta(k_1,k_2,k_3,k_4) = \tau_{\rm NL}(k_1,k_2,k_3)\sum_{i<j<l} P_\zeta(k_i)P_\zeta(k_j)[P_\zeta(k_{il})+P_\zeta(k_{jl})] + \frac{54}{25}g_{\rm NL}(k_1,k_2,k_3)\sum_{i<j<l} P_\zeta(k_i)P_\zeta(k_j)P_\zeta(k_l)\,. \tag{4.98}$$

4.6.3 *Physical Origin*

Once instrumental noise and foreground contamination are under control, it remains to explain the physical origin of a non-Gaussian signal. The plausible contributions are many but here we mention only a few.

First of all, due to cosmic variance even primordial perturbations with a perfectly Gaussian distribution would have a non-zero probability of producing a non-Gaussian statistics in the sky [121]. Apart from this statistical source, there do exist physical mechanisms of non-Gaussianity. Secondary mechanisms are the Sunyaev–Zel'dovich effect, gravitational lensing, and radio and infrared extragalactic sources.

After matter domination, perturbations enter a phase of gravitational amplification which deforms the primordial Gaussian distribution to a non-Gaussian one. This is the typical *non-linear* regime of the advanced evolutionary phase of cosmic structure formation, where the density contrast is $\delta\rho/\rho \gg 1$. Such type of non-Gaussianity has a specific form that can be obtained via analytic approximations and numerical simulations, is related with the problem of galaxy bias, and can be constrained by observations of the galaxy distribution [122–124].

Non-linearity is also the source of primary non-Gaussianity. The seed of matter cosmic structures and temperature fluctuations is, via (3.62), the primordial curvature perturbation ζ, which we have defined at linear order in (3.56). When higher-order perturbations are taken into account, they naturally generate a non-Gaussian signal.

4.6.3.1 Local Form

In momentum space, the bispectrum and trispectrum are given by (4.95) and (4.98), respectively. The form of the non-linear parameters depend on the model of primordial perturbations. In the simplest case [118, 120, 125, 126], one decomposes the Salopek–Bond non-linear curvature perturbation $\zeta_{\rm NL}(\mathbf{x})$ (3.51) into a Gaussian linear part ζ and a non-linear part:

$$\zeta_{\rm NL} = \zeta + \zeta^{\rm N} = \zeta + \frac{3}{5} f_{\rm NL}^{\rm local} \left(\zeta^2 - \langle\zeta^2\rangle\right) + \frac{9}{25} g_{\rm NL}^{\rm local} \zeta^3 \,, \tag{4.99}$$

where the non-linear parameters $f_{\rm NL}^{\rm local}$ and $g_{\rm NL}^{\rm local}$ are *constant*. By definition, $\langle\zeta_{\rm NL}\rangle = \langle\zeta\rangle = 0$. In the following, we set $g_{\rm NL} = 0$. Then, a direct calculation of the bispectrum shows that (Problem 4.5)

$$f_{\rm NL}(k_1, k_2, k_3) = f_{\rm NL}^{\rm local} \,. \tag{4.100}$$

For the trispectrum, similar steps lead to

$$\tau_{\rm NL}(k_1, k_2, k_3) = \left(\frac{6 f_{\rm NL}^{\rm local}}{5}\right)^2 \,. \tag{4.101}$$

The decomposition (4.99) is point-wise in position space and, for this reason, it is called the *local form* of non-Gaussianity. For a power-law scalar spectrum (4.62), the local bispectrum reads

$$B_\zeta^{\rm local}(k_1, k_2, k_3) = \frac{6}{5} f_{\rm NL}^{\rm local} A_\zeta^2 \sum_{\alpha<\beta} \frac{1}{(k_\alpha k_\beta)^{4-n_{\rm s}}} \,, \tag{4.102}$$

where $A_\zeta = 2\pi^2 A_{\rm s} \tau_0^{n_{\rm s}-1}$. Translating (4.99) in terms of the spherical harmonics decomposition (4.13), one can split the coefficients $a_{\ell m}$ into a linear and a non-linear part, $a_{\ell m}^{\rm NL} = a_{\ell m} + a_{\ell m}^{\rm N}$. Substituting (4.13) into (4.93) and using (4.122) and (4.94), we obtain (4.96) with constant $f_{\rm NL}^{\rm local}$. Using the full transfer function $\mathcal{T}_\ell$, one finds positive and negative acoustic peaks with twice the period of the angular spectrum [118]. Numerical and semi-analytic details of the shape of the bispectrum can be found in [118, 125, 127].

Expression (4.102) peaks at the *squeezed* limit where one of the edges of the triangle $(\mathbf{k}_1, \mathbf{k}_2, \mathbf{k}_3)$ collapses [128, 129]:

$$k_1 \simeq k_2 \gg k_3 \,, \qquad k_3 \approx 0 \,. \tag{4.103}$$

Sending, e.g., $k_3 \to 0$, by conservation of momenta one has $\mathbf{k}_1 \sim -\mathbf{k}_2$ and

$$B_\zeta^{\rm local}(k_1, k_1, k_3 \sim 0) \simeq \frac{12}{5} f_{\rm NL}^{\rm local} P_\zeta(k_1) P_\zeta(k_3) \,. \tag{4.104}$$

Measuring the bispectrum in this configuration, one can obtain an estimate of $f_{\rm NL}^{\rm local}$. In the local bispectrum, small- and large-scale modes are coupled together.

The squeezed limit can be understood in a fairly intuitive way in all models where the curvature perturbation ζ is constant at large scales [128, 130]. Consider the natural splitting (3.15) into a coarse-grained and a fine-grained perturbation. In the limit (4.103), $\zeta_{\mathbf{k}_3}$ is larger than the Hubble horizon and can be treated as constant in time. From (3.48), it is clear that

$$\zeta(\mathbf{x}_3) \overset{(4.103)}{\sim} \zeta_c(\mathbf{x}_3) \tag{4.105}$$

defines a new coordinate background $\mathbf{x}' \simeq [1 + \zeta_c(\mathbf{x}_3)]\mathbf{x}$ inside the horizon. In the new coordinates and up to linear order,

$$\zeta_q(\mathbf{x}') \simeq \zeta_q(\mathbf{x}) + (\mathbf{x}' - \mathbf{x}) \cdot \frac{d}{d\mathbf{x}}\zeta_q(\mathbf{x}) \simeq \zeta_q(\mathbf{x}) + \zeta_c(\mathbf{x}_3)\,\mathbf{x} \cdot \frac{d}{d\mathbf{x}}\zeta_q(\mathbf{x})\,. \tag{4.106}$$

If the linear perturbation $\zeta_q(\mathbf{x})$ is Gaussian, in the squeezed limit we have (Problem 4.6)

$$\langle \zeta(\mathbf{x}_1)\zeta(\mathbf{x}_2)\zeta(\mathbf{x}_3)\rangle \simeq -(n_s - 1)\xi_2^{(\zeta)}(0)\xi_2^{(\zeta)}(\varrho)\,. \tag{4.107}$$

Comparing this expression with (4.104), we finally obtain

$$\boxed{f_{\rm NL}^{\rm local} \simeq \frac{5}{12}(1 - n_s)\,.} \tag{4.108}$$

For spectra which are almost scale-invariant at large scales, the level of non-Gaussianity is very low, $f_{\rm NL} \ll 1$. Tensor modes produce an even lower signal.

4.6.3.2 Equilateral Form

Other models of primordial seeds predict a bispectrum which can be approximated by the *equilateral* form [131] $(i, j, l = 1, 2, 3)$

$$B_\zeta^{\rm equil}(k_1, k_2, k_3) = \frac{18 f_{\rm NL}^{\rm equil}}{5} A_\zeta^2 \left\{ -\sum_{i<j<l} \left[\frac{1}{(k_i k_j)^{4-n_s}} + \frac{2}{(k_i k_j k_l)^{2(4-n_s)/3}} \right] \right.$$
$$\left. + \sum_{i \neq j \neq l} \left[\frac{1}{(k_i k_j^2 k_l^3)^{(4-n_s)/3}} + \frac{1}{(k_i k_j k_l^4)^{(4-n_s)/3}} \right] \right\}, \tag{4.109}$$

where $f_{\rm NL}^{\rm equil}$ is constant. In the limit $k_1 \sim k_2 \sim k_3$, it coincides with (4.102). In the equilateral form, small- and very-large-scale modes are not coupled. The local and equilateral forms can be measured almost independently.

4.6.3.3 Orthogonal Form

Another *Ansatz* which is nearly orthogonal to the other two is the *orthogonal* form [132]

$$B_\zeta^{\rm ortho}(k_1,k_2,k_3) = \frac{18 f_{\rm NL}^{\rm ortho}}{5} A_\zeta^2 \left\{ -\sum_{i<j<l} \left[\frac{3}{(k_i k_j)^{4-n_{\rm s}}} + \frac{8}{(k_i k_j k_l)^{2(4-n_{\rm s})/3}} \right] \right.$$
$$\left. + \sum_{i\neq j\neq l} \left[\frac{3}{(k_i k_j^2 k_l^3)^{(4-n_{\rm s})/3}} + \frac{3}{(k_i k_j k_l^4)^{(4-n_{\rm s})/3}} \right] \right\}, \qquad (4.110)$$

where $f_{\rm NL}^{\rm ortho}$ is constant.

In the next chapter, we will discuss theoretical models predicting these three types of bispectrum.

4.6.4 *Current Estimates*

To date, we do not have significant constraints on the trispectrum but we have observational bounds on the non-linear parameter $f_{\rm NL}$ for the three types of primordial non-Gaussianity described above. At 68 % CL and combining temperature and E-mode polarization data, these estimates are [133]

$$\begin{aligned} f_{\rm NL}^{\rm local} &= 0.8 \pm 5.0\,, \\ f_{\rm NL}^{\rm equil} &= -4 \pm 43\,, \\ f_{\rm NL}^{\rm ortho} &= -26 \pm 21\,. \end{aligned} \qquad (4.111)$$

Bounds combining CMB data and large-scale structure surveys can change considerably depending on the data set, the estimator and the chosen model of galaxy power spectrum [134]. The progress in the determination of the level of non-Gaussianity has been remarkable since the beginning of precision cosmology. One can compare the 1σ errors in (4.111) by PLANCK with the final WMAP results, $\Delta f_{\rm NL}^{\rm local} \sim 20$ and $\Delta f_{\rm NL}^{\rm equil} \sim 140$, back to the first measurements by COBE where the error on local non-Gaussianity (the only one determined by that experiment) was 100 times larger, $\Delta f_{\rm NL}^{\rm local} \sim 600$.

Local non-Gaussianity can also be tested against a residual k-dependence in the non-linear parameter. This is done by estimating the index [135]

$$n_{f_{\rm NL}} := \frac{{\rm d}\ln |f_{\rm NL}|}{{\rm d}\ln k}\,. \qquad (4.112)$$

Observations do not disagree with the local model, since $-2.5 < n_{f_{\rm NL}} < 2.3$ [136]. Constraints on the trispectrum parameters can be found in [133, 136]. For $f_{\rm NL}^{\rm local} = 0$, the PLANCK 2015 bound on $g_{\rm NL}^{\rm local}$ in (4.99) is $g_{\rm NL}^{\rm local} = (-9.0 \pm 7.7) \times 10^4$ at 68 % CL.

Secondary and "post-primordial" effects easily generate non-Gaussianity but they can be subtracted from the maps. Primordial CMB temperature fluctuations are Gaussian within the present experimental confidence level. A detection of a non-Gaussian signal could constrain a number of high-energy models predicting large deviations from a Gaussian distribution. Several of these models will be analyzed in Chaps. 5 and 13.

4.7 Problems and Solutions

4.1 Two-point function. Derive (4.20) and (4.21).

Solution A direct calculation yields

$$\begin{aligned}
\langle a_{\ell m}^* a_{\ell' m'}\rangle &\overset{(4.19)}{=} \delta_{\ell'\ell}\delta_{m'm}\int_{-\infty}^{+\infty}\left(\prod_{\bar m=1}^{2\ell+1} d\alpha_{\ell,\bar m} f_\ell[\alpha_{\ell,\bar m}]\right)|a_{\ell m}|^2\\
&\overset{(4.15)}{=} \delta_{\ell'\ell}\delta_{m'm}\int_{-\infty}^{+\infty} d\alpha_{\ell,\bar m} f_\ell[\alpha_{\ell,\bar m}]\alpha_{\ell,\bar m}^2\\
&\overset{(3.70)}{=} C_\ell^{TT}\delta_{\ell'\ell}\delta_{m'm}\,.
\end{aligned} \tag{4.113}$$

In the second equality, the index $\bar m$ is silent and can pick any value in its range. The factor $\sqrt{2}$ in the definition (4.15) of the $\alpha_{\ell,\bar m}$'s was introduced to get this simple result. The temperature spectrum is

$$\begin{aligned}
\langle \Theta(\mathbf{e}_1)\Theta(\mathbf{e}_2)\rangle &\overset{(4.12)}{=} \sum_{\ell,\ell'=0}^{+\infty}\sum_{m=-\ell}^{\ell}\sum_{m'=-\ell'}^{\ell'}\langle a_{\ell m}a_{\ell' m'}\rangle\, Y_{\ell m}(\vartheta_1,\varphi_1)Y_{\ell' m'}(\vartheta_2,\varphi_2)\\
&= \sum_{\ell,\ell'=0}^{+\infty}\sum_{m=-\ell}^{\ell}\sum_{m'=-\ell'}^{\ell'}\langle a_{\ell m}^* a_{\ell' m'}\rangle\, Y_{\ell m}^*(\vartheta_1,\varphi_1)Y_{\ell' m'}(\vartheta_2,\varphi_2)\\
&\overset{(4.113)}{=} \sum_{\ell=0}^{+\infty} C_\ell^{TT}\sum_{m=-\ell}^{\ell} Y_{\ell m}^*(\vartheta_1,\varphi_1)Y_{\ell m}(\vartheta_2,\varphi_2)\\
&\overset{(4.6)}{=} \frac{1}{4\pi}\sum_{\ell=0}^{+\infty}(2\ell+1)C_\ell^{TT}P_\ell(\cos\vartheta_{12})\,.
\end{aligned} \tag{4.114}$$

4.2 Angular scales 1. Determine the angular, the multipole and the wave-number scale of anisotropies the size of the particle horizon at equality and decoupling for a flat universe.

Solution From (4.36), (4.37), (4.38) and Table 2.1, one gets

$$\theta_{\rm eq} \approx 0.6^\circ\,, \qquad \ell_{\rm eq} \approx 640\,, \qquad k \approx 0.04\,{\rm Mpc}^{-1}\,, \tag{4.115}$$

and

$$\theta_{\rm dec} \approx 1.3^\circ\,, \qquad \ell_{\rm dec} \approx 280\,, \qquad k \approx 0.02\,{\rm Mpc}^{-1}\,. \tag{4.116}$$

Anisotropies with $\ell \gg \ell_{\rm dec}$ depend more strongly on the causal astroparticle model describing photon-matter interaction. On the other hand, outside the last-scattering horizon the features of anisotropies are rather general and well described by linear perturbation theory.

4.3 Angular scales 2. Determine the comoving, the angular and the multipole scale at which anisotropies are blurred by effect of the finite thickness of the last-scattering surface. Assume $\textsc{k} = 0$.

Solution According to Sect. 4.1.2, the visibility function can be approximated by a Gaussian centered at $z_{\rm dec}$ and of width $\Delta z \approx 80$. During the period marked by the initial and final redshift

$$z_\pm = z_{\rm dec} \pm \frac{\Delta z}{2} \approx 1090 \pm 40\,,$$

photon-electron scattering washes out anisotropies smaller than the comoving particle horizon. Therefore, on the CMB sphere we cannot see fluctuations with wave-length smaller than the comoving scale

$$\chi(\Delta z) := \tau(z_-) - \tau(z_+)\,. \tag{4.117}$$

The calculation of comoving particle horizons at different epochs was done in Problem 2.6, where we saw that, at decoupling, we can ignore the cosmological constant but that radiation brings a correction factor (2.146) to the matter-dominated result. Then,

$$\begin{aligned}\chi(\Delta z) &\simeq 1.18[\tau_{\rm CDM}(z_-) - \tau_{\rm CDM}(z_+)] = \frac{2.36}{H_0}\left[\frac{1}{\sqrt{1+z_-}} - \frac{1}{\sqrt{1+z_+}}\right]\\ &\approx 12\,{\rm Mpc}\,.\end{aligned} \tag{4.118}$$

In terms of angular and multipole scales ((4.37) and (4.36)), we lose information on primordial fluctuations for

$$\theta \lesssim 0.047^\circ \approx 2.8' , \qquad \ell \gtrsim 7700 . \tag{4.119}$$

Experimental limitations practically increase (respectively, reduce) the lower limit of angular resolution (the upper limit of usable multipoles).

4.4 Linear tensor perturbations. Briefly discuss linear tensor adiabatic perturbations at decoupling. After matter domination, the anisotropic stress $\Pi_{\alpha\beta}$ can be neglected in the linearized Einstein equations. Ignore Λ.

Solution Neglecting Λ and for a matter-dominated universe, tensor perturbations are given by (3.36) and (3.37) with $D = 4$ and $p = 2/3$ ($\tau > 0$, $\nu = 3/2$). In the small wave-length limit,

$$\sqrt{k\tau} J_\nu(k\tau) \simeq -\sqrt{\frac{2}{\pi}} \cos(k\tau) , \qquad \sqrt{k\tau} Y_\nu(k\tau) \simeq -\sqrt{\frac{2}{\pi}} \sin(k\tau) . \tag{4.120}$$

Since primordial fluctuations are adiabatic, we choose the initial conditions at $\tau_{\rm dec} \approx 0$ as $h_{\lambda\mathbf{k}}(0) = \text{const} \neq 0$, $h'_{\lambda\mathbf{k}}(0) = 0$, which set $C_2 = 0$. From (3.31) and (2.187), each tensor mode is

$$h_\lambda(k\tau) = 3C_\lambda \sqrt{\frac{\pi}{2}} (k\tau)^{-3/2} J_{3/2}(k\tau) , \tag{4.121}$$

where we have chosen the normalization so that $h_\lambda(k\tau) \to C_\lambda$ outside the horizon. This solution is plotted in Fig. 4.23 together with the discarded unphysical solution $\propto (k\tau)^{-3/2} Y_{3/2}(k\tau)$, which diverges inside the horizon.

4.5 Local form of non-Gaussianity. Derive (4.100).

Solution The Fourier transform of the non-linear part of the Salopek–Bond perturbation (4.99) is

$$\zeta^{\rm N}_{\mathbf{k}} = \frac{3}{5} f^{\rm local}_{\rm NL} \left[-(2\pi)^3 \delta(\mathbf{k}) \langle \zeta^2 \rangle + \int \frac{d^3\mathbf{p}}{(2\pi)^3} \zeta_{\mathbf{p}} \zeta_{\mathbf{p}-\mathbf{k}} \right] .$$

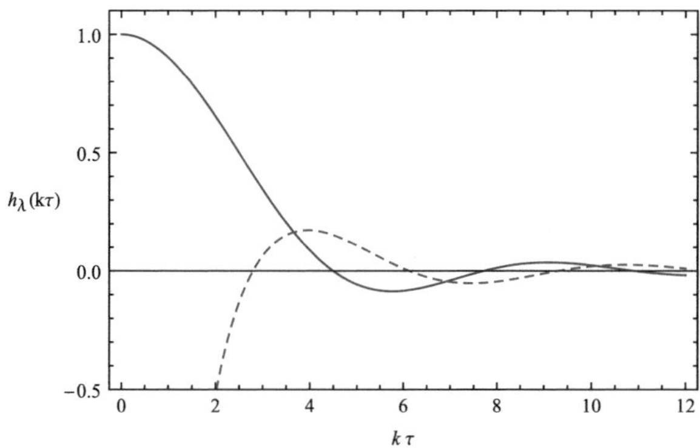

Fig. 4.23 Solutions of the tensor modes equation in terms of the Bessel function J (*continuous curve*, (4.121) with $C_\lambda = 1$) and Y (*dashed curve*). Adiabatic perturbations are described by the first, which goes to a non-zero constant outside the horizon, $k\tau \to 0$

The first term stems from the fact that the auto-correlation function is independent of $\mathbf{x}$. Since not all momenta can vanish at the same time, this piece can be thrown away. The second term enters into the three-point function, which at lowest order is

$$\begin{aligned}
\langle \zeta^{\rm NL}_{\mathbf{k}_1} \zeta^{\rm NL}_{\mathbf{k}_2} \zeta^{\rm NL}_{\mathbf{k}_3} \rangle &\simeq \langle \zeta_{\mathbf{k}_1} \zeta_{\mathbf{k}_2} \zeta^{\rm N}_{\mathbf{k}_3} \rangle + (\mathbf{k}_3 \leftrightarrow \mathbf{k}_2) + (\mathbf{k}_3 \leftrightarrow \mathbf{k}_1) \\
&= \frac{3}{5} f^{\rm local}_{\rm NL} \int \frac{\mathrm{d}^3\mathbf{p}}{(2\pi)^3} \langle \zeta_{\mathbf{k}_1} \zeta_{\mathbf{k}_2} \zeta_{\mathbf{p}} \zeta_{\mathbf{k}_3 - \mathbf{p}} \rangle \\
&\overset{(3.80)}{=} (2\pi)^3 \frac{3}{5} f^{\rm local}_{\rm NL} P_\zeta(k_1) P_\zeta(k_2) \\
&\quad \times \int \mathrm{d}^3\mathbf{p} \, [\delta(\mathbf{k}_1 + \mathbf{p}) \delta(\mathbf{k}_2 + \mathbf{k}_3 - \mathbf{p}) \\
&\quad + \delta(\mathbf{k}_2 + \mathbf{p}) \delta(\mathbf{k}_1 + \mathbf{k}_3 - \mathbf{p})] + (\mathbf{k}_3 \leftrightarrow \mathbf{k}_2) + (\mathbf{k}_3 \leftrightarrow \mathbf{k}_1) \\
&= (2\pi)^3 \delta(\mathbf{k}_1 + \mathbf{k}_2 - \mathbf{k}_3) \frac{3}{5} f^{\rm local}_{\rm NL} 2 P_\zeta(k_1) P_\zeta(k_2) \\
&\quad + (\mathbf{k}_3 \leftrightarrow \mathbf{k}_2) + (\mathbf{k}_3 \leftrightarrow \mathbf{k}_1) \, .
\end{aligned} \tag{4.122}$$

This yields (4.100) after comparing (3.78) and (4.95).

4.6 Squeezed limit. Derive (4.107) using (3.75).

Solution In the squeezed limit,

$$\begin{aligned}
\langle\zeta(\mathbf{x}_1)\zeta(\mathbf{x}_2)\zeta(\mathbf{x}_3)\rangle &\stackrel{(4.103)}{\sim} \langle\zeta_\mathrm{q}(\mathbf{x}_1')\zeta_\mathrm{q}(\mathbf{x}_1')\zeta_\mathrm{c}(\mathbf{x}_3)\rangle\\
&\stackrel{(4.106)}{\simeq} \left\langle\zeta_\mathrm{c}^2(\mathbf{x}_3)\mathbf{x}_1\cdot\frac{\mathrm{d}}{\mathrm{d}\mathbf{x}_1}\left[\zeta_\mathrm{q}(\mathbf{x}_1)\zeta_\mathrm{q}(\mathbf{x}_2)\right]\right\rangle\\
&\stackrel{(3.68)}{\simeq} \langle\zeta_\mathrm{c}^2(\mathbf{x}_3)\rangle_\mathrm{c}\,\mathbf{x}_1\cdot\frac{\mathrm{d}}{\mathrm{d}\mathbf{x}_1}\langle\zeta_\mathrm{q}(\mathbf{x}_1)\zeta_\mathrm{q}(\mathbf{x}_2)\rangle_\mathrm{q}\\
&= \xi_2^{(\zeta)}(0)\frac{\mathrm{d}}{\mathrm{d}\ln\varrho}\xi_2^{(\zeta)}(\varrho)\,,
\end{aligned}$$

where in the second line we exploited translation invariance and in the last line we used $\varrho = |\mathbf{x}_1 - \mathbf{x}_2|$ and $\partial\varrho/\partial\mathbf{x}_1 = \mathbf{x}_1/\varrho$. From (3.75),

$$\frac{\mathrm{d}\xi_2^{(\zeta)}(\varrho)}{\mathrm{d}\ln\varrho} = -(n_s - 1)\xi_2^{(\zeta)}(\varrho)\,, \qquad (4.123)$$

so that (4.107) holds.

References

1. R.A. Alpher, R.C. Hermann, Remarks on the evolution of the expanding universe. Phys. Rev. **75**, 1089 (1949)
2. A.A. Penzias, R.W. Wilson, A measurement of excess antenna temperature at 4080 Mc/s. Astrophys. J. **142**, 419 (1965)
3. R.H. Dicke, P.J.E. Peebles, P.G. Roll, D.T. Wilkinson, Cosmic black-body radiation. Astrophys. J. **142**, 414 (1965)
4. http://map.gsfc.nasa.gov
5. J.C. Mather et al., Measurement of the cosmic microwave background spectrum by the COBE FIRAS instrument. Astrophys. J. **420**, 439 (1994)
6. D.J. Fixsen, E.S. Cheng, J.M. Gales, J.C. Mather, R.A. Shafer, E.L. Wright, The cosmic microwave background spectrum from the full COBE/FIRAS data set. Astrophys. J. **473**, 576 (1996). [arXiv:astro-ph/9605054]
7. W. Hu, N. Sugiyama, Anisotropies in the cosmic microwave background: an analytic approach. Astrophys. J. **444**, 489 (1995). [arXiv:astro-ph/9407093]
8. W. Hu, N. Sugiyama, Toward understanding CMB anisotropies and their implication. Phys. Rev. D **51**, 2599 (1995). [arXiv:astro-ph/9411008]
9. W. Hu, N. Sugiyama, Small scale cosmological perturbations: an analytic approach. Astrophys. J. **471**, 542 (1996). [arXiv:astro-ph/9510117]
10. W. Hu, M.J. White, The damping tail of CMB anisotropies. Astrophys. J. **479**, 568 (1997). [arXiv:astro-ph/9609079]
11. http://www.astro.ucla.edu/~wright/spectrum.gif

12. R.A. Sunyaev, Ya.B. Zel'dovich, Microwave background radiation as a probe of the contemporary structure and history of the universe. Ann. Rev. Astron. Astrophys. **18**, 537 (1980)
13. M. Birkinshaw, The Sunyaev–Zel'dovich effect. Phys. Rep. **310**, 97 (1999). [arXiv:astro-ph/9808050]
14. W. Hu, J. Silk, Thermalization constraints and spectral distortions for massive unstable relic particles. Phys. Rev. Lett. **70**, 2661 (1993)
15. Ya.B. Zel'dovich, R.A. Sunyaev, The interaction of matter and radiation in a hot-model universe. Astrophys. Space Sci. **4**, 301 (1969)
16. D.H. Lyth, A.R. Liddle, *The Primordial Density Perturbation* (Cambridge University Press, Cambridge, 2009)
17. S. Weinberg, *Cosmology* (Oxford University Press, Oxford, 2008)
18. V. Mukhanov, *Physical Foundations of Cosmology* (Cambridge University Press, Cambridge, 2005)
19. L.F. Abbott, M.B. Wise, Anisotropy of the microwave background in the inflationary cosmology. Phys. Lett. B **135**, 279 (1984)
20. L.F. Abbott, M.B. Wise, Large-scale anisotropy of the microwave background and the amplitude of energy density fluctuations in the early universe. Astrophys. J. **282**, L47 (1984)
21. R. Fabbri, F. Lucchin, S. Matarrese, Multipole anisotropies of the cosmic background radiation and inflationary models. Astrophys. J. **315**, 1 (1987)
22. N. Bartolo, E. Komatsu, S. Matarrese, A. Riotto, Non-Gaussianity from inflation: theory and observations. Phys. Rep. **402**, 103 (2004). [arXiv:astro-ph/0406398]
23. J.R. Bond, G. Efstathiou, The statistics of cosmic background radiation fluctuations. Mon. Not. R. Astron. Soc. **226**, 655 (1987)
24. W. Hu, S. Dodelson, Cosmic microwave background anisotropies. Ann. Rev. Astron. Astrophys. **40**, 171 (2002). [arXiv:astro-ph/0110414]
25. M. Tegmark, M. Zaldarriaga, Current cosmological constraints from a 10 parameter CMB analysis. Astrophys. J. **544**, 30 (2000). [arXiv:astro-ph/0002091]
26. http://www.esa.int/Our_Activities/Space_Science/Planck
27. http://pla.esac.esa.int/pla/aio/index.html
28. http://planck.caltech.edu/index.html
29. R. Adam et al. [Planck Collaboration], Planck 2015 results. I. Overview of products and scientific results. Astron. Astrophys. **594**, A1 (2016). [arXiv:1502.01582]
30. J. Silk, M.L. Wilson, Residual fluctuations in the matter and radiation distribution after the decoupling epoch. Physica Scripta **21**, 708 (1980)
31. D. Scott, M. Srednicki, M.J. White, 'Sample variance' in small scale CMB anisotropy experiments. Astrophys. J. **421**, L5 (1994). [arXiv:astro-ph/9305030]
32. J.B. Peterson et al., Cosmic microwave background observations in the post-Planck era. arXiv:astro-ph/9907276
33. P.A.R. Ade et al. [Planck Collaboration], Planck 2015. XX. Constraints on inflation. Astron. Astrophys. **594**, A20 (2016). [arXiv:1502.02114]
34. A. Kogut et al., Dipole anisotropy in the COBE DMR first year sky maps. Astrophys. J. **419**, 1 (1993). [arXiv:astro-ph/9312056]
35. P.A.R. Ade et al. [Planck Collaboration], Planck 2013 results. XVI. Cosmological parameters. Astron. Astrophys. **571**, A16 (2014). [arXiv:1303.5076]
36. R.K. Sachs, A.M. Wolfe, Perturbations of a cosmological model and angular variations of the microwave background. Astrophys. J. **147**, 73 (1967) [Gen. Relat. Grav. **39**, 1929 (2007)]
37. M.J. White, W. Hu, The Sachs–Wolfe effect. Astron. Astrophys. **321**, 8 (1997). [arXiv:astro-ph/9609105]
38. E.R. Harrison, Fluctuations at the threshold of classical cosmology. Phys. Rev. D **1**, 2726 (1970)
39. Ya.B. Zel'dovich, A hypothesis, unifying the structure and the entropy of the universe. Mon. Not. R. Astron. Soc. **160**, 1P (1972)
40. S. Eidelman et al. [Particle Data Group], Review of particle physics. Phys. Lett. B **592**, 1 (2004). [arXiv:astro-ph/0406567]

41. A.D. Sakharov, The initial stage of an expanding universe and the appearance of a nonuniform distribution of matter. Zh. Eksp. Teor. Fiz. **49**, 345 (1965) [Sov. Phys. JETP **22**, 241 (1966)]
42. P.J.E. Peebles, J.T. Yu, Primeval adiabatic perturbation in an expanding universe. Astrophys. J. **162**, 815 (1970)
43. W. Hu, N. Sugiyama, J. Silk, The physics of microwave background anisotropies. Nature **386**, 37 (1997). [arXiv:astro-ph/9604166]
44. J. Silk, Fluctuations in the primordial fireball. Nature **215**, 1155 (1967)
45. J. Silk, Cosmic black-body radiation and galaxy formation. Astrophys. J. **151**, 459 (1968)
46. W.T. Hu, Wandering in the Background: A CMB Explorer. Ph.D. thesis, UC Berkeley, Berkeley (1995). [arXiv:astro-ph/9508126]
47. P.J.E. Peebles, The black-body radiation content of the universe and the formation of galaxies. Astrophys. J. **142**, 1317 (1965)
48. M.J. Rees, Cosmology and galaxy formation, in *The Evolution of Galaxies and Stellar Populations*, ed. by B.M. Tinsley, R.B. Larson (Yale University Observatory Publications, New Haven, 1977)
49. C.J. Hogan, A model of pregalactic evolution. Mon. Not. R. Astron. Soc. **188**, 781 (1979)
50. P.A.R. Ade et al. [Planck Collaboration], Planck 2015 results. XIII. Cosmological parameters. Astron. Astrophys. **594**, A13 (2016). [arXiv:1502.01589]
51. R.A. Sunyaev, Ya.B. Zel'dovich, The velocity of clusters of galaxies relative to the microwave background. The possibility of its measurement. Mon. Not. R. Astron. Soc. **190**, 413 (1980)
52. E. Komatsu et al., Seven-year Wilkinson Microwave Anisotropy Probe (WMAP) observations: cosmological interpretation. Astrophys. J. Suppl. **192**, 18 (2011). [arXiv:1001.4538]
53. M. Zemcov et al., First detection of the Sunyaev–Zel'dovich effect increment at $\lambda < 650\mu$m. Astron. Astrophys. **518**, L16 (2010). [arXiv:1005.3824]
54. M.J. Rees, D.W. Sciama, Large-scale density inhomogeneities in the universe. Nature **217**, 511 (1968)
55. N. Padmanabhan, C.M. Hirata, U. Seljak, D. Schlegel, J. Brinkmann, D.P. Schneider, Correlating the CMB with luminous red galaxies: the integrated Sachs–Wolfe effect. Phys. Rev. D **72**, 043525 (2005). [arXiv:astro-ph/0410360]
56. B.R. Granett, M.C. Neyrinck, I. Szapudi, An imprint of superstructures on the microwave background due to the integrated Sachs–Wolfe effect. Astrophys. J. **683**, L99 (2008). [arXiv:0805.3695]
57. J.P. Ostriker, E.T. Vishniac, Generation of microwave background fluctuations from nonlinear perturbations at the era of galaxy formation. Astrophys. J. **306**, L51 (1986)
58. A. Blanchard, J. Schneider, Gravitational lensing effect on the fluctuations of the cosmic background radiation. Astron. Astrophys. **184**, 1 (1987)
59. S. Cole, G. Efstathiou, Gravitational lensing of fluctuations in the microwave background radiation. Mon. Not. R. Astron. Soc. **239**, 195 (1989)
60. A. Lewis, A. Challinor, Weak gravitational lensing of the CMB. Phys. Rep. **429**, 1 (2006). [arXiv:astro-ph/0601594]
61. M. Kamionkowski, D.N. Spergel, N. Sugiyama, Small scale cosmic microwave background anisotropies as a probe of the geometry of the universe. Astrophys. J. **426**, L57 (1994). [arXiv:astro-ph/9401003]
62. P. de Bernardis et al. [Boomerang Collaboration], A flat universe from high-resolution maps of the cosmic microwave background radiation. Nature **404**, 955 (2000). [arXiv:astro-ph/0004404]
63. P. de Bernardis et al. [Boomerang Collaboration], Multiple peaks in the angular power spectrum of the cosmic microwave background: significance and consequences for cosmology. Astrophys. J. **564**, 559 (2002). [arXiv:astro-ph/0105296]
64. A. Balbi et al., Constraints on cosmological parameters from MAXIMA-1. Astrophys. J. **545**, L1 (2000); Erratum-ibid. **558**, L145 (2001). [arXiv:astro-ph/0005124]
65. A.T. Lee et al., A high spatial resolution analysis of the MAXIMA-1 cosmic microwave background anisotropy data. Astrophys. J. **561**, L1 (2001). [arXiv:astro-ph/0104459]

66. R. Stompor et al., Cosmological implications of the MAXIMA-1 high resolution cosmic microwave background anisotropy measurement. Astrophys. J. **561**, L7 (2001). [arXiv:astro-ph/0105062]
67. W. Hu, M. Fukugita, M. Zaldarriaga, M. Tegmark, CMB observables and their cosmological implications. Astrophys. J. **549**, 669 (2001). [arXiv:astro-ph/0006436]
68. R. Durrer, B. Novosyadlyj, S. Apunevych, Acoustic peaks and dips in the CMB power spectrum: observational data and cosmological constraints. Astrophys. J. **583**, 33 (2003). [arXiv:astro-ph/0111594]
69. A.A. Starobinsky, Cosmic background anisotropy induced by isotropic flat-spectrum gravitational-wave perturbations. Sov. Astron. Lett. **11**, 133 (1985)
70. M. Cortês, A.R. Liddle, P. Mukherjee, On what scale should inflationary observables be constrained? Phys. Rev. D **75**, 083520 (2007). [arXiv:astro-ph/0702170]
71. A. Kosowsky, M.S. Turner, CBR anisotropy and the running of the scalar spectral index. Phys. Rev. D **52**, 1739 (1995). [arXiv:astro-ph/9504071]
72. S. Pandolfi, A. Cooray, E. Giusarma, E.W. Kolb, A. Melchiorri, O. Mena, P. Serra, Harrison–Zel'dovich primordial spectrum is consistent with observations. Phys. Rev. D **81**, 123509 (2010). [arXiv:1003.4763]
73. P.A.R. Ade et al. [BICEP2 and Planck Collaborations], Joint analysis of BICEP2/*Keck Array* and *Planck* data. Phys. Rev. Lett. **114**, 101301 (2015). [arXiv:1502.00612]
74. P.A.R. Ade et al. [BICEP2 and Keck Array Collaborations], Improved constraints on cosmology and foregrounds from BICEP2 and Keck Array cosmic microwave background data with inclusion of 95 GHz band. Phys. Rev. Lett. **116**, 031302 (2016). [arXiv:1510.09217]
75. J.P. Zibin, D. Scott, M.J. White, Limits on the gravity wave contribution to microwave anisotropies. Phys. Rev. D **60**, 123513 (1999). [arXiv:astro-ph/9901028]
76. M.S. Turner, M. White, J.E. Lidsey, Tensor perturbations in inflationary models as a probe of cosmology. Phys. Rev. D **48**, 4613 (1993). [arXiv:astro-ph/9306029]
77. B. Allen, The stochastic gravity-wave background: sources and detection, in *Relativistic Gravitation and Gravitational Radiation*, ed. by J.-A. Marck, J.-P. Lasota (Cambridge University Press, Cambridge, 1997)
78. S. Chongchitnan, G. Efstathiou, Prospects for direct detection of primordial gravitational waves. Phys. Rev. D **73**, 083511 (2006). [arXiv:astro-ph/0602594]
79. A. Stewart, R. Brandenberger, Observational constraints on theories with a blue spectrum of tensor modes. JCAP **0808**, 012 (2008). [arXiv:0711.4602]
80. J. Aasi et al. [The LIGO Scientific Collaboration], Advanced LIGO. Class. Quantum Grav. **32**, 074001 (2015). [arXiv:1411.4547]
81. B.P. Abbott et al. [LIGO Scientific and Virgo Collaborations], GW150914: the Advanced LIGO detectors in the era of first discoveries. Phys. Rev. Lett. **116**, 131103 (2016). [arXiv:1602.03838]
82. https://www.elisascience.org
83. M.J. Rees, Polarization and spectrum of the primeval radiation in an anisotropic universe. Astrophys. J. **153**, L1 (1968)
84. M.M. Basko, A.G. Polnarev, Polarization and anisotropy of the relict radiation in an anisotropic universe. Mon. Not. R. Astron. Soc. **191**, 207 (1980)
85. M. Kaiser, Small-angle anisotropy of the microwave background radiation in the adiabatic theory. Mon. Not. R. Astron. Soc. **202**, 1169 (1983)
86. J.R. Bond, G. Efstathiou, Cosmic background radiation anisotropies in universes dominated by nonbaryonic dark matter. Astrophys. J. **285**, L45 (1984)
87. A.G. Polnarev, Polarization and anisotropy induced in the microwave background by cosmological gravitational waves. Astron. Zh. **62**, 1041 (1985)
88. D.D. Harari, M. Zaldarriaga, Polarization of the microwave background in inflationary cosmology. Phys. Lett. B **319**, 96 (1993). [arXiv:astro-ph/9311024]
89. D. Coulson, R.G. Crittenden, N.G. Turok, Polarization and anisotropy of the microwave sky. Phys. Rev. Lett. **73**, 2390 (1994). [arXiv:astro-ph/9406046]
90. R.G. Crittenden, D. Coulson, N.G. Turok, Temperature-polarization correlations from tensor fluctuations. Phys. Rev. D **52**, 5402 (1995). [arXiv:astro-ph/9411107]

91. M. Zaldarriaga, D.D. Harari, Analytic approach to the polarization of the cosmic microwave background in flat and open universes. Phys. Rev. D **52**, 3276 (1995). [arXiv:astro-ph/9504085]
92. M. Kamionkowski, A. Kosowsky, A. Stebbins, A probe of primordial gravity waves and vorticity. Phys. Rev. Lett. **78**, 2058 (1997). [arXiv:astro-ph/9609132]
93. U. Seljak, M. Zaldarriaga, Signature of gravity waves in polarization of the microwave background. Phys. Rev. Lett. **78**, 2054 (1997). [arXiv:astro-ph/9609169]
94. M. Zaldarriaga, U. Seljak, An all-sky analysis of polarization in the microwave background. Phys. Rev. D **55**, 1830 (1997). [arXiv:astro-ph/9609170]
95. M. Kamionkowski, A. Kosowsky, A. Stebbins, Statistics of cosmic microwave background polarization. Phys. Rev. D **55**, 7368 (1997). [arXiv:astro-ph/9611125]
96. D.N. Spergel, M. Zaldarriaga, CMB polarization as a direct test of inflation. Phys. Rev. Lett. **79**, 2180 (1997). [arXiv:astro-ph/9705182]
97. W. Hu, M.J. White, A CMB polarization primer. New Astron. **2**, 323 (1997). [arXiv:astro-ph/9706147]
98. A. Stebbins, Weak lensing on the celestial sphere. arXiv:astro-ph/9609149
99. F.J. Zerilli, Tensor harmonics in canonical form for gravitational radiation and other applications. J. Math. Phys. **11**, 2203 (1970)
100. E.T. Newman, R. Penrose, Note on the Bondi–Metzner–Sachs group. J. Math. Phys. **7**, 863 (1966)
101. J.N. Goldberg, A.J. MacFarlane, E.T. Newman, F. Rohrlich, E.C.G. Sudarshan, Spin-s spherical harmonics and $d_{\rm H}$. J. Math. Phys. **8**, 2155 (1967)
102. S. Dodelson, *Modern Cosmology* (Academic Press, San Diego, 2003)
103. D. Larson et al., Seven-year Wilkinson Microwave Anisotropy Probe (WMAP) observations: power spectra and WMAP-derived parameters. Astrophys. J. Suppl. **192**, 16 (2011). [arXiv:1001.4635]
104. P.A.R. Ade et al. [Planck Collaboration], Planck 2015 results. XV. Gravitational lensing. Astron. Astrophys. **594**, A15 (2016). [arXiv:1502.01591]
105. D. Hanson et al. [SPTpol Collaboration], Detection of B-mode polarization in the cosmic microwave background with data from the South Pole Telescope. Phys. Rev. Lett. **111**, 141301 (2013). [arXiv:1307.5830]
106. P.A.R. Ade et al. [POLARBEAR Collaboration], A measurement of the cosmic microwave background B-mode polarization power spectrum at sub-degree scales with POLARBEAR. Astrophys. J. **794**, 171 (2014). [arXiv:1403.2369]
107. http://camb.info
108. http://lambda.gsfc.nasa.gov/toolbox/tb_camb_form.cfm
109. http://cosmologist.info/cosmomc
110. http://www.class-code.net
111. http://montepython.net
112. N. Aghanim et al. [Planck Collaboration], Planck 2015 results. XI. CMB power spectra, likelihoods, and robustness of parameters. Astron. Astrophys. **594**, A11 (2016). [arXiv:1507.02704]
113. C.L. Bennett et al. [WMAP Collaboration], Nine-year Wilkinson Microwave Anisotropy Probe (WMAP) observations: final maps and results. Astrophys. J. Suppl. **208**, 20 (2013). [arXiv:1212.5225]
114. E. Komatsu et al., Non-Gaussianity as a probe of the physics of the primordial universe and the astrophysics of the low redshift universe. arXiv:0902.4759
115. L. Verde, R. Jimenez, M. Kamionkowski, S. Matarrese, Tests for primordial non-Gaussianity. Mon. Not. R. Astron. Soc. **325**, 412 (2001). [arXiv:astro-ph/0011180]
116. L. Wang, M. Kamionkowski, Cosmic microwave background bispectrum and inflation. Phys. Rev. D **61**, 063504 (2000). [arXiv:astro-ph/9907431]
117. A. Gangui, J. Martin, Cosmic microwave background bispectrum and slow roll inflation. Mon. Not. R. Astron. Soc. **313**, 323 (2000). [arXiv:astro-ph/9908009]

118. E. Komatsu, D.N. Spergel, Acoustic signatures in the primary microwave background bispectrum. Phys. Rev. D **63**, 063002 (2001). [arXiv:astro-ph/0005036]
119. W. Hu, Angular trispectrum of the cosmic microwave background. Phys. Rev. D **64**, 083005 (2001). [arXiv:astro-ph/0105117]
120. N. Kogo, E. Komatsu, Angular trispectrum of CMB temperature anisotropy from primordial non-Gaussianity with the full radiation transfer function. Phys. Rev. D **73**, 083007 (2006). [arXiv:astro-ph/0602099]
121. R. Scaramella, N. Vittorio, Non-Gaussian temperature fluctuations in the cosmic microwave background sky from a random Gaussian density field. Astrophys. J. **375**, 439 (1991)
122. J.N. Fry, Gravity, bias, the galaxy three-point correlation function. Phys. Rev. Lett. **73**, 215 (1994)
123. S. Matarrese, L. Verde, A.F. Heavens, Large-scale bias in the Universe: bispectrum method. Mon. Not. R. Astron. Soc. **290**, 651 (1997). [arXiv:astro-ph/9706059]
124. E. Sefusatti, E. Komatsu, The bispectrum of galaxies from high-redshift galaxy surveys: primordial non-Gaussianity and non-linear galaxy bias. Phys. Rev. D **76**, 083004 (2007). [arXiv:0705.0343]
125. A. Gangui, F. Lucchin, S. Matarrese, S. Mollerach, The three-point correlation function of the cosmic microwave background in inflationary models. Astrophys. J. **430**, 447 (1994). [arXiv:astro-ph/9312033]
126. L. Verde, L. Wang, A. Heavens, M. Kamionkowski, Large-scale structure, the cosmic microwave background, and primordial non-Gaussianity. Mon. Not. R. Astron. Soc. **313**, L141 (2000). [arXiv:astro-ph/9906301]
127. E. Komatsu, D.N. Spergel, B.D. Wandelt, Measuring primordial non-Gaussianity in the cosmic microwave background. Astrophys. J. **634**, 14 (2005). [arXiv:astro-ph/0305189]
128. J.M. Maldacena, Non-Gaussian features of primordial fluctuations in single field inflationary models. JHEP **0305**, 013 (2003). [arXiv:astro-ph/0210603]
129. D. Babich, P. Creminelli, M. Zaldarriaga, The shape of non-Gaussianities. JCAP **0408**, 009 (2004). [arXiv:astro-ph/0405356]
130. P. Creminelli, M. Zaldarriaga, Single field consistency relation for the 3-point function. JCAP **0410**, 006 (2004). [arXiv:astro-ph/0407059]
131. P. Creminelli, A. Nicolis, L. Senatore, M. Tegmark, M. Zaldarriaga, Limits on non-Gaussianities from WMAP data. JCAP **0605**, 004 (2006). [arXiv:astro-ph/0509029]
132. L. Senatore, K.M. Smith, M. Zaldarriaga, Non-Gaussianities in single field inflation and their optimal limits from the WMAP 5-year data. JCAP **1001**, 028 (2010). [arXiv:0905.3746]
133. P.A.R. Ade et al. [Planck Collaboration], Planck 2015 results. XVII. Constraints on primordial non-Gaussianity. Astron. Astrophys. **594**, A17 (2016). [arXiv:1502.01592]
134. F. De Bernardis, P. Serra, A. Cooray, A. Melchiorri, Constraints on primordial non-Gaussianity from WMAP7 and luminous red galaxies power spectrum and forecast for future surveys. Phys. Rev. D **82**, 083511 (2010). [arXiv:1004.5467]
135. C.T. Byrnes, S. Nurmi, G. Tasinato, D. Wands, Scale dependence of local f_{NL}. JCAP **1002**, 034 (2010). [arXiv:0911.2780]
136. J. Smidt, A. Amblard, C.T. Byrnes, A. Cooray, D. Munshi, CMB constraints on primordial non-Gaussianity from the bispectrum and trispectrum and a consistency test of single-field inflation. Phys. Rev. D **81**, 123007 (2010). [arXiv:1004.1409]

Chapter 5
Inflation

Le Réel, c'est l'impossible.
— Jacques Lacan, *Le Séminaire. Livre XI*, XIII.3

The Real is the impossible.

Contents

(continued)

G. Calcagni, *Classical and Quantum Cosmology*, Graduate Texts in Physics,
DOI 10.1007/978-3-319-41127-9_5

5.1 Problems of the Hot Big Bang Model

The hot big bang model is based upon some fundamental assumptions:

(i) The laws of physics verified today were valid also in primordial epochs.
(ii) The cosmological principle holds.
(iii) The initial conditions of the universe at the big bang $t_{\rm bb}$ are such that $\Omega(t_{\rm bb}) \sim 1$ and the universe is in thermal equilibrium at some temperature $T_{\rm bb} > 100\,{\rm MeV}$.
(iv) Large-scale cosmic structures (CMB anisotropies and galaxy distributions) formed from a primordial spectrum of almost Gaussian density fluctuations.

This standard cosmology has achieved remarkable results including the correct prediction of light elements abundances (big-bang nucleosynthesis) and the natural explanation of the CMB as a fossil imprint of the initial hot phase. We have already seen that the model contains at least two conceptual problems (the presence of a big-bang singularity and a very small cosmological constant), but there are actually other issues that deserve our attention.

5.1.1 Planck and GUT Scale

So far, we have followed a beaten path of knowledge that combined the cosmology of classical general relativity with ingredients of the Standard Model of particles. A problem in physics is almost invariably a point where the path stops and the explorer must resort to their own inventiveness and curiosity. Inevitably, in carving a new track one will follow false threads and meet dead ends, and often a certain amount of speculation will be involved. Such is the case while approaching the first instants of life of the Universe and energies we have not yet probed in accelerators.

For instance, in grand unification theories (GUT) the symmetry group of fundamental interactions is larger than the Standard Model group $SU(3) \otimes SU(2) \otimes U(1)$ and at high energies the strong and electromagnetic forces are unified. The minimal umbrella group is $SU(5)$ (Georgi–Glashow model [1, 2]), but there exist other proposals with larger groups such as $SO(10)$ [3, 4]. Typically, GUTs predict a finite lifetime for the proton [5–7]. The present experimental limit on the proton decay [8] almost rules out the $SU(5)$ model [9, 10], while $SO(10)$ is safe [6, 11, 12]. Regardless the chosen symmetry group, the characteristic energy scale at which symmetry breaking occurs is [13–18]

$$T_{\rm GUT} \approx 10^{15} - 10^{16}\,{\rm GeV}\,, \qquad z_{\rm GUT} \approx 10^{28} - 10^{29}\,. \tag{5.1}$$

At higher energies, not only should gravity be as important as the other forces, but the very concepts of gravity and matter might no longer be separable and a description in terms of particle fields might even fail. In Chap. 1 we introduced the Planck time $t_{\rm Pl}$, equation (1.3), which is interpreted as the time interval within which quantum fluctuations exist at Planck-length scale, equation (1.2). The other Planck quantities are obtained from these by dimensional arguments. From the point of view of cosmology, Planck time represents the time at which quantum effects are the size of the horizon, $H^{-1} \sim l_{\rm Pl}$. From what said above, it is conventional to identify the "beginning of time" or big bang at $t_{\rm bb} \sim t_{\rm Pl}$. One also expects that field theory breaks down at densities above $m_{\rm Pl}^4$. Beyond the Planck scale, only a theory of quantum gravity might be able to describe physical phenomena correctly. None of these features fit rigorously in a classical framework such as the hot big bang model of Chap. 2.

5.1.2 Flatness Problem

From (2.84), the deviation from the critical density parameter is proportional to the square of the comoving Hubble horizon:

$$|\Omega_{\rm K}| = |{\rm K}|r_H^2 \overset{(3.11)}{=} |{\rm K}|\,(\epsilon-1)^2\,\tau^2\,. \tag{5.2}$$

If we assume, as suggested by observations, that matter and radiation dominated the universe throughout most of its history, then the comoving Hubble horizon (2.137) increases with time. Consequently, any deviation from flat geometry are amplified. However, we know from (2.128) that today $\Omega_{\rm K0}$ is small, so that the curvature parameter must be fine tuned to a high degree near the big bang. More precisely, consider the ratio of the Hubble horizon at two different times $t_{\rm A}$ and $t_{\rm B}$ for a power-law cosmology. From (2.142),

$$\frac{r_{H,{\rm A}}}{r_{H,{\rm B}}} = \frac{\tau_{\rm A}}{\tau_{\rm B}} = \left(\frac{1+z_{\rm B}}{1+z_{\rm A}}\right)^{\frac{1}{p}-1} = \left(\frac{\rho_{\rm B}}{\rho_{\rm A}}\right)^{\frac{1-p}{2}}\,. \tag{5.3}$$

For a radiation-dominated universe ($p = 1/2$),

$$\frac{\tau_{\rm A}}{\tau_{\rm B}} = \frac{1+z_{\rm B}}{1+z_{\rm A}} = \left(\frac{\rho_{\rm B}}{\rho_{\rm A}}\right)^{\frac{1}{4}} = \frac{T_{\rm B}}{T_{\rm A}}\,, \tag{5.4}$$

while for a matter-dominated scenario ($p = 2/3$)

$$\frac{\tau_{\rm A}}{\tau_{\rm B}} = \left(\frac{1+z_{\rm B}}{1+z_{\rm A}}\right)^{\frac{1}{2}} = \left(\frac{\rho_{\rm B}}{\rho_{\rm A}}\right)^{\frac{1}{6}}\,. \tag{5.5}$$

Then,

$$|\Omega_{\rm K}| = |\Omega_{\rm K0}|\left[\frac{r_H(t)}{r_H(t_0)}\right]^2 = |\Omega_{\rm K0}|\left(\frac{\tau_{\rm eq}}{\tau_0}\right)^2\left(\frac{\tau}{\tau_{\rm eq}}\right)^2 \tag{5.6}$$

$$\overset{(5.3)}{=} |\Omega_{\rm K0}|\left(\frac{\tau_{\rm eq}}{\tau_0}\right)^2\left(\frac{1+z_{\rm eq}}{1+z}\right)^{\frac{2}{p}-2} \overset{(5.4)}{\simeq} \frac{1}{z^2}\,, \qquad z > z_{\rm eq}\,, \tag{5.7}$$

where in the last step we used (2.150) and (2.128) and radiation domination at early times. Already at radiation-matter equality, the curvature density should have been quite small, $|\Omega_{\rm K,eq}| \sim 10^{-6}$. Going backwards in the thermal history of the universe, this fine tuning is worsened. At energy scales of TeV order (the highest we can probe at the LHC),

$$|\Omega_{\rm K,LHC}| \sim 10^{-32}\,.$$

At the GUT scale (5.1), corresponding to $t \sim 10^{-38}-10^{-36}$ s, one has

$$\frac{\tau_{\rm GUT}}{\tau_{\rm eq}} \overset{(5.4)}{\approx} 10^{-26}-10^{-25}\,, \qquad |\Omega_{\rm K,GUT}| \sim 10^{-58}-10^{-56}\,. \tag{5.8}$$

At the Planck scale,

$$|\Omega_{\rm K,Pl}| \sim 10^{-64}\,.$$

We face a problem of initial conditions. The hot big bang model does not explain why the initial density of the universe was so close (up to one part over 10^{60} or less) to the critical density. The probability that Nature realized such a special initial condition is almost zero and it does not seem reasonable to accept that without a fight. An obvious way out is to assume that the universe has flat geometry,

$$\textsc{k} = 0\,.$$

In this case, however, we would like to have a theory of particles and geometry explaining why the flat case is elected above the others.

5.1.3 *Horizon Problem*

One of the main assumptions of the big bang model is that the universe is homogeneous. The cosmological principle justifies the assertion that the Universe was born at the initial time $t = t_{\rm bb}$ in a state of infinite energy (temperature). From CMB and large-scale observations, we know that the primordial universe was homogeneous and isotropic up to one part over 10^{-5}. This indicates a strong correlation of physical conditions in every region of the sky. The horizon at last scattering (4.116) subtends angular scales $\sim 1.3°$. Zones smaller than about one degree were in causal contact at the time when matter and radiation decoupled, while larger regions were outside the horizon $\tau_{\rm dec}$. These regions later entered the causal patch as soon as the growing horizon encompassed them. However, we observe temperature isotropy at scales far larger than $\tau_{\rm dec}$ and no causal process could have thermalized these ultra-horizon regions, since $\tau_{\rm dec} \ll \tau_0$. In power-law models with $0 < p < 1$, the presence of the initial singularity guarantees the existence of a *growing* cosmological horizon (equation (2.137)),

$$\dot{r}_H > 0\,,$$

distinguishing "in" and "out" regions. The big bang is also responsible for the fine tuning of the curvature energy density, but we have seen that the tuning is removed by a brute-force choice of geometry. In this respect, the horizon problem is truly endemic in big-bang expanding scenarios and more serious than the flatness problem.

5.1.4 Monopole Problem

Spontaneous symmetry breaking plays a major role in modern theoretical physics, from the particle Standard Model to condensed matter systems, from string theory to quantum gravity scenarios. Certain systems invariant under a discrete or continuous symmetry, and depending on one or more time-varying parameters, may undergo a phase transition when these parameters acquire a critical value. In this case, the symmetry is broken. An example is ferromagnetism. Above the Curie temperature, the spins of atoms in a ferromagnet are randomly oriented; a microscopic inspection of the material shows that its properties are invariant under spatial rotations. When the temperature drops below the critical value, the spins get the same orientation and the material acquires magnetic properties. Microscopically, there is a privileged direction and rotation invariance has been lost. This symmetry breaking is the transition from a high-entropy state to an ordered state.

Another example is the electroweak Higgs mechanism giving mass to the W and Z gauge bosons of the Standard Model. At even higher energies of order of (5.1), the mother group (e.g., $SU(5)$) is reduced to $SU(3) \otimes SU(2) \otimes U(1)$ and magnetic monopoles form. Monopoles are a particular type of structure generated via the *Kibble mechanism* [19]. During a cosmological phase transition, correlation lengths are limited by the particle horizon. Uncorrelated transitions take place at super-horizon scales, but at the border between causal regions false vacuum structures, called *topological defects*, are formed. Their geometry depends on the topology of the boundary. One can then have defects which are point-wise (*monopoles*), one-dimensional (*cosmic strings*), two-dimensional (*domain walls*), three-dimensional (*textures*), and so on, together with other formations such as Kibble gradients and fractal defects [20–22].[1]

The monopole problem is the excessive density of defects produced during the phase transition at a mass scale M [25–27]. A heuristic calculation yields (e.g., [28])

$$\Omega_{M0} > 10^{15},$$

incompatible with the observed quasi-critical density. This large monopole density can be avoided in certain models [29], and whenever one excludes grand unification. Therefore, the monopole and the flatness problems are less severe than the horizon problem.

[1]Cosmic strings can give rise to characteristic signatures in the CMB. See, e.g., [23, 24].

5.1.5 Primordial Seeds Problem

We have not yet identified *what* originated the primordial fluctuations. In doing so, we would like also to explain why the level of anisotropy is as low as in (4.69),

$$\frac{\delta T}{T_0} \sim 10^{-5},$$

and why the statistical distribution is Gaussian.

5.2 Inflationary Mechanism

If the exposition of the cosmological equations has been clear so far, it should be easy for the reader to devise a solution of the flatness and horizon problems. However, at the time of its proposal this solution was not so obvious and triggered a revolution in our way of thinking the universe [30–33].

We define as *inflation* any period of accelerating expansion,

$$\boxed{\ddot{a} > 0 \qquad \Leftrightarrow \qquad \epsilon < 1\,.} \tag{5.9}$$

From (2.168) and the discussion in Problem 2.9, this is equivalent to impose an equation of state with $w < -1/3$:

$$P < -\frac{1}{3}\rho\,. \tag{5.10}$$

In terms of exact solutions, the power-law solution (2.180) is inflationary when $p > 1$. The de Sitter background (2.172) realizes the prototype of extreme inflationary expansion:

$$a(t) = \mathrm{e}^{Ht}\,, \qquad H(t) = H\,, \qquad \epsilon(t) = 0\,. \tag{5.11}$$

Many approximate solutions will share almost all the qualitative features of de Sitter, including an *exponential expansion* of the scale factor, $a \sim \mathrm{e}^{Ht}$, and a *small first slow-roll parameter*, $\epsilon \ll 1$. For this reason, de Sitter cosmology has become almost a synonym of inflation. However, we must always keep in mind that de Sitter is an ideal setting and many realizations of inflation can, or actually must, deviate from it.

We assume that inflation lasted from some initial time $t_\mathrm{i} \gtrsim t_\mathrm{bb}$ until some time t_e, after which the universe was radiation-dominated. Inflation is characterized by an "improved" definition of the number of e-foldings (compare with (2.48)), which is the logarithm of the ratio of comoving Hubble horizons [34],

$$\mathcal{N} := \ln \frac{r_{H,\mathrm{i}}}{r_H} \,. \tag{5.12}$$

During inflation, the Hubble parameter is almost constant and $\mathcal{N}(\tau) \simeq \mathcal{N}_a$.

5.2.1 Solution of the Flatness Problem

Inflation solves the flatness problem almost by definition, since the comoving Hubble length *decreases* in time during an accelerated era (equation (3.9)):

$$\dot{r}_H = -\ddot{a}\, r_H^2 = -\frac{1}{a}(1-\epsilon) < 0 \,. \tag{5.13}$$

At primordial times, Ω naturally approaches the critical density even if $\Omega \neq 1$ as initial condition (Fig. 5.1). Because of the accelerated expansion, the universe inflates and its characteristic curvature scale is stretched in a short amount of time. After inflation, a local observer no longer feels any intrinsic curvature effect. This is the reason why we often ignored curvature so far.

We can estimate the amount of inflation sufficient to remove the fine-tuning problem of flatness. Suppose inflation starts soon after the Planck-scale big bang. As initial time, we set

$$t_\mathrm{i} \gtrsim t_\mathrm{bb} \sim t_\mathrm{Pl} \approx 10^{-44}\,\mathrm{s} \,, \tag{5.14}$$

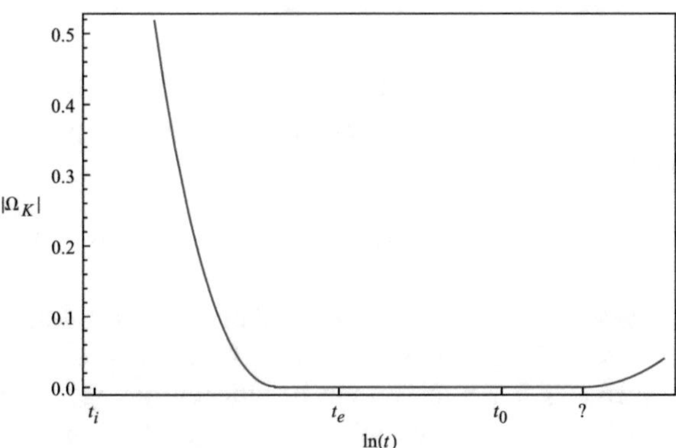

Fig. 5.1 Inflation and the flatness problem. Starting from a general initial condition $|\Omega_{\mathrm{Ki}}| \sim 1$, the curvature density parameter $|\Omega_\mathrm{K}|$ is strongly damped during inflation, $t_\mathrm{i} < t < t_\mathrm{e}$ (time axis arbitrary). Until today (t_0), the density parameter remains close to the critical value $\Omega_\mathrm{K} \sim 0$. Depending on the future evolution of the universe, the dark-energy era might end and curvature effects might eventually dominate

while in most models it is reasonable to assume that inflation ended at the GUT scale. Taking the upper limit of (5.1),

$$t_e \sim t_{\rm GUT} \approx 10^{-38}\,{\rm s}\,. \tag{5.15}$$

After the end of inflation, there may be a matter-dominated period called *reheating* that precedes the radiation era. For the time being, we assume it is so short we can ignore it, an approximation called *instantaneous reheating*. In this case, from (5.2), (5.12) and the lower energy bound of (5.8),

$$|\Omega_{K,i}| = |\Omega_{K,e}|e^{2\mathcal{N}_e} \simeq 10^{0.87\mathcal{N}_e - 58}\,.$$

To have a natural initial condition $|\Omega_{K,i}| \sim 1$, we need at least

$$\mathcal{N}_e \gtrsim 64\,. \tag{5.16}$$

This means that the expansion factor of the universe during inflation is at least $e^{64} \approx 10^{28}$.

5.2.2 Solution of the Horizon Problem

During inflation, the comoving Hubble length decreases as (5.13) and regions of the observable universe which were in causal contact disappear beyond the shrinking horizon. After the end of inflation, r_H increases and the external regions re-enter the causal patch. However, a post-inflationary observer would see these regions for the first time. This phenomenon is depicted in the upper panel of Fig. 5.2 for a comoving scale $\lambda_{\rm com} = \lambda/a$, which is constant by definition.

In the synchronous coordinate system (lower panel of Fig. 5.2), consider the physical scale λ inside the Hubble horizon at $t = t_i$. This scale can represent an inhomogeneity of wave-length λ, generated by some causal process within the horizon. Before inflation, both the scale $\lambda(t) \sim a(t)$ and the Hubble distance H^{-1} increase with time. During inflation, the scale $\lambda(t) \sim e^{Ht}$ increases exponentially but the Hubble distance almost remains constant. After the end of the accelerated expansion, the Hubble radius increases and begins to encompass nearby regions again, eventually englobing λ.

If the comoving horizon at the beginning of inflation is larger than the present one,

$$r_H(t_i) > r_H(t_0)\,,$$

then the horizon problem is solved because the regions we observe projected in the sky entered in causal contact before inflation, and they had time to reach thermal

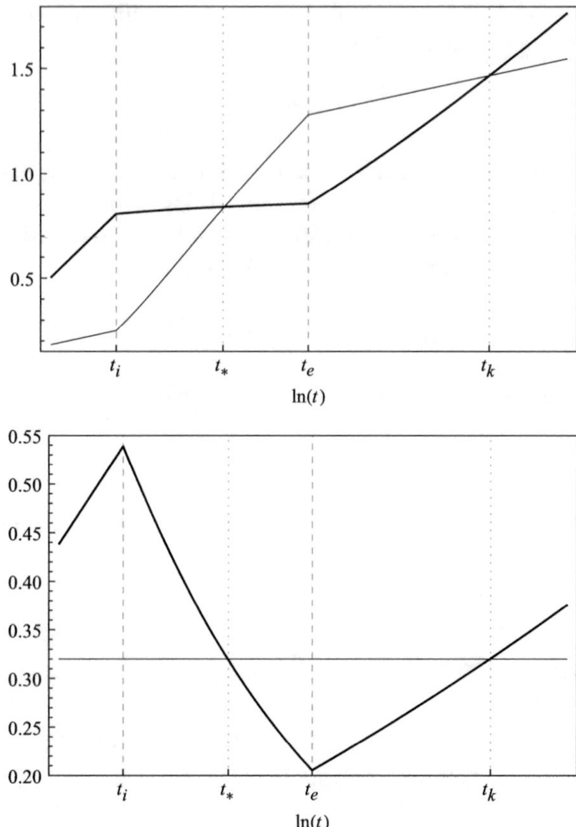

Fig. 5.2 Inflation and the horizon problem. The *upper panel* shows the time evolution of a physical scale or perturbation λ (*thin curve*) which starts inside the Hubble horizon R_H (*thick curve*). The curves are not smooth because we assume for simplicity a sudden begin and end of inflation at t_i and t_e, respectively. Before and after inflation, we assume a $0 < p < 1$ power-law evolution, $\ln\lambda \sim \ln a \sim p\ln t$ and $\ln R_H \sim \ln t + |\ln p|$, while for $t_\mathrm{i} < t < t_\mathrm{e}$ the scale factor expands quasi exponentially, $\ln\lambda \sim t$, $\ln R_H \sim$ const. The perturbation leaves the horizon at $t = t_*$ and re-enters after the end of inflation, at t_k. The same process in comoving coordinates is presented in the *bottom panel*

equilibrium before horizon exit. In instantaneous reheating scenarios, this places a lower bound on the number of e-foldings. For inflation ending at the GUT scale,

$$1 < \frac{r_H(t_\mathrm{i})}{r_H(t_0)} = \frac{r_H(t_\mathrm{i})}{r_H(t_\mathrm{e})}\frac{r_H(t_\mathrm{e})}{r_H(t_0)} = \mathrm{e}^{\mathcal{N}_\mathrm{e}}\frac{\tau_\mathrm{GUT}}{\tau_\mathrm{eq}}\frac{\tau_\mathrm{eq}}{\tau_0} \simeq 10^{0.43\mathcal{N}_\mathrm{e}-28}, \tag{5.17}$$

thus yielding again (5.16). To solve both the horizon and the flatness problem, inflation must have lasted more than 60 e-foldings.

In Sect. 3.1.3, we mentioned the fact that the particle horizon radius r_p is negative on inflationary backgrounds, so that it must be replaced by the Hubble horizon r_H

as an estimator of the size of the causal patch. Nevertheless, it still gives an intuitive physical insight into the causal properties of cosmological spacetimes. The power-law FLRW results (2.185) and (2.188) are sufficient to this aim. For $0 < p < 1$, the range of conformal time is the same as for proper time, $\tau \in (0, +\infty)$, so that the distance a signal can cover in the time interval $\Delta\tau = \tau - 0 = \tau$ since the singularity at $\tau = 0$ is no greater than $r_{\rm p} = c\tau$. For $p \geqslant 1$, conformal-time range is $\tau \in (-\infty, 0)$, meaning that for a given instant τ any finite distance $c\Delta\tau = c(\tau - \bar{\tau}) > 0$ can be covered by a signal travelling since the big bang at $\tau = -\infty$.

5.2.3 *Solution of the Monopole Problem*

During the exponential expansion of the inflationary period, monopoles and other relics are diluted and expelled from the horizon. Their observable density contribution is lowered to acceptable levels. This mechanism can work if the reheating temperature (see below) is not so high as to favour the thermal creation of new defects. In another scenario dubbed eternal inflation, which we shall discuss later, monopoles can actually inflate and constitute the seeds of other universes.

5.2.4 *Solution of the Primordial Seeds Problem*

Historically, inflation as a kinematical paradigm was invoked to solve the flatness, horizon and monopole problems [31–33, 35, 36]. Although CMB anisotropies were discovered later, it was soon recognized that the *dynamics* of the new mechanism could account for the observed density contrast $\delta\rho/\rho \sim 10^{-5} - 10^{-4}$ [37–42].

While (5.9) is a kinematical definition, (5.10) is a model-dependent dynamical statement which implicitly requires a knowledge of the macro- and microscopic properties of the cosmic fluid. The cosmic fluid should be constituted by one or more fields which are approximately homogeneous at the classical level but that naturally fluctuate upon quantization. The small-scale fluctuations of these fields are the primordial cosmic seeds we are looking for. In the next sections, we will illustrate this point by studying the classical and quantum dynamics of a real *scalar field*. The scalar field is perhaps the simplest and most intuitive realization of inflation, and by far the most popular and best studied.

5.3 Cold Big Bang

Consider the Friedmann equations (2.81) and (2.82) for a universe with no cosmological constant and no curvature term. What type of fluid dominated the first instants after the big bang? For simplicity, we can imagine that all forms of matter and radiation were originally enclosed in a single real scalar field ϕ, which later decayed into known particles and dark matter. The scalar is a "matter field" in this particle-field sense and the neutral spin-0 boson which it represents is called *inflaton*. The choice of a scalar field may be partially justified by supersymmetry, where scalars are abundant and play an important role. However, the inflaton does not have a precise place within field theory, like the other unknowns in cosmology (dark matter and dark energy).

5.3.1 Equation of State

We assume the following working hypotheses:

- The energy density of the inflaton dominates at the beginning, $\rho(t_\mathrm{i}) \simeq \rho_\phi$.
- The inflaton is minimally coupled with gravity, so that the total action is effectively given by (2.20) and (2.59), and the matter equation of motion is (2.66). In many models (*scalar-tensor theories*) a non-minimal coupling is allowed, but we will not consider this case until Sect. 7.4.
- The universe is flat, $\mathrm{K} = 0$.
- The inflaton is spatially homogeneous, $\phi(x) = \phi(t)$. From (2.79), this implies that the effective inflaton equation of state is

$$w_\phi = \frac{P_\phi}{\rho_\phi} = \frac{\frac{1}{2}\dot{\phi}^2 - V(\phi)}{\frac{1}{2}\dot{\phi}^2 + V(\phi)}\,. \tag{5.18}$$

The assumption of flat geometry is motivated both by simplicity (the dynamical equations being easier to solve) and by experiments, which place stringent bounds on the curvature of the universe [43]. Spectrum predictions different from the flat case are more difficult to accommodate in a natural way, since the location of the first CMB peak and of the Silk-damping drop-off are strongly affected by the parameters of curved models, both for a closed [44, 45] and an open universe [46–59]. Open models have recently experienced a raise of interest, due to their possible embedding in string theory [60, 61]; see Sect. 13.2.1.

Depending on the equation of state of ϕ, we can identify various regimes.

- In the *extreme slow-roll* (ESR) limit

$$\dot{\phi}^2 \ll V\,, \tag{5.19}$$

the inflaton behaves as an effective cosmological constant, $w_\phi \approx -1$. For this reason, the ESR limit is actually (flat) de Sitter. In this regime, one can neglect the scalar kinetic term in the first Friedmann equation and the second-derivative term in the scalar equation of motion:

$$H \simeq \sqrt{\frac{8\pi}{3m_{\rm Pl}^2}V}\,, \tag{5.20}$$

$$\dot{\phi} \simeq -\frac{V_{,\phi}}{3H}\,. \tag{5.21}$$

- The non-extreme *slow-roll* regime corresponds to (2.168c), $-1 < w_\phi < -1/3$. This ensures inflation, equation (5.9).
- In the kinetic fast-roll regime, or *kination* [62], the equation of state is $w_\phi \approx 1$ (stiff matter).

Then, the range of the barotropic index is

$$-1 < w_\phi < 1\,. \tag{5.22}$$

Notice, however, that relaxing the homogeneity condition extends this range. In fact, the energy density and pressure with gradients included are (equation (2.65))

$$\rho_\phi(x) = \tfrac{1}{2}\dot{\phi}^2 + V + \tfrac{1}{2}\partial_\alpha\phi\,\partial^\alpha\phi\,, \tag{5.23}$$

$$P_\phi(x) = \tfrac{1}{2}\dot{\phi}^2 - V - \tfrac{1}{6}\partial_\alpha\phi\,\partial^\alpha\phi\,. \tag{5.24}$$

In the *gradient-dominated* regime, $w_\phi \approx -1/3$, a value already included in (5.22); but in the *static* regime

$$\dot{\phi}^2 \ll \min[V,\,(\partial_\alpha\phi)^2]\,, \tag{5.25}$$

the barotropic index can be $w_\phi < -1$. Such values of w_ϕ are often associated with a phantom scenario where the kinetic term of the inflaton has the wrong sign. We have just seen that the static regime offers a more natural mechanism (see Sect. 7.4.4 for an alternative).

5.3.2 *Chaotic Inflation*

The old inflationary scenario, originally and independently proposed by Sato, Kazanas and Guth [30–33], was based on three fundamental assumptions. (i) Initially, the Universe expands in a high-temperature state of thermal equilibrium, during which the inflaton sits in a false vacuum (local maximum of the potential at $\phi = 0$). This is a homogeneous universe with the same value of temperature and energy density everywhere. (ii) Inflation ends with the decay of the false-vacuum state and the breaking of the action symmetry through a second-order phase transition. (iii) The decay of the unstable vacuum gives rises to the formation of bubbles where ϕ acquires the expectation values of the true degenerate vacuum. These bubbles expand at the speed of light, collide and eat up the false-vacuum regions. Linde [42] proposed an alternative and more natural model, called *chaotic inflation*, which has almost become a synonym of inflation since then.

As a quantum field operator, the inflaton will be denoted as $\hat{\phi}$. Initially, the expectation value $\langle\hat{\phi}\rangle$ is assumed to be strongly position-dependent, so that it spans a wide range of values (hence the name chaotic).[2] The most natural initial conditions for the inflaton should be defined as close as possible to the Planck time $t_{\rm i} = t_{\rm Pl}$, since this is the earliest time when we can still hope to make sense of initial conditions for classical fields on classical spacetimes. The Universe emerges from the Planck era with the scalar field shifted arbitrarily from the minima of its potential, so that the typical energy is

$$\rho_\phi(t_{\rm i}) \sim V(\phi_{\rm i}) \sim m_{\rm Pl}^4 \tag{5.26}$$

or $V(\phi_{\rm i}) \sim M_{\rm Pl}^4$. In regions where the field has appropriate initial conditions, inflation begins. These regions have a typical size of order of the Planck length $l_{\rm Pl}$. One can show that spatial gradients of the field do not prevent the onset of inflation at least in some regions [63].

The pre-inflationary Universe is highly inhomogeneous and anisotropic. The Planck scale acts as a coarse-graining or cut-off scale for inhomogeneities, so that the scalar field is approximately homogeneous in regions of volume

$$\mathcal{V}_{\rm i} \sim a_{\rm i}^3 \sim l_{\rm Pl}^3 ,$$

where, in the second member, we normalized the comoving volume to 1. These are the primordial analogues of the local patches of the separate universe approach. In some of the patches, the initial conditions are such that the slow-roll approximation is valid.

Assume a quadratic potential

$$V(\phi) = \tfrac{1}{2} m^2\phi^2 . \tag{5.27}$$

Obviously, the chaotic inflationary scenario can be applied to any well-motivated potential, but (5.27) is the simplest toy model of inflation and it well describes the dynamics near a local minimum. In the ESR regime, the dynamical equations are (5.20) and (5.21),

$$H \simeq \sqrt{\frac{4\pi}{3}}\frac{m}{m_{\rm Pl}}\phi , \qquad \dot{\phi} \simeq -\frac{m^2}{3}\frac{\phi}{H} \simeq -\frac{m_{\rm Pl}m}{\sqrt{12\pi}} ,$$

[2]Here the term "chaotic" loosely refers to the statistical distribution of the initial conditions, not to precise stochastic properties of chaos theory. In Chap. 6, we will see an example of chaotic evolution in the latter mathematical sense.

leading to a mild time dependence of ϕ,

$$\phi \simeq -\frac{m_{\rm Pl} m}{\sqrt{12\pi}}\, t\,,$$

and a scale factor

$$a \simeq a_{\rm i} {\rm e}^{H(t-t_{\rm i})} \simeq a_{\rm i} {\rm e}^{\frac{4\pi}{m_{\rm Pl}^2}(\phi_{\rm i}^2-\phi^2)}\,. \tag{5.28}$$

Observational bounds will later constrain the inflaton mass to be

$$m \approx 10^{-6}\, m_{\rm Pl}\,. \tag{5.29}$$

The initial condition (5.26) in the inflating regions requires $\phi_{\rm i} \sim 10^6 m_{\rm Pl}$. The end of inflation, as we have seen, can be placed at about the GUT scale, implying $\phi_{\rm e} \sim m_{\rm Pl} \ll \phi_{\rm i}$. From (5.28), the total number of e-foldings is

$$\mathcal{N}_{\rm e} \simeq \ln \frac{a_{\rm e}}{a_{\rm i}} \approx 10^{13}\,, \tag{5.30}$$

which is much larger than the minimum 60 e-folds (5.16). What is the physical size of a closed Universe after inflation? If the Universe at the big bang was a Planck-size sphere of radius $a_{\rm i} \sim l_{\rm Pl} \approx 10^{-35}$m, at $t_{\rm e}$ it is

$$a_{\rm e} \sim a_{\rm i} 10^{10^{12}} \approx 10^{10^{12}}\,{\rm m}\,. \tag{5.31}$$

In comparison, the linear size of the observable universe today is a tiny $\tau_0 \approx 10^{26}$m. Our causal patch is only a negligible part of a region which was very small in the primordial chaotic Universe. In this minute region, statistical fluctuations determined an appropriate initial mean value of the scalar field ϕ to trigger inflation. Thus, the cosmic principle and the slow-roll approximation combine naturally.

5.3.3 *Reheating*

The temperature of the scalar fluid during inflation does not need to be high. Then,

$$T_{\rm i} \sim T_{\rm e} \sim 0\,. \tag{5.32}$$

At the end of inflation, the universe is dominated by a non-relativistic scalar particle and we can assume a dust equation of state of the cosmic fluid, $w = 0$.

In cosmological and astrophysical surveys as well as in accelerators we have never observed effects which can be ascribed to a fundamental scalar particle of the typical mass of the inflaton. It is possible to explain this by our present technological limitations. However, a more plausible notion is that we do not see the boson ϕ because it decayed soon after inflation, thus generating all the matter (in cosmological jargon, matter and radiation) of the universe. This process is called *reheating* because it entails a rapid increase of temperature [64–79] (see [80] for a review). Reheating is typically divided into three phases.

1. In the first, called *preheating* [78, 81–83], the scalar field, after rolling slowly down its potential, oscillates around a local minimum and decays into massive bosons through a resonance. Gradually, oscillations are damped due to cosmological friction and the energy transfers to the new particles.
2. In the second phase, the new particles interact and decay into the actual matter constituents. This is the end of the cold big bang.
3. In the third stage, the final products are thermalized reaching the *reheating temperature* which, in the presence of supersymmetry, has an upper bound $T_{\rm reh} \lesssim 10^6 - 10^{10}\,{\rm GeV}$ [84, 85], corresponding to a cosmic age $t \sim 10^{-26} - 10^{-18}\,{\rm s}$. In this model, the lower limit for the redshift at the end of reheating is

$$z_{\rm reh} > 10^{23}\,. \tag{5.33}$$

From there on, the universe is described by the radiation-dominated hot big bang model of Chap. 2. Assuming instead the non-supersymmetric Standard-Model particle content and comparing the energy density at reheating with the one at horizon crossing (see equations (5.195) and (5.196) below), one obtains the more conservative upper bound

$$T_{\rm reh} \simeq \left(\frac{30}{\pi^2}\frac{\rho_{\rm reh}}{106.75}\right)^{\frac{1}{4}} < \left(\frac{30}{\pi^2}\frac{3H_*^2}{106.75\kappa^2}\right)^{\frac{1}{4}} < 7.6 \times 10^{15}\,{\rm GeV}\,. \tag{5.34}$$

In most of the models, the reheating process is explosive: this justifies, a posteriori, our initial assumption that the universe was radiation-dominated just after inflation.

To summarize, inflation might have taken place during the interval $10^{-43}\,{\rm s} < t < 10^{-30}\,{\rm s}$, followed by a reheating phase which ended at $t \sim 10^{-18}\,{\rm s}$. As anticipated, these numbers are somewhat speculative and depend significantly on assumptions involving semi-classical conjectures on quantum gravity and supersymmetric extensions of the Standard Model. Actually, the most conservative bound for exotic physics is dictated by the big-bang nucleosynthesis. The hot big bang model must start at $T \sim 10\,{\rm MeV}$ at the latest, in order to validate the standard calculations of light elements formation. The upper bound for the end of inflation is then

$$t_{\rm e} < 10^{-2}\,{\rm s}. \tag{5.35}$$

5.3.4 *Observable Inflation*

To calculate the observable amount of inflation [86], consider a comoving scale k crossing and re-entering the horizon at t_* and t_k, respectively during and after inflation. We can use $|\tau|$ instead of r_H in power-law cosmologies. In particular, $|\tau_k| = 1/k$. At horizon exit, $\mathcal{N}_*$ e-foldings have passed since the beginning of inflation. If we observe the scale k today, then $\tau_k < \tau_0$, and the portion of inflation which affected our local patch corresponds to the last e-foldings since horizon exit,

$$\mathcal{N}_k = \mathcal{N}_e - \mathcal{N}_* = \ln \frac{\tau_*}{\tau_e} . \qquad (5.36)$$

Since the horizon at exit and re-entry is the same,

$$1 = \frac{r_H(t_k)}{r_H(t_*)} = \frac{\tau_k}{\tau_0}\frac{\tau_0}{\tau_{eq}}\frac{\tau_{eq}}{\tau_{reh}}\frac{\tau_{reh}}{\tau_e}\frac{\tau_e}{\tau_*} .$$

Using (5.4) and (5.5),

$$\mathcal{N}_k = -\ln(k\tau_0) + \ln\left(\frac{\tau_0}{\tau_{eq}}\frac{\tau_{eq}}{\tau_{reh}}\frac{\tau_{reh}}{\tau_e}\right) = -\ln(k\tau_0) + \ln\left[\frac{\tau_0}{\tau_{eq}}\frac{T_{reh}}{T_{eq}}\left(\frac{\rho_e^{\frac{1}{4}}}{T_{reh}}\right)^{\frac{2}{3}}\right]$$

$$= -\ln(k\tau_0) + \ln\left[\frac{\tau_0}{\tau_{eq}}\frac{\left(T_{reh}^{max}\right)^{\frac{1}{3}} T_{GUT}^{\frac{2}{3}}}{T_{eq}}\frac{T_{reh}^{\frac{1}{3}}\rho_e^{\frac{1}{6}}}{\left(T_{reh}^{max}\right)^{\frac{1}{3}} T_{GUT}^{\frac{2}{3}}}\right] ,$$

where $T_{reh}^{max} = 10^{10}$ GeV is the upper bound of the reheating temperature and $T_{GUT} = 10^{15}$ GeV is the lower limit of the GUT scale. Therefore, from (2.150),

$$\boxed{\mathcal{N}_k \simeq 52 - \ln(k\tau_0) - \frac{2}{3}\ln\frac{10^{15}\,\text{GeV}}{\rho_e^{1/4}} - \frac{1}{3}\ln\frac{10^{10}\,\text{GeV}}{T_{reh}} .} \qquad (5.37)$$

This is the number of e-folds we can observe today, which is in general smaller (but see [87]) than the minimum amount of inflation (5.16) required to solve the horizon problem. In turn, $\mathcal{N}_k$ can be much smaller than the actual duration of inflation $\mathcal{N}_e$ (e.g., (5.30)).

Fix the perturbation wave-length to the horizon size today (largest observable scale), $k_{hor} = \tau_0^{-1}$. Shorter and shorter scales took more e-folds to cross the horizon, so that $\mathcal{N}_0 := \mathcal{N}_{k=k_{hor}}$ is the upper bound of "observable inflation." The minimum $\mathcal{N}_0$ occurs for the BBN bound $\rho_e^{1/4} \sim T_{reh} \sim 10$ MeV, $\mathcal{N}_0 \sim 17$. For instantaneous reheating at the GUT scale, $\rho_e^{1/4} \sim T_{reh} \sim T_{GUT}$, $\mathcal{N}_0 \sim 56$ (see Problem 5.1).

5.3.5 Timeline of the Early Universe

We have now to refine the timeline of the universe before BBN by adding the early cold big bang phase to the table in Problem 2.8 (Fig. 5.3). One can split the post-inflationary evolution into a matter-dominated era between the end of inflation and the end of reheating, and a radiation-dominated era between the end of reheating and matter-radiation equality. Table 5.1 summarizes the results of this section. The expansion ratio of the inflationary era is estimated via the lower bound (5.16),

$$\frac{r_H}{r_{H\mathrm{i}}} = \mathrm{e}^{\mathcal{N}_\mathrm{e}}\frac{r_H}{r_{H\mathrm{e}}} > 10^{28}\frac{r_H}{r_{H\mathrm{e}}},$$

while to calculate the ratio of comoving Hubble horizons after reheating we use

$$\frac{r_{H0}}{r_H} = \frac{\tau_0}{\tau_{\mathrm{eq}}}\frac{\tau_{\mathrm{eq}}}{\tau} \simeq 0.1\,z \simeq 10^9\frac{T}{1\,\mathrm{MeV}}.$$

Looking at the second column from the right in the table, one sees that the universe expanded more in the first 10^{-36} seconds than in the rest of its life. How much more, it greatly depends on the details of inflation: compare, for instance, (5.16) with (5.30).

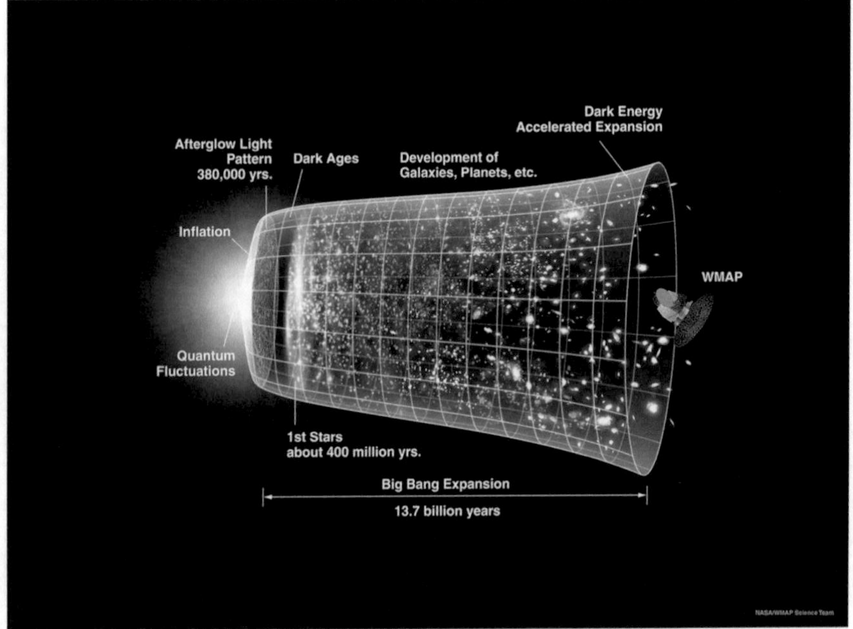

Fig. 5.3 Timeline of the universe (Credit: NASA/WMAP Science Team [88])

Table 5.1 Simplified history of the early universe from the cold big bang. The values of this table should be taken only as indicative

$\rho^{1/4}$ (GeV)	T (GeV)	t (s)	$r_H(t_0)/r_H(t)$	Event
$\sim 10^{19}$	~ 0	10^{-44}	$>10^{52}$	Cold big bang? **Inflation** begins at Planck scale?
$\sim 10^{15}$	~ 0	10^{-36}	10^{24}	Inflation ends at GUT scale? **Reheating** begins?
	$\sim 10^{10}$	10^{-7}	10^{22}	Reheating ends?
	10^{-2}	10^{-2}	10^{10}	Latest end of cold big bang model
	10^{-5}	200	10^{8}	BBN
	10^{-9}	10^{12}	100	Radiation-matter equality
	10^{-13}	10^{17}	1	Today

5.4 Scalar Field: Background Dynamics

In this section, we will describe the classical dynamics of a homogeneous, canonical scalar field. Let $D = 4$ and $\textsc{k} = 0$. The Friedmann and scalar field equations (2.169) and (2.171) are

$$H^2 = \frac{\kappa^2}{3}\left(\frac{\dot{\phi}^2}{2} + V\right), \tag{5.38}$$

$$0 = \ddot{\phi} + 3H\dot{\phi} + V_{,\phi}. \tag{5.39}$$

5.4.1 *Hamilton–Jacobi Formalism*

In Sect. 2.1.2, we briefly mentioned the possibility of using matter as an internal clock. For a scalar-field dominated universe, ϕ is the natural choice. Assume without loss of generality that $\dot{\phi} \geqslant 0$. From (2.135),

$$\dot{H} = -\frac{\kappa^2}{2}\dot{\phi}^2. \tag{5.40}$$

Thus, a canonical homogeneous scalar field in standard general relativity can never lead to super-inflation, since $\dot{H} < 0$. If we regard H as a function of ϕ, and if ϕ varies *monotonically* with time, then

$$H_{,\phi} = -\frac{\kappa^2}{2}\dot{\phi}. \tag{5.41}$$

Equation (5.38) then gives

$$H_{,\phi}^2 - \frac{3\kappa^2}{2}H^2 + \frac{\kappa^4}{2}V = 0\,. \tag{5.42}$$

The first slow-roll parameter can be recast in several different ways:

$$\epsilon = \frac{\kappa^2}{2}\frac{\dot{\phi}^2}{H^2} \tag{5.43a}$$

$$= \frac{2}{\kappa^2}\left(\frac{H_{,\phi}}{H}\right)^2 \tag{5.43b}$$

$$= \frac{\kappa^2}{2}\left(\frac{a}{a_{,\phi}}\right)^2, \tag{5.43c}$$

so that (5.42) becomes

$$V = \frac{1}{\kappa^2}(3-\epsilon)H^2\,. \tag{5.44}$$

Equation (5.43c) stems from (5.41) via

$$a_{,\phi}H_{,\phi} = -\frac{\kappa^2}{2}aH\,. \tag{5.45}$$

Equations (5.41) and (5.42) (or (5.44) and (5.45)) are the *Hamilton–Jacobi* equations [89, 90]. They hold only if $\dot{\phi}$ does not change sign and they break down, for instance, when the field oscillates around one of its minima. While the Hamilton–Jacobi formalism is not suited for reheating, it is useful during inflation, when the field is rolling far from the minima. In that case, it allows one to find the dynamical profiles of the model starting from the Hubble parameter $H(\phi)$: (i) replace $H(\phi)$ in (5.42) to obtain $V(\phi)$; (ii) integrate (5.41) to get $\phi(t)$ and, hence, $H(t)$ and $a(t)$. In general, it is not convenient to choose first the potential $V(\phi)$ because (5.44) is nonlinear; an exception is the exponential potential associated with power-law inflation (Problem 2.11), where ϵ is constant. Then, also H is exponential, $H(\phi) \propto \mathrm{e}^{-\phi/\phi_0}$.

5.4.2 *Slow-Roll Parameters*

The use of the slow-roll formalism [34, 91–93] simplifies the study of inflation; however, it can also be considered as an effective notation for some recurrent dimensionless combinations of cosmological quantities, without imposing any condition on their magnitude. We will keep calling these parameters "slow-roll" in this case, too. The most commonly used SR towers rely upon two different quantities, the geometric Hubble parameter H and the dynamical inflaton potential V. We will name these towers H-SR and V-SR, respectively, and explore some of their properties. Other SR towers can be constructed for particular cosmological scenarios or analyses [94–97].

5.4.2.1 *H*-SR Tower

The H-SR tower is defined as

$$\epsilon_0 := \epsilon\,, \qquad \epsilon_n := \prod_{i=1}^{n}\left[-\frac{\mathrm{d}\ln H^{(i)}}{\mathrm{d}\ln a}\right]^{\frac{1}{n}}, \qquad n \geqslant 1\,, \tag{5.46}$$

where (i) is the i-th ϕ derivative. The first three parameters, which are those appearing in all the main expressions for cosmological observables, are

$$\epsilon = \epsilon_0 = \frac{3\dot{\phi}^2}{\dot{\phi}^2 + 2V}\,, \tag{5.47}$$

$$\eta := \epsilon_1 = -\frac{\mathrm{d}\ln\dot{\phi}}{\mathrm{d}\ln a} = -\frac{\ddot{\phi}}{H\dot{\phi}}\,, \tag{5.48}$$

$$\xi^2 := \epsilon_2^2 = \frac{1}{H^2}\left(\frac{\ddot{\phi}}{\dot{\phi}}\right)^{\cdot} = \frac{\dddot{\phi}}{H^2\dot{\phi}} - \eta^2\,. \tag{5.49}$$

The conditions

$$\epsilon \ll 1\,, \qquad |\eta| \ll 1 \tag{5.50}$$

define the ESR regime. The second inequality is equivalent to assume solution (5.21). Formulæ truncated at the n-th power of these parameters will be referred to as "n-th order SR." The first-order SR is precisely the ESR approximation. If $|\eta| > 1$ or $|\xi| > 1$, inflation could still take place but it would soon end due to the rapid variation of ϵ. Noting that $\ddot{H} = -2H\dot{H}\eta$, we have

$$\dot{\epsilon} = 2H\epsilon(\epsilon - \eta)\,, \tag{5.51}$$

$$\dot{\eta} = H(\epsilon\eta - \xi^2)\,. \tag{5.52}$$

Differentiation with respect to the scalar field yields $\epsilon_{n,\phi} = \dot{\epsilon}_n/\dot{\phi}$. By (5.43a), the resulting prefactor $H/\dot{\phi}$ can be expressed as

$$\frac{H}{\dot{\phi}} = +\sqrt{\frac{\kappa^2}{2\epsilon}}\,, \tag{5.53}$$

where the plus sign has been chosen in order to have a slow rolling down the potential with $\dot{\phi} > 0$. This is always possible by a redefinition $\phi \to -\phi$.

A comment is in order. A slow-roll tower is dynamical (i.e., it does say something about the dynamics of the Hamilton–Jacobi equations) either when external constraints on the form and magnitude of the SR parameters are applied, or when distinct SR definitions are related through the Hamilton–Jacobi equations themselves. For example, the fundamental building block of the H-SR tower is the parameter ϵ; as long as the latter is not linked with the matter content (equation (2.166)), it is clear there will be no knowledge about the evolution of the system. However, when rewriting the H-parameters in terms of $\dot{\phi}$ through the first Hamilton–Jacobi equation, these parameters become dynamically informative. An interesting discussion on related issues can be found in [96].

5.4.2.2 V-SR Tower

The H-SR hierarchy is an elegant instrument of analysis coming from the Hamilton–Jacobi formulation of the equations of motion. However, often the natural starting point is not H but a reasonable inflaton potential V, for instance suggested by field-theory high-energy models. Given V, the Hubble parameter must be determined by the Hamilton–Jacobi equations, which are not always readily solvable. Therefore, it is convenient to define another SR tower and try to relate it to the original one, namely,

$$\epsilon_{V0} := \frac{1}{2\kappa^2}\left(\frac{V_{,\phi}}{V}\right)^2, \qquad \epsilon_{Vn} := \frac{1}{\kappa^2}\left[\frac{V^{(n+1)}V_{,\phi}^{n-1}}{V^n}\right]^{\frac{1}{n}}, \qquad n \geq 1\,, \tag{5.54}$$

where, again, we have introduced the first parameter separately. Therefore,

$$\epsilon_V := \epsilon_{V0}\,, \tag{5.55a}$$

$$\eta_V := \epsilon_{V1} = \frac{1}{\kappa^2}\frac{V_{,\phi\phi}}{V}\,, \tag{5.55b}$$

$$\xi_V^2 := \epsilon_{V2}^2 = \frac{1}{\kappa^4}\frac{V_{,\phi\phi\phi}V_{,\phi}}{V^2}\,, \tag{5.55c}$$

and their derivatives with respect to the scalar field are

$$\epsilon_{V,\phi} = -\frac{V_{,\phi}}{V}\left(2\epsilon_V - \eta_V\right), \tag{5.56}$$

$$\eta_{V,\phi} = -\frac{1}{2\epsilon_V}\frac{V_{,\phi}}{V}\left(2\epsilon_V\eta_V - \xi_V^2\right), \tag{5.57}$$

where

$$\frac{V_{,\phi}}{V} = -\sqrt{2\kappa^2 \epsilon_V}\,. \tag{5.58}$$

The conditions $\epsilon_V \ll 1$ and $|\eta_V| \ll 1$ are necessary to drop the kinetic term in (5.38) and the acceleration term in (5.39) but they are not sufficient. In general, the V-SR formalism requires a further assumption, namely (5.21), which is easy enough to be satisfied. This determines the minus sign in (5.58), provided $\dot{\phi} > 0$.

For a potential with a mass term $V_{,\phi\phi} = m^2$, the condition $|\eta_V| \ll 1$ is roughly equivalent to

$$\boxed{m^2 \ll H^2\,.} \tag{5.59}$$

Thus, the mass of the inflaton is expected to be much lighter than the Hubble energy during the accelerated expansion. Many models in supergravity and string cosmology face the challenge of respecting (5.59) without fine tuning. This "η-problem" will be discussed in Sect. 5.12.3.

5.4.2.3 Horizon-Flow Tower

The horizon-flow parameters [94, 95] are defined by

$$\epsilon_{H0} := \frac{H_{\rm inf}}{H}\,, \qquad \epsilon_{Hi} := \frac{{\rm d}\ln|\epsilon_{H\,i-1}|}{{\rm d}\mathcal{N}_a}\,, \qquad i \geq 1\,, \tag{5.60}$$

where $H_{\rm inf}$ is the Hubble rate at some chosen time and $\mathcal{N}_a = \ln(a/a_{\rm i})$ is the number of e-folds.[3] As shown in [96], these parameters (and others similarly defined) do not properly encode inflationary dynamics even if they provide a good algorithm for reconstructing the inflationary potentials. In fact, because of the absence of the $1/n$ power, definition (5.60) does not permit a power truncation similar to that of the traditional SR towers (5.46) and (5.54), unless one imposes a constraint such as $\partial^i_\phi H = 0$ for $i > i_{\max}$.

The evolution equation for the horizon-flow parameters is given by

$$\dot{\epsilon}_{Hi} = H\epsilon_{Hi}\epsilon_{H\,i+1}\,. \tag{5.61}$$

[3]Our definition counts $\mathcal{N}_a$ forward in time, in accordance with [94] where $\mathcal{N}_a(t_{\rm i}) = 0$ and goes up to $\mathcal{N}_a(t) > 0$. In [95], the "backward" definition is used, where $\mathcal{N}_k = \ln(a_{\rm e}/a)$ is the number of remaining e-folds at the time t before the end of inflation.

5.4.2.4 Relation Between SR Parameters

It is possible to map the three SR towers one to the other by some simple relations. Here we will restrict ourselves to the first three parameters. From (5.44), we get the exact relation

$$\epsilon_V = \epsilon \left(\frac{3-\eta}{3-\epsilon} \right)^2 . \tag{5.62}$$

Then, noting that

$$V_{,\phi} = \dot{\phi} H(\eta - 3) , \qquad V_{,\phi\phi} = H^2[3(\epsilon + \eta) - \eta^2 - \xi^2] , \tag{5.63}$$

one has

$$\eta_V = \frac{3(\epsilon + \eta) - \eta^2 - \xi^2}{3 - \epsilon} . \tag{5.64}$$

Finally, since $V_{,\phi\phi\phi} = -3(3\epsilon\eta + \xi^2)H^3 + O(\epsilon^3)$ we obtain, to first H-SR order,

$$\epsilon \simeq \epsilon_V , \qquad \eta \simeq \eta_V - \epsilon_V , \qquad \xi^2 \simeq \xi_V^2 - 3\epsilon_V\eta_V + 3\epsilon_V^2 . \tag{5.65}$$

These equations allow us to shift from one hierarchy to the other, according to the most convenient approach.

The horizon flow parameters are related to the first H-SR parameters by

$$\epsilon_{H1} = \epsilon , \qquad \epsilon_{H2} = 2(\epsilon - \eta) , \qquad \epsilon_{H2}\epsilon_{H3} = 2(2\epsilon^2 - 3\epsilon\eta + \xi^2) . \tag{5.66}$$

5.4.3 *Inflationary Attractor*

The predictive power of inflation depends on the behaviour of cosmological solutions with different initial conditions. If there exists an attractor such that the differences of these solutions rapidly vanish, then the inflationary and post-inflationary physics will generate observables which are independent of the initial conditions. Let $H_o(\phi) > 0$ be a generic expanding solution (denoted with the subscript o) of the Hamilton–Jacobi equations and consider a linear perturbation $\delta H(\phi)$ which does not reverse the sign of $\dot{\phi} > 0$. The linearized equation of motion (5.42) (V is fixed) is

$$2H_{o,\phi}\delta H_{,\phi} = 3\kappa^2 H_o \delta H ,$$

so that the perturbation is [89]

$$\delta H(\phi) = \delta H(\phi_o) \exp\left(\frac{3\kappa^2}{2}\int_{\phi_o}^{\phi} \mathrm{d}\phi \frac{H_o}{H_{o,\phi}}\right), \tag{5.67}$$

where we have integrated from some initial value ϕ_o. All linear perturbations are exponentially suppressed when the integrand is negative definite, which is the case during inflation as one can see from (5.41) and (5.53).

The number of e-foldings (5.12) can be written as an integral in the scalar field,

$$\begin{aligned}\mathcal{N} &= -\int_{r_{H\mathrm{i}}}^{r_H} \mathrm{d}\ln r_H = \int_{t_\mathrm{i}}^{t} \mathrm{d}t\, H(1-\epsilon)\\ &\stackrel{(5.53)}{=} \int_{\phi_\mathrm{i}}^{\phi} \mathrm{d}\phi\, [1-\epsilon(\phi)]\sqrt{\frac{\kappa^2}{2\epsilon(\phi)}}\,.\end{aligned} \tag{5.68}$$

In the SR approximation $\epsilon \simeq$ const, (5.43b) and (5.68) yield

$$\delta H(\phi) \simeq \delta H(\phi_o)\, \mathrm{e}^{-3(1-\epsilon)(\mathcal{N}-\mathcal{N}_o)}\,. \tag{5.69}$$

During the accelerated expansion, inflationary solutions all approach one another at exponential rate. This is known as "the inflationary attractor." The non-linear case is more complicated but the above result holds as soon as perturbations enter the linear regime.

In general, the *classical stability* of a solution $(a(t), \dot{a}(t), \rho_i(t))$ with various matter components ρ_i is checked by linearizing the background equations against a homogeneous perturbation

$$\delta X(t) := \begin{pmatrix} \delta a(t) \\ \delta \dot{a}(t) \\ \delta \rho_i(t) \end{pmatrix}, \tag{5.70}$$

and solving with respect to the latter. The set of linearized equations can be written in matrix form,

$$\dot{\delta X} = M\,\delta X\,, \tag{5.71}$$

where the entries M_{ij} of the matrix M are calculated on the background solution. The characteristic equation

$$\det(M - \lambda \mathbb{1}) = 0 \tag{5.72}$$

determines the eigenvalues λ and a solution is stable provided

$$\mathrm{Re}(\lambda) \leqslant 0\,. \tag{5.73}$$

When the eigenvalues are time dependent, they are interpreted as evaluated at a given time t [98].

To make sense physically, linear *inhomogeneous* perturbations must be defined on backgrounds which are classically stable in the above sense. In the case of inflationary solutions, we have just seen that they are all attractors. This means that we can use any of these solutions as a background whereon to calculate the primordial spectra.

5.5 Models of Inflation

Given a potential $V(\phi)$, we have seen how to obtain a complete cosmological profile and a set of SR parameters for the homogeneous scalar field $\phi(t)$. In preparation of calculating the primordial spectra, we summarize the recipe to obtain all the quantities appearing in cosmological observables.

1. Fix the inflaton potential $V(\phi)$ by phenomenological arguments or via a high-energy theory, for instance an effective field theory stemming from a suitable model of microscopic physics.
2. Determine the first V-SR parameters via (5.55) (or (5.65), if the model is an exact solution).
3. According to the chaotic scenario, the value of the inflaton ϕ_i at the onset of inflation is defined by

$$V(\phi_\mathrm{i}) = m_\mathrm{Pl}^4\,. \tag{5.74}$$

On the other hand, we can estimate the field value at the end of inflation by setting

$$\epsilon(\phi_\mathrm{e}) = 1\,. \tag{5.75}$$

One may adopt the alternative criterion $\eta_V(\phi_\mathrm{e}) = 1$, but the difference is small as long as the number of e-folds is large enough.

4. To first SR order, the total number of e-foldings (2.48) is

$$\mathcal{N}_\mathrm{e} \simeq \mathrm{sgn}(\dot{\phi})\sqrt{\frac{4\pi}{m_\mathrm{Pl}^2}}\int_{\phi_\mathrm{i}}^{\phi_\mathrm{e}} \frac{\mathrm{d}\phi}{\sqrt{\epsilon(\phi)}}\,, \tag{5.76}$$

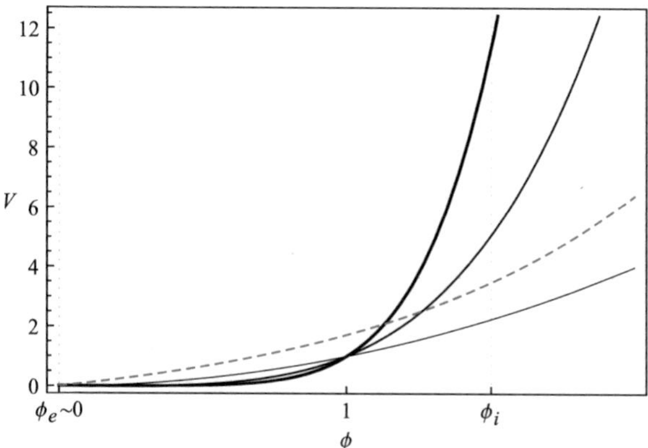

Fig. 5.4 Large-field inflationary models: potentials ϕ^n/n with $n = 2, 4, 6$ (increasing thickness) and exponential potential (*dashed curve*). Slow rolling begins at $\phi_{\rm i} \gg m_{\rm Pl}$ and terminates somewhere near the minimum at $\phi_{\rm e} \sim 0$

while at horizon crossing

$$\mathcal{N}_k \simeq \mathrm{sgn}(\dot{\phi})\sqrt{\frac{4\pi}{m_{\rm Pl}^2}}\int_{\phi_*}^{\phi_{\rm e}} \frac{\mathrm{d}\phi}{\sqrt{\epsilon(\phi)}} \,. \tag{5.77}$$

The value of the field ϕ_* at horizon crossing is obtained by inverting the last expression and fixing $\mathcal{N}_k$. Typically, $\mathcal{N}_k \approx 50-70$.

5. From ϕ_*, one obtains the value of the slow-roll parameters $\epsilon_* = \epsilon(\phi_*)$, $\eta_* = \eta(\phi_*)$, ..., when the cosmological scale k left the causal region. We observe the imprint of inflation at these scales in the CMB.

A standard classification [99] identifies three types of models when the potential has at least one global minimum[4]: large-field, small-field and multi-field. A more detailed charting of the inflationary models can be found in [100].

5.5.1 Large-Field Models

In *large-field* models, inflation starts with the scalar field displaced away from a minimum at the origin, $\phi_{\rm i} = \phi(t_{\rm i}) \neq 0$. From this configuration, the inflaton rolls down towards the minimum, $\phi_{\rm e} \ll \phi_{\rm i}$. Polynomial potentials of the form (2.61) with $\sigma_n > 0$ fall in this category. We consider two extreme examples: (2.182) and a monomial potential (Fig. 5.4).

[4]Classically, it is sufficient to have a local minimum, but when considering a quantum theory tunneling effects should be taken into account.

5.5.1.1 Exponential Potential

The exponential potential (2.182) in four dimensions,

$$V(\phi) = \frac{3p-1}{2}\phi_0^2\,\mathrm{e}^{-2\phi/\phi_0}\,, \tag{5.78}$$

corresponds to power-law inflation (2.180), which is an exact solution of the equations of motion discussed in Problem 2.11. In the flat case,

$$\phi_0 = \pm\sqrt{\frac{2p}{\kappa^2}}\,, \qquad \mathrm{K} = 0\,. \tag{5.79}$$

Inflation does not have a natural end, since all the SR parameters are constant:

$$\epsilon = \eta = \xi = \frac{1}{p}\,. \tag{5.80}$$

To terminate inflation, we have to force reheating by hand in the model. Despite this issue, called *graceful entry problem*, power-law inflation is important because it is not only the background around which inhomogeneous perturbations are often defined, but also the starting point for the construction of approximate solutions.

The potential (2.182) may arise in various settings beyond the cosmological standard model. In *Kaluza–Klein scenarios*, the Universe is allowed to have $D-1 > 3$ spatial dimensions, all but three of which are compactified to an unobservable size today. An example is an Einstein–Maxwell theory $S_6 = \int \mathrm{d}^4x\,\mathrm{d}^2y\sqrt{-\hat{g}}[c_6\hat{R} - (1/4)F_{AB}F^{AB} - 2\Lambda_6]$ defined on a six-dimensional spacetime with structure $\mathcal{M}_4\times S^2$, where $\mathcal{M}_4 \ni x$ is a four-dimensional Lorentzian manifold and $S^2 \ni y$ is the 2-sphere with a spacetime-dependent radius $L(x)$, frozen at some constant scale L_0 at late times [101]. Call $l(x) = L(x)/L_0$. The matrix $\hat{g}_{AB}$ is decomposed into a four-dimensional metric $\hat{g}_{\mu\nu}(x) = l^{-2}(x)g_{\mu\nu}(x)$ and an internal S^2 matrix $\hat{g}_{ab}(x,y) = l^2(x)g_{ab}(y)$. After integrating over the S^2 coordinates and by suitably tuning the constants, the effective four-dimensional action of the model is $S_4 = \mathrm{d}^4x\sqrt{-g}[R/(2\kappa^2) + \mathcal{L}_l]$, where the Lagrangian for the scalar l is of the form $\mathcal{L}_l = -(\partial l/l)^2 - V_0 l^{-2}(1-l^{-2})^2$. Defining $\phi := (2/\kappa)\ln l$, one obtains the effective potential $V(\phi) \propto \mathrm{e}^{-\kappa\phi}(1-\mathrm{e}^{-\kappa\phi})^2$, which reduces to (5.78) with $p = 2$ in the large-ϕ limit. The procedure can be extended to $\mathcal{M}_4 \times S^d$ and d compact dimensions, where the $(4+d)$-dimensional Lagrangian density is higher-order in curvature invariants, $\mathcal{L}_6 = O(R) + O(R^2)$ [102, 103]. The resulting effective potential for the scalar is $V(\phi) \propto \mathrm{e}^{-d\kappa\phi}(1-\mathrm{e}^{-2\kappa\phi})^2(1+c\mathrm{e}^{-2\kappa\phi})^{-2} \sim \mathrm{e}^{-d\kappa\phi}$. Still in the context of Kaluza–Klein compactification, the six-dimensional Nishino–Salam–Sezgin $\mathcal{N} = 2$ supergravity model [104–107] produces, when compactified to four dimensions, two mutually coupled scalar fields with exponential potentials [108–110]. None of these scenarios, however, can give sustainable inflation, since $p = O(1)$ typically. Other models leading to exponential potentials and acceleration will be discussed in Sect. 7.5.

5.5.1.2 Monomial Potential

The monomial potential ($n > 0$)

$$V(\phi) = \frac{\sigma_n}{n}\phi^n\,, \qquad \sigma_n > 0\,, \tag{5.81}$$

does not correspond to any exact background solution, except for the $H = \text{const}$, $\mathrm{K} = 1$ case (2.174), where $n = 2$. In the ESR approximation (5.20) and (5.21), the $\mathrm{K} = 0$ approximate solution features a Hubble parameter

$$H(\phi) \simeq \sqrt{\frac{8\pi\sigma_n}{3n\,m_{\rm Pl}^2}}\,\phi^{\frac{n}{2}}\,, \tag{5.82}$$

while integrating (5.21) with $H \sim \text{const} \sim \phi$, one has

$$\phi^{2-\frac{n}{2}} \simeq -\sqrt{\frac{n\sigma_n m_{\rm Pl}^2}{24\pi}}\,t\,. \tag{5.83}$$

Then,

$$a(t) \simeq a_{\rm i}\mathrm{e}^{\frac{8\pi}{n\,m_{\rm Pl}^2}[\phi_{\rm i}^2-\phi^2(t)]}\,. \tag{5.84}$$

The slow-roll parameters and field values at $t_{\rm i}$ and $t_{\rm e}$ read ($n > 1$)

$$\epsilon = \frac{n^2}{16\pi}\frac{m_{\rm Pl}^2}{\phi^2}\,, \tag{5.85a}$$

$$\eta = \frac{n(n-2)}{16\pi}\frac{m_{\rm Pl}^2}{\phi^2}\,, \tag{5.85b}$$

$$\phi_{\rm i}^2 = \left(\frac{nm_{\rm Pl}^4}{\sigma_n}\right)^{\frac{2}{n}}\,, \tag{5.85c}$$

$$\phi_{\rm e}^2 = \frac{n^2}{16\pi}m_{\rm Pl}^2 \ll \phi_{\rm i}^2\,. \tag{5.85d}$$

The total number of e-foldings is ($\dot\phi < 0$)

$$\mathcal{N}_{\rm e} = \frac{4\pi}{nm_{\rm Pl}^2}\left(\phi_{\rm i}^2-\phi_{\rm e}^2\right) = \frac{4\pi}{nm_{\rm Pl}^2}\left(\frac{nm_{\rm Pl}^4}{\sigma_n}\right)^{\frac{2}{n}} - \frac{n}{4}\,, \tag{5.85e}$$

while

$$\phi_*^2 = \frac{n(4\mathcal{N}_k + n)}{16\pi} m_{\rm Pl}^2, \tag{5.85f}$$

$$\epsilon_* = \frac{n}{4\mathcal{N}_k + n}, \tag{5.85g}$$

$$\eta_* = \frac{n-2}{4\mathcal{N}_k + n}. \tag{5.85h}$$

The time at which inflation ends is, from (5.83) and (5.85c),

$$t_{\rm e} \simeq \sqrt{\frac{24\pi}{n\sigma_n m_{\rm Pl}^2}\left(\frac{n m_{\rm Pl}^4}{\sigma_n}\right)^{\frac{2}{n}-\frac{1}{2}}}. \tag{5.86}$$

Inflation induced by a linear potential suffers from the same graceful exit problem as power-law inflation, since $\eta = -\epsilon = \text{const.}$

The linear term of an arbitrary potential can be reabsorbed by a field translation and it can be ignored without loss of generality. The $n = 1$ case will be resuscitated in Sect. 13.4.5 in the context of a fundamental theory of Nature (string theory), together with other potentials with non-integer $n = 2/5, 2/3, 4/5, 4/3$.

In the case of the quadratic potential ($n = 2$, $\sigma_2 = m^2$),

$$\phi_{\rm i} = \sqrt{2}\, m_{\rm Pl} \frac{m_{\rm Pl}}{m}, \tag{5.87a}$$

$$\phi_{\rm e} = \frac{m_{\rm Pl}}{\sqrt{4\pi}}, \tag{5.87b}$$

$$\mathcal{N}_{\rm e} \simeq 4\pi \frac{m_{\rm Pl}^2}{m^2}, \tag{5.87c}$$

$$\phi_*^2 \simeq \frac{\mathcal{N}_k}{2\pi} m_{\rm Pl}^2 \sim 10\, m_{\rm Pl}^2, \tag{5.87d}$$

$$\epsilon_* \simeq \frac{1}{2\mathcal{N}_k} \sim 10^{-2}, \tag{5.87e}$$

$$\eta_* = 0. \tag{5.87f}$$

The SR parameters are small. For $m = 10^{-6} m_{\rm Pl}$, one recovers the total number of e-foldings (5.30), $\mathcal{N}_{\rm e} \approx 10^{13}$, while inflation ends at $t_{\rm e} \approx 10^{13} t_{\rm Pl} \approx 10^{-31}$s. The minimum amount of e-folds $\mathcal{N}_{\rm e} \approx 60$ is obtained, from (5.87c), with a larger mass, $m \approx 0.5 m_{\rm Pl}$. The lighter the scalar field, the longer inflation.

The quadratic potential is the prototype of large-field models and one among the first proposals for chaotic inflation. Even if the potential energy V keeps below the Planck energy density, having field excursions $\Delta\phi \gg m_{\rm Pl}$ may pose an issue about the physical viability of these models (*trans-Planckian problem*). In fact, at scales comparable with or larger than the Planck mass, effective field theory can break down to give way to new degrees of freedom. To assess this possibility, it is necessary to explore these degrees of freedom in a fundamental theory (in jargon, to make a UV completion of the effective models of this chapter). That will be done in Chap. 13 (for the quadratic case, in Sect. 13.4.5).

For a quartic potential ($n = 4$) the numbers are similar. To get enough inflation in order to solve the horizon problem, $\sigma_4 \lesssim 10^{-2}$. Since $t_e \simeq 10\sigma_4^{-1/2} t_{\rm Pl}$, in that case inflation lasts only until $t_e \approx 10^{-42}$s. For $\sigma_4 = 10^{-16}$, one has $t_e \approx 10^{-35}$s. Quartic monomial inflation, however, is ruled out by data, while the quadratic potential is under strong pressure [43, 111].

5.5.2 *Small-Field Models*

In *small-field* models, the scalar field is initially very near a local maximum at the origin, $\phi_i \sim 0$, and rolls towards the minimum at $0 \neq \phi_e \gg \phi_i$. This case is sometimes called hilltop inflation because slow rolling takes place in a neighborhood of the local maximum. This type of potential is typically originated in spontaneous-symmetry-breaking scenarios (see Sect. 7.1.1).

A double-well (Mexican hat) example is shown in Fig. 5.5. Near the maximum, the typical potential can be approximated as

$$V(\phi) = \Lambda^4 \left[1 - \left(\frac{\phi}{2f}\right)^n\right], \tag{5.88}$$

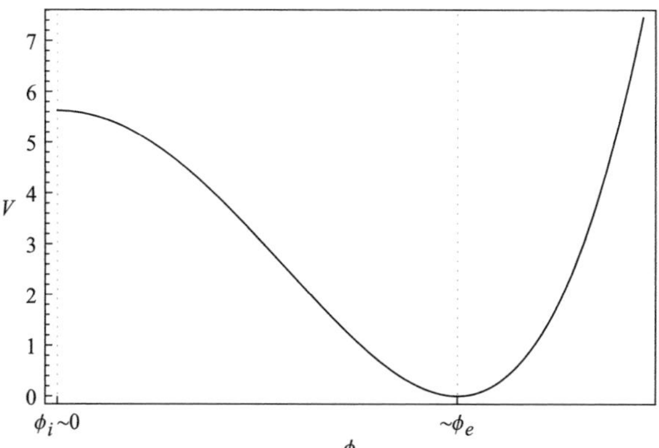

Fig. 5.5 Small-field inflationary model: double-well potential $V = \sigma_0 - \phi^2/2 + \sigma_4\phi^4/4$ with $\sigma_0, \sigma_4 > 0$. Slow rolling begins near the local maximum at $\phi_i \sim 0$ and terminates somewhere near the minimum at $\phi_e \gg M_{\rm Pl}$

where $\phi/(2f) \ll 1$ and Λ and f are mass scales. For $n \neq 2$, one has

$$\epsilon \simeq \frac{n^2}{8}\frac{M_{\rm Pl}^2}{f^2}\left(\frac{\phi}{2f}\right)^{2(n-1)}, \tag{5.89a}$$

$$\eta \simeq -\frac{n(n-1)}{4}\frac{M_{\rm Pl}^2}{f^2}\left(\frac{\phi}{2f}\right)^{n-2} \gg \epsilon\,, \tag{5.89b}$$

$$\phi_{\rm i} \simeq \left(1-64\pi^2\frac{M_{\rm Pl}^4}{\Lambda^4}\right)^{\frac{1}{n}} 2f\,, \tag{5.89c}$$

$$\phi_{\rm e} \simeq \left(\frac{8}{n^2}\frac{f^2}{M_{\rm Pl}^2}\right)^{\frac{1}{2(n-1)}} 2f\,, \tag{5.89d}$$

and

$$\mathcal{N}_{\rm e} \simeq \frac{4}{n(n-2)}\frac{f^2}{M_{\rm Pl}^2}\left(\frac{\phi_{\rm i}}{2f}\right)^{2-n}, \tag{5.89e}$$

$$\phi_*^{2-n} \simeq \frac{n(n-2)}{4}\frac{M_{\rm Pl}^2}{f^2}(2f)^{2-n}\mathcal{N}_k\,, \tag{5.89f}$$

$$\eta_* \simeq -\frac{n-1}{n-2}\frac{1}{\mathcal{N}_k} \gg \epsilon_*\,, \tag{5.89g}$$

where we used the reduced Planck mass for later convenience. The case $n = 2$ approximates, at small ϕ/f, the cosine potential

$$\boxed{V(\phi) = \frac{\Lambda^4}{2}\left(1+\cos\frac{\phi}{f}\right).} \tag{5.90}$$

This model is dubbed *natural inflation*, where the inflaton is a Nambu–Goldstone pseudo-scalar such as the axion and f is a spontaneous-symmetry-breaking mass scale called *decay constant* [112–114]. In the $\phi/f \ll 1$ limit,

$$\epsilon \simeq \frac{1}{8}\frac{M_{\rm Pl}^2}{f^2}\left(\frac{\phi}{f}\right)^2, \tag{5.91a}$$

$$\eta \simeq -\frac{1}{2}\frac{M_{\rm Pl}^2}{f^2}, \qquad |\eta| \gg \epsilon\,, \tag{5.91b}$$

$$\phi_{\rm i} \simeq \left(1 - 64\pi^2 \frac{M_{\rm Pl}^4}{\Lambda^4}\right)^{\frac{1}{2}} 2f\,, \tag{5.91c}$$

$$\phi_{\rm e} \simeq \sqrt{8}\frac{f^2}{M_{\rm Pl}}, \tag{5.91d}$$

and

$$\mathcal{N}_{\rm e} \simeq 2\frac{f^2}{M_{\rm Pl}^2} \ln \frac{\phi_{\rm e}}{\phi_{\rm i}}\,, \tag{5.91e}$$

$$\phi_* \simeq \sqrt{8}\frac{f^2}{M_{\rm Pl}} \exp\left(-\frac{M_{\rm Pl}^2}{f^2}\frac{\mathcal{N}_k}{2}\right). \tag{5.91f}$$

Supergravity and superstring realizations of (5.90) will be presented in Sects. 13.4.2, 13.4.3, and 13.4.4.

5.5.3 *Multi-field Inflation*

The dynamics of multi-field inflationary scenarios is considerably richer than for a single field. Simple models of multiple inflation feature two or several non-interacting scalar fields, which drive a series of inflationary periods [115–118] or just one period of accelerated expansion as in *assisted inflation* [119, 120]. In the latter case, even if each field has a potential too steep to roll slowly and cannot lead to inflation individually, their combined effect can drive acceleration. For instance, the scalar spectral index generated by one field with exponential potential (5.78) of slope p (equation (5.79)) is $n_{\rm s}-1 = -2/p$ (see (5.153) below), and it is close to scale invariance only if $p \gg 1$. On the other hand, a collection of n non-interacting scalar fields with exponential potentials of slope p_i is $n_{\rm s} - 1 = -2/(\sum_i p_i)$ and one can achieve scale invariance even if some of the slopes are too steep, $p_i < 1$ [119].[5] In the case of decoupled fields (no cross-terms in the total potential), they interact only through minimal gravitational coupling. More generally, the background equations of motion with an Einstein–Hilbert gravitational action and n homogeneous scalars

[5]Extensions of these models include the effect of curvature and a barotropic perfect fluid [121], Bianchi backgrounds [122] and polynomial potentials [123]. The stability of solutions as critical points in phase space was studied for decoupled [120, 124, 125] and coupled [126–128] exponential potentials, for decoupled inverse power-law potentials [125] and for general potentials with or without cross-interactions [129]. Late-time dark-energy scenarios can be found in [125, 130]. Multi-field inhomogeneous perturbations have also been considered at the linear [131] and non-linear level [132–134].

read

$$H^2 = \frac{\kappa^2}{3}\left(\frac{1}{2}\sum_{i=1}^{n}\dot{\phi}_i^2 + W\right), \qquad \ddot{\phi}_i + 3H\dot{\phi}_i + W_{,\phi_i} = 0\,, \tag{5.92}$$

where $W := W(\phi_1, \phi_2, \dots, \phi_n)$ is the potential with both self-interaction and interaction terms. These two equations can be combined to give $\dot{H} = -(1/2)\sum_i \dot{\phi}_i^2$. Multi-field assisted inflationary scenarios can be physically motivated within high-energy multi-dimensional theories. In fact, as we shall see in Chap. 13, they naturally emerge in Kaluza–Klein compactifications [123], M-theory and string theory.

In other settings, inflation is driven by one scalar field coupled with light or massless scalar spectators (e.g., an axion) [135–138]. These extra fields can play a variety of roles. For instance, curvature perturbations may be generated by a scalar companion of the inflaton, in which case the former is called *curvaton* [138–141]. Or, if the partner's mass is larger than that of the inflaton, it can effectively stop inflation in a natural way (Sect. 5.10.2). The potential of this *hybrid inflation* [142–146] is

$$V(\phi, \varphi) = \tfrac{1}{4}\lambda(\varphi^2 - M^2)^2 + \tfrac{1}{2}m^2\phi^2 + \tfrac{1}{2}\sigma\phi^2\varphi^2\,, \tag{5.93}$$

where λ and σ are the coupling constants of the auxiliary scalar field φ. A priori, the masses m and M range between 1 TeV (electroweak scale) and the Planck mass.

When the vacuum expectation value of the inflaton is greater than the critical value $\bar{\phi}^2 = \lambda M^2/\sigma$, the potential has a degenerate local minimum at $\langle\varphi\rangle = \varphi(t) = 0$ (Fig. 5.6). Then, inflation is driven by the effective potential

$$V_{\rm eff}(\phi) = \tfrac{1}{4}\lambda M^4 + \tfrac{1}{2}m^2\phi^2\,. \tag{5.94}$$

The dynamics is similar to a large-field model. The inflaton slowly rolls at the bottom of the $\varphi = 0$ trough, then at $|\phi| \lesssim |\bar{\phi}|$ it quickly falls into one of the

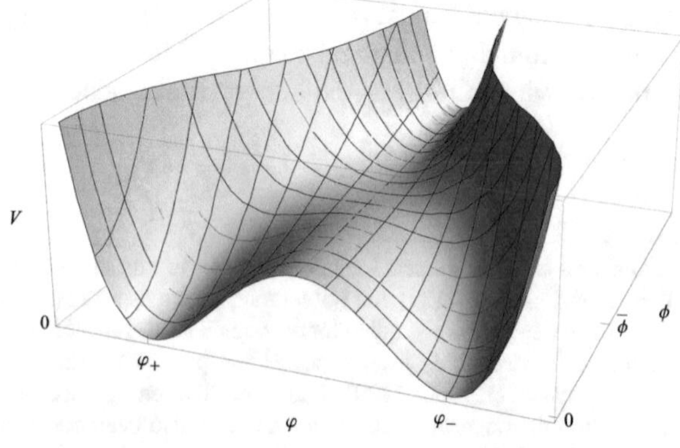

Fig. 5.6 Hybrid inflation potential: ϕ is the inflaton and φ an auxiliary scalar field

wells corresponding to the true minimum of the auxiliary field,

$$\varphi_\pm(\phi) = \pm M\sqrt{1 - \frac{\phi^2}{\bar{\phi}^2}}\,.$$

After rolling into either well, the scalar φ oscillates until the cosmic expansion damps its motion. If the vacuum energy associated with φ dominates the potential (5.94), then the phase transition stops inflation almost instantaneously.

The SR parameters in the false-vacuum (inflationary) regime (5.94) are

$$\epsilon \simeq 4\pi\eta^2\frac{\phi^2}{m_{\rm Pl}^2}\,, \qquad \eta \simeq \frac{1}{2\pi\lambda}\left(\frac{m m_{\rm Pl}}{M^2}\right)^2\,. \tag{5.95}$$

The vacuum-domination regime is equivalent to the condition

$$\epsilon \ll \eta\,. \tag{5.96}$$

Also, in this scenario inflation ends at $\bar{\phi}$, which is different, in general, from the value $\phi_{\rm e}$ obtained from the $\epsilon(\phi_{\rm e}) = 1$ condition (with some choices of the couplings magnitude, the end of inflation rather occurs at $\phi_{\rm e} > \bar{\phi}$). Also,

$$\phi_{\rm i}^2 \simeq \frac{1}{m^2}\left(2m_{\rm Pl}^4 - \frac{\lambda}{2}M^4\right)\,, \tag{5.97a}$$

$$\mathcal{N}_{\rm e} \simeq \frac{1}{2\eta}\ln\frac{\phi_{\rm i}^2}{\bar{\phi}^2}\,, \tag{5.97b}$$

$$\phi_* \simeq \bar{\phi}\,{\rm e}^{\eta\mathcal{N}_k}\,, \tag{5.97c}$$

$$\epsilon_* \simeq 4\pi\eta^2\frac{\lambda}{\sigma}\frac{M^2}{m_{\rm Pl}^2}{\rm e}^{2\eta\mathcal{N}_k}\,. \tag{5.97d}$$

The two-field potential is shown in Fig. 5.6. Typical values of the coupling constants are $\sigma \sim \lambda \sim 10^{-1}$, $m \sim 10^2$ GeV, $M \sim 10^{11}$ GeV [146]. Because of (5.96) and (5.153), hybrid models of this type predict blue-tilted scalar spectra and a very low tensor-to-scalar ratio. They are therefore ruled out experimentally [147].

5.6 First Glimpse of the Quantum Universe

So far in this book, we have treated fields and perturbations as classical objects, but quantum mechanics is expected to dominate the history of the universe at and near the big bang. What is the relation between the inflaton quantum fluctuations and classical perturbations spectra, and how did the transition between the quantum and the classical world happen?

5.6.1 Decoherence

Imagine we are able to describe the early universe as a quantum system where matter and gravitational degrees of freedom are encoded in field operators on a Hilbert space of physical states. This scenario is part of a broad area of investigation known under the name of *quantum cosmology*. In this section, we discuss a generic scalar field operator $\hat{\phi}$, eventually identified with the inflaton or a tensor polarization mode. In Chap. 10, we will quantize the Universe as a whole, which is the cosmological equivalent of the problem of how to quantize gravity. Here we anticipate that the Copenhagen interpretation of quantum mechanics (where macroscopic apparatus are treated as fundamentally classical in contrast with the quantum phenomena they observe) is replaced by the many-worlds or Everett interpretation [148–152], where the Universe itself is subject to quantum laws. Hence, the Universe is regarded as a superposition of quantum states and the space of states is spanned by all possible Universes.[6] In the Everett interpretation, it is assumed that the linear structure of a quantum theory is exact, so that equations such as the Schrödinger equation or the quantum constraints of canonical quantum gravity have universal validity and non-linear terms in the wave-functions are excluded.

Regardless of the interpretation, the canonical quantization of a bosonic field $\hat{\phi}$ on a curved manifold proceeds as in Minkowski spacetime. Namely, one defines a Fock space and a no-particle *vacuum state* $|0\rangle$ for the Heisenberg operator $\hat{\phi}$. The field is then expanded in creation and annihilation operators like the quantum harmonic oscillator. In the *multiverse* or "third" quantization, however, the vacuum state $|\Omega\rangle$ is not an eigenstate of $\hat{\phi}$. We can expand $|\Omega\rangle$ into a combination of eigenstates of $\hat{\phi}$,

$$|\Omega\rangle = \sum_i c_i|0\rangle_i\,, \qquad \hat{\phi}|0\rangle_i = \phi_i|0\rangle_i\,,$$

where the sum is over the ensemble of all possible universes. The probability to find a certain field distribution is $P_i[\phi_i] \sim |c_i|^2$. In usual quantum experiments, wave-functions lose phase correlations due to interaction with the environment and performing a measurement means to choose an eigenstate. This process is called *decoherence*, that is, a delocalization of interference terms or, in other words, a local suppression of interference. In collapse models, decoherence amounts to a global damping or destruction of interference (off-diagonal) terms in the density matrix associated with the eigenstates (see [152, 154] for reviews).

Within standard quantum theory, decoherence is a well-established and physically tested concept [155]. Quantum cosmology, however, lies on rather different grounds. Contrary to the laboratory set-up, choosing an eigenstate of $\hat{\phi}$ does not correspond to performing a measurement in a traditional sense, because there is no observer external to the quantum Universe. Here we face the same impasse we

[6] A third type of interpretation in quantum cosmology is Bohm interpretation [153].

met in relation with CMB correlators and multi-sky averages. To continue, we must assume that the universe we live in corresponds to a particular choice $i = I$ in the ensemble. The problem is to understand what mechanism forced the vacuum state into one of its eigenstate components. The issue of cosmic decoherence is still unclear in many respects, but some authors have considered it in great detail [156–164].

Here, we refrain from entertaining ourselves with this difficult subject but notice that the problem of classical-to-quantum transitions affects cosmology at different levels, not just the Universe as a whole. Let us either assume a second-quantized universe or admit that, in a multiverse scenario, some mechanism makes the scalar quantum field $\hat{\phi}$ lose its coherence and selects a given spectrum of cosmological perturbations. (In the rest of this chapter, the choice of interpretation is not important.) These perturbations are born as quantum fluctuations but the observer records them as a classical spectrum. At some point, a decoherence process must have taken place. In fact, we already have a hint of an answer. We have seen that "ultraviolet" and "infrared" scales are defined by the Hubble or particle horizon. The horizon signals the transition between small and large scales and a change of behaviour of classical perturbations, oscillating inside the horizon and frozen out at ultra-large scales. We can naturally regard the horizon as the watershed governing the transition of fluctuations between the quantum and the classical regime. Loosely speaking, the Hubble and particle horizons are the cosmological analogue of the laboratory measurement apparatus.

5.6.2 *From Quantum Fields to Classical Spectra*

An important assumption of inflationary dynamical models is that *the inflaton is in its vacuum state* $|0\rangle$ (corresponding to some specific realization $|0\rangle_I$ in the multiverse interpretation). We can imagine that all non-inflatonic fields adjust themselves in the state of minimum energy while $\hat{\phi}$ evolves. These fields, spatially averaged over a comoving region of the size of the Hubble horizon, constitute a classical background which defines the state $|0\rangle$.

Then, we reinterpret the homogeneous classical field $\phi(\tau)$, solution to the classical equations, as the expectation value of the quantum field $\hat{\phi}$ calculated on the vacuum state,

$$\langle\hat{\phi}(\tau,\mathbf{x})\rangle := \langle 0|\hat{\phi}(\tau,\mathbf{x})|0\rangle =: \phi(\tau)\,, \tag{5.98}$$

where we use conformal time to eventually match with the notation for classical perturbations. In Heisenberg picture, the quantum fluctuation $\delta\hat{\phi}$ of $\hat{\phi}$ is defined as

$$\hat{\phi}(\tau,\mathbf{x}) = \phi(\tau) + \delta\hat{\phi}(\tau,\mathbf{x})\,, \tag{5.99}$$

and has zero vacuum expectation value, $\langle\delta\hat{\phi}\rangle = 0$. At this point, we can quantize the fluctuation on a FLRW background. In general, the canonical field will be proportional to $\delta\hat{\phi}$ by some time-dependent factor:

$$\hat{u}(\tau,\mathbf{x}) = f(\tau)\delta\hat{\phi}(\tau,\mathbf{x})\,. \tag{5.100}$$

We can in fact be more specific and consider a field of the form

$$\hat{u}(\tau,\mathbf{x}) = a(\tau)\hat{\varphi}(\tau,\mathbf{x})\,, \tag{5.101}$$

where $\hat{\varphi}$ will represent either the inflaton fluctuation $\delta\hat{\phi}$ or a tensor polarization mode $\hat{h}_\lambda$. One imposes equal-time commutation relations to $\hat{u}$,

$$\left[\hat{u}(\tau,\mathbf{x}_1),\,\hat{u}(\tau,\mathbf{x}_2)\right] = 0 = \left[\hat{\Pi}_u(\tau,\mathbf{x}_1),\,\hat{\Pi}_u(\tau,\mathbf{x}_2)\right], \tag{5.102a}$$

$$\left[\hat{u}(\tau,\mathbf{x}_1),\,\hat{\Pi}_u(\tau,\mathbf{x}_2)\right] = \mathrm{i}\delta(\mathbf{x}_1,\mathbf{x}_2)\,, \tag{5.102b}$$

where $\hat{\Pi}_u = \hat{u}'$ is the conjugate momentum density. The fluctuation field can be expanded in spatial comoving Fourier modes,

$$\hat{u}(\tau,\mathbf{x}) = \int \frac{\mathrm{d}^3\mathbf{k}}{(2\pi)^3}\,\hat{u}_\mathbf{k}(\tau)\,\mathrm{e}^{\mathrm{i}\mathbf{k}\cdot\mathbf{x}}\,, \tag{5.103}$$

where the operator $\hat{u}_\mathbf{k}$ can be written as a combination of harmonic oscillators,

$$\hat{u}_\mathbf{k}(\tau) = u_k(\tau)a_\mathbf{k} + u_k^*(\tau)a_{-\mathbf{k}}^\dagger\,. \tag{5.104}$$

Here, $u_k(\tau)$ is a complex function of $k = |\mathbf{k}|$ and τ. The time-independent creation and annihilation operators $a_\mathbf{k}^\dagger$ and $a_\mathbf{k}$ are defined on the Fock space of normalized n-particle states $|n\rangle$ as

$$a_\mathbf{k}|0\rangle = 0\,, \qquad a_\mathbf{k}^\dagger|0\rangle = |1\rangle\,, \tag{5.105}$$

and satisfy canonical bosonic commutation relations,

$$[a_{\mathbf{k}_1},\,a_{\mathbf{k}_2}^\dagger] = \delta(\mathbf{k}_1,\mathbf{k}_2)\,, \tag{5.106}$$

$$[a_{\mathbf{k}_1},\,a_{\mathbf{k}_2}] = 0 = [a_{\mathbf{k}_1}^\dagger,\,a_{\mathbf{k}_2}^\dagger]\,. \tag{5.107}$$

Since $\hat{u}_\mathbf{k}^\dagger = \hat{u}_{-\mathbf{k}}$, the quantum operator $\hat{u}(t,\mathbf{x})$ is Hermitian. Plugging (5.104) into (5.102b) and using (5.106), one gets the *Wronskian*

$$u_k u_k^{*\prime} - u_k^* u_k' = \mathrm{i}\,. \tag{5.108}$$

The same quantization scheme is applied to gravitational waves [165, 166], in which case they are called gravitons.

The classical fluctuation $u_k(\tau)$ $(= u_{k,I}(\tau)$ in the multiverse scenario) is the eigenvalue of $\hat{u}_{\mathbf{k}}$ associated with the realization I. *This is the primordial cosmological perturbation whose spectrum we observe and measure*. Therefore,

$$\langle |\hat{u}_{\mathbf{k}}(\tau)|^2 \rangle := \langle 0 | \, \hat{u}_{\mathbf{k}} \hat{u}^\dagger_{\mathbf{k}} \, | 0 \rangle = |u_k(\tau)|^2 \,, \tag{5.109}$$

and the primordial spectrum is

$$\mathcal{P}_u(k) := \frac{k^3}{2\pi^2} |u_{k\tau \ll 1}(\tau_*)|^2 \,, \tag{5.110}$$

where in the right-hand side we have used the asymptotic form of the solution u_k at large scales and evaluated everything at horizon crossing (3.14): $|\tau_*| = 1/k$. The spectral index and its running, introduced in Sect. 4.4.2, can be obtained from (5.110).

5.6.3 Choice of Vacuum

In the absence of gravity, there exists a global inertial frame, the vacuum state of the free field $\hat{u}$ is uniquely determined by (5.105) and there is a clear distinction between positive- and negative-frequency modes. On curved spacetimes, however, the Poincaré group is no longer a symmetry group and there is neither a global Fermi frame nor a unique vacuum state [167]. An observer can build particle detectors and determine the local vacuum state $|0\rangle$ accordingly but for other, distant observers the state $|0\rangle$ will not correspond to their local vacuum $|\bar{0}\rangle$. Their detectors will measure the presence of particles:

$$\bar{a}_{\mathbf{k}} |\bar{0}\rangle = 0 \,, \qquad \bar{a}_{\mathbf{k}} |0\rangle \neq 0 \,. \tag{5.111}$$

This scenario is often described impressionistically by saying that particles are created from spacetime geometry. More precisely, the Fock vacuum (5.105) is regarded as the vacuum at some fixed time $\tau_{\rm i}$, annihilated by $a_{\mathbf{k}} = a_{\mathbf{k}}(\tau_{\rm i})$, while the vacuum at generic time τ is defined by the time-dependent creation and annihilation operators

$$a_{\mathbf{k}}(\tau) = v_{1k}(\tau) \, a_{\mathbf{k}} + v_{2k}(\tau) \, a^\dagger_{-\mathbf{k}} \,, \qquad a^\dagger_{\mathbf{k}}(\tau) = v^*_{1k}(\tau) \, a^\dagger_{\mathbf{k}} + v^*_{2k}(\tau) \, a_{-\mathbf{k}} \,. \tag{5.112}$$

These relations are the *Bogoliubov transformations* relating the operators of (5.105) with those of (5.111), $\bar{a}_{\mathbf{k}} = a_{\mathbf{k}}(\tau)$. The commutation relations (5.106) are preserved in time if, and only if,

$$|v_{1k}|^2 - |v_{2k}|^2 = 1\,.$$

Writing the field operator $\hat{u}_{\mathbf{k}}$ as the canonical harmonic oscillator in Heisenberg picture,

$$\hat{u}_{\mathbf{k}}(\tau) = a_{\mathbf{k}}(\tau) + a^\dagger_{-\mathbf{k}}(\tau)\,, \tag{5.113}$$

one recovers (5.104) provided

$$u_k = v_{1k} + v^*_{2k}\,.$$

The choice of vacuum $|0\rangle$ is an *Ansatz* (5.105) for $\hat{u}_{\mathbf{k}}$ at some fixed time $\tau_{\rm i}$. Physical considerations can guide us into its selection.

- *Adiabatic* or *Bunch–Davies vacuum* [167, 168]. This state is asymptotically Minkowski in the remote past/future, where all inertial observers would see no particles. The infinite-past time-like surface $\mathcal{I}^-$ at $\tau \to -\infty$ (defined rigorously in Sect. 6.1.1) corresponds to the "in" region bounded by the Hubble horizon, where the mode u_k is a positive-frequency plane wave. In the infinite-future time-like surface $\mathcal{I}^+$ at $\tau \to +\infty$, the mode u_k is a negative-frequency plane wave:

$$u_k \xrightarrow{\tau \to \pm\infty} \frac{\mathrm{e}^{\pm ik\tau}}{\sqrt{2k}}\,. \tag{5.114}$$

  In the absence of a cut-off scale at high energies, there is no reason to impose initial conditions at a finite scale $k = k(\tau_{\rm i})$. Then, the adiabatic vacuum at the infinite past is the unique choice compatible with the Wronskian condition (5.108).
- *Local* or *instantaneous Minkowski vacuum* [169]. It is a generalization of the adiabatic vacuum but imposed as a Cauchy problem at a finite initial time:

$$u_k(\tau_{\rm i}) = \frac{1}{\sqrt{2k}}\,, \qquad u'_k(\tau_{\rm i}) = \pm ik\, u_k(\tau_{\rm i})\,, \tag{5.115}$$

  where $\tau_{\rm i}$ is momentum-dependent and is determined by some new physics at high energies.
- *Minimal-energy vacuum* [170], so called because it minimizes the energy density [171]:

$$u_k(\tau_{\rm i}) = \frac{1}{\sqrt{2k^0}}\,, \qquad u'_k(\tau_{\rm i}) = \pm ik^0\, u_k(\tau_{\rm i})\,. \tag{5.116}$$

- *de Sitter vacua* [172–175]. This family of vacua was studied in [176–180] and, with particular reference to cosmology, in [181–185]. The family is defined by the Cauchy problem

$$u_k(\tau_\mathrm{i}) = -\frac{\tau_\mathrm{i} H}{\sqrt{2k}}\,\mathrm{e}^{\mathrm{i}\theta}\,, \qquad u_k'(\tau_\mathrm{i}) = -\mathrm{i}k\,u_k(\tau_\mathrm{i})\,, \tag{5.117}$$

where $\tau_\mathrm{i} = -\beta/k$ is momentum dependent, H is constant and θ and β are real parameters. Often these vacua are labelled by one complex parameter α, for which reason they are also called α-vacua. Since they correspond to states of minimum uncertainty [181, 182], they can be named *minimal uncertainty vacua.*

In the limit $\theta \to \beta \to +\infty$, (5.117) reduces to the infinite-past (positive frequency) adiabatic vacuum, while the local Minkowski vacuum is recovered for $\theta \to 0$ and $-\tau_\mathrm{i} H = \beta/a \to 1$ (here we used the horizon-crossing relation $k = aH$). In general, the α-vacua are all inequivalent and non-thermal, meaning that a detector in a de Sitter vacuum will not end up in thermal equilibrium.

The assumption that the Fourier modes $\hat{u}_\mathbf{k}$ have zero occupation number is necessary to guarantee a sufficiently long period of inflation, since a non-vacuum state with too many particles would drag and brake the accelerated expansion [186, 187].

5.6.4 Mukhanov–Sasaki Equation Revisited

A very general result of linear perturbation theory is that the equation of motion for u_k takes the form of the Mukhanov–Sasaki equation (3.32),

$$\boxed{u_k'' + (k^2 - M^2)u_k = 0\,,} \tag{5.118}$$

where

$$M^2 = (2-\epsilon)\mathcal{H}^2 - m^2(\tau) \tag{5.119}$$

and $m(\tau)$ is an effective mass whose details depend on the nature of the perturbation (scalar, tensor, choice of f in (5.100)) and on the dynamics (definition of the action, background Friedmann equations). During inflation or a power-law expansion, it is possible to write M^2 as

$$\boxed{M^2 = \frac{4\nu^2 - 1}{4\tau^2}\,,} \tag{5.120}$$

where ν is a constant. In this case, the solution of (5.118) was given by (3.36) in Sect. 3.2.3. Instead of the Bessel functions, one can express the solution also in terms of the Hankel functions of the first and second kind,

$$H_\nu^{(1)}(y) = J_\nu(y) + \mathrm{i}Y_\nu(y)\,, \qquad H_\nu^{(2)}(y) = J_\nu(y) - \mathrm{i}Y_\nu(y)\,, \tag{5.121}$$

which describe outgoing (advancing) and incoming cylindrical waves, respectively. Then,

$$u_k(\tau) = C_1\sqrt{k|\tau|}\,H_\nu^{(1)}(k|\tau|) + C_2\sqrt{k|\tau|}\,H_\nu^{(2)}(k|\tau|)\,. \tag{5.122}$$

Asymptotically (we assume $\nu > 0$),

$$H_\nu^{(1)}(y) \overset{y\ll 1}{\simeq} \mathrm{i}Y_\nu(y) \simeq -\mathrm{i}\frac{2^\nu\Gamma(\nu)}{\pi}y^{-\nu}\,,$$

$$H_\nu^{(2)}(y) \overset{y\ll 1}{\simeq} -\mathrm{i}Y_\nu(y) \simeq \mathrm{i}\frac{2^\nu\Gamma(\nu)}{\pi}y^{-\nu}\,,$$

$$H_\nu^{(1)}(y) \overset{y\gg 1}{\simeq} \sqrt{\frac{2}{\pi y}}\,\mathrm{e}^{-\mathrm{i}\frac{\pi}{2}\left(\nu+\frac{1}{2}\right)}\,\mathrm{e}^{\mathrm{i}y}\,,$$

$$H_\nu^{(2)}(y) \overset{y\gg 1}{\simeq} \sqrt{\frac{2}{\pi y}}\,\mathrm{e}^{\mathrm{i}\frac{\pi}{2}\left(\nu+\frac{1}{2}\right)}\,\mathrm{e}^{-\mathrm{i}y}\,.$$

At sub-horizon scales ($k|\tau| \gg 1$), the solution is a superposition of plane waves,

$$\begin{aligned} u_k(\tau) &\overset{k|\tau|\gg 1}{\simeq} C_1\sqrt{\frac{2}{\pi}}\,\mathrm{e}^{-\mathrm{i}\frac{\pi}{2}\left(\nu+\frac{1}{2}\right)}\,\mathrm{e}^{\mathrm{i}k|\tau|} + C_2\sqrt{\frac{2}{\pi}}\,\mathrm{e}^{\mathrm{i}\frac{\pi}{2}\left(\nu+\frac{1}{2}\right)}\,\mathrm{e}^{-\mathrm{i}k|\tau|} \\ &= C_+\,\mathrm{e}^{\mathrm{i}k|\tau|} + C_-\,\mathrm{e}^{-\mathrm{i}k|\tau|}\,. \end{aligned} \tag{5.123}$$

As we know, perturbations with small wave-lengths oscillate, in accordance with the fact that the quantum scalar field $\hat{u}$ locally behaves as a Minkowski harmonic oscillator. In turn, the quantum harmonic oscillator is defined through a choice of the vacuum state, which fixes the coefficients C_+ and C_-. With the Bunch–Davies infinite-past vacuum (during inflation, lower signs in (5.114)), for (5.123) one has

$$C_+ = \frac{1}{\sqrt{2k}}\,, \qquad C_- = 0\,. \tag{5.124}$$

The local Minkowski and minimal-energy vacua actually agree for the standard Mukhanov–Sasaki equation (5.118), where the dispersion relation of the field is

$$-(k^0)^2 + k^2 = 0\,. \tag{5.125}$$

In fact, the two branches of the local Minkowski vacuum (5.115) are

$$C_+ = 0\,, \quad C_- = \frac{e^{ik\tau_i}}{\sqrt{2k}}\,, \qquad \text{and} \qquad C_+ = \frac{e^{-ik\tau_i}}{\sqrt{2k}}\,, \quad C_- = 0\,,$$

while the minimal-energy state coincides with the previous one: initially and locally, the mode does not feel curvature effects. In models with modified dispersion relations [171], which can encode some quantum-gravity effects via a phenomenological cut-off (Sect. 5.10.3), the small-scale solution may differ from (5.123) and the local Minkowski and minimal-energy states are physically inequivalent. Notice also that, in general relativity and in de Sitter spacetime, a large class of initial states approach the adiabatic vacuum after a few e-foldings [188]. Therefore, at least in general-relativistic inflation the Bunch–Davies choice is very reasonable.

In the long wave-length limit, the appropriately normalized solution is

$$|u_{k\tau\ll 1}| = \frac{2^{\nu-2}}{\sqrt{k}}\frac{\Gamma(\nu)}{\Gamma(3/2)}|k\tau|^{\frac{1}{2}-\nu}\,. \tag{5.126}$$

In all cases of interest the constant ν is

$$\nu = \frac{3}{2} + \nu_\epsilon\,, \tag{5.127}$$

where $\nu_\epsilon = 0$ in de Sitter and $\nu_\epsilon \ll 1$ in the slow-roll regime. Therefore, (5.110) becomes

$$\mathcal{P}_u(k) = \frac{k^2}{4\pi^2}\left[2^{\nu_\epsilon}\frac{\Gamma\left(\frac{3}{2}+\nu_\epsilon\right)}{\Gamma\left(\frac{3}{2}\right)}\right]^2\,. \tag{5.128}$$

Recalling (5.101), one immediately gets $\mathcal{P}_\varphi = \mathcal{P}_u/a^2$. For power-law inflation (constant ϵ), the relation (3.11) between k and the Hubble parameter at horizon crossing is

$$k = aH(1-\epsilon)\,, \tag{5.129}$$

so that

$$\mathcal{P}_\varphi = (1-\epsilon)^2\left[2^{\nu_\epsilon}\frac{\Gamma\left(\frac{3}{2}+\nu_\epsilon\right)}{\Gamma\left(\frac{3}{2}\right)}\right]^2\left(\frac{H}{2\pi}\right)^2\,. \tag{5.130}$$

In pure de Sitter, (5.130) gives

$$\boxed{\mathcal{P}_\varphi = \left(\frac{H}{2\pi}\right)^2\,.} \tag{5.131}$$

The right-hand side is the square of the gravitational temperature associated with a FLRW spacetime:

$$\mathcal{T}_H := \frac{H}{2\pi} \,. \tag{5.132}$$

In particular, the de Sitter temperature is constant. We shall come back to this concept in Sect. 7.7.

Equation (5.131) is the leading-order term for the spectrum on a general background in slow-roll regime. Using the expansion

$$2^{\nu_\epsilon} \frac{\Gamma\left(\frac{3}{2}+\nu_\epsilon\right)}{\Gamma\left(\frac{3}{2}\right)} \simeq 1 + C\nu_\epsilon \,, \qquad \nu_\epsilon \ll 1 \,,$$

where $C = 2 - \ln 2 - \gamma_{\rm EM} \approx 0.73$ (recall that $\gamma_{\rm EM}$ is the Euler–Mascheroni constant), we get

$$\mathcal{P}_\varphi \simeq (1 - 2\epsilon + 2C\nu_\epsilon)\left(\frac{H}{2\pi}\right)^2 . \tag{5.133}$$

Below we will not consider next-to-leading slow-roll expressions such as this [189, 190], although they can be of interest in the future when high-precision cosmological tests become available.

We can rewrite (5.130) in another useful form. For power-law inflation, $a_* \propto |\tau_*|^{1/(\epsilon-1)} = k^{1/(1-\epsilon)}$ at horizon crossing (equation (2.187)), so that

$$\boxed{\mathcal{P}_\varphi(k) = A_\varphi (k\tau_0)^{-\frac{2\epsilon}{1-\epsilon}} \,,} \tag{5.134}$$

where A_φ is a dimensionless constant. The spectra in power-law backgrounds $a = t^p$ admit exact expressions in the SR parameters, which are all equal to $1/p$ [191]. Typically, however, one has inflation in mind and truncates these expressions to lowest or next-to-lowest SR order.

In all the above formulæ, we have omitted the subscript $*$ in slow-roll parameters, which are all evaluated at horizon crossing. Note that this computation did involve the Friedmann equations but only indirectly, through the effective mass term in the Mukhanov–Sasaki equation.

We can now revisit the issue of the quantum-to-classical transition of linear cosmological fluctuations. When omitting the negative-frequency part in (5.123), one is actually throwing away a large-scale decaying mode of the field operator in Heisenberg picture. The correlation between the decaying and non-decaying mode determines coherence, so that in this approximation one obtains "decoherence without decoherence," i.e., without invoking any interaction of the modes with an "environment" [181]. However, the presence of the cosmological horizon is crucial to determine the decaying, constant or growing nature of the modes. Beyond linear level, however, where different modes interact, the problem of decoherence forces itself again upon our attention. What one usually does, in this case, is to assume the Everett interpretation, start with the standard inflaton vacuum and turn it dynamically into a squeezed vacuum. This squeezed vacuum state can be written as a superposition of excited states and can decohere [192].

To summarize, once outside the horizon quantum fluctuations are frozen and become classical. One can reach the same conclusion by noting that the energy per quantum $\omega = k/a$ decreases by redshift, and by energy conservation the number of quanta $n_\omega \sim E/\omega$ for a given frequency (E is the Hamiltonian eigenvalue coming from the effective action of φ) must increase accordingly outside the horizon [193].

5.6.5 *Eternal Inflation*

The super-horizon modes of the scalar-field quantum fluctuations acquire the almost constant classical spectrum (5.131) with $\varphi = \delta\phi$, when the slow-roll approximation of the dynamics holds. Thus, during the time of a Hubble expansion $\delta t \sim H^{-1}$, the classical field evolution $\phi(t)$ is modified by jumps of size $\delta\phi \sim \pm\mathcal{T}_H$. If these random jumps dominate the dynamics, there will always be a non-vanishing probability that the field acquires values suitable to trigger a new inflationary era. Then, in an expanding region of Hubble size some sub-regions will end their accelerating expansion and reheating will begin, but in other sub-regions suitable initial conditions will be generated by the quantum fluctuations and new seedbeds of inflation will form. This scenario goes under the name of *eternal inflation* [194–203] (for updated references on the subject, see [204, 205]).

The broad picture is one of regions in accelerated expansion which, at some point, stop inflating and subsequently thermalize. Thermalized patches are separated by still inflating domains, which are continuously created by quantum fluctuations. The Universe is composed by thermalized and inflating regions. The comoving volume of inflating regions vanishes at $t \to +\infty$, while their physical volume grows exponentially: the Universe never thermalizes completely. Inflation, once started, reproduces itself *ad libitum* in the future. Specifying that the eternal process of reproduction is future-directed is important. If this scenario extended also to the infinite past, it would avoid the big-bang problem. We will see in Chap. 6 that this is not the case [206, 207] and eternal inflation does not extend indefinitely in the past. On the plus side, however, the graceful entry problem (Sect. 5.10.1) is alleviated in this framework [196, 197, 204, 208].

Many models of inflation admit fluctuation-dominated regimes. In the highly homogeneous universe of chaotic inflation, for instance, an infinite number of causally disconnected inflationary regions are expected to form. As we will see in Sect. 5.8.1, the dynamics of quantum fluctuations can be interpreted as a stochastic process, in particular a Brownian motion [37, 195, 209]. The distribution of reheating and inflating regions is characterized by a fractal geometry [142, 196, 197, 200, 202], which can be described by different choices of measure [202, 204, 210–217].

5.7 Cosmological Spectra

Once solved the horizon, flatness and monopole problems, inflation does not exhaust its task and plays a fundamental role in cosmic structure formation. Any pre-inflationary inhomogeneity and anisotropy is washed away by the primordial accelerated expansion, so that the origin of irregularities in the energy distribution of the universe must be traced back to events during or after inflation. It turns out that the observed anisotropies can be explained as being the Planck-size quantum fluctuations of the inflaton stretched at horizon scales. Tensor perturbations (gravitational waves) are generated via the same mechanism. By *coincidentia oppositorum*, the study of the large-scale structures is also the study of the smallest quantum scales and high-energy processes. Several orders of magnitude beyond the probing capability of accelerators is, at least potentially, cosmology.

The SR formalism gives good control over the theoretical shape and amplitude of cosmological perturbations. Here we shall restrict ourselves to the linear first-order approach [218–220], although it is possible to extend the discussion to second-order perturbations [221–223] and, as we have seen in Chap. 3, even to a non-perturbative, non-linear set-up.

The standard procedure to compute the perturbation spectra is: (i) Write the linearly perturbed metric in terms of gauge-invariant scalar and tensor quantities separately; (ii) Compute the effective action and the associated linearized equations of motion for a given background solution with constant or small SR parameters; solve the linearized equations with respect to the perturbations; (iii) Write the scalar and tensor perturbation amplitudes in terms of this solution; (iv) Since the observed fluctuations are originated at horizon crossing, the perturbation spectra are evaluated at this point (equation (3.14), $k|\tau_*| = 1$).

5.7.1 Gaussianity

We are in a position to understand why *linear* primordial fluctuations are Gaussian. Once chosen a realization $i = I$ for the universe, we can write the complex coefficient u_k as

$$u_k = |u_k| e^{i\vartheta_k} = \mathrm{Re}(u_k) + i\,\mathrm{Im}(u_k)\,. \tag{5.135}$$

By the cosmological principle, the distribution of each mode in (5.135) is isotropic, implying that the modes u_k as well as the spectrum $\mathcal{P}_u$ depend only upon the absolute value $k = |\mathbf{k}|$.

We use now the properties of random fields enunciated in Sect. 3.4. The Mukhanov–Sasaki equation (5.118) is linear in u_k. Then, the real and imaginary part of each mode behave like two independent harmonic oscillators for each k. In the vacuum state, the operators $\mathrm{Re}(u_k)a^\dagger_{\mathbf{k}}$ and $\mathrm{Im}(u_k)a^\dagger_{\mathbf{k}}$ have the same probability distribution, given by the ground-state wave eigenfunction of an harmonic oscillator (in this case, a Gaussian). The set of modes in (5.135) are then statistically independent and with the same distribution, and the probability distribution of the field u is Gaussian. Another way to see this is to note that the phases ϑ_k are mutually independent and randomly distributed in the interval $0 \leqslant \vartheta_k < 2\pi$. If the phase of each mode is random, then the central limit theorem guarantees that the (classical version of the) superposition (5.103) is Gaussian if the number of modes is large.

Since the power spectrum $\mathcal{P}_u(k)$ is continuous, the field u_k is ergodic, in agreement with the previous cosmic-choice assumption for any continuous transfer function describing the time evolution of the perturbation. We can summarize all these results as:

> ***Gaussianity.*** *In the linear approximation, inflationary fluctuations have a Gaussian probability distribution and are completely described by the power spectrum in momentum space. The statistical properties of the perturbations are evaluated in the ensemble of spatial points in the sky vault.*

The Gaussianity of the statistical distribution for the perturbations is a direct consequence of (i) neglecting second-order terms in the equations of motion and (ii) taking the cosmological principle for granted. Both are only approximations of the real world, although very good ones according to experiments. Going beyond the linear theory and accepting some deviations from perfect isotropy, as CMB probes indicate, small departures from the Gaussian distribution appear and generate new interesting features we shall explore later.

5.7.2 Linear Tensor Perturbations

The scalar field perturbation induces fluctuations on the metric and tensor perturbations are produced by back-reaction. In Sect. 3.2.3, we have seen that the Mukhanov–Sasaki equation for each tensor mode is (3.32) with

$$w_k = a h_{\lambda,k} \tag{5.136}$$

and (see equation (3.37)) [224]

$$\nu_\epsilon = \frac{\epsilon}{1-\epsilon}\,, \tag{5.137}$$

for power-law inflation with $\Lambda = 0$. To tie in with the quantum picture, one notices that (3.24) suggests to define the canonical field

$$u_k = \frac{w_k}{2\kappa}, \tag{5.138}$$

so that in pure de Sitter

$$\mathcal{P}_{h_\lambda} = 4\kappa^2 \left(\frac{H}{2\pi}\right)^2, \tag{5.139}$$

and our final result for the spectrum of tensor perturbations (4.63) is, to lowest SR order,

$$\boxed{\mathcal{P}_\mathrm{t} = 8\kappa^2 \left(\frac{H}{2\pi}\right)^2.} \tag{5.140}$$

In pure de Sitter, this is a constant. During inflation, quantum fluctuations of the gravitational field are pushed outside the horizon, where they remain frozen.

The tensor spectral index (4.64) can be calculated in two equivalent ways. For the lowest-order de Sitter expression (5.140), it is convenient to recast the momentum derivative into a time derivative,

$$\frac{\mathrm{d}}{\mathrm{d}\ln k} \simeq \frac{1}{H}\frac{\mathrm{d}}{\mathrm{d}t}, \tag{5.141}$$

and then act on (5.140) regarding H as non-constant, so that

$$\boxed{n_\mathrm{t} \simeq -2\epsilon\,.} \tag{5.142}$$

Alternatively, one uses (5.134) and obtains

$$n_\mathrm{t} = -\frac{2\epsilon}{1-\epsilon} = -\frac{2}{p-1}, \tag{5.143}$$

which is an exact expression in agreement with (5.142) for small ϵ. The inflationary spectrum of tensor perturbations in general relativity is almost scale invariant, always with a slight red tilt ($n_\mathrm{t} \lesssim 0$).

5.7.3 *Linear Scalar Perturbations*

Scalar perturbations are more involved than the tensor sector. The perturbed Einstein equations give a set of expressions for the inflaton fluctuation $\delta\phi$ and the gauge-invariant metric scalar perturbations Φ and Ψ. At large scales, one can ignore the anisotropic stress and set $\Phi = \Psi$, so that there are only two independent dynamical equations. Combining them together, one finds an equation for the canonical variable

$$u = z\mathcal{R} = a\delta\phi + \frac{\phi'}{\mathcal{H}}\Psi\,, \tag{5.144}$$

where $z := a\phi'/\mathcal{H}$ and $\mathcal{R}$ is the curvature perturbation (3.63). The exact (in the SR parameters) Mukhanov–Sasaki equation for the linear perturbation u_k reads [225–227]

$$u_k'' + \left(k^2 - \frac{z''}{z}\right)u_k = 0\,, \tag{5.145}$$

where

$$\frac{z''}{z} = \mathcal{H}^2\left(2 + 2\epsilon - 3\eta - 4\epsilon\eta + 2\epsilon^2 + \eta^2 + \xi^2\right)\,. \tag{5.146}$$

To understand where this equation comes from, we can either sketch its derivation from the linearized Einstein equations or employ a very efficient trick. Let us first briefly review the traditional approach.

The general scalar equation of motion is (2.66), $\Box\phi - V_{,\phi} = 0$. Splitting the conformal metric into a background and an inhomogeneous perturbation as in (3.52), the d'Alembertian is divided into a pure FLRW operator $a^2\tilde{\Box} = -\partial_\tau^2 - 2\mathcal{H}\partial_\tau$ and a contribution containing the metric back-reaction:

$$\Box = \tilde{\Box} + \frac{\nabla^2}{a^2} + \delta\Box\,. \tag{5.147}$$

Linearizing the equation of motion, we get (tilde omitted in background quantities)

$$\left(\partial_\tau^2 + 2\mathcal{H}\partial_\tau - \nabla^2 + a^2 V_{,\phi\phi}\right)\delta\phi - a^2(\delta\Box)\phi = 0\,. \tag{5.148}$$

If ϕ rolls slowly down its potential, in a zero-order approximation we can ignore the metric back-reaction and forget the last term $\delta\Box$. In this case, $\delta\phi$ is said to be a *test field*. In many circumstances, however, the metric back-reaction is not negligible. This happens, for instance, in eternal inflation or in certain high-energy models, and one should check the consistency of this assumption explicitly. First, one computes $(\delta\Box)\phi$, which turns out to be a function of the scalar perturbations of the metric. In longitudinal gauge and from (3.52),

$$\delta g^{00} = \frac{2\Phi}{a^2}\,, \qquad \delta\sqrt{-g} = a^4(\Phi - 3\Psi)\,;$$

plugging that into (2.67) and linearizing,

$$a^2(\delta\Box)\phi = a^2(\phi'' + 4\mathcal{H}\phi')\delta g^{00} + \phi'\left[a^2(\delta g^{00})' - \frac{1}{a^4}(\delta\sqrt{-g})' + \frac{4\mathcal{H}}{a^4}\delta\sqrt{-g}\right]$$
$$= \phi'(\Phi' + 3\Psi') - 2a^2\Phi V_{,\phi}\,. \tag{5.149}$$

In momentum space, (5.148) is identical to (3.26), with $u = a\delta\phi$. The equation of motion for Ψ yields (e.g., [193])

$$\Psi \simeq \epsilon\frac{\mathcal{H}}{\phi'}\delta\phi\,,$$

so that the mass contribution in (5.119) is

$$m^2 = a^2 V_{,\phi\phi} + 2a^2\epsilon\frac{\mathcal{H}}{\phi'}V_{,\phi} + O(\epsilon^2) \simeq \mathcal{H}^2(3\eta - 3\epsilon)\,,$$

where we recalled (5.63) and neglected the ϕ' term in (5.149). The effective mass term is of the form (5.120), with (5.127) and

$$\nu_\epsilon \simeq 2\epsilon - \eta \tag{5.150}$$

at linear SR order, in agreement with (5.146).

The two-line trick mentioned above is the following. One obtains a friction-free Mukhanov–Sasaki equation from (5.148) if $u_k \sim a\delta\phi$, which determines z in (5.144). At large scales, one can ignore Laplacian terms $\propto k^2$ in the linearized equations. Also, the comoving curvature perturbation is approximately constant (see (3.58) and (3.64)), so that at super-horizon scales $u_k'' \simeq z''\mathcal{R}_k$ and

$$u_k'' - \frac{z''}{z}u_k \simeq 0\,.$$

The only missing term, which can be readily added, is the Laplacian k^2 (with coefficient 1, as an inspection of (5.148) immediately shows). We have thus recovered (5.145) from the conservation equation of a gauge-invariant perturbation.

The scalar spectrum can be found by solving (5.145) or, to lowest SR order, directly from (5.148). For pedagogical purposes, the second option is clearer. If the SR approximation holds and ν_ϵ can be treated as a constant, then the only effect of metric back-reaction is to change the normalization of the power spectrum and we can ignore the details of the effective mass z''/z. This is tantamount to going to the flat gauge $\Phi = \Psi = 0$ and neglecting the contribution (5.149). Therefore, the spectrum of scalar field fluctuations in de Sitter is simply[7]

$$\mathcal{P}_{\delta\phi} = \left(\frac{H}{2\pi}\right)^2 . \tag{5.151}$$

The primordial scalar spectrum (4.57) is the spectrum of the curvature perturbation ζ. At large scales, ζ coincides with the comoving curvature perturbation (3.64). In flat gauge, $\mathcal{R} = (\mathcal{H}/\phi')\delta\phi$; taking (5.43a) into account, we get

$$\boxed{\mathcal{P}_\mathrm{s} = \frac{\kappa^2}{2}\frac{1}{\epsilon}\left(\frac{H}{2\pi}\right)^2 .} \tag{5.152}$$

Using (5.51) and (5.141), one obtains the scalar spectral index

$$\boxed{n_\mathrm{s} - 1 \simeq 2\eta - 4\epsilon .} \tag{5.153}$$

For an exact power-law expansion, the scalar index is the same as (5.143),

$$n_\mathrm{s} - 1 = -\frac{2\epsilon}{1-\epsilon} = -\frac{2}{p-1} . \tag{5.154}$$

The SR approximation guarantees that the scalar spectrum be almost scale invariant. To lowest order, the running (4.59) is quadratic in the SR parameters,

$$\alpha_\mathrm{s} \simeq 2(5\epsilon\eta - 4\epsilon^2 - \xi^2) , \tag{5.155}$$

where we used (5.51) and (5.52). Note that the SR parameters in all the observables (5.142), (5.152) and (5.153) are evaluated at horizon crossing. We omitted, and will do so again in similar expressions, the subscript * in ϵ_* and η_* everywhere.

In this section, we have used three types of approximation: linearity of perturbations, large-wave-length limit and extreme SR conditions. The latter allowed us to neglect the inflaton mass and metric back-reaction. This is no longer possible in the case of fast-roll inflation, where the SR approximation is not enough.

[7]For power-law inflation, the exact expression for $\mathcal{P}_{\delta\phi}$ is (5.130) with ν_ϵ given again by (5.137) [191].

5.7.4 Consistency Relations and Lyth Bound

The tensor-to-scalar ratio (4.66) is very small in inflationary models. This is because the scalar amplitude (5.152) is much larger than the tensor amplitude (5.140):

$$\boxed{r \simeq 16\epsilon\,.} \tag{5.156}$$

We can collect (5.142), (5.153) and (5.156) in the set of first-order *consistency equations*

$$n_{\rm t} \simeq -\frac{r}{8}\,, \tag{5.157}$$

$$\alpha_{\rm s} \simeq \frac{r}{16}\left[\frac{3}{4}r + 5(n_{\rm s}-1)\right] - 2\xi^2\,. \tag{5.158}$$

If

$$|\xi| \ll \min(\epsilon, |\eta|)\,, \tag{5.159}$$

then the set closes and the scalar running depends only on first-order observables. This does not happen in power-law inflation, where $|\xi| = O(\epsilon, |\eta|)$. The consistency equations relate cosmological observables in a way typical of inflationary scenarios, where the scalar and gravitational spectra have a common physical origin. They are a typical result of inflation which other models of structure formation cannot reproduce. Often, (5.157) is used in data analysis to lower the number of parameters in the model (Sect. 4.4.2). In doing so, one is implicitly assuming that inflation was realized by an ordinary scalar field.

Combining (5.76) and (5.156), one can express the excursion of the inflaton during $\mathcal{N}$ e-foldings as $\Delta\phi/M_{\rm Pl} \simeq \int_0^{\mathcal{N}} d\mathcal{N}'\,\sqrt{r/8}$. Since the tensor-to-scalar ratio is approximately constant during inflation,

$$\frac{\Delta\phi}{M_{\rm Pl}} \simeq \mathcal{N}\sqrt{\frac{r}{8}}\,. \tag{5.160}$$

The scales at which r is observed are the $2 < \ell \lesssim 100$ multipoles of the CMB spectrum, corresponding to modes which left the horizon during the last $\mathcal{N}_{\text{low-}\ell} \approx 4$ e-folds. The observed anisotropies in the sky have been produced in the last $\mathcal{N}_k \approx 60$ e-foldings, so that the excursion $\mathcal{N}_{\text{low-}\ell}$ sets both a lower bound on the total variation $\Delta\phi$, $8(\Delta\phi/M_{\rm Pl})^2 > \mathcal{N}_{\text{low-}\ell}^2 r$, and the upper limit $r < (\Delta\phi/M_{\rm Pl})^2/2$ for the tensor-to-scalar ratio, the *Lyth bound* [228]. A more refined estimate, which takes into account the experimental constraints on the scalar spectral index and the variation of $r(\mathcal{N})$ across all scales, sets $\mathcal{N}_{\min} \approx 30$ as an effective lower limit on the number of e-folds replacing $\mathcal{N}_{\text{low-}\ell}$ [229]:

$$r < \frac{8}{30^2}\left(\frac{\Delta\phi}{M_{\rm Pl}}\right)^2 \simeq 10^{-2}\left(\frac{\Delta\phi}{M_{\rm Pl}}\right)^2 . \tag{5.161}$$

On the other hand, the maximum observable tensor signal is set by $\mathcal{N}_k$,

$$r > \frac{8}{60^2}\left(\frac{\Delta\phi}{M_{\rm Pl}}\right)^2 \simeq 10^{-3}\left(\frac{\Delta\phi}{M_{\rm Pl}}\right)^2 . \tag{5.162}$$

Assuming a total excursion of order of the reduced Planck mass ($\Delta\phi \simeq M_{\rm Pl}$), we obtain the allowed range for the tensor-to-scalar ratio in single-field inflation:

$$10^{-3} < r < 10^{-2} . \tag{5.163}$$

A detection at the threshold (5.161) would signal an $O(M_{\rm Pl})$ field excursion. Conversely, models where the field variation is smaller than the Planck scale $M_{\rm Pl}$ are usually characterized by a low inflationary scale $V^{1/4}$ and produce a negligible gravitational spectrum, well below the range (5.163). We will see examples with $r \approx 0$ in Chap. 13

5.8 Non-Gaussianity

In Sect. 4.6.3, we have reviewed some of the mechanisms which can be responsible for a non-Gaussian statistical distribution of perturbations. Now that we have a model of the high-energy early universe, we can complete the discussion on non-linear perturbations. Before doing that, we mention just one more possible source of non-Gaussianity. It is the production of topological defects, which occurs during the whole history of the universe. Their evolution is highly non-linear and naturally generates non-Gaussian statistical distributions. Numerical simulations show that the combined effect of large populations of defects results in almost Gaussian perturbation spectra, in agreement with present observations. However, inflation is the main, if not the only, responsible of primordial fluctuations, so that the role of non-inflationary primordial sources is expected to be relatively marginal.

Within the inflationary paradigm, the standard way to generate non-Gaussian spectra is to consider self-interaction.[8] In general, even if one assumes a classical quadratic potential, quantum corrections to the effective action of the inflaton will

[8] Inflationary non-Gaussianity from self-interaction has been studied extensively in [230–248]. Non-Gaussianities are also generated, for instance, by the inclusion of higher-dimension operators in the inflaton Lagrangian [249], in warm inflation [250, 251], ghost inflation [252, 253] and when assuming that the inflaton does not sit in a vacuum state [186, 187, 254].

give rise to higher-order terms, for instance of the form (5.81). If the coupling constant is small, quantum fluctuations of the field $\phi = \langle\phi\rangle + \delta\phi$ are approximately Gaussian with variance $\sigma_\phi^2 := \langle|\delta\phi|^2\rangle$. However, if self-interaction is important we can no longer neglect second-order perturbations. For instance, for a monomial potential $V \propto \phi^n$, in the ESR regime given by (5.20) and (5.21) the comoving curvature perturbation in flat gauge is

$$\mathcal{R} \sim \frac{\mathcal{H}}{\phi'}\delta\phi \sim \phi\delta\phi = \langle\phi\rangle\,\delta\phi + \delta\phi^2\,,$$

and one can immediately see that the statistics is not Gaussian: $\langle\mathcal{R}\rangle \neq 0$, $\langle\mathcal{R}^3\rangle \neq 0$, and so on. This is precisely the non-linear mechanism discussed in Sect. 4.6.3, realized concretely by the inflaton. We already computed the non-linear parameter $f_{\rm NL}$ (4.108) in the local form of non-Gaussianity. That can be obtained also in the $\delta\mathcal{N}$ formalism, via the $\delta\mathcal{N}$ formula (3.51) [87]. Thinking $\mathcal{N}$ as a function of the scalar field and expanding $\delta\mathcal{N}$ in a Taylor series around the background trajectory $\phi(t)$,

$$\zeta_{\rm NL}(t,\mathbf{x}) = \delta\mathcal{N}(t,\mathbf{x}) \simeq \mathcal{N}_{,\phi}\delta\phi(t,\mathbf{x}) + \tfrac{1}{2}\mathcal{N}_{,\phi\phi}[\delta\phi(t,\mathbf{x})]^2 = \mathcal{N}_{,\phi}\delta\phi + \frac{\mathcal{N}_{,\phi\phi}}{2\mathcal{N}_{,\phi}^2}(\mathcal{N}_{,\phi}\delta\phi)^2,$$

where $\delta\phi(t,\mathbf{x}) := \phi(t,\mathbf{x}) - \phi(t)$. In the local-form case, comparing this expression with (4.99) we get

$$f_{\rm NL}^{\rm local} = \frac{5}{6}\frac{\mathcal{N}_{,\phi\phi}}{\mathcal{N}_{,\phi}^2}\,. \tag{5.164}$$

Noting that $\mathcal{N}$ is the backward-oriented number of e-folds,

$$\mathcal{N}_{,\phi} = -\frac{H}{\dot\phi} \stackrel{(5.53)}{=} -\sqrt{\frac{\kappa^2}{2\epsilon}}\,, \qquad \mathcal{N}_{,\phi\phi} = \mathcal{N}_{,\phi}\frac{\epsilon_{,\phi}}{2\epsilon} \stackrel{(5.51)}{=} \mathcal{N}_{,\phi}^2(\epsilon-\eta)\,, \tag{5.165}$$

one has $f_{\rm NL}^{\rm local} = (5/12)(2\epsilon - 2\eta) = (5/12)(1 - n_{\rm s} - 2\epsilon)$ from (5.153), which misses an extra term 2ϵ in brackets to get (4.108). This contribution comes from the back-reaction of the metric, which we ignored in (5.164). An improved calculation eventually yields the correct result in the squeezed limit.

5.8.1 *Stochastic Inflation*

Another framework wherein to study scalar-field non-linearities is *stochastic inflation* [198, 199, 230, 233, 255–262]. Stochastic inflation is an approximated method according to which the solutions of the equation of motion for the scalar perturbation in the long-wave-length limit $k \ll aH$ are connected to those in the $k \gg aH$ limit at the Hubble horizon. This approach allows one to model the coupling between large- and small-wave-length modes.

The scalar field (or any other derived scalar quantity φ, such as $\varphi = \delta\phi$) is separated into a "classical" or coarse-grained contribution φ_c, encoding all the modes outside the Hubble horizon, and a quantum or fine-grained part φ_q taking into account the in-horizon modes (equation (3.15)). Therefore, the classical part is the average of the scalar field on a comoving volume with radius $R = r_H/\varepsilon$, $0 < \varepsilon < 1$, larger than the shrinking Hubble radius $r_H = (aH)^{-1}$, while the oscillations of φ_q cancel one another in the average. With this decomposition, the equation of motion for φ becomes a Langevin equation with a stochastic noise source generated by the fine-grained contribution of the quantum fluctuations. If self-interaction is not negligible, the evolution of φ_c is non-linear even if the quantum fluctuations φ_q are exactly Gaussian.

Low frequencies are cut by a window function $W(|\mathbf{x} - \mathbf{x}'|/R)$ falling off at distances larger than R. Denoting with $\mathcal{W}$ the Fourier transform of W, we can choose

$$\mathcal{W}(k) = 1 - \frac{3}{kR} j_1(kR) .$$

Its asymptotic limits are

$$\lim_{kR\to 0} \mathcal{W} = 0 , \qquad \lim_{kR\to +\infty} \mathcal{W} = 1 .$$

We can approximate the window function to a Heaviside distribution,

$$\mathcal{W} \simeq \Theta(k - \varepsilon aH) , \tag{5.166}$$

where $0 < \varepsilon < 1$. The average of φ_c is performed over a volume $\sim R^3 = (r_H/\varepsilon)^3$.

Let φ be a test field and let us ignore metric perturbations. Within the horizon, the scalar-field modes are approximated by

$$\varphi_q(x) = \int \frac{d^3\mathbf{k}}{(2\pi)^3} \mathcal{W}(k) \left[\varphi_k(t) a_{\mathbf{k}} e^{i\mathbf{k}\cdot\mathbf{x}} + \varphi_k^*(t) a_{\mathbf{k}}^\dagger e^{-i\mathbf{k}\cdot\mathbf{x}}\right] , \tag{5.167}$$

while the long-wave-length modes φ_q have the same definition with $\Theta(k - \varepsilon aH)$ replaced by $\Theta(\varepsilon aH - k)$. Substituting (3.15) in the equation of motion

$$\left(\partial_t^2 + 3H\partial_t - \frac{\nabla^2}{a^2}\right) \varphi + V_{,\varphi} = 0 , \tag{5.168}$$

one gets

$$\left(\partial_t^2 + 3H\partial_t - \frac{\nabla^2}{a^2}\right) \varphi_c = -V_{,\varphi_c} + 3H\mu , \tag{5.169}$$

where $V_{,\varphi_c} = \partial V/\partial\varphi|_{\varphi=\varphi_c}$ is the average of $V_{,\varphi}$ over a comoving volume and

$$\mu(x) := -\frac{1}{3H}\left(\partial_t^2 + 3H\partial_t - \frac{\nabla^2}{a^2}\right)\varphi_q\,. \tag{5.170}$$

The noise term $\mu(x)$ stems from quantum fluctuations, which become classical at horizon exit and contribute to the background wherein φ_c lives. In this sense, one often refers to μ as the field back-reaction.

If the conditions of extreme slow-roll regime are satisfied, one can neglect the term $\ddot{\varphi}_c$ in (5.169) and obtain the *Langevin equation*

$$\dot{\varphi}_c(x) \simeq \frac{1}{3Ha^2}\nabla^2\varphi_c(x) - \frac{V_{,\varphi_c}}{3H} + \mu(x)\,. \tag{5.171}$$

We briefly mention that the Langevin equation is associated with the probability distribution $P[\varphi_c, t]$ of the value of the scalar field at time t at a given point $\mathbf{x}$ in the comoving volume. One can show that P obeys the *Fokker–Planck equation* [256, 263, 264]

$$\frac{\partial}{\partial t}P[\varphi_c, t] = -\frac{\partial}{\partial\varphi_c}\left\{\frac{1}{3H}\left(\frac{\nabla^2}{a^2}\varphi_c - V_{,\varphi_c}\right)P[\varphi_c, t]\right\} + \frac{\partial^2}{\partial\varphi_c^2}\frac{H^3}{8\pi^2}P[\varphi_c, t]\,, \tag{5.172}$$

presented here in Itô's version. Solutions to this equation determine the coupling among modes at horizon scale.

For a de Sitter universe ($H =$ const) the noise term, calculated from the equation of motion of the massless scalar (5.167), is [233]

$$\mu(x) = \mathrm{i}\varepsilon H^3 a\int\frac{\mathrm{d}^3\mathbf{k}}{(2\pi)^3}\frac{1}{\sqrt{2k^3}}\delta(k - \varepsilon aH)\left(a_{\mathbf{k}}\mathrm{e}^{-\mathrm{i}\mathbf{k}\cdot\mathbf{x}} - a_{\mathbf{k}}^\dagger\mathrm{e}^{\mathrm{i}\mathbf{k}\cdot\mathbf{x}}\right), \tag{5.173}$$

where $a = \exp(Ht)$. Its vacuum expectation value and two-point correlation function are

$$\langle\mu(x)\rangle = 0\,, \qquad \langle\mu(x)\mu(x')\rangle = -\frac{H^3}{4\pi^2}\delta(t - t')\frac{\sin(\varepsilon aH\varrho)}{\varepsilon aH\varrho},$$

where $\varrho = |\mathbf{x} - \mathbf{x}'|$. In particular,

$$\langle\mu(x)\mu(x')\rangle_{\mathbf{x}=\mathbf{x}'} = -\frac{H^3}{4\pi^2}\delta(t - t')\,. \tag{5.174}$$

The last equation characterizes μ as a white noise (infinitesimally short correlation time) but stochastic inflation can be extended to colored-noise models [265].

From (5.171), one can recognize two sources of non-linearity. The first is the self-interaction of the inflaton (e.g., if its potential contains terms such as ϕ^n, $n > 2$), while the second is the back-reaction of the field fluctuations on the background, encoded in the noise term. The statistical distribution of the field φ_c is non-Gaussian, even if quantum fluctuations are completely random. While fluctuations relevant for the visible universe are Gaussian, at scales much larger than the present Hubble horizon the field distribution can be more complicated and, in particular, non-Gaussian [137, 257]. Therefore, we expect a small non-Gaussian effect in the observable region, at least in the case of a single scalar field.

A calculation at first SR order recovers the non-linear parameter in the squeezed limit (4.108) via (5.153) [244]. A powerful approach obtaining this result in a reasonably easy and robust way is the space-gradient formalism [235, 266–268], a development of the separate universe method [269] (Sects. 3.1.4 and 3.3.1). It is convenient to leave the gauge unspecified until towards the end, and work with the metric

$$ds^2 = -N^2(t,\mathbf{x})dt^2 + a^2(t,\mathbf{x})dx_\alpha dx^\alpha\,, \tag{5.175}$$

where $N(t,\mathbf{x})$ and $a(t,\mathbf{x})$ are a locally defined lapse function (see Sect. 9.1.2) and scale factor, respectively. In synchronous gauge, $N = 1$. Physical quantities such as $H(t,\mathbf{x}) = \dot{a}/(Na)$ and the scalar field $\phi(t,\mathbf{x})$ are defined on an inhomogeneous background and evolve, separately in each homogeneous patch ("at each point"), through the dynamical equations once the initial conditions have been specified. Then, one can convert time derivatives into spatial gradients. For instance, at lowest order in the gradient expansion and at large scales,

$$\partial_\alpha H \simeq -\epsilon\frac{H^2}{\Pi}\partial_\alpha\phi\,, \qquad \partial_\alpha\Pi \simeq -\eta H\partial_\alpha\phi\,, \tag{5.176}$$

where

$$\Pi := \frac{\dot{\phi}}{N}\,, \qquad \epsilon = -\frac{\dot{H}}{NH^2}\,, \qquad \eta = -\frac{\dot{\Pi}}{NH\Pi}\,. \tag{5.177}$$

The non-linear extension of the Mukhanov–Sasaki variable (5.144) is

$$\mathcal{U}_\alpha := z\mathcal{R}_\alpha\,, \tag{5.178}$$

where

$$z := \frac{a(t,\mathbf{x})\Pi(t,\mathbf{x})}{H(t,\mathbf{x})} \tag{5.179}$$

and $\mathcal{R}_\alpha$ is the non-linear comoving curvature perturbation (3.44) (here we consider only spatial indices since $\mathcal{R}_0 \approx 0$ at lowest order). From (3.46),

$$\mathcal{U}_\alpha = a\left(\partial_\alpha\phi - \frac{\Pi}{H}\partial_\alpha \ln a\right). \tag{5.180}$$

At large scales, $\mathcal{R}_\alpha$ is conserved in time, so that differentiating twice with respect to t one gets

$$\ddot{\mathcal{U}}_\alpha - \frac{\ddot{z}}{z}\mathcal{U}_\alpha \simeq 0. \tag{5.181}$$

However, it is more convenient to keep the decaying mode implicitly dropped in the recursion of $\dot{\mathcal{U}}_\alpha = (\dot{z}/z)\mathcal{U}_\alpha$. Then, the equation of motion can be recast as

$$\ddot{\mathcal{U}}_\alpha + F\dot{\mathcal{U}}_\alpha - M^2\mathcal{U}_\alpha \simeq 0 \qquad \text{at large scales}, \tag{5.182}$$

where

$$F := NH - \frac{\dot{N}}{N}, \qquad M^2 := F\frac{\dot{z}}{z} + \frac{\ddot{z}}{z} = 2 + O(\epsilon). \tag{5.183}$$

In the linear approximation and in momentum space, (5.182) holds for $\mathcal{U}_\alpha \simeq \partial_\alpha u \to \mathrm{i}k_\alpha u_{\mathbf{k}}$. As in the linear case, the equation of motion for $\mathcal{U}_\alpha$ can be written as an equation for the coarse-grained part $\mathcal{U}_\alpha^{\rm c}$ sourced by a stochastic noise term. In momentum space, the coarse-grained part of the Mukhanov–Sasaki variable is $\mathcal{U}_{\alpha,\mathbf{k}}^{\rm c} = \mathcal{U}_{\alpha,\mathbf{k}}\mathcal{W}(kR)$. With ε sufficiently smaller than 1 ($R \gg r_H$), we can safely discard the k^2 term (second-order gradient) in the Mukhanov–Sasaki equation. Therefore, at large scales

$$\ddot{\mathcal{U}}_\alpha^{\rm c} + F\dot{\mathcal{U}}_\alpha^{\rm c} - M^2\mathcal{U}_\alpha^{\rm c} = \mu_\alpha, \tag{5.184a}$$

$$\mu_\alpha(t,\mathbf{x}) = \int\frac{\mathrm{d}^3\mathbf{k}}{(2\pi)^3}(\mathrm{i}k_\alpha)\mathrm{e}^{\mathrm{i}\mathbf{k}\cdot\mathbf{x}}[\ddot{\mathcal{W}} + \dot{\mathcal{W}}(2\partial_t + F)]u_k\beta_{\mathbf{k}} + \text{c.c.}, \tag{5.184b}$$

where c.c. stands for complex conjugate and $\beta_{\mathbf{k}}$ is a complex stochastic variable with ensemble average

$$\langle\beta_{\mathbf{k}}\beta_{\mathbf{k}'}^*\rangle = (2\pi)^3\delta(\mathbf{k}-\mathbf{k}'); \tag{5.185}$$

it simulates the continuous crossing of modes outside the horizon and their fueling of the coarse-grained part. As it stands, (5.184) is the non-linear extension of the Langevin equation (5.171) and it properly encodes the full stochastic contribution.

This would not be the case if we started from (5.181), where the velocity degree of freedom associated with the decaying mode has been absorbed [267].

Equation (5.184) can be expressed as a Langevin differential equation in the curvature perturbation (superscript c omitted):

$$\ddot{\mathcal{R}}_\alpha + \left(2\frac{\dot{z}}{z} + F\right)\dot{\mathcal{R}}_\alpha = -\frac{\mu_\alpha}{z}. \tag{5.186}$$

To lowest SR order, $\dot{z}/z \simeq NH$ and, neglecting the $\ddot{\mathcal{R}}_\alpha$ term, one gets

$$\dot{\mathcal{R}}_\alpha \simeq -\frac{\mu_\alpha}{(2NH + F)z}. \tag{5.187}$$

At first order in a perturbative gradient expansion,

$$\dot{\mathcal{R}}_\alpha^{(1)} \simeq -\frac{\mu_\alpha^{(1)}}{[(2NH + F)z]^{(0)}} \qquad \Rightarrow \qquad \dot{\mathcal{R}} \simeq -\frac{\mu}{(2NH + F)z}, \tag{5.188}$$

where we used the gradient expansion, $\mathcal{R}_\alpha^{(1)} \simeq \partial_\alpha \mathcal{R}$ and $\mu_\alpha^{(1)} \simeq \partial_\alpha \mu$ (Sect. 3.3.3) and in the second expression the superscript (0) has been omitted, so that z, N and H are defined on the homogeneous background. The power spectrum (5.152) is recovered from the solution of (5.188) (Problem 5.2).

At second order in the perturbation,

$$\dot{\mathcal{R}}_\alpha^{(2)} = -\mu_\alpha^{(1)}\{[(2NH + F)\bar{z}]^{-1}\}^{(1)} - \frac{\mu_\alpha^{(2)}}{z^{(0)}}. \tag{5.189}$$

The second term is higher order, $\mu_\alpha^{(2)} = O(\epsilon^2)$, and can be dropped. To calculate the first term, we fix the gauge so that the time coordinate coincides with the natural time variable during inflation, the logarithm of (the local value of the inverse of) the comoving Hubble radius [257, 267]:

$$t = \ln(aH).$$

Consistently, the right-hand side increases with time. Differentiating this expression with respect to t, we obtain

$$NH = \frac{1}{1-\epsilon},$$

so that, to lowest SR order, $\dot{N}/N \simeq -\dot{H}/H$ and $2NH + F \simeq 3$. We need also the relation

$$\begin{aligned}\partial_\alpha z &= \frac{\Pi}{H}\partial_\alpha a + \frac{a}{H}\partial_\alpha \Pi - \frac{a\Pi}{H^2}\partial_\alpha H \\ &\overset{(5.180)}{=} -\mathcal{U}_\alpha + a\partial_\alpha\phi + \frac{a}{H}\partial_\alpha \Pi - \frac{a\Pi}{H^2}\partial_\alpha H \\ &\overset{(5.176)}{\simeq} -\mathcal{U}_\alpha + a(1+\epsilon-\eta)\partial_\alpha\phi\,. \end{aligned} \tag{5.190}$$

Notice that in synchronous gauge one would have had to expand $Hz = a\dot{\phi}$ rather than z, thus loosing the term in ϵ crucial for the the final result. On surfaces of constant time, by definition $\partial_\alpha \tau = 0$, so that $\partial_\alpha a/a = -\partial_\alpha H/H$ [267]. Therefore, $\mathcal{U}_\alpha = (1-\epsilon)a\partial_\alpha\phi$ and, combining this with (5.190), we get

$$\partial_\alpha z = (2\epsilon - \eta)\mathcal{U}_\alpha = \tfrac{1}{2}(1-n_{\rm s})\mathcal{U}_\alpha\,.$$

Then,

$$\begin{aligned}\dot{\mathcal{R}}_\alpha^{(2)} &\simeq -(z^{-1})^{(1)}\frac{\mu_\alpha^{(1)}}{3} = -\left[-\frac{\partial^{-2}\partial^\beta\partial_\beta z^{(1)}}{z^{(0)}}\right]\frac{\mu_\alpha^{(1)}}{3z^{(0)}} \\ &\simeq -\tfrac{1}{2}(1-n_{\rm s})\mathcal{R}^{(1)}\dot{\mathcal{R}}_\alpha^{(1)}\,,\end{aligned}$$

from which

$$\mathcal{R}^{(2)} \simeq -\tfrac{1}{2}(1-n_{\rm s})\mathcal{R}^2\,. \tag{5.191}$$

Now we are ready to write down the curvature perturbation at second order,[9]

$$\mathcal{R}_{\rm NL} \simeq \mathcal{R} + \tfrac{1}{2}\mathcal{R}^{(2)} = \mathcal{R} - \tfrac{1}{4}(1-n_{\rm s})\mathcal{R}^2\,. \tag{5.192}$$

This expression should be compared with the local form of the non-linear comoving curvature perturbation,

$$\mathcal{R}_{\rm NL} \simeq \mathcal{R} - \frac{3}{5}f_{\rm NL}^{\rm local}\left(\mathcal{R}^2 - \langle\mathcal{R}^2\rangle\right)\,, \tag{5.193}$$

derived from (4.99) (the $-$ sign in front of $f_{\rm NL}^{\rm local}$ is because $\mathcal{R} \simeq -\zeta$ at large scales). Thus, upon computing the bispectrum, we have reobtained the non-linear parameter (4.108).

[9] In [270], various definitions of the second-order curvature perturbation are reviewed.

5.8.2 *Multi-field Non-Gaussianity*

In scenarios with dynamical scalar spectators during inflation, a non-Gaussian statistics arises when the light fields reach the bottom of the potential while their quantum fluctuations and the interaction with the inflaton generate non-linear effects [135–138, 271–276]. Also the curvaton scenario is characterized by appreciable non-Gaussianity [140, 141, 277–280]. In the presence of many light fields, the curvature perturbation can evolve after horizon exit due to the presence of isocurvature modes. Thus, one could expect a more pronounced non-Gaussianity statistics [237, 270, 281]. In general, however, non-Gaussianity is small in multi-field inflation, of which assisted inflation is a special case. Clear indications of this are given by analytical treatments (mainly, the $\delta\mathcal{N}$ formalism) of the multi-field bispectrum [87, 237, 270, 281–288] and trispectrum [100, 239, 240, 289, 290].

5.9 Observational Constraints on Inflation

5.9.1 *Temperature Spectra*

Since standard inflationary models produce fluctuations of the total density, the ensuing scenario of structure formations is adiabatic, in pleasant accordance with CMB and large-scale structure observations. Also, in Sect. 4.4.2 we saw the experimental bounds on the primordial spectral amplitudes and indices. These translate into constraints on the inflaton potential.

First, rewrite (5.152) in the ESR regime as

$$\mathcal{P}_{\rm s} = \frac{8}{3\epsilon}\frac{V}{m_{\rm Pl}^4}\,. \tag{5.194}$$

Assuming no tensor signal ($r = 0$), (4.69) fixes the normalization at the pivot scale $k_0 = 0.05\,\mathrm{Mpc}^{-1}$. Considering that $\epsilon < 1$, this corresponds to an upper bound on the inflaton potential at horizon crossing: $V^{1/4} < 6.5\times10^{16}\,\mathrm{GeV}$ at $k_0 = 0.05\,\mathrm{Mpc}^{-1}$. Including also a tensor signal and expressing ϵ in terms of r via the consistency relation (5.156), the PLANCK constraint (4.69) gives

$$V^{1/4} = \left[\frac{3}{128}\,r\mathcal{P}_{\rm s}(k_0)\right]^{1/4} m_{\rm Pl} \simeq 1.93\times10^{16}\left(\frac{r}{0.12}\right)^{1/4}\,\mathrm{GeV} \tag{5.195}$$

at $k_0 = 0.05\,\mathrm{Mpc}^{-1}$. Using the upper bound (4.73), one gets a 95 % CL upper bound for the Hubble parameter,

$$l_{\rm Pl}H < 7.2\times10^{-6} \qquad \text{at} \quad k_0 = 0.05\,\mathrm{Mpc}^{-1}, \tag{5.196}$$

while for (4.74) the coefficient is lowered to 5.5. The energy density at the end of inflation is at least 12 orders of magnitude smaller than the Planck density.

For the scalar potential (5.81) and from (5.85g), (5.85h), (5.153) and (5.156), we have

$$n_s - 1 \simeq -\frac{2n+4}{4\mathcal{N}_k + n}\,, \qquad r \simeq \frac{16n}{4\mathcal{N}_k + n}\,. \tag{5.197}$$

The scalar spectrum is always red tilted. For a given number of e-folds $\mathcal{N}_k$, these equations identify a map between monomial potentials ϕ^n and points in the (n_s, r) plane.

The border of large-field models ($0 \leqslant \eta < \epsilon$, $V_{,\phi\phi} > 0$) and small-field models ($\eta < 0 < \epsilon$, $V_{,\phi\phi} < 0$) corresponds to $n = 1$ ($V_{,\phi\phi} = 0$), giving [291]

$$r = -\frac{8}{3}(n_s - 1)\,.$$

The region spanned by large-field, small-field and hybrid models in the (n_s, r) plane are depicted in Fig. 5.7. The border between large-field and hybrid models is at $\epsilon = \eta$,

$$r = -8(n_s - 1)\,.$$

Figures 5.8 and 5.9 show the marginalized constraints at the 1σ and 2σ level for the scalar spectral index versus the tensor-to-scalar ratio [43, 111, 147]. This likelihood analysis, which can be performed via computer packages such as CosmoMC [292], is one of the so-called "top-down" approaches: one asks what the probability is that a theory predicting a given set of observables would realize the observed experimental data [293]. The Harrison–Zel'dovich spectrum is excluded at more

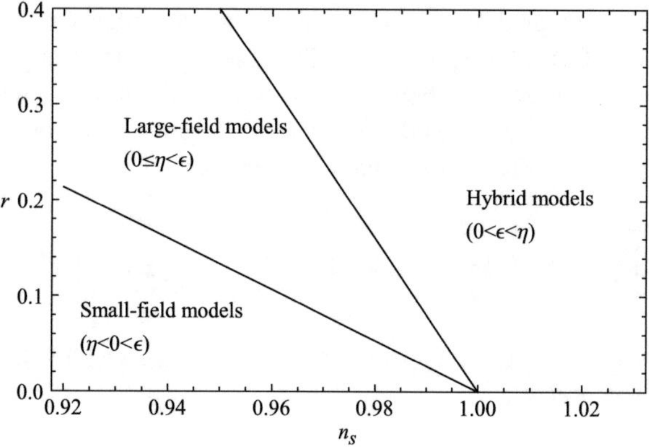

Fig. 5.7 Single-field models in the (n_s, r) plane

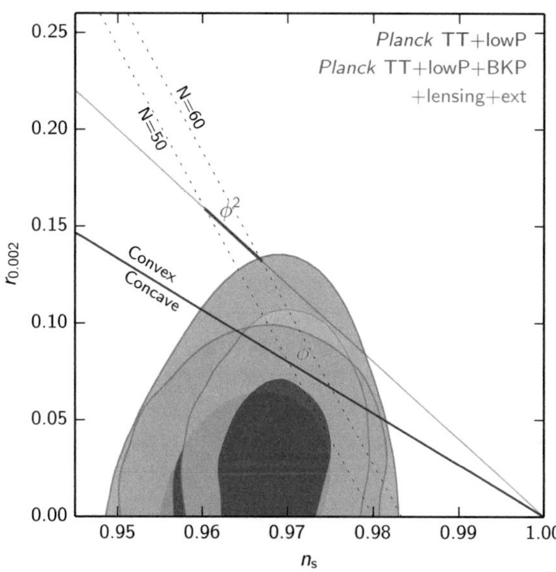

Fig. 5.8 Two-dimensional joint marginalized constraint (68 % and 95 % CL) on the primordial scalar index n_s and the tensor-to-scalar ratio $r_{0.002}$ at $k = 0.002\,\text{Mpc}^{-1}$ and for zero running, derived from the data combination of PLANCK and other data sets ("ext" includes BAO, supernovæ and H_0 observations, while BKP is the BICEP2+*Keck Array*+PLANCK joint likelihood on *B* modes). These contours change when the running is included; see Fig. 4.16 (Credit: [43], reproduced with permission ©ESO)

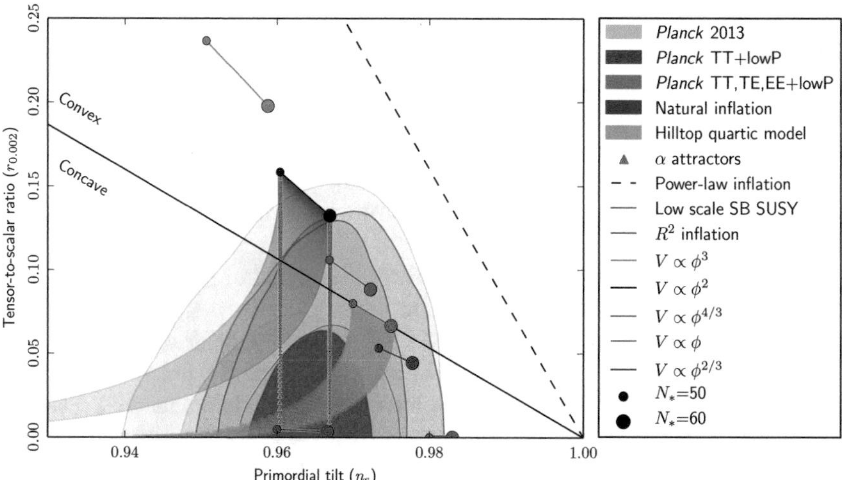

Fig. 5.9 Joint marginalized constraints in the (n_s, $r_{0.002}$) plane from PLANCK in combination with other data sets, compared with the theoretical predictions of selected inflationary models. Hilltop quartic models have a potential $V \propto 1 - (\phi/\phi_0)^4$ (Credit: [111], reproduced with permission ©ESO)

than 99 % CL, and so are monomial potentials with large n. According to data, the quartic potential ϕ^4 is ruled out and even the quadratic potential is under tight pressure. Small-single-field models such as natural inflation and higher-derivative gravity seem to be favoured. The PLANCK constraint $f \gtrsim 6.9 M_{\rm Pl}$ [33] disfavours the original interpretation of natural inflation as a pseudo-Nambu–Goldstone boson, where $f \lesssim M_{\rm Pl}$. However, a multi-axion generalization of the same scenario can easily produce a effective super-Planckian f [114].

The formulæ (5.142) and (5.153) express the spectral indices $n_{\rm t}$ and $n_{\rm s}$ in terms of the slow-roll parameters. If the accelerated expansion is rapid enough ($\epsilon, \eta \ll 1$), perturbations of cosmological scale exit the horizon soon one after another. Since, with good approximation, the physical conditions are about the same when small and large scales cross the horizon, the fluctuation spectra will be invariant. The fact that inflation does not predict perfect scale invariance is a great success. The zero-order Harrison–Zel'dovich scalar spectrum $n_{\rm s} = 1$ is ruled out by observations and it is necessary to know at least the first-SR-order expressions of the spectral index.

Inflationary models with different Lagrangians often produce similar observables $n_{\rm s}$ and r, so that a single type of observation cannot distinguish among inequivalent theories. The combination of various experiments and their improving precision can break this degeneracy [294].

5.9.2 *Polarization*

Cosmic variance forbids an accurate determination of the primordial gravitational-wave spectrum at large scales, where the latter has the largest amplitude,[10] while small scales are more affected by the uncertainty on cosmological parameters. Polarization maps can help to decompose the tensor contribution from the scalar one. Since the inflaton is a scalar field, no B-modes arise from density perturbations to linear order. Inflationary gravitational waves are not directly detectable but they leave a BB imprint. Magnetic-type polarization is produced by vector and tensor modes but it is negligible in both cases. In particular, polarization from gravitational waves is quite weak due to the small tensor-to-scalar ratio predicted in standard inflationary models.

As for the temperature spectra, the polarization signal of topological defects is quite distinct from that of inflation. Due to vector modes, the signal of the B-type polarization is enhanced to the order of $1\,\mu$K at small angular scales [296]. The position and amplitude of the observed temperature and TE-polarization peaks, together with the upper bounds on BB, already rule out defect models. The role of B polarization will be relevant to constrain other scenarios alternative to standard inflation and non-inflationary seeds generating a strong gravitational-wave signal or vorticity.

[10] A preheating phase can enhance the tensor signal after inflation, thus enhancing the expectation for detection of a stochastic gravitational-wave background [295].

An even more interesting situation would be determined by a non-zero cosmological contribution to the cross-correlation spectral C_ℓ^{TB} and C_ℓ^{EB}, showing that the primordial perturbation spectrum has definite handedness.

Parity violations may arise from several physical mechanisms. In scalar-tensor theories and other non-minimally coupled models, a scalar field Φ can interact with the electromagnetic field strength $F_{\mu\nu}$ via terms of the form $\mathcal{L}_{\Phi,F} \propto f(\Phi)\epsilon^{\mu\nu\sigma\tau}F_{\mu\nu}F_{\sigma\tau}$. If $f(\Phi) = \Phi$, the field Φ is a pseudo-scalar [297–300]. A non-minimal axion inflaton in the same class of theories can couple with gravity through interactions such as the Pontryagin term $\mathcal{L}_{\Phi,R} \propto f(\Phi)\epsilon^{\mu\nu\sigma\tau}R^{\rho}{}_{\lambda\mu\nu}R^{\lambda}{}_{\rho\sigma\tau}$. Non-trivial C_ℓ^{TB} and C_ℓ^{EB} spectra are generated by both types of terms [301–303] as well as by others with gravity-electromagnetic axial coupling [304]. Another intriguing possibility, arising in models of quantum gravity, is that gravity itself may be chiral. In metric formalism, this means that one can identify left- and right-handed graviton modes $h_{\rm L,R} = (h_+ \pm i h_\times)/\sqrt{2}$ coupling differently with matter, via some effective Newton constants $G_{\rm L,R}$. The physical origin of such a situation [305–309] can be understood in first-order formulation (see Chap. 9), where $G_{\rm L,R}$ depend on the Barbero–Immirzi parameter. We postpone a discussion of axion inflation in string theory to Sect. 13.4.

5.9.3 Non-Gaussianity

Current experimental bounds on the non-linear parameter $f_{\rm NL}$ (Sect. 4.6.4) are compatible with the small level of non-Gaussianity predicted by inflation. In the local model, this is (4.108),

$$f_{\rm NL}^{\rm local} \simeq \frac{5}{12}(1-n_{\rm s})\,. \tag{5.198}$$

We might regard this equation as a lowest-SR-order consistency relation joining the set (5.157) and (5.158) [247, 310].[11]

One should note that the post-inflationary era greatly enhances non-Gaussianity, up to $f_{\rm NL}^{\rm post} = O(1)$ [311–313][12] or even $f_{\rm NL}^{\rm post} = O(100)$ from a suitable preheating phase [314, 315]. In addition to the post-inflationary contribution, one must consider also angular averaging. The total observed $f_{\rm NL}$ is thus $f_{\rm NL}^{\rm obs} = O(1) + f_{\rm NL}$ rather than the bare inflationary result (4.108) [316]. Therefore, the non-linear effect of standard

[11] A small non-Gaussian component, proportional to the tensor amplitude and spectral index $n_{\rm t}$, also comes from the three-point functions involving the graviton zero-mode [245]. Since the tensor amplitude is much smaller than the scalar one, we can neglect this term with respect to the scalar bispectrum.

[12] This is due to the fact that, at second order in perturbation theory, the longitudinal gauge condition $\Phi - \Psi = 0$ is modified as $\Phi^{(2)} - \Psi^{(2)} = 4\Psi^2$ at large scales, thus providing a non-trivial second-order correction to the Sachs–Wolfe effect [222].

single-field inflation, if the SR approximation holds, is always sub-dominant. Fortunately, we are able to discriminate between primordial and post-inflationary contributions and separate the signals according to their features. The constraints quoted in Sect. 4.6.4 are on the first type of contribution.

5.10 Unsolved and New Problems

The big bang and the cosmological constant still constitute unsolved issues. Inflation neither resolves the big-bang singularity nor explains the fine tuning of Λ. In parallel, it solves four puzzles of the hot big bang model, but at a price. It is almost the rule in theoretical physics that a good solution to a problem entails the emerging of other problems. This happens because an answer to a question often is just another, more instructive way to ask the same question.

5.10.1 Graceful Entry Problem

To develop the idea of inflation, we have first considered a kinematical and, then, a dynamical model in a FLRW spacetime, Sects. 5.2 and 5.3.1. Then, we set the initial conditions via the pre-inflationary chaotic scenario, which is far from being FLRW. Linear inhomogeneous perturbations fall short of justifying the homogeneous approximation in the face of the extremely inhomogeneous initial conditions. The question remains whether inflation is a likely event in a spacetime with general metric or not. Sometimes, this is called *graceful entry problem*. The question of the naturalness of inflation in cosmological models with non-FLRW initial conditions has been debated in early papers [317–322].

In [317, 318], it was conjectured that all expanding universes with a positive cosmological constant tend asymptotically to de Sitter. There are obvious counter-examples to this, for instance a closed universe collapsing before inflation. However, under certain hypotheses this became a "no-hair" theorem for Bianchi universes [320], later generalized to inhomogeneous cosmologies [321]. The name suggests that a universe endowed with an arbitrarily complicated structure ("hair") would lose its memory (get "bald") in an exponentially fast time scale. With due caution, the no-hair theorem can be stated as follows.

No-hair theorem. *Given a spacetime* $(\mathcal{M}, g)$ *such that:*

1. *initially, it is expanding;*
2. *it is endowed with a a positive cosmological constant* Λ*;*
3. *it is described by an energy-momentum tensor obeying the strong and dominant energy conditions;*
4. *the spatial Ricci curvature scalar* ${}^{(3)}R$ *is non-positive in "sufficiently" large regions.*

Then, these regions evolve towards de Sitter spacetime.

The statement requires some comments. The first assumption, which we would translate as $H(t_i) > 0$ in a cosmological setting, adapts the theorem to the physical scenario. The presence of a cosmological constant is essential. For an inflationary universe, Λ mimics the slow-rolling inflaton with effective equation of state $w \approx -1$. The energy-momentum tensor described in point 3 represents the contribution of all matter fields except the inflaton, the latter corresponding to the cosmological constant. Point 4 is somewhat delicate. It is possible to require that space has negative or zero curvature ${}^{(3)}R$ at all points, corresponding to an open or flat universe. In this case, one is excluding the existence of regions collapsing into black holes, but these regions do not interfere with inflation if they are sufficiently small in comparison with expanding volumes.

The theorem, unfortunately, does not really demonstrate the naturalness of inflation. In fact, a cosmological constant is much less than a (set of) dynamical slow-rolling field(s), even if the two notions blur in the extreme slow-roll regime. In part, however, it justifies the use of the FLRW metric *during* inflation and strongly suggests that an accelerated expansion is a rather general and model-independent feature of the early universe.

The problem of initial conditions is relaxed in eternally inflating models (Sect. 5.6.5).

5.10.2 Graceful Exit Problem

As already stressed by Guth, the old inflationary scenario suffered from a *graceful exit problem*: if the formation rate of the true-vacuum bubbles is greater than the expansion rate of the universe, then the phase transition is very fast and inflation does not begin. Conversely, if the vacuum decay is too slow the post-inflationary universe becomes highly inhomogeneous and, in some of its parts, the phase transition is not completed.

Although the bubble scenario has been replaced by chaotic inflation, the exit problem has been persisting in other forms. The simplest models of chaotic inflation do not contain in themselves the physics necessary to limit the acceleration period. For instance, all models with constant $\epsilon < 1$ never cease to inflate ($t_e = \infty$) and they need to be supplemented by an ad hoc engineered reheating mechanism.

A simple solution preserving the advantages of chaotic inflation is to introduce another scalar field φ whose decay determines the end of inflation. Such a field configuration can be realized in hybrid inflation, already described in Sect. 5.5.3.

5.10.3 Trans-Planckian Problem

In most of the above models, the number of e-foldings of accelerated expansion is such that all observable cosmological scales were sub-Planckian at the onset of inflation. Since we expect that the quantum-field-theory description breaks down at these scales, one might question the validity of the models themselves. Inflation, however, does work, so it is plausible that this trans-Planckian problem [323] is due to the qualitative nature of the set of initial conditions. Another possibility is that the latter will be determined in a satisfactory way only in a robust theory of high-energy physics. At Planck scale, exotic physics should modify the effective action of the inflation. For instance, this can happen in the form of a non-standard dispersion relation

$$-(k^0)^2 + |\mathbf{k}|^2 = 0 \quad \longrightarrow \quad -(k^0)^2 + f(k^0, \mathbf{k})|\mathbf{k}|^2 = 0\,. \tag{5.199}$$

The resulting inflationary predictions can drastically change and observations can place constraints on phenomenology at the Planck scale [171, 182–185, 324–337].

5.10.4 Naturalness or Model-Building Problem

We saw that the parameters of common inflationary potentials require a certain level of tuning. For instance, according to (5.29) the mass in a quadratic potential should be six orders of magnitude smaller than the Planck mass. This is a consequence of having a small CMB temperature fluctuation $\delta T/T \sim 10^{-5}$. Sometimes, this is perceived as a fine-tuning issue.

The *naturalness* of a theory can be assessed by different criteria. The first, formulated by Dirac, establishes that a physical theory is "natural" when all its dimensionless parameters are of order 1. This condition, however, is too restrictive because it would label as unnatural very successful models such as the electroweak theory (fine structure constant $\alpha \sim 10^{-2}$, electron Yukawa coupling $\sim 10^{-5}$, and so on).

Another criterion was introduced by 't Hooft [338] and defines a theory as natural if a new symmetry is acquired when sending to zero all the dimensionless parameters in the action. Consequently, all the parameters associated with this approximate symmetry are stable against radiative corrections, so that they do not receive significant (i.e., $O(1)$) contributions from higher-order quantum effects [339].

The second criterion or its adaptations[13] can remove the fine-tuning issue related to the inflationary potential but $V(\phi)$ depends on the chosen high-energy model. The problem is that observations are fit by a number of early-universe "natural," more or less convincing theories, from supersymmetric particle physics to string theory,

[13]In Polchinski's formulation [340], a field theory is natural if all masses are forbidden by symmetries. The Standard Model is not natural due to the Higgs mass but, interestingly, this naturalness argument suggests that a new symmetry should appear at electroweak energies.

from loop quantum cosmology to modified gravity, and so on. A task for the present generation of physicists is to remove the theoretical degeneracy. Much of this book is devoted to this problem.

5.11 The Inflaton and Particle Physics

The construction of a physical picture of the early universe would be incomplete if one could not embed the inflaton in a quantum field theory of particles. This entails three types of questions: (i) whether inflation can be triggered by fields different from a real scalar, (ii) whether the inflaton, in whatever incarnation, can be identified with a specific particle field, and (iii) how to apply quantum-field-theory techniques in order to improve our understanding of the inflationary era. In this section, we comment on the first two questions, leaving the third aside.[14]

5.11.1 *Not Only Scalars*

We have seen that inflation, a generic mechanism producing a sufficiently long period of accelerated expansion, can be realized dynamically by a real scalar field. One can ask oneself if other fields can perform a similar job. For instance, a vector field on cosmological backgrounds [349, 350] is an alternative candidate to play the role of the inflaton [351–359].[15] The simplest example is that of an Abelian massive vector with Lagrangian

$$\mathcal{L} = -\frac{1}{4}F_{\mu\nu}F^{\mu\nu} - \frac{1}{2}\left(m^2 - \frac{R}{6}\right)A_\mu A^\mu\,, \tag{5.200}$$

where $F_{\mu\nu} = 2\nabla_{[\mu}A_{\nu]}$ is the vector field strength. One can envisage more complicated self-interactions, so that A_μ is non-gauge; these interactions can arise, for instance, from a spontaneous breaking of the $U(1)$ symmetry. Rescaling the field as $B_\mu = A_\mu/a$ on an FLRW background, the field equations $\nabla^\mu F_{\mu\nu} - m^2 A_\mu = 0$ become

$$\ddot{B}_\alpha + 3H\dot{B}_\alpha + m^2 B_\alpha = 0\,, \qquad B_0 = 0\,. \tag{5.201}$$

---

[14]Examples of applications of quantum field theory are the effective field theory approach [341] and the study of inflationary infrared divergences and loop contributions to cosmological observables (see the review [342] and later papers [343–348]).

[15]Vector fields have also been considered as curvaton or auxiliary fields [360–366] and in the context of dark energy [367–373].

The non-minimal term in (5.200) is required in order to make the dynamics compatible with the slow-roll approximation.

Compared with the scalar case, some extra difficulties arise. First, the Einstein equations are not satisfied by a homogeneous field because the spatial part of the stress-energy tensor contains off-diagonal components which are of the same order of magnitude as the diagonal components. Intuitively, this happens because the vector field selects a preferred direction. The stress-energy tensor can be made diagonal statistically by a large number of randomly oriented vector fields or, exactly, by a triplet of orthogonal fields $A_\alpha^{(i)} = \phi\delta_\alpha^i$, where $i = 1, 2, 3$ and ϕ is a scalar mode. Second, perturbation theory is more complicate because of the coupling between scalar, vector and tensor sectors already at the linear level. In the limit of a large number of random fields, this problem is relaxed. Third, perturbation analyses possibly indicate the presence of an unstable mode from the longitudinal component of the field, and hence an instability.

Vector inflation can be generalized to p-forms [374–380]. Contrary to the vector case (1-form), 3-forms are consistent with an FLRW background (their deploy in cosmology is not new [381]); also, a non-minimal coupling is not necessary to support an inflating background. Unstable modes, apparently, persist also for $p \geqslant 2$ but in some regimes they can be avoided. Finally, both inflation and dark energy can be realized by a fermionic field [382–391], even via a condensation mechanism [392–398]. The level of fine tuning entailed in these models and required to match with observations is still under inspection.

5.11.2 Higgs Inflation

p-forms are often invoked as alternative inflationary fields on the ground that, since we have observed only one fundamental scalar in Nature, we should keep an open mind to other types of particles. Yet, scalars maintain a certain theoretical fascination. In supersymmetric scenarios and models with spontaneous symmetry breaking, a pseudo-Goldstone boson can emerge (the axion, which made a brief appearance in Sect. 5.5.2) as the driver of natural inflation [112, 113, 143, 399]. In Chap. 13, we shall see how the inflaton can be embedded in string theory as a modulus field. Here we mention a more minimalistic possibility.

In the non-supersymmetric Standard Model, there is only one fundamental field, the Higgs. Its origin from spontaneous symmetry breaking in the electroweak sector will be very briefly sketched in Sect. 7.1.1. By itself, the Higgs cannot sustain inflation. In fact, from (5.195), and considering an inflaton field scale $\phi \sim m_{\rm Pl}$, the mass of the inflaton is close to the GUT scale, $m \sim 10^{13}$ GeV. Therefore, one cannot identify ϕ with the much lighter minimally coupled Higgs field h. The recent discovery at LHC of the Higgs neutral boson found the value $m_h \approx 125$ GeV [400–402]. The coupling of the quartic potential term is also quite different, being it $\lambda \sim 10^{-13}$ for the inflaton and $\lambda \sim 10^{-1}$ for the Higgs [403].

Interestingly, by allowing a *non-minimal coupling* between the Standard Model and gravity the Higgs can act as the inflaton. This type of coupling (in contrast with the minimal one, where matter interacts with gravity only via the determinant of the metric $\sqrt{-g}$) arises naturally in the renormalization of the theory on curved backgrounds [167] and can be further justified [404, 405]. Advances in precision cosmology eventually revived the interest in early attempts to identify the Higgs and the inflaton [118, 406] within a realistic particle-physics scenario [407–431]. The total Lagrangian is

$$\mathcal{L} = \frac{R}{2\kappa^2} + \frac{\xi}{2} h^\dagger h R - \frac{1}{2} D_\mu h^\dagger D^\mu h - \frac{\lambda}{4} (h^\dagger h - v^2)^2 + \mathcal{L}_{\rm SM} \,, \tag{5.202}$$

where ξ are λ are coupling constants, h is the Higgs doublet, D_μ is the $SU(2) \otimes U(1)$ covariant derivative, v is the symmetry-breaking scale of the Standard Model, and $\mathcal{L}_{\rm SM}$ is the Lagrangian of all the other Standard-Model particles. For instance, for $1 \ll \xi \ll 10^4$, at the tree level viable inflation is achieved by tuning $\xi \sim 10^4 \sqrt{\lambda} \sim 10^4 m_h / v$, where in the last step we defined the Higgs effective mass

$$m_h^2 = 2\lambda v^2 \,. \tag{5.203}$$

Radiative corrections and a one-loop renormalization-group improvement already modify the constraints on the parameters [408, 412, 416, 418], while in a renormalization-group analysis at 2-loop level the scalar spectral index $n_{\rm s}$ is shown to be correlated to the Higgs mass [411, 415]. Higgs inflation has been embedded also in supersymmetric extensions of the Standard Model [432] and in supergravity (Sect. 5.12).

A Higgs mass of about $m_h \approx 125\,\text{GeV}$ leads, in general, to a red-tilted scalar spectrum ($n_{\rm s} \approx 0.97$), a small negative running ($\alpha_{\rm s} = -O(10^{-4})$) and a tiny tensor-to-scalar ratio ($r \sim 10^{-3}$), all compatible within 68 % CL with PLANCK observations [147, 433].

The inflaton Higgs field takes values $h \sim m_{\rm Pl} / \sqrt{\xi} \gg E$ above the cut-off $E = m_{\rm Pl} / \xi \sim H$ of the effective theory (which is non-renormalizable, due to gravity); this highlights a unitarity problem [413, 414, 420–425, 427, 428]. The latter can also be seen as a fine-tuning problem, akin to that affecting ad hoc models of inflation. The actual cut-off, however, depends on the background, and the argument about unitarity is true only for small field values. The model can be rescued [425, 430] by introducing new physics beyond the Standard Model, which spoils the original simplicity of the Higgs-inflation idea [427].

Another possible resolution is to take a different prescription for the Lagrangian allowing for non-minimal couplings of derivative type ("new Higgs inflation") [434, 435]. These couplings replace the second and third terms in (5.202) in the form

$$\frac{\xi}{2} h^\dagger h R - \frac{1}{2} D_\mu h^\dagger D^\mu h \to -\frac{1}{2} \left(g^{\mu\nu} - \beta^2 G^{\mu\nu} \right) D_\mu h^\dagger D_\nu h \,, \tag{5.204}$$

where β is a real constant and $G^{\mu\nu} = R^{\mu\nu} - g^{\mu\nu}R/2$ is Einstein's tensor (2.24). Viable inflation is obtained if $\beta \simeq 10^8\lambda^{1/4}m_{\rm Pl}^{-1}$, so that $\beta^{-1} \sim 10^9 - 10^{10}$ TeV for the Higgs values of λ. Again, the spectral scalar index is within the observational bounds, $n_s \approx 0.97$. Unitarity bounds seem to be respected throughout the evolution of the universe (but see [427]).

A second proposal to address the unitarity issue is to couple the Higgs with an extra scalar field [436] but this is not a genuine completion of Higgs inflation, since it is the extra field to act as the inflaton [437]. A fourth option is to resort to Palatini formalism, where the affine connection is independent of the tetrad and the cut-off is higher by a factor of $\sqrt{\xi}$ [428]. Yet another model, built with the same purpose of achieving unitarity, does not feature derivative non-minimal couplings between the Higgs and gravity but it prescribes a non-canonical kinetic term $(1 + \xi h^2)\partial_\mu h\partial^\mu h$ [438–440], which can easily arise in supergravity. The ensuing cosmological consequences can be discriminated from those of the other implementations [431].

Without introducing new matter degrees of freedom and complicating an originally parsimonious model, one could turn the attention to another medicine for unitarity, namely, a UV improvement of the gravitational sector (for instance, asymptotic safety [427] or loop quantum gravity [428]). A fully satisfactory realization of the inflationary paradigm within particle physics and cosmology might lie in the realm of quantum gravity. By no means, there is no conclusive indication that the latter will be a necessary ingredient of a full theory of the early universe. Yet, as the Higgs example shows, research is in progress in many directions.

5.12 Supersymmetry and Supergravity

The inflaton can be embedded also in supergravity (SUGRA), the supersymmetric formulation of covariant general relativity. Models of SUGRA inflation are important for several reasons. First and foremost, the quest for supersymmetry at LHC will give, in the very near future, crucial information about the energy scale of the theory, even in the least exciting case where no supersymmetric partners of the Standard-Model particles were detected. Since supergravity scenarios are among the most promising fundamental theories, it is interesting to see whether and how inflation can be realized therein. Second, supersymmetry is one of the key ingredients of string theory and SUGRA is the low-energy limit of that framework (Sect. 12.2). However, not all potentials of SUGRA inflation can originate from strings and we should differentiate between pure SUGRA models and string models. Here we concentrate on the former, postponing a discussion of string cosmology to Chap. 13. Vacua in supergravity will be discussed in Sect. 7.1.3.

Since the machinery of supersymmetry will play a key role in the cosmological constant problem and will be essential to formulate string theory, we will delve upon it in some detail.

5.12.1 Global Supersymmetry

Let us first recall the main idea of $\mathcal{N} = 1$ supersymmetry in $D = 4$ Minkowski spacetime. The Poincaré algebra of translation generators (hat symbol of operators omitted) $P^\mu = \int \mathrm{d}\mathbf{x}\, T^{\mu 0}$ and Lorentz generators $J_{\mu\nu}$,

$$[P_\mu, P_\nu] = 0\,, \tag{5.205a}$$

$$[P_\mu, J_{\nu\rho}] = \mathrm{i}(\eta_{\mu\rho}P_\nu - \eta_{\mu\nu}P_\rho)\,, \tag{5.205b}$$

$$[J_{\mu\nu}, J_{\sigma\rho}] = \mathrm{i}(\eta_{\mu\rho}J_{\nu\sigma} - \eta_{\nu\rho}J_{\mu\sigma} + \eta_{\nu\sigma}J_{\mu\rho} - \eta_{\mu\sigma}J_{\nu\rho})\,, \tag{5.205c}$$

is augmented by a set of operators $Q_a = \int \mathrm{d}\mathbf{x}\, J^0_a$ and $Q^\dagger_b = \int \mathrm{d}\mathbf{x}\, J^{0\dagger}_b$ called supercharges, stemming from a conserved supercurrent vector J^μ_a where the indices a, b run from 1 to 2. Supercharges are two-component Weyl spinors obeying the set of commutation relations

$$[Q_a, P_\mu] = 0\,, \qquad [Q_a, J_{\mu\nu}] = (\sigma_{\mu\nu})_a^{\ c} Q_c \tag{5.206}$$

$$\{Q_a, Q^\dagger_b\} = 2(\sigma^\mu)_{ab}P_\mu\,, \qquad \{Q_a, Q_b\} = 0 = \{Q^\dagger_a, Q^\dagger_b\}\,, \tag{5.207}$$

where $\{A, B\} = AB + BA$ is the anti-commutator and $(\sigma^{\mu\nu})_a^{\ c} = [(\sigma^\mu)_{ab}(\bar\sigma^\nu)^{bc} - (\sigma^\nu)_{ab}(\bar\sigma^\mu)^{bc}]/4$ are the generators of the special linear algebra $sl(2,\mathbb{C})$. In this expression, σ^μ are the 2×2 Pauli matrices

$$\sigma^0 = -\mathbb{1}_2\,, \quad \sigma^1 = \begin{pmatrix} 0 & 1 \\ 1 & 0 \end{pmatrix}, \quad \sigma^2 = \begin{pmatrix} 0 & -\mathrm{i} \\ \mathrm{i} & 0 \end{pmatrix}, \quad \sigma^3 = \begin{pmatrix} 1 & 0 \\ 0 & -1 \end{pmatrix}, \tag{5.208}$$

and $\bar\sigma^\mu = (\sigma^0, -\sigma^\alpha)$. The same supercharge algebra can be rewritten in terms of a four-component Majorana spinor, in which case one uses the 4×4 Dirac matrices γ^μ in the Weyl basis,

$$\gamma^0 = \begin{pmatrix} 0 & \sigma^0 \\ \sigma^0 & 0 \end{pmatrix}, \qquad \gamma^\alpha = \begin{pmatrix} 0 & \sigma^\alpha \\ -\sigma^\alpha & 0 \end{pmatrix}, \tag{5.209}$$

obeying the Clifford algebra $\{\gamma^\mu, \gamma^\nu\} = -2\eta^{\mu\nu}\mathbb{1}_4$.

Taking the trace of (5.207) over spinor indices, only the $\mu = 0$ component survives to yield the Hamiltonian $H = P^0$:

$$H = \frac{1}{4}\sum_a \{Q_a, Q^\dagger_a\}\,. \tag{5.210}$$

Global supersymmetry is the requirement that the supercharges annihilate the Fock vacuum, $Q_a\,|0\rangle = 0 = Q^\dagger_a\,|0\rangle$, so that the vacuum expectation value of the

Hamiltonian is

$$\rho_{\rm vac} = \langle 0|\,H\,|0\rangle = 0\,, \tag{5.211}$$

to all orders in perturbation theory.

A second way to recast this result is to consider the simplest analogue of (7.1), the Wess–Zumino model [441–444]. Its on-shell matter content is a scalar multiplet made of a real scalar, a real pseudo-scalar and a Majorana spinor. It is convenient to work in superspace, tagged by the 4-vector x^μ and four anti-commuting complex coordinates θ^a, $\theta^\dagger_{\dot a}$ [443, 444]. The latter are Grassmann variables such that all their anti-commutators vanish: $\{\theta,\theta\} = \{\theta^\dagger,\theta^\dagger\} = \{\theta,\theta^\dagger\} = 0$. Integration over Grassmann coordinates can be defined with a measure $\mathrm{d}^2\theta = -(1/4)\mathrm{d}\theta^a\mathrm{d}\theta^b\epsilon_{ab}$ (and its conjugate; ϵ_{ab} is the two-index Levi-Civita symbol) and obeys some simple rules. For instance, $\int \mathrm{d}^2\theta = 0 = \int \mathrm{d}^2\theta\,\theta$, $\int \mathrm{d}^2\theta\,\theta\theta = 1$, and so on (notice the engineering dimensions $[\theta] = -1/2$, $[\mathrm{d}\theta] = 1/2$). On superspace, one can define superfields such as the scalar $\Phi(x,\theta,\theta^\dagger)$. The Wess–Zumino particle fields can be grouped into a chiral superfield $\Phi(y,\theta)$, which does not depend on $\theta^\dagger$ except through the combination $y^\mu := x^\mu + \mathrm{i}\theta\sigma^\mu\theta^\dagger$. Using the so-called Fierz identities for Grassmann variables (tabulated in, e.g., [445]), one can show that

$$\begin{aligned}\Phi(y,\theta) &= \phi(y) + \sqrt{2}\theta\,\psi(y) + \theta\theta\,F(y)\\ &= \phi(x) + \mathrm{i}\theta\sigma^\mu\theta^\dagger\,\partial_\mu\phi(x) + \frac{1}{4}\theta\theta\theta^\dagger\theta^\dagger\,\Box\phi(x) + \sqrt{2}\theta\,\psi(x)\\ &\quad -\frac{\mathrm{i}}{\sqrt{2}}\theta\theta\,\partial_\mu\psi(x)\sigma^\mu\theta^\dagger + \theta\theta\,F(x)\,,\end{aligned} \tag{5.212}$$

where omitted spinor indices are contracted via ϵ_{ab}, ϕ (often called A in the literature) is a complex field encoding both the proper and pseudo real scalar particles, ψ^a is a complex left-handed Weyl spinor and F is a scalar auxiliary field. The anti-chiral superfield $\Phi^\dagger$ is obtained by Hermitian conjugation. By our choice of units, $[\Phi] = 1 = [\phi]$, $[\psi] = 1/2$ and $[F] = 2$. The particle fields ϕ and ψ have their usual scaling in four dimensions.

The Wess–Zumino action $S_{\rm WZ} = \int \mathrm{d}^4x\,\mathcal{L}_{\rm WZ}$ is invariant under supergauge transformations:

$$\mathcal{L}_{\rm WZ} = \int \mathrm{d}^2\theta\mathrm{d}^2\theta^\dagger\,\Phi^\dagger\Phi - \left[\int \mathrm{d}^2\theta\,W(\Phi) + \mathrm{H.c.}\right], \tag{5.213a}$$

$$W(\Phi) = \frac{1}{2}m\Phi^2 + \frac{1}{3}\lambda\Phi^3\,, \tag{5.213b}$$

where m is the superfield mass, λ is the interaction coupling and "H.c." stands for Hermitian conjugate. The functional $W(\Phi)$ is called *superpotential* and has dimension $[W] = 3$. Then, $[m] = 1$, $[\lambda] = 0$ and $[\mathcal{L}_{\rm WZ}] = 4$ (the Lagrangian is a four-dimensional mass density).

Let us consider only the bosonic part of the model and ignore the fermion ψ in (5.212). Expanding the Lagrangian (5.213) and eliminating the auxiliary field F by its equations of motion, one eventually obtains

$$\mathcal{L}_{\rm WZ} = -\partial_\mu\phi\partial^\mu\phi^* - V(\phi,\phi^*) + \mathcal{L}[\psi,\phi]\,, \tag{5.214a}$$

$$V(\phi,\phi^*) = \left|\frac{\partial W}{\partial \Phi}\right|^2_{\Phi=\phi} = |m\phi + \lambda\phi^2|^2\,. \tag{5.214b}$$

The Wess–Zumino model can be generalized to an arbitrary number of chiral superfields Φ^i with generic holomorphic superpotential $W(\Phi^i)$, independent of the complex conjugate of the superfields, and a generic kinetic term $K(\Phi^i,\Phi^{j\dagger})$ called *Kähler potential*. Indices in the argument of the functionals are omitted from now on:

$$S = \int d^4x\,\mathcal{L}\,,\qquad \mathcal{L} = \int d^2\theta d^2\theta^\dagger\, K(\Phi,\Phi^\dagger) - \left[\int d^2\theta\, W(\Phi) + \text{H.c.}\right]. \tag{5.215}$$

In terms of particle fields, this action gives rise to what is known as a non-linear sigma model [446, 447]:

$$\mathcal{L} = -\mathcal{G}_{ij}\partial_\mu\phi^i\partial^\mu\phi^{j\dagger} - V(\phi,\phi^\dagger)\,,\qquad V(\phi,\phi^\dagger) = \sum_i \left|\frac{\partial W}{\partial \Phi^i}\right|^2_{\Phi=\phi}\,, \tag{5.216}$$

where we ignored Yang–Mills gauge fields, the dimensionless Hermitian matrix

$$\mathcal{G}_{ij} := \left.\frac{\partial^2 K(\Phi,\Phi^\dagger)}{\partial\Phi^i\partial\Phi^{j\dagger}}\right|_{\Phi=\phi} \tag{5.217}$$

is called *Kähler metric* and the associated complex Riemannian manifold is a *Kähler manifold*. The ϕ^i act as coordinates on the Kähler manifold. To summarize, the units we have used in this sub-section are

$$[\Phi] = [\phi] = 1\,,\qquad [\mathcal{G}_{ij}] = 0\,,\qquad [K] = 2\,,\qquad [W] = 3\,,\qquad [V] = 4\,. \tag{5.218}$$

The reader is warned that we will change convention from the end of the next sub-section on.

5.12.2 Supergravity

When gravity is included, supersymmetry becomes a local gauge symmetry [448–450]. In $D = 4$, $\mathcal{N} = 1$ supergravity, instead of (5.215) one has [446, 451–454]

$$\mathcal{L} = -\frac{3}{\kappa^2}\int \mathrm{d}^2\theta \mathrm{d}^2\theta^\dagger\, E \exp\left[-\frac{\kappa^2}{3}K(\Phi, \Phi^\dagger)\right] - \left[\int \mathrm{d}^2\theta\, \tilde{E}\, W(\Phi) + \mathrm{H.c.}\right], \tag{5.219}$$

where E and $\tilde{E}$ are, respectively, the determinant of superspace and the chiral determinant [448, 455–458]. Other choices for the kinetic term are possible [459], but for the sake of the main argument we limit our attention to (5.219). Again, we drop contributions (called D-terms) generated by Yang–Mills gauge fields [453, 460–462]. In spacetime, the bosonic part of the Lagrangian is $\mathcal{L} = \sqrt{-g}\,\mathcal{L}_{\mathrm{SUGRA}}$, where

$$\boxed{\mathcal{L}_{\mathrm{SUGRA}} = \frac{R}{2\kappa^2} - \mathcal{G}_{ij}\partial_\mu\phi^i\partial^\mu\phi^{j\dagger} - V(\phi, \phi^\dagger)\,,} \tag{5.220a}$$

$$V = \mathrm{e}^{\kappa^2 K}\left[\mathcal{G}^{ij}D_iW(D_jW)^\dagger - 3\kappa^2|W|^2\right]_{\Phi=\phi}. \tag{5.220b}$$

The first term in (5.220a) is the Einstein–Hilbert Lagrangian, $\mathcal{G}^{ij}$ is the inverse of the Kähler metric and

$$D_iW := \frac{\partial W}{\partial\Phi^i} + \kappa^2\frac{\partial K}{\partial\Phi^i}W\,. \tag{5.221}$$

The contribution

$$F^2 := \mathrm{e}^{\kappa^2 K}\mathcal{G}^{ij}D_iW(D_jW)^\dagger \tag{5.222}$$

is called F-term. Supersymmetry is locally preserved as long as

$$D_iW = 0\,, \tag{5.223}$$

corresponding to field configurations such that V is stationary.

In Minkowski spacetime and in the limit $\kappa^2 \to 0$ (vanishing G), the expressions (5.219), (5.220a) and (5.220b) reduce to (5.215) and (5.216), which do not depend on Newton's constant. At this point, we change units to make all the rest of the formulæ throughout the book simpler. We rescale $\Phi \to \kappa^{-1}\Phi$, $K \to \kappa^{-2}K$ and $W \to \kappa^{-3}W$, so that

$$[\Phi] = [\mathcal{G}_{ij}] = [W] = [K] = 0\,, \qquad [V] = 4\,, \tag{5.224}$$

while equations (5.220b) and (5.221) are

$$\boxed{\kappa^4 V = \mathrm{e}^K \left[\mathcal{G}^{ij} D_i W (D_j W)^\dagger - 3|W|^2\right]_{\Phi=\phi}\,, \qquad D_i W = \frac{\partial W}{\partial \Phi^i} + \frac{\partial K}{\partial \Phi^i} W.} \tag{5.225}$$

Notice that the Wess–Zumino potential becomes

$$W = \frac{1}{2} M \Phi^2 + \frac{1}{3} \lambda \Phi^3\,, \qquad M := \frac{m}{\kappa}\,. \tag{5.226}$$

5.12.3 *η*-problem

To realize inflation in supergravity, it is not sufficient to reproduce one of the potentials discussed in this chapter. The inflaton mass must be light enough to drive acceleration without fine tuning the parameters of the Kähler and superpotentials. However, this is not always possible due to the *η-problem* [146, 463].

Consider first the single-superfield case. The action (5.215) with (5.225) is invariant under the transformations

$$K(\Phi, \Phi^\dagger) \to K(\Phi, \Phi^\dagger) + f(\Phi) + f^\dagger(\Phi^\dagger)\,, \qquad W(\Phi) \to \mathrm{e}^{f(\Phi)} W(\Phi)\,, \tag{5.227}$$

provided $\int \mathrm{d}^2\theta f^\dagger(\Phi^\dagger) = 0 = \int \mathrm{d}^2\theta^\dagger f(\Phi)$. In particular, the metric (5.217) is determined up to shifts of the Kähler potential by a holomorphic function f. We can make a holomorphic field redefinition $\Phi_{\rm old} \to \Phi = h(\Phi_{\rm old})$ so that $\mathcal{G}^{ij} = \delta^{ij}$ at $\Phi = 0$ (canonical kinetic term) and then cancel the purely holomorphic part of K via (5.227), so that

$$K = |\Phi|^2 + \dots\,, \qquad \mathrm{e}^K = 1 + |\Phi|^2 + \dots\,, \tag{5.228}$$

up to higher-order terms. Unfortunately, the $\exp K \propto \exp|\phi|^2$ term dominates the potential with its steepness and (5.59) is violated. Expanding (5.225) around the origin, one has $V(\phi) = V(0)(1 + |\phi|^2) + \dots$ and the second slow-roll parameter is (as ϕ is dimensionless, there is no κ^{-2} prefactor here)

$$\eta_V = \frac{V_{,\phi\phi^*}}{V} = 1 + \dots\,. \tag{5.229}$$

In order to cancel the $O(1)$ term and obtain sufficient slow-roll, one must fine tune the contributions "…" stemming from $|D_\phi W|^2 - 3|W|^2$ [464–469].

In the absence of further input this is unlikely to happen, especially in the multi-superfield case where there are many independent parameters involved. SUGRA models typically feature a number of scalar fields called *moduli* that we will meet quite often in string theory (Sects. 12.1.4 and 12.3.5). Field redefinitions and the Kähler transformation (5.227) make the metric diagonal at the origin ($\mathcal{G}^{ij}(\Phi^i = 0) = \delta^{ij}$) and the above argument is repeated without changes for each individual field ϕ^i.

The origin of the η-problem can be intuitively traced back to the fact that the inflaton is just one among many moduli. If one assumes (or proves) that all the other moduli are stabilized (i.e., sit at their minimum) when inflation occurs, then their masses will be much larger than the Hubble scale during this epoch, $m_i^2 \gg H^2$. However, any such stabilization mechanism will, in general, stabilize also the modulus we dub as the inflaton. Therefore, as long as we have a dynamical inflaton, either there are also dynamical moduli with $m_i^2 \ll H^2$ (which may lead to the moduli problem of Sect. 13.2.4 and spoil inflation) or, if the moduli are stabilized beforehand, the inflaton has a large mass $m_\varphi^2 = O(H^2)$.

5.12.4 *Inflation in Supergravity*

There are ways out of the η-problem, but with caveats. One is to enforce some symmetry in the SUGRA action that relaxes the fine tuning of the parameters [463]. Another, which we will explore in this sub-section, is to tailor the super- and Kähler potentials to obtain inflationary potentials with a plausible parameter range. A disadvantage of this approach is its ad hoc nature and a lack of justification, from first principles, in the choice of K and W. This issue will be partly addressed in string cosmology (Chap. 13).

A very important class of supergravity scenarios with applications to inflation and to the cosmological constant problem (Sect. 7.1.3) are the *no-scale models* [470–477]. Let $\Phi^i = T, S^s, \Phi^c$ be a set of scalar fields constituting the bosonic sector of a chiral multiplet. The superpotential and Kähler potential we consider are

$$\boxed{W = W_1(\Phi) + W_2(S), \qquad K = -3\ln|T + T^* - h^2(\Phi, \Phi^*)| + \tilde{K}(S, S^*),} \tag{5.230}$$

where h^2 and $\tilde{K}$ are real functions of, respectively, Φ and S (and their complex conjugate). Notice that the superpotential does not depend on T, so that the TT^* component of the Kähler metric $\mathcal{G}$ exactly cancels the second term in (5.225). Then, one can show that the potential (5.225) is

$$V = \frac{e^{\tilde{K}}}{|T+T^*-h^2|^3}\left[\mathcal{G}^{s_1s_2}D_{s_1}W(D_{s_2}W)^* + \frac{\kappa^2}{3}|T+T^*-h^2|f^{c_1c_2}\frac{\partial W_1}{\partial\Phi^{c_1}}\left(\frac{\partial W_1}{\partial\Phi^{c_2}}\right)^*\right], \qquad (5.231)$$

where

$$(f^{c_1c_2})^{-1} = f_{c_1c_2} := \frac{\partial^2 h^2}{\partial(\Phi^{c_1})^*\partial\Phi^{c_2}}\,. \qquad (5.232)$$

The independence of the exponential factor $\exp K$ from $\mathrm{Im}T$ is the key to solve the η-problem when identifying the inflaton with this field. In Sect. 12.3.5, we will see how some (but not all) no-scale models can emerge in string theory.

No-scale models have flat directions and they can provide an embedding for inflation [478–481]. Having seen a class of Kähler potentials for a generic W, we present a non-exhaustive list of some four-dimensional supergravity models with specific inflationary potentials. Some are no-scale models, others are not. We consider the bosonic part of three chiral multiplets, a "matter" scalar Φ and two "moduli" S and T (all dimensionless, while the inflaton $[\varphi]=1$), the first called dilaton and the second volume modulus. This nomenclature will find a rationale in Chap. 12. After a rescaling, the inflaton can be identified either with the matter field or with the volume modulus T. The second case will be analyzed also in Sect. 13.3.2 in the context of string theory. The reason why one usually demands the dilaton to sit at its minimum during inflation, as well as a discussion of scenarios where acceleration is driven precisely by S, can be found in Sect. 13.7.5.

- *Quadratic potential* $V(\varphi)=\frac{1}{2}M^2\varphi^2$, stemming from the two-superfield model $W=mS\Phi$ [482] and [482, 483]

$$K = |S|^2 - \frac{1}{2}(\Phi-\Phi^*)^2 - \frac{\zeta}{3}|S|^4, \qquad (5.233)$$

  where ζ is a constant and $\kappa\varphi=\sqrt{2}\mathrm{Re}\Phi$. Both $\mathrm{Im}\Phi$ and S are sub-dominant with respect to $\varphi\gg m_{\mathrm{Pl}}$ during inflation and their vacuum is stable thanks to the ζ term (see below).
- *Polinomial potential* $V(\varphi)=b_0+b_2\varphi^2+b_3\varphi^3+b_4\varphi^4$, stemming (with $b_0=0$) from the single-superfield Wess–Zumino model (5.226)–(5.214) [484] (but this case has global supersymmetry, since the potential is (5.216) here) or from its two-superfield generalization with $K=c_1(\Phi-\Phi^*)+c_2(\Phi-\Phi^*)^2+|S|^2$ [485]. Then, $\kappa\varphi=\mathrm{Re}\Phi$ and $\mathrm{Im}\Phi=0=S$. The same potential (but with $b_0\neq 0$ and $b_3=0$) arises with (5.233) and $W=\sqrt{b_0}S(1-\mathrm{const}\times\Phi^2)$ [486].

- *Non-minimally coupled inflation*, which includes Higgs inflation and related models with $\mathcal{L} = (\xi/2)\varphi^2 R - (\nabla\varphi)^2/2 - (\lambda/4)\varphi^4$ [487–494] as well as conformally-coupled two-field models [495].
- *Starobinsky inflation* [496–504] is a minimally-coupled single-field model with potential

$$V(\varphi) = \frac{3}{4}M^4\left(1 - e^{-\sqrt{\frac{2}{3}}\kappa\varphi}\right)^2 . \tag{5.234}$$

Computing the slow-roll parameters, one finds that

$$n_s \simeq 1 - \frac{2}{\mathcal{N}_k} \approx 0.967\,, \qquad r \simeq \frac{12}{\mathcal{N}_k^2} \approx 3\times 10^{-3}\,, \tag{5.235}$$

where we calculated the observables at $\mathcal{N}_k = 60$ e-foldings. After the release of the first PLANCK data, Starobinsky inflation has been among the most favoured models of early acceleration, due to the low tensor-to-scalar ratio well below the upper bounds reported in Sect. 4.4.2.

A realization of the Starobinsky potential has the Wess–Zumino superpotential (5.226) with [505, 506]

$$K = -3\ln\left(T + T^* - \frac{|\Phi|^2}{3}\right) . \tag{5.236}$$

After a certain field redefinition $\chi = \chi(\Phi)$ and taking the vacuum expectations values $\langle \mathrm{Im}T\rangle = 0$, $\langle \mathrm{Re}T\rangle = m^2/(6\lambda^2) \neq 0$ and $\langle \mathrm{Im}\chi\rangle = 0$ (at the minimum of the potential corrected by non-perturbative effects), the inflaton is identified with $\kappa\varphi = \mathrm{Re}\chi$. Stability of the minimum along the non-inflaton T-direction requires to break the no-scale structure of (5.236) by adding terms which we will see below.

Another two-superfield model corresponding to Starobinsky inflation has [507–509]

$$W = 3mS(T-1)\,, \qquad K = -3\ln\left(T + T^* - \frac{1}{3}|S|^2 + \frac{\zeta}{3}|S|^4\right) . \tag{5.237}$$

The modulus S sits at the minimum $\langle S\rangle = 0$ and the resulting potential is (5.234), where $\kappa\varphi = \sqrt{2}\mathrm{Re}T$. If $\zeta = 0$ [507] the S direction has a tachyonic instability.

The Kähler potentials in (5.236) and (5.237) are no-scale SUGRA scenarios. Other no-scale models with different Kähler potentials, and where both the matter and volume-modulus field can play the role of the inflaton, are presented in [506, 510–513], where supersymmetry breaking in no-scale inflationary scenarios is also discussed. On the other hand, non–no-scale SUGRA models of Starobinsky inflation and an implementation in the SUGRA version of $f(R)$ theories (Sect. 7.5.2) can be found, respectively, in [509, 514, 515] and [516, 517].

Notice, however, that higher-order corrections (in the field strengths) to the SUGRA Lagrangian can reintroduce the η-problem [515].

A different origin of Starobinsky inflation, from a quadratic Ricci-curvature term, will be given in Sect. 7.5.4, while a string-theory realization can be found in Sect. 13.4.5.

- *Goncharov–Linde inflation* [518, 519] is based on the potential

$$V(\varphi) = \frac{M^4}{9}\left[4 - \tanh^2\left(\sqrt{\frac{3}{2}}\kappa\varphi\right)\right]\tanh^2\left(\sqrt{\frac{3}{2}}\kappa\varphi\right) \tag{5.238}$$

and it generates the cosmological observables

$$n_s \simeq 1 - \frac{2}{\mathcal{N}_k} \approx 0.967\,, \qquad r \simeq \frac{4}{3\mathcal{N}_k^2} \approx 4\times 10^{-4} \tag{5.239}$$

for $\mathcal{N}_k = 60$ e-foldings. The tensor-to-scalar is even smaller than Starobinsky's. This scenario was originally formulated from the single-superfield model [518, 519]

$$W = \frac{m^2}{3}\frac{\sinh^2(\sqrt{3}\Phi)}{\cosh(\sqrt{3}\Phi)}\,, \qquad K = -\frac{1}{2}(\Phi - \Phi^*)^2\,, \tag{5.240}$$

where $\varphi = \mathrm{Re}[\chi(\Phi)]$ for some suitable field redefinition of Φ. Equation (5.238) can be recovered also from [520]

$$W = \frac{m}{3\sqrt{3}}Z^2(1 - Z^2)\,, \qquad K = -3\ln\left[1 - |Z|^2 - \frac{4}{9}\frac{(Z - Z^*)^2}{1 - |Z|^2}\right], \tag{5.241}$$

where

$$Z := \frac{T - 1}{T + 1}\,. \tag{5.242}$$

Kähler potentials of the form $K = -3\ln(1 - |Z|^2 - |h|^2)$, where $h = h(Z, S, \Phi)$ is a generic term, are of no-scale type if $h = (T + 1)^{-1}A(S, \Phi)$ for an arbitrary function A. In fact, from (5.242) and after a Kähler transformation (5.227) with $f(T) = -3\ln[(T + 1)/\sqrt{2}]$, one has

$$K = -3\ln(1 - |Z|^2 - |h|^2) \to -3\ln\left(T + T^* - |h'|^2\right)\,, \qquad h' := \frac{(T + 1)h}{\sqrt{2}}\,; \tag{5.243}$$

we recover (5.230) if $\partial_T h' = 0$.

It is also possible to reproduce any potential $V(\varphi)$ or certain parametric classes $V_{(\alpha)}(\varphi)$. The above potentials are thus recovered in a more general setting:

- *Generic potentials* $V(\varphi)$ [479, 483, 521, 522]. For an arbitrary holomorphic function $f(\Phi)$, one takes the superpotential

$$W = Sf(\Phi) \tag{5.244}$$

and either the Kähler potential (5.233) or

$$K = -3\ln\left[1 - \frac{1}{3}|S|^2 + \frac{1}{6}(\Phi - \Phi^*)^2 + \frac{\zeta}{3}|S|^4\right]. \tag{5.245}$$

In both cases, freezing both S and $\mathrm{Im}\Phi$ at their minimum where they vanish, the potential for the canonically normalized inflaton $\kappa\varphi = \sqrt{2}\mathrm{Re}\Phi$ takes the form

$$V(\varphi) = \kappa^4 \left| f\left(\frac{\kappa\varphi}{\sqrt{2}}\right)\right|^2. \tag{5.246}$$

This is a dynamical attractor, since the trajectory $S = 0 = \mathrm{Im}\Phi$ is stable thanks to the higher-order term $|S|^4$. In fact, the mass of S in the vacuum trajectory can be shown to be $m_S^2 = V_{,SS}|_{S=0=\mathrm{Im}\Phi} = 4\zeta H^2 + f_{,\varphi}^2$ for (5.233) and $m_S^2 = 2(6\zeta - 1)H^2 + f_{,\varphi}^2$ for (5.245). In order for $S = 0$ to be a stable minimum, one needs $m_S^2 > 0$, which is achieved for any f when $\zeta \geqslant 0, 1/6$, respectively. Moreover, to avoid that S excitations spoil inflation, the mass must be greater than the Hubble energy, which is guaranteed if $\zeta > 1/4$.

  Generic non-minimally coupled models are obtained for a different choice of K.

- *α-attractors* [520, 523–532]. The members of this class of models are called α-attractors because they are all comprised between a finite vertical line in the (n_s, r) plane parametrized by $\alpha > 0$ and because their properties for $\alpha \lesssim O(1)$ are not very sensitive to the shape of the potential, which can be of the form $V_{(\alpha)}(\varphi) \propto \mathcal{F}^2[\tanh(\kappa\varphi/\sqrt{6\alpha})]$ for a wide range of choices of $\mathcal{F}$. Here we take $\mathcal{F}(Z) = Z$.

  The simplest α-attractors are the so-called T-models [483]. The total Lagrangian is $\mathcal{L} = R/(2\kappa^2) + \mathcal{L}_\alpha$, where the Einstein–Hilbert Lagrangian is augmented by the contribution of a non-canonical real-valued scalar field with a quadratic potential (here ϕ and m are dimensionless):

$$\kappa^4\mathcal{L}_\alpha = -\frac{\kappa^2}{2}\frac{\partial_\mu\phi\partial^\mu\phi}{1 - \phi^2/(6\alpha)} - \frac{m^2}{2}\phi^2, \tag{5.247}$$

where $\alpha > 0$. With the field redefinition $\phi = \sqrt{6\alpha}\tanh(\kappa\varphi/\sqrt{6\alpha})$, we can transform (5.247) into the canonical Lagrangian

$$\mathcal{L}_\alpha = -\frac{1}{2}\partial_\mu\varphi\partial^\mu\varphi - 3\alpha M^4\left(\tanh\frac{\kappa\varphi}{\sqrt{6\alpha}}\right)^2, \tag{5.248}$$

where $M^2 = m/\kappa^2$. The inflationary observables are [524]

$$n_s \simeq 1 - \frac{2}{\mathcal{N}_k}, \qquad r \simeq \frac{12\alpha}{\mathcal{N}_k(\mathcal{N}_k + 3\alpha/2)}. \tag{5.249}$$

Just like the scalar spectral index, the scalar spectral amplitude does not depend on α and, for $\mathcal{N}_k = 60$, it agrees with observations when $m = O(10^{-5})$.

The T-models (5.247)–(5.248) can be derived from specific super- and Kähler potentials. Up to stabilization terms we will ignore, two possibilities are [524, 526]

$$W = \sqrt{3\alpha}\, m\, S\, Z(1 - Z^2)^{\frac{3\alpha-1}{2}}, \tag{5.250}$$

$$K = -3\alpha \ln\left(1 - |Z|^2 - \frac{1}{3\alpha}|S|^2\right), \tag{5.251}$$

and [495]

$$W = \sqrt{3\alpha}\, m\, S\, Z(1 - Z^2), \tag{5.252}$$

$$K = -3\ln\left[1 - |Z|^2 + \frac{\alpha-1}{2}\frac{(Z - Z^*)^2}{1 - |Z|^2} - \frac{1}{3}|S|^2\right], \tag{5.253}$$

where Z is given by (5.242). The potential $V(S, Z)$, calculated from (5.225), has a minimum at $S = 0 = Z$. After stabilizing the field S at $S = 0$, or taking a nilpotent field $S^2 = 0$, the potential (5.225) is the one in (5.247) with $\mathrm{Re}Z = \phi/\sqrt{6\alpha}$. The geometric interpretation of $\alpha \propto |R_K|^{-1}$ is that it is the inverse of the negative curvature R_K of the Kähler manifold. The case $\alpha \to +\infty$ of a flat Kähler manifold is also meaningful [528].

For special values of α, one recovers the predictions of several of the earliest models of inflation. For $\alpha \gtrsim 10^3$ (and exactly, in the limit $\alpha \to +\infty$), this is the usual model of quadratic inflation ((5.197) with $n = 2$ and for large $\mathcal{N}_k$) in tension with PLANCK data, at the border with the 2σ-level region in the (n_s, r) plane. The 1σ region is entered for $\alpha \lesssim 40$. For $\alpha = 1$ and $\alpha = 1/9$, one hits the point in the (n_s, r) plane corresponding, respectively, to (5.235) (Starobinsky inflation; compare (5.251) and (5.253) for $\alpha = 1$ with (5.237) and (5.243)) and (5.239) (Goncharov–Linde inflation). Starobinsky's potential is recovered exactly with $W = S\mathcal{F}(Z)(1 - Z^2)$ for $\mathcal{F}(Z) \propto Z/(1 + Z)$.

Since $V(S = 0, Z = 0) = 0$ at the minimum, these models preserve supersymmetry. Supersymmetry can be broken and the vanishing minimum can be lifted to some small value by introducing new parameters in the superpotential W [495]. This can be done in such ways that either the inflaton potential is modified but the cosmological predictions are not (for $\alpha \sim 1$ and $\alpha \gg 1$), or the form of the inflaton potential remains the same up to an additive constant, $V(\varphi) \to V(\varphi)+V_0$ [495]. In either case, α-attractors with broken supersymmetry on a de Sitter vacuum can realize models of quintessential inflation (Sect. 7.3.6).

Another class of α-attractors (called E-models) consists in exponential models such that [525]

$$V(\varphi) = 3\alpha M^4 \left(1 - \mathrm{e}^{-\sqrt{\frac{2}{3\alpha}}\kappa\varphi}\right)^2 . \tag{5.254}$$

$\alpha = 1$ is precisely Starobinsky's model (5.234). The potential (5.254) stems from the super- and Kähler potential

$$W = \alpha m^2 \sqrt{6T}\,(1 + \sqrt{3}\,S)(T \ln T - T + 1), \tag{5.255}$$

$$K = -3 \ln \left(T + T^* + \frac{\alpha - 1}{2} \frac{(T - T^*)^2}{T + T^*}\right) + |S|^2 . \tag{5.256}$$

Again, one can deform the superpotential to break supersymmetry on a de Sitter vacuum. More two-field α-attractors coming from different combinations of Kähler and superpotentials can be found in [520].

Two classes of single-field models can be devised for

$$K = -3\alpha \ln(T + T^*) \tag{5.257}$$

and for a suitable α-dependent superpotential $W(T)$ which gives rise either to attractors for $\alpha > 1$ [528] or to attractors for $\alpha < 1$ [529]. After the imaginary part of the volume modulus is settled in the minimum $\mathrm{Im}T = 0$ and for $T = \exp[-\sqrt{2/(3\alpha)}\kappa\varphi] \ll 1$, the potential at large φ reads

$$V(\varphi) \simeq V_0 - V_1\, \mathrm{e}^{-\sqrt{\frac{2}{3\alpha}}\kappa\varphi} . \tag{5.258}$$

For $\alpha = 1/9$, the α-attractors of [529] exactly reproduce the Goncharov–Linde superpotential in (5.241).

For simplicity, we have presented single-field SUGRA potentials, but one can generalize to multi-field situations where not all non-inflatonic degrees of freedom have been stabilized in advance. In a no-scale model partially motivated by string theory, it was checked that non-Gaussianity can be low enough to appease observations [511, 512].

Finally, we mention the case where the main contribution to the inflaton energy density is not the F-term (5.222) but a D-term, $V = F^2 - 3\mathrm{e}^K|W|^2 \to F^2 - 3\mathrm{e}^K|W|^2 + D^2$ [533–537]. This contribution, which we have ignored so far, is generated by a gauge symmetry and is of the form $D^2 \propto \mathrm{Re}(f_{ab}^{-1})D^aD^b$, where f_{ab} is the gauge kinetic function, $D^a = \Phi^i(t^a)_i^{\ j}\partial_j K + \xi^a$, t^a are the generators of the gauge group and ξ^a are constants which are non-zero only for $U(1)$. If the D-term dominates over the F-term responsible for the η-problem, it is possible to achieve the slow-roll condition $\eta_V \ll 1$ without fine tuning. We do not delve into D-term inflation here as we will dedicate most of Chap. 13 to related string scenarios.

5.13 Problems and Solutions

5.1 Observable inflation. Determine how many e-foldings before the end of inflation the modes $k_0 = 0.002\,\mathrm{Mpc}^{-1}$, $k_0' = 0.05\,\mathrm{Mpc}^{-1}$ (pivot scales in observations) and $k_{\rm dec} = 0.02\,\mathrm{Mpc}^{-1}$ (size of the horizon at decoupling, equation (4.116)), crossed the horizon. Assume instantaneous reheating at the GUT scale, $\rho_{\rm e}^{1/4} \sim T_{\rm reh} \sim T_{\rm GUT}$.

Solution With instantaneous reheating at the GUT scale,

$$\mathcal{N}_k = \mathcal{N}_0 - \ln(k\tau_0) \simeq 56 - \ln(k\tau_0)\,. \qquad (5.259)$$

The pivot scale $k_0 = 0.002\,\mathrm{Mpc}^{-1}$ crossed the horizon $-\ln(k_0\tau_0) \approx 3.6 \sim 4$ e-foldings before $k_{\rm hor}$, so that $\mathcal{N}_{k=k_0} \approx 60$, while the mode $k_0' = 0.05\,\mathrm{Mpc}^{-1}$ crossed the horizon less than half e-folding before $k_{\rm hor}$. For the scale corresponding to the Hubble radius at decoupling, $-\ln(k_{\rm dec}\tau_0) \approx 1.3 \sim 1$ and $\mathcal{N}_{k=k_{\rm dec}} \approx 57$.

5.2 Inflationary scalar spectrum. Recover the power spectrum (5.152) from (5.188) in the quasi-de Sitter approximation. Use synchronous gauge, $N = 1$. The linear comoving curvature perturbation $\mathcal{R}$ at large scales is obtained by integrating $\mu/(3Hz)$ from the initial time $t_{\rm i}$ to t and then sending both extrema to infinity. The asymptotic-past limit $t_{\rm i} \to -\infty$ sets the initial conditions, while the asymptotic-future limit $t \to +\infty$ corresponds to the large-scale regime $k\tau \ll 1$. Use a Gaussian window function

$$\mathcal{W}(k) = \exp(-k^2R^2/2)\,. \qquad (5.260)$$

Solution The calculation is rather simple in the quasi-de Sitter (ESR) regime, where $a \simeq \exp(Ht)$ and both H and ϵ are constant; then, by virtue of (5.43a), $z = a\sqrt{2\epsilon/\kappa^2}$ and ϵ can be taken outside the time integral. Also, at large scales the scalar mode u_k

is given by (5.126), with $\nu = 0$ in quasi-de Sitter:

$$u_k \simeq \frac{aH}{\sqrt{2k^3}}, \tag{5.261}$$

so that

$$\begin{aligned}\mathcal{R} &\simeq -\frac{1}{3H} \lim_{(t_i,t)\to(-\infty,+\infty)} \int_{t_i}^{t} dt' \frac{\mu}{z} \\ &\overset{(5.261)}{\simeq} -\frac{1}{3}\sqrt{\frac{\kappa^2}{2\epsilon}} \int \frac{d^3\mathbf{k}}{(2\pi)^3} \frac{e^{i\mathbf{k}\cdot\mathbf{x}}}{\sqrt{2k^3}} \beta_{\mathbf{k}} \int_{-\infty}^{+\infty} dt\, \frac{1}{a}[\ddot{\mathcal{W}} + \dot{\mathcal{W}}(2\partial_t + H)]a + \text{c.c.}\,.\end{aligned}$$

From the window function (5.260), $\dot{\mathcal{W}} \simeq (kR)^2 H\mathcal{W}$ and $\ddot{\mathcal{W}} \simeq [(kR)^2 - 2](kR)^2 H^2 \mathcal{W}$; since $\dot{a} = Ha$, the integral is

$$H^2 \int_{-\infty}^{+\infty} dt\,[(kR)^2 + 1](kR)^2 \mathcal{W} = H\mathcal{W}[(kR)^2 + 3]\Big|_{-\infty}^{+\infty} = 3H\,,$$

where we used $R(t) = (\varepsilon H)^{-1}\exp(-Ht)$. Therefore,

$$\mathcal{R}(\mathbf{x}) = -\sqrt{\frac{\kappa^2 H^2}{2\epsilon}} \int \frac{d^3\mathbf{k}}{(2\pi)^3} \frac{e^{i\mathbf{k}\cdot\mathbf{x}}}{\sqrt{2k^3}} \beta_{\mathbf{k}} + \text{c.c.}\,. \tag{5.262}$$

By definition, the power spectrum is $\mathcal{P}_s = k^3 P_s/(2\pi^2)$, where (Problem 3.5)

$$\langle|\mathcal{R}(\mathbf{x})|^2\rangle =: \int \frac{d^3\mathbf{k}}{(2\pi)^3} P_s(k)\,.$$

Using the stochastic relation (5.185), one obtains the scalar spectrum (5.152).

5.3 Patch cosmology. Consider a modified effective Friedmann equation of the form

$$H^{2-\theta} = \beta_\theta \rho\,, \tag{5.263}$$

where $-1 \leqslant \theta \leqslant 1$ is a constant and $\beta_\theta > 0$ is a coupling with energy dimension $[\beta_\theta] = -2-\theta$. Assume a standard continuity equation and a nearly homogeneous scalar field as matter content, with power-law potential (5.81). Calculate the number of e-foldings, the scalar spectrum, its index and the index running for a test field on de Sitter (lowest-order slow-roll expressions).

(continued)

Express the scalar index as a function of the number of e-foldings, as done in Sect. 5.5.1.2. Do the inflationary predictions change much with respect to the standard case $\theta = 0$?

Sketch of solution We follow [538–540], where the reader can find more details. As functions of the slow-roll parameter, the scalar observables read

$$\mathcal{P}_s = \frac{6\beta_\theta}{(2-\theta)25\pi^2}\frac{H^{2+\theta}}{2\epsilon}, \tag{5.264a}$$

$$n_s - 1 = (2\eta - 4\epsilon) + O(\epsilon^2), \tag{5.264b}$$

$$\alpha_s = 2\left[2(\theta-2)\,\epsilon^2 + 5\epsilon\eta - \xi^2\right]. \tag{5.264c}$$

For the scalar potential (5.81), we have in particular

$$n_s - 1 = -\frac{\sigma_n^{1-q}}{6(n\beta_\theta)^q}\frac{n(6q-4)+4}{\phi^{2+(q-1)n}}, \tag{5.265}$$

where $q := 2/(2-\theta)$. Setting $\epsilon = 1$ at the end of inflation, one gets $\phi_e^{n(q-1)+2} \simeq qn^{3-q}/(6\beta_\theta^q\sigma_n^{q-1})$ and the number of e-foldings (5.85e) is modified as

$$\mathcal{N}_e = \frac{3\beta_\theta^q\sigma_n^{q-1}}{n^q[n(q-1)+2]}\phi^{2+n(q-1)} - \frac{qn}{2[n(q-1)+2]}, \tag{5.266}$$

which is valid for $n \neq 2/(1-q)$. Then, (5.265) reads

$$n_s - 1 = -\frac{2n(3q-2)+4}{2\mathcal{N}_e(nq-n+2)+nq} = \frac{2(n+2)+2\theta(n-1)}{\mathcal{N}_e(4-2\theta+n\theta)+n}. \tag{5.267}$$

For a quartic potential ($n = 4$), we have $n_s - 1 = 3/[\mathcal{N}_e + 2/(2+\theta)]$. Since the correction $2/3 \leqslant 2/(2+\theta) \leqslant 2$ is much smaller than the number of e-foldings $\mathcal{N}_e \approx 60$, the inflationary predictions do not change much.

The effective Friedmann equation (5.263) arise in high-energy regimes of certain braneworld inflationary scenarios which we shall revisit in Sect. 13.7.1.

References

1. H. Georgi, S.L. Glashow, Unity of all elementary particle forces. Phys. Rev. Lett. **32**, 438 (1974)
2. H. Georgi, H.R. Quinn, S. Weinberg, Hierarchy of interactions in unified gauge theories. Phys. Rev. Lett. **33**, 451 (1974)
3. H. Georgi, The state of the art—Gauge theories. AIP Conf. Proc. **23**, 575 (1975)

4. H. Fritzsch, P. Minkowski, Unified interactions of leptons and hadrons. Ann. Phys. (N.Y.) **93**, 193 (1975)
5. A.J. Buras, J.R. Ellis, M.K. Gaillard, D.V. Nanopoulos, Aspects of the grand unification of strong, weak and electromagnetic interactions. Nucl. Phys. B **135**, 66 (1978)
6. G. Lazarides, Q. Shafi, C. Wetterich, Proton lifetime and fermion masses in an $SO(10)$ model. Nucl. Phys. B **181**, 287 (1981)
7. J.R. Ellis, D.V. Nanopoulos, S. Rudaz, A phenomenological comparison of conventional and supersymmetric GUTs. Nucl. Phys. B **202**, 43 (1982)
8. Y. Hayato et al. [Super-Kamiokande Collaboration], Search for proton decay through $p \rightarrow \bar{\nu}K^+$ in a large water Cherenkov detector. Phys. Rev. Lett. **83**, 1529 (1999). [arXiv:hep-ex/9904020]
9. H. Murayama, A. Pierce, Not even decoupling can save minimal supersymmetric $SU(5)$. Phys. Rev. D **65**, 055009 (2002). [arXiv:hep-ph/0108104]
10. B. Bajc, P. Fileviez Perez, G. Senjanović, Proton decay in minimal supersymmetric $SU(5)$. Phys. Rev. D **66**, 075005 (2002). [arXiv:hep-ph/0204311]
11. K.S. Babu, J.C. Pati, F. Wilczek, Fermion masses, neutrino oscillations, and proton decay in the light of SuperKamiokande. Nucl. Phys. B **566**, 33 (2000). [arXiv:hep-ph/9812538]
12. B. Dutta, Y. Mimura, R.N. Mohapatra, Neutrino mixing predictions of a minimal $SO(10)$ model with suppressed proton decay. Phys. Rev. D **72**, 075009 (2005). [arXiv:hep-ph/0507319]
13. S. Dimopoulos, H. Georgi, Softly broken supersymmetry and SU(5). Nucl. Phys. B **193**, 150 (1981)
14. S. Dimopoulos, S. Raby, F. Wilczek, Supersymmetry and the scale of unification. Phys. Rev. D **24**, 1681 (1981)
15. N. Sakai, Naturalnes in supersymmetric GUTS. Z. Phys. C **11**, 153 (1981)
16. W.J. Marciano, G. Senjanović, Predictions of supersymmetric grand unified theories. Phys. Rev. D **25**, 3092 (1982)
17. M.B. Einhorn, D.R.T. Jones, The weak mixing angle and unification mass in supersymmetric SU(5). Nucl. Phys. B **196**, 475 (1982)
18. L.E. Ibáñez, G.G. Ross, Low-energy predictions in supersymmetric grand unified theories. Phys. Lett. B **105**, 439 (1981)
19. T.W.B. Kibble, Some implications of a cosmological phase transition. Phys. Rep. **67**, 183 (1980)
20. M.B. Hindmarsh, T.W.B. Kibble, Cosmic strings. Rep. Prog. Phys. **58**, 477 (1995). [arXiv:hep-ph/9411342]
21. A. Vilenkin, E.P.S. Shellard, *Cosmic Strings and Other Topological Defects* (Cambridge University Press, Cambridge, 2000)
22. D. Tong, TASI lectures on solitons. arXiv:hep-th/0509216
23. R.J. Danos, R.H. Brandenberger, G. Holder, Signature of cosmic strings wakes in the CMB polarization. Phys. Rev. D **82**, 023513 (2010). [arXiv:1003.0905]
24. N. Bevis, M. Hindmarsh, M. Kunz, J. Urrestilla, CMB power spectra from cosmic strings: predictions for the Planck satellite and beyond. Phys. Rev. D **82**, 065004 (2010). [arXiv:1005.2663]
25. Ya.B. Zel'dovich, M.Y. Khlopov, On the concentration of relic magnetic monopoles in the universe. Phys. Lett. B **79**, 239 (1978)
26. J. Preskill, Cosmological production of superheavy magnetic monopoles. Phys. Rev. Lett. **43**, 1365 (1979)
27. J. Preskill, Magnetic monopoles. Ann. Rev. Nucl. Part. Sci. **34**, 461 (1984)
28. V. Mukhanov, *Physical Foundations of Cosmology* (Cambridge University Press, Cambridge, 2005)
29. S. Weinberg, *Cosmology* (Oxford University Press, Oxford, 2008)
30. K. Sato, Cosmological baryon-number domain structure and the first order phase transition of a vacuum. Phys. Lett. B **99**, 66 (1981)

31. D. Kazanas, Dynamics of the universe and spontaneous symmetry breaking. Astrophys. J. **241**, L59 (1980)
32. A.H. Guth, Inflationary universe: a possible solution to the horizon and flatness problems. Phys. Rev. D **23**, 347 (1981)
33. K. Sato, First-order phase transition of a vacuum and the expansion of the universe. Mon. Not. R. Astron. Soc. **195**, 467 (1981)
34. A.R. Liddle, P. Parsons, J.D. Barrow, Formalising the slow-roll approximation in inflation. Phys. Rev. D **50**, 7222 (1994). [arXiv:astro-ph/9408015]
35. A.D. Linde, A new inflationary universe scenario: a possible solution of the horizon, flatness, homogeneity, isotropy and primordial monopole problems. Phys. Lett. B **108**, 389 (1982)
36. A. Albrecht, P.J. Steinhardt, Cosmology for grand unified theories with radiatively induced symmetry breaking. Phys. Rev. Lett. **48**, 1220 (1982)
37. J.M. Bardeen, P.J. Steinhardt, M.S. Turner, Spontaneous creation of almost scale-free density perturbations in an inflationary universe. Phys. Rev. D **28**, 679 (1983)
38. A.A. Starobinsky, Dynamics of phase transition in the new inflationary universe scenario and generation of perturbations. Phys. Lett. B **117**, 175 (1982)
39. S.W. Hawking, The development of irregularities in a single bubble inflationary universe. Phys. Lett. B **115**, 295 (1982)
40. A.D. Linde, Scalar field fluctuations in expanding universe and the new inflationary universe scenario. Phys. Lett. B **116**, 335 (1982)
41. A.H. Guth, S.-Y. Pi, Fluctuations in the new inflationary universe. Phys. Rev. Lett. **49**, 1110 (1982)
42. A.D. Linde, Chaotic inflation. Phys. Lett. B **129**, 177 (1983)
43. P.A.R. Ade et al. [Planck Collaboration], Planck 2015 results. XIII. Cosmological parameters. Astron. Astrophys. **594**, A13 (2016). [arXiv:1502.01589]
44. A.D. Linde, Can we have inflation with $\Omega > 1$? JCAP **0305**, 002 (2003). [arXiv:astro-ph/0303245]
45. S. del Campo, R. Herrera, Extended closed inflationary universes. Class. Quantum Grav. **22**, 2687 (2005). [arXiv:gr-qc/0505084]
46. M. Kamionkowski, D.N. Spergel, N. Sugiyama, Small scale cosmic microwave background anisotropies as a probe of the geometry of the universe. Astrophys. J. **426**, L57 (1994). [arXiv:astro-ph/9401003]
47. J.R. Gott, Creation of open universes from de Sitter space. Nature **295**, 304 (1982)
48. D.H. Lyth, E.D. Stewart, Inflationary density perturbations with $\Omega < 1$. Phys. Lett. B **252**, 336 (1990)
49. B. Ratra, P.J.E. Peebles, CDM cosmogony in an open universe. Astrophys. J. **432**, L5 (1994)
50. A. Kashlinsky, I.I. Tkachev, J. Frieman, Microwave background anisotropy in low Ω_0 inflationary models and the scale of homogeneity in the universe. Phys. Rev. Lett. **73**, 1582 (1994). [arXiv:astro-ph/9405024]
51. N. Sugiyama, J. Silk, The imprint of Ω on the cosmic microwave background. Phys. Rev. Lett. **73**, 509 (1994). [arXiv:astro-ph/9406026]
52. M. Kamionkowski, B. Ratra, D.N. Spergel, N. Sugiyama, CBR anisotropy in an open inflation, CDM cosmogony. Astrophys. J. **434**, L1 (1994). [arXiv:astro-ph/9406069]
53. B. Ratra, P.J.E. Peebles, Inflation in an open universe. Phys. Rev. D **52**, 1837 (1995)
54. M. Bucher, A.S. Goldhaber, N. Turok, An open universe from inflation. Phys. Rev. D **52**, 3314 (1995). [arXiv:hep-ph/9411206]
55. K. Yamamoto, M. Sasaki, T. Tanaka, Large angle CMB anisotropy in an open universe in the one bubble inflationary scenario. Astrophys. J. **455**, 412 (1995). [arXiv:astro-ph/9501109]
56. A.D. Linde, Inflation with variable Ω. Phys. Lett. B **351**, 99 (1995). [arXiv:hep-th/9503097]
57. A.D. Linde, Toy model for open inflation. Phys. Rev. D **59**, 023503 (1999). [arXiv:hep-ph/9807493]
58. A.D. Linde, M. Sasaki, T. Tanaka, CMB in open inflation. Phys. Rev. D **59**, 123522 (1999). [arXiv:astro-ph/9901135]

59. S. del Campo, R. Herrera, Extended open inflationary universes. Phys. Rev. D **67**, 063507 (2003). [arXiv:gr-qc/0303024]
60. B. Freivogel, M. Kleban, M. Rodríguez Martínez, L. Susskind, Observational consequences of a landscape. JHEP **0603**, 039 (2006). [arXiv:hep-th/0505232]
61. D. Yamauchi, A. Linde, A. Naruko, M. Sasaki, T. Tanaka, Open inflation in the landscape. Phys. Rev. D **84**, 043513 (2011). [arXiv:1105.2674]
62. M. Joyce, Electroweak baryogenesis and the expansion rate of the universe. Phys. Rev. D **55**, 1875 (1997). [arXiv:hep-ph/9606223]
63. J.H. Kung, R.H. Brandenberger, Chaotic inflation as an attractor in initial-condition space. Phys. Rev. D **42**, 1008 (1990)
64. A.D. Dolgov, A.D. Linde, Baryon asymmetry in the inflationary universe. Phys. Lett. B **116**, 329 (1982)
65. L.F. Abbott, E. Fahri, M.B. Wise, Particle production in the new inflationary cosmology. Phys. Lett. B **117**, 29 (1982)
66. A.D. Dolgov, D.P. Kirilova, Production of particles by a variable scalar field. Sov. J. Nucl. Phys. **51**, 172 (1990)
67. J.H. Traschen, R.H. Brandenberger, Particle production during out-of-equilibrium phase transitions. Phys. Rev. D **42**, 2491 (1990)
68. L. Kofman, A.D. Linde, A.A. Starobinsky, Reheating after inflation. Phys. Rev. Lett. **73**, 3195 (1994). [arXiv:hep-th/9405187]
69. Y. Shtanov, J.H. Traschen, R.H. Brandenberger, Universe reheating after inflation. Phys. Rev. D **51**, 5438 (1995). [arXiv:hep-ph/9407247]
70. D. Boyanovsky, H.J. de Vega, R. Holman, D.S. Lee, A. Singh, Dissipation via particle production in scalar field theories. Phys. Rev. D **51**, 4419 (1995). [arXiv:hep-ph/9408214]
71. D.I. Kaiser, Post inflation reheating in an expanding universe. Phys. Rev. D **53**, 1776 (1996). [arXiv:astro-ph/9507108]
72. D. Boyanovsky, M. D'Attanasio, H.J. de Vega, R. Holman, D.S. Lee, Linear versus nonlinear relaxation: consequences for reheating and thermalization. Phys. Rev. D **52**, 6805 (1995). [arXiv:hep-ph/9507414]
73. S.Yu. Khlebnikov, I.I. Tkachev, Classical decay of inflaton. Phys. Rev. Lett. **77**, 219 (1996). [arXiv:hep-ph/9603378]
74. S.Yu. Khlebnikov, I.I. Tkachev, The universe after inflation: the wide resonance case. Phys. Lett. B **390**, 80 (1997). [arXiv:hep-ph/9608458]
75. T. Prokopec, T.G. Roos, Lattice study of classical inflaton decay. Phys. Rev. D **55**, 3768 (1997). [arXiv:hep-ph/9610400]
76. S.Yu. Khlebnikov, I.I. Tkachev, Resonant decay of Bose condensates. Phys. Rev. Lett. **79**, 1607 (1997). [arXiv:hep-ph/9610477]
77. D. Boyanovsky, D. Cormier, H.J. de Vega, R. Holman, A. Singh, M. Srednicki, Scalar field dynamics in Friedman–Robertson–Walker spacetimes. Phys. Rev. D **56**, 1939 (1997). [arXiv:hep-ph/9703327]
78. L. Kofman, A.D. Linde, A.A. Starobinsky, Towards the theory of reheating after inflation. Phys. Rev. D **56**, 3258 (1997). [arXiv:hep-ph/9704452]
79. B.A. Bassett, S. Liberati, Geometric reheating after inflation. Phys. Rev. D **58**, 021302(R) (1998); Erratum-ibid. **60**, 049902(E) (1999). [arXiv:hep-ph/9709417]
80. B.A. Bassett, S. Tsujikawa, D. Wands, Inflation dynamics and reheating. Rev. Mod. Phys. **78**, 537 (2006). [arXiv:astro-ph/0507632]
81. P.B. Greene, L. Kofman, A.D. Linde, A.A. Starobinsky, Structure of resonance in preheating after inflation. Phys. Rev. D **56**, 6175 (1997). [arXiv:hep-ph/9705347]
82. B.A. Bassett, D.I. Kaiser, R. Maartens, General relativistic preheating after inflation. Phys. Lett. B **455**, 84 (1999). [arXiv:hep-ph/9808404]
83. G.N. Felder, L. Kofman, A.D. Linde, Instant preheating. Phys. Rev. D **59**, 123523 (1999). [arXiv:hep-ph/9812289]
84. M. Kawasaki, K. Kohri, T. Moroi, Big-bang nucleosynthesis and hadronic decay of long-lived massive particles. Phys. Rev. D **71**, 083502 (2005). [arXiv:astro-ph/0408426]

85. K. Kohri, T. Moroi, A. Yotsuyanagi, Big-bang nucleosynthesis with unstable gravitino and upper bound on the reheating temperature. Phys. Rev. D **73**, 123511 (2006). [arXiv:hep-ph/0507245]
86. A.R. Liddle, S.M. Leach, How long before the end of inflation were observable perturbations produced? Phys. Rev. D **68**, 103503 (2003). [arXiv:astro-ph/0305263]
87. D.H. Lyth, A.R. Liddle, *The Primordial Density Perturbation* (Cambridge University Press, Cambridge, 2009)
88. http://map.gsfc.nasa.gov
89. D.S. Salopek, J.R. Bond, Nonlinear evolution of long wavelength metric fluctuations in inflationary models. Phys. Rev. D **42**, 3936 (1990)
90. A.G. Muslimov, On the scalar field dynamics in a spatially flat Friedmann universe. Class. Quantum Grav. **7**, 231 (1990)
91. P.J. Steinhardt, M.S. Turner, Prescription for successful new inflation. Phys. Rev. D **29**, 2162 (1984)
92. A.R. Liddle, D.H. Lyth, COBE, gravitational waves, inflation and extended inflation. Phys. Lett. B **291**, 391 (1992). [arXiv:astro-ph/9208007]
93. E.W. Kolb, S.L. Vadas, Relating spectral indices to tensor and scalar amplitudes in inflation. Phys. Rev. D **50**, 2479 (1994). [arXiv:astro-ph/9403001]
94. D.J. Schwarz, C.A. Terrero-Escalante, A.A. García, Higher order corrections to primordial spectra from cosmological inflation. Phys. Lett. B **517**, 243 (2001). [arXiv:astro-ph/0106020]
95. W.H. Kinney, Inflation: flow, fixed points, and observables to arbitrary order in slow roll. Phys. Rev. D **66**, 083508 (2002). [arXiv:astro-ph/0206032]
96. A.R. Liddle, Inflationary flow equations. Phys. Rev. D **68**, 103504 (2003). [arXiv:astro-ph/0307286]
97. E. Ramírez, A.R. Liddle, Stochastic approaches to inflation model building. Phys. Rev. D **71**, 123510 (2005). [arXiv:astro-ph/0502361]
98. E.J. Copeland, M.R. Garousi, M. Sami, S. Tsujikawa, What is needed of a tachyon if it is to be the dark energy? Phys. Rev. D **71**, 043003 (2005). [arXiv:hep-th/0411192]
99. S. Dodelson, W.H. Kinney, E.W. Kolb, Cosmic microwave background measurements can discriminate among inflation models. Phys. Rev. D **56**, 3207 (1997). [arXiv:astro-ph/9702166]
100. L. Alabidi, D.H. Lyth, Inflation models and observation. JCAP **0605**, 016 (2006). [arXiv:astro-ph/0510441]
101. C. Wetterich, Kaluza–Klein cosmology and the inflationary universe. Nucl. Phys. B **252**, 309 (1985)
102. Q. Shafi, C. Wetterich, Inflation with higher dimensional gravity. Phys. Lett. B **152**, 51 (1985)
103. Q. Shafi, C. Wetterich, Inflation from higher dimensions. Nucl. Phys. B **289**, 787 (1987)
104. H. Nishino, E. Sezgin, Matter and gauge couplings of $N = 2$ supergravity in six dimensions. Phys. Lett. B **144**, 187 (1984)
105. A. Salam, E. Sezgin, Chiral compactification on Minkowski $\times S^2$ of $N = 2$ Einstein–Maxwell supergravity in six dimensions. Phys. Lett. B **147**, 47 (1984)
106. S. Randjbar-Daemi, A. Salam, E. Sezgin, J.A. Strathdee, An anomaly-free model in six dimensions. Phys. Lett. B **151**, 351 (1985)
107. I.G. Koh, H. Nishino, Towards realistic $D = 6$, $N = 2$ Kaluza–Klein supergravity on coset $\mathrm{E}_7/\mathrm{SO}(12) \times \mathrm{Sp}(1)$ with chiral fermions. Phys. Lett. B **153**, 45 (1985)
108. K.-i. Maeda, H. Nishino, Cosmological solutions in $D = 6$, $N = 2$ Kaluza–Klein supergravity: Friedmann universe without fine tuning. Phys. Lett. B **154**, 358 (1985)
109. K.-i. Maeda, H. Nishino, An attractor universe in six-dimensional $N = 2$ supergravity Kaluza–Klein theory. Phys. Lett. B **158**, 381 (1985)
110. J.J. Halliwell, Classical and quantum cosmology of the Salam–Sezgin model. Nucl. Phys. B **286**, 729 (1987)
111. P.A.R. Ade et al. [Planck Collaboration], Planck 2015. XX. Constraints on inflation. Astron. Astrophys. **594**, A20 (2016). [arXiv:1502.02114]
112. K. Freese, J.A. Frieman, A.V. Olinto, Natural inflation with pseudo Nambu–Goldstone bosons. Phys. Rev. Lett. **65**, 3233 (1990)

113. F.C. Adams, J.R. Bond, K. Freese, J.A. Frieman, A.V. Olinto, Natural inflation: particle physics models, power-law spectra for large-scale structure, and constraints from COBE. Phys. Rev. D **47**, 426 (1993). [arXiv:hep-ph/9207245]
114. J.E. Kim, H.P. Nilles, M. Peloso, Completing natural inflation. JCAP **0501**, 005 (2005). [arXiv:hep-ph/0409138]
115. M. Sasaki, E.D. Stewart, A general analytic formula for the spectral index of the density perturbations produced during inflation. Prog. Theor. Phys. **95**, 71 (1996). [arXiv:astro-ph/9507001]
116. A.A. Starobinsky, Multicomponent de Sitter (inflationary) stages and the generation of perturbations. Pis'ma Zh. Eksp. Teor. Fiz. **42**, 124 (1985) [JETP Lett. **42**, 152 (1985)]
117. J. Silk, M.S. Turner, Double inflation. Phys. Rev. D **35**. 419 (1987)
118. D.S. Salopek, J.R. Bond, J.M. Bardeen, Designing density fluctuation spectra in inflation. Phys. Rev. D **40**, 1753 (1989)
119. A.R. Liddle, A. Mazumdar, F.E. Schunck, Assisted inflation. Phys. Rev. D **58**, 061301 (1998). [arXiv:astro-ph/9804177]
120. K.A. Malik, D. Wands, Dynamics of assisted inflation. Phys. Rev. D **59**, 123501 (1999). [arXiv:astro-ph/9812204]
121. A.A. Coley, R.J. van den Hoogen, Dynamics of multi-scalar-field cosmological models and assisted inflation. Phys. Rev. D **62**, 023517 (2000). [arXiv:gr-qc/9911075]
122. J.M. Aguirregabiria, A. Chamorro, L.P. Chimento, N.A. Zuccalá, Assisted inflation in Friedmann–Robertson–Walker and Bianchi spacetimes. Phys. Rev. D **62**, 084029 (2000). [arXiv:gr-qc/0006108]
123. P. Kanti, K.A. Olive, Realization of assisted inflation. Phys. Rev. D **60**, 043502 (1999). [arXiv:hep-ph/9903524]
124. Z.K. Guo, Y.S. Piao, Y.Z. Zhang, Cosmological scaling solutions and multiple exponential potentials. Phys. Lett. B **568**, 1 (2003). [arXiv:hep-th/0304048]
125. S.A Kim, A.R. Liddle, S. Tsujikawa, Dynamics of assisted quintessence. Phys. Rev. D **72**, 043506 (2005). [arXiv:astro-ph/0506076]
126. Z.K. Guo, Y.S. Piao, R.G. Cai, Y.Z. Zhang, Cosmological scaling solutions and cross-coupling exponential potential. Phys. Lett. B **576**, 12 (2003). [arXiv:hep-th/0306245]
127. A. Collinucci, M. Nielsen, T. Van Riet, Scalar cosmology with multi-exponential potentials. Class. Quantum Grav. **22**, 1269 (2005). [arXiv:hep-th/0407047]
128. J. Hartong, A. Ploegh, T. Van Riet, D.B. Westra, Dynamics of generalized assisted inflation. Class. Quantum Grav. **23**, 4593 (2006). [arXiv:gr-qc/0602077]
129. G. Calcagni, A.R. Liddle, Stability of multifield cosmological solutions. Phys. Rev. D **77**, 023522 (2008). [arXiv:0711.3360]
130. S. Tsujikawa, General analytic formulae for attractor solutions of scalar-field dark energy models and their multifield generalizations. Phys. Rev. D **73**, 103504 (2006). [arXiv:hep-th/0601178]
131. J.c. Hwang, H. Noh, Cosmological perturbations with multiple scalar fields. Phys. Lett. B **495**, 277 (2000). [arXiv:astro-ph/0009268]
132. D. Langlois, F. Vernizzi, Nonlinear perturbations of cosmological scalar fields. JCAP **0702**, 017 (2007). [arXiv:astro-ph/0610064]
133. G.I. Rigopoulos, E.P.S. Shellard, B.J.W. van Tent, Nonlinear perturbations in multiple-field inflation. Phys. Rev. D **73**, 083521 (2006). [arXiv:astro-ph/0504508]
134. M. Sasaki, T. Tanaka, Super-horizon scale dynamics of multi-scalar inflation. Prog. Theor. Phys. **99**, 763 (1998). [arXiv:gr-qc/9801017]
135. T.J. Allen, B. Grinstein, M.B. Wise, Non-gaussian density perturbations in inflationary cosmologies. Phys. Lett. B **197**, 66 (1987)
136. L. Kofman, D.Yu. Pogosyan, Nonflat perturbations in inflationary cosmology. Phys. Lett. B **214**, 508 (1988)
137. S. Mollerach, S. Matarrese, A. Ortolan, F. Lucchin, Stochastic inflation in a simple two-field model. Phys. Rev. D **44**, 1670 (1991)

138. A.D. Linde, V. Mukhanov, Non-Gaussian isocurvature perturbations from inflation. Phys. Rev. D **56**, 535 (1997). [arXiv:astro-ph/9610219]
139. D.H. Lyth, D. Wands, Generating the curvature perturbation without an inflaton. Phys. Lett. B **524**, 5 (2002). [arXiv:hep-ph/0110002]
140. D.H. Lyth, C. Ungarelli, D. Wands, The primordial density perturbation in the curvaton scenario. Phys. Rev. D **67**, 023503 (2003). [arXiv:astro-ph/0208055]
141. N. Bartolo, S. Matarrese, A. Riotto, Non-Gaussianity in the curvaton scenario. Phys. Rev. D **69**, 043503 (2004). [arXiv:hep-ph/0309033]
142. A.D. Linde, Eternal extended inflation and graceful exit from old inflation without Jordan–Brans–Dicke. Phys. Lett. B **249**, 18 (1990)
143. A.D. Linde, Axions in inflationary cosmology. Phys. Lett. B **259**, 38 (1991)
144. A.D. Linde, Hybrid inflation. Phys. Rev. D **49**, 748 (1994). [arXiv:astro-ph/9307002]
145. S. Mollerach, S. Matarrese, F. Lucchin, Blue perturbation spectra from inflation. Phys. Rev. D **50**, 4835 (1994). [arXiv:astro-ph/9309054]
146. E.J. Copeland, A.R. Liddle, D.H. Lyth, E.D. Stewart, D. Wands, False vacuum inflation with Einstein gravity. Phys. Rev. D **49**, 6410 (1994). [arXiv:astro-ph/9401011]
147. S. Tsujikawa, J. Ohashi, S. Kuroyanagi, A. De Felice, Planck constraints on single-field inflation. Phys. Rev. D **88**, 023529 (2013). [arXiv:1305.3044]
148. J.A. Wheeler, W.H. Zurek (eds.), *Quantum Theory and Measurement* (Princeton University Press, Princeton, 1983)
149. H. Everett, "Relative state" formulation of quantum mechanics. Rev. Mod. Phys. **29**, 454 (1957)
150. J.B. Hartle, Quantum mechanics of individual systems. Am. J. Phys. **36**, 704 (1968)
151. B.S. DeWitt, R.N. Graham (eds.), *The Many-Worlds Interpretation of Quantum Mechanics* (Princeton University Press, Princeton, 1973)
152. W.H. Zurek, Decoherence, einselection, and the quantum origins of the classical. Rev. Mod. Phys. **75**, 715 (2003)
153. N. Pinto-Neto, G. Santos, W. Struyve, Quantum-to-classical transition of primordial cosmological perturbations in de Broglie–Bohm quantum theory. Phys. Rev. D **85**, 083506 (2012). [arXiv:1110.1339]
154. E. Joos, H.D. Zeh, C. Kiefer, D. Giulini, J. Kupsch, I.-O. Stamatescu, *Decoherence and the Appearance of a Classical World in Quantum Theory* (Springer, Berlin, 2003)
155. M. Schlosshauer, Experimental motivation and empirical consistency in minimal no-collapse quantum mechanics. Ann. Phys. (N.Y.) **321**, 112 (2006). [arXiv:quant-ph/0506199]
156. H.D. Zeh, Emergence of classical time from a universal wave function. Phys. Lett. A **116**, 9 (1986)
157. C. Kiefer, Continuous measurement of mini-superspace variables by higher multipoles. Class. Quantum Grav. **4**, 1369 (1987)
158. M. Sakagami, Evolution from pure states into mixed states in de Sitter space. Prog. Theor. Phys. **79**, 442 (1988)
159. J.J. Halliwell, Decoherence in quantum cosmology. Phys. Rev. D **39**, 2912 (1989)
160. T. Padmanabhan, Decoherence in the density matrix describing quantum three-geometries and the emergence of classical spacetime. Phys. Rev. D **39**, 2924 (1989)
161. J.P. Paz, S. Sinha, Decoherence and back reaction: the origin of the semiclassical Einstein equations. Phys. Rev. D **44**, 1038 (1991)
162. C. Kiefer, Decoherence in quantum electrodynamics and quantum gravity. Phys. Rev. D **46**, 1658 (1992)
163. C. Kiefer, Topology, decoherence, and semiclassical gravity. Phys. Rev. D **47**, 5414 (1993). [arXiv:gr-qc/9306016]
164. I.G. Moss, *Quantum Theory, Black Holes and Inflation* (Wiley, Chichester, 1996)
165. L.P. Grishchuk, Amplification of gravitational waves in an isotropic universe. Zh. Eksp. Teor. Fiz. **67**, 825 (1974) [Sov. Phys. JETP **40**, 409 (1975)]
166. L.H. Ford, L. Parker, Quantized gravitational wave perturbations in Robertson–Walker universes. Phys. Rev. D **16**, 1601 (1977)

167. N.D. Birrell, P.C.W. Davies, *Quantum Fields in Curved Space* (Cambridge University Press, Cambridge, 1982)
168. T.S. Bunch, P.C.W. Davies, Quantum field theory in de Sitter space: renormalization by point splitting. Proc. R. Soc. Lond. A **360**, 117 (1978)
169. R.H. Brandenberger, Quantum fluctuations as the source of classical gravitational perturbations in inflationary universe models. Nucl. Phys. B **245**, 328 (1984)
170. M.R. Brown, C.R. Dutton, Energy-momentum tensor and definition of particle states for Robertson–Walker space-times. Phys. Rev. D **18**, 4422 (1978)
171. J. Martin, R.H. Brandenberger, The trans-Planckian problem of inflationary cosmology. Phys. Rev. D **63**, 123501 (2001). [arXiv:hep-th/0005209]
172. N.A. Chernikov, E.A. Tagirov, Quantum theory of scalar fields in de Sitter space-time. Ann. Poincaré Phys. Theor. A **9**, 109 (1968)
173. E.A. Tagirov, Consequences of field quantization in de Sitter type cosmological models. Ann. Phys. (N.Y.) **76**, 561 (1973)
174. J. Géhéniau, C. Schomblond, Fonctions de Green dans l'univers de de Sitter. Acad. R. Belg. Bull. Cl. Sci. **54**, 1147 (1968)
175. C. Schomblond, P. Spindel, Uniqueness conditions for the $\Delta^1(x, y)$ propagator of the scalar field in the de Sitter universe. Ann. Poincaré Phys. Theor. A **25**, 67 (1976)
176. E. Mottola, Particle creation in de Sitter space. Phys. Rev. D **31**, 754 (1985)
177. B. Allen, Vacuum states in de Sitter space. Phys. Rev. D **32**, 3136 (1985)
178. B. Allen, A. Folacci, Massless minimally coupled scalar field in de Sitter space. Phys. Rev. D **35**, 3771 (1987)
179. R. Bousso, A. Maloney, A. Strominger, Conformal vacua and entropy in de Sitter space. Phys. Rev. D **65**, 104039 (2002). [arXiv:hep-th/0112218]
180. M. Spradlin, A. Volovich, Vacuum states and the S-matrix in dS/CFT. Phys. Rev. D **65**, 104037 (2002). [arXiv:hep-th/0112223]
181. D. Polarski, A.A. Starobinsky, Semiclassicality and decoherence of cosmological perturbations. Class. Quantum Grav. **13**, 377 (1996). [arXiv:gr-qc/9504030]
182. U.H. Danielsson, Note on inflation and trans-Planckian physics. Phys. Rev. D **66**, 023511 (2002). [arXiv:hep-th/0203198]
183. U.H. Danielsson, Inflation, holography and the choice of vacuum in de Sitter space. JHEP **0207**, 040 (2002). [arXiv:hep-th/0205227]
184. N. Kaloper, M. Kleban, A. Lawrence, S. Shenker, L. Susskind, Initial conditions for inflation. JHEP **0211**, 037 (2002). [arXiv:hep-th/0209231]
185. U.H. Danielsson, On the consistency of de Sitter vacua. JHEP **0212**, 025 (2002). [arXiv:hep-th/0210058]
186. J. Lesgourgues, D. Polarski, A.A. Starobinsky, Quantum to classical transition of cosmological perturbations for nonvacuum initial states. Nucl. Phys. B **497**, 479 (1997). [arXiv:gr-qc/9611019]
187. J. Martin, A. Riazuelo, M. Sakellariadou, Nonvacuum initial states for cosmological perturbations of quantum-mechanical origin. Phys. Rev. D **61**, 083518 (2000). [arXiv:astro-ph/9904167]
188. R.H. Brandenberger, C.T. Hill, Energy-density fluctuations in de Sitter space. Phys. Lett. B **179**, 30 (1986)
189. E.D. Stewart, D.H. Lyth, A more accurate analytic calculation of the spectrum of cosmological perturbations produced during inflation. Phys. Lett. B **302**, 171 (1993). [arXiv:gr-qc/9302019]
190. E.J. Copeland, E.W. Kolb, A.R. Liddle, J.E. Lidsey, Reconstructing the inflaton potential: perturbative reconstruction to second order. Phys. Rev. D **49**, 1840 (1994). [arXiv:astro-ph/9308044]
191. D.H. Lyth, E.D. Stewart, The curvature perturbation in power law (e.g. extended) inflation. Phys. Lett. B **274**, 168 (1992)
192. C. Kiefer, D. Polarski, Why do cosmological perturbations look classical to us? Adv. Sci. Lett. **2**, 164 (2009). [arXiv:0810.0087]

193. A. Riotto, Inflation and the theory of cosmological perturbations. arXiv:hep-ph/0210162
194. P.J. Steinhardt, Natural inflation, in *The Very Early Universe*, ed. by G.W. Gibbons, S.W. Hawking, S.T.C. Siklos (Cambridge University Press, Cambridge, 1983)
195. A. Vilenkin, Birth of inflationary universes. Phys. Rev. D **27**, 2848 (1983)
196. A.D. Linde, Eternally existing selfreproducing chaotic inflationary universe. Phys. Lett. B **175**, 395 (1986)
197. A.D. Linde, Eternal chaotic inflation. Mod. Phys. Lett. A **01**, 81 (1986)
198. A.A. Starobinsky, Stochastic de sitter (inflationary) stage in the early universe. Lect. Notes Phys. **246**, 107 (1986)
199. A.S. Goncharov, A.D. Linde, V.F. Mukhanov, The global structure of the inflationary universe. Int. J. Mod. Phys. A **2**, 561 (1987)
200. M. Aryal, A. Vilenkin, The fractal dimension of the inflationary universe. Phys. Lett. B **199**, 351 (1987)
201. K. Nakao, Y. Nambu, M. Sasaki, Stochastic dynamics of new inflation. Prog. Theor. Phys. **80**, 1041 (1988)
202. A.D. Linde, D.A. Linde, A. Mezhlumian, From the big bang theory to the theory of a stationary universe. Phys. Rev. D **49**, 1783 (1994). [arXiv:gr-qc/9306035]
203. A.D. Linde, D.A. Linde, A. Mezhlumian, Do we live in the center of the world? Phys. Lett. B **345**, 203 (1995). [arXiv:hep-th/9411111]
204. S. Winitzki, *Eternal Inflation* (World Scientific, Singapore, 2009)
205. A.D. Linde, Towards a gauge invariant volume-weighted probability measure for eternal inflation. JCAP **0706**, 017 (2007). [arXiv:0705.1160]
206. A. Borde, Geodesic focusing, energy conditions and singularities. Class. Quantum Grav. **4**, 343 (1987)
207. A. Vilenkin, Did the universe have a beginning? Phys. Rev. D **46**, 2355 (1992)
208. F. Helmer, S. Winitzki, Self-reproduction in k-inflation. Phys. Rev. D **74**, 063528 (2006). [arXiv:gr-qc/0608019]
209. A. Vilenkin, Quantum fluctuations in the new inflationary Universe. Nucl. Phys. B **226**, 527 (1983)
210. A. Vilenkin, Making predictions in eternally inflating universe. Phys. Rev. D **52**, 3365 (1995). [arXiv:gr-qc/9505031]
211. A. Vilenkin, Unambiguous probabilities in an eternally inflating universe. Phys. Rev. Lett. **81**, 5501 (1998). [arXiv:hep-th/9806185]
212. V. Vanchurin, A. Vilenkin, S. Winitzki, Predictability crisis in inflationary cosmology and its resolution. Phys. Rev. D **61**, 083507 (2000). [arXiv:gr-qc/9905097]
213. J. Garriga, A. Vilenkin, Prescription for probabilities in eternal inflation. Phys. Rev. D **64**, 023507 (2001). [arXiv:gr-qc/0102090]
214. J. Garriga, D. Schwartz-Perlov, A. Vilenkin, S. Winitzki, Probabilities in the inflationary multiverse. JCAP **0601**, 017 (2006). [arXiv:hep-th/0509184]
215. R. Easther, E.A. Lim, M.R. Martin, Counting pockets with world lines in eternal inflation. JCAP **0603**, 016 (2006). [arXiv:astro-ph/0511233]
216. R. Bousso, Holographic probabilities in eternal inflation. Phys. Rev. Lett. **97**, 191302 (2006). [arXiv:hep-th/0605263]
217. R. Bousso, B. Freivogel, I-S. Yang, Eternal inflation: the inside story. Phys. Rev. D **74**, 103516 (2006). [arXiv:hep-th/0606114]
218. H. Kodama, M. Sasaki, Cosmological perturbation theory. Prog. Theor. Phys. Suppl. **78**, 1 (1984)
219. V.F. Mukhanov, H.A. Feldman, R.H. Brandenberger, Theory of cosmological perturbations. Phys. Rep. **215**, 203 (1992)
220. J.M. Bardeen, Gauge-invariant cosmological perturbations. Phys. Rev. D **22**, 1882 (1980)
221. S. Matarrese, S. Mollerach, M. Bruni, Relativistic second-order perturbations of the Einstein-de Sitter universe. Phys. Rev. D **58**, 043504 (1998). [arXiv:astro-ph/9707278]
222. V. Acquaviva, N. Bartolo, S. Matarrese, A. Riotto, Gauge-invariant second-order perturbations and non-Gaussianity from inflation. Nucl. Phys. B **667**, 119 (2003). [arXiv:astro-ph/0209156]

223. K.A. Malik, D. Wands, Evolution of second-order cosmological perturbations. Class. Quantum Grav. **21**, L65 (2004). [arXiv:astro-ph/0307055]
224. L.F. Abbott, M.B. Wise, Constraints on generalized inflationary cosmologies. Nucl. Phys. B **244**, 541 (1984)
225. V.F. Mukhanov, Gravitational instability of the universe filled with a scalar field. Pis'ma Zh. Eksp. Teor. Fiz. **41**, 402 (1985) [JETP Lett. **41**, 493 (1985)]
226. M. Sasaki, Large scale quantum fluctuations in the inflationary universe. Prog. Theor. Phys. **76**, 1036 (1986)
227. V.F. Mukhanov, Quantum theory of cosmological perturbations in R^2 gravity. Phys. Lett. B **218**, 17 (1989)
228. D.H. Lyth, What would we learn by detecting a gravitational wave signal in the cosmic microwave background anisotropy? Phys. Rev. Lett. **78**, 1861 (1997). [arXiv:hep-ph/9606387]
229. D. Baumann, L. McAllister, A microscopic limit on gravitational waves from D-brane inflation. Phys. Rev. D **75**, 123508 (2007). [arXiv:hep-th/0610285]
230. A. Ortolan, F. Lucchin, S. Matarrese, Non-Gaussian perturbations from inflationary dynamics. Phys. Rev. D **38**, 465 (1988)
231. H.M. Hodges, G.R. Blumenthal, L.A. Kofman, J.R. Primack, Nonstandard primordial fluctuations from a polynomial inflation potential. Nucl. Phys. B **335**, 197 (1990)
232. T. Falk, R. Rangarajan, M. Srednicki, The angular dependence of the three point correlation function of the cosmic microwave background radiation as predicted by inflationary cosmologies. Astrophys. J. **403**, L1 (1993). [arXiv:astro-ph/9208001]
233. I. Yi, E.T. Vishniac, Inflationary stochastic dynamics and the statistics of large-scale structure. Astrophys. J. Suppl. Ser. **86**, 333 (1993)
234. I. Yi, E.T. Vishniac, Simple estimate of the statistics of large scale structure. Phys. Rev. D **48**, 950 (1993)
235. G.I. Rigopoulos, E.P.S. Shellard, B.J.W. van Tent, Simple route to non-Gaussianity in inflation. Phys. Rev. D **72**, 083507 (2005). [arXiv:astro-ph/0410486]
236. D. Seery, J.E. Lidsey, Primordial non-Gaussianities in single field inflation. JCAP **0506**, 003 (2005). [arXiv:astro-ph/0503692]
237. D.H. Lyth, Y. Rodríguez, The inflationary prediction for primordial non-Gaussianity. Phys. Rev. Lett. **95**, 121302 (2005). [arXiv:astro-ph/0504045]
238. X. Chen, M.x. Huang, S. Kachru, G. Shiu, Observational signatures and non-Gaussianities of general single field inflation. JCAP **0701**, 002 (2007). [arXiv:hep-th/0605045]
239. D. Seery, J.E. Lidsey, M.S. Sloth, The inflationary trispectrum. JCAP **0701**, 027 (2007). [arXiv:astro-ph/0610210]
240. X. Chen, M.x. Huang, G. Shiu, The inflationary trispectrum for models with large non-Gaussianities. Phys. Rev. D **74**, 121301 (2006). [arXiv:hep-th/0610235]
241. F. Arroja, K. Koyama, Non-Gaussianity from the trispectrum in general single field inflation. Phys. Rev. D **77**, 083517 (2008). [arXiv:0802.1167]
242. L. Wang, M. Kamionkowski, The cosmic microwave background bispectrum and inflation. Phys. Rev. D **61**, 063504 (2000). [arXiv:astro-ph/9907431]
243. A. Gangui, J. Martin, Cosmic microwave background bispectrum and slow roll inflation. Mon. Not. R. Astron. Soc. **313**, 323 (2000). [arXiv:astro-ph/9908009]
244. A. Gangui, F. Lucchin, S. Matarrese, S. Mollerach, The three-point correlation function of the cosmic microwave background in inflationary models. Astrophys. J. **430**, 447 (1994). [arXiv:astro-ph/9312033]
245. J.M. Maldacena, Non-Gaussian features of primordial fluctuations in single field inflationary models. JHEP **0305**, 013 (2003). [arXiv:astro-ph/0210603]
246. D. Babich, P. Creminelli, M. Zaldarriaga, The shape of non-Gaussianities. JCAP **0408**, 009 (2004). [arXiv:astro-ph/0405356]
247. P. Creminelli, M. Zaldarriaga, Single field consistency relation for the 3-point function. JCAP **0410**, 006 (2004). [arXiv:astro-ph/0407059]

248. L. Senatore, K.M. Smith, M. Zaldarriaga, Non-Gaussianities in single field inflation and their optimal limits from the WMAP 5-year data. JCAP **1001**, 028 (2010). [arXiv:0905.3746]
249. P. Creminelli, On non-Gaussianities in single-field inflation. JCAP **0310**, 003 (2003). [arXiv:astro-ph/0306122]
250. S. Gupta, A. Berera, A.F. Heavens, S. Matarrese, Non-Gaussian signatures in the cosmic background radiation from warm inflation. Phys. Rev. D **66**, 043510 (2002). [arXiv:astro-ph/0205152]
251. S. Gupta, Dynamics and non-Gaussianity in the weak-dissipative warm inflation scenario. Phys. Rev. D **73**, 083514 (2006). [arXiv:astro-ph/0509676]
252. N. Arkani-Hamed, P. Creminelli, S. Mukohyama, M. Zaldarriaga, Ghost inflation. JCAP **0404**, 001 (2004). [arXiv:hep-th/0312100]
253. K. Izumi, S. Mukohyama, Trispectrum from ghost inflation. JCAP **1006**, 016 (2010). [arXiv:1004.1776]
254. A. Gangui, J. Martin, M. Sakellariadou, Single field inflation and non-Gaussianity. Phys. Rev. D **66**, 083502 (2002). [arXiv:astro-ph/0205202]
255. S.-J. Rey, Dynamics of inflationary phase transition. Nucl. Phys. B **284**, 706 (1987)
256. A. Hosoya, M. Morikawa, K. Nakayama, Stochastic dynamics of scalar field in the inflationary universe. Int. J. Mod. Phys. A **4**, 2613 (1989)
257. D.S. Salopek, J.R. Bond, Stochastic inflation and nonlinear gravity. Phys. Rev. D **43**, 1005 (1991)
258. I. Yi, E.T. Vishniac, Stochastic analysis of the initial condition constraints on chaotic inflation. Phys. Rev. D **47**, 5280 (1993)
259. A.A. Starobinsky, J. Yokoyama, Equilibrium state of a self-interacting scalar field in the de Sitter background. Phys. Rev. D **50**, 6357 (1994). [arXiv:astro-ph/9407016]
260. J. Martin, M. Musso, Solving stochastic inflation for arbitrary potentials. Phys. Rev. D **73**, 043516 (2006). [arXiv:hep-th/0511214]
261. J. Martin, M. Musso, Reliability of the Langevin pertubative solution in stochastic inflation. Phys. Rev. D **73**, 043517 (2006). [arXiv:hep-th/0511292]
262. F. Kühnel, D.J. Schwarz, Large-scale suppression from stochastic inflation. Phys. Rev. Lett. **105**, 211302 (2010). [arXiv:1003.3014]
263. H. Haken, *Synergetics* (Springer, Berlin, 1978)
264. H. Risken, *The Fokker–Planck Equation* (Springer, Berlin, 1984)
265. S. Matarrese, M.A. Musso, A. Riotto, Influence of super-horizon scales on cosmological observables generated during inflation. JCAP **0405**, 008 (2004). [arXiv:hep-th/0311059]
266. G.I. Rigopoulos, E.P.S. Shellard, The separate universe approach and the evolution of nonlinear superhorizon cosmological perturbations. Phys. Rev. D **68**, 123518 (2003). [arXiv:astro-ph/0306620]
267. G.I. Rigopoulos, E.P.S. Shellard, Non-linear inflationary perturbations. JCAP **0510**, 006 (2005). [arXiv:astro-ph/0405185]
268. D.H. Lyth, K.A. Malik, M. Sasaki, A general proof of the conservation of the curvature perturbation. JCAP **0505**, 004 (2005). [arXiv:astro-ph/0411220]
269. D. Wands, K.A. Malik, D.H. Lyth, A.R. Liddle, A new approach to the evolution of cosmological perturbations on large scales. Phys. Rev. D **62**, 043527 (2000). [arXiv:astro-ph/0003278]
270. D.H. Lyth, Y. Rodríguez, Non-Gaussianity from the second-order cosmological perturbation. Phys. Rev. D **71**, 123508 (2005). [arXiv:astro-ph/0502578]
271. P.J.E. Peebles, An isocurvature cold dark matter cosmogony. I. A worked example of evolution through inflation. Astrophys. J. **510**, 523 (1999). [arXiv:astro-ph/9805194]
272. P.J.E. Peebles, An isocurvature cold dark matter cosmogony. II. Observational tests. Astrophys. J. **510**, 531 (1999). [arXiv:astro-ph/9805212]
273. F. Bernardeau, J.-P. Uzan, Non-Gaussianity in multi-field inflation. Phys. Rev. D **66**, 103506 (2002). [arXiv:hep-ph/0207295]
274. F. Bernardeau, J.-P. Uzan, Inflationary models inducing non-Gaussian metric fluctuations. Phys. Rev. D **67**, 121301 (2003). [arXiv:astro-ph/0209330]

275. M. Zaldarriaga, Non-Gaussianities in models with a varying inflaton decay rate. Phys. Rev. D **69**, 043508 (2004). [arXiv:astro-ph/0306006]
276. K. Enqvist, A. Väihkönen, Non-Gaussian perturbations in hybrid inflation. JCAP **0409**, 006 (2004). [arXiv:hep-ph/0405103]
277. D.H. Lyth, Non-Gaussianity and cosmic uncertainty in curvaton-type models. JCAP **0606**, 015 (2006). [arXiv:astro-ph/0602285]
278. K.A. Malik, D.H. Lyth, A numerical study of non-Gaussianity in the curvaton scenario. JCAP **0609**, 008 (2006). [arXiv:astro-ph/0604387]
279. M. Sasaki, J. Valiviita, D. Wands, Non-Gaussianity of the primordial perturbation in the curvaton model. Phys. Rev. D **74**, 103003 (2006). [arXiv:astro-ph/0607627]
280. K. Enqvist, S. Nurmi, Non-Gaussianity in curvaton models with nearly quadratic potential. JCAP **0510**, 013 (2005). [arXiv:astro-ph/0508573]
281. D. Seery, J.E. Lidsey, Primordial non-Gaussianities from multiple-field inflation. JCAP **0509**, 011 (2005). [arXiv:astro-ph/0506056]
282. G.I. Rigopoulos, E.P.S. Shellard, B.J.W. van Tent, Large non-Gaussianity in multiple-field inflation. Phys. Rev. D **73**, 083522 (2006). [arXiv:astro-ph/0506704]
283. D.H. Lyth, I. Zaballa, A bound concerning primordial non-Gaussianity. JCAP **0510**, 005 (2005). [arXiv:astro-ph/0507608]
284. G.I. Rigopoulos, E.P.S. Shellard, B.J.W. van Tent, Quantitative bispectra from multifield inflation. Phys. Rev. D **76**, 083512 (2007). [arXiv:astro-ph/0511041]
285. I. Zaballa, Y. Rodríguez, D.H. Lyth, Higher order contributions to the primordial non-Gaussianity. JCAP **0606**, 013 (2006). [arXiv:astro-ph/0603534]
286. F. Vernizzi, D. Wands, Non-Gaussianities in two-field inflation. JCAP **0605**, 019 (2006). [arXiv:astro-ph/0603799]
287. L. Alabidi, Non-Gaussianity for a two component hybrid model of inflation. JCAP **0610**, 015 (2006). [arXiv:astro-ph/0604611]
288. T. Battefeld, R. Easther, Non-Gaussianities in multi-field inflation. JCAP **0703**, 020 (2007). [arXiv:astro-ph/0610296]
289. D. Seery, J.E. Lidsey, Non-Gaussianity from the inflationary trispectrum. JCAP **0701**, 008 (2007). [arXiv:astro-ph/0611034]
290. C.T. Byrnes, M. Sasaki, D. Wands, The primordial trispectrum from inflation. Phys. Rev. D **74**, 123519 (2006). [arXiv:astro-ph/0611075]
291. W.H. Kinney, Constraining inflation with cosmic microwave background polarization. Phys. Rev. D **58**, 123506 (1998). [arXiv:astro-ph/9806259]
292. http://cosmologist.info/cosmomc.
293. A. Aguirre, M. Tegmark, Multiple universes, cosmic coincidences, and other dark matters. JCAP **0501**, 003 (2005). [arXiv:hep-th/0409072]
294. D.A. Easson, B.A. Powell, Identifying the inflaton with primordial gravitational waves. Phys. Rev. Lett. **106**, 191302 (2011). [arXiv:1009.3741]
295. M. Bastero-Gil, J. Macias-Pérez, D. Santos, Nonlinear metric perturbation enhancement of primordial gravitational waves. Phys. Rev. Lett. **105**, 081301 (2010). [arXiv:1005.4054]
296. U. Seljak, U.L. Pen, N. Turok, Polarization of the microwave background in defect models. Phys. Rev. Lett. **79**, 1615 (1997). [arXiv:astro-ph/9704231]
297. N. Barnaby, R. Namba, M. Peloso, Phenomenology of a pseudo-scalar inflaton: naturally large non-Gaussianity. JCAP **1104**, 009 (2011). [arXiv:1102.4333]
298. L. Sorbo, Parity violation in the cosmic microwave background from a pseudoscalar inflaton. JCAP **1106**, 003 (2011). [arXiv:1101.1525]
299. N. Barnaby, E. Pajer, M. Peloso, Gauge field production in axion inflation: consequences for monodromy, non-Gaussianity in the CMB, and gravitational waves at interferometers. Phys. Rev. D **85**, 023525 (2012). [arXiv:1110.3327]
300. N. Bartolo, S. Matarrese, M. Peloso, M. Shiraishi, Parity-violating and anisotropic correlations in pseudoscalar inflation. JCAP **1501**, 027 (2015). [arXiv:1411.2521]
301. A. Lue, L. Wang, M. Kamionkowski, Cosmological signature of new parity-violating interactions. Phys. Rev. Lett. **83**, 1506 (1999). [arXiv:astro-ph/9812088]

302. R. Jackiw, S.-Y. Pi, Chern–Simons modification of general relativity. Phys. Rev. D **68**, 104012 (2003). [arXiv:gr-qc/0308071]
303. M. Pospelov, A. Ritz, C. Skordis, Pseudoscalar perturbations and polarization of the cosmic microwave background. Phys. Rev. Lett. **103**, 051302 (2009). [arXiv:0808.0673]
304. B. Feng, H. Li, M. Li, X. Zhang, Gravitational leptogenesis and its signatures in CMB. Phys. Lett. B **620**, 27 (2005). [arXiv:hep-ph/0406269]
305. S. Mercuri, Fermions in Ashtekar–Barbero connections formalism for arbitrary values of the Immirzi parameter. Phys. Rev. D **73**, 084016 (2006). [arXiv:gr-qc/0601013]
306. C.R. Contaldi, J. Magueijo, L. Smolin, Anomalous CMB polarization and gravitational chirality. Phys. Rev. Lett. **101**, 141101 (2008). [arXiv:0806.3082]
307. S. Mercuri, Modifications in the spectrum of primordial gravitational waves induced by instantonic fluctuations. Phys. Rev. D **84**, 044035 (2011). [arXiv:1007.3732]
308. L. Bethke, J. Magueijo, Inflationary tensor fluctuations, as viewed by Ashtekar variables, their imaginary friends. Phys. Rev. D **84**, 024014 (2011). [arXiv:1104.1800]
309. L. Bethke, J. Magueijo, Chirality of tensor perturbations for complex values of the Immirzi parameter. Class. Quantum Grav. **29**, 052001 (2012). [arXiv:1108.0816]
310. A. Gruzinov, Consistency relation for single scalar inflation. Phys. Rev. D **71**, 027301 (2005). [arXiv:astro-ph/0406129]
311. N. Bartolo, S. Matarrese, A. Riotto, Enhancement of non-Gaussianity after inflation. JHEP **0404**, 006 (2004). [arXiv:astro-ph/0308088]
312. N. Bartolo, S. Matarrese, A. Riotto, Evolution of second-order cosmological perturbations and non-Gaussianity. JCAP **0401**, 003 (2004). [arXiv:astro-ph/0309692]
313. N. Bartolo, S. Matarrese, A. Riotto, Gauge-invariant temperature anisotropies and primordial non-Gaussianity. Phys. Rev. Lett. **93**, 231301 (2004). [arXiv:astro-ph/0407505]
314. K. Enqvist, A. Jokinen, A. Mazumdar, T. Multamäki, A. Väihkönen, Non-Gaussianity from preheating. Phys. Rev. Lett. **94**, 161301 (2005). [arXiv:astro-ph/0411394]
315. N. Barnaby, J.M. Cline, Non-Gaussian and nonscale-invariant perturbations from tachyonic preheating in hybrid inflation. Phys. Rev. D **73**, 106012 (2006). [arXiv:astro-ph/0601481]
316. N. Bartolo, E. Komatsu, S. Matarrese, A. Riotto, Non-Gaussianity from inflation: theory and observations. Phys. Rep. **402**, 103 (2004). [arXiv:astro-ph/0406398]
317. S.W. Hawking, I.G. Moss, Supercooled phase transitions in the very early universe. Phys. Lett. B **110**, 35 (1982)
318. G.W. Gibbons, S.W. Hawking, Cosmological event horizons, thermodynamics, and particle creation. Phys. Rev. D **15**, 2738 (1977)
319. A.A. Starobinsky, Isotropization of arbitrary cosmological expansion given an effective cosmological constant. Pis'ma Zh. Eksp. Teor. Fiz. **37**, 55 (1983) [JETP Lett. **37**, 66 (1983)]
320. R.M. Wald, Asymptotic behavior of homogeneous cosmological models in the presence of a positive cosmological constant. Phys. Rev. D **28**, 2118 (1983)
321. L.G. Jensen, J.A. Stein-Schabes, Is inflation natural? Phys. Rev. D **35**, 1146 (1987)
322. M. Bruni, S. Matarrese, O. Pantano, A local view of the observable universe. Phys. Rev. Lett. **74**, 1916 (1995). [arXiv:astro-ph/9407054]
323. R.H. Brandenberger, J. Martin, Trans-Planckian issues for inflationary cosmology. Class. Quantum Grav. **30**, 113001 (2013). [arXiv:1211.6753]
324. R.H. Brandenberger, J. Martin, The robustness of inflation to changes in super-Planck-scale physics. Mod. Phys. Lett. A **16**, 999 (2001). [arXiv:astro-ph/0005432]
325. A.A. Starobinsky, Robustness of the inflationary perturbation spectrum to trans-Planckian physics. Pis'ma Zh. Eksp. Teor. Fiz. **73**, 415 (2001) [JETP Lett. **73**, 371 (2001)]. [arXiv:astro-ph/0104043]
326. J. Martin, R.H. Brandenberger, The Corley–Jacobson dispersion relation and trans-Planckian inflation. Phys. Rev. D **65**, 103514 (2002). [arXiv:hep-th/0201189]
327. R.H. Brandenberger, J. Martin, On signatures of short distance physics in the cosmic microwave background. Int. J. Mod. Phys. A **17**, 3663 (2002). [arXiv:hep-th/0202142]
328. R. Easther, B.R. Greene, W.H. Kinney, G. Shiu, Generic estimate of trans-Planckian modifications to the primordial power spectrum in inflation. Phys. Rev. D **66**, 023518 (2002). [arXiv:hep-th/0204129]

329. G.L. Alberghi, R. Casadio, A. Tronconi, Trans-Planckian footprints in inflationary cosmology. Phys. Lett. B **579**, 1 (2004). [arXiv:gr-qc/0303035]
330. J. Martin, R.H. Brandenberger, Dependence of the spectra of fluctuations in inflationary cosmology on trans-Planckian physics. Phys. Rev. D **68**, 063513 (2003). [arXiv:hep-th/0305161]
331. S. Cremonini, Effects of quantum deformations on the spectrum of cosmological perturbations. Phys. Rev. D **68**, 063514 (2003). [arXiv:hep-th/0305244]
332. S. Koh, S.P. Kim, D.J. Song, Gravitational wave spectrum in inflation with nonclassical states. JHEP **0412**, 060 (2004). [arXiv:gr-qc/0402065]
333. S. Shankaranarayanan, L. Sriramkumar, Trans-Planckian corrections to the primordial spectrum in the infrared and the ultraviolet. Phys. Rev. D **70**, 123520 (2004). [arXiv:hep-th/0403236]
334. L. Sriramkumar, T. Padmanabhan, Initial state of matter fields and trans-Planckian physics: can CMB observations disentangle the two? Phys. Rev. D **71**, 103512 (2005). [arXiv:gr-qc/0408034]
335. R.H. Brandenberger, J. Martin, Back-reaction and the trans-Planckian problem of inflation revisited. Phys. Rev. D **71**, 023504 (2005). [arXiv:hep-th/0410223]
336. R. Easther, W.H. Kinney, H. Peiris, Observing trans-Planckian signatures in the cosmic microwave background. JCAP **0505**, 009 (2005). [arXiv:astro-ph/0412613]
337. R. Brandenberger, X.m. Zhang, The trans-Planckian problem for inflationary cosmology revisited. arXiv:0903.2065
338. G. 't Hooft, Naturalness, chiral symmetry, and spontaneous chiral symmetry breaking. NATO Adv. Study Inst. Ser. B Phys. **59**, 135 (1980)
339. R. Barbieri, Looking beyond the standard model: the supersymmetric option. Riv. Nuovo Cim. **11N4**, 1 (1988)
340. J. Polchinski, Effective field theory and the Fermi surface, in *Recent Directions in Particle Theory—From Superstrings and Black Holes to the Standard Model*, ed. by J. Harvey, J. Polchinski (World Scientific, Singapore, 1993). [arXiv:hep-th/9210046]
341. C. Cheung, P. Creminelli, A.L. Fitzpatrick, J. Kaplan, L. Senatore, The effective field theory of inflation. JHEP **0803**, 014 (2008). [arXiv:0709.0293]
342. D. Seery, Infrared effects in inflationary correlation functions. Class. Quantum Grav. **27**, 124005 (2010) [arXiv:1005.1649]
343. C.T. Byrnes, M. Gerstenlauer, A. Hebecker, S. Nurmi, G. Tasinato, Inflationary infrared divergences: geometry of the reheating surface versus δN formalism. JCAP **1008**, 006 (2010). [arXiv:1005.3307]
344. C.P. Burgess, R. Holman, L. Leblond, S. Shandera, Breakdown of semiclassical methods in de Sitter space. JCAP **1010**, 017 (2010). [arXiv:1005.3551]
345. Y. Urakawa, T. Tanaka, Infrared divergence divergence does not affect the gauge-invariant curvature perturbation. Phys. Rev. D **82**, 121301 (2010). [arXiv:1007.0468]
346. Y. Urakawa, T. Tanaka, Natural selection of inflationary vacuum required by infra-red regularity and gauge-invariance. Prog. Theor. Phys. **125**, 1067 (2011). [arXiv:1009.2947]
347. M. Gerstenlauer, A. Hebecker, G. Tasinato, Inflationary correlation functions without infrared divergences. JCAP **1106**, 021 (2011). [arXiv:1102.0560]
348. W. Xue, X. Gao, R. Brandenberger, IR divergences in inflation and entropy perturbations. JCAP **1206**, 035 (2012). [arXiv:1201.0768]
349. Y. Hosotani, Exact solution to the Einstein–Yang–Mills equation. Phys. Lett. B **147**, 44 (1984)
350. D.V. Galt'sov, M.S. Volkov, Yang–Mills cosmology. Cold matter for a hot universe. Phys. Lett. B **256**, 17 (1991)
351. L.H. Ford, Inflation driven by a vector field. Phys. Rev. D **40**, 967 (1989)
352. A.B. Burd, J.E. Lidsey, Analysis of inflationary models driven by vector fields. Nucl. Phys. B **351**, 679 (1991)
353. M.C. Bento, O. Bertolami, P. Vargas Moniz, J.M. Mourão, P.M. Sá, On the cosmology of massive vector fields with $SO(3)$ global symmetry. Class. Quantum Grav. **10**, 285 (1993). [arXiv:gr-qc/9302034]

354. A. Golovnev, V. Mukhanov, V. Vanchurin, Vector inflation. JCAP **0806**, 009 (2008). [arXiv:0802.2068]
355. K. Bamba, S. Nojiri, S.D. Odintsov, Inflationary cosmology and the late-time accelerated expansion of the universe in nonminimal Yang–Mills-$F(R)$ gravity and nonminimal vector-$F(R)$ gravity. Phys. Rev. D **77**, 123532 (2008). [arXiv:0803.3384]
356. K. Dimopoulos, M. Karčiauskas, D.H. Lyth, Y. Rodríguez, Statistical anisotropy of the curvature perturbation from vector field perturbations. JCAP **0905**, 013 (2009). [arXiv:0809.1055]
357. B. Himmetoglu, C.R. Contaldi, M. Peloso, Instability of anisotropic cosmological solutions supported by vector fields. Phys. Rev. Lett. **102**, 111301 (2009). [arXiv:0809.2779]
358. A. Golovnev, V. Mukhanov, V. Vanchurin, Gravitational waves in vector inflation. JCAP **0811**, 018 (2008). [arXiv:0810.4304]
359. E. Dimastrogiovanni, N. Bartolo, S. Matarrese, A. Riotto, Non-Gaussianity and statistical anisotropy from vector field populated inflationary models. Adv. Astron. **2010**, 752670 (2010). [arXiv:1001.4049]
360. K. Dimopoulos, Can a vector field be responsible for the curvature perturbation in the universe? Phys. Rev. D **74**, 083502 (2006). [arXiv:hep-ph/0607229]
361. K. Dimopoulos, Supergravity inspired vector curvaton. Phys. Rev. D **76**, 063506 (2007). [arXiv:0705.3334]
362. K. Dimopoulos, M. Karčiauskas, Non-minimally coupled vector curvaton. JHEP **0807**, 119 (2008). [arXiv:0803.3041]
363. S. Yokoyama, J. Soda, Primordial statistical anisotropy generated at the end of inflation. JCAP **0808**, 005 (2008). [arXiv:0805.4265]
364. S. Kanno, M. Kimura, J. Soda, S. Yokoyama, Anisotropic inflation from vector impurity. JCAP **0808**, 034 (2008). [arXiv:0806.2422]
365. M. Karčiauskas, K. Dimopoulos, D.H. Lyth, Anisotropic non-Gaussianity from vector field perturbations. Phys. Rev. D **80**, 023509 (2009). [arXiv:0812.0264]
366. M.-a. Watanabe, S. Kanno, J. Soda, Inflationary universe with anisotropic hair. Phys. Rev. Lett. **102**, 191302 (2009). [arXiv:0902.2833]
367. M. Novello, S.E. Perez Bergliaffa, J. Salim, Nonlinear electrodynamics and the acceleration of the universe. Phys. Rev. D **69**, 127301 (2004). [arXiv:astro-ph/0312093]
368. V.V. Kiselev, Vector field as a quintessence partner. Class. Quantum Grav. **21**, 3323 (2004). [arXiv:gr-qc/0402095]
369. C. Armendáriz-Picón, Could dark energy be vector-like? JCAP **0407**, 007 (2004). [arXiv:astro-ph/0405267]
370. H. Wei, R.-G. Cai, Interacting vector-like dark energy, the first and second cosmological coincidence problems. Phys. Rev. D **73**, 083002 (2006). [arXiv:astro-ph/0603052]
371. C.G. Boehmer, T. Harko, Dark energy as a massive vector field. Eur. Phys. J. C **50**, 423 (2007). [arXiv:gr-qc/0701029]
372. J. Beltrán Jiménez, A.L. Maroto, Cosmic vector for dark energy. Phys. Rev. D **78**, 063005 (2008). [arXiv:0801.1486]
373. T. Koivisto, D.F. Mota, Vector field models of inflation and dark energy. JCAP **0808**, 021 (2008). [arXiv:0805.4229]
374. C. Germani, A. Kehagias, P-nflation: generating cosmic inflation with p-forms. JCAP **0903**, 028 (2009). [arXiv:0902.3667]
375. T. Kobayashi, S. Yokoyama, Gravitational waves from p-form inflation. JCAP **0905**, 004 (2009). [arXiv:0903.2769]
376. T.S. Koivisto, D.F. Mota, C. Pitrou, Inflation from N-forms and its stability. JHEP **0909**, 092 (2009). [arXiv:0903.4158]
377. T.S. Koivisto, N.J. Nunes, Three-form cosmology. Phys. Lett. B **685**, 105 (2010). [arXiv:0907.3883]
378. C. Germani, A. Kehagias, Scalar perturbations in p-nflation: the 3-form case. JCAP **0911**, 005 (2009). [arXiv:0908.0001]
379. T.S. Koivisto, N.J. Nunes, Inflation and dark energy from three-forms. Phys. Rev. D **80**, 103509 (2009). [arXiv:0908.0920]

380. T.S. Koivisto, F.R. Urban, Three-magnetic fields. Phys. Rev. D **85**, 083508 (2012) [arXiv:1112.1356]
381. T. Banks, Relaxation of the cosmological constant. Phys. Rev. Lett. **52**, 1461 (1984)
382. C. Armendáriz-Picón, P.B. Greene, Spinors, inflation, and nonsingular cyclic cosmologies. Gen. Relat. Grav. **35**, 1637 (2003). [arXiv:hep-th/0301129]
383. B. Saha, T. Boyadjiev, Bianchi type I cosmology with scalar and spinor fields. Phys. Rev. D **69**, 124010 (2004). [arXiv:gr-qc/0311045]
384. M.O. Ribas, F.P. Devecchi, G.M. Kremer, Fermions as sources of accelerated regimes in cosmology. Phys. Rev. D **72**, 123502 (2005). [arXiv:gr-qc/0511099]
385. B. Saha, Spinor field and accelerated regimes in cosmology. Grav. Cosmol. **12**, 215 (2006). [arXiv:gr-qc/0512050]
386. B. Saha, Nonlinear spinor field in Bianchi type-I cosmology: inflation, isotropization, and late time acceleration. Phys. Rev. D **74**, 124030 (2006)
387. C.G. Böhmer, D.F. Mota, CMB anisotropies and inflation from non-standard spinors. Phys. Lett. B **663**, 168 (2008). [arXiv:0710.2003]
388. M.O. Ribas, F.P. Devecchi, G.M. Kremer, Cosmological model with non-minimally coupled fermionic field. Europhys. Lett. **81**, 19001 (2008). [arXiv:0710.5155]
389. C.G. Böhmer, Dark spinor inflation: theory primer and dynamics. Phys. Rev. D **77**, 123535 (2008). [arXiv:0804.0616]
390. Y.-F. Cai, J. Wang, Dark energy model with spinor matter and its quintom scenario. Class. Quantum Grav. **25**, 165014 (2008). [arXiv:0806.3890]
391. D. Gredat, S. Shankaranarayanan, Modified scalar and tensor spectra in spinor driven inflation. JCAP **1001**, 008 (2010). [arXiv:0807.3336]
392. D.G. Caldi, A. Chodos, Cosmological neutrino condensates. arXiv:hep-ph/9903416
393. T. Inagaki, X. Meng, T. Murata, Dark energy problem in a four fermion interaction model. arXiv:hep-ph/0306010
394. F. Giacosa, R. Hofmann, M. Neubert, A model for the very early universe. JHEP **0802**, 077 (2008). [arXiv:0801.0197]
395. S. Alexander, T. Biswas, The cosmological BCS mechanism and the big bang singularity. Phys. Rev. D **80**, 023501 (2009). [arXiv:0807.4468]
396. S. Alexander, T. Biswas, G. Calcagni, Cosmological Bardeen–Cooper–Schrieffer condensate as dark energy. Phys. Rev. D **81**, 043511 (2010); Erratum-ibid. D **81**, 069902(E) (2010). [arXiv:0906.5161]
397. N.J. Popławski, Cosmological constant from quarks and torsion. Ann. Phys. (Berlin) **523**, 291 (2011). [arXiv:1005.0893]
398. J.M. Weller, Fermion condensate from torsion in the reheating era after inflation. Phys. Rev. D **88**, 083511 (2013). [arXiv:1307.2423]
399. D.H. Lyth, A. Riotto, Particle physics models of inflation and the cosmological density perturbation. Phys. Rep. **314**, 1 (1999). [arXiv:hep-ph/9807278]
400. G. Aad et al. [ATLAS Collaboration], Observation of a new particle in the search for the Standard Model Higgs boson with the ATLAS detector at the LHC. Phys. Lett. B **716**, 1 (2012). [arXiv:1207.7214]
401. S. Chatrchyan et al. [CMS Collaboration], Observation of a new boson at a mass of 125 GeV with the CMS experiment at the LHC. Phys. Lett. B **716**, 30 (2012). [arXiv:1207.7235]
402. D. Carmi, A. Falkowski, E. Kuflik, T. Volansky, J. Zupan, Higgs after the discovery: a status report. JHEP **1210**, 196 (2012). [arXiv:1207.1718]
403. K.A. Olive et al. [Particle Data Group], Review of particle physics. Chin. Phys. C **38**, 090001 (2014)
404. J.J. van der Bij, Can gravity make the Higgs particle decouple? Acta Phys. Polon. B **25**, 827 (1994)
405. J.J. van der Bij, Can gravity play a role at the electroweak scale? Int. J. Phys. **1**, 63 (1995). [arXiv:hep-ph/9507389]
406. J.L. Cervantes-Cota, H. Dehnen, Induced gravity inflation in the standard model of particle physics. Nucl. Phys. B **442**, 391 (1995). [arXiv:astro-ph/9505069]

407. F.L. Bezrukov, M. Shaposhnikov, The Standard Model Higgs boson as the inflaton. Phys. Lett. B **659**, 703 (2008). [arXiv:0710.3755]
408. A.O. Barvinsky, A.Yu. Kamenshchik, A.A. Starobinsky, Inflation scenario via the Standard Model Higgs boson and LHC. JCAP **0811**, 021 (2008). [arXiv:0809.2104]
409. F. Bezrukov, D. Gorbunov, M. Shaposhnikov, On initial conditions for the hot big bang. JCAP **0906**, 029 (2009). [arXiv:0812.3622]
410. J. García-Bellido, D.G. Figueroa, J. Rubio, Preheating in the standard model with the Higgs inflaton coupled to gravity. Phys. Rev. D **79**, 063531 (2009). [arXiv:0812.4624]
411. A. De Simone, M.P. Hertzberg, F. Wilczek, Running inflation in the Standard Model. Phys. Lett. B **678**, 1 (2009). [arXiv:0812.4946]
412. F.L. Bezrukov, A. Magnin, M. Shaposhnikov, Standard Model Higgs boson mass from inflation. Phys. Lett. B **675**, 88 (2009). [arXiv:0812.4950]
413. C.P. Burgess, H.M. Lee, M. Trott, Power-counting and the validity of the classical approximation during inflation. JHEP **0909**, 103 (2009). [arXiv:0902.4465]
414. J.L.F. Barbón, J.R. Espinosa, Naturalness of Higgs inflation. Phys. Rev. D **79**, 081302 (2009). [arXiv:0903.0355]
415. F. Bezrukov, M. Shaposhnikov, Standard model Higgs boson mass from inflation: two loop analysis. JHEP **0907**, 089 (2009). [arXiv:0904.1537]
416. A.O. Barvinsky, A.Yu. Kamenshchik, C. Kiefer, A.A. Starobinsky, C.F. Steinwachs, Asymptotic freedom in inflationary cosmology with a non-minimally coupled Higgs field. JCAP **0912**, 003 (2009). [arXiv:0904.1698]
417. T.E. Clark, B. Liu, S.T. Love, T. ter Veldhuis, Standard model Higgs boson-inflaton and dark matter. Phys. Rev. D **80**, 075019 (2009). [arXiv:0906.5595]
418. A.O. Barvinsky, A.Yu. Kamenshchik, C. Kiefer, A.A. Starobinsky, C.F. Steinwachs, Higgs boson, renormalization group, and naturalness in cosmology. Eur. Phys. J. C **72**, 2219 (2012). [arXiv:0910.1041]
419. A.O. Barvinsky, A.Yu. Kamenshchik, C. Kiefer, C.F. Steinwachs, Tunneling cosmological state revisited: origin of inflation with a nonminimally coupled standard model Higgs inflaton. Phys. Rev. D **81**, 043530 (2010). [arXiv:0911.1408]
420. R.N. Lerner, J. McDonald, Higgs inflation and naturalness. JCAP **1004**, 015 (2010). [arXiv:0912.5463]
421. M. Atkins, X. Calmet, On the unitarity of linearized general relativity coupled to matter. Phys. Lett. B **695**, 298 (2011). [arXiv:1002.0003]
422. C.P. Burgess, H.M. Lee, M. Trott, Comment on Higgs inflation and naturalness. JHEP **1007**, 007 (2010). [arXiv:1002.2730]
423. M.P. Hertzberg, On inflation with non-minimal coupling. JHEP **1011**, 023 (2010). [arXiv:1002.2995]
424. D.I. Kaiser, Conformal transformations with multiple scalar fields. Phys. Rev. D **81**, 084044 (2010). [arXiv:1003.1159]
425. F. Bezrukov, A. Magnin, M. Shaposhnikov, S. Sibiryakov, Higgs inflation: consistency and generalisations. JHEP **1101**, 016 (2011). [arXiv:1008.5157]
426. L.A. Popa, A. Caramete, Cosmological constraints on Higgs boson mass. Astrophys. J. **723**, 803 (2010). [arXiv:1009.1293]
427. M. Atkins, X. Calmet, Remarks on Higgs inflation. Phys. Lett. B **697**, 37 (2011). [arXiv:1011.4179]
428. F. Bauer, D.A. Demir, Higgs–Palatini inflation and unitarity. Phys. Lett. B **698**, 425 (2011). [arXiv:1012.2900]
429. R.N. Lerner, J. McDonald, Distinguishing Higgs inflation and its variants. Phys. Rev. D **83**, 123522 (2011). [arXiv:1104.2468]
430. F. Bezrukov, D. Gorbunov, M. Shaposhnikov, Late and early time phenomenology of Higgs-dependent cutoff. JCAP **1110**, 001 (2011). [arXiv:1106.5019]
431. L.A. Popa, Observational consequences of the standard model Higgs inflation variants. JCAP **1110**, 025 (2011). [arXiv:1107.3436]

432. K. Nakayama, F. Takahashi, Higgs mass and inflation. Phys. Lett. B **707**, 142 (2012). [arXiv:1108.3762]
433. P.A.R. Ade et al. [Planck Collaboration], Planck 2013 results. XXII. Constraints on inflation. Astron. Astrophys. **571**, A22 (2014). [arXiv:1303.5082]
434. C. Germani, A. Kehagias, New model of inflation with nonminimal derivative coupling of Standard Model Higgs boson to gravity. Phys. Rev. Lett. **105**, 011302 (2010). [arXiv:1003.2635]
435. C. Germani, A. Kehagias, Cosmological perturbations in the new Higgs inflation. JCAP **1005**, 019 (2010); Erratum-ibid. **1006**, E01 (2010). [arXiv:1003.4285]
436. G.F. Giudice, H.M. Lee, Unitarizing Higgs inflation. Phys. Lett. B **694**, 294 (2011). [arXiv:1010.1417]
437. R.N. Lerner, J. McDonald, Unitarity-violation in generalized Higgs inflation models. JCAP **1211**, 019 (2012). [arXiv:1112.0954]
438. R.N. Lerner, J. McDonald, Unitarity-conserving Higgs inflation model. Phys. Rev. D **82**, 103525 (2010). [arXiv:1005.2978]
439. K. Nakayama, F. Takahashi, Higgs chaotic inflation in standard model and NMSSM. JCAP **1102**, 010 (2011). [arXiv:1008.4457]
440. K. Kamada, T. Kobayashi, M. Yamaguchi, J.'i. Yokoyama, Higgs G-inflation. Phys. Rev. D **83**, 083515 (2011). [arXiv:1012.4238]
441. J. Wess, B. Zumino, Supergauge transformations in four dimensions. Nucl. Phys. B **70**, 39 (1974)
442. J. Wess, B. Zumino, A Lagrangian model invariant under supergauge transformations. Phys. Lett. B **49**, 52 (1974)
443. A. Salam, J.A. Strathdee, Supergauge transformations. Nucl. Phys. B **76**, 477 (1974)
444. S. Ferrara, J. Wess, B. Zumino, Supergauge multiplets and superfields. Phys. Lett. B **51**, 239 (1974)
445. J.D. Lykken, Introduction to supersymmetry. arXiv:hep-th/9612114
446. E. Cremmer, J. Scherk, The supersymmetric non-linear σ-model in four dimensions and its coupling to supergravity. Phys. Lett. B **74**, 341 (1978)
447. B. Zumino, Supersymmetry and Kähler manifolds. Phys. Lett. B **87**, 203 (1979)
448. R.L. Arnowitt, P. Nath, B. Zumino, Superfield densities and action principle in curved superspace. Phys. Lett. B **56**, 81 (1975)
449. D.Z. Freedman, P. van Nieuwenhuizen, S. Ferrara, Progress toward a theory of supergravity. Phys. Rev. D **13**, 3214 (1976)
450. S. Deser, B. Zumino, Consistent supergravity. Phys. Lett. B **62**, 335 (1976)
451. E. Cremmer, B. Julia, J. Scherk, P. van Nieuwenhuizen, S. Ferrara, L. Girardello, Super-Higgs effect in supergravity with general scalar interactions. Phys. Lett. B **79**, 231 (1978)
452. E. Cremmer, B. Julia, J. Scherk, S. Ferrara, L. Girardello, P. van Nieuwenhuizen, Spontaneous symmetry breaking and Higgs effect in supergravity without cosmological constant. Nucl. Phys. B **147**, 105 (1979)
453. R. Barbieri, S. Ferrara, D.V. Nanopoulos, K.S. Stelle, Supergravity, R invariance and spontaneous supersymmetry breaking. Phys. Lett. B **113**, 219 (1982)
454. E. Witten, J. Bagger, Quantization of Newton's constant in certain supergravity theories. Phys. Lett. B **115**, 202 (1982)
455. J. Wess, J. Bagger, *Supersymmetry and Supergravity* (Princeton University Press, Princeton, 1992)
456. J. Wess, B. Zumino, Superspace formulation of supergravity. Phys. Lett. B **66**, 361 (1977)
457. J. Wess, B. Zumino, Superfield Lagrangian for supergravity. Phys. Lett. B **74**, 51 (1978)
458. M. Müller, The density multiplet in superspace. Z. Phys. C **16**, 41 (1982)
459. N.-P. Chang, S. Ouvry, X. Wu, $N = 1$ supergravity with nonminimal coupling: a class of models. Phys. Rev. Lett. **51**, 327 (1983)
460. E. Cremmer, S. Ferrara, L. Girardello, A. Van Proeyen, Coupling supersymmetric Yang–Mills theories to supergravity. Phys. Lett. B **116**, 231 (1982)

461. E. Cremmer, S. Ferrara, L. Girardello, A. Van Proeyen, Yang–Mills theories with local supersymmetry: Lagrangian, transformation laws and super-Higgs effect. Nucl. Phys. B **212**, 413 (1983)
462. J.A. Bagger, Coupling the gauge-invariant supersymmetric non-linear sigma model to supergravity. Nucl. Phys. B **211**, 302 (1983)
463. E.D. Stewart, Inflation, supergravity, and superstrings. Phys. Rev. D **51**, 6847, (1995). [arXiv:hep-ph/9405389]
464. D.V. Nanopoulos, K.A. Olive, M. Srednicki, K. Tamvakis, Primordial inflation in simple supergravity. Phys. Lett. B **123**, 41 (1983)
465. G.B. Gelmini, D.V. Nanopoulos, K.A. Olive, Finite temperature effects in primordial inflation. Phys. Lett. B **131**, 53 (1983)
466. A.D. Linde, Primordial inflation without primordial monopoles. Phys. Lett. B **132**, 317 (1983)
467. B.A. Ovrut, P.J. Steinhardt, Supersymmetry and inflation: a new approach. Phys. Lett. B **133**, 161 (1983)
468. R. Holman, P. Ramond, G.G. Ross, Supersymmetric inflationary cosmology. Phys. Lett. B **137**, 343 (1984)
469. B.A. Ovrut, P.J. Steinhardt, Inflationary cosmology and the mass hierarchy in locally supersymmetric theories. Phys. Rev. Lett. **53**, 732 (1984)
470. E. Cremmer, S. Ferrara, C. Kounnas, D.V. Nanopoulos, Naturally vanishing cosmological constant in $N = 1$ supergravity. Phys. Lett. B **133**, 61 (1983)
471. J.R. Ellis, A.B. Lahanas, D.V. Nanopoulos, K. Tamvakis, No-scale supersymmetric standard model. Phys. Lett. B **134**, 429 (1984)
472. J.R. Ellis, C. Kounnas, D.V. Nanopoulos, Phenomenological $SU(1,1)$ supergravity. Nucl. Phys. B **241**, 406 (1984)
473. J.R. Ellis, C. Kounnas, D.V. Nanopoulos, No-scale supersymmetric GUTs. Nucl. Phys. B **247**, 373 (1984)
474. J.R. Ellis, C. Kounnas, D.V. Nanopoulos, No-scale supergravity models with a Planck mass gravitino. Phys. Lett. B **143**, 410 (1984)
475. N. Dragon, M.G. Schmidt, U. Ellwanger, Sliding scales in minimal supergravity. Phys. Lett. B **145**, 192 (1984)
476. R. Barbieri, E. Cremmer, S. Ferrara, Flat and positive potentials in $N = 1$ supergravity. Phys. Lett. B **163**, 143 (1985)
477. A.B. Lahanas, D.V. Nanopoulos, The road to no-scale supergravity. Phys. Rep. **145**, 1 (1987)
478. G. Gelmini, C. Kounnas, D.V. Nanopoulos, Primordial inflation with flat supergravity potentials. Nucl. Phys. B **250**, 177 (1985)
479. A.S. Goncharov, A.D. Linde, A simple realisation of the inflationary Universe scenario in $SU(1,1)$ supergravity. Class. Quantum Grav. **1**, L75 (1984)
480. J.R. Ellis, K. Enqvist, D.V. Nanopoulos, K.A. Olive, M. Srednicki, SU(N,1) inflation. Phys. Lett. B **152**, 175 (1985); Erratum-ibid. B **156**, 452 (1985)
481. H. Murayama, H. Suzuki, T. Yanagida, J. Yokoyama, Chaotic inflation and baryogenesis in supergravity. Phys. Rev. D **50**, 2356 (1994). [arXiv:hep-ph/9311326]
482. M. Kawasaki, M. Yamaguchi, T. Yanagida, Natural chaotic inflation in supergravity. Phys. Rev. Lett. **85**, 3572 (2000). [arXiv:hep-ph/0004243]
483. R. Kallosh, A. Linde, New models of chaotic inflation in supergravity. JCAP **1011**, 011 (2010). [arXiv:1008.3375]
484. D. Croon, J. Ellis, N.E. Mavromatos, Wess–Zumino inflation in light of Planck. Phys. Lett. B **724**, 165 (2013). [arXiv:1303.6253]
485. K. Nakayama, F. Takahashi, T.T. Yanagida, Polynomial chaotic inflation in the Planck era. Phys. Lett. B **725**, 111 (2013). [arXiv:1303.7315]
486. M. Kawasaki, M. Yamaguchi, Supersymmetric topological inflation model. Phys. Rev. D **65**, 103518 (2002). [arXiv:hep-ph/0112093]
487. M.B. Einhorn, D.R.T. Jones, Inflation with non-minimal gravitational couplings in supergravity. JHEP **1003**, 026 (2010). [arXiv:0912.2718]

488. S. Ferrara, R. Kallosh, A. Linde, A. Marrani, A. Van Proeyen, Jordan frame supergravity and inflation in the NMSSM. Phys. Rev. D **82**, 045003 (2010). [arXiv:1004.0712]
489. H.M. Lee, Chaotic inflation in Jordan frame supergravity. JCAP **1008**, 003 (2010). [arXiv:1005.2735]
490. S. Ferrara, R. Kallosh, A. Linde, A. Marrani, A. Van Proeyen, Superconformal symmetry, NMSSM, and inflation. Phys. Rev. D **83**, 025008 (2011). [arXiv:1008.2942]
491. I. Ben-Dayan, M.B. Einhorn, Supergravity Higgs inflation and shift symmetry in electroweak theory. JCAP **1012**, 002 (2010). [arXiv:1009.2276]
492. K. Nakayama, F. Takahashi, General analysis of inflation in the Jordan frame supergravity. JCAP **1011**, 039 (2010). [arXiv:1009.3399]
493. M. Arai, S. Kawai, N. Okada, Higgs inflation in minimal supersymmetric $SU(5)$ GUT. Phys. Rev. D **84**, 123515 (2011). [arXiv:1107.4767]
494. R. Kallosh, A. Linde, Superconformal generalization of the chaotic inflation model $\frac{\lambda}{4}\phi^4 - \frac{\xi}{2}\phi^2 R$. JCAP **1306**, 027 (2013). [arXiv:1306.3211]
495. R. Kallosh, A. Linde, Universality class in conformal inflation. JCAP **1307**, 002 (2013). [arXiv:1306.5220]
496. A.A. Starobinsky, A new type of isotropic cosmological models without singularity. Phys. Lett. B **91**, 99 (1980)
497. V.F. Mukhanov, G.V. Chibisov, Quantum fluctuation and nonsingular universe. Pis'ma Zh. Eksp. Teor. Fiz. **33**, 549 (1981) [JETP Lett. **33**, 532 (1981)]
498. J.D. Barrow, A.C. Ottewill, The stability of general relativistic cosmological theory. J. Phys. A **16**, 2757 (1983)
499. A.A. Starobinskiĭ, The perturbation spectrum evolving from a nonsingular initially de Sitter cosmology and the microwave background anisotropy. Pis'ma Astron. Zh. **9**, 579 (1983) [Sov. Astron. Lett. **9**, 302 (1983)]
500. B. Whitt, Fourth-order gravity as general relativity plus matter. Phys. Lett. B **145**, 176 (1984)
501. L.A. Kofman, A.D. Linde, A.A. Starobinsky, Inflationary universe generated by the combined action of a scalar field and gravitational vacuum polarization. Phys. Lett. B **157**, 361 (1985)
502. A.A. Starobinsky, H.-J. Schmidt, On a general vacuum solution of fourth-order gravity. Class. Quantum Grav. **4**, 695 (1987)
503. K.-i. Maeda, Inflation as a transient attractor in R^2 cosmology. Phys. Rev. D **37**, 858 (1988)
504. K.-i. Maeda, J.A. Stein-Schabes, T. Futamase, Inflation in a renormalizable cosmological model and the cosmic no-hair conjecture. Phys. Rev. D **39**, 2848 (1989)
505. J. Ellis, D.V. Nanopoulos, K.A. Olive, No-scale supergravity realization of the Starobinsky model of inflation. Phys. Rev. Lett. **111**, 111301 (2013); Erratum-ibid. **111**, 129902 (2013). [arXiv:1305.1247]
506. J. Ellis, D.V. Nanopoulos, K.A. Olive, Starobinsky-like inflationary models as avatars of no-scale supergravity. JCAP **1310**, 009 (2013). [arXiv:1307.3537]
507. S. Cecotti, Higher derivative supergravity is equivalent to standard supergravity coupled to matter. Phys. Lett. B **190**, 86 (1987)
508. S. Cecotti, S. Ferrara, M. Porrati, S. Sabharwal, New minimal higher derivative supergravity coupled to matter. Nucl. Phys. B **306**, 160 (1988)
509. R. Kallosh, A. Linde, Superconformal generalizations of the Starobinsky model. JCAP **1306**, 028 (2013). [arXiv:1306.3214]
510. J.A. Casas, Baryogenesis, inflation and superstrings, in *International Europhysics Conference on High Energy Physics*, ed. by D. Lellouch, G. Mikenberg, E. Rabinovici (Springer, Berlin, 1999). [arXiv:hep-ph/9802210]
511. J. Ellis, M.A.G. García, D.V. Nanopoulos, K.A. Olive, A no-scale inflationary model to fit them all. JCAP **1408**, 044 (2014). [arXiv:1405.0271]
512. J. Ellis, M.A.G. García, D.V. Nanopoulos, K.A. Olive, Two-field analysis of no-scale supergravity inflation. JCAP **1501**, 010 (2015). [arXiv:1409.8197]
513. J. Ellis, M.A.G. García, D.V. Nanopoulos, K.A. Olive, Phenomenological aspects of no-scale inflation models. JCAP **1510**, 003 (2015). [arXiv:1503.08867]

514. W. Buchmuller, V. Domcke, K. Kamada, The Starobinsky model from superconformal D-term inflation. Phys. Lett. B **726**, 467 (2013). [arXiv:1306.3471]
515. F. Farakos, A. Kehagias, A. Riotto, On the Starobinsky model of inflation from supergravity. Nucl. Phys. B **876**, 187 (2013). [arXiv:1307.1137]
516. S.V. Ketov, A.A. Starobinsky, Embedding $R + R^2$ inflation in supergravity. Phys. Rev. D **83**, 063512 (2011). [arXiv:1011.0240]
517. S.V. Ketov, A.A. Starobinsky, Inflation and non-minimal scalar-curvature coupling in gravity and supergravity. JCAP **1208**, 022 (2012). [arXiv:1203.0805]
518. A.S. Goncharov, A.D. Linde, Chaotic inflation of the universe in supergravity. Zh. Eksp. Teor. Fiz. **86**, 1594 (1984) [JETP **59**, 930 (1984)]
519. A.S. Goncharov, A.D. Linde, Chaotic inflation in supergravity. Phys. Lett. B **139**, 27 (1984)
520. R. Kallosh, A. Linde, Planck, LHC, and α-attractors. Phys. Rev. D **91**, 083528 (2015). [arXiv:1502.07733]
521. R. Kallosh, A. Linde, T. Rube, General inflaton potentials in supergravity. Phys. Rev. D **83**, 043507 (2011). [arXiv:1011.5945]
522. R. Kallosh, A. Linde, K.A. Olive, T. Rube, Chaotic inflation and supersymmetry breaking. Phys. Rev. D **84**, 083519 (2011). [arXiv:1106.6025]
523. S. Ferrara, R. Kallosh, A. Linde, M. Porrati, Minimal supergravity models of inflation. Phys. Rev. D **88**, 085038 (2013). [arXiv:1307.7696]
524. R. Kallosh, A. Linde, D. Roest, Superconformal inflationary α-attractors. JHEP **1311**, 198 (2013). [arXiv:1311.0472]
525. S. Cecotti, R. Kallosh, Cosmological attractor models and higher curvature supergravity. JHEP **1405**, 114 (2014). [arXiv:1403.2932]
526. R. Kallosh, A. Linde, D. Roest, Large field inflation and double α-attractors. JHEP **1408**, 052 (2014). [arXiv:1405.3646]
527. R. Kallosh, A. Linde, Escher in the sky. C. R. Phys. **16**, 914 (2015). [arXiv:1503.06785]
528. D. Roest, M. Scalisi, Cosmological attractors from α-scale supergravity. Phys. Rev. D **92**, 043525 (2015). [arXiv:1503.07909]
529. A. Linde, Single-field α-attractors. JCAP **1505**, 003 (2015). [arXiv:1504.00663]
530. J.J.M. Carrasco, R. Kallosh, A. Linde, D. Roest, Hyperbolic geometry of cosmological attractors. Phys. Rev. D **92**, 041301 (2015). [arXiv:1504.05557]
531. J.J.M. Carrasco, R. Kallosh, A. Linde, Cosmological attractors and initial conditions for inflation. Phys. Rev. D **92**, 063519 (2015). [arXiv:1506.00936]
532. J.J.M. Carrasco, R. Kallosh, A. Linde, α-attractors: Planck, LHC and dark energy. JHEP **1510**, 147 (2015). [arXiv:1506.01708]
533. J.A. Casas, C. Muñoz, Inflation from superstrings. Phys. Lett. B **216**, 37 (1989)
534. J.A. Casas, J.M. Moreno, C. Muñoz, M. Quirós, Cosmological implications of an anomalous U(1): inflation, cosmic strings and constraints on superstring parameters. Nucl. Phys. B **328**, 272 (1989)
535. P. Binetruy, G.R. Dvali, D-term inflation. Phys. Lett. B **388**, 241 (1996). [arXiv:hep-ph/9606342]
536. E. Halyo, Hybrid inflation from supergravity D-terms. Phys. Lett. B **387**, 43 (1996). [arXiv:hep-ph/9606423]
537. T. Matsuda, Successful D-term inflation with moduli. Phys. Lett. B **423**, 35 (1998). [arXiv:hep-ph/9705448]
538. G. Calcagni, Slow-roll parameters in braneworld cosmologies. Phys. Rev. D **69**, 103508 (2004). [arXiv:hep-ph/0402126]
539. G. Calcagni, S. Tsujikawa, Observational constraints on patch inflation in noncommutative spacetime. Phys. Rev. D **70**, 103514 (2004). [arXiv:astro-ph/0407543]
540. G. Calcagni, Braneworld Cosmology and Noncommutative Inflation. Ph.D. thesis, Parma University, Parma (2005). [arXiv:hep-ph/0503044]

Chapter 6
Big-Bang Problem

Vaccha, the speculative view that the world is eternal... that the world is not eternal... that the world is finite... that the world is infinite is a thicket of views, a wilderness of views, a contortion of views, a vacillation of views, a fetter of views. It is beset by suffering, by vexation, by despair, and by fever, and it does not lead to disenchantment, to dispassion, to cessation, to peace, to direct knowledge, to enlightenment, to Nibbāna.

— *Aggivacchagotta Sutta, Majjhima Nikāya* 72.14 [1]

Contents

The big-bang problem is an open issue in theoretical physics. In this and the next chapters, we will review some of the proposals, few of which successful or completely satisfactory, advanced to solve it. Let us first explain why the big bang is a problem at the classical level. We ask ourselves the following question:

Is the big-bang singularity typical in the inflationary universe? (6.1)

G. Calcagni, *Classical and Quantum Cosmology*, Graduate Texts in Physics,
DOI 10.1007/978-3-319-41127-9_6

We shall give a positive answer thanks to a set of theorems valid in *classical* spacetimes (Sect. 6.2). These theorems determine the existence of singularities in certain spacetimes, but they do not yield any information about their structure. Classical singularities can be classified (Sect. 6.1.3) and their details studied by standard techniques. In Sect. 6.3, the example of the chaotic BKL singularity in anisotropic universes is presented in some detail; it will be further developed in Sect. 13.9.2.

6.1 Spacetimes and Singularities

In Chap. 2, we have discussed several cosmological solutions of very general character. In particular, power-law solutions $a(t) \propto t^p$ are instrumental in understanding the inflationary expansion and the history of the universe after reheating. They are singular at the big bang,

$$\lim_{t \to t_{\rm bb}} a(t) = 0\,, \tag{6.2}$$

where $t_{\rm bb} \sim 0$ is the proper-time big bang event. At this point, the laws of physics break down and the history of the universe we worked out at such a painstaking length turns out to be based on ill-defined initial conditions. Inspecting the Friedmann and continuity equations, one sees that the energy density blows up in the past for ordinary matter fields. One might hope that this pathological behaviour occurs only in special situations. After all, de Sitter and power-law cosmologies are idealizations of the real world and one might have inflation work out on a more complicated but non-singular background $a(t)$, perhaps to be analyzed via numerical methods. However, particular non-singular solutions would rely on particular initial conditions, while general results are desirable to assess "how often" these regular scenarios are realized. Unfortunately, the big-bang singularity is intrinsic not only to the Einstein equations, but also to other gravitational dynamics on a wide class of Lorentzian manifolds. This is the actual reason why the big bang is a serious problem and not just a glitch of special classes of solutions: it is not special at all, and is even expected.

6.1.1 Globally Hyperbolic Spacetimes

Before stating the singularity theorems, we need a few definitions (see, e.g., [2] and the pedagogical introduction [3]). Since we wish to go beyond pure or perturbed FLRW spacetimes, covariant formalism shall be used.

Many of the following concepts have already been employed, such as the notion of *spacetime*. Given a Lorentzian manifold $(\mathcal{M}, g)$, a spacetime is the maximal manifold $\mathcal{M}$ on which the metric $g_{\mu\nu}$ is smooth. We also assumed that all spacetimes we have been dealing with where *time-orientable*, which is the case if there exists a non-vanishing time-like vector field t^μ. This is equivalent to state that all tangent spaces $T_x\mathcal{M}$ can be time-oriented consistently. Time-like vectors $v^\mu \in T_x\mathcal{M}$ are *future-directed* if they have the same orientation defined by t^μ. In other words, $t^\mu v_\mu < 0$. They are *past-directed* when the have opposite orientation.

The Lorentz classification of vectors, curves and surfaces also played a role in, for example, the discussion of the energy conditions (Sect. 2.2.1). Let $\sigma \in I \subseteq \mathbb{R}$ be a real parameter and $\gamma^\mu : I \to \mathcal{M}$ a curve. A curve γ^μ is *space-like*, *time-like* or *null* if its tangent vector $\mathrm{d}\gamma^\mu(\sigma)/\mathrm{d}\sigma$ is, respectively, space-like (positive norm), time-like (negative norm) or null (zero norm) for all $\sigma \in I$. Also, a *future-directed* (respectively, *past-directed*) *causal curve* is a curve γ^μ such that the tangent vector $\mathrm{d}\gamma^\mu(\sigma)/\mathrm{d}\sigma$ is either null or time-like future-oriented (respectively, past-oriented) for all σ. For a time-like geodesic (the observer), the parameter $\sigma = t$ is proper time, while for a null geodesic it is called affine parameter. The concepts of "past" and "future" of an event at point $x \in \mathcal{M}$ can be refined progressively. The *chronological future* of $x \in \mathcal{M}$ is the set $\mathcal{I}^+(x)$ of all points to which x can be connected by a future-directed time-like curve. If the curve is causal, the set of points is called the *causal future* of x and is denoted by $\mathcal{J}^+(x)$. Replacing future-directed with past-directed curves, one obtains the chronological past $\mathcal{I}^-(x)$ and causal past $\mathcal{J}^-(x)$ of x. The difference between the causal and the chronological future of $x \in \mathcal{M}$ is the *future light cone* $\mathcal{E}^+(x) = \mathcal{J}^+(x) \setminus \mathcal{I}^+(x)$. The *past light cone* is $\mathcal{E}^-(x) = \mathcal{J}^-(x) \setminus \mathcal{I}^-(x)$. In particular, for a time-oriented manifold the causal past or future is the closure of the chronological past or future, $\mathcal{J}^\pm(x) = \overline{\mathcal{I}^\pm(x)}$. For every pair $x, y \in \mathcal{M}$, the set $\mathcal{J}^-(x) \cap \mathcal{J}^+(y)$ is the set of all points that can be reached both from x along a past-oriented time-like curve and from y along a future-oriented time-like curve.

Globally, a Lorentzian manifold is *future* (respectively, *past*) *causally simple* if $\mathcal{E}^\pm(x) = \partial\mathcal{I}^\pm(x) \neq \emptyset$ for all $x \in \mathcal{M}$, where $\partial\mathcal{I}^\pm(x)$ is the boundary of $\mathcal{I}^\pm(x)$. Light cones and particle horizons in causally simple spacetimes are shown in Fig. 6.1. Furthermore, $(\mathcal{M}, g)$ is *causal* if there are no closed causal curves, and is *stably causal* if there exists a global time function $t : \mathcal{M} \to \mathbb{R}$ such that $\nabla_\mu t$ is time-like. In particular, stably causal manifolds are time-orientable and do not admit closed time-like curves, so that they are causal. Stable causality is a stronger requirement than causality.

We are finally able to define global hyperbolicity [4, 5], which is an important ingredient of singularity theorems and of the canonical formulation of classical gravity. A Lorentzian manifold $(\mathcal{M}, g)$ is *globally hyperbolic* if:

(i) it is causal, and
(ii) $\mathcal{J}^+(x) \cap \mathcal{J}^-(y)$ is compact for all $x, y \in \mathcal{M}$.

See Fig. 6.2. We can give an alternative definition of global hyperbolicity. This version is more restrictive and hence less preferable, but we mention it because it is widely found in the literature. First, we need a few more definitions. A smooth

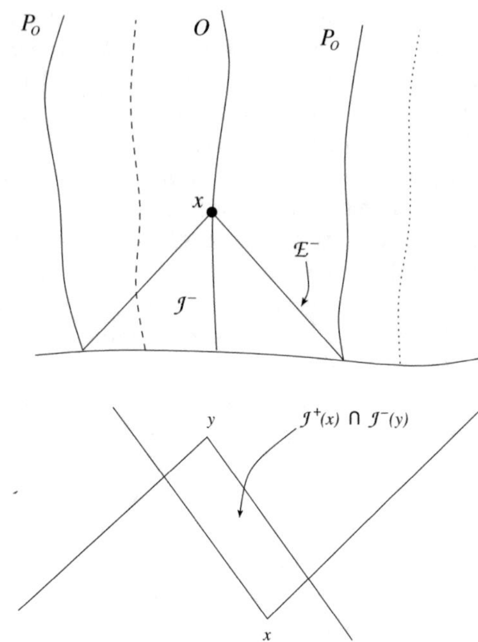

Fig. 6.1 Past light cone $\mathcal{E}^-$ of an observer O at point x in a causally simple spacetime. Particles (*dashed world-line*) inside the particle horizon $\mathcal{P}_O$ have been observed by O, while those outside (*dotted world-line*) not yet. The *bottom curve* is the space-like past infinity

Fig. 6.2 The intersection $\mathcal{J}^+(x) \cap \mathcal{J}^-(y)$ of the causal future and past of any two points in a globally hyperbolic manifold is a compact set

future-directed causal curve $\gamma^\mu : (a, b) \to \mathcal{M}$ (with possibly $a = -\infty$ or $b = +\infty$) is *future-inextendible* if $\lim_{\sigma \to b} \gamma^\mu(\sigma)$ does not exist. A *past-inextendible* past-directed curve is defined analogously. Also, the *past domain of dependence* of $\Sigma \subset \mathcal{M}$ is the set $\mathcal{D}^-(\Sigma)$ of all points $x \in \mathcal{M}$ such that any future-inextendible causal curve starting at x intersects Σ. The *future domain of dependence* $\mathcal{D}^+(\Sigma)$ is defined analogously. The *domain of dependence* of Σ is the set $\mathcal{D}(\Sigma) = \mathcal{D}^-(\Sigma) \cup \mathcal{D}^+(\Sigma)$. Then, a Lorentzian manifold is globally hyperbolic if

(i) it is stably causal, and
(ii) for any $a \in \mathbb{R}$, the time slice $\Sigma_a = t^{-1}(a)$ defined as the pre-image of the global time function t is such that $\mathcal{D}(\Sigma_a) = \mathcal{M}$.

Global hyperbolicity can also be characterized by Geroch splitting theorem. Let $\Sigma \subset \mathcal{M}$ be a *Cauchy hypersurface* in $\mathcal{M}$, i.e., a sub-set intersected exactly once by every inextendible time-like curve in $\mathcal{M}$. Roughly speaking, the future (and past) development of $\mathcal{M}$ can be pre- (retro-)dicted from data on Σ. Then, we have [6, 7]:

Geroch splitting theorem. *Any globally hyperbolic spacetime* $(\mathcal{M}, g)$ *admits a smooth space-like Cauchy hypersurface* Σ *and, then, it is diffeomorphic to* $\mathbb{R} \times \Sigma$.

In other words, a global time function t can be chosen in such a way that each surface of constant t is a Cauchy surface Σ_t and the spacetime topology is $\mathcal{M} \cong \mathbb{R} \times \Sigma$. The set $\{\Sigma_t\}$ is called a *foliation* of spacetime.

Cosmology obviously requires a time orientation and a foliation; in previous chapters we tacitly assumed that spacetime is globally hyperbolic. One can also extend the concepts of "open" and "closed" universe to non-FLRW spacetimes. A generic Lorentzian manifold is *open* if it contains no compact (i.e., without boundary) edgeless surfaces where any two points cannot be connected by time-like curves. An alternative but more intuitive definition makes use of a compact space-like surface. According to this definition, universes previously called "flat" are open. A spacetime is *closed* if all its space-like surfaces are compact.

6.1.2 Focusing Theorems

In the presence of a singularity, we expect that all time-like or null geodesics will focus (i.e., converge) at a point in the past or future. Any time-like vector t^μ obeys the *Raychaudhuri equation* (Problem 6.1),

$$\dot{\theta} = \omega_{\mu\nu}\omega^{\mu\nu} - \sigma_{\mu\nu}\sigma^{\mu\nu} - R_{\mu\nu}t^\mu t^\nu - \frac{1}{D-1}\theta^2 + \nabla_\mu \dot{t}^\mu \,. \tag{6.3}$$

Let $t^\mu = u^\mu$ be a unit *time-like congruence*, i.e., a family of integral curves such that[1]

$$u^\nu \nabla_\nu u^\mu = 0 = u_\mu \nabla_\nu u^\mu \,, \qquad u^2 := u_\mu u^\mu = -1 \,. \tag{6.4}$$

When an observer with velocity u^μ follows a time-like congruence, its acceleration $a^\mu = \dot{u}^\mu = u^\nu \nabla_\nu u^\mu$ is identically zero. Also the last term of (6.3) vanishes. Convergence of congruences is governed by the *focusing theorem*, which we now state for time-like curves.

Focusing theorem 1. *Let $(\mathcal{M}, g)$ be a spacetime obeying the time-like convergence condition (2.38), $R_{\mu\nu}u^\mu u^\nu \geqslant 0$, where u^μ is a time-like congruence of past-directed geodesics such that its volume expansion satisfies $\theta \leqslant \theta_0 < 0$ and the vorticity $\omega_{\mu\nu}$ vanishes. Then, the congruence must blow up ($\theta \to -\infty$) at proper time no greater than $t_0 = (D-1)/|\theta_0|$ in the past.*

Proof If vorticity vanishes and the time-like convergence condition holds, the Raychaudhuri equation for the family of geodesics implies that

$$\dot{\theta} \leqslant -\frac{1}{D-1}\theta^2 \,.$$

[1]In Chap. 2, we used the symbol u^μ for a generic unit time-like vector. Here and in the following, we reserve it for congruences.

Integrating from $\theta(t=0) \leqslant \theta_0$ to $\theta(t)$, we get the lower bound

$$\frac{1}{\theta} \geqslant \frac{1}{\theta_0} + \frac{t}{D-1}, \tag{6.5}$$

where the parameter t is positive and increasing to the past. This means that the congruence must converge to a caustic ($\theta \to -\infty$) at proper time no greater than $t_0 = (D-1)/|\theta_0|$ in the past. □

The theorem implies that all geodesics leaving a point will eventually reconverge after a finite time. Replacing the expansion condition with $\theta \geqslant \theta_0 > 0$, one gets the same result for congruences focusing in the future. Employing the Raychaudhuri equation for null geodesics, one can prove the same theorem almost verbatim for focusing null congruences, with proper time t replaced by an affine parameter σ.[2]

A stronger version of the focusing theorem assumes that the time-like convergence condition is violated everywhere by at least a minimum amount.

Focusing theorem 2. *Let $(\mathcal{M}, g)$ be a Lorentzian manifold obeying the convergence condition*

$$R_{\mu\nu}u^\mu u^\nu \leqslant -\frac{\beta^2}{D-1} < 0, \tag{6.6}$$

where $\beta > 0$ and u^μ is a unit time-like congruence of past-directed geodesics such that its volume expansion satisfies $\theta \leqslant \theta_0 < -\beta$ and the vorticity $\omega_{\mu\nu}$ vanishes. Then, the congruence must converge to a caustic at proper time no greater than

$$t_0 = -\frac{D-1}{\beta}\text{arccoth}\frac{\theta_0}{\beta} > 0. \tag{6.7}$$

Proof From (6.3) and in the absence of shear, the congruence of time-like geodesics obeys

$$\dot{\theta} \geqslant \frac{1}{D-1}(\beta^2 - \theta^2).$$

Integrating as before,

$$\theta \leqslant \beta \coth\left[\frac{\beta(t-t_0)}{D-1}\right],$$

[2]Null congruences will be defined and employed in Sect. 7.7.

so that

$$\theta < -\beta \qquad \text{for} \qquad t < t_0 \,. \tag{6.8}$$

The congruence diverges into the past faster than in the previous case, by the amount given in (6.8). This concludes the proof. □

The typical applications of the last result are inflationary universes.

6.1.3 Classifications of Singularities

Let us clarify what we mean by "singularity." A *space-like singularity* is a point where matter is concentrated with infinite density; time-like geodesics, which represent motions of free particles, cannot be extended beyond that point. A *time-like singularity* is a region of infinite curvature where null geodesics (motions of light rays) end [8].

Then, a spacetime is time-like (respectively, null) *geodesically incomplete* if there exist time-like (null) incomplete geodesics. In this case, spacetime is said to be *singular*. The requirement of maximal extension in the definition of spacetime is to avoid the obvious definition of a regular spacetime as a manifold with the set of singularity points removed. The big bang is both a past space- and time-like singularity because it marks the beginning (more precisely, the boundary) of time and space, respectively. It is also a *naked* singularity, since there is no event horizon to screen it. Note that these singularities are all physical and should be discriminated from coordinate singularities, that can be removed by a suitable recharting of the manifold.

Singularities can be organized in more refined systems. For instance, one speaks about curvature singularities if at least one of the components of the Riemann or Weyl tensors diverges or is discontinuous at a point [9]. If it is the Weyl tensor to be ill behaved, then the singularity is conformal. According to which curvature invariant displays the worst behaviour, the classification is further specialized. Also, if the pathological components are bounded, the curvature singularity is oscillatory, otherwise it is divergent; depending on the model, the big bang can be either (Sect. 6.3).

Moreover, if objects can fall into the singularity without being torn apart by tidal forces, the singularity is called *weak*; if they do not remain intact, it is called *strong* [9]. These definitions can be made more precise by invoking only geometric quantities, thus guaranteeing their validity irrespectively of the physical properties of the bodies [10–12]. In general, physical singularities are strong. The big bang is, by definition, a strong curvature singularity.

While the singularity theorems below guarantee geodesic incompleteness under a set of sufficient conditions, other pathological situations may arise from different necessary requirements and they must be dealt with separately. For the sake of completeness, we should mention that in the context of cosmology the big bang may be not the only singularity one encounters in the evolution of the universe [13–16]. It is possible, in fact, that certain matter configurations violating the dominant energy condition (2.56) lead to *sudden future singularities* (or type II future singularities), where the fluid pressure P and the Ricci scalar R (as well as its time derivatives) blow up in a finite proper time t_* since the big bang [17–20]. The scale factor $a(t_*)$, the Hubble parameter $H(t_*)$ and the energy density $\rho(t_*)$ remain finite, although one has infinite acceleration:

$$a(t_*),\ H(t_*),\ \rho(t_*) < \infty\,, \qquad P, \ddot{a} \to -\infty\,.$$

This can even happen not too far from today, with $t_* - t_0 \sim 10\,\text{Myr}$ [21], or even when the DEC is preserved [15]. Sudden future singularities do not entail geodesic incompleteness and the evolution of the universe may continue through them. They are not, in fact, of strong type and the surge of tidal forces at t_* does not destroy all objects [22].

A more serious case of sudden singularity is the *big rip* (or type I future singularity), where the scale factor, the energy density and pressure of the fluid all go to infinity [23–26]:

$$a,\ H,\ \rho,\ -P,\ -\ddot{a} \to +\infty\,.$$

This is caused by an exotic matter with effective equation of state with $w < -1$. The universe neither collapses into a big crunch (the singularity at the end of a contraction phase) nor expands forever, but it is torn apart (about $\sim 20\,\text{Gyr}$ from now) due to the increase of the energy $\rho \propto a^{-3(1+w)}$ with the expansion.

Big rip and sudden future singularities can be brought about not only by non-conventional matter components but also by modifications of general relativity (e.g., [27]). Exotic equations of state, where $P(\rho) \not\propto \rho$ and the energy density and pressure are not proportional to each other, can be responsible for other types of future singularities, where $\rho, -P \to +\infty$ for finite scale factor (big freeze, type III) [14, 28, 29] or finite a, ρ and P but divergent derivatives of H (big brake, type IV) [14, 30, 31]. Like the big rip, the big freeze is a strong singularity, while the big brake is a weak one.

Much of future singularities depends on the details of the matter content and on the nature of dark energy. It is not clear how a complete theory of quantum gravity would fare with respect to future extremal regimes of curvature and energy and one might argue that the big-bang problem is a more pressing business to conclude. This chapter, therefore, concentrates on the latter, leaving further discussion on the fate of the universe to Chap. 7.

6.2 Singularity Theorems

The presence of the big bang in various scenarios is argued by a series of theorems developed by Penrose, Hawking and Geroch in the 1960s [32–38] (and their later extensions [39–41]) and by Borde, Vilenkin and Guth in the 1990s [42–47]. They hold for *classical* general relativity and some also describe other singularities such as those in the interior of black holes.

6.2.1 Hawking–Penrose Theorems

The strategy of singularity theorems is to follow the evolution of a geodesic (that is, an observer) and see if we can do so indefinitely to the infinite past. All these theorems share the same structure of hypotheses:

- a local convergence or energy condition;
- a global condition on the causal structure;
- an assumption that gravity is strong enough in some regions to force the convergence of geodesics therein.

Playing with the strength of each condition, one can obtain different types of singularity theorems.[3]

Hawking–Penrose theorem. *Let* $(\mathcal{M}, g)$ *be a spacetime such that:*

1. *it obeys the time-like convergence condition (2.38),* $R_{\mu\nu}t^\mu t^\nu \geqslant 0$;
2. *it is globally hyperbolic;*
3. *the volume expansion satisfies* $\theta \leqslant \theta_0 < 0$ *on a time slice* Σ_t.

Then $(\mathcal{M}, g)$ *is singular, i.e., no past-directed time-like geodesics orthogonal to* Σ_t *can be extended to proper time greater than* $t_0 = (D-1)/|\theta_0|$ *to the past of* Σ_t.

Sketch of proof Consider a point $x \in \mathcal{M}$ on a time slice Σ_t. The first focusing theorem prescribes that all geodesics from x reconverge to another point y in the time slice $\Sigma_{t'}$ at finite distance $t'-t$ from Σ_t. This implies that there is no geodesic of maximal length. Such a geodesic exists in globally hyperbolic spacetimes which are geodesically complete. We are thus forced to conclude that $(\mathcal{M}, g)$ is geodesically incomplete. □

Applying the theorem to an expanding universe, this situation describes a singularity in the past but it is straightforward to extend it to future singularities, replacing the expansion condition with $\theta \geqslant \theta_0 > 0$. In both cases, after the finite amount of proper time t_0 a free-falling observer ceases to be represented by a point

[3]We refer the interested reader to [4] for a complete review of older results and [2] for a concise proof of the following theorem by Hawking and Penrose [38].

of the Lorentzian manifold. A similar claim holds when the time-like convergence condition is replaced by (6.6) [45].

Closed time-like curves announce a breakdown of causality and are, perhaps, more objectionable than singularities. Therefore, the requirement of their non-existence, implicit in condition 2 of the theorem, seems reasonable. One might ask, however, if their presence would prevent the formation of singularities. This is actually not the case [36, 37] as, for instance, in a closed expanding or contracting universe.

The singularity theorem applies to the region of the Schwarzschild solution inside the event horizon (neutral non-rotating black holes), to the Milne universe (where, however, geodesic incompleteness is not associated with a curvature singularity) and to non-accelerating FLRW expanding spacetimes. In the latter case, one can check the theorem's conditions either by looking at (2.70) in $D = 4$ ($R_{00} = 3H^2(\epsilon - 1) > 0$) or by employing the Einstein equations to get (2.39) and (2.57). If $\rho > 0$, then the SEC is

$$w \geqslant -\frac{1}{3} + \frac{2\Lambda}{3\rho\kappa^2},$$

which is valid for ordinary matter when $\Lambda \leqslant 0$ or when Λ is positive but small. A small positive cosmological constant, as observed today, would be negligible in the extreme high-curvature regime near the big bang [38], so that it can be ignored. The singularity at $t = 0$ is a physical singularity where the Ricci tensor diverges; from (2.70) and (2.71),

$$R_{00} \to +\infty, \qquad R_{\alpha\beta} \to +\infty.$$

Notice that the theorem does *not* apply, in particular, to the following cases:

- Minkowski spacetime, since there is no time slice with expansion bounded away from zero.
- Anti-de Sitter spacetime ($\Lambda < 0$), which is not globally hyperbolic.
- de Sitter spacetime and expanding inflationary FLRW spacetimes, neither of which satisfy the time-like convergence condition.

A comment about the last result is in order. So far, we have regarded de Sitter as an idealization of realistic inflationary cosmologies where the Hubble parameter H is approximately constant. However, it is important to realize how misleading it can be to confuse the mathematical de Sitter spacetime with realistic models of the early universe. In fact, for flat de Sitter the scale factor is $a = e^{Ht}$, the big bang is in the infinite past at $t \to -\infty$ but the Ricci tensor and scalar are all finite there. From (2.70), (2.71) and (2.72) with constant H and $\textsc{k} = 0$,

$$R_{00} = -3H^2 < 0, \qquad R_{\alpha\beta} \to 0^+, \qquad R = 12H^2.$$

On the other hand, for power-law inflation ($p \gg 1$) the big bang is at $t \to 0$, where

$$R_{00} \to -\infty\,, \qquad R_{\alpha\beta} \to 0^+\,, \qquad R \to +\infty\,.$$

Thus, the singularity in pure de Sitter is not a curvature singularity. It is actually a coordinate singularity from the point of view of full de Sitter spacetime (four-dimensional hyperboloid embedded in five-dimensional Minkowski), which is geodesically complete [4]. For power-law cosmologies, it is a curvature singularity and only in this case we call it "big bang."

The Hawking–Penrose theorem is not decisive as far as the big bang issue is concerned. We have just seen that it does not hold for de Sitter and inflationary backgrounds. While de Sitter is geodesically complete, we cannot reach the same conclusion for inflationary spacetimes. The continuity equation (2.76) shows that, for $\rho > 0$ and $w > -1$, the energy density increases backwards in time indefinitely. In fact, another singularity theorem suggests that the big-bang singularity is typical also in inflationary cosmologies. A closed *trapped surface* $\mathcal{T}$ is a compact space-like two-surface such that the two families of null geodesics meeting the surface orthogonally converge locally at $\mathcal{T}$. Trapped surfaces can be regarded as regions where gravity is so strong that even light rays cannot escape. They arise during black-hole formation but also in the early universe when matter is extremely dense. Then, we have the following [32, 33]

Penrose–Hawking theorem. *Let $(\mathcal{M}, g)$ be a spacetime such that:*

1. *it obeys the null convergence condition (2.34), $R_{\mu\nu}n^\mu n^\nu \geqslant 0$;*
2. *there is a non-compact Cauchy surface;*
3. *there is a closed trapped surface.*

Then $(\mathcal{M}, g)$ is null geodesically incomplete.

The theorem does not distinguish between future and past singularities, but we can see that it finds a physical application in the very early universe. Let us examine the three hypotheses of the theorem for a universe with $\Lambda = 0$ and a perfectly homogeneous FLRW background. Einstein equations are assumed. First, the NEC (2.54) is satisfied for inflationary fluids realized by ordinary field theories, where $\rho > 0$ and $w > -1$. In fact, from (2.65)

$$R_{\mu\nu}n^\mu n^\nu = \kappa^2 T_{\mu\nu}n^\mu n^\nu = \kappa^2 (n^\mu \partial_\mu \phi) \geqslant 0\,,$$

which is a background-independent result. One can see this also by choosing a null vector $n^\mu = (1, n^\alpha)$, where $n_\alpha n^\alpha = 1$. Then, from (2.70) and (2.71)

$$R_{\mu\nu}n^\mu n^\nu = -(D-2)\dot{H} + \frac{2\textsf{K}}{a^2}\,,$$

which is positive definite if the universe does not super-accelerate and the curvature term is positive or negligible. Second, FLRW spacetimes are globally hyperbolic and, by Geroch splitting theorem, flat and open universes are naturally foliated by non-compact Cauchy surfaces Σ_t. Third, there are various ways to prove the existence of trapped surfaces in the early universe [4, 33]. We recall one in Problem 6.2.

Other singularity theorems can be tailored for closed universes [39, 41] and by replacing the null energy condition with the WEC and/or the SEC on average [39, 40, 42]. In particular, if the strong energy condition is valid on the average along all complete causal geodesics, then closed universes in which Einstein equations hold have an incomplete time-like or null geodesic [40]; suitable focusing theorems allow one to further extend these results [42].

6.2.2 Borde–Vilenkin Theorems

The eternally inflating scenario of Sect. 5.6.5 attracted particular interest [44–49]. Given a point $x \in \mathcal{B}$ in an inflating region $\mathcal{B} \subset \mathcal{M}$, there exists (with non-zero probability) another point $y \in \mathcal{B}$ in the future of x at a given finite geodesic distance. Assuming that the boundaries of thermalized regions expand at (almost) the speed of light, the whole region $\mathcal{I}^-(y) \setminus \mathcal{I}^-(x)$ would be non-thermal. Realistically, however, thermalized regions will form almost surely (i.e., with probability 1) in an infinite spacetime volume. This scenario satisfies all three hypotheses of the following theorem.

Borde–Vilenkin theorem. *Let $(\mathcal{M}, g)$ be a spacetime such that:*

1. *it obeys the null convergence condition $R_{\mu\nu}n^\mu n^\nu \geqslant 0$;*
2. *it is open;*
3. *there is at least one point x such that for some point y to the future of x the volume of $\mathcal{I}^-(y) \setminus \mathcal{I}^-(x)$ is finite.*

Then $(\mathcal{M}, g)$ is null geodesically incomplete.

The local convergence condition of this theorem is the same as before. On the other hand, the restriction on the global causal structure of spacetime is much milder. The third condition is necessary for inflation to be future-eternal [43]. As a result, almost all points in the inflating regions will have a singularity in their past.

The global condition on the geometry of spacetime can be relaxed to include also closed universes. In turn, we need a causality condition prescribing that no light cone can wrap around to "swallow" the universe. Let $(\mathcal{M}, g)$ be a stably causal Lorentzian manifold. A past light cone $\mathcal{E}^-$ is called *localized* if from every spacetime point $x \notin \mathcal{E}^-$ there is at least one past-directed time-like curve that does not intersect the cone. In particular, localized light cones cannot be compact. Minkowski, Schwarzschild, de Sitter, flat and open FLRW spacetimes and some closed FLRW spacetimes all have only localized past light cones. Examples of spacetimes with non-localized light cones are given in [45].

Open and closed spacetimes obeying a set of rather general conditions are null geodesically incomplete [45]:

Borde theorem. *Let $(\mathcal{M}, g)$ be a spacetime such that:*

1. *it obeys the null convergence condition $R_{\mu\nu}n^\mu n^\nu \geqslant 0$;*
2. *it is stably causal and past causally simple;*
3. *all past light cones are localized;*
4. *it contains one of the following:*

   a. *a point with a reconverging past light cone (i.e., such that $\theta < 0$ along every past-directed null geodesic in the light cone), or*
   b. *a past-trapped surface, or*
   c. *a point x such that for some point y to the future of x the volume of $\overline{\mathcal{I}^-(y) \setminus \mathcal{I}^-(x)}$ is finite.*

Then $(\mathcal{M}, g)$ is null geodesically incomplete.

This theorem modifies some of the requirements of the Borde–Vilenkin and the Penrose–Hawking theorem. With respect to the latter and the Hawking–Penrose theorem, it is also more specific about the location of the would-be singularity (in the past).

Semi-classical considerations show that the null convergence condition might be violated in eternally inflating scenarios where quantum fluctuations of the energy-momentum tensor can produce super-acceleration [48]. However, there is actually a much stronger singularity theorem for universes which expand sufficiently fast.

6.2.3 *Borde–Guth–Vilenkin Theorem*

First, we need a new kinematical, local definition of the volume expansion or Hubble parameter. Consider a Lorentzian manifold $(\mathcal{M}, g)$ and an observer O described by the unit time-like or null geodesic $v^\mu(\sigma) = \mathrm{d}x^\mu/\mathrm{d}\sigma$, embedded in the unit velocity field $u^\mu(\sigma)$ such that (6.4) holds. This congruence may represent, but not necessarily, a flux of real particles whose world-lines u^μ cross the observer's path with zero proper acceleration (Fig. 6.3). Define $\gamma := -u_\mu v^\mu \geqslant 0$. In the time-like case, this is the usual special-relativistic Lorentz factor between v^μ and u^μ, while in the null case $\gamma = \mathrm{d}t/\mathrm{d}\sigma$ (here the parameter σ is chosen to increase to the future), where t is time measured by comoving observers.

At proper time (or affine parameter) σ and $\sigma + \delta\sigma$ the observer meets, respectively, the world-line $u^\mu(\sigma)$ and $u^\mu(\sigma + \delta\sigma)$. Furthermore, define the unit space-like vector

$$w^\mu := \frac{v^\mu - \gamma u^\mu}{\sqrt{\gamma^2 + v^2}}, \qquad w^2 = 1, \qquad u_\mu w^\mu = 0. \tag{6.9}$$

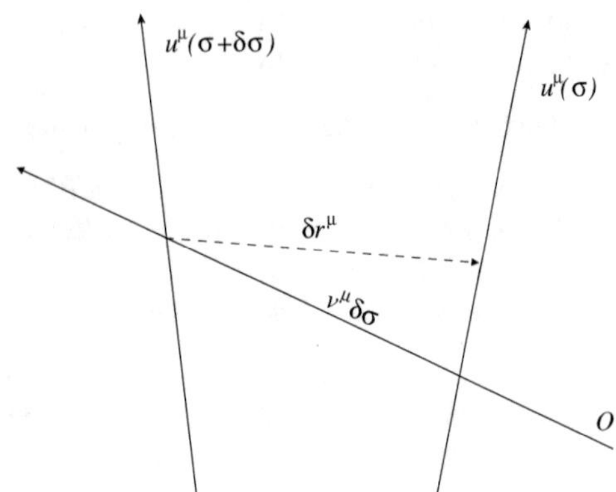

Fig. 6.3 Set-up of Borde–Guth–Vilenkin theorem: an observer O in a velocity field

The space-like vector $\delta r^\mu = \sqrt{\gamma^2+v^2}w^\mu\,\delta\sigma$ is orthogonal to the two world-lines and joins them at equal times in their own rest frame. The distance between the two particles is $\delta r := |\delta r^\mu| = \sqrt{\gamma^2+v^2}\,\delta\sigma$.

The generalized Hubble parameter is the radial component of the projected gradient of u^μ along the radial direction:

$$\mathcal{H} := w_\mu w^\nu \nabla_\nu u^\mu\,. \tag{6.10}$$

Its interpretation is straightforward. The relative velocity between the two test particles is $\delta u^\mu = \dot{u}^\mu\delta\sigma$, where $\dot{} := v^\mu\nabla_\mu$. Its radial component is $\delta u_r := w_\mu\delta u^\mu$. Then, $\mathcal{H} = \delta u_r/\delta r$ is the variation of the radial relative velocity of two particles with respect to their distance at time σ, computed in the rest frame of one of the particles. Substituting (6.9) in (6.10), we get

$$\mathcal{H} = -\frac{\dot{\gamma}}{\gamma^2+v^2} = \dot{F}\,, \tag{6.11}$$

where we used $\dot{v}^\mu = 0$ and

$$F = \begin{cases} \gamma^{-1}\,, & v^2 = 0 \quad \text{(null geodesic)} \\ \operatorname{arctanh}\left(\gamma^{-1}\right)\,, & v^2 = -1 \quad \text{(time-like geodesic)} \end{cases}. \tag{6.12}$$

Notice that the volume expansion $\theta = \nabla_\mu u^\mu$ of the congruence is nothing but the average over all spatial directions of $\mathcal{H}$. In fact, the average of $w_\mu w^\nu$ is $\int d\Omega_{D-2} w_\mu w^\nu = \Omega_{D-2}\delta^\nu_\mu$, where Ω_{D-2} is given in (3.101). Then,

$$\frac{1}{\Omega_{D-2}}\int d\Omega_{D-2}\mathcal{H} = \frac{\theta}{D-1}\,. \tag{6.13}$$

Therefore, $\mathscr{H}$ is the inhomogeneous and anisotropic generalization of the Hubble parameter H.

We define the *averaged expansion condition* as

$$\mathscr{H}_{\rm av} := \frac{1}{\sigma_{\rm e}-\sigma_{\rm i}}\int_{\sigma_{\rm i}}^{\sigma_{\rm e}} d\sigma\, \mathscr{H}(\sigma) > 0\,, \tag{6.14}$$

where $\sigma_{\rm e}$ and $\sigma_{\rm i}$ are two reference times such that $\sigma_{\rm e}$ is in the future of $\sigma_{\rm i}$. This excludes long (but not occasional) contracting phases in the past and it does not require the expansion to be accelerating. The last theorem we review takes advantage of this general condition [47, 49]:

Borde–Guth–Vilenkin (BGV) theorem. *Let $(\mathcal{M}, g)$ be a spacetime where a congruence u^μ can be continuously defined along any past-directed time-like or null geodesic v^μ. Let u^μ obey the averaged expansion condition (6.14) for almost any v^μ. Then $(\mathcal{M}, g)$ is geodesically past-incomplete.*

Proof Consider two reference times $\sigma_{\rm e} > \sigma_{\rm i}$ ($\sigma_{\rm e}$ is in the future of $\sigma_{\rm i}$). From (6.11), one has

$$\int_{\sigma_{\rm i}}^{\sigma_{\rm e}} d\sigma\, \mathscr{H}(\sigma) = F(\gamma_{\rm e}) - F(\gamma_{\rm i})\,.$$

If the averaged expansion condition holds, then the function $F > 0$ increases forward in time ($d\gamma/d\sigma < 0$). In particular,

$$0 < \mathscr{H}_{\rm av} = \frac{F(\gamma_{\rm e}) - F(\gamma_{\rm i})}{\sigma_{\rm e}-\sigma_{\rm i}} = \frac{C}{\sigma_{\rm e}-\sigma_{\rm i}}\,,$$

where $0 < C < +\infty$ is a positive finite constant. Thus, the proper or affine length of almost any past-directed geodesic is finite and spacetime is geodesically past-incomplete. □

The averaged expansion condition might not be valid for all observers (for some, the universe might have contracted during most of its history), hence the specification "for almost any v^μ" in the theorem. In particular, the momentum carried by almost any past-directed causal geodesic is blue-shifted by an infinite amount in a finite proper time (or affine parameter) interval.

6.2.4 An Undecided Issue

Singularity theorems can be useful but tricky and one should exercise caution in their interpretation.

On one hand, their failure on a given Lorentzian manifold does not prevent the manifold to be geodesically incomplete. For instance, the Hawking–Penrose theorem does not apply to inflation. If we assume that the universe inflates from the very beginning, then the theorem lefts question (6.1) unanswered. However, one can envisage a scenario where the universe does not actually begin to expand in acceleration, thus satisfying the theorem just before inflation. Moreover, the other theorems clearly show that geodesic incompleteness is manifest under a variety of local and global conditions. Therefore, in most circumstances a negative result is due to an unsuitable set of hypotheses.

On the other hand, it is not obvious that the past singularity found in geodesically incomplete manifolds is the big bang. In fact, the theorems only demonstrate the existence of incomplete geodesics but they do not guarantee that this feature is associated with a spacetime singularity, nor are they precise about where in the past the would-be singularity lies. In other words, to be identified with the cosmic big bang, geodesic incompleteness should hold for *all* observers with end-points at the *same* time slice $t_0 = t_{\rm bb}$. There is evidence, nonetheless, that at least in the case of eternal inflation almost all observers satisfy the last condition of the Borde–Vilenkin theorem, albeit there is no information about the location of all these singularities. Things get better when the focusing condition (6.8) and the existence of a global Cauchy hypersurface are assumed, so that *all* time-like geodesics emanating orthogonally from the Cauchy surface terminate at t_0 [45].

The situation is all the more unclear in inhomogeneous, non-FLRW cosmologies such as those found in the initial conditions of chaotic inflation, where the perfect-fluid approximation is almost certainly too optimistic. One would have to look at dissipative and interacting relativistic fluids [50], which might or might not violate the hypotheses of the theorems (such as the energy conditions and the existence of trapped surfaces). Anyway, the BGV theorem applies to a fairly wide class of cosmological models, without any assumption about homogeneity, isotropy and the energy conditions. Provided the average expansion condition (6.14) holds, time-like and null geodesic incompleteness is guaranteed also in extremely inhomogeneous scenarios. This is an important achievement but, unfortunately, inconclusive like its predecessors: from geodesic incompleteness one cannot infer that inflating universes have a unique beginning. On top of that, there are particular scenarios which avoid also the BGV theorem. Based on arguments inspired by geodesic completeness of full de Sitter, a geodesically complete eternally inflating spacetime was constructed in [51, 52].

To summarize, the answer to question (6.1) is: "At the classical level, Yes, there was a big bang…maybe. And there are exceptions." The big-bang problem exists as soon as we have a big bang and the evidence collected so far is that likely there was one. This fact begs for a solution, but there is another reason why we should prepare ourselves for this task. The singularity theorems show that singularities plague spacetime manifolds under rather general and reasonable conditions, for instance in the formation of black holes or in inflationary cosmologies. Maybe it is not possible to always avoid geodesic incompleteness, or to avoid all singularities with the same mechanism, but a general solution to the problem (universal in its qualitative features) would touch upon the very nature of geometry and spacetime inside black holes and in the early universe and would have deep consequences in our way of thinking Nature.

6.2.4.1 Bouncing and Cyclic Universes

From a philosophical perspective, the big bang has raised many questions about the nature of time and its birth, leading to alternative scenarios where the initial singularity is replaced by a finite event (a *bounce*) or a series of bounce events (*cyclic universe*). A bounce is a moment of the evolution of the universe where, after a period of contraction, the geometry acquires a minimum volume configuration where energy density, pressure and curvature do not diverge. After the bounce, the universe expands. At the semi-classical level, the structure of the perturbations generated through a single bounce can be more complicated than the standard one in a monotonically expanding universe; for example, vector modes cannot be neglected during the contracting phase in contrast to their decaying behaviour in the post-big-bang phase.

In other models, the Universe experiences a cyclic succession of expansions and contractions in which a single bounce is just a transitory phase in a wider process of evolution. This idea is embedded in several ancient cosmogonies (including Hindu and Theravāda Buddhist traditions[4]) but also modern science tried to implement it [56] (see [57–64] for early attempts).

Since the big bang is typical of cosmological models in general relativity, the existence of special bouncing or cyclic solutions does not provide a genuine solution to the big-bang problem. This led to the study of cosmologies with standard gravity and alternative forms of matter, such as fermion condensates [65], or with a negative cosmological constant [66–72]. In the latter case, the dynamics is driven by a negative Λ via the standard Friedmann equation (in $D = 4$)

$$H^2 = \frac{\kappa^2}{3}\rho - \frac{|\Lambda|}{3}, \tag{6.15}$$

where ρ is made of ordinary dust matter and radiation. The universe has no beginning and undergoes linear asymmetric cycles with constant period due to entropy exchange between different matter species. Entropy increases monotonically from cycle to cycle and so does the scale factor. The average expansion mimics one with acceleration, thus giving rise to a scenario of emergent cyclic inflation.

A cyclic universe can be realized also in models beyond general relativity, including higher-derivative gravity and scalar-tensor theories [56] and multi-scale spacetimes where the dynamics takes place in a scale-dependent multi-fractal

[4]For Theravāda cosmology and the creation-destruction myth, see Dīgha Nikāya 27.10-31 [53], Majjhima Nikāya 28.7,12 [1], Aṅguttara Nikāya 4:156, 7:66 [54]. A comprehensive account can be found in the *Visuddhimagga*, XIII 13, 28–65 [55].

geometry [73]. In the latter case, one has a universe with a finite past but an infinite number of cycles, which are log-periodic and asymmetric by construction of the measure. Another possibility, in the presence of a cosmological constant, is to join cycles of expansions from the big bang to the indefinite de Sitter acceleration by identifying the conformal 3-surface representing the big bang with the conformal infinity of the past cycle. This scenario of conformal cyclic cosmology [74–77] gives rise to ring structures that are still under search in CMB maps [78–80].

It should be stressed that in many cyclic models the average expansion condition (6.14) holds and $\mathscr{H}_{\rm av} > 0$ for a null geodesic over the cycles. Thus, the BGV theorem implies that also cyclic scenarios are, in general, geodesically incomplete. Rather than model-dependent resolutions of the problem, one might look at altogether different frameworks where classical gravity is heavily modified by quantum effects and big-bang avoidance is a robust outcome of quantization. In Chaps. 10, 11, and 13 we shall see candidate models claiming such a feature.

6.3 BKL Singularity

Instead of attempting to address the big-bang issue in quantum gravity, let us assume, for the time being, that gravity is purely classical and that the initial singularity exists and is unavoidable. The analytic or semi-analytic behaviour of solutions to the Einstein's equations near the singularity [81, 82] has been studied first in a series of papers [83–93] whose results are collected in the major works [94, 95] and later in [96–106]. Numerical methods can also be employed [107] (see [108, 109] for reviews on numerical analyses of singularities).

6.3.1 *Tetrads and Bianchi Models*

Let $(\mathcal{M}, g)$ be a D-dimensional spacetime manifold $\mathcal{M}$ locally equipped with a metric $g_{\mu\nu}$. The tangent space $T_x\mathcal{M}$ is isomorphic to Minkowski space and we can define the one-to-one map $e : \mathcal{M} \to T_x\mathcal{M}$ which sends tensor fields from the manifold to the Minkowskian tangent space. This map, generally called *vielbein* (*tetrad* or *vierbein* in the four-dimensional case), is a local reference system for spacetime, physically representing the gravitational field. Its relation with the metric $g_{\mu\nu}$ is summarized by the following formulæ:[5]

$$g_{\mu\nu} = \eta_{ab} e^a{}_\mu e^b{}_\nu \,, \qquad e^a{}_\mu e_a{}^\nu = \delta^\nu_\mu \,, \qquad e_a{}^\mu e^b{}_\mu = \delta^b_a \,, \tag{6.16}$$

[5]The co-tetrad $e_a{}^\mu$ is often denoted as $\omega_a{}^\mu$ in the literature.

where both Greek and Latin indices run from 0 to $D-1$ and transform, respectively, under general coordinate and local Lorentz transformations. The gravitational field $e^a{}_\mu$ is the transformation matrix between local Minkowski coordinates $X^a = X^a(x)$ and arbitrary coordinates x^μ,

$$e^a{}_\mu = \frac{\partial X^a}{\partial x^\mu} \,. \tag{6.17}$$

Vielbein fields incorporate all the metric properties of spacetime but the converse is not true. In fact, due to manifest local Lorentz invariance, there are infinitely many realizations of the local basis reproducing the same metric tensor. This is also the reason why there are more components in the vielbeins than in the metric field, the difference being exactly $D(D-1)/2$, which is the number of free parameters of the $SO(1, D-1)$ group representing Lorentz transformations on the Minkowski tangent space.

A priori, the homogeneous and isotropic assumption of FLRW backgrounds is too restrictive in the neighborhood of the big bang, but the choice of too general a background might be intractable. As in many other situations, it is customary to achieve a compromise and choose some background encoding, say, anisotropies but not inhomogeneities. Denoting with Latin indices i, j the spatial components of the tangent space, on a *homogeneous* manifold $\mathcal{M}$ the triad $e^i{}_\alpha = e^i{}_\alpha(t)$ is space-independent. We define $D-1$ $(D-1)$-dimensional constant frame vectors $\mathbf{l}^i$ and $D-1$ scalars a_i such that (no sum over the inert index i)

$$e^i{}_\alpha = a_i(t)\, l^i_\alpha \,. \tag{6.18}$$

Then, the metric in synchronous time can be written as

$$d^2s = -dt^2 + g_{\alpha\beta} dx^\alpha dx^\beta = -dt^2 + a_i^2(t)\delta_{ij} l^i_\alpha l^j_\beta dx^\alpha dx^\beta \,. \tag{6.19}$$

Three-dimensional homogeneous spaces can be divided according to the *Bianchi classification* of three-dimensional real Lie algebras [110–112]. The complete list of spaces is discussed in [94, 113]; here we mention just a few special cases.

- *Bianchi I*: homogeneous spacetimes with zero spatial curvature. Isotropic examples are Euclidean, Minkowski and flat FLRW spacetimes.
- *Bianchi V*: special cases are homogeneous spacetimes with constant negative spatial curvature. An isotropic example is open FLRW spacetimes.
- *Bianchi IX*: special cases are homogeneous spacetimes with constant positive spatial curvature. An isotropic example is closed FLRW spacetimes.

Some line elements, vacuum and perfect-fluid solutions of Bianchi II, III, IV, VI_h, VI_0 and VII_h can be found in [114].

6.3.2 Kasner Metric

Anisotropic Bianchi I models are important for cosmology. A particular flat metric is

$$ds^2 = -dt^2 + \sum_{i=1}^{D-1} a_i^2(t)(dx^i)^2. \tag{6.20}$$

It describes a homogeneous flat universe characterized by $D-1$ scale factors $a_i(t)$, one for each spatial direction expanding at different rates $H_i = \dot{a}_i/a_i$. In the isotropic (FLRW) case, $a_i(t) = a(t)$ for all $i = 1, \ldots, D-1$. Since $l^i_\alpha = \delta^i_\alpha$, Latin and Greek indices are interchangeable.

A vacuum solution in $D > 3$ is the *Kasner metric* [115], where

$$\boxed{a_i(t) = t^{p_i}} \tag{6.21}$$

up to a normalization constant. In fact, the only non-vanishing Levi-Civita components in flat Bianchi I are

$$\Gamma^0_{ij} = H_i g_{ij}, \qquad \Gamma^j_{i0} = H_i \delta^j_i, \tag{6.22}$$

and the Ricci tensor reads[6]

$$R_{00} = -\sum_{i=1}^{D-1}(H_i^2 + \dot{H}_i) = -\sum_i \frac{\ddot{a}_i}{a_i}, \tag{6.23}$$

$$R_{ij} = \left(\dot{H}_i + H_i \sum\nolimits_k H_k\right) g_{ij}. \tag{6.24}$$

Plugging in the profile (6.21), the vacuum equations $R_{\mu\nu} = 0$ are solved if

$$\boxed{\sum_i p_i = 1, \qquad \sum_i p_i^2 = 1.} \tag{6.25}$$

Therefore, the Kasner exponents p_i obey the conditions (6.25), the first defining a hyperplane and the second a hypersphere S^{D-2}. The intersection of the Kasner plane and sphere determines the space of solutions p_i as a hypersphere S^{D-3}. The covariant volume has a singularity linear in t, since by the first Kasner condition

[6] Sum and product indices range from 1 to $D-1$ unless stated otherwise. Sums or products with subscript $i < j$ run both on j and on $i < j$.

$$\sqrt{-g} = \prod_i a_i(t) = t\,. \tag{6.26}$$

The exponents can always be ordered as $p_1 \leqslant p_2 \leqslant \cdots \leqslant p_{D-1} \leqslant 1$ but at least one inequality must be strict. In fact, isotropic power-law cosmology is *not* a vacuum solution ($R_{\mu\nu} = 0$), since the Kasner conditions cannot be satisfied simultaneously for $p_i = p$, for all i. Notice also that the Riemann tensor $R_{\mu\nu\lambda\sigma}$ vanishes on the solution only when $p_{D-1} = 1$ and $p_i = 0$ for all $i = 1, \ldots, D-2$; after a coordinate transformation, this is actually Minkowski. Barring this case (which we will ignore from now on), one can show that at least one Kasner exponent is always negative: squaring the first Kasner condition in (6.25) and using the second, one obtains

$$\sum_{i<j} p_i p_j = 0\,. \tag{6.27}$$

Thus, at least one direction is contracting although the total spatial volume increases. In three spatial dimensions, there is exactly one non-positive Kasner exponent and the range is

$$\boxed{-\tfrac{1}{3} \leqslant p_1 < 0\,, \qquad 0 < p_2 \leqslant \tfrac{2}{3}\,, \qquad \tfrac{2}{3} \leqslant p_3 < 1\,.} \tag{6.28}$$

They are equal in pairs only when $(p_1, p_2, p_3) = (-1/3,\, 2/3,\, 2/3)$ (or $(p_1, p_2, p_3) = (0, 0, 1)$, which we disregard).

In the presence of matter, the Kasner metric is no longer a solution. However, near the singularity it is still a good approximation. In fact, consider a perfect fluid with world-line u^μ and constant barotropic index w. The volume expansion is $\theta = \sum_i H_i$, so that in the absence of shear the continuity equation (2.49) and the spatial components of (2.52) become

$$\dot\rho + \left(\sum_i H_i\right)\rho(1+w) = 0\,, \qquad u_i\dot\rho + \rho\left(1+\frac{1}{w}\right)\dot u_i = 0\,. \tag{6.29}$$

The solutions are

$$\rho \propto \left(u_0 \prod_i a_i\right)^{-(1+w)}, \qquad \rho \propto u_i^{-\left(1+\frac{1}{w}\right)}, \tag{6.30}$$

where in the second equation the index of u_i is inert. The second equation states that the covariant components u_i all have the same magnitude. The largest contravariant component near the singularity ($t \to 0$) is $u^{D-1} = u_{D-1}/a_{D-1}^2$, so that

$$u_0 = \sqrt{1 + u_i u^i} \sim \frac{u_{D-1}}{a_{D-1}}.$$

Then, solving for u_{D-1} in (6.30) and combining the two results one has

$$\rho \sim \left(\prod_{i \neq D-1} a_i\right)^{-\frac{1+w}{1-w}} = t^{-\frac{1+w}{1-w}(1-p_{D-1})},$$

$$u_i \sim t^{\frac{w}{1-w}(1-p_{D-1})}, \qquad u_0 \sim t^{\frac{w}{1-w}(1-p_{D-1})-p_{D-1}},$$

where we assumed $w \neq 1$. For matter obeying the dominant energy condition (2.56) (radiation, dust, scalar field and so on), the energy density diverges at $t = 0$ except in the special case $p_{D-1} = 1$, as anticipated. From (2.40), and remembering that $p_1 < 0$,

$$|T_0^0| \simeq (1+w)\rho u_0^2 \sim t^{-(1+p_{D-1})} \sim T_{D-1}^{D-1},$$

$$T_1^1 \simeq w\rho \sim t^{-\frac{1+w}{1-w}(1-p_{D-1})},$$

$$T_i^i \simeq (1+w)\rho u_i u^i \sim t^{-(1+2p_i-p_{D-1})}, \qquad i = 2, \dots, D-2.$$

The $t = 0$ singularity in the energy-momentum tensor is milder than for the Ricci tensor components (6.23) and (6.24), which scale as $R_\mu^{\ \nu} \sim t^{-2}$. We conclude that curvature terms dominate near the big bang and it is not restrictive to neglect matter.

6.3.3 *Generalized Kasner Metric*

The non-flat generalized Kasner metric [81] is a special case of (6.19) where $a_i = t^{p_i}$. The Ricci tensor reads

$$R_0^0 = \sum_i (H_i^2 + \dot{H}_i) = \sum_i \frac{\ddot{a}_i}{a_i}, \tag{6.31}$$

$$R_i^j = R_\alpha^\beta l_i^\alpha l_\beta^j = \left(\dot{H}_i + H_i \sum\nolimits_k H_k\right)\delta_i^j + {}^{(D-1)}R_i^j, \tag{6.32}$$

where ${}^{(D-1)}R_i^j$ is the Ricci tensor of spatial slices, built with the purely spatial connection $\Gamma_{\alpha\beta}^\gamma(l)$. One can check whether the Kasner profile (6.21) is an approximate vacuum solution in the limit $t \to 0$. This happens if the spatial curvature term ${}^{(D-1)}R_i^j$ can be neglected with respect to the rest, that is, when it does not grow faster than

t^{-2} near the singularity. It turns out that

$$^{(D-1)}R_i^j \sim t^{-2(1-p_{ijk})}, \tag{6.33}$$

where

$$p_{ijk} := 2p_i + \sum_{m\neq i,j,k} p_m = 1 + p_i - p_j - p_k\,, \qquad i \neq j \neq k \neq i\,. \tag{6.34}$$

In the stability region of the parameter space

$$p_{ijk} > 0\,, \tag{6.35}$$

the spatial curvature is negligible and one has the desired result. In $D \geqslant 11$ dimensions, there always exists a region wherein the inequality (6.35) is satisfied, but for $D \leqslant 10$ this is not possible unless one contradicts the Kasner conditions [97].[7] In that case, one has to impose some constraints on the functions l^i_α (and their derivatives, which are zero in this case) so that (6.35) is enforced.

For instance, in $D = 4$ one has $p_{ijk} = 2p_i$ and the dominant term goes as $t^{-2(1-2p_1)}$. Define

$$\lambda^{(i)} := \frac{1}{\mathcal{V}_0}\,\mathbf{l}^i\cdot(\nabla\times\mathbf{l}^i)\,, \tag{6.36}$$

where $(\nabla\times\mathbf{l}^i)^\alpha = \epsilon^{\alpha\beta\gamma}\partial_\beta l^i_\gamma$ is the curl of the vector $\mathbf{l}^i$. The comoving (coordinate) spatial volume is $\mathcal{V}_0 = |\det l| = \mathbf{l}^1\cdot(\mathbf{l}^2\times\mathbf{l}^3)$. The spatial curvature is [81] (no sum over indices)

$$^{(3)}R_i^i = \frac{1}{2(a_1a_2a_3)^2}\left[\lambda^{(i)2}a_i^4 - (\lambda^{(j)}a_j^2 - \lambda^{(k)}a_k^2)^2\right], \qquad i \neq j \neq k \neq i\,, \tag{6.37}$$

while off-diagonal components vanish. For each component, the dominant term is

$$\frac{\lambda^{(1)2}a_1^2}{2(a_2a_3)^2} \sim \lambda^{(1)2}t^{-2(1-2p_1)}\,,$$

which is negligible in R_i^j only if

$$\lambda^{(1)} = 0\,. \tag{6.38}$$

This is the additional constraint advertized in the general case.

[7] In nine spatial dimensions, the maximum p_{ijk} is zero at $p_1 = p_2 = p_3 = -1/3$, $p_4 = \cdots = p_9 = 1/3$.

In Bianchi I spaces, all the $\lambda^{(i)}$ vanish. In Bianchi VIII models, the $\lambda^{(i)}$ are constant and, all but one, with the same sign. Without loss of generality, one can choose $-\lambda^{(1)} = \lambda^{(2)} = \lambda^{(3)} = 1$. The volume of spatial slices is infinite [94]. In Bianchi IX models, all $\lambda^{(i)}$ are constant and of the same sign; we can set $\lambda^{(1)} = \lambda^{(2)} = \lambda^{(3)} = 1$. Then, the four-dimensional metric (6.19) represents a "twisted" ellipsoid [116] of volume $\mathcal{V} = 16\pi^2 \prod_i a_i$ [94] (here $\mathcal{V}_0 = 1$), i.e., a closed universe whose spatial hypersurfaces have non-constant positive curvature. When all a_i are equal (closed FLRW), the curvature is constant.

6.3.4 *Mixmaster Dynamics*

Since the constants $\lambda^{(i)}$ do not vanish in Bianchi VIII and Bianchi IX spaces, the generalized Kasner metric with condition (6.38) is an approximate solution. We can imagine a Bianchi VIII or Bianchi IX universe where the Kasner metric (6.20) plays the role of a Bianchi I flat background, while spatial curvature terms are regarded as homogeneous perturbations. Bianchi VIII and IX are geodesically incomplete [117, 118] and at $t = 0$ there is a physical (curvature) singularity.

Fix $D = 4$. Recasting (6.31), (6.32) and (6.37) in terms of the time variable

$$d\tau := \frac{dt}{\prod_i a_i}, \tag{6.39}$$

the vacuum equations $R_\mu^{\ \nu} = 0$ read (here $\mathcal{H}_i := a_i'/a_i = a_i^{-1}da_i/d\tau$)

$$0 = \sum_{i=1}^{3} \mathcal{H}_i' - 2\sum_{i<j} \mathcal{H}_i\mathcal{H}_j, \tag{6.40}$$

$$0 = 2\mathcal{H}_i' + a_i^4 - \left(\lambda^{(j)}a_j^2 - \lambda^{(k)}a_k^2\right)^2, \qquad i \neq j \neq k \neq i, \tag{6.41}$$

which can be combined into the first-order expression

$$4\sum_{i<j} \mathcal{H}_i\mathcal{H}_j = \sum_i a_i^4 - 2\sum_{i<j} \lambda^{(i)}\lambda^{(j)}(a_ia_j)^2. \tag{6.42}$$

These equations can be studied semi-analytically in a neighborhood of the singularity. Initial conditions are set to be Kasner-like at the infinite future,

$$\mathcal{H}_i(\tau = +\infty) = p_i, \qquad p_1 < 0.$$

As time evolves (backwards towards the singularity), the perturbation along the $i = 1$ direction grows, while those corresponding to positive Kasner exponents decay.

Then, at late times, $\tau \sim \ln t$ and (6.41) can be approximated as

$$\mathcal{H}_1' + \tfrac{1}{2}a_1^4 \simeq 0\,, \qquad \mathcal{H}_{2,3}' - \tfrac{1}{2}a_1^4 \simeq 0\,,$$

whose positive solutions are

$$a_1(\tau) = \sqrt{\frac{2|p_1|}{\cosh(2|p_1|\tau)}}\,, \qquad a_{2,3}(\tau) = \mathrm{e}^{(p_{2,3}-|p_1|)\tau}\sqrt{\cosh(2|p_1|\tau)}\,. \tag{6.43}$$

Near the singularity ($\tau \to -\infty$),

$$a_1(\tau) \sim \mathrm{e}^{|p_1|\tau}\,, \qquad a_{2,3}(\tau) \sim \mathrm{e}^{(p_{2,3}-2|p_1|)\tau}\,, \qquad t \sim \frac{2|p_1|}{1-2|p_1|}\mathrm{e}^{(1-2|p_1|)\tau}\,,$$

so that asymptotically one has another *Kasner epoch*

$$a_i \sim t^{\tilde{p}_i}, \tag{6.44}$$

where

$$\boxed{\tilde{p}_1 = \frac{|p_1|}{1-2|p_1|} > 0\,, \qquad \tilde{p}_2 = \frac{p_2-2|p_1|}{1-2|p_1|} < 0\,, \qquad \tilde{p}_3 = \frac{p_3-2|p_1|}{1-2|p_1|} > 0\,.} \tag{6.45}$$

The new coefficients still obey the Kasner conditions (6.25). Note that the volume in the new epoch is $\mathcal{V} = \prod_i a_i \sim (1-2|p_1|)t$.

6.3.4.1 Kasner Epochs

To summarize the physical picture moving towards the singularity, one starts at large t with a Kasner epoch where space contracts along the $\alpha = i = 1$ direction. By effect of the homogeneous perturbation, a_1 and a_2 reach, respectively, a minimum and a maximum value, at which there is a transition to another Kasner epoch with contracting direction $\alpha = i = 2$. The previously increasing perturbation $\propto a_1^4$ now dies away, while the perturbation $\propto \tilde{a}_2^4$ drives the universe to another Kasner epoch, and so on. p_3, the greatest of the two positive powers in one era, remains positive also in the next.

It is convenient to parametrize the Kasner exponents with a parameter $u > 1$ (Fig. 6.4):

$$p_1(u) = -\frac{u}{1+u+u^2} = p_1\left(\frac{1}{u}\right), \tag{6.46a}$$

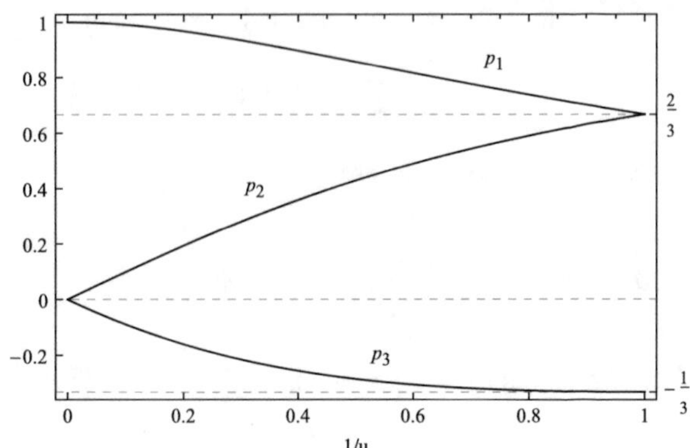

Fig. 6.4 Kasner exponents $p_i(u)$ in a four-dimensional Bianchi I universe

$$p_2(u) = \frac{1+u}{1+u+u^2} = p_3\left(\frac{1}{u}\right), \tag{6.46b}$$

$$p_3(u) = \frac{u(1+u)}{1+u+u^2} = p_2\left(\frac{1}{u}\right), \tag{6.46c}$$

where the rightmost-hand sides show how to extend the parametrization to $u \leqslant 1$. Comparing (6.45) and (6.46), the transition rule (in terms of the parameter of the first epoch) can be recast as

$$\tilde{p}_1(u) = p_2(u-1), \qquad \tilde{p}_2(u) = p_1(u-1), \qquad \tilde{p}_3(u) = p_3(u-1). \tag{6.47}$$

Let $u = u_0 > 1$ be the initial value at $\tau = +\infty$. Realistic scenarios will not obey a Kasner power law exactly, so that the only meaningful case is when u_0 is irrational. We can decompose u_0 into its integer part $n_1 = \lfloor u_0 \rfloor$ (or floor function: the greatest integer number n_1 smaller than u_0) and a remainder $1 > q_0 \in \mathbb{R} \setminus \mathbb{Q}$:

$$u_0 = \lfloor u_0 \rfloor + q_0 .$$

The negative exponent will bounce back and forth between the $i = 1$ and $i = 2$ directions for a certain number $k_1 = 1 + n_1$ of iterations of the parameter

$$u_n = u_0 - n = n_1 + q_0 - n, \qquad n = 0, 1, \dots, n_1 = k_1 - 1. \tag{6.48}$$

When the integer part is exhausted and $u_{k_1-1} = q_0 < 1$, either p_1 or p_2 is negative and p_3 becomes the smallest positive exponent. This is the beginning of the last Kasner epoch and u_{k_1} signals the end of a *Kasner era* (sequence of Kasner epochs), i.e., a period of time when one of the directions evolves monotonically.

From (6.43), the maximum value of a_1 in a given Kasner epoch is $a_{\max} = \sqrt{2|p_1|}$. Since $|p_1|$ decreases in successive epochs, the maximal expansion of one direction decreases in a given era. More precisely, consider the Kasner profile parametrized by

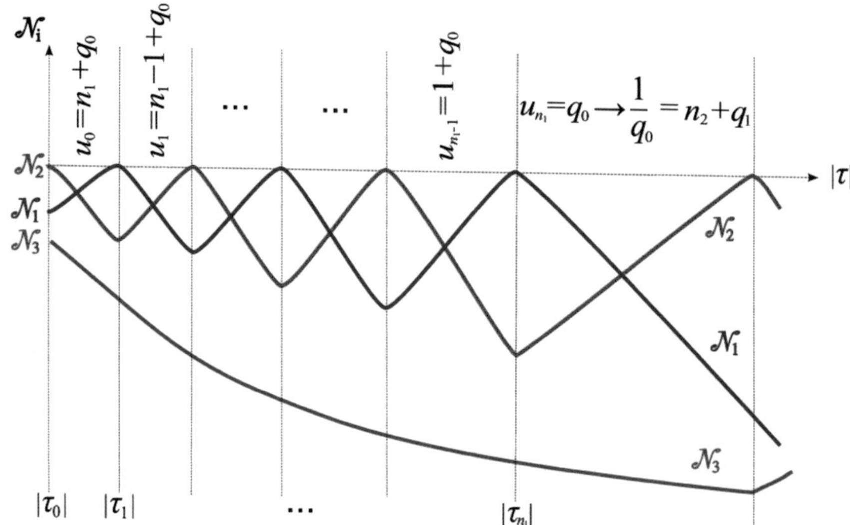

Fig. 6.5 Kasner era made of n_1 Kasner epochs. Time flows backwards towards the singularity to the right. The symbols are explained in the text (Source: adaptation of "Kasner epochs" by Lantonov – Own work. Licensed under Creative Commons Attribution-Share Alike 3.0 via Wikimedia Commons [119])

the parameter u defined in the preceding epoch. The maximum is $\tilde{a}_{\max} = \sqrt{2|\tilde{p}_1(u)|}$. In the next epoch, the maximum value is $\tilde{\tilde{a}}_{\max} = \sqrt{2|\tilde{p}_2(u)|}$, so that

$$\frac{\tilde{\tilde{a}}_{\max}}{\tilde{a}_{\max}} = \sqrt{\frac{|\tilde{p}_2(u)|}{|\tilde{p}_1(u)|}} = \sqrt{\frac{|p_1(u-1)|}{|p_2(u-1)|}} = \sqrt{\frac{u-1}{u}} < 1\,. \tag{6.49}$$

The evolution towards the singularity can be represented as in Fig. 6.5 for the first Kasner era with n_1 epochs. We define the anisotropic e-foldings

$$\mathcal{N}_i := \ln a_i\,, \qquad \sum_{i=1}^{3} \mathcal{N}_i = -|\tau|\,. \tag{6.50}$$

In the second expression, we noticed that near the singularity the change in the normalization of the volume $\prod_i a_i$ between one epoch and another is negligible for large $|\tau|$ (small t), so that we can define a global time variable $\tau = \ln t < 0$ and neglect the gradual decrease (6.49), setting $\mathcal{N}_i^{\max} \approx 0$. In particular, $\mathcal{N}_i \leqslant 0$ for all i. In Fig. 6.5, we denote with $|\tau_n|$ the beginning of the $(n+1)$-st epoch. All the minima of $\mathcal{N}_{1,2}$ can be written as

$$\mathcal{N}_{i,n} = \tau_n \delta_n\,, \qquad i = 1, 2\,,$$

where $0 < \delta_n < 1$, while from (6.50)

$$\mathcal{N}_{3,n} = \tau_n(1 - \delta_n) .$$

The discrete evolution of the number of e-foldings is governed by

$$\frac{\mathcal{N}_{i,n+1} - \mathcal{N}_{i,n}}{\tau_{n+1} - \tau_n} = p_i(u_n) . \tag{6.51}$$

During the epoch starting at τ_n and ending at τ_{n+1}, one of the functions $\mathcal{N}_{i,n}$ (say, $\mathcal{N}_{1,n}$) increases from $\mathcal{N}_{1,n} = \tau_n\delta_n$ to $\mathcal{N}_{1,n+1} = 0$, while the other ($\mathcal{N}_{2,n}$) decreases from $\mathcal{N}_{2,n} = 0$ to $\mathcal{N}_{2,n+1} = \tau_{n+1}\delta_{n+1}$. Combining these expressions with (6.51), one obtains

$$|\tau_{n+1}|\delta_{n+1} = \frac{1 + u_n}{u_n}|\tau_n|\delta_n = \frac{1 + u_0}{u_n}|\tau_0|\delta_0 .$$

Therefore, the oscillation amplitudes $|\mathcal{N}_{\alpha,n}|$ increase during the whole Kasner era. The ratio of the amplitude of the last oscillation with respect to the first is proportional to the "length" of the era if the latter is long,

$$\frac{|\mathcal{N}_{i,k_1-1}|}{|\mathcal{N}_{i,0}|} = \frac{k_1 + q_0}{1 + q_0} \simeq k_1 , \qquad i = 1, 2 .$$

The oscillation amplitude of the scale factors $a_{1,2}$ grows as a power law, $a_{1,2}(\tau_{k_1-1}) = [a_{1,2}(\tau_0)]^{k_1}$.

Still from (6.51), the duration of the Kasner epochs is

$$|\tau_{n+1}| = |\tau_n|\left(1 + \frac{1 + u_n + u_n^2}{u_n}\delta_n\right) = |\tau_0|\left[1 + (n+1)\left(n + \frac{1 + u_n + u_n^2}{u_n}\right)\delta_0\right] .$$

The time length of the epochs increases within the era, $|\tau_{n+1}| - |\tau_n| > |\tau_n| - |\tau_{n-1}|$, while the total time duration of the era is

$$|\tau_{k_1}| - |\tau_0| = k_1\left(k_1 + q_0 + \frac{1}{q_0}\right)|\tau_0|\delta_0 .$$

Here $|\tau_{k_1}|$ is the beginning of the next era. Notice that it is consistent to assume instantaneous transitions between Kasner epochs, since the time duration of transition periods, from (6.43), is long when $|p_1| \ll 1$ ($u \gg 1$) and proportional to u. This should be compared with $|\tau_{n+1}| - |\tau_n| \simeq u_n|\tau_n|\delta_n \gg u_n$.

6.3.4.2 Kasner Eras and Chaos

After the end of the first Kasner era, according to the reparametrization $u \to 1/u$ in (6.46) the negative exponent will bounce between the directions $i = 3$ and $i = 1$ or (as in the example of Fig. 6.5) between $i = 3$ and $i = 2$. The function $\mathcal{N}_3$ increases from its minimum value $\mathcal{N}_{3,k_1} = \tau_{k_1}(1 - \delta_{k_1})$, which sets the initial amplitude for the new series of (wider) oscillations:

$$\tilde{\delta}_0|\tilde{\tau}_0| := |\mathcal{N}_{3,k_1}| = \left[\frac{1}{\delta_0} + k_1(k_1 + q_0) - 1\right]|\tau_0|\delta_0 > |\tau_0|\delta_0 \,. \tag{6.52}$$

If u_0 is irrational, there is an infinite sequence of Kasner eras. Asymptotically, both the order of the transitions between Kasner epochs and the "length" of a Kasner era (number of Kasner epochs in it) are governed by a random process. In fact, the initial value of the parameter u for the second era is $1/q_0 = n_2 + q_1$ and its length is n_2. From this conclusion and (6.48), the iterative map governing the discrete evolution of the parameter u_n can be written as

$$u_{n+1} = \begin{cases} u_n - 1 \,, & u_n > 2 \\ (u_n - 1)^{-1} \,, & 1 < u_n < 2 \end{cases} . \tag{6.53}$$

By induction, the length k_i of successive eras are given by the expansion of u_0 as a continuous fraction:

$$u_0 = k_1 + \cfrac{1}{k_2 + \cfrac{1}{k_3 + \dots}} \,. \tag{6.54}$$

The values of the sequence $\{k_m\}$ are distributed according to stochastic laws [94] which can be studied with the methods of fractal geometry and chaos theory [120–127]. In this sense, the singularity is *chaotic*. This chaotic behaviour has an agile representation in terms of Misner variables [82, 128]:

$$\Omega := -\tfrac{1}{3}\ln(a_1 a_2 a_3) \,, \qquad a_{1,2} =: \mathrm{e}^{-\Omega+\beta_+\pm\sqrt{3}\beta_-} \,, \qquad a_3 =: \mathrm{e}^{-\Omega-2\beta_+} \,. \tag{6.55}$$

One can show that the potential associated with the Bianchi IX cosmology is

$$U \propto -\det(g_{ij})\,{}^{(3)}R \propto \mathrm{e}^{-4\Omega}\Big\{\mathrm{e}^{-8\beta_+} - 4\mathrm{e}^{-2\beta_+}\cosh(2\sqrt{3}\,\beta_-) + 2\mathrm{e}^{4\beta_+}\left[\cosh(4\sqrt{3}\,\beta_-) - 1\right]\Big\} \,. \tag{6.56}$$

Anisotropies are encoded in the two variables $\beta_\pm$ defined by (6.55). The potential (6.56) is depicted in Fig. 6.6. The dynamics of the universe going towards the

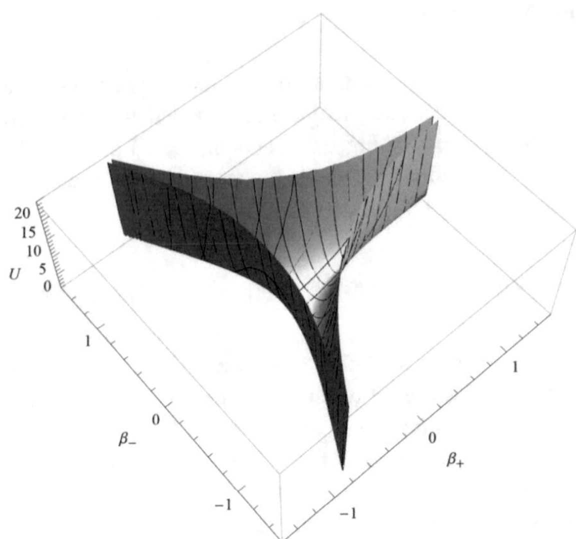

Fig. 6.6 The potential (6.56) of the Bianchi IX mixmaster universe

singularity is described by a particle falling down a potential U with exponentially steep walls and three sharp corners. As soon as the particle enters a corner and gets reflected by one of the walls, a Kasner epoch ends and is followed by another with different expansion. The infinite number of reflections gives rise to a type of chaotic dynamics known as a cosmological *billiard*, for which a wide toolbox of sophisticated techniques is available [82, 128–135] (see [136, 137] for reviews). These tools have gradually superseded the patchwork-like intuitive method followed in this section and in the original papers on this type of singularities. We will describe classical billiards in more detail in Sect. 13.9.2.

The natural stochastic character of the solution has an important consequence. Given a set of "initial" conditions at a time $t > 0$, after a few eras evolving towards the singularity the system loses memory of the chosen initial conditions and admits a chaotic description. Thus, the solution is stable in a qualitative sense, inasmuch as the oscillatory regime will be realized in general. However, the details of specific solutions will be highly sensitive to the initial conditions.

This big-bang model based on a Bianchi VIII or Bianchi IX metric is know as the *Belinsky–Khalatnikov–Lifshitz* (BKL) *singularity*, or also as the *mixmaster universe*.[8]

[8]This name was given by Misner [82] after a famous mechanical kitchen mixer, produced by an American brand of electric home appliances. Reference to this tool is due to the fact that, after a large number of eras, all parts of the whole universe are in causal contact with one another, along all directions: the texture of a cream or a dough is homogenized after enough mixing cycles. In fact, the BKL model was a candidate solution to the horizon problem well before inflation was proposed. This can be roughly seen from (2.188) and the discussion at the end of Sect. 5.2.2. In

The BKL singularity is space-like: the metric becomes singular in the spatial dimensions while approaching the big bang. There is also a time-like version with a similar chaotic behaviour, which we will not discuss [138, 139].

6.3.5 BKL Conjecture

There is good evidence that the mixmaster behaviour is a general property of solutions of the Einstein equations near the singularity, for fairly arbitrary "initial" conditions (boundary conditions on a hypersurface in the future of the singularity). The chaotic singularity is not a peculiar feature of Bianchi VIII and Bianchi IX models in vacuum. Bianchi I and Bianchi IX universes with Yang–Mills non-Abelian fields also evolve chaotically, although in a qualitatively different way [140–142]. Going beyond general relativity and the Einstein–Hilbert action, the mixmaster behaviour on Bianchi IX anisotropic backgrounds is present in scalar-tensor theories [143].

The mixmaster universe can be realized also in Bianchi I [144], Bianchi II [145] and Bianchi VI_0 general relativity [146–148] with a magnetic field and a perfect fluid with $|w| \leqslant 1$. However, it is not a global attractor and there exist exceptional solutions which do not display the oscillatory regime near the singularity.

There are other counter-examples of models which do not oscillate. Particular forms of matter (a massless scalar field or a stiff fluid, $w = 1$) do lead to a homogeneous singularity but spoil the mixmaster behaviour [149, 150]. For $D \geqslant 10 + 1$ general relativity in vacuum, the oscillatory chaotic approach to the singularity disappears and the general solution is monotonic, contrary to what happens for $4 \leqslant D \leqslant 10$. Therefore, the structure of the singularity is qualitatively different in theories with 10 or more spatial dimensions [97–105]. This result has a beautiful mathematical description in terms of the Kac–Moody algebra AE_{D-1} defining the BKL oscillatory structure: for $D-1 < 10$, AE_{D-1} is hyperbolic, while for $D-1 \geqslant 10$ it is not [106]. We will comment in Sect. 12.4, and see more in detail in Sect. 13.9.2, that p-forms with $p > 0$ can induce a chaotic evolution in a number of situations where it would otherwise be suppressed, including in $D = 11$. String theory will have to say much about that.

Near the singularity, higher-order curvature terms should become important in the effective action, so that it is natural to extend the BKL analysis to these models. The mixmaster universe is also a solution in higher-order $D = 4$ gravity with $(\text{Riemann})^2$ contributions in the action [151–154] but it is non-generic because unstable; on the other hand, there exists a general (stable) solution which is monotonic, isotropic and non-chaotic. In higher dimensions, it seems unlikely that

---

a given Kasner era, τ plays the role of conformal time along the direction expanding (forward in time) monotonically. Inflation, as we have seen, does much more than addressing the horizon problem.

the mixmaster universe be realized in fourth-order gravity, since the Kasner metric is not a Bianchi I solution [155].

Furthermore, one can abandon homogeneous spacetimes and consider less restrictive backgrounds. The vacuum BKL model can be generalized to the presence of matter and to inhomogeneous (non-diagonal) metrics as well, where $\lambda^{(i)} = \lambda^{(i)}(\mathbf{x})$ are functions of spatial coordinates. In both cases, the only qualitative difference is that the direction of the oscillation axes changes from one epoch to another [94, 95, 104, 105]; the BKL singularity is confirmed with other methods for inhomogeneous backgrounds with barotropic fluids [156]. Generic spacetimes with no special symmetries have been studied via the orthonormal frame formalism in four dimensions [157] and with Iwasawa variables [158] and further evidence was collected that the mixmaster, oscillatory, homogeneous universe is typically realized near the singularity in $4 \leqslant D \leqslant 10$ [157, 159].

We end up with the following conclusion. In a variety of gravitational set-ups with different total actions and dimensionality, homogeneous (Bianchi) metrics are a good local approximation of generic spacetimes near the singularity. The dynamics is either monotonic (in particular, Kasner-like) and non-chaotic, or oscillatory and (modulo exceptional cases) chaotic. The *BKL conjecture* elevates the lesson of all these models to a paradigm. Intuitively, it states that spatially different points decouple from one another near the big-bang singularity. More precisely:

> ***BKL conjecture.*** *Near a singularity, almost all spacetimes become spatially homogeneous at every point and time derivatives dominate over spatial derivatives. The approach towards the singularity is either monotonic or oscillatory in some of the quantities describing the system.*

The sentence "almost all spacetimes" takes into account the existence of exceptional cases, while the generic formula "in some of the quantities describing the system" encompasses both traditional mixmaster scenarios (where spatial directions oscillate) and others where oscillations occur in other variables (for instance, shear and color stress in Einstein–Yang–Mills models [141]).

Stronger versions of the conjecture also require that matter be negligible near the singularity (in this case, the singularity is described by *vacuum* Bianchi models) and specify the most general Bianchi metric near the singularity to be of type IX. Since there are several exceptions to both these requirements, the weaker formulation is preferable.

The approach in Iwasawa variables is also convenient because it unravels a surprising feature of oscillatory singularities, with which we conclude the discussion on the BKL universe. A question which we did not pose so far is what happens *at* the big bang: Is there any geometric structure left at the singularity? When the dynamical evolution is monotonic, the answer is positive. For instance, at least one direction of a Bianchi I Kasner universe expands when approaching the big bang and one still has the asymptotic notion of which direction expands or contracts in the limit $\tau \to -\infty$. If the approach is oscillatory, however, we would expect chaos to erase any information of the metric in the same limit. This is not the case and the asymptotic geometry at the cosmological singularity, dubbed *partially framed flag* [158], is non-trivial.

6.4 Problems and Solutions

6.1 Raychaudhuri equation. Derive the Raychaudhuri equation (6.3) using (2.43) and (3.88) for a time-like vector t^μ.

Solution From (2.43),

$$(\nabla_\nu t_\mu)(\nabla^\mu t_\nu) = \sigma_{\mu\nu}\sigma^{\mu\nu} - \omega_{\mu\nu}\omega^{\mu\nu} + \frac{1}{D-1}\theta^2\,, \tag{6.57}$$

where we used (2.44), (2.45) and the following relations:

$$\sigma_{\mu\nu}h^{\mu\nu} + \sigma^{\mu\nu}h_{\mu\nu} = 0, \quad \sigma_{\mu\nu}\dot{t}^\nu t^\mu + \sigma^{\mu\nu}\dot{t}_\nu t_\mu = 0, \quad \omega_{\mu\nu}\dot{t}^\nu t^\mu + \omega^{\mu\nu}\dot{t}_\nu t_\mu = \dot{t}^\nu t_\nu \dot{t}_\mu t^\mu.$$

Also, from (3.88) one gets

$$2\nabla_{[\mu}\nabla_{\nu]}t^\mu = R_{\mu\nu}t^\mu\,. \tag{6.58}$$

Contracting with t^ν,

$$R_{\mu\nu}t^\mu t^\nu = t^\nu\nabla_\mu\nabla_\nu t^\mu - \dot{\theta} = \nabla_\mu \dot{t}^\mu - (\nabla_\mu t^\nu)(\nabla_\nu t^\mu) - \dot{\theta}\,;$$

plugging this into (6.57), we obtain (6.3). On a FLRW background, both shear and vorticity vanish and one gets (2.82) provided the Einstein equations hold.

6.2 Trapped surfaces. Show the existence of trapped surfaces (Sect. 6.2.1) in a FLRW spacetime.

Solution After the coordinate redefinition

$$\varrho = \int \frac{dr}{\sqrt{1-\textsc{k}r^2}} = \begin{cases} \text{arcsinh}\, r\,, & \textsc{k} = -1 \\ r\,, & \textsc{k} = 0 \\ \arcsin r\,, & \textsc{k} = +1 \end{cases}\,,$$

the spatial FLRW line element (2.2b) in four dimensions becomes

$$\gamma_{\alpha\beta}\mathrm{d}x^\alpha \mathrm{d}x^\beta = \mathrm{d}\varrho^2 + r^2(\varrho)\mathrm{d}\Omega_2^2 \,.$$

Consider a sphere $\mathcal{T} = S^2$ of comoving radius ϱ in the slice Σ_t. The two families of past-directed null geodesics orthogonal to S^2 intersect the slice $\Sigma_{t'}$ for $t' < t$ in two two-spheres of radius

$$\varrho_\pm(t') = \varrho \pm \int_t^{t'} \frac{\mathrm{d}t''}{a(t'')} = \varrho \pm [r_\mathrm{p}(t') - r_\mathrm{p}(t)] \,,$$

where r_p is, as usual, the particle-horizon radius. The surface area of a two-sphere of radius ϱ is $A = 4\pi a^2 r^2(\varrho)$, so that both families of null geodesics will be converging into the past if $\dot{A}[t, \varrho_\pm(t)] > 0$. This happens if

$$aH + \frac{a}{r}\frac{\mathrm{d}r}{\mathrm{d}\varrho}\dot{\varrho} = \frac{1}{r_H} \pm \frac{1}{r}\frac{\mathrm{d}r}{\mathrm{d}\varrho} = \frac{1}{r_H} \pm \sqrt{\frac{1}{\varrho^2} - \mathrm{K}} > 0 \,,$$

thus yielding the condition

$$\frac{r_H^2}{\varrho^2} - \mathrm{K} r_H^2 < 1 \,. \tag{6.59}$$

Notice now that the Friedmann equation (2.81) with $\Lambda = 0$ can be written as

$$\frac{1}{r_H^2} = \frac{1}{r_\mathrm{S}^2} - \mathrm{K} \,, \tag{6.60}$$

where

$$r_\mathrm{S} := \sqrt{\frac{3}{\kappa^2 \rho}} \begin{cases} > r_H \,, & \mathrm{K} = -1 \\ = r_H \,, & \mathrm{K} = 0 \\ < r_H \,, & \mathrm{K} = +1 \end{cases} \tag{6.61}$$

is the comoving *Schwarzschild length* of the universe filled with a perfect fluid with positive energy density ρ. Equation (6.59) becomes

$$\varrho > r_\mathrm{S} \,. \tag{6.62}$$

In each time slice, there is a trapped surface of size greater than the Schwarzschild length.

References

1. *The Middle Length Discourses of the Buddha. A Translation of the Majjhima Nikāya*, transl. by Bhikkhu Ñāṇamoli and Bhikkhu Bodhi (Wisdom, Somerville, 1995)
2. J. Natário, Relativity and singularities —A short introduction for mathematicians. Resenhas **6**, 309 (2005). [arXiv:math/0603190]
3. S.W. Hawking, Nature of space and time. arXiv:hep-th/9409195
4. S.W. Hawking, G.F.R. Ellis, *The Large Scale Structure of Space-Time* (Cambridge University Press, Cambridge, 1973)
5. A.N. Bernal, M. Sánchez, Globally hyperbolic spacetimes can be defined as causal instead of strongly causal. Class. Quantum Grav. **24**, 745 (2007). [arXiv:gr-qc/0611138]
6. R.P. Geroch, Domain of dependence. J. Math. Phys. **11**, 437 (1970)
7. A.N. Bernal, M. Sánchez, On smooth Cauchy hypersurfaces and Geroch's splitting theorem. Commun. Math. Phys. **243**, 461 (2003). [arXiv:gr-qc/0306108]
8. K. Tomita, On inhomogeneous cosmological models containing space-like and time-like singularities alternately. Prog. Theor. Phys. **59**, 1150 (1978)
9. G.F.R. Ellis, B.G. Schmidt, Singular space-times. Gen. Relat. Grav. **8**, 915 (1977)
10. F.J. Tipler, Singularities in conformally flat spacetimes. Phys. Lett. A **64**, 8 (1977)
11. C.J.S. Clarke, A. Królak, Conditions for the occurence of strong curvature singularities. J. Geom. Phys. **2**, 127 (1985)
12. A. Królak, Towards the proof of the cosmic censorship hypothesis. Class. Quantum Grav. **3**, 267 (1986)
13. S. Cotsakis, I. Klaoudatou, Future singularities of isotropic cosmologies. J. Geom. Phys. **55**, 306 (2005). [arXiv:gr-qc/0409022]
14. S. Nojiri, S.D. Odintsov, S. Tsujikawa, Properties of singularities in (phantom) dark energy universe. Phys. Rev. D **71**, 063004 (2005). [arXiv:hep-th/0501025]
15. C. Cattoën, M. Visser, Necessary and sufficient conditions for big bangs, bounces, crunches, rips, sudden singularities, and extremality events. Class. Quantum Grav. **22**, 4913 (2005). [arXiv:gr-qc/0508045]
16. L. Fernández-Jambrina, R. Lazkoz, Classification of cosmological milestones. Phys. Rev. D **74**, 064030 (2006). [arXiv:gr-qc/0607073]
17. J.D. Barrow, Sudden future singularities. Class. Quantum Grav. **21**, L79 (2004). [arXiv:gr-qc/0403084]
18. K. Lake, Sudden future singularities in FLRW cosmologies. Class. Quantum Grav. **21**, L129 (2004). [arXiv:gr-qc/0407107]
19. J.D. Barrow, More general sudden singularities. Class. Quantum Grav. **21**, 5619 (2004). [arXiv:gr-qc/0409062]
20. J.D. Barrow, C.G. Tsagas, New isotropic and anisotropic sudden singularities. Class. Quantum Grav. **22**, 1563 (2005). [arXiv:gr-qc/0411045]
21. M.P. Dąbrowski, T. Denkiewicz, M.A. Hendry, How far is it to a sudden future singularity of pressure? Phys. Rev. D **75**, 123524 (2007). [arXiv:0704.1383]
22. L. Fernández-Jambrina, R. Lazkoz, Geodesic behaviour of sudden future singularities. Phys. Rev. D **70**, 121503 (2004). [arXiv:gr-qc/0410124]
23. R.R. Caldwell, A phantom menace? Phys. Lett. B **545**, 23 (2002). [arXiv:astro-ph/9908168]
24. A.A. Starobinsky, Future and origin of our universe: modern view. Grav. Cosmol. **6**, 157 (2000). [arXiv:astro-ph/9912054]
25. B. McInnes, The dS/CFT correspondence and the big smash. JHEP **0208**, 029 (2002). [arXiv:hep-th/0112066]
26. R.R. Caldwell, M. Kamionkowski, N.N. Weinberg, Phantom energy and cosmic doomsday. Phys. Rev. Lett. **91**, 071301 (2003). [arXiv:astro-ph/0302506]
27. Y. Shtanov, V. Sahni, New cosmological singularities in braneworld models. Class. Quantum Grav. **19**, L101 (2002). [arXiv:gr-qc/0204040]

28. S. Nojiri, S.D. Odintsov, Final state and thermodynamics of dark energy universe. Phys. Rev. D **70**, 103522 (2004). [arXiv:hep-th/0408170]
29. H. Štefančić, Expansion around the vacuum equation of state: sudden future singularities and asymptotic behavior. Phys. Rev. D **71**, 084024 (2005). [arXiv:astro-ph/0411630]
30. V. Gorini, A. Kamenshchik, U. Moschella, V. Pasquier, Tachyons, scalar fields, and cosmology. Phys. Rev. D **69**, 123512 (2004). [arXiv:hep-th/0311111]
31. A. Kamenshchik, C. Kiefer, B. Sandhöfer, Quantum cosmology with big-brake singularity. Phys. Rev. D **76**, 064032 (2007). [arXiv:0705.1688]
32. R. Penrose, Gravitational collapse and space-time singularities. Phys. Rev. Lett. **14**, 57 (1965)
33. S.W. Hawking, Occurrence of singularities in open universes. Phys. Rev. Lett. **15**, 689 (1965)
34. S.W. Hawking, The occurrence of singularities in cosmology. Proc. R. Soc. Lond. A **294**, 511 (1966)
35. S.W. Hawking, The occurrence of singularities in cosmology. II. Proc. R. Soc. Lond. A **295**, 490 (1966)
36. R.P. Geroch, Singularities in closed universes. Phys. Rev. Lett. **17**, 445 (1966)
37. S.W. Hawking, The occurrence of singularities in cosmology. III. Causality and singularities. Proc. R. Soc. Lond. A **300**, 187 (1967)
38. S.W. Hawking, R. Penrose, The singularities of gravitational collapse and cosmology. Proc. R. Soc. Lond. A **314**, 529 (1970)
39. F.J. Tipler, General relativity and conjugate ordinary differential equations. J. Diff. Equ. **30**, 165 (1978)
40. F.J. Tipler, Energy conditions and spacetime singularities. Phys. Rev. D **17**, 2521 (1978)
41. G.J. Galloway, Curvature, causality and completeness in space-times with causally complete spacelike slices. Math. Proc. Camb. Philos. Soc. **99**, 367 (1986)
42. A. Borde, Geodesic focusing, energy conditions and singularities. Class. Quantum Grav. **4**, 343 (1987)
43. A. Vilenkin, Did the universe have a beginning? Phys. Rev. D **46**, 2355 (1992)
44. A. Borde, A. Vilenkin, Eternal inflation and the initial singularity. Phys. Rev. Lett. **72**, 3305 (1994). [arXiv:gr-qc/9312022]
45. A. Borde, Open and closed universes, initial singularities and inflation. Phys. Rev. D **50**, 3692 (1994). [arXiv:gr-qc/9403049]
46. A. Borde, A. Vilenkin, Singularities in inflationary cosmology: a review. Int. J. Mod. Phys. D **5**, 813 (1996). [arXiv:gr-qc/9612036]
47. A. Borde, A.H. Guth, A. Vilenkin, Inflationary spacetimes are incomplete in past directions. Phys. Rev. Lett. **90**, 151301 (2003). [arXiv:gr-qc/0110012]
48. A. Borde, A. Vilenkin, Violation of the weak energy condition in inflating spacetimes. Phys. Rev. D **56**, 717 (1997). [arXiv:gr-qc/9702019]
49. A.H. Guth, Eternal inflation and its implications. J. Phys. A **40**, 6811 (2007). [arXiv:hep-th/0702178]
50. D. Langlois, F. Vernizzi, Nonlinear perturbations for dissipative and interacting relativistic fluids. JCAP **0602**, 014 (2006). [arXiv:astro-ph/0601271]
51. A. Aguirre, S. Gratton, Steady-state eternal inflation. Phys. Rev. D **65**, 083507 (2002). [arXiv:astro-ph/0111191]
52. A. Aguirre, S. Gratton, Inflation without a beginning: a null boundary proposal. Phys. Rev. D **67**, 083515 (2003). [arXiv:gr-qc/0301042]
53. *The Long Discourses of the Buddha. A Translation of the Dīgha Nikāya*, transl. by M. Walshe (Wisdom, Somerville, 1995)
54. *The Numerical Discourses of the Buddha. A Translation of the Aṅguttara Nikāya*, transl. by Bhikkhu Bodhi (Wisdom, Somerville, 2012)
55. B. Buddhaghosa, *Visuddhimagga — The Path of Purification*, transl. by Bhikkhu Ñāṇamoli (Buddhist Publication Society, Kandy, 2010), pp. 404–414
56. M. Novello, S.E. Perez Bergliaffa, Bouncing cosmologies. Phys. Rep. **463**, 127 (2008). [arXiv:0802.1634]

57. R.C. Tolman, On the problem of the entropy of the Universe as a whole. Phys. Rev. **37**, 1639 (1931)
58. R.C. Tolman, On the theoretical requirements for a periodic behaviour of the Universe. Phys. Rev. **38**, 1758 (1931)
59. G. Lemaître, L'univers en expansion. Ann. Soc. Sci. Bruxelles A **53**, 51 (1933)
60. R.C. Tolman, *Relativity, Thermodynamics and Cosmology* (Clarendon Press, Oxford, 1934)
61. M.J. Rees, The collapse of the universe: an eschatological study. Observatory **89**, 193 (1969)
62. R.H. Dicke, P.J.E. Peebles, The big bang cosmology – enigmas and nostrums, in [63]
63. S.W. Hawking, W. Israel (eds.), *General Relativity: An Einstein Centenary Survey* (Cambridge University Press, Cambridge, 1979)
64. Ya.B. Zel'dovich, I.D. Novikov, *Relativistic Astrophysics*. The Structure and Evolution of the Universe, vol. 2 (University of Chicago Press, Chicago, 1983)
65. S. Alexander, T. Biswas, Cosmological BCS mechanism and the big bang singularity. Phys. Rev. D **80**, 023501 (2009). [arXiv:0807.4468]
66. T. Biswas, Emergence of a cyclic universe from the Hagedorn soup. arXiv:0801.1315
67. T. Biswas, S. Alexander, Cyclic inflation. Phys. Rev. D **80**, 043511 (2009). [arXiv:0812.3182]
68. T. Biswas, A. Mazumdar, Inflation with a negative cosmological constant. Phys. Rev. D **80**, 023519 (2009). [arXiv:0901.4930]
69. T. Biswas, A. Mazumdar, A. Shafieloo, Wiggles in the cosmic microwave background radiation: echoes from nonsingular cyclic inflation. Phys. Rev. D **82**, 123517 (2010). [arXiv:1003.3206]
70. T. Biswas, T. Koivisto, A. Mazumdar, Could our universe have begun with $-\Lambda$? arXiv:1105.2636
71. T. Biswas, T. Koivisto, A. Mazumdar, Phase transitions during cyclic inflation and non-Gaussianity. Phys. Rev. D **88**, 083526 (2013). [arXiv:1302.6415]
72. W. Duhe, T. Biswas, Emergent cyclic inflation, a numerical investigation. Class. Quantum Grav. **31**, 155010 (2014). [arXiv:1306.6927]
73. G. Calcagni, Multi-scale gravity and cosmology. JCAP **1312**, 041 (2013). [arXiv:1307.6382]
74. R. Penrose, Before the big bang: an outrageous new perspective and its implications for particle physics. Conf. Proc. C **060626**, 2759 (2006)
75. R. Penrose, *Cycles of Time: An Extraordinary New View of the Universe* (Bodley Head, London, 2010)
76. V.G. Gurzadyan, R. Penrose, Concentric circles in WMAP data may provide evidence of violent pre-big-bang activity. arXiv:1011.3706
77. E. Newman, A fundamental solution to the CCC equations. Gen. Relat. Grav. **46**, 1717 (2014). [arXiv:1309.7271]
78. V.G. Gurzadyan, R. Penrose, On CCC-predicted concentric low-variance circles in the CMB sky. Eur. Phys. J. Plus **128**, 22 (2013). [arXiv:1302.5162]
79. A. DeAbreu, D. Contreras, D. Scott, Searching for concentric low variance circles in the cosmic microwave background. JCAP **1512**, 031 (2015). [arXiv:1508.05158]
80. D. An, K.A. Meissner, P. Nurowski, Ring-type structures in the Planck map of the CMB. arXiv:1510.06537
81. E.M. Lifshitz, I.M. Khalatnikov, Investigations in relativistic cosmology. Adv. Phys. **12**, 185 (1963)
82. C.W. Misner, Mixmaster universe. Phys. Rev. Lett. **22**, 1071 (1969)
83. E.M. Lifshitz, I.M. Khalatnikov, Problems of relativistic cosmology. Usp. Fiz. Nauk **80**, 391 (1963) [Sov. Phys. Usp. **6**, 495 (1964)]
84. V.A. Belinskiĭ, I.M. Kahalatnikov, A general solution of the gravitational equations with a simultaneous fictitious singularity. Zh. Eksp. Teor. Fiz. **49**, 1000 (1965) [Sov. Phys. JETP **22**, 694 (1966)]
85. L.P. Grishchuk, A.G. Doroshkevich, I.D. Novikov, Anisotropy of the early stages of cosmological expansion and of relict radiation. Zh. Eksp. Teor. Fiz. **55**, 2281 (1968) [Sov. Phys. JETP **28**, 1210 (1969)]

86. V.A. Belinskiĭ, I.M. Khalatnikov, On the nature of the singularities in the general solutions of the gravitational equations. Zh. Eksp. Teor. Fiz. **56**, 1701 (1969) [Sov. Phys. JETP **29**, 911 (1969)]
87. E.M. Lifshitz, I.M. Khalatnikov, Oscillatory approach to singular point in the open cosmological model. Pis'ma Zh. Eksp. Teor. Fiz. **11**, 200 (1970) [JETP Lett. **11**, 123 (1970)]
88. V.A. Belinskiĭ, E.M. Lifshitz, I.M. Khalatnikov, Oscillatory approach to the singular point in relativistic cosmology. Usp. Fiz. Nauk **102**, 463 (1970) [Sov. Phys. Usp. **13**, 745 (1971)]
89. E.M. Lifshitz, I.M. Lifshitz, I.M. Khalatnikov, Asymptotic analysis of oscillatory mode of approach to a singularity in homogeneous cosmological models. Zh. Eksp. Teor. Fiz. **59**, 322 (1970) [Sov. Phys. JETP **32**, 173 (1971)]
90. I.M. Khalatnikov, E.M. Lifshitz, General cosmological solution of the gravitational equations with a singularity in time. Phys. Rev. Lett. **24**, 76 (1970)
91. V.A. Belinskiĭ, E.M. Lifshitz, I.M. Khalatnikov, The oscillatory mode of approach to a singularity in homogeneous cosmological models with rotating axes. Zh. Eksp. Teor. Fiz. **60**, 1969 (1971) [Sov. Phys. JETP **33**, 1061 (1971)]
92. V.A. Belinskiĭ, E.M. Lifshitz, I.M. Khalatnikov, Construction of a general cosmological solution of the Einstein equation with a time singularity. Zh. Eksp. Teor. Fiz. **62**, 1606 (1972) [Sov. Phys. JETP **35**, 838 (1972)]
93. V.A. Belinskiĭ, I.M. Khalatnikov, E.M. Lifshitz, On the problem of the singularities in the general cosmological solution of the Einstein equations. Phys. Lett. A **77**, 214 (1980)
94. V.A. Belinskiĭ, I.M. Khalatnikov, E.M. Lifshitz, Oscillatory approach to a singular point in the relativistic cosmology. Adv. Phys. **19**, 525 (1970)
95. V.A. Belinskiĭ, I.M. Khalatnikov, E.M. Lifshitz, A general solution of the Einstein equations with a time singularity. Adv. Phys. **31**, 639 (1982)
96. I.M. Khalatnikov, E.M. Lifshitz, K.M. Khanin, L.N. Shchur, Ya.G. Sinai, On the stochasticity in relativistic cosmology. J. Stat. Phys. **38**, 97 (1985)
97. J. Demaret, M. Henneaux, P. Spindel, Non-oscillatory behavior in vacuum Kaluza–Klein cosmologies. Phys. Lett. B **164**, 27 (1985)
98. J. Demaret, J.L. Hanquin, M. Henneaux, P. Spindel, A. Taormina, The fate of the mixmaster behaviour in vacuum inhomogeneous Kaluza–Klein cosmological models. Phys. Lett. B **175**, 129 (1986)
99. R.T. Jantzen, Symmetry and variational methods in higher-dimensional theories. Phys. Rev. D **34**, 424 (1986); Symmetry and variational methods in higher-dimensional theories: Errata and addendum. Phys. Rev. D **35**, 2034 (1987)
100. Y. Elskens, M. Henneaux, Chaos in Kaluza–Klein models. Class. Quantum Grav. **4**, L161 (1987)
101. Y. Elskens, M. Henneaux, Ergodic theory of the mixmaster model in higher space-time dimensions. Nucl. Phys. B **290**, 111 (1987)
102. Y. Elskens, Ergodic theory of the mixmaster universe in higher space-time dimensions. II. J. Stat. Phys. **48**, 1269 (1987)
103. A. Hosoya, L.G. Jensen, J.A. Stein-Schabes, The critical dimension for chaotic cosmology. Nucl. Phys. B **283**, 657 (1987)
104. J. Demaret, Y. De Rop, M. Henneaux, Chaos in non-diagonal spatially homogeneous cosmological models in spacetime dimensions $\leqslant 10$. Phys. Lett. B **211**, 37 (1988)
105. J. Demaret, Y. De Rop, M. Henneaux, Are Kaluza–Klein models of the universe chaotic? Int. J. Theor. Phys. **28**, 1067 (1989)
106. T. Damour, M. Henneaux, B. Julia, H. Nicolai, Hyperbolic Kac–Moody algebras and chaos in Kaluza–Klein models. Phys. Lett. B **509**, 323 (2001). [arXiv:hep-th/0103094]
107. B.K. Berger, D. Garfinkle, E. Strasser, New algorithm for mixmaster dynamics. Class. Quantum Grav. **14**, L29 (1997). [arXiv:gr-qc/9609072]
108. B.K. Berger, Numerical approaches to spacetime singularities. Living Rev. Relat. **5**, 1 (2002)
109. D. Garfinkle, Numerical simulations of general gravitational singularities. Class. Quantum Grav. **24**, S295 (2007). [arXiv:0808.0160]

110. L. Bianchi, Sugli spazii a tre dimensioni che ammettono un gruppo continuo di movimenti. On the three-dimensional spaces which admit a continuous group of motions. Soc. Ital. Sci. Mem. Mat. **11**, 267 (1898) [Gen. Relat. Grav. **33**, 2157 (2001); Gen. Relat. Grav. **33**, 2171 (2001)]
111. A. Krasiński et al., The Bianchi classification in the Schücking–Behr approach. Gen. Relat. Grav. **35**, 475 (2003)
112. W. Kundt, The spatially homogeneous cosmological models. Gen. Relat. Grav. **35**, 491 (2003)
113. L.D. Landau, E.M. Lifshitz, *The Classical Theory of Fields* (Butterworth–Heinemann, London, 1980)
114. L. Hsu, J. Wainwright, Self-similar spatially homogeneous cosmologies: orthogonal perfect fluid and vacuum solutions. Class. Quantum Grav. **3**, 1105 (1986)
115. E. Kasner, Geometrical theorems on Einstein's cosmological equations. Am. J. Math. **43**, 217 (1921)
116. D.L. Wiltshire, An introduction to quantum cosmology, in *Cosmology: The Physics of the Universe*, ed. by B. Robson, N. Visvanathan, W.S. Woolcock (World Scientific, Singapore, 1996). [arXiv:gr-qc/0101003]
117. H. Ringström, Curvature blow up in Bianchi VIII and IX vacuum spacetimes. Class. Quantum Grav. **17**, 713 (2000). [arXiv:gr-qc/9911115]
118. H. Ringström, The Bianchi IX attractor. Ann. Henri Poincaré **2**, 405 (2001). [arXiv:gr-qc/0006035]
119. http://commons.wikimedia.org/wiki/File:Kasner_epochs.svg#mediaviewer/File:Kasner_epochs.svg
120. J.D. Barrow, Chaos in the Einstein equations. Phys. Rev. Lett. **46**, 963 (1981); Erratum-ibid. **46**, 1436 (1981)
121. J.D. Barrow, Chaotic behavior in general relativity. Phys. Rep. **85**, 1 (1982)
122. D.F. Chernoff, J.D. Barrow, Chaos in the mixmaster universe. Phys. Rev. Lett. **50**, 134 (1983)
123. P. Halpern, Chaos in the long-term behavior of some Bianchi-type VIII models. Gen. Relat. Grav. **19**, 73 (1987)
124. N.J. Cornish, J.J. Levin, The mixmaster universe is chaotic. Phys. Rev. Lett. **78**, 998 (1997). [arXiv:gr-qc/9605029]
125. N.J. Cornish, J.J. Levin, The mixmaster universe: a chaotic Farey tale. Phys. Rev. D **55**, 7489 (1997). [arXiv:gr-qc/9612066]
126. A.E. Motter, P.S. Letelier, Mixmaster chaos. Phys. Lett. A **285**, 127 (2001). [arXiv:gr-qc/0011001]
127. A.E. Motter, Relativistic chaos is coordinate invariant. Phys. Rev. Lett. **91**, 231101 (2003). [arXiv:gr-qc/0305020]
128. C.W. Misner, Quantum cosmology. I. Phys. Rev. **186**, 1319 (1969)
129. C.W. Misner, Minisuperspace, in *Magic Without Magic*, ed. by J.R. Klauder (Freeman, San Francisco, 1972)
130. D.M. Chitré, Investigation of Vanishing of a Horizon for Bianchi Type IX (The Mixmaster Universe). Ph.D. thesis, University of Maryland, College Park (1972)
131. N.L. Balazs, A. Voros, Chaos on the pseudosphere. Phys. Rep. **143**, 109 (1986)
132. A. Csordás, R. Graham, P. Szépfalusy, Level statistics of a noncompact cosmological billiard. Phys. Rev. A **44**, 1491 (1991)
133. R. Graham, R. Hübner, P. Szépfalusy, G. Vattay, Level statistics of a noncompact integrable billiard. Phys. Rev. A **44**, 7002 (1991)
134. R. Benini, G. Montani, Frame independence of the inhomogeneous mixmaster chaos via Misner–Chitré-like variables. Phys. Rev. D **70**, 103527 (2004). [arXiv:gr-qc/0411044]
135. J.M. Heinzle, C. Uggla, N. Rohr, The cosmological billiard attractor. Adv. Theor. Math. Phys. **13**, 293 (2009). [arXiv:gr-qc/0702141]
136. G. Montani, M.V. Battisti, R. Benini, G. Imponente, Classical and quantum features of the mixmaster singularity. Int. J. Mod. Phys. A **23**, 2353 (2008). [arXiv:0712.3008]
137. M. Henneaux, D. Persson, P. Spindel, Spacelike singularities and hidden symmetries of gravity. Living Rev. Relat. **11**, 1 (2008)

138. S.L. Parnovsky, Gravitation fields near the naked singularities of the general type. Physica A **104**, 210 (1980)
139. E. Shaghoulian, H. Wang, Timelike BKL singularities and chaos in AdS/CFT. Class. Quantum Grav. **33**, 125020 (2016). [arXiv:1601.02599]
140. B.K. Darian, H.P. Kunzle, Axially symmetric Bianchi I Yang–Mills cosmology as a dynamical system. Class. Quantum Grav. **13**, 2651 (1996). [arXiv:gr-qc/9608024]
141. J.D. Barrow, J.J. Levin, Chaos in the Einstein–Yang–Mills equations. Phys. Rev. Lett. **80**, 656 (1998). [arXiv:gr-qc/9706065]
142. Y. Jin, K.-i. Maeda, Chaos of Yang–Mills field in class A Bianchi spacetimes. Phys. Rev. D **71**, 064007 (2005). [arXiv:gr-qc/0412060]
143. R. Carretero-González, H.N. Núñez-Yépez, A.L. Salas-Brito, Evidence of chaotic behavior in Jordan–Brans–Dicke cosmology. Phys. Lett. A **188**, 48 (1994)
144. V.G. LeBlanc, Asymptotic states of magnetic Bianchi I cosmologies. Class. Quantum Grav. **14**, 2281 (1997)
145. V.G. LeBlanc, Bianchi II magnetic cosmologies. Class. Quantum Grav. **15**, 1607 (1998)
146. V.G. LeBlanc, D. Kerr, J. Wainwright, Asymptotic states of magnetic Bianchi VI_0 cosmologies. Class. Quantum Grav. **12**, 513 (1995)
147. B.K. Berger, Comment on the chaotic singularity in some magnetic Bianchi VI_0 cosmologies. Class. Quantum Grav. **13**, 1273 (1996). [arXiv:gr-qc/9512005]
148. M. Weaver, Dynamics of magnetic Bianchi VI_0 cosmologies. Class. Quantum Grav. **17**, 421 (2000). [arXiv:gr-qc/9909043]
149. V.A. Belinskiĭ, I.M. Khalatnikov, Effect of scalar and vector fields on the nature of the cosmological singularity. Zh. Eksp. Teor. Fiz. **63**, 1121 (1972) [Sov. Phys. JETP **36**, 591 (1973)]
150. L. Andersson, A.D. Rendall, Quiescent cosmological singularities. Commun. Math. Phys. **218**, 479 (2001). [arXiv:gr-qc/0001047]
151. J.D. Barrow, H. Sirousse-Zia, Mixmaster cosmological model in theories of gravity with a quadratic Lagrangian. Phys. Rev. D **39**, 2187 (1989); Erratum-ibid. D **41**, 1362 (1990)
152. J.D. Barrow, S. Cotsakis, Chaotic behaviour in higher-order gravity theories. Phys. Lett. B **232**, 172 (1989)
153. S. Cotsakis, J. Demaret, Y. De Rop, L. Querella, Mixmaster universe in fourth-order gravity theories. Phys. Rev. D **48**, 4595 (1993)
154. J. Demaret, L. Querella, Hamiltonian formulation of Bianchi cosmological models in quadratic theories of gravity. Class. Quantum Grav. **12**, 3085 (1995). [arXiv:gr-qc/9510065]
155. N. Deruelle, On the approach to the cosmological singularity in quadratic theories of gravity: the Kasner regimes. Nucl. Phys. B **327**, 253 (1989)
156. N. Deruelle, D. Langlois, Long wavelength iteration of Einstein's equations near a spacetime singularity. Phys. Rev. D **52**, 2007 (1995). [arXiv:gr-qc/9411040]
157. C. Uggla, H. van Elst, J. Wainwright, G.F.R. Ellis, The past attractor in inhomogeneous cosmology. Phys. Rev. D **68**, 103502 (2003). [arXiv:gr-qc/0304002]
158. T. Damour, S. de Buyl, Describing general cosmological singularities in Iwasawa variables. Phys. Rev. D **77**, 043520 (2008). [arXiv:0710.5692]
159. D. Garfinkle, Numerical simulations of generic singuarities. Phys. Rev. Lett. **93**, 161101 (2004). [arXiv:gr-qc/0312117]

Chapter 7
Cosmological Constant Problem

What shall we use to fill the empty spaces?
— Roger Waters (Pink Floyd), *The Wall*

Contents

(continued)

G. Calcagni, *Classical and Quantum Cosmology*, Graduate Texts in Physics,
DOI 10.1007/978-3-319-41127-9_7

A suspicion shared by some, including the author, is that a satisfactory solution of the cosmological constant problem would shed much light on the puzzle of quantum gravity. Quantum field theory, vacuum fluctuations, the microscopic degrees of freedom of gravity and their coarse graining all converge to this Pandora's box in ways that still fascinate even the most consummate expert. About three quarters of the content of the universe is something whose intimate nature is utterly unknown. Hundreds of explanations, theories, models, conjectures have been put forward without successfully convincing the scientist. This chapter is an account, with neither sad nor happy ending, of the problem and of some of the efforts dedicated to its solution.

7.1 The Problem in Field Theory

7.1.1 Spontaneous Symmetry Breaking and Dynamical Λ

We saw in Sect. 2.5.3 that the cosmological constant can be interpreted as the contribution of field vacuum fluctuations to the energy density of the universe. In quantum field theory, the physics of the vacuum is regulated by the evolution of particle fields determined by their self and mutual interactions. Therefore, it is expected that Λ becomes *dynamical* in realistic models [1, 2]. This is naturally achieved by the mechanism of *spontaneous symmetry breaking* from a "false" to a "true" vacuum, where vacuum configurations corresponding to maxima of a classical potential are unstable and the system decays to vacua of lower energy. We saw instances of this in hybrid inflation (Sect. 5.5.3) and, indirectly, in the Higgs Lagrangian (5.202).

The Goldstone model gives the standard example of spontaneous symmetry breaking. Consider a complex scalar ϕ with Lagrangian

$$\mathcal{L} = -\partial_\mu\phi\partial^\mu\phi^\dagger - V(\phi,\phi^\dagger)\,, \qquad V(\phi,\phi^\dagger) = V_0 - \mu^2|\phi|^2 + \lambda|\phi|^4\,, \tag{7.1}$$

where $\mu^2, \lambda > 0$ and V_0 are constants and $|\phi|^2 = \phi^\dagger\phi$. The system is invariant under a $U(1)$ global transformation $\phi \to e^{i\alpha}\phi$, where $0 \leqslant \alpha < 2\pi$ is a real, spacetime-independent parameter. The vacuum structure reflects this continuous symmetry. The double-well (or Mexican-hat) potential V has a local maximum (a false vacuum) at $\phi = 0$, where $V = V_0$, and an infinite number of global minima at

$$\phi = \phi_\vartheta := \frac{\mu}{\sqrt{2\lambda}}\, e^{i\vartheta}\,, \tag{7.2}$$

labelled by $0 \leqslant \vartheta < 2\pi$ and where $V = V_{\rm min} := V_0 - \mu^4/(4\lambda)$. The system is said to have a $U(1)$-symmetric true vacuum state. The $U(1)$ symmetry is "spontaneously broken" when the scalar falls into one of the minima. We can choose, for example, $\vartheta = 0$ and the minimum $\phi_0 = \mu/\sqrt{2\lambda}$. To describe the system near the ground state, it is convenient to parametrize the field with real scalars σ and η,

$$\phi(x) = \frac{1}{\sqrt{2}}[v + \sigma(x) + i\eta(x)]\,, \tag{7.3}$$

where $v = \sqrt{2}\phi_0$, and expand the Lagrangian (7.1) accordingly:

$$\mathcal{L} = -\tfrac{1}{2}(\partial\sigma)^2 - \tfrac{1}{2}m_\sigma^2\sigma^2 - \tfrac{1}{2}(\partial\eta)^2 - U(\sigma,\eta)\,. \tag{7.4}$$

Up to interaction terms $U(\sigma,\eta)$, one can single out the free part and recognize σ as a massive scalar with standard kinetic term and mass $m_\sigma^2 := 2\lambda v^2 = 2\mu^2$, and η as a massless scalar called *Nambu–Goldstone boson*. In general, Nambu–Goldstone bosons arise whenever a continuous symmetry is spontaneously broken, a result known as Goldstone's theorem [3–5]. The model can be generalized to an Abelian $U(1)$ gauge symmetry as well as to the non-Abelian local gauge symmetries of the Standard Model [6–9]. In the Abelian case, the Lagrangian (7.1) is extended to include a vector boson A_μ with field strength $F_{\mu\nu} := \partial_\mu A_\nu - \partial_\nu A_\mu$ (for instance, the electromagnetic field),

$$\mathcal{L} = -\tfrac{1}{4}F_{\mu\nu}F^{\mu\nu} - D_\mu\phi(D^\mu\phi)^\dagger - V(\phi)\,, \tag{7.5}$$

where $D_\mu = \partial_\mu + ieA_\mu$ is the covariant derivative with Abelian coupling (electric charge) e. The system is invariant under a local $U(1)$ gauge transformation

$$\phi(x) \to \phi'(x) = e^{i\alpha(x)}\phi(x)\,, \quad A_\mu(x) \to A'_\mu(x) = A_\mu(x) - e^{-1}\partial_\mu\alpha(x)\,, \tag{7.6}$$

where now α is a spacetime-dependent scalar. The vacuum structure (7.2) is unchanged, since $A_\mu = 0$ in a Lorentz-invariant vacuum. The field decomposition leading to (7.4) follows the same steps as above except for two differences: the gauge vector acquires a mass $m_A^2 = (ev)^2/2$ and the Nambu–Goldstone boson η is not an independent degree of freedom and can be removed by a suitable gauge transformation (7.6) (unitary gauge). Then, (7.4) is replaced by

$$\mathcal{L} = -\tfrac{1}{4}F_{\mu\nu}F^{\mu\nu} + \tfrac{1}{2}m_A^2 A_\mu A^\mu - \tfrac{1}{2}(\partial\sigma)^2 - \tfrac{1}{2}m_\sigma^2\sigma^2 - W(\sigma, A)\,, \tag{7.7}$$

up to some interactions W which we do not write here. The generation of a mass for a gauge boson through spontaneous symmetry breaking is called *Higgs mechanism*. Generalizing to non-Abelian groups, the same mechanism applies to the electroweak sector, where $\sigma = h$ is the Higgs boson with mass (5.203). The resulting Lagrangian is (5.202) with $g_{\mu\nu} = \eta_{\mu\nu}$.

We may ask now whether the Higgs particle accounts for the observed cosmological constant [1, 10, 11]. In the semi-classical sense of (2.112), the cosmological constant is the contribution of vacuum quantum fluctuations of gravity and matter fields. These fluctuations sum up to a term of the form (2.113), which adds to a purely classical cosmological constant Λ_0 to give the total vacuum energy

$$\rho_{\rm vac} = \langle\rho\rangle + \rho_{\Lambda_0} = \frac{\kappa^2\langle\rho\rangle + \Lambda_0}{\kappa^2} =: \frac{\Lambda}{\kappa^2} = \rho_\Lambda\,. \tag{7.8}$$

The cosmological vacuum contribution of electroweak interactions is, to first approximation, $\langle\rho\rangle \simeq V_{\rm min}$ and the vacuum energy density today reads

$$\rho_{\rm vac}^{(0)} \simeq \left(\frac{\Lambda_0}{\kappa^2} + V_0\right) - \frac{m_h^2 v^2}{8}\,. \tag{7.9}$$

The vacuum expectation value v can be determined from its relation $v^2 = (\sqrt{2}\,G_{\rm F})^{-1}$ with the Fermi constant $G_{\rm F}$, which is measured from the muon lifetime [12]: $v^2 \approx 6.06\times 10^4\,{\rm GeV}^2$. (From this and the LHC value $m_h \approx 125\,{\rm GeV}$, one can get the quartic coupling $\lambda = m_h^2/(2v^2) \approx 0.13$, which is more difficult to detect directly [13].) Assuming that V_0 exactly cancels the classical cosmological constant (or that $\Lambda_0 = 0 = V_0$ identically), one can estimate (7.9) as

$$|\rho_{\rm vac}^{(0)}| \simeq 1.2\times 10^8\,{\rm GeV}^4 \simeq 10^{-68} m_{\rm Pl}^4\,, \tag{7.10}$$

to be compared with the observed contribution

$$\boxed{\rho_\Lambda \simeq 10^{-48}\,{\rm GeV}^4 \simeq 10^{-124} m_{\rm Pl}^4\,.} \tag{7.11}$$

Recall that the Higgs cannot play the role of the inflaton as it stands, since its quartic interaction is too strong. In Sect. 5.11.2, we saw that a non-minimal coupling with gravity can lead to a viable scenario. However, the principal reason why the Higgs physics may work for inflation but not for the cosmological constant problem is that early- and late-time accelerating periods are characterized by very different energy scales. Equation (7.11) is separated from the GUT scale typical of inflation by about 12 – 16 orders of magnitude.

7.1.2 *Zero-Point Energy and Higher Loops*

The vacuum energy density (or zero-point energy) is the eigenvalue of the Hamiltonian when acting upon the physical vacuum state, i.e., the partition-function contribution $Z[0]$ of all bubble diagrams; the latter are Feynman diagrams with no external legs [14]. One can try to fill the gap of about 56 orders of magnitude between (7.10) and (7.11) by refining the argument leading to (7.9), valid only at the tree level in perturbation theory. The first correction comes for one-loop bubble diagrams which, in the absence of interactions, amount to one bubble diagram per free field. In Sect. 2.5.3, we identified $\rho_{\rm vac}$ with ρ_Λ in (2.123), implicitly regarding Λ as the sum of a classical part Λ_0 plus the vacuum contribution. Here we reconsider that calculation with more details.

For a free field of mass m_i, the zero-point energy is an infinite superposition of harmonic oscillators, given by an integral over all frequencies $\omega(\mathbf{p}) = \sqrt{|\mathbf{p}|^2 + m_i^2}$. In four dimensions,

$$\rho^{(1)}_{\mathrm{vac},i} = \frac{N_i}{2}\int_{-\infty}^{+\infty}\frac{\mathrm{d}^3\mathbf{p}}{(2\pi)^3}\sqrt{|\mathbf{p}|^2+m_i^2} = \frac{N_i}{(2\pi)^2}\int_0^{+\infty}\mathrm{d}p\,p^2\sqrt{p^2+m_i^2}\,, \tag{7.12}$$

where $p = |\mathbf{p}|$ and $|N_i|$ is the number of one-particle states, positive for bosonic fields (real scalar: $N = 1$; complex scalar and real massless vector: $N = 2$; real massive vector: $N = 3$) and negative for fermions (Majorana spinor: $N = -2$, Dirac massive spinor: $N = -4$). Equation (7.12) can be easily checked for a massive scalar field $\hat{\phi}$ on Minkowski background by using the expansion in creation and annihilation operators of Sect. 5.6.2 for the Bunch–Davies vacuum (5.114),

$$\hat{\phi}(t,\mathbf{x}) = \frac{1}{(2\pi)^3}\int\frac{\mathrm{d}^3\mathbf{p}}{\sqrt{2\omega(\mathbf{p})}}\left[a_{\mathbf{p}}\,\mathrm{e}^{-\mathrm{i}\omega(\mathbf{p})\,t+\mathrm{i}\mathbf{p}\cdot\mathbf{x}} + a^\dagger_{\mathbf{p}}\,\mathrm{e}^{\mathrm{i}\omega(\mathbf{p})\,t-\mathrm{i}\mathbf{p}\cdot\mathbf{x}}\right]. \tag{7.13}$$

Taking the vacuum expectation value of the energy-density (5.23), $2\rho_\phi = (\partial_t\hat{\phi})^2 + \partial_\alpha\hat{\phi}\partial^\alpha\hat{\phi} + m^2\hat{\phi}^2$, one obtains (7.12) with $N = 1$. The check for other fields is similar [14].

The zero-point energy $\rho^{(1)}_{\text{vac},i}$ is divergent. To get a finite expression, one can introduce an ultraviolet momentum cut-off $p_{\max}$ much larger than m_i. For each species, we can then evaluate the integral (7.12) (with domain $(0, p_{\max})$) exactly by an analytic continuation to D dimensions, then expanding for large $p_{\max}$ and setting $D = 4$. To keep the expression for the energy density dimensionally correct in $D = 4$, we also multiply by a mass scale M. For each species,

$$\begin{aligned}
&\frac{M^{4-D}}{2}\int_{-p_{\max}}^{p_{\max}}\frac{d^{D-1}\mathbf{p}}{(2\pi)^{D-1}}\sqrt{|\mathbf{p}|^2+m^2}\\
&=\frac{M^{4-D}}{2}\int d\Omega_{D-2}\int_0^{p_{\max}}\frac{dp}{(2\pi)^{D-1}}\,p^{D-2}\sqrt{p^2+m^2}\\
&=\frac{M^{4-D}(4\pi)^{\frac{1-D}{2}}}{\Gamma\left(\frac{D-1}{2}\right)}\int_0^{p_{\max}}dp\,p^{D-2}\sqrt{p^2+m^2}\\
&=\frac{M^{4-D}(4\pi)^{\frac{1-D}{2}}}{(D-1)\Gamma\left(\frac{D-1}{2}\right)}\,m p_{\max}^{D-1}F\left(-\frac{1}{2},\frac{D-1}{2};\frac{D+1}{2};-\frac{p_{\max}^2}{m^2}\right) &(7.14)\\
&\overset{D=4}{=}\frac{1}{16\pi^2}\left[p_{\max}^4+p_{\max}^2m^2+\frac{m^4}{4}\ln\left(\frac{\sqrt{e}m^2}{4p_{\max}^2}\right)\right]+O\left(\frac{1}{p_{\max}}\right), &(7.15)
\end{aligned}$$

where we integrated the solid angle in spherical coordinates to get the area of the hypersphere S^{D-2}, $\Omega_{D-2} = 2\pi^{(D-1)/2}/\Gamma[(D-1)/2]$, and F (also denoted as ${}_2F_1$) is the hypergeometric function. Summing over all species i and neglecting all mass contributions, the result is [15]

$$\rho^{(1)}_{\text{vac}} = \sum_i \rho^{(1)}_{\text{vac},i} \sim O(1)p_{\max}^4\,. \tag{7.16}$$

For a Planck-scale cut-off (2.124),

$$\rho^{(1)}_{\text{vac}} \sim 10^{76}\,\text{GeV}^4 \sim m_{\text{Pl}}^4\,, \tag{7.17}$$

an abyss of 124 orders with respect to the observed value (7.11), much worse than the tree-level estimate (7.10).

What went amiss? First of all, in (7.15) the cut-off $p_{\max}$ has been regarded as physical, but it is easy to see that the first two terms in the expansion break Lorentz invariance. In fact, computing the expectation value of the pressure (5.24), one verifies that $\langle P\rangle \neq -\langle\rho\rangle$. It is then more natural to employ a regularization scheme preserving Lorentz invariance. One such scheme is dimensional regularization. Instead of setting $D = 4$ in (7.14), we first send the cut-off to infinity and then expand with respect to the parameter $\epsilon = D - 4 \ll 1$:

$$\begin{aligned}
&\lim_{p_{\max}\to+\infty} \frac{M^{4-D}(4\pi)^{\frac{1-D}{2}}}{(D-1)\Gamma\left(\frac{D-1}{2}\right)} m p_{\max}^{D-1} F\left(-\frac{1}{2}, \frac{D-1}{2}; \frac{D+1}{2}; -\frac{p_{\max}^2}{m^2}\right) \\
&= -\frac{M^4\Gamma\left(-\frac{D}{2}\right)}{2(4\pi)^{\frac{D}{2}}}\left(\frac{m}{M}\right)^D \\
&= \frac{m^4}{(8\pi)^2}\left[\frac{2}{\epsilon} - \frac{3}{2} - \gamma_{\rm EM} - \ln(4\pi) + \ln\left(\frac{m^2}{M^2}\right)\right] + O(\epsilon)\,,
\end{aligned} \tag{7.18}$$

where $\gamma_{\rm EM} \approx 0.577$ (Sect. 3.2.3). In the "non-minimal subtraction" renormalization scheme, one adds to this expression some counter-terms exactly canceling the divergent term $1/\epsilon$ as well as, for convenience, mass-independent contributions inside the square brackets. Thus, summing over all species, the regularized and renormalized vacuum energy reads [16]

$$\rho_{\rm vac}^{(1)} = \sum_i \frac{N_i m_i^4}{(8\pi)^2} \ln\left(\frac{m_i^2}{M^2}\right)\,. \tag{7.19}$$

In contrast with (7.15), this expression is cut-off independent and its size is determined both by the mass of the particles and by the renormalization scale M. The latter may be chosen as the average between the graviton energy $\sim H_0$ and the energy of the photons coming from supernovæ, $M \sim 10^{-44} m_{\rm Pl}$. Taking the massive particle content of the minimal Standard Model after electroweak symmetry breaking, one obtains an estimate of the zero-point vacuum density [16]:

$$|\rho_{\rm vac}^{(1)}| \approx 2\times 10^8\,{\rm GeV}^4 \sim 10^{-68} m_{\rm Pl}^4\,, \tag{7.20}$$

of the same order of magnitude as (7.10). One might hope that the contribution of field interactions would ameliorate the situation, but this is not the case. Using the non-perturbative method of the Gaussian effective potential [17–19], one can show that interactions leave the mass dependence of (7.19) substantially unchanged [14]:

$$\rho_{\rm vac}^{\rm int} = \sum_i \frac{N_i m_i^4}{(8\pi)^2}\left[\frac{1}{2} - \ln\left(\frac{m_i^2}{M^2}\right)\right]\,.$$

As a last stand, we could take the tidal forces of gravity into account: the effect is negligible [14, 20]. One can see this also by a heuristic argument. Including gravity and assuming that contributions up to one loop are exactly canceled by Λ_0, two-loop diagrams are the next. One can consider a bath of virtual pairs of massive particles–anti-particles with Compton wave-length $\lambda = \hbar/m$ with uniform number-density distribution $n(\lambda) \propto \lambda^{-3}$. The gravitational interaction of a pair of such particles is $V(\lambda) \sim Gm^2/\lambda$, yielding an energy density [21]

$$\rho_{\text{vac}} \sim V(\lambda)\, n(\lambda) \sim \frac{m^6}{m_{\text{Pl}}^2}\,, \tag{7.21}$$

not an improvement. Thus, $\rho_\Lambda \neq \rho_{\text{vac}}$ in the non-supersymmetric Standard Model.

7.1.3 Supersymmetry and Supergravity

Supersymmetry itself, rather than loop corrections in the Standard Model, may lower the hierarchy gap to acceptable values. Here we pick up the thread left at Sect. 5.12, where we introduced the superpotential $W = W(\phi^i)$ and the Kähler potential $K = K(\phi^i, \phi^{i\dagger})$ for the fields ϕ^i acting as coordinates on a Kähler manifold.

As a matter of fact, the simplest implementation of supersymmetry overshoots the target. Thanks to the exact cancellation of vacuum diagrams order by order, the zero-point energy-momentum density is zero [22]. In models on Minkowski spacetime, global supersymmetry is realized by values of the fields where the superpotential is stationary, $\partial_i W = 0$, so that

$$\langle \rho \rangle = V_{\min} = 0\,. \tag{7.22}$$

One reaches the same conclusion by noting that the zero-point vacuum energy (7.16) with (7.12) is predicted to be exactly zero when combining the components of super-multiplets, which all share the same mass. For the Wess–Zumino multiplet (5.212) (two real scalars and one Majorana fermion), $\sum_i N_i = 1 + 1 - 2 = 0$. A massive vector multiplet has a real massive vector, a real scalar and two Majorana spinors, giving again $\sum_i N_i = 3 + 1 - 2 - 2 = 0$; and so on.

Getting zero instead of something does not solve the Λ problem but sets the stage for an interesting line of attack. Global supersymmetry is broken at some scale (2.125) above TeV and the magical cancellations leading to (5.211) and (7.22) no longer take place. Equation (7.16) with $p_{\max} = M_{\text{SUSY}}$ yields $\rho_{\text{vac}}^{(1)} \gtrsim 10^{-64}\, m_{\text{Pl}}$. According to the crude estimate (7.21), loop corrections with broken supersymmetry slightly reduce the vacuum contribution to

$$\rho_{\text{vac}} \sim 10^{-96} m_{\text{Pl}}^4\,. \tag{7.23}$$

This lowers the gap with the value (7.11) down to 30 orders of magnitude, still a serious fine tuning.

In supergravity, the vacuum structure changes. Supersymmetry becomes local and is encoded in the condition (5.223). Now the degeneracy of the gauge-invariant globally supersymmetric vacuum (7.22) is broken by gravity and the minima of the potential are negative semi-definite:

$$V_{\rm min} = -3\kappa^4|W|^2 {\rm e}^K \leqslant 0\,. \tag{7.24}$$

All these vacuum solutions are stable but, in general, only one of them corresponds to the Minkowski-spacetime vacuum (7.22), provided we fine tune W so that it vanishes for such a solution. This special vacuum, however, is *not* the state with lowest energy [23].

Conversely, one can find non-supersymmetric solutions ($D_iW \neq 0$) such that $V = 0$, but these solutions will not be stable in general. Still, there exist special classes of Kähler potentials giving rise to positive-definite potentials with a non-supersymmetric minimum $V_{\rm min} = 0$ without fine tuning. These are the no-scale models introduced in Sect. 5.12.4 [24]. The matrices $\mathcal{G}$ and f in (5.217) and (5.232) determine the kinetic terms in the Lagrangian density (5.220a) and are therefore positive definite. This implies that the potential (5.231) is positive semi-definite and has an absolute minimum for any field configuration such that

$$V = 0 \qquad \Leftrightarrow \qquad W_1 = {\rm const}\,, \quad D_sW = 0\,. \tag{7.25}$$

This vacuum state fixes the values of the fields Φ^c and S^s but not of T. The name of these models stems from their featuring only $m_{\rm Pl}$ as a mass scale, while the hierarchy of all the other masses is determined dynamically from the Planck mass by spontaneous symmetry breaking. If W_2 and K did not depend on all scalars S^s, then for some value $\tilde{s}$ of the index s one would have $0 = D_{\tilde{s}}W = \partial W/\partial S^{\tilde{s}} + W\partial\tilde{K}/\partial S^{\tilde{s}} = W\partial\tilde{K}/\partial S^{\tilde{s}}$, which would imply that $W = 0$. Then, the supersymmetry condition (5.223) would be trivially obeyed for all i in the vacuum state (7.25). On the other hand, assuming that the superpotential and the Kähler potential depend on all scalars S^s, one has $W \neq 0$ and the vacuum (7.25) does break supersymmetry, since for indices c (5.223) is violated: $D_cW = W\partial K/\partial\Phi^c = 3W|T + T^\dagger - h|^{-1}\partial h/\partial\Phi^c \neq 0$.

To summarize, no-scale models predict a non-supersymmetric configuration with vanishing vacuum energy. The cosmological constant problem is not solved, first of all because one needs to generate a mechanism to lift the minimum of the potential to some non-zero value compatible with (7.11). Secondly, one is simply assuming (5.230) with exactly those coefficients and field dependencies. Not only are these structures chosen ad hoc at the classical level, but they are not expected to survive when loop corrections are taken into account. We will come back to these issues in Chap. 12, where we will see how no-scale models of the form (5.230) and related lifting mechanisms can be naturally generated in string theory (Sect. 12.3.5).

7.2 Other Versions of the Problem and Strategies

Tree- and loop-level calculations in the minimal Standard Model of particles and inclusion of supersymmetry confirm the existence of a hierarchy or fine-tuning problem. This is sometimes called the "old" cosmological constant problem, to distinguish it from the "new" problem asking why ρ_Λ is about the same order of magnitude as the total matter density $\rho_{\rm m}$ (Sect. 2.6). To these issues, one should add the coincidence problem as well as the following observations.

7.2.1 *Broken Symmetries*

Regardless of its value, by itself the cosmological constant is a bizarre object from the perspective of general relativity as a field theory [24]. Normally, in a field theory invariant under a group $\mathbb{G}$ one can find vacuum solutions symmetric with respect to some sub-group $\mathbb{G}' \in \mathbb{G}$ without fine tuning the parameters. Examples are provided by the $U(1)$-invariant vacuum expectation values of doublet scalars in the $SU(2) \otimes U(1)$ electroweak model, or by supergravity models with partial breaking of supersymmetry.

In the case of plain, non-supersymmetric classical gravity, $\mathbb{G}$ includes the diffeomorphism group of general coordinate transformations. Fixing a background solution $g_{\mu\nu} = \tilde{g}_{\mu\nu}$ of the Einstein equations (2.23) always breaks diffeomorphism invariance, but some residual symmetry may be left. When $\Lambda = 0$, the Minkowski metric $g_{\mu\nu} = \eta_{\mu\nu}$ is a vacuum solution of the Einstein equations (2.23), $R_{\mu\nu} - g_{\mu\nu}R/2 = 0$. This vacuum solution carries with it the symmetries of the global Fermi frame, the Poincaré group of proper Lorentz transformations and translations. The presence of a non-vanishing cosmological constant forbids such a vacuum solution (the solution now is de Sitter or anti-de Sitter spacetime, depending on the sign of Λ) and, in particular, breaks invariance under spatial and time translations. Solutions preserving translations would then require a fine tuning on Λ. To put it differently, one may look for translation-invariant solutions where the gravitational and all matter fields ϕ_n are constant. These would obey the equations of motion $\partial\mathcal{L}/\partial\phi_n = 0$ and $\partial(\sqrt{-g}\,\mathcal{L})/\partial g_{\mu\nu} = 0$ but the latter is incompatible with the solution $\mathcal{L} = -\rho_\Lambda = -\Lambda/\kappa^2$, unless the constant Λ is exactly set to zero.

In yet other words, the cosmological constant is related to a symmetry of the matter sector (e.g., [25, 26]): in the absence of gravity, a constant shift $\rho_0 = \text{const}$ of the matter Lagrangian

$$\mathcal{L}_{\rm m} \to \mathcal{L}_{\rm m} + \rho_0 \qquad \Rightarrow \qquad T_\mu^{\ \nu} \to T_\mu^{\ \nu} + \rho_0\delta_\mu^\nu \tag{7.26}$$

would leave the equations of motion $\nabla_\nu T_\mu^{\ \nu} = 0$ invariant. This is not the case when a dynamical gravitational field with a non-trivial action is introduced. In particular, the Einstein's equations (2.23) are not invariant under (7.26) (the cosmological constant is shifted by $\Lambda \to \Lambda - \kappa^2 \rho_0$) and we cannot impose (7.26) as a symmetry of the system. At the level of particle physics, this restriction has no natural explanation.

7.2.2 *The 4π Puzzle*

Perhaps no other great question in theoretical physics admits as many formulations as the cosmological constant problem. As a final piece of information, we recast the problem in terms of an astounding numerical coincidence purely based on classical cosmology [27–29].

Consider the Hubble volume in $N = 1$ gauge

$$\mathcal{V}_H = \frac{4\pi}{3} R_H^3 = \frac{4\pi}{3H^3} \,. \tag{7.27}$$

To measure the duration of a cosmological era marked by some initial and final time $t_{1,2}$ (which is which depends on the sign of cosmic acceleration), we note that the number of comoving modes in the comoving volume $\mathcal{V}_{\rm com} = \mathcal{V}_H/a^3$ crossing the Hubble horizon with wave-numbers $a(t_1)H(t_1) = k_1 \leqslant k \leqslant a(t_2)H(t_2) = k_2$ is

$$\mathcal{N}_{12} = \int_{k_1}^{k_2} \frac{{\rm d}^3\mathbf{k}}{(2\pi)^3} \mathcal{V}_{\rm com} = \frac{2}{3\pi} \ln \frac{r_{H,2}}{r_{H,1}} \,, \tag{7.28}$$

which is nothing but the improved number of e-foldings (5.12) up to a numerical prefactor. Consider now the cosmic history from inflation until today. The evolution of the proper Hubble horizon is schematically depicted in Fig. 7.1. During both inflation and late-time acceleration, H is approximately constant. The end of inflation and the beginning of radiation and dust domination is marked by point X, while the beginning of the dark-energy era is at point Y. Modes with $k < k_-$ and $k > k_+$ exit the horizon only once, respectively at point P and Q, and never re-enter. Modes with $k_- < k < k_+$ exit twice, at A and C, re-entering at some point B during radiation or dust domination. We are interested in counting the modes $k \in [k_-, k_+]$ which re-entered the Hubble patch during the radiation-matter era XY. We call this number, characteristic of our universe, $\mathcal{N}_{\rm c} := \mathcal{N}_{\rm XY}$. It can be calculated from the condition ${\rm d}k_-/{\rm d}a = {\rm d}(aH)/{\rm d}a|_{\rm Y} = 0$ (the tangent to the curve at Y has unit slope) and the Friedmann equation (2.81) in the absence of curvature ($\textsc{k} = 0$) and with radiation, dust matter and a cosmological constant Λ. Denoting the scale factor at Y normalized at radiation-matter equality as $\alpha = a_{\rm Y}/a_{\rm eq}$, the slope condition at Y states that $\alpha = \alpha(\sigma)$ is solution to the quartic equation $2\sigma^4\alpha^4 - \alpha - 2 = 0$, where $\sigma^4 := \rho_\Lambda/\rho_{\rm eq}$. Using the Friedmann equation, the result is

$$\mathcal{N}_{\rm c} = \frac{1}{6\pi} \ln \left[c(\sigma) \frac{\rho_{\rm reh}}{\rho_\Lambda} \right], \tag{7.29}$$

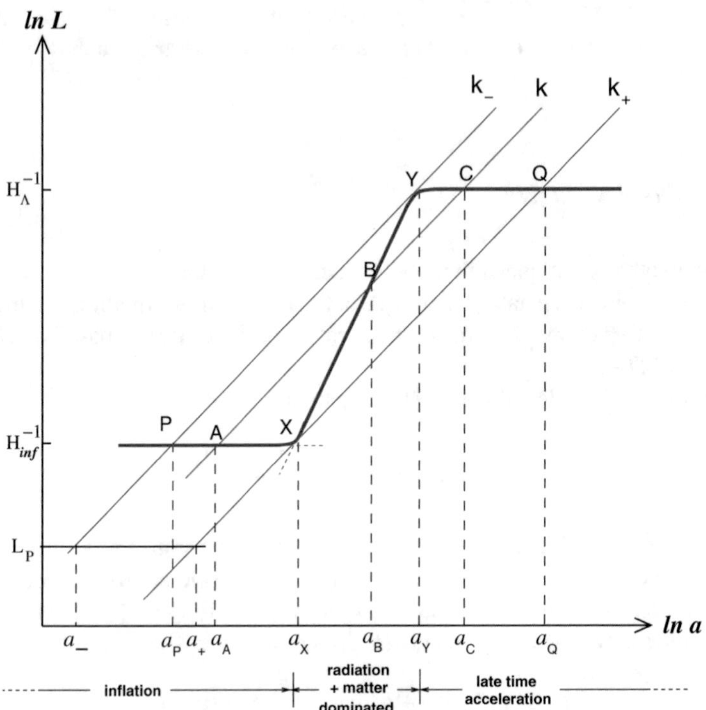

Fig. 7.1 Schematic history of the universe from inflation to late-time acceleration. Here $L_P = l_{\text{Pl}}$ is the Planck length, $H_{\text{inf}} = H_e$ is the Hubble parameter at the end of inflation and $H_\Lambda^2 = \Lambda/3$. The rest of the symbols are explained in the text (Source: [27], ©2014 World Scientific)

where $c(\sigma) = 2[\alpha(\sigma) + 2]/[3\alpha(\sigma) + 4]^2$ and ρ_{reh} is the energy density at the end of reheating (for instantaneous reheating, $\rho_{\text{reh}} = \rho_e$). With $\sigma \approx 2.6 \times 10^{-3}$ (Sect. 2.5.3), one has $c \approx 10^{-3}$. Assuming the energy at the end of inflation around the GUT scale, $\rho_{\text{reh}}^{1/4} \approx 10^{15}$ GeV (Sect. 5.9.1), we also have $\rho_{\text{reh}}/\rho_\Lambda \approx 10^{108}$. Combining everything and taking into account experimental uncertainties, we obtain

$$\mathcal{N}_c = 4\pi \tag{7.30}$$

up to about one part over 10^3. The result is robust with respect to changes of ρ_{reh} of about one order of magnitude, which is roughly the energy excursion between instantaneous reheating and less efficient scenarios.

Such a precise determination of this number has no explanation in standard cosmology: there is no apparent reason why the duration of the radiation-dust era should be of exactly 4π e-foldings! Barring a fortuitous coincidence or exotic reheating scenarios, one may wonder whether a yet undiscovered fundamental principle is being enforced. According to this principle, the cosmological constant is completely determined by the thermal properties of the universe at the end of inflation and at radiation-matter equality, as well as by the duration $\mathcal{N}_\mathrm{c}$ of the subsequent thermal history:

$$\Lambda = \kappa^2 \rho_\mathrm{reh}\, c(\sigma)\, \mathrm{e}^{-6\pi \mathcal{N}_\mathrm{c}}. \tag{7.31}$$

It is plausible that, in any theory explaining the relations (7.30) and (7.31), gravitational and matter degrees of freedom should emerge in a unified way.

7.2.3 UV or IR Problem? Strategies for a Solution

The introduction of supersymmetry in Sect. 7.1.3 attempted to address the cosmological constant problem in two steps: first, by modifying the dynamics of the system, from the Einstein–Hilbert action to the SUGRA action. And, second, by interpreting Λ as the vacuum energy of the quantum theory. This is an instance where the cosmological constant problem is regarded as due to some unknown physics above a certain energy scale (in this case, the scale M_SUSY of supersymmetry breaking). However, there is also a possibility that the root of the problem essentially lies in the infrared rather than in the ultraviolet. For example, if we do not want to abandon local perturbative quantum field theory, we can revisit the untested assumption that it is valid at arbitrary large scales and introduce an IR cut-off of order of the Hubble horizon. This act has its inspiration in the thermodynamics of black holes, where the Bekenstein–Hawking entropy-area law [30–33] indicates that, in the presence of a black hole, ordinary quantum field theory overcounts the number of degrees of freedom within a given volume and ceases to be valid at large scales [34] (we will see more of this in Sect. 7.7.3). Similarly, in the presence of a cosmological horizon it turns out that a UV cut-off must be related to an IR cut-off in such a way that the fine-tuning problem of Sect. 7.1 disappears, even if ρ_Λ maintains its interpretation as the vacuum energy of a quantum field theory [35].

However, on one hand the IR cut-off is not sufficient to explain the observed value of Λ and, on the other hand, some UV modification of physics (quantum gravity?) may still be needed to explain the fundamentals of the overcounting of degrees of freedom. In this and many other cases, it is therefore difficult, or perhaps artificial, to reduce solutions of the cosmological constant problem to purely UV or purely IR effects. This will become most apparent when surveying the landscape of explanations offered in this chapter and in Chaps. 11 and 13.

Having seen that the cosmological constant cannot be easily interpreted as the vacuum of a particle field theory, we should look towards other directions. Intuition, prejudice or experience may lead to very different approaches, sometimes rediscovered or reformulated in independent frameworks. We limit the discussion to some broad classes of proposals.

(A) Without abandoning the particle-physics perspective, the cosmological constant can still be the vacuum energy of an effective field theory but in scenarios more sophisticated that those outlined in Sects. 7.1.1, 7.1.2 and 7.1.3, as, for instance, in string theory.
(B) The energy-density contribution attributed to the cosmological constant and the observed acceleration come from dynamical matter fields (Sect. 7.3). Their full dynamics, not just their vacuum state, is tuned via the free parameters of the model so that to reproduce the observed energy density and equation of state. This dynamical approach does not explain why the number in (7.11) is realized in Nature, but it attempts to combine a reasonable theoretical framework with a modest tuning of the free parameters of the model. It also aims to alleviate another aspect of the cosmological problem when Λ is exactly constant, the coincidence problem discussed in Sect. 2.5.3. Extensions to many-field configurations are also envisaged.
(C) The culprit is neither the vacuum nor the dynamics of a matter field but geometry itself (Sects. 7.4 and 7.5.2). This can be made possible if we modify the Einstein–Hilbert action (2.18) by changing the coupling between gravity and matter or by including higher-order curvature invariants, or both. The goal is the same as in class (B), namely, to find a region in the parameter space which accommodates all observations related to the cosmological constant without fine tuning.
(D) More complicated matter dynamics, spacetime backgrounds, or deeper modifications of geometry going beyond the inclusion of higher-derivative terms may give rise to scenarios which combine some of the features of the previous classes of models (Sect. 7.6). This general case also includes analogue-gravity (Sect. 7.6.4) or emergent-gravity scenarios (Sect. 7.7).
(E) A yet undiscovered fundamental principle is in action.

Other classifications are possible [36, 37] but the present one covers virtually all models where the measured cosmological constant is effectively determined by the dynamics of geometric and matter degrees of freedom. It was recognized rather early that matter fields produce a time-dependent contribution tuning with Λ, both at the classical and at the quantum level [38–40], with particular emphasis on non-minimally coupled scalars [41–55]. In all these cases, Λ effectively decays in time from possibly large values, but there are other types of predictions. In one class of models, the cosmological constant is identically zero, which can be obtained by requiring the vanishing of graviton and scalar-particle creation in radiation-dominated universe [56], via Kaluza–Klein compactification [57], via spontaneous symmetry breaking in modified gravity with non-minimal coupling [58] and in some quantum-gravity models (Sect. 10.2.4). Still in the same category, we may include the possibility where Λ is not zero but so strongly suppressed that it vanishes for all practical purposes (a Euclidean quantum-gravity example is [59]). With the discovery in the late 1990s of the acceleration of the present universe, models with exactly $\Lambda = 0$ turned out to be, at best, incomplete.

In general, all the above strategies and their specific incarnations can suffer from two drawbacks when cosmological models are formulated. First, they involve a constraint on the parameter space of the model, not a *prediction* of the value (7.11). Therefore, none of the solutions of the cosmological constant problem falling in the above categories imposes itself as *the* solution to the problem. Second, sometimes they lack a solid theoretical background explaining from first principles why an action should have such and such form with such and such free parameters. In a way, these questions are akin to the one we asked for the inflaton: Where does Λ come from? This issue may be tackled by facing the Λ problem in a candidate fundamental theory of Nature, such as string theory or one of the many quantum-gravity approaches. This will be the subject of Sects. 9.4, 10.2.4 and 13.1.2, with some interesting twist also regarding the issue of the value.

It might also be the case that asking for a unique and unequivocal prediction of the value (7.11) is an unreasonable request, even in a fundamental theory. Whenever quantum mechanics or chaos make their appearance, exactly deterministic predictions become nonsensical and one is entitled only to talk about probabilities. Canonical quantum cosmology (Sects. 9.4 and 10.2.4) and the string landscape (Sect. 13.1.2) are precisely examples of that.[1] To some, this attitude might look like an admission of defeat, considering that it almost waves away an unknown that constitutes 70 % of the observed universe. However, one cannot rule it out only on the grounds of subjective taste.

On a positive note, there are at least two theories (asymptotic safety, Sect. 11.2; causal sets, Sect. 11.6) offering a prediction for Λ in agreement with (7.11), while another (emergent gravity, Sect. 7.7) has the ambition to address the 4π puzzle by first principles.

7.3 Quintessence

Inspired by the concept of dynamical vacuum in quantum field theory, it is tempting to consider scenarios with a varying cosmological constant, replacing $\Lambda \to \Lambda(t)$ with no kinetic term [61–66] (a number of time profiles are catalogued in [36]). The presence of derivatives for Λ in the action, making it a dynamical field, gives rise to a more flexible dynamics. After early applications with a 3-form [39] and a non-minimally coupled scalar [41, 42, 50, 51], the attention moved to a minimally coupled scalar field weakly coupled with the rest of matter [67–81]. As it constitutes a fifth contribution to the energy density of the universe apart from baryons, neutrinos, radiation and dark matter, this scalar has been dubbed *quintessence*

[1] An early attempt to explain the cosmological constant in perturbative quantum gravity is [60], where the bare Λ is compensated by one-loop terms. This case is still of deterministic type, the cosmological constant being driven to zero by quantum effects.

[71]. Due to inhomogeneous fluctuations, quintessence clusters gravitationally (e.g., [82]). This can leave an imprint both on the CMB spectrum and on the evolution of large-scale structures, which allows one to place constraints via independent data sets.

A major assumption in quintessence models is that the cosmological constant is set exactly to zero and that only dynamics contributes to acceleration. In this respect, the cosmological constant problem in its earliest version is simply assumed to be solved, as the mechanism leading to $\Lambda = 0$ is left unspecified.

7.3.1 *Tracking, Freezing and Thawing*

The general-relativity action (2.20) with $\Lambda = 0$ is assumed. The energy density and pressure of quintessence are given by (2.79), while the scalar equation of motion, the continuity equation for matter and the first Friedmann equation are, respectively, (5.39) also written as

$$\dot{\rho}_\phi + 3H(1+w_\phi)\rho_\phi = 0\,, \tag{7.32}$$

and

$$\dot{\rho} + 3H(1+w)\rho = 0\,, \qquad H^2 = \frac{\kappa^2}{3}(\rho + \rho_\phi)\,, \tag{7.33}$$

where $\rho = \rho_{\rm m} + \rho_{\rm r}$ is the contribution of matter and radiation with effective barotropic index $w = P/\rho$. The energy density of quintessence can be integrated for any potential,

$$\rho_\phi(a) = \rho_\phi(\bar{a}) \exp\left[-3\int_{\bar{a}}^{a} \frac{da'}{a'}(1+w_\phi)\right],$$

where $\bar{a}$ is some reference scale factor. In the range (5.22), for w_ϕ approximately constant one has

$$\rho_\phi(a) \simeq \rho_\phi(\bar{a})\left(\frac{a}{\bar{a}}\right)^{-\xi}, \qquad 0 < \xi = 3(1+w_\phi) < 6\,. \tag{7.34}$$

Solutions of this form with ξ constant are called *scaling* [76, 77, 79, 83]. Since $\dot{\rho}_\phi = -3H\dot{\phi}^2$, scaling solutions are such that the kinetic energy of the scalar is a fixed fraction of the total energy, $\dot{\phi}^2/\rho_\phi = \xi/3$. In the extreme slow-roll regime, $\xi \sim 0$, while during kination $\xi \sim 6$.

Scaling behaviour occurs as a response to one or more cosmological fluids [67, 68]. For a viable evolution, the energy density of the field should be tuned to decrease at a lower rate than matter ($\rho_{\rm m} \sim a^{-3}$) and radiation ($\rho_{\rm r} \sim a^{-4}$) in order to dominate at late times. To classify all possible cases, it is convenient to assume a constant barotropic index w for the fluid and consider the ratio

$$\frac{\rho_\phi}{\rho} \propto a^{3(w-w_\phi)}. \tag{7.35}$$

Tracking solutions are those with $w = w_\phi$. In this case, quintessence "tracks" in parallel the evolution of matter or radiation energy density, at least for part of the cosmic history.

Scaling profiles alleviate the coincidence problem since they are cosmological attractors: the evolution of the universe is fairly insensitive to the initial conditions. The problem may be claimed to be solved only if no other fine tuning, for instance on the mass scale of the potential, is introduced.

If $w_\phi \leqslant w$, the scaling condition (7.34) occurs when [79]

$$\boxed{\Gamma := \frac{\eta_V}{2\epsilon_V} = \frac{V_{,\phi\phi}V}{V_{,\phi}^2} \geqslant 1\,, \qquad w_\phi \leqslant w\,,} \tag{7.36}$$

where η_V and ϵ_V are defined in (5.55) and the equality holds for tracking solutions. In fact, defining $X := (1+w_\phi)/(1-w_\phi) = \dot\phi^2/(2V)$ and $X^{(p)} := \mathrm{d}^p \ln X/(\mathrm{d}\ln a)^p$, one sees that

$$\Gamma = 1 + \frac{1}{2(1+w_\phi)}\left\{w - w_\phi - (1+w-2w_\phi)\frac{X^{(1)}}{6+X^{(1)}} - \frac{4X^{(2)}}{[6+X^{(1)}]^2}\right\}, \tag{7.37}$$

and the scaling condition is nothing but $w_\phi \approx$ const (hence $X \approx$ const, $\Gamma \approx$ const). The constancy of w_ϕ in the scaling regime determines the value of w_ϕ via (7.37). Dropping almost vanishing terms in $X^{(1)}$ and $X^{(2)}$, one finds $w_\phi \sim [w - 2(\Gamma - 1)]/(2\Gamma - 1)$, valid in the deep radiation- or dust-domination regime. Setting $w = 0$ does not give the precise value of w_ϕ today, as the approximation $w_\phi \approx$ const is too coarse to characterize the interpolating solution at late times, since scaling solutions are not, in general, exact solutions of the equations of motion. To get the correct w_ϕ, one can either integrate the equations of motion numerically for a given potential or perturb the tracker solution to get semi-analytic formulæ for $w_\phi(\Gamma)$ [84, 85] which permit to find model-independent constraints on the barotropic index w_0 of Λ. In particular, for scaling solutions $-1.19 < w_0 = w_\phi|_{\rm today} < -0.95$ [86].

The scaling condition can also be re-expressed in terms of the density parameter Ω_ϕ (2.83). Noting that $\dot\phi^2/H^2 = (3/\kappa^2)\Omega_\phi(1+w_\phi)$, using (7.32) and the definition of X the following identity holds:

$$\frac{V_{,\phi}}{V} = \pm 3\sqrt{\frac{\kappa^2(1+w_\phi)}{3\Omega_\phi}}\left[1 + \frac{X^{(1)}}{6}\right], \tag{7.38}$$

where the $\pm$ sign holds when, respectively, $\dot{\phi} \lessgtr 0$ (these manipulations are valid only in regimes where ϕ is monotonic). Away from the extreme slow-roll regime $w_\phi \approx -1$, $|V_{,\phi}/V| \propto 1/\sqrt{\Omega_\phi}$. The density parameter Ω_ϕ increases in time if $w_\phi < w$ (quintessence dominates at late times), so that the left-hand side must decrease if one wants to avoid a fine tuning of the initial conditions. In fact, if the left- and right-hand side of (7.38) had opposite evolution, there would be a time after which, due to strong cosmological friction, the field freezes at some constant value $\phi = \phi_\mathrm{f}$ and behaves just like a cosmological constant. This *creeping* or *freezing* regime [80, 87, 88] dominates the late expansion of the universe. At that point, however, one would have to tune the initial conditions such that $\kappa^2 V(\phi_\mathrm{f})$ acquires the observed value H_0^2, which would lead us back to the Λ problem. To avoid this situation, we require that $|V_{,\phi}/V|$ decreases in time. Since $(V_{,\phi}/V)_{,\phi} = (V_{,\phi}/V)^2(\Gamma - 1)$, this condition is satisfied when $\Gamma > 1$. When $w_\phi > w$, the scaling condition is $1 - (1 - w)/[2(3 + w)] < \Gamma < 1$ [77, 79].

Realistic solutions should admit time-varying equations of state for quintessence and matter and, hence, different scaling regimes at different epochs (Problems 7.1 and 7.2 and Figs. 7.2 and 7.3). The late acceleration of the universe is observed, for instance, via standard candles such as type I supernovæ, objects with a known luminosity-distance relation. Their proper distance is determined by (2.143) and (2.156) as a function of the redshift or, equivalently, of the scale factor. It is therefore useful to parametrize the barotropic index as a function of a. Equation (7.38) can be recast as

$$\frac{\mathrm{d}w_\phi}{\mathrm{d}\ln a} = \frac{\dot{w}_\phi}{H} = -3(1 - w_\phi^2) \pm \sqrt{\frac{3\Omega_\phi}{\kappa^2}}(1 - w_\phi)\sqrt{1 + w_\phi}\,\frac{V_{,\phi}}{V}\,. \tag{7.39}$$

For a given potential, (7.39) evaluated today and (2.120) allow one to plot a theoretical point in the (w_0, w_a) plane and to compare it with observations. Other parametrizations than (2.120) are possible [84, 85, 88–91]. One which has attracted much attention is related to the so-called generalized Chaplygin gas, i.e., any cosmological fluid obeying and equation of state of the form $P = w_0/\rho^\beta$, where $\beta \neq -1$ [92–96]. The special case $\beta = 1$ [92, 93] is the original model by Chaplygin [97]. This class of equations of state can be realized, for a specific reconstructed potential, by a homogeneous scalar field [92] or by a complex scalar [93, 94]. Replacing P in the continuity equation (7.32), one obtains $\rho_\phi(a)$ and hence $w_\phi(a) = P_\phi(a)/\rho_\phi(a) = w_0/\rho_\phi^{\beta+1}(a) = w_0[(1 + w_0)a^{-3(1+\beta)} - w_0]^{-1}$ [94].

A scaling field slows down near the present as it starts to feel the friction induced by its own dominance of the energy density, hence its barotropic index decreases with time ($\dot{w}_\phi < 0$). In this setting, quintessence goes through a creeping regime where $w_\phi \approx -1$ before entering the scaling behaviour [79]. The duration and end of the freezing regime depend on the parameters of the model. Whenever the scaling regime is reached *after* the present epoch, some fine tuning is required, albeit less severe than for a pure cosmological constant.

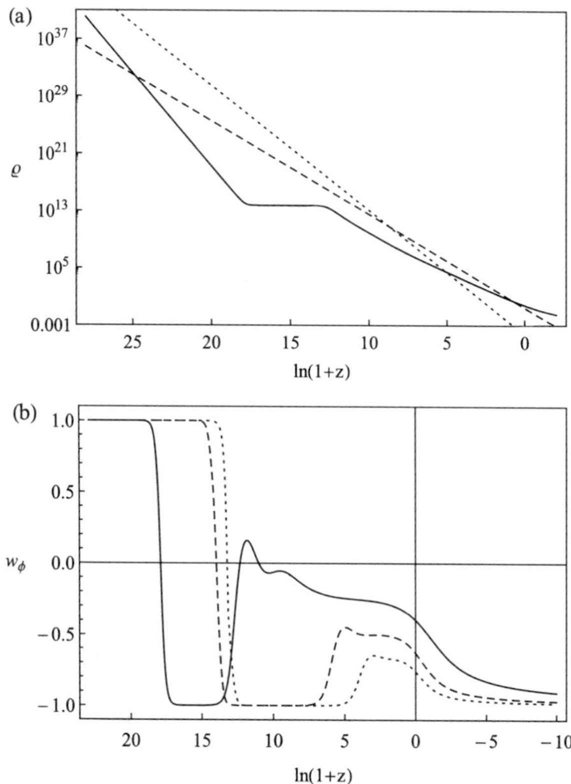

Fig. 7.2 (**a**) Quintessence (*solid curve*), radiation (*dotted curve*) and matter (*dashed curve*) energy density as a function of redshift for a potential $V = M^{4+n}\phi^{-n}$ with $n = 6$, $M \approx 5 \times 10^6$ GeV, for a scaling overshooting solution with initial conditions $\phi_{\rm i} = \phi(z = 10^{12}) = 3.5 \times 10^{-6} m_{\rm Pl}$ and $\phi'_{\rm i} = 3.5 \times 10^{-3} m_{\rm Pl}$, where $' = \partial_N$. In the dimensionless variables of Problems 7.1 and 7.2, the mass and initial conditions read $A \approx 60$, $y_{\rm i} = 10^{-5}$ and $y'_{\rm i} = 10^{-2}$. (**b**) Quintessence barotropic index w_ϕ for the $n = 6$ model (*solid curve*) compared with the output for the $n = 1$ (*dotted curve*, $M \approx 2$ keV, $A \approx 0.38$) and $n = 2$ (*dashed curve*, $M \approx 27$ MeV, $A \approx 0.54$) models with same initial conditions. Today, the barotropic index is $w_\phi(z = 0, n = 1) \approx -0.76$, $w_\phi(z = 0, n = 2) \approx -0.63$, $w_\phi(z = 0, n = 6) \approx -0.40$

Equation (7.39) predicts that the closer w_ϕ to -1, the closer is $\dot{w}_\phi$ to zero and the more indistinguishable is quintessence from a pure cosmological constant. According to observations, the present universe is under strong acceleration (w_ϕ very close to -1). Naively, one might try to explain this fact by requiring quintessence to roll slowly at late times and to obey the SR conditions

$$\epsilon_V \ll 1\,, \qquad |\eta_V| \ll 1\,. \tag{7.40}$$

(Incidentally, this would guarantee the stability of the solution according to the inflationary analysis in Sect. 5.4.3.) The dynamics of quintessence, however, is more

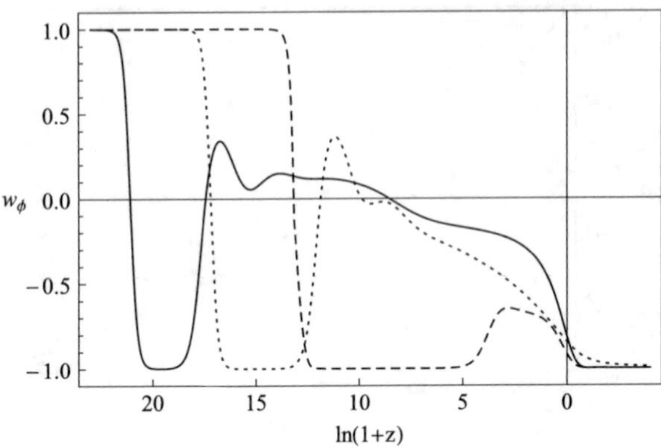

Fig. 7.3 Quintessence barotropic index w_ϕ for the inverse exponential model (7.49) (*dotted curve*, $V_0 = 0.28\rho_{\rm crit,0}$, $y_{\rm i} = 10^{-2}$) and the SUGRA-inspired model (7.50) with $n = 1$ (*dashed curve*, $M \approx 2\,{\rm keV}$, $y_{\rm i} = 10^{-4}$) and $n = 11$ (*solid curve*, $M \approx 6 \times 10^{10}\,{\rm GeV}$, $y_{\rm i} = 10^{-3}$), all with $y_{\rm i}' = 10^{-2}$ (in the dimensionless variables of Problems 7.1 and 7.2). Today, the barotropic index is, respectively $w_\phi(z=0) \approx -0.85, -0.92, -0.82$

difficult to assess that of inflation, since it is not potential dominated and (7.40) does not imply the strong SR approximation (5.19). The SR conditions (7.40) should be applied with care. When they hold, $w_\phi \lesssim -0.99$ [88, 89]. Moreover, while the inflaton is the only content of the early universe, dark energy is the main component in the present time but it does not dominate the other fluids completely. These two factors make a numerical analysis an indispensable tool.

The problem is that solutions with scaling regimes typically predict a negative barotropic index today, but not too close to -1 for sensible values of the parameters in the potential or for the simplest potentials. On the other hand, one can get closer to -1 by allowing a rapid variation of the barotropic index at low redshifts, which is in disagreement with present observations. Therefore, the simplest scaling quintessence scenarios are not satisfactory solutions to the Λ problem because an effective barotropic index $w_\phi \approx -1$ implies a fine tuning in the models.

To overcome these difficulties, one may look for attractor solutions indistinguishable from a cosmological constant, i.e., field models such that $w_\phi \approx -1$ at late times for a modest tuning of the initial conditions. A creeping field is one which sits static at low energy density until the density of other components drops low enough for it to become dynamical and, then, starts to move near the present epoch ("thawing out" [87–89]). When this happens, the barotropic index of the creeper increases away from -1 ($\dot{w}_\phi > 0$). Potentials with tracking regimes always feature a freezing era and, depending on its onset and ending, it can lead to different observable predictions for the same potential. Thus, creeping solutions can be found both for potentials which do not support observable tracking and in scaling models where the field is frozen until it reaches the attractor in our recent past. In all these cases, some fine tuning persists although less severe than for a pure cosmological constant. Thawing solutions, which can be found for most of the potentials below [84, 89], do not fare better, since the field undergoes a SR phase (7.40) which does simplify the dynamics but is set by special initial conditions.

7.3.2 Periodic and Power-Law Potentials

This tension can be appreciated by looking at possible candidates for the quintessence potential. In general, the fine tuning manifests itself either in the initial conditions of models with potentials more or less motivated by particle physics or in the specific choice of purely phenomenological potentials. The difference in the energy scale of the problem makes viable inflaton models unsuitable as dark energy, inasmuch as their parameters require severe fine tuning [78].

For instance, a simple quadratic potential under the SR conditions (7.40) is unfit, since the field would be practically massless, $m = \sqrt{V_{,\phi\phi}} \lesssim \sqrt{\kappa^2 V} \sim \sqrt{3}H_0 \sim 10^{-33}$ eV. The field can acquire a large mass from perturbative radiative corrections unless some symmetry prevents it ("symmetry protection"). Such is the case for an ultra-light pseudo-Nambu–Goldstone boson with potential (5.90) [69, 98–102]. The effective mass is then $m \sim \Lambda^2/f$ and, setting $f \sim m_{\rm Pl}$, one would need the rather reasonable value $\Lambda \sim 10^{-3}$ eV. Near the local maximum at $\phi = 0$, the field rolls down slowly, an example of "hilltop" dynamics [103]. From (5.91b), it is clear that the second SR condition in (7.40) is not satisfied unless $f \gtrsim m_{\rm Pl}$; this does not necessarily signal a trans-Planckian problem and viable late-time acceleration can be obtained even when $f \lesssim m_{\rm Pl}$ with some theoretical justification [102]. The potential (5.90) arises also for the axion in string and M-theory [104] (Sect. 13.4) and in extended ($\mathcal{N} \geq 2$) gauged supergravity, with f of order of the Planck mass [105].

Neither periodic nor quadratic potentials yield scaling solutions. The power law (5.81) does give stable scaling profiles if $n > 2(3+w)/(1-w)$ but, in this case, $w_\phi > w$ and there cannot be acceleration when the field rolls down the potential. Still, in the oscillatory phase at the minimum at $\phi = 0$ the virial theorem produces a mean equation of state [106]. In fact, if the oscillation frequency is much larger than the expansion of the universe H, for a potential $V \propto \phi^n$ the average over the oscillations of the kinetic energy $\mathcal{K}_\phi = \dot{\phi}^2/2$ is $n/2$ times the average of V, so that

$$\langle w_\phi \rangle = \frac{\langle \mathcal{K}_\phi \rangle - \langle V \rangle}{\langle \mathcal{K}_\phi \rangle + \langle V \rangle} = \frac{n-2}{n+2} \,. \tag{7.41}$$

Consequently, the mean expansion rate of the quintessence energy density is $\rho_\phi \propto a^{-3(1+\langle w_\phi \rangle)} = a^{-6n/(n+2)}$. Equation (7.41) is valid classically but oscillations are bound to stop due to quantum particle production. The latter does not occur if one assumes that quintessence couples very weakly with the rest of matter. If $n \ll 1$, one can reach a cosmological-constant regime while having a tracking attractor at early times. Unfortunately, in order to have $\rho_\phi/\rho_{\rm m} = O(1)$ at present times one should tune this ratio initially to the same levels as (7.11).

7.3.3 Exponential and Hyperbolic Potentials

Take also the exponential potential [51, 68, 72, 73, 107] $V = V_0 e^{\lambda\kappa\phi}$, where solutions with $\lambda \to -\lambda$ are physically identical upon changing the sign of ϕ. The usual inflationary attractor ($\lambda^2 < 3(1+w)$, $w_\phi = -1 + \lambda^2/3$) cannot work as dark energy without fine tuning, since in this case the energy density ρ_ϕ would have a sizable contribution at nucleosynthesis, thus impeding regular structure formation [72, 83]. On the other hand, there are tracking attractors for $\lambda^2 > 3(1+w)$ [83]. Since $w_\phi = w \geqslant 0$ and $\epsilon_V = \eta_V/2 = \lambda^2/2$, the resulting universe does not accelerate in the infinite future but, as a transient regime, today it does ($w_\phi(z=0) \approx -0.50$) under some parameter tuning, although not enough to account for the observed acceleration.

A much better scenario is achieved with a sum of exponentials [108–113]:

$$V(\phi) = V_0 e^{\lambda_0\kappa\phi} + V_1 e^{\lambda_1\kappa\phi}, \tag{7.42}$$

where the sum can be extended to more terms. If $\lambda_0 > 5.5$ and $0 < \lambda_1 < 0.8$, the evolution of ρ_ϕ is such that the BBN bound is not violated, but observations on late-time acceleration further require that $\lambda_0 > 17$ and $\lambda_1 \lesssim 0.1$ [86]. At early times the field follows the tracking attractor, while at late times one hits the inflationary attractor. Today, in the transition between the two regimes where we are approaching the de Sitter phase but ρ_ϕ does not dominate yet, w_ϕ can be close to -1 for a large range of parameters. If $\lambda_1 < 0$, instead of rolling indefinitely the field falls into a local minimum, where it acts as an effective cosmological constant.

Potentials of the type (7.42) arise in extended gauged supergravity in four dimensions [114, 115] and in the *dimensional reduction* of $D = 11$ and $D = 10$ supergravity, which, in turn, can be regarded as the low-energy limit of M- and string theory (Sect. 12.3.3). Dimensional reduction is the compactification of D-dimensional spacetime as $\mathcal{M} = M^4 \times \mathcal{C}$, where the $(D-4)$-dimensional manifold $\mathcal{C}$ is called internal space. The values of the coefficients V_i and λ_i depend on the geometry and topology of the internal space. Models based on spherical and toroidal compactifications with static internal space [116–119] predict parameters that do not lie in the range suitable for dark energy and the BBN bound is violated. Other types of dimensional reductions are more successful, *in primis* flux compactifications (Sect. 12.3.9).

Yet another phenomenological possibility is [120, 121]

$$V(\phi) = U(\phi)\, e^{\lambda\kappa\phi}, \tag{7.43}$$

where U is chosen so that the potential exhibits a local minimum. Examples are $U(\phi) = V_0 + V_1(\phi - \phi_0)^\alpha$ and $U(\phi) = V_0 + V_1/[V_2 + (\phi - \phi_0)^2]$, the latter arising in string theory where ϕ governs the separation of branes via a Yukawa-like interaction [122]. With natural choices of the parameters α and V_i, the field may remain trapped in the minimum, thus behaving as an effective Λ at late times and giving $w_\phi \approx -1$ today, similarly to the model (7.42).

Another class of models, with hybrid properties between the exponential potential at early times and the power-law potential $V \propto \phi^n$ at late times, is [123]

$$V(\phi) = V_0 \left[\cosh\left(\frac{2\lambda}{n}\kappa\phi\right) - 1\right]^{\frac{n}{2}} \tag{7.44}$$

and other variations on the theme such as $V(\phi) \propto (\cosh\kappa\phi)^\beta$, $(\sinh\kappa\phi)^\beta$ [124, 125] and the cosh types arising in extended gauged supergravity [105, 114]. All these cases avoid the fine tuning of a pure power law or exponential but reintroduce it from the backdoor, since (7.44) is an ad hoc choice with some crucial assumptions on the field dependence, the smallness of n and the absence of interactions with matter.

7.3.4 *Inverse Power-Law Potential*

Apart from a positive power law and the exponentials, a potential with exact scaling solutions is the inverse power law [67, 68]

$$V(\phi) = \frac{M^{4+n}}{\phi^n}, \qquad M > 0, \qquad n > 0. \tag{7.45}$$

Such potential may arise in dynamical supersymmetry breaking scenarios, as in super-chromodynamics and its generalizations [126–130]. In that case, at the classical level one realizes an effective action of the form (5.216) with a flat Kähler potential $K = \phi\phi^\dagger$ and a superpotential $W = M^{2+n/2}\phi^{1-n/2}$, where $n > 2$ depends on the Dynkin indices of the gauge group. It is, however, difficult to construct a realistic model with global supersymmetry. In fact, while curvature corrections do not disturb the tracking evolution and radiative quantum corrections to the superpotential cancel out, the Kähler potential is not protected by the same symmetries and the flat potential $K = \phi\phi^\dagger$ receives low-energy corrections which modify the kinetic term of the field. In general, these modifications spoil the scaling behaviour and make (7.45) hard to embed in a realistic supersymmetric theory [130]. We temporarily ignore this issue and concentrate on the phenomenology.

The SR conditions (7.40) become $\epsilon_V = n^2/(2\kappa^2\phi^2) \ll 1$ and $\eta_V = n(n+1)/(\kappa^2\phi^2) \ll 1$, obeyed if $\phi \gg n m_{\rm Pl}$. For $\phi \sim m_{\rm Pl}$, the mass scale in (7.45) is of order

$$M \sim 0.7 \sim (10^{-1} H_0^2 m_{\rm Pl}^{n+2})^{\frac{1}{n+4}}. \tag{7.46}$$

Using (2.11) and (2.12), for $n = 2$ one has $M = O(10)$ MeV, while $M = O(10^6)$ GeV for $n = 6$. For the power-law expansion (2.93), the solution of (5.39) is

$$\phi = \left[\frac{2(6p + 3pn - n)}{M^{4+n}n(2+n)^2}\right]^{-\frac{1}{2+n}} t^{\frac{2}{2+n}} =: \phi_0 \left(\frac{a}{\bar{a}}\right)^{\frac{2}{p(2+n)}} . \tag{7.47}$$

The power-law expansion (2.93) is also the background solution of the continuity equation in (7.33) for $\rho \propto a^{2/p}$ and, approximately, of (7.33) when $\rho_\phi/\rho \ll 1$. Consequently, $w_\phi = -1 + 2n/[3p(2+n)] = -1 + n(1+w)/(2+n)$, the ratio (7.35) $\rho_\phi/\rho \propto a^{4/[p(2+n)]}$ increases in time and eventually quintessence dominates the expansion of the universe. This scaling solution, with $w > w_\phi$ and $\Gamma = 1 + 1/n > 1$, is stable [68, 77]. However, it requires some tuning (either ρ_ϕ very low initially or $n \geqslant 5$) as, otherwise, today we would still be in the creeping regime, the scaling attractor being in the future. Moreover, the barotropic index w_ϕ in realistic scenarios (Fig. 7.2) is somewhat smaller than the theoretical value $w_\phi = -2/(2+n)$ for exact scaling with dust but it is not very close to -1 for $n \geqslant 1$. Therefore, this model is unsuitable to explain the present acceleration unless $n < 1$. In that case, the observational bound is [86]

$$n < 0.075 . \tag{7.48}$$

In Problems 7.1 and 7.2 we work out in detail the numerical solution for the inverse power-law potential (7.45) with $n = 6$. This example is experimentally non-viable, since the predicted barotropic index today is about $w_\phi \approx -0.40$. Still, it gives a fair idea of the phenomenology of quintessence (Fig. 7.2) and is a practical example of how to integrate the cosmological equations of motion numerically.

7.3.5 Other Potentials

Phenomenological potentials with slightly better properties are the sum of inverse powers, $V = \sum_i V_i \phi^{-n_i}$ [79] or the inverse exponential [76, 79]

$$V(\phi) = V_0 e^{\frac{\gamma}{\kappa\phi}} - V_1 , \tag{7.49}$$

where γ is a constant and V_1 is either V_0 or zero. For $\phi > 0$, at early times ($\kappa\phi \ll 1$) the potential is very steep with large SR parameters $2\epsilon_V \simeq \eta_V \sim 1/(\kappa\phi)^4 \gg 1$, which produces a tracking behaviour during radiation domination. At late times ($\kappa\phi \gg 1$), (7.49) approaches the inverse power law (7.45) with $n = 1$ and $4\epsilon_V \simeq \eta_V \sim 2/(\kappa\phi)^2 \ll 1$. This model does not need fine-tuning of the initial conditions, as the field can lie anywhere on the flat part of the potential. However, one has to fix the value of V_0 by hand in order to reproduce the observed dark energy density, which of course does not explain the cosmological constant. Observationally it is not favoured either [131], since $w_\phi(z = 0) \approx -0.85$ (Fig. 7.3). Here is another negative example illustrating the difficulties of the problem.

Other potentials for quintessence come from supergravity. The fact that, in the inverse power-law model (7.45), $\phi \sim m_{\rm Pl}$ at low redshift indicates that SUGRA corrections are somewhat unavoidable. The resulting effective potential for the scalar field is not positive definite [130, 132]: plugging the flat Kähler potential $K = \phi^2$ (here ϕ is real) and the superpotential $W = M^{2+n/2}\phi^{1-n/2}$ into (5.225) yields $V(\phi) = e^{(\kappa\phi)^2}M^{4+n}\phi^{-n}[(n-2)^2/4 - (n+1)(\kappa\phi)^2 + (\kappa\phi)^4]$. The potential (7.45) is recovered in the limit $\kappa \to 0$. All SUGRA corrections, including the exponential one, are relevant only at late times, when $\kappa\phi = O(1)$, and they do not affect the early history of the universe. However, the second term in square brackets is negative definite and leads to an inconsistent late-time evolution. Various ad hoc choices of the Kähler potential lead to an effective potential V of the form [130, 132–138]:

$$V(\phi) = M^{4-\alpha}\phi^\alpha\, e^{\lambda(\kappa\phi)^\beta}. \tag{7.50}$$

In one of these models, with a string-inspired non-flat Kähler potential, the quintessence field is one of the compactification moduli (see Sect. 12.3.5) and the parameters in (7.50) are $\lambda = 1/2$, $\beta = 2$ and $\alpha \leqslant -11$ in order to get a realistic scale hierarchy [130, 132]. The barotropic index can be seen in Fig. 7.3; today, $w_\phi(z = 0) \approx -0.82$. A lower w_ϕ can be obtained for low n (for instance, $w_\phi(z = 0) \approx -0.92$ for $n = 1$ and $w_\phi(z = 0) \approx -0.88$ for $n = 2$) but these cases have no solid theoretical background.

An index closer to $w_\phi(z = 0) \approx -1$ can be obtained for other Kähler potentials such that V acquires a non-zero local minimum. In one such case [133], on the side of the minimum with $\kappa\phi \ll 1$ (increasing ϕ), $V \sim \phi^{-2/3}$, while on the other side ($|\kappa\phi| \ll 1$, decreasing ϕ) V takes the asymptotic form (7.50) with $\lambda = (3/2)^\beta$, $\alpha = 2/3$ and $\beta = 4/3$, $V \sim \phi^{2/3}\exp[(3\kappa\phi/2)^{4/3}]$.

Traditional supersymmetry breaking should happen at energy scales $\sqrt{\langle F\rangle} \sim 10^{10}\,$GeV. In order not to increase the vacuum energy density to unacceptable levels, the second contribution in (5.225) should compensate the F-term, $W \sim \langle F\rangle\kappa^{-1} \sim m_{3/2}\kappa^{-2}$, where $m_{3/2}$ is the gravitino mass. As the energy scale of the superpotential in these models is much smaller than $\langle F\rangle\kappa^{-1}$, such corrections can spoil the quintessence scenario [133]. One way out may be offered by string theory, in $D \neq 4$ models where supersymmetry is unbroken (see [139] and Sect. 13.1.2). Another possibility is to reconsider the problem in extended gauged supergravities, where the scale of supersymmetry breaking is much smaller [105, 115].

7.3.6 *Quintessence and the Inflaton*

Both quintessence and the inflaton are, at some stage of their evolution, slowly rolling real scalar fields. The culprit of both early- and late-time acceleration could therefore be the same [68, 107], a possibility named "quintessential inflation" [136, 140–145]. The potential is tailored to sustain inflation at early times and acceleration at late times, with a sequence of kination-, radiation- and matter-domination periods in between. If the scalar field is dynamical throughout the evolution of the universe, its potential cannot have an absolute minimum at early times; to this class of profiles there belong *runaway potentials* with a minimum at infinity, $V(\phi \to \infty) = 0$. Also, the scalar field should not decay at the end of inflation, which forbids the inflaton to be directly coupled with other fields (in particle-physics jargon, such a field is called "sterile"). A mechanism alternative to standard reheating must be invoked [140]. In fact, particles can be produced gravitationally due to changes in the spacetime metric [146, 147] or via preheating, where particle production occurs via oscillations of the scalar field when it has a weak non-minimal coupling with matter [145, 148].

Identifying the inflaton with quintessence is not straightforward: any such model possesses a number of finely tuned features, whose ultimate source is the η-problem [144, 149]. The latter can be stated as follows. As the inflaton potential is very flat, in order to reach the much lower quintessence energy scales the curvature $V_{,\phi\phi}$ of the potential must be large towards the end of inflation. However, this implies a large slow-roll parameter η_V and a spoiling of scale invariance in the CMB power spectrum. Among the various requirements, the period of late-time acceleration must be triggered very recently. The absence of tracking behaviour during matter domination makes these cosmologies sensitive to the initial conditions. Difficulties persist if general relativity is extended to a scalar-tensor theory [150]. Quintessential inflation, however, may be possible with the introduction of further ingredients (such as non-trivial measures [151], extra dimensions [152] or a curvaton field [144]). A variant of the scenario may be realized by a complex scalar field, the real part being the inflaton and the imaginary one quintessence [153].

The η-problem will reappear in all its strength in some important models of supergravity and string cosmology (Sects. 13.2.4 and 13.5). String theory has an arsenal of tools that can solve this issue as well as the fine-tuning problem of the initial conditions. This will permit to revive the idea of quintessential inflation and to realize it in concrete models presented in Chap. 13.

7.3.7 Summary

Quintessence can solve the coincidence problem but, at best, it only restates the old Λ problem. When the latter is relaxed, some non-negligible amount of parameter tuning is often present. Models which avoid fine tuning of initial conditions are, in general, constrained by observations into parameter regimes with no theoretical motivation. These findings are a practical illustration of the obstacles cosmologists have been facing in their attempts to address the cosmological constant problem.

7.4 Scalar-Tensor Theories

A dynamically damped effective Λ can also arise with non-minimally coupled and multiple scalar fields [41–55]. A scalar field is minimally coupled to gravity when the only interaction terms are via the covariant derivatives $\nabla[g]$ and the metric determinant weight $\sqrt{-g}$. Non-minimal couplings between scalar fields and curvature invariants give rise to a wide class of models known as *scalar-tensor* theories [154–169].

Their key characteristic is that fundamental constants of Nature are neither fundamental nor constant. For example, in the model

$$\boxed{\bar{S}=\int \mathrm{d}^D x\,\sqrt{-\bar{g}}\left[\frac{F(\bar{\phi})}{2\kappa^2}\,\bar{R}-\frac{1}{2}\omega(\bar{\phi})\,\bar{g}^{\mu\nu}\,\partial_\mu\bar{\phi}\partial_\nu\bar{\phi}-U(\bar{\phi})+h(\bar{\phi})\bar{\mathcal{L}}_{\rm m}\right]}\,, \tag{7.51}$$

a non-canonical real scalar is coupled both with the Ricci scalar and with the matter Lagrangian (in the next sub-section we will explain the bars). Then, both the gravitational coupling $G_*(\bar{\phi}) := G/F(\bar{\phi})$ and the matter masses $m^2(\bar{\phi}) := m^2 h(\bar{\phi})$ are effectively spacetime-dependent. The special case

$$F=\kappa\bar{\phi}\,,\qquad \omega=\frac{\omega_{\rm JBD}}{\kappa\bar{\phi}}\,,\qquad U=0\,,\qquad D=4\,, \tag{7.52}$$

where $\omega_{\rm JBD}$ is a constant, is known as *Jordan–Brans–Dicke theory* [154, 155, 157, 158]. General relativity is recovered in the limit $\omega_{\rm JBD}\to\infty$, provided the trace of the matter energy-momentum tensor does not vanish, $\bar{T}\neq 0$ [170–172]. The equations of motion for the action (7.51) are

$$\left(\bar{R}_{\mu\nu}-\frac{1}{2}\bar{g}_{\mu\nu}\bar{R}\right)F=\kappa^2\left(h\bar{T}_{\mu\nu}+\bar{T}^{\bar{\phi}}_{\mu\nu}\right)+\bar{\nabla}_\mu\bar{\nabla}_\nu F-\bar{g}_{\mu\nu}\bar{\Box}F\,, \tag{7.53a}$$

$$\omega\bar{\Box}\bar{\phi}-U_{,\bar{\phi}}+\frac{1}{2}\omega_{,\bar{\phi}}\bar{\partial}^\mu\bar{\phi}\partial_\mu\bar{\phi}+F_{,\bar{\phi}}\frac{\bar{R}}{2\kappa^2}+h_{,\bar{\phi}}\bar{\mathcal{L}}_{\rm m}=0\,, \tag{7.53b}$$

where $\bar{T}^{\bar{\phi}}_{\mu\nu}=\omega\partial_\mu\bar{\phi}\partial_\nu\bar{\phi}+\bar{g}_{\mu\nu}\bar{\mathcal{L}}_{\bar{\phi}}$ is the energy-momentum tensor of the scalar and the matter energy-momentum tensor is covariantly conserved, $\bar{\nabla}_\mu\bar{T}^{\mu\nu}=0$. Contracting the indices we obtain the trace equation

$$\left(\frac{D}{2}-1\right)F\bar{R}=-\kappa^2\left(h\bar{T}+\bar{T}^{\bar{\phi}}\right)+(D-1)\bar{\Box}F\,. \tag{7.54}$$

7.4.1 Motivations

A non-minimal coupling $F \propto \bar{\phi}^2$ (with $\omega = 1$) naturally appears in quantum field theory on classical curved backgrounds. When quantum corrections to the minimally-coupled scalar-field action are taken into account, virtual loop processes of the scalar field produce divergent mass counter-terms, which are exactly canceled by a $\bar{\phi}^2\bar{R}$ contribution [20, 173–177]. The minimal coupling $\bar{\phi}^2\bar{R}$ is thus required by mass renormalization.

A scalar-tensor action can arise also in more exotic situations, for instance in Kaluza–Klein compactifications [178, 179] of supergravity where the D-dimensional metric is block diagonal, $g_{MN} = (\bar{g}_{\mu\nu}, \phi_{ab})$, $\mu,\nu = 0,1,2,3$, $a,b = 4,\dots,D-1$, and $\bar{\phi} = \det\phi_{ab}$ [180, 181]. Then, the typical potential in the Einstein frame (see (7.55) below) is of exponential type (Sect. 7.3.3). The dilaton of string theory (Sects. 12.1.5 and 13.7.5) is also governed, at lowest order in the string scale, by the action (7.51) except in the matter sector, where the coupling is not universal and one cannot reduce it to an overall function $h(\bar{\phi})$. The equations of motion are easily extended to this case by replacing $h\bar{\mathcal{L}}_{\rm m} \to \sum_i h_i\bar{\mathcal{L}}_{\rm m}^{(i)}$.

7.4.2 Conformal Transformations

The metric $\bar{g}_{\mu\nu}$ is denoted with a bar to highlight that the action (7.51) is presented in a metric frame where gravity is non-minimally coupled to the scalar field. If $F(\bar{\phi}) \neq 1$ and $h(\bar{\phi}) = 1$, $\bar{\phi}$ is minimally coupled to matter fields ψ^i and the set of variables $(\bar{g}_{\mu\nu}, \bar{\phi}, \psi^i)$ is called *Jordan frame*. There exists, however, the *Einstein frame* [155] where $F(\phi) = 1$ and the scalar field is minimally coupled with the D-dimensional metric $g_{\mu\nu}$:

$$S = \int d^D x\sqrt{-g}\left[\frac{R}{2\kappa^2} - \frac{1}{2}g^{\mu\nu}\partial_\mu\phi\partial_\nu\phi - V(\phi) + h\Omega^{-D}\mathcal{L}_{\rm m}\right], \tag{7.55}$$

where $\mathcal{L}_{\rm m} = \bar{\mathcal{L}}_{\rm m}[\Omega^{-2}g_{\mu\nu}, \psi^i]$. The two frames are related by a *conformal* (or *Weyl*) *transformation* of the metric (Problems 7.3, 7.4 and 7.5)

$$\boxed{g_{\mu\nu} := \Omega^2\,\bar{g}_{\mu\nu}\,,} \tag{7.56}$$

where

$$\boxed{\Omega = \Omega(\phi) = \{F[\bar{\phi}(\phi)]\}^{\frac{1}{D-2}}} \tag{7.57}$$

is a function of spacetime coordinates which is expressed in terms of the scalar field and $\phi = \phi(\bar{\phi})$ is given by (7.155). Thus, $U = \Omega^D V$. The equations of motion are a modification of the usual ones in the matter sector:

$$R_{\mu\nu} - \frac{1}{2} g_{\mu\nu} R = \kappa^2 \left(hT_{\mu\nu} + T^{\phi}_{\mu\nu} \right) , \tag{7.58a}$$

$$\Box\phi - V_{,\phi} = -QhT - h_{,\phi}\Omega^{-D}\mathcal{L}_{\rm m} , \tag{7.58b}$$

where

$$Q := -\frac{\Omega_{,\phi}}{\Omega} \tag{7.59}$$

and, calling $L_{\rm m}[\Omega^{-2} g_{\mu\nu}, \psi^i] = \Omega^{-D}\sqrt{-g}\mathcal{L}_{\rm m}$, we used $\delta L_{\rm m}/\delta\phi = (\partial L_{\rm m}/\partial \bar{g}^{\mu\nu}) \times\delta\bar{g}^{\mu\nu}/\delta\phi = (\partial L_{\rm m}/\partial g^{\mu\nu}) g^{\mu\nu}\delta \ln \Omega^2/\delta\phi = \sqrt{-g}QT$. Noting also that

$$T_{\mu\nu} = -\frac{2}{\sqrt{-g}}\frac{\delta L_{\rm m}}{\delta g^{\mu\nu}} = -\Omega^{2-D}\frac{2}{\sqrt{-\bar{g}}}\frac{\delta(\sqrt{-\bar{g}}\bar{\mathcal{L}}_{\rm m})}{\delta \bar{g}^{\mu\nu}} = \Omega^{2-D}\bar{T}_{\mu\nu} , \tag{7.60}$$

from (7.149) one has

$$\nabla_\mu (hT^{\mu\nu}) = QhT\nabla^\nu\phi . \tag{7.61}$$

The total energy-momentum tensor is conserved, $\nabla_\mu(hT^{\mu\nu} + T^{\mu\nu}_\phi) = 0$, as required by general covariance.

In general, a theory can be conformally invariant if no dimensionful coupling appears in the action. As the Newton constant is dimensionful, a Weyl transformation will map an action to another with different characteristics. For instance, if $\Omega(\phi) = \cosh(\sqrt{\kappa^2/6}\phi)$ in $D = 4$, the scalar field in the Jordan frame has $\omega(\phi) = 1$ (which might be regarded as a canonical kinetic term, were it not for the FR term which can be integrated by parts to give extra field derivatives). Thus, starting from a cosmological constant $U = \Lambda/\kappa^2$ in one frame we can get a hyperbolic potential $V \propto \Omega^{-D}$ in the other frame, similar to those considered in Sect. 7.3.3. When combined with a field redefinition $\phi \to \bar{\phi}(\phi)$, different potentials U and form factors ω in the Jordan frame can lead to a variety of potentials for the scalar field in the Einstein frame.

On a FRW background with a perfect fluid, (7.53) become

$$F\left[\left(\frac{D}{2} - 1\right)\bar{H}^2 + \frac{\textsf{K}}{\bar{a}^2}\right] = \frac{\kappa^2}{D-1}\left(h\,\bar{\rho} + \bar{\rho}_{\bar{\phi}}\right) - H\dot{F} , \tag{7.62a}$$

$$\left(\omega + \frac{D-1}{D-2}\frac{F^2_{,\bar{\phi}}}{\kappa^2 F}\right)[\ddot{\bar{\phi}} + (D-1)\bar{H}\dot{\bar{\phi}}] + U_{,\bar{\phi}} - \frac{D}{D-2}\frac{F_{,\bar{\phi}}}{F}U$$

$$+\frac{1}{2}\left[\omega_{,\bar{\phi}}+\omega\frac{F_{,\bar{\phi}}}{F}+\frac{2(D-1)}{D-2}\frac{F_{,\bar{\phi}}F_{,\bar{\phi}\bar{\phi}}}{F\kappa^2}\right]\dot{\bar{\phi}}^2$$

$$+\frac{h}{D-2}\frac{F_{,\bar{\phi}}}{F}[(D-1)\bar{P}-\bar{\rho}]-h_{,\bar{\phi}}\bar{P}=0\,, \tag{7.62b}$$

where we used the trace equation (7.54),

$$-\left(\frac{D}{2}-1\right)F\frac{\bar{R}}{\kappa^2}=h[(D-1)\bar{P}-\bar{\rho}]+\left(\frac{D}{2}-1\right)\omega\dot{\bar{\phi}}^2-DU$$
$$+\frac{D-1}{\kappa^2}[F_{,\bar{\phi}}\ddot{\bar{\phi}}+F_{,\bar{\phi}}(D-1)\bar{H}\dot{\phi}+F_{,\bar{\phi}\bar{\phi}}\dot{\bar{\phi}}^2].$$

The continuity equation for matter is

$$\dot{\bar{\rho}}+(D-1)\bar{H}(\bar{\rho}+\bar{P})=0\,. \tag{7.63}$$

It is important to stress that, in all these equations, dots represent derivatives with respect to synchronous time $\bar{t}$, related to time in the Einstein frame by

$$\mathrm{d}t=\Omega\mathrm{d}\bar{t}\,,\qquad a=\Omega\bar{a}\,. \tag{7.64}$$

Both relations in (7.64) stem from (7.56) and the relation between line elements $\mathrm{d}s^2=\Omega^2\mathrm{d}\bar{s}^2$.

In the Einstein frame, the first Friedmann equation is (2.74) with $\Lambda=0$ and total energy density $\rho+\rho_\phi$. The continuity equation (2.76) and the scalar-field equation (2.80) are augmented by source terms,

$$\dot{\rho}+(D-1)H(\rho+P)=-[(D-1)P-\rho]Q\dot{\phi}\,, \tag{7.65}$$

$$\ddot{\phi}+(D-1)H\dot{\phi}+V_{,\phi}=[(D-1)P-\rho]Q\,, \tag{7.66}$$

as per (7.61) and (7.58b). Here dots are derivatives with respect to t. Comparing (7.63) and (7.65) and taking into account (7.64), it follows that

$$\Omega H=\bar{H}+\frac{\mathrm{d}\Omega}{\mathrm{d}t}\,,\qquad \rho=\Omega^{-D}\bar{\rho}\,,\qquad P=\Omega^{-D}\bar{P}\,. \tag{7.67}$$

The change of frame realized by a conformal transformation does not entail a coordinate transformation, hence it is not a diffeomorphism. The debate about "which frame is the physical one" has been ongoing for some time [182–185]. In the original definition of the Jordan–Brans–Dicke theory, $h=1$ and the conformal transformation (7.56) connects the Jordan frame to a frame where measurement units are changed [158]. On the other hand, setting $h\Omega^{-D}=1$ in (7.55) would imply that effective masses are constant in both frames, in which case the frame without bars can be regarded as an Einstein frame with fixed units [185]. The answer to the above question is conditioned by which pair one chooses: (Jordan frame: frame with bars where $h = 1$)–(Einstein frame with running units), (frame with bars where $1 \neq h \neq \Omega^D$)–(Einstein frame with running units) or (frame with bars where $h = \Omega^D$)–(Einstein frame with fixed units). For definiteness, we consider the first pair, i.e., the Jordan frame ($h = 1$) automatically connected to the Einstein frame with running units (from now on, Einstein frame in short).

There are some evident differences between the Jordan and the Einstein frame: (i) the two frames have different field equations and solutions; (ii) only in the Einstein frame both the scalar-field Hamiltonian and the ADM energy [186] are positive semi-definite [182, 187]; (iii) in the Jordan frame, the energy-momentum tensor for matter is conserved, while in the Einstein frame it is not (equation (7.61)). Similarly, the scalar-field equation (7.58b) in the Einstein frame has a source term; (iv) by definition, in the Jordan frame matter is minimally coupled with the scalar field: therefore, massive particles with D-velocity $\bar{u}^\mu = \Omega u^\mu$ follow the geodesic equation $\bar{u}^\mu \bar{\nabla}_\mu \bar{u}^\nu = 0$ in the Jordan frame but they do not in the Einstein frame, where $u^\mu \nabla_\mu u^\nu = (u^\mu u^\nu + g^{\mu\nu})\partial_\mu \Omega / \Omega$ (here we used (7.149)); (v) conformal transformations encode a change of metric units, as the length of space and time intervals, as well as the norm of vectors, are clearly affected. Notwithstanding, the light-cone causal structure of spacetime is maintained, since time-like, space-like and null vectors are such with respect to both metrics.

At the classical level, for $h \neq \Omega^D$ the physics of the background is the same: it is simply described differently, with a different interpretation of the phenomena. Time and length intervals are rescaled as $\mathrm{d}x^\mu \to \mathrm{d}\bar{x}^\mu = \Omega \mathrm{d}x^\mu$, masses rescale as $m \to \bar{m} = \Omega^{-1} m$ and measurements (based on time or length or mass ratios) are unaffected [184, 185, 188–191]. This is no longer true in the Einstein frame with fixed units, where $h = \Omega^D$ and masses are constant as in the frame without bars.

Due to the mass rescaling, the gravitational coupling in the Einstein frame acquires a field dependence. From this, one might naively conclude that the strong equivalence principle is violated in the Einstein frame but respected in the Jordan frame. (The strong equivalence principle requires that all matter fields gravitate in the same way independently of the location in spacetime.) However, in $D = 4$ a post-Newtonian calculation shows that the gravitational coupling measured in a Cavendish experiment is [192, 193]

$$G_{\text{eff}} = \frac{G}{F}\left[1 + \frac{F_{,\bar{\phi}}^2}{2\omega\kappa^2 F + 3F_{,\bar{\phi}}^2}\right] \tag{7.68}$$

in both frames [191]. The strong equivalence principle is thus violated in both frames, since the effective Newton's coupling depends on ϕ. The weak equivalence principle (stating that all forms of neutral matter couple in the same way to gravity, i.e., test bodies fall with the same acceleration in a gravitational field independently of their mass and composition) holds classically but is violated in general due to quantum effects [194–197]. Notice that (7.68) differs from the coupling $G_* = G/F = G/\Omega^2$ by less than one part over 10^4 today [198].

Another example of the classical physical equivalence of the two frames is provided by the flat FRW metric, conformally equivalent to the Minkowski metric. In one case cosmology is described by cosmic expansion, in the other by the dynamics of a scalar field. What ultimately matters is the output of measurements, which is the same in both frames despite the widely different theoretical explanation. The redshift of a light signal is given by the ratio of the frequency of the signal measured at a source S *in units of the observer's reference frame*, divided by the frequency measured by a distant observer O. Using the geodesic equation, $1+\bar{z} = \bar{\nu}_S/\bar{\nu}_O = \bar{a}_O/\bar{a}_S$ in the Jordan frame. In the Einstein frame, the units in the observer's and in the source reference frame do not coincide, since matter is non-minimally coupled to gravity. Therefore, one should rescale the source frequency ν_S in the observer units, yielding $1+z = (\Omega_O^{-1}/\Omega_S^{-1})\nu_S/\nu_O = (\Omega_O^{-1}a_O)/(\Omega_S^{-1}a_S) = \bar{a}_O/\bar{a}_S = 1+\bar{z}$. The observed redshift is then conformal-frame independent [188–190] and so is the magnitude-redshift relation [189, 190].

7.4.3 Perturbations, Quantum Theory and Extended Inflation

In cosmological linear perturbation theory and in the absence of isocurvature perturbations, the Jordan and Einstein frames are physically equivalent, as the amplitudes of density perturbations have the same momentum dependence [199, 200]. The reason is that the Weyl transformation (7.56) is linearized in perturbation theory, so that quantized fields are rescaled by background factors. In particular, the linear curvature perturbation ζ on uniform density hypersurfaces is conformally invariant under (7.56) either in the presence of radiation only or provided the entropy perturbation between matter and the scalar ϕ vanishes [191]. In single-field inflation, where acceleration is driven by the scalar mode non-minimally coupled to gravity, a stronger result actually holds: the fully non-linear comoving curvature perturbation is conformally invariant [201, 202],

$$\mathcal{R}_{\rm NL} = \bar{\mathcal{R}}_{\rm NL}\,. \tag{7.69}$$

On a cosmological background, there is no difference between inflationary observables calculated in either frame to leading order in the SR parameters [203, 204]. This is no longer the case when other matter components are present apart from the non-minimally coupled mode [205], such as in multi-field scenarios [202, 206–208].

Related to perturbation theory is the fact that, in general, quantization and the field-dependent conformal transformation do not commute and the two frames are physically inequivalent when ϕ or $g_{\mu\nu}$ (or both) are treated at the quantum level [55, 183, 185, 209–214] (see also Sect. 8.1). Scalar-graviton interactions must be transformed away to allow for a standard field quantization, which points towards the Einstein frame as a preferred choice.

Thanks to the physical equivalence between the two frames at the leading perturbative level, non-minimally coupled single-field inflationary scenarios are the simplest. There, it is sufficient to apply most of the standard results of Chap. 5 in the Einstein frame and consider the dynamics generated by the potential V; a full analysis in the Jordan frame is also possible [215, 216]. In Sect. 5.11.2, we already discussed the case where the scalar field is the Higgs. Models of extended inflation [217, 218] predict, in general, a low tensor-to-scalar ratio. In terms of the number of e-foldings, for the Jordan–Brans–Dicke parametrization $n_s - 1 \simeq -2/\mathcal{N}_k$ and $r = 4(3 + 2\omega_{\rm JBD})/\mathcal{N}_k^2$, which results in the upper bound $\omega_{\rm JBD} < 11.5$ at 68% CL (much larger at the $2-3\sigma$ level) [219]. The prediction (5.235) of Starobinsky inflation corresponds to $\omega_{\rm JBD} = 0$.

7.4.4 Cosmological Constant Problem

How can scalar-tensor theories of gravity address the cosmological constant problem? By itself, the non-minimal coupling cannot suppress large Λ values to small ones. In fact, let $U = \bar{\rho}_\Lambda = \Lambda/\kappa^2$ be the energy density of the cosmological constant in the Jordan frame. In the Einstein frame in four dimensions, the Λ term is $\rho_\Lambda = \Omega^{-4}\bar{\rho}_\Lambda$ and, if the conformal factor is, say, $\Omega = e^{c\kappa\phi}$, then the cosmological constant seems to be exponentially suppressed at late times ($c\phi \to +\infty$). However, actual measurements of the energy density are performed with respect to some standard units $\rho_{\rm ref}$, which scale as $\rho_{\rm ref} = \Omega^{-4}\bar{\rho}_{\rm ref}$ with respect to the units in the Jordan frame. The ratio $\rho_\Lambda/\rho_{\rm ref} = \bar{\rho}_\Lambda/\bar{\rho}_{\rm ref}$ is therefore unchanged [185].

The scalar ϕ, however, can act as the quintessence field when endowed with a non-trivial potential [198, 220–239]. Scaling solutions exist for exponential, inverse- and (contrary to minimally coupled quintessence) also positive-power-law potentials U in the Jordan frame [221, 227, 233] and, in general, for $U(\phi) \propto [F(\phi)]^c$, where $c = O(1)$ is a constant [223, 225]. Just like for the inflaton, theoretical models can be compared with observations to reduce the parameter space or, conversely, one can use experiments to reconstruct the functions in (7.51) [198, 230, 238]. The interest in these models (dubbed extended or, more generally, coupled quintessence) is in their ability to generate scalar-field potentials with peculiar properties and an effective barotropic index smaller than or close to -1. This possibility [238] arises via the non-minimal coupling with dark matter in the Einstein frame [191, 240]. Let us set $D = 4$, no curvature and a dust component, $\rho = \rho_{\rm m}$, $P = 0$. In the Einstein frame, the dark-matter energy density is $\rho_{\rm m} = \Omega^{-4}\bar{\rho}_{\rm m}$. An observer assuming the standard Einstein dynamics would parametrize this contribution as in (2.88), so that the first Friedmann equation would read

$$H^2 = \frac{\kappa^2}{3}\left[\rho_{\rm m0}\left(\frac{a_0}{a}\right)^3 + \rho_\phi^{\rm eff}\right], \qquad \rho_\phi^{\rm eff} := \rho_\phi + \rho_{\rm m} - \rho_{\rm m0}\left(\frac{a_0}{a}\right)^3. \tag{7.70}$$

From (7.63), (7.64) and (7.67),

$$\rho_{\rm m} = \Omega^{-4}\bar{\rho}_{\rm m} = \Omega^{-4}\bar{\rho}_{\rm m0}\left(\frac{\bar{a}_0}{\bar{a}}\right)^3 = \frac{\Omega_0}{\Omega}\rho_{\rm m0}\left(\frac{a_0}{a}\right)^3. \tag{7.71}$$

Combining (7.65) and (7.66), one has $\partial_t(\rho_{\rm m}+\rho_\phi)+3H(\rho_{\rm m}+\rho_\phi+P_\phi)=0$. Comparing this equation with the effective dark-energy continuity equation $\dot{\rho}_\phi^{\rm eff} + 3H(1+w_{\rm eff})\rho_\phi^{\rm eff}=0$ and plugging in (7.71), one obtains the effective barotropic index

$$w_{\rm eff} := \frac{P_\phi}{\rho_\phi^{\rm eff}} = w_\phi \frac{\rho_\phi}{\rho_\phi^{\rm eff}} = \frac{w_\phi}{1+\left(\frac{\Omega_0}{\Omega}-1\right)\frac{\rho_{\rm m0}}{\rho_\phi}\left(\frac{a_0}{a}\right)^3}. \tag{7.72}$$

Therefore, if Ω decreases in time then $\Omega_0/\Omega < 1$, the denominator is smaller than 1 and $w_{\rm eff} < w_\phi$. For a given choice of parameters in the potential, this effect lowers the barotropic index of tracking solutions which, as we have seen in the minimally-coupled cases, fail to reach $w_\phi \approx -1$ today. For a slow-rolling field, $w_\phi \approx -1$ and $w_{\rm eff} \lesssim -1$. In this case, however, one would not fall back into freezing scenarios where the initial conditions are fine tuned. The intuitive reason is that gravity now induces an effective potential which is non-trivial even when the potential U is almost flat. In the ordinary Klein–Gordon equation $0 = \ddot{\phi}+3H\dot{\phi}+V_{,\phi} \simeq \ddot{\phi}+3H\dot{\phi}$, flatness implies a slowly-varying or almost constant scalar profile. On the other hand, the $\bar{R}$ curvature term in (7.53b) (or the source in the Einstein-frame counterpart (7.66)) fuels the dynamics of ϕ even if V is constant, a mechanism dubbed "R-boost" [241]. Consequently, for a given set of different initial conditions on the field value, and requiring $w_0 \approx -1$ and $\Omega_{\Lambda,0} \approx 0.7$, extended quintessence spans a wide range of magnitudes for the initial field density, contrary to the minimal-coupling case where the density converges to the single value $\Omega_{\Lambda,0}$ [236]. Thus, the scalar-tensor mechanism can produce an effective equation of state with ultra-negative barotropic index $w_{\rm eff} < -1$ without invoking ghosts, phantom matter [242] or any other type of unstable fields (we saw another mechanism, based on gradients only, in Sect. 5.3.1).

7.4.5 Experimental Bounds and Chameleon Mechanism

Scalar-tensor gravity is subject to stringent experimental bounds coming from post-Newtonian solar-system tests, equivalence-principle tests and cosmology [243]. For instance, in the Jordan–Brans–Dicke parametrization solar-system tests require the global bound [244]

$$\omega_{\rm JBD} > 4\times 10^4 \qquad (95\%\ {\rm CL})\,. \tag{7.73}$$

For a constant Q, (7.59) gives $\Omega = \mathrm{e}^{-Q\phi}$ and, since in this case

$$3 + 2\omega_{\text{JBD}} = \frac{1}{2Q^2}, \tag{7.74}$$

for massless quintessence (7.73) corresponds to the constraint $|Q| < 1.25 \times 10^{-3} l_{\text{Pl}}$ on the coupling with dark matter in (7.65) and (7.66). The model is thus very close to standard general relativity throughout the evolution of the universe. More generally, at late times the scalar field ϕ is driven towards a minimum of the Einstein-frame coupling $h(\phi)\,\Omega^{-D}(\phi)$ with matter, so that after radiation domination the system converges to general relativity [245, 246]. A scalar with a non-trivial potential and a large effective mass m can satisfy local constraints provided its interaction range $\sim 1/m$ is sufficiently short. The value and the variation of Newton's coupling G, however, is different in clusters of over-dense regions than at cosmological scales. Spatial differences between an over-density and the background are typically small, $\delta G/G \sim 10^{-6}$; time differences are even less appreciable, $\dot{G}/G \lesssim 10^{-20} - 10^{-19}\,\mathrm{s}^{-1}$ outside and inside an over-density, respectively [247]. As it turns out, G may approximately be locally constant in the solar system even if the scalar field continues evolving. Local bounds should therefore be applied with some caution to the cosmic evolution.

Another effect to take into account is the *chameleon mechanism* [248, 249]. Let $D = 4$ and $h = 1$. The non-minimal coupling with matter induces an effective potential $V_{\text{eff}} = V(\phi) + \Omega^{-4}\rho$ in the Einstein frame. Calling $\tilde{\rho} := \Omega^{-3}\rho$ the conserved energy density of non-relativistic matter (dust, $P \approx 0$), for a constant Q the effective potential reads

$$V_{\text{eff}}(\phi, \tilde{\rho}) = V(\phi) + \mathrm{e}^{Q\phi}\,\tilde{\rho}\,. \tag{7.75}$$

The scalar-field effective mass $m^2_{\text{eff}}(\tilde{\rho}) = V_{\text{eff},\phi\phi}$ depends on the energy density of the environment. In general, m_{eff} increases with $\tilde{\rho}$ and it is possible that the scalar field is massive enough in regions of high density to respect local constraints on the equivalence principle and fifth-force effects, while at cosmological scales such effects become important. Therefore, quintessence may not have yet been detected in Earth-based experiments because of the high density of the environment. Inside a compact object, the scalar ϕ acquires a constant value minimizing the effective potential (7.75), with $\tilde{\rho}$ being the density in the interior of the object. Assuming a spherical body of radius r_0, the field grows within a thin shell of width Δr_0 below the surface, while at distances $r > r_0$ the field $\phi \sim \mathrm{e}^{-m_{\text{env}} r}/r$ behaves as a massive scalar with mass m_{env} determined by the energy density $\tilde{\rho}_{\text{env}}$ of the surrounding environment. At $r \gg r_0$, ϕ tends to the value minimizing $V_{\text{eff}}(\phi, \tilde{\rho}_{\text{env}})$. In other words, the fifth force mediated by the scalar field is mainly generated by the thin shell and is consequently much weaker than it would be if the whole bulk of the body contributed.

Since the ratio $\Delta r_0/r_0$ is inversely proportional to the Newtonian potential of the body, massive objects such as planets obey the thin-shell condition $\Delta r_0/r_0 \ll 1$.

For instance, the chameleon mechanism affects laboratory measurements of G because $\Delta r_0/r_0 < 10^{-3/2}$ for test masses in a vacuum chamber on Earth. Also, preservation of equivalence-principle bounds yield, for Earth, $\Delta r_\oplus/r_\oplus < 10^{-7}$. In this case, one can show that the scalar field is short-ranged in the atmosphere (for an inverse power-law potential, $m_{\rm atm}^{-1} \lesssim 1-10\,{\rm mm}$) and free already at solar-system scales: $m_{\rm sol}^{-1} \lesssim 1-10\,{\rm AU}$ (Astronomical Units), while at cosmological scales $m_{\rm cosmo}^{-1} \lesssim 10^{-1}-10^3\,{\rm pc}$. In the presence of a thin shell, one can show that the effectively measured coupling is

$$Q_{\rm eff} \simeq 3Q\frac{\Delta r_0}{r_0}\,, \tag{7.76}$$

which is greatly suppressed near Earth. This parameter replaces Q in all expressions related to experiments, including (7.74). Therefore, scalar-tensor models can be made compatible with local and solar-system tests even if they predict a large $Q \gtrsim O(1)$.

7.5 Higher-Order and Higher-Derivative Gravity Models

The scalar field appearing in quintessence and extended quintessence models is an extra degree of freedom independent of the gravitational sector. When a scalar-tensor model arises from dimensional reduction, ϕ is indeed related to geometry, as a residual manifestation of the higher-dimensional metric. Another way to generate a spin-0 degree of freedom from geometry is to replace the Einstein–Hilbert action with a higher-order expression in the curvature invariants. Although these theories of modified gravity can ultimately be recast as multi-scalar-tensor models and, hence, do not really solve the cosmological constant problem, it is attractive to identify the agent of present cosmic acceleration with gravity itself [250, 251].

7.5.1 *Motivation and Ghosts*

Consider a quartic Lagrangian containing the curvature invariants R, $R_{\mu\nu}R^{\mu\nu}$ and $R_{\mu\nu\sigma\tau}R^{\mu\nu\sigma\tau}$ [252–254]:

$$\mathcal{L}_g = \frac{1}{2\kappa^2}\left(R + \alpha R^2 + \beta R_{\mu\nu}R^{\mu\nu} + \gamma R_{\mu\nu\sigma\tau}R^{\mu\nu\sigma\tau}\right), \tag{7.77}$$

where α, β and γ are dimensionful constants. In four dimensions only, the Gauss–Bonnet term

$$\mathcal{L}_{\rm GB} := R^2 - 4R_{\mu\nu}R^{\mu\nu} + R_{\mu\nu\sigma\tau}R^{\mu\nu\sigma\tau} \tag{7.78}$$

is topological and does not contributes to the equations of motion. This allows one, in $D = 4$ and at the classical level, to rewrite (7.77) only in terms of the Ricci scalar and the Ricci tensor ($\gamma = 0$).

Counter-terms of the kind (7.77) are introduced, in four dimensions, in the renormalization of matter quantum fields living in a fixed, classical, curved spacetime [252]. They appear, for instance, in vacuum loop diagrams of an interacting scalar field and conspire to cancel the ensuing divergences [177]. On a FLRW background, the renormalized energy-momentum tensor of matter acquires curvature corrections which lead to an anomalous trace $g^{\mu\nu}\langle T_{\mu\nu}\rangle = b_1\Box R + b_2 R_{\sigma\tau}R^{\sigma\tau} + b_3 R^2$ [255] and to an effective quartic Lagrangian.

In perturbative quantum gravity (Sect. 8.2), the Einstein–Hilbert Lagrangian $\mathcal{L} = R/(2\kappa^2)$ receives loop corrections in the curvature invariants, coming from the self-interactions of the graviton [176, 256]. Up to quadratic order (one loop), the gravitational Lagrangian is of the form (7.77) [176, 256–266]. Fourth-order quantum gravity (often called Stelle's theory [257, 258]) is indeed renormalizable but at the price of loosing S-matrix unitarity, due to the presence of ghost modes [257, 258]. A ghost is, by definition, a field whose kinetic term is unbounded from below (roughly speaking, it has the "wrong sign;" see Sect. 11.8.1). It implies a macroscopic instability which, if not healed, would lead to a breakdown of the theory, a violation of unitarity and wild particle creation in a time interval of cosmological length. The same problem arises in a generalization of Stelle's theory to D-dimensional spacetimes. Let

$$\begin{aligned}\mathcal{L}_{\rm poly} &= \alpha_2 R + \alpha_4 R^2 + \beta_4 R_{\mu\nu}R^{\mu\nu} + \ldots + \alpha_X R^{\frac{X}{2}} + \beta_X (R_{\mu\nu}R^{\mu\nu})^{\frac{X}{4}}\\ &\quad + \gamma_X (R_{\mu\nu\sigma\tau}R^{\mu\nu\sigma\tau})^{\frac{X}{4}} + \delta_X R\,\Box^{\frac{X}{2}-2} R + \ldots \end{aligned} \tag{7.79}$$

be a polynomial Lagrangian with at most X derivatives of the metric, where the first dots include a finite number of extra terms of order $< X$ in the derivatives of the metric tensor and the second ones indicate a finite number of operators $O(R^2\Box^{X/2-3}R)$ with the same number of derivatives but higher powers of the curvature. This theory is renormalizable when $X = D$ (Sect. 8.2) but it also contains a ghost.

On backgrounds with constant Ricci curvature $R = R_0$ (such as Minkowski and de Sitter), a generic theory $f(R, R_{\mu\nu}R^{\mu\nu}, R_{\mu\nu\sigma\tau}R^{\mu\nu\sigma\tau})$ can be expanded up to quadratic order as an Einstein–Hilbert action plus a Λ, an R^2 and a Weyl-squared term $C_{\mu\nu\sigma\tau}C^{\mu\nu\sigma\tau}$, where $C_{\mu\nu\sigma\tau}$ is the Weyl tensor. There are eight degrees of freedom associated with this class of actions, including (7.77): two of them are the usual polarization modes of the spin-2 graviton of general relativity, one corresponds to a spin-0 massive scalar coming from the R^2 term and five more to a massive spin-2 particle with negative-definite kinetic energy (a ghost), coming from the Weyl-squared term [257, 258, 267–269]. The massive scalar is a ghost only if the graviton is a ghost, which is not the case if the leading curvature term is Einstein gravity.

On these backgrounds with $R = R_0$, the Weyl ghost can be eliminated if some quite restrictive conditions on the Lagrangian are satisfied [270]. These strong selection rules in the space of parameters establish whether a given scalar-tensor or modified gravity model has ghosts, classical instabilities (when the squared propagation speed of inhomogeneous perturbative modes is negative definite) and super-luminal modes (i.e., modes propagating faster than light, which can cause breaking of causality and an ill-posed Cauchy problem).

No-ghost conditions, classical-stability conditions and sub-luminal conditions can be generalized to FLRW backgrounds (where the Ricci curvature is time dependent) and to actions whose higher-order term is either a function of the Gauss–Bonnet invariant (7.78) [271, 272] or the Gauss–Bonnet term non-minimally coupled with a scalar field [273–275]. Even when they can be tuned to produce a viable history of the universe [276, 277], ghosts, super-luminal modes and classical instabilities arise in a number of models for general initial conditions and during cosmologically relevant periods. Gauss–Bonnet actions are possibly inadequate, by themselves, to relax the fine tuning of the cosmological constant problem.

Higher-order gravity admits non-singular cosmological solutions [278–281], but they are not typical and there are classes of solutions which contain singularities [254, 282, 283]. Moreover, although changing Einstein's equations would change the energy condition of the singularity theorems of Chap. 6 [284], the Borde–Guth–Vilenkin theorem of Sect. 6.2.3 is independent of the dynamics as long as the background is inflationary. Therefore, modified-gravity models do not help to solve the big-bang problem. Yet, modulo stability issues, they do have cosmological applications to inflation [285] and to quintessence: acceleration is a typical dynamical feature.

Another modification of standard general relativity consists in including higher covariant derivatives $\Box^n R$, $n = 0, \ldots, N$ [268, 286–288]. Terms of the form $R\Box^n R$ can be generated by loop corrections that include massive matter fields, in the limit of scales much larger than the mass of the fields [286]. In general, higher-order and higher-derivative actions can arise as effective theories valid up to the Planck scale, after integrating out the microscopic degrees of freedom of a more fundamental description of gravity. They also admit an implementation in supergravity [289, 290].

7.5.2 General $f(R)$ Action

In quantum field theory, the inclusion of operators made only of the Ricci scalar R is inconsistent, since one should take into account all tensor invariants that contribute to the propagator. Nevertheless, to illustrate the main cosmological properties of higher-order gravity it may be useful to consider the simple case where the Lagrangian only depends on R [291–296]. More general scenarios of dark energy include $f(R, R_{\mu\nu}R^{\mu\nu})$ terms [297]; these give similar results, both in ordinary [295] and Palatini formalism [298, 299]. The action we start from is thus

$$\boxed{\bar{S} = \int d^D x \sqrt{-\bar{g}} \left[\frac{f(\bar{R})}{2\kappa^2} + \bar{\mathcal{L}}_m\right].} \tag{7.80}$$

The bars have been introduced to stress that this theory is classically equivalent to the scalar-tensor action (7.51) with Jordan-frame functions

$$F = \frac{\partial f}{\partial \bar{R}}, \qquad U = \frac{F\bar{R} - f}{2\kappa^2}, \qquad \omega = 0. \tag{7.81}$$

In fact, the system (7.80) is on-shell (i.e., dynamically) identical to the one given by the Lagrangian $\mathcal{L} = f(\bar{\phi}) + (\bar{R} - \bar{\phi})F$, by varying with respect to the field F. There is a scalar mode $\bar{\phi} = \bar{R}$ hidden in (7.80) and two more ($\phi_2 = \bar{R}_{\mu\nu}\bar{R}^{\mu\nu}$, $\phi_3 = \bar{R}_{\mu\nu\sigma\tau}\bar{R}^{\mu\nu\sigma\tau}$) in the action (7.77) [269]. By expanding the action at second order in the perturbations on a given background, the kinetic term of the graviton tensor modes is $F h^\alpha{}_\beta \Box h^\beta{}_\alpha$. Taking into account both graviton [300] and scalar perturbations [301, 302], $f(R)$ theories are free of ghosts and other instabilities only if the constraints

$$F = \frac{\partial f}{\partial \bar{R}} > 0, \qquad F' = \frac{\partial^2 f}{\partial \bar{R}^2} > 0, \tag{7.82}$$

are respected on the chosen background before the future de Sitter attractor (if present) is reached.

The equations of motion in the Jordan frame are

$$F\bar{R}_{\mu\nu} - \tfrac{1}{2} f\, \bar{g}_{\mu\nu} - \bar{\nabla}_\mu \bar{\nabla}_\nu F + \bar{g}_{\mu\nu} \bar{\Box} F = \kappa^2 \bar{T}_{\mu\nu}. \tag{7.83}$$

It is more convenient to map this model [294, 295, 303–305] conformally into the Einstein-frame action (7.55) with $h = 1$ and, setting $\omega = 0$ in (7.155),

$$\kappa\phi = \sqrt{(D-1)(D-2)}\, \ln \Omega = \sqrt{\frac{D-1}{D-2}}\, \ln F. \tag{7.84}$$

In particular, $Q = -1/\sqrt{(D-1)(D-2)}$. For general f, the potential in the Einstein frame is given by (7.81),

$$V = \frac{(F\bar{R} - f)F^{-\frac{D}{D-2}}}{2\kappa^2}. \tag{7.85}$$

If $f = f(\bar{R}, \bar{\Box}\bar{R}, \ldots, \bar{\Box}^N \bar{R})$, then (7.81) is generalized to $F = \sum_{n=0}^N \bar{\Box}^n \partial f / \partial \varphi_n$ and the theory contains N independent scalar modes $\varphi_n = \bar{\Box}^n \bar{R}$ [286–288]. This is an example of the extra degrees of freedom typically appearing in higher-derivative theories.

7.5.3 Palatini Formulation

For any given scalar-tensor action with fixed form factors F, ω and U, there exist two inequivalent definitions of the variational principle for gravity.[2] In the second-order formalism we used so far, the dynamical degrees of freedom are encoded in the metric $g_{\mu\nu}$ and the action is varied with respect to infinitesimal fluctuations $\delta g^{\mu\nu}$. In Chap. 9 we will introduce another possibility, the Palatini formalism, where the metric and the connection $\Gamma^{\sigma}_{\mu\nu}$ are treated as independent variables. While the two procedures yield the same dynamics for standard general relativity or in vacuum, they are mutually exclusive in the case of modified gravity with matter [250, 292, 297, 307]. For instance, the Jordan-frame equations of motion (7.83) are replaced by

$$F\bar{R}_{\mu\nu} - \tfrac{1}{2}f\,\bar{g}_{\mu\nu} = \kappa^2\bar{T}_{\mu\nu}\,, \qquad \nabla^{(\Gamma)}_{\rho}\left(\sqrt{-\bar{g}}\,F\,\bar{g}_{\mu\nu}\right) = 0\,, \tag{7.86}$$

in Palatini formalism. The Jordan–Brans–Dicke theory classically equivalent to (7.80) has $\omega_{\rm JBD} = 0$ in the metric formalism but $\omega_{\rm JBD} = -3/2$ in Palatini formalism [308, 309]. The resulting dynamics is somewhat different but both cases are well below the lower bound (7.73) and the chameleon mechanism is generally required [310–314].

7.5.4 Form of $f(R)$

Various functionals f have been proposed, including (bars are omitted from now on) the monomial $f = R^n$ [283, 295, 315, 316], the polynomial [291, 317]

$$f = R + c_n R^n\,, \tag{7.87}$$

trigonometric, exponential and hyperbolic functions [283, 318, 319] and logarithmic functions such as $f = \ln R$ [320] and $f \propto R + R^n(\ln R)^m$ [321], the case $n = 1 = m$ arising from one-loop quantum corrections in a $\phi^2 R$ scalar field theory [322]. Positive powers of the Ricci scalar are important at high curvature and play a role in the early universe and inflation [283, 285, 303, 305, 323–325]. Special attention has been devoted to the polynomial case (7.87) with $n = 2$ [283, 294, 297, 303, 326, 327],

$$f = R + c_2 R^2\,, \tag{7.88}$$

[2] When the form factors are left unspecified, the two formulations can be mapped one onto the other by a redefinition of ω [306].

for which the Einstein–frame potential is

$$V = \frac{1}{8c_2} e^{\frac{(D-4)\kappa\phi}{\sqrt{(D-1)(D-2)}}} \left(1 - e^{-\sqrt{\frac{D-2}{D-1}}\kappa\phi}\right)^2, \tag{7.89}$$

the generalization of (5.234) to D dimensions. Here we see the origin of Starobinsky inflation, the model of accelerating dynamics discussed in Sect. 5.12.4: it is the Einstein–frame counterpart of the $f(R)$ model (7.88) and acceleration is fueled by the quadratic curvature correction. The Lagrangian (7.88) will find another possible motivation by quantum gravity in Sect. 11.2.2.

For $f(R)$ given by (7.87) and a dust fluid ($w = 0$), in the Jordan frame the effective barotropic index of the universe defined by $a = t^{2/[3(1+w_\Lambda)]}$ is $w_\Lambda = -1 + 2(2-n)/[3(n-1)(2n-1)]$ [328]. Therefore, the larger n the closer w_Λ to -1. A similar conclusion, but with different n dependence, holds also in first-order formalism, where $w_\Lambda = -1 + (w+1)/n$ in the Jordan frame in the presence of a perfect fluid with equation of state $P = w\rho$ [320].

By themselves, positive powers of the Ricci scalar are tightly constrained throughout the history of the universe [316, 329]. On the other hand, negative powers of R dominate in low-curvature regimes and can lead to late-time acceleration [330, 331]. A problem with inverse powers of R is that the Minkowski metric is not a vacuum solution of the equations of motion, which would invalidate standard quantum field theory in local inertial frames. The condition that f be analytic in R (i.e., finite at $R = 0$) seems therefore fundamental [332]. However, both de Sitter and anti-de Sitter spacetimes are vacuum solutions and, with some care, quantum field theory can be defined thereon [20].

The polynomial (7.87) with $n < 0$ suffers from instabilities in metric formalism [333] and do not reproduce the correct evolution of the universe [334–336]. These issues can be overcome either by adopting the Palatini formalism [337–340] or by assuming, in metric formalism, a phenomenological combination of both positive and negative powers [341, 342],

$$f = R + c_n(R - R_0)^n + d_l(R - R_1)^{-l}, \tag{7.90}$$

where $R_{0,1}$ are fixed curvature scales and $n, l > 0$. This model can fit observations of the late-time acceleration. In particular, a viable Lagrangian with $n = 0$ and $R_0 = 0 = R_1$ is

$$f(R) = R - \mu R_2 \left(1 - \left|\frac{R_2}{R}\right|^l\right), \qquad l > 0, \tag{7.91}$$

where $\mu > 0$ is a dimensionless constant and R_2 is some curvature scale. The corresponding potential (7.81) in the Jordan frame is, in $D = 4$, of the form $U(\phi) = V_0(\mu, R_0)[1 - A(\mu, l)(1 - e^{2\kappa\phi/\sqrt{6}})^{l/(l+1)}]$. Since $Q < 0$ but $\dot{\phi} > 0$, then $\dot{\Omega}/\Omega = -Q\dot{\phi} > 0$ and, according to (7.72), $w_{\rm eff} > w_\phi$. Then, one cannot mimic a phantom barotropic index but the de Sitter attractor can be reached nonetheless [239]. Via the chameleon mechanism, (7.91) is compatible with equivalence-principle tests provided [239, 314]

$$l > 1.8 \,. \tag{7.92}$$

Another model compatible with observations is [313]

$$f(R) = R - \frac{bcH_0^2}{d + (cH_0^2/R)^n} \,, \tag{7.93}$$

where $n = 1, 2, 3$ and b, c and d lie in a certain range [343].

The quest for viable modified gravity actions is a broad subject; we refer the reader to the papers cited above, a sample literature such as [343–356] and the reviews [357–359]. Here we just mention that most of the simplest models with inverse powers, logarithms or exponentials are ruled out because they fail to obey certain requirements. Having solutions with an ordinary matter era with $a \propto t^{2/3}$ and an adequate late-time acceleration constrains the derivatives $\partial_R f$ and $\partial_R^2 f$, which single out a general classification for the possible cosmic evolutions of these models [351]. Linear perturbation theory [302, 360–366] further restricts the shape of $f(R)$ via the observation of large-scale structures and the CMB [239, 356]. To date, the base ΛCDM model is compatible with observations and there is no compelling evidence pointing towards a dynamical dark energy, including that from a modification of general relativity [367].

Viable $f(R)$ profiles such as (7.91) and (7.93) can be tailored nonetheless and the fine tuning of a pure cosmological constant can be relaxed. The modern challenge to accept is then twofold: observationally, to distinguish viable $f(R)$ or other higher-order gravity models from a pure Λ term and, theoretically, to give phenomenology a solid motivation. Whether quantum-gravity approaches or string theory can provide such a motivation will be a subject of the next chapters.

7.5.5 Horndeski Theory and Extensions

Attention has been devoted also to Horndeski theory, the most general four-dimensional scalar-tensor Lagrangian which is higher-order in the derivatives of the metric and of the scalar field but giving rise to equations of motion which are at most second order [219, 368–370]. Ghost modes are therefore easily avoided. Horndeski's general action includes scalar-tensor theories with second-order kinetic terms, $f(R)$ models, k-essence [371–375] (where the covariant scalar-field Lagrangian $\mathcal{L}(X, \phi)$ is higher order in the kinetic term $X = \nabla_\mu\phi\nabla^\mu\phi$) and the Galileon model (higher-order scalar kinetic terms but second-order field equations [376, 377]) and it is equivalent to generalized Galileon theory [358, 359, 378, 379].

A yet more general class of models produces equations of motion with higher-order derivatives but physical degrees of freedom obey second-order dynamics [380–387]. In this "healthy" extension of Horndeski theory, freedom from ghosts can be checked via a canonical analysis.

7.6 Other Approaches

Quintessence, scalar-tensor theories and higher-order gravity are all considered classic approaches to dark energy. However, there exist other scenarios where the cosmological constant problems are tackled under difference premisses, sometimes conservative, sometimes involving a deeper conceptual leap beyond standard general relativity. In this section we mention only a few, leaving the discussion of other candidates motivated by quantum gravity or string theory for later.

7.6.1 Varying Couplings

Scalar-tensor theories are a special case of models where all the couplings are spacetime dependent via a Lorentz scalar field. The phenomenology of such models can be tightly constrained by cosmological as well as atomic and particle-physics experiments. For instance, in ordinary electrodynamics the fine-structure constant $\alpha = e^2/(\hbar c)$ depends on the electron charge e, the Planck constant $\hbar$ and the speed of light c. When one or more of these constants are promoted to coordinate-dependent parameters, one effectively obtains a time-space varying α, for which there are various observational data sets. A possibility for varying α is to keep $\hbar$ and c constant while allowing for a non-constant electric charge, $e \to e(x)$, as in Bekenstein's model [388, 389]. In other scenarios, the speed of light is made spacetime dependent [390]:

$$c \to c(x)\,. \tag{7.94}$$

Varying-speed-of-light (VSL) theories simply correspond to frameworks where units are adapted with the scales in the dynamics (and, in particular, chosen such that c varies) [391]. Time and space units are redefined so that differentials scale as $\mathrm{d}t \to [f(x)]^a \mathrm{d}t$, $\mathrm{d}x^i \to [f(x)]^b \mathrm{d}x^i$, where f is a function, a and b are constants and local Lorentz invariance of the line element requires $c(x) \propto [f(x)]^{b-a}$. In particular, when $b = 0$ one formally reabsorbs c in the coordinate $x^0 = \int \mathrm{d}t\, c(t)$, which scales as a length. With this coordinate, all equations can be made general-covariant and gauge invariant (in a word, formally identical to the usual ones) provided some conditions are met. For instance, the field strength of the Abelian electromagnetic field A is of the form $F_{\mu\nu} = (\hbar c/e)\{\partial_\mu[eA_\nu/(\hbar c)] - \partial_\nu[eA_\mu/(\hbar c)]\}$ and explicit c dependence disappears if $e \propto \hbar(c)\, c$. Therefore, VSL models also require, in general, a varying electric charge and a varying Planck's constant. They are locally Lorentz invariant under transformations which look like the usual ones but with a varying c. Defining the scalar field $\chi := \chi_0 \ln(c/c_0)$, the total action for a minimal version of VSL can be written as

$$S = \int \mathrm{d}^4x \sqrt{-g} \left[\frac{\mathrm{e}^{4\chi/\chi_0}}{2\kappa^2} R + \mathrm{e}^{\chi/\chi_0} (\mathcal{L}_\mathrm{m} + \mathcal{L}_F) - \frac{\omega_c}{2} \partial_\mu \chi \partial^\mu \chi \right], \tag{7.95}$$

where χ_0 and ω_c are constants. The theory is similar to a scalar-tensor model and the two are equivalent only when $e \propto \hbar(c)\, c = \text{const}$. A varying c can have an impact in the history of the early universe and provide a conceptual alternative to standard inflation. In fact, the comoving particle horizon is modified as $r_\mathrm{p} = \int_0^t \mathrm{d}t'\, c(t')/a(t')$ and, depending on the profile $c(t)$, acceleration and nearly scale invariance can be sustained by a purely geometric effect [392]. For the same reason, the late-time evolution can be affected by a dynamical cosmological constant with energy density $\rho_\Lambda = c^2(t)\Lambda/\kappa^2$ [82, 393–395]. Due to the similarity with scalar-tensor models and a lack of a compelling theoretical motivation, varying-coupling models meet with about the same limitations as any other phenomenological explanation of dark energy. However, if varying couplings were to arise in the effective semi-classical limit of a fundamental theory of quantum geometry and particle physics, these models might become strong contenders in explaining the evolution of the universe.

7.6.2 *Void Models*

So far, the cosmological principle detailed in Sect. 2.1 has been the main cornerstone whereupon to found a cosmological theory of the universe. Inhomogeneities have been taken into account in the form of perturbations around a homogeneous background, where the energy density of each constituent i is a well-defined average ρ_i over sufficiently large volumes. The cosmological principle is an important statistical assumption and, if the universe was strongly inhomogeneous, the mean values ρ_i would be of little physical significance. Homogeneity has been experimentally checked up to Hubble-horizon scales but deviations from a FLRW background become more apparent at very large scales, where super-clusters of galaxies thread along the outskirts of giant under-dense regions. Thus, such clusters find themselves surrounded by large voids.

According to the Copernican principle, embedded in the cosmological principle, the observed isotropy must be the same as seen at any other point in the universe. However, we have no other reference point than the Earth and one might consider a non-FLRW background where the observer occupies a somewhat special place with respect to the surrounding inhomogeneities. In void models [396–401], we live close to the centre (to avoid a strong dipole effect) of an under-dense region where the Hubble parameter is larger than in the surroundings. The faster expansion of the under-dense region produces a velocity gradient between the motion of inner and outer galaxies. Objects in the near outer region are, for a given redshift z, at a larger distance from us than in an FLRW universe and, hence, have a lower apparent magnitude. An observer assuming a homogeneous and isotropic background would interpret the data as nearby galaxies strongly redshifting away in acceleration.

A concrete example of void is given by the Lemaître–Tolman–Bondi (LTB) model [402–407]. The local FLRW patch with cosmological constant is replaced by a spherically symmetric neighborhood with $\Lambda = 0$, filled with dust matter and inhomogeneous in the radial direction, the centre being under-dense. LTB regions can be patched together through FLRW junctions [408, 409] to obtain a more realistic large-scale "Swiss cheese" model of voids. If we lived at (or near [406]) the centre of one such void region, we could still fit the isotropic set of redshift-dependent cosmological observables and explain late-time acceleration without invoking a non-vanishing cosmological constant [398–400, 410–424], as long as the void is not too large (size $\lesssim$ 1.5 Gpc [418]). In fact, from (2.147) and (2.154) one can Taylor expand the luminosity distance $d_L = \chi[\tau(z)]$ around $z = 0$ and compute the series coefficients as functions of $\Omega_{\rm m0}$ and $\Omega_{\Lambda 0}$. Repeating the same procedure on an LTB background and matching the coefficients of the Taylor expansion, one finds an effective $\Omega_{\Lambda 0}$ contribution determined by the parameters of the LTB metric.

The assumption that, by sheer coincidence, we should live very close ($\lesssim$ 20 Mpc [422]) to the centre of an under-dense region introduces a fine tuning in void models. Still, their validity as a viable alternative to dark energy is ultimately assessed by experiments. There are three main classes of models.

For a local (or small, or minimal) void [399, 400, 416], the under-dense region has typical redshift size $z \sim 0.07$–0.1 ($\sim$ 200–300 h^{-1} Mpc). Large-redshift supernovæ data rule out this class, unless the outer regions are curved [421]. An off-centre observer can detect *TB* and *EB* correlations with a characteristic multipole dependence, generated by gravitational lensing from the inhomogeneous matter distribution. However, the signal is rather small and requires high-precision polarization experiments [425].

Scenarios with large voids can be roughly divided into two categories, one where the void has a compensating over-dense outer shell ($z \sim 0.5 - 1.5$, size $\sim$ 1–2 h^{-1} Gpc) and another where the void has no compensating shell and approaches FLRW only asymptotically (size less sharply defined). With respect to small voids, large-void scenarios are statistically less favoured by standard structure formation [426], although some large voids are indeed observed. The details of specific models as well as of independent data analyses may differ to the point where it is not completely clear whether large-void models can be rendered compatible with CMB, baryon acoustic oscillations, BBN, large-scale structure and type I supernovæ observations [421] or not [418, 419, 422, 423, 427].

Even if, as it seems now likely, void scenarios cannot reconcile the tension between different data sets, they constitute an important reminder that, no matter how fascinating the cosmological constant puzzle, there is still a possibility that (part of) its solution lies within traditional physics.

7.6.3 Unimodular Gravity

By definition, in unimodular gravity some of the components of the metric $g_{\mu\nu}$ are non-dynamical. This can be implemented in three ways: preserving (in full or almost) general covariance but demanding $\sqrt{-g}$ to be non-dynamical [24, 428]; as a partial gauge fixing of general coordinate transformations preserving the volume (so that $\sqrt{-g}$ transforms as a scalar rather than a density) [429–453]; or replacing the volume weight $\sqrt{-g}$ with a scalar with some internal symmetries and no kinetic term [454–457].

The case where the determinant of the metric is a trivial scalar $\sqrt{-g} = 1$ is particularly simple. The variation of the action (2.20) with respect to the metric is now done with $\delta\sqrt{-g} = 0$ instead of (2.21). Since $g_{\mu\nu}\delta g^{\mu\nu} = 0$, only the traceless part of (2.23) survives as a dynamical equation:

$$R_{\mu\nu} - \frac{1}{D} g_{\mu\nu} R = \kappa^2 \left(T_{\mu\nu} - \frac{1}{D} g_{\mu\nu} T \right). \tag{7.96}$$

At the classical level, the dynamics of this model of gravity is equivalent to general relativity: applying the covariant derivative to (7.96) and using the Bianchi identity (2.32), one obtains the trace (2.26), where Λ is an *arbitrary integration constant*. Plugging (2.26) into (7.96), one obtains the Einstein equations (2.23). When $\sqrt{-g}$ is not globally fixed to 1, the dynamics is that of a scalar-tensor theory.

A justification for this version of unimodular gravity comes from the theory of the massless spin-2 graviton on Minkowski spacetime [441, 447]. In general, classical and quantum stability of the graviton is achieved if its dynamics is invariant under *transverse diffeomorphisms* or transverse Fierz–Pauli symmetry, defined by (3.90) with

$$\partial^\mu \xi_\mu = 0\,. \tag{7.97}$$

The name transverse stems from the fact that $k^\mu \xi_\mu = 0$ in momentum space and the covector ξ_μ is orthogonal to the graviton's direction of propagation. Here one is assuming that Minkowski spacetime is described in Cartesian coordinates, so that the covariant derivatives with respect to the metric $\eta_{\mu\nu}$ are plain derivatives (as in (3.90) and (7.97)) and $\eta = \det \eta_{\mu\nu} = -1$. (Generalization to an arbitrary coordinate system is possible [449, 450].) The most general quadratic action respecting transverse diffeomorphism invariance is

$$S_h = \int d^D x \sqrt{-\eta}\, \mathcal{L}_h = \int d^D x\, \mathcal{L}_h\,, \tag{7.98}$$

$$\mathcal{L}_h = \frac{1}{2}\left(\partial_\mu h^{\mu\tau} \partial_\nu h^\nu{}_\tau - \frac{1}{2} \partial_\mu h^{\nu\tau} \partial^\mu h_{\nu\tau} - \alpha \partial^\mu h \partial^\nu h_{\mu\nu} + \frac{\beta}{2} \partial_\mu h \partial^\mu h \right), \tag{7.99}$$

where α and β are constants and $h = \eta_{\mu\nu}h^{\mu\nu}$. Apart from the graviton, the theory also contains a spin-0 particle, which can be removed by extending the symmetry group in two alternative ways. In the first, one drops the condition (7.97) and thus moves from transverse to full diffeomorphisms, in which case one obtains the Fierz–Pauli Lagrangian [458] at the linear level ((7.99) with $\alpha = 1 = \beta$) and Einstein's gravity at the non-linear level. In the second, one imposes the Weyl symmetry (the infinitesimal version of the conformal transformation (7.56))

$$\tilde{h}_{\mu\nu} = h_{\mu\nu} + \Phi\eta_{\mu\nu}, \tag{7.100}$$

where Φ is a scalar. The trace h can thus be changed arbitrarily and, in particular, set to zero, so that the scalar mode is removed from the spectrum. To have $S_{\tilde{h}} = S_h$ and invariance under both transverse diffeomorphisms and conformal symmetry, we need $\alpha = 2/D$ and $\beta = (D+2)/D^2$ in (7.99).

When adding non-linear terms, one combines the background metric and the graviton into the metric (here $[h_{\mu\nu}] = 1$, while in (3.1) $h_{\mu\nu}$ is dimensionless)

$$g_{\mu\nu} = \eta_{\mu\nu} + \sqrt{2\kappa^2}h_{\mu\nu}\,. \tag{7.101}$$

The metric $\hat{g}_{\mu\nu} := |\det g/\det\eta|^{-1/D}g_{\mu\nu}$ is a tensor density such that $\sqrt{-\hat{g}} = \sqrt{-\eta}$, equal to 1 in the Cartesian reference frame. This is the non-linear version of the traceless condition $h = 0$. At the non-linear level [441, 449, 450], the resulting theory in the Cartesian frame is unimodular gravity:

$$S_{\hat{g}} = \frac{1}{2\kappa^2}\int \mathrm{d}^D x\, R[\hat{g}]\,, \tag{7.102}$$

while in terms of $g_{\mu\nu}$ one has an extra kinetic term for the metric. The equations of motion are (7.96) with $g_{\mu\nu}$ replaced by $\hat{g}_{\mu\nu}$. The key difference between general relativity and unimodular gravity is in the choice of the symmetry group.

There are some theories which realize the unimodular constraint $\sqrt{-g} = 1$ from quite different first principles, for instance by regarding gravity as an emergent phenomenon born out of more fundamental degrees of freedom or by assuming a radically altered structure of spacetime geometry. We will describe these approaches in Sects. 7.7 and 11.6, respectively. While the classical dynamics of general relativity and unimodular gravity is the same, there are differences in the quantum theories which depend on the (and disappear in some) formulation of the unimodular paradigm [451, 452].

In all these scenarios, the cosmological constant Λ is an integration constant associated with the total D-volume $\mathcal{V} = \int \mathrm{d}^D x\sqrt{-g} = \int \mathrm{d}^D x$. An integration constant can be set at one's leisure without incurring into the conceptual problem of the fine tuning of a physical parameter in an action. This constant is unrelated to the zero-point energy of matter [450]. At the quantum level, the vacuum zero-point energy of matter $\mathcal{V}\rho_{\mathrm{vac}}$ does not gravitate as in general relativity, since the minimal coupling with the metric is trivial here. Vacuum bubble diagrams are thereby factored out of the partition function, just as in quantum field theory on Minkowski spacetime.

Although the magnitude of the vacuum energy as prescribed by particle physics becomes irrelevant for the cosmological constant problem, a small value such as (7.11) may be natural in quantum unimodular gravity. In Hamiltonian formalism, the wave-function of the Universe becomes a superposition of states with different values of the cosmological constant, which allows for a probabilistic reinterpretation of the Λ problem (see Sect. 10.2.4). In fact, the Euclidean partition function features a functional integration over all possible values of Λ [437, 438, 444]:

$$Z = \int \mathrm{d}\mu(\Lambda) \int [\mathcal{D}g][\mathcal{D}\phi]\, \mathrm{e}^{-S_g[g,\phi,\Lambda]} = \int \mathrm{d}\mu(\Lambda)\, \mathrm{e}^{-S_{\rm eff}[\bar{g},\bar{\phi},\Lambda]}\,, \tag{7.103}$$

where $\mu(\Lambda)$ is a generic smooth quantum measure weight with an a priori uniform distribution (i.e., the point $\Lambda = 0$ is neither preferred nor disfavoured), the measures in square brackets are functional measures in terms of the metric and a generic matter scalar field, and $\bar{g}$ and $\bar{\phi}$ are the background fields minimizing the effective action. In vacuum, the background metric is de Sitter, implying in $D = 4$ that

$$Z = \int \mathrm{d}\mu(\Lambda)\, \exp\left(-\frac{3\kappa^2}{8\Lambda}\right). \tag{7.104}$$

The contribution $\Lambda = 0$ dominates exponentially.

7.6.4 *Analogue Gravity and Condensates*

A yet different perspective regards the cosmological constant problem as the reflection of infrared effects. In this case, Λ is determined by macroscopic, "collective" degrees of freedom describing effective regimes where fundamental physics intervenes only indirectly. This limit is often called hydrodynamical, as it entails a long-wave-length coarse graining of the microscopic structure. Such situation often happens in condensed-matter systems, which in fact provide a stimulating inspiration to understand gravity in regimes usually inaccessible by modern experiments. For instance, the excitations in certain superfluid systems propagate on an effective background which closely resembles a Lorentzian curved spacetime in a strong-curvature regime. This motivated the study of a number of models of *analogue gravity*, where the tools of condensed-matter problems are applied to astrophysical or cosmological situations [459, 460]. One such example will be reported in Chap. 14.

In many of these models, a formal analogy with superfluid phases indicates that Λ is an emergent phenomenon [461–463]. More precisely, the cosmological constant neither is the energy of the vacuum of a field theory nor is dynamically determined by a basic scalar field. In some cases, it is related to the grand canonical energy of the vacuum [464]. In others, it corresponds to the expectation value of a condensate, whose dynamics governs the energy scale of the problem. A *condensate* is a composite object resulting from the binding together of some fundamental degrees of freedom into an energetically favourable configuration. We mention three possible condensate scenarios.

(i) In loop quantum gravity and group field theory, the condensate might be composed by the gravitational degrees of freedom themselves, under some extreme conditions. These cases can provide an IR resolution or alleviation of the cosmological constant problem but rooted in a UV modification of gravity. We postpone a brief discussion of this possibility, which radically differs from the next two as far as physical interpretation and mathematical tools are concerned, to Sects. 9.4 and 11.5.
(ii) Gluon condensate. By thermodynamical arguments and without invoking quantum gravity, one can devise models where Λ is determined by an extensive thermodynamic variable characterized by a Lorentz-invariant quantum vacuum which is "self-sustained," i.e., supporting an equilibrium state even in the absence of an external environment [465, 466]. Such variable can be realized by the fields of the Standard Model without introducing exotic particles or modifying general relativity. An example is the gluon condensate $q := \langle F^a_{\mu\nu} F_a^{\mu\nu} \rangle$, i.e., the vacuum expectation value of the squared QCD (quantum chromodynamics) field strength [467]. While $\Lambda(q) = \Lambda(0) = 0$ in perfect vacuum, a small non-zero value can be obtained in the presence of thermal matter, also leading to breaking of Lorentz invariance. The effective cosmological constant term is time dependent, it dominates after the cold-dark-matter era and it is naturally small (the energy density of a condensate is typically much smaller than the free vacuum energy of the fields it is made of). Thus, both the coincidence problem and the old cosmological constant problem are at least relaxed.
(iii) Fermion condensate forming in the early universe [468–474]. One starts with ordinary general relativity plus matter. The latter is described by the fermionic Dirac action with a four-fermion interaction term:

$$S_\psi = -\frac{\mathrm{i}}{2}\int_{\mathcal{M}} \mathrm{d}^4x\, e \left(\bar\psi \gamma^a e_a{}^\mu \nabla_\mu \psi + \text{c.c.} + \frac{J_{5a} J_5^a}{M^2}\right), \tag{7.105}$$

where $e_a{}^\mu$ is the gravitational field introduced in Sect. 6.3.1,[3] $e = \det e_a{}^\mu$, ψ and $\bar\psi$ are a spinor and its conjugate, "c.c." indicates the complex conjugate of the first term, $J_5^a = \bar\psi\gamma_5\gamma^a\psi$ is the axial fermionic current and M is a mass scale. The interaction term can be decomposed into a scalar interaction

[3]First-order formalism is mandatory in order to couple fermions with gravity consistently.

$(\bar{\psi}\psi)^2$ plus other contributions which can be ignored for simplicity. Thus, the interaction reduces to

$$S_{\rm int} = \int d^4x\, e\, \frac{(\bar{\psi}\psi)^2}{M^2} = \int d^4x\, e \left[(\bar{\psi}\psi)\Delta - \frac{M^2}{4}\Delta^2\right], \tag{7.106}$$

where in the second equality we have introduced the auxiliary scalar Δ, which acts as a mass term for the fermions. On a FLRW background, a non-zero value for the auxiliary field would signal a cosmological fermionic condensation. The quantum theory associated with this system is encoded into the path integral $Z = \int[\mathcal{D}\Delta][\mathcal{D}\psi][\mathcal{D}\bar{\psi}]\, e^{iS_\psi} =: \int[\mathcal{D}\Delta]\, e^{iS_{\rm eff}} \simeq e^{iS_{\rm eff}}\big|_{\rm sp}$, where one integrates over the fields, defines the effective action $S_{\rm eff}$ and approximates the functional integral by the saddle point (mean-field approximation). The effective action $S_{\rm eff}$ can be evaluated by performing the Gaussian integrals in terms of the fermionic coordinates and leads to a non-perturbative potential $V_{\rm eff}(\Delta)$ for the fermion gap. At equilibrium, the system lies at the minimum of the potential and the gap equation $\partial V_{\rm eff}/\partial\Delta = 0$ determines the mass $M(\Delta)$ as a function of Δ and, in turn, Δ as a spacetime function such that M remains constant. Introducing a chemical potential μ, which corresponds to having a non-zero number density n of fermions, the total effective energy density can be shown to be [472]

$$\rho_{\rm gap} = V_{\rm min} + \mu n = \frac{\Delta^2}{32\pi^2}\left(\Delta^2 - 8\mu^2\right)\left(2N + 3 + 2\ln\Delta^2\right), \tag{7.107}$$

where M has been reabsorbed by the gap equation and $N = O(10^2)$ is a regularization free parameter. (The four-fermion interaction is non-renormalizable in Minkowski spacetime and the divergence cannot be eliminated. A standard approach is to interpret the regularization parameter in terms of a finite physical cut-off scale. On curved spacetimes, the issue of renormalization is still open.) The energy density (7.107) is called a "gap" because it constitutes a non-perturbative hiatus in the energy spectrum with respect to the perturbative vacuum $\rho = 0$. Via the gap equation, the gap size $\rho_{\rm gap}$ is governed by M.

An analysis of the effective barotropic index $w_{\rm eff} := -1 - (1/3)d\ln|\rho_{\rm gap}|/d\ln a$ shows that the model can lead to late-time acceleration. At primordial energy scales $\Delta \sim \mu \gg M$, $\rho_{\rm gap}$ is negative and ensures the existence of a non-singular bounce point where $\rho_{\rm gap}$ precisely cancels the energy density of ordinary matter [471]. After the bounce, the gap energy density redshifts away faster than regular matter and remains sub-dominant as compared to ordinary matter and radiation. However, once $\Delta \sim \mu \sim M$, we gradually fall into a regime where $\Delta \sim M \gg \mu$ and $\rho_{\rm gap} \propto 4\pi^2M^2\Delta^2 + \Delta^4 \sim M^4$. Once the matter energy density drops down to $\rho_{\rm m} \sim M^4$, we enter the present dark-energy-dominated quasi-de Sitter phase. The details of the scenario and the size of the effects depend not only on the behaviour of the chemical potential across the early inflationary stage and the subsequent reheating era [474], but also on the issues of regularization and renormalization, which have not been fully assessed yet. In particular, a very small value $M = O(1)$ meV of the mass parameter is required. Therefore, one is indeed assuming that a bare $O(1)$ cosmological constant miraculously cancels by virtue of some mechanism: the old cosmological constant problem is not solved. It is relaxed, however, since the cosmological constant is strongly suppressed and the level of fine-tuning is reduced to just one part over 100, via the choice of N.

7.7 Emergent Gravity

The condensate models briefly described in Sect. 7.6.4 regard Λ as an emergent phenomenon within frameworks where gravity is a fundamental force. One step further is to consider gravity itself as an effective long-range interaction. An often-invoked analogy is that of hydrodynamics as a macroscopic model emerging from molecular physics. The graviton can be conceived as a composite particle or the collective excitation of some underlying fundamental degrees of freedom. This collective mode, born from the statistical description of an altogether different microscopic system, arises through some sort of phase transition. In this purely classical context, not only the cosmological constant [475] but the whole gravitational sector emerges from other degrees of freedom and there is no meaning in quantizing gravity per se.

Emergent gravity [476, 477] is a special class of analogue gravity models with widely and wildly different incarnations. For instance, the linearized graviton can be obtained as a Goldstone boson associated with the breaking of vacuum degeneracy in interacting fermionic or scalar field theories, or from a spin-2 particle in flat space [477–480]. However, it is difficult to recover the fully non-linear gravitational dynamics from these models. Another idea is to get an *effective metric* from some approximation of a classical field theory and an effective dynamics for the metric from low-energy quantum corrections. A scalar field theory in Minkowski spacetime produces, when linearized around a given background ϕ_0, an effective metric given by $\sqrt{-g}g^{\mu\nu} := \partial^2\mathcal{L}/[\partial(\partial_\mu\phi)\partial(\partial_\nu\phi)]|_{\phi=\phi_0}$ [481, 482]. A dynamics for the effective metric is then generated automatically by the quantum effective action, which induces an Einstein–Hilbert term for g at the one-loop level [253, 481, 482]. Other corrections can modify the effective equations of motion in a way deviating from simple Einstein gravity, by adding higher-order "curvature" and non-minimal coupling terms.

If gravity was an emergent force, it should be possible to describe its large-scale dynamics by macroscopic effective degrees of freedom, while near singularities and in high-curvature regimes the microscopic structure of the fundamental theory (whatever it may be) should become apparent. A similar transition appears in thermodynamics: its fundamental substratum is quantum statistical mechanics but its range of application encompasses macroscopic bodies. Perhaps surprisingly, this example is not just an analogy. To each vector in spacetimes with horizons, one can assign thermodynamic potentials: entropy S and temperature $\mathcal{T}$ [30–33, 483–486], energy E, free energy F, enthalpy Q,[4] and so on. These potentials obey thermodynamical laws, which govern the dynamics within the horizon. In the language of field theory, the potentials are Noether charges and the thermodynamical laws are the conservation laws for the associated Noether currents. In the context of emergent gravity, however, Noether charges are derived before the currents.

In general, these thermodynamical laws are equivalent to Einstein's equations and reproduce the dynamics of gravity and matter.[5] This result holds for generic backgrounds with a horizon [487–495] and for specific metrics. Examples of spacetimes with a global horizon are those with spherical symmetry (black holes) [496–499] and the FLRW universe [499–505] (where the Hubble horizon is associated with the de Sitter temperature (5.132)). The range of applicability is much wider than where a global horizon is available; in general, local horizons form around any observer in constant acceleration. Moreover, the form of the potentials is dictated by the type of gravity one wishes to obtain in the equations of motion. Modulo a difference of utmost importance for the cosmological constant problem, in this way one obtains the dynamics of the Einstein–Hilbert action as well as of higher-order actions such as $f(R)$, Gauss–Bonnet and Lanczos–Lovelock gravity. Here we limit the discussion to ordinary Einstein gravity.

In all these constructions, spacetime is a pre-existing manifold with metric $g_{\mu\nu}$ and spanned by coordinate charts, as in usual differential geometry; the metric of such spacetimes, however, is not dynamical and the equations of motion are derived from thermodynamical laws which are assumed to be fundamental.

One such formulation by Padmanabhan [490–495, 506] elects the horizon of a local Rindler observer, on which we now make a short digression.

7.7.1 *Rindler Observer and Null Congruences*

Consider D-dimensional Minkowski spacetime charted by coordinates $(T, X, \mathbf{x}_\perp)$, where we singled out one spatial coordinate X. In a Rindler frame, time T and the coordinate X are redefined as

$$T = \varrho \sinh(gt)\,, \qquad X = \varrho \cosh(gt)\,, \tag{7.108}$$

[4]For vanishing chemical potential, the enthalpy is $E - F = TS$ and measures the difference between the energy and the free energy of a finite-temperature system.

[5]An ultra-simplified instance of this mechanism is the recovery of the Friedmann equations with $\textsc{k} = 0 = \Lambda$ from Newtonian and thermodynamics considerations. Let $D = 4$ and consider an expanding ball of volume $\mathcal{V} = 4\pi a^3/3$ filled with energy (mass) $E = M = \rho\mathcal{V}$. Assuming the hypothesis of adiabatic expansion (no change in entropy, $\mathrm{d}S = 0$), the first law of thermodynamics $\mathrm{d}E + P\,\mathrm{d}\mathcal{V} = 0$ is equivalent to the continuity equation $\dot\rho + 3H(\rho + P) = 0$. On the other hand, the first Friedmann equation can be interpreted as an energy conservation equation $m\dot a^2/2 - GmM/a = 0$, where the first term is a kinetic energy of a small mass m at distance a from the observer in the uniform medium and the second term is Newton's potential.

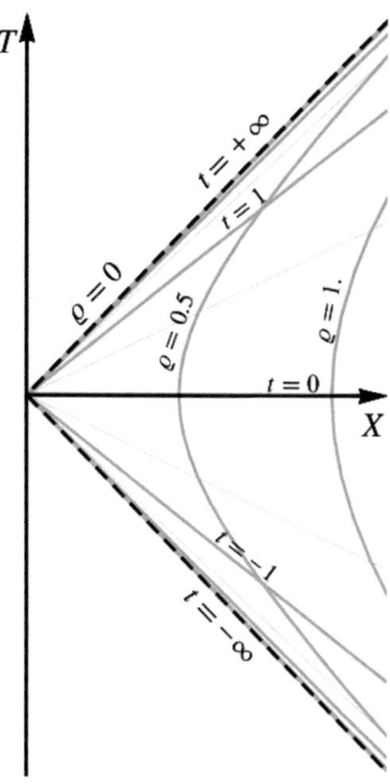

Fig. 7.4 Two-dimensional Rindler chart (7.108) (with $g=1$) in coordinates (ϱ,t) plotted against the two-dimensional local inertial frame in Cartesian coordinates (X,T). The *dashed line* is the $X>0$ portion of the light cone $X^2-T^2=0$ of the local inertial observer at $(X,T)=(0,0)$. It corresponds to the Rindler horizon (7.110). Some $\varrho=$ const and $t=$ const lines are shown for reference. Rindler observers correspond to $\varrho=$ const hyperbolæ

where g is a constant. The line element is then

$$ds^2 = -dT^2 + dX^2 + d\mathbf{x}_\perp^2 = -g^2\varrho^2 dt^2 + d\varrho^2 + d\mathbf{x}_\perp^2 . \tag{7.109}$$

The two charts are depicted in Fig. 7.4 in the (X,T) plane. A *Rindler observer* is at rest in the Rindler frame and thus corresponds to a hyperbola with constant ϱ. This observer has constant proper acceleration and g is the proper acceleration at $\varrho=1$.

In the following, we set $\mathbf{x}_\perp = 0$. Since there can be no incoming information from outside the portion of light cone enveloping the Rindler frame, this null surface acts as a horizon for the Rindler observer. The light cone is defined by $X^2-T^2=\varrho^2=0$ or, equivalently, by $T=\pm X$, which corresponds to the two null hypersurfaces $t=\pm\infty$ (infinite future and past). Therefore, the *Rindler horizon* is the locus of points

$$\mathcal{H}_\mathrm{R}(\varrho,t) = \{(\varrho,t)\,|\,\varrho=0\} = \{(\varrho,t)\,|\,t=\pm\infty\} . \tag{7.110}$$

The proper acceleration of a Rindler observer increases as the horizon is approached asymptotically. One can show that $\mathcal{T} = g/(2\pi)$ is the Davies–Unruh temperature of the horizon [507].

Presently, we move to the covariant setting of a $D = 4$ curved spacetime. In the neighborhood of any point P, one can define a local inertial frame centered at P and spanned by the Cartesian coordinates $(T, X, \mathbf{x}_\perp)$. A *local* Rindler observer is then a Rindler observer charted in this neighborhood, which is called local Rindler frame. If u^μ is the four-velocity of a local Rindler observer, its acceleration with respect to the corresponding free-falling observer will be $\mathrm{a}^\mu = \dot{u}^\mu = u^\nu \nabla_\nu u^\mu$. The Davies–Unruh temperature associated with this observer is then

$$\mathcal{T} = \frac{N\mathrm{a}}{2\pi}\,, \tag{7.111}$$

where N is the lapse function and $\mathrm{a} = \sqrt{|\mathrm{a}_\mu \mathrm{a}^\mu|}$. Notice that $\mathcal{T}$ is a local quantity.

One can attach a local Rindler frame to any point in a generic null surface $\partial\mathcal{V}$. Therefore, locally $\partial\mathcal{V}$ can be mapped to the Rindler horizon (7.110) of a local Rindler frame. This null surface is globally described by a family of freely propagating light rays, a *null congruence*. An element n^μ of a null congruence[6] is a vector normal to $\partial\mathcal{V}$ such that

$$n^\mu \nabla_\mu n^\nu = k n^\nu\,, \qquad n^2 := n_\mu n^\mu = 0\,, \tag{7.112}$$

where k is a scalar. In a local Rindler frame, $k = g$ is a constant; if $k = 0$, one has a null *geodesic* congruence [508]. We also take a parametrization of the null vectors $n^\mu(\sigma) = \mathrm{d}x^\mu/\mathrm{d}\sigma$, where the affine parameter σ introduced in Sect. 6.1.1 is defined by $n^\mu \nabla_\mu \sigma = 1$:

$$n^\mu \nabla_\mu = \frac{\mathrm{d}}{\mathrm{d}\sigma}\,. \tag{7.113}$$

The expansion of the congruence is $\theta := [\nabla_\mu, n^\mu]$, so that by definition $\nabla_\mu n^\mu = k + \theta$. All these expressions can be compared with their counterparts (6.4), (2.47) and (2.46) for a time-like congruence.

[6] In Chap. 2 we used the symbol n^μ for a generic null vector but here it will denote a congruence.

7.7.2 Dynamics

Having a horizon at hand, one can proceed with the construction of thermodynamical potentials within or at the horizon. In particular, for any null congruence n^μ in $\partial\mathcal{V}$, the quantity

$$\mathcal{Q} := \nabla_\mu n^\nu \nabla_\nu n^\mu - (\nabla_\mu n^\mu)^2 \tag{7.114}$$

can be rigorously shown to be the heat or enthalpy density of $\partial\mathcal{V}$ [495]. Equation (7.114) only depends on the derivatives of the congruence because it is required to be invariant under covariantly constant translations of the null vectors, just like the entropy and enthalpy of an elastic solid are invariant under translations. On the other hand, translation invariance is broken in the presence of matter and the heat density for matter $T_{\mu\nu}n^\mu n^\nu$ is quadratic in the null vectors, where $T_{\mu\nu}$ is a symmetric tensor. The analogy with elastic solids (constancy of the heat coefficient, leading to $\nabla^\mu T_{\mu\nu} = 0$) and the fact that, for a perfect fluid, the heat density would simply be $T_{\mu\nu}n^\mu n^\nu = \rho + P$ immediately identify $T_{\mu\nu}$ with the energy-momentum tensor of matter fields.

The total heat of the null surface $\partial\mathcal{V}$ is the total heat density integrated over the surface and constitutes a conserved charge associated with the bulk $\mathcal{V}$ whose boundary is $\partial\mathcal{V}$. If $\partial\mathcal{V}$ is spanned by embedding coordinates y^i and has induced metric γ_{ij} $(i,j = 1,2)$, the total heat is [495]

$$\mathcal{Q}[n] := \frac{1}{8\pi}\int_{\sigma_1}^{\sigma_2} d\sigma \int_{\partial\mathcal{V}} d^2y\,\sqrt{\gamma}\,\left(\mathcal{Q} + \kappa^2 T_{\mu\nu}n^\mu n^\nu\right). \tag{7.115}$$

Demanding that the potential has an extremum for *all* Rindler observers in spacetime yields the field equations

$$\frac{\delta\mathcal{Q}}{\delta n^\mu} = 0\,, \tag{7.116}$$

where the variation is taken with respect to an arbitrary null vector not necessarily belonging to $\partial\mathcal{V}$.

To calculate (7.116), we notice that $R_{\mu\nu}n^\mu n^\nu = n^\nu[\nabla_\mu,\nabla_\nu]n^\mu = -\mathcal{Q} - \nabla_\mu(\theta n^\mu)$. The last term is a total covariant derivative and contributes with a vanishing boundary term in (7.115), so that the integrand can be rewritten as $\mathcal{Q} + \kappa^2 T_{\mu\nu}n^\mu n^\nu \to (-R_{\mu\nu} + \kappa^2 T_{\mu\nu})n^\mu n^\nu$. Equation (7.116) is then equivalent to [490, 495]

$$\left(R_{\mu\nu} - \kappa^2 T_{\mu\nu}\right) n^\nu = 0\,. \tag{7.117}$$

Since this equation is valid for all n^ν in spacetime, we can contract (7.117) with any n^μ. The expression $(R_{\mu\nu} - \kappa^2 T_{\mu\nu})\,n^\mu n^\nu = 0$ is invariant if we add a term $fg_{\mu\nu}$ in the bracket, where f is a function. Using the continuity equation (2.31) and the contracted Bianchi identity (2.32) on $(R_{\mu\nu}+fg_{\mu\nu}-\kappa^2T_{\mu\nu})\,n^\mu n^\nu=0$, we find that $f=-R/2+\Lambda$, where Λ is an *arbitrary* constant. The equations of motion can be rewritten as

$$\boxed{\left(G_{\mu\nu}+\Lambda g_{\mu\nu}-\kappa^2T_{\mu\nu}\right)\,n^\mu n^\nu=0\,.} \tag{7.118}$$

Just like, when applied to all free-falling observers, the equivalence principle determines how gravity interacts with matter in any local inertial frame, so the maximization (7.116) of the total heat for all local Rindler observers establishes the dynamics of gravity.

Since the metric $g_{\mu\nu}$ is not varied to obtain Einstein's equations, the theory may be regarded as a special case of unimodular gravity (Sect. 7.6.3) where all the metric components are non-dynamical. The underlying construction is more involved and differently motivated, since in the present case gravity is not modified in its symmetry structure but is assumed to be emergent from other degrees of freedom.

Contrary to (2.23), the equations of motion (7.118) are invariant under the shift symmetry (7.26) because $\rho_0 g_{\mu\nu}n^\mu n^\nu=0$. Therefore, Padmanabhan's theory is not just a reformulation of general relativity in the language of thermodynamics. The fact that we obtained unimodular gravity rather than Einstein's seems a rather general characteristic of the thermodynamic derivation of the equations of motion [487], through the use of null hypersurfaces.

7.7.3 Holographic Equipartition

Equation (7.118) can be obtained from the entropy S associated with a three-dimensional bulk $\mathcal{V}$ rather than the enthalpy of its boundary [490, 492]. In particular, one can devise a natural definition for the total energy E in $\mathcal{V}$ which is functionally equal to the heat content of $\partial\mathcal{V}$ [495]. The model is therefore a concrete example of *holography* [30, 31, 34, 486, 509, 510], a principle according to which, for a wide class of gravitational and field theories, the dynamics in the bulk is determined by the physics (a thermodynamical one, in the present case) at the boundary.[7]

[7] Without invoking thermodynamical arguments, the equations of motion (7.118) can be obtained in metric formalism by splitting the Einstein–Hilbert Lagrangian $\mathcal{L}_{\rm EH}=\mathcal{L}_{\rm bulk}+\mathcal{L}_{\rm sur}$ into a bulk and a surface term and, then, varying only the surface term with respect to special variations of the metric encoding a normal displacement to a null surface [511–513]. This is possible thanks to the holographic relation $\sqrt{-g}\mathcal{L}_{\rm sur}=-(D/2-1)^{-1}\partial_\sigma[g_{\mu\nu}\partial(\sqrt{-g}\mathcal{L}_{\rm bulk})/\partial(\partial_\sigma g_{\mu\nu})]$. The same procedure is generalizable to an arbitrary gravitational Lagrangian [514]. The lack of a thermodynamical interpretation in metric formalism, however, does not explain why gravity is holographic.

To make holography more apparent, we specify surface and bulk degrees of freedom. Let $\mathcal{A} = \int_{\partial\mathcal{V}} d^2y\,\sqrt{\gamma}$ be the area of a generic two-surface $\partial\mathcal{V}$ (not necessarily null) bounding a region with three-volume $\mathcal{V}$. The boundary of a spatial region $\mathcal{V}$ is compact, while $\partial\mathcal{V}$ is non-compact in the case of a Rindler horizon. The number of degrees of freedom on $\partial\mathcal{V}$ is defined by counting the number of elementary areas $l_{\rm Pl}^2$ in $\mathcal{A}$:

$$N_{\partial\mathcal{V}} := \frac{\mathcal{A}}{l_{\rm Pl}^2}\,. \tag{7.119}$$

This reproduces the entropy-area law for black holes [30, 32, 33, 165], as

$$S = \frac{N_{\partial\mathcal{V}}}{4} = \frac{\mathcal{A}}{4l_{\rm Pl}^2}\,. \tag{7.120}$$

Averaging the local temperature (7.111) over the surface, one gets the average temperature of $\partial\mathcal{V}$:

$$\mathcal{T}_{\partial\mathcal{V}} := \frac{1}{\mathcal{A}}\int_{\partial\mathcal{V}} d^2y\,\sqrt{\gamma}\,\mathcal{T}\,. \tag{7.121}$$

The number of effective degrees of freedom in the bulk $\mathcal{V}$ are then defined as minus the total gravitational energy E in the bulk divided by the thermodynamical energy:

$$N_{\mathcal{V}} := -\frac{E}{\frac{1}{2}k_{\rm B}\mathcal{T}_{\partial\mathcal{V}}}\,,\qquad E := \int_{\mathcal{V}} d^3x\,\sqrt{h}\,\rho_{\rm TK}\,, \tag{7.122}$$

where $k_{\rm B}$ is Boltzmann constant, h is the determinant of the induced metric $h_{\mu\nu}$ on $\mathcal{V}$ and

$$\rho_{\rm TK} := 2N\left(T_{\mu\nu} - \tfrac{1}{2}g_{\mu\nu}T\right)u^\mu u^\nu \tag{7.123}$$

is called Tolman–Komar energy density [515, 516]. The total energy E is defined as the volume integral of the Tolman–Komar energy density. For a perfect fluid (2.40) with $N = 1$, $\rho_{\rm TK} = \rho + 3P = (1 + 3w)\rho$ is negative definite for a barotropic index $w < -1/3$ and positive definite for ordinary matter ($w > -1/3$). The number of bulk degrees of freedom takes positive values only for matter driving acceleration.

By simple manipulations of the Noether currents of the system, one can show that [495]

$$2\int_{\mathcal{V}} d^3x\,h_{\mu\nu}\mathcal{L}_\xi\pi^{\mu\nu} = \frac{1}{2}\,k_{\rm B}\mathcal{T}_{\partial\mathcal{V}}\left(N_{\partial\mathcal{V}} - N_{\mathcal{V}}\right)\,, \tag{7.124}$$

where

$$\pi^{\mu\nu} = \frac{1}{2\kappa^2}\sqrt{h}\,(K^{\mu\nu} - h^{\mu\nu}K) \tag{7.125}$$

is the momentum conjugate to $h_{\mu\nu}$,

$$K_{\mu\nu} := \tfrac{1}{2}\mathcal{L}_u h_{\mu\nu} = \tfrac{1}{2}\left(u^\rho\nabla_\rho h_{\mu\nu} + h_{\nu\rho}\nabla_\mu u^\rho + h_{\rho\mu}\nabla_\nu u^\rho\right) \tag{7.126}$$

is the *second fundamental form*, $K := K_\mu^{\ \mu} = h^{\mu\nu}K_{\mu\nu}$ and $\mathcal{L}_\xi$ is the Lie derivative with respect to the evolution vector $\xi^\mu = Nu^\mu$ corresponding to observers with four-velocity u^μ normal to the constant-t hypersurface. For later use, we notice that

$$\begin{aligned}
h_{\mu\nu}\mathcal{L}_\xi\pi^{\mu\nu} &= \mathcal{L}_\xi(h_{\mu\nu}\pi^{\mu\nu}) - \pi^{\mu\nu}\mathcal{L}_\xi h_{\mu\nu}\\
&= \frac{1}{2\kappa^2}\left[-2\mathcal{L}_\xi(\sqrt{h}\,K) - 2N(2\kappa^2)\pi^{\mu\nu}K_{\mu\nu}\right]\\
&= -\frac{N}{\kappa^2}\left(\sqrt{h}\,u^\mu\partial_\mu K + K\sqrt{h}\,h^{\mu\nu}K_{\mu\nu} + 2\kappa^2\pi^{\mu\nu}K_{\mu\nu}\right)\\
&= -\frac{N}{\kappa^2}\sqrt{h}\left(u^\mu\partial_\mu K + K_{\mu\nu}K^{\mu\nu}\right).
\end{aligned} \tag{7.127}$$

The left-hand side of (7.124) vanishes for any static spacetime, where an observer would experience *holographic equipartition* [517–519]:

$$N_\mathcal{V} = N_{\partial\mathcal{V}}\,. \tag{7.128}$$

Thermodynamically, it corresponds to the relation $E = 2\mathcal{T}S$. In this case, the counting of degrees of freedom in the bulk and at the boundary coincide. For a general dynamical spacetime, the difference between $N_\mathcal{V}$ and $N_{\partial\mathcal{V}}$ governs the rate of change of gravitational momentum and the dynamical evolution. Note that, from what said below (7.123), holographic equipartition can be achieved only in the presence of dark-energy types of matter.

Equation (7.124) is foliation dependent but is valid for all foliations of spacetime, hence it carries the same dynamical information as Einstein's equations. We can check this for de Sitter cosmology and an FLRW spacetime filled with a perfect fluid. In the first case, one has holographic equipartition and $N_{\partial\mathcal{V}} = N_\mathcal{V}$. We take $\mathcal{V} = \mathcal{V}_H$ to be the $N = 1$ Hubble volume (7.27), so that the boundary surface has an area $\mathcal{A} = 4\pi R_H^2 = 4\pi H^{-2}$ and the temperature is (5.132). With the Boltzmann constant reinstated,

$$\mathcal{T}_{\partial\mathcal{V}} = \mathcal{T} = \mathcal{T}_H = \frac{H}{2\pi k_\mathrm{B}}\,. \tag{7.129}$$

Therefore,

$$N_{\partial\mathcal{V}} = \frac{2(4\pi)^2}{\kappa^2 H^2}\,, \qquad N_{\mathcal{V}} = -\frac{4\pi\mathcal{V}_H}{H}(\rho+3P) = -\frac{(4\pi)^2}{3H^4}(\rho+3P)\,, \tag{7.130}$$

with $P = -\rho$ and we obtain the first Friedmann equation (2.81) for $\textsc{k} = 0$ and a pure cosmological constant. The FLRW case is more involved. Comoving observers follow geodesics with zero acceleration and they do not see a Davies–Unruh temperature. Therefore, $N_{\partial\mathcal{V}} = 0$ and

$$\frac{1}{2}k_{\rm B}\mathcal{T}_{\partial\mathcal{V}}\,(N_{\partial\mathcal{V}} - N_{\mathcal{V}}) = E = (\rho+3P)\mathcal{V}\,, \tag{7.131}$$

for any three-volume $\mathcal{V} = \int d^3x\,\sqrt{h}$. On an FLRW background,

$$K_{\mu\nu} = \frac{H}{N}h_{\mu\nu}\,, \qquad K = \frac{3H}{N}\,, \tag{7.132}$$

so that the left-hand side of (7.124) in synchronous gauge has

$$2h_{\mu\nu}\mathcal{L}_\xi\pi^{\mu\nu} \stackrel{(7.127)}{=} -\frac{6}{\kappa^2}\sqrt{h}(\dot{H}+H^2)\,. \tag{7.133}$$

Combining (7.131) and (7.133), we get the second Friedmann equation (2.82) without Λ term,

$$\dot{H}+H^2 = \frac{\ddot{a}}{a} = -\frac{\kappa^2}{6}(\rho+3P)\,. \tag{7.134}$$

The presence of a preferred time foliation permits to recast the cosmological dynamics in a way similar but inequivalent to the background-independent equation (7.124) [520, 521]:

$$\frac{d\mathcal{V}_H}{dt} = l_{\rm Pl}^2\,(N_{\partial\mathcal{V}} - N_{\mathcal{V}})\,, \tag{7.135}$$

which corresponds to (7.134) under the replacements (7.27), (7.129) and (7.130). Here, the degrees of freedom of the spacetime bulk emerge during the dynamical evolution: as the Hubble horizon and the causal patch $\mathcal{V}_H$ expand, more bulk modes become accessible and achieve holographic equipartition.

7.7.4 Cosmological Constant Problem

Having set the main ingredients of the theory, we turn to the Λ problem [25–29, 495, 521]. The first step towards its solution is (7.118): as in any classical unimodular gravity model, the cosmological constant in Padmanabhan's theory is an integration constant. The most direct consequence of this fact is that gravity is now insensitive to the value of the cosmological constant. Therefore, if we could find a fundamental principle fixing the cosmological constant to the observed value (2.118), the transformation (7.26) would not change it: the shift $\rho_\Lambda \to \rho_\Lambda + \rho_0$ in the matter sector would be reabsorbed by an opposite effect in the gravitational sector.

The second step is to identify such principle and, with it, to explain the 4π puzzle (7.30). Recall that $\mathcal{N}_{\rm c}$ is the number of modes which became accessible to our causal patch $\mathcal{V}_H$ during the transient phase XY (radiation-dust era) between two quasi-de Sitter regimes. According to the emergent-gravity scenario of (7.128), this number should coincide with the number of degrees of freedom populating the Hubble sphere $\partial\mathcal{V}_H$. On the other hand, since the line k_- in Fig. 7.1 is at 45°, the expansion rate of the era XY is the same as of the inflationary era PX, $a_{\rm X}/a_{\rm P} = a_{\rm Y}/a_{\rm X}$. But 4π is precisely the number of degrees of freedom of the boundary of an elementary Planck ball, $N_{\partial\mathcal{V}_{\rm Pl}} = (4\pi l_{\rm Pl}^2)/l_{\rm Pl}^2 = 4\pi$. Therefore, a possible solution of the 4π puzzle and of the Λ problem could reside in a realization of a theoretically and experimentally consistent Planck-scale inflationary regime within the emergent-gravity paradigm, so that

$$\mathcal{N}_{\rm c} = N_{\partial\mathcal{V}_{\rm Pl}}\,. \tag{7.136}$$

Another hint that this theory may be on the right track is the following. Whenever one has a thermodynamical description of a system, the latter must be constituted by microscopic degrees of freedom. Therefore, in this particular incarnation of emergent gravity one is still entitled to look for "quanta of geometry," although these will not correspond to graviton modes. A natural bridge between thermodynamics and a microscopic description of these fundamental degrees of freedom is provided by statistical mechanics. In general, one can define a partition function Z via a density of energy states $\varrho(E)$ such that the thermodynamical potentials such as S and the free energy F are recovered. In the present scenario, we have defined the surface degrees of freedom as proportional to the area $\mathcal{A}$ of the surface $\partial\mathcal{V}$ which, according to the entropy-area law, is proportional to the entropy S. In order to reproduce the entropy-area law, this density of states must be exponential,

$$\varrho(E) \simeq \exp\left(4\pi\frac{E^2}{m_{\rm Pl}^2}\right), \tag{7.137}$$

to lowest order in an $E/m_{\rm Pl}$ expansion [522, 523]. Then, the entropy reads $S(E) = \ln\varrho = 4\pi E^2/m_{\rm Pl}^2 \to S(\mathcal{A}) = \mathcal{A}/(4l_{\rm Pl}^2) = N_{\partial\mathcal{V}}/4$. This result is valid in all spacetimes with a hypersurface with infinite redshift, such as a black hole. Roughly speaking, the event horizon "stretches" virtual high-energy field excitations (representing the quantum interactions of matter with geometry) to sub-Planckian energies and allows them to become real modes, which then populate thermodynamical energy levels. Via (7.137), one can show that the enforcement of the entropy-area law implies the presence of a minimal area $4\pi l_{\rm Pl}^2$, so that below the Planck scale spacetime ceases to have the ordinary properties of a continuum [524].[8] Since the cosmological constant (7.31) can be interpreted, in this context, as an energy times a density of states, there is an indication that

$$\mathcal{N}_{\rm c} \propto N_{\partial\mathcal{V}}\,,$$

similar but not quite equal to (7.136). We still lack the details of the theory at the Planck scale to make this correspondence, and a true solution to the old Λ problem, more precise.

Note that, in general, a holographic universe prefers a small or vanishing cosmological constant. The argument is simple and is based on the entropy-area law (7.120) applied to the whole universe. For a homogeneous and isotropic universe, the area $\mathcal{A}$ in question is the surface $\mathcal{A} = 4\pi/H^2$ of the Hubble horizon enclosing the causal region, so that for a de Sitter patch $S = 3\pi m_{\rm Pl}^2/\Lambda$. The Boltzmann probability distribution goes as the exponential of the entropy [527]:

$$P(\Lambda) \propto \exp[S(\Lambda)] = \exp\left(\frac{3\pi m_{\rm Pl}^2}{\Lambda}\right), \tag{7.138}$$

which is peaked at $\Lambda = 0$. Therefore, it is both conceivable and consistent that a solution of the 4π puzzle (which implies a small observed cosmological constant) relies on holography.

7.8 Problems and Solutions

7.1 Numerical cosmology and quintessence 1. Write the first Friedmann equation in (7.33), the scalar-field equation (5.39) and the barotropic index $w_\phi = P_\phi/\rho_\phi$ in a way both convenient and stable for a numerical code, i.e., in the vector form $\vec{X}' = \vec{f}(N, y, z)$, where $\vec{X} = (z, y)^{\rm t}$, $z \propto \kappa\phi'$, $y \propto \kappa\phi$ and $\vec{f}$ is a two-component vector. Here a prime denotes a derivative with respect to the number of e-foldings $N := \ln a/a_0$, where a_0 is the scale factor today. All quantities should be dimensionless.

[8] The presence of a minimal length in Padmanabhan's theory can be inferred also by other arguments independent of (7.137) [525, 526].

Solution Since the cosmological evolution spans many orders of magnitude in synchronous time, for numerical work it is useful to switch to the number of e-folds $N = -\ln(1+z)$, which also coincides (up to a sign) with the redshift. In particular, near the end of inflation and the beginning of reheating $N_{\rm reh} \lesssim -53$ ($z < 10^{23}$, equation (2.164)), at big-bang nucleosynthesis $N_{\rm BBN} \approx -21$ ($z \approx 10^9$, equation (2.164)), at matter-radiation equality $N_{\rm eq} \approx -8$ ($z \approx 3400$, equation (2.90)), at recombination $N_{\rm rec} \approx -7$ ($z \approx 1100$, equation (2.131)), at dark-energy domination $N_{\rm acc} \approx -0.7$ ($z \approx 1$, equation (2.126)).

In this way, an exponential time span is rendered linear and possible precision issues are prevented. First, we rewrite the Friedmann equation in (7.33) in terms of dimensionless densities. Let $\rho_{\rm crit,0} = 3H_0^2/\kappa^2$ be the critical density today and $\varrho = \rho/\rho_{\rm crit,0}$ a density normalized to the present value of the critical density, not to be confused with the density parameter (2.83). Then, $\varrho = \varrho_{\rm m} + \varrho_{\rm r}$, where (see (2.87) and (2.88))

$$\varrho_{\rm r} = \Omega_{\rm r,0} e^{-4N}, \qquad \varrho_{\rm m} = \Omega_{\rm m,0}\, e^{-3N}. \tag{7.139}$$

Calling $x := H^2/H_0^2$, $y := \kappa\phi/\sqrt{3}$ and $U := V/\rho_{\rm crit,0}$, the first Friedmann equation reads

$$x = \varrho_{\rm m} + \varrho_{\rm r} + \varrho_y, \tag{7.140a}$$

$$\varrho_y = x\frac{(y')^2}{2} + U(y), \tag{7.140b}$$

where we used $\dot{y} = Hy'$. Since $\ddot{y} = H^2[y'' + x'y'/(2x)]$, using (7.140) the equation of motion (5.39) for quintessence reads

$$y' = z, \qquad z' = -\frac{1}{2x}\left[(3\varrho_{\rm m} + 2\varrho_{\rm r})z + 2(3zU + U_{,y})\right], \tag{7.141}$$

where

$$x = 2\frac{\varrho_{\rm m} + \varrho_{\rm r} + U}{2 - z^2}. \tag{7.142}$$

The right-hand sides of (7.141) are only functions of N, y and z, as requested. The barotropic index is

$$w_\phi = w_y = \frac{1}{\varrho_y}\left[x\frac{(y')^2}{2} - U(y)\right]. \tag{7.143}$$

7.2 Numerical cosmology and quintessence 2. Write a portable code for a general U. Integrate (7.141) numerically for the inverse power-law potential (7.45) with $n = 6$ and plot ϱ_y and w_y for a scaling solution.

Solution The present density parameters have values $\Omega_{\rm m,0} \approx 0.31$ and $\Omega_{\rm r,0} \approx 5 \times 10^{-5}$, where we used (2.12), (2.96) and (2.99). (As an input in the numerical code, we include more digits but all results are unaffected by small changes in the matter and radiation densities.) Next, the equations of motion should be integrated *forwards* in time from before matter-radiation equality. We choose the initial instant $N_{\rm i} = -28$ ($z \approx 10^{12}$), an epoch between the end of reheating and the start of nucleosynthesis, but modulo convergence issues one can adapt the code to start just at the end of inflation. At $N = N_{\rm i}$, the energy density of quintessence should be comparable with or lower than the contribution of radiation. Assuming a small initial velocity, this translates into the relation $\varrho_{y,\rm i} \simeq U \lesssim \varrho_{\rm r,i} = \varrho_{\rm r}(N_{\rm i})$. In dimensionless variables and constants, the potential (7.45) reads $U = A/y^n$, where $A = M^{4+n}(\kappa^2/3)^{n/2+1}/H_0^2$. From (7.46), we get the order of magnitude that A should have in order to obtain a viable solution: $A \sim \Omega_\phi \sim 1$. Therefore,

$$y_{\rm i} \gtrsim \left(\frac{A}{\varrho_{\rm r,i}}\right)^{\frac{1}{n}} \sim 10^{-\frac{45}{n}}. \tag{7.144}$$

For $n = 6$, we get a rough lower limit $y_{\rm i} \sim 10^{-8}$. Any value around or above this will do; we set then $y_{\rm i} = 10^{-5}$.

The initial velocity may be set depending on whether we want a scaling or a creeping asymptotic solution. For scaling attractors, we can consider two qualitative cases. In the first one (overshooting), the initial energy density $\varrho_{y,\rm i}$ is much greater than the tracker value. The energy density ϱ_y then drops quickly as the solution overshoots the scaling attractor, which is reached after a transient regime with $w_y \approx$ const. In the second case (undershooting), the field starts in a freezing regime with a negligible kinetic energy, $(y_{\rm i}')^2 \ll U(y_{\rm i})$, and ϱ_y is stuck at a constant value. At late times, it thaws and reaches the scaling attractor. We concentrate on the overshooting case without loss of generality, setting $y_{\rm i}' = 10^{-2}$ (positive, since the field increases when it rolls down its potential). One can create an undershooting example by fixing $y_{\rm i}' = 0$.

Having fixed the initial values $y_{\rm i}$ and $y_{\rm i}'$, our numerical code should adjust the normalization A of the potential to yield the solutions which satisfy the boundary condition today, $x(N = 0) = 1$. Although the problem is defined by boundary conditions at the present time, it is unwise to attempt to integrate backwards from them, for instance by choosing $y_{\rm i}$ freely while $y_{\rm i}'$ is constrained by the Friedmann equation at $N = 0$. This is because strong attractors become repellers in backwards integration, leading to a rapid growth of numerical instabilities. Instead, we integrate forwards using a shooting method to adjust the initial conditions to obtain the correct present properties. The shooting method transforms a boundary-value problem into an initial-value problem. It consists in having A run over a sensible range, in the present case around values $0.1 < A < 100$. The code runs a loop over A. For each value of A starting from 0.1, the solution $y_{\rm num}(N)$ is found and plugged into (7.142) at $N = 0$ to check the value $x_{\rm num}(0)$. If the latter is away from 1 above the given tolerance (about one part over 10^3, according to the experimental error in (2.128)), then A is increased by one step (typically, between 10^{-3} and 10^{-1}) and the code is rerun until a normalization is reached such that $|x_{\rm num}(0) - 1| < 10^{-3}$.

Let us implement this strategy in MATHEMATICA. First, we specify the constants $\sqrt{\kappa^2/3}$ (kap3), H_0 (H0), $\Omega_{\rm r,0}$ (OMr0) and $\Omega_{\rm m,0}$ (OMm0) associating to them a numerical value:

```
kap3 = Sqrt[8 * Pi/(3 * (1.22 * 10^19)^2)];

H0 = 2.133 * 0.68 * 10^-42;

OMm0 = 0.1426 * 0.68^-2;

OMr0 = 2.469 * 10^-5 * 0.68^-2;
```

Second, we write down all equations valid for any potential U (dependent on two parameters A and n) and its y derivative Upr: the radiation and matter density parameters rhor and rhom, equation (7.141) (eq1, x and eq2) and equations (7.140b) (rhoQ) and (7.143) (wQ), everything as a function of the number of e-foldings ne:

```
rhor[ne_] :=  OMr0 * Exp[-4 * ne]

rhom[ne_] :=  OMm0 * Exp[-3 * ne]

eq1[ne_] :=  y'[ne] - z[ne]

                  rhom[ne] + rhor[ne] + U[y, ne, A, n]
x[ne_, U_, n_] := ------------------------------------
                         1 - z[ne]^2/2

                                             1
eq2[ne_, U_, Upr_, n_] :=  z'[ne] + ---------------- ((3 * rhom[ne] + 2
                                     2 * x[ne, U, n]
  *rhor[ne]) * z[ne] + 2 * (3 * z[ne] * U[y, ne, A, n] + Upr[y, ne, A, n]))

                     rhom[ne] + rhor[ne] + U[Y, ne, A, n]
X[ne_, U_, A_, n_] := ------------------------------------
                          1 - (Y'[ne])^2/2

                   X[ne, U, Anum, n](Y'[ne])^2
rhoQ[ne_, U_, n_] := --------------------------- + U[Y, ne, Anum, n]
                               2

                  ( X[ne, U, Anum, n] * (Y'[ne])^2
wQ[ne_, U_, n_] := ( ------------------------------
                  (              2

  - U[Y, ne, Anum, n])) / rhoQ[ne, U, n]
```

We defined x twice, once (function x, included in the equations of motion) exactly as in (7.142) and another (function X) with $z \to y'$, the second form being the one to be evaluated on the numerical solution Y. Notice that the symbols U and Upr denote functions in the right-hand sides but variables in the left-hand sides, so that eventually one can plug in different potentials by specifying explicit functional forms. We do it now for an inverse power law:

```
Un[y_, ne_, A_, n_] :=  A * y[ne]^-n

Uprn[y_, ne_, A_, n_] :=  -n * A * y[ne]^(-n-1)

mass[n_] :=  (H0^2 * Anum/kap3^(2 + n))^(1/(4 + n))
```

where the last expression relates M to A. At this point, we create a loop with the Do command for the $n = 6$ potential:

```
Do[

 Y = NDSolveValue[{eq1[ne] == 0, eq2[ne, Un, Uprn, 6] == 0, y[-28] == 10^-5,

  z[-28] == 0.01}, y, {ne, -28, 10}, SolveDelayed -> True];

 If[X[0, Un, A, 6] > 1 - 10^-3 && X[0, Un, A, 6] < 1 + 10^-3, Anum = A;

  Print[{Anum"=A", mass[6]"GeV=M", X[0, Un, 6]"=x(0)"}]; Break[]],

 {A, 59, 60, .01}]
```

At each new iteration with a given value $59 < A < 60$ (the range is guessed a priori), NDSolveValue is the numerical integration of the differential equations of motion with given initial conditions at $N = -28$ (in this example, for the tracking overshooting solution depicted in Fig. 7.2a), up to some future time $N = 10$. The output is the interpolating function Y[ne]. The option SolveDelayed -> True avoids singularities in the solved form of the equations. The conditional If checks whether $|x_{\rm num}(0)-1| < 10^{-3}$. If the convergence criterion is not met, the loop begins anew with a higher value of A, increased by a step $\Delta A = 0.01$. Otherwise, Break[] interrupts the loop and Print gives some output values, in the above example the normalization A, the mass M of the potential in GeV and $x_{\rm num}(0)$: {59.75 = A, 5.12595×10^6 GeV = M, 0.999011 = x(0)}.

It is easy to plot ρ_y = rhoQ and w_ϕ = wQ as in Fig. 7.3 and check that $w_\phi(z = 0)$ = wQ[0, Un, 6] is -0.39951.

7.3 Conformal transformations 1. Consider the action (7.55) with $\mathcal{L}_m = 0$ and the conformal transformation (7.56), where $\Omega = \Omega(x)$ is a generic scalar function. Define

$$\mathcal{O}_\mu := \partial_\mu \ln \Omega \,, \quad \mathcal{O}_{\mu\nu} := \bar{\nabla}_\mu \mathcal{O}_\nu = \bar{\nabla}_\mu \bar{\nabla}_\nu \ln \Omega \,, \quad \mathcal{O} := \mathcal{O}_\mu^{\ \mu} = \bar{\Box} \ln \Omega \,, \tag{7.145}$$

where

$$\bar{\Box} = \Omega^2 \Box - (D-2)\mathcal{O}_\mu \bar{\partial}^\mu \tag{7.146}$$

and indices are raised and lowered with the Jordan metric. Show how the linear Riemann invariants and (7.55) transform under (7.56).

Solution The inverse metric, measure factor and affine connection (2.15) transform as

$$g^{\mu\nu} = \Omega^{-2} \bar{g}^{\mu\nu} \,, \tag{7.147}$$

$$\sqrt{-g} = \Omega^D \sqrt{-\bar{g}} \,, \tag{7.148}$$

$$\Gamma^\rho_{\mu\nu} = \bar{\Gamma}^\rho_{\mu\nu} + \left[2\delta^\rho_{(\mu} \mathcal{O}_{\nu)} - \bar{g}_{\mu\nu} \mathcal{O}^\rho \right] . \tag{7.149}$$

From (7.148) one obtains (7.146), while (7.149) and (2.16) yield

$$\begin{aligned}
R^\rho_{\ \mu\sigma\nu} &= \bar{R}^\rho_{\ \mu\sigma\nu} + r^\rho_{\ \mu\sigma\nu} , \\
r^\rho_{\ \mu\sigma\nu} &:= 2\delta^\rho_{[\nu} \mathcal{O}_{\sigma]\mu} + 2\bar{g}_{\mu[\sigma} \mathcal{O}^\rho_{\ \nu]} + 2\left(\delta^\rho_{[\sigma} \mathcal{O}_{\nu]} \mathcal{O}_\mu + \bar{g}_{\mu[\nu} \mathcal{O}_{\sigma]} \mathcal{O}^\rho + \delta^\rho_{[\nu} \bar{g}_{\sigma]\mu} \mathcal{O}_\tau \mathcal{O}^\tau \right) , \\
R_{\mu\nu} &= \bar{R}_{\mu\nu} + r_{\mu\nu} , \\
r_{\mu\nu} &:= r^\rho_{\ \mu\rho\nu} = (2-D)\mathcal{O}_{\mu\nu} - \bar{g}_{\mu\nu} \mathcal{O} + (D-2)\left(\mathcal{O}_\mu \mathcal{O}_\nu - \bar{g}_{\mu\nu} \mathcal{O}_\tau \mathcal{O}^\tau \right) , \\
\Omega^2 R &= \bar{R} - (D-1)[2\mathcal{O} + (D-2)\mathcal{O}_\mu \mathcal{O}^\mu] .
\end{aligned} \tag{7.150}$$

After integrating by parts and throwing away boundary terms, the last expression yields

$$\int d^D x \sqrt{-g}\, R = \int d^D x \sqrt{-\bar{g}}\, \Omega^{D-2} \left[\bar{R} + (D-1)(D-2)\mathcal{O}_\mu \mathcal{O}^\mu \right] ,$$

so that (7.55) becomes

$$\bar{S} = \int d^D x \sqrt{-\bar{g}}\, \Omega^{D-2} \left[\frac{\bar{R}}{2\kappa^2} + \frac{(D-1)(D-2)}{2\kappa^2} \mathcal{O}_\mu \mathcal{O}^\mu - \frac{1}{2} \bar{\partial}_\mu \phi \bar{\partial}^\mu \phi - \Omega^2 V(\phi) \right] \tag{7.151}$$

in the Jordan frame.

7.4 Conformal transformations 2. (a) Find $\Omega(\phi)$ explicitly in (7.151) such that the kinetic term of the scalar field is canonical. (b) Without specifying Ω, find a field redefinition $\phi \to \bar{\phi}$ such that the kinetic term reads $-(\omega/2)\bar{\partial}_\mu \bar{\phi} \bar{\partial}^\mu \bar{\phi}$, where ω is an arbitrary function of ϕ.

Solution

(a) Since $\mathcal{O}_\mu = (\Omega_{,\phi}/\Omega)\partial_\mu \phi$, the second and third terms in (7.151) can be combined into

$$\Omega^{D-2} \left[\frac{(D-1)(D-2)}{2\kappa^2} \left(\frac{\Omega_{,\phi}}{\Omega} \right)^2 - \frac{1}{2} \right] \bar{\partial}_\mu \phi \bar{\partial}^\mu \phi \,. \tag{7.152}$$

The coefficient equals $-1/2$ if the following non-linear differential equation is satisfied:

$$\frac{(D-1)(D-2)}{\kappa^2} \Omega_{,\phi}^2 + \Omega^{4-D} - \Omega^2 = 0\,.$$

This equation is exactly solved by the profile

$$\Omega(\phi) = \left\{ \cosh \left[\frac{D-2}{2} \sqrt{\frac{\kappa^2}{(D-1)(D-2)}} \phi \right] \right\}^{\frac{2}{D-2}} . \tag{7.153}$$

Then, the action in the Jordan frame is

$$\bar{S} = \int d^D x \sqrt{-\bar{g}} \left[\frac{1}{2\kappa^2} \Omega^{D-2}(\phi) \bar{R} - \frac{1}{2} \bar{\partial}_\mu \phi \bar{\partial}^\mu \phi - U(\phi) \right], \tag{7.154}$$

where $U(\phi) = \Omega^D(\phi) V(\phi)$.

(b) Imposing the relation

$$\Omega^{D-2} \left[\frac{(D-1)(D-2)}{2\kappa^2} \left(\frac{\Omega_{,\bar{\phi}}}{\Omega} \right)^2 - \frac{1}{2} \left(\frac{d\phi}{d\bar{\phi}} \right)^2 \right] \bar{\partial}_\mu \bar{\phi} \bar{\partial}^\mu \bar{\phi} = -\frac{\omega}{2} \bar{\partial}_\mu \bar{\phi} \bar{\partial}^\mu \bar{\phi} \,,$$

one gets

$$\phi = \int d\bar{\phi}\sqrt{\frac{\omega}{\Omega^{D-2}} + \frac{(D-1)(D-2)}{\kappa^2}\left(\frac{\Omega_{,\bar{\phi}}}{\Omega}\right)^2}$$
$$= \int d\bar{\phi}\sqrt{\frac{\omega}{F} + \frac{(D-1)(D-2)}{4\kappa^2}\left(\frac{F_{,\bar{\phi}}}{F}\right)^2}, \qquad (7.155)$$

where $F(\bar{\phi}) = \Omega^{D-2}(\bar{\phi})$.

7.5 Conformal transformations 3. Let

$$\Omega(\phi) = \frac{1}{\cosh\left(\sqrt{\kappa^2/6}\,\phi\right)} \qquad (7.156)$$

in (7.151) with $D = 4$. Find a transformation of the scalar field $\phi \to \bar{\phi}$ such that the kinetic term of $\bar{\phi}$ is canonical. Rewrite $\bar{S}$ in terms of $\bar{\phi}$.

Solution Since $\Omega_{,\phi}/\Omega = -\sqrt{\kappa^2/6}\tanh(\sqrt{\kappa^2/6}\,\phi) = -\sqrt{\kappa^2/6}\sqrt{1-\Omega^2}$, expression (7.152) in four dimensions becomes

$$\left(\frac{3}{\kappa^2}\Omega_{,\phi}^2 - \frac{1}{2}\Omega^2\right)\bar{\partial}_\mu\phi\bar{\partial}^\mu\phi = -\frac{1}{2}\Omega^4\bar{\partial}_\mu\phi\bar{\partial}^\mu\phi\,.$$

Defining the field

$$\bar{\phi} := \int d\phi\,\Omega^2(\phi) = \sqrt{\frac{6}{\kappa^2}}\tanh\left(\sqrt{\frac{\kappa^2}{6}}\phi\right), \qquad \bar{\phi}^2 < \frac{6}{\kappa^2}, \qquad (7.157)$$

the action (7.151) with $D = 4$ becomes

$$\bar{S} = \int d^4x\sqrt{-\bar{g}}\left[\frac{1}{2\kappa^2}f(\bar{\phi})\bar{R} - \frac{1}{2}\bar{\partial}_\mu\bar{\phi}\bar{\partial}^\mu\bar{\phi} - U(\bar{\phi})\right], \qquad (7.158)$$

where

$$f(\bar{\phi}) = 1 - \frac{\kappa^2}{6}\bar{\phi}^2, \qquad U(\bar{\phi}) = \Omega^4[\phi(\bar{\phi})]\,V[\phi(\bar{\phi})]\,. \qquad (7.159)$$

The energy-momentum tensor of this model has finite matrix entries at every order of renormalized perturbation theory [173].

References

1. A.D. Linde, Is the cosmological constant a constant? Pis'ma Zh. Eksp. Teor. Fiz. **19**, 320 (1974) [JETP Lett. **19**, 183 (1974)]
2. A.D. Linde, Phase transitions in gauge theories and cosmology. Rep. Prog. Phys. **42**, 389 (1979)
3. Y. Nambu, Quasi-particles and gauge invariance in the theory of superconductivity. Phys. Rev. **117**, 648 (1960)
4. J. Goldstone, Field theories with "superconductor" solutions. Nuovo Cim. **19**, 154 (1961)
5. J. Goldstone, A. Salam, S. Weinberg, Broken symmetries. Phys. Rev. **127**, 965 (1962)
6. F. Englert, R. Brout, Broken symmetry and the mass of gauge vector mesons. Phys. Rev. Lett. **13**, 321 (1964)
7. P.W. Higgs, Broken symmetries, massless particles and gauge fields. Phys. Lett. **12**, 132 (1964)
8. P.W. Higgs, Broken symmetries and the masses of gauge bosons. Phys. Rev. Lett. **13**, 508 (1964)
9. G.S. Guralnik, C.R. Hagen, T.W.B. Kibble, Global conservation laws and massless particles. Phys. Rev. Lett. **13**, 585 (1964)
10. J. Dreitlein, Broken symmetry and the cosmological constant. Phys. Rev. Lett. **33**, 1243 (1974)
11. S.A. Bludman, M.A. Ruderman, Induced cosmological constant expected above the phase transition restoring the broken symmetry. Phys. Rev. Lett. **38**, 255 (1977)
12. D.B. Chitwood et al. [MuLan Collaboration], Improved measurement of the positive muon lifetime and determination of the Fermi constant. Phys. Rev. Lett. **99**, 032001 (2007). [arXiv:0704.1981]
13. T. Plehn, M. Rauch, Quartic Higgs coupling at hadron colliders. Phys. Rev. D **72**, 053008 (2005). [arXiv:hep-ph/0507321]
14. J. Martin, Everything you always wanted to know about the cosmological constant problem (but were afraid to ask). C. R. Phys. **13**, 566 (2012). [arXiv:1205.3365]
15. Ya.B. Zel'dovich, The cosmological constant and the theory of elementary particles. Sov. Phys. Usp. **11**, 381 (1968)
16. J.F. Koksma, T. Prokopec, The cosmological constant and Lorentz invariance of the vacuum state. arXiv:1105.6296
17. P.M. Stevenson, Gaussian effective potential. I. Quantum mechanics. Phys. Rev. D **30**, 1712 (1984).
18. P.M. Stevenson, Gaussian effective potential. II. $\lambda\phi^4$ field theory. Phys. Rev. D **32**, 1389 (1985).
19. P.M. Stevenson, R. Tarrach, The return of $\lambda\phi^4$. Phys. Lett. B **176**, 436 (1986).
20. N.D. Birrell, P.C.W. Davies, *Quantum Fields in Curved Space* (Cambridge University Press, Cambridge, 1982)
21. Ya.B. Zel'dovich, Cosmological constant and elementary particles. Pis'ma Zh. Eksp. Teor. Fiz. **6**, 883 (1967) [JETP Lett. **6**, 316 (1967)]
22. B. Zumino, Supersymmetry and the vacuum. Nucl. Phys. B **89**, 535 (1975)
23. S. Weinberg, Does gravitation resolve the ambiguity among supersymmetry vacua? Phys. Rev. Lett. **48**, 1776 (1982)
24. S. Weinberg, The cosmological constant problem. Rev. Mod. Phys. **61**, 1 (1989)
25. T. Padmanabhan, Dark energy: mystery of the millennium. AIP Conf. Proc. **861**, 179 (2006). [arXiv:astro-ph/0603114]
26. T. Padmanabhan, Dark energy and gravity. Gen. Relat. Grav. **40**, 529 (2008). [arXiv:0705.2533]
27. T. Padmanabhan, H. Padmanabhan, Cosmological constant from the emergent gravity perspective. Int. J. Mod. Phys. D **23**, 1430011 (2014). [arXiv:1404.2284]

28. T. Padmanabhan, The physical principle that determines the value of the cosmological constant. arXiv:1210.4174
29. H. Padmanabhan, T. Padmanabhan, CosMIn: the solution to the cosmological constant problem. Int. J. Mod. Phys. D **22**, 1342001 (2013). [arXiv:1302.3226]
30. J.D. Bekenstein, Black holes and entropy. Phys. Rev. D **7**, 2333 (1973)
31. J.D. Bekenstein, Generalized second law of thermodynamics in black hole physics. Phys. Rev. D **9**, 3292 (1974)
32. S.W. Hawking, Particle creation by black holes. Commun. Math. Phys. **43**, 199 (1975); Erratum-ibid. **46**, 206 (1976)
33. S.W. Hawking, Black holes and thermodynamics. Phys. Rev. D **13**, 191 (1976)
34. G. 't Hooft, Dimensional reduction in quantum gravity, in *Salamfestschrift*, ed. by A. Ali, J. Ellis, S. Randjbar-Daemi (World Scientific, Singapore, 1993). [arXiv:gr-qc/9310026]
35. A.G. Cohen, D.B. Kaplan, A.E. Nelson, Effective field theory, black holes, and the cosmological constant. Phys. Rev. Lett. **82**, 4971 (1999). [arXiv:hep-th/9803132]
36. V. Sahni, A.A. Starobinsky, The case for a positive cosmological Λ-term. Int. J. Mod. Phys. D **9**, 373 (2000). [arXiv:astro-ph/9904398]
37. S. Nobbenhuis, Categorizing different approaches to the cosmological constant problem. Found. Phys. **36**, 613 (2006). [arXiv:gr-qc/0411093]
38. E. Mottola, Particle creation in de Sitter space. Phys. Rev. D **31**, 754 (1985)
39. T. Banks, Relaxation of the cosmological constant. Phys. Rev. Lett. **52**, 1461 (1984)
40. L.F. Abbott, A mechanism for reducing the value of the cosmological constant. Phys. Lett. B **150**, 427 (1985)
41. M. Endō, T. Fukui, The cosmological term and a modified Brans–Dicke cosmology. Gen. Relat. Grav. **8**, 833 (1977)
42. A.D. Dolgov, An attempt to get rid of the cosmological constant, in *The Very Early Universe*, ed. by G.W. Gibbons, S.W. Hawking, S.T.C. Siklos (Cambridge University Press, Cambridge, 1983)
43. Y. Fujii, Origin of the gravitational constant and particle masses in a scale-invariant scalar-tensor theory. Phys. Rev. D **26**, 2580 (1982)
44. L.H. Ford, Quantum instability of de Sitter spacetime. Phys. Rev. D **31**, 710 (1985)
45. O. Bertolami, Time-dependent cosmological term. Nuovo Cim. B **93**, 36 (1986)
46. O. Bertolami, Brans–Dicke cosmology with a scalar field dependent cosmological term. Fortsch. Phys. **34**, 829 (1986)
47. L.H. Ford, Cosmological-constant damping by unstable scalar fields. Phys. Rev. D **35**, 2339 (1987).
48. R.D. Peccei, J. Solà, C. Wetterich, Adjusting the cosmological constant dynamically: cosmons and a new force weaker than gravity. Phys. Lett. B **195**, 183 (1987)
49. S.M. Barr, Attempt at a classical cancellation of the cosmological constant. Phys. Rev. D **36**, 1691 (1987)
50. C. Wetterich, Cosmologies with variable Newton's "constant". Nucl. Phys. B **302**, 645 (1988)
51. C. Wetterich, Cosmology and the fate of dilatation symmetry. Nucl. Phys. B **302**, 668 (1988)
52. W.-M. Suen, C.M. Will, Damping of the cosmological constant by a classical scalar field. Phys. Lett. B **205**, 447 (1988)
53. Y. Fujii, Saving the mechanism of a decaying cosmological constant. Mod. Phys. Lett. A **04**, 513 (1989)
54. E.T. Tomboulis, Dynamically adjusted cosmological constant and conformal anomalies. Nucl. Phys. B **329**, 410 (1990)
55. Y. Fujii, T. Nishioka, Model of a decaying cosmological constant. Phys. Rev. D **42**, 361 (1990)
56. L. Parker, Cosmological constant and absence of particle creation. Phys. Rev. Lett. **50**, 1009 (1983)
57. V.A. Rubakov, M.E. Shaposhnikov, Extra space-time dimensions: towards a solution to the cosmological constant problem. Phys. Lett. B **125**, 139 (1983)
58. I. Antoniadis, N.C. Tsamis, On the cosmological constant problem. Phys. Lett. B **144**, 55 (1984)

59. S.G. Rajeev, Why is the cosmological constant small? Phys. Lett. B **125**, 144 (1983)
60. T.R. Taylor, G. Veneziano, Quenching the cosmological constant. Phys. Lett. B **228**, 311 (1989)
61. M. Özer, M.O. Taha, A solution to the main cosmological problems. Phys. Lett. B **171**, 363 (1986)
62. M. Özer, M.O. Taha, A model of the universe free of cosmological problems. Nucl. Phys. B **287**, 776 (1987)
63. T.S. Olson, T.F. Jordan, Ages of the Universe for decreasing cosmological constants. Phys. Rev. D **35**, 3258 (1987)
64. K. Freese, F.C. Adams, J.A. Frieman, E. Mottola, Cosmology with decaying vacuum energy. Nucl. Phys. B **287**, 797 (1987)
65. M. Reuter, C. Wetterich, Time evolution of the cosmological "constant". Phys. Lett. B **188**, 38 (1987)
66. J.M. Overduin, F.I. Cooperstock, Evolution of the scale factor with a variable cosmological term. Phys. Rev. D **58**, 043506 (1998). [arXiv:astro-ph/9805260]
67. P.J.E. Peebles, B. Ratra, Cosmology with a time variable cosmological constant. Astrophys. J. **325**, L17 (1988)
68. B. Ratra, P.J.E. Peebles, Cosmological consequences of a rolling homogeneous scalar field. Phys. Rev. D **37**, 3406 (1988)
69. K. Coble, S. Dodelson, J.A. Frieman, Dynamical Λ models of structure formation. Phys. Rev. D **55**, 1851 (1997). [arXiv:astro-ph/9608122]
70. M.S. Turner, M.J. White, CDM models with a smooth component. Phys. Rev. D **56**, 4439 (1997). [arXiv:astro-ph/9701138]
71. R.R. Caldwell, R. Dave, P.J. Steinhardt, Cosmological imprint of an energy component with general equation of state. Phys. Rev. Lett. **80**, 1582 (1998). [arXiv:astro-ph/9708069]
72. P.G. Ferreira, M. Joyce, Structure formation with a selftuning scalar field. Phys. Rev. Lett. **79**, 4740 (1997). [arXiv:astro-ph/9707286]
73. P.G. Ferreira, M. Joyce, Cosmology with a primordial scaling field. Phys. Rev. D **58**, 023503 (1998). [arXiv:astro-ph/9711102]
74. G. Huey, L. Wang, R. Dave, R.R. Caldwell, P.J. Steinhardt, Resolving the cosmological missing energy problem. Phys. Rev. D **59**, 063005 (1999). [arXiv:astro-ph/9804285]
75. S.M. Carroll, Quintessence and the rest of the world: suppressing long-range interactions. Phys. Rev. Lett. **81**, 3067 (1998). [arXiv:astro-ph/9806099]
76. I. Zlatev, L. Wang, P.J. Steinhardt, Quintessence, cosmic coincidence, and the cosmological constant. Phys. Rev. Lett. **82**, 896 (1999). [arXiv:astro-ph/9807002]
77. A.R. Liddle, R.J. Scherrer, Classification of scalar field potentials with cosmological scaling solutions. Phys. Rev. D **59**, 023509 (1999). [arXiv:astro-ph/9809272]
78. C.F. Kolda, D.H. Lyth, Quintessential difficulties. Phys. Lett. B **458**, 197 (1999). [arXiv:hep-ph/9811375]
79. P.J. Steinhardt, L. Wang, I. Zlatev, Cosmological tracking solutions. Phys. Rev. D **59**, 123504 (1999). [arXiv:astro-ph/9812313]
80. L. Wang, R.R. Caldwell, J.P. Ostriker, P.J. Steinhardt, Cosmic concordance and quintessence. Astrophys. J. **530**, 17 (2000). [arXiv:astro-ph/9901388]
81. A. de la Macorra, G. Piccinelli, General scalar fields as quintessence. Phys. Rev. D **61**, 123503 (2000). [arXiv:hep-ph/9909459]
82. E.J. Copeland, M. Sami, S. Tsujikawa, Dynamics of dark energy. Int. J. Mod. Phys. D **15**, 1753 (2006). [arXiv:hep-th/0603057]
83. E.J. Copeland, A.R. Liddle, D. Wands, Exponential potentials and cosmological scaling solutions. Phys. Rev. D **57**, 4686 (1998). [arXiv:gr-qc/9711068]
84. T. Chiba, Slow-roll thawing quintessence. Phys. Rev. D **79**, 083517 (2009); Erratum-ibid. D **80**, 109902(E) (2009). [arXiv:0902.4037]
85. T. Chiba, Equation of state of tracker fields. Phys. Rev. D **81**, 023515 (2010). [arXiv:0909.4365]

86. T. Chiba, A. De Felice, S. Tsujikawa, Observational constraints on quintessence: thawing, tracker, and scaling models. Phys. Rev. D **87**, 083505 (2013). [arXiv:1210.3859]
87. R.R. Caldwell, E.V. Linder, Limits of quintessence. Phys. Rev. Lett. **95**, 141301 (2005). [arXiv:astro-ph/0505494]
88. E.V. Linder, Paths of quintessence. Phys. Rev. D **73**, 063010 (2006). [arXiv:astro-ph/0601052]
89. R.J. Scherrer, A.A. Sen, Thawing quintessence with a nearly flat potential. Phys. Rev. D **77**, 083515 (2008). [arXiv:0712.3450]
90. P.S. Corasaniti, E.J. Copeland, Model independent approach to the dark energy equation of state. Phys. Rev. D **67**, 063521 (2003). [arXiv:astro-ph/0205544]
91. D.K. Hazra, S. Majumdar, S. Pal, S. Panda, A.A. Sen, S.P. Trivedi, Post-Planck dark energy constraints. Phys. Rev. D **91**, 083005 (2015). [arXiv:1310.6161]
92. A.Yu. Kamenshchik, U. Moschella, V. Pasquier, An alternative to quintessence. Phys. Lett. B **511**, 265 (2001). [arXiv:gr-qc/0103004]
93. N. Bilic, G.B. Tupper, R.D. Viollier, Unification of dark matter and dark energy: the inhomogeneous Chaplygin gas. Phys. Lett. B **535**, 17 (2002). [arXiv:astro-ph/0111325]
94. M.C. Bento, O. Bertolami, A.A. Sen, Generalized Chaplygin gas, accelerated expansion and dark energy matter unification. Phys. Rev. D **66**, 043507 (2002). [arXiv:gr-qc/0202064]
95. M.C. Bento, O. Bertolami, A.A. Sen, Generalized Chaplygin gas model: dark energy – dark matter unification and CMBR constraints. Gen. Relat. Grav. **35**, 2063 (2003). [arXiv:gr-qc/0305086]
96. A.A. Sen, R.J. Scherrer, Generalizing the generalized Chaplygin gas. Phys. Rev. D **72**, 063511 (2005). [arXiv:astro-ph/0507717]
97. S. Chaplygin, On gas jets. Sci. Mem. Moscow Univ. Math. **21**, 1 (1904)
98. N. Weiss, Possible origins of a small, nonzero cosmological constant. Phys. Lett. B **197**, 42 (1987)
99. C.T. Hill, D.N. Schramm, J.N. Fry, Cosmological structure formation from soft topological defects. Comments Nucl. Part. Phys. **19**, 25 (1989)
100. J.A. Frieman, C.T. Hill, R. Watkins, Late-time cosmological phase transitions: particle-physics models and cosmic evolution. Phys. Rev. D **46**, 1226 (1992)
101. J.A. Frieman, C.T. Hill, A. Stebbins, I. Waga, Cosmology with ultralight pseudo Nambu–Goldstone bosons. Phys. Rev. Lett. **75**, 2077 (1995). [arXiv:astro-ph/9505060]
102. L.J. Hall, Y. Nomura, S.J. Oliver, Evolving dark energy with $w \neq -1$. Phys. Rev. Lett. **95**, 141302 (2005). [arXiv:astro-ph/0503706]
103. S. Dutta, R.J. Scherrer, Hilltop quintessence. Phys. Rev. D **78**, 123525 (2008). [arXiv:0809.4441]
104. K. Choi, String or M theory axion as a quintessence. Phys. Rev. D **62**, 043509 (2000). [arXiv:hep-ph/9902292]
105. R. Kallosh, A.D. Linde, S. Prokushkin, M. Shmakova, Supergravity, dark energy and the fate of the universe. Phys. Rev. D **66**, 123503 (2002). [arXiv:hep-th/0208156]
106. M.S. Turner, Coherent scalar-field oscillations in an expanding universe. Phys. Rev. D **28**, 1243 (1983)
107. B. Spokoiny, Deflationary universe scenario. Phys. Lett. B **315**, 40 (1993). [arXiv:gr-qc/9306008]
108. T. Barreiro, E.J. Copeland, N.J. Nunes, Quintessence arising from exponential potentials. Phys. Rev. D **61**, 127301 (2000). [arXiv:astro-ph/9910214]
109. C. Rubano, P. Scudellaro, On some exponential potentials for a cosmological scalar field as quintessence. Gen. Relat. Grav. **34**, 307 (2002). [arXiv:astro-ph/0103335]
110. A.A. Sen, S. Sethi, Quintessence model with double exponential potential. Phys. Lett. B **532**, 159 (2002). [arXiv:gr-qc/0111082]
111. M. Gutperle, R. Kallosh, A.D. Linde, M/string theory, S-branes and accelerating universe. JCAP **0307**, 001 (2003). [arXiv:hep-th/0304225]
112. I.P. Neupane, Accelerating cosmologies from exponential potentials. Class. Quantum Grav. **21**, 4383 (2004). [arXiv:hep-th/0311071]

113. L. Järv, T. Mohaupt, F. Saueressig, Quintessence cosmologies with a double exponential potential. JCAP **0408**, 016 (2004). [arXiv:hep-th/0403063]
114. C.M. Hull, The minimal couplings and scalar potentials of the gauged $N = 8$ supergravities. Class. Quantum Grav. **2**, 343 (1985)
115. R. Kallosh, A.D. Linde, S. Prokushkin, M. Shmakova, Gauged supergravities, de Sitter space and cosmology. Phys. Rev. D **65**, 105016 (2002). [arXiv:hep-th/0110089]
116. H. Lü, C.N. Pope, p-brane solitons in maximal supergravities. Nucl. Phys. B **465**, 127 (1996). [arXiv:hep-th/9512012]
117. I.V. Lavrinenko, H. Lü, C.N. Pope, Fiber bundles and generalized dimensional reduction. Class. Quantum Grav. **15**, 2239 (1998). [arXiv:hep-th/9710243]
118. M.S. Bremer, M.J. Duff, H. Lü, C.N. Pope, K.S. Stelle, Instanton cosmology and domain walls from M theory and string theory. Nucl. Phys. B **543**, 321 (1999). [arXiv:hep-th/9807051]
119. S.W. Hawking, H.S. Reall, Inflation, singular instantons and eleven-dimensional cosmology. Phys. Rev. D **59**, 023502 (1999). [arXiv:hep-th/9807100]
120. A. Albrecht, C. Skordis, Phenomenology of a realistic accelerating universe using only Planck scale physics. Phys. Rev. Lett. **84**, 2076 (2000). [arXiv:astro-ph/9908085]
121. C. Skordis, A. Albrecht, Planck scale quintessence and the physics of structure formation. Phys. Rev. D **66**, 043523 (2002). [arXiv:astro-ph/0012195]
122. G.R. Dvali, S.H.H. Tye, Brane inflation. Phys. Lett. B **450**, 72 (1999). [arXiv:hep-ph/9812483]
123. V. Sahni, L. Wang, New cosmological model of quintessence and dark matter. Phys. Rev. D **62**, 103517 (2000). [arXiv:astro-ph/9910097]
124. L.P. Chimento, A.S. Jakubi, Scalar field cosmologies with perfect fluid in Robertson–Walker metric. Int. J. Mod. Phys. D **5**, 71 (1996). [arXiv:gr-qc/9506015]
125. L.A. Ureña-López, T. Matos, New cosmological tracker solution for quintessence. Phys. Rev. D **62**, 081302 (2000). [arXiv:astro-ph/0003364]
126. T.R. Taylor, G. Veneziano, S. Yankielowicz, Supersymmetric QCD and its massless limit: an effective lagrangian analysis. Nucl. Phys. B **218**, 493 (1983)
127. I. Affleck, M. Dine, N. Seiberg, Dynamical supersymmetry breaking in four dimensions and its phenomenological implications. Nucl. Phys. B **256**, 557 (1985)
128. P. Binétruy, Models of dynamical supersymmetry breaking and quintessence. Phys. Rev. D **60**, 063502 (1999). [arXiv:hep-ph/9810553]
129. A. Masiero, M. Pietroni, F. Rosati, SUSY QCD and quintessence. Phys. Rev. D **61**, 023504 (2000). [arXiv:hep-ph/9905346]
130. P. Brax, J. Martin, Robustness of quintessence. Phys. Rev. D **61**, 103502 (2000). [arXiv:astro-ph/9912046]
131. P.-Y. Wang, C.-W. Chen, P. Chen, Confronting tracker field quintessence with data. JCAP **1202**, 016 (2012). [arXiv:1108.1424]
132. P. Brax, J. Martin, Quintessence and supergravity. Phys. Lett. B **468**, 40 (1999). [arXiv:astro-ph/9905040]
133. E.J. Copeland, N.J. Nunes, F. Rosati, Quintessence models in supergravity. Phys. Rev. D **62**, 123503 (2000). [arXiv:hep-ph/0005222]
134. P. Brax, J. Martin, A. Riazuelo, Exhaustive study of cosmic microwave background anisotropies in quintessential scenarios. Phys. Rev. D **62**, 103505 (2000). [arXiv:astro-ph/0005428]
135. P. Brax, J. Martin, A. Riazuelo, Quintessence with two energy scales. Phys. Rev. D **64**, 083505 (2001). [arXiv:hep-ph/0104240]
136. S.C.C. Ng, N.J. Nunes, F. Rosati, Applications of scalar attractor solutions to cosmology. Phys. Rev. D **64**, 083510 (2001). [arXiv:astro-ph/0107321]
137. P.S. Corasaniti, E.J. Copeland, Constraining the quintessence equation of state with SnIa data and CMB peaks. Phys. Rev. D **65**, 043004 (2002). [arXiv:astro-ph/0107378]
138. S.A. Bludman, Tracking quintessence would require two cosmic coincidences. Phys. Rev. D **69**, 122002 (2004). [arXiv:astro-ph/0403526]

139. E. Witten, The cosmological constant from the viewpoint of string theory. arXiv:hep-ph/0002297
140. P.J.E. Peebles, A. Vilenkin, Quintessential inflation. Phys. Rev. D **59**, 063505 (1999). [arXiv:astro-ph/9810509]
141. M. Giovannini, Spikes in the relic graviton background from quintessential inflation. Class. Quantum Grav. **16**, 2905 (1999). [arXiv:hep-ph/9903263]
142. M. Peloso, F. Rosati, On the construction of quintessential inflation models. JHEP **9912**, 026 (1999). [arXiv:hep-ph/9908271]
143. K. Dimopoulos, J.W.F. Valle, Modeling quintessential inflation. Astropart. Phys. **18**, 287 (2002). [arXiv:astro-ph/0111417]
144. K. Dimopoulos, Curvaton hypothesis and the η-problem of quintessential inflation, with and without branes. Phys. Rev. D **68**, 123506 (2003). [arXiv:astro-ph/0212264]
145. A.H. Campos, H.C. Reis, R. Rosenfeld, Preheating in quintessential inflation. Phys. Lett. B **575**, 151 (2003). [arXiv:hep-ph/0210152]
146. L.H. Ford, Gravitational particle creation and inflation. Phys. Rev. D **35**, 2955 (1987)
147. M. Joyce, T. Prokopec, Turning around the sphaleron bound: electroweak baryogenesis in an alternative postinflationary cosmology. Phys. Rev. D **57**, 6022 (1998). [arXiv:hep-ph/9709320]
148. G.N. Felder, L. Kofman, A.D. Linde, Inflation and preheating in nonoscillatory models. Phys. Rev. D **60**, 103505 (1999). [arXiv:hep-ph/9903350]
149. E.J. Copeland, A.R. Liddle, D.H. Lyth, E.D. Stewart, D. Wands, False vacuum inflation with Einstein gravity. Phys. Rev. D **49**, 6410 (1994). [arXiv:astro-ph/9401011]
150. R.A. Frewin, J.E. Lidsey, On identifying the present day vacuum energy with the potential driving inflation. Int. J. Mod. Phys. D **2**, 323 (1993). [arXiv:astro-ph/9312035]
151. A.B. Kaganovich, Field theory model giving rise to "quintessential inflation" without the cosmological constant and other fine tuning problems. Phys. Rev. D **63**, 025022 (2001). [arXiv:hep-th/0007144]
152. G. Huey, J.E. Lidsey, Inflation, braneworlds and quintessence. Phys. Lett. B **514**, 217 (2001). [arXiv:astro-ph/0104006]
153. R. Rosenfeld, J.A. Frieman, A simple model for quintessential inflation. JCAP **0509**, 003 (2005). [arXiv:astro-ph/0504191]
154. P. Jordan, Formation of the stars and development of the universe. Nature **164**, 637 (1949)
155. M. Fierz, Über die physikalische Deutung der erweiterten Gravitationstheorie P. Jordans. Helv. Phys. Acta **29**, 128 (1956)
156. P. Jordan, Zum gegenwärtigen Stand der Diracschen kosmologischen Hypothesen. Z. Phys. **157**, 112 (1959)
157. C. Brans, R.H. Dicke, Mach's principle and a relativistic theory of gravitation. Phys. Rev. **124**, 925 (1961)
158. R.H. Dicke, Mach's principle and invariance under transformation of units. Phys. Rev. **125**, 2163 (1962)
159. P.G. Bergmann, Comments on the scalar-tensor theory. Int. J. Theor. Phys. **1**, 25 (1968)
160. K. Nordtvedt, Post-Newtonian metric for a general class of scalar-tensor gravitational theories and observational consequences. Astrophys. J. **161**, 1059 (1970)
161. R.V. Wagoner, Scalar-tensor theory and gravitational waves. Phys. Rev. D **1**, 3209 (1970)
162. S. Deser, Scale invariance and gravitational coupling. Ann. Phys. (N.Y.) **59**, 248 (1970)
163. J. O'Hanlon, Intermediate-range gravity: a generally covariant model. Phys. Rev. Lett. **29**, 137 (1972)
164. Y. Fujii, Scalar-tensor theory of gravitation and spontaneous breakdown of scale invariance. Phys. Rev. D **9**, 874 (1974)
165. J.D. Bekenstein, Exact solutions of Einstein-conformal scalar equations. Ann. Phys. (N.Y.) **82**, 535 (1974)
166. P. Minkowski, On the spontaneous origin of Newton's constant. Phys. Lett. B **71**, 419 (1977)
167. V. Canuto, S.H. Hsieh, P.J. Adams, Scale-covariant theory of gravitation and astrophysical applications. Phys. Rev. Lett. **39**, 429 (1977)

168. A. Zee, Broken-symmetric theory of gravity. Phys. Rev. Lett. **42**, 417 (1979)
169. Y. Fujii, J.M. Niedra, Solutions of a cosmological equation in the scale invariant scalar-tensor theory of gravitation. Prog. Theor. Phys. **70**, 412 (1983)
170. J.L. Anderson, Scale invariance of the second kind and the Brans–Dicke scalar-tensor theory. Phys. Rev. D **3**, 1689 (1971)
171. N. Banerjee, S. Sen, Does Brans–Dicke theory always yield general relativity in the infinite ω limit? Phys. Rev. D **56**, 1334 (1997)
172. V. Faraoni, The $\omega \rightarrow \infty$ limit of Brans–Dicke theory. Phys. Lett. A **245**, 26 (1998). [arXiv:gr-qc/9805057]
173. C.G. Callan, S.R. Coleman, R. Jackiw, A new improved energy-momentum tensor. Ann. Phys. (N.Y.) **59**, 42 (1970)
174. D.Z. Freedman, I.J. Muzinich, E.J. Weinberg, On the energy-momentum tensor in gauge field theories. Ann. Phys. (N.Y.) **87**, 95 (1974)
175. D.Z. Freedman, E.J. Weinberg, The energy-momentum tensor in scalar and gauge field theories. Ann. Phys. (N.Y.) **87**, 354 (1974)
176. G. 't Hooft, M.J.G. Veltman, One loop divergencies in the theory of gravitation. Ann. Poincaré Phys. Theor. A **20**, 69 (1974)
177. L.S. Brown, J.C. Collins, Dimensional renormalization of scalar field theory in curved space-time. Ann. Phys. (N.Y.) **130**, 215 (1980)
178. T. Kaluza, Zum Unitätsproblem in der Physik. Sitz.-ber. Kgl. Preuss. Akad. Wiss. **1921**, 966 (1921)
179. O. Klein, Quantentheorie und fünfdimensionale Relativitätstheorie. Z. Phys. **37**, 895 (1926) [Surveys High Energy Phys. **5**, 241 (1986)]
180. A.H. Chamseddine, $N = 4$ supergravity coupled to $N = 4$ matter and hidden symmetries. Nucl. Phys. B **185**, 403 (1981)
181. P.G.O. Freund, Kaluza–Klein cosmologies. Nucl. Phys. B **209**, 146 (1982)
182. G. Magnano, L.M. Sokołowski, Physical equivalence between nonlinear gravity theories and a general-relativistic self-gravitating scalar field. Phys. Rev. D **50**, 5039 (1994). [arXiv:gr-qc/9312008]
183. V. Faraoni, E. Gunzig, P. Nardone, Conformal transformations in classical gravitational theories and in cosmology. Fund. Cosmic Phys. **20**, 121 (1999). [arXiv:gr-qc/9811047]
184. É.É. Flanagan, The conformal frame freedom in theories of gravitation. Class. Quantum Grav. **21**, 3817 (2004). [arXiv:gr-qc/0403063]
185. V. Faraoni, S. Nadeau, (Pseudo)issue of the conformal frame revisited. Phys. Rev. D **75**, 023501 (2007). [arXiv:gr-qc/0612075]
186. R.L. Arnowitt, S. Deser, C.W. Misner, The dynamics of general relativity, in *Gravitation: An Introduction to Current Research*, ed. by L. Witten (Wiley, New York, 1962). [arXiv:gr-qc/0405109]
187. L.M. Sokołowski, Uniqueness of the metric line element in dimensionally reduced theories. Class. Quantum Grav. **6**, 59 (1989)
188. C. Armendariz-Picón, Predictions and observations in theories with varying couplings. Phys. Rev. D **66**, 064008 (2002). [arXiv:astro-ph/0205187]
189. R. Catena, M. Pietroni, L. Scarabello, Einstein and Jordan reconciled: a frame-invariant approach to scalar-tensor cosmology. Phys. Rev. D **76**, 084039 (2007). [arXiv:astro-ph/0604492]
190. N. Deruelle, M. Sasaki, Conformal equivalence in classical gravity: the example of "veiled" general relativity. Springer Proc. Phys. **137**, 247 (2011). [arXiv:1007.3563]
191. T. Chiba, M. Yamaguchi, Conformal-frame (in)dependence of cosmological observations in scalar-tensor theory. JCAP **1310**, 040 (2013). [arXiv:1308.1142]
192. K. Nordtvedt, Equivalence principle for massive bodies. II. Theory. Phys. Rev. **169**, 1017 (1968)
193. T. Damour, G. Esposito-Farèse, Tensor multiscalar theories of gravitation. Class. Quantum Grav. **9**, 2093 (1992)

194. Y.M. Cho, Violation of equivalence principle in Brans–Dicke theory. Class. Quantum Grav. **14**, 2963 (1997)
195. L. Hui, A. Nicolis, Equivalence principle for scalar forces. Phys. Rev. Lett. **105**, 231101 (2010). [arXiv:1009.2520]
196. C. Armendariz-Picón, R. Penco, Quantum equivalence principle violations in scalar-tensor theories. Phys. Rev. D **85**, 044052 (2012). [arXiv:1108.6028]
197. F. Nitti, F. Piazza, Scalar-tensor theories, trace anomalies, and the QCD frame. Phys. Rev. D **86**, 122002 (2012). [arXiv:1202.2105]
198. G. Esposito-Farèse, D. Polarski, Scalar-tensor gravity in an accelerating universe. Phys. Rev. D **63**, 063504 (2001). [arXiv:gr-qc/0009034]
199. N. Makino, M. Sasaki, The density perturbation in the chaotic inflation with non-minimal coupling. Prog. Theor. Phys. **86**, 103 (1991)
200. R. Fakir, S. Habib, W. Unruh, Cosmological density perturbations with modified gravity. Astrophys. J. **394**, 396 (1992)
201. J. Weenink, T. Prokopec, Gauge invariant cosmological perturbations for the nonminimally coupled inflaton field. Phys. Rev. D **82**, 123510 (2010). [arXiv:1007.2133]
202. J.-O. Gong, J.-c. Hwang, W.-I. Park, M. Sasaki, Y.-S. Song, Conformal invariance of curvature perturbation. JCAP **1109**, 023 (2011). [arXiv:1107.1840]
203. D.I. Kaiser, Primordial spectral indices from generalized Einstein theories. Phys. Rev. D **52**, 4295 (1995). [arXiv:astro-ph/9408044]
204. D.I. Kaiser, Frame independent calculation of spectral indices from inflation. arXiv:astro-ph/9507048
205. G. Domènech, M. Sasaki, Conformal frame dependence of inflation. JCAP **1504**, 022 (2015). [arXiv:1501.07699]
206. J. White, M. Minamitsuji, M. Sasaki, Curvature perturbation in multi-field inflation with non-minimal coupling. JCAP **1207**, 039 (2012). [arXiv:1205.0656]
207. J. White, M. Minamitsuji, M. Sasaki, Non-linear curvature perturbation in multi-field inflation models with non-minimal coupling. JCAP **1309**, 015 (2013). [arXiv:1306.6186]
208. T. Qiu, J.-Q. Xia, Perturbations of single-field inflation in modified gravity theory. Phys. Lett. B **744**, 273 (2015). [arXiv:1406.5902]
209. M.J. Duff, Inconsistency of quantum field theory in curved space-time, in *Quantum Gravity 2*, ed. by C.J. Isham, R. Penrose, D.W. Sciama (Oxford University Press, Oxford, 1981)
210. S.P. de Alwis, Quantization of a theory of 2*D* dilaton gravity. Phys. Lett. B **289**, 278 (1992). [arXiv:hep-th/9205069]
211. R. Fakir, S. Habib, Quantum fluctuations with strong curvature coupling. Mod. Phys. Lett. A **08**, 2827 (1993)
212. E. Elizalde, S. Naftulin, S.D. Odintsov, The renormalization structure and quantum equivalence of 2*D* dilaton gravities. Int. J. Mod. Phys. A **9**, 933 (1994). [arXiv:hep-th/9304091]
213. D. Grumiller, W. Kummer, D.V. Vassilevich, Dilaton gravity in two dimensions. Phys. Rep. **369**, 327 (2002). [arXiv:hep-th/0204253]
214. D. Grumiller, W. Kummer, D.V. Vassilevich, Positive specific heat of the quantum corrected dilaton black hole. JHEP **0307**, 009 (2003). [arXiv:hep-th/0305036]
215. E. Komatsu, T. Futamase, Complete constraints on a nonminimally coupled chaotic inflationary scenario from the cosmic microwave background. Phys. Rev. D **59**, 064029 (1999). [arXiv:astro-ph/9901127]
216. S. Tsujikawa, B. Gumjudpai, Density perturbations in generalized Einstein scenarios and constraints on nonminimal couplings from the cosmic microwave background. Phys. Rev. D **69**, 123523 (2004). [arXiv:astro-ph/0402185]
217. D. La, P.J. Steinhardt, Extended inflationary cosmology. Phys. Rev. Lett. **62**, 376 (1989); Erratum-ibid. **62**, 1066 (1989)
218. D. La, P.J. Steinhardt, E.W. Bertschinger, Prescription for successful extended inflation. Phys. Lett. B **231**, 231 (1989)
219. S. Tsujikawa, J. Ohashi, S. Kuroyanagi, A. De Felice, Planck constraints on single-field inflation. Phys. Rev. D **88**, 023529 (2013). [arXiv:1305.3044]

220. C. Wetterich, An asymptotically vanishing time-dependent cosmological “constant”. Astron. Astrophys. **301**, 321 (1995) [arXiv:hep-th/9408025]
221. J.-P. Uzan, Cosmological scaling solutions of nonminimally coupled scalar fields. Phys. Rev. D **59**, 123510 (1999). [arXiv:gr-qc/9903004]
222. T. Chiba, Quintessence, the gravitational constant, and gravity. Phys. Rev. D **60**, 083508 (1999). [arXiv:gr-qc/9903094]
223. L. Amendola, Scaling solutions in general nonminimal coupling theories. Phys. Rev. D **60**, 043501 (1999). [arXiv:astro-ph/9904120]
224. F. Perrotta, C. Baccigalupi, S. Matarrese, Extended quintessence. Phys. Rev. D **61**, 023507 (1999). [arXiv:astro-ph/9906066]
225. R. de Ritis, A.A. Marino, C. Rubano, P. Scudellaro, Tracker fields from nonminimally coupled theory. Phys. Rev. D **62**, 043506 (2000). [arXiv:hep-th/9907198]
226. L. Amendola, Coupled quintessence. Phys. Rev. D **62**, 043511 (2000). [arXiv:astro-ph/9908023]
227. D.J. Holden, D. Wands, Selfsimilar cosmological solutions with a nonminimally coupled scalar field. Phys. Rev. D **61**, 043506 (2000). [arXiv:gr-qc/9908026]
228. N. Bartolo, M. Pietroni, Scalar-tensor gravity and quintessence. Phys. Rev. D **61**, 023518 (2000). [arXiv:hep-ph/9908521]
229. O. Bertolami, P.J. Martins, Nonminimal coupling and quintessence. Phys. Rev. D **61**, 064007 (2000). [arXiv:gr-qc/9910056]
230. B. Boisseau, G. Esposito-Farèse, D. Polarski, A.A. Starobinsky, Reconstruction of a scalar-tensor theory of gravity in an accelerating universe. Phys. Rev. Lett. **85**, 2236 (2000). [arXiv:gr-qc/0001066]
231. V. Faraoni, Inflation and quintessence with nonminimal coupling. Phys. Rev. D **62**, 023504 (2000). [arXiv:gr-qc/0002091]
232. C. Barceló, M. Visser, Scalar fields, energy conditions and traversable wormholes. Class. Quantum Grav. **17**, 3843 (2000). [arXiv:gr-qc/0003025]
233. C. Baccigalupi, S. Matarrese, F. Perrotta, Tracking extended quintessence. Phys. Rev. D **62**, 123510 (2000). [arXiv:astro-ph/0005543]
234. S. Sen, T.R. Seshadri, Self interacting Brans–Dicke cosmology and quintessence. Int. J. Mod. Phys. D **12**, 445 (2003). [arXiv:gr-qc/0007079]
235. D.F. Torres, Quintessence, superquintessence, and observable quantities in Brans–Dicke and nonminimally coupled theories. Phys. Rev. D **66**, 043522 (2002). [arXiv:astro-ph/0204504]
236. S. Matarrese, C. Baccigalupi, F. Perrotta, Approaching Λ without fine-tuning. Phys. Rev. D **70**, 061301 (2004). [arXiv:astro-ph/0403480]
237. S.M. Carroll, A. De Felice, M. Trodden, Can we be tricked into thinking that w is less than -1? Phys. Rev. D **71**, 023525 (2005). [arXiv:astro-ph/0408081]
238. L. Perivolaropoulos, Crossing the phantom divide barrier with scalar tensor theories. JCAP **0510**, 001 (2005). [arXiv:astro-ph/0504582]
239. S. Tsujikawa, K. Uddin, S. Mizuno, R. Tavakol, J.'i. Yokoyama, Constraints on scalar-tensor models of dark energy from observational and local gravity tests. Phys. Rev. D **77**, 103009 (2008). [arXiv:0803.1106]
240. S. Das, P.S. Corasaniti, J. Khoury, Superacceleration as signature of dark sector interaction. Phys. Rev. D **73**, 083509 (2006). [arXiv:astro-ph/0510628]
241. M. Bruni, S. Matarrese, O. Pantano, A local view of the observable universe. Phys. Rev. Lett. **74**, 1916 (1995). [arXiv:astro-ph/9407054]
242. R.R. Caldwell, A phantom menace? Phys. Lett. B **545**, 23 (2002). [arXiv:astro-ph/9908168]
243. C.M. Will, The confrontation between general relativity and experiment. Living Rev. Relat. **17**, 4 (2014)
244. B. Bertotti, L. Iess, P. Tortora, A test of general relativity using radio links with the Cassini spacecraft. Nature **425**, 374 (2003)
245. T. Damour, K. Nordtvedt, General relativity as a cosmological attractor of tensor scalar theories. Phys. Rev. Lett. **70**, 2217 (1993)

246. T. Damour, K. Nordtvedt, Tensor-scalar cosmological models and their relaxation toward general relativity. Phys. Rev. D **48**, 3436 (1993)
247. T. Clifton, D.F. Mota, J.D. Barrow, Inhomogeneous gravity. Mon. Not. R. Astron. Soc. **358**, 601 (2005). [arXiv:gr-qc/0406001]
248. J. Khoury, A. Weltman, Chameleon fields: awaiting surprises for tests of gravity in space. Phys. Rev. Lett. **93**, 171104 (2004). [arXiv:astro-ph/0309300]
249. J. Khoury, A. Weltman, Chameleon cosmology. Phys. Rev. D **69**, 044026 (2004). [arXiv:astro-ph/0309411]
250. S. Capozziello, Curvature quintessence. Int. J. Mod. Phys. D **11**, 483 (2002). [arXiv:gr-qc/0201033]
251. S. Capozziello, S. Carloni, A. Troisi, Quintessence without scalar fields. Recent Res. Dev. Astron. Astrophys. **1**, 625 (2003). [arXiv:astro-ph/0303041]
252. R. Utiyama, B.S. DeWitt, Renormalization of a classical gravitational field interacting with quantized matter fields. J. Math. Phys. **3**, 608 (1962)
253. A.D. Sakharov, Vacuum quantum fluctuations in curved space and the theory of gravitation. Dokl. Akad. Nauk SSSR **177**, 70 (1967) [Gen. Relat. Grav. **32**, 365 (2000)]
254. T.V. Ruzmaikina, A.A. Ruzmaikin, Quadratic corrections to the Lagrangian density of the gravitational field and the singularity. Zh. Eksp. Teor. Fiz. **57**, 680 (1969) [Sov. Phys. JETP **30**, 372 (1970)]
255. P.C.W. Davies, S.A. Fulling, S.M. Christensen, T.S. Bunch, Energy-momentum tensor of a massless scalar quantum field in a Robertson–Walker universe. Ann. Phys. (N.Y.) **109**, 108 (1977)
256. B.S. DeWitt, Quantum theory of gravity. II. The manifestly covariant theory. Phys. Rev. **162**, 1195 (1967)
257. K.S. Stelle, Renormalization of higher-derivative quantum gravity. Phys. Rev. D **16**, 953 (1977)
258. K.S. Stelle, Classical gravity with higher derivatives. Gen. Relat. Grav. **9**, 353 (1978)
259. J. Julve, M. Tonin, Quantum gravity with higher derivative terms. Nuovo Cim. B **46**, 137 (1978)
260. A. Salam, J.A. Strathdee, Remarks on high-energy stability and renormalizability of gravity theory. Phys. Rev. D **18**, 4480 (1978)
261. E.S. Fradkin, A.A. Tseytlin, Renormalizable asymptotically free quantum theory of gravity. Nucl. Phys. B **201**, 469 (1982)
262. D.G. Boulware, G.T. Horowitz, A. Strominger, Zero-energy theorem for scale-invariant gravity. Phys. Rev. Lett. **50**, 1726 (1983)
263. N.H. Barth, S.M. Christensen, Quantizing fourth-order gravity theories: the functional integral. Phys. Rev. D **28**, 1876 (1983)
264. I.L. Buchbinder, S.D. Odintsov, I.L. Shapiro, *Effective Action in Quantum Gravity* (IOP, Bristol, 1992)
265. M. Asorey, J.L. López, I.L. Shapiro, Some remarks on high derivative quantum gravity. Int. J. Mod. Phys. A **12**, 5711 (1997). [arXiv:hep-th/9610006]
266. F.d.O. Salles, I.L. Shapiro, Do we have unitary and (super)renormalizable quantum gravity below Planck scale? Phys. Rev. D **89**, 084054 (2014). [arXiv:1401.4583]
267. A. Hindawi, B.A. Ovrut, D. Waldram, Consistent spin-two coupling and quadratic gravitation. Phys. Rev. D **53**, 5583 (1996). [arXiv:hep-th/9509142]
268. A. Hindawi, B.A. Ovrut, D. Waldram, Nontrivial vacua in higher derivative gravitation. Phys. Rev. D **53**, 5597 (1996). [arXiv:hep-th/9509147]
269. T. Chiba, Generalized gravity and ghost. JCAP **0503**, 008 (2005). [arXiv:gr-qc/0502070]
270. A. Núñez, S. Solganik, Ghost constraints on modified gravity. Phys. Lett. B **608**, 189 (2005). [arXiv:hep-th/0411102]
271. S.M. Carroll, A. De Felice, V. Duvvuri, D.A. Easson, M. Trodden, M.S. Turner, Cosmology of generalized modified gravity models. Phys. Rev. D **71**, 063513 (2005). [arXiv:astro-ph/0410031]

272. A. De Felice, M. Hindmarsh, M. Trodden, Ghosts, instabilities, and superluminal propagation in modified gravity models. JCAP **0608**, 005 (2006). [arXiv:astro-ph/0604154]
273. S. Kawai, M.-a. Sakagami, J. Soda, Instability of 1-loop superstring cosmology. Phys. Lett. B **437**, 284 (1998). [arXiv:gr-qc/9802033]
274. S. Kawai, J. Soda, Evolution of fluctuations during graceful exit in string cosmology. Phys. Lett. B **460**, 41 (1999). [arXiv:gr-qc/9903017]
275. G. Calcagni, B. de Carlos, A. De Felice, Ghost conditions for Gauss–Bonnet cosmologies. Nucl. Phys. B **752**, 404 (2006). [arXiv:hep-th/0604201]
276. T. Koivisto, D.F. Mota, Cosmology and astrophysical constraints of Gauss–Bonnet dark energy. Phys. Lett. B **644**, 104 (2007). [arXiv:astro-ph/0606078]
277. T. Koivisto, D.F. Mota, Gauss–Bonnet quintessence: background evolution, large scale structure and cosmological constraints. Phys. Rev. D **75**, 023518 (2007). [arXiv:hep-th/0609155]
278. H. Nariai, On the removal of initial singularity in a big-bang universe in terms of a renormalized theory of gravitation. I. Examination of the present status and a new approach. Prog. Theor. Phys. **46**, 433 (1971)
279. H. Nariai, K. Tomita, On the removal of initial singularity in a big-bang universe in terms of a renormalized theory of gravitation. II. Criteria for obtaining a physically reasonable model. Prog. Theor. Phys. **46**, 776 (1971)
280. P.C.W. Davies, Singularity avoidance and quantum conformal anomalies. Phys. Lett. B **68**, 402 (1977)
281. M.V. Fischetti, J.B. Hartle, B.L. Hu, Quantum effects in the early universe. I. Influence of trace anomalies on homogeneous, isotropic, classical geometries. Phys. Rev. D **20**, 1757 (1979)
282. K. Tomita, T. Azuma, H. Nariai, On anisotropic and homogeneous cosmological models in the renormalized theory of gravitation. Prog. Theor. Phys. **60**, 403 (1978)
283. J.D. Barrow, A.C. Ottewill, The stability of general relativistic cosmological theory. J. Phys. A **16**, 2757 (1983)
284. S.W. Hawking, R. Penrose, The singularities of gravitational collapse and cosmology. Proc. R. Soc. Lond. A **314**, 529 (1970)
285. A.A. Starobinsky, A new type of isotropic cosmological models without singularity. Phys. Lett. B **91**, 99 (1980)
286. S. Gottlöber, H.-J. Schmidt, A.A. Starobinsky, Sixth-order gravity and conformal transformations. Class. Quantum Grav. **7**, 893 (1990)
287. H.-J. Schmidt, Variational derivatives of arbitrarily high order and multi-inflation cosmological models. Class. Quantum Grav. **7**, 1023 (1990)
288. D. Wands, Extended gravity theories and the Einstein–Hilbert action. Class. Quantum Grav. **11**, 269 (1994). [arXiv:gr-qc/9307034]
289. S. Cecotti, Higher derivative supergravity is equivalent to standard supergravity coupled to matter. Phys. Lett. B **190**, 86 (1987)
290. S. Cecotti, S. Ferrara, M. Porrati, S. Sabharwal, New minimal higher derivative supergravity coupled to matter. Nucl. Phys. B **306**, 160 (1988)
291. B.N. Breizman, V.Ts. Gurovich, V.P. Sokolov, The possibility of setting up regular cosmological solutions. Zh. Eksp. Teor. Fiz. **59**, 288 (1970) [Sov. Phys. JETP **32**, 155 (1971)]
292. H.A. Buchdahl, Non-linear Lagrangians and cosmological theory. Mon. Not. R. Astron. Soc. **150**, 1 (1970)
293. H. Nariai, On a phenomenological modification of Einstein's gravitational Lagrangian. Prog. Theor. Phys. **51**, 613 (1974)
294. P. Teyssandier, P. Tourrenc, The Cauchy problem for the $R + R^2$ theories of gravity without torsion. J. Math. Phys. **24**, 2793 (1983)
295. G. Magnano, M. Ferraris, M. Francaviglia, Nonlinear gravitational Lagrangians. Gen. Relat. Grav. **19**, 465 (1987)
296. H.-J. Schmidt, Comparing self-interacting scalar fields and $R + R^3$ cosmological models. Astron. Nachr. **308**, 183 (1987). [arXiv:gr-qc/0106035]
297. H.A. Buchdahl, Quadratic Lagrangians and Palatini's device. J. Phys. A **12**, 1229 (1979)

298. G. Allemandi, A. Borowiec, M. Francaviglia, Accelerated cosmological models in Ricci squared gravity. Phys. Rev. D **70**, 103503 (2004). [arXiv:hep-th/0407090]
299. A. Borowiec, M. Ferraris, M. Francaviglia, I. Volovich, Universality of Einstein equations for the Ricci squared Lagrangians. Class. Quantum Grav. **15**, 43 (1998). [arXiv:gr-qc/9611067]
300. A. Núñez, S. Solganik, The content of $f(R)$ gravity. arXiv:hep-th/0403159
301. V. Faraoni, Solar system experiments do not yet veto modified gravity models. Phys. Rev. D **74**, 023529 (2006). [arXiv:gr-qc/0607016]
302. Y.-S. Song, W. Hu, I. Sawicki, Large scale structure of $f(R)$ gravity. Phys. Rev. D **75**, 044004 (2007). [arXiv:astro-ph/0610532]
303. K.-i. Maeda, Inflation as a transient attractor in R^2 cosmology. Phys. Rev. D **37**, 858 (1988)
304. K.-i. Maeda, Towards the Einstein–Hilbert action via conformal transformation. Phys. Rev. D **39**, 3159 (1989)
305. J.D. Barrow, S. Cotsakis, Inflation and the conformal structure of higher-order gravity theories. Phys. Lett. B **214**, 515 (1988)
306. T. Koivisto, H. Kurki-Suonio, Cosmological perturbations in the Palatini formulation of modified gravity. Class. Quantum Grav. **23**, 2355 (2006). [arXiv:astro-ph/0509422]
307. M. Ferraris, M. Francaviglia, I. Volovich, The universality of vacuum Einstein equations with cosmological constant. Class. Quantum Grav. **11**, 1505 (1994). [arXiv:gr-qc/9303007]
308. G.J. Olmo, The gravity Lagrangian according to solar system experiments. Phys. Rev. Lett. **95**, 261102 (2005). [arXiv:gr-qc/0505101]
309. T.P. Sotiriou, $f(R)$ gravity and scalar-tensor theory. Class. Quantum Grav. **23**, 5117 (2006). [arXiv:gr-qc/0604028]
310. B. Li, J.D. Barrow, Cosmology of $f(R)$ gravity in metric variational approach. Phys. Rev. D **75**, 084010 (2007). [arXiv:gr-qc/0701111]
311. I. Navarro, K. Van Acoleyen, $f(R)$ actions, cosmic acceleration and local tests of gravity. JCAP **0702**, 022 (2007). [arXiv:gr-qc/0611127]
312. T. Faulkner, M. Tegmark, E.F. Bunn, Y. Mao, Constraining $f(R)$ gravity as a scalar-tensor theory. Phys. Rev. D **76**, 063505 (2007). [arXiv:astro-ph/0612569]
313. W. Hu, I. Sawicki, Models of $f(R)$ cosmic acceleration that evade solar system tests. Phys. Rev. D **76**, 064004 (2007). [arXiv:0705.1158]
314. S. Capozziello, S. Tsujikawa, Solar system and equivalence principle constraints on $f(R)$ gravity by the chameleon approach. Phys. Rev. D **77**, 107501 (2008). [arXiv:0712.2268]
315. I.W. Roxburgh, Nonlinear Lagrangian theories of gravity. Gen. Relat. Grav. **8**, 219 (1977)
316. T. Clifton, J.D. Barrow, The power of general relativity. Phys. Rev. D **72**, 103005 (2005). [arXiv:gr-qc/0509059]
317. H. Nariai, Gravitational instability of regular model-universes in a modified theory of general relativity. Prog. Theor. Phys. **49**, 165 (1973)
318. P. Zhang, Testing gravity against the early time integrated Sachs–Wolfe effect. Phys. Rev. D **73**, 123504 (2006). [arXiv:astro-ph/0511218]
319. G. Cognola, E. Elizalde, S. Nojiri, S.D. Odintsov, L. Sebastiani, S. Zerbini, Class of viable modified $f(R)$ gravities describing inflation and the onset of accelerated expansion. Phys. Rev. D **77**, 046009 (2008). [arXiv:0712.4017]
320. G. Allemandi, A. Borowiec, M. Francaviglia, Accelerated cosmological models in first-order nonlinear gravity. Phys. Rev. D **70**, 043524 (2004). [arXiv:hep-th/0403264]
321. S. Nojiri, S.D. Odintsov, Modified gravity with $\ln R$ terms and cosmic acceleration. Gen. Relat. Grav. **36**, 1765 (2004). [arXiv:hep-th/0308176]
322. L.H. Ford, D.J. Toms, Dynamical symmetry breaking due to radiative corrections in cosmology. Phys. Rev. D **25**, 1510 (1982)
323. A.A. Starobinsky, H.-J. Schmidt, On a general vacuum solution of fourth-order gravity. Class. Quantum Grav. **4**, 695 (1987)
324. K.-i. Maeda, J.A. Stein-Schabes, T. Futamase, Inflation in a renormalizable cosmological model and the cosmic no-hair conjecture. Phys. Rev. D **39**, 2848 (1989)
325. S. Cotsakis, P.J. Saich, Power-law inflation and conformal transformations. Class. Quantum Grav. **11**, 383 (1994)

326. H. Weyl, *Space, Time, and Matter* (Dover, Mineola, 1952)
327. B. Whitt, Fourth-order gravity as general relativity plus matter. Phys. Lett. B **145**, 176 (1984)
328. S. Capozziello, F. Occhionero, L. Amendola, The phase-space view of inflation II: fourth order models. Int. J. Mod. Phys. D **1**, 615 (1993)
329. S. Capozziello, S. Nojiri, S.D. Odintsov, A. Troisi, Cosmological viability of $f(R)$-gravity as an ideal fluid and its compatibility with a matter dominated phase. Phys. Lett. B **639**, 135 (2006). [arXiv:astro-ph/0604431]
330. S.M. Carroll, V. Duvvuri, M. Trodden, M.S. Turner, Is cosmic speed-up due to new gravitational physics? Phys. Rev. D **70**, 043528 (2004). [arXiv:astro-ph/0306438]
331. T. Chiba, $1/R$ gravity and scalar-tensor gravity. Phys. Lett. B **575**, 1 (2003). [arXiv:astro-ph/0307338]
332. G.A. Vilkovisky, Effective action in quantum gravity. Class. Quantum Grav. **9**, 895 (1992)
333. A.D. Dolgov, M. Kawasaki, Can modified gravity explain accelerated cosmic expansion? Phys. Lett. B **573**, 1 (2003). [arXiv:astro-ph/0307285]
334. G.J. Olmo, Post-Newtonian constraints on $f(R)$ cosmologies in metric and Palatini formalism. Phys. Rev. D **72**, 083505 (2005). [arXiv:gr-qc/0505135]
335. L. Amendola, D. Polarski, S. Tsujikawa, Are $f(R)$ dark energy models cosmologically viable? Phys. Rev. Lett. **98**, 131302 (2007). [arXiv:astro-ph/0603703]
336. L. Amendola, D. Polarski, S. Tsujikawa, Power-laws $f(R)$ theories are cosmologically unacceptable. Int. J. Mod. Phys. D **16**, 1555 (2007). [arXiv:astro-ph/0605384]
337. D.N. Vollick, $1/R$ curvature corrections as the source of the cosmological acceleration. Phys. Rev. D **68**, 063510 (2003). [arXiv:astro-ph/0306630]
338. X. Meng, P. Wang, Modified Friedmann equations in R^{-1}-modified gravity. Class. Quantum Grav. **20**, 4949 (2003). [arXiv:astro-ph/0307354]
339. X. Meng, P. Wang, Cosmological evolution in $1/R$-gravity theory. Class. Quantum Grav. **21**, 951 (2004). [arXiv:astro-ph/0308031]
340. M. Amarzguioui, Ø. Elgarøy, D.F. Mota, T. Multamäki, Cosmological constraints on $f(R)$ gravity theories within the Palatini approach. Astron. Astrophys. **454**, 707 (2006). [arXiv:astro-ph/0510519]
341. S. Nojiri, S.D. Odintsov, Modified gravity with negative and positive powers of the curvature: unification of the inflation and of the cosmic acceleration. Phys. Rev. D **68**, 123512 (2003). [arXiv:hep-th/0307288]
342. A.W. Brookfield, C. van de Bruck, L.M.H. Hall, Viability of $f(R)$ theories with additional powers of curvature. Phys. Rev. D **74**, 064028 (2006). [arXiv:hep-th/0608015]
343. Á. de la Cruz-Dombriz, P.K.S. Dunsby, S. Kandhai, D. Sáez-Gómez, Theoretical and observational constraints of viable $f(R)$ theories of gravity. Phys. Rev. D **93**, 084016 (2016). [arXiv:1511.00102]
344. G. Cognola, E. Elizalde, S. Nojiri, S.D. Odintsov, S. Zerbini, One-loop $f(R)$ gravity in de Sitter universe. JCAP **0502**, 010 (2005). [arXiv:hep-th/0501096]
345. J.A.R. Cembranos, The Newtonian limit at intermediate energies. Phys. Rev. D **73**, 064029 (2006). [arXiv:gr-qc/0507039]
346. T. Koivisto, Matter power spectrum in $f(R)$ gravity. Phys. Rev. D **73**, 083517 (2006). [arXiv:astro-ph/0602031]
347. S. Capozziello, V.F. Cardone, A. Troisi, Dark energy and dark matter as curvature effects. JCAP **0608**, 001 (2006). [arXiv:astro-ph/0602349]
348. S. Nojiri, S.D. Odintsov, M. Sami, Dark energy cosmology from higher-order, string-inspired gravity and its reconstruction. Phys. Rev. D **74**, 046004 (2006). [arXiv:hep-th/0605039]
349. S. Nojiri, S.D. Odintsov, Modified $f(R)$ gravity consistent with realistic cosmology: from matter dominated epoch to dark energy universe. Phys. Rev. D **74**, 086005 (2006). [arXiv:hep-th/0608008]
350. S. Nojiri, S.D. Odintsov, Modified gravity and its reconstruction from the universe expansion history. J. Phys. Conf. Ser. **66**, 012005 (2007). [arXiv:hep-th/0611071]
351. L. Amendola, R. Gannouji, D. Polarski, S. Tsujikawa, Conditions for the cosmological viability of $f(R)$ dark energy models. Phys. Rev. D **75**, 083504 (2007). [arXiv:gr-qc/0612180]

352. I. Sawicki, W. Hu, Stability of cosmological solutions in $f(R)$ models of gravity. Phys. Rev. D **75**, 127502 (2007). [arXiv:astro-ph/0702278]
353. S. Fay, S. Nesseris, L. Perivolaropoulos, Can $f(R)$ modified gravity theories mimic a ΛCDM cosmology? Phys. Rev. D **76**, 063504 (2007). [arXiv:gr-qc/0703006]
354. S.A. Appleby, R.A. Battye, Do consistent $F(R)$ models mimic general relativity plus Λ? Phys. Lett. B **654**, 7 (2007). [arXiv:0705.3199]
355. A.A. Starobinsky, Disappearing cosmological constant in $f(R)$ gravity. JETP Lett. **86**, 157 (2007). [arXiv:0706.2041]
356. S. Tsujikawa, Observational signatures of $f(R)$ dark energy models that satisfy cosmological and local gravity constraints. Phys. Rev. D **77**, 023507 (2008). [arXiv:0709.1391]
357. A. De Felice, S. Tsujikawa, $f(R)$ theories. Living Rev. Relat. **13**, 3 (2010).
358. S. Tsujikawa, Modified gravity models of dark energy. Lect. Notes Phys. **800**, 99 (2010). [arXiv:1101.0191]
359. T. Clifton, P.G. Ferreira, A. Padilla, C. Skordis, Modified gravity and cosmology. Phys. Rep. **513**, 1 (2012). [arXiv:1106.2476]
360. J.-c. Hwang, Cosmological perturbations in generalized gravity theories: formulation. Class. Quantum Grav. **7**, 1613 (1990)
361. J.-c. Hwang, Perturbations of the Robertson–Walker space: multicomponent sources and generalized gravity. Astrophys. J. **375**, 443 (1991)
362. J.-c. Hwang, H. Noh, Cosmological perturbations in generalized gravity theories. Phys. Rev. D **54**, 1460 (1996)
363. J.-c. Hwang, H. Noh, Gauge-ready formulation of the cosmological kinetic theory in generalized gravity theories. Phys. Rev. D **65**, 023512 (2002). [arXiv:astro-ph/0102005]
364. R. Bean, D. Bernat, L. Pogosian, A. Silvestri, M. Trodden, Dynamics of linear perturbations in $f(R)$ gravity. Phys. Rev. D **75**, 064020 (2007). [arXiv:astro-ph/0611321]
365. L. Pogosian, A. Silvestri, Pattern of growth in viable $f(R)$ cosmologies. Phys. Rev. D **77**, 023503 (2008); Erratum-ibid. D **81**, 049901 (2010). [arXiv:0709.0296]
366. S. Tsujikawa, K. Uddin, R. Tavakol, Density perturbations in $f(R)$ gravity theories in metric and Palatini formalisms. Phys. Rev. D **77**, 043007 (2008). [arXiv:0712.0082]
367. P.A.R. Ade et al. [Planck Collaboration], Planck 2015 results. XIV. Dark energy and modified gravity. Astron. Astrophys. **594**, A14 (2016). [arXiv:1502.01590]
368. G.W. Horndeski, Second-order scalar-tensor field equations in a four-dimensional space. Int. J. Theor. Phys. **10**, 363 (1974)
369. C. Charmousis, E.J. Copeland, A. Padilla, P.M. Saffin, General second-order scalar-tensor theory and self-tuning. Phys. Rev. Lett. **108**, 051101 (2012). [arXiv:1106.2000]
370. R. Kase, S. Tsujikawa, Cosmology in generalized Horndeski theories with second-order equations of motion. Phys. Rev. D **90**, 044073 (2014). [arXiv:1407.0794]
371. T. Chiba, T. Okabe, M. Yamaguchi, Kinetically driven quintessence. Phys. Rev. D **62**, 023511 (2000). [arXiv:astro-ph/9912463]
372. C. Armendáriz-Picón, V.F. Mukhanov, P.J. Steinhardt, Dynamical solution to the problem of a small cosmological constant and late-time cosmic acceleration. Phys. Rev. Lett. **85**, 4438 (2000). [arXiv:astro-ph/0004134]
373. C. Armendáriz-Picón, V.F. Mukhanov, P.J. Steinhardt, Essentials of k-essence. Phys. Rev. D **63**, 103510 (2001). [arXiv:astro-ph/0006373]
374. A. Melchiorri, L. Mersini, C.J. Ödman, M. Trodden, The state of the dark energy equation of state. Phys. Rev. D **68**, 043509 (2003). [arXiv:astro-ph/0211522]
375. S. Tsujikawa, M. Sami, A unified approach to scaling solutions in a general cosmological background. Phys. Lett. B **603**, 113 (2004). [arXiv:hep-th/0409212]
376. D.B. Fairlie, J. Govaerts, Universal field equations with reparametrization invariance. Phys. Lett. B **281**, 49 (1992). [arXiv:hep-th/9202056]
377. A. Nicolis, R. Rattazzi, E. Trincherini, Galileon as a local modification of gravity. Phys. Rev. D **79**, 064036 (2009). [arXiv:0811.2197]
378. C. Deffayet, X. Gao, D.A. Steer, G. Zahariade, From *k*-essence to generalised Galileons. Phys. Rev. D **84**, 064039 (2011). [arXiv:1103.3260]

379. T. Kobayashi, M. Yamaguchi, J.'i. Yokoyama, Generalized G-inflation: inflation with the most general second-order field equations. Prog. Theor. Phys. **126**, 511 (2011). [arXiv:1105.5723]
380. J. Gleyzes, D. Langlois, F. Piazza, F. Vernizzi, Healthy theories beyond Horndeski. Phys. Rev. Lett. **114**, 211101 (2015). [arXiv:1404.6495]
381. C. Lin, S. Mukohyama, R. Namba, R. Saitou, Hamiltonian structure of scalar-tensor theories beyond Horndeski. JCAP **1410**, 071 (2014). [arXiv:1408.0670]
382. J. Gleyzes, D. Langlois, F. Piazza, F. Vernizzi, Exploring gravitational theories beyond Horndeski. JCAP **1502**, 018 (2015). [arXiv:1408.1952]
383. X. Gao, Hamiltonian analysis of spatially covariant gravity. Phys. Rev. D **90**, 104033 (2014). [arXiv:1409.6708]
384. A. De Felice, S. Tsujikawa, Inflationary gravitational waves in the effective field theory of modified gravity. Phys. Rev. D **91**, 103506 (2015). [arXiv:1411.0736]
385. S. Tsujikawa, Possibility of realizing weak gravity in redshift space distortion measurements. Phys. Rev. D **92**, 044029 (2015). [arXiv:1505.02459]
386. F. Arroja, N. Bartolo, P. Karmakar, S. Matarrese, The two faces of mimetic Horndeski gravity: disformal transformations and Lagrange multiplier. JCAP **1509**, 051 (2015). [arXiv:1506.08575]
387. F. Arroja, N. Bartolo, P. Karmakar, S. Matarrese, Cosmological perturbations in mimetic Horndeski gravity. JCAP **1604**, 042 (2016). [arXiv:1512.09374]
388. J.D. Bekenstein, Fine structure constant: is it really a constant? Phys. Rev. D **25**, 1527 (1982)
389. J.D. Bekenstein, Fine structure constant variability, equivalence principle and cosmology. Phys. Rev. D **66**, 123514 (2002). [arXiv:gr-qc/0208081]
390. J. Magueijo, New varying speed of light theories. Rep. Prog. Phys. **66**, 2025 (2003). [arXiv:astro-ph/0305457]
391. J. Magueijo, Covariant and locally Lorentz invariant varying speed of light theories. Phys. Rev. D **62**, 103521 (2000). [arXiv:gr-qc/0007036]
392. J. Magueijo, Speedy sound and cosmic structure. Phys. Rev. Lett. **100**, 231302 (2008). [arXiv:0803.0859]
393. A. Albrecht, J. Magueijo, A time varying speed of light as a solution to cosmological puzzles. Phys. Rev. D **59**, 043516 (1999). [arXiv:astro-ph/9811018]
394. J.D. Barrow, J. Magueijo, Solutions to the quasi-flatness and quasilambda problems. Phys. Lett. B **447**, 246 (1999). [arXiv:astro-ph/9811073]
395. B.A. Bassett, S. Liberati, C. Molina-París, M. Visser, Geometrodynamics of variable-speed-of-light cosmologies. Phys. Rev. D **62**, 103518 (2000). [arXiv:astro-ph/0001441]
396. K. Tomita, Bulk flows and CMB dipole anisotropy in cosmological void models. Astrophys. J. **529**, 26 (2000). [arXiv:astro-ph/9905278]
397. K. Tomita, Distances and lensing in cosmological void models. Astrophys. J. **529**, 38 (2000). [arXiv:astro-ph/9906027]
398. M.-N. Célérier, Do we really see a cosmological constant in the supernovae data? Astron. Astrophys. **353**, 63 (2000). [arXiv:astro-ph/9907206]
399. K. Tomita, A local void and the accelerating universe. Mon. Not. R. Astron. Soc. **326**, 287 (2001). [arXiv:astro-ph/0011484]
400. K. Tomita, Analyses of type Ia supernova data in cosmological models with a local void. Prog. Theor. Phys. **106**, 929 (2001). [arXiv:astro-ph/0104141]
401. M.-N. Célérier, The accelerated expansion of the universe challenged by an effect of the inhomogeneities. A review. New Adv. Phys. **1**, 29 (2007). [arXiv:astro-ph/0702416]
402. G. Lemaître, The expanding universe. Ann. Soc. Sci. Bruxelles Ser. I Sci. Math. Astron. Phys. A **53**, 51 (1933) [Gen. Relat. Grav. **29**, 641 (1997)]
403. R.C. Tolman, Effect of imhomogeneity on cosmological models. Proc. Natl. Acad. Sci. **20**, 169 (1934) [Gen. Relat. Grav. **29**, 935 (1997)]
404. H. Bondi, Spherically symmetrical models in general relativity. Mon. Not. R. Astron. Soc. **107**, 410 (1947)
405. M. Hossein Partovi, B. Mashhoon, Toward verification of large-scale homogeneity in cosmology. Astrophys. J. **276**, 4 (1984)

406. N.P. Humphreys, R. Maartens, D.R. Matravers, Anisotropic observations in universes with nonlinear inhomogeneity. Astrophys. J. **477**, 47 (1997). [arXiv:astro-ph/9602033]
407. N. Mustapha, C. Hellaby, G.F.R. Ellis, Large-scale inhomogeneity versus source evolution: can we distinguish them observationally? Mon. Not. R. Astron. Soc. **292**, 817 (1997). [arXiv:gr-qc/9808079]
408. S. Khakshournia, R. Mansouri, Dynamics of general relativistic spherically symmetric dust thick shells. Gen. Relat. Grav. **34**, 1847 (2002). [arXiv:gr-qc/0308025]
409. T. Biswas, A. Notari, 'Swiss-cheese' inhomogeneous cosmology and the dark energy problem. JCAP **0806**, 021 (2008). [arXiv:astro-ph/0702555]
410. J.W. Moffat, Cosmic microwave background, accelerating universe and inhomogeneous cosmology. JCAP **0510**, 012 (2005). [arXiv:astro-ph/0502110]
411. J.W. Moffat, Late-time inhomogeneity and acceleration without dark energy. JCAP **0605**, 001 (2006). [arXiv:astro-ph/0505326]
412. H. Alnes, M. Amarzguioui, O. Grøn, Inhomogeneous alternative to dark energy? Phys. Rev. D **73**, 083519 (2006). [arXiv:astro-ph/0512006]
413. D.J.H. Chung, A.E. Romano, Mapping luminosity-redshift relationship to Lemaître–Tolman–Bondi cosmology. Phys. Rev. D **74**, 103507 (2006). [arXiv:astro-ph/0608403]
414. K. Enqvist, T. Mattsson, The effect of inhomogeneous expansion on the supernova observations. JCAP **0702**, 019 (2007). [arXiv:astro-ph/0609120]
415. K. Enqvist, Lemaître–Tolman–Bondi model and accelerating expansion. Gen. Relat. Grav. **40**, 451 (2008). [arXiv:0709.2044]
416. S. Alexander, T. Biswas, A. Notari, D. Vaid, Local void vs dark energy: confrontation with WMAP and type Ia supernovae. JCAP **0909**, 025 (2009). [arXiv:0712.0370]
417. J. García-Bellido, T. Haugbølle, Confronting Lemaître–Tolman–Bondi models with observational cosmology. JCAP **0804**, 003 (2008). [arXiv:0802.1523]
418. J. García-Bellido, T. Haugbølle, Looking the void in the eyes—the kinematic Sunyaev–Zeldovich effect in Lemaître–Tolman–Bondi models. JCAP **0809**, 016 (2008). [arXiv:0807.1326]
419. J.P. Zibin, A. Moss, D. Scott, Can we avoid dark energy? Phys. Rev. Lett. **101**, 251303 (2008). [arXiv:0809.3761]
420. S. February, J. Larena, M. Smith, C. Clarkson, Rendering dark energy void. Mon. Not. R. Astron. Soc. **405**, 2231 (2010). [arXiv:0909.1479]
421. T. Biswas, A. Notari, W. Valkenburg, Testing the void against cosmological data: fitting CMB, BAO, SN and H_0. JCAP **1011**, 030 (2010). [arXiv:1007.3065]
422. A. Moss, J.P. Zibin, D. Scott, Precision cosmology defeats void models for acceleration. Phys. Rev. D **83**, 103515 (2011). [arXiv:1007.3725]
423. M. Zumalacárregui, J. García-Bellido, P. Ruiz-Lapuente, Tension in the void: cosmic rulers strain inhomogeneous cosmologies. JCAP **1210**, 009 (2012). [arXiv:1201.2790]
424. R. de Putter, L. Verde, R. Jimenez, Testing LTB void models without the cosmic microwave background or large scale structure: new constraints from galaxy ages. JCAP **1302**, 047 (2013). [arXiv:1208.4534]
425. H. Goto, H. Kodama, The gravitational lensing effect on the CMB polarisation anisotropy in the Λ-LTB model. Prog. Theor. Phys. **125**, 815 (2011). [arXiv:1101.0476]
426. P. Hunt, S. Sarkar, Constraints on large scale inhomogeneities from WMAP-5 and SDSS: confrontation with recent observations. Mon. Not. R. Astron. Soc. **401**, 547 (2010). [arXiv:0807.4508]
427. P.A.R. Ade et al. [Planck Collaboration], Planck intermediate results. XIII. Constraints on peculiar velocities. Astron. Astrophys. **561**, A97 (2014). [arXiv:1303.5090]
428. M. Henneaux, C. Teitelboim, The cosmological constant and general covariance. Phys. Lett. B **222**, 195 (1989)
429. J.L. Anderson, D. Finkelstein, Cosmological constant and fundamental length. Am. J. Phys. **39**, 901 (1971)
430. J. Rayski, The problems of quantum gravity. Gen. Relat. Grav. **11**, 19 (1979)

431. J.J. van der Bij, H. van Dam, Y.J. Ng, The exchange of massless spin-two particles. Physica A **116**, 307 (1982)
432. A. Zee, Remarks on the cosmological constant paradox, in *High Energy Physics: Proceedings of the 20th Orbis Scientiae, 1983*, ed. by S.L. Mintz, A. Perlmutter (Plenum, New York, 1985)
433. W. Buchmüller, N. Dragon, Einstein gravity from restricted coordinate invariance. Phys. Lett. B **207**, 292 (1988)
434. W. Buchmüller, N. Dragon, Gauge fixing and the cosmological constant. Phys. Lett. B **223**, 313 (1989)
435. W.G. Unruh, Unimodular theory of canonical quantum gravity. Phys. Rev. D **40**, 1048 (1989)
436. W.G. Unruh, R.M. Wald, Time and the interpretation of canonical quantum gravity. Phys. Rev. D **40**, 2598 (1989)
437. Y.J. Ng, H. van Dam, Possible solution to the cosmological constant problem. Phys. Rev. Lett. **65**, 1972 (1990)
438. Y.J. Ng, H. van Dam, Unimodular theory of gravity and the cosmological constant. J. Math. Phys. **32**, 1337 (1991)
439. A.N. Petrov, On the cosmological constant as a constant of integration. Mod. Phys. Lett. A **06**, 2107 (1991)
440. E. Álvarez, Can one tell Einstein's unimodular theory from Einstein's general relativity? JHEP **0503**, 002 (2005). [arXiv:hep-th/0501146]
441. E. Álvarez, D. Blas, J. Garriga, E. Verdaguer, Transverse Fierz–Pauli symmetry. Nucl. Phys. B **756**, 148 (2006). [arXiv:hep-th/0606019]
442. E. Álvarez, A.F. Faedo, J.J. López-Villarejo, Ultraviolet behavior of transverse gravity. JHEP **0810**, 023 (2008). [arXiv:0807.1293]
443. B. Fiol, J. Garriga, Semiclassical unimodular gravity. JCAP **1008**, 015 (2010). [arXiv:0809.1371]
444. L. Smolin, Quantization of unimodular gravity and the cosmological constant problems. Phys. Rev. D **80**, 084003 (2009). [arXiv:0904.4841]
445. E. Álvarez, R. Vidal, Weyl transverse gravity (WTDiff) and the cosmological constant. Phys. Rev. D **81**, 084057 (2010). [arXiv:1001.4458]
446. D. Blas, M. Shaposhnikov, D. Zenhäusern, Scale-invariant alternatives to general relativity. Phys. Rev. D **84**, 044001 (2011). [arXiv:1104.1392]
447. E. Álvarez, The weight of matter. JCAP **1207**, 002 (2012). [arXiv:1204.6162]
448. A. Eichhorn, On unimodular quantum gravity. Class. Quantum Grav. **30**, 115016 (2013). [arXiv:1301.0879]
449. C. Barceló, R. Carballo-Rubio, L.J. Garay, Unimodular gravity and general relativity from graviton self-interactions. Phys. Rev. D **89**, 124019 (2014). [arXiv:1401.2941]
450. C. Barceló, R. Carballo-Rubio, L.J. Garay, Absence of cosmological constant problem in special relativistic field theory of gravity. arXiv:1406.7713
451. A. Padilla, I.D. Saltas, A note on classical and quantum unimodular gravity. Eur. Phys. J. C **75**, 561 (2015). [arXiv:1409.3573]
452. R. Bufalo, M. Oksanen, A. Tureanu, How unimodular gravity theories differ from general relativity at quantum level. Eur. Phys. J. C **75**, 477 (2015). [arXiv:1505.04978]
453. A. Basak, O. Fabre, S. Shankaranarayanan, Cosmological perturbation of unimodular gravity and general relativity are identical. Gen. Relat. Grav. **48**, 123 (2016). [arXiv:1511.01805]
454. E.I. Guendelman, A.B. Kaganovich, Principle of nongravitating vacuum energy and some of its consequences. Phys. Rev. D **53**, 7020 (1996)
455. E.I. Guendelman, A.B. Kaganovich, Gravitational theory without the cosmological constant problem. Phys. Rev. D **55**, 5970 (1997). [arXiv:gr-qc/9611046]
456. E.I. Guendelman, Scale invariance, new inflation and decaying lambda terms. Mod. Phys. Lett. A **14**, 1043 (1999). [arXiv:gr-qc/9901017]
457. E.I. Guendelman, A.B. Kaganovich, Dynamical measure and field theory models free of the cosmological constant problem. Phys. Rev. D **60**, 065004 (1999). [arXiv:gr-qc/9905029]
458. M. Fierz, W. Pauli, On relativistic wave equations for particles of arbitrary spin in an electromagnetic field. Proc. R. Soc. Lond. A **173**, 211 (1939)

459. G.E. Volovik, Superfluid analogies of cosmological phenomena. Phys. Rep. **351**, 195 (2001). [arXiv:gr-qc/0005091]
460. C. Barceló, S. Liberati, M. Visser, Analogue gravity. Living Rev. Relat. **14**, 3 (2011)
461. G.E. Volovik, Vacuum energy and cosmological constant: view from condensed matter. J. Low Temp. Phys. **124**, 25 (2001). [arXiv:gr-qc/0101111]
462. G.E. Volovik, Cosmological constant and vacuum energy. Ann. Phys. (Berlin) **14**, 165 (2005). [arXiv:gr-qc/0405012]
463. G.E. Volovik, Vacuum energy: myths and reality. Int. J. Mod. Phys. D **15**, 1987 (2006). [arXiv:gr-qc/0604062]
464. G. Jannes, G.E. Volovik, The cosmological constant: a lesson from the effective gravity of topological Weyl media. JETP Lett. **96**, 215 (2012). [arXiv:1108.5086]
465. F.R. Klinkhamer, G.E. Volovik, Self-tuning vacuum variable and cosmological constant. Phys. Rev. D **77**, 085015 (2008). [arXiv:0711.3170]
466. F.R. Klinkhamer, G.E. Volovik, Dynamic vacuum variable and equilibrium approach in cosmology. Phys. Rev. D **78**, 063528 (2008). [arXiv:0806.2805]
467. F.R. Klinkhamer, G.E. Volovik, Gluonic vacuum, q-theory, and the cosmological constant. Phys. Rev. D **79**, 063527 (2009). [arXiv:0811.4347]
468. D.G. Caldi, A. Chodos, Cosmological neutrino condensates. arXiv:hep-ph/9903416
469. T. Inagaki, X. Meng, T. Murata, Dark energy problem in a four fermion interaction model. arXiv:hep-ph/0306010
470. F. Giacosa, R. Hofmann, M. Neubert, A model for the very early universe. JHEP **0802**, 077 (2008). [arXiv:0801.0197]
471. S. Alexander, T. Biswas, The cosmological BCS mechanism and the big bang singularity. Phys. Rev. D **80**, 023501 (2009). [arXiv:0807.4468]
472. S. Alexander, T. Biswas, G. Calcagni, Cosmological Bardeen–Cooper–Schrieffer condensate as dark energy. Phys. Rev. D **81**, 043511 (2010); Erratum-ibid. D **81**, 069902(E) (2010). [arXiv:0906.5161]
473. N.J. Popławski, Cosmological constant from quarks and torsion. Ann. Phys. (Berlin) **523**, 291 (2011). [arXiv:1005.0893]
474. J.M. Weller, Fermion condensate from torsion in the reheating era after inflation. Phys. Rev. D **88**, 083511 (2013). [arXiv:1307.2423]
475. S. Finazzi, S. Liberati, L. Sindoni, Cosmological constant: a lesson from Bose–Einstein condensates. Phys. Rev. Lett. **108**, 071101 (2012). [arXiv:1103.4841]
476. B.L. Hu, Can spacetime be a condensate? Int. J. Theor. Phys. **44**, 1785 (2005). [arXiv:gr-qc/0503067]
477. L. Sindoni, Emergent models for gravity: an overview of microscopic models. SIGMA **8**, 027 (2012). [arXiv:1110.0686]
478. H.C. Ohanian, Gravitons as Goldstone bosons. Phys. Rev. **184**, 1305 (1969)
479. D. Atkatz, Dynamical method for generating the gravitational interaction. Phys. Rev. D **17**, 1972 (1978)
480. S. Deser, Gravity from self-interaction redux. Gen. Relat. Grav. **42**, 641 (2010). [arXiv:0910.2975]
481. C. Barceló, S. Liberati, M. Visser, Analog gravity from field theory normal modes? Class. Quantum Grav. **18**, 3595 (2001). [arXiv:gr-qc/0104001]
482. C. Barceló, M. Visser, S. Liberati, Einstein gravity as an emergent phenomenon? Int. J. Mod. Phys. D **10**, 799 (2001). [arXiv:gr-qc/0106002]
483. P.C.W. Davies, Scalar particle production in Schwarzschild and Rindler metrics. J. Phys. A **8**, 609 (1975)
484. W.G. Unruh, Notes on black hole evaporation. Phys. Rev. D **14**, 870 (1976)
485. G.W. Gibbons, S.W. Hawking, Action integrals and partition functions in quantum gravity. Phys. Rev. D **15**, 2752 (1977)
486. J.D. Bekenstein, Universal upper bound on the entropy-to-energy ratio for bounded systems. Phys. Rev. D **23**, 287 (1981)

487. T. Jacobson, Thermodynamics of space-time: the Einstein equation of state. Phys. Rev. Lett. **75**, 1260 (1995). [arXiv:gr-qc/9504004]
488. T. Padmanabhan, Gravity and the thermodynamics of horizons. Phys. Rep. **406**, 49 (2005). [arXiv:gr-qc/0311036]
489. C. Eling, R. Guedens, T. Jacobson, Non-equilibrium thermodynamics of spacetime. Phys. Rev. Lett. **96**, 121301 (2006). [arXiv:gr-qc/0602001]
490. T. Padmanabhan, A. Paranjape, Entropy of null surfaces and dynamics of spacetime. Phys. Rev. D **75**, 064004 (2007). [arXiv:gr-qc/0701003]
491. D. Kothawala, T. Padmanabhan, Thermodynamic structure of Lanczos–Lovelock field equations from near-horizon symmetries. Phys. Rev. D **79**, 104020 (2009). [arXiv:0904.0215]
492. T. Padmanabhan, A physical interpretation of gravitational field equations. AIP Conf. Proc. **1241**, 93 (2010). [arXiv:0911.1403]
493. T. Padmanabhan, D. Kothawala, Lanczos–Lovelock models of gravity. Phys. Rep. **531**, 115 (2013). [arXiv:1302.2151]
494. K. Parattu, B.R. Majhi, T. Padmanabhan, Structure of the gravitational action and its relation with horizon thermodynamics and emergent gravity paradigm. Phys. Rev. D **87**, 124011 (2013). [arXiv:1303.1535]
495. T. Padmanabhan, General relativity from a thermodynamic perspective. Gen. Relat. Grav. **46**, 1673 (2014). [arXiv:1312.3253]
496. T. Padmanabhan, Classical and quantum thermodynamics of horizons in spherically symmetric space-times. Class. Quantum Grav. **19**, 5387 (2002). [arXiv:gr-qc/0204019]
497. A. Paranjape, S. Sarkar, T. Padmanabhan, Thermodynamic route to field equations in Lanczos–Lovelock gravity. Phys. Rev. D **74**, 104015 (2006). [arXiv:hep-th/0607240]
498. D. Kothawala, S. Sarkar, T. Padmanabhan, Einstein's equations as a thermodynamic identity: the cases of stationary axisymmetric horizons and evolving spherically symmetric horizons. Phys. Lett. B **652**, 338 (2007). [arXiv:gr-qc/0701002]
499. M. Akbar, R.-G. Cai, Thermodynamic behavior of field equations for $f(R)$ gravity. Phys. Lett. B **648**, 243 (2007). [arXiv:gr-qc/0612089]
500. A.V. Frolov, L. Kofman, Inflation and de Sitter thermodynamics. JCAP **0305**, 009 (2003). [arXiv:hep-th/0212327]
501. R.-G. Cai, S.P. Kim, First law of thermodynamics and Friedmann equations of Friedmann–Robertson–Walker universe. JHEP **0502**, 050 (2005). [arXiv:hep-th/0501055]
502. G. Calcagni, de Sitter thermodynamics and the braneworld. JHEP **0509**, 060 (2005). [arXiv:hep-th/0507125]
503. M. Akbar, R.-G. Cai, Friedmann equations of FRW universe in scalar-tensor gravity, $f(R)$ gravity and first law of thermodynamics. Phys. Lett. B **635**, 7 (2006). [arXiv:hep-th/0602156]
504. M. Akbar, R.-G. Cai, Thermodynamic behavior of Friedmann equations at the apparent horizon of the FRW universe. Phys. Rev. D **75**, 084003 (2007). [arXiv:hep-th/0609128]
505. R.-G. Cai, L.-M. Cao, Unified first law and thermodynamics of the apparent horizon in the FRW universe. Phys. Rev. D **75**, 064008 (2007). [arXiv:gr-qc/0611071]
506. T. Padmanabhan, The atoms of space, gravity and the cosmological constant. Int. J. Mod. Phys. D **25**, 1630020 (2016). [arXiv:1603.08658]
507. T. Padmanabhan, Thermodynamics of horizons: a comparison of Schwarzschild, Rindler and de Sitter spacetimes. Mod. Phys. Lett. A **17**, 923 (2002). [arXiv:gr-qc/0202078]
508. T.M. Adamo, C.N. Kozameh, E.T. Newman, Null geodesic congruences, asymptotically flat space-times and their physical interpretation. Living Rev. Relat. **15**, 1 (2012).
509. L. Susskind, The world as a hologram. J. Math. Phys. **36**, 6377 (1995). [arXiv:hep-th/9409089]
510. R. Bousso, A covariant entropy conjecture. JHEP **9907**, 004 (1999). [arXiv:hep-th/9905177]
511. T. Padmanabhan, The holography of gravity encoded in a relation between entropy, horizon area and action for gravity. Gen. Relat. Grav. **34**, 2029 (2002). [arXiv:gr-qc/0205090]
512. T. Padmanabhan, Holographic gravity and the surface term in the Einstein–Hilbert action. Braz. J. Phys. **35**, 362 (2005). [arXiv:gr-qc/0412068]

513. T. Padmanabhan, A new perspective on gravity and the dynamics of space-time. Int. J. Mod. Phys. D **14**, 2263 (2005). [arXiv:gr-qc/0510015]
514. A. Mukhopadhyay, T. Padmanabhan, Holography of gravitational action functionals. Phys. Rev. D **74**, 124023 (2006). [arXiv:hep-th/0608120]
515. R.C. Tolman, On the use of the energy-momentum principle in general relativity. Phys. Rev. **35**, 875 (1930)
516. A. Komar, Covariant conservation laws in general relativity. Phys. Rev. **113**, 934 (1959)
517. T. Padmanabhan, Entropy of static spacetimes and microscopic density of states. Class. Quantum Grav. **21**, 4485 (2004). [arXiv:gr-qc/0308070]
518. T. Padmanabhan, Equipartition of energy in the horizon degrees of freedom and the emergence of gravity. Mod. Phys. Lett. A **25**, 1129 (2010). [arXiv:0912.3165]
519. T. Padmanabhan, Surface density of spacetime degrees of freedom from equipartition law in theories of gravity. Phys. Rev. D **81**, 124040 (2010). [arXiv:1003.5665]
520. T. Padmanabhan, Emergence and expansion of cosmic space as due to the quest for holographic equipartition. arXiv:1206.4916
521. T. Padmanabhan, Emergent perspective of gravity and dark energy. Res. Astron. Astrophys. **12**, 891 (2012). [arXiv:1207.0505]
522. T. Padmanabhan, Quantum structure of space-time and black hole entropy. Phys. Rev. Lett. **81**, 4297 (1998). [arXiv:hep-th/9801015]
523. T. Padmanabhan, Event horizon: magnifying glass for Planck length physics. Phys. Rev. D **59**, 124012 (1999). [arXiv:hep-th/9801138]
524. M. Arzano, G. Calcagni, Black-hole entropy and minimal diffusion. Phys. Rev. D **88**, 084017 (2013). [arXiv:1307.6122]
525. D. Kothawala, T. Padmanabhan, Entropy density of spacetime as a relic from quantum gravity. Phys. Rev. D **90**, 124060 (2014). [arXiv:1405.4967]
526. D. Kothawala, T. Padmanabhan, Entropy density of spacetime from the zero point length. Phys. Lett. B **748**, 67 (2015). [arXiv:1408.3963]
527. P. Hořava, D. Minic, Probable values of the cosmological constant in a holographic theory. Phys. Rev. Lett. **85**, 1610 (2000). [arXiv:hep-th/0001145]

Chapter 8
The Problem of Quantum Gravity

Keating: Why do I stand up here? Anybody?
Charlie: To feel taller.
Keating: No! (Keating rings the bell on his desk with his foot.) Thank you for playing, Mr. Dalton. I stand upon my desk to remind yourself that we must constantly look at things in a different way.

— Tom Schulman, *Dead Poets Society* (1989)

Contents

In Chap. 6, we examined the big-bang problem in general relativity and in generic expanding cosmologies. All the arguments therein are classical and assume that spacetime can be described by a Lorentzian manifold, including at arbitrarily small scales. A fascinating possibility is that classical singularities are resolved in a quantum setting. In fact, at the classical level a singularity may emerge because there is no geometrical obstruction for matter to be compressed in arbitrarily small regions of space and for spacetimes to be distorted to arbitrarily high curvature. However, if gravity was a quantum interaction, at sufficiently small scales or large curvature a continuum manifold picture might be invalid and replaced by a very different structure preventing the formation of pathological configurations. In particular, the hope is that "quanta of geometry" would not allow themselves to be packed into regions smaller than a certain critical size, as it happens in loop quantum gravity. Or, like in string theory, if one was able to set geometry on equal footing with particle

G. Calcagni, *Classical and Quantum Cosmology*, Graduate Texts in Physics,
DOI 10.1007/978-3-319-41127-9_8

matter fields and to unify all forces of nature in the dynamics of extended objects, all ultraviolet divergences would be smoothed out because there would be no point-wise interactions. Or, again, a certain "minimal length scale" might be implemented to achieve a similar result.

There are great differences between these scenarios. Technically, their formalization in a rigorous framework turns out to be a formidable task and one may ask whether we need to quantize gravity at all and, if we make the attempt, why it is so difficult. This is the subject of the present chapter.

8.1 Do We Need to Quantize Gravity?

The resolution of issues such as the big-bang and the cosmological constant problems could lie in a fully quantum realm but, to be pedantic, we do not observe phenomena that, as far as we can tell, necessarily require to quantize gravity. Then, we should seriously contemplate the eventuality that this is not the right track to follow and that gravity is an intrinsically classical interaction [1, 2]. In such a case, only matter fields would be quantized (as dictated by experimental evidence) and embedded in a classical, general relativistic background. Then, the Einstein equations would exactly be of the form (2.112), which we repeat here:

$$G_{\mu\nu} = \kappa^2 \langle T_{\mu\nu} \rangle \,. \tag{8.1}$$

Angular brackets denote the expectation value of the total matter energy-momentum tensor on a quantum state. At this point, we should start devising experiments to detect deviations of Lorentzian quantum field theory or quantum mechanics from the predictions based on a flat Minkowski background (e.g., by taking the Schrödinger equation in curved space [3]). So, how far can we maintain the view that gravity is classical? Are there arguments against that?

According to the Einstein equations (8.1), a matter state $|m, g\rangle$ cannot be the linear combination of matter eigenstates $|m_i, g_i\rangle$ describing a mass m_i in a classical metric g_i, because there is only one spacetime for all states. Therefore, the superposition principle for matter wave-functions does not hold. In doing so, we face a problem concerning the theory of measurement. Imagine a matter state $|m, g\rangle$ such that the total mass of matter is either m_1 or m_2, with probability $1/2$. Write the linear superposition

$$|m, g\rangle = \frac{|m_1, g_1\rangle + |m_2, g_2\rangle}{\sqrt{2}} \,.$$

If the wave-function $|m, g\rangle$ collapsed during the measurement of the gravitational field, as in the Copenhagen interpretation, an experiment would measure either m_1 or m_2 and the right-hand side of the Einstein equations (8.1) would not be covariantly conserved [4].[1] If, instead, we adopted the Everett interpretation and assumed that $|m, g\rangle$ does not collapse, then (8.1) would predict that any measurement on matter gives a mass equal to the average $(m_1 + m_2)/2$. This is not the case experimentally [6], thus favouring the hypothesis that (8.1) is an approximation and gravity is intrinsically quantum. The result is not conclusive because one can avoid the above problems by devising more complicated ways to couple classical gravity with quantum matter fields, for instance by changing the representation of matter quantum operators and the Hilbert space of the theory [7, 8]. In general, due to back-reaction of the gravitational field, quantum mechanics governing particles on a classical curved background is non-linear. This can be subject to experimental tests [9].

A theoretical objection against classical gravity is the following [10].[2] Consider an action $S[g, \phi]$ dependent on the metric and some matter fields generically denoted as ϕ. Consider also a suitably regular field redefinition

$$\bar{g}_{\mu\nu} = \bar{g}_{\mu\nu}(g_{\mu\nu}, \phi)\,, \qquad \bar{\phi} = \bar{\phi}(g_{\mu\nu}, \phi)\,,$$

so that the action

$$\bar{S}[\bar{g}, \bar{\phi}] = S[g, \phi]$$

describes the same physics as S at the classical level. At the quantum level, if all fields (including $g_{\mu\nu}$) are quantized the two theories are equivalent on shell (i.e., after using the equations of motion) order by order in perturbation theory, although they differ as far as individual Feynman diagrams and off-shell S-matrix elements are concerned. This is because the on-shell S-matrix is invariant under field redefinitions. However, if gravity is purely classical only matter fields are quantized and the two theories are physically inequivalent. The intuitive reason is that internal graviton lines, which are essential to maintain the on-shell equivalence, are now absent. An example is the minimally-coupled massless scalar field theory

$$S[g, \phi] = \int \mathrm{d}^4x \sqrt{-g} \left(\frac{R}{2\kappa^2} - \frac{1}{2} g^{\mu\nu} \partial_\mu \phi \partial_\nu \phi \right) . \tag{8.2}$$

[1] Another *Gedankenexperiment* along similar lines is given in [5].

[2] Much of the criticism of [10] was actually directed towards quantum field theories in curved space even as approximated models, but one can take the milder view that (8.1) has a limited validity, depending on the circumstances [4, 11].

At the one-loop level, UV divergences are removed if one adds a certain counter-term ΔS [12]. The classical theory (8.2) is equivalent to the non-minimal action

$$\bar{S}[\bar{g},\bar{\phi}] = \int d^4x \sqrt{-\bar{g}} \left[\bar{R} \left(\frac{1}{2\kappa^2} - \frac{\bar{\phi}^2}{12} \right) - \frac{1}{2} \bar{g}^{\mu\nu} \partial_\mu \bar{\phi} \partial_\nu \bar{\phi} \right] \tag{8.3}$$

via a conformal transformation (Problem 8.1). One-loop finiteness of this theory requires a counter-term $\Delta\bar{S}$. When graviton internal lines are taken into account, on shell we have $\Delta S = \Delta\bar{S}$. However, when only the scalar field is quantized one finds that $\Delta S \neq \Delta\bar{S}$ [10].

The conclusion is that the same classical theory could be written in infinitely many different ways and one would have to invoke a criterion selecting one frame among all the others. This may be problematic, but the existence of such a criterion is not altogether unreasonable: for instance, one could impose positivity of energy and choose the Einstein frame as the frame where the fundamental theory is defined [13].

Other objections have been moved to rule out classical gravity but, after careful scrutiny, to date none of them has been found to be strong enough. Classical gravity can be made plausible under certain interpretations and applications of quantum mechanics and, ultimately, the issue lies with experimental evidence [14]. The beautiful economy of thought of unification theories is very appealing, also because it has been instrumental in the discovery of new physics in the past. Hereafter the working hypothesis is maintained that gravity is quantum, but one should keep an open mind about other possibilities as in Sect. 7.7.

8.2 Perturbative Quantum Gravity

The most natural attempt to unify the gravitational interaction and particle physics consists in regarding gravity as a particle field and quantizing it. This can be done either in canonical formalism [15] or via the perturbative background-field method in covariant quantization [16–18]. In the second case, one treats the graviton as a small perturbation on a background,

$$g_{\mu\nu} = \tilde{g}_{\mu\nu} + h_{\mu\nu} \,. \tag{8.4}$$

The background is typically chosen as Minkowski, $\tilde{g}_{\mu\nu} = \eta_{\mu\nu}$. The squared Planck length plays the role of coupling constant, $\kappa^2 \propto G \sim l_{\rm Pl}^{D-2}$. Its engineering dimension (2.19), however, is positive for spacetime dimensionality larger than or equal to 3 and the ensuing perturbation theory is not renormalizable, meaning that the S-matrix diverges at a given finite order.

A standard power-counting argument is sufficient to understand qualitatively the relation between coupling dimensionality and renormalization properties of a theory [19, 20]. Consider the classical action

$$S=\int_{\mathcal{M}} \mathrm{d}^D x\mathcal{L}\,,\qquad \mathcal{L}=\sum_{d\geqslant 0}\alpha_d\mathcal{O}_d\,, \tag{8.5}$$

where α_d and $\mathcal{O}_d$ are, respectively bare coupling constants and local operators with scaling dimension

$$[\alpha_d]=D-d\,,\qquad [\mathcal{O}_d]=d\,. \tag{8.6}$$

Here D is the topological dimension of the manifold $\mathcal{M}$ whereon the field theory lives. More generally, it is the scaling dimension of the measure defining the action. The expressions (8.6) are imposed so that S is dimensionless in $\hbar=1$ units.

Operators are classified according to their scaling dimension:

- An operator $\mathcal{O}_d$ is *relevant* if $d<D$. Then $[\alpha_d]>0$.
- An operator $\mathcal{O}_d$ is *marginal* if $d=D$. Then $[\alpha_d]=0$.
- An operator $\mathcal{O}_d$ is *irrelevant* if $d>D$. Then $[\alpha_d]<0$.

For example, in the scalar field theory given by (2.59), (2.60), (2.61) and (2.62), in four dimensions the operators ϕ^n with $n\leqslant 3$ are relevant, the kinetic and quartic terms are marginal, while the operators ϕ^n with $n\geqslant 5$ are irrelevant.

When constructing perturbation theory, one must take into account all possible gauge-inequivalent interactions $\tilde{\alpha}_d\tilde{\mathcal{O}}_d$ order by order in the effective low-energy action. Some couplings diverge ($\tilde{\alpha}_d\to\infty$) when the regulator in the given regularization scheme is removed. However, if the operator $\tilde{\mathcal{O}}_d$ associated with one of these couplings is already present at the tree level ($\tilde{\mathcal{O}}_d=\mathcal{O}_d$ for some d), one can absorb the divergence into an effective coupling which is defined to be finite when the regulator is removed:

$$\alpha_d\mathcal{O}_d+\tilde{\alpha}_d\mathcal{O}_d=:\alpha_d^{\rm eff}\mathcal{O}_d\,,\qquad |\alpha_d^{\rm eff}|<\infty\,.$$

Contrary to the bare coupling α_d, the effective coupling is what one measures physically. If this procedure works order by order, the theory is said to be *perturbatively renormalizable* and, hence, physically predictive. This means that the number of physical couplings we must measure at any perturbative order is finite. In renormalizable theories, high-momentum modes only shift the bare values of the couplings and high-energy effects (i.e., of heavy particles with mass above the energy cut-off) are under complete control [21]. This result is known as the decoupling theorem.

The way operators enter the effective action is governed by the dimension of the bare couplings. In the renormalization group picture, the physical action stems from the bare action by integrating out momentum modes greater than a certain energy cut-off scale M, and then removing the cut-off. In terms of the dimensionless constants

$$\lambda_d = \alpha_d M^{d-D}, \qquad [\lambda_d] = 0\,, \tag{8.7}$$

the operators scale as

$$\int \mathrm{d}^D x\, \alpha_d \mathcal{O}_d \sim \lambda_d \left(\frac{p}{M}\right)^{d-D}, \tag{8.8}$$

where p is the proper (physical) momentum. Therefore:

- Relevant operators are important at low energies ($p/M \ll 1$).
- Marginal operators are equally important at all scales. Their detailed behaviour is not obvious and they can be, case by case, either marginally relevant or irrelevant.
- Irrelevant operators become important in the ultraviolet ($p/M \gg 1$) but, contrary to what the name suggests, some of them can also alter macroscopic physics. If they are important also for the IR physics, they are called dangerous.

Since there is a finite number of relevant operators and, typically, also of marginal operators, macroscopic physics is described only by few observables.

If divergences are present, they correspond to local operators of dimensionality increasing with the order of the perturbative expansion. Suppose the bare action S contains only relevant operators; then, only a finite number of relevant operators (those which do not appear in S) will enter the effective action and any divergence will be absorbed in the finite number of couplings $\{\alpha_d\}$. For instance, in electromagnetism the electron mass and charge have non-negative dimension in natural units and one can formally absorb the divergences just in these two coupling constants, which are then determined by experiments. Conversely, if even one irrelevant operator appears in S one can construct new irrelevant operators at each order. Explicit calculations can determine whether their couplings are finite or not. If they diverge, the perturbative approach loses predictive power because we can absorb all the divergences only by adding an infinite number of operators to the action.

A theory is said to be *power-counting renormalizable* if

$$\boxed{[\alpha_d] \geqslant 0} \tag{8.9}$$

for all couplings α_d. This condition is not sufficient to guarantee that the theory be renormalizable in the sense of the full renormalization group flow, but it provides a good guiding principle in many situations. If a model is not power-counting renormalizable, then it will likely be non-renormalizable unless remarkable divergence cancellations happen. In the case of gravity, these cancellations do happen [22] and explicit calculations are necessary to settle the issue.

The condition that a theory have a good UV behaviour in the absence of irrelevant operators can be understood by looking at the *superficial degree of divergence* of a Feynman diagram. Consider a one-particle-irreducible (1PI) Feynman subgraph with L loops, I internal propagators and V vertices. The superficial degree of divergence δ is the canonical dimension of all these contributions: given a UV energy cut-off M, the divergent part of the diagram scales as M^δ. If $\delta = 0$, one has at most superficial logarithmic divergences and the theory is power-counting renormalizable. When $\delta < 0$ for every sub-diagram in a Feynman graph, the graph is convergent; if only a finite number of Feynman diagrams diverge superficially, the theory is power-counting *super-renormalizable*.

We can count divergences in the case of the scalar field theory (2.59), (2.60), (2.61) and (2.62). Each loop integral over momenta gives $[\mathrm{d}^D p] = D$, while the propagator $\tilde{G}(p^2) = -1/(p^2 + m^2)$ has $[\tilde{G}(p^2)] = -2$. For the scalar field theory (2.59), (2.60), (2.61) and (2.62), interaction vertices do not carry dimensionality and, overall, $\delta = DL - 2I$. Since $I \geqslant L$, the maximum superficial degree of divergence can be $L(D-2)$. L is the number of independent momenta, given by I minus the number of relations they satisfy among themselves: these are $V - 1$ (one for each vertex, up to the total momentum conservation), so that

$$L = I - V + 1\,. \tag{8.10}$$

This result is often called Euler's theorem for graphs. With only mass and a $\sigma_n \phi^n$ interaction, for each vertex there are n lines and we get $nV = N + 2I$, where N is the number of external legs in the diagram. Replacing L and I with these expressions, one obtains

$$\delta = DL - 2I = D - [\sigma_n]V - \left(\frac{D}{2} - 1\right)N\,, \tag{8.11}$$

where we used (2.62). This formula can be derived also by dimensional arguments (Problem 8.2).

If N is the maximum power in (2.61), the superficial degree of divergence is $\delta = [\sigma_N](1 - V)$. For the theory to be power-counting renormalizable, it must be $[\sigma_N] \geqslant 0$, implying

$$N \leqslant \frac{2D}{D-2}\,.$$

In two dimensions, δ does not depend on the number of external legs (N is unconstrained) and, the greater the number of vertices, the more convergent the diagram. In four dimensions, the ϕ^4 theory is renormalizable while higher powers of ϕ are responsible for an infinite number of divergent diagrams. In general, δ is bounded by the dimension of operators which already appear in the bare action.

The coupling constant governing the perturbative expansion of graviton field theory is $l_{\rm Pl}^2$. This is not immediately obvious from the Einstein–Hilbert action, but we already have almost all the ingredients sufficient to understand this result. Let us revert to units where $[h_{\mu\nu}] = 0$. As we have seen in Chaps. 3 and 5, the quadratic kinetic term of the action (3.24) for $h_{\mu\nu}$ is canonical if one defines the graviton as the field (5.138), $u \sim h/\sqrt{G} \sim m_{\rm Pl}h$, with scaling dimension $[u] = 1$ (the polarization indices λ are omitted here). The next-to-leading terms are irrelevant operators. In fact, borrowing (9.65) and (9.66) from Chap. 9, one can see that the classical Lagrangian for u contains not only the kinetic term $K^2 \sim \dot{u}^2$ but also interactions of the form

$$\frac{1}{G}h(\partial h)^2 \sim l_{\rm Pl}^2 u(\partial u)^2\,, \qquad \frac{1}{G}(h\partial h)^2 \sim l_{\rm Pl}^2(u\partial u)^2\,, \tag{8.12}$$

where ∂ is a spatial derivative. Contrary to the polynomial scalar field theory of the previous example, also vertices contribute to the superficial degree of divergence, each with a factor of 2. In particular, for any 1PI Feynman diagram there can be up to $2V$ extra factors of internal momenta, so that $\delta \leqslant DL+2(V-I) = (D-2)L+2$. Apart from this difference, the power-counting argument applies also to gravity, which is not renormalizable due to the presence of the dangerously irrelevant operators (8.12). Intuitively, the cause why these operators are dangerous despite the small value of the Planck length is the strong equivalence principle: the coupling strength with gravity is the same for all forms of matter and energy, including gravity itself. As soon as a curved background notably affects a matter quantum field theory, it does so also in the gravitational sector and at all spacetime scales.

Looking more in detail at the perturbative diagrammatic expansion, it turns out that, in the absence of supersymmetry, the ultraviolet properties of the theory are in agreement with the power-counting argument. Pure vacuum Einstein gravity in four dimensions is perturbatively non-renormalizable at two loops, meaning that its S-matrix diverges at that order on shell [23–25]. This is due to the presence of dimension-six operators made of three Riemann invariants. In vacuum, the only non-vanishing contribution is

$$\tilde{\mathcal{O}}_6 = R^{\mu\nu}{}_{\rho\sigma}R^{\rho\sigma}{}_{\tau\lambda}R^{\tau\lambda}{}_{\mu\nu}\,, \qquad \tilde{\alpha}_6 \to \infty\,. \tag{8.13}$$

Inclusion of matter does not improve the scenario. On the contrary, divergences appear already at the one-loop level in the presence of a massive scalar field [12],[3] Dirac fermions [26] and Yang–Mills fields [27].

Stelle's higher-order polynomial theory (7.79), on the other hand, is renormalizable. An elementary power counting shows that the maximal superficial degree of divergence of a Feynman graph is $\delta = D-(D-X)(V-I) = D+(D-X)(L-1)$. For $X = D$, the theory is renormalizable since the maximal divergence is $\delta = D$. Then, all the infinities can be absorbed in the operators already present in the Lagrangian (7.79). Unfortunately, as we saw in Sect. 7.5.1, the propagator contains at least one ghost (i.e., a state of negative norm) that marks a violation of unitarity.

[3] A non-minimal coupling between gravity and the scalar can cancel some of the divergences but not all [12].

8.2.1 Supergravity

In Sect. 5.12, we briefly introduced $\mathcal{N} = 1$ supergravity models in four dimensions. Generalization to $\mathcal{N}$ supercharges is also possible, in which case (5.207) becomes $\{Q_a^A, Q_{bB}^\dagger\} = 2(\sigma^\mu)_{ab}P_\mu\delta_B^A$ and $\{Q_a^A, Q_b^B\} = \epsilon_{ab}X^{AB}$, where $A = 1, \dots, \mathcal{N}$, ϵ_{ab} is the two-index Levi-Civita symbol and the scalars X^{AB} are called central charges. We do not enter into details and only quote some results obtained in SUGRA scenarios.

The minimal $D = 4$, $\mathcal{N} = 1$ supergravity model is non-renormalizable at three (respectively, one) loops in the absence (presence) of matter [28]. $\mathcal{N} = 8$ extended supergravity fares better. In fact, in four dimensions and in vacuum this theory is finite at the four-loop level and promising evidence justifies the conjecture that $\mathcal{N} = 8$ SUGRA be perturbatively finite [29–37], although there is disagreement on such an optimistic view [38] and the theory may have a divergence at seven loops [39–41]. In any case, the improved UV behaviour of the theory is due both to supersymmetry and to a class of diagrammatic cancellations typical of gravity which lower its degree of divergence [22]. These cancellations are not accounted for by traditional power-counting arguments and can be made manifest only in certain representations of the quantum amplitudes.

8.2.2 Effective Field Theory

Quantum fluctuations involve all energy scales. For this reason, loop corrections to tree-level interactions carry information on all such scales, as they involve integration on all momenta p. This information is confined to the shifting of a finite number of bare parameters in renormalizable theories, while in non-renormalizable ones high-energy scales affect infinitely many parameters and, consequently, most types of experimental measurements. Such is the case of gravity, as we saw in this chapter. However, it is possible to treat it as a consistent *effective* field theory (e.g., [42]) even when the usual perturbative expansions are non-predictive [43–47] (see [48, 49] for reviews). To achieve this result, one reorganizes the theory by separating the known low-energy quantum effects from the unknown high-energy contributions via a momentum expansion (i.e., grouping terms with the same number of derivatives). The graviton is nothing but the residual low-energy degrees of freedom of the effective theory.

Under this perspective, the series "$R^2 + R^3 + R^4 + \dots$" of Riemann curvature operators is regarded as an expansion in a low-curvature, low-energy regime. By definition, higher-order operators are sub-dominant at such scales and the series can be truncated at any given order without the necessity to know the form of the full expansion (if any) or to worry about ghosts (which would be artifacts of the truncation). For instance, the quartic Lagrangian (7.77) would be interpreted as an effective quantum description of gravity valid at scales much larger than the Planck scale, where the parameters α, β and γ are all $O(l_{\text{Pl}}^2) = O(\hbar)$.

A quantum field theory of gravity is thus obtained at low curvature and low energies, where classical dynamics is modified by well-defined and well-behaved quantum corrections. A fully calculable example is the one-loop correction to the Newtonian potential between two masses $m_{1,2}$ at distance r [45, 50]:

$$\boxed{\Phi(r) \simeq -\frac{Gm_1m_2}{r}\left[1+3\frac{G(m_1+m_2)}{r}+\frac{41}{10\pi}\frac{l_{\rm Pl}^2}{r^2}\right],} \tag{8.14}$$

where the middle term [51–53] is the classical relativistic correction. The $O(r^{-3})$ term is expected to arise in the low-curvature, low-energy limit of quantum gravity, whatever the form of the theory at high energies.

For consistency, solutions to higher-order equations of motion should be truncated at lowest order in the couplings (α, β and γ in the example (7.77)), while solutions which are non-perturbative in such couplings must be discarded [54, 55]. When applied to the models of Sect. 7.5, this prescription virtually excludes most of their interesting applications to cosmology, since corrections to the Einstein–Hilbert dynamics are negligible at inflationary as well as dark-energy scales. Such corrections scale as $\sim (\rho/\rho_{\rm Pl})^n$, with $n > 0$ and $\rho \ll \rho_{\rm Pl}$ (see (10.69) and below (11.101)). Therefore, the problem of quantum gravity is not whether gravity admits a consistent quantum description (*it does*, at an effective level) but to understand the deep UV structure of the theory. Getting access to the high-curvature, high-energy limit would also pave the way to phenomenology beyond the standard cold big-bang ΛCDM model and, hopefully, to the resolution of its problems.

8.2.3 *Resummed Quantum Gravity*

An approach complementary to effective field theory is *resummed quantum gravity* [56–62]. The latter is based on the Feynman expansion of perturbative quantum gravity (where the graviton $h_{\mu\nu}$ is in the decomposition (8.4) of the metric) with matter fields [16] and resums all the propagators *á la* Dyson [63] in the loop expansion of the theory. For instance, for a massive scalar particle in four dimensions the propagator is modified as

$$\tilde{G}(p^2) = -\frac{1}{p^2+m^2} \to \bar{G}(p^2) = -\frac{e^{B_g''(p)}}{p^2+m^2}, \qquad B_g''(p) = -\frac{\kappa^2 p^2}{8\pi^2}\ln\left(\frac{p^2}{m^2}+1\right), \tag{8.15}$$

and loop diagrams of the graviton self-energy with matter loops feature $\bar{G}(p^2)$ instead of the bare scalar propagator $\tilde{G}(p^2)$. Remarkably [58, 60], the inclusion of all radiative corrections at all orders and the rearrangement of the IR divergences tame the UV divergences that appear in the finite-loop calculation where resummation is not performed [12] and counter-terms are not added. In other words, after the IR resummation, an eventual UV regulator can be removed and the UV regime remains finite order by order in the loop expansion.

In this approach, the graviton propagator with matter loops is strongly damped in the UV. At one loop in the resummed series and to leading order in the momentum p, the gauge-invariant graviton propagator acquires a correction of the form

$$-\frac{1}{p^2} \to -\frac{1}{p^2+p^4/\gamma^2}\,, \tag{8.16}$$

where γ depends on the masses of the matter fields [56]. Without resummation, γ would be infinity and the one-loop propagator would diverge [12, 57] but, for the Standard-Model masses, $\gamma \approx 0.21 m_{\rm Pl}$ [60]. When Fourier transforming (8.16), the corresponding correction to Newton's potential is [56]

$$\boxed{\Phi(r) \simeq -\frac{G m_1 m_2}{r}\,(1-\mathrm{e}^{-\gamma r})\,.} \tag{8.17}$$

Note the difference with respect to the result (8.14). In the effective-field-theory case, the final expression for Newton's potential is valid at large distances $\gamma r = O(1) r/l_{\rm Pl} \gg 1$, i.e., at low energies (momenta lower than the Planck mass); the corrective IR term in (8.14) is perturbative in $1/r$. On the other hand, the resummed-gravity approach is focussed on the UV regime of the theory and (8.17) is the one-loop result valid at all distances $r > 1/\gamma$, even near the Planck length. The relationship between the effective-field-theory and the resummed-gravity approximation is similar to that between, respectively, the use of chiral perturbation theory for soft strong interactions at large distance and the use of perturbative quantum chromodynamics for hard strong interactions [58]. Both approaches to perturbative quantum gravity recover the classical potential for $r \ll l_{\rm Pl}$ and they should agree at some intermediate scale.

An interesting property of (8.17) is that Newton's potential vanished in the UV and, apparently, the gravitational interaction becomes weakly couple. This stems from the one-loop propagator (8.16) and the associated effective Newton's coupling

$$G_{\rm eff}(p) = \frac{G_0}{1+p^2/\gamma^2}\,. \tag{8.18}$$

This is no longer a constant in the momentum and its value can change with the probed energy scale. We will come back to this important point in Sect. 11.2.1.

8.3 Approaches to Quantum Gravity

The lack of a UV-finite perturbative formulation of quantum gravity, the non-conclusive nature of finiteness proofs in the $\mathcal{N} = 8$ supersymmetric extension (together with the fact that we have not observed supersymmetry yet), the limitations of the effective-field-theory approach and some criticism [64, 65] on the decomposition (8.4) are all motivations for turning our attention to other proposals related to quantum gravity.[4] We can roughly divide them in three categories.

In the first, gravity is on a completely different status with respect to matter fields. A conservative but not very satisfactory point of view is that it may be just a classical force which does not admit a quantum formulation, contrary to the other interactions. A somewhat more rewarding possibility is that gravity is not a fundamental interaction but an emergent phenomenon, as detailed in Sect. 7.7.

The second class of frameworks is constituted by the "theories of everything," which aim to describe both gravity and matter in a unified, consistent framework. Among the theories of everything, we count:

- Supergravity, where all fields are components of a super-multiplet.
- String theory [67–69], which we will meet in Chap. 12. The fundamental objects are strings, propagating in a D-dimensional target spacetime. Different vibration modes of a string correspond to particles; gravity emerges from the modes of closed strings. The end-points of open strings live on lower-dimensional extended objects called Dp-branes, whose dynamics is highly non-perturbative. There exist several formulations of the theory, most of which supersymmetric. Self-consistency requirements impose the number of spatial dimensions to be larger than three: $D-1 = 25$ in the bosonic string and $D-1 = 9$ in the supersymmetric versions. The low-energy limit of the supersymmetric string is supergravity. Non-perturbative dualities between all these theories suggest that they are just different aspects of a bigger and yet elusive framework, exotically called M-theory (Sect. 12.4).
- Group field theory [70–77] (Sect. 11.5). Field theories on group manifolds are higher-dimensional generalizations of matrix models, characterized by Feynman diagrams having the combinatorial structure of D-dimensional simplicial complexes. The Feynman amplitudes of these theories are spin-foam models, based on the same simplicial complexes, so that group field theories potentially represent a truly unified framework for both loop quantum gravity and simplicial approaches. The interpretation is that group field theories define the covariant dynamics of quantum spacetime as a sum over Feynman diagrams, interpreted as discrete spacetimes of arbitrary topology, each weighted by a purely algebraic and discrete version of a quantum-gravity path integral.

---

[4] For a nice historical overview of ideas in quantum gravity, we refer the reader to [66].

All these theories have several problems. On one hand, they include a number of ingredients which are phenomenologically unnecessary (e.g., supersymmetry, large or infinite particle spectra, extra spatial dimensions) or yet unclear (e.g., individuation of physical observables, definition of semi-classical and continuum limits). These ingredients can be accommodated to fit with experiments but the task is challenging. Both the study of the theoretical structure of these theories and their predictive power advance rather slowly despite the concentrated efforts by a large part of the scientific community.[5]

On the other hand, it is difficult to devise experiments unravelling a fundamentally quantum structure of the gravitational force. This is due to an inevitable combination of technological limitations and the way we are presently asking the questions. The third group of theories, which we introduce now, is no exception [72]. There, unification is abandoned[6] in favour of tackling the less ambitious but still difficult problem of how to quantize the gravitational sector alone. The way general relativity is fused with quantum mechanics notably differs from the theories of everything, mainly because greater emphasis is placed on diffeomorphisms (in string theory, these symmetries are not fundamental and stem from a much larger gauge group). To this category there belong:

- *Loop quantum gravity* (LQG), where gravity is quantized in canonical formalism and quanta of space are described by states called spin networks [64, 78–80]. The spectra of area and volume operators are discrete and bounded from below. A path-integral evolution of spin networks is *spin foams* [81–84] (Sect. 11.4).
- *Asymptotic safety*, also known as quantum Einstein gravity [85–90], a non-perturbative renormalization group approach realizing asymptotic safety in a traditional field-theory setting (Sect. 11.2).
- *Causal dynamical triangulations* (CDT) [91–93], a Lorentzian path-integral formulation of quantum gravity where the integral is performed over piecewise flat 4-geometries (Sect. 11.3). This is an example of approach where spacetime geometry is fundamentally continuous but discretized on a lattice to regularize infinities; the limit of zero lattice spacing is eventually taken.
- *Causal sets* [94–99], where the texture of spacetime is fundamentally discrete (Sect. 11.6).

Other approaches (*non-commutative spacetimes*, *non-local gravity* and models with dimensional flow) will be discussed in Sects. 11.7, 11.8 and 11.9.

Contrary to supergravity and string theory, in group field theory and the rest of the approaches the notion of "quantizing gravity" is replaced by that of "quantizing spacetime." In the first case, (8.4) singles out a particular background whereupon

[5]To give two instances, it has become apparent that string theory is not a model of Nature but, rather, a framework wherein to construct such a model, just in the same way "quantum field theory" stands to the Standard Model of particles. Group field theory is much less developed than any of the other models and its status as a viable theory of everything is unclear.

[6]Notice, however, that there exist arguments against non-unified frameworks of quantum-gravity (Sect. 13.2.2).

the perturbation $h_{\mu\nu}$ is the graviton particle. On the other hand, in any of the above non-perturbative theories the decomposition (8.4) is somewhat artificial and the fundamental objects are quanta of spacetime itself rather than particle quanta.

8.4 Problems and Solutions

8.1 Conformal transformations. Find the transformation mapping (8.2) into (8.3).

Solution From Problems 7.3, 7.4 and 7.5, it is easy to see that

$$\bar{g}_{\mu\nu} = \Omega^2 g_{\mu\nu}\,, \qquad \Omega^2 = \frac{1}{\cosh^2\left(\sqrt{\kappa^2/6}\,\phi\right)}\,, \qquad \bar{\phi} = \sqrt{\frac{6}{\kappa^2}}\tanh\left(\sqrt{\frac{\kappa^2}{6}}\phi\right).$$

8.2 Power-counting renormalizability. Derive (8.11) with a dimensional argument.

Solution A diagram with N external lines can be generated by a $\sigma_N \phi^N$ term, so that its scaling dimension is $[\sigma_N]$. On the other hand, with only the ϕ^n interaction term available the divergent part of the diagram scales as $\sigma_n^V M^\delta$, where M is the UV cut-off. Therefore, we have $[\sigma_N] = [\sigma_n]V + \delta$ and (8.11) is recovered.

References

1. C. Møller, The energy-momentum complex in general relativity and related problems, in *Les Théories Relativistes de la Gravitation*, ed. by A. Lichnerowicz, M.A. Tonnelat (CNRS, Paris, 1962)
2. L. Rosenfeld, On quantization of fields. Nucl. Phys. **40**, 353 (1963)
3. L. Parker, Path integrals for a particle in curved space. Phys. Rev. D **19**, 438 (1979)
4. L.H. Ford, Gravitational radiation by quantum systems. Ann. Phys. (N.Y.) **144**, 238 (1982)
5. R.M. Wald, *General Relativity* (The University of Chicago Press, Chicago, 1984)
6. D.N. Page, C.D. Geilker, Indirect evidence for quantum gravity. Phys. Rev. Lett. **47**, 979 (1981)
7. B. Mielnik, Generalized quantum mechanics. Commun. Math. Phys. **37**, 221 (1974)
8. T.W.B. Kibble, S. Randjbar-Daemi, Nonlinear coupling of quantum theory and classical gravity. J. Phys. A **13**, 141 (1980)

9. S. Carlip, Is quantum gravity necessary? Class. Quantum Grav. **25**, 154010 (2008). [arXiv:0803.3456]
10. M.J. Duff, Inconsistency of quantum field theory in curved space-time, in *Quantum Gravity 2*, ed. by C.J. Isham, R. Penrose, D.W. Sciama (Oxford University Press, Oxford, 1981)
11. J.P. Paz, S. Sinha, Decoherence and back reaction: the origin of the semiclassical Einstein equations. Phys. Rev. D **44**, 1038 (1991)
12. G. 't Hooft, M.J.G. Veltman, One loop divergencies in the theory of gravitation. Ann. Poincaré Phys. Theor. A **20**, 69 (1974)
13. V. Faraoni, E. Gunzig, P. Nardone, Conformal transformations in classical gravitational theories and in cosmology. Fund. Cosm. Phys. **20**, 121 (1999). [arXiv:gr-qc/9811047]
14. S. Boughn, Nonquantum gravity. Found. Phys. **39**, 331 (2009). [arXiv:0809.4218]
15. B.S. DeWitt, Quantum theory of gravity. I. The canonical theory. Phys. Rev. **160**, 1113 (1967)
16. R.P. Feynman, Quantum theory of gravitation. Acta Phys. Polon. **24**, 697 (1963) [Reprinted in *100 Years of Gravity and Accelerated Frames*, ed. by J.-P. Hsu, D. Fine (World Scientific, Singapore, 2005)]
17. B.S. DeWitt, Quantum theory of gravity. II. The manifestly covariant theory. Phys. Rev. **162**, 1195 (1967)
18. B.S. DeWitt, Quantum theory of gravity. III. Applications of the covariant theory. Phys. Rev. **162**, 1239 (1967)
19. P. Ramond, *Field Theory: A Modern Primer* (Westview Press, Boulder, 1997)
20. S. Weinberg, *The Quantum Theory of Fields*, vol. I (Cambridge University Press, Cambridge, 1995)
21. T. Appelquist, J. Carazzone, Infrared singularities and massive fields. Phys. Rev. D **11**, 2856 (1975)
22. Z. Bern, J.J. Carrasco, D. Forde, H. Ita, H. Johansson, Unexpected cancellations in gravity theories. Phys. Rev. D **77**, 025010 (2008). [arXiv:0707.1035]
23. M.H. Goroff, A. Sagnotti, Quantum gravity at two loops. Phys. Lett. B **160**, 81 (1985)
24. M.H. Goroff, A. Sagnotti, The ultraviolet behavior of Einstein gravity. Nucl. Phys. B **266**, 709 (1986)
25. A.E.M. van de Ven, Two loop quantum gravity. Nucl. Phys. B **378**, 309 (1992)
26. S. Deser, P. van Nieuwenhuizen, Nonrenormalizability of the quantized Dirac–Einstein system. Phys. Rev. D **10**, 411 (1974)
27. S. Deser, H.-S. Tsao, P. van Nieuwenhuizen, One-loop divergences of the Einstein–Yang–Mills system. Phys. Rev. D **10**, 3337 (1974)
28. S. Deser, J.H. Kay, K.S. Stelle, Renormalizability properties of supergravity. Phys. Rev. Lett. **38**, 527 (1977)
29. Z. Bern, L.J. Dixon, R. Roiban, Is $N = 8$ supergravity ultraviolet finite? Phys. Lett. B **644**, 265 (2007). [arXiv:hep-th/0611086]
30. Z. Bern, J.J. Carrasco, L.J. Dixon, H. Johansson, D.A. Kosower, R. Roiban, Three-loop superfiniteness of $N = 8$ supergravity. Phys. Rev. Lett. **98**, 161303 (2007). [arXiv:hep-th/0702112]
31. Z. Bern, J.J.M. Carrasco, L.J. Dixon, H. Johansson, R. Roiban, Manifest ultraviolet behavior for the three-loop four-point amplitude of $N = 8$ supergravity. Phys. Rev. D **78**, 105019 (2008). [arXiv:0808.4112]
32. N. Arkani-Hamed, F. Cachazo, J. Kaplan, What is the simplest quantum field theory? JHEP **1009**, 016 (2010). [arXiv:0808.1446]
33. Z. Bern, J.J. Carrasco, L.J. Dixon, H. Johansson, R. Roiban, The ultraviolet behavior of $N = 8$ supergravity at four loops. Phys. Rev. Lett. **103**, 081301 (2009). [arXiv:0905.2326]
34. R. Kallosh, The ultraviolet finiteness of $N = 8$ supergravity. JHEP **1012**, 009 (2010). [arXiv:1009.1135]
35. N. Beisert, H. Elvang, D.Z. Freedman, M. Kiermaier, A. Morales, S. Stieberger, $E_{7(7)}$ constraints on counterterms in $N = 8$ supergravity. Phys. Lett. B **694**, 265 (2010). [arXiv:1009.1643]

36. R. Kallosh, $E_{7(7)}$ symmetry and finiteness of $N = 8$ supergravity. JHEP **1203**, 083 (2012). [arXiv:1103.4115]
37. R. Kallosh, $N = 8$ counterterms and $E_{7(7)}$ current conservation. JHEP **1106**, 073 (2011). [arXiv:1104.5480]
38. T. Banks, Arguments against a finite $N = 8$ supergravity. arXiv:1205.5768
39. M.B. Green, J.G. Russo, P. Vanhove, String theory dualities and supergravity divergences. JHEP **1006**, 075 (2010). [arXiv:1002.3805]
40. J. Björnsson, M.B. Green, 5 loops in $24/5$ dimensions. JHEP **1008**, 132 (2010). [arXiv:1004.2692]
41. J. Björnsson, Multi-loop amplitudes in maximally supersymmetric pure spinor field theory. JHEP **1101**, 002 (2011). [arXiv:1009.5906]
42. S. Weinberg, Effective field theory, past and future. Proc. Sci. **CD09**, 001 (2009). [arXiv:0908.1964]
43. J.F. Donoghue, Leading quantum correction to the Newtonian potential. Phys. Rev. Lett. **72**, 2996 (1994). [arXiv:gr-qc/9310024]
44. J.F. Donoghue, General relativity as an effective field theory: the leading quantum corrections. Phys. Rev. D **50**, 3874 (1994). [arXiv:gr-qc/9405057]
45. N.E.J. Bjerrum-Bohr, J.F. Donoghue, B.R. Holstein, Quantum gravitational corrections to the nonrelativistic scattering potential of two masses. Phys. Rev. D **67**, 084033 (2003); Erratum-ibid. D **71**, 069903(E) (2005). [arXiv:hep-th/0211072]
46. N.E.J. Bjerrum-Bohr, J.F. Donoghue, B.R. Holstein, Quantum corrections to the Schwarzschild and Kerr metrics. Phys. Rev. D **68**, 084005 (2003); Erratum-ibid. D **71**, 069904(E) (2005). [arXiv:hep-th/0211071]
47. N.E.J. Bjerrum-Bohr, J.F. Donoghue, P. Vanhove, On-shell techniques and universal results in quantum gravity. JHEP **1402**, 111 (2014). [arXiv:1309.0804]
48. J.F. Donoghue, The effective field theory treatment of quantum gravity. AIP Conf. Proc. **1483**, 73 (2012). [arXiv:1209.3511]
49. J.F. Donoghue, B.R. Holstein, Low energy theorems of quantum gravity from effective field theory. J. Phys. G **42**, 103102 (2015). [arXiv:1506.00946]
50. I.B. Khriplovich, G.G. Kirilin, Quantum long-range interactions in general relativity. Zh. Eksp. Teor. Fiz. **125**, 1219 (2004) [JETP **98**, 1063 (2004)]. [arXiv:gr-qc/0402018]
51. A. Einstein, L. Infeld, B. Hoffmann, The gravitational equations and the problem of motion. Ann. Math. **39**, 65 (1938)
52. A. Eddington, G. Clark, The problem of n bodies in general relativity theory. Proc. R. Soc. Lond. A **166**, 465 (1938)
53. Y. Iwasaki, Quantum theory of gravitation vs. classical theory. Fourth-order potential. Prog. Theor. Phys. **46**, 1587 (1971)
54. J.Z. Simon, Higher-derivative Lagrangians, nonlocality, problems, and solutions. Phys. Rev. D **41**, 3720 (1990)
55. J.Z. Simon, Stability of flat space, semiclassical gravity, and higher derivatives. Phys. Rev. D **43**, 3308 (1991)
56. B.F.L. Ward, Quantum corrections to Newton's law. Mod. Phys. Lett. A **17**, 2371 (2002). [arXiv:hep-ph/0204102]
57. B.F.L. Ward, Are massive elementary particles black holes? Mod. Phys. Lett. A **19**, 143 (2004). [arXiv:hep-ph/0305058]
58. B.F.L. Ward, Massive elementary particles and black holes. JCAP **0402**, 011 (2004). [arXiv:hep-ph/0312188]
59. B.F.L. Ward, Planck scale cosmology in resummed quantum gravity. Mod. Phys. Lett. A **23**, 3299 (2008). [arXiv:0808.3124]
60. B.F.L. Ward, An estimate of Λ in resummed quantum gravity in the context of asymptotic safety. Phys. Dark Univ. **2**, 97 (2013)
61. B.F.L. Ward, Running of the cosmological constant and estimate of its value in quantum general relativity. Mod. Phys. Lett. A **30**, 1540030 (2015). [arXiv:1412.7417]

62. B.F.L. Ward, Einstein–Heisenberg consistency condition interplay with cosmological constant prediction in resummed quantum gravity. Mod. Phys. Lett. A **30**, 1550206 (2015). [arXiv:1507.00661]
63. D.R. Yennie, S.C. Frautschi, H. Suura, The infrared divergence phenomena and high-energy processes. Ann. Phys. (N.Y.) **13**, 379 (1961)
64. C. Rovelli, *Quantum Gravity* (Cambridge University Press, Cambridge, 2007)
65. T.P. Singh, T. Padmanabhan, Notes on semiclassical gravity. Ann. Phys. (N.Y.) **196**, 296 (1989)
66. C. Rovelli, Notes for a brief history of quantum gravity, in *Recent Developments in Theoretical and Experimental General Relativity, Gravitation and Relativistic Field Theories*, ed. by V.G. Gurzadyan, R.T. Jantzen, R. Ruffini (World Scientific, Singapore, 2002). [arXiv:gr-qc/0006061]
67. J. Polchinski, *String Theory* (Cambridge University Press, Cambridge, 1998)
68. B. Zwiebach, *A First Course in String Theory* (Cambridge University Press, Cambridge, 2009)
69. K. Becker, M. Becker, J.H. Schwarz, *String Theory and M-Theory* (Cambridge University Press, Cambridge, 2007)
70. L. Freidel, Group field theory: an overview. Int. J. Theor. Phys. **44**, 1769 (2005). [arXiv:hep-th/0505016]
71. D. Oriti, The group field theory approach to quantum gravity, in [72]. [arXiv:gr-qc/0607032]
72. D. Oriti (ed.), *Approaches to Quantum Gravity* (Cambridge University Press, Cambridge, 2009)
73. D. Oriti, The microscopic dynamics of quantum space as a group field theory, in [74]. [arXiv:1110.5606]
74. G.F.R. Ellis, J. Murugan, A. Weltman (eds.), *Foundations of Space and Time* (Cambridge University Press, Cambridge, 2012)
75. D. Oriti, Group field theory as the second quantization of loop quantum gravity. Class. Quantum Grav. **33**, 085005 (2016). [arXiv:1310.7786]
76. A. Baratin, D. Oriti, Ten questions on group field theory (and their tentative answers). J. Phys. Conf. Ser. **360**, 012002 (2012). [arXiv:1112.3270]
77. S. Gielen, L. Sindoni, Quantum cosmology from group field theory condensates: a review. SIGMA **12**, 082 (2016). [arXiv:1602.08104]
78. T. Thiemann, *Modern Canonical Quantum General Relativity* (Cambridge University Press, Cambridge, 2007); Introduction to modern canonical quantum general relativity. arXiv:gr-qc/0110034
79. T. Thiemann, Quantum gravity: from theory to experimental search. Lect. Notes Phys. **631**, 412003 (2003). [arXiv:gr-qc/0210094]
80. A. Ashtekar, J. Lewandowski, Background independent quantum gravity: a status report. Class. Quantum Grav. **21**, R53 (2004). [arXiv:gr-qc/0404018]
81. D. Oriti, Spacetime geometry from algebra: spin foam models for non-perturbative quantum gravity. Rep. Prog. Phys. **64**, 1489 (2001). [arXiv:gr-qc/0106091]
82. A. Perez, Spin foam models for quantum gravity. Class. Quantum Grav. **20**, R43 (2003). [arXiv:gr-qc/0301113]
83. C. Rovelli, A new look at loop quantum gravity. Class. Quantum Grav. **28**, 114005 (2011). [arXiv:1004.1780]
84. A. Perez, The spin-foam approach to quantum gravity. Living Rev. Relat. **16**, 3 (2013)
85. M. Niedermaier, The asymptotic safety scenario in quantum gravity: an introduction, Class. Quantum Grav. **24**, R171 (2007). [arXiv:gr-qc/0610018]
86. M. Niedermaier, M. Reuter, The asymptotic safety scenario in quantum gravity. Living Rev. Relat. **9**, 5 (2006)
87. M. Reuter, F. Saueressig, Functional renormalization group equations, asymptotic safety and quantum Einstein gravity. arXiv:0708.1317
88. A. Codello, R. Percacci, C. Rahmede, Investigating the ultraviolet properties of gravity with a Wilsonian renormalization group equation. Ann. Phys. (N.Y.) **324**, 414 (2009). [arXiv:0805.2909]

89. D.F. Litim, Renormalisation group and the Planck scale. Philos. Trans. R. Soc. Lond. A **369**, 2759 (2011). [arXiv:1102.4624]
90. M. Reuter, F. Saueressig, Asymptotic safety, fractals, and cosmology. Lect. Notes Phys. **863**, 185 (2013). [arXiv:1205.5431]
91. R. Loll, The emergence of spacetime, or, quantum gravity on your desktop. Class. Quantum Grav. **25**, 114006 (2008). [arXiv:0711.0273]
92. J. Ambjørn, J. Jurkiewicz, R. Loll, Causal dynamical triangulations and the quest for quantum gravity, in [74]. [arXiv:1004.0352]
93. J. Ambjørn, A. Görlich, J. Jurkiewicz, R. Loll, Nonperturbative quantum gravity. Phys. Rep. **519**, 127 (2012). [arXiv:1203.3591]
94. R.D. Sorkin, Spacetime and causal sets, in *Relativity and Gravitation: Classical and Quantum*, ed. by J.C. D'Olivo, E. Nahmad-Achar, M. Rosenbaum, M.P. Ryan, L.F. Urrutia, F. Zertuche (World Scientific, Singapore, 1991)
95. D.D. Reid, Introduction to causal sets: an alternate view of spacetime structure. Can. J. Phys. **79**, 1 (2001). [arXiv:gr-qc/9909075]
96. J. Henson, The causal set approach to quantum gravity, in [72]. [arXiv:gr-qc/0601121]
97. J. Henson, Discovering the discrete universe. arXiv:1003.5890
98. S. Surya, Directions in causal set quantum gravity, in *Recent Research in Quantum Gravity*, ed. by A. Dasgupta (Nova Science, Hauppauge, 2011). [arXiv:1103.6272]
99. F. Dowker, Introduction to causal sets and their phenomenology. Gen. Relat. Grav. **45**, 1651 (2013)

Chapter 9
Canonical Quantum Gravity

Tirem-me daqui a metafísica!
— Fernando Pessoa, *Lisbon Revisited (1923)*

Spare me metaphysics!

Contents

Letting aside temporarily the possibility to unify all forces in a single consistent framework, a question one can ask oneself is whether quantum mechanics can resolve a particular singularity, such as the big bang or that inside a black hole. With this goal in mind, we examine an approach to gravity based on the Hamiltonian formalism and its cosmological applications. Section 9.1 is devoted to classical canonical gravity, where the so-called Schwinger, Ashtekar–Barbero and ADM variables are defined. Quantization in ADM variables is discussed in Sect. 9.2, where the Wheeler–DeWitt equation is introduced for the first time. This equation will be the starting point for tackling the big-bang problem in Chap. 10.

G. Calcagni, *Classical and Quantum Cosmology*, Graduate Texts in Physics,
DOI 10.1007/978-3-319-41127-9_9

9.1 Canonical Variables in General Relativity

For the purpose of quantization, the Lagrangian formalism can be used in path-integral approaches, but the Hamiltonian formalism is more transparent regarding the choice of classical canonical variables (from which quantum operators are then constructed). Historically, the first attempt to carry out the program of Hamiltonian quantization was via the second-order *Arnowitt–Deser–Misner* (ADM) *formalism* [1–3], but eventually it has become clear that first-order *Ashtekar–Barbero variables* [4–6] are more convenient. We will present the Hamiltonian analysis in first-order formalism, later sketching the results for the ADM decomposition. Throughout this chapter, we specialize to four spacetime dimensions and do not include supersymmetry (see [7–10] for supergravity Hamiltonian formalism, [11–16] for its applications to quantum cosmology and Sect. 13.9.3 for supersymmetric canonical quantum cosmology in string theory).

9.1.1 First-Order Formalism and Parity

The fundamental object of first-order gravity is the local frame, or gravitational field, or tetrad $e^a{}_\mu$, which we introduced in Sect. 6.3.1. Since *parity symmetry* is by definition a frame symmetry, we will briefly discuss it here before moving to Hamiltonian formalism. Chosen a frame $e^a{}_\mu(x)$ in the internal (i.e., tangent) space at a point x in a spacetime manifold $\mathcal{M}$, one can consider an internal Lorentz $\Lambda^b{}_a$:

$$e^a{}_\mu \to e^a{}_\mu{}' = \Lambda_b{}^a e^b{}_\mu \,. \tag{9.1}$$

This transformation may be *proper* or *improper*, according to the sign of the Jacobian determinant

$$\det \Lambda = \pm 1 \,. \tag{9.2}$$

Improper transformations ($\det \Lambda = -1$) include time reversal and spatial reflections. In four dimensions, there are four such reflections: those flipping only one spatial direction and one flipping all of them. An even number of reflections corresponds to a proper Lorentz rotation and, for the same reason, an odd combination of time reversal and reflections is just a time reversal plus a rotation. In the following, we shall define a parity transformation by a single spatial reflection.

In general relativity, parity is a well-defined operation on the internal space tangent at every point of the manifold $\mathcal{M}$: it changes the orientation of a frame.

Since, in general, it is not possible to define a global frame, i.e., a spatially constant tetrad, parity is a local symmetry. Conversely, in homogeneous spacetimes one can choose one and the same frame for every point in $\mathcal{M}$ or along particle geodesics: this is the so-called *Fermi frame* [17–20]. In this case, which includes the obvious examples of Newtonian physics, (quantum) field theory and FLRW cosmology, parity flips the orientation of all the copies of the same frame at all points in $\mathcal{M}$ and the whole universe is mirrored in its reflected image. The notion of parity in Minkowski or FLRW spacetimes is a special case of the one in general relativity.

An immediate consequence is that reflection operations can be rigorously defined only in internal space, so that parity properties should apply only to the projected tensors. For instance, one can ask whether a vector v^μ is a polar or pseudo-vector only after considering its projection $v^a := e^a_{\ \mu} v^\mu$ on the local frame. The vielbein internal basis changes orientation under reflection, so that if v^a is a polar internal vector, then one simply says that "v^μ is a polar vector." In this sense, "internal parity" and "parity" are synonyms.

Consider a proper scalar field $\phi(x)$. By definition, it is invariant under internal Lorentz transformations, including parity. On the other hand, a *pseudo-scalar* $\varphi(x)$ changes sign under reflection and can be defined as the weight-zero scalar

$$\varphi(x) := \varepsilon\phi(x)\,, \qquad \varepsilon := \frac{{}^{(4)}e}{|{}^{(4)}e|} = \pm 1\,, \tag{9.3}$$

where ${}^{(4)}e := \det(e_a^{\ \mu})$ is the determinant of the tetrad. φ is a frame-dependent object and, by definition, it makes sense only after choosing a frame. An internal Lorentz transformation $x \to x' = \Lambda x$ yields

$$\varphi'(x) = \varphi(x') = (\det\Lambda)\,\varphi(x)\,. \tag{9.4}$$

An internal scalar is not necessarily a spacetime scalar (any spacetime tensor $T^{\mu\nu\cdots}$ is a collection of internal scalars) but a spacetime (pseudo-)scalar is also an internal (pseudo-)scalar. Also, while a shift $\phi \to \phi + \phi_0$ of a proper scalar with a constant is still a scalar field, a shifted pseudo-scalar is no longer a pseudo-scalar, as $(\varphi+\varphi_0)' = -\varphi + \varphi_0$, unless $\varphi_0 = \varepsilon\phi_0$ be a frame-dependent constant.

Equation (9.4) is commonly adopted as the definition of pseudo-scalars. The quantity ε is a constant whose presence is typically "hidden" in pseudo-tensors, because the choice of frame (and its orientation) is often left implicit.

Let us consider a proper internal totally anti-symmetric tensor η_{abcd}. Under reflections, it transforms as

$$\eta'_{a'b'c'd'} = \Lambda^a_{\ a'}\Lambda^b_{\ b'}\Lambda^c_{\ c'}\Lambda^d_{\ d'}\eta_{abcd}.$$

On the other hand, the fully anti-symmetric tensor

$$\epsilon_{abcd} := \varepsilon\eta_{abcd} \tag{9.5}$$

is an internal pseudo-tensor, i.e.,

$$\epsilon'_{a'b'c'd'} = (\det \Lambda)\Lambda^a{}_{a'}\Lambda^b{}_{b'}\Lambda^c{}_{c'}\Lambda^d{}_{d'}\epsilon_{abcd}\,.$$

If

$$\epsilon_{0123} = -\epsilon_{3210} = 1 \tag{9.6}$$

and the other usual permutation rules follow, this is the *Levi-Civita symbol*, which can be characterized either as an internal pseudo-tensor or as a spacetime pseudo-scalar. The *Levi-Civita tensor* differs from the "symbol" by an extra weight factor $|^{(4)}e|$.

The determinant of the tetrad can be expressed as

$$^{(4)}e = \frac{1}{4!}\epsilon_{\mu\nu\rho\sigma}\epsilon^{abcd}e_a{}^{\mu}e_b{}^{\nu}e_c{}^{\rho}e_d{}^{\sigma} = \epsilon^{abcd}e_a{}^0e_b{}^1e_c{}^2e_d{}^3\,.$$

A property of the symbol with n indices is

$$\epsilon_{a_1a_2\cdots a_n}\epsilon^{a_1a_2\cdots a_n} = n!\,,$$

that is, the number of permutations of a set with n elements. Also,

$$\epsilon_{a_1\cdots a_k a_{k+1}\cdots a_n}\epsilon^{a_1\cdots a_k b_{k+1}\cdots b_n} = k!(n-k)!\delta^{b_{k+1}}_{[a_{k+1}}\cdots\delta^{b_n}_{a_n]}\,,$$

where anti-symmetrization on multiple indices is done by subtracting odd index permutations to even permutations and dividing by the factorial of the number of indices. For $n = 3$, we get two relations which we shall use later:

$$\epsilon_{ijk}\epsilon^{ij'k'} = 2\delta^{j'}_{[j}\delta^{k'}_{k]}\,, \qquad \epsilon_{ijk}\epsilon^{ijk'} = 2\delta^{k'}_k\,. \tag{9.7}$$

9.1.1.1 Volume Form

The Einstein–Hilbert action for gravity has a counterpart in first-order Palatini formalism. Let us discuss the terms in (2.18) one by one, ignoring the cosmological constant.

In second-order formalism, the volume element is $\sqrt{-g}\,\mathrm{d}^4x$. In terms of the gravitational field, $g = -[^{(4)}e]^2$. In the real field $\mathbb{R}$, one has $\sqrt{-g} = |^{(4)}e|$ and we can define the volume form

$$e^a \wedge e^b \wedge e^c \wedge e^d := |^{(4)}e|\epsilon^{abcd}\mathrm{d}^4x\,, \tag{9.8}$$

where $e^a := e^a{}_{\mu}\mathrm{d}x^{\mu}$ and (9.8) defines the *wedge product* $\wedge$. The volume form is a pseudo-tensor.

Before moving on, a remark is in order. The first-order Hamiltonian formalism is actually an independent formulation of general relativity. On one hand, in this case the existence of a metric is not even required. A metric $g_{\mu\nu}$ is well defined only if ${}^{(4)}e \neq 0$.[1] Here we assume that the gravitational sector is *non-degenerate*, so that it is possible to define a Lagrangian theory. On the other hand, one can define the volume form with ${}^{(4)}e$ rather than with its absolute value as in (9.8). This step is far from being innocuous for at least two reasons. First, it changes the parity properties of fields and canonical variables in the Hamiltonian formulation. Second, in view of an operational definition of physical volume and quantum mechanical measurements, the orientation of a frame should not make any difference in such measurements. This is all the more important in general relativity, as there is no universal frame and different observers can choose their own orientation. Then, one can show that only (9.8) is a meaningful choice for the volume form; we shall give an explicit argument later (Problem 9.1).

9.1.1.2 Lorentz Spin Connection and Curvature

According to the *Palatini formulation*, we introduce the Lorentz-valued *spin connection* $\omega^{ab}{}_{\mu}$ as an independent field. Its *curvature* is

$$R_{\mu\nu}{}^{ab} := 2\partial_{[\mu}\omega^{ab}{}_{\nu]} + 2\omega^{a}{}_{c[\mu}\omega^{cb}{}_{\nu]}\,, \tag{9.9}$$

where Latin indices are lowered via the internal metric. The curvature is invariant under the exchange of the index pairs $\mu\nu$ and ab. At this stage, the spin connection is unrelated to the Levi-Civita connection $\Gamma^{\sigma}_{\mu\nu}$ but the analogue of the Riemann tensor is simply

$$R_{\mu\nu}{}^{\rho\sigma} = e_a{}^{\rho} e_b{}^{\sigma} R_{\mu\nu}{}^{ab}\,. \tag{9.10}$$

Quite often in the literature, the language of differential forms is used to write in compact notation expressions with explicit spacetime components. One defines

$$\omega^{ab} := \omega^{ab}{}_{\mu}\mathrm{d}x^{\mu} \tag{9.11}$$

and the *exterior derivative*

$$\mathrm{d}\omega^{ab} := \partial_{\mu}\omega^{ab}{}_{\nu}\mathrm{d}x^{\mu} \wedge \mathrm{d}x^{\nu}\,. \tag{9.12}$$

[1] Thanks to this, first-order quantization is fit for describing quantum gravitational systems at scales where the concept of metric and smooth spacetime can break down.

The curvature of ω is

$$R^{ab} = \mathrm{d}\omega^{ab} + \omega^{ac} \wedge \omega_c{}^b = \tfrac{1}{2} R^{ab}{}_{\mu\nu} \mathrm{d}x^\mu \wedge \mathrm{d}x^\nu = \tfrac{1}{2} e^c \wedge e^d R^{ab}{}_{cd} . \tag{9.13}$$

9.1.1.3 Torsion

All spacetimes we have considered so far were manifolds with vanishing *torsion tensor* $T^a{}_{\mu\nu}$, defined in the generalized Bianchi identity

$$R^a{}_{b[\mu\nu} e^b{}_{\sigma]} = D(\omega)_{[\mu} T^a{}_{\nu\sigma]} \neq 0 , \tag{9.14}$$

where D_μ is the covariant derivative operator made with the spin connection ω^{ab}. Intuitively, torsion characterizes how tangent planes (moving frames) twist around a geodesic. Certain fields (such as β in Problem 9.1) or fermions [21–24] can act as sources for torsion.[2] In that case, one writes the Lorentz spin connection as a torsionless part $\bar{\omega}^{ab}$ (the Ricci connection, which obeys the homogeneous structure equation (9.49) we will see later) plus the *contortion* 1-form $\mathcal{K}^{ab}$ [27]:

$$\omega^{ab}{}_\mu = \bar{\omega}^{ab}{}_\mu + \mathcal{K}^{ab}{}_\mu , \tag{9.15}$$

where the contortion tensor

$$\mathcal{K}^{ab}{}_\mu = e^a{}_\nu e^b{}_\rho \mathcal{K}^{\nu\rho}{}_\mu , \qquad \mathcal{K}^{\nu\rho}{}_\mu = -\mathcal{K}^{\rho\nu}{}_\mu , \tag{9.16}$$

is related to the torsion $T^\nu{}_{\rho\mu} = -T^\nu{}_{\mu\rho}$ by

$$\mathcal{K}^\nu{}_{\rho\mu} = \tfrac{1}{2}(T^\nu{}_{\rho\mu} - T_\rho{}^\nu{}_\mu - T_\mu{}^\nu{}_\rho) . \tag{9.17}$$

It is particularly convenient to split torsion into its irreducible components in accordance with the Lorentz group [28–31]:

$$T_{\mu\nu\rho} = \frac{1}{3}\left(T_\nu g_{\mu\rho} - T_\rho g_{\mu\nu}\right) - \frac{1}{6}\,\epsilon_{\mu\nu\rho\sigma} S^\sigma + q_{\mu\nu\rho} , \tag{9.18}$$

where

$$T_\mu := T^\nu{}_{\mu\nu} \tag{9.19}$$

[2]Spinors cannot be coupled directly to the metric but they are easily accommodated in the vielbein formulation [25, 26].

is the *trace vector*,

$$S_\mu := \epsilon_{\nu\rho\sigma\mu} T^{\nu\rho\sigma} \tag{9.20}$$

is the *pseudo-trace axial vector* and the anti-symmetric tensor $q_{\mu\nu\rho}$ is such that $q^{\nu}{}_{\rho\nu} = 0 = \epsilon^{\mu\nu\rho\sigma} q_{\mu\nu\rho}$.

9.1.1.4 Hilbert–Palatini Action

The *Hilbert–Palatini action* can be obtained formally by replacing the Ricci scalar in the Einstein–Hilbert action (2.18) with the spin connection curvature scalar $R(e,\omega) := e_a^{\ \mu} e_b^{\ \nu} R^{ab}{}_{\mu\nu}$:

$$S_{\rm HP} = \frac{1}{2\kappa^2} \int d^4x\, |{}^{(4)}e|\, e_a^{\ \mu} e_b^{\ \nu} R^{ab}{}_{\mu\nu} = \frac{1}{4\kappa^2} \int \epsilon_{abcd} e^a \wedge e^b \wedge R^{cd}, \tag{9.21}$$

which is a proper scalar. Often the notation $*R^{ab} = \epsilon_{abcd} R^{cd}/2$ is used.

9.1.1.5 Holst and Nieh–Yan Terms

The Hilbert–Palatini action can be generalized by adding terms which do not affect the classical dynamics. These terms will be involved in the theory of canonical quantum gravity and loop quantum cosmology.

The first of these terms was introduced by Holst [32] and it reads

$$S_{\rm H} := \frac{1}{4\kappa^2\gamma} \int d^4x\, |{}^{(4)}e|\, e_a^{\ \mu} e_b^{\ \nu} \epsilon^{ab}{}_{cd} R_{\mu\nu}{}^{cd} = \frac{1}{2\kappa^2\gamma} \int e^a \wedge e^b \wedge R_{ab}, \tag{9.22}$$

where γ is a real constant called the *Barbero–Immirzi parameter* [6, 33, 34]. Often the notation $\beta := -1/\gamma$ is used. The Holst term violates parity at the level of the action but the classical dynamics is unchanged. In fact, the integrand in (9.22) vanishes because of the Bianchi cyclic identity $R^a{}_{b[\mu\nu} e^b{}_{\sigma]} = 0$. This identity is also called *second Cartan structure equation* and is obtained when solving the dynamics. For this reason the Holst term, which is not topological, is said to vanish "on half shell," i.e., when half of the dynamical equations have been taken into account.

This situation is unusual in Lagrangian theories, where one would expect that non-dynamical terms be topological. Indeed, the Holst modification is not completely general and it constitutes only one of the two terms composing a well-known topological density called the Nieh–Yan 4-form [23, 35–39]:

$$\begin{aligned} S_{\rm NY} &= -\frac{1}{4\kappa^2\gamma}\int \mathrm{d}^4x\,|^{(4)}e|\left(\epsilon^{\mu\nu\rho\sigma}\eta_{ab}T^a{}_{\mu\nu}T^b{}_{\rho\sigma} - e_a{}^{\mu}e_b{}^{\nu}\epsilon^{ab}{}_{cd}R_{\mu\nu}{}^{cd}\right)\\ &= -\frac{1}{2\kappa^2\gamma}\int \left(T^a\wedge T_a - e^a\wedge e^b\wedge R_{ab}\right)\\ &= -\frac{1}{2\kappa^2\gamma}\int \mathrm{d}\left(e_a\wedge T^a\right). \end{aligned} \tag{9.23}$$

The Nieh–Yan density is linear in the curvature and contains a torsion-torsion term. Contrary to the Holst term, it is a total divergence, as the last line shows.

Topological invariants become dynamical when their coupling constant is promoted to a field. In Problem 9.1, we study the general case where $-1/\gamma = \beta \to \beta(x)$ [40–44], since it is an example of a gravitational theory with torsion which is dynamically equivalent to a torsion-free model.

9.1.2 Hamiltonian Analysis

At the classical level, we can ignore the Holst or Nieh–Yan term and consider torsion-free gravity minimally coupled with a real scalar field (bars omitted):

$$S = \int \mathrm{d}t\,L = \int \mathrm{d}^4x\,|^{(4)}e|\left[\frac{1}{2\kappa^2}e_a{}^{\mu}e_b{}^{\nu}R_{\mu\nu}{}^{ab} - \frac{1}{2}\partial_\mu\phi\partial^\mu\phi - V(\phi)\right]. \tag{9.24}$$

Here the field ϕ is left arbitrary and we study the dynamics as a constrained Hamiltonian system.[3]

In torsion-free (pseudo-)Riemannian manifolds, the requirement that the spin connection realizes parallel transport of angles and lengths translates into the *compatibility equation*. In second-order formalism, it states that the metric is covariantly constant, $\nabla_\rho g_{\mu\nu} = 0$.

Systems endowed with gauge symmetries are characterized by a phase-space structure defined by a set of relations called *constraints*. Different phase-space points describe the same solution if they are related by a gauge transformation. Also, some of the canonical variables are not dynamical and the phase space is larger than the space of physical degrees of freedom. Both these occurrences are translated into geometric surfaces by the constraints. Constraint surfaces are subsets of phase-space points whose intersections correspond to dynamical solutions. Examples of constrained systems are the parametrized non-relativistic point particle

[3]The scalar ϕ can be identified with the Barbero–Immirzi field, in which case $\phi = \widetilde{\beta}$ is a pseudo-scalar and $V(\phi) = 0$. One can check that the Hamiltonian analysis of (9.24) is consistent with the one starting from the fundamental action (9.136) (Problems 9.1 and 9.2).

[3], the relativistic particle [45, 46], Yang–Mills theory on curved space and, last but not least, gravity.

Following the Dirac procedure [47–49], we calculate the first- and second-class constraints of the theory (9.24). The latter can be easily solved and the system turns out to be characterized by a set of first-class constraints which reflect rotational and diffeomorphism gauge freedom.

We assume spacetime $\mathcal{M}$ to be globally hyperbolic, so that we can define a Cauchy surface Σ, the normal u^μ to the surface and the first fundamental form (2.41). Let $t^\mu = t^\mu(x)$ be the *time-flow vector* field on $\mathcal{M} \ni x$ satisfying $t^\mu \nabla_\mu t = t^\mu \partial_\mu t = 1$. Neither t nor t^μ can be interpreted in terms of physical measurements of time, since the metric is, in fact, an unknown dynamical field. The time-flow vector field generates a one-parameter group of diffeomorphisms, known as embedding diffeomorphisms, $x_t : \mathbb{R} \times \Sigma \to \mathcal{M}$, defined as $x_t(\mathbf{x}) := x(t, \mathbf{x})$. This allows to represent spacetime as a smooth deformation of the three-dimensional Cauchy surfaces Σ into a one-parameter family of three-dimensional Cauchy surfaces Σ_t. These are described by the parametric equations $x_t^\mu = x_t^\mu(\mathbf{x})$, where t denotes the hypersurface at different "times."

A general parametrization can be obtained by introducing the normal and tangential components of the vector field $t^\mu(x)$ with respect to Σ. Namely, we define

$$N := -t_\mu u^\mu \,, \qquad N^\mu := h^\mu{}_\nu t^\nu \,, \tag{9.25}$$

respectively called the *lapse function* and the *shift vector*. As a consequence, we have

$$t^\mu(t, \mathbf{x}) = \partial_t x^\mu(t, \mathbf{x})|_{x(t,\mathbf{x}) = x_t(\mathbf{x})} = N(t, \mathbf{x})\, u^\mu(t, \mathbf{x}) + N^\mu(t, \mathbf{x}) \,.$$

By acting with a Wigner boost on the local basis, we can rotate it so that its zeroth component is made parallel, at each point of Σ, to the normal vector u_μ, i.e., $u_\mu = e^0{}_\mu$, implying that the local boost parameter $e^i{}_0$ vanishes at each point of Σ:

$$e^i{}_0 = 0 \,.$$

The requirement that this particular choice of the orientation of the local basis be preserved along the evolution fixes the so-called *Schwinger* or *time gauge*, the net result being that the action no longer depends on the boost parameters. Also, the local symmetry group is reduced from the initial $SO(1, 3)$ to $SO(3)$, which encodes the spatial rotational symmetry. It can be demonstrated that fixing the time gauge into the action does not affect the consistency of the canonical analysis, this procedure being equivalent to a canonical gauge fixing.

9.1.2.1 Step 1: Decomposing the Action in the (3 + 1) Splitting

The action (9.24) can be written as follows (all quantities are torsion free):

$$
\begin{aligned}
S &= \frac{1}{2\kappa^2}\int \mathrm{d}t\mathrm{d}^3\mathbf{x}\,|^{(4)}e|\left(2e_0^{\ \mu}e_i^{\ \nu}R_{\mu\nu}^{\ \ \ 0i} + e_i^{\ \mu}e_j^{\ \nu}R_{\mu\nu}^{\ \ \ ij}\right)\\
&\quad -\int \mathrm{d}t\mathrm{d}^3\mathbf{x}\,|^{(4)}e|\left(\frac{1}{2}\partial_\mu\phi\partial^\mu\phi + V\right)\\
&= \frac{1}{2\kappa^2}\int \mathrm{d}t\mathrm{d}^3\mathbf{x}\,|^{(4)}e|\left[2\frac{t^\mu - N^\mu}{N}e_i^{\ \nu}R_{\mu\nu}^{\ \ \ 0i} + e_i^{\ \alpha}e_j^{\ \gamma}R_{\alpha\gamma}^{\ \ \ ij}\right]\\
&\quad -\int \mathrm{d}t\mathrm{d}^3\mathbf{x}\,|^{(4)}e|\left[\frac{1}{2}(h^{\mu\nu} - u^\mu u^\nu)\partial_\mu\phi\partial_\nu\phi + V\right]\\
&= \frac{1}{\kappa^2}\int \mathrm{d}t\mathrm{d}^3\mathbf{x}\,|e|\left\{\frac{N}{2}e_i^\alpha e_j^\gamma\left({}^{(3)}R_{\alpha\gamma}^{\ \ \ ij} + 2K^i_{[\alpha}K^j_{\gamma]}\right) - N^\alpha e_i^{\gamma}\,{}^{(3)}R_{\alpha\gamma}^{\ \ \ 0i}\right.\\
&\quad \left. + e_i^\alpha\left[\dot{K}^i_\alpha + \omega^{0i}_{\ \ \mu}\partial_\alpha t^\mu - \partial_\alpha\left(t\cdot\omega^i\right) - \omega^i_{\ k\alpha}(t\cdot\omega^k) + (t\cdot\omega^i_{\ k})K^k_\alpha\right]\right\}\\
&\quad + \int \mathrm{d}t\mathrm{d}^3\mathbf{x}\,|e|\left[\frac{1}{2N}(\dot\phi - N^\alpha\partial_\alpha\phi)^2 - N\left(\frac{1}{2}\partial_\alpha\phi\partial^\alpha\phi + V\right)\right],
\end{aligned}
$$

where $\dot{} = t^\mu\partial_\mu$, ${}^{(4)}e = Ne$, $e = \det e^i_\alpha$ is the determinant of the triad, ${}^{(3)}R_{\alpha\gamma}^{\ \ \ ij}$ is the curvature of the spin connection $\omega^{ij}_{\ \ \alpha}$ with all indices spatial and

$$K^i_\alpha := \omega^{0i}_{\ \ \alpha} \tag{9.26}$$

is the *extrinsic curvature*. The following notation has been used: $t\cdot\omega^i = t^\mu\omega^{0i}_{\ \ \mu}$, $t\cdot\omega^i_{\ k} = t^\mu\omega^i_{\ k\mu}$. Greek indices $\alpha,\gamma,\dots$ from the beginning of the alphabet and Latin indices $i,j,\dots$ from the middle of the alphabet run from 1 to 3 and denote, respectively, components transforming under spatial diffeomorphisms and local spatial rotations. The three-dimensional Levi-Civita symbol is defined as in (9.6) and we will often make use of the relation (9.7). The absolute value of e is often omitted in the literature; here we keep it in order to avoid confusion when studying how the action and canonical variables transform under parity.

Now, remembering the definition of the Lie derivative operator (3.39) on a vector, one has

$$
\begin{aligned}
S = \frac{1}{\kappa^2}\int \mathrm{d}t\mathrm{d}^3\mathbf{x}\,|e|&\left\{e_i^\alpha\left[\mathcal{L}_t K^i_\alpha - D_\alpha\left(t\cdot\omega^i\right) + (t\cdot\omega^i_{\ k})K^k_\alpha\right]\right.\\
&\left. -2N^\alpha e_i^\gamma D_{[\alpha}K^i_{\gamma]} + \frac{N}{2}e_i^\alpha e_j^\gamma\left({}^{(3)}R_{\alpha\gamma}^{\ \ \ ij} + 2K^i_{[\alpha}K^j_{\gamma]}\right)\right\} + S_\phi\,,
\end{aligned}
\tag{9.27}
$$

where

$$S_\phi = \int \mathrm{d}t\mathrm{d}^3\mathbf{x}\,|e|\left[\frac{1}{2N}(\mathcal{L}_t\phi - N^\alpha\partial_\alpha\phi)^2 - N\left(\frac{1}{2}\partial_\alpha\phi\partial^\alpha\phi + V\right)\right]. \tag{9.28}$$

We introduced the $so(3)$-valued torsion-free covariant derivative D_α, which can be written as

$$D_\alpha v^i = \partial_\alpha v^i + \epsilon^i{}_{jk}\Gamma^j_\alpha v^k \tag{9.29}$$

on a gauge vector, where

$$\Gamma^i_\alpha := -\frac{1}{2}\,\epsilon^i{}_{jk}\omega_\alpha{}^{jk}, \qquad \omega_\alpha{}^{jk} = -\epsilon_i{}^{jk}\Gamma^i_\alpha. \tag{9.30}$$

Notice the minus sign, which stems from the definition (9.6). In much of the literature the other convention $\epsilon^{0123} = 1$ with upper indices is in use and the minus signs are absorbed into the Levi-Civita symbol.

The curvature of Γ is defined as

$$R^i_{\alpha\gamma} := 2\partial_{[\alpha}\Gamma^i_{\gamma]} + \epsilon^i{}_{jk}\Gamma^j_\alpha\Gamma^k_\gamma = -\frac{1}{2}\,\epsilon^i{}_{jk}\,{}^{(3)}R_{\alpha\gamma}{}^{jk}. \tag{9.31}$$

Later we will discover that $\omega^{ij}{}_\alpha$ is the Ricci spin connection, which depends on the triad field. The canonical analysis will eventually show that it is not an independent variable but we must keep it as such for the time being.

9.1.2.2 Step 2: Defining Canonical Variables and Symplectic Structure

The next step is the definition of the momenta conjugate to the fundamental variables. Since the Lagrangian is singular, we expect a set of primary constraints to appear. In particular, the only non-vanishing momenta are those conjugate to K^i_α and ϕ:

$$K^i_\alpha : \qquad E^\alpha_i := \kappa^2\frac{\delta S}{\delta\mathcal{L}_t K^i_\alpha} = |e|e^\alpha_i, \tag{9.32a}$$

$$\phi : \qquad \Pi_\phi := \frac{\delta S}{\delta\mathcal{L}_t\phi} = \frac{|e|}{N}(\dot\phi - N^\alpha\partial_\alpha\phi). \tag{9.32b}$$

All the others vanish identically, i.e.,

$$e^\alpha_i : \qquad \mathcal{P}^i_\alpha = 0, \tag{9.33a}$$

$$\Gamma^i_\alpha : \qquad \Pi^\alpha_i = 0, \tag{9.33b}$$

$$t\cdot\omega^i:\qquad \Pi_i=0\,,\tag{9.33c}$$

$$t\cdot\omega^{ij}:\qquad \Pi_{ij}=0\,,\tag{9.33d}$$

$$N^\alpha:\qquad \pi_\alpha=0\,,\tag{9.33e}$$

$$N:\qquad \pi^0=0\,.\tag{9.33f}$$

The field E_i^α is a weight-1 proper vector called *densitized triad* or, in analogy with Maxwell theory, *gravielectric field*. The pair (K,E) is called Schwinger variables [50].

Once a set of canonical fields and conjugate momenta (Q_n,P_n) is established, one can define the equal-time *Poisson bracket* between two functionals f and g as

$$\begin{aligned}&\left\{f[Q_n(t,\mathbf{x}),P_n(t,\mathbf{x})],g[Q_n(t,\mathbf{x}'),P_n(t,\mathbf{x}')]\right\}\\ &:=\sum_n\left[\frac{\delta f}{\delta Q_n(t,\mathbf{y})}\frac{\delta g}{\delta P_n(t,\mathbf{y})}-\frac{\delta f}{\delta P_n(t,\mathbf{y})}\frac{\delta g}{\delta Q_n(t,\mathbf{y})}\right].\end{aligned}\tag{9.34}$$

This writing is only formal for functionals wherein spatial derivatives act on the canonical variables. In that case, one can define *smeared functionals*

$$F(t):=\int d^3\mathbf{x}\Lambda_f(\mathbf{x})f[Q_n(t,\mathbf{x}),P_n(t,\mathbf{x})]\,,$$

$$G(t):=\int d^3\mathbf{x}'\Lambda_g(\mathbf{x}')g[Q_n(t,\mathbf{x}'),P_n(t,\mathbf{x}')]\,,$$

where $\Lambda_{f,g}$ are arbitrary space-dependent tensors, and calculate the Poisson bracket $\{F,G\}$.

The phase space is equipped with the *symplectic structure*

$$\left\{K_\alpha^i(t,\mathbf{x}),E_j^\beta(t,\mathbf{x}')\right\}=\kappa^2\delta_\alpha^\beta\delta_j^i\delta(\mathbf{x},\mathbf{x}')\,,\tag{9.35a}$$

$$\{\phi(t,\mathbf{x}),\Pi_\phi(t,\mathbf{x}')\}=\delta(\mathbf{x},\mathbf{x}')\,,\tag{9.35b}$$

$$\left\{e_i^\alpha(t,\mathbf{x}),\mathcal{P}_\beta^j(t,\mathbf{x}')\right\}=\delta_\beta^\alpha\delta_i^j\delta(\mathbf{x},\mathbf{x}')\,,\tag{9.35c}$$

$$\left\{\Gamma_\alpha^i(t,\mathbf{x}),\Pi_j^\beta(t,\mathbf{x}')\right\}=\delta_\alpha^\beta\delta_j^i\delta(\mathbf{x},\mathbf{x}')\,,\tag{9.35d}$$

$$\{t\cdot\omega^i(t,\mathbf{x}),\Pi_k(t,\mathbf{x}')\}=\delta_k^i\delta(\mathbf{x},\mathbf{x}')\,,\tag{9.35e}$$

$$\{t\cdot\omega^{ij}(t,\mathbf{x}),\Pi_{kl}(t,\mathbf{x}')\}=\delta_{[k}^i\delta_{l]}^j\delta(\mathbf{x},\mathbf{x}')\,,\tag{9.35f}$$

$$\{N(t,\mathbf{x}),\pi^0(t,\mathbf{x}')\}=\delta(\mathbf{x},\mathbf{x}')\,,\tag{9.35g}$$

$$\{N^\alpha(t,\mathbf{x}),\pi_\beta(t,\mathbf{x}')\}=\delta_\beta^\alpha\delta(\mathbf{x},\mathbf{x}')\,.\tag{9.35h}$$

9.1.2.3 Step 3: Primary Constraints

As one can immediately note, in none of the above conjugate momenta except Π_ϕ is there the temporal Lie derivative of any of the fundamental variables. Therefore, in principle, all momenta but one should be considered as *primary constraints*, that is to say, relations encoding the fact that not all the canonical variables are independent. Thus, the following set of primary constraints has to be imposed:

$$^{(K)}C_i^\alpha := E_i^\alpha - |e|e_i^\alpha \approx 0\,, \tag{9.36a}$$

$$^{(e)}C_\alpha^i := \mathcal{P}_\alpha^i \approx 0\,, \tag{9.36b}$$

$$^{(\Gamma)}C_i^\alpha := \Pi_i^\alpha \approx 0\,, \tag{9.36c}$$

$$C_i := \Pi_i \approx 0\,, \tag{9.36d}$$

$$C_{ij} := \Pi_{ij} \approx 0\,, \tag{9.36e}$$

$$C^{(N)} := \pi^0 \approx 0\,, \tag{9.36f}$$

$$C_\alpha := \pi_\alpha \approx 0\,, \tag{9.36g}$$

where the symbol $\approx$ indicates *weak equalities* valid only on the constraint surface, i.e., the manifold in phase space defined by the imposition of the primary constraints. Expressions which vanish weakly do not vanish identically throughout phase space. In contrast, *strong equalities* $=$ hold on all phase space.

9.1.2.4 Step 4: Dirac Hamiltonian

Having calculated the conjugate momenta, we can now perform a Legendre transformation and extract the canonical Hamiltonian. Since the latter is not uniquely determined because of the primary constraints (generically denoted as C_m), we write the *Dirac Hamiltonian*:

$$\begin{aligned}
H_\mathrm{D} &= \int \mathrm{d}^3\mathbf{x} \left(\frac{1}{\kappa^2} E_i^\alpha \mathcal{L}_t K_\alpha^i + \Pi_\phi \mathcal{L}_t \phi + \lambda^m C_m \right) - L \\
&= \int \mathrm{d}^3\mathbf{x} \Bigg\{ N^\alpha \mathcal{H}_\alpha + N\mathcal{H} + \frac{1}{\kappa^2} E_i^\alpha \left[D_\alpha \left(t \cdot \omega^i \right) - (t \cdot \omega^i{}_k) K_\alpha^k \right] \\
&\quad + {}^{(K)}\lambda_\alpha^i\, {}^{(K)}C_i^\alpha + {}^{(e)}\lambda_i^\alpha\, {}^{(e)}C_\alpha^i + {}^{(\Gamma)}\lambda_\alpha^i\, {}^{(\Gamma)}C_i^\alpha + \lambda^i C_i \\
&\quad + \lambda^{ij} C_{ij} + \lambda C^{(N)} + \lambda^\alpha C_\alpha \Bigg\}\,,
\end{aligned}$$

where λ^m are arbitrary functions,

$$\boxed{\mathcal{H}_\alpha := \frac{2}{\kappa^2} E_i^\gamma D_{[\alpha} K^i_{\gamma]} + \Pi_\phi \partial_\alpha \phi} \tag{9.37}$$

is the *super-momentum* and

$$\boxed{\mathcal{H} := \frac{1}{2\kappa^2 |e|} E_i^\alpha E_j^\gamma \left(\epsilon^{ij}{}_k R^k_{\alpha\gamma} - 2K^i_{[\alpha} K^j_{\gamma]} \right) + \mathcal{H}_\phi} \tag{9.38a}$$

is the *super-Hamiltonian* (or scalar constraint),[4] where

$$\boxed{\mathcal{H}_\phi := \frac{1}{2|e|} \Pi_\phi^2 + \frac{1}{2} |e| \partial_\alpha \phi \partial^\alpha \phi + |e| V(\phi) \,.} \tag{9.38b}$$

For autonomous systems (no explicit time dependence), the time evolution of a functional $f[e, K, \Gamma, N, \dots]$ of the canonical variables is governed by the Hamilton equation

$$\dot{f} = \{f, H_\mathrm{D}\} \,. \tag{9.39}$$

9.1.2.5 Step 5: Secondary and First-Class Constraints

The primary constraints (9.36) have been imposed at a given time t. As a consistency requirement, the Dirac canonical procedure calculates the Poisson brackets between the primary constraints and the Dirac Hamiltonian. In fact, the constraints (9.36) must be (weakly) constant in time,

$$\dot{C}_m = \{C_m, H_\mathrm{D}\} \approx 0 \,. \tag{9.40}$$

For example, a momentum which is zero on a hypersurface Σ_t must vanish also on any other surface $\Sigma_{t'}$. If (9.40) is not realized on the primary surface for some value of the Lagrange multipliers λ_m, then it must be constrained to vanish. In this way, *secondary constraints* are generated which determine the secondary constraint surface on the phase space [51].

[4]The context should be clear enough to avoid confusion with the Hubble parameter $\mathcal{H}$ in conformal time.

The Poisson brackets between the Dirac Hamiltonian and the first two primary constraints (9.36a) and (9.36b) do not generate any secondary constraint. In fact, they can be set to zero by choosing the Lagrange multiplier ${}^{(K)}\lambda_i^\alpha$ suitably.

For the other primary constraints, one gets

$$\kappa^2\{C_i, H_\mathrm{D}\} = \mathcal{B}_i := D_\alpha E_i^\alpha\,, \tag{9.41}$$

$$\begin{aligned}\kappa^2\{{}^{(\Gamma)}C_i^\alpha, H_\mathrm{D}\} &= \epsilon_{ijk}E^{\alpha j}\left[t\cdot\omega^k - N^\gamma K_\gamma^k - E^{\gamma k}\partial_\gamma\left(\frac{N}{|e|}\right)\right]\\ &\quad + N^\alpha\epsilon_{ijk}E^{\gamma j}K_\gamma^k - \frac{N}{|e|}\epsilon_i{}^{jk}D_\gamma\left(E_j^\alpha E_k^\gamma\right)\,,\end{aligned} \tag{9.42}$$

$$\kappa^2\{C_{ij}, H_\mathrm{D}\} = K_{\alpha[j}E_{i]}^\alpha\,, \tag{9.43}$$

$$\{C_\alpha, H_\mathrm{D}\} = -\mathcal{H}_\alpha\,, \tag{9.44}$$

$$\{C^{(N)}, H_\mathrm{D}\} = -\mathcal{H}\,. \tag{9.45}$$

Note that all these equations are to be regarded as a shorthand for relations involving smeared functionals. For instance, define the smeared super-momentum and super-Hamiltonian as

$$d[N^\alpha] := \int \mathrm{d}^3\mathbf{x}\, N^\alpha\mathcal{H}_\alpha\,, \qquad H[N] := \int \mathrm{d}^3\mathbf{x}\, N\mathcal{H}\,, \tag{9.46}$$

and the smeared constraints

$$C[\lambda^\alpha] := \int \mathrm{d}^3\mathbf{x}'\,\lambda^\alpha C_\alpha\,, \qquad C[\lambda] := \int \mathrm{d}^3\mathbf{x}'\,\lambda C^{(N)}\,.$$

Then, the last two equations are actually (time dependence omitted)

$$\begin{aligned}\{C[\lambda^\alpha], H_\mathrm{D}\} &= \{C[\lambda^\alpha], d[N^\beta]\}\\ &= 0 - \int \mathrm{d}^3\mathbf{y}\frac{\delta}{\delta\pi_\alpha(\mathbf{y})}\int \mathrm{d}^3\mathbf{x}'\lambda^\alpha(\mathbf{x}')\pi_\alpha(\mathbf{x}')\\ &\quad\times\frac{\delta}{\delta N^\alpha(\mathbf{y})}\int \mathrm{d}^3\mathbf{x}N^\beta(\mathbf{x})\mathcal{H}_\beta(\mathbf{x})\\ &= -\int \mathrm{d}^3\mathbf{y}\mathrm{d}^3\mathbf{x}'\mathrm{d}^3\mathbf{x}\lambda^\alpha(\mathbf{x}')\delta(\mathbf{x}',\mathbf{y})\delta(\mathbf{x},\mathbf{y})\mathcal{H}_\alpha(\mathbf{x})\\ &= -\int \mathrm{d}^3\mathbf{x}\lambda^\alpha(\mathbf{x})\mathcal{H}_\alpha(\mathbf{x})\\ &= -d[\lambda^\alpha]\,,\end{aligned} \tag{9.47}$$

$$\{C[\lambda], H_\mathrm{D}\} = -H[\lambda]\,. \tag{9.48}$$

To keep notation light, we will always write constraint formulæ in their non-smeared version.

Since the spin connection is torsion-free, the triad obeys the *homogeneous structure equation*

$$D_{[\alpha}e^i_{\gamma]} = 0\,. \tag{9.49}$$

This expression is an equation of motion in the Lagrangian approach. Its derivation in time-gauge canonical formalism is not obvious since it comes from a second-class constraint which, in this case, is identically zero. Without fixing the time gauge, this second-class constraint is non-trivial and vanishes if (9.49) holds weakly.

As a consequence of (9.49), Γ^i_α is the spatial torsion-free $so(3)$ spin connection

$$2\Gamma^i_\alpha[E] = \epsilon^i{}_{jk}E^{\gamma j}\nabla_\alpha E^k_\gamma\,, \tag{9.50}$$

where the covariant derivative ∇_α contains the Christoffel symbol, in turn expressed as a function of E^α_i (see [52] for the explicit expression). While in second-order formalism (9.50) is the definition of the spin connection, in first-order formalism the spin connection is not an independent variable and (9.50) is a result of the constraint analysis. In particular, this implies that the Lagrange multiplier ${}^{(\Gamma)}\lambda^i_\alpha$ is determined by the equation of motion of $\Gamma^i_\alpha[E]$.

On the other hand, the dynamical equations of the canonical variables $t\cdot\omega^i$, $t\cdot\omega^{ij}$, N and N^α are completely arbitrary, since each of their Poisson brackets with the full Dirac Hamiltonian is proportional to the associated Lagrange multiplier λ_m (the same is true for the equations of motion of e^α_i and Γ^i_α but, as argued above, their Lagrange multipliers are no longer arbitrary). Therefore, at this point a useful simplification of the canonical system of constraints can be naturally provided and consists in treating the above sub-set of canonical variables directly as Lagrange multipliers. This *gauge fixing* could have been done at the very beginning by discriminating, by an educated guess, dynamical variables from Lagrange multipliers. However, the Dirac procedure does not give us any hint about this classification ab initio, so that we have preferred to follow the general analysis and to arrive at this conclusion after having calculated the set of primary and secondary constraints.

Equation (9.41) is solved because of the compatibility equation; then, $D_\alpha E^\alpha_i = 0$ strongly. Equations (9.42), (9.43), (9.44) and (9.45) do not contain Lagrange multipliers λ_m and do not vanish on the primary surface. Hence they have to be considered as secondary constraints.

Equation (9.42) has been arranged to isolate three terms. The first includes some of the new Lagrange multipliers and can be made to vanish by definition. The second is proportional to (9.43) and it vanishes weakly. The third term is strongly equal to zero as it is nothing but the compatibility equation. Overall, (9.42) is redundant with other constraints and it will be ignored from now on.

The left-hand side of (9.43) suggests to define

$$\boxed{\mathcal{R}_i := \epsilon_{ij}{}^{k} K^j_\alpha E^\alpha_k \,.} \tag{9.51}$$

Equations (9.43), (9.44) and (9.45) (i.e., (9.37), (9.38) and (9.51)) must be imposed to vanish weakly, in which case they are called *first-class constraints*. In general, first-class functions of canonical variables are those whose Poisson brackets with every constraint vanish weakly. The Poisson bracket of two first-class constraints, which we omit to report here, is a linear superposition of first-class constraints. Moreover, it is a general result of constrained Hamiltonian systems that first-class primary constraints generate small gauge transformations [51] (see Problem 9.4). The "Dirac conjecture" postulates that also first-class secondary constraints are gauge generators.

Functions which are not first class (and so admit a non-zero Poisson bracket with at least one constraint) are *second class*. After the gauge fixing, we are left with no second-class constraints.

To summarize, the initial complicated system of constraints has been reduced to a set of seven first-class constraints,

$$\boxed{\begin{aligned} \mathcal{R}_i &\approx 0\,,\\ \mathcal{H}_\alpha &\approx 0\,,\\ \mathcal{H} &\approx 0\,, \end{aligned}} \tag{9.52}$$

which reflect the gauge freedom of the physical system. In fact, $\mathcal{R}_i \approx 0$ and $\mathcal{H}_\alpha \approx 0$ establish, respectively, invariance under rotations of the local spatial basis and spacetime diffeomorphisms within the three-surfaces Σ. The super-momentum and super-Hamiltonian constraints correspond, respectively, to the 0α and 00 components of Einstein's equations. The Hamiltonian constraint *both* encodes invariance under time reparametrizations *and* generates the dynamics (time evolution) of the system. Symmetry and dynamics are thus entangled.

A standard counting of degrees of freedom (d.o.f.) shows that, before any gauge fixing,

$$2\times\begin{pmatrix}\text{\# of physical}\\ \text{d.o.f.}\end{pmatrix} = \begin{pmatrix}\text{\# of canonical}\\ \text{variables}\end{pmatrix} - \begin{pmatrix}\text{\# of 2nd-class}\\ \text{constraints}\end{pmatrix} -2\times\begin{pmatrix}\text{\# of 1st-class}\\ \text{constraints}\end{pmatrix}.$$

The number of second-class constraints is even, as it must be in order for them to be completely solved. Once the Dirac bracket [51] is defined for the system, the second-class constraints can be considered as strong equations and the original variables are reintroduced in the theory by using a simple identity.

After gauge fixing, the system is completely described by the $so(3)$-valued extrinsic curvature K^i_α and its momentum E^α_i, together with the field ϕ and its momentum Π_ϕ, for a total of $9 + 9 + 1 + 1 = 20$ canonical variables. Then, the physical degrees of freedom on the phase space are

$$20 - 0 - 2 \times 7 = 6,$$

specifically four corresponding to the two polarizations of the graviton and two associated with the scalar field.

9.1.3 Ashtekar–Barbero Variables

The Ashtekar–Barbero variables are defined as

$$A^i_\alpha := \gamma K^i_\alpha + \Gamma^i_\alpha, \tag{9.53}$$

$$P^\alpha_i := \frac{1}{\gamma} E^\alpha_i, \tag{9.54}$$

where γ is the Barbero–Immirzi parameter and the internal indices $i, j, \dots$ live in the $su(2)$ Lie algebra. For quantization purposes, it is desirable that (9.54) defines a one-to-one mapping between the triad and the momentum P^α_i. This requires γ to be a frame-independent constant. In Problem 9.3, it is shown that the symplectic structure is non-canonical if one tries to define the pair (A, P) with the Barbero–Immirzi field $\beta(x)$ instead of $\beta_0 = -1/\gamma$. β_0 can arise as the expectation value of the Barbero–Immirzi field.

The symplectic structure in the new variables is

$$\boxed{\begin{aligned} \left\{A^i_\alpha(t,\mathbf{x}), P^\gamma_j(t,\mathbf{x}')\right\} &= \kappa^2 \delta^\gamma_\alpha \delta^i_j \delta(\mathbf{x},\mathbf{x}'), \\ \{\phi(t,\mathbf{x}), \Pi_\phi(t,\mathbf{x}')\} &= \delta(\mathbf{x},\mathbf{x}'), \end{aligned}} \tag{9.55}$$

while the other brackets vanish. Since (9.53) and (9.54) define a canonical transformation of variables (a symplectomorphism), one can apply it directly to the constraints without repeating the canonical analysis.

The first-class constraints can be easily rearranged. Equations (9.41) and (9.51) are, respectively, the polar (boost) and axial part of the covariant Gauss tensor, governing frame Lorentz transformations (small gauge transformations). The strong and weak equations $\mathcal{B}_i = 0$ and $\mathcal{R}_i \approx 0$ can be combined into the *Gauss constraint*

$\mathcal{G}_i := \mathcal{R}_i + \mathcal{B}_i/\gamma$ (Problem 9.4):

$$\boxed{\mathcal{G}_i := \mathcal{D}_\alpha P_i^\alpha = \partial_\alpha P_i^\alpha + \epsilon_{ij}{}^k A_\alpha^j P_k^\alpha\,,} \tag{9.56}$$

where $\mathcal{D}$ is the covariant derivative with respect to A_α^i.

The curvature of the Ashtekar–Barbero connection is

$$F_{\alpha\beta}^i := 2\partial_{[\alpha}A_{\beta]}^i + \epsilon^i{}_{jk}A_\alpha^j A_\beta^k\,. \tag{9.57}$$

This is related to (9.31) by the formula

$$R_{\alpha\beta}^i = F_{\alpha\beta}^i + \gamma\epsilon^i{}_{jk}K_\alpha^j K_\beta^k - 2\gamma D_{[\alpha}K_{\beta]}^i\,. \tag{9.58}$$

Ignoring the matter part, (9.37) can be rewritten as

$$\begin{aligned}
\kappa^2\mathcal{H}_\alpha &= 2P_i^\beta D_{[\alpha}A_{\beta]}^i - 2P_i^\beta D_{[\alpha}\Gamma_{\beta]}^i \\
&= P_i^\beta F_{\alpha\beta}^i + \epsilon^i{}_{jk}P_i^\beta A_\alpha^j A_\beta^k - 2\gamma\epsilon^i{}_{jk}P_i^\beta K_{[\alpha}^j A_{\beta]}^k - P_i^\beta R_{\alpha\beta}^i - \epsilon^i{}_{jk}P_i^\beta \Gamma_\alpha^j \Gamma_\beta^k \\
&\approx P_i^\beta F_{\alpha\beta}^i - \gamma K_\alpha^i \mathcal{R}_i\,,
\end{aligned}$$

where in the third (weak) equality we made use of the Bianchi identity $E_i^\gamma R_{\alpha\gamma}^i = 0$ and of the rotation constraint (9.51). Then, dropping the last term the super-momentum constraint is

$$\boxed{\mathcal{H}_\alpha \approx \frac{1}{\kappa^2}P_i^\beta F_{\alpha\beta}^i + \Pi_\phi\partial_\alpha\phi\,.} \tag{9.59}$$

Exploiting the compatibility condition and the rotation constraint, the super-Hamiltonian constraint is, up to a weakly vanishing term,

$$\boxed{\mathcal{H} \approx \frac{\gamma^2 P_i^\alpha P_j^\beta}{2\kappa^2|e|}\left[\epsilon^{ij}{}_k F_{\alpha\beta}^k - 2(1+\gamma^2)K_{[\alpha}^i K_{\beta]}^j\right] + \mathcal{H}_\phi\,,} \tag{9.60}$$

where

$$|e| = \sqrt{|\gamma \det P|}\,. \tag{9.61}$$

The Hamiltonian constraint is non-linear in the connection (via the curvature) and in the extrinsic curvature.

Note that A^i_α does not have definite parity and that the way improper Lorentz transformations act on the Hamiltonian constraint is not transparent. This is a problem intrinsic to the choice of the action.

Loop quantum gravity makes extensive use of the Ashtekar–Barbero connection and its conjugate momentum, which allow to reformulate gravity almost as a gauge theory and facilitate a well-defined quantization. The main difference with respect to gauge theories is in the fact that the Hamiltonian constraint is not only dynamical but it also generates time reparametrizations. Moreover, it contains quadratic extrinsic curvature terms, which decouple from the theory when

$$\gamma = \mp\mathrm{i}\,. \tag{9.62}$$

In this case, the Ashtekar connection A is complex and it is said to be *(anti-)self-dual*. The (anti-)self-dual theory is greatly simplified but, unfortunately, it turns out that it is difficult to quantize, the reason being that the gauge group becomes the non-compact group $SL(2,\mathbb{C})$. Therefore, most of the progress in LQG is based upon the real Ashtekar–Barbero connection.

9.1.4 ADM Variables

The ADM variables were the first to be used in canonical quantization [53, 54]. As we have seen, the foliation into spatial hypersurfaces Σ_t defines the lapse function and shift vector (9.25). The corresponding metric is

$$g_{\mu\nu} = \begin{pmatrix} -N^2 + N^\gamma N_\gamma & N_\alpha \\ N_\beta & h_{\alpha\beta} \end{pmatrix}, \tag{9.63}$$

whose inverse is

$$g^{\mu\nu} = \frac{1}{N^2}\begin{pmatrix} -1 & N^\alpha \\ N^\beta & N^2 h^{\alpha\beta} - N^\alpha N^\beta \end{pmatrix}. \tag{9.64}$$

The geometric meaning of the metric components is the same as before. Given the normal $u^\mu = (-N,0,0,0)^\mu$ to the hypersurface Σ_t, the lapse N relates coordinate time t with proper time $\sigma = \int \mathrm{d}t\, N$ on curves orthogonal to Σ_t. The shift vector vanishes in comoving coordinates. The projected spatial components of the second fundamental form (7.126) are the extrinsic curvature $K_{\alpha\beta} = K_{\alpha\beta}(t,\mathbf{x}) = e^i_{(\alpha}K_{\beta)i}$ of the hypersurface Σ_t with respect to the embedding manifold $\mathcal{M}$:

$$K_{\alpha\beta} = \frac{1}{N}\left[\frac{1}{2}\dot{h}_{\alpha\beta} - \nabla_{(\alpha}N_{\beta)}\right]. \tag{9.65}$$

In terms of (9.65), the Einstein–Hilbert action becomes

$$S = \frac{1}{2\kappa^2}\int \mathrm{d}t\mathrm{d}^3\mathbf{x}\,\sqrt{h}\,N\left[{}^{(3)}R + K_{\alpha\beta}K^{\alpha\beta} - K^2\right] + S_\phi\,, \tag{9.66}$$

where $K := K_\alpha^{\ \alpha}$ is the trace of the extrinsic curvature and h is the determinant of the 3-metric. This result can be found either via a direct computation or from (9.27). We have implicitly thrown away the York–Gibbons–Hawking boundary term $S_\partial \propto \int_{\partial\mathcal{M}} \mathrm{d}^3\mathbf{x}\,\sqrt{h}\,K$ mentioned in Sect. 2.2 [55, 56].[5] In a quantum setting, however (as when looking for instantonic solutions in Euclidean signature), the boundary piece may be important since the classical equations of motion do not hold. It is customary to add its opposite in the definition of the fundamental action, so that it cancels away and (9.66) remains the same.

Define the DeWitt "metric of metrics" (or super-metric) [54]

$$\mathcal{G}_{\alpha\beta\gamma\delta} := h_{\alpha(\gamma}h_{\delta)\beta} - \tfrac{1}{2}h_{\alpha\beta}h_{\gamma\delta}\,, \tag{9.67}$$

and its inverse

$$\mathcal{G}^{\alpha\beta\gamma\delta} := h^{\alpha(\gamma}h^{\delta)\beta} - h^{\alpha\beta}h^{\gamma\delta}\,.$$

Note that the inverse of $\mathcal{G}$ is not $\mathcal{G}$ with indices raised by $h^{\alpha\beta}$. The canonical momenta in the gravity sector are defined as

$$\pi^{\alpha\beta} := \frac{\delta S}{\delta\dot{h}_{\alpha\beta}} = \frac{1}{2\kappa^2}\sqrt{h}\,\mathcal{G}^{\alpha\beta\gamma\delta}K_{\gamma\delta} = \frac{1}{2\kappa^2}\sqrt{h}\,(K^{\alpha\beta} - h^{\alpha\beta}K)\,, \tag{9.68a}$$

$$\pi^0 := \frac{\delta S}{\delta\dot{N}} = 0\,, \tag{9.68b}$$

$$\pi^\alpha := \frac{\delta S}{\delta\dot{N}_\alpha} = 0\,. \tag{9.68c}$$

Here we used the time derivative $\dot{} = t^\mu\nabla_\mu$ rather than the Lie derivative $\mathcal{L}_t$, but this difference is immaterial due to the simple form of the action (9.66). Equation (9.68a) is the spatial version of (7.125).

The total Dirac Hamiltonian is

$$H_\mathrm{D} = \int \mathrm{d}^3\mathbf{x}(N^\alpha\mathcal{H}_\alpha + N\mathcal{H} + \pi^0\dot{N} + \pi^\alpha\dot{N}_\alpha)\,, \tag{9.69}$$

[5] A derivation of the York–Gibbons–Hawking boundary term can be found in [57]. Within first-order formalism, the boundary term is discussed in [58, 59].

where

$$\boxed{\mathcal{H}_\alpha = -2\nabla^\beta \pi_{\alpha\beta} + \Pi_\phi \partial_\alpha \phi\,,} \tag{9.70}$$

and

$$\mathcal{H} = \frac{2\kappa^2}{\sqrt{h}} \pi^{\alpha\beta} G_{\alpha\beta\gamma\delta} \pi^{\gamma\delta} - \frac{\sqrt{h}}{2\kappa^2}\, {}^{(3)}R + \mathcal{H}_\phi\,. \tag{9.71}$$

More explicitly, the last expression is

$$\boxed{\mathcal{H} = \frac{\sqrt{h}}{2\kappa^2} \left[K^{\alpha\beta} K_{\alpha\beta} - K^2 - {}^{(3)}R \right] + \mathcal{H}_\phi\,.} \tag{9.72}$$

The classical constraints are the primary relations

$$\pi^0 \approx 0\,, \qquad \pi^\alpha \approx 0\,, \tag{9.73}$$

and the momentum and Hamiltonian constraints

$$\boxed{\mathcal{H}_\alpha \approx 0\,, \qquad \mathcal{H} \approx 0\,.} \tag{9.74}$$

The latter are first-class constraints (Problem 9.5). After imposing second-class constraints and skimming out Lagrange multipliers, (9.69) is reduced to

$$H_\mathrm{D} = \int \mathrm{d}^3\mathbf{x}\, (N^\alpha \mathcal{H}_\alpha + N\mathcal{H})\,. \tag{9.75}$$

9.2 Wheeler–DeWitt Equation

Once the phase space of a Hamiltonian system has been identified, it can be quantized according to the Dirac prescription. The first step is extremely important and consists in finding suitable phase-space coordinates (Q, P) (e.g., Ashtekar–Barbero or ADM variables). Not all possible choices will lead to a well-defined quantization. In the case of gravity, for example, the ADM variables turn out to be a natural but also ill-suited choice, while the Ashtekar–Barbero connection, although less intuitive from the point of view of classical metric spacetimes, allows for greater progress.

The second step is to identify an algebra of functionals $f[P,Q]$ with well-defined Poisson brackets, closed under Poisson brackets (meaning that the bracket $\{f,g\}$ of two elements f and g belong to the algebra) and separating distinct points in phase space (i.e., if $Q \sim Q'$ and $P \sim P'$ under a small gauge transformation, then $f[Q',P'] = f[Q,P]$). The third step is to find an irreducible representation of the algebra on a Hilbert space, that is to say, a mapping $f \to \hat{f}$ from elements of the algebra to operators on a Hilbert space. Poisson brackets are mapped into commutators, complex conjugation to adjointness, and so on. Only at this point can one deal with the quantum constraints.

We will not follow this detailed agenda but we will introduce some basic concepts first in ADM and then in Ashtekar–Barbero variables, with special reference to quantum cosmology in the next chapter. The Planck constant $\hbar$ is restored.

9.2.1 Superspace and Quantization

The DeWitt metric (9.67) can be regarded as the metric of a six-dimensional manifold with pseudo-Riemannian structure [54]. Indices can be collected in pairs to define the metric $\mathcal{G}_{AB} = \mathcal{G}_{(\alpha\beta)(\gamma\delta)}$, $A = 1,\ldots,6$, and the elements h^A as $h^\alpha = h_{\alpha\alpha}$, $\alpha = 1,2,3$, $h^4 = 2^{-1/2}h_{23}$, $h^5 = 2^{-1/2}h_{31}$, $h^6 = 2^{-1/2}h_{12}$. In the Minkowski case $h_{\alpha\beta} = \delta_{\alpha\beta}$, it is easy to solve the characteristic equation $\det(\mathcal{G} - \lambda\mathbb{1}) = 0$ and to find that $\mathcal{G}$ has signature $(-,+,\cdots,+)$. This can be extended to a metric of fields accounting for all the physical degrees of freedom of the problem. In the case of a matter scalar field, for example, one can define the block-diagonal metric

$$\mathbb{G} := \frac{1}{\sqrt{h}}\begin{pmatrix} 4\kappa^2\mathcal{G}^{AB} & 0 \\ 0 & 1 \end{pmatrix}, \qquad (9.76)$$

which appears in the Hamiltonian constraint. $\mathbb{G}$ is defined over the infinite-dimensional space $\mathrm{Riem}_\phi(\Sigma_t)$ of all Riemannian 3-metrics $h_{\alpha\beta}(t,\mathbf{x})$ and scalar-field configurations $\phi(t,\mathbf{x})$ at each point $\mathbf{x} \in \Sigma_t \subset \mathcal{M}$ in a spatial hypersurface with fixed topology:

$$\mathrm{Riem}_\phi(\Sigma_t) := \left\{h_{\alpha\beta}(t,\mathbf{x}),\ \phi(t,\mathbf{x})\,|\,\mathbf{x}\in\Sigma_t\right\}. \qquad (9.77)$$

Since diffeomorphisms define equivalence classes of physical geometries, the actual configuration space is the so-called *superspace*[6]

$$\mathcal{S}(\Sigma_t) := \mathrm{Riem}_\phi(\Sigma_t) \setminus \mathrm{Diff}_0(\Sigma_t), \qquad (9.78)$$

[6]This has nothing to do with the superspace of supersymmetry of Sect. 5.12.

where Diff_0 is the space of maps realizing infinitesimal foliation-preserving diffeomorphisms.

The superspace (9.78) should run over all the canonical coordinates Q^A (conjugate to their momenta P_A), including the lapse and the shift. We have not included N and N_α because they factor out as Lagrange multipliers. It is instructive to see this explicitly. Consider the extended superspace $\tilde{\mathcal{S}} = \mathcal{S} \cup \{N(t,\mathbf{x}),\, N_\alpha(t,\mathbf{x})\,|\,\mathbf{x} \in \Sigma_t\}$. In the Dirac quantization, $\tilde{\mathcal{S}}$ is regarded as the Hilbert space whereon we define functionals $\Psi[h_{\alpha\beta}, \phi, N, N_\alpha]$ of the canonical variables. The classical momenta are promoted to functional derivative operators

$$P^A \to \hat{P}^A := -\mathrm{i}\hbar \frac{\delta}{\delta Q_A}\,. \tag{9.79}$$

In particular,

$$\pi^{\alpha\beta} \to \hat{\pi}^{\alpha\beta} := -\mathrm{i}\hbar \frac{\delta}{\delta h_{\alpha\beta}}\,, \tag{9.80a}$$

$$\Pi_\phi \to \hat{\Pi}_\phi := -\mathrm{i}\hbar \frac{\delta}{\delta \phi}\,, \tag{9.80b}$$

$$\pi^0 \to \hat{\pi}^0 := -\mathrm{i}\hbar \frac{\delta}{\delta N}\,, \tag{9.80c}$$

$$\pi^\alpha \to \hat{\pi}^\alpha := -\mathrm{i}\hbar \frac{\delta}{\delta N_\alpha}\,. \tag{9.80d}$$

The quantum version of a constraint $C(Q,P)$, denoted with a hat, is a composite operator acting on the Hilbert space of functionals Ψ. It is made of the above momentum operators and of multiplication operators on $\mathcal{S}$ corresponding to the canonical variables:

$$h_{\alpha\beta} \to \hat{h}_{\alpha\beta} := h_{\alpha\beta}\,, \qquad \phi \to \hat{\phi} := \phi\,. \tag{9.80e}$$

The classical Poisson bracket is replaced by the commutator bracket as in standard quantum mechanics,

$$\{\cdot,\cdot\} \to -\frac{\mathrm{i}}{\hbar}[\cdot,\cdot]\,. \tag{9.81}$$

Each non-smeared constraint equation

$$\hat{C}\Psi = \hat{C}[\hat{Q},\hat{P}]\Psi = 0 \tag{9.82}$$

is an infinite set of relations, one for each spatial point in the foliation Σ_t. The way the operators (9.80) are arranged in the definition of the quantum constraints gives rise to inequivalent orderings. The *operator ordering* problem is a source of ambiguity which, typically, cannot be removed in canonical approaches. In general, a classical constraint is of the form

$$\mathcal{C} = \sum_n \mathcal{C}_n\,, \qquad \mathcal{C}_n = \prod_m A_{n,m}\,,$$

where $A_{n,m}$ are functions of the canonical variables. Take, for instance, terms like $A^l B$. When quantizing, we have an infinite number of options, as many as the ways to resolve the classical identity:

$$A^l B \to \begin{cases} \widehat{A^l}\hat{B} \\ (\hat{A})^l\hat{B} \\ \hat{B}(\hat{A})^l \\ (\hat{A})^{l-q}\hat{B}(\hat{A})^q \\ \dots \end{cases} . \tag{9.83}$$

Sometimes, independent arguments can select one or more choices over the others; below we will see an example.

The functional obeying the quantum version of the constraints is often called the *wave-function of the Universe*. The primary constraints (9.73) are

$$\hat{\pi}^0\Psi = -i\hbar\frac{\delta\Psi}{\delta N} = 0\,, \qquad \pi^\alpha = -i\hbar\frac{\delta\Psi}{\delta N_\alpha} = 0\,, \tag{9.84}$$

implying that Ψ does not depend on any of the non-dynamical variables:

$$\Psi = \Psi[h_{\alpha\beta}, \phi]\,.$$

From now on, we drop $\tilde{\mathcal{S}}$ and refer to $\mathcal{S}$ as the superspace.

Choosing derivative operators to act to the right, the momentum constraint is

$$\hat{\mathcal{H}}_\alpha\Psi = -i\hbar\left(\partial_\alpha\phi\frac{\delta}{\delta\phi} - 2\nabla^\beta\frac{\delta}{\delta h^{\alpha\beta}}\right)\Psi = 0\,. \tag{9.85}$$

One can show that this condition is satisfied if Ψ is invariant under coordinate transformations $\mathbf{x} \to \mathbf{x}'$ in the hypersurface Σ_t [60]. This is in agreement with the definition of superspace (9.78) and the classical interpretation of the momentum constraint as the generator of diffeomorphisms on Σ_t.

The quantum dynamics is governed by the Hamiltonian constraint, which has come to be known as the *Wheeler–DeWitt* (WDW) *equation* [54, 61]. With derivative operators acting to the right of $\mathcal{G}$, it reads

$$\hat{\mathcal{H}}\Psi = \left[-\frac{\hbar^2}{\sqrt{h}}\left(2\kappa^2\mathcal{G}_{\alpha\beta\gamma\delta}\frac{\delta^2}{\delta h_{\alpha\beta}\delta h_{\gamma\delta}} + \frac{1}{2}\frac{\delta^2}{\delta\phi^2}\right) + U(h_{\alpha\beta}, \phi)\right]\Psi = 0\,, \tag{9.86}$$

where the effective potential is

$$\boxed{U(h_{\alpha\beta},\phi)=\sqrt{h}\left[-\frac{{}^{(3)}R}{2\kappa^2}+\frac{1}{2}h^{\alpha\beta}\partial_\alpha\phi\partial_\beta\phi+V(\phi)\right].}\tag{9.87}$$

With a different operator ordering where $\mathcal{G}$ is in between the momenta π^A [54, 62], the WDW equation becomes

$$\boxed{\hat{\mathcal{H}}\Psi=\left[-\frac{1}{2}\hbar^2\nabla_A\nabla^A+U(h^A,\phi)\right]\Psi=0\,,}\tag{9.88}$$

where

$$\nabla_A\nabla^A:=\frac{1}{\sqrt{-\mathbb{G}}}\partial_A\left(\sqrt{-\mathbb{G}}\,\mathbb{G}^{AB}\partial_B\right)\tag{9.89}$$

is the Laplace–Beltrami operator in the super-metric. The operator ordering ambiguity is removed if one requires invariance of the WDW equation under a conformal transformation of the super-metric $\mathbb{G}$ and the lapse function [63, 64]. Then, in (9.88) one must add an $O(\hbar^2)$ operator proportional to the scalar curvature $\mathcal{R}$ of the n-dimensional superspace [54], with conformal coupling:

$$\nabla_A\nabla^A\to\nabla_A\nabla^A-\frac{n-2}{4(n-1)}\mathcal{R}\,.\tag{9.90}$$

We should bear in mind that gravity is not perturbatively renormalizable when the three-metric $h_{\alpha\beta}$ is elected as the fundamental field, so that we expect that the quantum constraints in ADM variables be invalid, or receive corrections, when sufficiently small volumes are considered. Therefore, the interpretation of Ψ should be as the wave-function of a *large* universe. The adjectives "small" and "large" are intentionally used in a loose sense. The understanding is that these universes are large compared to Planck-size patches but small enough to justify a quantum treatment.

9.2.2 *Semi-classical States*

In quantum mechanics and quantum field theory on classical spacetimes, the *path integral* or *transition amplitude* [65, 66] gives, after squaring, the probability for a given quantum state at the initial time t_i to evolve into another state at the final time t_f. A general-covariant, background independent theory of gravitation does not have a preferred time coordinate by which to define "initial" and "final" but, nevertheless, it may be possible to define a notion of path integral. In particular, the most general solution of the WDW equation on a spatial hypersurface Σ with induced metric $h_{\alpha\beta}$ and a generic matter content ϕ_h can be formally written as the path integral [56, 67]

$$\Psi[h_{\alpha\beta}, \phi_h] = \int [\mathcal{D}g][\mathcal{D}\phi]\, \mathrm{e}^{\frac{\mathrm{i}}{\hbar}S[g_{\mu\nu},\phi]}, \tag{9.91}$$

where the functional integration is over all Lorentzian 4-geometries and matter fields acquiring the values $h_{\alpha\beta}$ and ϕ_h on Σ. The measure is, in general, ill-defined and the integral diverges. One can restrict the sum to geometries with Euclidean signature and perform a Wick rotation $t \to -\mathrm{i}t$, so that one replaces $\mathrm{i}S$ with the Euclidean action I, $S \to \pm\mathrm{i}I$, and the integrand is $\mathrm{e}^{\pm I/\hbar}$ (the sign depends on various considerations [68, 69]). However, this does not guarantee the convergence of expression (9.91), which is too general to be manageable. In order to find semi-classical solutions of the WDW equation (9.88), we can employ the *Wentzel–Kramers–Brillouin* (WKB) *approximation* [70–73]. This approximation finds applications in quantum mechanics, inflationary cosmology and cosmological perturbation theory [74–77] and in the WDW equation [78–82].

Classically, the WKB method can be illustrated as follows. Consider an ordinary differential equation in one dimension,

$$\sum_{m=0}^{M} c_m(t)\, \partial_t^m f(t) = 0\,, \qquad c_M(t) = c \ll 1\,, \qquad t \geqslant t_0\,.$$

We look for perturbative solutions of the form

$$f(t) \simeq f_{\mathrm{WKB}} = \exp\left[\frac{1}{\varepsilon}\sum_{n=0}^{N} \varepsilon^n f_n(t)\right], \qquad N \geqslant M-1\,, \tag{9.92}$$

in the formal limit $\varepsilon \to 0$. "Formal" means that, eventually, the expansion parameter remains finite and both c and ε can be set to be $O(1)$.

If $\sqrt{\varepsilon} \sim 1/\Delta t$ is the inverse of a large time interval Δt over which the system evolves slowly, the expansion (9.92) is called *adiabatic*. For example, plugging this *Ansatz* up to order $N = 1$ into the second-order equation

$$[c\partial_t^2 + \omega^2(t)]f = 0\,,$$

where $\omega^2 \in \mathbb{R}$ can also be negative, one finds

$$0 \simeq \frac{c}{\varepsilon^2}\left(\dot{f}_0 + \varepsilon\dot{f}_1\right)^2 + \frac{c}{\varepsilon}\left(\ddot{f}_0 + \varepsilon\ddot{f}_1\right) + \omega^2(t) = \frac{c}{\varepsilon^2}\dot{f}_0^2 + \frac{c}{\varepsilon}\left(\ddot{f}_0 + 2\dot{f}_0\dot{f}_1\right) + \omega^2(t)\,,$$

where we have neglected the sub-leading term $\ddot{f}_1 + \dot{f}_1^2$. The leading-order term must match ω^2. Setting $c = \varepsilon^2$, this yields

$$\dot{f}_0^2 = -\omega^2(t) \quad \Rightarrow \quad f_0(t) = \pm\mathrm{i}\int_{t_0}^{t} \mathrm{d}t'\omega(t'),$$

while the next-to-leading term vanishes if

$$\dot{f}_1 = -\frac{\ddot{f}_0}{2\dot{f}_0} \quad \Rightarrow \quad f_1(t) = -\tfrac{1}{2}\ln\omega(t),$$

up to integration constants. Collecting these results, the approximated WKB solution is

$$f_{\mathrm{WKB}}(t) = \frac{1}{\sqrt{\omega(t)}}\left[b_+ \mathrm{e}^{\frac{\mathrm{i}}{\sqrt{c}}\int_{t_0}^{t}\mathrm{d}t'\omega(t')} + b_- \mathrm{e}^{-\frac{\mathrm{i}}{\sqrt{c}}\int_{t_0}^{t}\mathrm{d}t'\omega(t')}\right], \tag{9.93}$$

where $b_\pm$ are constants. Depending on the sign of ω^2, at any time slice t the WKB solution can be regarded as a superposition of plane waves or exponentials.

In quantum mechanical systems, the expansion parameter is $\sqrt{\varepsilon} \sim \hbar$ and pure-phase WKB solutions are regarded as *semi-classical*. In this approximation, (9.91) is peaked at a classical trajectory $(h_{\alpha\beta}(x), \phi_h(x))$ and one can neglect integration over all possible 4-geometries and field configurations. Consider the Hamiltonian constraint (9.88) and the trial wave-function

$$\Psi = \sum_n \Psi_n = \sum_n \Psi_{0,n}\mathrm{e}^{\frac{1}{\hbar}(\mathrm{i}S_n - I_n)}, \tag{9.94}$$

where $\Psi_{0,n}$ are slowly-varying complex amplitudes and S_n and I_n are real-valued functionals. We can assume that each mode Ψ_n obeys the WDW equation separately. Then, denoting with ∇ covariant gradients in superspace (for a superspace scalar, $\nabla_A = \partial_A$),

$$\begin{aligned} 0 &= \left(-\tfrac{1}{2}\hbar^2\nabla^2 + U\right)\Psi_n \\ &= -\tfrac{1}{2}\nabla\cdot\left[\hbar^2\nabla\Psi_{0,n} + \hbar\Psi_{0,n}\nabla(\mathrm{i}S_n - I_n)\right]\mathrm{e}^{\frac{1}{\hbar}(\mathrm{i}S_n - I_n)} \\ &= \mathrm{e}^{\frac{1}{\hbar}(\mathrm{i}S_n - I_n)}\left[\Psi_{0,n}\left(\tfrac{1}{2}X_1 + \mathrm{i}X_2\right) - \hbar X_3 - \tfrac{1}{2}\hbar^2\nabla^2\Psi_{0,n}\right], \end{aligned} \tag{9.95}$$

where the dot denotes the scalar product in superspace and

$$X_1 = (\nabla S_n)^2 - (\nabla I_n)^2 + 2U, \tag{9.96a}$$

$$X_2 = \nabla S_n \cdot \nabla I_n, \tag{9.96b}$$

$$X_3 = \nabla\Psi_{0,n}\cdot\nabla(\mathrm{i}S_n - I_n) + \tfrac{1}{2}\Psi_{0,n}\nabla^2(\mathrm{i}S_n - I_n). \tag{9.96c}$$

Ignoring the $O(\hbar^2)$ term, the solution of (9.95) is found after setting

$$X_1 = 0\,, \qquad X_2 = 0\,, \qquad X_3 = 0\,.$$

Equation (9.96c) determines the real and imaginary part of $\Psi_{0,n}$ once S_n and I_n are known. These are fixed by (9.96a) and (9.96b).

In analogy with the quantum mechanical particle in a potential, if $I_n = 0$ one can interpret the single mode $\Psi_n \sim \mathrm{e}^{\mathrm{i}S_n/\hbar}$ as the wave-function of a post-big-bang, sufficiently large universe in the classical oscillatory regime with Lorentzian action S_n. If $S_n = 0$, the exponential wave-function $\Psi_n \sim \mathrm{e}^{-I_n/\hbar}$ represents a universe in the classically forbidden region across the potential, reachable through a tunneling process described by the Euclidean action I_n. This suggests that we can go to classically admissible regions in phase space if superspace gradients of the imaginary part of the action are negligible with respect to the real part,

$$(\nabla S_n)^2 \gg (\nabla I_n)^2\,. \tag{9.97}$$

Dropping the subscript n, we then consider a single semi-classical WKB mode of (9.94) of the form[7]

$$\Psi = \Psi_0\,\mathrm{e}^{\frac{\mathrm{i}}{\hbar}S}\,. \tag{9.98}$$

At this point we face a problem. In general, having dropped the exponential factor $\mathrm{e}^{-I/\hbar}$, this state is not square integrable. This implies that it does not belong to the physical Hilbert space of square-integrable distributions and that the probabilistic interpretation associated with wave-functions breaks down. For the usual Schrödinger equation of non-relativistic quantum mechanical systems, thanks to the time-evolution operator $-\mathrm{i}\hbar\partial_t$ one can define Gaussian states, wave-packets with a probability density $|\Psi|^2$ peaked about a classical trajectory. In this sense, Gaussian states are regarded as semi-classical. Then, one can easily relate Gaussian and semi-classical WKB states, thus showing explicitly that semi-classical WKB states are approximations of square-integrable functionals [80, 83]. In gravity this is not possible in a forthright way, since there is no explicit time dependence in the quantum Hamiltonian constraint. WKB wave-functions annihilate the Hamiltonian constraint and are thus "energy eigenstates with zero energy." Therefore, it is not clear in which sense semi-classical WKB states of quantum gravity are approximations of physical states. In fact, there seems to be no physical reason why $|\Psi|^2$ should be zero at the boundary of superspace. On the other hand, WKB states can give sensible descriptions of what a semi-classical universe could look like and, a posteriori, it is reasonable to regard them as approximations

[7]Very often in the literature, WKB wave-functions are *defined* as (9.98) ($I \equiv 0$), in which case "WKB state" and "semi-classical WKB state" are one and the same thing. Here we keep the distinction.

of yet-unknown (but hopefully existing) states in the full theory. In the absence of tools performing the appropriate checks, the use of WKB states in quantum gravity is ambiguous and requires a certain dose of faith, as well as caution in accepting the ensuing results. In Sect. 11.5.2, we will see a remarkable example where such tools are indeed available: in group field theory, WKB-like states can be constructed as approximations of known normalizable states.

Having said that, let us study (9.98). The expression $X_1 = 0$ becomes the Einstein–Hamilton–Jacobi equation [78, 84]

$$(\nabla S)^2 + 2U = \mathbb{G}^{AB}\nabla_A S \nabla_B S + 2U = 0\,. \tag{9.99}$$

Comparing this expression with the classical constraint, one can identify the classical momenta π_A with the gradient of the real action S,

$$\pi_A = \partial_A S\,. \tag{9.100}$$

Sometimes these are called Hamilton–Jacobi momenta. Define now the vector field

$$\frac{\mathrm{d}}{\mathrm{d}s} := \nabla^A S \nabla_A\,. \tag{9.101}$$

Taking the superspace derivative of (9.99) and plugging in (9.100), one gets the geodesic equation

$$\begin{aligned} 0 &= \partial^A S \partial_A \partial_C S + \tfrac{1}{2}\partial_A S \partial_B S \partial_C \mathbb{G}^{AB} + \partial_C U \\ &= \frac{\mathrm{d}\pi_C}{\mathrm{d}s} + \tfrac{1}{2}\pi_A \pi_B \partial_C \mathbb{G}^{AB} + \partial_C U\,. \end{aligned} \tag{9.102}$$

The parameter s in (9.101) is called *WKB time* [85]. The signature of the DeWitt metric already suggested the possibility to choose an intrinsic time coordinate in superspace [86]. This definition depends on the choice of a solution $h_{\alpha\beta}$ of the Einstein equations but not on the choice of coordinates. However, the definition of the vacuum state of the theory does depend on coordinate transformations, as it happens in quantum field theories on curved spacetime [80].

Equation $X_3 = 0$ in the classically allowed region (9.97) simply becomes

$$\frac{\mathrm{d}\Psi_0}{\mathrm{d}s} + \frac{1}{2}\Psi_0 \nabla^2 S = 0\,, \tag{9.103}$$

yielding

$$\Psi_0 = \mathcal{A}\exp\left(-\frac{1}{2}\int \mathrm{d}s \nabla^2 S\right), \tag{9.104}$$

where $\mathcal{A} \in \mathbb{C}$ is an integration constant. Summarizing, the semi-classical WKB state (9.98) is

$$\Psi = \mathcal{A} \exp\left(\frac{\mathrm{i}}{\hbar} S - \frac{1}{2} \int \mathrm{d}s \nabla^2 S\right). \tag{9.105}$$

This wave-function is characterized by the precise correlation (9.100) between conjugate variables [80, 83]. It is the analogue of a wave-packet peaked about a classical particle trajectory in ordinary quantum mechanics. However, a superposition of WKB states is no longer semi-classical and, as in the case of inflationary perturbations in Sect. 5.6.1, one could invoke a decoherence process [87].

Semi-classical states can be useful to understand how to track down quantum-gravity effects in the empirical world. However, classical gravity is already too weak to affect particle physics in ordinary laboratory set-ups and WDW canonical quantum gravity corrections are even smaller. To give a quantum mechanical example, these corrections alter the Schrödinger equation and, in particular, modify the spectral lines of hydrogen-type atoms; the effect is unobservable [81]. In this respect, as we have already emphasized in early occasions, cosmology of the early universe and black-hole physics are generally regarded as more promising test grounds for quantum-gravity models.

9.2.3 *Boundary Conditions*

The Wheeler–DeWitt equation does not encode a unique quantum dynamics and one must specify boundary conditions for the wave-function of the Universe (9.91), fixing a pre-classical geometry. Various boundary conditions have been put forward, chiefly the *Hartle–Hawking no-boundary* proposal [88] and the *Vilenkin tunneling* proposal [89], among others [90–92].

The Hartle–Hawking no-boundary proposal assumes that the fundamental formulation of (9.91) is in Euclidean signature and that Lorentzian spacetimes emerge from Riemannian quantum configurations only in certain cases. Second, the path integral in (9.91) is estimated over complex 4-geometries such that the Lorentzian universe emerges from compact Riemannian configurations [62, 68, 88, 93–105]. Thus, the "initial" boundary of the path integral is a Riemannian manifold with no boundary at all, while the final one is the present universe. A typical initial configuration is half the 4-sphere S^4, "sewn" to de Sitter spacetime [89]. Contrary to initial expectations, the no-boundary prescription does not fix the solution Ψ_{HH} of the WDW equation uniquely [96, 97]. A density matrix approach describing a microcanonical ensemble of cosmological models generalizes the vacuum state of the Hartle–Hawking proposal to a quasi-thermal ensemble [106–108].

The Vilenkin tunneling proposal pictures the nucleation of a bubble universe from nothing by quantum tunneling [68, 69, 89, 98, 99, 109–113]. The Lorentzian path integral (9.91) is summed over compact 4-geometries with the restriction to solutions Ψ_{V} with only "outgoing" modes at singular boundaries of superspace. For a WKB mode $\sim e^{\mathrm{i}S_n}$, the "outgoing" condition is with respect to WKB time (9.101)

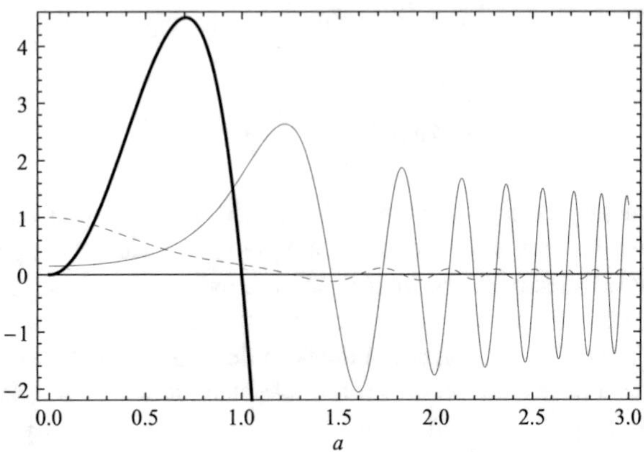

Fig. 9.1 The no-boundary (*thin solid curve*) and the real part of the tunneling (*dashed curve*) wave-functions in a cosmological simplified setting described in Sect. 10.2.1, for $\kappa = 1 = H$. The no-boundary wave-function is damped through the superspace potential barrier U (*thick curve*), while the other tunnels through it

and requires that the superspace vectors $-\nabla S_n$ point out of superspace at the boundary. Lorentzian configurations beginning from degenerate $(3+1)$-geometries are thus excluded, since the nucleation point is non-singular; still, allowed solutions can end at a singularity. A cosmological example [89, 109] illustrates the point. The solution of the Friedmann equation (2.81) for a closed universe ($\textsf{K} = 1$) and a cosmological constant is (2.111). The universe bounces at $t = 0$ after an infinite period of contraction. The cosmological constant is, more generally, the false vacuum of a scalar field (i.e., a local maximum of its potential), which later decays into true vacuum. Analytic continuation $t \to -it$ leads to the Hawking–Moss instanton [56, 114]

$$a(t) = H^{-1} \cos(Ht) , \tag{9.106}$$

where $|t| < \pi/(2H)$ and which describes the four-sphere S^4 of radius H^{-1}.[8] As in quantum mechanics, the solution of the classical Euclidean equations represents tunneling from a barrier at $t = 0$. In this simplified setting, the path integral (9.91) is peaked about the de Sitter instanton and the wave-function $\Psi_{\rm V}$ tunnels through the superspace potential barrier U. In the Hartle–Hawking proposal, on the other hand, the wave-function $\Psi_{\rm HH}$ is damped through the barrier, the universe does bounce at $t = 0$ and the evolution is symmetric in time. Also, while the no-boundary prescription yields real wave-functions, the tunneling proposal in general does not. Figure 9.1 shows the general behaviour of $\Psi_{\rm HH}$ and ${\rm Re}(\Psi_{\rm V})$, in a typical cosmological model which we will describe in Sect. 10.2.1.

[8] For the inquisitive reader, we notice that the profile (9.106) is periodic with period $1/\mathcal{T}_H = 2\pi/H$. Its inverse is precisely the de Sitter temperature (5.132) giving the size of a field fluctuation. An explanation of this fact can be found in [115, 116].

9.3 Some Features of Loop Quantum Gravity

Modulo some important differences, a superspace can be defined also for Ashtekar–Barbero variables. In this case, geometry is not necessarily associated with a metric and the first-order superspace can include also a degenerate sector of triads and connections not corresponding to metric structures. Moreover, the quotient space is taken over the group $SU(2)$ of small gauge transformations (this is isomorphic to the universal cover of $SO(3)$, $\mathrm{Spin}(3) \cong SU(2)$).

The connection itself cannot be promoted to a well-defined operator; only the *holonomy* of the connection can be consistently quantized. The holonomy along an oriented path e is an element of the $SU(2)$ group defined as

$$h_e := \mathcal{P}\exp\left[\int_e \mathrm{d}\lambda\, e^\alpha(\lambda)A^i_\alpha(\lambda)\tau_i\right], \tag{9.107}$$

where $e^\alpha(\lambda) = \mathrm{d}x^\alpha/\mathrm{d}\lambda$ is the tangent vector along the path parametrized by λ and τ_i, $i = 1, 2, 3$, are the three generators of the $su(2)$ algebra in irreducible j-representation (of dimension $2j + 1$). Expanding the exponential, the path ordering $\mathcal{P}$ cancels the factor $1/n!$ in the Taylor series, so that for a path of length λ_e

$$h_e = \mathbb{1} + \sum_{n=1}^{+\infty}\int_0^{\lambda_e}\mathrm{d}\lambda_1\int_{\lambda_1}^{\lambda_e}\mathrm{d}\lambda_2\dots\int_{\lambda_{n-1}}^{\lambda_e}\mathrm{d}\lambda_n\prod_{k=1}^{n}\left[e^\alpha(\lambda_k)A^i_\alpha(\lambda_k)\tau_i\right]. \tag{9.108}$$

Holonomies describe how the connection is transported along any given path. If the path is closed, h_e is called a *loop*. In the original formulation of LQG, the kinematical Hilbert space was spanned by a basis of loops, hence the name of the theory.

For technical convenience, in the quantization of the Hamiltonian constraint one can fix the representation to be the fundamental one, i.e., $j = 1/2$. This choice may be justified also by theoretical considerations,[9] which however are not compelling and can be bypassed. In this representation, one introduces the Pauli matrices (5.208), so that the generators are

$$\tau_i = \frac{\sigma_i}{2\mathrm{i}}, \qquad \tau_i\tau_j = \frac{\epsilon_{ij}{}^k\tau_k}{2} - \mathbb{1}_2\frac{\delta_{ij}}{4}, \tag{9.109}$$

[9]In the high-j case, the Hamiltonian constraint is a difference equation of higher-than-second order. This may lead to an enlargement of the physical Hilbert space and, as a consequence, to the presence of solutions with incorrect large-volume limit [117]. Even if this were not the case, there is evidence (in $2 + 1$ dimensions the proof is actually complete) that LQG has a well-defined continuum limit to quantum field theory only in the fundamental representation of the gauge group [118].

where $\mathbb{1}_2$ is the 2×2 identity matrix. We will not fix the spin label j unless stated otherwise.

The quadratic Casimir invariant C_2 in representation j is given by the algebraic relation

$$\sum_i \tau_i \tau_i = -C_2(j)\mathbb{1}_2 = -j(j+1)\mathbb{1}_2 \,. \tag{9.110}$$

The variable conjugate to the holonomy is not the triad by its *flux* through an arbitrary two-surfaces Σ:

$$E[\Sigma] = \int_\Sigma \mathrm{d}^2\mathbf{y}\, E^i_\alpha n^\alpha \tau_i \,, \tag{9.111}$$

where n^α is the normal to Σ. These are interpreted as elementary areas.

The spatial volume of a given space can be expressed as

$$\mathcal{V} = \int \mathrm{d}^3\mathbf{x}\sqrt{|\det E|} \,. \tag{9.112}$$

The determinant (9.61) of the densitized triad appears in the Hamiltonian constraint (9.60) as an *inverse* power. In quantizing inverse powers of the volume, one faces the problem that naive inverse operators $(\hat{\mathcal{V}})^{-s}$, with $s > 0$, are not well-defined. As a standard procedure, one rewrites inverse powers of the densitized triad using classical Poisson brackets involving positive powers of $\mathcal{V}$ [119, 120]. For instance, in (9.60) we want to rewrite the term

$$\frac{E^\alpha_i E^\gamma_j}{2\kappa^2|e|}\epsilon^{ij}{}_k F^k_{\alpha\gamma} \,.$$

Starting from (9.8) and the property (9.7), one can prove Thiemann identity

$$\begin{aligned}
\epsilon^{ijk}\frac{E^\alpha_i E^\gamma_j}{\sqrt{|\det E|}} &= 2\epsilon^{\alpha\gamma\beta}\frac{\mathcal{V}^{1-q}}{q}\frac{\delta\mathcal{V}^q}{\delta E^\beta_k} \\
&= \frac{4}{\kappa^2\gamma q l_0}\mathcal{V}^{1-q}\epsilon^{\alpha\gamma\beta}\sum_{i'} e^{i'}_\beta \mathrm{tr}(\tau_k h_{i'}\{h^{-1}_{i'}, \mathcal{V}^q\}) \,,
\end{aligned} \tag{9.113}$$

where we introduced an ambiguity parameter $q \in \mathbb{R}$ and in the second equality we used the expression $h_k \delta h^{-1}_k / \delta A^i_\alpha = -l_0 e^\alpha_k \tau_i$, where l_0 is the length of the holonomy path.

The field strength $F^k_{\alpha\gamma}$ can be manipulated as follows. One considers the loop $h_{\Box_{ij}} := h_i h_j h^{-1}_i h^{-1}_j$ on a closed oriented square path whose edges have length l_0 and are labelled by spin indices i, j, i, j. Then, from (9.108) for a closed path and applying Stokes theorem one gets [121, 122]

$$F^k_{\alpha\gamma} = -2e^i_\alpha e^j_\gamma \lim_{l_0\to 0} \frac{\mathrm{tr}(\tau_k h_{\Box_{ij}})}{l_0^2}\,. \tag{9.114}$$

The field strength is a local function of spacetime coordinates and is obtained in the limit where the area of the plaquette $\Box_{ij}$ tends to zero. Overall,

$$\frac{E^\alpha_i E^\gamma_j}{2\kappa^2|e|}\epsilon^{ij}{}_k F^k_{\alpha\gamma} = -\frac{4}{\kappa^4\gamma}\lim_{l_0\to 0}\frac{\mathcal{V}^{1-q}}{q l_0^3}\sum_{i,j,k,i'}\mathrm{tr}(\tau_k h_{\Box_{ij}})\epsilon^{iji'}\mathrm{tr}(\tau_k h_{i'}\{h_{i'}^{-1},\,\mathcal{V}^q\})\,. \tag{9.115}$$

The kinematical states of LQG are spin networks, graphs in an embedding space whose edges e are labeled by spin quantum numbers j_e. Edges meet at nodes in sets of three; at each node, the labels of group elements are governed by a map $SU(2)\otimes SU(2) \longrightarrow SU(2)$, called intertwiner. Physical states are based upon the kinematical Hilbert space of the theory and are annihilated by the quantum constraints. We leave this analysis to specialized literature [52, 123, 124], limiting this section to the barest details we shall need when discussing the cosmological version of the LQG quantization scheme (Sect. 10.3).

A key feature of loop quantum gravity, which is not an assumption but a consequence of the full theory, is that *the spectrum of the area operator is bounded from below* by the Planck scale. In particular, a direct calculation shows that the lowest area eigenvalue on any gauge-invariant state is [124, 125]

$$\tilde{\Delta}_{\rm Pl} := 2\sqrt{3}\pi\gamma\ell_{\rm Pl}^2\,. \tag{9.116}$$

This result can be roughly illustrated as follows. A given two-dimensional surface Σ is divided into elementary cells Σ_e on which a flux $E_i[\Sigma_e] = \int_{\Sigma_e} \mathrm{d}y_1\mathrm{d}y_2 n_\alpha(y)E^\alpha_i(y)$ is defined, where $y_{1,2}$ are coordinates over Σ_e and n^α is the normal to the surface. The classical area of Σ is given by the surface integral of the induced two-metric and the sum of the areas of the Σ_e. Each area is determined by the fluxes through it, $\mathcal{A}_e = \gamma^{-1}\sqrt{E_i[\Sigma_e]E^i[\Sigma_e]}$. Upon quantization, each Σ_e is pierced by the edge of an “adapted” spin network (i.e., the edge starts at Σ_e, or is embedded in Σ_e, or it has empty intersection with Σ_e). The quantum number determines the spectrum $\mathcal{A}_e$ of the area operator $\hat{\mathcal{A}}$ associated with an elementary plaquette intersecting only one edge e:

$$\mathcal{A}_e = \gamma\ell_{\rm Pl}^2\sqrt{j_e(j_e+1)}\,. \tag{9.117}$$

The geometrical size of the plaquette changes only when the latter intersects another edge, thus increasing in quantum jumps. Therefore, the action of two flux operators over an adapted edge gives the quadratic Casimir invariant of the algebra $su(2)$, which takes discrete values and is non-zero in the fundamental representation

$j_e = 1/2$. Multiplying (9.117) by a solid-angle factor of 4π yields (9.116). Although this is only a kinematical result, it makes one hope that physical singularities such as those found inside black holes and at the birth of the Universe be resolved in the quantum setting of LQG.

The value of γ can be constrained by the computation of the entropy of non-rotating black-hole isolated horizons [126–128]. Matching the LQG result with Bekenstein–Hawking's semi-classical area law fixes the Barbero–Immirzi parameter to be [128]

$$\gamma \approx 0.2375 \,. \tag{9.118}$$

This is the only available determination of γ and it would be highly desirable to obtain further input from LQG calculations performed in independent physical settings. In loop quantum cosmology, the Barbero–Immirzi parameter is left arbitrary, its value affecting (but in a mild way) only the energy scale of the quantum bounce. Note, however, that the black-hole area law is recovered also in group field theory, a powerful Lagrangian extension of canonical quantum gravity (Sect. 11.5), but for any value of the Barbero–Immirzi parameter [129]. The difference is due to the fact that in group field theory one uses large superpositions of graphs (condensate states), while in LQG one picks a very special choice of states, the eigenstates of the black-hole horizon area.

Loop quantum gravity and other approaches of discrete gravity have often inspired the following naive objection. One starts from a theory defined on a smooth manifold where classical general relativity applies. Then, one quantizes and hopes to obtain something like a continuum limit for geometry. Is this not a snake biting its tail? The answer is No and is exactly the same as for ordinary quantum mechanics. When we say that "a system is quantized" we do not mean that a classical physical system is transformed into something else. Rather, quantization means to define a set of operators on a Hilbert space (or a tower of spaces, if we consider also the kinematical spaces instrumental to the construction of the physical Hilbert space) in a certain representation. Everything which happens before we complete the construction of these operators and Hilbert space is a mathematical framework. For instance, the embedding manifold of LQG is just an auxiliary object with no physical significance. *Only after quantizing* can we start talking about physics, because the assumption is that the interaction we are studying is intrinsically quantum.

9.4 Cosmological Constant Problem

Canonical quantum gravity in the ordinary Wheeler–DeWitt quantization faces the cosmological constant problem from a probabilistic point of view. Results in this direction can be obtained in a cosmological setting, where the number of degrees of freedom of the full theory is reduced and the problem becomes tractable. We still need to introduce many ingredients of this construction and we postpone the discussion to Sect. 10.2.4. Here we concentrate on Λ-related aspects of loop quantum gravity and its modifications. At the time of writing, loop quantum gravity has not been fully formulated in the presence of a cosmological constant, which deforms the internal group to a quantum group and therefore complicates the structure of the theory; attempts to relate such quantum-group structure to the one in spin foams are fairly recent [130, 131]. In parallel, proposals to solve or relax the Λ problem in LQG have been preliminary but, interestingly, none of them invokes symmetry reduction to a cosmological background. Results in this direction have not been supported by much study [132–134] and are to be considered as speculative but they provide a case, among recent others, of cross-fertilization between quantum gravity and ideas taken from condensed matter physics. This section can be skipped on a first reading.

9.4.1 Chern–Simons State

Spin-network solutions of the constraints with a cosmological constant have been sketched in [132]. Their small number suggests that the state of the Universe should have been initially one with $\Lambda = 0$. Such indication is not incompatible with the current notion that well-defined solutions of the constraints with a cosmological constant are difficult to find. LQG solutions of the full theory are, in fact, incompletely understood in the presence of a Λ term.

For simplicity, we work in complex Ashtekar variables, such that the connection is (9.53) with $\gamma = -\mathrm{i}$. In particular, in the absence of matter, the classical Hamiltonian constraint (9.60) simplifies to $\mathcal{H} = -[P_i^\alpha P_j^\beta/(2\kappa^2|e|)]\epsilon^{ij}{}_k F^k_{\alpha\beta}$, which is remindful of a gauge theory. A formal solution of the quantum constraints in vacuum is the Chern–Simons (or Kodama) state [135–139]:

$$\Psi_{\rm CS}[A] = C\exp\left(\frac{3m_{\rm Pl}^2}{2\Lambda}\int_{\mathcal{M}_3} Y_{\rm CS}\right) \tag{9.119}$$

in Lorentzian spacetime, where A is the self-dual connection, $\mathcal{M}_3$ is the three-dimensional spatial sub-manifold and $Y_{\rm CS}$ is the Chern–Simons form [140]

$$\begin{aligned} Y_{\rm CS} &:= \frac{1}{2}\mathrm{tr}\left(A\wedge \mathrm{d}A + \frac{2}{3}A\wedge A\wedge A\right) \\ &= \mathrm{d}^3x\,\epsilon^{\alpha\beta\gamma}\left(A^i_\alpha\partial_\beta A_{\gamma i} + \tfrac{1}{3}\epsilon_{ijk}A^i_\alpha A^j_\beta A^k_\gamma\right). \end{aligned} \tag{9.120}$$

The trace is taken in the adjoint representation, where the $su(2)$ algebra generators are $f_{ijk} = \epsilon_{ijk}$. The normalization factor $\mathcal{N}$ is independent of both the triad and the connection but it can depend on the topology, as it will be soon demanded.

The Chern–Simons state is not only a WKB state, but also a genuine ground state of the theory, inasmuch as by adding a matter sector its weakly-coupled excitations reproduce standard quantum field theory on de Sitter background [141], while linearizing the quantum theory one recovers long-wave-length gravitons on de Sitter [139]. However, there are several issues associated with this state. $\Psi_{\rm CS}$ is solution to the constraints only for a special operator ordering such that the super-Hamiltonian is not Hermitian. Moreover, it depends on the self-dual Ashtekar connection, not on the real Ashtekar–Barbero connection (9.53) by which the actual theory is defined. This makes the whole construction of the state (9.119) rather formal, even when re-expressed in terms of holonomies. On top of that, $\Psi_{\rm CS}$ is not normalizable. This is expected because states satisfying the constraints are generally non-normalizable in the inner product on the kinematical Hilbert space. An immediate consequence is that it does not make sense to calculate the probability $|\Psi_{\rm CS}|^2$ that the Universe realizes a configuration with $\Lambda = 0$, a datum which could have been compared with the arguments about spin-network solutions [132]. Last, $\Psi_{\rm CS}$ violates CPT symmetry, which may result in negative energy levels and a Lorentz violation [142].

Eventually, these problems may be solvable. Hermiticity is a requirement only for excited states in a quantum mechanical theory and it is not necessary in vacuum gravity, where one is interested only in the ground state. Next, in the self-dual formulation there are many perturbative self-consistency checks which can be performed on the mathematical structure emerging from the Chern–Simons state [138, 143]. Further, the latter can be generalized to a real Immirzi parameter (real connection) [144–146]. This state solves all the quantum constraints, is δ-normalizable (i.e., $\langle\Psi|\Psi\rangle = \delta$) just like momentum eigenstates in quantum mechanics (since de Sitter space is unbounded, the corresponding wave-function cannot be normalizable as $\langle\Psi|\Psi\rangle = 1$) and invariant under large-gauge and CPT transformations (it violates CP and T separately). Last, many key features of quantum mechanics and quantum field theory were discovered even before a Fock or Hilbert space were defined. Therefore, although there is no rigorously known kinematical Hilbert space for self-dual gravity with Lorentzian signature, one can still work in complex variables and make generic predictions about the quantum theory.

The Chern–Simons wave-function can give a useful insight in non-perturbative properties of the theory. Since it is a WKB state, even if one does not believe it to lie in the physical Hilbert space of the full theory, at some level a true quantum state close to the de Sitter ground state must be approximated by $\Psi_{\rm CS}$ reasonably well. Taking these caveats on board, one may illustrate how to relate the Λ problem to the topological structure of the theory.

As mentioned in Sect. 7.2.1, the full gauge group of gravity $\mathbb{G}$ is made of diffeomorphisms, but it contains also the group $\mathcal{G}$ of *small* as well as *large gauge transformations* in internal space. In the case of the complex connection formulation [4, 5], $\mathcal{G} = SU(2)$. Under a local gauge transformation, the Ashtekar connection transforms as

$$A \to A' = gAg^{-1} - g^{-1}\mathrm{d}g\,, \qquad g(\mathbf{x}) \in \mathcal{G}\,. \tag{9.121}$$

Let $\mathcal{G}_0 \subset \mathcal{G}$ be the sub-group of small gauge transformations, local transformations continuously connected to the identity. Its elements are of the form $g_0 = \exp[-\mathrm{i}\tau_i\theta^i(\mathbf{x})]$, where τ_i are the generators of the $su(2)$ algebra and $\theta^i(\mathbf{x})$ are some functions on a spatial slice of the four-manifold. Pure gauge configurations $A^o = g^{-1}\mathrm{d}g$ are equivalent to the flat gauge $A = 0$. Invariance of the exponent in (9.119) under small gauge transformations requires the integration manifold $\mathcal{M}_3$ to have no boundary. It turns out that $\mathcal{M}_3$ must have constant curvature in the semi-classical limit, selecting the 3-sphere S^3 as the only classically viable alternative [143].

On the other hand, large gauge transformations are unitary transformations $g(\mathbf{x})$ which are not homotopic to the identity but tend to the identity at large $\mathbf{x}$, $\lim_{|\mathbf{x}|\to\infty} g(\mathbf{x}) = \mathbb{1}$. All directions in Euclidean space are identified at infinity, so that g is a mapping from $\mathcal{M}_3 \cong S^3$ to S^3. The different ways in which the sphere S^3 can be continuously mapped onto itself are summarized by the homotopy group $\pi_3(S^3) = \mathbb{Z}$. Therefore, each disconnected component $\mathcal{G}_n \subset \mathcal{G}$ is labelled by an integer n, called *winding number*, generated out of a pure gauge connection A^o in that component,

$$w(A^o) = \frac{I(A^o)}{24\pi^2} = \frac{1}{24\pi^2}\int_{S^3} \mathrm{tr}(A^o \wedge A^o \wedge A^o), \tag{9.122}$$

where I is the Cartan–Maurer invariant [147]. The winding number characterizes how many times $g(\mathbf{x})$ winds around the non-contractible 3-sphere in the $SU(2)$ internal space as $\mathbf{x}$ ranges over S^3 in space. If $g = g_0 \in \mathcal{G}_0$, the winding number is zero, while all large gauge transformations fall into homotopy classes g_n generated by the $n = 1$ transformation: $g_n(\mathbf{x}) = [g_1(\mathbf{x})]^n$. Then, the quotient $\mathcal{G}/\mathcal{G}_0$ is isomorphic to the homotopy sequence of integer-valued winding configurations $\{(g_1)^n \,|\, n \in \mathbb{Z}\}$.

This is the origin of topological θ vacua in quantum Yang–Mills theory [148, 149] and gravity [150, 151]. On each disconnected component of $\mathcal{G}$, there lives an independent physical sector (an inequivalent quantum theory) with ground state $\Psi[A] = \langle A|n\rangle$. When promoted to a quantum operator, a unitary large gauge transformation jumps from a sector to another, acting as $\hat{g}_1|n\rangle = |n+1\rangle$. By definition, physical observables are invariant under the full gauge group, so that $\hat{g}_1$ must commute with the Hamiltonian and they share the same basis of eigenstates, chosen so that the eigenvalue of $\hat{g}_1$ on a state is a pure phase $\mathrm{e}^{\mathrm{i}\theta}$, $0 \leqslant \theta \leqslant 2\pi$. In other words, under a gauge transformation as in (9.121), $\Psi[A'] = \Psi[A]$ if g is small ($g = g_0$), while $\Psi[A'] = \mathrm{e}^{\mathrm{i}n\theta}\Psi[A]$ if $g = g_n$ is large. This implies that the quantum vacuum is a superposition of states $|\theta\rangle = \sum_n \mathrm{e}^{\mathrm{i}n\theta}|n\rangle$.

In the case of quantum chromodynamics, the θ sector appears as parity-breaking terms for the gauge field and the fermions. In particular, the classical Yang–Mills Lagrangian

$$\mathcal{L}_F = -\frac{1}{4}\sum_i F^i_{\mu\nu}F^{i\,\mu\nu}\,, \qquad F^i_{\mu\nu} := \partial_\mu A^i_\nu - \partial_\nu A^i_\nu + f^i_{\ jk}A^j_\mu A^k_\nu\,, \tag{9.123}$$

receives a correction of the form

$$\mathcal{L}_F \to \mathcal{L}_F - \frac{\tilde{\theta}}{32\pi^2}\sum_i F^i_{\mu\nu}\tilde{F}^{i\,\mu\nu}, \tag{9.124}$$

where $\tilde{F}^{i\,\mu\nu} = \epsilon^{\mu\nu\rho\sigma}F^i_{\rho\sigma}/2$. Classical dynamics is unaffected, since the new term is topological, i.e., it can be written as a total derivative. At the quantum level, however, topological terms do change the dynamics. But then (9.124) creates a problem, since there is no experimental evidence that chromodynamics violates the CP symmetry. To solve this "strong CP problem," the QCD partition function is augmented by a $U(1)$ symmetry, namely, the Peccei–Quinn symmetry corresponding to invariance under a rotation by the θ angle [152, 153]. This procedure corresponds to promoting the $\tilde{\theta}$ parameter to a field θ, a light particle called the *axion* [154, 155]. The Lagrangian (9.124) is modified by $\tilde{\theta} \to \tilde{\theta} - \theta$ and augmented by a kinetic term $\propto \partial_\mu\theta\partial^\mu\theta$. Instanton effects spontaneously break the $U(1)$ symmetry associated with the axion by introducing a potential with a minimum at $\theta \simeq \tilde{\theta}$, thus canceling or suppressing the CP-violating term.

For self-dual gravity, something similar happens. At the level of the action, a θ sector is, on solutions to the field equations, realized by adding the self-dual Pontryagin term $\mathcal{L}_{\tilde{\theta}} \propto \mathrm{i}\tilde{\theta}R^\rho{}_{\sigma\mu\nu}\tilde{R}_\rho{}^{\sigma\mu\nu}$, which is topological. In the classical super-Hamiltonian, there appear parity-violating terms at linear and quadratic order in $\tilde{\theta}$ [156]. At the quantum level, the Chern–Simons form transforms non-trivially under large gauge transformations, as $\int Y_{\rm CS} \to \int Y_{\rm CS} + 4\pi^2 w$:

$$\Psi_{\rm CS}[A] \to \Psi_{\rm CS}[A'] = \mathrm{e}^{\mathrm{i}n\theta}\Psi_{\rm CS}[A]\,, \qquad n \in \mathbb{Z}\,, \tag{9.125}$$

where

$$\mathrm{i}\theta = \frac{6\pi^2 m_{\rm Pl}^2}{\Lambda}\,. \tag{9.126}$$

The θ parameter is indeed a phase in the Euclidean case, where instantonic solutions are considered. Invariance of the state under large gauge transformations can be achieved by setting the normalization constant in (9.119) to be $C = \mathrm{e}^{-\mathrm{i}w(A^o)\theta}$ [143, 157]. The θ dependence is absorbed in the state normalization but reappears in the inner product, giving inequivalent quantum probabilities.

9.4.2 Λ as a Condensate?

At this point, in analogy with the Peccei–Quinn mechanism in QCD, we modify the theory and promote Λ to an evolving functional $\Lambda(A)$ [133, 134]. The Peccei–Quinn mechanism (and its pseudo Nambu–Goldstone boson) has been proposed to explain the observed cosmological constant in quantum field theory [158–160], but in the present context it is directly related to gravitational quantum degrees of freedom. The operation $\Lambda \to \Lambda(A)$ is recognizable, thanks to (9.126), as a deformation of the topological sector of the quantum theory: $\theta \to \theta(A)$. A large gauge transformation (9.125) in the Chern–Simons wave-function is regarded as a $U(1)$ rotation, so that by allowing θ to vary as a function of the connection we explicitly break the $U(1)$ symmetry. Deforming θ leads to an altogether new theory, which reduces to ordinary loop quantum gravity when the deformation is very weak (perturbative limit). An expansion of the new ground state and of the total Hamiltonian in a perturbative parameter must yield, at leading order, just LQG with a cosmological constant. This is sufficient to justify the application of the deformation procedure to a constraint and a to state of the usual theory, obtaining some new constraint annihilating a new state. In other words, one modifies the super-Hamiltonian constraint by adding a counter-term breaking gauge invariance. The Chern–Simons state will also be deformed because, according to (9.125) and (9.126), Λ enters in the phase of the wave-function. As the Chern–Simons state lives in the connection space, we assume that Λ in the deformed state Ψ_* depends only on A and not on the triad operator.

The new state Ψ_* turns out to be compatible with all the constraints only in two cases: either $\Lambda(A)$ is invariant under small gauge transformations or $\det E = 0$, leading to a *degenerate* sector of gravity with rk$E \leqslant 2$ (rk denotes the rank of the triad). The special case rk$E = 1$ gives rise to an intriguing structure. Classically, this configuration has been studied by Jacobson [161]. Geometry amounts to a collection of parallel gravitational lines where the triad E_i^α has rank 1 and vanishes elsewhere. Spatial diffeomorphism invariance can be partially fixed by spreading the gravitational lines along the z direction. Then, the only non-zero component of the electric field is E_i^z. Solving the classical Gauss constraint and fixing the gauge, one ends up with $A_z^i = 0$, $A_a^3 = A_a^3(x^a)$, $A_a^{i\neq 3} = A_a^{i\neq 3}(t,z)$, $E_{i\neq 3}^z = 0 = E_i^a$ and $E_3^z = E_3^z(x^a)$, where $a = 1, 2$ and the last equality stems from the Gauss constraint and the equation of motion coming from the super-Hamiltonian:

$$\dot{A}_\alpha^i = \mathrm{i}\epsilon^i{}_{jk} E^{\beta j}\left(F_{\alpha\beta}^k + \frac{\Lambda}{2}\epsilon_{\alpha\beta\gamma}E^{\gamma k}\right), \tag{9.127}$$

where the presence of a cosmological constant is, at this stage, irrelevant. The residual freedom in the choice of the transverse coordinates x^a can be used to fix $E_3^z = 1$. The equation of motion (9.127) for the transverse-transverse components of the connection is $\dot{A}_a^i = -\mathrm{i}\epsilon^{i3}{}_j\partial_z A_a^j$, which can be written as a $(1+1)$-dimensional

Dirac equation at every point x^a in the transverse plane [133, 134]:

$$\gamma^0\dot{\psi} + \gamma^z\partial_z\psi = 0\,, \tag{9.128}$$

where

$$\psi := \begin{pmatrix} \mathrm{i}A_1^1 \\ A_2^1 \\ A_1^2 \\ \mathrm{i}A_2^2 \end{pmatrix} \tag{9.129}$$

and γ^μ are the Dirac matrices in the Dirac basis (compare with (5.209))

$$\gamma^0 = \begin{pmatrix} \sigma^0 & 0 \\ 0 & -\sigma^0 \end{pmatrix}, \qquad \gamma^\alpha = \begin{pmatrix} 0 & \sigma^\alpha \\ -\sigma^\alpha & 0 \end{pmatrix}. \tag{9.130}$$

The causal structure of the degenerate sector is, from (9.128), that of a *world-sheet*, a $(1+1)$-dimensional spacetime. Rotating back in the target space, (9.128) can be put in covariant form as $\gamma^\mu\partial_\mu\psi = 0$. One can take several differently oriented gravitational lines and patch them together at their boundaries [161]. The emerging classical picture would be that of two-dimensional world-sheets interacting at the edges, where the gravitational field may be non-degenerate. However, at the classical level one lacks a model for this interaction, as well as a physical interpretation of the world-sheet network and fermionic degrees of freedom.

Both naturally emerge at the quantum level when the deformed constraints act upon the Chern–Simons state. To see this, consider that we are working in a semi-classical approximation, so that the counter-term in the quantum Hamiltonian constraint can be interpreted as a deformation also of the classical equation of motion (9.127), which reads

$$\dot{A}^i_\alpha = \mathrm{i}\epsilon^i{}_{jk}E^{\beta j}\left[F^k_{\alpha\beta} + \frac{\Lambda}{2}\epsilon_{\alpha\beta\gamma}E^{\gamma k} - \epsilon_{\alpha\beta\gamma}\int_{\mathcal{M}_3} Y_{\rm CS}\frac{\delta\ln\Lambda(A)}{\delta A_{\gamma k}}\right]. \tag{9.131}$$

In Jacobson's degenerate sector, this equation can be manipulated as in the classical case. To proceed, one has to make an *Ansatz* for the functional $\Lambda(A)$. We choose $\Lambda(\psi) = \Lambda_0\exp(\xi^{\rm t}\psi)$, where $\Lambda_0 = O(1)$ is a constant and $\xi^{\rm t} = (\xi_1, \xi_2, \xi_3, \xi_4)$ depends on A and is, for the time being, arbitrary. Then, (9.131) yields $\gamma^0\dot{\psi} + \gamma^z\partial_z\psi + \mathrm{i}m\gamma^x\xi = 0$, where ψ is given by (9.129) and $m := -\mathrm{i}\psi^{\rm t}\gamma^x\gamma^z\partial_z\psi$ is an effective mass. Imposing the condition $\xi = -\gamma^x\psi$, the field ψ obeys the Dirac equation

$$\gamma^0\dot{\psi} + \gamma^z\partial_z\psi + \mathrm{i}m\psi = 0\,. \tag{9.132}$$

Classically the mass vanishes, as $\xi^{\mathrm{t}}\psi = 0$. However, after quantization the field components become Grassmann variables and the symmetry-breaking mechanism described at the beginning of this section comes into effect. In order to get quantities with a well-defined Lorentz structure, ψ^{t} should actually be related to $\bar{\psi} := \psi^\dagger\gamma^0$. This is realized if $\psi = \psi_c := -\mathrm{i}\gamma^y\psi^*$, i.e., if ψ is equal to its charge conjugate (Majorana fermion). Then, the connection components must satisfy the conditions $A_1^2 = A_2^{1*}$, $A_2^2 = A_1^{1*}$. In particular, $\xi^{\mathrm{t}}\psi = \psi^{\mathrm{t}}\gamma^x\psi = -\bar{\psi}\gamma^5\gamma^z\psi =: -j_5^z$, where $\gamma^5 := \mathrm{i}\gamma^0\gamma^x\gamma^y\gamma^z$ (see Problem 9.1) and Λ becomes

$$\Lambda = \Lambda_0 \exp(-j_5^z)\,. \tag{9.133}$$

The cosmological constant encodes the imprint of an axial vector current j_5^α, associated with a chiral transformation $\psi \to e^{\mathrm{i}\theta\gamma^5}\psi$ and not conserved in the presence of the effective mass

$$m = -\mathrm{i}\bar{\psi}\gamma^5\partial_z\psi\,. \tag{9.134}$$

This is yet another reminder that the deformation process affects the topological sector and the CP symmetry of the theory.

Before applying (9.133) to the cosmological constant problem, its physical interpretation must be sharpened. Equations (9.132) and (9.134) are the starting point of a scenario which has loop quantum gravity as an effective limit. One begins with a spinor field ψ in two dimensions and canonically quantizes it as a fermion. Expand it in discrete one-dimensional momentum space, $\psi = \sum_{k,\sigma} e^{ikz}(2\mathcal{E}_k)^{-1/2}(c_{k\sigma}u_{k\sigma} + c_{-k\sigma}^\dagger v_{k\sigma})$, where $\sigma = \pm$ is the spin, $\mathcal{E}_k$ is the energy, c and $c^\dagger$ are annihilation and creation operators obeying a fermionic algebra $\{c_{k\sigma}, c_{k'\sigma'}^\dagger\} = \delta_{kk'}\delta_{\sigma\sigma'}$ and u and v are spinorial functions. Due to (9.134), there appear four-fermion non-local (i.e., dependent on non-coincident points) correlations in the Hamiltonian, of the form

$$\sum_{\sigma,k,k'} V_{kk'}c_{k\sigma}^\dagger c_{k'\sigma'}^\dagger c_{k\sigma}c_{k'\sigma'}\,, \tag{9.135}$$

where $V_{kk'}$ is a function of the momenta determined by the effective mass (9.134). Spin models of the form (9.135), generically called Fermi-liquid theories, are employed in condensed matter physics to describe fermionic systems with many-body interactions at sufficiently low temperature (see, e.g., [162–164] for introductory reviews). In the mean-field approximation $V_{kk'} = g$ =const, the interaction is effectively independent of the momenta. This is the Bardeen–Cooper–Schrieffer (BCS) model of superconductivity [165–170]. Fermions with opposite spin can interact non-locally in N pairs at a given energy level, lower than the Fermi energy E_{F} of the free-fermion sea. Let E_{pair} be the binding energy of the two-body bound state in vacuum. At weak coupling (small binding energy $E_{\mathrm{pair}} \ll E_{\mathrm{F}}$), Cooper pairs are weakly bound by an attractive potential and may overlap, forming a condensate with superconducting properties. This lowest-energy state $|\mathrm{BCS}\rangle$ is regarded as the true vacuum of the theory, whose energy is separated from the perturbative-vacuum energy by a mass gap $\Delta \sim e^{-1/g}$. At strong coupling (large binding energy $E_{\rm pair} \gg E_{\rm F}$), pairs are tighter and do not overlap, so that they can be treated as a gas of non-interacting bosons, which is expected to undergo a Bose–Einstein condensation (BEC) [171–177]. The transition between the two regimes (called BEC-BCS crossover [178, 179]) is smooth and well described by a weak-field approximation.

One can check that the gravitational analogues of the BCS wave-function $|\mathrm{BCS}\rangle$ and mass gap Δ are the Chern–Simons state and the cosmological constant. As we will see in Chap. 12, a world-sheet is invariant under conformal transformations and one can build a conformal field theory (CFT) on it. The quantum degenerate sector of the present gravity model is described by world-sheets interacting at their boundary. Indeed, one can place a CFT on each of these world-sheets. However, conformal invariance is broken at the boundary by a marginal operator encoding the interaction, which results in a deformed CFT. It turns out that this deformed CFT is fermionic in nature, as the classical picture suggests, and describes a Fermi liquid. On the other hand, its geometric and algebraic structures reproduce those of a *quantum* (or *framed*) *spin network* (e.g., [139]), which is the natural environment wherein to embed the Chern–Simons state. This and other pieces of evidence lead to the conjecture that the degenerate sector has a direct interpretation in terms of framed spin networks, so that the $(1+1)$-dimensional theory holographically generates the *full* four-dimensional spacetime via quantum interactions.

The edges of quantum spin networks are two-dimensional tubular surfaces and vertices are promoted to punctured two-spheres. BCS levels are mapped to edges of quantum spin networks, interacting at nodes via a BCS coupling. The BCS interaction is the cornerstone of the construction of three-dimensional geometry from quantum spin networks. Classically, it describes scattering of Jacobson electric lines at their end-points; these world-sheets, patched together at their edges, span the three-dimensional space, giving rise to the geometric sector which was lost in the free-field picture. The screening charges at a given node have an intuitive picture as the sites activated in an area measurement, i.e., when a classical area intersects the spin network. In condensed matter physics, the Fermi sea is a ground state of uncorrelated electron pairs whose Fermi energy is higher that the BCS pair-correlated state. In quantum gravity, pair correlation can be regarded as a process of quantum decoherence. An abstract spin network is the gravity counterpart of an unexcited Fermi sea. As soon as an area measurement is performed on the state, N edges (as many as the number of Cooper pairs) of a given node are activated and the system relaxes to a lower-energy vacuum corresponding to the selection of one of the area eigenstates in a wave superposition. In a sense, measuring quantum geometry means counting Cooper pairs. This picture [134] is actually more general than the BCS case and gravity possibly allows for a dual description in terms of a more generic non-locally interacting Fermi liquid.

A natural $3+1$ embedding structure arises from the quantum interaction of electric lines, but it is still an open issue how to precisely recover the usual structure of LQG, i.e., the bosonic statistics of connection variables living in $3+1$ dimensions from the fermions describing the $1 + 1$ quantum degenerate sector. Perhaps, the physics of the BEC-BCS crossover could play an important role. If this scenario could be clarified, one would benefit from a possible relaxation of the cosmological constant problem. In fact, (9.133) can address the Λ smallness problem in terms of a condensate with vacuum expectation value (with respect to the deformed Chern–Simons state Ψ_*) $\langle j_5^z \rangle = O(10^2)$. In the perturbative regime (small values of the connection, $|\langle j_5^z \rangle| \ll 1$), $\langle \Lambda \rangle \simeq \Lambda_0(1 - \langle j_5^z \rangle) = O(1)$. In the non-perturbative regime, the effective mass becomes important and the cosmological constant, supposing $\langle j_5^z \rangle$ to be positive definite for large connection values, becomes exponentially small. In this case, the smallness of the cosmological constant would be regarded as a large-scale non-perturbative quantum mechanism similar to quark confinement.

9.5 Problems and Solutions

9.1 Barbero–Immirzi field and torsion 1. Consider the Hilbert–Palatini action (9.21) augmented by the Nieh–Yan term (9.23), but with $-1/\gamma = \beta \to \beta(x)$ now inside the integral:

$$\begin{aligned} S &= S_{\mathrm{HP}} + S_{\mathrm{NY},\beta} \\ &= \frac{1}{2\kappa^2} \int \mathrm{d}^4x \, |{}^{(4)}e| \, e_a^{\ \mu} e_b^{\ \nu} \\ &\quad \times \left[R_{\mu\nu}^{\ \ \ ab} + \frac{\beta}{2} \left(\epsilon^{\tau\lambda\rho\sigma} \eta_{ab} T^{\mu}_{\ \ \tau\lambda} T^{\nu}_{\ \ \rho\sigma} - \epsilon^{ab}_{\ \ \ cd} R_{\mu\nu}^{\ \ \ cd} \right) \right]. \end{aligned} \tag{9.136}$$

Plug (9.15), (9.17) and (9.18) into the action and obtain an expression in terms of the irreducible torsion components. Vary the action with respect to these components and pull back the resulting equations into the action to obtain an effective action. Is β a scalar o a pseudo-scalar field?

Solution The Hilbert–Palatini piece becomes

$$\begin{aligned} S_{\mathrm{HP}} &= \frac{1}{2\kappa^2} \int \mathrm{d}^4x \, |{}^{(4)}e| \, e_a^{\ \mu} e_b^{\ \nu} \left(\bar{R}_{\mu\nu}^{\ \ \ ab} + \mathcal{K}^a_{\ c\mu} \mathcal{K}^{cb}_{\ \ \ \nu} - \mathcal{K}^a_{\ c\nu} \mathcal{K}^{cb}_{\ \ \ \mu} \right) \\ &= \frac{1}{2\kappa^2} \int \mathrm{d}^4x \, |{}^{(4)}e| \left[e_a^{\ \mu} e_b^{\ \nu} \bar{R}_{\mu\nu}^{\ \ \ ab} + \frac{1}{24} S_\mu S^\mu - \frac{2}{3} T_\mu T^\mu + \frac{1}{2} q_{\mu\nu\rho} q^{\mu\nu\rho} \right], \end{aligned} \tag{9.137}$$

where in the first line we have dropped out a total divergence and quantities with an overbar are torsion-free. In particular, $\bar{\nabla}_\mu$ is the torsionless and metric-compatible covariant derivative. The Nieh–Yan contribution is rather simple, as a number of cancellations take place between the Holst and the torsion-torsion term:

$$S_{\mathrm{NY},\beta} = \frac{1}{4\kappa^2}\int \mathrm{d}^4x\, |^{(4)}e|\beta\bar{\nabla}_\mu S^\mu\,. \tag{9.138}$$

By varying the action with respect to the irreducible components of torsion S^μ, T^ν and $q^{\rho\sigma\tau}$, we obtain, respectively,

$$\partial_\mu\beta - \tfrac{1}{6}S_\mu = 0\,, \qquad T_\mu = 0\,, \qquad q_{\mu\nu\rho} = 0\,. \tag{9.139}$$

After reinserting the solutions above into (9.136), we get the effective action

$$S_{\mathrm{eff}} = \frac{1}{2\kappa^2}\int \mathrm{d}^4x\, |^{(4)}e| \left(e_a{}^\mu e_b{}^\nu \bar{R}_{\mu\nu}{}^{ab} - \frac{3}{2}\partial_\mu\beta\partial^\mu\beta \right)\,. \tag{9.140}$$

Therefore, the system is equivalent to Hilbert–Palatini torsion-free gravity plus a canonical massless scalar field

$$\tilde{\beta} := \sqrt{\frac{3}{2\kappa^2}}\,\beta\,, \qquad [\tilde{\beta}] = 1\,. \tag{9.141}$$

Notice that S_{eff} preserves parity because β is a pseudo-scalar. To see this, project the first equation in (9.139), $\partial_a\beta := e_a^\mu\partial_\mu\beta = S_a/6$. ∂_a is a polar internal vector, but the pseudo-vector component of the torsion $S_a = \epsilon_{abcd}T^{bcd}$ is defined via the Levi-Civita symbol. Therefore, β is a pseudo-scalar. If we had chosen the volume form with $^{(4)}e$ instead of $|^{(4)}e|$, the classical decomposition of torsion according to the Lorentz group would have lost its meaning.

Another way to reach the same conclusion is by adding fermions to the action. Then, the Barbero–Immirzi field obeys a massless Klein–Gordon equation with a fermionic axial bilinear as a source [44]:

$$\Box\beta \propto \bar{\psi}\gamma^5\psi\,, \tag{9.142}$$

where the matrix γ^5 is defined as

$$\gamma^5 = \frac{\mathrm{i}}{4!}\epsilon_{abcd}\gamma^a\gamma^b\gamma^c\gamma^d\,. \tag{9.143}$$

Since a parity operation would change only one of the γ^a's, γ^5 changes sign and hence it is a pseudo-scalar. As spinors are internal scalars, this yields the desired result. With the volume form without absolute value of the determinant, one would have been obliged to change the definition of γ^5, but this would ultimately spoil the usual notion of parity in quantum field theory. Since the physics on Minkowski spacetime is a collection of local experiments conducted in a given frame, the notion of parity must be consistent. We are then forced to assume (9.8) and it then follows that this choice is the only self-consistent one when matter is included in the theory.

9.2 Barbero–Immirzi field and torsion 2. Perform the Hamiltonian analysis of $S = S_{\mathrm{HP}} + S_{\mathrm{NY},\beta}$ given by (9.137) and (9.138). Set $\kappa^2 = 1$. Fix the tensor $q_{\mu\nu\rho}$ to zero since it is non-dynamical and does not contribute to the torsion tensor, as it was clear from (9.139). Setting $q_{\mu\nu\rho} = 0$ does not affect the generality of the formulation and it has the advantage of simplifying the canonical analysis, which is rather involved for 3–tensors.

Solution Following step by step the calculation of Sect. 9.1.2, the action $S = S_{\mathrm{HP}} + S_{\mathrm{NY},\beta}$ can be written as follows:

$$
\begin{aligned}
S &= \frac{1}{2}\int \mathrm{d}t\,\mathrm{d}^3\mathbf{x}\,|{}^{(4)}e|\,\Big(2e_0^{\ \mu}e_i^{\ \nu}R_{\mu\nu}^{\ \ \ 0i} + e_i^{\ \mu}e_j^{\ \nu}R_{\mu\nu}^{\ \ \ ij} \\
&\quad -\frac{1}{2}S^\mu\partial_\mu\beta + \frac{1}{24}S_\mu S^\mu - \frac{2}{3}T_\mu T^\mu\Big) \\
&= \frac{1}{2}\int \mathrm{d}t\,\mathrm{d}^3\mathbf{x}\,|{}^{(4)}e|\,\Big[2\frac{t^\mu - N^\mu}{N}\,e_i^{\ \nu}R_{\mu\nu}^{\ \ \ 0i} + e_i^{\ \alpha}e_j^{\ \gamma}R_{\alpha\gamma}^{\ \ \ ij} \\
&\quad +(h^{\mu\nu} - u^\mu u^\nu)\Big(-\frac{1}{2}S_\mu\partial_\nu\beta + \frac{1}{24}S_\mu S_\nu - \frac{2}{3}T_\mu T_\nu\Big)\Big] \\
&= \int \mathrm{d}t\,\mathrm{d}^3\mathbf{x}\,|e|\,\Big\{\frac{N}{2}e_i^\alpha e_j^\gamma\left({}^{(3)}R_{\alpha\gamma}^{\ \ \ ij} + 2K^i_{[\alpha}K^j_{\gamma]}\right) - N^\alpha e_i^{\gamma}\,{}^{(3)}R_{\alpha\gamma}^{\ \ \ 0i} \\
&\quad + e_i^\alpha\left[\dot{K}^i_\alpha + \omega^{0i}_{\ \ \mu}\partial_\alpha t^\mu - \partial_\alpha\left(t\cdot\omega^i\right) - \omega^i_{\ k\alpha}(t\cdot\omega^k) + (t\cdot\omega^i_{\ k})K^k_\alpha\right] \\
&\quad + \frac{1}{4}(u\cdot S)\dot{\beta} - \frac{1}{4}(u\cdot S)N^\alpha\partial_\alpha\beta - \frac{N}{4}S^\alpha\partial_\alpha\beta \\
&\quad + N\left[\frac{1}{3}(u\cdot T)^2 - \frac{1}{48}(u\cdot S)^2 + \frac{1}{48}S_\alpha S^\alpha - \frac{1}{3}T_\alpha T^\alpha\right]\Big\}
\end{aligned}
$$

$$= \int \mathrm{d}t\, \mathrm{d}^3\mathbf{x}\, |e| \Big\{ e_i^\alpha \left[\mathcal{L}_t K_\alpha^i - D_\alpha \left(t \cdot \omega^i \right) + (t \cdot \omega^i{}_k) K_\alpha^k \right]$$

$$-2N^\alpha e_i^\gamma D_{[\alpha} K_{\gamma]}^i + \frac{N}{2} e_i^\alpha e_j^\gamma \left({}^{(3)}R_{\alpha\gamma}{}^{ij} + 2K_{[\alpha}^i K_{\gamma]}^j \right)$$

$$+\frac{1}{4}(u \cdot S)\mathcal{L}_t \beta - \frac{1}{4}(u \cdot S) N^\alpha \partial_\alpha \beta - \frac{N}{4} S^\alpha \partial_\alpha \beta$$

$$+N \left[\frac{1}{3}(u \cdot T)^2 - \frac{1}{48}(u \cdot S)^2 + \frac{1}{48} S_\alpha S^\alpha - \frac{1}{3} T_\alpha T^\alpha \right] \Big\},$$

where we have omitted the bars for torsionless geometrical objects, we fixed $q_{\mu\nu\rho}$ to zero and the following notation have been used: $u \cdot S = u^\mu S_\mu$, $u \cdot T = u^\mu T_\mu$.

The only non-zero momenta are the conjugate to K_α^i (equation (9.32a)) and to β:

$$\beta : \qquad \Pi_{(\beta)} := \frac{\delta S}{\delta \mathcal{L}_t \beta} = \frac{1}{4} |e| u \cdot S . \tag{9.144}$$

All the other momenta vanish identically, as given by (9.33) and

$$u \cdot S : \qquad \Pi^{(S)} = 0 , \tag{9.145a}$$

$$u \cdot T : \qquad \Pi^{(T)} = 0 , \tag{9.145b}$$

$$S^\alpha : \qquad \Pi_\alpha^{(S)} = 0 , \tag{9.145c}$$

$$T^\alpha : \qquad \Pi_\alpha^{(T)} = 0 . \tag{9.145d}$$

The phase space is equipped with the symplectic structure (9.35) with ϕ replaced by β and $\Pi_\phi \to \Pi_{(\beta)}$, plus the Poisson brackets

$$\left\{ u \cdot S(t, \mathbf{x}), \Pi^{(S)}(t, \mathbf{x}') \right\} = \delta(\mathbf{x}, \mathbf{x}') , \tag{9.146a}$$

$$\left\{ u \cdot T(t, \mathbf{x}), \Pi^{(T)}(t, \mathbf{x}') \right\} = \delta(\mathbf{x}, \mathbf{x}') , \tag{9.146b}$$

$$\left\{ S^\alpha(t, \mathbf{x}), \Pi_\gamma^{(S)}(t, \mathbf{x}') \right\} = \delta_\gamma^\alpha \delta(\mathbf{x}, \mathbf{x}') , \tag{9.146c}$$

$$\left\{ T^\alpha(t, \mathbf{x}), \Pi_\gamma^{(T)}(t, \mathbf{x}') \right\} = \delta_\gamma^\alpha \delta(\mathbf{x}, \mathbf{x}') . \tag{9.146d}$$

The primary constraints are (9.36) and

$$\mathcal{C}_{(\beta)} := \Pi_{(\beta)} - \frac{1}{4} |e| u \cdot S \approx 0 , \tag{9.147a}$$

$$C^{(b)} := \Pi^{(b)} \approx 0 , \qquad b = S, T , \tag{9.147b}$$

$$C_\alpha^{(b)} := \Pi_\alpha^{(b)} \approx 0 , \qquad b = S, T . \tag{9.147c}$$

Note that, contrary to an ordinary scalar field, the momentum of the Barbero–Immirzi field defines a primary constraint, as it does not contain Lie derivatives. The Dirac Hamiltonian reads

$$\begin{aligned} H_{\mathrm{D}} &= \int \mathrm{d}^3\mathbf{x}\left(E^\alpha_i \mathcal{L}_t K^i_\alpha + \Pi_{(\beta)}\mathcal{L}_t\beta + \lambda^m C_m\right) - L \\ &= \int \mathrm{d}^3\mathbf{x}\left\{N^\alpha \mathcal{H}_\alpha + N\mathcal{H} + E^\alpha_i\left[D_\alpha\left(t\cdot\omega^i\right) - (t\cdot{\omega^i}_k)K^k_\alpha\right]\right. \\ &\quad + {}^{(K)}\lambda^{\ i}_\alpha\, {}^{(K)}C^\alpha_i + \lambda C_{(\beta)} + {}^{(e)}\lambda^\alpha_i\, {}^{(e)}C^i_\alpha + {}^{(\Gamma)}\lambda^{\ i}_\alpha\, {}^{(\Gamma)}C^\alpha_i + \lambda^i C_i + \lambda^{ij}C_{ij} \\ &\quad \left. + \lambda C^{(N)} + \lambda^\alpha C_\alpha + \sum_{b=S,T}\left[\lambda^{(b)}C^{(b)} + \lambda^\alpha_{(b)}C^{(b)}_\alpha\right]\right\}, \end{aligned} \tag{9.148}$$

where the super-momentum and super-Hamiltonian are, respectively,

$$\mathcal{H}_\alpha := 2E^\gamma_i D_{[\alpha}K^i_{\gamma]} + \Pi_{(\beta)}\partial_\alpha\beta \tag{9.149}$$

and

$$\begin{aligned} \mathcal{H} := {}& \frac{1}{2|e|}E^\alpha_i E^\gamma_j\left({\epsilon^{ij}}_k R^k_{\alpha\gamma} - 2K^i_{[\alpha}K^j_{\gamma]}\right) + \frac{1}{3|e|}\Pi^2_{(\beta)} \\ & + |e|\left[\frac{1}{3}T_\alpha T^\alpha - \frac{1}{48}S_\alpha S^\alpha - \frac{1}{3}(u\cdot T)^2 + \frac{1}{4}S^\alpha\partial_\alpha\beta\right]. \end{aligned} \tag{9.150}$$

The analysis of the constraints runs as in the torsion-free case, with slight modifications. The Poisson bracket between the Dirac Hamiltonian and the primary constraint (9.147a) does not generate any secondary constraints and it can be set to zero by suitably choosing the Lagrange multiplier $\lambda^{(S)}$. The secondary constraints are (9.41), (9.42), (9.43), (9.44) and (9.45) and

$$\{C^{(S)}, H_{\mathrm{D}}\} = 0\,, \tag{9.151}$$

$$\{C^{(S)}_\alpha, H_{\mathrm{D}}\} = -\frac{1}{4}N|e|\left(\partial_\alpha\beta - \frac{1}{6}S_\alpha\right), \tag{9.152}$$

$$\{C^{(T)}_\alpha, H_{\mathrm{D}}\} = -\frac{2}{3}N|e|T_\alpha\,, \tag{9.153}$$

$$\{C^{(T)}, H_{\mathrm{D}}\} = -\frac{2}{3}N|e|u\cdot T\,. \tag{9.154}$$

Equation (9.151) is a consequence of the fact that $u\cdot S$ disappears from the Dirac Hamiltonian, so that its momentum is preserved by the Hamiltonian flow. Then we can set $\Pi^{(S)} = 0$ strongly, as it vanishes initially. The dynamical equations of the canonical variables S^α, T^α and $u\cdot T$ are completely arbitrary, since they depend only on the associated Lagrange multipliers λ_m. Therefore, they can be considered as new Lagrange multipliers, which are easily calculated in order for the secondary constraints (9.152), (9.153) and (9.154) to vanish:

$$u \cdot T = 0\,, \qquad T_\alpha = 0\,, \qquad S_\alpha = 6\partial_\alpha\beta\,, \tag{9.155}$$

in agreement, after projection, with (9.139). Accordingly, β must be a pseudo-scalar. Plugging these expressions into (9.150), we get

$$\mathcal{H} = \frac{1}{2|e|}E_i^\alpha E_j^\gamma\left(\epsilon^{ij}{}_k R^k_{\alpha\gamma} - 2K^i_{[\alpha}K^j_{\gamma]}\right) + \frac{1}{3|e|}\Pi^2_{(\beta)} + \frac{3}{4}|e|\,\partial_\alpha\beta\partial^\alpha\beta\,. \tag{9.156}$$

Notice that the system possesses a shift symmetry $\beta \to \beta + \beta_0$ which is absent in the Holst case. Four out of the six physical degrees of freedom correspond to the two polarizations of the graviton, while the remaining two are associated with the pseudo-scalar field β.

9.3 Barbero–Immirzi field. Recast the first-class constraints of the previous problem in terms of the variables

$$\tilde{A}^i_\alpha := -\frac{1}{\beta}K^i_\alpha + \Gamma^i_\alpha\,, \qquad \tilde{P}^\alpha_i := -\beta E^\alpha_i\,, \tag{9.157}$$

where β is the Barbero–Immirzi field. Show that the symplectic structure generated by $\tilde{A}$ and $\tilde{P}$ is not canonical.

Solution Equation (9.56) remains formally the same. On the other hand, (9.58) with non-constant β becomes

$$R^i_{\alpha\gamma} = F^i_{\alpha\gamma} - \frac{1}{\beta^2}\epsilon^i{}_{jk}K^j_\alpha K^k_\gamma + \frac{2}{\beta}D_{[\alpha}K^i_{\gamma]} + \frac{2}{\beta^2}K^i_{[\alpha}\partial_{\gamma]}\beta\,.$$

Then, the super-momentum (9.149) can be rewritten as

$$\mathcal{H}_\alpha \approx \tilde{\Pi}_{(\beta)}\partial_\alpha\beta + P^\gamma_i F^i_{\alpha\gamma} + \frac{1}{\beta^2}K^i_\alpha P^\gamma_i\partial_\gamma\beta\,, \tag{9.158}$$

where

$$\tilde{\Pi}_{(\beta)} := \Pi_{(\beta)} + \frac{1}{\beta}E^\alpha_i K^i_\alpha\,. \tag{9.159}$$

The weak equality stems, as before, from the use of the rotation constraint (9.51). The super-Hamiltonian is

$$\mathcal{H} \approx \frac{P_i^\alpha P_j^\gamma}{2\beta^2|e|}\left[\epsilon^{ij}{}_k F^k_{\alpha\gamma} - 2\left(1+\frac{1}{\beta^2}\right)K^i_{[\alpha}K^j_{\gamma]}\right]$$
$$+\frac{1}{3|e|}\left(\tilde{\Pi}_{(\beta)} + \frac{1}{\beta^2}P_i^\alpha K^i_\alpha\right)^2 + \frac{3}{4}|e|\,\partial_\alpha\beta\partial^\alpha\beta\,. \qquad (9.160)$$

Although the constraints include couplings between $\partial\beta$ and the other variables, this does not signal a transition from the Einstein to a Jordan frame, since the physical metric is still the same.

It is immediately clear that the naive generalization (9.157) of the Ashtekar–Barbero connection does not lead to a canonical algebra [180]. Due to the mixing of matter and gravitational degrees of freedom in A, the symplectic structure in the new variables is non-canonical:

$$\{\tilde{A}^i_\alpha(t,\mathbf{x}), \tilde{\Pi}_{(\beta)}(t,\mathbf{x}')\} = -\frac{1}{\beta}E^\gamma_j\frac{\delta\Gamma^i_\alpha}{\delta E^\gamma_j} \neq 0\,. \qquad (9.161)$$

There is another way to state this result. The rotation constraint and the saturated compatibility condition combine into the Gauss constraint (9.56). Taking the Poisson bracket of the Gauss constraint with itself, one can see that the algebra of gauge rotations does not close.

One should justify the definition (9.53) and explain the relation between the constant β_0 and the Barbero–Immirzi field. If β_0 was regarded as the expectation value of the β field, for instance at a local minimum of an effective potential, then its parity properties would be irrelevant as long as the state in the inner product is an eigenstate of the parity operator. This is indeed the case, as the expectation value of a quantum field is always tacitly defined as an operation which fixes the orientation of local frames. On the other hand, if β_0 was interpreted as the asymptotic value of β at a suitable boundary of the ambient manifold, then β_0 would be a pseudo-constant. However, β_0 is a proper constant and the second interpretation cannot hold.

9.4 Small gauge transformations. Let $\Lambda^j(\mathbf{x})$ be an internal 3-vector defining the *generating functional*

$$\mathcal{G}[\Lambda^j] := \int \mathrm{d}^3\mathbf{x}\,\Lambda^j\mathcal{G}_j\,, \qquad (9.162)$$

called smeared Gauss constraint. Show that $\mathcal{G}[\Lambda]$ generates infinitesimal gauge transformations of the gravielectric field and of the Ashtekar–Barbero connection. For a variable X, such transformations are given by $\delta X = \{X, \mathcal{G}\}$.

Solution Taking the Poisson bracket between P_i^α and the smeared constraint $\mathcal{G}$, we have

$$\delta P_i^\alpha = \{P_i^\alpha(\mathbf{x}), \mathcal{G}[\Lambda^j]\} \overset{(9.56)}{=} \left\{P_i^\alpha(\mathbf{x}), \int \mathrm{d}^3\mathbf{x}' \Lambda^j \epsilon_{jk}{}^l A_\beta^k P_l^\beta\right\}$$

$$\overset{(9.55)}{=} -\kappa^2 \delta_\beta^\alpha \delta_i^k \int \mathrm{d}^3\mathbf{x} \Lambda^j \epsilon_{jk}{}^l P_l^\beta \delta(\mathbf{x}, \mathbf{x}') = \kappa^2 \Lambda^j(\mathbf{x}) \epsilon_{ij}{}^l P_l^\alpha(\mathbf{x}) . \quad (9.163)$$

Therefore, the (smeared) Gauss constraint generates gauge rotations of the gravi-electric field. The Poisson bracket with the connection reads

$$\delta A_\alpha^i = \frac{1}{\kappa^2} \{A_\alpha^i(\mathbf{x}), \mathcal{G}[\Lambda^j]\} = \frac{1}{\kappa^2} \left\{A_\alpha^i(\mathbf{x}), \int \mathrm{d}^3\mathbf{x}' \left(-P_j^\beta \partial_\beta \Lambda^j + \Lambda^j \epsilon_{jk}{}^l A_\beta^k P_l^\beta\right)\right\}$$

$$= \delta_\alpha^\beta \left(-\delta_j^i \partial_\beta \Lambda^j + \delta_l^i \Lambda^j \epsilon_{jk}{}^l A_\beta^k\right) = \left(-\partial_\alpha \Lambda^i + \epsilon_{jk}{}^i A_\alpha^k \Lambda^j\right)$$

$$= -\mathcal{D}_\alpha \Lambda^i . \quad (9.164)$$

This is the expected form of a small gauge transformation of the connection.

9.5 Smeared constraints. The total Hamiltonian in ADM variables (9.69) can be written as

$$H_\mathrm{D} = H[N] + d[N^\alpha] + \int \mathrm{d}^3\mathbf{x}(\pi^0 \dot{N} + \pi^\alpha \dot{N}_\alpha) , \quad (9.165)$$

where the first two terms are the smeared constraints (9.46). In order to show that the super-momentum and super-Hamiltonian constraints are first class, we must check that

$$\dot{d}[N^\alpha] = \{d[N^\alpha], H_\mathrm{D}\} \approx 0 , \qquad \dot{H}[N] = \{H[N], H_\mathrm{D}\} \approx 0 .$$

Calculate these Poisson brackets. Can the result give a hint about why gravity is difficult to quantize?

Sketch of solution: The only non-trivial Poisson brackets to compute are

$$\{d[N^\alpha], H[N]\} , \qquad \{d[N^\alpha], d[M^\beta]\} , \qquad \{H[N], H[M]\} .$$

The smeared constraints are convenient because derivatives can be transferred to the smearing functions N and N^α. In all the above brackets, non-derivative terms commute. With a calculation similar to that of (9.47), one can see that [49]

$$\{d[N^\alpha], H[N]\} = H[N^\beta \partial_\beta N] \approx 0\,, \tag{9.166a}$$

$$\{d[N^\alpha], d[M^\beta]\} = d[N^\gamma \partial_\gamma M^\alpha - M^\gamma \partial_\gamma N^\alpha] \approx 0\,, \tag{9.166b}$$

$$\{H[N], H[M]\} = d[h^{\alpha\beta}(N\partial_\beta M - M\partial_\beta N)] \approx 0\,. \tag{9.166c}$$

All these constraints are satisfied on the secondary constraint surface. Equations (9.166a) and (9.166b) specify how the smeared Hamiltonian and momentum constraints transform under spatial diffeomorphisms. The last expression is the only equation featuring the canonical variable $h^{\alpha\beta}$ in the smearing function. When quantizing the system, canonical variables (and, in particular, $h^{\alpha\beta}$) become operators on a Hilbert space, but, from (9.166c), it is clear that the commutator $[\hat{H}, \hat{H}]\Psi$ on a wave-functional Ψ will vanish only after a suitable operator ordering. The same obstruction occurs in the Ashtekar–Barbero phase space, but it can be overcome. The curious reader might enjoy the fact that the commutator of the Hamiltonian constraint with itself is strongly zero in the so-called ultra-local gravity [181–183].

References

1. R.L. Arnowitt, S. Deser, C.W. Misner, Dynamical structure and definition of energy in general relativity. Phys. Rev. **116**, 1322 (1959)
2. R.L. Arnowitt, S. Deser, C.W. Misner, Canonical variables for general relativity. Phys. Rev. **117**, 1595 (1960)
3. R.L. Arnowitt, S. Deser, C.W. Misner, The dynamics of general relativity, in *Gravitation: An Introduction to Current Research*, ed. by L. Witten (Wiley, New York, 1962). [arXiv:gr-qc/0405109]
4. A. Ashtekar, New variables for classical and quantum gravity. Phys. Rev. Lett. **57**, 2244 (1986)
5. A. Ashtekar, New Hamiltonian formulation of general relativity. Phys. Rev. D **36**, 1587 (1987)
6. J.F. Barbero, Real Ashtekar variables for Lorentzian signature space-times. Phys. Rev. D **51**, 5507 (1995). [arXiv:gr-qc/9410014]
7. S. Deser, J.H. Kay, K.S. Stelle, Hamiltonian formulation of supergravity. Phys. Rev. D **16**, 2448 (1977)
8. E.S. Fradkin, M.A. Vasiliev, Hamiltonian formalism, quantization and *S* matrix for supergravity. Phys. Lett. B **72**, 70 (1977)
9. M. Pilati, The canonical formulation of supergravity. Nucl. Phys. B **132**, 138 (1978)
10. P.D. D'Eath, The canonical quantization of supergravity. Phys. Rev. D **29**, 2199 (1984); Erratum-ibid. D **32**, 1593 (1985)
11. P.D. D'Eath, *Supersymmetric Quantum Cosmology* (Cambridge University Press, Cambridge, 2005)
12. P. Vargas Moniz, *Quantum Cosmology – The Supersymmetric Perspective*. Lect. Notes Phys. **803**, 1 (2010); Lect. Notes Phys. **804**, 1 (2010)
13. T. Damour, P. Spindel, Quantum supersymmetric cosmology and its hidden Kac–Moody structure. Class. Quantum Grav. **30**, 162001 (2013). [arXiv:1304.6381]
14. T. Damour, P. Spindel, Quantum supersymmetric Bianchi IX cosmology. Phys. Rev. D **90**, 103509 (2014). [arXiv:1406.1309]

15. P. Vargas Moniz, Supersymmetric quantum cosmology: a 'Socratic' guide. Gen. Relat. Grav. **46**, 1618 (2014)
16. P. Vargas Moniz, Quantum cosmology: meeting SUSY. Springer Proc. Math. Stat. **60**, 117 (2014)
17. E. Fermi, Sopra i fenomeni che avvengono in vicinanza di una linea oraria. Atti Accad. Naz. Lincei Cl. Sci. Fis. Mat. Nat. Rend. **31**, 21-51-101 (1922)
18. B. Bertotti, Fermi's coordinates and the principle of equivalence, in *Enrico Fermi*, ed. by C. Bernardini, L. Bonolis (Springer, Berlin, 2004)
19. F.K. Manasse, C.W. Misner, Fermi normal coordinates and some basic concepts in differential geometry. J. Math. Phys. **4**, 735 (1963)
20. N. Ashby, B. Bertotti, Relativistic effects in local inertial frames. Phys. Rev. D **34**, 2246 (1986)
21. L. Freidel, D. Minic, T. Takeuchi, Quantum gravity, torsion, parity violation and all that. Phys. Rev. D **72**, 104002 (2005). [arXiv:hep-th/0507253]
22. A. Perez, C. Rovelli, Physical effects of the Immirzi parameter. Phys. Rev. D **73**, 044013 (2006). [arXiv:gr-qc/0505081]
23. S. Mercuri, Fermions in Ashtekar–Barbero connection formalism for arbitrary values of the Immirzi parameter. Phys. Rev. D **73**, 084016 (2006). [arXiv:gr-qc/0601013]
24. M. Bojowald, R. Das, Fermions in loop quantum cosmology and the role of parity. Class. Quantum Grav. **25**, 195006 (2008). [arXiv:0806.2821]
25. S. Deser, P. van Nieuwenhuizen, Nonrenormalizability of the quantized Dirac–Einstein system. Phys. Rev. D **10**, 411 (1974)
26. E. Cartan, *Leçons sur la Théorie des Spineurs*, vol. II (Hermann, Paris, 1938)
27. F.W. Hehl, P. von der Heyde, G.D. Kerlick, J.M. Nester, General relativity with spin and torsion: foundations and prospects. Rev. Mod. Phys. **48**, 393 (1976)
28. I.L. Buchbinder, S.D. Odintsov, I.L. Shapiro, *Effective Action in Quantum Gravity* (IOP, Bristol, 1992)
29. J.D. McCrea, Irreducible decompositions of nonmetricity, torsion, curvature and Bianchi identities in metric-affine spacetimes. Class. Quantum Grav. **9**, 553 (1992)
30. J.A. Helayel-Neto, A. Penna-Firme, I.L. Shapiro, Conformal symmetry, anomaly and effective action for metric-scalar gravity with torsion. Phys. Lett. B **479**, 411 (2000). [arXiv:gr-qc/9907081]
31. S. Capozziello, G. Lambiase, C. Stornaiolo, Geometric classification of the torsion tensor in space-time. Ann. Phys. (Berlin) **10**, 713 (2001). [arXiv:gr-qc/0101038]
32. S. Holst, Barbero's Hamiltonian derived from a generalized Hilbert–Palatini action. Phys. Rev. D **53**, 5966 (1996). [arXiv:gr-qc/9511026]
33. G. Immirzi, Real and complex connections for canonical gravity. Class. Quantum Grav. **14**, L177 (1997). [arXiv:gr-qc/9612030]
34. C. Rovelli, T. Thiemann, Immirzi parameter in quantum general relativity. Phys. Rev. D **57**, 1009 (1998). [arXiv:gr-qc/9705059]
35. H.T. Nieh, M.L. Yan, An identity in Riemann–Cartan geometry. J. Math. Phys. **23**, 373 (1982)
36. O. Chandía, J. Zanelli, Topological invariants, instantons and chiral anomaly on spaces with torsion. Phys. Rev. D **55**, 7580 (1997). [arXiv:hep-th/9702025]
37. H.T. Nieh, A torsional topological invariant. Int. J. Mod. Phys. A **22**, 5237 (2007)
38. S. Mercuri, From the Einstein–Cartan to the Ashtekar–Barbero canonical constraints, passing through the Nieh–Yan functional. Phys. Rev. D **77**, 024036 (2008). [arXiv:0708.0037]
39. G. Date, R.K. Kaul, S. Sengupta, Topological interpretation of Barbero–Immirzi parameter. Phys. Rev. D **79**, 044008 (2009). [arXiv:0811.4496]
40. L. Castellani, R. D'Auria, P. Frè, *Supergravity and Superstrings: A Geometric Perspective*, vol. 1 (World Scientific, Singapore, 1991)
41. V. Taveras, N. Yunes, Barbero–Immirzi parameter as a scalar field: K-inflation from loop quantum gravity? Phys. Rev. D **78**, 064070 (2008). [arXiv:0807.2652]
42. A. Torres-Gomez, K. Krasnov, Remarks on Barbero–Immirzi parameter as a field. Phys. Rev. D **79**, 104014 (2009). [arXiv:0811.1998]

43. G. Calcagni, S. Mercuri, Barbero–Immirzi field in canonical formalism of pure gravity. Phys. Rev. D **79**, 084004 (2009). [arXiv:0902.0957]
44. S. Mercuri, V. Taveras, Interaction of the Barbero–Immirzi field with matter and pseudoscalar perturbations. Phys. Rev. D **80**, 104007 (2009). [arXiv:0903.4407]
45. C. Teitelboim, Quantum mechanics of the gravitational field. Phys. Rev. D **25**, 3159 (1982)
46. J.B. Hartle, K.V. Kuchař. Path integrals in parametrized theories: the free relativistic particle. Phys. Rev. D **34**, 2323 (1986)
47. P.A.M. Dirac, Generalized Hamiltonian dynamics. Can. J. Math. **2**, 129 (1950)
48. P.A.M. Dirac, Generalized Hamiltonian dynamics. Proc. R. Soc. Lond. A **246**, 326 (1958)
49. P.A.M. Dirac, The theory of gravitation in Hamiltonian form. Proc. R. Soc. Lond. A **246**, 333 (1958)
50. J. Schwinger, Quantized gravitational field. Phys. Rev. **130**, 1253 (1963)
51. M. Henneaux, C. Teitelboim, *Quantization of Gauge Systems* (Princeton University Press, Princeton, 1994)
52. T. Thiemann, *Modern Canonical Quantum General Relativity* (Cambridge University Press, Cambridge, 2007); Introduction to modern canonical quantum general relativity. arXiv:gr-qc/0110034
53. J.A. Wheeler, Geometrodynamics and the issue of the final state, in *Relativity, Groups and Topology*, ed. by C. DeWitt, B.S. DeWitt (Gordon and Breach, New York, 1964)
54. B.S. DeWitt, Quantum theory of gravity. I. The canonical theory. Phys. Rev. **160**, 1113 (1967)
55. J.W. York, Role of conformal three-geometry in the dynamics of gravitation. Phys. Rev. Lett. **28**, 1082 (1972)
56. G.W. Gibbons, S.W. Hawking, Action integrals and partition functions in quantum gravity. Phys. Rev. D **15**, 2752 (1977)
57. A. Guarnizo, L. Castañeda, J.M. Tejeiro, Boundary term in metric $f(R)$ gravity: field equations in the metric formalism. Gen. Relat. Grav. **42**, 2713 (2010). [arXiv:1002.0617]
58. A. Ashtekar, *Lectures on Non-perturbative Canonical Gravity* (World Scientific, Singapore, 1991)
59. A. Ashtekar, J. Engle, D. Sloan, Asymptotics and Hamiltonians in a first order formalism. Class. Quantum Grav. **25**, 095020 (2008). [arXiv:0802.2527]
60. P.W. Higgs, Integration of secondary constraints in quantized general relativity. Phys. Rev. Lett. **1**, 373 (1958); Erratum-ibid. **3**, 66 (1959)
61. J.A. Wheeler, Superspace and the nature of quantum geometrodynamics, in *Battelle Rencontres: 1967 Lectures in Mathematics and Physics*, ed. by C. DeWitt, J.A. Wheeler (Benjamin, New York, 1968)
62. S.W. Hawking, D.N. Page, Operator ordering and the flatness of the universe. Nucl. Phys. B **264**, 185 (1986)
63. J.J. Halliwell, Derivation of the Wheeler–DeWitt equation from a path integral for minisuperspace models. Phys. Rev. D **38**, 2468 (1988)
64. I. Moss, Quantum cosmology and the self-observing universe. Ann. Poincaré Phys. Theor. A **49**, 341 (1988)
65. R.P. Feynman, A.R. Hibbs, *Quantum Mechanics and Path Integrals: Emended Edition* (Dover, Mineola, 2010)
66. M. Chaichian, A. Demichev, *Path Integrals in Physics* (IOP, Bristol, 2001)
67. S.W. Hawking, The path-integral approach to quantum gravity, in *General Relativity: An Einstein Centenary Survey*, ed. by S.W. Hawking, W. Israel (Cambridge University Press, Cambridge, 1979)
68. A.D. Linde, Quantum creation of an inflationary universe. Zh. Eksp. Teor. Fiz. **87**, 369 (1984) [Sov. Phys. JETP **60**, 211 (1984)]; Quantum creation of the inflationary universe. Lett. Nuovo Cim. **39**, 401 (1984)
69. A. Vilenkin, Wave function discord. Phys. Rev. D **58**, 067301 (1998). [arXiv:gr-qc/9804051]
70. H. Jeffreys, On certain approximate solutions of linear differential equations of the second order. Proc. Lond. Math. Soc. **s2-23**, 428 (1925)

71. G. Wentzel, Eine Verallgemeinerung der Quantenbedingungen für die Zwecke der Wellenmechanik. Z. Phys. A **38**, 518 (1926)
72. H.A. Kramers, Wellenmechanik und halbzahlige Quantisierung. Z. Phys. A **39**, 828 (1926)
73. L. Brillouin, La mécanique ondulatoire de Schrödinger; une méthode générale de résolution par approximations successives. C. R. Acad. Sci. **183**, 24 (1926)
74. D.S. Salopek, J.R. Bond, J.M. Bardeen, Designing density fluctuation spectra in inflation. Phys. Rev. D **40**, 1753 (1989)
75. M. Nagasawa, J. Yokoyama, Phase transitions triggered by quantum fluctuations in the inflationary universe. Nucl. Phys. B **370**, 472 (1992)
76. J. Martin, D.J. Schwarz, WKB approximation for inflationary cosmological perturbations. Phys. Rev. D **67**, 083512 (2003). [arXiv:astro-ph/0210090]
77. R. Casadio, F. Finelli, M. Luzzi, G. Venturi, Improved WKB analysis of cosmological perturbations. Phys. Rev. D **71**, 043517 (2005). [arXiv:gr-qc/0410092]
78. U.H. Gerlach, Derivation of the ten Einstein field equations from the semiclassical approximation to quantum geometrodynamics. Phys. Rev. **177**, 1929 (1969)
79. T. Banks, TCP, quantum gravity, the cosmological constant and all that.... Nucl. Phys. B **249**, 332 (1985)
80. T.P. Singh, T. Padmanabhan, Notes on semiclassical gravity. Ann. Phys. (N.Y.) **196**, 296 (1989)
81. C. Kiefer, T.P. Singh, Quantum gravitational corrections to the functional Schrödinger equation. Phys. Rev. D **44**, 1067 (1991)
82. S.P. Kim, New asymptotic expansion method for the Wheeler–DeWitt equation. Phys. Rev. D **52**, 3382 (1995). [arXiv:gr-qc/9511038]
83. J.J. Halliwell, Correlations in the wave function of the universe. Phys. Rev. D **36**, 3626 (1987)
84. A. Peres, On Cauchy's problem in general relativity – II. Nuovo Cim. **26**, 53 (1962)
85. H.D. Zeh, Time in quantum gravity. Phys. Lett. A **126**, 311 (1988)
86. S.P. Kim, Quantum mechanics of conformally and minimally coupled Friedmann–Robertson–Walker cosmology. Phys. Rev. D **46**, 3403 (1992)
87. C. Kiefer, Continuous measurement of mini-superspace variables by higher multipoles. Class. Quantum Grav. **4**, 1369 (1987)
88. J.B. Hartle, S.W. Hawking, Wave function of the Universe. Phys. Rev. D **28**, 2960 (1983)
89. A. Vilenkin, Creation of universes from nothing. Phys. Lett. B **117**, 25 (1982)
90. W.-M. Suen, K. Young, Wave function of the Universe as a leaking system. Phys. Rev. D **39**, 2201 (1989)
91. H.D. Conradi, H.D. Zeh, Quantum cosmology as an initial value problem. Phys. Lett. A **154**, 321 (1991)
92. H.-D. Conradi, Initial state in quantum cosmology. Phys. Rev. D **46**, 612 (1992)
93. S.W. Hawking, The quantum state of the universe. Nucl. Phys. B **239**, 257 (1984)
94. D.N. Page, Density matrix of the Universe. Phys. Rev. D **34**, 2267 (1986)
95. G.W. Gibbons, L.P. Grishchuk, What is a typical wave function for the universe? Nucl. Phys. B **313**, 736 (1989)
96. J.J. Halliwell, J. Louko, Steepest-descent contours in the path-integral approach to quantum cosmology. I. The de Sitter minisuperspace model. Phys. Rev. D **39**, 2206 (1989)
97. J.J. Halliwell, J. Louko, Steepest-descent contours in the path-integral approach to quantum cosmology. II. Microsuperspace. Phys. Rev. D **40**, 1868 (1989)
98. J.J. Halliwell, J. Louko, Steepest-descent contours in the path-integral approach to quantum cosmology. III. A general method with applications to anisotropic minisuperspace models. Phys. Rev. D **42**, 3997 (1990)
99. J.J. Halliwell, J.B. Hartle, Integration contours for the no-boundary wave function of the universe. Phys. Rev. D **41**, 1815 (1990)
100. L.P. Grishchuk, L.V. Rozhansky, Does the Hartle–Hawking wavefunction predict the universe we live in? Phys. Lett. B **234**, 9 (1990)
101. G.W. Lyons, Complex solutions for the scalar field model of the Universe. Phys. Rev. D **46**, 1546 (1992)

102. A. Lukas, The no boundary wave-function and the duration of the inflationary period. Phys. Lett. B **347**, 13 (1995). [arXiv:gr-qc/9409012]
103. J.B. Hartle, S.W. Hawking, T. Hertog, No-boundary measure of the Universe. Phys. Rev. Lett. **100**, 201301 (2008). [arXiv:0711.4630]
104. J.B. Hartle, S.W. Hawking, T. Hertog, Classical universes of the no-boundary quantum state. Phys. Rev. D **77**, 123537 (2008). [arXiv:0803.1663]
105. J.B. Hartle, S.W. Hawking, T. Hertog, No-boundary measure in the regime of eternal inflation. Phys. Rev. D **82**, 063510 (2010). [arXiv:1001.0262]
106. A.O. Barvinsky, A.Yu. Kamenshchik, Cosmological landscape from nothing: some like it hot. JCAP **0609**, 014 (2006). [arXiv:hep-th/0605132]
107. A.O. Barvinsky, A.Yu. Kamenshchik, Thermodynamics via creation from nothing: limiting the cosmological constant landscape. Phys. Rev. D **74**, 121502 (2006). [arXiv:hep-th/0611206]
108. A.O. Barvinsky, Why there is something rather than nothing: cosmological constant from summing over everything in Lorentzian quantum gravity. Phys. Rev. Lett. **99**, 071301 (2007). [arXiv:0704.0083]
109. A. Vilenkin, Quantum creation of universes. Phys. Rev. D **30**, 509 (1984)
110. A. Vilenkin, Quantum origin of the universe. Nucl. Phys. B **252**, 141 (1985)
111. A. Vilenkin, Boundary conditions in quantum cosmology. Phys. Rev. D **33**, 3560 (1986)
112. A. Vilenkin, Quantum cosmology and the initial state of the Universe. Phys. Rev. D **37**, 888 (1988)
113. A. Vilenkin, Approaches to quantum cosmology. Phys. Rev. D **50**, 2581 (1994). [arXiv:gr-qc/9403010]
114. S.W. Hawking, I.G. Moss, Supercooled phase transitions in the very early universe. Phys. Lett. B **110**, 35 (1982)
115. T. Padmanabhan, Cosmological constant: the weight of the vacuum. Phys. Rep. **380**, 235 (2003). [arXiv:hep-th/0212290]
116. M. Spradlin, A. Strominger, A. Volovich, de Sitter space, in *Unity from Duality: Gravity, Gauge Theory and Strings*, ed. by C. Bachas, A. Bilal, M. Douglas, N. Nekrasov, F. David (Springer, Berlin, 2002). [arXiv:hep-th/0110007]
117. K. Vandersloot, Hamiltonian constraint of loop quantum cosmology. Phys. Rev. D **71**, 103506 (2005). [arXiv:gr-qc/0502082]
118. A. Perez, Regularization ambiguities in loop quantum gravity. Phys. Rev. D **73**, 044007 (2006). [arXiv:gr-qc/0509118]
119. T. Thiemann, Anomaly-free formulation of non-perturbative, four-dimensional Lorentzian quantum gravity. Phys. Lett. B **380**, 257 (1996). [arXiv:gr-qc/9606088]
120. T. Thiemann, Quantum spin dynamics (QSD). Class. Quantum Grav. **15**, 839 (1998). [arXiv:gr-qc/9606089]
121. C. Rovelli, L. Smolin, Knot theory and quantum gravity. Phys. Rev. Lett. **61**, 1155 (1988)
122. C. Rovelli, L. Smolin, Loop space representation of quantum general relativity. Nucl. Phys. B **331**, 80 (1990)
123. C. Rovelli, *Quantum Gravity* (Cambridge University Press, Cambridge, 2007)
124. A. Ashtekar, J. Lewandowski, Background independent quantum gravity: a status report. Class. Quantum Grav. **21**, R53 (2004). [arXiv:gr-qc/0404018]
125. A. Ashtekar, J. Lewandowski, Quantum theory of gravity I: area operators. Class. Quantum Grav. **14**, A55 (1997). [arXiv:gr-qc/9602046]
126. A. Ashtekar, J.C. Baez, A. Corichi, K. Krasnov, Quantum geometry and black hole entropy. Phys. Rev. Lett. **80**, 904 (1998). [arXiv:gr-qc/9710007]
127. A. Ashtekar, J.C. Baez, K. Krasnov, Quantum geometry of isolated horizons and black hole entropy. Adv. Theor. Math. Phys. **4**, 1 (2000). [arXiv:gr-qc/0005126]
128. K.A. Meissner, Black hole entropy in loop quantum gravity. Class. Quantum Grav. **21**, 5245 (2004). [arXiv:gr-qc/0407052]
129. D. Oriti, D. Pranzetti, L. Sindoni, Horizon entropy from quantum gravity condensates. Phys. Rev. Lett. **116**, 211301 (2016). [arXiv:1510.06991]

130. M. Dupuis, F. Girelli, Observables in loop quantum gravity with a cosmological constant. Phys. Rev. D **90**, 104037 (2014). [arXiv:1311.6841]
131. V. Bonzom, M. Dupuis, F. Girelli, E.R. Livine, Deformed phase space for 3d loop gravity and hyperbolic discrete geometries. arXiv:1402.2323
132. R. Gambini, J. Pullin, Does loop quantum gravity imply $\Lambda = 0$? Phys. Lett. B **437**, 279 (1998). [arXiv:gr-qc/9803097]
133. S.H.S. Alexander, G. Calcagni, Quantum gravity as a Fermi liquid. Found. Phys. **38**, 1148 (2008). [arXiv:0807.0225]
134. S.H.S. Alexander, G. Calcagni, Superconducting loop quantum gravity and the cosmological constant. Phys. Lett. B **672**, 386 (2009). [arXiv:0806.4382]
135. S.B. Treiman, R. Jackiw, B. Zumino, E. Witten, *Current Algebra and Anomalies* (World Scientific, Singapore, 1985), p. 258
136. H. Kodama, Specialization of Ashtekar's formalism to Bianchi cosmology. Prog. Theor. Phys. **80**, 1024 (1988)
137. H. Kodama, Holomorphic wave function of the Universe. Phys. Rev. D **42**, 2548 (1990)
138. R. Gambini, J. Pullin, *Loops, Knots, Gauge Theories and Quantum Gravity* (Cambridge University Press, Cambridge, 1996)
139. L. Smolin, Quantum gravity with a positive cosmological constant. arXiv:hep-th/0209079
140. S.S. Chern, J. Simons, Characteristic forms and geometric invariants. Ann. Math. **99**, 48 (1974)
141. L. Smolin, C. Soo, The Chern–Simons invariant as the natural time variable for classical and quantum cosmology. Nucl. Phys. B **449**, 289 (1995). [arXiv:gr-qc/9405015]
142. E. Witten, A note on the Chern–Simons and Kodama wave functions. arXiv:gr-qc/0306083
143. C. Soo, Wave function of the universe and Chern-Simons perturbation theory. Class. Quantum Grav. **19**, 1051 (2002). [arXiv:gr-qc/0109046]
144. A. Randono, Generalizing the Kodama state. I. Construction. arXiv:gr-qc/0611073
145. A. Randono, Generalizing the Kodama state. II. Properties and physical interpretation. arXiv:gr-qc/0611074
146. A. Randono, A mesoscopic quantum gravity effect. Gen. Relat. Grav. **42**, 1909 (2010). [arXiv:0805.2955]
147. S. Weinberg, *The Quantum Theory of Fields*, vol. II (Cambridge University Press, Cambridge, 1997)
148. C.G. Callan, R.F. Dashen, D.J. Gross, The structure of the gauge theory vacuum. Phys. Lett. B **63**, 334 (1976)
149. R. Jackiw, C. Rebbi, Vacuum periodicity in a Yang–Mills quantum theory. Phys. Rev. Lett. **37**, 172 (1976)
150. S. Deser, M.J. Duff, C.J. Isham, Gravitationally induced CP effects. Phys. Lett. B **93**, 419 (1980)
151. A. Ashtekar, A.P. Balachandran, S. Jo, The CP problem in quantum gravity. Int. J. Mod. Phys. A **4**, 1493 (1989)
152. R.D. Peccei, H.R. Quinn, CP conservation in the presence of pseudoparticles. Phys. Rev. Lett. **38**, 1440 (1977)
153. R.D. Peccei, H.R. Quinn, Constraints imposed by CP conservation in the presence of pseudoparticles. Phys. Rev. D **16**, 1791 (1977)
154. F. Wilczek, Problem of strong P and T invariance in the presence of instantons. Phys. Rev. Lett. **40**, 279 (1978)
155. S. Weinberg, A new light boson? Phys. Rev. Lett. **40**, 223 (1978)
156. M. Montesinos, Self-dual gravity with topological terms. Class. Quantum Grav. **18**, 1847 (2001). [arXiv:gr-qc/0104068]
157. R. Paternoga, R. Graham, Triad representation of the Chern–Simons state in quantum gravity. Phys. Rev. D **62**, 084005 (2000). [arXiv:gr-qc/0003111]
158. N. Weiss, Possible origins of a small, nonzero cosmological constant. Phys. Lett. B **197**, 42 (1987)

159. J.A. Frieman, C.T. Hill, R. Watkins, Late time cosmological phase transitions: particle physics models and cosmic evolution. Phys. Rev. D **46**, 1226 (1992)
160. J.A. Frieman, C.T. Hill, A. Stebbins, I. Waga, Cosmology with ultralight pseudo Nambu–Goldstone bosons. Phys. Rev. Lett. **75**, 2077 (1995). [arXiv:astro-ph/9505060]
161. T. Jacobson, $1 + 1$ sector of $3 + 1$ gravity. Class. Quantum Grav. **13**, L111 (1996); Erratum-ibid. **13**, 3269 (1996). [arXiv:gr-qc/9604003]
162. G.E. Volovik, Superfluid analogies of cosmological phenomena. Phys. Rep. **351**, 195 (2001). [arXiv:gr-qc/0005091]
163. G.E. Volovik, Field theory in superfluid He-3: what are the lessons for particle physics, gravity, and high temperature superconductivity? Proc. Natl. Acad. Sci. **96**, 6042 (1999). [arXiv:cond-mat/9812381]
164. G.E. Volovik, *The Universe in a Helium Droplet* (Clarendon Press, Oxford, 2003)
165. J. Bardeen, L.N. Cooper, J.R. Schrieffer, Microscopic theory of superconductivity. Phys. Rev. **106**, 162 (1957)
166. J. Bardeen, L.N. Cooper, J.R. Schrieffer, Theory of superconductivity. Phys. Rev. **108**, 1175 (1957)
167. N.N. Bogoliubov, On a new method in the theory of superconductivity. Nuovo Cim. **7**, 794 (1958)
168. J. Polchinski, Effective field theory and the Fermi surface, in *Recent Directions in Particle Theory*, ed. by J.A. Harvey, J.G. Polchinski (World Scientific, Singapore, 1993). [arXiv:hep-th/9210046]
169. G. Sierra, Conformal field theory and the exact solution of the BCS Hamiltonian. Nucl. Phys. B **572**, 517 (2000). [arXiv:hep-th/9911078]
170. M. Asorey, F. Falceto, G. Sierra, Chern–Simons theory and BCS superconductivity. Nucl. Phys. B **622**, 593 (2002). [arXiv:hep-th/0110266]
171. S.N. Bose, Plancks Gesetz und Lichtquantenhypothese. Z. Phys. A **26**, 178 (1924)
172. A. Einstein, Quantentheorie des einatomigen idealen Gases. Sitz.-ber. Kgl. Preuss. Akad. Wiss. **1924**, 261 (1924)
173. A. Einstein, Quantentheorie des einatomigen idealen Gases. Zweite Abhandlung. Sitz.-ber. Kgl. Preuss. Akad. Wiss. **1925**, 3 (1925)
174. F. Dalfovo, S. Giorgini, L.P. Pitaevskii, S. Stringari, Theory of Bose–Einstein condensation in trapped gases. Rev. Mod. Phys. **71**, 463 (1999)
175. A.J. Leggett, Bose–Einstein condensation in the alkali gases: some fundamental concepts. Rev. Mod. Phys. **73**, 307 (2001)
176. C.J. Pethick , H. Smith, *Bose–Einstein Condensation in Dilute Gases* (Cambridge University Press, Cambridge, 2002)
177. L.P. Pitaevskii, S. Stringari, *Bose–Einstein Condensation* (Clarendon Press, Oxford, 2003)
178. S. Giorgini, L.P. Pitaevskii, S. Stringari, Theory of ultracold atomic Fermi gases. Rev. Mod. Phys. **80**, 1215 (2008). [arXiv:0706.3360]
179. M. Randeria, E. Taylor, BCS-BEC crossover and the unitary Fermi gas. Ann. Rev. Cond. Matter Phys. **5**, 209 (2014). [arXiv:1306.5785]
180. G.A. Mena Marugán, Extent of the Immirzi ambiguity in quantum general relativity. Class. Quantum Grav. **19**, L63 (2002). [arXiv:gr-qc/0203027]
181. C.J. Isham, Some quantum field theory aspects of the superspace quantization of general relativity. Proc. R. Soc. Lond. A **351**, 209 (1976)
182. M. Henneaux, Zero Hamiltonian signature spacetimes. Bull. Soc. Math. Belg. **31**, 47 (1979)
183. C. Teitelboim, The Hamiltonian structure of space-time, in *General Relativity and Gravitation*, ed. by A. Held, vol. 1 (Plenum, New York, 1980)

Chapter 10
Canonical Quantum Cosmology

Not only is the universe stranger than we imagine, it is stranger than we can imagine.

—Sir Arthur Eddington

Contents

A complete background-independent quantum theory of gravity is a hard goal to achieve and the formal discussion of any of the proposals in this direction lies beyond the scope of this book. However, in order to understand certain techniques it is often useful to specialize to a simple background and study the quantum system thereon (Sect. 10.1). *Symmetry reduction* is done not only for didactic purposes but also as a (hopefully temporary) necessity in active research. For instance, it is natural

G. Calcagni, *Classical and Quantum Cosmology*, Graduate Texts in Physics,
DOI 10.1007/978-3-319-41127-9_10

to consider the big-bang and cosmological constant problems in an expanding cosmological background. In such a setting, one does not employ the full machinery of general theories of quantum gravity and a number of simplifications take place. We apply the procedure of symmetry reduction to canonical gravity and quantum cosmological scenarios, first in the Wheeler–DeWitt approach in ADM variables (Sect. 10.2) and then within loop quantum cosmology (Sect. 10.3).

10.1 Mini-superspace

The space $\mathcal{S}$ of geometric and matter configurations is rather formal and cannot be handled easily also due to the fact that it is infinite dimensional. However, if one restricts the theory to homogeneous geometries, the number of degrees of freedom becomes finite and one can construct a quantum model on a finite-dimensional *mini-superspace* [1, 2]. This reduction has no robust justification a priori and it is generally agreed that mini-superspace quantization is only a toy model of a quantum theory of geometry. In this case, the symmetry reduction of the theory is performed at the *kinematical* level, that is, before solving the quantum constraints. In fact, we first restrict the classical theory to a homogeneous background and then quantize:

$$\text{classical theory} \quad \to \quad \text{symmetry reduction} \quad \to \quad \text{quantization}\,.$$

This is in implicit violation of Heisenberg uncertainty principle, since one is simultaneously freezing inhomogeneous modes out of the wave-function Ψ.

On the other hand, a complete quantum canonical theory of gravity should implement quantization in a background-independent way and *then* specialize, if desired, to a homogeneous cosmological setting:

$$\text{classical theory} \quad \to \quad \text{quantization} \quad \stackrel{?}{\to} \quad \text{symmetry reduction}\,.$$

The question mark on the last arrow symbolizes the fact that, in many approaches,[1] the limit to a continuum (smooth manifold) and the operation of symmetry reduction are non-trivial and require intermediate steps. In several cases, the formulation of the theory is incomplete and these steps are still unknown. Therefore, in general, mini-superspace models can capture only some of the qualitative features of the cosmology of the full theory. A complete check of that can be done only if the underlying theory is under control and full solutions can be constructed. In LQG, this seems to be the case (Sect. 10.3.7). In Sect. 11.5, we will present a second-quantized extension of the canonical theory where symmetry reduction and

[1] For instance, in discrete gravity models geometric variables pick countable labels which characterize simplicial complexes. These complexes are mathematical objects quite distinct from manifolds, but they can approximate smooth manifolds under certain limits.

quantization may commute. However, in general they do not [3, 4] and we shall see a concrete example of this in Sect. 11.3.2. Ignoring for the moment possible embeddings of canonical quantum gravity into a more complete theory, we can exploit the knowledge we have of the full canonical theory to get a hint of the level at which we can trust the mini-superspace results.

Having so cautioned the reader, we proceed to analyze the canonical variables and first-class constraints in a cosmological setting, beginning with Wheeler–DeWitt quantization [1, 2]. We limit the discussion to FLRW models in vacuum or with a minimally coupled scalar field [5], having inflation in mind [6–10].[2] We will then discuss loop quantum cosmology and cosmological perturbations. Mini-superspace FLRW scenarios and their generalization to Bianchi models are usually sensible and give a very stimulating insight into the problem of the big bang in a quantum universe, especially in the loop quantization scheme. However, it has become progressively clear that many of the properties of quantum geometry become accessible only when inhomogeneous perturbations and, consequently, the full structure of the constraint algebra, have been included in the picture. In this case, the properties of the quantum system do change, unless the back-reaction of anisotropic or inhomogeneous modes (loosely speaking, the "production of gravitons") be negligible [3, 4, 24–27].

10.1.1 Classical FLRW Hamiltonian

The symmetry reduction takes place at the level of the classical action. From this, one can define mini-superspace canonical variables and obtain the first-class constraints. To get the flavour of what happens, we first perform the symmetry reduction directly on the classical constraints and then rederive the mini-superspace constraints from the symmetry-reduced action.

On homogeneous backgrounds, one can always choose the shift vector N^α to vanish. Then, the extrinsic curvature is (7.132), $K_{\alpha\beta} = Hh_{\alpha\beta}/N$, and the spatial Ricci curvature is ${}^{(3)}R = 6\textsc{k}/a^2$. The momentum constraint (9.70) vanishes identically, so that the only constraint left is the super-Hamiltonian (9.71), which can be written as

$$a^3\left[\frac{1}{2}\frac{\dot{\phi}^2}{N^2}+V(\phi)-\frac{3}{\kappa^2}\left(\frac{H^2}{N^2}+\frac{\textsc{k}}{a^2}\right)\right]=0\,. \tag{10.1}$$

This is nothing but the first Friedmann equation (2.81) with $N \neq 1$.

Let us now recover (10.1) from the action. On an ideal FLRW background, open and flat universes have infinite spatial volume and the super-Hamiltonian

[2]Early papers considering a massless scalar field are [11, 12]. The case with non-minimal couplings has also been studied [6, 7, 13–21]. Adding higher-order curvature terms does not change the results much [8, 22, 23].

constraint is formally ill-defined because it entails a divergent integration of a spatially constant quantity over a comoving spatial slice Σ,

$$\int_{\Sigma} d^3\mathbf{x}\,\sqrt{h} = +\infty\,.$$

To make the integral finite, it is customary to define the constraint on a freely chosen finite region of size $\mathcal{V} = a^3\mathcal{V}_0$, where $\mathcal{V}_0$ is the corresponding comoving volume:

$$\int_{\Sigma} d^3\mathbf{x}\,\sqrt{h} \to \int_{\Sigma(\mathcal{V}_0)} d^3\mathbf{x}\,\sqrt{h} =: a^3\mathcal{V}_0 < +\infty\,. \tag{10.2}$$

If the universe is closed, $\mathcal{V}_0$ represents its comoving volume and is therefore physical. In the open and flat cases, the *fiducial volume* $\mathcal{V}_0$ is arbitrary and has no physical meaning, so that it should not appear in physical observables eventually. We will face this issue in loop quantum cosmology.

From (9.66),

$$\begin{aligned}
S &= \mathcal{V}_0 \int dt\, Na^3 \left[\frac{3}{\kappa^2}\left(\frac{\textsf{K}}{a^2} - \frac{H^2}{N^2}\right) + \frac{1}{2}\frac{\dot\phi^2}{N^2} - V(\phi)\right] \\
&= \mathcal{V}_0 \int dt\, N \left[\frac{3a}{\kappa^2}\left(\textsf{K} - \frac{\dot a^2}{N^2}\right) + a^3\left(\frac{1}{2}\frac{\dot\phi^2}{N^2} - V\right)\right]\,.
\end{aligned} \tag{10.3}$$

The momenta conjugate to the variables N, a and ϕ are $\pi^0 = 0$ and

$$p_{(a)} := \frac{\delta S}{\delta \dot a} = -\frac{6\mathcal{V}_0}{\kappa^2}\frac{a\dot a}{N}\,, \qquad p_\phi := \frac{\delta S}{\delta\dot\phi} = \mathcal{V}_0 a^3 \frac{\dot\phi}{N}\,, \tag{10.4}$$

with commutation relations

$$\{a, p_{(a)}\} = 1\,, \qquad \{\phi, p_\phi\} = 1\,. \tag{10.5}$$

The Dirac Hamiltonian is

$$H_{\rm D} = \pi^0\dot N + p_{(a)}\dot a + p_\phi\dot\phi - L = \pi^0\dot N + N\mathcal{H}\,,$$

where, in agreement with (10.1),

$$\mathcal{H} = \frac{1}{2\mathcal{V}_0 a^3}\left[-\frac{\kappa^2}{6}a^2 p_{(a)}^2 + p_\phi^2 + \mathcal{U}(a,\phi)\right]\,, \tag{10.6a}$$

and

$$\mathcal{U}(a,\phi) := 2\mathcal{V}_0 a^3 U(a,\phi)\,, \qquad U(a,\phi) = \mathcal{V}_0 a\left(a^2 V - \frac{3\textsf{K}}{\kappa^2}\right)\,. \tag{10.6b}$$

Notice that the mini-superspace scalar constraint differs with respect to the Hamiltonian constraint of the full theory by the integrated volume factor. To keep the notation simple, we will use the same symbol $\mathcal{H}$ in both cases.

The Hamilton equations are (Problem 10.1)

$$\dot{p}_{(a)} = \{p_{(a)}, H_\mathrm{D}\} = 2\mathcal{V}_0 N a^2 \left[\kappa^2\left(\frac{\dot{\phi}^2}{N^2} - V\right) - 3\frac{H^2}{N^2}\right], \tag{10.7}$$

$$\dot{p}_\phi = \{p_\phi, H_\mathrm{D}\} = -N\mathcal{V}_0 a^3 V_{,\phi}\,. \tag{10.8}$$

These are, respectively, the second Friedmann equation and the scalar-field equation of motion: check these expressions for $N = 1$ with (2.170) and (2.171), noting that $\dot{p}_{(a)} = -(6\mathcal{V}_0/\kappa^2)a^2(2H^2 + \dot{H})$ and $\dot{p}_\phi = \mathcal{V}_0 a^3(\ddot{\phi} + 3H\dot{\phi})$.

In preparation for a comparison with the results of loop quantum cosmology, we note that, at the classical level, one can define other canonical coordinates (pairs of conjugate variables) in phase space and recast the Hamiltonian constraint (10.6) accordingly. For instance, consider the pair

$$b := -\frac{\gamma\kappa^2}{6v_0}\frac{p_{(a)}}{a^{1+2n}}\,, \qquad v := \frac{3v_0}{(1+n)\gamma\kappa^2}\,a^{2(1+n)}\,, \tag{10.9}$$

where v_0 is a constant. The case $n = -1/2$ corresponds to the simple scale-factor variables. From (10.5),

$$\{b, v\} = 1\,, \tag{10.10}$$

while the Hamiltonian constraint (10.6) is

$$\mathcal{H} = \frac{1}{2\mathcal{V}_0}\left(\frac{v_0}{v}\right)^{\frac{3}{2(1+n)}}\left[-\frac{2(1+n)^2\kappa^2}{3}(bv)^2 + p_\phi^2 + \mathcal{U}(v,\phi)\right], \tag{10.11a}$$

$$\mathcal{U} = 2\mathcal{V}_0^2\left(\frac{v}{v_0}\right)^{\frac{2}{1+n}}\left[\left(\frac{v}{v_0}\right)^{\frac{1}{1+n}} V(\phi) - \frac{3\textsc{K}}{\kappa^2}\right]. \tag{10.11b}$$

10.2 Wheeler–DeWitt Quantum Cosmology

In the traditional WDW quantization, one promotes b and v to operators and the Poisson bracket (10.10) to a commutator. This begins the construction of the mini-superspace quantum theory:

$$[\hat{b}, \hat{v}] = \mathrm{i}\,. \tag{10.12}$$

Since we are in $\hbar = 1$ units,

$$\kappa^2 = 8\pi l_{\rm Pl}^2 = \frac{8\pi}{m_{\rm Pl}^2} = \frac{1}{M_{\rm Pl}^2}\,. \tag{10.13}$$

At this point, we can choose two unitarily equivalent representations for the states of the Hilbert space: one where the volume operator is diagonal or one where $\hat{b}$ is diagonal. We consider first the v representation, where

$$\hat{v} := v\,, \qquad \hat{b} := \mathrm{i}\frac{\partial}{\partial v}\,. \tag{10.14}$$

In the scale-factor representation, $\hat{p}_{(a)} = -\mathrm{i}\partial_a$ and $[\hat{p}_{(a)}, \hat{a}] = -\mathrm{i}$. In general, the kinematical Hilbert space $\mathcal{H}_{\rm kin}^{\rm g}$ of the gravity sector is spanned by the basis $\{|v\rangle\}$ of eigenstates of the multiplicative operator $\hat{v}$, which can be rendered orthonormal:

$$\langle v|v'\rangle = \delta_{vv'}\,. \tag{10.15}$$

For the matter sector, one chooses a standard Schrödinger quantization with a natural representation of the Hilbert space $\mathcal{H}_{\rm kin}^{\phi}$, the space of square-integrable functions on $\mathbb{R}$, on which $\hat{\phi}$ acts by multiplication and $\hat{p}_\phi$ by derivation,

$$\hat{\phi} := \phi\,, \qquad \hat{p}_\phi := -\mathrm{i}\frac{\partial}{\partial \phi}\,. \tag{10.16}$$

An orthonormal basis is given by

$$\langle \phi|\phi'\rangle = 2\pi\,\delta(\phi-\phi')\,, \tag{10.17}$$

where the 2π factor is for later convenience (Sect. 11.4). The Hilbert space of the coupled system is then just the tensor product $\mathcal{H}_{\rm kin}^{\rm g}\otimes\mathcal{H}_{\rm kin}^{\phi}$.

Symmetry reduction of the Wheeler–DeWitt equation (9.88) or quantization of the classical constraint (10.11) with the ordering $\hat{v}\hat{b}\hat{v}\hat{b}$ yield the same result:

$$\hat{\mathcal{H}}\Psi[v,\phi] = \frac{1}{2\mathcal{V}_0}\left(\frac{v_0}{v}\right)^{\frac{3}{2(1+n)}}\left[\frac{2(1+n)^2\kappa^2}{3}\frac{\partial^2}{(\partial\ln v)^2} - \frac{\partial^2}{\partial\phi^2} + \mathcal{U}\right]\Psi[v,\phi] = 0\,. \tag{10.18}$$

The mini-superspace Hamiltonian constraint (10.18) is a sort of "Schrödinger equation without time," i.e., with no $\mathrm{i}\hbar\partial_t$ term. The role of physical time in general relativity is played by some suitably chosen internal clock. In particular, the WDW equation is hyperbolic and can be written as a second-order differential equation on a two-dimensional configuration-space manifold. To get a cleaner expression, we rewrite the square bracket in terms of the number of e-foldings $\mathcal{N} = \ln a$ and set $n = -1/2$:

$$\left[\frac{\partial^2}{\partial\mathcal{N}^2} - \frac{6}{\kappa^2}\frac{\partial^2}{\partial\phi^2} + \frac{6\mathcal{U}}{\kappa^2}\right]\Psi[\mathcal{N},\phi] = 0\,. \tag{10.19}$$

10.2.1 de Sitter Solutions and Probability of Inflation

The type of information we would like to extract from the solution Ψ of the WDW equation is about the probability in the space of initial conditions that inflation be realized and whether the big-bang singularity is resolved. To get an idea of how these solutions behave, we invoke the mini-superspace WKB approximation (9.94) [13, 24, 28–32], valid only for long-wave-length gravitational modes [33].[3] The further requirement of decoherence excludes significant interference between different WKB solutions [24, 29, 39–41]. In particular, rather than considering a superposition of expanding and contracting branches of a quantum universe, we super-select only one of them (the expanding one, which we observe). Finally, we assume that the wave-function Ψ slowly varies with respect to ϕ, which is tantamount to considering a de Sitter background.

In this setting, the configuration space becomes one-dimensional. After rescaling $S \to \mathcal{V}_0 S$ in (9.98) and plugging (9.98) into (10.19), (9.99) becomes

$$\partial_{\mathcal{N}} S \simeq \pm\sqrt{\frac{6\mathcal{U}}{\mathcal{V}_0^2\kappa^2}} = \pm\sqrt{\frac{12}{\kappa^2}}\,\mathrm{e}^{2\mathcal{N}}\sqrt{\mathrm{e}^{2\mathcal{N}}V - \frac{3\textsc{k}}{\kappa^2}}\,,$$

so that

$$S \simeq \pm\sqrt{\frac{12}{\kappa^2}}\frac{1}{3V}\left(\mathrm{e}^{2\mathcal{N}}V - \frac{3\textsc{k}}{\kappa^2}\right)^{\frac{3}{2}}. \tag{10.20}$$

Using this result and (9.103), we get

$$0 = \partial_{\mathcal{N}} S\,\partial_{\mathcal{N}}\Psi_0 + \frac{1}{2}\Psi_0\partial_{\mathcal{N}}^2 S = \partial_{\mathcal{N}} S\left[\partial_{\mathcal{N}}\Psi_0 + \left(1 + \frac{1}{2}\frac{\mathrm{e}^{2\mathcal{N}}V}{\mathrm{e}^{2\mathcal{N}}V - 3\textsc{k}/\kappa^2}\right)\Psi_0\right],$$

giving, via (9.105), the semi-classical limit

$$\Psi[a,\phi] \simeq \frac{\mathcal{A}(\phi)}{a(a^2V - 3\textsc{k}/\kappa^2)^{1/4}}\exp\left[\pm\frac{\mathrm{i}}{3V}\sqrt{\frac{12}{\kappa^2}}\left(a^2V - \frac{3\textsc{k}}{\kappa^2}\right)^{\frac{3}{2}}\right] \tag{10.21}$$

for $a^2V - 3\textsc{k}/\kappa^2 > 0$, while the limit in the classically forbidden region $a^2V - 3\textsc{k}/\kappa^2 < 0$ is

$$\Psi[a,\phi] \simeq \frac{\mathcal{A}(\phi)}{a(3\textsc{k}/\kappa^2 - a^2V)^{1/4}}\exp\left[\pm\frac{1}{3V}\sqrt{\frac{12}{\kappa^2}}\left(\frac{3\textsc{k}}{\kappa^2} - a^2V\right)^{\frac{3}{2}}\right]. \tag{10.22}$$

[3] Other related methods were developed in [34–38].

If the scalar potential or cosmological constant is positive, $V > 0$, then exponential solutions exist only for closed universes with small enough scale factor. There exist oscillatory semi-classical solutions (with progressively damped amplitude) for a flat or open geometry or for a closed geometry with sufficiently large scale factor. In this case, the Hamilton–Jacobi momenta (9.100) give the classical equations in the extreme slow-roll approximation.

The signs in (10.20), (10.21) and (10.22) and the coefficients of linear combinations of solutions depend on the choice of boundary conditions. The closed case $\textsf{K} = 1$ is of special interest because one can construct solutions obeying the no-boundary or tunneling conditions discussed in Sect. 9.2.3. Analytic solutions exist for a pure de Sitter background (2.111), where $\kappa^2 V = \Lambda = 3H^2$ is the cosmological constant. In that case, one neglects the field derivative term in (10.19) and the potential is

$$\frac{6\mathcal{U}}{\kappa^2} = \left(\frac{6\mathcal{V}_0}{\kappa^2}\right)^2 a^4(H^2a^2 - 1)\,. \tag{10.23}$$

To solve the WDW equation exactly, one exploits the ambiguity in the ordering choice of $\hat{a}$ and $\hat{p}_{(a)}$ factors and augments the second-order derivative in the scale factor in (10.19) by a friction term, $\partial^2/\partial a^2 \to \partial^2/\partial a^2 - a^{-1}\partial/\partial a$. Then, imposing $\Psi[a = \infty] = 0$, one finds the Hartle–Hawking wave-function [42–44]

$$\Psi_{\rm HH}[a] = \mathcal{A}\frac{\mathrm{Ai}[z(a)]}{\mathrm{Ai}[z(0)]}, \tag{10.24}$$

where $\mathcal{A}$ is a normalization constant, Ai is the Airy function (solution of $\partial_z^2\Psi - z\Psi = 0$) and $z(a) = [3\mathcal{V}_0/(H^2\kappa^2)]^{2/3}(1 - H^2a^2)$. Setting $\mathcal{V}_0 = 1$ and using the asymptotic expansions of Ai,

$$\mathrm{Ai}(z) \overset{z\to+\infty}{\simeq} \frac{1}{2\sqrt{\pi}} z^{-\frac{1}{4}} \exp\left(-\frac{2}{3}z^{\frac{3}{2}}\right),$$

$$\mathrm{Ai}(z) \overset{z\to-\infty}{\simeq} \frac{1}{\sqrt{\pi}}(-z)^{-\frac{1}{4}} \cos\left[\frac{2}{3}(-z)^{\frac{3}{2}} - \frac{\pi}{4}\right],$$

one obtains the approximate form of $\Psi_{\rm HH}$ in the semi-classical and classically forbidden ranges [5, 45]:

$$\Psi_{\rm HH}[a] \sim \frac{1}{(H^2a^2-1)^{1/4}} \cos\left[\frac{2}{H^2\kappa^2}(H^2a^2-1)^{\frac{3}{2}} - \frac{\pi}{4}\right], \qquad H^2a^2 > 1\,, \tag{10.25a}$$

$$\Psi_{\rm HH}[a] \sim \frac{1}{(1-H^2a^2)^{1/4}} \exp\left[-\frac{2}{H^2\kappa^2}(1-H^2a^2)^{\frac{3}{2}}\right], \qquad H^2a^2 < 1\,. \tag{10.25b}$$

The wave-function in the tunneling proposal is complex-valued [42]:

$$\Psi_{\rm V}[a] = \mathcal{A}\frac{{\rm Ai}[z(a)] + {\rm i}\,{\rm Bi}[z(a)]}{{\rm Ai}[z(0)] + {\rm i}\,{\rm Bi}[z(0)]}, \tag{10.26}$$

where Bi is the second solution of the Airy equation, with asymptotic limits

$$\begin{aligned}
{\rm Bi}(z) &\stackrel{z\to+\infty}{\simeq} \frac{1}{\sqrt{\pi}}z^{-\frac{1}{4}}\exp\left(\frac{2}{3}z^{\frac{3}{2}}\right), \\
{\rm Bi}(z) &\stackrel{z\to-\infty}{\simeq} \frac{1}{\sqrt{\pi}}(-z)^{-\frac{1}{4}}\cos\left[\frac{2}{3}(-z)^{\frac{3}{2}} + \frac{\pi}{4}\right].
\end{aligned}$$

Thus,

$$\Psi_{\rm V}[a] \sim \frac{1}{(H^2a^2-1)^{1/4}}\exp\left[-\frac{2{\rm i}}{H^2\kappa^2}(H^2a^2-1)^{\frac{3}{2}}\right], \qquad H^2a^2 > 1\,, \tag{10.27a}$$

$$\Psi_{\rm V}[a] \sim \frac{1}{(1-H^2a^2)^{1/4}}\exp\left[\frac{2}{H^2\kappa^2}(1-H^2a^2)^{\frac{3}{2}}\right], \qquad H^2a^2 < 1\,. \tag{10.27b}$$

Figure 9.1 shows the wave-function (10.24) and the real part of the wave-function (10.26).

We now turn to the problem of whether WDW quantum cosmology predicts sufficiently long inflation. For the tunneling proposal, it is easy to argue for an affirmative answer [43]. Let us revert to a scalar-field cosmology in the ESR approximation, so that $H^2 \simeq \kappa^2 V(\phi)/3$. The probability density function $P(\phi_{\rm i})$ of the initial state of the Universe is the nucleation probability given by the ratio of the squared wave-function at the classical turning point $a = H^{-1}$ and at $a = 0$, $P(\phi_{\rm i}) \sim |\Psi[H^{-1},\phi]/\Psi[0,\phi_{\rm i}]|^2 \sim |\Psi[0,\phi_{\rm i}]|^{-2} \propto \exp[4/(H^2\kappa^2)]$, where we used (10.25b) and (10.27b). Thus,

$$P_{\rm HH}(\phi_{\rm i}) = \exp\left[\frac{12}{\kappa^4 V(\phi_{\rm i})}\right], \qquad P_{\rm V}(\phi_{\rm i}) = \exp\left[-\frac{12}{\kappa^4 V(\phi_{\rm i})}\right]. \tag{10.28}$$

It is worth stressing that the main difference between the probability densities of the no-boundary and tunneling proposals amounts to just a sign only in the de Sitter mini-superspace approximation. The Hartle–Hawking wave-function is the path-integrated exponential of the Euclidean action $I < 0$ under standard Wick rotation $t \to -{\rm i}t$, while Vilenkin's wave-function is a Lorentzian path integral with Lorentzian action S:

$$\Psi_{\rm HH} \sim \int {\rm e}^{-I}, \qquad \Psi_{\rm V} \sim \int {\rm e}^{{\rm i}S}.$$

For the de Sitter solution, in the tunneling regime these correspond to probability densities

$$P_{\rm HH} \sim e^{-I}, \qquad P_{\rm V} \sim e^{-|I|}.$$

Both cases are different from Linde's $t \to +it$ prescription [46], leading to $\Psi_{\rm L} \sim \int e^{+I}$ and $P_{\rm L} \sim e^{+I} \sim P_{\rm V}$. The fact that $I < 0$ in the de Sitter case has sometimes led to confusion between Linde's and Vilenkin's prescriptions [47].

In the tunneling case, $P_{\rm V}$ is maximal but finite at large V, corresponding to large values of ϕ for the monomial potentials of chaotic inflation. Large-field potentials may then be regarded as typical in realizations of the quantum universe, although the tunneling wave-function can describe viable small-field inflation as well (see [48] for the case of natural inflation). The no-boundary proposal has the opposite behaviour and its probability distribution is strongly peaked at small potential values, where it diverges (this is not a problem per se, as P is a probability density). Therefore, $P_{\rm HH}$ is not normalizable unless the scalar field has a finite range $\phi \in [\phi_1, \phi_2]$. However, maximal probability would correspond to the absolute minimum of the potential, a condition incompatible with inflationary dynamics. This result [43] can be refined for a monomial potential (e.g., [49]) but it holds also for small-field potentials [50, 51]. Later studies, however, reassessed the way conditional probabilities are weighted in quantum cosmology, apparently reconciling the Hartle–Hawking no-boundary proposal with sustainable inflation [52–54]. The same conclusion can be reached by another mechanism, by including one-loop corrections to the effective action [56–59], also in the case of a non-minimally coupled scalar field [55, 56, 60, 61]. In general, the modified probability density function is peaked at configurations favouring eternal inflation in the no-boundary case, and standard inflation in the tunneling case.[4]

10.2.2 Massless Scalar Field and Group Averaging

The WKB solutions are useful tools to study the structure of the quantum theory, but it would be desirable to construct the full Hilbert space of physical states in exactly solvable models. A case of particular interest in this respect is that of a flat geometry with a massless scalar field, $V(\phi) = 0 = \textsc{k}$.

[4]The probability issue can be tackled also from other perspectives, some of which are adopted when facing the cosmological constant problem in the context of eternal inflation (Sects. 5.6.5 and 10.2.1). Gibbons and Turok proposed a classical measure for classical cosmological trajectories [62] which, apparently, disfavours inflation [62]. In fact, in the simplest case of a single slow-rolling scalar field, this classical probability of having $\mathcal{N}$ e-foldings of inflation contains a damping factor $e^{-3\mathcal{N}}$ which makes long inflationary periods ($\mathcal{N} \sim 50-60$) highly improbable. There are several caveats and criticisms regarding this measure [63–65], including about the way initial conditions are imposed and the fact that they are set in regimes where quantum effects should be taken into account.

For later comparison with loop quantum cosmology, we take the b representation where (10.14) is replaced by the multiplicative operator $\hat{b} := b$ and

$$\hat{v} := -\mathrm{i}\frac{\partial}{\partial b}\,. \tag{10.29}$$

Then, keeping the inverse v factor in (10.11) to the left and the ordering $(bv)^2 \to \hat{b}\hat{v}\hat{b}\hat{v}$, the Hamiltonian constraint operator reduces to

$$\boxed{\hat{C}\Psi[y,\phi] := -\left(\Theta + \partial_\phi^2\right)\Psi[y,\phi] = 0\,,} \tag{10.30}$$

where

$$\Theta := -\partial_y^2\,, \qquad y := \sqrt{\frac{3}{2(1+n)^2\kappa^2}}\ln\frac{b}{b_0}\,, \tag{10.31}$$

and b_0 is an integration constant. A first consequence of the absence of a potential is that the operator annihilating Ψ commutes with the parity operator changing the orientation of the triad, so that both wave-functions encoding either orientation are solutions. Since physical observables should not depend on the triad orientation, we can impose Ψ to be even under parity in the v representation, which is equivalent to $\Psi[-b] = -\Psi[b]$ in b representation (this can be seen from the Fourier transform connecting the two representations). Here, however, we have to restrict the sign of b so that (10.31) be well defined. Therefore, (10.30) implicitly requires a prior fixing of the frame orientation.

A second consequence is that the scalar field ϕ appears only in the second-derivative term, while the operator Θ is ϕ-independent. The WDW equation takes the form of a massless Klein–Gordon equation where ϕ plays the role of time. In fact, the absence of a potential guarantees that ϕ is monotonic throughout the whole evolution of the wave-function, thus constituting a reliable internal clock.

The operator Θ is self-adjoint and positive definite on the space of square-integrable functions on the real line; its eigenfunctions are $e_\omega(y) = \mathrm{e}^{\pm\mathrm{i}\omega y}$ with eigenvalues ω^2 and obey the orthonormality relation

$$\int_{-\infty}^{+\infty}\frac{\mathrm{d}y}{2\pi}\,e_\omega^*(y)\,e_{\omega'}(y) = \delta(\omega,\omega') \tag{10.32}$$

and the completeness relation

$$\int_{-\infty}^{+\infty}\mathrm{d}y\,e_\omega(y)\,\Psi[y,\phi] = 0 \quad \forall\,\omega \qquad \Leftrightarrow \qquad \Psi[y,\phi] = 0\,. \tag{10.33}$$

Any solution Ψ can be decomposed into a positive-frequency and a negative-frequency part. The *group averaging* procedure [66–73] is one way to find this result [74]. One considers the sub-space $\mathcal{S}$ of rapidly decreasing functionals $\psi(y,\phi)$, dense in the Hilbert space of square-integrable functionals over the mini-superspace. These are not physical states unless they satisfy the scalar constraint (10.30). The latter is implemented by the Schwinger representation (i.e., a one-parameter integral) of the formal operator $2\pi\delta(\hat{C})$:

$$\Psi[y,\phi] = \Psi[\psi] = \int_{-\infty}^{+\infty} d\alpha\, e^{i\alpha\hat{C}}\psi(y,\phi)\,. \tag{10.34}$$

The first equality stresses that physical states can be regarded as distributions on $\mathcal{S}$: Ψ is the average of ψ over the volume of the one-dimensional group $\mathbb{R}$ generated by the unitary operator $e^{i\alpha\hat{C}}$, hence the name of "group averaging." Expanding this expression in Fourier modes ω and k (eigenvalues of ∂_y and ∂_ϕ), the integral in α becomes the Dirac delta distribution $2\pi\delta(\omega^2-k^2) = 2\pi[\delta(|\omega|+k)+\delta(|\omega|-k)]/(2|k|)$. Indicating as $\psi(\omega,k)$ the Fourier transform of $\psi(y,\phi)$ and performing the integral in k, one obtains

$$\Psi[y,\phi] = \Psi_+[y,\phi]+\Psi_-[y,\phi]\,, \tag{10.35a}$$

$$\Psi_\pm[y,\phi] := \int_{-\infty}^{+\infty}\frac{d\omega}{2\pi}\,e^{-i\omega y}e^{\pm i|\omega|\phi}\psi_\pm(\omega)\,, \tag{10.35b}$$

where $\psi_\pm(\omega) = \psi(\omega,\pm|\omega|)/(2|\omega|)$. The auxiliary Hilbert space $\mathcal{S}$ is thus decomposed into positive-, null- and negative-frequency sub-spaces, $\mathcal{S} = \mathcal{S}_+\oplus\mathcal{S}_0\oplus\mathcal{S}_-$. In turn, each positive- and negative-frequency wave-function is decomposed into "left-moving" and "right-moving" components having support on the negative (respectively, positive) ω-axis:

$$\Psi_\pm[y,\phi] = \Psi_{\pm,L}[y_\pm]+\Psi_{\pm,R}[y_\mp]\,, \tag{10.36a}$$

$$\Psi_{\pm,L}[y_\pm] := \int_{-\infty}^{+\infty}\frac{d\omega}{2\pi}\,e^{-i\omega y_\pm}\theta(-\omega)\psi_\pm(\omega) = \int_{-\infty}^{0}\frac{d\omega}{2\pi}\,e^{-i\omega y_\pm}\psi_\pm(\omega)\,, \tag{10.36b}$$

$$\Psi_{\pm,R}[y_\mp] := \int_{-\infty}^{+\infty}\frac{d\omega}{2\pi}\,e^{-i\omega y_\mp}\theta(\omega)\psi_\pm(\omega) = \int_{0}^{+\infty}\frac{d\omega}{2\pi}\,e^{-i\omega y_\mp}\psi_\pm(\omega)\,, \tag{10.36c}$$

where $y_\pm = y\pm\phi$ and θ is the Heaviside step function: $\theta(x) = 0$ for $x<0$ and $\theta(x) = 1$ for $x \geqslant 1$. Individually, $\Psi_\pm$ obey the "square root" of the WDW equation, which is a first-order Schrödinger-type equation:

$$\mp\, i\partial_\phi\Psi_\pm = \sqrt{\Theta}\,\Psi_\pm\,. \tag{10.37}$$

Let $\phi = \phi_0$ be the initial internal time where the system starts its evolution with wave-function $\Psi[y, \phi_0]$. Then, (10.37) evolves the state into the solution

$$\Psi_\pm[y, \phi] = \mathrm{e}^{\pm\mathrm{i}\sqrt{\Theta}(\phi-\phi_0)}\Psi_\pm[y, \phi_0]\,. \tag{10.38}$$

From (10.34), one would be tempted to define the scalar product

$$(\Psi_1, \Psi_2) \sim \int \mathrm{d}\alpha \int \mathrm{d}\alpha' \langle\psi_1|\mathrm{e}^{-\mathrm{i}\alpha'\hat{C}}\mathrm{e}^{\mathrm{i}\alpha\hat{C}}|\psi_2\rangle\,,$$

where $\langle\cdot|\cdot\rangle$ is the scalar product in $\mathcal{S}$, but this quantity is clearly divergent (perform one integration, or just invoke invariance of the Haar measure over the group manifold to factorize the infinite group volume). Instead, the physical inner product in the general group averaging procedure is defined as

$$(\Psi_1, \Psi_2) := \int \mathrm{d}\alpha \langle\psi_1|\mathrm{e}^{\mathrm{i}\alpha\hat{C}}|\psi_2\rangle\,. \tag{10.39}$$

In the case under examination, the Hilbert space of physical states is naturally equipped with the inner product on constant ϕ slices:

$$(\Psi_1, \Psi_2) := -\mathrm{i}\int \mathrm{d}y \left(\Psi_1^*\partial_\phi\Psi_2 - \Psi_2\partial_\phi\Psi_1^*\right)\big|_{\phi=\phi_0}\,. \tag{10.40}$$

In terms of the cross-products of negative- and positive-frequency parts,

$$(\Psi_{1\pm}, \Psi_{2\pm}) = \int_{-\infty}^{+\infty}\frac{\mathrm{d}\omega}{2\pi}\left[\pm_2\mathrm{e}^{\mathrm{i}(\mp_1|\omega|\pm_2|\omega|)\phi} \pm_1 \mathrm{e}^{\mathrm{i}(\mp_2|\omega|\pm_1|\omega|)\phi}\right] \times\psi_{1\pm}^*(\omega)|\omega|\psi_{2\pm}(\omega) \tag{10.41}$$

and one has four terms:

$$(\Psi_{1+}, \Psi_{2+}) = 2\int_{-\infty}^{+\infty}\frac{\mathrm{d}\omega}{2\pi}\,\psi_{1+}^*(\omega)|\omega|\psi_{2+}(\omega)\,, \tag{10.42a}$$

$$(\Psi_{1-}, \Psi_{2-}) = -2\int_{-\infty}^{+\infty}\frac{\mathrm{d}\omega}{2\pi}\,\psi_{1-}^*(\omega)|\omega|\psi_{2-}(\omega)\,, \tag{10.42b}$$

$$(\Psi_{1+}, \Psi_{2-}) = 2\mathrm{i}\int_{-\infty}^{+\infty}\frac{\mathrm{d}\omega}{2\pi}\,\sin(2|\omega|\phi)\,\psi_{1+}^*(\omega)|\omega|\psi_{2-}(\omega)\,, \tag{10.42c}$$

$$(\Psi_{1-}, \Psi_{2+}) = 2\mathrm{i}\int_{-\infty}^{+\infty}\frac{\mathrm{d}\omega}{2\pi}\,\sin(2|\omega|\phi)\,\psi_{1-}^*(\omega)|\omega|\psi_{2+}(\omega)\,. \tag{10.42d}$$

The inner product (10.40) is formally equivalent to the one for a relativistic particle but other choices are possible [75–77]. For instance, the "non-relativistic" form

$$(\Psi_1, \Psi_2)_{\text{nrel}} := \int \mathrm{d}y\, \Psi_1^*[y, \phi_0]\, \Psi_2[y, \phi_0] \qquad (10.43)$$

corresponds to defining the functionals $\Psi_\pm$ over the auxiliary Hilbert spaces $\mathcal{S}'_\pm \ni 2|\omega|\psi_\pm(\omega)$ rather than on $\mathcal{S}_\pm$. The two theories equipped with the inner products (10.40) and (10.43) lead to unitarily equivalent representations of the algebra of Dirac observables, so that there is no preferred choice from the point of view of physics.

The wave-function (10.35) is non-normalizable with respect to the scalar product (10.40) because the eigenvalue $\omega = 0$ is in the continuous part of the spectrum of the Θ operator. In fact, the sum of the terms (10.42) is not positive definite. The problem can be avoided if there exists a super-selection rule allowing us to consider only the positive- or the negative-frequency sector. This rule can be realized by identifying a complete set of Dirac observables which preserve the two sectors separately [12, 74, 78]. In the present case, the self-adjoint operators realizing these Dirac observables can be chosen to be $\hat{p}_\phi = -\mathrm{i}\partial_\phi$ and the self-adjoint part of the volume operator $\hat{v} = -\mathrm{i}\partial_b$ [78]. Therefore, we can elect the positive-frequency wave-functions Ψ_+ to be the actual physical states. Setting $\Psi_- = 0$, the inner product (10.40) reduces to (10.42a). Moreover, also the left- and right-moving sectors are preserved and they can be studied separately. This means that the expanding and contracting cosmological branches never interfere with each other.

10.2.3 Quantum Singularity

What happens to the big bang singularity? The original hope when the WDW formalism was proposed was that, in general, the wave-function solving the Hamiltonian constraint is $\Psi = 0$ at the big bang, so that the probability for the universe to hit the initial singularity is zero [1]. To solutions with this property, there would correspond a set of semi-classical effective equations of motion solved by an effective, non-singular scale factor. Although these solutions exist (e.g., [38]), the problem is the same as in classical cosmology: How "typical" are they? Classical singularity theorems show that, roughly speaking, non-singular solutions are the exception to the rule. In quantum mini-superspace models, we do not have such powerful results to achieve the same conclusion, but the fact that both the no-boundary and tunneling wave-functions discussed in Sect. 10.2.1 are non-vanishing at $a = 0$ is not encouraging. At small scales, however, the WKB approximation breaks down and spacetime becomes "fuzzy," a regime where singularities could possibly be smeared. This is roughly what we would expect in a quantum-mechanical universe where Heisenberg uncertainty principle applies to quanta of matter and spacetime. It is difficult to make these concepts more precise in second-order formalism.

The issue of the singularity is still unsettled in WDW cosmology. We begin by reviewing the cons, some general arguments suggesting that the big bang is not resolved, then moving onto the pros. Consider the massless scalar model of the previous section and a classical trajectory $(\phi_{\rm cl}, p_{\phi,\rm cl}, y_{\rm cl}, v_{\rm cl})$. When followed backwards or forwards in time, the classical system falls into a big-bang singularity. In fact, the classical expanding and contracting solutions in synchronous time are, up to integration constants,

$$\phi_{\rm cl\pm}(t) = \pm\sqrt{\frac{2}{3\kappa^2}}\,\ln t\,, \qquad a_{\rm cl\pm}(t) = \left(\pm\sqrt{\frac{3\kappa^2}{2}}\,t\right)^{\frac{1}{3}}. \tag{10.44}$$

This can be recast as a solution in phase space, $\phi - \bar\phi \propto \ln v$, where $\bar\phi$ is a constant. Also, ϕ is monotonic and can be used as an internal clock throughout the whole evolution.

Take now a Gaussian semi-classical physical state (positive frequency, subscripts $+$ omitted) peaked at this trajectory at the time $\phi = \phi_0$:

$$\Psi_{\rm sc}[y, \phi_0] = \int_{-\infty}^{+\infty} \frac{d\omega}{2\pi}\, e^{-i\omega y} e^{i|\omega|(\phi_0-\phi_{\rm cl})} \psi_{\rm sc}(\omega)\,, \qquad \psi_{\rm sc}(\omega) = e^{-\frac{(\omega-\omega_{\rm cl})^2}{2\sigma}}\,, \tag{10.45}$$

where the variance σ is constant and $\omega_{\rm cl} = p_{\phi,\rm cl}$ is one of the two branches. Equation (10.38) determines univocally the state $\Psi[y, \phi]$, which is simply (10.45) with ϕ_0 replaced by ϕ. Therefore, a semi-classical state always remains such all the way through the classical trajectory and up to the big-bang or big-crunch singularity. We can conclude that, at least *in the massless model*, the big-bang problem also affects the WDW quantum evolution.

The same result holds actually for *any* physical state, not only semi-classical. Since the left- and right-moving sectors are super-selected, they must be considered separately rather than in the joint expression (10.184).[5] The expectation value of $|\hat v|$ on (a dense sub-set of) physical states at a time ϕ for each sector is (Problem 10.2)

$$\langle|\hat v|\rangle_{\rm L} = (\Psi_{+,\rm L}, \hat v\Psi_{+,\rm L}) = \mathcal{V}_{\rm L} e^{\kappa_0\phi}\,, \tag{10.46}$$

$$\langle|\hat v|\rangle_{\rm R} = (\Psi_{+,\rm R}, \hat v\Psi_{+,\rm R}) = \mathcal{V}_{\rm R} e^{-\kappa_0\phi}\,, \tag{10.47}$$

where $\mathcal{V}_{\rm L,R}$ are two positive constants. Consider first the left sector. The expectation value of the volume goes from zero at $\phi = -\infty$ to infinity at the infinite future; this corresponds to a quantum big-bang singularity. On the other hand, for the right sector one starts with an infinitely large universe and eventually encounters a big-crunch singularity at $\phi = +\infty$.

In the presence of a potential, it is possible to obtain quantum wave-functions that vanish at $a = 0$ and avoid the big bang [79]. This result depends on the form of

[5] The reason is that the frame orientation has been fixed at the very beginning.

the potential $V(\phi)$ near the classical singularity but it is not difficult to find specific examples.

A final answer may demand further proof beyond the FLRW mini-superspace isotropic approximation. A natural direction to look at are Bianchi IX wave-functions and the quantum version of the mixmaster universe [2, 80–82]. For a small-enough realistic universe, it is not possible to follow a semi-classical evolution down to the singularity and one does not expect any counterpart to the classical infinite series of oscillations. Nonetheless, one can study the evolution of wave packets and the expectation values of anisotropic variables [83–90], especially with the methods of quantum billiards [86, 88, 89, 91–93]. Quantum theories can be chaotic in a precise sense [94, 95]. For diagonal Hamiltonians, simple dynamical variables are represented by pseudo-random matrices with many non-zero entries. The expectation values of these variables tend to equilibrium values which are independent of the initial conditions. The quantum mixmaster universe belongs to this class of models, so that the chaotic nature of the BKL singularity is thus preserved, *mutatis mutandis*, at the quantum level. This is not the case in string theory, where chaos and quantum effects at ultra-microscopic scales could eventually solve the big-bang problem (Sect. 13.9.3).

10.2.4 Cosmological Constant and the Multiverse

The probability (10.28) to have inflation also gives valuable information about the cosmological constant problem. Paradoxically, here the tunneling proposal fares worse than the no-boundary one: what was a virtue of the former and a (solvable) problem for the latter turns out to be the other way around when applying the same formulæ to the cosmological constant. Interpreting $\rho_\Lambda = \Lambda/\kappa^2$ today as the effective contribution generated by all matter fields near their ground states, we assume that matter is non-dynamical, such as a scalar field without kinetic term or a 3-form field [96–99] in four dimensions, and replace V in (10.28) with this effective cosmological constant. The tunneling proposal predicts large values of Λ as the most probable:

$$P_{\rm V}(\Lambda) = \exp\left(-\frac{3m_{\rm Pl}^2}{2\pi\Lambda}\right). \tag{10.48}$$

A similar conclusion (probability density peaked away from $\Lambda = 0$) is reached in tunneling models where matter fields compensate a negative cosmological constant to produce a positive effective cosmological constant [33]. On the other hand, for the no-boundary proposal one gets [100–104]:

$$P_{\rm HH}(\Lambda) = \exp\left(\frac{3m_{\rm Pl}^2}{2\pi\Lambda}\right). \tag{10.49}$$

Thus, the most probable configuration for the Hartle–Hawking wave-function is $\Lambda = 0$, in regions where fields are near their ground state.[6] Notice the similarity with the probability distribution (7.138) obtained in generic holographic theories of gravity.

However, a Λ-dependent normalization of the wave-function may erase the effect or, in other words, the latter is only a normalization artifact. One cannot check this point in canonical theories, which are linear in the wave-function. In some loose sense, this issue is related to the absence of interactions between wave-functions describing different universes. Turning on these interactions is tantamount to going beyond quantum mechanics and promoting wave-function to fields. This is a second-quantization procedure but, since the degrees of freedom of a single universe are already fields, eventually to be quantized, such a framework is sometimes called of "third quantization" [44, 105–124], recently revived also in the context of group field cosmology [125–131]. While in canonical schemes geometry is fully dynamical but the topology of the Universe is fixed by construction, in third quantization the Universe is allowed to change topology and, in particular, to branch into disconnected components. This interacting multiverse scenario obeys a set of quantum rules which greatly vary from model to model. Various early realizations of a branching multiverse seemed to solve or alleviate the cosmological constant problem: predictions on the most probable value of Λ range from $\Lambda = 0$ (with $0 < \Lambda \ll 1$ requiring some amount of fine tuning) [106, 107, 112, 115][7] to small Λ [105, 111], as in the single-universe probability (10.49) [5, 100]. Most of these results, however, are obtained in Euclidean gravity [106, 107, 111, 112, 115] and Lorentzian signature can even spoil the peak at $\Lambda = 0$ in the probability density, making the cosmological constant in third-quantized scenarios a random variable [116]. Also, as recalled in Sect. 7.2.3, nowadays we would like to obtain a very small but non-zero cosmological constant, a requirement in tension with models with $\Lambda = 0$.

To circumvent these problems (while creating others), one can exploit the fact that in eternal inflation there are infinitely many inflating regions, each with a different value of the true vacuum of the scalar field. Combining this picture with some anthropic arguments [135–139], one may conceive that it is highly probable that we live in one such region with a scale reproducing the observed value of the cosmological constant. As seen in Sect. 7.1.1, the inflationary mechanism turns out to be too short a blanket to solve both the inflationary and Λ problems at the same time and cosmic inhomogeneities should then be explained by an alternative to (or modification of) inflation. Another problem with this model is that the probability

[6]The same conclusion might be reached also in scalar-tensor models, both for Vilenkin's and Hartle–Hawking's boundary conditions [16, 19, 20], albeit with some recent reservations [21].

[7]This probabilistic enhancement of the $\Lambda = 0$ configuration, known as Coleman mechanism, has been extended also to Einstein–Yang–Mills dynamics [132] and scalar-tensor theories [133, 134]. The latter are not favoured with respect to general relativity.

measure is highly speculative and an endless source of debate (see, e.g., [140–145] and bibliography in Sect. 5.6.5).

The main message one may get from these efforts is that, in general, quantum cosmology offers a probabilistic interpretation of the cosmological constant rather than explaining why Λ takes the value (7.11). In our realization of the universe, (7.11) holds but Λ could have taken *any* other value among the most probable ones. In the best-case scenario, no fine tuning is needed to achieve extremely small values such as (7.11). Similar considerations hold for quantum models lying outside the canonical or the third-quantization framework, as in proposals based on decoherence [146].

10.2.5 Perturbations and Inflationary Observables

Inflationary perturbations and the associated spectra allow us to track down quantum corrections and confront them with the observed CMB power spectrum. Although the outcome of this procedure is a constraint on the free parameters of the models rather than an actual prediction, as a minimal present-day achievement we can at least state that quantum cosmology WDW models are compatible with observations.

10.2.5.1 Perturbations

When inhomogeneities are switched on, the FLRW mini-superspace framework breaks down and one should consider the full Dirac Hamiltonian (9.75) for gravity and a scalar field. Since the super-momentum and super-Hamiltonian constraints are non-linear in the canonical variables, the problem quickly becomes intractable unless one resorts to some approximations. To obtain the inflationary spectra, linear perturbation theory is sufficient. For a closed universe, a full treatment begins with the expansion of scalar, vector and tensor perturbations in spherical harmonics over the 3-sphere [147], thus extending mini-superspace to the infinite number of degrees of freedom of superspace. Eventually, the expansion in spherical harmonics can be combined with the WKB approximation to find solutions of the WDW and diffeomorphism equations [28, 39, 43, 147–149]. In this section, however, we ignore both vector modes and the metric back-reaction $\delta g_{\mu\nu}$, in which case the scalar is regarded as a test field. In the standard cosmological model, back-reaction does not affect the power spectrum at lowest order in perturbation theory and in the slow-roll truncation. This suffices for our purposes also in WDW quantum cosmology. The scalar perturbation is decomposed into Fourier modes,

$$\delta\phi(t,\mathbf{x}) = \sum_{\mathbf{k}} \delta\phi_k(t)\,\mathrm{e}^{\mathrm{i}\mathbf{k}\cdot\mathbf{x}}, \tag{10.50}$$

where we assumed spatial slices to be compact ($\textsc{k} = 1$) and that each Fourier mode depends on $k = |\mathbf{k}|$. Replacing $\phi(t)$ with $\phi(t,\mathbf{x})$ in the WDW equation (10.19) for a quadratic potential $V(\phi) = \frac{1}{2}m^2\phi^2$, the mini-superspace is augmented by the infinity of modes $\delta\phi_k$. The wave-function $\Psi[\mathcal{N}, \phi, \{\delta\phi_k\}_k]$ can be factorized as a background plus the rest, $\Psi[\mathcal{N}, \phi, \{\delta\phi_k\}_k] = \Psi_0[\mathcal{N}, \phi] \prod_{k>0} \Psi_k[\mathcal{N}, \phi, \delta\phi_k]$. In doing so, one drops self-interaction terms which are consistently negligible in first-order perturbation theory. Eventually, ignoring the curvature term one obtains [28, 147]

$$\left[\frac{\partial^2}{\partial\mathcal{N}^2} - \frac{6}{\kappa^2}\frac{\partial^2}{\partial\delta\phi_k^2} + \mathrm{e}^{6\mathcal{N}}\left(\frac{6H}{\kappa^2}\right)^2 + \frac{6}{\kappa^2}\left(\mathrm{e}^{6\mathcal{N}}m^2 + \mathrm{e}^{4\mathcal{N}}k^2\right)\delta\phi_k^2\right]\psi_k \simeq 0\,, \tag{10.51}$$

where $\psi_k[\mathcal{N}, \delta\phi_k] = \Psi_0[\mathcal{N}, \phi]\Psi_k[\mathcal{N}, \phi, \delta\phi_k]$ and the ϕ dependence is omitted because we used the slow-roll approximation $6H^2/\kappa^2 \simeq m^2\phi^2$ to express the background potential in terms of the Hubble parameter.

Noting that $\mathcal{N}$ and $\delta\phi_k$ correspond, respectively, to slow- and fast-evolving variables, at this point one can make a Born–Oppenheimer approximation on the solution [150, 151]. The latter is written as

$$\psi_k[\mathcal{N}, \delta\phi_k] = \exp[\mathrm{i}S(\mathcal{N}, \delta\phi_k)] \tag{10.52}$$

and the functional S is expanded in $m_{\rm Pl}^2 = l_{\rm Pl}^{-2} = 8\pi/\kappa^2$: $S = m_{\rm Pl}^2 S_0 + S_1 + m_{\rm Pl}^{-2} S_2 + \dots$. Plugging the *Ansatz* (10.52) into (10.51) and expanding, the $O(m_{\rm Pl}^4)$ and $O(m_{\rm Pl}^2)$ terms imply $S_0 = \pm\mathrm{e}^{3\mathcal{N}}H/(4\pi)$ (the equivalent of (10.20)), while at the next two orders one finds

$$\psi_k^{(0)}[\mathcal{N}, \delta\phi_k] := A(\mathcal{N})\,\mathrm{e}^{\mathrm{i}S_1(\mathcal{N}, \delta\phi_k)}\,, \tag{10.53}$$

$$\psi_k^{(1)}[\mathcal{N}, \delta\phi_k] := B(\mathcal{N})\psi_k^{(0)}[\mathcal{N}, \delta\phi_k]\,\mathrm{e}^{\mathrm{i}m_{\rm Pl}^{-2}S_2(\mathcal{N}, \delta\phi_k)}\,, \tag{10.54}$$

where A and B are chosen to match the amplitudes in the WKB approximation.

10.2.5.2 Observables

The wave-functions $\psi_k^{(0)}$ and $\psi_k^{(1)}$ can be found semi-analytically [152–154]. From the explicit solutions, one can calculate the two-point correlation function

$$P_\phi^{(n)}(k) := \langle\psi_k^{(n)}|\,|\delta\phi_k|^2\,|\psi_k^{(n)}\rangle \tag{10.55}$$

of the scalar perturbation order by order. This quantity is directly related to the imprint (5.110) of inhomogeneous fluctuations in the cosmic microwave background, when approximated to the long-wave-length limit $k \ll k_*$ and evaluated at $k = k_*$. This is the n-th order power spectrum

$$\mathcal{P}_{\rm s}^{(n)}(k) := \frac{k^3}{2\pi^2}P_\phi^{(n)}(k \ll k_*)\big|_{k=k_*}\,. \tag{10.56}$$

The lowest-order result $\mathcal{P}_{\rm s}^{(0)}$ coincides with the classical one (5.152).

The next-to-lowest-order expression is the standard one times a quantum correction:

$$\mathcal{P}_{\rm s}(k) \simeq \mathcal{P}_{\rm s}^{(1)}(k) = \mathcal{P}_{\rm s}^{(0)}(k) C_k^2 . \tag{10.57}$$

For $\psi_k^{(1)}$, there are actually two solutions with the same boundary condition, giving rise to quantum corrections with opposite sign and slightly different size:[8]

$$(C_k^-)^2 \simeq \left[1 - \frac{9.2}{(k/k_0)^3}(l_{\rm Pl}H)^2\right]^{-3}\left[1 - \frac{40.1}{(k/k_0)^3}(l_{\rm Pl}H)^2\right]^2$$
$$= 1 - \frac{52.6}{(k/k_0)^3}(l_{\rm Pl}H)^2 + O\left(\frac{l_{\rm Pl}^4 H^4}{k^6}\right), \tag{10.58}$$

$$(C_k^+)^2 \simeq \left[1 - \frac{11.5}{(k/k_0)^3}(l_{\rm Pl}H)^2\right]^{-3}\left[1 + \frac{1.7}{(k/k_0)^3}(l_{\rm Pl}H)^2\right]^2$$
$$= 1 + \frac{38.0}{(k/k_0)^3}(l_{\rm Pl}H)^2 + O\left(\frac{l_{\rm Pl}^4 H^4}{k^6}\right), \tag{10.59}$$

where k_0 is some reference scale. We can write the leading Wheeler–DeWitt quantum correction as

$$(C_k^\pm)^2 \simeq 1+\delta_{\rm WDW}^\pm(k)+O(\delta_{\rm WDW}^2), \qquad \delta_{\rm WDW}^\pm(k) := \pm 50\left(\frac{k_0}{k}\right)^3 (l_{\rm Pl}H)^2. \tag{10.60}$$

The approximation scheme used to derive (10.58) and (10.59) breaks down in the limit $C_k \to 0$ and the critical k at which that happens should not be taken as a physical threshold. While $C_k \to 1$ in the small-scale limit ($k \to \infty$), at large scales ($k \ll k_*$) the quantum-corrected power spectrum acquires a mild scale dependence which suppresses ($-$) or enhances ($+$) the signal with respect to the standard result. The imaginary part of the $-$ solution is discontinuous; if one demands continuity, then the prediction of the model is a power enhancement, as in loop quantum cosmology (Sect. 10.3.9). In the case of suppression of the spectrum, a similar effect happens also in other models where a bounce is present [155, 156] or geometry is quantized, such as non-commutative and string inflation [157–159]. It might seem counter-intuitive that quantum gravity affects large scales more than small scales. However, large-scale perturbations left the horizon before (and hence re-entered after) smaller-scale fluctuations and they were longer exposed to quantum effects.

From the power spectrum and using (5.51) and (5.141), we get the scalar spectral index (4.58) and its running (4.59). Since at horizon crossing $\mathrm{d}/\mathrm{d}\ln k \simeq \mathrm{d}/(H\mathrm{d}t)$

[8] Numerical coefficients differ from those in [154], where the rescaled Planck mass $\tilde{M}_{\rm Pl}^2 = (3\pi/2)m_{\rm Pl}^2$ is used instead of (1.4).

and $H \simeq$ const, we obtain

$$\frac{d\delta^{\pm}_{\rm WDW}}{d\ln k} \simeq -3\delta^{\pm}_{\rm WDW} \tag{10.61}$$

and [154, 160]

$$n_{\rm s} - 1 \simeq 2\eta - 4\epsilon - 3\delta^{\pm}_{\rm WDW}\,, \tag{10.62}$$

$$\alpha_{\rm s} \simeq 2\left(5\epsilon\eta - 4\epsilon^2 - \xi^2\right) + 9\delta^{\pm}_{\rm WDW}\,, \tag{10.63}$$

where we dropped higher-order terms in the combined $\delta_{\rm WDW}$/slow-roll expansion. The slow-roll part of these expression reproduces the classical results (5.153) and (5.155). Using the slow-roll tower (5.55) and relations (5.65), equations (10.62) and (10.63) are recast as

$$n_{\rm s} - 1 \simeq -6\epsilon_V + 2\eta_V - 3\delta^{\pm}_{\rm WDW}\,, \tag{10.64}$$

$$\alpha_{\rm s} \simeq -24\epsilon_V^2 + 16\epsilon_V\eta_V - 2\xi_V^2 + 9\delta^{\pm}_{\rm WDW}\,. \tag{10.65}$$

The scalar power spectrum expanded to all orders in the perturbation wave-number about a pivot scale k_0 is the generalization of (4.67),

$$\ln \mathcal{P}_{\rm s}(k) = \ln \mathcal{P}_{\rm s}(k_0) + [n_{\rm s}(k_0) - 1]x + \frac{\alpha_{\rm s}(k_0)}{2}x^2 + \sum_{m=3}^{\infty}\frac{\alpha_{\rm s}^{(m)}(k_0)}{m!}x^m\,, \tag{10.66}$$

where $x := \ln(k/k_0)$. As the order of the observables

$$\alpha_{\rm s}^{(m)} := \frac{d^{m-2}\alpha_{\rm s}}{(d\ln k)^{m-2}} = O(\epsilon^m) + (-3)^m\delta^{\pm}_{\rm WDW} \tag{10.67}$$

increases, the classical part becomes smaller and smaller but the leading-order quantum correction survives. At some order m, the quantum correction dominates over the standard part. Taking (10.67) into account, (10.66) can be recast as

$$\ln \mathcal{P}_{\rm s}(k) \simeq \ln \mathcal{P}_{\rm s}^{(0)}(k) - \delta^{\pm}_{\rm WDW}(k_0)\left[1 - \left(\frac{k_0}{k}\right)^3\right]. \tag{10.68}$$

10.2.5.3 Experimental Bounds

The typical energy scale during inflation is estimated to be about the grand-unification scale, $H \sim 10^{15}$ GeV, corresponding to an energy density $\rho_{\rm infl} \sim H^2/l_{\rm Pl}^2 \sim 10^{68}\,{\rm GeV}^4$. In contrast, classical gravity is believed to break down at distances shorter than the Planck length $l_{\rm Pl}$, i.e., at energies above $m_{\rm Pl} \sim 10^{19}$ GeV. The ratio between the inflationary and Planck energy density is very small,

$$\frac{\rho_{\rm infl}}{\rho_{\rm Pl}} \sim (l_{\rm Pl}H)^2 \sim 10^{-8}\,, \tag{10.69}$$

and quantum corrections are expected to be of the same order of magnitude or lower, well below any reasonable experimental sensitivity threshold. WDW quantum cosmology realizes precisely this type of corrections: their size is set by the energy scale of inflation.

To make this argument more precise, we choose the reference scale k_0 in (10.60). One possibility is $k_0 = k_{\rm min} \sim 1.4 \times 10^{-4}\,{\rm Mpc}^{-1}$, the largest observable scale. Via (4.38), one can re-express $\delta_{\rm WDW}$ in terms of spherical multipoles. The lowest early-universe contribution to the CMB spectrum is the quadrupole $\ell = 2$, so that $k/k_{\rm min} = \ell/\ell_{\rm min} = \ell/2$. A more generous estimate for the quantum correction is obtained by replacing $k_{\rm min}$ by the pivot scale $k_0 \gg k_{\rm min}$ of a CMB experiment, which we adopt from now on. From the PLANCK bound (5.196) for the Hubble parameter at $k_0 = 0.002\,{\rm Mpc}^{-1}$, the WDW quantum correction is constrained to be

$$|\delta^{\pm}_{\rm WDW}(k_0)| < 3 \times 10^{-9}\,. \tag{10.70}$$

With $k_{\rm min}$ instead of k_0, the quantum correction is further suppressed, $|\delta^{\pm}_{\rm WDW}(k_0)| < 7 \times 10^{-13}$. As anticipated, quantum corrections are too small to be detected. Their dependence on the inflationary energy scale is crucial for this result. Another reason is that, at large scales, cosmic variance is the leading source of error. Quantum-gravity effects should be compared with the error bars due to cosmic variance (4.34) with respect to the classical spectrum $\mathcal{P}_{\rm s}^{(0)}(\ell)$. The latter is determined up to the normalization $\mathcal{P}_{\rm s}(\ell_0)$, so that the region in the $(\ell, \mathcal{P}_{\rm s}(\ell)/\mathcal{P}_{\rm s}(\ell_0))$ plane affected by cosmic variance is roughly delimited by the two curves

$$\frac{\mathcal{P}_{\rm s}^{(0)}(\ell) \pm \sqrt{\sigma^2_{\mathcal{P}_{\rm s}^{(0)}}(\ell)}}{\mathcal{P}_{\rm s}^{(0)}(\ell_0)} = \left(1 \pm \sqrt{\frac{2}{2\ell+1}}\right)\frac{\mathcal{P}_{\rm s}^{(0)}(\ell)}{\mathcal{P}_{\rm s}^{(0)}(\ell_0)}\,, \tag{10.71}$$

where we take the classical spectrum as a reference. The WDW-corrected spectrum is given by (10.68) with the classical spectrum (4.67) truncated at second order, since slow-roll parameters are at most of order $\epsilon = O(10^{-2})$. To plot the WDW spectrum, we only need to plug in values for the scalar index and its running. These are yielded by (10.64) and (10.65) for a given potential.

The quadratic potential $V(\phi) \propto \phi^2$ is under strong experimental pressure, but it illustrates well the effect of quantum-gravity corrections. In that case, from (5.85a) and (5.85b),

$$\epsilon_V = \frac{2}{\kappa^2\phi^2}\,, \qquad \eta_V = \epsilon_V\,, \qquad \xi_V^2 = 0\,.$$

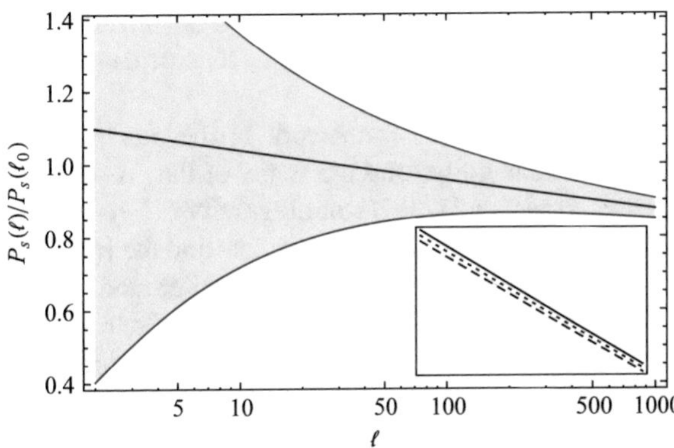

Fig. 10.1 Log-linear plot of the Wheeler–DeWitt primordial scalar spectrum (10.68) for a quadratic inflaton potential, with $\epsilon_V(k_0) = 0.009$ and for the pivot wave-number $k_0 = 0.002\,\mathrm{Mpc}^{-1}$, corresponding to $\ell_0 = 29$. The *shaded region*, delimited by the two curves (10.71), is affected by cosmic variance. The inset shows the negligible difference between the standard "classical" spectrum $\mathcal{P}_s^{(0)}(\ell)$ (*dotted line*) and the spectrum $\mathcal{P}_s(\ell)$ with Wheeler–DeWitt quantum correction $\delta^+_{\rm WDW}$ (*solid line*, enhanced spectrum) and $\delta^-_{\rm WDW}$ (*dashed line*, suppressed spectrum), at $2 < \ell < 3$. Quantum corrections are magnified to $|\delta^{\pm}_{\rm WDW}(k_0)| = 10^{-5}$ in order to show their effect (Credit: [160])

This allows one to reduce the slow-roll parameters to just one. A realistic theoretical value for ϵ_V at the pivot scale is $\epsilon_V(k_0) = 0.009$. As shown in Fig. 10.1, WDW quantum corrections are extremely small even in the most conservative estimate and they are completely drowned by cosmic variance. The loop quantization inspired by loop quantum gravity will produce corrections potentially much larger that those of WDW cosmology. The intuitive reason will be given in Sect. 10.3.8, while a comparison with experiments along the same lines above is in Sect. 10.3.9.

10.3 Loop Quantum Cosmology

Loop quantum gravity aims to quantize the gravitational interaction in a rigorous and consistent way. At the classical level, it relies on the Hamiltonian formalism in Ashtekar–Barbero canonical variables and features the real Barbero–Immirzi constant γ. While the understanding of physical consequences implied by full LQG is still an open research topic, insight has been gained by studying the theory in symmetry-reduced spacetimes. In particular, *loop quantum cosmology* (LQC) opens up the possibility of resolving the singularities that plague classical cosmological spacetimes [161–166]. For a spatially flat FLRW background with a massless scalar field, the model can be analyzed rigorously in terms of physical observables [167]. Although the symmetry reduction is performed at the classical level as in standard mini-superspace quantization, the techniques used in LQC closely follow those of loop quantum gravity. Consequently, LQC quantization is inequivalent to the standard Wheeler–DeWitt quantization and, thus, it can lead to significantly different predictions [166].

In fact, most of the conceptual framework of the previous sections is rather general and its applications go beyond the scope of the ADM formalism. Having reviewed some basic aspects of loop quantum gravity in Sect. 9.3, we can proceed with the LQC construction of kinematical operators and the Hamiltonian constraint on a spatially-flat, homogeneous and isotropic background. Then, the big-bang problem is faced at a fourfold level: (a) looking at the spectrum of the kinematical volume operator, (b) studying how zero-volume physical modes decouple from the Hamiltonian evolution, (c) computing the expectation value of the volume operator on physical states and (d) showing that the effective cosmological dynamics displays a bounce rather than a singularity. All these approaches agree. The interest in loop quantum cosmology, however, is not exhausted by the big-bang problem and modifications to the standard inflationary picture may arise. Although the basic idea of inflation as a dynamical scalar field remains the same, geometric quantum corrections do change cosmological perturbation theory and the observable spectra. Again, we will limit the discussion to a minimally coupled scalar field, although scalar-tensor and $f(R)$ systems can also be contemplated (Problems 7.1, 7.2, 7.3, 7.4, 7.5 and [168–170]).

10.3.1 Classical FLRW Variables and Constraints

As in the ADM case, the symplectic structure is reduced at the classical level. We consider only the $\textsc{k} = 0$ case. We already had occasion to mention that the volume of a spatially flat universe is infinite and, to render the volume integral (9.112) well defined, we need to consider a patch of the universe with finite fiducial comoving volume $\mathcal{V}_0$. To the fiducial patch there correspond a fiducial triad ${}^o e_i^\alpha$ and co-triad ${}^o e_\alpha^i$ with fixed orientation. The spin connection Γ_α^i vanishes for spatially flat slices, so that in this background the symmetric Ashtekar–Barbero connection is just the extrinsic curvature, which is expressed as

$$A_\alpha^i = \tilde{c}\,{}^o e_\alpha^i\,, \qquad \tilde{c} := c\mathcal{V}_0^{-1/3} = \gamma\dot{a}\,. \tag{10.72}$$

The relation $c \propto \dot{a}$ is valid only at the classical level and it will be modified later in the case of the effective quantum dynamics. The symmetry-reduced conjugate momentum is

$$E_i^\alpha = \tilde{p}|{}^o e|\,{}^o e_i^\alpha\,, \qquad \tilde{p} := p\mathcal{V}_0^{-2/3} = a^2\,. \tag{10.73}$$

The definitions of $c = \tilde{c}\mathcal{V}_0^{1/3}$ and $p = \tilde{p}\mathcal{V}_0^{2/3}$ are suggested by the holonomy and flux expressions (9.107) and (9.111) in the full theory.

In view of implementing the various quantization ambiguities and parametrizations, it is convenient to define a new pair of variables as [78, 167]

$$b = \frac{\bar{\mu}c}{2}, \qquad v = \frac{6}{(1+n)\gamma\kappa^2}\frac{p}{\bar{\mu}}, \tag{10.74}$$

where $\bar{\mu}$ is an arbitrary dimensionless function of a. For a power law,

$$\bar{\mu} = \left(\frac{p_*}{p}\right)^n = \left(\frac{a_*}{a}\right)^{2n}, \tag{10.75}$$

where $n \in \mathbb{R}$ and p_* and a_* are, respectively, constants of dimension $[p_*] = -2$ and $[a_*] = 0$. In particular, $v_0 = 2\mathcal{V}_0^{2(1+n)/3} p_*^{-n} = 2\mathcal{V}_0^{2/3} a_*^{-2n}$ in (10.9) and v is dimensionless. It is easy to show that b and v are canonically conjugate and (10.10) holds (notice that $\{c, p\} = \gamma\kappa^2/3$).

The mini-superspace Hamiltonian constraint in these variables is (10.11). When $n = 0$, it is expressed directly in terms of the symmetry-reduced connection and densitized triad:

$$\mathcal{H} = -\frac{3}{\gamma^2\kappa^2}\sqrt{p}\,c^2 + \mathcal{H}_\phi\,. \tag{10.76}$$

The goal of this section is to recast $\mathcal{H}$ as a functional of holonomies and positive powers of the triad.

In the $\textsc{k} = 0$ FLRW case, comoving spatial slices are flat and the tangent vector in (9.107) is the same at every point along the edge e. In particular, the holonomy of oriented length l_0 along the i-th direction is

$$h_i = \exp(l_0\mathcal{V}_0^{-1/3}c\tau_i) =: \exp(\mu c\tau_i)\,, \tag{10.77}$$

where the path ordering $\mathcal{P}$ in (9.107) becomes trivial and

$$\mu := \frac{l_0}{\mathcal{V}_0^{1/3}} \tag{10.78}$$

is the ratio between the holonomy length and the size of the comoving volume. Defining $\lambda := \mu/\bar{\mu}$ as the path length in units of $\bar{\mu}$, we can express the holonomy (10.77) in terms of the new variables as

$$h_i = \exp(2\lambda b\tau_i) = \mathbb{1}_2\cos\lambda b + 2\tau_i\,\sin\lambda b\,, \tag{10.79}$$

where we employed the identity in (9.109) valid in the fundamental representation of the algebra. Using (10.79) and $\tau_i\tau_j\tau_i = \tau_j/4$, one can show that

$$\mathrm{tr}(\tau_k h_{\Box_{ij}}) = -\tfrac{1}{2}\epsilon_{ijk}\sin^2(2\lambda b)\,. \tag{10.80}$$

This equation will be needed to compute the field strength (9.114) as in (9.115) and, from that, the Hamiltonian.

We have already mentioned the issue of writing inverse powers of the densitized triad using classical relations. In particular, the formula

$$\{b, v^l\} = l v^{l-1} \tag{10.81}$$

is valid for arbitrary values of l. Via (10.79) and (10.81), one can express powers of v as

$$\begin{aligned} v^{l-1} &= \frac{1}{3l\lambda}\sum_i \mathrm{tr}(\tau_i h_i \{h_i^{-1}, v^l\}) \\ &= \frac{1}{l\lambda}(\cos\lambda b\,\{\sin\lambda b, v^l\} - \sin\lambda b\,\{\cos\lambda b, v^l\})\,, \end{aligned} \tag{10.82}$$

where we explicitly wrote the sum over indices when summation convention is not clear, used (9.110) and took the trace in the $j = 1/2$ representation (for which $C_2 = j(j+1) = 3/4$).

For a flat homogeneous background, the scalar constraint (9.60) becomes (Problem 10.3)

$$\begin{aligned} \mathcal{H} = &\frac{12}{\gamma^3\kappa^4 q}\lim_{l_0\to 0}\frac{1}{l_0^3}\left[\frac{(1+n)\gamma\kappa^2}{3}\frac{p_*^n}{2}\right]^{\frac{3}{2(1+n)}} v^{\frac{3(1-q)}{2(1+n)}}\sin^2 2\lambda b \\ &\times[\sin\lambda b\,\{\cos\lambda b, v^{\frac{3q}{2(1+n)}}\} - \cos\lambda b\,\{\sin\lambda b, v^{\frac{3q}{2(1+n)}}\}] + \mathcal{H}_\phi\,. \end{aligned} \tag{10.83}$$

The gravitational sector is only a function of b and v. The scalar-field part only contains volume factors,

$$\mathcal{H}_\phi = \frac{p_\phi^2}{2p^{3/2}} + p^{3/2}V(\phi)\,. \tag{10.84}$$

The other constraints vanish. The constraint (10.83) should be compared with the WDW Hamiltonian (10.11) in the same variables. These expressions are classical and, therefore, physically equivalent. However, the following quantization will make the LQC Hamiltonian crucially deviate from the WDW model.

For later purposes, we note that the full dynamics is encoded in $\mathcal{H}$ if we set the lapse function $N = 1$. A different gauge choice at the classical level can simplify the Hamiltonian $H_\mathrm{D} = N\mathcal{H}$ significantly. In particular, setting $N = \sqrt{|\det E|}$ one avoids the need to rewrite inverse powers of the volume in terms of Poisson brackets:

$$\begin{aligned} H_{\rm D} &= -N\frac{1}{2\gamma^2\kappa^2}\frac{E_i^\alpha E_j^\gamma}{\sqrt{|\det E|}}\epsilon^{ij}{}_k F^k_{\alpha\gamma} + N\mathcal{H}_\phi \\ &\overset{(9.114)}{=} \frac{1}{\gamma^2\kappa^2}\lim_{l_0\to 0}|e|^2\epsilon^{ij}{}_k\frac{\mathrm{tr}(\tau_k h_{\square_{ij}})}{l_0^2} + N\mathcal{H}_\phi \\ &\overset{(10.80)}{=} -\frac{3}{4\gamma^2\kappa^2}\lim_{l_0\to 0}\frac{\bar{\mu}^2}{l_0^2}v^2\sin^2 2\lambda b + N\mathcal{H}_\phi\,. \end{aligned} \tag{10.85}$$

10.3.2 Quantization and Inverse-Volume Spectrum

In the quantization procedure (9.81) of loop quantum cosmology, the classical Poisson bracket is replaced by the commutator bracket as in standard quantum mechanics. As mentioned earlier, the connection operator does not have a well-defined action on the kinematical Hilbert space of LQG, which suggests not to define a connection-dependent operator $\hat{b}$. Nonetheless, the other elementary variable v can be promoted to the self-adjoint operator (10.29), $v \to \hat{v} := -\mathrm{i}\partial_b$. It is easy to check that the functionals

$$e_v(b) = \mathrm{e}^{\mathrm{i}vb} =: \langle b|v\rangle \tag{10.86}$$

are the eigenfunctions in b-representation of $\hat{v}$. In Dirac notation, we can write down the action of (10.29) on the states $|v\rangle$ as

$$\hat{v}|v\rangle = v|v\rangle\,. \tag{10.87}$$

These eigenstates form a basis $\{|v\rangle\}$ of the kinematical Hilbert space. On this basis, the action of a holonomy operator of (dimensionless) edge v' can be constructed from the basic operator $\widehat{\mathrm{e}^{\mathrm{i}v'b}}$, which acts simply as a translation:

$$\widehat{\mathrm{e}^{\mathrm{i}v'b}}|v\rangle = |v+v'\rangle\,. \tag{10.88}$$

Comparing with (10.87), one sees that the holonomy operator increases the volume of the universe by attaching edges to the symmetry-reduced, simplified spin network. What is the size of an edge? Eigenvalues $v = O(1)$ correspond to areas of order of the Planck area $\kappa^2 = 8\pi l_{\rm Pl}^2$, so that the holonomy edge is of order of the Planck length. Note also that, contrary to the full theory, due to the symmetry reduction the spectrum of flux-area-volume operators is continuous and labelled by the real parameter v. However, all eigenstates are normalizable with respect to the inner product (10.15) for the Hilbert space of square-integrable functions on the quantum configuration space, including the state with zero eigenvalue. In this sense, the spectrum of volume and dreibein operators is discrete [162, 165, 166, 171–173]. This has an important consequence for the spectrum of inverse-volume operators because, de facto, v can be regarded as a parameter on natural numbers. Since the normalization in (10.15) has a Kronecker delta rather than a Dirac distribution, one is dealing with a non-separable Hilbert space.

We consider the range of l being $0 < l < 1$ in (10.82), so that v^{l-1} is an inverse power of the volume. The ambiguity parameter l determines the initial slope of the effective geometrical density. To preserve coordinate invariance when quantizing geometrical densities before symmetry reduction, l must be discrete, $l_m = 1 - (2m)^{-1}$, $m \in \mathbb{N}$ [174, 175]. Hence one can select the bound

$$\tfrac{1}{2} \leqslant l < 1\,, \tag{10.89}$$

which is also favoured phenomenologically [175]; a natural choice is $l = 3/4$.

To quantize the system in terms of elementary operators, we fix the length of the holonomy to be such that $\mu = \bar{\mu}$ and we consider their symmetric ordering so that (10.82) (with $\lambda = 1$) becomes the self-adjoint operator

$$\widehat{|v|^{l-1}} = \frac{\mathrm{i}}{l}\left[\widehat{\cos b}\,\widehat{|v|^l}\,\widehat{\sin b} - \widehat{\sin b}\,\widehat{|v|^l}\,\widehat{\cos b}\right] = \frac{1}{2l}\left[\widehat{\mathrm{e}^{-\mathrm{i}b}}\,\widehat{|v|^l}\,\widehat{\mathrm{e}^{\mathrm{i}b}} - \widehat{\mathrm{e}^{\mathrm{i}b}}\,\widehat{|v|^l}\,\widehat{\mathrm{e}^{-\mathrm{i}b}}\right], \tag{10.90}$$

where $\widehat{|v|^l} = |\hat{v}|^l$. The absolute value of $\hat{v}$ is taken in order for the eigenvalues of $\widehat{|v|^{l-1}}$ to be real. The basis states $|v\rangle$ are also the eigenstates of the operator (10.90),

$$\boxed{\widehat{|v|^{l-1}}\,|v\rangle = \frac{1}{2l}\left(|v+1|^l - |v-1|^l\right)|v\rangle\,.} \tag{10.91}$$

Remarkably, the spectrum of this operator is bounded from above. In fact, the maximum eigenvalue is obtained when the negative term in (10.91) vanishes, i.e., for $v = 1$:

$$|v|^{l-1}_{\max} = \frac{2^{l-1}}{l}\,. \tag{10.92}$$

Classically, as $\hbar \to 0$ and the Planck length goes to zero, the value (10.92) diverges. In general, however, the inverse volume cannot go arbitrarily to infinity towards a singular geometric configuration. Here we have the first evidence that, contrary to ADM quantum cosmology, in LQC the big-bang problem could be avoided. This happens because we chose the holonomy rather than the connection as the fundamental variable.

Therefore, the volume spectrum (10.87) does admit the $v = 0$ eigenvalue and its inverse fails to be a densely defined operator. In spite of that, the inverse-volume operator (10.90) is regular (with zero eigenvalue) at $v = 0$: volume singularities are healed by purely geometrical effects. However, (10.91) and (10.92) are only kinematical relations and they do not guarantee that similar properties be realized at the dynamical level. In particular, nothing has been said about the quantum dynamical evolution of the system: Does it stop at a big-bang singularity at $v = 0$? To check this, we need to consider physical states or, in other words, the action of the quantum Hamiltonian constraint on a wave-function.

10.3.3 Mini-superspace Parametrization

While in LQG the area spectrum is bounded from below by the minimal area (9.116), due to symmetry reduction the same property is not shared by loop quantum cosmology. Nevertheless, one may draw inspiration from the full theory and *assume* that the kinematical area of any loop inside the comoving volume $\mathcal{V}_0$ is bounded by the area gap for the gauge-invariant states which are likely to be realized in a homogeneous context. This value is twice the area gap (9.116) [176],

$$\Delta_{\rm Pl} := 2\tilde{\Delta}_{\rm Pl} = 4\sqrt{3}\pi\gamma l_{\rm Pl}^2\,, \tag{10.93}$$

so that

$$(al_0)^2 \geqslant \Delta_{\rm Pl}\,. \tag{10.94}$$

This step is rather speculative inasmuch as it borrows a result of the background-independent framework and forces it into the symmetry-reduced model. It is necessary, however, because the quantum scalar constraint in mini-superspace would be singular if one maintained the limit $l_0 \to 0$.

If the inequality (10.94) is saturated (smallest possible holonomy path), then the comoving cell area is also the comoving area gap, that is, the smallest non-vanishing eigenvalue of the area operator measuring comoving surfaces.[9] In particular,

$$\mu^2 = \frac{l_0^2}{\mathcal{V}_0^{2/3}} \geqslant \frac{\Delta_{\rm Pl}}{p} = \left(\frac{p_*}{\Delta_{\rm Pl}}\bar{\mu}\right)^{\frac{1}{n}}. \tag{10.95}$$

Taking the equality in the second step, one has $\mu = \bar{\mu}$ if $p_* = \Delta_{\rm Pl}$ and

$$n = \frac{1}{2}\,, \qquad \mu = \bar{\mu} = \sqrt{\frac{\Delta_{\rm Pl}}{p}}\,, \tag{10.96}$$

[9] Big-bang nucleosynthesis can place a bound over the smallest physical area [177].

a choice known as *improved quantization scheme* or improved dynamics [78, 167, 178]. The set $\{|v\rangle\}$ becomes the eigenstate basis of the volume operator itself, $v \propto p^{3/2} = \mathcal{V}$. From the point of view of spin-network dynamics, the Universe grows by changing a graph with all spin labels fixed to $j = 1/2$. As the Universe expands, the comoving area gap shrinks to zero and the geometry is better and better described by classical general relativity, while near the putative big bang quantum effects become important.

Originally, the variables p and c were used instead of v and b, corresponding to $\bar{\mu} = 1$ ($n = 0$). In this *old quantization scheme*, the states $|v\rangle = |\mu\rangle$ coincide with the basis eigenstates of the momentum operator $\hat{p}$, with eigenvalues $v \propto p$ [166, 179]. Here the Universe grows by changing the spin label j on a fixed graph. This case leads to severe restrictions on the matter sector if the wave-functions solving the Hamiltonian constraint are required to be normalizable and to reproduce the classical limit at large scales [180]. Also for such reason, the improved quantization scheme seems to be the most natural and, as we are going to see very soon, the most reasonable in a purely homogeneous context. However, later motivations lead us to keep n, p_* and the other free parameters of the model as general as possible. In this case, p_* is some physical squared length determined by the theory which may differ from the mass gap $\Delta_{\rm Pl}$.

10.3.4 Quantum Hamiltonian Constraint

If we assume that holonomy plaquettes cannot be shrunk indefinitely, then we must replace the limit $l_0 \to 0$ in (10.83) with $l_0 \to \mathcal{V}_0^{1/3}\mu$. With this substitution, the quantum Hamiltonian operator corresponding to (10.83) is well-defined. When $\mu = \bar{\mu}$,

$$\hat{\mathcal{H}} = -4\,\widehat{\sin 2b}\,\hat{C}\,\widehat{\sin 2b} + \hat{\mathcal{H}}_\phi\,, \tag{10.97}$$

where

$$\begin{aligned}\hat{C} &= \frac{3i}{8q\mathcal{V}_0\gamma^3\kappa^4}\left[\frac{(1+n)\gamma\kappa^2}{3}\right]^{\frac{3(1+2n)}{2(1+n)}}\left(\frac{p_*^n}{2}\right)^{-\frac{3}{2(1+n)}}\widehat{|v|^{\frac{3(1+2n-q)}{2(1+n)}}}\\&\quad\times\left[\widehat{\cos b}\,\widehat{|v|^{\frac{3q}{2(1+n)}}}\,\widehat{\sin b}-\widehat{\sin b}\,\widehat{|v|^{\frac{3q}{2(1+n)}}}\,\widehat{\cos b}\right]\\&= \frac{3}{16q\mathcal{V}_0\gamma^3\kappa^4}\left[\frac{(1+n)\gamma\kappa^2}{3}\right]^{\frac{3(1+2n)}{2(1+n)}}\left(\frac{p_*^n}{2}\right)^{-\frac{3}{2(1+n)}}\widehat{|v|^{\frac{3(1+2n-q)}{2(1+n)}}}\\&\quad\times\left[\widehat{e^{-ib}}\,\widehat{|v|^{\frac{3q}{2(1+n)}}}\,\widehat{e^{ib}}-\widehat{e^{ib}}\,\widehat{|v|^{\frac{3q}{2(1+n)}}}\,\widehat{e^{-ib}}\right],\end{aligned} \tag{10.98}$$

and we used the relation $p = [(1+n)\gamma\kappa^2 p_*^n v/6]^{1/(1+n)}$. The states $|v\rangle$ are eigenstates of $\hat{C}$,

$$\hat{C}|v\rangle = c_v|v\rangle\,, \tag{10.99}$$

with eigenvalues

$$c_v = \frac{3}{16q\mathcal{V}_0\gamma^3\kappa^4}\left[\frac{(1+n)\gamma\kappa^2}{3}\right]^{\frac{3(1+2n)}{2(1+n)}}\left(\frac{p_*^n}{2}\right)^{-\frac{3}{2(1+n)}}$$
$$\times|v|^{\frac{3(1+2n)}{2(1+n)}}\left(\left|1+\frac{1}{v}\right|^{\frac{3q}{2(1+n)}}-\left|1-\frac{1}{v}\right|^{\frac{3q}{2(1+n)}}\right). \tag{10.100}$$

The quantization leading to (10.97) is also known as *polymeric* and is based on the replacement (formal in general, rigorous in the case of LQC) of a classically small canonical variable x with a trigonometric function, $x \to \sin x$. Then,

$$\hat{\mathcal{H}}|v\rangle = c_{v+2}|v+4\rangle - (c_{v+2}+c_{v-2})|v\rangle + c_{v-2}|v-4\rangle + \hat{\mathcal{H}}_\phi|v\rangle\,. \tag{10.101}$$

Taking $\hat{\mathcal{H}}^\dagger$, one obtains the same expression, thus showing that the Hamiltonian constraint is real-valued. In fact, $\hat{C}^\dagger = \hat{C}$, independently of the parametrization (choice of n and q). It is more delicate to show that the Hamiltonian constraint is self-adjoint. This property depends on the choice of measure in the physical inner product (see [181], Sect. IIIB) and it is also important for numerical purposes (solutions of non-self-adjoint constraints may lead to unstable modes [182]). In order to have a self-adjoint extension, the constraint must be positive definite and symmetric under the chosen inner product (which can be achieved also by a state redefinition) [183]. The details of the proof depend on those of the constraint [184], but the qualitative features of the effective dynamics are not greatly affected by this property [185].

A physical state is a linear superposition of the eigenstates $|v\rangle$ over the discrete variable v (recall (10.15)),[10]

$$\Psi = \sum_v \Psi_v|v\rangle\,, \tag{10.102}$$

and such that the coefficients Ψ_v obey the LQC version of the Wheeler–DeWitt equation:

$$\boxed{c_{v+2}\Psi_{v+4} - (c_{v+2}+c_{v-2})\Psi_v + c_{v-2}\Psi_{v-4} + \langle v|\hat{\mathcal{H}}_\phi|v\rangle\Psi_v = 0\,.} \tag{10.103}$$

[10] A precise description of physical states as bras in a rigged Hilbert space is given in [166].

Upon quantization, the scalar-field Hamiltonian is

$$\hat{\mathcal{H}}_\phi = \frac{1}{2}\widehat{p^{-3/2}}\widehat{p_\phi^2} + \widehat{p^{3/2}}V(\hat{\phi}) . \tag{10.104}$$

The operator $\hat{\phi}$ acts multiplicatively on the states $|v\rangle$, so that

$$\rho := p^{-3/2}\langle v|\hat{\mathcal{H}}_\phi|v\rangle = \frac{\nu p_\phi^2}{2p^3} + V , \tag{10.105}$$

where the correction function ν is

$$\begin{aligned} \nu &:= \langle v|\widehat{p^{3/2}}\widehat{p^{-3/2}}|v\rangle = (\langle v|\widehat{v^{1-l}}\widehat{v^{l-1}}|v\rangle)^{\frac{6}{(1-l)\sigma}} \\ &= \left[\frac{v}{2l}\left(\left|1+\frac{1}{v}\right|^l - \left|1-\frac{1}{v}\right|^l\right)\right]^{\frac{6}{(1-l)\sigma}} , \end{aligned} \tag{10.106}$$

and

$$\sigma = 4(1+n) . \tag{10.107}$$

The first departing point from the Wheeler–DeWitt equation in ADM variables is that (10.103) is a difference, rather than differential, equation in v. Secondly, while in ADM quantization the wave-function of the universe vanishes at the big bang, here it is undetermined. In fact, (10.103) fixes Ψ_v recursively for all v except $v = 0$. In that case, the coefficient $\langle v|\hat{\mathcal{H}}_\phi|v\rangle$ vanishes on the zero-volume state if the potential V is non-singular there and so does the coefficient $c_2 + c_{-2} = 0$ (because $|v| = |-v|$). Therefore, Ψ_0 decouples from the evolution automatically, the wave-function at either side of the classical big bang singularity are uniquely determined and one "jumps across" the singularity, which is effectively removed from the quantum evolution [186]. Restricting the support of the wave-function onto a discrete lattice without $\lambda = 0$ picks out a separable sub-space of the originally non-separable Hilbert space.

In this sense, the big-bang problem receives an answer different from the one possibly expected by the questioner. In classical cosmology, the issue is whether the "universe" (matter density, metric) is finite at the beginning of time. In the ADM quantization, it is whether the probability of having a classical big-bang configuration is zero or not. Here, the problem is simply cut off from the range of questions we can ask about the quantum dynamics of the system.

Furthermore, there exists a semi-classical limit where the effective equations of motion show how the big bang is replaced by a bounce with finite energy density. Before considering the effective dynamics, however, it is instructive to further compare (10.103) with the WDW equation (10.18). The states annihilated by the scalar constraint are in the v-representation and, in fact, Ψ_v was denoted as $\Psi[v,\phi]$ in the ADM theory. In (10.103), the length of the discrete steps is fixed but in the limit where it vanishes one can show that the WDW equation is recovered. This statement is not immediately obvious because, despite the use of the same variables, $\hat{\mathcal{H}}$ has been engineered in two quite different ways. The b-representation is more convenient for the comparison.

To this purpose, we choose the gauge $N = a^3$ [78] and quantize the Hamiltonian (10.85). In the absence of a scalar potential and up to an overall factor 1/2, the operator ordering can be arranged so that

$$\left[\frac{3}{2\gamma^2\kappa^2}(\sin 2b\,\partial_b)^2 - \partial_\phi^2\right]\Psi[b,\phi] = 0\,. \tag{10.108}$$

Because one has a discrete one-dimensional lattice in v space and the Fourier transform in b-space has support on the interval $b \in (0, \pi/2)$ [78], one can define

$$\Theta := -\partial_z^2\,, \qquad z := \sqrt{\frac{\gamma^2\kappa^2}{6}}\ln\tan b\,, \tag{10.109}$$

so that we get

$$\boxed{\hat{C}\Psi[z,\phi] := -\left(\Theta + \partial_\phi^2\right)\Psi[z,\phi] = 0\,.} \tag{10.110}$$

This expression is formally identical to the WDW equation (10.30) and the ensuing quantization follows step by step that of Sect. 10.2.2. A key difference, however, is that invariance of the wave-function under parity (frame re-orientation) is not gauge-fixed ab initio and physical states are required to satisfy $\Psi_+[-z,\phi] = -\Psi_+[z,\phi]$. It follows that the left- and right-moving sectors are not super-selected and must be considered together [78]. In particular, we can write

$$\Psi_+[z,\phi] = \Psi_{+,\mathrm{L}}[z_+] + \Psi_{+,\mathrm{R}}[z_-] = \xi(z_+) - \xi(z_-)\,, \tag{10.111}$$

where $z_\pm = z \pm \phi$ and ξ is some function.

This fact is crucial for the resolution of the big-bang singularity. The volume operator (10.29) in the z variable is

$$\hat{v} = -\mathrm{i}\sqrt{\frac{2\gamma^2\kappa^2}{3}}\cosh\left[\sqrt{\frac{6}{\gamma^2\kappa^2}}\,z\right]\partial_z =: -\mathrm{i}v_*\cosh(\kappa_0 z)\,\partial_z\,, \tag{10.112}$$

and one has

$$\begin{aligned}(\Psi_{1+}, |\hat{v}|\Psi_{2+}) = 2v_* \int \mathrm{d}z\,\big[&(\partial_z\Psi_{1+,\mathrm{L}}^*)\cosh(\kappa_0 z)(\partial_z\Psi_{2+,\mathrm{L}})\\ &+(\partial_z\Psi_{1+,\mathrm{R}}^*)\cosh(\kappa_0 z)(\partial_z\Psi_{2+,\mathrm{R}})\big]\big|_{\phi=\phi_0}.\end{aligned} \tag{10.113}$$

The relative sign of the two terms is positive since we are taking the expectation value of $|\hat{v}|$, which coincides with that of $\hat{v}$ on the left sector and with that of $-\hat{v}$ on the right sector. At any internal time ϕ and on any physical state (Problem 10.4),

$$\langle|\hat{v}|\rangle = (\Psi_+, |\hat{v}|\Psi_+) = \mathcal{V}_* \cosh(\kappa_0 \phi)\,, \tag{10.114}$$

where $\mathcal{V}_* > 0$ is the minimal volume at the bounce. Equation (10.114) completes the proof that the big-bang singularity is avoided in mini-superspace loop quantum cosmology. Further evidence comes from noting that matter energy density has an absolute upper bound on the whole physical Hilbert space [78]. We can reach the same quantitative conclusion, albeit not as robustly, when looking at the effective dynamics on semi-classical states.

10.3.5 Models with Curvature or a Cosmological Constant

In the previous sections, we ignored the contribution both of the intrinsic curvature $\Gamma^i_\alpha = (\textsc{k}/2)\delta^i_\alpha$ and of a cosmological constant Λ. Here, we sketch scenarios where the universe is not flat ($\textsc{k} = \pm 1$) and $\Lambda \neq 0$.

10.3.5.1 Closed Universe

The case of a universe with positive-definite spin connection was studied in [178, 187–193]. Due to the extra term in the connection, the form of the classical Hamiltonian constraint (10.76) as a function of c (classically related to metric variables as $c = \gamma\dot{a} + \textsc{k}$ with $\textsc{k} = 1$) is modified by the replacement $c^2 \to c(c - 1) + (1 + \gamma^2)/4$, up to fiducial-volume factors. In the classical Friedmann equation, this replacement corresponds to $H^2 \to H^2 + 1/a^2$. The quantum constraint and the resulting difference equation are modified accordingly. There is no arbitrariness in the fiducial volume $\mathcal{V}_0$, since it can be identified with the total volume of the Universe, which is finite and well defined. Then, the choice of elementary holonomy is more natural than in the flat case and, locally, one can distinguish between the group structure of $SU(2)$ and that of $SO(3)$ [190]. As in the flat case, the constraint operator is essentially self-adjoint [190] and the singularity at $v = 0$ is removed from the quantum evolution [178, 187, 190]. However, instead of a single-bounce event one now has a cyclic model [178]. This happens because the classical and quantum scalar constraints have both contracting and expanding branches coexisting in closed-universe solutions, while these branches correspond to distinct solutions in the flat case.

10.3.5.2 Open Universe

Loop quantum cosmology of an open universe [181, 193, 194] is slightly more delicate to deal with. In contrast with the flat and closed cases $\textsc{k} = 0, 1$, the spin connection is non-diagonal, so that also the connection is non-diagonal and it has two (rather than one) dynamical components $c(t)$ and $c_2(t)$. The Gauss constraint eventually fixes $c_2(t) = 1$ and one ends up with the same number of degrees of freedom as usual. The volume of the universe is infinite as in the flat case and a fiducial volume must be defined. The classical Hamiltonian constraint is (10.76) with $c^2 \to c^2 - \mathcal{V}_0^{2/3}\gamma^2$. The quantum constraint is constructed after defining a suitable holonomy loop; the bounce still takes place and the $v = 0$ big-bang state factors out of the dynamics.

10.3.5.3 $\Lambda \neq 0$

Another generalization is to add a cosmological constant term, positive [184, 187, 192] or negative [183, 191, 194, 195]. The finite-difference equations of these models have been studied in relation to the self-adjoint property.

For $\Lambda > 0$, below a critical value Λ_* (of order of the Planck energy) the Hamiltonian constraint operator admits many self-adjoint extensions, each with a discrete spectrum. Above Λ_*, the operator is essentially self-adjoint but there are no physically interesting states in the Hilbert space of the model [184].

For $\Lambda < 0$, the scalar constraint is essentially self-adjoint and its spectrum is discrete [183] (while for $\Lambda = 0$ it is continuous and with support on the positive real line), also when $\textsc{k} = -1$ [194]. As in the $\Lambda = 0$, $\textsc{k} = 1$ case, the universe undergoes cycles of bounces [195].

10.3.6 Homogeneous Effective Dynamics

Effective cosmological equations of motion are derived from the expression of the Hamiltonian constraint on a semi-classical state $|\Psi_{\rm sc}\rangle$. The latter is typically decomposed into a gravitational and a matter sector, $|\Psi_{\rm sc}\rangle = \sum_{A,B} |{\rm grav}\rangle_A \otimes |{\rm mat}\rangle_B$. In general, geometrical and matter operators do not act separately on physical states because solutions to the Hamiltonian constraint already incorporate correlations between the two sectors. Therefore, operators on such states are in general complicated, entangled observables. However, on a semi-classical state geometrical and matter operators commute and they can be treated separately.

The semi-classical states $|\Psi_{\rm sc}\rangle$ are peaked around some point in the classical phase space [74, 196] and they can be defined also in the group-averaging formalism [197]. One then computes the expectation value of the Hamiltonian constraint operator thereon, using an appropriate inner product. Accordingly, for the gravitational part of the Hamiltonian operator (10.97) we approximate its expectation value as

$\langle\Psi_{\rm sc}|\widehat{\sin}(\bar{\mu}c)\,\hat{C}\,\widehat{\sin}(\bar{\mu}c)|\Psi_{\rm sc}\rangle\simeq c_v\sin^2(\bar{\mu}c)$ and we may write

$$\langle\Psi_{\rm sc}|\hat{\mathcal{H}}|\Psi_{\rm sc}\rangle\simeq-\frac{3}{\kappa^2}\,\alpha\sqrt{p}\,\frac{\sin^2(\bar{\mu}c)}{\gamma^2\bar{\mu}^2}+p^{3/2}\rho=0\,, \tag{10.115}$$

where the correction function α is

$$\alpha:=\frac{4\mathcal{V}_0\gamma^2\kappa^2}{3}\frac{\bar{\mu}^2}{\sqrt{p}}C_v=\frac{\sigma}{12q}v\left(\left|1+\frac{1}{v}\right|^{\frac{6q}{\sigma}}-\left|1-\frac{1}{v}\right|^{\frac{6q}{\sigma}}\right), \tag{10.116}$$

σ is given in (10.107) and the range of the ambiguity q is

$$0<q\leqslant 1\,. \tag{10.117}$$

When $\alpha=1$ and the matter sector is a massless free scalar field, (10.115) is exact [198]. In general, however, the evolution of a finitely spread semi-classical state will produce quantum fluctuations leading to additional corrections to (10.115) [199, 200]. Assuming that the semi-classical wave-packet of the Universe does not spread appreciably, we can stick with (10.115) also in the presence of a non-trivial scalar potential. Then, the matter energy density ρ is given by (10.105).

The Hamilton equation of motion $\dot{p}=\{p,\langle N\mathcal{H}\rangle\}$ for the densitized triad gives the Hubble parameter

$$H:=\frac{\dot{p}}{2p}=\alpha\,\frac{\sin(2\bar{\mu}c)}{2\gamma\bar{\mu}\sqrt{p}}\,. \tag{10.118}$$

In the classical limit, $c\to\gamma\dot{a}$ and the right-hand side tends to $\dot{a}/a$ for small $\bar{\mu}c$. Equation (10.115) yields

$$\alpha\sin^2(\bar{\mu}c)=\frac{\rho}{\rho_*}\,, \tag{10.119}$$

where

$$\rho_*:=\frac{3}{\gamma^2\kappa^2\bar{\mu}^2p}\,. \tag{10.120}$$

Combining (10.119) and (10.118), one gets the Friedmann equation

$$\boxed{H^2=\frac{\kappa^2}{3}\rho\left(\alpha-\frac{\rho}{\rho_*}\right).} \tag{10.121}$$

The equation of motion $\dot{\phi} = \{\phi, \langle N\mathcal{H}\rangle\}$ yields (compare with (10.4))

$$p_\phi = p^{3/2}\frac{\dot{\phi}}{N\nu}\,, \tag{10.122}$$

while $\dot{p}_\phi = \{p_\phi, \langle N\mathcal{H}\rangle\}$ leads to the effective Klein–Gordon equation

$$\boxed{\ddot{\phi} + \left(3H - \frac{\dot{\nu}}{\nu}\right)\dot{\phi} + \nu V_{,\phi} = 0\,.} \tag{10.123}$$

As $\nu \geqslant 0$ has a maximum at $v = 1$ and then decreases down to unity for large v, the friction term in (10.123) changes sign during the evolution of the universe, the first stage being of super-acceleration.

Setting $\alpha = 1 = \nu$ in the equations of motion (10.121) and (10.123), one ignores inverse-volume corrections. On the other hand, in the limit $\sin(2\bar{\mu}c) \to 2\bar{\mu}c$ one neglects holonomy corrections and the second term in (10.121) is dropped.

The left-hand side of (10.121) is positive definite and, if $\rho > 0$ ($\alpha > 0$ if $n > -1$), the energy density is *bounded from above*:

$$\boxed{\rho \leqslant \alpha\rho_*\,.} \tag{10.124}$$

When $n \neq 1/2$, $\rho_* \propto a^{2(2n-1)}$ varies with time and, at some point during the history of the universe, it can be made arbitrarily small, thus loosing physical meaning as an absolute lower bound. This is avoided in the improved quantization (10.96), where the critical density is constant:

$$\rho_* = \frac{3}{\gamma^2\kappa^2 p_*}\,. \tag{10.125}$$

For the particular choice $p_* = \Delta_{\rm Pl}$, the critical density is less than half the Planck density,

$$\rho_* = \frac{\sqrt{3}}{32\pi^2\gamma^3}m_{\rm Pl}^4 \approx 0.41\, m_{\rm Pl}^4\,. \tag{10.126}$$

The numerical prefactor depends on (10.93) and (10.94) and it could change in a more complete formulation of the model, but not in a way leading to qualitative differences. If the ambiguity q is set equal to 1, then $\alpha = 1$ and the lower bound (10.124) is the constant (10.126) [167, 201, 202]. Thus, also at the level of the effective dynamics the big-bang singularity is removed.

The homogeneous effective dynamics describes the evolution of sharply peaked semi-classical states down to Planck densities. In these regimes, quantum fluctuations of the canonical variables due to their relative uncertainties might become large enough to spoil the effective equations of motion. However, this is not the case [203]. Provided the fiducial volume $\mathcal{V} = a^3\mathcal{V}_0$ is chosen to be much larger than a Planck volume, states that are initially sharply peaked will remain so during their evolution, even through the bounce. Conversely, effective equations may be less reliable for states which are not sharply peaked, or when the fiducial volume is too small. Barring this possibility, effective homogeneous dynamics is an excellent approximation of the exact dynamics of homogeneous LQC [167, 197]. Of course, the system is fully quantum and local quantum fluctuations have not disappeared. However, canonical variables are average quantities and, in the large-$\mathcal{V}_0$ limit, they are insensitive to local fluctuations. These homogeneous fields behave similarly to the center of mass of a many-particle body. While individual atoms experience quantum fluctuations, quantum back-reaction effects are negligible on the center of mass when the number of atoms composing the body is large.

10.3.6.1 Inverse-Volume Corrections

We discuss now the correction functions α and ν from the point of view of their asymptotic limits, later stressing an interpretational issue.

On a semi-classical state, the eigenvalues of $\widehat{|v|^{l-1}}$ are approximated by the classical variable v^{l-1} itself. Consistently, the classical limit corresponds to a large-volume approximation where $v \gg 1$, while in the near-Planck regime ("small volumes"; the reason for quotation marks will be clear soon) $v \ll 1$. Since

$$v = \frac{12\sqrt{3}}{\sigma}\frac{p_*}{\Delta_{\rm Pl}}\left(\frac{p}{p_*}\right)^{\frac{\sigma}{4}}, \tag{10.127}$$

"near the Planck scale" ($v \ll 1$) the correction functions read

$$\alpha \simeq v^{2-\frac{6q}{\sigma}} =: \alpha_1 \delta_{\rm inv}^{-q_\alpha}, \qquad \nu \simeq v^{\frac{6(2-l)}{(1-l)\sigma}} =: \nu_1 \delta_{\rm inv}^{-q_\nu}, \tag{10.128}$$

where

$$q_\alpha = 1 - \frac{3q}{\sigma}, \qquad \alpha_1 = \left(\frac{12\sqrt{3}}{\sigma}\frac{p_*}{\Delta_{\rm Pl}}\right)^{2q_\alpha}, \tag{10.129}$$

$$q_\nu = \frac{3(2-l)}{(1-l)\sigma}, \qquad \nu_1 = \left(\frac{12\sqrt{3}}{\sigma}\frac{p_*}{\Delta_{\rm Pl}}\right)^{2q_\nu}, \tag{10.130}$$

and

$$\boxed{\delta_{\rm inv} := \left(\frac{p_*}{p}\right)^{\frac{\sigma}{2}} = \left(\frac{a_*}{a}\right)^{\sigma}.} \tag{10.131}$$

From the calculation leading to α and ν, it is clear that the "natural" choice of the ambiguities l and q can be set at the middle of their range:[11]

$$l = \frac{3}{4}, \qquad q = \frac{1}{2}. \tag{10.132}$$

In the mini-superspace parametrization, the old quantization scheme corresponds to $\sigma = 4$ and

$$q_\alpha = \frac{5}{8}, \qquad \alpha_1 = 3^{\frac{15}{8}} = O(10), \tag{10.133}$$

$$q_\nu = \frac{15}{4}, \qquad \nu_1 = 3^{\frac{45}{4}} = O(10^5), \tag{10.134}$$

while the improved scheme has $\sigma = 6$ and

$$q_\alpha = \frac{3}{4}, \qquad \alpha_1 = 2^{3/2}3^{3/4} \approx 6, \tag{10.135}$$

$$q_\nu = \frac{5}{2}, \qquad \nu_1 = 2^5 3^{5/2} \approx 500. \tag{10.136}$$

In homogeneous models with $n = 0$, the duration of this regime depends on the spin representation of the holonomies, small j implying a very short super-inflationary period and, actually, almost no intermediate stage between the discrete quantum regime and the continuum classical limit [175]. Since small-j representations are theoretically favoured, this may constitute a problem. It will be relaxed in a different parametrization when inhomogeneities are taken into account.

In the quasi-classical limit (large volumes), (10.116) and (10.106) can be approximated as

$$\boxed{\alpha \simeq 1 + \alpha_0 \delta_{\rm inv}, \qquad \nu \simeq 1 + \nu_0 \delta_{\rm inv},} \tag{10.137}$$

[11]Different parameter choices have been made in the literature. For instance, the same sequence of steps we reviewed was followed in [172, 173] (arbitrary j, $l = 1/2$, $\sigma = 4$, $\alpha = 1$), [204] (arbitrary j and l, $\sigma = 4$, $\alpha = 1$), [167] ($j = 1/2$, $l = 1/2$, $q = 1$, $\sigma = 6$), [205] (arbitrary j and l, $q = 1$, $\sigma = 6$), and partly [180] ($j = 1/2$, arbitrary l, $q = 1$, arbitrary σ). All these papers use the area gap $\tilde{\Delta}_{\rm Pl}$ (10.93) rather than $\Delta_{\rm Pl}$.

where

$$\alpha_0 = \frac{(3q-\sigma)(6q-\sigma)}{6^4}\left(\frac{\Delta_{\rm Pl}}{p_*}\right)^2, \qquad \nu_0 = \frac{\sigma(2-l)}{6^3}\left(\frac{\Delta_{\rm Pl}}{p_*}\right)^2. \tag{10.138}$$

For the natural choice (10.132), the old and the improved quantization schemes in mini-superspace parametrization correspond, respectively, to

$$\sigma = 4\,, \qquad \alpha_0 = \tfrac{5}{2^5 3^4} \approx 0.002\,, \qquad \nu_0 = \tfrac{5}{6^3} \approx 0.02\,, \tag{10.139}$$

and

$$\sigma = 6\,, \qquad \alpha_0 = \tfrac{1}{96} \approx 0.01\,, \qquad \nu_0 = \tfrac{5}{144} \approx 0.03\,. \tag{10.140}$$

Taking $q = 1$ instead, one gets a negative $\alpha_0 = -1/648$ for $\sigma = 4$ and $\alpha_0 = 0$ for $\sigma = 6$.

Although one can resort to different quantization schemes, equations (10.128) and (10.137) maintain the same structure. The coefficients σ, q_α and q_ν are robust in the choice of the parameters, inasmuch as their order of magnitude does not change appreciably [205]. All these parameters can be set to their "natural" values, which are dictated by the form of the Hamiltonian and other considerations.

Now we examine an interpretation issue related to any parametrization in pure mini-superspace. For a compact universe (e.g., a sphere or a torus) the fiducial volume defined in (10.2) is identified with the total physical volume, which is given by the theory; no issue arises in this case. Otherwise, the fiducial volume in a non-compact setting is an arbitrary quantity and should not appear in physical observables. However, both $\bar\mu$ and other volume-dependent objects do manifest themselves in a number of cases, from the effective equations of motion to cosmological spectra. In particular, $\mathcal{V}$ appears in the correction function (10.131) as $\delta_{\rm inv} \sim a^{-\sigma} \sim \mathcal{V}^{-\sigma/3}$. To make $\delta_{\rm inv}$ dimensionless, one can use the Planck length $l_{\rm Pl}$ to write

$$\delta_{\rm inv} \sim \left(\frac{l_{\rm Pl}^3}{\mathcal{V}_0}\right)^{\frac{\sigma}{3}} a^{-\sigma}\,. \tag{10.141}$$

Physically, the parameter σ is related to how the number of plaquettes of an underlying discrete state changes with respect to the volume as the universe expands. This is a phenomenological prescription for the area of holonomy plaquettes, but ideally it should be an input from the full theory [206]. For phenomenology at the current level of precision, the most significant parameter among $\{\alpha_0, \nu_0, \sigma\}$ is σ, which is not as much affected by different choices of the mini-superspace scheme.

Since $\delta_{\rm Pl}$ is $\mathcal{V}_0$-dependent, inverse-volume corrections are not strictly meaningful in a pure mini-superspace treatment. Casting the problem into another way, the arbitrariness of $\mathcal{V}_0$ corresponds to the freedom to perform a conformal rescaling of non-closed FLRW metrics. The ratio $a_*/a = \sqrt{p_*/p}$ appearing in correction functions, as well as statements such as "the quasi-classical regime takes place at $a < a_*$," seem to violate the conformal invariance $a \to Aa$, since a_* is supposed to be some parameter fixed once and for all by the underlying theory. One could interpret $\mathcal{V}_0$ as a regulator and send $\mathcal{V}_0 \to \infty$ at the end of calculations, so that in the quasi-classical limit there are no inverse-volume corrections at all. However, the full theory does contain these corrections and they should appear also in a cosmological setting. This highlights that something is missing in the theoretical construction of the homogeneous LQC effective dynamics. To get a clearer picture, we should include inhomogeneous perturbations. In fact, in the presence of inhomogeneities there is no conformal freedom in the volume choice, which depends on a local scale factor. In general, the strength of inverse-volume corrections for a given perturbation mode is expected to depend on the scale of the problem, which is measured by the wave-length of a given mode relative to the Planck size [207].

10.3.6.2 Models with $\textsc{k} \neq 0$ and $\Lambda \neq 0$

The flat effective dynamical model has been extended to cases with curvature and a cosmological constant.

For a closed universe, $\textsc{k} = 1$, as mentioned in Sect. 10.3.5 there is no fiducial-volume problem and inverse-volume corrections make sense also in a pure homogeneous and isotropic setting. The cyclic bounces appearing in the dynamics of the difference evolution equation [178] exist also at the effective level [189, 191, 192]; in particular, the big crunch of classical closed universes can be avoided [188]. The bounce persists in an open universe, $\textsc{k} = -1$ [181]. In general, all past and future strong curvature singularities are resolved in $\textsc{k} = \pm 1$ isotropic models; for the closed model, weak singularities in the past evolution may also be resolved [193].

There is evidence that a cosmological constant, if suitably tuned, does not spoil the singularity resolution. When $\Lambda > 0$ and $\textsc{k} = 1$, the bounce is preserved if the cosmological constant is sufficiently small [192]. Above a certain critical value, however, periodic oscillations take place. When $\Lambda < 0$, recollapse of the universe is possible, even cyclically [191, 195]. Whichever the sign of the cosmological constant, the effective Friedmann equation is (10.121), with the critical density ρ_* shifted by a constant, Λ-dependent term.

10.3.6.3 Probability of Inflation

In Sect. 10.2.1, we asked whether inflation is "probable" in WDW quantization. The same question has been posed in LQC, relying on the Gibbons–Turok classical measure. The suppression factor $\sim \mathrm{e}^{-3\mathcal{N}}$ is still present if one ignores holonomy corrections [208] but, in general, the bounce fixes the ambiguities in the measure present in the classical case. It turns out that the probability to have observationally viable inflation is very close to 1 [209]. We do not elaborate further on this point, as the measure problem in cosmology is still under scrutiny.

10.3.7 Singularity Resolved?

To summarize, the avoidance of the big-bang singularity in *homogeneous* and isotropic LQC is confirmed in four ways:

1. At the kinematical level, via the spectrum (10.91) and (10.92) of the inverse volume operator.
2. By the finite-difference quantum Hamiltonian constraint (10.103).
3. By the expectation value (10.114) of the volume operator on physical states in the massless case.
4. Through the effective dynamical equations (10.124), (10.125) and (10.126) in the improved quantization scheme.

The big bang is replaced by a bounce at $H = 0$, where the energy density is about half the Planck energy. Relaxing the condition of isotropy does not spoil the bounce, as shown in Bianchi I and Gowdy models [176, 210–214],[12] Bianchi II [216] and Bianchi IX cosmology [217]. In general, strong singularities are effectively excluded in purely homogeneous LQC [193, 218]. In accordance with the disappearance of the singularity, the chaotic behaviour of the mixmaster universe is drastically changed by LQC quantum effects and *the BKL singularity is removed* [219–221]. As the billiard ball representing the dynamics falls down the potential U from a Kasner regime towards the classical singularity, it starts bouncing off the walls as in the classical case (6.56). However, at sufficiently small volumes the BKL sequence of Kasner eras comes to an end because quantum gravity becomes important and the shape of the potential is modified. The walls are now of finite height, beyond which U falls off smoothly to negative values approaching zero (where $a_i = 0$) from below. Thus, the zero-volume limit corresponds to a local maximum of the potential, not to a global minimum. Also spin-foam cosmology, the mini-superspace path-integral version of LQC, agrees that the singularity $a = 0$ disappears from the quantum dynamics (Sect. 11.4).

These results are encouraging but insufficient to establish a definitive solution of the big-bang issue. The mini-superspace quantization is characterized by a number of ambiguities encoding our ignorance about the details of the full theory and of the underlying state [200]. Even choosing the improved quantization scheme, the critical density $\alpha\rho_*$ may be non-constant if $q \neq 1$, in which case one would lose the neat bounce interpretation.

These ambiguities cannot easily be dismissed as artifacts of the mini-superspace quantization. It is possible to obtain cosmology directly from the full quantum

[12]Gowdy spacetime is globally hyperbolic and has the spatial topology of a 3-torus [215].

theory by reducing its degrees of freedom at the level of kinematical quantum states. One fixes a gauge in the kinematical Hilbert space of LQG where the triad (hence the metric) is diagonal. Spin networks with $SU(2)$ labels are therefore reduced to a three-dimensional cubic lattice whose links are labelled by elements of a subgroup $U(1)$ of $SU(2)$. Each 3-valent node is therefore characterized by the elements of a sub-group $U(1) \otimes U(1) \otimes U(1)$ of $SU(2) \otimes SU(2) \otimes SU(2)$. Since one has a grid with different intertwiners at each point, these reduced states represent inhomogeneous geometries; and since one has different spin labels on the links associated with the three directions at each node, the states are also anisotropic. The homogeneous kinematical sector is thus obtained when all the intertwiners are equal, while the anisotropic sector is recovered when all spins are equal. This procedure is known as *quantum reduced loop gravity* [222–225]. The effective Friedmann equation is obtained by picking a semi-classical reduced state, peaked at large spin quantum numbers and calculating the expectation value of the Hamiltonian. The spread of the state gives rise to quantum corrections. These results depend on three regulators ϵ_α, representing the length of the links along the α-th direction; in the isotropic case, the regulators $\epsilon_\alpha = \epsilon$ are all equal. Sending $\epsilon_\alpha \to 0$ produces the correct classical limit of Bianchi I cosmology, plus corrections. On the other hand, fixing $\epsilon = \bar{\mu}$ one obtains the dynamics (10.115) of homogeneous loop quantum cosmology [225]. Clearly, the quantization ambiguities of canonical formalism as well as the uncertainty on the function $\bar{\mu}(a)$ still remain.

Therefore, from the point of view of the full theory the mini-superspace loop quantization of gravity is less of a toy model than expected (an intriguing confirmation of this result will be discussed in Sect. 11.5.2). But as far as empirical cosmology is concerned, it is not enough. On one hand, loop quantum gravity could resolve the big-bang singularity but, on the other hand, we are far from a complete control of the full theory. An important step forward consists in extending the mini-superspace (regardless of whether the symmetry reduction is done before or after quantization) to include inhomogeneities. In this case, although the notion of a fixed background is still present and one must assume a number of approximations, one can better appreciate the interplay of the degrees of freedom of the full theory and check whether the bounce is an artifact of homogeneity or is a robust feature of the theory. Moreover, the study of inhomogeneous perturbations is essential to extract early-universe cosmological observables.

This is the subject of the remainder of the chapter. As a spoiler for the reader, we anticipate that no clear-cut answer is yet available in the state of the art. In the so-called deformed-algebra approach, the bounce picture is even put on trial by the appearance of something interpreted as an effective signature change of spacetime, suggesting that the universe "dissolves" at Planckian scales rather than bouncing off a minimal size. However, there is no general consensus on this issue. The deformed-algebra method is still under inspection and other formalisms such as the dressed-metric approach and hybrid quantization do not see any signature change.

10.3.8 Lattice Refinement: Quantum Corrections Revisited

Until now, we have not given any motivation for taking $\mu = \bar{\mu} \propto p^{-n}$. This is the next subject and it resides in a framework which does not enjoy all the symmetries of a purely FLRW background.

In loop quantum gravity, the classical continuum of general relativity is replaced by the appearance of discrete spatial structures. It is often expected that the discreteness scale is determined by the Planck length $l_{\rm Pl}$ but, if discreteness is fundamental, its scale must be set by the dynamical parameters of some underlying spin-network state. The scaling of the plaquette in the area law (9.117) is determined by the Planck length for dimensional reasons, but the actual size is given by the spin quantum number. Its values in a specific physical situation have to be derived from the LQG dynamical equations, a task which remains extremely difficult to date. However, given the form in which j_e appears in the dynamical equations, its implications for physics can be understood in certain phenomenological situations, such as in cosmology. Then, instead of using the spin labels j_e, it is useful to refer to an elementary quantum-gravity length scale L, which needs not be exactly the Planck length.

The scale L naturally arises if translation invariance is broken, e.g., by the presence of clustering matter or inhomogeneous perturbations. The comoving volume $\mathcal{V}_0$ of the system can be discretized as a *lattice* whose $\mathcal{N}$ *cells* or patches are nearly isotropic, have characteristic comoving volume l_0^3 and correspond to the vertices of the spin network associated with $\mathcal{V}_0$. The proper size of a cell is

$$L^3 := a^3 l_0^3 = \frac{\mathcal{V}}{\mathcal{N}} . \qquad (10.142)$$

To calculate the curvature at the lattice sites within $\mathcal{V}_0$, we need to specify closed holonomy paths around such points. As we have seen in Sect. 9.3, a generic holonomy plaquette is given by the composition of elementary holonomies over individual plaquettes. Therefore, we can set the length of the elementary holonomy to be that of the characteristic lattice cell. In other words, the elementary loops of comoving size l_0 we have talked about until now define the cells' walls, while in a pure FLRW background there is only one cell of volume $\mathcal{V}_0$ (the number $\mathcal{N}$ is arbitrary). We naturally identify the previously ad hoc function $\bar{\mu}(p)$ as the ratio of the cell-to-lattice size, under the requirement that the lattice be *refined* in time:

$$\bar{\mu} = \mu = \mathcal{N}^{-\frac{1}{3}} . \qquad (10.143)$$

The patch size l_0^3 is independent of the size of the fiducial region, since both $\mathcal{V}_0$ and $\mathcal{N}$ scale in the same way when the size of the region is changed. Physical predictions should not feature the region one chooses unless one is specifically asking region-dependent questions (such as: What is the number of vertices in a given volume?).

Lattice refinement addresses the fiducial-volume issue mentioned in Sects. 10.1.1 and 10.3.6. Fluxes $E[\Sigma]$ are determined by the inhomogeneous spin-network quantum state of the full theory associated with a given patch [204]. This implies that the number of vertices of the underlying physical state would change when changing the fiducial volume $a^3\mathcal{V}_0$. Therefore, there is no scaling ambiguity in the equations of motion [204, 226], although *the physical observables will depend on the choice of spin-network state*.

The spin-network state described by the lattice can be (and usually is) excited by the action of the Hamiltonian operator on the spin vertices, increasing their number and changing their edge labels [227, 228]. This process has not yet been established univocally in the full theory, so that it is convenient to parametrize the number of vertices as in (10.142) [206], where the length $L(t)$ is state dependent and, *by assumption*, coordinate independent; its time dependence is inherited from the state itself. Since the kinematical Hilbert space is usually factorized into gravitational and matter sectors, the problem arises of how to define a natural clock when matter does not enter the definition of a (purely geometrical) spin network. This issue requires a much deeper understanding of the theory. Thus, as unsatisfactory as the free scale (10.142) may be, we take it as a phenomenological ingredient in the present formulation of inhomogeneous LQC.

The general form (10.75) of $\bar{\mu}(p)$ is obtained if $L(t)$ scales as

$$L \propto a^{3(1-2n)}\,. \tag{10.144}$$

Homogeneous models adopting (10.95) feature holonomies which depend on triad variables; in other words, curvature components are constrained by the area operator although this does not appear in the full constraint. This does not seem justified in a purely homogeneous setting. On the other hand, in inhomogeneous models the dependence of the parameter $\bar{\mu}$ on p is implemented at the state (rather than operatorial) level, in closer conformity with the full theory [206]. This partly explains why inverse-volume corrections should depend only on the densitized triad. A better understanding of this fact is gained in a more complete derivation of these corrections within lattice refinement, showing that they naturally arise in a moment expansion of flux operators [229].

10.3.8.1 Lattice Refinement and Inhomogeneous Perturbations

The patches of volume L^3 find a most natural representation in inhomogeneous cosmologies within the separate universe picture. Regions of size L^3 are generated by an underlying discrete state and correspond to quantum degrees of freedom absent classically. The discrete nature of the state implies that inhomogeneities are unavoidable and no perfectly homogeneous geometry can exist. Given these inhomogeneities and their scale provided by the state, one can reinterpret them in a classical context, making use of the separate universe approach. There, the volume $\mathcal{V}$ can be regarded as a region of the universe where inhomogeneities are non-zero but small. This region is coarse-grained into smaller regions of volume L^3, each centered at some point $\mathbf{x}$, wherein the universe is FLRW and described by a "local" scale factor $a(t, \mathbf{x}) = a_{\mathbf{x}}(t)$. The difference between scale factors separated by the typical perturbation wave-length $|\mathbf{x}' - \mathbf{x}| \sim \lambda \ll \mathcal{V}^{1/3}$ defines a spatial gradient interpreted as a metric perturbation. In a perfectly homogeneous context, $L^3 \sim \mathcal{V}$ and there is no sensible notion of dividing $\mathcal{V}$ into cells; this is tantamount to stating that only the fiducial volume enters the quantum corrections and the observables. On the other hand, in an inhomogeneous universe the quantity L^3 carries a time dependence which, in turn, translates into a momentum dependence. The details of the cell sub-division (number of cells per unit volume) are intimately related to the structure of small-scale perturbations and their spectrum. Thus, lattice refinement naturally fits into the cosmological perturbation analysis. As long as perturbations are linear and almost scale invariant, the size of the volume within which the study is conducted is totally irrelevant.

10.3.8.2 Critical Density and Quantum Corrections

Under the replacement $\mathcal{V}_0 \to l_0^3$ everywhere, also the definition of the classical canonical pair (c, p) changes: $c = l_0 \gamma \dot{a}$, $p = l_0^2 a^2$. From (10.120), (10.143) and (10.142), the critical density is

$$\rho_* = \frac{3}{\gamma^2 \kappa^2} \left(\frac{\mathcal{N}}{\mathcal{V}} \right)^{2/3} = \frac{3}{\gamma^2 \kappa^2 L^2}. \tag{10.145}$$

In all quantization schemes but the improved one ($n = 1/2$), the patch size L is dynamical and ρ_* is not constant. In any case, the critical density is a number density which depends neither on the size of the fiducial volume nor on coordinates explicitly, so that it is physically well-defined even outside the improved quantization scheme. Holonomy corrections are defined as

$$\delta_{\rm hol} := \frac{\rho}{\rho_*}, \tag{10.146}$$

and the Friedmann equation (10.121) can be written as $H^2 = (\kappa^2/3)\rho(\alpha - \delta_{\rm hol})$, where the Hubble parameter (10.118) is

$$H = \alpha \frac{\sin(2\gamma L H_{\rm cl})}{2\gamma L}, \qquad H_{\rm cl} := \frac{\dot{a}}{a}. \tag{10.147}$$

The maximum density (size of H) is obtained when $\sin(2\gamma L H_{\rm cl}) = 1$, i.e., for $L H_{\rm cl} = \pi/(4\gamma) \approx 3.3$ In terms of the classical energy density $\rho = 3H_{\rm cl}^2/\kappa^2$, (10.146) can be quantified as

$$\delta_{\rm hol} \sim (\gamma H_{\rm cl} L)^2. \tag{10.148}$$

Intuitively, holonomy corrections become large when the Hubble scale $H_{\rm cl}^{-1} = a/\dot{a} \sim \gamma L$ is of the size of the discreteness scale, certainly an extreme regime in cosmology. In fact, the relative size of the quantum effect at maximal density is very large, since for $\alpha = 1$

$$\frac{H_{\rm cl} - H}{H_{\rm cl}} = 1 - \frac{\sin(2\gamma L H_{\rm cl})}{2\gamma L H_{\rm cl}} = 1 - \frac{2}{\pi} \approx 30\,\% \,. \tag{10.149}$$

Similar modifications occur for the quantum correction $\delta_{\rm inv}$. In inverse-volume as well as in holonomy corrections, one refers to elementary building blocks of a discrete state, respectively, the plaquette areas and the edge lengths. A pure mini-superspace quantization makes use of macroscopic parameters, such as the volume of some fiducial region, and fluxes are calculated on comoving areas $\sim \mathcal{V}_0^{2/3}$. In a purely homogeneous universe, the only way to write down (10.131) is $\delta_{\rm inv} \propto (l_{\rm Pl}/\mathcal{V}^{1/3})^{\sigma}$, which is volume dependent. On the other hand, in the lattice interpretation of loop quantum cosmology one uses the microscopic volume of a cell and both holonomy plaquettes and fluxes are defined on comoving areas $\sim l_0^2$. In particular, the individual cell area L^2 is nothing but the expectation value of the flux operator $\hat{E}[\Sigma]$ through a surface Σ on a semi-classical state. Inverse-volume corrections are determined by the inhomogeneous state through the patch size L:

$$\delta_{\rm inv} = \left(\frac{l_{\rm Pl}}{L}\right)^{\tilde{\sigma}} , \tag{10.150}$$

with some phenomenological parameter $\tilde{\sigma} > 0$ [229]. The latter is not the parameter σ determined by (10.107); $n = 1/2$ will not imply $\tilde{\sigma} = 6$. The inverse-volume correction (10.150) does not depend on holonomies due to the use of the Thiemann identity [229]. We saw that inverse powers of L cannot be quantized to a densely defined operator because the spectrum of the volume contains 0. Inverse volumes appear in the dynamics via the Hamiltonian constraint (of both gravity and matter, as in kinetic matter terms) and are an unavoidable consequence of spatial discreteness in loop quantum gravity. This requires to recast their classical expressions via Poisson brackets, which in turn feature derivatives by L. Quantum discreteness then replaces classical continuous derivatives by finite-difference quotients. For example, the expression $(2\sqrt{L})^{-1} = \partial\sqrt{L}/\partial L$ would become $(\sqrt{L+l_{\rm Pl}} - \sqrt{L-l_{\rm Pl}})/(2l_{\rm Pl})$, strongly differing from $(2\sqrt{L})^{-1}$ when $L \sim l_{\rm Pl}$. For larger L, corrections are perturbative and of order $l_{\rm Pl}/L$, so that in general this type of inverse-volume quantum corrections are encoded by the ratio (10.150). In terms of energy densities,

$$\frac{\rho_*}{\rho_{\rm Pl}} \sim \left(\frac{l_{\rm Pl}}{L}\right)^2 \lesssim 1 \,. \tag{10.151}$$

In practice, the actual size of LQC effects on the inflationary spectrum is well below the over-optimistic upper bound (10.151) but above the naive estimate (10.69) and the WDW effect (10.70). The non-local nature of LQG effects prevents the formation of singularities one would typically find classically. The physical interpretation of inverse-volume corrections stems exactly from the same mechanism: classically divergent quantities such as inverse powers of volumes remain finite due to intrinsically quantum effects. Loosely speaking, quanta of geometry cannot be compressed too densely and they determine the onset of a repulsive force at Planck scale [230, 231] which is responsible for the various corrections to the dynamics.

There are indications that holonomy corrections are not significant in the energy regime of inflation, but only at near-Planckian densities [167]. This is suggested by effective equations for certain matter contents with a dominant kinetic energy [198, 232]. In order to compare inverse-volume and holonomy corrections, we notice from (10.150) and (10.148) that

$$\delta_{\rm inv} \sim \left(\gamma l_{\rm Pl} H \delta_{\rm hol}^{-1/2}\right)^{\tilde{\sigma}} . \qquad (10.152)$$

For a universe of causal size $H^{-1} \sim l_{\rm Pl}$, inverse-volume corrections are considerable and behave very differently from what is normally expected for quantum gravity. Holonomy corrections are small for small densities, but inverse-volume corrections may still be large because they are magnified by an inverse power of $\delta_{\rm hol}$. As the energy density decreases in an expanding universe, holonomy corrections fall to small values, while inverse-volume corrections increase. For instance, in an inflationary regime with a typical energy scale of $\rho \sim 10^{-10}\rho_{\rm Pl}$, we can use (10.152) with $\tilde{\sigma} = 4$ to write $\delta_{\rm hol} \sim 10^{-9}/\sqrt{\delta_{\rm inv}}$. Small holonomy corrections of size $\delta_{\rm hol} < 10^{-6}$ then require inverse-volume correction larger than $\delta_{\rm inv} > 10^{-6}$. This interplay of holonomy and inverse-volume corrections can make loop quantum cosmology testable because it leaves only a finite window for consistent parameter values, rather than just providing Planckian upper bounds. It also shows that inverse-volume corrections become dominant for sufficiently small densities, although they are eventually suppressed as the density decreases.

10.3.8.3 Lattice-Refinement Parametrization

In the lattice refinement picture, (10.150) replaces the total lattice fiducial volume $\mathcal{V}$ with the "patch" (i.e., cell) volume L^3 [233]. This means that one makes the formal replacement $\mathcal{V} \to \mathcal{V}/\mathcal{N}$ everywhere in mini-superspace expressions, which can also be justified as follows.

While writing down the semi-classical Hamiltonian with inverse-volume and holonomy corrections, one is at a non-dynamical quantum-geometric level. At this kinematical level, internal time is taken at a fixed value but the geometry still varies on the whole phase space. In this setting, we must keep $\mathcal{N}$ fixed to some constant $\mathcal{N}_0$ while formulating the constraint as a composite operator. Since the vertex density does not depend on the choice of fiducial volume, it is physically reasonable to expect the $\mathcal{N}_0$ factor to be hidden in the kinematical quantity $a_* = \sqrt{p_*}$. The net result is the Hamiltonian constraint operator of the old formulation of loop quantum cosmology [162, 179] not taking into account any refinement, corresponding to $n = 0$ and $\tilde{\sigma} = \sigma$.

However, when one solves the constraint or uses it for effective equations, one has to bring in the dynamical nature of $\mathcal{N}$ from an underlying full state. This is the motivation for promoting $\mathcal{N}$ to a time-dependent quantity, a step which captures operator as well as state properties of the effective dynamics. For some stretches of time, one can choose to use the scale factor a as the time variable and to represent $\mathcal{N}(a)$ as a power law,

$$\mathcal{N} = \mathcal{N}_0 a^{6n} . \tag{10.153}$$

Then, from (10.143) one gets (10.75). Overall, quantum corrections are of the form (10.150),

$$\delta_{\rm inv} = \left(l_{\rm Pl}^3 \frac{\mathcal{N}}{\mathcal{V}}\right)^{\frac{\tilde{\sigma}}{3}} = \left(l_{\rm Pl}^3 \frac{\mathcal{N}_0}{\mathcal{V}_0}\right)^{\frac{\tilde{\sigma}}{3}} a^{(2n-1)\tilde{\sigma}} , \tag{10.154}$$

where $\tilde{\sigma} > 0$. This equation cannot be obtained in a pure mini-superspace setting. The parameter a plays two roles, one as a dynamical geometric quantity and the other as internal time. The parametrization (10.153) as a power law of the scale factor is simply a way to encode the qualitative (yet robust) phenomenology of the theory. The general viewpoint is similar to mean-field approximations which model effects of underlying degrees of freedom by a single, physically motivated function.

Comparing with the earlier mini-superspace parametrization, (10.154) gives $\sigma = (1 - 2n)\tilde{\sigma}$. Since $\partial\mathcal{N}/\partial\mathcal{V} \geqslant 0$, one has $n \geqslant 0$: the number of vertices $\mathcal{N}$ must not decrease with the volume and it is constant for $n = 0$. Also, $l_0 \propto a^{1-2n}$ is the geometry as determined by the state; in a discrete geometrical setting, this has a lower non-zero bound which requires $n \leqslant 1/2$. In particular, for $n = 1/2$ we have a constant patch volume as in the improved mini-superspace quantization scheme [167]. In contrast with the mini-superspace parametrization (10.107), in the effective parametrization of (10.154) we have $\sigma = 0$ for the improved quantization scheme $n = 1/2$. The range of n is then

$$0 < n \leqslant \tfrac{1}{2} . \tag{10.155}$$

The critical density (10.145), $\rho_* \propto a^{6(2n-1)}$, is constant for $n = 1/2$.

The exponent $\tilde{\sigma}$ in (10.153) can be taken as a positive integer. In fact, the correction function $\delta_{\rm inv}$ depends on flux values, corresponding to p for the isotropic background. Since p changes sign under orientation reversal but the operators are parity invariant, only even powers of p can appear, giving $\tilde{\sigma} = 4$ as the smallest value. Therefore we set $\tilde{\sigma} \geqslant 4$.

To summarize, σ is a time-independent parameter given by the quasi-classical theory and with range

$$\boxed{\sigma \geqslant 0\,.} \tag{10.156}$$

The lattice parametrization replaces the one for homogeneous LQC. In fact, strictly speaking, the use of one parametrization instead of the other is not a matter of choice. A perfectly homogeneous FLRW background is an idealization of reality which, in some applications, may turn out to be untenable. The study of cosmological perturbations is an example in this respect. In that case, the lattice refinement parametrization is not only useful, but also required for consistency when inverse-volume corrections are considered.

In the near-Planckian regime, the small-j problem in the homogeneous parametrization is reinterpreted and relaxed in terms of the lattice embedding. The volume spectrum depends on the quadratic Casimir in j representation: $\bar{\mu}^{-n} \sim L^2 \sim \sqrt{C_2(j)} \sim j$. A higher-$j$ effect can be obtained as a refinement of the lattice (smaller $\bar{\mu}$) [234], thus allowing for long enough super-acceleration. A change in $\bar{\mu}(p)$ can be achieved by varying the comoving volume. This is an arbitrary operation in pure FLRW, while in inhomogeneous models $\bar{\mu}$ is a physical quantity related to the number of vertices of the underlying reduced spin-network state. As long as we lack a calculation of this effect in the full theory, we will not be able to predict the duration of the small-volume regime.

In the quasi-classical regime in the lattice parametrization (10.156), σ may be different in α and ν for an inhomogeneous model but we assume that the background equations (10.137) are valid also in the perturbed case. The coefficients α_0 and ν_0 become arbitrary but *positive* parameters. In fact, from the explicit calculations of inverse-volume operators and their spectra in exactly isotropic models and for regular lattice states in the presence of inhomogeneities [165, 172, 205], correction functions implementing inverse-volume corrections approach the classical value always from above:

$$\boxed{\alpha_0 \geqslant 0\,, \qquad \nu_0 \geqslant 0\,.} \tag{10.157}$$

10.3.9 Perturbations and Inflationary Observables

The original formulation of LQC mainly dealt with the quantization of homogeneous spaces but efforts have been made to incorporate inhomogeneities at the quantum level as well as within an effective dynamics. The goal is to identify characteristic observational signatures allowing one to place bounds on the model.

Currently, there are three main approaches to the problem of LQC perturbations. In the *deformed-algebra* (or effective-constraints) *approach* [226, 235–237], one encodes quantum corrections in effective constraints. In *hybrid quantization* [238–243], it is assumed that the main effects of quantum geometry are encoded in the homogeneous background, which is quantized polymerically as in Sects. 10.3.4 and 10.3.6, while inhomogeneities are Fock quantized in a standard way [212]. This gauge-invariant formalism aims to capture the quantum dynamics in a regime between full quantum gravity and quantum field theory on curved spacetimes and it served as a basis where to deal with inhomogeneous Gowdy cosmology [244–247]. In the *dressed metric approach*, classical constraints are solved for gauge-invariant modes before quantizing. The quantization is of hybrid type and in the geometry sector a "dressed" metric encodes the quantum corrections [248–251]. At the time of writing, most of the LQC approaches to inhomogeneities are still under active inspection and there is ongoing debate concerning both their mathematical aspects and, whenever available, the differences in their physical predictions [237]. The reader is invited to look at the literature and thread carefully on this ground, which is still at the research frontier. It is with this mind attitude that, to obtain the dynamics of inhomogeneities, we briefly report on the deformed-algebra scheme.

10.3.9.1 Perturbations with Deformed Algebra

In standard cosmology, it is common to perturb the action or the Einstein equations (Chap. 3). However, it is also possible to apply perturbation theory to the constraints in Hamiltonian formalism; the two methods agree [252]. In canonical quantum cosmology, the second option is forced upon us by the structure of the theory and constitutes the first step of the following strategy. In the classical theory with a set of constraints C_α, closure of the constraint algebra $\{C_\alpha, C_\beta\} = f_{\alpha\beta}{}^{\gamma} C_\gamma$ is guaranteed by general covariance. For instance, the perturbed smeared constraints in ADM variables automatically obey the canonical algebra (9.166); in Ashtekar–Barbero variables, we have seven first-class constraints (9.52): the super-Hamiltonian, the three components of the diffeomorphism constraint and the three components of the Gauss constraint generating infinitesimal $su(2)$ gauge transformations in the internal space. However, when quantum corrections are included the constraint algebra can present anomalies, terms which do not let the algebra close. In order to guarantee closure of the algebra, one must include counter-terms order by order in perturbation theory that cancel the anomalies. LQG effects can be captured by effective constraints, where quantum corrections are inserted by hand in all the places and shapes expected from the full theory. Since one is solving the quantum-corrected diffeomorphism and Gauss constraints in parallel with the super-Hamiltonian, the theory is fully gauge and diffeomorphism invariant, but not with respect to the standard classical transformations. The structure of spacetime itself is deformed by quantum effects, via the effective constraints. Gauge transformations belonging to a deformed algebra no longer correspond to ordinary coordinate transformations on a manifold. Thus, in order to take the new gauge structure into account, one should rely only on gauge-invariant perturbations. This procedure (first quantize the classical system, then cast it in gauge-invariant variables) is the core of the effective-dynamics approach and it gives rise to the somewhat surprising possibility that LQC quantum corrections be large even during inflation.

One may wonder whether one would get the same results by fixing the gauge *before* quantizing (examples of this possibility are the partial gauge-fixing at the classical level of [207, 253, 254] and early papers on hybrid quantization). In principle, the answer is No. Gauge fixing and quantization do not commute because quantization deeply affects the very notion of gauge invariance. Moreover, whenever gauge-ready variables are not constructed after quantizing, one will generally obtain a physically different quantum system. This is one of the reasons why the effective-dynamics approach (which defines gauge invariance from deformed effective constraints, i.e., supposedly at the quantum level) differs from other proposals.[13]

In the case of inhomogeneous LQC, the seven first-class constraints (9.52) display three types of quantum corrections. These are not limited to inverse-volume and holonomy corrections but can include higher-moment terms generating higher derivatives in the effective constraints [256–260]. The setting is an inflationary era driven by a slowly rolling scalar field, the only difference with respect to the standard case being the presence of the quantum corrections. Ignoring moment corrections, the background equations of motion are (10.121) (with $\rho_* \to \infty$ when holonomy corrections are dropped) and (10.123). As in standard general relativity, linear perturbations can be decomposed into scalar, vector and tensor modes which can be studied independently. Triad and connection components are separated into a flat FLRW background and an inhomogeneous perturbation,

$$E_i^\alpha = a^2\delta_i^\alpha + \delta E_i^\alpha\,, \qquad A_\alpha^i = c\delta_\alpha^i + \left(\delta\Gamma_\alpha^i + \gamma\delta K_\alpha^i\right), \tag{10.158}$$

where the curvature and triad perturbations are canonically conjugate:

$$\{\delta K_\alpha^i(\mathbf{x}), \delta E_j^\beta(\mathbf{x}')\} = \kappa^2\delta_\alpha^\beta\delta_j^i\delta(\mathbf{x},\mathbf{x}')\,. \tag{10.159}$$

The form of the perturbations depends on the sector one considers. In turn, perturbed effective dynamics changes according to whether one includes only holonomy corrections, inverse-volume corrections, moments, or all. After some early works based on toy models [173, 175, 261–266], the constraint algebra has been closed explicitly for holonomy corrections only (for scalar [255, 267, 268], vector [235, 269] and tensor modes [226]), inverse-volume corrections only (for scalar [236, 270], vector [235] and tensor modes [226]), and both at the same time [226, 235, 271]. Contrary

[13]Eventually, the discrepancy might be not so severe. The dynamics of [207, 253] is the same found in [255] once recast in the longitudinal gauge. This match is not sufficient to have a complete proof of equivalence because the longitudinal gauge is legal only in the absence of inverse-volume corrections.

to the WDW case, ignoring back-reaction of the metric and considering just a perturbed test scalar is undesirable, because back-reaction contributes to the actual form of quantum gauge transformations and hence of the gauge-invariant variables. This can lead to an incomplete treatment in partial disagreement with the full gauge-invariant equations. The inflationary tensor spectrum and its index are known for holonomy corrections only [272–279], inverse-volume corrections only [229, 280–283] and both corrections simultaneously [284], while the scalar spectrum and index have been computed in the presence of inverse-volume [229, 282, 283] and small holonomy corrections [285] separately. We concentrate on the inverse-volume case, since one needs the full set of first-order plus non-Gaussianity inflationary observables in order to make a stronger comparison with observations. At the end of the chapter, we will sketch the status of LQC spectra with holonomy corrections.

Since the near-Planckian regime is intrinsically non-perturbative, one can safely trust linear perturbation theory only in the quasi-classical limit, where a consistent closure of the effective constraint algebra has been verified explicitly. One defines the smeared effective Hamiltonian constraint with large-volume inverse-volume corrections as

$$H[N] = \int \mathrm{d}^3\mathbf{x}\, N[\alpha(E)\mathcal{H}_g + \nu(E)\mathcal{H}_\Pi + \xi(E)\mathcal{H}_\nabla + \mathcal{H}_V]\,, \tag{10.160}$$

where α, ν and ξ are arbitrary functions of the densitized triad and different contributions $\mathcal{H}_{g,\Pi,\nabla,V}$ pertain to the gravitational sector and to the scalar field kinetic, gradient and potential terms, respectively. Similarly, one considers the perturbed Gauss and diffeomorphism constraints. The resulting perturbed equations contain counter-terms which fix the functions α, ν and ξ and guarantee anomaly cancellation [235, 236]. These counter-terms have been computed in the quasi-classical limit where (10.137) holds and they depend only on the three parameters α_0, ν_0 and σ. A consistency condition further reduces the parameter space to two dimensions:

$$\alpha_0\left(\frac{\sigma}{6}-1\right) - \nu_0\left(\frac{\sigma}{6}+1\right)\left(\frac{\sigma}{3}-1\right) = 0\,. \tag{10.161}$$

Notably, when $\sigma = 6$ this equation is satisfied only if $\nu_0 = 0$, which is not possible in a pure mini-superspace calculation (see (10.138)). Further tension with the mini-superspace parametrization with $\sigma = 6$ arises from inflationary and de Sitter background solutions, which exist for $0 \lesssim \sigma \lesssim 3$ [282].

Once a closed constraint algebra and the full set of linearized perturbed equations are obtained, one can study the dynamics of the perturbations. Inverse-volume (and holonomy) corrections suppress vector perturbations even faster than in classical cosmology [235] and we can ignore them as usual. Scalar fluctuations Ψ and $\delta\phi$ in the metric and in the scalar field generate the gauge-invariant curvature perturbation on comoving hypersurfaces,

$$\mathcal{R} = \Psi + \frac{\mathcal{H}}{\phi'}\left(1 - \frac{\sigma\nu_0}{6}\delta_{\rm inv}\right)\delta\phi\,.$$

At large scales, this quantity is conserved thanks to a delicate cancellation of counter-terms [282]:

$$\mathcal{R}' = \left[1 + \left(\frac{\alpha_0}{2} + 2\nu_0\right)\delta_{\rm inv}\right]\frac{2\mathcal{H}}{\kappa^2\phi'^2}\nabla^2\Psi\,.$$

Because of this property, one can argue (and also rigorously show) that the Mukhanov–Sasaki scalar variable $u = z_{\rm inv}\mathcal{R}$, where

$$z_{\rm inv} := \frac{a\phi'}{\mathcal{H}}\left[1 + \left(\frac{\alpha_0}{2} - \nu_0\right)\delta_{\rm inv}\right], \tag{10.162}$$

obeys the simple dynamical equation in momentum space

$$u_k'' + \left(s_{\rm inv}^2 k^2 - \frac{z_{\rm inv}''}{z_{\rm inv}}\right)u_k = 0\,, \tag{10.163}$$

where

$$s_{\rm inv}^2 = 1 + \chi\delta_{\rm inv}\,, \qquad \chi := \frac{\sigma\nu_0}{3}\left(\frac{\sigma}{6} + 1\right) + \frac{\alpha_0}{2}\left(5 - \frac{\sigma}{3}\right), \tag{10.164}$$

is the (squared) propagation speed of the perturbation.

Tensor observables can be calculated analogously and display the same type of corrections. In terms of the transverse traceless part of the perturbed 3-metric, the triad and curvature perturbations are

$$\delta E_i^\beta = -\frac{1}{2}a^2 h_i^\beta\,, \qquad \delta K_\beta^i = \frac{1}{2}\left(\frac{1}{\alpha}h_i^{\beta\prime} + \frac{c}{\gamma}h_i^\beta\right). \tag{10.165}$$

Defining h_k as the Fourier transform of the tensor polarization mode h_λ and

$$w_k := a_{\rm inv}h_k\,, \qquad a_{\rm inv} := a\left(1 - \frac{\alpha_0}{2}\delta_{\rm inv}\right), \tag{10.166}$$

the Mukhanov–Sasaki equation for the individual tensor mode h_k is [226, 281, 282]

$$w_k'' + \left(\alpha^2 k^2 - \frac{a_{\rm inv}''}{a_{\rm inv}}\right)w_k = 0\,. \tag{10.167}$$

This is formally identical to the scalar Mukhanov–Sasaki equation and the analysis is exactly the same up to the substitutions $z_{\rm inv} \to a_{\rm inv}$, $\chi \to 2\alpha_0$.

10.3.9.2 Observables

According to the inflationary paradigm, observables are expanded in terms of the slow-roll parameters. Asymptotic solutions to the Mukhanov–Sasaki equation at large scales, evaluated at horizon crossing, yield the scalar spectrum, the scalar index and its running to lowest order in the SR parameters:

$$\mathcal{P}_{\rm s} = \frac{k^3}{2\pi^2 z_{\rm inv}^2}\left\langle |u_{k\ll\mathcal{H}}|^2\right\rangle\Big|_{k|\tau|=1} = \frac{\kappa^2}{8\pi^2}\frac{\mathcal{H}^2}{a^2\epsilon}(1+\gamma_{\rm s}\delta_{\rm inv})\,, \tag{10.168}$$

$$n_{\rm s}-1 = 2\eta-4\epsilon+\sigma\gamma_{n_{\rm s}}\delta_{\rm inv}\,, \tag{10.169}$$

$$\alpha_{\rm s} = 2(5\epsilon\eta-4\epsilon^2-\xi^2)+\sigma(4\tilde\epsilon-\sigma\gamma_{n_{\rm s}})\delta_{\rm inv}\,, \tag{10.170}$$

where

$$\gamma_{\rm s} := \nu_0\left(\frac{\sigma}{6}+1\right)+\frac{\sigma\alpha_0}{2\epsilon}-\frac{\chi}{\sigma+1}\,,$$

$$\gamma_{n_{\rm s}} := \alpha_0-2\nu_0+\frac{\chi}{\sigma+1}\,,$$

$$\tilde\epsilon := \alpha_0\left(\frac{\sigma}{2}+2\epsilon-\eta\right)+\nu_0\left(\frac{\sigma}{6}-1\right)\epsilon\,.$$

One can notice a large-scale enhancement of power via the term

$$\delta_{\rm inv}\sim a^{-\sigma}\sim|\tau|^{\sigma}\sim k^{-\sigma}\,. \tag{10.171}$$

If large enough, quantum corrections dominate and $\alpha_{\rm s}\simeq\sigma f_{\rm s}\delta_{\rm inv}$ to lowest SR order, where

$$f_{\rm s} := \frac{\sigma[3\alpha_0(13\sigma-3)+\nu_0\sigma(6+11\sigma)]}{18(\sigma+1)}\,.$$

Bounds on the scalar running turn out to be the main constraint on the parameters. Due to cosmic variance, there is an intrinsic uncertainty in the determination of the power spectrum at large scales (small multipoles ℓ), which should be compared with the strength of the typical signal from quantum corrections.

From (10.167), one derives the tensor spectrum and its index:

$$\mathcal{P}_{\rm t} = \frac{4\kappa^2}{\pi^2}\frac{k^3}{a_{\rm inv}^2}\left\langle |w_{k\ll\mathcal{H}}|^2\right\rangle\big|_{k|\tau|=1} = \frac{2\kappa^2}{\pi^2}\frac{\mathcal{H}^2}{a^2}(1+\gamma_{\rm t}\delta_{\rm inv})\,, \tag{10.172}$$

$$n_{\rm t} = -2\epsilon-\sigma\gamma_{\rm t}\delta_{\rm inv}\,,\qquad \gamma_{\rm t} := \frac{\sigma-1}{\sigma+1}\alpha_0\,. \tag{10.173}$$

Finally, a modified consistency relation between scalar and tensor perturbations holds:

$$r = -8\{n_{\rm t} + [n_{\rm t}(\gamma_{\rm t} - \gamma_{\rm s}) + \sigma\gamma_{\rm t}]\delta_{\rm inv}\}\,. \tag{10.174}$$

10.3.9.3 Experimental Bounds

As in Sect. 10.2.5, one rewrites the observables in terms of the potential-dependent slow-roll parameters (5.54); the resulting lengthy expressions can be found in [229]. Since $\alpha_{\rm s}^{(m)}(k_0) \simeq (-1)^m\sigma^{m-1}f_{\rm s}\delta_{\rm inv}(k_0)$, the scalar spectrum (10.66) becomes

$$\begin{aligned}\mathcal{P}_{\rm s}(k) \simeq \mathcal{P}_{\rm s}(k_0)\exp\Bigg\{&[n_{\rm s}(k_0) - 1]x + \frac{\alpha_{\rm s}(k_0)}{2}x^2\\ &+f_{\rm s}\delta_{\rm inv}(k_0)\left[x\left(1 - \frac{1}{2}\sigma x\right) + \frac{1}{\sigma}(e^{-\sigma x} - 1)\right]\Bigg\}\,,\end{aligned} \tag{10.175}$$

where $x = \ln(k/k_0)$. This is the expression to be used in numerical analyses and when comparing the LQC signal with cosmic variance. Before doing so, we notice the existence of a theoretical upper bound on the quantum correction $\delta_{\rm LQC} := \alpha_0\delta_{\rm inv}$. (We recall that ν_0 is not independent and it can be removed from parameter space via (10.161), except in the case $\sigma = 3$ which can be treated separately.) For the validity of the linear expansion of the perturbation formulæ where the $O(\delta_{\rm inv})$ truncation has been systematically implemented, we require that $\delta_{\rm LQC}(k) = \delta_{\rm LQC}(k_0)(k_0/k)^\sigma = \delta_{\rm LQC}(k_0)(\ell_0/\ell)^\sigma < 1$ for all wave-numbers relevant to the CMB anisotropies. For the pivot scale $\ell_0 = 29$, the quadrupole $\ell = 2$ gives the bound

$$\delta_{\rm LQC}(k_0) < \delta_{\rm LQC}^{\max} = 14.5^{-\sigma}\,. \tag{10.176}$$

One can choose an inflationary potential and, for any given choice of σ, find an upper bound for $\delta_{\rm LQC}$. For instance, for a quadratic potential and using the 7-year WMAP data combined with large-scale structure, the Hubble constant measurement from the Hubble Space Telescope, Supernovae type Ia and BBN, one finds an experimental upper bound that depends on the value of σ [229, 283]. The 95%-confidence-level upper limits of δ constrained by observations for the quadratic potential with $k_0 = 0.002\ {\rm Mpc}^{-1}$ are

$$\delta_{\rm LQC}^{\max} = 0.27,\ 3.5\times10^{-2},\ 6.8\times10^{-5},\ 4.3\times10^{-7}\,, \qquad \sigma = 0.5, 1, 2, 3\,. \tag{10.177}$$

For $\sigma = 3$, the parameter $\delta = \nu_0\delta_{\rm inv}$ is used instead. For $\sigma \lesssim 1$, the signal can be greater than cosmic variance at large scales (Fig. 10.2). All the cases in (10.177) are compatible with the theoretical upper limit (10.176) but $\delta_{\rm LQC}(k_0) \sim \delta_{\rm LQC}^{\max}$ for $\sigma = 0.5$. More recent constraints on other potentials are about $2-3$ orders of magnitude

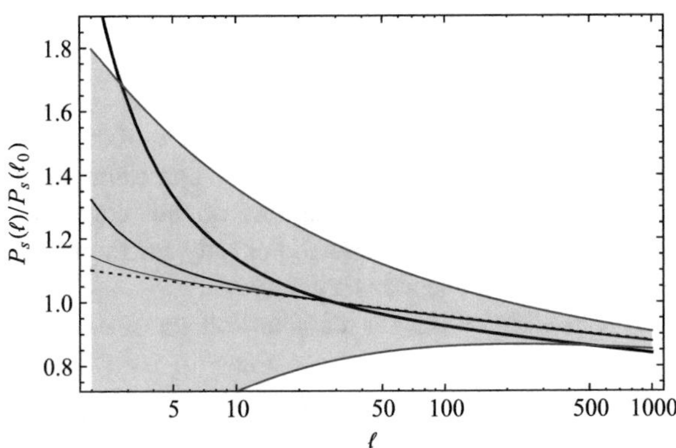

Fig. 10.2 Log-linear plot of the LQC primordial scalar spectrum (10.175) with inverse-volume quantum corrections for a quadratic inflaton potential, with $\epsilon_V(k_0) = 0.009$ and for the pivot wave-number $k_0 = 0.002\,\mathrm{Mpc}^{-1}$, corresponding to $\ell_0 = 29$. The classical case is represented by the *dotted line*, while *solid curves* correspond to $\sigma = 1, 1.5, 2$ (decreasing thickness). The *shaded region* is affected by cosmic variance (Credit: [160])

stronger [286]. For instance, given the same pivot scale, $\delta_{\mathrm{LQC}}^{\mathrm{max}} \sim 10^{-3} - 10^{-5}$ for $\sigma = 0.5$.

We conclude by reviewing the path which led to the above observational implications of quantum gravity. Corrections to the general relativistic dynamics are expected to arise in different ways. For instance, loop corrections are always present in perturbative graviton field theory, which can be captured in effective actions with higher-curvature corrections to the Einstein–Hilbert action. The additional terms change the Newtonian potential as well as the cosmological dynamics. However, in currently observable regimes the curvature scale is very small and one expects only tiny corrections of (dimensionless) size at most $l_{\mathrm{Pl}}H$. In such cases, exemplified by (10.69), tests of quantum gravity are possible at best indirectly, for instance if it provides concrete and sufficiently constrained models for inflation. So far, however, models do not appear tight enough. The same type of modifications arises in WDW quantum cosmology, predicting corrections of the size of (10.70), $|\delta_{\mathrm{WDW}}| < 3\times10^{-9}$. The LQC quantum corrections are not governed by the energy scale of inflation but by some quantum-state scale. In background-independent frameworks such as loop quantum gravity, stronger modifications of the theory are possible since the usual covariant continuum dynamics is generalized, and entirely new effects may be contemplated. In LQG, gauge transformations as well as the dynamics are generated by constraint equations. Since the latter are modified with respect to the classical constraints, gauge transformations change and a new spacetime structure becomes apparent. This global deformation of the classical geometry is the ultimate responsible for possibly sizable inverse-volume quantum effects of the form (10.151). The stark contrast between WDW and LQC quantum corrections arising in these scenarios highlights how sensitive the physics is to the quantization scheme and to the choice of canonical variables.

10.3.9.4 Holonomy Corrections

Another type of quantum effect in the dynamics, holonomy corrections, is realized in a highly non-linear fashion, from the exponentiation h_e of curvature components. Just as in the case of inverse-volume corrections, the constraint algebra is deformed by quantum effects and gauge transformations do not correspond to standard diffeomorphisms. This has important effects not only for the inflationary spectra, but also for the fate of the homogeneous bounce.

The Mukhanov–Sasaki equations for scalar and tensor modes are [226, 255, 268]

$$u'' - \left(s_{\rm hol}^2\nabla^2 + \frac{z_{\rm hol}''}{z_{\rm hol}}\right)u = 0\,, \qquad w'' - \left(s_{\rm hol}^2\nabla^2 + \frac{a_{\rm hol}''}{a_{\rm hol}}\right)w = 0\,, \qquad (10.178)$$

where the effective propagation speeds and background functions $z_{\rm hol}$ and $a_{\rm hol}$ read

$$s_{\rm hol}^2 := \cos(2\gamma L H_{\rm cl}) = 1 - 2\frac{\rho}{\rho_*}\,, \qquad z_{\rm hol} := \frac{\phi'}{H}\,, \qquad a_{\rm hol} := \frac{a}{|s_{\rm hol}|}\,, \qquad (10.179)$$

and H is given by (10.147) with $\alpha = 1$. These expressions should be compared with their inverse-volume counterparts (10.162), (10.163), (10.164), (10.166) and (10.167).

Although the full set of equations of motions for perturbations with holonomy corrections is known, cosmological observational signatures of holonomy effects have been studied for the tensor sector alone. The phenomenology of the scalar sector has not been developed in sufficient detail to compare with experiments as in the WDW and inverse-volume LQC cases. With respect to the inverse-volume case, the analysis of the spectra is complicated by the analytic form of holonomy corrections. In general, tensor modes are amplified during the bounce. However, after the bounce these modes are enhanced by inflationary expansion later than in the classical case and the spectrum is thus *suppressed* at low multipoles, as

$$\mathcal{P}_{\rm t} \overset{k\to 0}{\sim} k^2\,, \qquad (10.180)$$

on a de Sitter background. It also shows an oscillatory pattern, progressively damped towards small scales. The gravitational spectrum is notoriously difficult to detect by itself and information from the scalar spectrum (which, from (10.178), is expected to behave similarly to the tensor one) is needed, also to determine whether the large-scale suppression is observable past the cosmic-variance noise. Recent results invoking the PLANCK upper bounds on B-modes show that inflation with holonomy corrections might be excluded observationally [279]. This does not rule out the deformed-algebra approach because the interplay with inverse-volume corrections should also be taken into account when comparing the model with experiments.

10.3.9.5 Non-Gaussianity

The effect of quantum corrections goes beyond linear perturbation theory and higher-order observables can be calculated. As the perturbative level increases, the statistics of inhomogeneous fluctuations deviates from the Gaussian one and odd-order correlation functions acquire non-vanishing values. In particular, the bispectrum (three-point correlation function of the curvature perturbation) can be constrained by observations.

Both WDW and LQC quantum cosmology with inverse-volume corrections predict a small scalar index $n_s - 1$, given by (10.62) and (10.169). In Sect. 4.6.3, we saw that any inflationary model with such a prediction implies a negligible non-Gaussianity in the squeezed limit (4.108) [287]. One can therefore conclude that no appreciable WDW or inverse-volume-corrected LQC non-Gaussian signal of local form can be detected [160]. A detailed calculation of the LQC momentum-dependent bispectrum for inverse-volume corrections reaches the same conclusion [288].

10.3.10 Inflation in Other Approaches

Compared with the effective constraints, the hybrid quantization approach [240–243] is more closely related to the dressed-metric framework [249–251]. In the former case, the gauge-invariant Mukhanov–Sasaki variables are determined at the classical level; perturbations are afterwards quantized on a Fock space. Making a Born–Oppenheimer *Ansatz* for physical states (valid, for instance, if the latter are semi-classical with a sharp peak), the Mukhanov–Sasaki equation for, say, the scalar perturbation u_k is

$$u_k'' + \left(k^2 - \frac{\langle \hat{z}'' \rangle}{\langle \hat{z} \rangle}\right) u_k = 0\,, \tag{10.181}$$

where $\langle z \rangle$ is the expectation value on the chosen vacuum of a suitable background operator. The Mukhanov equation in the dressed-metric framework is formally the same as (10.181) but with a different operator $\hat{z}$ (and possibly different choices of vacuum). In this case, $\hat{z}$ contains the scale-factor operator $\hat{a}$ whose expectation value $\langle \hat{a} \rangle = a$ is solution to the modified Friedmann equations. $\langle \hat{z} \rangle = z$ is approximated with the expectation value at which $\hat{z}$ is peaked. The two approaches yield qualitatively the same spectra, which show a suppression of power at large scales [242, 251].

The inflationary spectra have been computed also in the partial gauge-fixing scenario of [253, 254] and it is possible to suppress the tensor-to-scalar ratio with respect to the standard result. Although, as discussed in Sect. 10.3.9.1, the quantum theory after gauge fixing is different from the other approaches, it is a full loop quantization of both the background and the perturbations, in the sense that it makes use of the well-known standard LQC quantization in each patch in the separate-universe approach.

At intermediate and large scales, there is agreement between the tensor spectrum in the deformed-algebra approach in the absence of inverse-volume corrections (or when holonomy corrections dominate) and the dressed-metric approach [289]. For wave-numbers $k < \sqrt{\kappa^2\rho_*} = O(l_{\rm Pl}^{-1})$, the form of all these spectra is quantitatively the same and is independent of unknown quantum-geometry parameters (there is a dependence on the inflaton initial conditions). Therefore, at large scales the effective-dynamics and dressed-metric approaches predict suppressed power spectra and are in mutual agreement.[14] However, in the deep ultraviolet (i.e., at small scales, large wave-number k) the effective-dynamics spectra have a different asymptotic behaviour with respect to the other three formalisms. This is expected from the different way the classical system is quantized and raises an interpretational issue near Planckian scales, which we now examine.

10.3.11 Is There a Bounce?

Let us go back to the issue of singularity resolution. In homogeneous LQC, the big bang is replaced by a bounce; this result can be found in so many different ways that its robustness seems beyond question. Is this feature preserved in the presence of inhomogeneities? In approaches with gauge fixing or where quantum gauge invariance stems directly from classical gauge invariance, the answer is in the affirmative. However, in the deformed-algebra framework gauge invariance is implemented at the quantum level, which can result in a dramatic rewriting of the behaviour of the Universe near Planckian densities.

The propagation speed of perturbations is never super-luminal ($|s_{\rm hol}^2| \leqslant 1$) but it does change sign near the bouncing point [255, 271, 278]. This marks a possible instability and a change of effective spacetime signature at near-Planckian scales [290, 291], in a super-inflationary early era. This happens because $s_{\rm hol}^2$ appears in the right-hand side of the deformed commutator $\{H, H\}$, (9.166c). In general, this constraint is deformed as

$$\{H[N], H[M]\} = d[\beta h^{\alpha\gamma}(N\partial_\gamma M - M\partial_\gamma N)]\,, \tag{10.182}$$

where β is a function of the phase-space variables. When $\beta = +1$, one recovers the constraint (9.166c) of general relativity in Lorentzian signature, while when $\beta = -1$, one has the algebra of Riemannian gravity (spacetime with Euclidean signature) [292]. The deformation of LQC with inverse-volume and holonomy corrections is, respectively, $\beta = s_{\rm inv}^2$ and $\beta = s_{\rm hol}^2$; only in the latter case β changes sign [268].

[14] It is not clear whether the gauge-fixed quantization via the separate-universe approach [253, 254] makes the same prediction of a suppression of power at large scales.

The same structure (10.182) appears also in the loop quantization of spherically symmetric spacetimes [237, 291, 293–296] and in (2+1)-dimensional loop quantum gravity [297, 298]. In the case of LQC holonomy corrections, at the critical energy density $\rho = \rho_*$ one has $\beta = s_{\text{hol}}^2 \approx -1$, while at low energies $\beta = s_{\text{hol}}^2 \approx +1$. Going backwards in time from today, β changes sign at some point before reaching the bounce density, which poses a conceptual problem: Is the homogeneous bounce reached at all or does spacetime geometry change so much as to invalidate the mini-superspace approximation?

A change of sign in front of the Laplacian of the Mukhanov–Sasaki equations is not a novelty in cosmology. The same effect occurs, for instance, in higher-order gravity where the Gauss–Bonnet curvature invariant is non-minimally coupled with a scalar field (see [299] and references therein). In that case, this change of sign is simply interpreted as a classical instability of the perturbations on the FLRW background affecting cosmological spacetime scales. Although ghost instabilities and super-luminal propagation are problematic and can be avoided by a restriction of the parameter space, the nature of the spacetime wherein perturbation modes propagate remains purely classical and Lorentzian. The higher-order terms of Gauss–Bonnet theory, in fact, do not lead to deformations of the constraint algebra.

In LQC, however, the change in the perturbation equations is a direct consequence of the deformation (10.182) of the constraint algebra of gravity and, hence, of a deformation of the classical spacetime structure. The type of field equations changes from hyperbolic to elliptic for all modes simultaneously. Moreover, the manifold on which physical fields are defined has no causal structure at high curvature. These are the main reasons why, in the present context, such an effect is interpreted as a signature change of spacetime rather than a simple perturbative instability. This effect is not a transition to classical Euclidean space, since $\beta = -1$ is realized only on one hypersurface. Rather, it is a change in the type of partial differential equations.

Within the effective-dynamics approach, there are choices of counter-terms which avoid the signature change but still imply the existence of a space-like surface where the propagation speed of perturbations vanishes and initial conditions must be set [271]. There is also the possibility that inverse-volume and moment terms may compensate these deformations so that no signature change takes place. However, these corrections would not cancel holonomy effects exactly and are presently unknown. Overall, although the deformation (10.182) happens on different backgrounds in the deformed-algebra approach, the resolution of classical gravitational singularities via a Lorentzian mechanism is not as evident as in homogeneous cosmology. This preliminary conclusion is reinforced in full $(3+1)$-dimensional LQG, where inverse-volume operators are well-defined but unbounded from above on zero-volume eigenstates (including the big bang) [300, 301].

The Laplacian modification of the Mukhanov–Sasaki equation does not take place in the hybrid quantization and dressed-metric approaches. Since gauge invariants are defined with respect to the classical gauge transformations, no deformation of the constraint algebra (10.182) arises which could give rise to a signature change. In fact, the propagation speed of the perturbations in (10.181) is always equal to 1

[240–243]. As we already mentioned, both the hybrid quantization and the dressed-metric frameworks are different ways of quantizing the same classical system and, as long as they are self-consistent, they should be regarded as physically inequivalent but equally valid alternatives to the effective-dynamics setting. In the gauge-fixed approach [207, 253, 254], the factor in front of the kinetic term does change sign. However, for sub-Planckian modes the derivative term in the Mukhanov–Sasaki equation is always sub-leading with respect to the effective mass. Since this framework cannot be applied to trans-Planckian modes, where the kinetic term would be important, it does not give any conclusive evidence about signature change.

At present, the signature change remains a possibility, although its appearance, not just in the effective-dynamics approach but also from operator computations of off-shell constraint algebras, may be very generic [237, 302, 303]. If it was confirmed that spacetime changes from Lorentzian to Euclidean when reaching the critical density from below, this would not mean that singularities are not resolved in loop quantum cosmology. One could no longer talk of a bounce in the sense of Lorentzian cosmology, but there might be some tunneling or topology change as in WDW quantum cosmology, to be translated into some general mechanism in the full theory more subtle than the requirement of having well-defined inverse-volume operators. It may turn out that there is a mechanism within loop quantum gravity by which signature change could be avoided after all. Even if this were the case, however, one could not show it within mini-superspace models because one must have access to temporal and spatial variations. This is the reason why such models do not seem to be completely understood at high density.

10.4 Problems and Solutions

10.1 Classical FLRW Hamilton equations. Derive (10.7).

Solution A direct calculation yields

$$\dot{p}_{(a)} = \{p_{(a)}, H_{\rm D}\} = N\left[\frac{3p_\phi^2}{2\mathcal{V}_0 a^4} - 3\mathcal{V}_0 a^2 V + \frac{3}{\kappa^2}\mathcal{V}_0 \textsc{K} - \frac{\kappa^2 p_{(a)}^2}{12\mathcal{V}_0 a^2}\right]$$

$$= \mathcal{V}_0 N a^2\left[\frac{3\dot{\phi}^2}{2N^2} - 3V + \frac{3}{\kappa^2}\frac{\textsc{K}}{a^2} - \frac{3}{\kappa^2}\frac{H^2}{N^2}\right] = 2\mathcal{V}_0 N a^2\left[\kappa^2\left(\frac{\dot{\phi}^2}{N^2} - V\right) - 3\frac{H^2}{N^2}\right].$$

10.2 Volume expectation value. Noting that $\partial_\phi \Psi_{\pm,\mathrm{L}} = \pm\partial_y \Psi_{\pm,\mathrm{L}}$ and $\partial_\phi \Psi_{\pm,\mathrm{R}} = \mp\partial_y \Psi_{\pm,\mathrm{R}}$ for any state, rewrite the inner product (10.40). From there, derive (10.46) and (10.47).

Solution After integration by parts, (10.40) becomes

$$(\Psi_{1+}, \Psi_{2+}) = 2\mathrm{i} \int \mathrm{d}y \left(\Psi_{2+,\mathrm{L}} \partial_y \Psi^*_{1+,\mathrm{L}} - \Psi_{2+,\mathrm{R}} \partial_y \Psi^*_{1+,\mathrm{R}}\right) \big|_{\phi=\phi_0} . \tag{10.183}$$

Writing the volume operator (10.29) as

$$\hat{v} = -\mathrm{i}\sqrt{\frac{3}{2(1+n)^2\kappa^2 b_0^2}}\, \mathrm{e}^{-\sqrt{\frac{2\kappa^2}{3}}(1+n)\,y}\partial_y =: -\mathrm{i}v_* \mathrm{e}^{-\kappa_0 y}\partial_y ,$$

one has

$$\begin{aligned}(\Psi_{1+}, \hat{v}\Psi_{2+}) &= 2v_* \int \mathrm{d}y \left[(\partial_y \Psi^*_{1+,\mathrm{L}})\mathrm{e}^{-\kappa_0 y}(\partial_y \Psi_{2+,\mathrm{L}})\right.\\ &\qquad \left.-(\partial_y \Psi^*_{1+,\mathrm{R}})\mathrm{e}^{-\kappa_0 y}(\partial_y \Psi_{2+,\mathrm{R}})\right] \big|_{\phi=\phi_0}\\ &= (\Psi_{1+,\mathrm{L}}, |\hat{v}|\Psi_{2+,\mathrm{L}}) - (\Psi_{1+,\mathrm{R}}, |\hat{v}|\Psi_{2+,\mathrm{R}}) .\end{aligned} \tag{10.184}$$

The relative $-$ sign comes from the fact that $\hat{v}$ leaves the left sector invariant and the right sector anti-invariant. Physically, what matters is the expectation value of $|\hat{v}|$. Finally, we have

$$\begin{aligned}\langle|\hat{v}|\rangle_\mathrm{L} &= (\Psi_{+,\mathrm{L}}, \hat{v}\Psi_{+,\mathrm{L}}) = 2v_* \int \mathrm{d}y |\partial_y \Psi_{+,\mathrm{L}}[y_+]|^2 \mathrm{e}^{-\kappa_0 y}\\ &= 2v_* \int \mathrm{d}y_+ |\partial_{y_+} \Psi_{+,\mathrm{L}}[y_+]|^2 \mathrm{e}^{-\kappa_0 (y_+ - \phi)} =: \mathcal{V}_\mathrm{L} \mathrm{e}^{\kappa_0 \phi} ,\\ \langle|\hat{v}|\rangle_\mathrm{R} &= (\Psi_{+,\mathrm{R}}, \hat{v}\Psi_{+,\mathrm{R}}) = 2v_* \int \mathrm{d}y |\partial_y \Psi_{+,\mathrm{R}}[y_-]|^2 \mathrm{e}^{-\kappa_0 y}\\ &= 2v_* \int \mathrm{d}y_- |\partial_{y_-} \Psi_{+,\mathrm{R}}[y_-]|^2 \mathrm{e}^{-\kappa_0 (y_- + \phi)} =: \mathcal{V}_\mathrm{R} \mathrm{e}^{-\kappa_0 \phi} .\end{aligned}$$

10.3 Classical super-Hamiltonian in LQC. Derive (10.83).

Solution A direct calculation yields

$$
\begin{aligned}
\mathcal{H} &= -\frac{1}{2\gamma^2\kappa^2}\frac{E_i^\alpha E_j^\gamma}{\sqrt{|\det E|}}\epsilon^{ij}{}_k F_{\alpha\gamma}^k + \mathcal{H}_\phi \\
&\overset{(9.115)}{=} \frac{4}{\gamma^3\kappa^4}\lim_{l_0\to 0}\frac{\mathcal{V}^{1-q}}{q l_0^3}\sum_{i,j,k,i'}\mathrm{tr}(\tau_k h_{\square_{ij}})\epsilon^{iji'}\,\mathrm{tr}(\tau_k h_{i'}\{h_{i'}^{-1},\,\mathcal{V}^q\}) + \mathcal{H}_\phi \\
&\overset{(10.80)}{=} -\frac{4}{\gamma^3\kappa^4}\lim_{l_0\to 0}\frac{\mathcal{V}^{1-q}}{q l_0^3}\sin^2 2\lambda b\sum_k \mathrm{tr}(\tau_k h_k\{h_k^{-1},\,\mathcal{V}^q\}) + \mathcal{H}_\phi \\
&\overset{(10.82)}{=} \frac{12}{\gamma^3\kappa^4 q}\lim_{l_0\to 0}\frac{1}{l_0^3}\left[\frac{(1+n)\gamma\kappa^2}{3}\frac{p_*^n}{2}\right]^{\frac{3}{2(1+n)}} v^{\frac{3(1-q)}{2(1+n)}}\sin^2 2\lambda b \\
&\quad\times[\sin\lambda b\,\{\cos\lambda b,\,v^{\frac{3q}{2(1+n)}}\} - \cos\lambda b\,\{\sin\lambda b,\,v^{\frac{3q}{2(1+n)}}\}] + \mathcal{H}_\phi .
\end{aligned}
$$

10.4 Expectation value of the volume operator in LQC. Derive (10.114).

Solution A direct calculation yields

$$
\begin{aligned}
\langle|\hat{v}|\rangle &= (\Psi_+,|\hat{v}|\Psi_+) \\
&= 2v_*\int \mathrm{d}z\left(|\partial_z\Psi_{+,\mathrm{L}}[z_+]|^2 + |\partial_z\Psi_{+,\mathrm{R}}[z_-]|^2\right)\cosh(\kappa_0 z) \\
&= 2v_*\int \mathrm{d}z_+|\partial_{z_+}\xi|^2\cosh[\kappa_0(z_+ - \phi)] \\
&\quad +2v_*\int \mathrm{d}z_-|\partial_{z_-}\xi|^2\cosh[\kappa_0(z_- + \phi)] \\
&= \left[4v_*\int \mathrm{d}z|\partial_z\xi|^2\cosh(\kappa_0 z)\right]\cosh(\kappa_0\phi) \\
&=: \mathcal{V}_*\cosh(\kappa_0\phi) .
\end{aligned}
$$

References

1. B.S. DeWitt, Quantum theory of gravity. I. The canonical theory. Phys. Rev. **160**, 1113 (1967)
2. C.W. Misner, Quantum cosmology. I. Phys. Rev. **186**, 1319 (1969)
3. K.V. Kuchař, M.P. Ryan, Can mini superspace quantization be justified? in *Gravitational Collapse and Relativity*, ed. by H. Sato, T. Nakamura (World Scientific, Singapore, 1986)
4. K.V. Kuchař, M.P. Ryan, Is minisuperspace quantization valid?: Taub in mixmaster. Phys. Rev. D **40**, 3982 (1989)
5. S.W. Hawking, The quantum state of the universe. Nucl. Phys. B **239**, 257 (1984)
6. S.P. Kim, Quantum mechanics of conformally and minimally coupled Friedmann–Robertson–Walker cosmology. Phys. Rev. D **46**, 3403 (1992)
7. I.G. Moss, W.A. Wright, Wave function of the inflationary universe. Phys. Rev. D **29**, 1067 (1984)
8. S.W. Hawking, Z.C. Wu, Numerical calculations of minisuperspace cosmological models. Phys. Lett. B **151**, 15 (1985)
9. U. Carow, S. Watamura, Quantum cosmological model of the inflationary universe. Phys. Rev. D **32**, 1290 (1985)
10. C. Kiefer, Wave packets in minisuperspace. Phys. Rev. D **38**, 1761 (1988)
11. D.J. Kaup, A.P. Vitello, Solvable quantum cosmological model and the importance of quantizing in a special canonical frame. Phys. Rev. D **9**, 1648 (1974)
12. W.F. Blyth, C.J. Isham, Quantization of a Friedmann universe filled with a scalar field. Phys. Rev. D **11**, 768 (1975)
13. R. Brout, G. Horwitz, D. Weil, On the onset of time and temperature in cosmology. Phys. Lett. B **192**, 318 (1987)
14. L. Liu, C.-G. Huang, The quantum cosmology in the Brans–Dicke theory. Gen. Relat. Grav. **20**, 583 (1988)
15. D.N. Page, Minisuperspaces with conformally and minimally coupled scalar fields. J. Math. Phys. **32**, 3427 (1991)
16. Z.H. Zhu, Boundary conditions in quantum cosmology in the Brans–Dicke theory. Chin. Phys. Lett. **9**, 273 (1992)
17. C. Kiefer, E.A. Martínez, On time and the quantum to classical transition in Jordan–Brans–Dicke quantum gravity. Class. Quantum Grav. **10**, 2511 (1993). [arXiv:gr-qc/9306029]
18. J.E. Lidsey, Scale factor duality and hidden supersymmetry in scalar-tensor cosmology. Phys. Rev. D **52**, 5407 (1995). [arXiv:gr-qc/9510017]
19. Z.-H. Zhu, Y.-Z. Zhang, X.-P. Wu, On the cosmological constant in quantum cosmology of the Brans–Dicke theory. Mod. Phys. Lett. A **13**, 1333 (1998)
20. Z.-H. Zhu, Cosmic wave functions with the Brans–Dicke theory. Chin. Phys. Lett. **17**, 856 (2000)
21. D.-i. Hwang, H. Sahlmann, D.-h. Yeom, The no-boundary measure in scalar-tensor gravity. Class. Quantum Grav. **29**, 095005 (2012). [arXiv:1107.4653]
22. S.W. Hawking, J.C. Luttrell, Higher derivatives in quantum cosmology: (I). The isotropic case. Nucl. Phys. B **247**, 250 (1984)
23. P.F. González-Díaz, On the wave function of the universe. Phys. Lett. B **159**, 19 (1985)
24. T.P. Singh, T. Padmanabhan, Notes on semiclassical gravity. Ann. Phys. (N.Y.) **196**, 296 (1989)
25. S. Sinha, B.L. Hu, Validity of the minisuperspace approximation: an example from interacting quantum field theory. Phys. Rev. D **44**, 1028 (1991)
26. F.D. Mazzitelli, Midisuperspace-induced corrections to the Wheeler–DeWitt equation. Phys. Rev. D **46**, 4758 (1992). [arXiv:hep-th/9203072]
27. A. Ishikawa, T. Isse, The stability of the minisuperspace. Mod. Phys. Lett. A **08**, 3413 (1993). [arXiv:gr-qc/9308004]
28. C. Kiefer, Continuous measurement of mini-superspace variables by higher multipoles. Class. Quantum Grav. **4**, 1369 (1987)

29. J.J. Halliwell, Correlations in the wave function of the universe. Phys. Rev. D **36**, 3626 (1987)
30. R. Brout, On the concept of time and the origin of the cosmological temperature. Found. Phys. **17**, 603 (1987)
31. R. Brout, G. Venturi, Time in semiclassical gravity. Phys. Rev. D **39**, 2436 (1989)
32. D.P. Datta, Geometric phase in vacuum instability: applications in quantum cosmology. Phys. Rev. D **48**, 5746 (1993). [arXiv:gr-qc/9306028]
33. T. Banks, TCP, quantum gravity, the cosmological constant and all that.... Nucl. Phys. B **249**, 332 (1985)
34. S.P. Kim, New asymptotic expansion method for the Wheeler–DeWitt equation. Phys. Rev. D **52**, 3382 (1995). [arXiv:gr-qc/9511038]
35. S.P. Kim, Classical spacetime from quantum gravity. Class. Quantum Grav. **13**, 1377 (1996). [arXiv:gr-qc/9601049]
36. C. Bertoni, F. Finelli, G. Venturi, The Born–Oppenheimer approach to the matter-gravity system and unitarity. Class. Quantum Grav. **13**, 2375 (1996). [arXiv:gr-qc/9604011]
37. S.P. Kim, Problem of unitarity and quantum corrections in semiclassical quantum gravity. Phys. Rev. D **55**, 7511 (1997). [arXiv:gr-qc/9611040]
38. S.P. Kim, Quantum potential and cosmological singularities. Phys. Lett. A **236**, 11 (1997). [arXiv:gr-qc/9703065]
39. J.J. Halliwell, Decoherence in quantum cosmology. Phys. Rev. D **39**, 2912 (1989)
40. C. Kiefer, Decoherence in quantum electrodynamics and quantum gravity. Phys. Rev. D **46**, 1658 (1992)
41. J.P. Paz, S. Sinha, Decoherence and back reaction: the origin of the semiclassical Einstein equations. Phys. Rev. D **44**, 1038 (1991)
42. A. Vilenkin, Boundary conditions in quantum cosmology. Phys. Rev. D **33**, 3560 (1986)
43. A. Vilenkin, Quantum cosmology and the initial state of the Universe. Phys. Rev. D **37**, 888 (1988)
44. A. Vilenkin, Approaches to quantum cosmology. Phys. Rev. D **50**, 2581 (1994). [arXiv:gr-qc/9403010]
45. J.B. Hartle, S.W. Hawking, Wave function of the Universe. Phys. Rev. D **28**, 2960 (1983)
46. A.D. Linde, Quantum creation of an inflationary universe. Zh. Eksp. Teor. Fiz. **87**, 369 (1984) [Sov. Phys. JETP **60**, 211 (1984)]; Quantum creation of the inflationary universe. Lett. Nuovo Cim. **39**, 401 (1984)
47. A. Vilenkin, Wave function discord. Phys. Rev. D **58**, 067301 (1998). [arXiv:gr-qc/9804051]
48. G. Calcagni, C. Kiefer, C.F. Steinwachs, Quantum cosmological consistency condition for inflation. JCAP **1410**, 026 (2014). [arXiv:1405.6541]
49. D.L. Wiltshire, An introduction to quantum cosmology, in *Cosmology: The Physics of the Universe*, ed. by B. Robson, N. Visvanathan, W.S. Woolcock (World Scientific, Singapore, 1996). [arXiv:gr-qc/0101003]
50. L.P. Grishchuk, L.V. Rozhansky, Does the Hartle–Hawking wavefunction predict the universe we live in? Phys. Lett. B **234**, 9 (1990)
51. A. Lukas, The no boundary wave-function and the duration of the inflationary period. Phys. Lett. B **347**, 13 (1995). [arXiv:gr-qc/9409012]
52. J.B. Hartle, S.W. Hawking, T. Hertog, No-boundary measure of the Universe. Phys. Rev. Lett. **100**, 201301 (2008). [arXiv:0711.4630]
53. J.B. Hartle, S.W. Hawking, T. Hertog, Classical universes of the no-boundary quantum state. Phys. Rev. D **77**, 123537 (2008). [arXiv:0803.1663]
54. J.B. Hartle, S.W. Hawking, T. Hertog, No-boundary measure in the regime of eternal inflation. Phys. Rev. D **82**, 063510 (2010). [arXiv:1001.0262]
55. A.O. Barvinsky, A.Yu. Kamenshchik, C. Kiefer, C.F. Steinwachs, Tunneling cosmological state revisited: origin of inflation with a non-minimally coupled standard model Higgs inflaton. Phys. Rev. D **81**, 043530 (2010). [arXiv:0911.1408]
56. A.O. Barvinsky, A.Yu. Kamenshchik, 1-loop quantum cosmology: the Normalizability of the Hartle-Hawking wave function and the probability of inflation. Class. Quantum Grav. **7**, L181 (1990)

57. A.O. Barvinsky, Unitarity approach to quantum cosmology. Phys. Rep. **230**, 237 (1993)
58. A.O. Barvinsky, Reduction methods for functional determinants in quantum gravity and cosmology. Phys. Rev. D **50**, 5115 (1994). [arXiv:gr-qc/9311023]
59. A.O. Barvinsky, A.Yu. Kamenshchik, C. Kiefer, Effective action and decoherence by fermions in quantum cosmology. Nucl. Phys. B **552**, 420 (1999). [arXiv:gr-qc/9901055]
60. A.O. Barvinsky, A.Yu. Kamenshchik, Quantum scale of inflation and particle physics of the early universe. Phys. Lett. B **332**, 270 (1994). [arXiv:gr-qc/9404062]
61. A.O. Barvinsky, A.Yu. Kamenshchik, Effective equations of motion and initial conditions for inflation in quantum cosmology. Nucl. Phys. B **532**, 339 (1998). [arXiv:hep-th/9803052]
62. G.W. Gibbons, N. Turok, The measure problem in cosmology. Phys. Rev. D **77**, 063516 (2008). [arXiv:hep-th/0609095]
63. A.D. Linde, Inflationary cosmology. Lect. Notes Phys. **738**, 1 (2008). [arXiv:0705.0164]
64. J.S. Schiffrin, R.M. Wald, Measure and probability in cosmology. Phys. Rev. D **86**, 023521 (2012). [arXiv:1202.1818]
65. A. Kaya, Comments on the canonical measure in cosmology. Phys. Lett. B **713**, 1 (2012). [arXiv:1203.2807]
66. A. Higuchi, Quantum linearization instabilities of de Sitter space-time. II. Class. Quantum Grav. **8**, 1983 (1991)
67. A. Higuchi, Linearized quantum gravity in flat space with toroidal topology. Class. Quantum Grav. **8**, 2023 (1991)
68. N.P. Landsman, Rieffel induction as generalized quantum Marsden–Weinstein reduction. J. Geom. Phys. **15**, 285 (1995). [arXiv:hep-th/9305088]
69. D. Marolf, The spectral analysis inner product for quantum gravity, in *Proceedings of the Seventh Marcel Grossman Meeting on General Relativity*, ed. by R. Ruffini, M. Keiser (World Scientific, Singapore, 1994). [arXiv:gr-qc/9409036]
70. A. Ashtekar, J. Lewandowski, D. Marolf, J. Mourão, T. Thiemann, Quantization of diffeomorphism invariant theories of connections with local degrees of freedom. J. Math. Phys. **36**, 6456 (1995). [arXiv:gr-qc/9504018]
71. D. Marolf, Refined algebraic quantization: systems with a single constraint. arXiv:gr-qc/9508015
72. J.B. Hartle, D. Marolf, Comparing formulations of generalized quantum mechanics for reparametrization-invariant systems. Phys. Rev. D **56**, 6247 (1997). [arXiv:gr-qc/9703021]
73. A. Ashtekar, L. Bombelli, A. Corichi, Semiclassical states for constrained systems. Phys. Rev. D **72**, 025008 (2005). [arXiv:gr-qc/0504052]
74. A. Ashtekar, T. Pawłowski, P. Singh, Quantum nature of the big bang: an analytical and numerical investigation. I. Phys. Rev. D **73**, 124038 (2006). [arXiv:gr-qc/0604013]
75. J.J. Halliwell, M.E. Ortiz, Sum-over-histories origin of the composition laws of relativistic quantum mechanics and quantum cosmology. Phys. Rev. D **48**, 748 (1993). [arXiv:gr-qc/9211004]
76. J.J. Halliwell, J. Thorwart, Decoherent histories analysis of the relativistic particle. Phys. Rev. D **64**, 124018 (2001). [arXiv:gr-qc/0106095]
77. G. Calcagni, S. Gielen, D. Oriti, Two-point functions in (loop) quantum cosmology. Class. Quantum Grav. **28**, 125014 (2011). [arXiv:1011.4290]
78. A. Ashtekar, A. Corichi, P. Singh, Robustness of key features of loop quantum cosmology. Phys. Rev. D **77**, 024046 (2008). [arXiv:0710.3565]
79. A. Kamenshchik, C. Kiefer, B. Sandhöfer, Quantum cosmology with big-brake singularity. Phys. Rev. D **76**, 064032 (2007). [arXiv:0705.1688]
80. S.W. Hawking, J.C. Luttrell, The isotropy of the universe. Phys. Lett. B **143**, 83 (1984)
81. W.A. Wright, I.G. Moss, The anisotropy of the universe. Phys. Lett. B **154**, 115 (1985)
82. P. Amsterdamski, Wave function of an anisotropic universe. Phys. Rev. D **31**, 3073 (1985)
83. T. Furusawa, Quantum chaos of mixmaster universe. Prog. Theor. Phys. **75**, 59 (1986)
84. T. Furusawa, Quantum chaos of mixmaster universe. II. Prog. Theor. Phys. **76**, 67 (1986)
85. B.K. Berger, Quantum chaos in the mixmaster universe. Phys. Rev. D **39**, 2426 (1989)

86. R. Graham, P. Szépfalusy, Quantum creation of a generic universe. Phys. Rev. D **42**, 2483 (1990)
87. B.K. Berger, Application of Monte Carlo simulation methods to quantum cosmology. Phys. Rev. D **48**, 513 (1993)
88. R. Graham, Chaos and quantum chaos in cosmological models. Chaos Solitons Fractals **5**, 1103 (1995). [arXiv:gr-qc/9403030]
89. R. Benini, G. Montani, Inhomogeneous quantum mixmaster: from classical toward quantum mechanics. Class. Quantum Grav. **24**, 387 (2007). [arXiv:gr-qc/0612095]
90. E. Calzetta, Chaos, decoherence and quantum cosmology. Class. Quantum Grav. **29**, 143001 (2012). [arXiv:1205.1841]
91. G. Montani, M.V. Battisti, R. Benini, G. Imponente, Classical and quantum features of the mixmaster singularity. Int. J. Mod. Phys. A **23**, 2353 (2008). [arXiv:0712.3008]
92. A. Csordás, R. Graham, P. Szépfalusy, Level statistics of a noncompact cosmological billiard. Phys. Rev. A **44**, 1491 (1991)
93. R. Graham, R. Hübner, P. Szépfalusy, G. Vattay, Level statistics of a noncompact integrable billiard. Phys. Rev. A **44**, 7002 (1991)
94. A. Peres, Ergodicity and mixing in quantum theory. I. Phys. Rev. A **30**, 504 (1984)
95. M. Feingold, N. Moiseyev, A. Peres, Ergodicity and mixing in quantum theory. II. Phys. Rev. A **30**, 509 (1984)
96. S. Weinberg, The cosmological constant problem. Rev. Mod. Phys. **61**, 1 (1989)
97. M.J. Duff, P. van Nieuwenhuizen, Quantum inequivalence of different field representations. Phys. Lett. B **94**, 179 (1980)
98. A. Aurilia, H. Nicolai, P.K. Townsend, Hidden constants: the θ parameter of QCD and the cosmological constant of $N = 8$ supergravity. Nucl. Phys. B **176**, 509 (1980)
99. M. Henneaux, C. Teitelboim, The cosmological constant as a canonical variable. Phys. Lett. B **143**, 415 (1984)
100. E. Baum, Zero cosmological constant from minimum action. Phys. Lett. B **133**, 185 (1983)
101. S.W. Hawking, The cosmological constant is probably zero. Phys. Lett. B **134**, 403 (1984)
102. M.J. Duff, The cosmological constant is possibly zero, but the proof is probably wrong. Phys. Lett. B **226**, 36 (1989)
103. M.J. Duncan, L.G. Jensen, Four-forms and the vanishing of the cosmological constant. Nucl. Phys. B **336**, 100 (1990)
104. Z.C. Wu, The cosmological constant is probably zero, and a proof is possibly right. Phys. Lett. B **659**, 891 (2008). [arXiv:0709.3314]
105. T. Banks, Prolegomena to a theory of bifurcating universes: a nonlocal solution to the cosmological constant problem or little lambda goes back to the future. Nucl. Phys. B **309**, 493 (1988)
106. S.R. Coleman, Why there is nothing rather than something: a theory of the cosmological constant. Nucl. Phys. B **310**, 643 (1988)
107. S.B. Giddings, A. Strominger, Baby universes, third quantization and the cosmological constant. Nucl. Phys. B **321**, 481 (1989)
108. M. McGuigan, Third quantization and the Wheeler–DeWitt equation. Phys. Rev. D **38**, 3031 (1988)
109. A. Hosoya, M. Morikawa, Quantum field theory of the Universe. Phys. Rev. D **39**, 1123 (1989)
110. V.A. Rubakov, P.G. Tinyakov, Gravitational instantons and creation of expanding universes. Phys. Lett. B **214**, 334 (1988)
111. V.A. Rubakov, On third quantization and the cosmological constant. Phys. Lett. B **214**, 503 (1988)
112. I.R. Klebanov, L. Susskind, T. Banks, Wormholes and the cosmological constant. Nucl. Phys. B **317**, 665 (1989)
113. W. Fischler, L. Susskind, A wormhole catastrophe. Phys. Lett. B **217**, 48 (1989)
114. M. McGuigan, Universe creation from the third-quantized vacuum. Phys. Rev. D **39**, 2229 (1989)

115. J. Preskill, Wormholes in spacetime and the constants of nature. Nucl. Phys. B **323**, 141 (1989)
116. W. Fischler, I.R. Klebanov, J. Polchinski, L. Susskind, Quantum mechanics of the googolplexus. Nucl. Phys. B **327**, 157 (1989)
117. Y.-M. Xiang, L. Liu, Third quantization of a solvable model in quantum cosmology in Brans–Dicke theory. Chin. Phys. Lett. **8**, 52 (1991)
118. H.J. Pohle, Coherent states and Heisenberg uncertainty relation in a third-quantized minisuperspace. Phys. Lett. B **261**, 257 (1991)
119. K. Kuchař, Time and interpretations of quantum gravity, in *Proceedings of the Fourth Canadian Conference on General Relativity and Relativistic Astrophysics*, ed. by G. Kunstatter et al. (World Scientific, Singapore, 1992) [Int. J. Mod. Phys. Proc. Suppl. D **20**, 3 (2011)]
120. C. Isham, Canonical quantum gravity and the problem of time, in *Integrable Systems, Quantum Groups, and Quantum Field Theories*, ed. by L.A. Ibort, M.A. Rodríguez (Kluwer, Dordrecht, 1993). [arXiv:gr-qc/9210011]
121. S. Abe, Fluctuations around the Wheeler–DeWitt trajectories in third-quantized cosmology. Phys. Rev. D **47**, 718 (1993)
122. T. Horiguchi, Uncertainty relation in a third-quantized universe. Phys. Rev. D **48**, 5764 (1993)
123. M.A. Castagnino, A. Gangui, F.D. Mazzitelli, I.I. Tkachev, Third quantization, decoherence and the interpretation of quantum gravity in minisuperspace. Class. Quantum Grav. **10**, 2495 (1993)
124. L.O. Pimentel, C. Mora, Third quantization of Brans–Dicke cosmology. Phys. Lett. A **280**, 191 (2001). [arXiv:gr-qc/0009026]
125. S. Gielen, D. Oriti, Discrete and continuum third quantization of gravity, in *Quantum Field Theory and Gravity*, ed. by F. Finster et al. (Springer, Basel, 2012). [arXiv:1102.2226]
126. G. Calcagni, S. Gielen, D. Oriti, Group field cosmology: a cosmological field theory of quantum geometry. Class. Quantum Grav. **29**, 105005 (2012). [arXiv:1201.4151]
127. Y. Ohkuwa, Y. Ezawa, Third quantization of $f(R)$-type gravity. Class. Quantum Grav. **29**, 215004 (2012). [arXiv:1203.1361]
128. M. Faizal, Super-group field cosmology. Class. Quantum Grav. **29**, 215009 (2012). [arXiv:1209.2346]
129. Y. Ohkuwa, Y. Ezawa, Third quantization of $f(R)$-type gravity II: general $f(R)$ case. Class. Quantum Grav. **30**, 235015 (2013). [arXiv:1210.4719]
130. M. Faizal, Absence of black holes information paradox in group field cosmology. Int. J. Geom. Methods Mod. Phys. **11**, 1450010 (2014). [arXiv:1301.0224]
131. S. Gielen, D. Oriti, L. Sindoni, Cosmology from group field theory formalism for quantum gravity. Phys. Rev. Lett. **111**, 031301 (2013). [arXiv:1303.3576]
132. A. Hosoya, W. Ogura, Wormhole instanton solution in the Einstein–Yang–Mills system. Phys. Lett. B **225**, 117 (1989)
133. F.S. Accetta, A. Chodos, B. Shao, Wormholes and baby universes in scalar-tensor gravity. Nucl. Phys. B **333**, 221 (1990)
134. L.J. Garay, J. García-Bellido, Jordan–Brans–Dicke quantum wormholes and Coleman's mechanism. Nucl. Phys. B **400**, 416 (1993). [arXiv:gr-qc/9209015]
135. J. García-Bellido, A.D. Linde, D.A. Linde, Fluctuations of the gravitational constant in the inflationary Brans–Dicke cosmology. Phys. Rev. D **50**, 730 (1994). [arXiv:astro-ph/9312039]
136. A. Vilenkin, Predictions from quantum cosmology. Phys. Rev. Lett. **74**, 846 (1995). [arXiv:gr-qc/9406010]
137. J. García-Bellido, A.D. Linde, Stationarity of inflation and predictions of quantum cosmology. Phys. Rev. D **51**, 429 (1995). [arXiv:hep-th/9408023]
138. A. Vilenkin, Making predictions in eternally inflating universe. Phys. Rev. D **52**, 3365 (1995). [arXiv:gr-qc/9505031]
139. H. Martel, P.R. Shapiro, S. Weinberg, Likely values of the cosmological constant. Astrophys. J. **492**, 29 (1998). [arXiv:astro-ph/9701099]
140. J. Garriga, A. Vilenkin, On likely values of the cosmological constant. Phys. Rev. D **61**, 083502 (2000). [arXiv:astro-ph/9908115]

141. S. Weinberg, A priori probability distribution of the cosmological constant. Phys. Rev. D **61**, 103505 (2000). [arXiv:astro-ph/0002387]
142. S. Weinberg, The cosmological constant problems. arXiv:astro-ph/0005265
143. M.L. Graesser, S.D.H. Hsu, A. Jenkins, M.B. Wise, Anthropic distribution for cosmological constant and primordial density perturbations. Phys. Lett. B **600**, 15 (2004). [arXiv:hep-th/0407174]
144. B. Feldstein, L.J. Hall, T. Watari, Density perturbations and the cosmological constant from inflationary landscapes. Phys. Rev. D **72**, 123506 (2005). [arXiv:hep-th/0506235]
145. J. Garriga, A. Vilenkin, Anthropic prediction for Λ and the Q catastrophe. Prog. Theor. Phys. Suppl. **163**, 245 (2006). [arXiv:hep-th/0508005]
146. C. Kiefer, F. Queisser, A.A. Starobinsky, Cosmological constant from decoherence. Class. Quantum Grav. **28**, 125022 (2011). [arXiv:1010.5331]
147. J.J. Halliwell, S.W. Hawking, Origin of structure in the Universe. Phys. Rev. D **31**, 1777 (1985)
148. S. Wada, Quantum cosmological perturbations in pure gravity. Nucl. Phys. B **276**, 729 (1986); Erratum-ibid. B **284**, 747 (1987)
149. T. Vachaspati, A. Vilenkin, Uniqueness of the tunneling wave function of the Universe. Phys. Rev. D **37**, 898 (1988)
150. C. Kiefer, *Quantum Gravity* (Oxford University Press, Oxford, 2012)
151. C. Kiefer, T.P. Singh, Quantum gravitational corrections to the functional Schrödinger equation. Phys. Rev. D **44**, 1067 (1991)
152. C. Kiefer, M. Krämer, Quantum gravitational contributions to the CMB anisotropy spectrum. Phys. Rev. Lett. **108**, 021301 (2012). [arXiv:1103.4967]
153. C. Kiefer, M. Krämer, Can effects of quantum gravity be observed in the cosmic microwave background? Int. J. Mod. Phys. D **21**, 1241001 (2012). [arXiv:1205.5161]
154. D. Bini, G. Esposito, C. Kiefer, M. Krämer, F. Pessina, On the modification of the cosmic microwave background anisotropy spectrum from canonical quantum gravity. Phys. Rev. D **87**, 104008 (2013). [arXiv:1303.0531]
155. Y.-S. Piao, B. Feng, X. Zhang, Suppressing CMB quadrupole with a bounce from contracting phase to inflation. Phys. Rev. D **69**, 103520 (2004). [arXiv:hep-th/0310206]
156. Z.-G. Liu, Z.-K. Guo, Y.-S. Piao, Obtaining the CMB anomalies with a bounce from the contracting phase to inflation. Phys. Rev. D **88**, 063539 (2013). [arXiv:1304.6527]
157. G. Calcagni, S. Tsujikawa, Observational constraints on patch inflation in noncommutative spacetime. Phys. Rev. D **70**, 103514 (2004). [arXiv:astro-ph/0407543]
158. S. Tsujikawa, R. Maartens, R. Brandenberger, Non-commutative inflation and the CMB. Phys. Lett. B **574**, 141 (2003). [arXiv:astro-ph/0308169]
159. Y.-S. Piao, S. Tsujikawa, X. Zhang, Inflation in string inspired cosmology and suppression of CMB low multipoles. Class. Quantum Grav. **21**, 4455 (2004). [arXiv:hep-th/0312139]
160. G. Calcagni, Observational effects from quantum cosmology. Ann. Phys. (Berlin) **525**, 323 (2013); Erratum-ibid. **525**, A165 (2013). [arXiv:1209.0473]
161. M. Bojowald, Loop quantum cosmology I: kinematics. Class. Quantum Grav. **17**, 1489 (2000). [arXiv:gr-qc/9910103]
162. M. Bojowald, Loop quantum cosmology. II: volume operators. Class. Quantum Grav. **17**, 1509 (2000). [arXiv:gr-qc/9910104]
163. M. Bojowald, Loop quantum cosmology III: Wheeler–DeWitt operators. Class. Quantum Grav. **18**, 1055 (2001). [arXiv:gr-qc/0008052]
164. M. Bojowald, Loop quantum cosmology IV: discrete time evolution. Class. Quantum Grav. **18**, 1071 (2001). [arXiv:gr-qc/0008053]
165. M. Bojowald, Inverse scale factor in isotropic quantum geometry. Phys. Rev. D **64**, 084018 (2001). [arXiv:gr-qc/0105067]
166. A. Ashtekar, M. Bojowald, J. Lewandowski, Mathematical structure of loop quantum cosmology. Adv. Theor. Math. Phys. **7**, 233 (2003). [arXiv:gr-qc/0304074]
167. A. Ashtekar, T. Pawłowski, P. Singh, Quantum nature of the big bang: improved dynamics. Phys. Rev. D **74**, 084003 (2006). [arXiv:gr-qc/0607039]

168. X. Zhang, Y. Ma, Extension of loop quantum gravity to $f(R)$ theories. Phys. Rev. Lett. **106**, 171301 (2011). [arXiv:1101.1752]
169. X. Zhang, Y. Ma, Loop quantum $f(R)$ theories. Phys. Rev. D **84**, 064040 (2011). [arXiv:1107.4921]
170. X. Zhang, Y. Ma, Nonperturbative loop quantization of scalar-tensor theories of gravity. Phys. Rev. D **84**, 104045 (2011). [arXiv:1107.5157]
171. M. Bojowald, Loop quantum cosmology. Living Rev. Relat. **11**, 4 (2008)
172. M. Bojowald, Quantization ambiguities in isotropic quantum geometry. Class. Quantum Grav. **19**, 5113 (2002). [arXiv:gr-qc/0206053]
173. M. Bojowald, Inflation from quantum geometry. Phys. Rev. Lett. **89**, 261301 (2002). [arXiv:gr-qc/0206054]
174. T. Thiemann, Quantum spin dynamics (QSD): V. Quantum gravity as the natural regulator of the Hamiltonian constraint of matter quantum field theories. Class. Quantum Grav. **15**, 1281 (1998). [arXiv:gr-qc/9705019]
175. M. Bojowald, J.E. Lidsey, D.J. Mulryne, P. Singh, R. Tavakol, Inflationary cosmology and quantization ambiguities in semiclassical loop quantum gravity. Phys. Rev. D **70**, 043530 (2004). [arXiv:gr-qc/0403106]
176. A. Ashtekar, E. Wilson-Ewing, Loop quantum cosmology of Bianchi I models. Phys. Rev. D **79**, 083535 (2009). [arXiv:0903.3397]
177. M. Bojowald, R. Das, R.J. Scherrer, Dirac fields in loop quantum gravity and big bang nucleosynthesis. Phys. Rev. D **77**, 084003 (2008). [arXiv:0710.5734]
178. A. Ashtekar, T. Pawłowski, P. Singh, K. Vandersloot, Loop quantum cosmology of $k = 1$ FRW models. Phys. Rev. D **75**, 024035 (2007). [arXiv:gr-qc/0612104]
179. M. Bojowald, Isotropic loop quantum cosmology. Class. Quantum Grav. **19**, 2717 (2002). [arXiv:gr-qc/0202077]
180. W. Nelson, M. Sakellariadou, Lattice refining LQC and the matter Hamiltonian. Phys. Rev. D **76**, 104003 (2007). [arXiv:0707.0588]
181. K. Vandersloot, Loop quantum cosmology and the $k = -1$ Robertson–Walker model. Phys. Rev. D **75**, 023523 (2007). [arXiv:gr-qc/0612070]
182. M. Bojowald, P. Singh, A. Skirzewski, Coordinate time dependence in quantum gravity. Phys. Rev. D **70**, 124022 (2004). [arXiv:gr-qc/0408094]
183. W. Kamiński, J. Lewandowski, The flat FRW model in LQC: the self-adjointness. Class. Quantum Grav. **25**, 035001 (2008). [arXiv:0709.3120]
184. W. Kamiński, T. Pawłowski, The LQC evolution operator of FRW universe with positive cosmological constant. Phys. Rev. D **81**, 024014 (2010). [arXiv:0912.0162]
185. K. Vandersloot, Hamiltonian constraint of loop quantum cosmology. Phys. Rev. D **71**, 103506 (2005). [arXiv:gr-qc/0502082]
186. M. Bojowald, Absence of singularity in loop quantum cosmology. Phys. Rev. Lett. **86**, 5227 (2001). [arXiv:gr-qc/0102069]
187. M. Bojowald, K. Vandersloot, Loop quantum cosmology, boundary proposals, and inflation. Phys. Rev. D **67**, 124023 (2003). [arXiv:gr-qc/0303072]
188. P. Singh, A. Toporensky, Big crunch avoidance in $k = 1$ semiclassical loop quantum cosmology. Phys. Rev. D **69**, 104008 (2004). [arXiv:gr-qc/0312110]
189. J.E. Lidsey, D.J. Mulryne, N.J. Nunes, R. Tavakol, Oscillatory universes in loop quantum cosmology and initial conditions for inflation. Phys. Rev. D **70**, 063521 (2004). [arXiv:gr-qc/0406042]
190. Ł. Szulc, W. Kamiński, J. Lewandowski, Closed FRW model in loop quantum cosmology. Class. Quantum Grav. **24**, 2621 (2007). [arXiv:gr-qc/0612101]
191. M. Bojowald, R. Tavakol, Recollapsing quantum cosmologies and the question of entropy. Phys. Rev. D **78**, 023515 (2008). [arXiv:0803.4484]
192. J. Mielczarek , O. Hrycyna, M. Szydłowski, Effective dynamics of the closed loop quantum cosmology. JCAP **0911**, 014 (2009). [arXiv:0906.2503]
193. P. Singh, F. Vidotto, Exotic singularities and spatially curved loop quantum cosmology. Phys. Rev. D **83**, 064027 (2011). [arXiv:1012.1307]

194. Ł. Szulc, Open FRW model in loop quantum cosmology. Class. Quantum Grav. **24**, 6191 (2007). [arXiv:0707.1816]
195. E. Bentivegna, T. Pawłowski, Anti-de Sitter universe dynamics in LQC. Phys. Rev. D **77**, 124025 (2008). [arXiv:0803.4446]
196. P. Singh, K. Vandersloot, Semiclassical states, effective dynamics and classical emergence in loop quantum cosmology. Phys. Rev. D **72**, 084004 (2005). [arXiv:gr-qc/0507029]
197. A. Corichi, E. Montoya, Coherent semiclassical states for loop quantum cosmology. Phys. Rev. D **84**, 044021 (2011). [arXiv:1105.5081]
198. M. Bojowald, Large scale effective theory for cosmological bounces. Phys. Rev. D **75**, 081301(R) (2007). [arXiv:gr-qc/0608100]
199. M. Bojowald, H.H. Hernández, A. Skirzewski, Effective equations for isotropic quantum cosmology including matter. Phys. Rev. D **76**, 063511 (2007). [arXiv:0706.1057]
200. M. Bojowald, Quantum nature of cosmological bounces. Gen. Relat. Grav. **40**, 2659 (2008). [arXiv:0801.4001]
201. P. Singh, Loop cosmological dynamics and dualities with Randall–Sundrum braneworlds. Phys. Rev. D **73**, 063508 (2006). [arXiv:gr-qc/0603043]
202. P. Singh, K. Vandersloot, G.V. Vereshchagin, Nonsingular bouncing universes in loop quantum cosmology. Phys. Rev. D **74**, 043510 (2006). [arXiv:gr-qc/0606032]
203. C. Rovelli, E. Wilson-Ewing, Why are the effective equations of loop quantum cosmology so accurate? Phys. Rev. D **90**, 023538 (2014). [arXiv:1310.8654]
204. M. Bojowald, Loop quantum cosmology: recent progress. Pramana **63**, 765 (2004). [arXiv:gr-qc/0402053]
205. M. Bojowald, H.H. Hernández, M. Kagan, A. Skirzewski, Effective constraints of loop quantum gravity. Phys. Rev. D **75**, 064022 (2007). [arXiv:gr-qc/0611112]
206. M. Bojowald, Loop quantum cosmology and inhomogeneities. Gen. Relat. Grav. **38**, 1771 (2006). [arXiv:gr-qc/0609034]
207. E. Wilson-Ewing, Holonomy corrections in the effective equations for scalar mode perturbations in loop quantum cosmology. Class. Quantum Grav. **29**, 085005 (2012). [arXiv:1108.6265]
208. C. Germani, W. Nelson, M. Sakellariadou, On the onset of inflation in loop quantum cosmology. Phys. Rev. D **76**, 043529 (2007). [arXiv:gr-qc/0701172]
209. A. Ashtekar, D. Sloan, Probability of inflation in loop quantum cosmology. Gen. Relat. Grav. **43**, 3619 (2011). [arXiv:1103.2475]
210. D.-W. Chiou, Loop quantum cosmology in Bianchi type I models: analytical investigation. Phys. Rev. D **75**, 024029 (2007). [arXiv:gr-qc/0609029]
211. D.-W. Chiou, K. Vandersloot, Behavior of nonlinear anisotropies in bouncing Bianchi I models of loop quantum cosmology. Phys. Rev. D **76**, 084015 (2007). [arXiv:0707.2548]
212. M. Martín-Benito, L.J. Garay, G.A. Mena Marugán, Hybrid quantum Gowdy cosmology: combining loop and Fock quantizations. Phys. Rev. D **78**, 083516 (2008). [arXiv:0804.1098]
213. G.A. Mena Marugán, M. Martín-Benito, Hybrid quantum cosmology: combining loop and Fock quantizations. Int. J. Mod. Phys. A **24**, 2820 (2009). [arXiv:0907.3797]
214. M. Martín-Benito, G.A. Mena Marugán, E. Wilson-Ewing, Hybrid quantization: from Bianchi I to the Gowdy model. Phys. Rev. D **82**, 084012 (2010). [arXiv:1006.2369]
215. R.H. Gowdy, Vacuum space-times with two parameter spacelike isometry groups and compact invariant hypersurfaces: topologies and boundary conditions. Ann. Phys. (N.Y.) **83**, 203 (1974)
216. A. Ashtekar, E. Wilson-Ewing, Loop quantum cosmology of Bianchi type II models. Phys. Rev. D **80**, 123532 (2009). [arXiv:0910.1278]
217. E. Wilson-Ewing, Loop quantum cosmology of Bianchi type IX models. Phys. Rev. D **82**, 043508 (2010). [arXiv:1005.5565]
218. P. Singh, Are loop quantum cosmos never singular? Class. Quantum Grav. **26**, 125005 (2009). [arXiv:0901.2750]
219. M. Bojowald, G. Date, Quantum suppression of the generic chaotic behavior close to cosmological singularities. Phys. Rev. Lett. **92**, 071302 (2004). [arXiv:gr-qc/0311003]

220. M. Bojowald, G. Date, K. Vandersloot, Homogeneous loop quantum cosmology: the role of the spin connection. Class. Quantum Grav. **21**, 1253 (2004). [arXiv:gr-qc/0311004]
221. M. Bojowald, G. Date, G.M. Hossain, The Bianchi IX model in loop quantum cosmology. Class. Quantum Grav. **21**, 3541 (2004). [arXiv:gr-qc/0404039]
222. E. Alesci, F. Cianfrani, A new perspective on cosmology in loop quantum gravity. Europhys. Lett. **104**, 10001 (2013). [arXiv:1210.4504]
223. E. Alesci, F. Cianfrani, Quantum-reduced loop gravity: cosmology. Phys. Rev. D **87**, 083521 (2013). [arXiv:1301.2245]
224. E. Alesci, F. Cianfrani, C. Rovelli, Quantum-reduced loop gravity: relation with the full theory. Phys. Rev. D **88**, 104001 (2013). [arXiv:1309.6304]
225. E. Alesci, F. Cianfrani, Quantum reduced loop gravity: semiclassical limit. Phys. Rev. D **90**, 024006 (2014). [arXiv:1402.3155]
226. M. Bojowald, G.M. Hossain, Loop quantum gravity corrections to gravitational wave dispersion. Phys. Rev. D **77**, 023508 (2008). [arXiv:0709.2365]
227. T. Thiemann, Quantum spin dynamics (QSD). Class. Quantum Grav. **15**, 839 (1998). [arXiv:gr-qc/9606089]
228. C. Rovelli, L. Smolin, The physical Hamiltonian in nonperturbative quantum gravity. Phys. Rev. Lett. **72**, 446 (1994). [arXiv:gr-qc/9308002]
229. M. Bojowald, G. Calcagni, S. Tsujikawa, Observational test of inflation in loop quantum cosmology. JCAP **1111**, 046 (2011). [arXiv:1107.1540]
230. M. Bojowald, Quantum geometry and its implications for black holes. Int. J. Mod. Phys. D **15**, 1545 (2006). [arXiv:gr-qc/0607130]
231. A. Ashtekar, Quantum space-times. Fund. Theories Phys. **165**, 163 (2010). [arXiv:0810.0514]
232. M. Bojowald, W. Nelson, D. Mulryne, R. Tavakol, The high-density regime of kinetic-dominated loop quantum cosmology. Phys. Rev. D **82**, 124055 (2010). [arXiv:1004.3979]
233. M. Bojowald, Consistent loop quantum cosmology. Class. Quantum Grav. **26**, 075020 (2009). [arXiv:0811.4129]
234. M. Bojowald, D. Cartin, G. Khanna, Lattice refining loop quantum cosmology, anisotropic models and stability. Phys. Rev. D **76**, 064018 (2007). [arXiv:0704.1137]
235. M. Bojowald, G.M. Hossain, Cosmological vector modes and quantum gravity effects. Class. Quantum Grav. **24**, 4801 (2007). [arXiv:0709.0872]
236. M. Bojowald, G.M. Hossain, M. Kagan, S. Shankaranarayanan, Anomaly freedom in perturbative loop quantum gravity. Phys. Rev. D **78**, 063547 (2008). [arXiv:0806.3929]
237. A. Barrau, M. Bojowald, G. Calcagni, J. Grain, M. Kagan, Anomaly-free cosmological perturbations in effective canonical quantum gravity. JCAP **1505**, 051 (2015). [arXiv:1404.1018]
238. M. Fernández-Méndez, G.A. Mena Marugán, J. Olmedo, Hybrid quantization of an inflationary universe. Phys. Rev. D **86**, 024003 (2012). [arXiv:1205.1917]
239. M. Fernández-Méndez, G.A. Mena Marugán, J. Olmedo, Hybrid quantization of an inflationary model. The flat case. Phys. Rev. D **88**, 044013 (2013). [arXiv:1307.5222]
240. L. Castelló Gomar, M. Fernández-Méndez, G.A. Mena Marugán, J. Olmedo, Cosmological perturbations in hybrid loop quantum cosmology: Mukhanov–Sasaki variables. Phys. Rev. D **90**, 064015 (2014). [arXiv:1407.0998]
241. L. Castelló Gomar, M. Martín-Benito, G.A. Mena Marugán, Gauge-invariant perturbations in hybrid quantum cosmology. JCAP **1506**, 045 (2015). [arXiv:1503.03907]
242. D. Martín de Blas, J. Olmedo, Primordial power spectra for scalar perturbations in loop quantum cosmology. JCAP **1606**, 029 (2016). [arXiv:1601.01716]
243. L. Castelló Gomar, M. Martín-Benito, G.A. Mena Marugán, Quantum corrections to the Mukhanov–Sasaki equations. Phys. Rev. D **93**, 104025 (2016). [arXiv:1603.08448]
244. L.J. Garay, M. Martín-Benito, G.A. Mena Marugán, Inhomogeneous loop quantum cosmology: hybrid quantization of the Gowdy model. Phys. Rev. D **82**, 044048 (2010). [arXiv:1005.5654]
245. D. Brizuela, G.A. Mena Marugán, T. Pawłowski, Big bounce and inhomogeneities. Class. Quantum Grav. **27**, 052001 (2010). [arXiv:0902.0697]

246. M. Martín-Benito, D. Martín de Blas, G.A. Mena Marugán, Matter in inhomogeneous loop quantum cosmology: the Gowdy T^3 model. Phys. Rev. D **83**, 084050 (2011). [arXiv:1012.2324]
247. M. Martín-Benito, D. Martín de Blas, G.A. Mena Marugán, Approximation methods in loop quantum cosmology: from Gowdy cosmologies to inhomogeneous models in Friedmann–Robertson–Walker geometries. Class. Quantum Grav. **31**, 075022 (2014). [arXiv:1307.1420]
248. I. Agulló, A. Ashtekar, W. Nelson, Quantum gravity extension of the inflationary scenario. Phys. Rev. Lett. **109**, 251301 (2012). [arXiv:1209.1609]
249. I. Agulló, A. Ashtekar, W. Nelson, Extension of the quantum theory of cosmological perturbations to the Planck era. Phys. Rev. D **87**, 043507 (2013). [arXiv:1211.1354]
250. I. Agulló, A. Ashtekar, W. Nelson, The pre-inflationary dynamics of loop quantum cosmology: confronting quantum gravity with observations. Class. Quantum Grav. **30**, 085014 (2013). [arXiv:1302.0254]
251. I. Agulló, N.A. Morris, Detailed analysis of the predictions of loop quantum cosmology for the primordial power spectra. Phys. Rev. D **92**, 124040 (2015). [arXiv:1509.05693]
252. D. Langlois, Hamiltonian formalism and gauge invariance for linear perturbations in inflation. Class. Quantum Grav. **11**, 389 (1994)
253. E. Wilson-Ewing, Lattice loop quantum cosmology: scalar perturbations. Class. Quantum Grav. **29**, 215013 (2012). [arXiv:1205.3370]
254. E. Wilson-Ewing, Separate universes in loop quantum cosmology: framework and applications. Int. J. Mod. Phys. D **25**, 1642002 (2016). [arXiv:1512.05743]
255. T. Cailleteau, J. Mielczarek, A. Barrau, J. Grain, Anomaly-free scalar perturbations with holonomy corrections in loop quantum cosmology. Class. Quantum Grav. **29**, 095010 (2012). [arXiv:1111.3535]
256. M. Bojowald, A. Skirzewski, Effective equations of motion for quantum systems. Rev. Math. Phys. **18**, 713 (2006). [arXiv:math-ph/0511043]
257. M. Bojowald, A. Skirzewski, Quantum gravity and higher curvature actions. Int. J. Geom. Methods Mod. Phys. **4**, 25 (2007). [arXiv:hep-th/0606232]
258. M. Bojowald, S. Brahma, E. Nelson, Higher time derivatives in effective equations of canonical quantum systems. Phys. Rev. D **86**, 105004 (2012). [arXiv:1208.1242]
259. M. Bojowald, B. Sandhöfer, A. Skirzewski, A. Tsobanjan, Effective constraints for quantum systems. Rev. Math. Phys. **21**, 111 (2009). [arXiv:0804.3365]
260. M. Bojowald, A. Tsobanjan, Effective constraints for relativistic quantum systems. Phys. Rev. D **80**, 125008 (2009). [arXiv:0906.1772]
261. S. Tsujikawa, P. Singh, R. Maartens, Loop quantum gravity effects on inflation and the CMB. Class. Quantum Grav. **21**, 5767 (2004). [arXiv:astro-ph/0311015]
262. G.M. Hossain, Primordial density perturbation in effective loop quantum cosmology. Class. Quantum Grav. **22**, 2511 (2005). [arXiv:gr-qc/0411012]
263. G. Calcagni, M. Cortês, Inflationary scalar spectrum in loop quantum cosmology. Class. Quantum Grav. **24**, 829 (2007). [arXiv:gr-qc/0607059]
264. E.J. Copeland, D.J. Mulryne, N.J. Nunes, M. Shaeri, Super-inflation in loop quantum cosmology. Phys. Rev. D **77**, 023510 (2008). [arXiv:0708.1261]
265. M. Artymowski, Z. Lalak, Ł. Szulc, Loop quantum cosmology: holonomy corrections to inflationary models. JCAP **0901**, 004 (2009). [arXiv:0807.0160]
266. M. Shimano, T. Harada, Observational constraints on a power spectrum from super-inflation in loop quantum cosmology. Phys. Rev. D **80**, 063538 (2009). [arXiv:0909.0334]
267. T. Cailleteau, A. Barrau, Gauge invariance in loop quantum cosmology: Hamilton–Jacobi and Mukhanov-Sasaki equations for scalar perturbations. Phys. Rev. D **85**, 123534 (2012). [arXiv:1111.7192]
268. T. Cailleteau, A. Barrau, J. Grain, F. Vidotto, Consistency of holonomy-corrected scalar, vector and tensor perturbations in loop quantum cosmology. Phys. Rev. D **86**, 087301 (2012). [arXiv:1206.6736]
269. J. Mielczarek, T. Cailleteau, A. Barrau, J. Grain, Anomaly-free vector perturbations with holonomy corrections in loop quantum cosmology. Class. Quantum Grav. **29**, 085009 (2012). [arXiv:1106.3744]

270. M. Bojowald, G.M. Hossain, M. Kagan, S. Shankaranarayanan, Gauge invariant cosmological perturbation equations with corrections from loop quantum gravity. Phys. Rev. D **79**, 043505 (2009); Erratum-ibid. D **82**, 109903(E) (2010). [arXiv:0811.1572]
271. T. Cailleteau, L. Linsefors, A. Barrau, Anomaly-free perturbations with inverse-volume and holonomy corrections in loop quantum cosmology. Class. Quantum Grav. **31**, 125011 (2014). [arXiv:1307.5238]
272. A. Barrau, J. Grain, Holonomy corrections to the cosmological primordial tensor power spectrum. arXiv:0805.0356
273. J. Mielczarek, Gravitational waves from the big bounce. JCAP **0811**, 011 (2008). [arXiv:0807.0712]
274. J. Grain, A. Barrau, Cosmological footprints of loop quantum gravity. Phys. Rev. Lett. **102**, 081301 (2009). [arXiv:0902.0145]
275. J. Mielczarek, Tensor power spectrum with holonomy corrections in loop quantum cosmology. Phys. Rev. D **79**, 123520 (2009). [arXiv:0902.2490]
276. J. Mielczarek, T. Cailleteau, J. Grain, A. Barrau, Inflation in loop quantum cosmology: dynamics and spectrum of gravitational waves. Phys. Rev. D **81**, 104049 (2010). [arXiv:1003.4660]
277. J. Grain, A. Barrau, T. Cailleteau, J. Mielczarek, Observing the big bounce with tensor modes in the cosmic microwave background: phenomenology and fundamental loop quantum cosmology parameters. Phys. Rev. D **82**, 123520 (2010). [arXiv:1011.1811]
278. L. Linsefors, T. Cailleteau, A. Barrau, J. Grain, Primordial tensor power spectrum in holonomy corrected Ω loop quantum cosmology. Phys. Rev. D **87**, 107503 (2013). [arXiv:1212.2852]
279. B. Bolliet, A. Barrau, J. Grain, S. Schander, Observational exclusion of a consistent quantum cosmology scenario. Phys. Rev. D **93**, 124011 (2016). [arXiv:1510.08766]
280. E.J. Copeland, D.J. Mulryne, N.J. Nunes, M. Shaeri, The gravitational wave background from super-inflation in loop quantum cosmology. Phys. Rev. D **79**, 023508 (2009). [arXiv:0810.0104]
281. G. Calcagni, G.M. Hossain, Loop quantum cosmology and tensor perturbations in the early universe. Adv. Sci. Lett. **2**, 184 (2009). [arXiv:0810.4330]
282. M. Bojowald, G. Calcagni, Inflationary observables in loop quantum cosmology. JCAP **1103**, 032 (2011). [arXiv:1011.2779]
283. M. Bojowald, G. Calcagni, S. Tsujikawa, Observational constraints on loop quantum cosmology. Phys. Rev. Lett. **107**, 211302 (2011). [arXiv:1101.5391]
284. J. Grain, T. Cailleteau, A. Barrau, A. Gorecki, Fully loop-quantum-cosmology-corrected propagation of gravitational waves during slow-roll inflation. Phys. Rev. D **81**, 024040 (2010). [arXiv:0910.2892]
285. J. Mielczarek, Inflationary power spectra with quantum holonomy corrections. JCAP **1403**, 048 (2014). [arXiv:1311.1344]
286. T. Zhu, A. Wang, K. Kirsten, G. Cleaver, Q. Sheng, Q. Wu, Inflationary spectra with inverse-volume corrections in loop quantum cosmology and their observational constraints from Planck 2015 data. JCAP **1603**, 046 (2016). [arXiv:1510.03855]
287. P. Creminelli, M. Zaldarriaga, Single field consistency relation for the 3-point function. JCAP **0410**, 006 (2004). [arXiv:astro-ph/0407059]
288. L.-F. Li, R.-G. Cai, Z.-K. Guo, B. Hu, Non-Gaussian features from the inverse volume corrections in loop quantum cosmology. Phys. Rev. D **86**, 044020 (2012). [arXiv:1112.2785]
289. B. Bolliet, J. Grain, C. Stahl, L. Linsefors, A. Barrau, Comparison of primordial tensor power spectra from the deformed algebra and dressed metric approaches in loop quantum cosmology. Phys. Rev. D **91**, 084035 (2015). [arXiv:1502.02431]
290. M. Bojowald, G.M. Paily, Deformed general relativity and effective actions from loop quantum gravity. Phys. Rev. D **86**, 104018 (2012). [arXiv:1112.1899]
291. M. Bojowald, G.M. Paily, A no-singularity scenario in loop quantum gravity. Class. Quantum Grav. **29**, 242002 (2012). [arXiv:1206.5765]

292. S.A. Hojman, K. Kuchař, C. Teitelboim, Geometrodynamics regained. Ann. Phys. (N.Y.) **96**, 88 (1976)
293. M. Bojowald, J.D. Reyes, R. Tibrewala, Nonmarginal Lemaitre–Tolman–Bondi-like models with inverse triad corrections from loop quantum gravity. Phys. Rev. D **80**, 084002 (2009). [arXiv:0906.4767]
294. J.D. Reyes, Spherically Symmetric Loop Quantum Gravity: Connections to 2-Dimensional Models and Applications to Gravitational Collapse. Ph.D. thesis, Pennsylvania State University, University Park (2009)
295. M. Bojowald, G.M. Paily, J.D. Reyes, Discreteness corrections and higher spatial derivatives in effective canonical quantum gravity. Phys. Rev. D **90**, 025025 (2014). [arXiv:1402.5130]
296. M. Bojowald, S. Brahma, Signature change in 2-dimensional black-hole models of loop quantum gravity. arXiv:1610.08850
297. A. Perez, D. Pranzetti, On the regularization of the constraints algebra of quantum gravity in $2+1$ dimensions with non-vanishing cosmological constant. Class. Quantum Grav. **27**, 145009 (2010). [arXiv:1001.3292]
298. A. Henderson, A. Laddha, C. Tomlin, Constraint algebra in loop quantum gravity reloaded. I. Toy model of a $U(1)^3$ gauge theory. Phys. Rev. D **88**, 044028 (2013). [arXiv:1204.0211]
299. G. Calcagni, B. de Carlos, A. De Felice, Ghost conditions for Gauss–Bonnet cosmologies. Nucl. Phys. B **752**, 404 (2006). [arXiv:hep-th/0604201]
300. J. Brunnemann, T. Thiemann, On (cosmological) singularity avoidance in loop quantum gravity. Class. Quantum Grav. **23**, 1395 (2006). [arXiv:gr-qc/0505032]
301. J. Brunnemann, T. Thiemann, Unboundedness of triad-like operators in loop quantum gravity. Class. Quantum Grav. **23**, 1429 (2006). [arXiv:gr-qc/0505033]
302. M. Bojowald, S. Brahma, U. Büyükçam, F. D'Ambrosio, Hypersurface-deformation algebroids and effective space-time models. arXiv:1610.08355
303. M. Bojowald, S. Brahma, Signature change in loop quantum gravity: general midisuperspace models and dilaton gravity. arXiv:1610.08840

Chapter 11
Cosmology of Quantum Gravities

Magna et spatiosa res est sapientia; vacuo illi loco opus est; de divinis humanisque discendum est, de praeteritis de futuris, de caducis de aeternis, de tempore. De quo uno vide quam multa quaerantur: primum an per se sit aliquid; deinde an aliquid ante tempus sit sine tempore; cum mundo coeperit an etiam ante mundum quia fuerit aliquid, fuerit et tempus. [...] Quamcumque partem rerum humanarum divinarumque conprenderis, ingenti copia quaerendorum ac discendorum fatigaberis. Haec tam multa, tam magna ut habere possint liberum hospitium, supervacua ex animo tollenda sunt. Non dabit se in has angustias virtus; laxum spatium res magna desiderat. Expellantur omnia, totum pectus illi vacet.
— Seneca, *Ad Lucilium Epistularum Moralium*, XI, 88, 33–35

Wisdom is a great and spacious thing; it needs plenty of free space. It teaches us about the divine and the human, the past and the future, the transient and the eternal, and about time. See how many issues arise just about the latter: First, whether time is anything by itself; then, if anything existed prior to time and without time; if time began with the world or, since something must have existed before the world, if also time existed before the world. [...] Whatever part you embrace of human and divine sciences, you will have to make a great effort in studying a vast number of things. These are so many and so important that, in order for them to have free shelter in your soul, you will have to remove all superfluous things. Virtue does not confine itself to narrow quarters; great things wish large space. Let us expel everything else from our breast and make it empty to make room for virtue.

G. Calcagni, *Classical and Quantum Cosmology*, Graduate Texts in Physics,
DOI 10.1007/978-3-319-41127-9_11

Contents

A theory of everything explaining all the cosmological puzzles is an attractive utopia. It should explain how an expanding universe can arise from the chaotic foam of quantum geometry; whether the big-bang singularity of classical general relativity is replaced by some configuration without infinities; what is the fate of other gravitational singularities such as those met in black holes; what caused the early inflationary expansion, and what the late one; how the degrees of freedom of the Standard Model of particles emerged; what is the cosmological constant and why it acquired such a small value as observed today; and so on.

In the lack of a concrete realization of a theory of everything, one should be able to ask the same questions in non-unified quantum-gravity scenarios while keeping exploring all the possibilities with an open mind. To this end, it is useful to compare the approaches to quantum gravity listed in Sect. 8.3 and see what they can say about cosmology and its problems. We will do so in this chapter, summarizing here the salient points of each framework:

- *Asymptotic safety* [1–45] (Sect. 11.2). Reviews are [46–51]. Various types of $f(R)$ actions are naturally produced. The big-bang problem is not solved. The cosmological constant problem is reformulated but not quite solved in the functional renormalization approach, while in the resummed-quantum-gravity approach to asymptotic safety an actual prediction for Λ is given and is in agreement with the observed value, although the observed equation of state is not reproduced [52].

- *Causal dynamical triangulations* [53–75] (Sect. 11.3). Reviews are [76–79]. A de Sitter universe naturally emerges as the semi-classical limit of full non-perturbative, background-independent quantum gravity. Neither the Λ nor the big-bang problem are addressed.
- *Spin foams* [80–95] (reviews are [96–99]) and their symmetry-reduced cosmological version [100–110] (Sect. 11.4). The big bang is removed in an LQC-related mini-superspace context. The Λ problem is not solved.
- *Group field theory* [111–141] (Sect. 11.5). Reviews are [142–148]. As in causal dynamical triangulation, a cosmological limit is obtained from quantum gravity under certain approximations. Semi-classical cosmological equations of motions arise from the full, background-independent quantum theory via a non-perturbative condensation mechanism. Interestingly, the homogeneous and isotropic dynamics can match with that of LQC in the improved quantization scheme of the canonical theory.
- *Causal sets* [149–180] (Sect. 11.6). Reviews are [181–186]. A prediction for the cosmological constant is found which is in agreement with the observed value (2.118); however, its robustness against inhomogeneities is not clear and there might be some residual fine tuning. The big-bang problem seems to be removed. There arises the possibility to have an early de Sitter stage of purely geometric origin. So far, none of these features have been embedded in a fully realistic cosmological history, which is difficult to extract since the main building blocks of the theory are pre-geometric and the notion of a spacetime continuum is not fundamental but, rather, emergent from a discrete ordered structure.
- In Sects. 11.7, 11.8 and 11.9, we will overview other approaches: non-commutative spacetimes, non-local gravity and models with dimensional flow.

11.1 Hausdorff and spectral dimension

In preparation for a comparison among these theories (Sect. 11.9), we introduce two notions of spacetime dimension. We recall that the *Hausdorff dimension* $d_{\rm H}$ of a set is the scaling of volumes defined thereon. One can proceed by covering the set with a minimum number of balls of different size and then send their radii to zero. For a smooth manifold of topological dimension D, this reduces to a local, operational definition of the Hausdorff dimension $d_{\rm H}$ as the scaling law for the volume $\mathcal{V}^{(D)}(R)$ of a D-ball of radius R:

$$d_{\rm H} := \frac{{\rm d}\ln \mathcal{V}^{(D)}(R)}{{\rm d}\ln R}\,. \tag{11.1}$$

For Euclidean space, $\mathcal{V}^{(D)} \propto R^D$ and the Hausdorff and topological dimension coincide, $d_{\rm H} = D$. For a Riemannian manifold, the same result holds only in the limit $R \to 0$, lest curvature or topological effects become important. For a Lorentzian manifold, one computes the Hausdorff dimension either of spatial hypersurfaces or of the whole spacetime with imaginary time; the result, in the limit of vanishing radius, is still $d_{\rm H} = D$. For a generic spacetime coming from some effective limit of a quantum-gravity model, the Hausdorff dimension can vary depending on the scale (in which case one must not take the limit $R \to 0$ in (11.1)) and takes non-integer values.

A different geometric indicator is the *spectral dimension* $d_{\rm S}$. Its most common interpretation is based upon the diffusion equation method. One places a classical test particle on the geometry one wants to probe, at some initial point x' (suitably defined also in discrete geometries) and let it diffuse pushed around by geometry itself. Then one asks what the probability P is to find the particle at point x after some diffusion time T. However, in covariant settings it is difficult to interpret T as an actual time variable. The time coordinate t is either omitted (when diffusing on spatial hypersurfaces) or put on equal footing as spatial coordinates (when considering the Euclideanized version of the geometry under inspection). Therefore, it is advisable to modify the interpretation of the diffusing process and regard the latter as a probing of the geometry with a resolution $1/T \to 1/\ell$. The length scale ℓ represents the minimal detectable separation between points. By definition, the heat kernel $P(x, x'; \ell)$ is the solution of the running (or diffusion) equation

$$\left[\frac{\partial}{\partial \ell^2} - \mathcal{K}_x\right] P(x, x'; \ell) = 0 \,, \qquad P(x, x'; 0) = f(x)\, \delta(x - x') \,, \tag{11.2}$$

where $\mathcal{K}_x$ is the kinetic operator characteristic of the geometry and acting on the x dependence of P. For a smooth manifold, $\mathcal{K}_x$ is the Euclidean version of the covariant Laplace–Beltrami operator $\Box$, i.e., the curved Laplacian ∇^2. Effective continuous quantum geometries can induce strong deviations from (11.2) via changes in the initial condition, in the diffusion operator $\partial/\partial\ell^2$, in $\mathcal{K}_x$ (which can be a higher-order or even a non-local derivative operator on a continuum, or a finite-difference operator if the setting is discrete) and by the presence of source terms [41, 187, 188]. The initial condition at $\ell = 0$ (where the weight factor is $f = 1/\sqrt{|g|}$ for a Riemannian geometry) expresses the fact that, at infinite resolution, the probe is a point-wise particle exactly localized at x'.

The trace $\mathcal{P}(\ell) := \int d^D x\, P(x, x; \ell)$ is called return probability. The spectral dimension of spacetime is then defined as the scaling of $\mathcal{P}$ with respect to ℓ,[1]

$$d_{\rm S}(\ell) := -\frac{d \ln \mathcal{P}(\ell)}{d \ln \ell} \,. \tag{11.3}$$

[1]In order to make sense of $d_{\rm S}$ as an indicator of the physical geometry, one should ensure that the diffusion process or resolution-dependent probing be well defined. If the solution P of (11.2) was regarded as a probability density function, as in transport theory, then it should be positive semi-definite and normalized to 1. However, in many approaches to quantum gravity P becomes negative for certain values of ℓ and x, in which case the interpretation of (11.2) becomes problematic, even if the return probability $\mathcal{P}$ is positive-definite. In fact, if there were no operationally sensible procedure by which the test particle could be found "somewhere" with a certain probability, then no clear geometric and physical meaning could be attached to (11.3). The negative probabilities problem can be fixed either by modifying (11.2) [41] or by adopting a quantum-field-theory viewpoint [189], where P is an amplitude rather than a probability.

As for the Hausdorff dimension, to ensure that large-scale curvature and topological effects do not vitiate the result,[2] one should limit the attention to a local definition of d_S, in the limit $\ell \to 0$. However, in any situation where we expect a running of the spectral dimension one should not take this limit and, instead, ignore both curvature and topological effects whenever possible. For instance, on a Riemannian manifold one ignores the curvature and just considers D-dimensional Euclidean space. The solution of (11.2) is the Gaussian $P(x, x'; \ell) = (4\pi\ell^2)^{-D/2} \exp[-|x - x'|^2/(4\ell^2)]$, the return probability is a simple power law $\mathcal{P}(\ell) \propto \ell^{-D}$ and the spectral dimension is $d_S = D = d_H$.

11.2 Asymptotic Safety

Although graviton perturbation theory suffers from divergences, the gravitational interaction can be quantized as a field theory non-perturbatively via the functional renormalization approach. This is the aim of asymptotically safe gravity. Let λ_i be the essential couplings of a field theory, i.e., those combinations of the parameters in the action which are invariant under field redefinitions. Asymptotic safety is a condition imposed on λ_i such that they approach a fixed point as the momentum-energy scale k above which all quantum fluctuations are integrated out goes to infinity [1].[3] As in a perturbatively renormalizable theory, there is a finite number of essential couplings and divergences in the UV are supposedly removed.

If the λ_i's have energy dimension d_i, then $\bar{\lambda}_i(k) = k^{-d_i}\lambda_i(k)$ are dimensionless and remain so under an infinitesimal change in the energy scale, so that their variation can depend only on a combination β_i of the other couplings:

$$\frac{\mathrm{d}\bar{\lambda}_i}{\mathrm{d}\ln k} = \beta_i\left(\{\bar{\lambda}_j\}\right). \tag{11.4}$$

For given initial conditions, the solution of the Gell-Mann–Low equation (11.4) describes the trajectory $\bar{\lambda}_i(k)$ in the space of couplings. Different initial conditions correspond to different theories. An ultraviolet *non-Gaussian fixed point* (NGFP) in the trajectory is defined by

$$\lim_{k\to\infty} \bar{\lambda}_i(k) = \bar{\lambda}_i^* \neq 0\,, \qquad \beta_i(\{\bar{\lambda}_j^*\}) = 0\,, \tag{11.5}$$

[2]For instance, in the diffusion interpretation on a sphere the particle can come back to x' more easily than on a plane and, if we wait too long ($T \to +\infty$), the return probability tends to a constant. In the resolution interpretation, waiting too long means taking too low a resolution ($1/\ell \to 0$), so that the sphere cannot be distinguished from a point. In both cases, $d_S \to 0$ instead of $d_S \to D$.

[3]In this section, we reserve the symbol k for this cut-off.

where $\bar{\lambda}_i^*$ are finite constants not necessarily small (hence, perturbation theory may not apply). In general, any theory possesses the Gaussian fixed point $\bar{\lambda}_i^* = 0$, corresponding to the free limit; the existence of a non-Gaussian fixed point is much less trivial. The main assumption of asymptotic safety is that there exist initial conditions such that the fixed point (11.5) lies in the trajectory solving (11.4). If it does not, then in general $\bar{\lambda}_i \to \infty$ at high energies and the theory may develop a singularity in physical observables.

11.2.1 Framework

For a given k, effective asymptotic-safety spacetime and dynamics are described by an effective action $\Gamma_k[g_{\mu\nu}]$. The metric is subject to quantum fluctuations and, depending on the resolution $1/\ell(k)$ of the microscope probing the geometry, one observes different properties of spacetime. The change of the microscope amounts to a coarse-graining procedure. Denote as $\langle g_{\mu\nu}\rangle_k$ the metric averaged over Wick-rotated spacetime volumes of linear size $\ell(k)$. Then, $\langle g_{\mu\nu}\rangle_k$ is solution of

$$\frac{\delta\Gamma_k}{\delta g_{\mu\nu}}\left[\langle g_{\mu\nu}\rangle_k\right] = 0\,. \tag{11.6}$$

Letting k take any non-negative value, one ends up with a continuous family of actions and of field equations (11.6), all valid simultaneously. From the scale dependence of Γ_k, one can extract the solution $\langle g_{\mu\nu}\rangle_k$ at any scale from the UV to the IR. Although $\langle g_{\mu\nu}\rangle_k$ is typically a smooth classical metric, its change with the scale, regulated by the renormalization group (RG) flow, makes effective asymptotic-safety spacetime a non-smooth multi-scale (in particular, fractal [34]) object with possibly very "irregular" geometry.

In the absence of matter and in the so-called Einstein–Hilbert truncation, the effective action is the usual one but with scale-dependent couplings:

$$\Gamma_k = \frac{1}{16\pi G_k}\int d^Dx\sqrt{-g}\,(R-2\Lambda_k)\,. \tag{11.7}$$

A non-trivial non-Gaussian fixed point $(\bar{G}_*, \bar{\Lambda}_*)$ exists in the plane of the dimensionless couplings $\bar{G}_k = k^{D-2}G_k$ and $\bar{\Lambda}_k = k^{-2}\Lambda_k$ [2, 16]. In the IR perturbative regime near the Gaussian fixed point, $G_k = G_0 + O(k^2)$ and $\Lambda_k = \Lambda_0 + \nu G_0 k^4 + O(k^6)$, where $\nu = O(1)$ is a constant. In terms of the combined quantities $\bar{\Lambda}_{\rm T} := (4\nu\Lambda_0 G_0)^{1/2}$, $k_{\rm T} := [\Lambda_0/(\nu G_0)]^{1/4}$ and $\bar{G}_{\rm T} := \Lambda_{\rm T}/(2\nu)$, in four dimensions

$$\bar{G}_k = \bar{G}_{\rm T}\left(\frac{k}{k_{\rm T}}\right)^2 + O(k^4), \qquad \bar{\Lambda}_k = \frac{1}{2}\bar{\Lambda}_{\rm T}\left[\left(\frac{k_{\rm T}}{k}\right)^2 + \left(\frac{k}{k_{\rm T}}\right)^2\right] + O(k^4)\,. \tag{11.8}$$

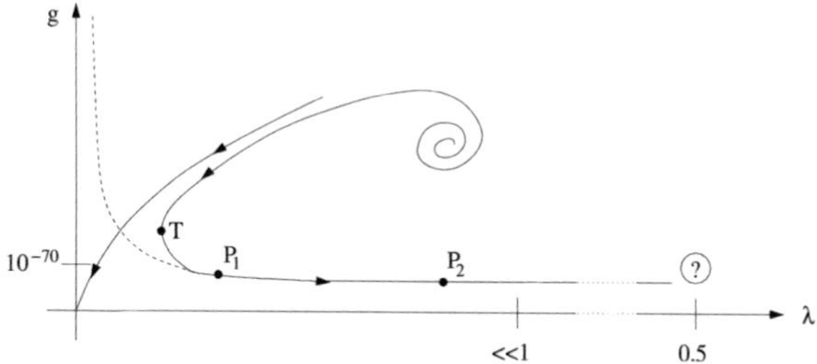

Fig. 11.1 Renormalization-group flow on the plane of the dimensionless couplings $g = \bar{G}_k$ and $\lambda = \bar{\Lambda}_k$ described in the text, for asymptotically safe quantum gravity with Einstein–Hilbert effective action. Other classes of trajectories are possible but this is the only one realized by Nature (positive Newton's and cosmological constant in the IR). Classical general relativity corresponds to the segment between P_1 and P_2. The *question mark* indicates a singularity point where the Einstein–Hilbert approximation breaks down. The separatrix from other trajectories is also shown (Source: [18])

Here k_T represents the scale at which the turning point T in Fig. 11.1 is passed. Overall, the gravitational coupling decreases from the UV to the IR along the RG flow, while the cosmological constant (if positive) increases. Both acquire very small values near the Gaussian fixed point $(\bar{G}, \bar{\Lambda}) = (0, 0)$ at the origin of the plot [18].

The scale dependence of the average metric is related to the one of the running cosmological constant Λ_k. In fact, the field equations read $R_{\mu\nu}[\langle g\rangle_k] = [2/(2-D)]\Lambda_k \langle g_{\mu\nu}\rangle_k$. Let k_0 be an arbitrary reference scale (typically in the infrared, $k_0/k \gg 1$) and assume that the cosmological constant scales according to a function F,

$$\Lambda_k = F(k^2)\,\Lambda_{k_0}\,, \tag{11.9}$$

where F depends on k^2 by the requirement of Lorentz invariance. We also assume that F stays positive throughout the flow between k and k_0. Then, from $F^{-1}\,R^{\mu}{}_{\nu}[\langle g\rangle_k] = [2/(2-D)]\Lambda_{k_0}\,\delta^{\mu}_{\nu}$ and the scaling property $R^{\mu}{}_{\nu}[c\,g] = c^{-1}\,R^{\mu}{}_{\nu}[g]$ of the Riemann tensor ($c > 0$ is a constant), one gets

$$\langle g_{\mu\nu}\rangle_k = \frac{1}{F(k^2)}\langle g_{\mu\nu}\rangle_{k_0}\,, \qquad \langle g^{\mu\nu}\rangle_k = F(k^2)\langle g^{\mu\nu}\rangle_{k_0}\,. \tag{11.10}$$

The function F acquires different asymptotic forms depending on the regime. In the far IR, $F \simeq 1$ by definition ($k \to k_0 \to +\infty$). In an intermediate semi-classical regime near the Gaussian fixed point, $F \simeq k^4$. In the deep UV, at the NGFP one has $\Lambda_k \simeq k^2\bar{\Lambda}_*$ and asymptotic safety requires that $F \simeq k^2$ [21]. All in all, we can write

$$F \simeq |k|^{\delta} \tag{11.11}$$

in asymptotic regimes, with $\delta = 0, 4, 2$ in the IR, semi-classical limit and UV, respectively. The semi-classical limit may change depending on the truncation and on the presence of matter [37], but the main scaling argument is unaltered.

In classical geometries, Γ_k can be qualitatively interpreted as encoding dynamical variables which are coarse-grained on regions of spacetime of size L [190]. In this context, in the simplest classical non-compact case $\ell \sim k^{-1}$ and the resolution is directly identified with k. This is the sense in which Γ_k acts as a "microscope" with resolving power $\ell(k)$. In quantum gravity not only is the functional form $\ell(k)$ more complicated a priori, but the relation between resolving power, proper distances and IR cut-off k is much subtler. In a general quantum manifold with fractal properties, the k-dependence of lengths is an example of the typical effect of momentum dependence of measurements on multi-fractals [191]. There appear situations where ℓ can decrease arbitrarily even when there exists a non-vanishing minimum resolution: the limit $k \to +\infty$ is no longer sufficient to probe arbitrarily small proper distances because the effective quantum geometry "shrinks" faster in the UV limit. This is a direct consequence of the scaling (11.10) of the average metric and, more generally, of the running of the couplings in the effective action. Lengths can also be measured by a macroscopic observer with average metric $\langle g_{\mu\nu}\rangle_{k_0}$, who does see a finite minimal length corresponding to the minimal resolution [22, 24].

The cut-off scale k is often identified with the physical momentum p, although sometimes other identifications are more convenient (see below and [41]). With the identification $k = p$ and ignoring intermediate regimes, one can interpolate the IR and UV regions via the graviton effective dispersion relation $p^2 + p^4/\gamma^2 = 0$ and the effective Newton's constant $G_{\rm eff}(p) = G_0/(1 + p^2/\gamma^2)$. For small momenta, $G_{\rm eff}(p) = G_0 + O(p^2)$, while for large momenta (i.e., near the NGFP) $G_{\rm eff}(p) \simeq (G_0\gamma^2)p^{-2}$. The numerical coefficient γ is such that $G_0\gamma^2 = \bar{G}_*$. Remarkably, this is precisely the behaviour of (8.16) and (8.18) found in Sect. 8.2.3. A direct calculation yields $\bar{G}_* \approx 0.0442$ [192].

In Sect. 7.1.2, we have seen that the zero-point energy density of each degree of freedom of the i-th matter field yields the contribution (7.12) to the cosmological constant, which stems from the four-dimensional integral

$$\rho^{(1)}_{{\rm vac},i} \propto \int_{-\infty}^{+\infty} \frac{{\rm d}^4 p}{(2\pi)^4}\, \frac{p_0^2}{p^2 + m_i^2}\,. \tag{11.12}$$

In the resummed quantum-gravity framework, the propagator is replaced by its dressed version (8.15). Summing over all matter particles with N_i degrees of freedom (negative for fermions), up to an overall constant now one has [52]

$$\rho^{(1)}_{\rm vac} \propto \sum_i N_i \int_{-\infty}^{+\infty} \frac{{\rm d}^4 p}{(2\pi)^4}\, \frac{p_0^2\, {\rm e}^{B''_g(p)}}{p^2 + m_i^2} \simeq -\frac{m_{\rm Pl}^4}{64} \sum_i \frac{N_i}{[\ln(\pi m_{\rm Pl}^2/m_i^2)]^2}\,. \tag{11.13}$$

To encode the running of Newton's constant in (11.13), one should replace the Planck mass with the one coming from (8.18),

$$m_{\rm Pl}^2(p) = \left(1 + \frac{p^2}{\gamma^2}\right) m_{\rm Pl}^2 , \tag{11.14}$$

where $1/\gamma^2 \approx 22.635/m_{\rm Pl}^2$. This replacement does not have an important effect on the logarithm in (11.13) and can be done just in the numerator $m_{\rm Pl}^4$:

$$\rho_\Lambda(p) \simeq -\frac{m_{\rm Pl}^4(p)}{64} \sum_i \frac{N_i}{[\ln(\pi m_{\rm Pl}^2/m_i^2)]^2} . \tag{11.15}$$

For the particle content of the Standard Model, the sum value is $\sum_i N_i/[\ln(\pi m_{\rm Pl}^2/m_i^2)]^2 \approx -9.194 \times 10^{-3}$.

Thus, $\Lambda(p) = 8\pi\rho_\Lambda(p)/m_{\rm Pl}^2(p) \simeq \Lambda_0 + \bar{\Lambda}_* p^2$, where $\bar{\Lambda}_* \approx 0.0817$ [52]. This estimate changes with the number and species of particles and, in turn, can provide a constraint on the admissible supersymmetric extensions of the Standard Model.

Therefore, the existence of a non-Gaussian fixed point is confirmed independently by the functional renormalization approach and by resummed quantum gravity. It is more difficult to compare the actual numerical values of $\bar{G}_*$ and $\bar{\Lambda}_*$ in the two frameworks. In the case of resummed gravity, $(\bar{G}_*, \bar{\Lambda}_*) \approx (0.0442, 0.0817)$ [52], but in the functional renormalization approach the values depend on the assumptions on the truncation scheme and on the matter content and $\bar{G}_*, |\bar{\Lambda}_*| = O(0.1) - O(1)$ [5, 37, 39, 43, 45].

Despite the pending issue of the NGFP value, a posteriori we recognize resummed quantum gravity as another explicit realization of asymptotic safety because $(\bar{G}_*, \bar{\Lambda}_*) \neq (0, 0)$ [25]. This formalism has also the advantage of being gauge invariant, thus avoiding the issues of gauge dependence that inevitably arise in the truncation scheme of the functional renormalization approach.

11.2.2 Cosmology

The scale dependence of the couplings in the average effective action Γ_k is inherited by the Einstein equations of this theory. For a FLRW background, the cut-off $k = k(t)$ can only depend on time and, dimensionally, the simplest identification is with the Hubble horizon, $k \propto H(t)$ [26]. Then, the RG-improved cosmological dynamics is encoded into two equation. One is the usual first Friedmann equation except for the replacements $\Lambda \to \Lambda(t)$ and $G \to G(t)$, where the precise time dependence is determined by integrating the RG equations numerically. In four dimensions,

$$H^2 = \frac{8\pi G(t)}{3}\rho - \frac{\textsc{k}}{a^2} + \frac{\Lambda(t)}{3} , \qquad \rho_\Lambda = \frac{\Lambda(t)}{8\pi G(t)} . \tag{11.16}$$

The Bianchi identity is not automatically solved when the energy-momentum tensor is covariantly conserved:

$$\dot{\rho} + 3H(\rho + P) = -\frac{\dot{\Lambda} + 8\pi\rho\dot{G}}{8\pi G}, \tag{11.17}$$

where ρ and P are the energy density and pressure of a perfect fluid. Energy is exchanged between matter and geometry. If the energy-momentum tensor is conserved, then (11.17) splits into $\dot{\rho} + 3H(\rho + P) = 0$ and $\dot{\Lambda} + 8\pi\rho\dot{G} = 0$, giving the general solution

$$\rho = -\frac{1}{8\pi}\frac{\dot{\Lambda}}{\dot{G}}, \qquad a \propto \left(-\frac{\dot{G}}{\dot{\Lambda}}\right)^{\frac{1}{3(1+w)}}, \tag{11.18}$$

valid for a constant barotropic index $w = P/\rho$. Such a dynamics with varying G and varying Λ is not just a phenomenological model with ad hoc functions $G(t)$ and $\Lambda(t)$ [193], since the latter are determined by the RG evolution.

In particular, one has a concrete quantum-gravity example of an alternative acceleration mechanism without scalar fields, solving the horizon and the entropy problem [6, 26, 51]. Near the non-Gaussian fixed point (11.20), the equations of motion with flat background ($\textsc{k} = 0$) admit the power-law asymptotic solution $a \propto t^{\alpha}$, $\rho \propto t^{-4}$, where

$$\alpha = \frac{2}{3(1+w)(1-\Omega_\Lambda^*)} \tag{11.19}$$

and $\Omega_\Lambda^* = \rho_\Lambda^*/\rho_{\rm crit}^*$ is evaluated at the fixed point. Therefore, near the NGFP $k \propto H \propto t^{-1}$ and

$$\Lambda(t) \propto t^{-2}, \qquad G(t) \propto t^2 \tag{11.20}$$

in four dimensions. From (11.17), one can find that $P_\Lambda/\rho_\Lambda = 4/(3\alpha) - 1$: the running cosmological constant (11.20) obeys the usual equation of state $P_\Lambda \simeq -\rho_\Lambda$ only for $\alpha \gg 1$.

The UV asymptotic behaviour (11.20) is readily recovered in resummed quantum gravity, by identifying $p \sim 1/t$ in the energy density (11.15),

$$\rho_\Lambda(t) \simeq \left(1 + \frac{1}{\gamma^2 t^2}\right)^2 m_{\rm Pl}^4 \left\{-\frac{1}{64}\sum_i \frac{N_i}{[\ln(\pi m_{\rm Pl}^2/m_i^2)]^2}\right\}, \tag{11.21}$$

and taking the limit $\gamma t \propto t/t_{\rm Pl} \ll 1$.

The deep UV regime takes place at energy scales $k \propto H \sim m_{\rm Pl}$, corresponding to times prior to $t_* \sim \alpha t_{\rm Pl}$. Starting with radiation ($w = 1/3$), for $0 < t < t_*$ the universe accelerates ($\alpha > 1$) provided $1/2 < \Omega_\Lambda^* \lesssim 1$. At times $t > t_*$, inflation stops and there follows a classical evolution with $a \propto t^{1/2}$. The primordial power spectrum is generated by matter fluctuations, which are almost scale invariant near the NGFP. In fact, at the UV fixed point the $D = 4$ graviton propagator in momentum space scales as $\tilde{G}(p^2) \simeq -1/p^4$ for $p^2 \gg m_{\rm Pl}^2$, amounting to $G(x-x') \sim \ln|x-x'|^2$ in position space [7]. In particular, the graviton two-point correlation function on a cosmological spatial slice is (up to tensor indices) $\langle h(t,\mathbf{x})h(t,\mathbf{x}')\rangle \sim \ln|\mathbf{x}-\mathbf{x}'|^2$, so that $\langle \delta R(t,\mathbf{x})\delta R(t,\mathbf{x}')\rangle \sim |\mathbf{x}-\mathbf{x}'|^{-4}$ for the curvature fluctuation $\delta R \sim \partial^2 h$. Back to momentum space, this corresponds to a scale-invariant power spectrum [6].

Unfortunately, the amplitude of tensor modes produced during this stage is too large, since it is governed by the dimensionless combination $GH^2 \propto G_* \Lambda_* = O(10^{-3})$ at the NGFP [36]. To fill the gap of seven orders of magnitude between this value and the present constraints on the tensor amplitude (from (4.66) and (4.69), $A_{\rm t} \sim 0.1 A_{\rm s} \sim 10^{-10}$), one must abandon the Einstein–Hilbert truncation and consider higher-order curvature terms, for instance of the Starobinsky type (7.88) [42] or encoded in a generic functional $f(R)$ [32]. The Taylor coefficients of f are determined at the NGFP for any given truncation order.

An $f(R)$ effective action with limit (7.88) can also arise from the Einstein–Hilbert truncation when making the scale identification [36]

$$k^2 \propto R\,. \tag{11.22}$$

Near the NGFP, $G_k \simeq G_* k^{-2} \propto R^{-1}$, so that $R/G_k \sim R^2$. In the semi-classical perturbative regime near the Gaussian fixed point, the intermediate limit $f(R) \simeq R + bR^2$ is obtained. This may provide a theoretical justification to Starobinsky inflation. A more precise determination of the effective action, still by the identification (11.22), comes from the analytic expressions of G_k and Λ_k solving the linearized flow equations around the NGFP [35]:

$$\mathcal{L} = R^2 + bR^2 \left(\frac{R}{R_0}\right)^{\beta} \cos\left[\omega \ln\left(\frac{R}{R_0}\right)\right], \tag{11.23}$$

where $b > 0$ and $\beta < 0$. The logarithmic oscillations arise from the interplay of the pair of complex-conjugate critical exponents $\theta_1 = \theta_2^* = -\beta + 2ib$ governing the spiral approach to the fixed point (Fig. 11.1). The simplest cosmological solution with acceleration is de Sitter.

In general, inflation can be sustained by higher-order curvature terms one can add to the Einstein–Hilbert truncation [31]. It is also possible to introduce matter in the form of a real scalar field, in which case one obtains an ordinary inflationary scenario where the RG quantum corrections modify the constraints on viable potentials [33, 38, 40].

The big-bang problem still persists, as there is a singularity at $t = 0$. In the functional renormalization approach, the cosmological constant problems are not solved either, since today's value (2.118) corresponds to the IR semi-classical limit of the running $\bar{\Lambda}$, which is tuned by extremely small values of $\bar{G}_{\rm T} \sim \bar{\Lambda}_{\rm T} \sim 10^{-60}$. However, once the emergence of classical spacetime is achieved in a robust way, the smallness of Λ is a natural consequence of the fact that the RG trajectory realized in Nature spends several orders of magnitude of RG time $\ln(k/k_{\rm T})$ near the Gaussian fixed point $(\bar{G}, \bar{\Lambda}) = (0, 0)$ [20]. Thus, the old Λ problem is reformulated by asking why the perturbative semi-classical regime of such a trajectory is fine tuned to this degree [18]. Another interesting trajectory runs along the $\bar{\Lambda} = 0$ line from a non-Gaussian fixed point $(\bar{G}, \bar{\Lambda}) = (\bar{G}_* \neq 0, 0)$ directly to the Gaussian fixed point [44]. In this case, the cosmological constant is zero at all scales and, as in any "overshooting" approach to the old Λ problem, the small value of the dark-energy density must be explained with something more than the vacuum energy.

In the resummed-propagator approach to asymptotic safety, it is possible to make a more precise prediction for the value of the effective cosmological constant today. This estimate relies on a transition between the deep UV regime where (11.21) holds and the classical regime where the standard cosmological dynamics applies. The transition is marked by the critical time $t_* = \alpha t_{\rm Pl}$. Numerical integration of the RG-improved cosmological equations of the functional renormalization approach points towards a value $\alpha \approx 25$ [26]. The same value is also obtained in resummed quantum gravity with the help of Heisenberg uncertainty principle [194]. In the classical regime, the vacuum equation of state is evaluated at energies smaller than the Planck mass. One takes the view that the energy density and pressure of the cosmological constant and matter fields are not separately conserved. Instead, one should consider the full continuity equation

$$\dot{\rho} + \dot{\rho}_\Lambda + 3H[(\rho + \rho_\Lambda) + (P + P_\Lambda)] = 0\,. \tag{11.24}$$

To maintain the usual equation of state $w_\Lambda = -1$ that is observed by experiments, $\rho_\Lambda + P_\Lambda = 0$ and $\rho_\Lambda(t)$ is determined dynamically by Einstein's equations. If the cosmological constant is small enough during cosmic evolution, then with good approximation $\dot{\rho} + 3H(\rho + P) \approx 0$ for matter and radiation separately. At times $t_* < t < t_{\rm eq}$, the universe is dominated by radiation and, by virtue of the standard Friedmann equation, $a \propto t^{1/2}$ (equation (2.78)). For $t > t_{\rm eq}$, dust matter dominates and $a \propto t^{2/3}$. In both eras, $\rho_\Lambda = -\rho + 3H^2/\kappa^2 \propto t^{-2}$, so that integrating (11.24) between t_* and t_0 (today), we have

$$\rho_\Lambda(t_0) = \rho_\Lambda(t_*)\left(\frac{t_*}{t_0}\right)^2. \tag{11.25}$$

The age of the universe is given by (2.14), $t_0 \approx 13.8\,{\rm Gyr}$, so that $t_*/t_0 \approx 3.097 \times 10^{-60}$. The key information provided by resummed quantum gravity is the value of $\rho_\Lambda(t_*)$, which is determined by (11.21) at the transition point t_*: $\rho_\Lambda(t_*) \approx 1.54 \times 10^{-4} m_{\rm Pl}^4$. Combining all this information into (11.25), one finally obtains the attractive result

$$\rho_\Lambda(t_0) \approx 1.5 \times 10^{-123} m_{\rm Pl}^4 \approx (2.4 \times 10^{-3}\text{eV})^4\,, \tag{11.26}$$

quite close to the observed value (2.118). A posteriori, one can check that the contribution of ρ_Λ is small enough from the time of the big-bang nucleosynthesis onwards up to the recent acceleration phase [52]. Therefore, the above approximation on the equations of state is consistent. Since $w_\Lambda = -1$, (11.26) solves the old cosmological constant problem without disrupting the evolution of the universe in an unwanted way.

11.3 Causal Dynamical Triangulations

In analogy with quantum mechanics and quantum field theory, quantization of geometry can be defined as the sum over histories

$$Z = \int [\mathcal{D}g]\, e^{iS[g]}\,, \tag{11.27}$$

where we set $\hbar = 1$, $[\mathcal{D}g]$ is the functional integration measure (performed with certain boundary conditions) of the space of equivalence classes of Lorentzian D-dimensional metrics $g_{\mu\nu}$ under diffeomorphisms and S is the gravitational action. We saw examples of path integrals for gravity in (7.103) and (9.91). For simplicity, we have factored out and ignored the contribution of matter fields. The path integral Z depends on the coupling constants in the Lorentzian action S. If S is chosen to be the Einstein–Hilbert action with cosmological term, the space of couplings is two-dimensional and spanned by all possible values of G and Λ. The presence of all allowed geometries in the path creates quantum interference in the evolution of the metric $g_{\mu\nu}(x)$ via the dynamics of S.

We had already occasion to mention that the path-integral approach is technically difficult due to the divergences hidden in (11.27). To overcome this problem, one can discretize the geometries and thus regularize the functional integration, in a way preserving the local causal structure of inertial frames. Such is the goal of causal dynamical triangulations (CDT).

11.3.1 Framework

The most obvious cause of concern regarding (11.27) is the oscillatory integrand e^{iS}. Going to imaginary time $t \to it$ produces a formally convergent factor e^{-S_E}, where S_E is the Euclideanized classical action; the Euclideanized path integral is then called partition function. The Euclidean analytic continuation of gravity can

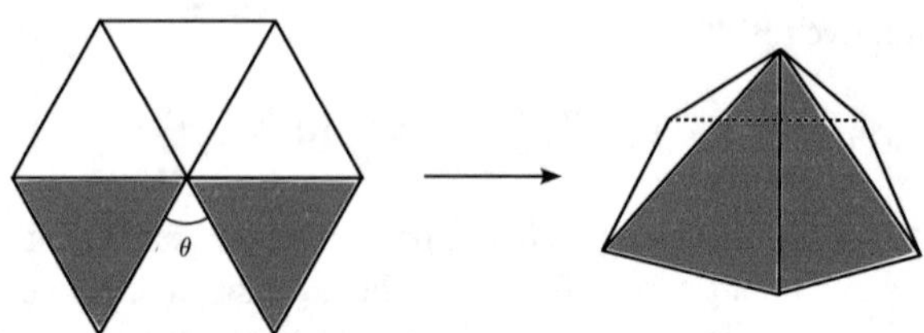

Fig. 11.2 Triangulation of a Riemannian disk with positive curvature (Reprinted figure with permission from [70]. ©2013 by the American Physical Society)

give a wide overview of possible quantum phenomena, as seen in Chaps. 7 and 9, but its viability as a physical approach has never been established conclusively. One can appreciate this in Euclidean dynamical triangulations (EDT) [195–203], which is also helpful to introduce some of the building blocks of CDT.

The *triangulation* of a given smooth manifold with metric $g_{\mu\nu}$ is a collection of *D-simplices*, pieces of a D-dimensional flat space. A 2-simplex is a triangle, a 3-simplex a tetrahedron. Denoting with ℓ_i the length of the edges of a D-simplex, the *continuum limit* is defined by $\ell_i \to 0$ for all edges of all simplices in a triangulation. In the case of EDT, the continuum limit is a Riemannian manifold, the metric has positive signature and the D-simplices are chunks of Euclidean space. Simplices are usually assumed to be equilateral, $\ell_i = \ell$ for all i. The meeting point of different edges is called a vertex. Curvature in a triangulation is introduced by gluing D-simplices together according to different angles. Take the $D = 2$ example of a disk, which is triangulated by six equilateral triangles (Fig. 11.2). Removing one triangle and gluing (i.e., identifying) the free edges, one realizes a disk with positive curvature, characterized by a deficit angle (in this case, $\theta = \pi/3$). Generalizing to D dimensions and many simplices, one gets triangulations whose geometry is intrinsically expressed in terms of the edge length ℓ and deficit angles θ_i, with no reference to an external embedding space. The picture of a triangulation embedded in a Riemannian manifold is therefore unnecessary and one can also construct arbitrary triangulations not associated with a smooth manifold in the continuum limit. By virtue of a set of manifold conditions, however, one can guarantee that the triangulation looks like a D-dimensional space everywhere.

A triangulation is said to be dynamical when a classical action for it is considered. The classical dynamics is described by *Regge calculus* [204–208], a discrete approximation of general relativity. When the quantum theory is considered, the path integral (11.27) over continuous geometries becomes a sum over triangulations. The action S is discretized according to Regge calculus and the parameter space can be studied numerically via Monte Carlo simulations. Operationally, different triangulations with the same topology are realized by creating and destroying vertices with a set of combinatorial moves [55]. It is found that $D = 4$ EDT possess neither a large-scale nor a continuum limit reproducing general relativity.

Contrary to EDT, causal dynamical triangulations do not lose the information on the causal structure of the original continuous system. At a first step, geometry is Lorentzian and D-simplices are pieces of Minkowski spacetime. This requires that the squared lengths ℓ_i^2 of their edges can also take null or negative values, and that the concurrence of time- and space-like edges at each vertex respect the local light-cone structure of inertial frames with the correct Lorentzian signature $(-,+,\cdots,+)$. In the simplest formulation of CDT, there is also a global notion of causality thanks to the introduction of a preferred foliation in each path-integral history, thus ensuring the presence of a global, discrete proper time. This foliation into $D = 2$ strips (or $D = 3$ slabs, and so on) is the piecewise discrete realization of global hyperbolicity. There is good evidence, anyway, that the main features of CDT (in particular, the de Sitter phase we shall describe below) do not depend on the existence of a preferred foliation in the triangulation [69, 70].

Each strip (or slab, and so on) of fixed width $\Delta t = 1$ is constituted by an arbitrary sequence of triangles (tetrahedra) pointing up or down. The requirements of local and global causality lead, however, to restrictions on the type of simplices allowed. These are classified according to the number (i,j) of vertices in one spatial slice and the next (or the previous: in the following we omit to mention the time-reversed simplices (j,i)). For CDT with foliation, in $D = 1 + 1$ dimensions triangles can have only one space-like edge and are therefore all of type (1,2), while in $D = 2 + 1$ dimensions tetrahedra can have either two non-contiguous space-like edges (type (2,2) tetrahedra) or one space-like face (type (3,1) tetrahedra), while no edge can be light-like (Fig. 11.3). In $D = 3 + 1$, four-simplices are only of type (4,1) (one space-like tetrahedral face, i.e., six out of ten edges are space-like) and (3,2) (one space-like triangle and one edge, while the other six edges are time-like); see Fig. 11.4. Without foliation, there are more classes of simplices allowed.

Therefore, one can parametrize equilateral CDT with one space-like squared length $\ell_\mathrm{s}^2 > 0$ and a time-like one $\ell_\mathrm{t}^2 = -\alpha\ell_\mathrm{s}^2 < 0$, where α is a dimensionless constant called asymmetry parameter. In foliated CDT, $\alpha > 0$ corresponds to Lorentzian signature and $\alpha < -7/12$ to the Euclidean one. In numerical codes, the edge length is fixed to $\ell_\mathrm{s} = 1$.

For actual numerical calculations, one must Wick rotate the system, but this does not lead to EDT because the causal structure is preserved through the length labels of the edges. The discrete partition function of foliated CDT is thus

$$Z_\mathrm{CDT} = \sum_T \frac{1}{\mathrm{Aut}(T)}\,\mathrm{e}^{-S_\mathrm{E}^\mathrm{Regge}(T)}, \tag{11.28}$$

the sum of causal triangulations T (piecewise globally hyperbolic geometries) whose spatial sections are further restricted to have fixed $(D-1)$-sphere S^{D-1} compact topology. The factor $1/\mathrm{Aut}(T)$ is the analogue of the path-integral measure $[\mathcal{D}g]$ and Aut is the order of the automorphism group of T.[4] The action $S_\mathrm{E}^\mathrm{Regge}$ is the Wick-rotated Regge action, which depends on the couplings G and Λ, on α and on

[4] An automorphism of a graph is an edge-preserving permutation of vertices. The set of automorphisms of a graph forms a group called automorphism group.

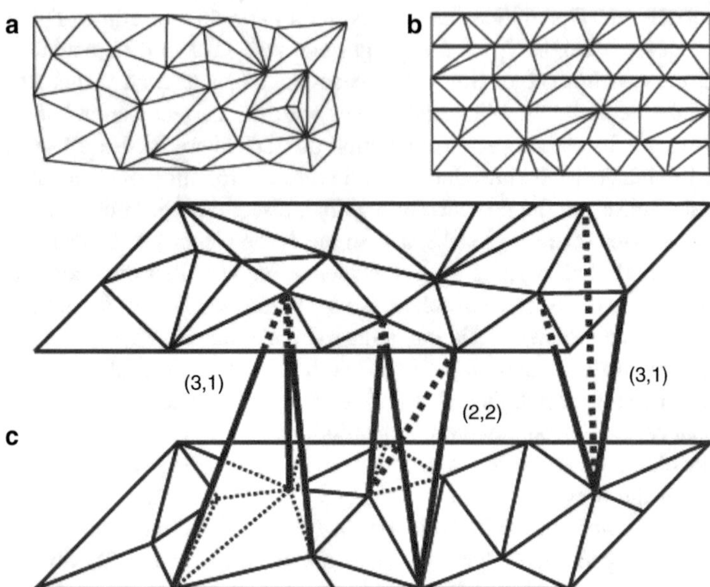

Fig. 11.3 (**a**) Two-dimensional EDT, consisting of equilateral triangles with Euclidean signature. (**b**) Two-dimensional foliated CDT, consisting of equilateral triangles with one space-like (*blue, thick lines*) and two time-like (*red, thin lines*) edges. In both cases, curvature is present at all interior vertices where the number of concurrent edges is not six. Two-dimensional graphs constitute a flattening out of a curved surface and, therefore, cannot respect the equilateral length assignment. (**c**) A piece of foliated $D = 2+1$ triangulation (Reprinted figure with permission from [70]. ©2013 by the American Physical Society)

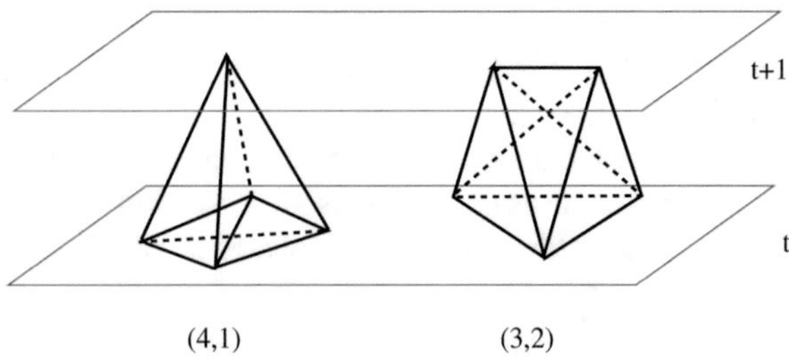

Fig. 11.4 Four-simplices of type (4,1) and (3,2), described in the text (Reprinted figure with permission from [59]. ©2005 by the American Physical Society)

the weighted volumes of vertices and allowed (i,j) simplices of each triangulation. Volumes are nothing but the counting of the number of each type of simplex and are combined together according to the combinatorial structure of T. Euclideanization is achieved by rotating the asymmetry parameter $\alpha \to -\alpha$ by -180 degrees in the complex plane $(\mathrm{Re}\,\alpha,\ \mathrm{Im}\,\alpha)$, under certain restrictions on the range of α.

The form of the action in $D = 2, 3, 4$ can be found in [53, 55], to which we refer for the details. In four-dimensional foliated CDT, it reads

$$S_{\rm E}^{\rm Regge}(T) = -c_0(G,\alpha)\,N_0(T) + c_4(G,\Lambda,\alpha)\,N_4(T) + \Delta(\alpha)\,[N_{41}(T) - 6N_0(T)]\,, \tag{11.29}$$

where N_0 is the number of vertices of the triangulation T and $N_4 = N_{41} + N_{32}$ is the total number of 4-simplices, in turn split into those of type (4,1) and (3,2). The constant c_0 is proportional to $1/G$ with α-dependent coefficient, c_4 is an α-dependent linear combination of Λ and $1/G$ and the parameter Δ depends on α; their explicit form can be found in [79]. Since $\Delta = 0$ when $\alpha_{\rm E} = -\alpha = 1$, it encodes the asymmetry between the lengths of spatial and time-like edges.

We now briefly describe the setting and main results of numerical simulations. The cosmological constant Λ is usually kept fixed, which corresponds to fixing the four-volume $\langle \mathcal{V}_4\rangle \sim \Lambda^{-1}$ averaged over paths in the path integral (that is, over different computer-generated triangulations). The four-volume is $\int {\rm d}^4x\sqrt{|g|}$ in the continuum limit, while in simplicial gravity it is $\mathcal{V}_4 = \ell_{\rm s}^4(\sqrt{8\alpha+3}\,N_{41} + \sqrt{12\alpha+7}\,N_{32})$. Fixing the volume is therefore equivalent to constrain the number of 4-simplices of type (4,1) and (3,2) but, since $N_{41} \propto N_{32} \propto N_4$ for a given α, this is tantamount to fixing the total number of 4-simplices.

The overall picture that emerges from the simulations is that there are different phases of geometry depending on the values of the parameters [59]. Although each individual triangulation in the partition function obeys the manifold conditions (corresponding, in the continuum limit, to summing over smooth manifolds with metric $g_{\mu\nu}$), the geometry coming out of the sum over histories can be highly non-classical. Such is the case in phases A and B of the phase-space diagram of the theory (Fig. 11.5).

The *branched-polymer phase* (phase A) dominates at sufficiently large c_0 (small Newton's constant G) above some critical value $\bar{c}_0$. The time evolution of the three-volume is characterized by an irregular sequence of maxima of variable size and minima of the order of the smallest three-volume $\ell_{\rm s}^3$ (i.e., one single vertex). The lifetime of these spacetime configurations with macroscopic spatial extension is very short, $\Delta t \approx 3$. These mutually disconnected "lumps" of space are realizations of branched polymers, which dominate the path for this choice of parameters. Branched polymers [209] belong to a class of random fractals which include tree graphs and random combs. These objects have $d_{\rm H} = 2$, while their spectral dimension is bounded both from above and from below, according to the relation $2d_{\rm H}/(d_{\rm H}+1) \leqslant d_{\rm S} \leqslant d_{\rm H}$ [210–214]. Hence, the geometry of phase A is far from being Riemannian.

The *crumpled phase* (phase B) occurs for small $c_0 < \bar{c}_0$ and small asymmetry parameter $0 \leqslant \Delta < \bar{\Delta}$. Here, spacetime has a vanishing temporal extension since it is limited to a large-volume configuration at a thin slice $\Delta t = 2$, where all 4-simplices are concentrated. Everywhere else along the time evolution, the three-volume stays close to its minimum $\ell_{\rm s}^3$ (just one vertex), which in the continuum

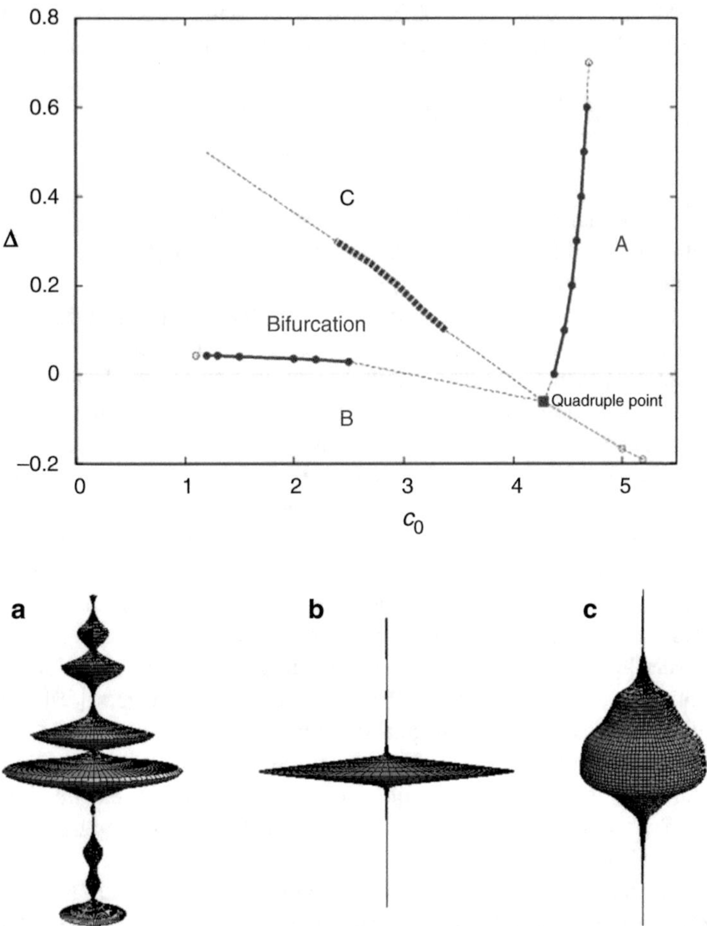

Fig. 11.5 *Upper plot*: Phase diagram of four-dimensional causal dynamical triangulations in the parameter space spanned by the gravitational coupling $c_0 \propto 1/G$ and the asymmetry parameter Δ. "Bifurcation" is phase D. *Thick points* mark where numerical simulations have been conducted at the phase transitions, while the *dashed lines* converging to the quadruple point and extending to the outer regions are an extrapolation. The jump in the order parameter $N_{41} - 6N_0$ (conjugate to Δ according to (11.29)) decreases for decreasing c_0, until the B $\leftrightarrow$ D (bifurcation) phase transition has an end-point moving towards the left of the phase diagram. *Lower graphics*: snapshots of typical spacetimes in phases A ($N_4 = 45{,}500$, $c_0 = 5.6$, $\Delta = 0$, $t_{\rm tot} = 20$), B ($N_4 = 22{,}250$, $c_0 = 1.6$, $t_{\rm tot} = 20$) and C ($N_4 = 91{,}100$, $c_0 = 2.2$, $\Delta = 0.6$, $t_{\rm tot} = 40$). At each snapshot, taken at intervals $\Delta t = 1$, the circumference is proportional to the spatial three-volume $\mathcal{V}_3(t)$, while the surface is an interpolation between adjacent spatial volumes, to give the reader the feeling of a continuous evolution (Source: adaptation from [59, 73])

limit means that the universe has zero spatial extension. The spacetime at the thin slice is not just a short-lived three-dimensional universe, since it has large-volume but almost no extension. In fact, there the number of vertices is very small while the number of simplices concurring at each vertex is extremely high. In particular, one can jump from one tetrahedron to any other in just a few steps. The geometry is highly correlated and displays a very high, or possibly infinite, Hausdorff dimension $d_{\rm H}$, and does not correspond to any classical smooth configuration.

Within the *bifurcation phase* (phase D or C_b) that separates phase B from C, periodic clusters of volume form around singular vertices and the metric of spacetime spontaneously changes signature from Lorentzian (in phase C, where an effective metric is meaningful) to Euclidean [71, 73–75].

Both phase A and B have an analogue in Euclidean dynamical triangulations [195, 196, 199, 200, 203] but phase C and the bifurcation phase are typical only of CDT. The intuitive reason is that in EDT all edges have the same length and $\Delta = 0$, but on the $\Delta = 0$ line one can only access to phases A and B with numerical methods. The phase transition A $\leftrightarrow$ B is first order: the number of vertices N_0 abruptly jumps from one type of geometry to the other, with no smooth interpolating configurations in between [201, 202]. This explains why EDT fails to have a continuum limit, for which existence of a second-order phase transition is required. On the other hand, in CDT it is still not clear how the continuum limit is approached. While the A $\leftrightarrow$ C transition is first order [65, 68] and the B $\leftrightarrow$ D transition is second order [67, 68, 75],[5] the order of the transition between the bifurcation phase and phase C is under study. A first- or second-order transition would imply that the change of spacetime signature is, respectively, discontinuous or continuous. Recent results give preliminary indications that the B $\leftrightarrow$ C transition may be second or higher order [74].

11.3.2 Cosmology

In phase C (or $C_{\rm dS}$), a semi-classical limit is reached non-perturbatively. Here the geometry is macroscopically four-dimensional [55–57, 59] and can be identified, at large scales, with a de Sitter universe [61, 63, 66, 70]. The typical result is shown in the bottom right of Fig. 11.5. The number of simplices is fixed to some $N_4 = O(10^4)$; the vertical axis is the time span $t \in [0, t_{\rm tot}]$.

To check the geometric properties of this spacetime, one can study its dimensionality and the time evolution of spatial hypersurfaces. Let $d_{\rm H}$ be the Hausdorff dimension of the average geometry. For a classical spacetime, time intervals are expected to scale as $N_4^{1/d_{\rm H}} \propto \mathcal{V}_4^{1/d_{\rm H}}$ and spatial volumes as $N_4^{(d_{\rm H}-1)/d_{\rm H}} \propto \mathcal{V}_4^{(d_{\rm H}-1)/d_{\rm H}}$. For the output geometry to correspond to a macroscopic semi-classical spacetime, it should be $d_{\rm H} \approx 4 = D$ with good approximation. One check consists in setting $d_{\rm H} = 4$ from the start, rescaling spatial volumes and time as the dimensionless

[5]In [67, 68], phase D had not been discovered yet and it was indicated as "phase C."

quantities

$$\mathcal{V}_3 \to \frac{\mathcal{V}_3}{\mathcal{V}_4^{3/4}}, \qquad t \to \frac{t}{\mathcal{V}_4^{1/4}}, \tag{11.30}$$

and comparing the rescaled spatial slices for simulations with different N_4, for instance by measuring the volume-volume correlator $\langle \mathcal{V}_3(0)\, \mathcal{V}_3(\bar{t})\rangle := t_{\rm tot}^{-2} \sum_{j=1}^{t_{\rm tot}} \mathcal{V}_3(j)\, \mathcal{V}_3(j+\bar{t})$. It turns out that the profiles of the volume-volume correlator as a function of $\bar{t}$ match. Alternatively, one can let $d_{\rm H}$ free and fit it to match the volume correlators; indeed, the best fit for $d_{\rm H}$ is 4 within the given accuracy. As said above, the rescaling (11.30) is typical of a classical spacetime, which means that it is appropriate to call the geometry under study macroscopic. A further suggestive result is that the average geodesic distance $\langle r\rangle$ between points in the spatial volume $\mathcal{V}_3(t)$ scales, for any given t, as $\langle r\rangle \sim \mathcal{V}_3^{1/3}$. This scaling and the strong matching of the rescaling (11.30) indicate that what is produced by the simulations are genuine, macroscopic, four-dimensional spacetimes whose spatial volumes are three-dimensional. Another geometric indicator is the spectral dimension. At large scales, $d_{\rm S} \simeq 4$, while at small scales there is uncertain numerical evidence on the values $d_{\rm S} \simeq 2$ [58, 59] and $d_{\rm S} \simeq 3/2$ [72].

This is not sufficient, by itself, to conclude that we are observing the emergence of a smooth manifold, since there exist very "irregular" geometries with integer Hausdorff and spectral dimension. However, at large scales the geometry matches de Sitter spacetime. First, the Euclidean effective action for a large number N_3 of tetrahedra (i.e., large spatial volumes) is [57, 59, 61]

$$S_{\rm E}^{\rm eff} \propto \sum_{t=0}^{t_{\rm tot}} \left\{ b_1 \frac{[N_3(t+1) - N_3(t)]^2}{N_3(t)} + b_2 N_3^{1/3}(t) - \lambda N_3(t) \right\}, \tag{11.31}$$

where $b_{1,2}$ are constants and the parameter λ is determined so that $\sum_{t=0}^{t_{\rm tot}} N_3(t) = N_4(t)$. Calling the three-volume $N_3(t) \propto \mathcal{V}_3(t) =: a^3(t)$ and taking arbitrarily small time steps, up to an overall minus sign we obtain the action of Euclidean closed FLRW cosmology in synchronous gauge with scale factor a and cosmological constant, evaluated within a finite time interval:

$$S_{\rm E}^{\rm eff} \simeq -S_{\rm E}^{\rm FLRW} \propto \int_0^{t_{\rm tot}} dt\, a^3 \left(\frac{3}{\kappa^2} \frac{\dot{a}^2}{a^2} + \frac{1}{a^2} - \lambda \right). \tag{11.32}$$

Thus, at large scales and in this region of parameter space, one does get cosmology from background-independent quantum gravity. Furthermore, $\langle \mathcal{V}_3(t)\rangle$ is very well fitted by the classical three-volume of Euclidean de Sitter space $\mathcal{V}_3 = a^3(t)$ with $a(t) \propto \cos(bt)$, for some constant b related to b_1. The linear size of the universe is $O(10) l_{\rm Pl}$ [63]. Small quantum fluctuations accompany this configuration.

The embedding of the CDT semi-classical limit in a realistic history of the universe is still missing. The cosmological constant is fixed throughout the numerical simulations, which therefore cannot say anything about the Λ problems. The typical de Sitter universe is about 10 times larger than a Planck volume, implying that such configuration may be of interest only for primordial cosmology, for instance as the birth of an inflationary universe from quantum physics. How such birth and the UV properties of this geometry can accommodate the big-bang problem remains to be seen. Also, the CDT de Sitter universe has a finite life span and one should study how to extend its evolution to the future, possibly with the inclusion of matter degrees of freedom. Despite all these open issues, CDT gives encouraging hints that, indeed, cosmology can naturally come from quantum gravity.

To summarize, in CDT one starts from a fully background-independent, non-perturbative formulation of quantum gravity and obtains various geometric phases. In general, the geometry of quantum spacetime does not correspond to any continuous manifold. Under such premises, it is highly non-trivial to show the existence of a semi-classical limit. However, the theory does possess a phase where at large scales the geometry naturally tends to de Sitter spacetime, a stable, classical background of Planckian size. Therefore, accelerating cosmology can emerge at early times from quantum gravity non-perturbatively. At microscopic scales in the same phase, the spectral dimension of spacetime is appreciably lower than 4, signalling a highly non-classical UV geometric regime. This also gives an explicit illustration of the fact, mentioned in Sect. 10.1, that quantization and mini-superspace symmetry reduction do not commute: if they did, (11.32) would be valid at all scales, not only in the large-volume limit. On the contrary, at small scales geometry is radically different and the semi-classical limit (11.32) breaks down.

11.4 Spin Foams

A disadvantage of canonical quantization is the loss of general covariance. A complete covariant definition of the quantum dynamics may be achieved by the path-integral version of LQG going under the name of spin foams.

11.4.1 Framework

In Sect. 9.2.2, we briefly introduced the concept of path integral. In quantum mechanics and quantum field theory, this is related to the probability for a state specified at some initial configuration $\Sigma_{\rm i}$ (initial time $t_{\rm i}$ or field value $\phi_{\rm i}$) to evolve into a state with configuration $\Sigma_{\rm f}$ (final time $t_{\rm f}$ or field value $\phi_{\rm f}$). A notion of global time is employed in foliated CDT but it is absent in a general-covariant, background independent theory of gravitation such as the one symbolically represented by the path integral (9.91). In particular, in canonical formalism the dynamics is not encoded in an evolution variable but, rather, in the solutions to the quantum constraints. This can be achieved either via the group averaging detailed in

Sect. 10.2.2 or by a relational time variable that can be found from the available degrees of freedom of gravity and matter, a procedure known as *deparametrization*. The interpretation of the "initial" and "final" states is modified accordingly.

Just as LQG states are graphs labelled by group elements or group representations, spin-foam histories are 2-complexes labelled by the same type of data. A *2-complex* is a collection of faces, edges and vertices, triangulating a four-dimensional geometry bounded by an "initial" and a "final" spin-network state $|\Sigma_\mathrm{i}\rangle$ and $|\Sigma_\mathrm{f}\rangle$. These boundary states are solutions to the quantum first-class constraints. One usually takes the dual of the triangulation, constructed with the surfaces orthogonal to the faces of the 4-simplices. Then, the labels of edges and vertices of a spin network (respectively, half-integer spin numbers and intertwiners) are now attached to the faces and edges of the dual triangulation (a "foam" of spin labels); a dual vertex corresponds to a 4-simplex. In the continuum classical limit, the overall configuration can be heuristically regarded as a four-dimensional manifold with boundary, describing the "evolution" of the geometry Σ_i into Σ_f. At the quantum level, the transition amplitude is the physical inner product $Z = \langle\Sigma_\mathrm{f}|\Sigma_\mathrm{i}\rangle$ between the two states, in the case of LQG obtained by summing over all possible spin labels and dual triangulations with the same boundary. The sum over dual triangulations is called *vertex expansion*, since the n-th term of the series contains n vertices (i.e., n 4-simplices in the original triangulation). A simplex is a chunk of flat Euclideanized spacetime, so that the vertex expansion might be roughly interpreted (but see Sect. 11.5.1) as an approximation around flat spacetime. It well captures both the classical and quantum features of the continuum limit [94, 95]. The spin-foam models currently under the most intense scrutiny are the Engle–Pereira–Rovelli–Livine (EPRL) amplitude [85, 86, 89, 90] and its variant the Freidel–Krasnov (FK) amplitude [84, 87, 88, 91]. Asymptotically for large spin labels in the unitary representations of the gauge group, both EPRL and FK models reproduce the path integral $Z \simeq \exp(\mathrm{i}S_\mathrm{Regge})$ for the Regge action, of which we saw a foliated example in (11.29) [92, 93]. Therefore, spin foams possess not only a continuum limit, but also the correct semi-classical one of simplicial gravity. In this sense, CDT can be regarded as an asymptotic approximated limit of spin foams.

11.4.2 Cosmology

A path-integral formulation of quantum cosmology also exists, both for the old Wheeler–DeWitt quantization [215–217] and for LQC [100–110]. The dynamics of homogeneous and isotropic LQC is recast in a form that formally resembles the spin-foam formulation of the full theory; in particular, it is expressed in terms of a vertex expansion. This cosmological model can exemplify issues and conceptual points that one would eventually have to address in the spin-foam formulation of quantum gravity and can suggest solutions to the same by providing a testing ground for novel techniques.

The covariant construction of two-point functions is particularly simple because of the close analogy between flat cosmology with a massless scalar field and the relativistic point particle in quantum mechanics [216, 218]. In fact, not only do both systems have the same number of degrees of freedom, but they are also formally identical. To see this, we recall that the parametrized form of the particle action is

$$S=\int_{\tau'}^{\tau''}\mathrm{d}\tau\left(p_\mu\frac{\mathrm{d}x^\mu}{\mathrm{d}\tau}-N\mathcal{H}_0\right),\qquad \mathcal{H}_0=p_\mu p^\mu+m^2\,, \tag{11.33}$$

which is invariant under reparametrizations $\tau\to f(\tau)$ of the world-line. $\mathcal{H}_0$ is the Hamiltonian, which for a massless particle in $1+1$ dimensions reads $\mathcal{H}_0=p_x^2-p_t^2$ and can be quantized as $\hat{\mathcal{H}}_0=-(\partial_x^2-\partial_t^2)$. On the other hand, the space of physical states in canonical quantum cosmology is obtained by solving the Hamiltonian constraint

$$\hat{C}\,\Psi[v,\phi]=-\left(\Theta-\hat{p}_\phi^2\right)\,\Psi[v,\phi]=0\,, \tag{11.34}$$

corresponding to (10.30) for Wheeler–DeWitt cosmology and (10.110) for LQC. The two expressions only differ in the explicit form of the operator Θ, which we do not need to use here. In Sects. 10.2.2 and 10.3.4, the space of solution states was reduced in terms of two sectors satisfying, respectively, $(\hat{p}_\phi\mp\sqrt{\Theta})\Psi_\pm=0$, resulting from the fact that $\hat{p}_\phi$ is a Dirac observable for the system governed by the above dynamics. This allowed us to perform an explicit splitting of the space of physical states into positive- and negative-frequency states.

Now, the form of the constraint (11.34) is that of a relativistic particle in 1+1 dimensions (with a discretized spatial coordinate in the case of LQC), where Θ plays the role of a second-order spatial derivative (up to a mass terms) and the scalar field ϕ of the time coordinate t. The volume eigenstates $|v\rangle$ with normalization (10.15) are the counterpart of the position states $|x\rangle$. The conjugate momentum p_ϕ is the analogue of the energy E. Defining

$$|p_\phi\rangle:=\int\frac{\mathrm{d}\phi}{2\pi}\,\sqrt{2|p_\phi|}\,\mathrm{e}^{\mathrm{i}p_\phi\phi}\,|\phi\rangle\,, \tag{11.35}$$

one verifies

$$\hat{p}_\phi|p_\phi\rangle=p_\phi|p_\phi\rangle\,,\qquad p_\phi\in\mathbb{R}\,, \tag{11.36}$$

as well as the positive-definite normalization

$$\langle p_\phi|p'_\phi\rangle=2|p_\phi|\,\delta(p_\phi-p'_\phi)\,. \tag{11.37}$$

Alternatively, one may remove the absolute value in the definition (11.35) to obtain the indefinite normalization

$$\langle p_\phi | p'_\phi \rangle = 2p_\phi\, \delta(p_\phi - p'_\phi)\,. \tag{11.38}$$

Just as in the particle case, the definition of different two-point functions is fully characterized by three choices: (i) the canonical inner product, via the choice of canonical representation of the physical quantum states in the ϕ space, depending on which normalization between (11.37) and (11.38) is selected; (ii) whether one restricts the attention to either positive- or negative-frequency sectors of the Hilbert space, or works in the full Hilbert space (super-selection); (iii) the integration range in proper time (lapse) in the group-averaging representation of the two-point function, or, equivalently, the class of histories summed over in the sum-over-histories (i.e., path-integral) formulation of the same. Integrals over the full real line give, in general, solutions of the constraint equation and, thus, true inner products for the canonical theory. Integrals of the positive semi-axis only give Green's functions (propagators) for the same constraint equation, i.e., solutions of the constraint equation in the presence of a delta source.

After making the choices (i)–(iii), one can proceed to define various two-point functions for quantum cosmology and to give their spin-foam representation in terms of a sum over histories. As an example, consider a "relativistic" representation in terms of states

$$|v,\phi\rangle := |v,\phi;+\rangle + |v,\phi;-\rangle\,, \tag{11.39}$$

where the kets

$$|v,\phi;\pm\rangle = \int \frac{\mathrm{d}p_\phi}{2|p_\phi|}\, \mathrm{e}^{\pm \mathrm{i} p_\phi \phi}\, \delta(p_\phi \mp \sqrt{\Theta})\, |v,p_\phi\rangle \tag{11.40}$$

are not orthogonal at fixed ϕ. Choosing the positive-definite normalization (11.37) and the corresponding resolution of the identity in momentum space, these states satisfy the completeness relation

$$\mathbb{1} = \mathrm{i}\sum_v \left(|v,\phi;+\rangle\, \overleftrightarrow{\partial}_\phi\, \langle v,\phi;+| - |v,\phi;-\rangle\, \overleftrightarrow{\partial}_\phi\, \langle v,\phi;-| \right), \tag{11.41}$$

where $\overleftrightarrow{\partial} = \overrightarrow{\partial} - \overleftarrow{\partial}$. The sum becomes an integral in the WDW case. The inner product is then positive definite. Super-selecting the positive-frequency sector, only the first term in the resolution of the identity (11.41) would survive.

The inner product between two states $|v_{\rm i},\phi_{\rm i}\rangle$ and $|v_{\rm f},\phi_{\rm f}\rangle$ can be defined rigorously in the group averaging procedure (Sect. 10.2.2). In the relativistic representation (11.39) with both frequency sectors, the inner product is [105, 219][6]

$$G_{\rm H}(v_{\rm f},\phi_{\rm f};v_{\rm i},\phi_{\rm i}) := \int_{-\infty}^{+\infty} d\alpha\, \langle v_{\rm f},\phi_{\rm f}|e^{i\alpha\hat{\mathcal{C}}}|v_{\rm i},\phi_{\rm i}\rangle\,. \tag{11.42}$$

This two-point function governs the evolution of the universe from one volume-matter configuration to another. For the relativistic particle, the analogue of (11.42) is the Hadamard function, hence the subscript H. The Hadamard function has a sum-over-histories representation obtained by fixing the reparametrization invariance of the action (11.33) in proper-time gauge $dN/d\tau = 0$:

$$G_{\rm H}(x'',t'';x',t') = \int_{-\infty}^{+\infty} d\tau\; g(x'',t'';\tau|x',t';0)\,, \tag{11.43}$$

where $g(x'',t'';\tau|x'',t';0)$ is a non-relativistic transition amplitude for the Hamiltonian $\mathcal{H}_0$, in (proper) time τ. Using a different ("non-relativistic") canonical representation, one ends up with another type of two-point function, named Newton–Wigner, that can be described also via deparametrization [100, 102, 105]. Group averaging and the deparametrized framework lead to distinct vertex expansions, which correspond to different perturbative expressions of the same two-point function.

The Hadamard function (11.42) can be recast in the vertex expansion. One observes that $\hat{\mathcal{C}}$ consists of two pieces that act on the kinematical Hilbert spaces $\mathcal{H}_{\rm kin}^{\rm g}$ and $\mathcal{H}_{\rm kin}^{\phi}$, respectively, so that for (11.42) the group-averaged inner product takes the form

$$G_{\rm H}(v_{\rm f},\phi_{\rm f};v_{\rm i},\phi_{\rm i}) =: \int_{-\infty}^{+\infty} d\alpha\; A_{\rm H}(\Delta\phi;\alpha)A_{\Theta}(v_{\rm f},v_{\rm i};\alpha)\,,$$

where $\Delta\phi = \phi_{\rm f} - \phi_{\rm i}$ and

$$A_{\Theta}(v_{\rm f},v_{\rm i};\alpha) := \langle v_{\rm f}|e^{-i\alpha\Theta}|v_{\rm i}\rangle\,, \qquad A_{\rm H}(\Delta\phi;\alpha) := \int_{-\infty}^{+\infty}\frac{dp_\phi}{2\pi}\, e^{i\alpha p_\phi^2}e^{ip_\phi\Delta\phi}\,.$$

[6] Restriction of the integration interval to the positive semi-axis $\alpha \in [0,+\infty)$ leads to the definition of the Feynman Green's function $G_{\rm F}$. While two-point functions such as $G_{\rm H}$ define canonical inner products, Green's functions such as $G_{\rm F}$ are not solutions of the Hamiltonian constraint equation but they are true transition amplitudes (propagators), in the sense that they take into account the relative ordering in the "time" variable ϕ labelling the states (and thus defining a background-independent notion of "in" and "out") [105]. In other words, true transition amplitudes propagate solutions of the constraint equation into other solutions.

A_Θ is recast in the manner of Feynman by splitting up the "time" interval α into N parts of equal length $\epsilon = \alpha/N$ and introducing decompositions of the identity on $\mathcal{H}^{\rm g}_{\rm kin}$:

$$A_\Theta(v_{\rm f}, v_{\rm i}; \alpha) = \sum_{\bar{v}_{N-1},\dots,\bar{v}_1} \langle v_{\rm f}|{\rm e}^{-{\rm i}\epsilon\Theta}|\bar{v}_{N-1}\rangle\langle\bar{v}_{N-1}|{\rm e}^{-{\rm i}\epsilon\Theta}|\bar{v}_{N-2}\rangle \dots \langle\bar{v}_1|{\rm e}^{-{\rm i}\epsilon\Theta}|v_{\rm i}\rangle\,.$$

One then reorganizes the sum by characterizing each possible history $(v_{\rm i}, \bar{v}_1, \dots, \bar{v}_{N-2}, \bar{v}_{N-1}, v_{\rm f})$ by the number of volume transitions M that occur. In the limit $N \to \infty$ and performing the integration over α and, then, the one over p_ϕ, the final result has the cosine typical of the Hadamard two-point function [105]:

$$G_{\rm H}(v_{\rm f}, \phi_{\rm f}; v_{\rm i}, \phi_{\rm i}) = \sum_{M=0}^{+\infty} \sum_{\substack{v_{M-1},\dots,v_1\\ v_m\neq v_{m+1}}} \Theta_{v_{\rm f}v_{M-1}} \dots \Theta_{v_2v_1}\Theta_{v_1v_{\rm i}} \prod_{k=1}^{p} \frac{1}{(n_k-1)!}$$
$$\times \left(\frac{\partial}{\partial\Theta_{w_kw_k}}\right)^{n_k-1} \sum_{m=1}^{p} \frac{1}{\sqrt{\Theta_{w_mw_m}}} \frac{\cos(\sqrt{\Theta_{w_mw_m}}\Delta\phi)}{\prod_{\substack{j=1\\ j\neq m}}^{p}(\Theta_{w_mw_m}-\Theta_{w_jw_j})}, \tag{11.44a}$$

where

$$\Theta_{v_kv_l} := \langle v_k|\Theta|v_l\rangle \tag{11.44b}$$

are the matrix elements of Θ and the p distinct values appearing in $(v_{\rm i}, v_1, \dots, v_{M-1}, v_{\rm f})$ are denoted by $w_1, \dots, w_p$ with multiplicities n_i, so that $n_1 + \dots + n_p = M + 1$. This last step can only be done formally if the variable v has a continuous range. The expression (11.44a), which is rigorous for LQC, is understood to hold for WDW theory in this formal sense.

The purpose of the example (11.44a) is to illustrate how the form of the path integral depends on the various steps (i)–(iii); other possibilities are available in the literature. Here we are mainly interested in quoting the cosmological applications of the theory. To this purpose, we revert to a generic setting where a spin-foam sum has been properly derived.

To extract phenomenology, one must make a concrete choice for the boundary states $|\Sigma_{\rm i}\rangle = |v_{\rm i}, \phi_{\rm i}\rangle$ and $|\Sigma_{\rm f}\rangle = |v_{\rm f}, \phi_{\rm f}\rangle$. Assuming $|\Sigma_{\rm i}\rangle$ and $|\Sigma_{\rm f}\rangle$ to be coherent states and taking the leading term in the vertex expansion, the classical dynamics is recovered in the large-volume limit [103, 106, 110]. This is a weaker result than that in CDT, since here the path integral is calculated in the symmetry reduction of mini-superspace. However, one should appreciate that, away from the classical limit, the support of the LQC transition amplitude does not include $a = 0$ [110]. The canonical and path-integral approaches to FLRW loop quantum cosmology agree on the removal of the big-bang singularity.

11.5 Group Field Theory

In any canonical scheme such as LQG, while geometry is fully dynamical the topology of the universe is fixed by construction. In general, however, one may ask whether it is possible to build a quantum theory inclusive of topology change or, in other words, if one can envisage an interacting multiverse scenario obeying a set of quantum rules. In its general lines, this is the field (or, more improperly, "third") quantization approach mentioned in Sect. 10.2.4.

Beside the issue of topology change, the main difficulty faced by LQG is the complete definition of the quantum dynamics and the proof that the resulting theory contains Einstein's gravity in an appropriate limit. A tentative but complete definition of the quantum dynamics of spin-network states is obtained, via spin-foam models, by embedding LQG states into the larger framework of group field theory (GFT), in turn strictly related to tensor models [220, 221].

11.5.1 Framework

GFTs are quantum field theories on group manifolds. Instead of Lorentzian tensor fields $\phi_{\mu\nu\dots}(x)$ on a D-dimensional spacetime manifold, one has a complex-valued object $\varphi(g) := \varphi(g_1,\dots,g_D)$ dependent on D elements g_I ($I = 1,\dots,D$) of a given Lie group $\mathbb{G}$, the local gauge group of gravity. Gauge invariance of vertices is expressed by the property

$$\varphi(g_1,\dots,g_D) = \varphi(g_1 h,\dots,g_D h) \qquad \forall h \in \mathbb{G}\,, \tag{11.45}$$

i.e., the field $\varphi : \mathbb{G}^D \to \mathbb{C}$ is invariant under right multiplication of all its arguments g_I with the same group element h. The classical dynamics is governed by the action

$$\boxed{S_{\rm GFT}[\varphi,\varphi^*] = \int_{\mathbb{G}} d^D g \left\{ \int_{\mathbb{G}} d^D g'\, \varphi^*(g)\mathcal{K}(g,g')\,\varphi(g') + V[\varphi(g),\varphi^*(g)]\right\}\,,} \tag{11.46}$$

where the kinetic operator $\mathcal{K}$ is a non-local operator on $\mathbb{G}\otimes\mathbb{G}$ and the potential V is a non-linear interaction of the fields; choices of $\mathcal{K}$ and V fix the model. Variation of (11.46) with respect to φ^* yields the classical equation of motion (group domain in the integral omitted)

$$\int d^D g'\,\mathcal{K}(g,g')\varphi(g') + \frac{\delta V}{\delta\varphi^*(g)} = 0\,, \tag{11.47}$$

while varying with respect to φ gives the conjugate of (11.47).

The classical field $\varphi(g)$ is interpreted as the D-valent vertex of a spin network, with group labels $g_1, \dots, g_D$ attached to the D links. Each g_I is then the holonomy (parallel transport) of the connection along the I-th link, (11.45) being the gauge transformations acting on the vertex. To each vertex in a spin network there corresponds a $(D-1)$-simplex (in $D = 4$, a tetrahedron) in the dual simplicial complex. In this representation, $\varphi(g)$ is a $(D-1)$-simplex whose D $(D-2)$-faces are labelled by the g_I's and (11.45) is called closure constraint, since it can be shown to be equivalent to the requirement that the faces (four triangles, for a tetrahedron) close to form the simplex. The interaction term V in the action describes how $(D-1)$-simplices are glued together along their faces to form a D-simplex.

GFTs are the direct second quantization of spin networks [146]. The quantum scalar field $\hat{\varphi}$ can be expanded in terms of creation and annihilation operators on a Fock space. A choice of operator ordering is necessary in the interaction term, since $\hat{\varphi}$ and $\hat{\varphi}^\dagger$ do not commute and obey the algebra

$$\left[\hat{\varphi}(g), \hat{\varphi}^\dagger(g')\right] = \mathbb{1}_{\mathbb{G}}(g, g'), \qquad \left[\hat{\varphi}(g), \hat{\varphi}(g')\right] = 0 = \left[\hat{\varphi}^\dagger(g), \hat{\varphi}^\dagger(g')\right], \tag{11.48}$$

where $\mathbb{1}_{\mathbb{G}}$ is the identity operator on $\mathbb{G}$ compatible with (11.45). For a compact group, $\mathbb{1}_{\mathbb{G}}(g, g') := \int_{\mathbb{G}} \mathrm{d}h \prod_{I=1}^{D} \delta\left(g_I h g_I'^{-1}\right)$, where $\mathrm{d}h$ is the Haar measure of the group such that the group volume is normalized to 1, $\int_{\mathbb{G}} \mathrm{d}h = 1$.

The Fock states of GFT closely resemble spin networks or their dual and they represent certain geometries according to the choice of the group. In the continuum limit of differential manifolds, for Riemannian four-dimensional geometries one has $\mathbb{G} = SO(4)$, while for covariant Lorentzian four-dimensional models $\mathbb{G} = SL(2, \mathbb{C})$. When GFT is constructed as the generalization of loop quantum gravity, the group is $\mathbb{G} = SU(2)$ and the geometry described by the states is three-dimensional and spatial. In this case, the connection is the Ashtekar–Barbero connection and the elements of the group are the holonomies (9.107) along the edges of the tetrahedra.

The GFT Hilbert space is very similar to the standard Hilbert space of loop quantum gravity, although there are some technical differences.[7] These are not very important in what follows and we will stick with the LQG terminology in $D = 4$. The Fock vacuum $|\emptyset\rangle$ is, by definition, annihilated by $\hat{\varphi}$, $\hat{\varphi}|\emptyset\rangle = 0$, and corresponds to a "no-spacetime" configuration where no quantum-geometry degree of freedom is present and all area and volume operators have vanishing expectation value. It is normalized to 1 by convention, $\langle\emptyset|\emptyset\rangle = 1$. The one-particle GFT state $|g\rangle := \hat{\varphi}^\dagger(g)|\emptyset\rangle$ is interpreted as the creation of a four-valent spin-network vertex or of its dual tetrahedron with labels $g_1, \dots, g_4$. The labels of all the vertices of a spin network with $\mathcal{N}$ vertices are specified by constructing an $\mathcal{N}$-particle state.

[7]Contrary to LQG spin networks, the combinatorial structure of GFT states is not embedded in an abstract space; no cylindrical equivalence conditions are imposed; states associated to different graphs have a different scalar product.

Just like in LQG, one can define area and volume operators on the kinematical Hilbert space with different spectra. For instance, the square of the total area $\langle\hat{A}_I\rangle^2$ associated with all I-th faces in a given state will differ, in general, from the sum $\langle\widehat{A_I^2}\rangle$ of squared areas associated with each I-th face, where angular brackets denote the expectation value on the state of the operators

$$\hat{A}_I := \gamma\kappa^2 \int \mathrm{d}^4g\,\hat{\varphi}^\dagger(g)\sqrt{-\int \mathrm{d}^4g'\,\mathcal{K}(g_I,g')\,\hat{\varphi}(g')}\,, \tag{11.49a}$$

$$\widehat{A_I^2} := -(\gamma\kappa^2)^2 \int \mathrm{d}^4g\,\hat{\varphi}^\dagger(g) \int \mathrm{d}^4g'\,\mathcal{K}(g_I,g')\,\hat{\varphi}(g')\,, \tag{11.49b}$$

where γ is the Barbero–Immirzi parameter. The minus signs are such that expectation values are positive definite in the case of gravity in three spatial dimensions. Contrary to LQG, the number of spin-network vertices $\mathcal{N}$ is an observable in GFT, given by the expectation value of the operator

$$\hat{\mathcal{N}} := \int \mathrm{d}^4g\,\hat{\varphi}^\dagger(g)\,\hat{\varphi}(g)\,, \qquad \langle\hat{\mathcal{N}}\rangle =: \mathcal{N}\,. \tag{11.50}$$

The elements $B_I \in \mathfrak{g}$ of the Lie algebra associated with the group $\mathbb{G}$ are the coordinates of the GFT momentum space, related to group space via the momentum transform [122, 123, 129, 222, 223]

$$\tilde{\varphi}(B_1,\ldots,B_4) := \int \mathrm{d}^4g \prod_{I=1}^{4} e_{g_I}(B_I)\,\varphi(g_1,\ldots,g_4)\,, \tag{11.51}$$

where $\{e_g(B)\}$ is a basis of "plane waves" on $\mathbb{G}$. Since $\mathbb{G}$ is non-Abelian and $gh \neq hg$ for any two group elements, the product of two elements of the basis is *non-commutative*: $e_g(B) \star e_h(B) = e_{gh}(B) \neq e_{hg}(B) = e_h(B) \star e_g(B)$. The bivectors $(B_I)^{\alpha\beta}$ can be directly related to the flux variables of Sect. 9.3 and represent the oriented areas, spanned by the triads e_i^α, of the face I of the dual tetrahedron labelled by the g_I's. Therefore, the one-particle momentum-space Fock state $\hat{\tilde{\varphi}}^\dagger(B)|\emptyset\rangle$ fully determines the metric of one tetrahedron. In quantum mechanics, it would correspond to the momentum particle state $|\mathbf{p}\rangle$ with $\mathbf{p}$ being a vector multiplet with components p_I.

The quantum dynamics of the theory can be studied with the various methods available in quantum field theory, adapted to group manifolds. The GFT path integral or partition function can be written perturbatively as a series of Feynman diagrams. Decomposing the interaction $V = \sum_n \lambda_n V_n$ as a sum of polynomials of order n, one has

$$Z_{\mathrm{GFT}} := \int [\mathcal{D}\varphi][\mathcal{D}\varphi^*]\,\mathrm{e}^{-S_{\mathrm{GFT}}[\varphi,\varphi^*]} = \sum_\sigma \frac{\prod_n \lambda_n^{\mathcal{N}_n(\sigma)}}{\mathrm{Aut}(\sigma)}\,\mathcal{A}_\sigma\,, \tag{11.52}$$

where σ are cellular complexes dual to the GFT Feynman diagrams, $\mathcal{N}_n(\sigma)$ is the number of vertices with n legs in the Feynman diagram dual to σ, Aut(σ) is the order of the automorphisms of σ and $\mathcal{A}_\sigma$ is the Feynman amplitude assigned to σ. This expression can be compared with (11.28). The dynamics of GFT is defined by the superposition of interaction processes (creation and annihilation) of spin-network vertices, forming complexes of arbitrary topology; thus, topology is naturally made dynamical. The Feynman amplitudes $\mathcal{A}_\sigma$ of the theory can be represented as spin-foam models and, therefore, encode four-dimensional spacetime geometries. The spin-foam vertex expansion is nothing but the GFT perturbative expansion in powers of the GFT coupling constant in V; in this case, however, its interpretation as an expansion around flat spacetime is uncertain, since the GFT vacuum is highly non-geometric.

General relativity is not a sub-sector of the theory but a regime obtained wherever a continuum and classical limit of the pre-geometric fundamental construct exists. Gravity is therefore an emergent phenomenon within group field theory, as we are going to see now.

Matter fields can be added to the picture, either as emergent degrees of freedom of the theory arising as perturbations around background dynamical solutions [117–120] or by hand as new coordinates in an extension of the group manifold $\mathbb{G}^D = \mathbb{G} \otimes \cdots \otimes \mathbb{G}$ [116, 121]. In the second case and for a real scalar field, the GFT field becomes $\varphi(g) \to \varphi(g, \phi)$, where the ϕ dependence is rendered dimensionless. The generalization of the action (11.46), of the right-hand side of the commutation relations (11.48) and of the rest of the theory is straightforward.

11.5.2 Cosmology

We had several occasions to remark that loop quantum cosmology is not the cosmological limit of full LQG but, rather, a mini-superspace model (plus perturbations) employing LQG techniques. One of the advantages of GFT over LQG is the possibility to extract cosmology directly from the full quantum theory [130, 132–135, 139–141, 148]. The ensuing model warrants a thorough comparison with the findings of homogeneous LQC and has offered important consistency checks that the known features of LQC do belong to the complete theory and are not artifacts of the symmetry reduction before quantization. As a companion to CDT, GFT allows one to face the problem of how to obtain cosmological dynamical equations from a background-independent, non-perturbative formulation of quantum gravity.

A problem in GFT, not present in CDT, in obtaining the double limit of semi-classicality and the continuum is that the fundamental pre-geometric discrete structure contains different information with respect to gravity on a fixed topology. The corresponding method for defining this double limit is therefore more involved that those described in the preceding sections, albeit not overly so.

The physical interpretation of the continuum limit is somewhat delicate. We want to translate statements regarding differential manifolds (homogeneity in $D - 1 = 3$, for instance) into the language of simplicial complexes, but to do so it should be made possible to embed any such complex into a smooth continuous geometry. If a limit where the complex describes a differentiable spatial hypersurface exists, each tetrahedron is nearly flat. This *flatness condition* consists in having the triads e_i^α, associated with the bivectors B_I labelling the faces of each tetrahedron, almost constant. Since the four B_I's must close into a tetrahedron ($\sum_I B_I = 0$), for each tetrahedron one can reparametrize this information by three linearly independent bivectors defined by the *simplicity constraints*

$$(B_i)^{\alpha\beta} = \epsilon_i^{\ jk} e_j^\alpha e_k^\beta \,. \tag{11.53}$$

Consequently, one can approximate the integral of the triad along the edges of the tetrahedron T by its value $e_i^\alpha(x_{\mathrm{v}(T)})$ at the point $x_{\mathrm{v}(T)}$ where the dual spin-network vertex $\mathrm{v}(T)$ associated with T is located in the abstract embedding. Then, for each tetrahedron one has triad data that can be converted into information on the metric $g_{\alpha\beta}(x_{\mathrm{v}(T)}) = \delta_{ij} e_\alpha^i(x_{\mathrm{v}(T)})\, e_\beta^j(x_{\mathrm{v}(T)})$ at each point, and a description in terms of continuous fields is made possible. Retroactively, this allows one to compute the curvature of the geometry as a whole and compare it with the linear size of each tetrahedron. The flatness condition is thus nothing but the requirement that the tetrahedra be much smaller than the overall curvature radius of the embedding geometry. It is not an assumption imposed on the model a priori but, rather, a self-consistency check to be done at the end of the calculation. If this check failed, then the interpretation of the resulting quantum state in terms of continuum geometry would lose robustness or even break down.

With this caveat on board, let us construct GFT quantum states in the Fock space capable of describing homogeneous semi-classical geometries with a continuum limit. We begin with the vacuum theory and the double requirement of continuity and semi-classicality.

Consider an n-particle complex given by the superposition of different discrete geometries:

$$|\Omega_n\rangle = \prod_{j=0}^{n} \hat{\xi}_j |\emptyset\rangle \,, \qquad \hat{\xi}_j := \int \mathrm{d}^4 g\, \xi_j(g)\, \hat{\varphi}^\dagger(g) \,,$$

where the operators $\hat{\xi}_j$ are composed of creation operators summed over all possible group configurations. The group weights ξ_j are left-invariant under gauge transformations, $\xi_j(g_1, \dots, g_4) = \xi_j(hg_1, \dots, hg_4)$ for all $h \in \mathbb{G}$. Intuitively, a continuum limit is approximated by states with an infinite number of particles:

$$|\Psi_{\mathrm{cont}}\rangle = \lim_{\mathcal{N}\to+\infty} \sum_{n=0}^{\mathcal{N}} b_n\, |\Omega_n\rangle \,.$$

This most generic state corresponds to an arbitrarily non-classical geometry, which may or may not be well defined depending on whether the formal limit $\mathcal{N} \to +\infty$ can actually be performed. To achieve semi-classicality, one should also be able to replace operators $\hat{\varphi}$ in the quantum dynamics with a classical group field $\xi(g)$ representing their expectation value on the state. Note that the constituents of the GFT field remain quantum even if the field itself is taken to a classical limit. This means that, at small scales, the system is in a quantum regime where individual "molecules" or "atoms" interact quantum mechanically, while at large scales it will admit a classical description in terms of collective or "hydrodynamical" degrees of freedom. In the LQG interpretation, the scale separating the microscopic and macroscopic regimes is the Planck scale.

In fact, a configuration with (i) an infinite number of particles with a concrete $\mathcal{N} \to +\infty$ limit and (ii) realizing semi-classicality but with fully quantum elements is nothing but a condensate, of which we saw examples in Sects. 7.6.4 and 9.4.2. GFT condensates are known for quantum cosmology [130, 132–137] and for spherically-symmetric quantum geometries with an horizon [138]. Here we consider cosmology and the case where the condensate is of elementary building blocks, i.e., tetrahedra. Let

$$\hat{\xi} := \int \mathrm{d}^4 g\, \xi(g)\, \hat{\varphi}^\dagger(g) \tag{11.54}$$

be a single-tetrahedron operator with group weight ξ (called σ in the original literature) left-invariant under gauge transformations. By changing (11.54), one can obtain other types of states where the building blocks are composite ("molecules" or bound states such as pairs of tetrahedra). According to the lore of condensed matter, a condensate of tetrahedra is represented by the gauge-invariant kinematical state

$$|\xi\rangle := A\, \mathrm{e}^{\hat{\xi}} |\emptyset\rangle\,, \tag{11.55}$$

where A is a normalization constant. If, as in realistic situations, the state lies in the Fock space so that it does not define an inequivalent representation of the same quantum system, one can choose A so that $\langle\xi|\xi\rangle = 1$:

$$A = \exp\left[-\frac{1}{2}\int \mathrm{d}^4 g\, |\xi(g)|^2\right]. \tag{11.56}$$

An easy calculation shows that $|\xi\rangle$ is a *coherent state*, that is, an eigenstate of the annihilation operator $\hat{\varphi}$ with eigenvalue ξ,

$$\hat{\varphi}(g)|\xi\rangle = \xi(g)|\xi\rangle\,. \tag{11.57}$$

The metric can be reconstructed from ξ when working in the space of bivectors via the momentum transform (11.51).

Expression (11.55) defines a non-perturbative vacuum on the kinematical Fock space, on which the GFT field acquires a non-trivial expectation value: $\langle\xi|\hat{\varphi}|\xi\rangle = \xi \neq 0$. In condensed-matter physics, expanding the field $\varphi = \xi + \delta\varphi$ around its non-zero vacuum expectation value and truncating the equations of motion up to some order in the fluctuations $\delta\varphi$ is called *mean-field approximation*. To spell this approximation in precise terms, we consider the single-particle condensate (11.55). Full quantum dynamics is given by the infinite tower of constraints $\langle\xi|\hat{\mathcal{O}}\hat{\mathcal{C}}|\xi\rangle = 0$, where

$$\hat{\mathcal{C}} := \int \mathrm{d}^4 g'\, \mathcal{K}(g,g')\hat{\varphi}(g') + \frac{\delta\hat{V}}{\delta\hat{\varphi}^\dagger(g)} \tag{11.58}$$

is the quantum version of the classical equation of motion (11.47) and $\hat{\mathcal{O}}[\hat{\varphi},\hat{\varphi}^\dagger]$ is an arbitrary operator of the field and its Hermitian conjugate. Exact solutions to the quantum dynamics are able to solve all these conditions simultaneously. Approximated solutions can be found by imposing only the first of such constraints, with $\mathcal{O} = \mathbb{1}$. The expectation value of the quantum equation of motion (11.58) on $|\xi\rangle$ is the analogue of the Gross–Pitaevskii equation for Bose–Einstein condensation [224–226]. Taking a normal ordering in $\hat{V}$ such that all $\hat{\varphi}^\dagger$ are to the left of all the $\hat{\varphi}$, from (11.57) one has

$$0 = \langle\xi|\hat{\mathcal{C}}|\xi\rangle = \int \mathrm{d}^4 g'\, \mathcal{K}(g,g')\xi(g') + \left.\frac{\delta V}{\delta\varphi^*(g)}\right|_{\varphi=\xi}. \tag{11.59}$$

Solutions $\xi(g)$ of this equation give, when plugged into (11.55), approximate physical states.

The scalar weight $|\xi|$ is interpreted as a probability distribution on the space of homogeneous geometries. It is not a wave-function of the quantum geometry in the canonical sense, since (11.59) is non-linear in general (just like the Gross–Pitaevskii equation, which is a "non-linear Schrödinger equation" for the condensate wave-function). In WDW and loop quantum cosmology, a wave-function Ψ describes a single quantum Universe with fixed topology. In mini-superspace spin-foam cosmology, the quantum Universe is represented by a path integral analogous to that for an individual quantum particle. In GFT cosmology (or GFC in short), the scalar ξ is a highly quantum object, the interpretation of a continuum geometry and the semi-classical limit being recovered only by the macroscopic, large-scale collective behaviour of this many-particle ensemble. In the case of molecular condensates, ξ also encodes the correlation between different quanta.

In a homogeneous manifold, all points of space carry the same information on the metric or the connection. In a classical dual simplicial complex where the flatness condition holds for each individual tetrahedron, the equivalent of points of space are tetrahedra and their metric information is carried by their group or algebra labels. The redundancy required by homogeneity is thus achieved by asking that all the building blocks of the combinatorial structure be in the same microscopic configuration. This does not mean that the Hilbert space has been reduced to states with certain symmetry requirements, which would correspond to an ad hoc mini-superspace symmetry reduction. Rather, we have considered special states in the full Hilbert space and we have imposed that the quantum distribution $\xi(g)$ of geometric data g is identical for all the quanta. Accordingly, and in this precise sense, the GFT condensate (11.55), composed by an infinite number of tetrahedra all with the same wave-function, represents a quantum continuous geometry which is homogeneous at the scale of the quanta (the Planck scale). At ultra-macroscopic scales where only global properties are apparent, by construction the geometry is classical and homogeneous.

Note that the definition of semi-classicality employed here differs from the one from the point of view of spacetime geometry. In canonical quantum gravity, semi-classicality means peakedness of the wave-function of the Universe around some classical phase-space point. In GFT, this characteristic is additional to all the rest of the construction and should be imposed, if desired, after reaching the "classical" (in the sense of the mean-field approximation) hydrodynamical description in terms of a distribution on mini-supespace. This distribution, which enters final hydrodynamical equations, is formally equal to the quantum weight ξ but, at this effective level, it encodes information on the collective behaviour of the system rather than on the individual tetrahedra.

Inhomogeneities could be included in different ways, most of which are still under exploration. In general, at the microscopic level, inhomogeneities would correspond to deviations from the simple condensate states. Some possibilities include (i) quasi-particle fluctuations over the condensate, (ii) states in which a non-negligible number of fundamental GFT quanta do not aggregate to the condensate and (iii) multi-condensate states, i.e., states in which several large sub-sets of the GFT atoms condense in different states, each characterized by a different wave-function.

Another scenario, which amounts to an effective description of inhomogeneities, is the following. Once the hydrodynamical description of the quantum system is given, one can also interpret it as the purely statistical approximation of the dynamics of *classical* building blocks; the weight ξ is then regarded as a statistical probability distribution. These constituents might be reconstructed from the classical fluid by regarding any hydrodynamical observable (for instance, the total area associated with the state) as the composition of elementary observables (the area of each element). The statistically reconstructed constituents are not the fundamental GFT quanta. In fact, condensed quanta yield exactly the same distribution of geometries, but the reconstructed effective elements may be defined to differ in the value of their geometric observables (the value of their area) even if they are homogeneous individually.[8] The resulting effective geometry captures the properties

[8]Starting from the fundamental level, this can be realized by grouping a different number $\mathcal{N}$ of fundamental quanta in the effective building blocks.

of the universe at mesoscopic scales intermediate between the fundamental one and the global picture and it is interpreted as an inhomogeneous universe [137].

This picture is strikingly coincident with the separate-universe approach: the global geometry is described as a collection of patches of continuous and homogeneous space and inhomogeneities are the correlations among these patches. The effective constituents are the equivalent of the homogeneous patches and, in order for the mesoscopic regime to take place, their size turns out to be a few orders of magnitude larger than the Planck scale. The level of inhomogeneity depends, among other details, on the shape of the mean field $\xi(g)$. If the latter is sharply peaked around one particular set of group labels, then inhomogeneities are expected to be small, since the excursion of the geometric data between different patches is considerably limited. This aspect of the theory is still under development.

To summarize, the GFC condensate (11.55) is characterized by three scales. At the Planck scale, the geometry is quantum but continuous and homogeneous. At macroscopic scales many orders of magnitude larger than the Planck scale or than the present Hubble horizon, the geometry is classical (in the sense of the mean-field approximation), continuous and homogeneous. Some extra ingredient might be necessary to extend this notion of classicality with the one of canonical gravity (there, a classical spacetime is represented by a point in the classical phase space). In the effective approach to inhomogeneities, the transition between the micro- and macroscopic regimes is set by a mesoscopic scale a few orders of magnitude larger than the Planck scale, at which inhomogeneities are incorporated as an effect of statistical fine graining of the hydrodynamical description.[9]

To study a concrete model of quantum dynamics, one must make a choice of operators in (11.59). Renormalization analyses indicate that finiteness of the theory requires the kinetic operator $\mathcal{K}$ to be the Laplacian Δ_g on the group manifold [124, 127, 128, 131]. At first, we will ignore the matter part and assume that non-linear interactions are negligible, $V = M^2|\varphi(g)|^2$, where M is a dimensionless constant. The dynamical equation to solve is thus

$$\left(\sum_{I=1}^{4}\Delta_{g_I}+M^2\right)\xi(g)=0\,. \tag{11.60}$$

We now use the representation of $SU(2)$ in a neighborhood of the identity in terms of the generators (9.109), where $g_I(\pi_I)=\sqrt{1-\vec{\pi}_I\cdot\vec{\pi}_I}\mathbb{1}_2+2\vec{\tau}\cdot\vec{\pi}_I$ is a 2×2 matrix and the four three-vectors $\vec{\pi}_I$ are elements of the Lie algebra $su(2)$ such that $|\vec{\pi}_I|\leqslant 1$. The Laplacian with these coordinates reads (index I omitted everywhere)

$$\Delta_g\xi[\pi(g)]=(\delta^{ij}-\pi^i\pi^j)\partial_i\partial_j\xi(\pi)-3\pi^i\partial_i\xi(\pi)\,. \tag{11.61}$$

[9]The reader should consider this scale hierarchy *cum grano salis*, since the techniques and interpretations in GFC are still under intense study at the time of writing.

Thanks to gauge invariance and the closure condition of the group elements on the tetrahedron, one can manipulate (11.60) with the replacement (11.61) to express it only in terms of the elements $\vec{\pi}_{\mathcal{I}}$ of the first three links (dual faces). The $\vec{\pi}_{\mathcal{I}}$'s can be combined into the matrix invariants $\pi_{\mathcal{I}\mathcal{J}} := \vec{\pi}(g_{\mathcal{I}}g_4^{-1}) \cdot \vec{\pi}(g_{\mathcal{J}}g_4^{-1})$, where $\mathcal{I}, \mathcal{J} = 1, 2, 3$, $|\pi_{\mathcal{I}\mathcal{J}}| \leqslant 1$ and $\pi_{\mathcal{I}\mathcal{I}} \geqslant 0$. For isotropic states in which ξ depends only on the diagonal components $\pi_{\mathcal{I}\mathcal{I}}$, one gets

$$\sum_{\mathcal{I}=1}^{3}\left[2\pi_{\mathcal{I}\mathcal{I}}(1-\pi_{\mathcal{I}\mathcal{I}})\frac{\partial^2\xi}{\partial\pi_{\mathcal{I}\mathcal{I}}^2}+(3-4\pi_{\mathcal{I}\mathcal{I}})\frac{\partial\xi}{\partial\pi_{\mathcal{I}\mathcal{I}}}\right.$$
$$\left.+\sum_{\mathcal{J}\neq\mathcal{I}}\sqrt{1-\pi_{\mathcal{I}\mathcal{I}}}\sqrt{1-\pi_{\mathcal{J}\mathcal{J}}}\pi_{\mathcal{I}\mathcal{J}}\frac{\partial^2\xi}{\partial\pi_{\mathcal{I}\mathcal{I}}\partial\pi_{\mathcal{J}\mathcal{J}}}\right]+\frac{M^2}{4}\xi=0\,. \qquad (11.62)$$

Note that the assumption of isotropy $\xi = \xi(\pi_{11}, \pi_{22}, \pi_{33})$ does not amount to a classical symmetry reduction of the theory as in Chap. 10, since it is performed *after* quantization. Therefore, homogeneity is an ingredient recovered after taking the continuum limit of a special but fully quantum state, while isotropy is imposed only as a useful requirement to find analytic solutions of the GFT condensate giving rise to a FLRW cosmological background.

Compatibility with isotropy requires the last term in square brackets in (11.62) to vanish, implying that $\xi = \sum_{\mathcal{I}} \xi_{\mathcal{I}}(\pi_{\mathcal{I}\mathcal{I}})$ for some functions $\xi_{\mathcal{I}}$. For simplicity, one can further assume that the diagonal components are all equal, $\pi_{\mathcal{I}\mathcal{I}} = \chi$ for all $\mathcal{I}$, so that (11.62) is finally recast as [133]

$$2\chi(1-\chi)\frac{\mathrm{d}^2\xi(\chi)}{\mathrm{d}\chi^2}+(3-4\chi)\frac{\mathrm{d}\xi(\chi)}{\mathrm{d}\chi}+m\,\xi(\chi)=0\,, \qquad m:=\frac{M^2}{12}\,, \qquad (11.63)$$

where $0 \leqslant \chi \leqslant 1$. To give an interpretation to $\vec{\pi}$ and χ, we notice that, assuming that the connection remains approximately constant along a dual link with length l_0, the holonomy thereon is $g \simeq \exp(l_0\vec{\omega} \cdot \vec{\tau})$, where $\omega^i = e^\alpha A^i_\alpha$. Expanding as in (10.79), one has $g = \cos(l_0|\vec{\omega}|/2)\,\mathbb{1}_2 + 2\vec{\tau} \cdot (\vec{\omega}/|\vec{\omega}|)\sin(l_0|\vec{\omega}|/2)$, leading to the identification $\vec{\pi} = (\vec{\omega}/|\vec{\omega}|)\sin(l_0|\vec{\omega}|/2)$ and of χ as the square of the sine of the connection:

$$\chi=\sin^2\left(\frac{l_0|\vec{\omega}|}{2}\right)=:\sin^2\left(\frac{\bar{\mu}c}{2}\right). \qquad (11.64)$$

In the second step, we used a notation reminiscent of the cosmological setting of (10.72) and (10.74), where we encoded the information on the holonomy length and on the macroscopic fiducial volume into a parameter $\bar{\mu} = \bar{\mu}(a)$, whose time evolution is parametrized by the scale factor. In accordance with the lattice-refinement interpretation, it is natural to regard $\bar{\mu}$ as the number (10.143) of elementary cells per fiducial volume $\mu = l_0 \mathcal{V}_0^{-1/3} = \mathcal{N}^{-1/3}$. Near the identity,

$\vec{\pi} \simeq \vec{\omega}/2$ and $\sqrt{\chi} \simeq \bar{\mu}c/2$ is proportional to the connection c at low curvature. At the classical level for $\textsc{k} = 0$, $c \propto \dot{a}$, so that the low-curvature classical limit is

$$\chi \propto (a\bar{\mu}H)^2 \ll 1 \,. \tag{11.65}$$

The general solution of (11.63) is a linear combination of the associated Legendre functions of the first and second kind (here $P_\ell^m = (-1)^m P_{\ell,-m}$),

$$\begin{aligned}\xi(\chi) &= \xi_m(\chi)\\ &:= \left(\frac{1-\chi}{\chi}\right)^{1/4}\left[A_1 P^{\frac{1}{2}}_{\frac{1}{2}(\sqrt{1+2m}-1)}(2\chi-1) + A_2 Q^{\frac{1}{2}}_{\frac{1}{2}(\sqrt{1+2m}-1)}(2\chi-1)\right],\end{aligned} \tag{11.66}$$

which can be compared with (10.25) and (10.27) of WDW quantum cosmology. The solution (11.66), where $A_{1,2}$ are constants, is always normalizable with respect to the group measure. Its behaviour depends on the sign and value of m but, in the most general case,

$$\xi(\chi) \stackrel{\chi\ll 1}{\sim} \frac{1}{\sqrt{\chi}} \,. \tag{11.67}$$

The probability density $\xi(\chi)$ diverges when the connection variable χ tends to zero. Therefore, the general isotropic vacuum solution is infinitely peaked at small curvature, meeting the expectation that, in the continuum limit, tetrahedra of a classical geometry are nearly flat (spatially constant triad and connection). This is an important self-consistency check of the theory.

The exact vacuum solutions of (11.63) are well defined also in high-curvature regimes where $\chi \approx 1$ and the flatness condition fails; away from $\chi \approx 0$, they can even have finite oscillatory maxima. Moreover, a special class of solutions diverges also at $\chi = 1$. These regimes and special solutions are not unphysical but, according to the above discussion on the embedding picture, they do not admit a simple geometric interpretation in the language of continuous smooth manifolds. This situation is strongly remindful of what happens in LQC, where a non-classical dynamics is effectively described by equations on a continuum even if there is no underlying smooth manifold structure. It is in this sense that the Universe described by $\xi(\chi)$, although semi-classical, retains many quantum features, contrary to the WKB wave-functions of canonical quantum cosmology which represent conventional semi-classical geometries for all values of their arguments (scale factor a and matter field ϕ).

Including a matter scalar field, the kinetic operator reads

$$\mathcal{K} = \delta(g^{-1}g')\delta(\phi-\phi')\left(\sum_{I=1}^{4}\Delta_{g_I} + 12\mathcal{E}^2\partial_\phi^2 + M\right), \tag{11.68}$$

where $\mathcal{E}^2$ is some constant whose sign will be chosen later in relation with the classical equations of motion. The analytic extension of all the following equations to $\mathcal{E}^2 < 0$ is straightforward. Equation (11.60) becomes

$$\left(\sum_{I=1}^{4} \Delta_{g_I} + 12\mathcal{E}^2\partial_\phi^2 + M^2\right)\xi(g,\phi) = 0\,. \tag{11.69}$$

Under the isotropy condition, a one-parameter family of exact solutions in factorized form is

$$\xi(\chi,\phi) = \xi_{\beta m}(\chi)\,\left(B_+\mathrm{e}^{\mathrm{i}\Omega\phi} + B_-\mathrm{e}^{-\mathrm{i}\Omega\phi}\right), \qquad \Omega := \frac{\sqrt{(1-\beta)m}}{\mathcal{E}},$$

where $B_\pm$ are constant coefficients and β is real. When $\beta = 1$ and $\beta = 0$, one gets the special solutions $\xi(\chi,\phi) = \xi_m(\chi)\,(B_1 + B_2\phi)$ and $\xi(\chi,\phi) = (C_1 + C_2\sqrt{1/\chi - 1})(B_+\mathrm{e}^{\mathrm{i}\Omega\phi} + B_-\mathrm{e}^{-\mathrm{i}\Omega\phi})$, respectively ($B_{1,2}$ and $C_{1,2}$ are constants).

Another and no less important check is the recovery of the continuum classical equations of motion, but it may be possible to get even more: the modified Friedmann equation of homogeneous and isotropic LQC. At first, we take a WKB *Ansatz* of the form

$$\xi_{\mathrm{WKB}}(\chi,\phi) = \mathcal{A}(\chi,\phi)\,\mathrm{e}^{[\mathrm{i}S(\chi,\phi)-I(\chi,\phi)]/l_{\mathrm{Pl}}^2}\,. \tag{11.70}$$

A purely semi-classical approximation with $\mathcal{A} = \mathrm{const}$, $S = S(\chi)$ and $I = 0$ is incompatible with the exact solutions [133], which do not oscillate rapidly in the semi-classical region $\chi \ll 1$. In general, the functions $\mathcal{A}$, S and I must be tuned so that the continuum and classical geometric interpretation with $\xi \sim \chi^{-1/2}$ be given in vacuum by the simultaneous limits $l_{\mathrm{Pl}} \to 0$ and $\chi \to 0$. Incidentally, the problem of the normalization of WKB states discussed in Sect. 9.2.2 is solved in GFT, since the general exact solution of (11.69) is known.

From now on, we look for solutions of the form $S = S(\chi,\phi)$ and $I = 0$; damping terms can be included in $\mathcal{A}$. Plugging (11.70) into (11.63), expanding the constant m as $m = m_4 l_{\mathrm{Pl}}^{-4} + m_2 l_{\mathrm{Pl}}^{-2} + m_0$ and separating order by order in l_{Pl}, one obtains the WKB equations

$$0 = -2\chi(1-\chi)(S_{,\chi})^2 - \mathcal{E}^2(S_{,\phi})^2 + m_4\,, \tag{11.71a}$$

$$0 = 2\chi(1-\chi)S_{,\chi\chi} + 4\chi(1-\chi)S_{,\chi}\frac{\mathcal{A}_{,\chi}}{\mathcal{A}} + (3-4\chi)S_{,\chi} + \mathcal{E}^2\left(S_{,\phi\phi} + 2S_{,\phi}\frac{\mathcal{A}_{,\phi}}{\mathcal{A}}\right) - \mathrm{i}m_2\,, \tag{11.71b}$$

$$0 = 2\chi(1-\chi)\frac{\mathcal{A}_{,\chi\chi}}{\mathcal{A}} + (3-4\chi)\frac{\mathcal{A}_{,\chi}}{\mathcal{A}} + \mathcal{E}^2\frac{\mathcal{A}_{,\phi\phi}}{\mathcal{A}} + m_0\,. \tag{11.71c}$$

Setting $m_4 = 0$ in the first equation, we get

$$2\chi(1-\chi) = -\mathcal{E}^2 \left(\frac{S_{,\phi}}{S_{,\chi}}\right)^2. \tag{11.72}$$

Ignoring matter one has $2\chi(1-\chi) \approx 0$ which, consistently with the above analysis, admits solutions both in the continuum geometric limit $\chi \approx 0$ and away from it where $\chi \approx 1$.

As in the usual Hamilton–Jacobi formalism, we identify $\partial_\chi S \propto p_\chi$ and $\partial_\phi S \propto p_\phi$ with the semi-classical momenta. Classically, from (10.3) in $N = 1$ gauge they would correspond to

$$p_\chi \sim p_{\bar{\mu}^2 \dot{a}^2} \sim \frac{a}{\bar{\mu}^2 H}, \qquad p_\phi \sim a^3 \dot{\phi}. \tag{11.73}$$

Two main results stem from (11.72): (A) the purely classical limit fixes the behavior of $\bar{\mu}$ and (B) the limit of LQC effective dynamics is also recovered and confirms (A) [134].

(A) Plugging (11.65) into (11.71a), one has $H^2 \propto \mathcal{E}^2\dot{\phi}^2 - m_4 a^{-6} + (a\bar{\mu})^{-2}$, which is, assuming $\mathcal{E}^2 > 0$, the standard Friedmann equation for a massless scalar field and two extra contributions. One is a stiff matter term which can be removed by setting $m_4 = 0$. The other is a curvature term $\textsc{k} = 1$ if $\bar{\mu} = 1$ or a cosmological constant if $\bar{\mu} \propto a^{-1}$. The first possibility is excluded because a curvature term could only come from the classical connection $c = \gamma\dot{a} + 1$ and also because, if we want to embed LQC in group field cosmology and to identify the GFC function $\bar{\mu}$ with the LQC function (10.75), restrictions on the LQC matter content forbid a constant $\bar{\mu}$. The other choice is more interesting, but we will see that $\mathcal{E}^2 > 0$ does not lead to Lorentzian LQC. Also, Wick-rotating the above equation to compensate for a positive $\mathcal{E}^2$ ($H^2 \to -H^2$, $\dot{\phi}^2 \to -\dot{\phi}^2$) would give a negative cosmological constant.

We therefore turn to another derivation of the classical equation of motion. Taking the extreme regime $\chi \approx 0$, we now make the expansion $\chi(1-\chi) \simeq \chi$ in (11.72) and get

$$(a\bar{\mu})^{-2} \propto -\mathcal{E}^2\dot{\phi}^2. \tag{11.74}$$

If we take $\mathcal{E}^2 < 0$, the right-hand side is the scalar field energy density plus a cosmological constant. The left-hand side is H^2 only if $\bar{\mu} \propto 1/\dot{a}$. For the inverse power law (10.75) and an expanding universe, this condition is verified if $a \propto t^{1/(1-2n)}$ for $n < 1/2$, or if $a \propto e^{Ht}$ when H is constant in the improved quantization scheme $n = 1/2$. Although both cases rely on a specific form of the scale factor, the second is more realistic in the presence of a cosmological constant Λ, which is bound to dominate over matter asymptotically (de Sitter attractor, Sect. 5.10.1). Preliminary calculations confirm that only interactions, which we ignored here, could generate a Λ term when the scalar is in slow rolling. Remarkably, the choice

$$\bar{\mu} \propto \frac{1}{a} \tag{11.75}$$

is the one of the improved quantization scheme of Sect. 10.3.3. In the canonical theory, we saw how other quantization choices are compatible with the classical limit. If we want to embed LQC in group field cosmology and identify the GFC function $\bar{\mu}$ with the LQC function (10.75), and if we demand the classical limit of the GFC dynamics to be the Einstein-gravity Friedmann equation in de Sitter approximation, then one can conclude that the quantization ambiguity of the canonical theory is partially removed in this model of GFT. Of the functions $\bar{\mu}$ decreasing with the scale factor, only (11.75) gives rise to the classical Friedmann equation. At the level of homogeneous and isotropic cosmology, any other quantization choice of the canonical theory would make the embedding of LQC into GFC inconsistent with the classical limit of general relativity.

(B) One can also obtain the Friedmann equation (10.119) (equivalent to (10.121)) of loop quantum cosmology for general χ. Observing that $4\chi(1-\chi) = \sin^2(\bar{\mu}c)$, (11.72) and (10.119) agree provided $\mathcal{E}^2 < 0$ and the Hamilton–Jacobi momentum p_χ be

$$p_\chi \propto \sqrt{\frac{\alpha}{\nu}\frac{a^2}{\bar{\mu}}}, \tag{11.76}$$

where α and ν are the inverse-volume LQC corrections of the gravity and matter sectors. Equation (11.76) and the functions α and ν have not been derived from first principles, but the characteristic structure of LQC dynamics is indeed reproduced. The classical limit $\alpha, \nu \to 1$ agrees with (11.73) only if $\bar{\mu} \propto 1/\dot{a}$, consistently with (11.74).

The WKB approximation may be too limited, or even inadequate [133], to describe GFT condensates, but there are other ways to recover the improved dynamics of LQC. One of them makes the same assumptions on the kinetic term of GFT (the group Laplacian up to a constant) and on the potential (quadratic if the constant in the Laplacian is zero; vanishing if such constant is non-zero) [137]. With $\hat{\mathcal{O}} = \int \mathrm{d}^4 g\, \hat{\varphi}^\dagger(g)$ and recalling the operators (11.49) and (11.50), the constraint $\langle\xi|\hat{\mathcal{O}}\hat{\mathcal{C}}|\xi\rangle = 0$ can be written as

$$\sum_I \langle\widehat{A_I^2}\rangle - (\gamma\kappa^2 M)^2\langle\hat{\mathcal{N}}\rangle = 0\,, \tag{11.77}$$

which is an alternative form of (11.59) and (11.60). The first expectation value is proportional to the square of the fiducial area $a^2\mathcal{V}_0^{2/3}$ but the proportionality coefficient is also important: for consistency, in the low-curvature limit one should recover the Poisson brackets (classical algebra) between area and holonomy variables. In GFT, these variables are given by the expectation values $\mathrm{i}\gamma\kappa^2\langle\int \mathrm{d}^4 g\, \hat{\varphi}^\dagger\partial_i\hat{\varphi}\rangle \propto a^2$

and $\langle \int d^4g\, \pi^i \hat{\varphi}^\dagger \hat{\varphi} \rangle / \mathcal{N} \propto \sin(\mathcal{N}^{-1/3} c/2)$, where we omitted numerical coefficients. In particular, the second expectation value is an extensive quantity and, therefore, scales as $\mathcal{N}$; the overall scaling of the right-hand side is thus independent of $\mathcal{N}$. Therefore, the only way to compensate the $\mathcal{N}$ dependence in the holonomy variable in the low-curvature limit (and hence to have a Poisson bracket $\{c, a^2\} \simeq$ const) is to attach a factor $\mathcal{N}^{1/3}$ to the area variable a^2. In the homogeneous limit, this implies that $\langle \hat{A}_I \rangle \simeq \mathcal{N}^{1/3} a^2 \mathcal{V}_0^{2/3} \simeq \sqrt{\mathcal{N} \langle \widehat{A_I^2} \rangle}$. Assuming all I contributions in (11.77) are equal, the latter reads $4\mathcal{N}^{-1/3} a^4 \mathcal{V}_0^{4/3} - (\gamma \kappa^2 M)^2 \mathcal{N} = 0$, implying that

$$\mathcal{N} \propto a^3 .$$

This is nothing but the volume–number-of-vertices scaling in the improved quantization scheme of LQC.

We have thus found that the effective dynamics of the isotropic condensate state of group field theory with Laplacian kinetic operator may be equivalent to that of homogeneous and isotropic LQC in the improved quantization scheme (11.75) where the area of elementary holonomies is constant. This constitutes a somewhat surprising example of a cosmological model of quantum gravity where the operations of mini-superspace symmetry reduction and quantization possibly commute (see the diagrams at the beginning of Sect. 10.1). Group field theory can have the potential of deriving homogeneous and isotropic LQC from a fundamental quantum setting. In doing so, at least part of the quantization ambiguities unavoidable in the Hamiltonian formalism are removed for a given GFC model. The lattice-refinement interpretation of Sect. 10.3.8 is also naturally embedded in GFT: the number of vertices $\mathcal{N}$ corresponds to the expectation value of the number operator $\hat{\mathcal{N}}$, the number of particles of the condensate for a given macroscopic volume.

Evidence has been piling up that LQC at large can be fully derived from GFT but these results need some extensions. The Laplacian Δ_g and the mass term used above do not correspond to GFT models based on popular spin-foam amplitudes such as the EPRL model, and one can conceive more involved kinetic operators $\mathcal{K}$ and interactions V. Regarding the kinetic term, a generalization of the EPRL model has indeed been considered in [139, 140], where LQC-like Friedmann equations with a bounce were derived from a condensate. Non-trivial interactions V have just begun to be implemented in models with the Δ_g [141], while there are still no studies for the EPRL Laplacian. Non-linear interactions are carriers of topology change and can produce an effective cosmological constant term as well as non-minimally coupled effective models when matter is included [227]. Moreover, the mean-field approximation should be controlled by computing fluctuations around ξ and checking that they remain small. If they did, then the quantum solutions above would remain peaked at small curvature. This check would also be crucial for the study of cosmological inhomogeneities and their comparison with LQC perturbations. Cosmological perturbations, in fact, would be naturally interpreted as phonons, fluctuations propagating in the condensate.

The GFC approach will likely have something to say about the cosmological constant problem. The peak (11.67) in the probability density can be translated into one for Λ, since in the classical limit (where the standard Friedmann equation holds), in the improved-dynamics prescription (11.75) and for negligible or nearly constant matter energy density one has $\chi \simeq H^2 \propto \Lambda$ and

$$\xi(\Lambda) \sim \frac{1}{\sqrt{\Lambda}}. \tag{11.78}$$

This peak is less pronounced than the exponential probabilities found in WDW quantum cosmology, but it is perhaps better motivated, as it does not rely on a mini-superspace quantization.

The resonance between the ideas exposed here and in Sect. 9.4.2 might not be accidental, either. The equation of motion (9.131) and the corresponding Hamiltonian constraint have interaction terms that could find a justification in the second quantization of GFT. The issues with the Chern–Simons state disappear in GFT, where we deal with normalizable states of the full theory. It is quite conceivable that the condensate of the Ashtekar connection of the rudimentary model of Sect. 9.4.2 might be the approximation of a GFT condensate on the Fock space in group variables. To date, this possibility has not been investigated yet.

11.6 Causal Sets

A spacetime with a discrete texture may avoid the infinities plaguing general relativity and quantum field theory. Loop quantum gravity is an example where such type of geometry apparently removes gravitational singularities. Causal sets [149] is another theory with fundamental discreteness devised for the same motivation. Its development is still at an early stage but one of its most intriguing features is an actual prediction for the value (2.118) of the cosmological constant.

11.6.1 Framework

The type of discreteness considered in causal sets is such that any bounded region of spacetime is made of a finite (rather than just countable) number of points. This means that measuring any D-dimensional spacetime volume $\mathcal{V}$ is tantamount to counting the N points within $\mathcal{V}$. To match units, we assume that an average number α^2 of points occupy a volume of Planck size $t_{\rm Pl} l_{\rm Pl}^{D-1} = m_{\rm Pl}^{-D}$:

$$N \approx \alpha^2 m_{\rm Pl}^D \mathcal{V}, \tag{11.79}$$

where $\alpha > 0$ is a free real parameter and the symbol "$\approx$" will be clarified shortly. If $\alpha = O(1)$, then the Planck scale $l_{\rm Pl}$ is also the discreteness scale l_* characterizing the theory, and Planck volumes are elementary. If $\alpha \ll 1$, then $l_* \gg l_{\rm Pl}$ and the scale at which discreteness effects become apparent is larger than the Planck size.

A second assumption is that this set of points has an ordering, which reproduces the causal structure of continuous Lorentzian geometries whenever these can be recovered after some coarse-graining or large-scale approximation on the set. The most general set will not possess a macroscopic light-cone continuum structure, which will therefore be a byproduct of the fundamental ordering. The appealing economy of the recipe "order and number" has the potential of unifying topology, geometry and causal structure.

11.6.1.1 Kinematics and Poisson Statistics

These properties can be made rigorous and embodied, with no reference to any embedding, in a causal set (or *causet*) $\mathcal{C}$, defined by a partial-order relation $\preceq$ and the following rules:

1. Reflexivity: for all $x \in \mathcal{C}$, $x \preceq x$.
2. Transitivity: for all $x, y, z \in \mathcal{C}$, $x \preceq y \preceq z \Rightarrow x \preceq z$.
3. Anti-symmetry (or non-circularity, or acyclicity): for all $x, y \in \mathcal{C}$, $x \preceq y \preceq x \Rightarrow x = y$. In the continuum approximation, this property forbids closed time-like curves.
4. Local finiteness: for all $x, y \in \mathcal{C}$, $\mathrm{Card}([x, y]) < \infty$, where Card is the cardinality (the number of points) of the "interval" set $[x, y] := \{z \in \mathcal{C} \,|\, x \preceq z \preceq y\}$, i.e., all the ordered points comprised between x and y.

In particular, whenever a Lorentzian-manifold limit exists, the relation $x \prec y$ corresponds to the continuum statement $x \in \mathcal{J}^-(y)$, i.e., the point x belongs to the causal past of y (Sect. 6.1.1). The partial order $\preceq$ is therefore the discrete equivalent of the time ordering. The order is said to be partial because not all points are related by $\prec$. In the continuum, two such unrelated points are separated by a space-like vector.

Before considering any dynamics, one should understand the kinematical properties of causets, namely, the circumstances under which concepts such as length, geodesic, spacetime dimensionality and continuum can emerge. The latter is often studied by sprinkling (i.e., randomly generating) a finite set of points in the portion of a Lorentzian manifold $\mathcal{M}$ and then imposing an order relation reproducing the light-cone structure of the given continuum geometry. One then attempts to classify the type of causal sets compatible with this embedding. The intuitive conjecture, by now extensively verified, underlying this procedure is that a causet $\mathcal{C}$ admits a continuum approximation $\mathcal{M}$ if it can be generated by sprinkling on $\mathcal{M}$.

In order to preserve both Lorentz invariance and the point-volume relation (11.79), the random sprinkling must be done via a Poisson process. The probability that the random variable N take the value n is given by the Poisson distribution of n

points into a spacetime volume $\mathcal{V}$:

$$P(N = n) = \frac{1}{n!} (\rho\mathcal{V})^n e^{-\rho\mathcal{V}}, \tag{11.80}$$

where ρ is the sprinkling density. By this definition, the expected value of N is thus $N = \rho\mathcal{V}$ and its variance is $\sigma_N^2 = \langle N^2 \rangle = \rho\mathcal{V} = N$. Consequently, any determination of N will carry a statistical uncertainty ΔN given by the standard deviation

$$\Delta N := \sigma_N = \sqrt{N}, \qquad \Delta\mathcal{V} = \frac{\sqrt{\mathcal{V}}}{\alpha m_{\rm Pl}^{D/2}}. \tag{11.81}$$

Such fluctuations, often called Poisson noise, are purely kinematical and do not entail any dynamics. Thus, the relation (11.79) is not an exact equality but a statistical one, valid up to fluctuations of the size (11.81).

The Poisson distribution (11.80) is manifestly covariant since it depends only on the D-volume element $\mathcal{V}$ and on the sprinkling density ρ. This does not ensure that the resulting causet be covariant but, for a Minkowski embedding, it can be proven that individual realizations of the Poisson process are indeed Lorentz invariant [164]. A Lorentz transformation on a causet would change the relative position of the elements but not the properties of their statistical distribution. Also, no finite-valency graph (i.e., such that each node has a finite number of neighbors) can be associated with a sprinkling of Minkowski spacetime consistent with Lorentz invariance. A regular lattice or the graphs used in spin foams and CDT, for instance, do not satisfy this property. The intuitive reason is that a regular distribution of points can be boosted to a distribution of over-dense and under-dense regions, thus changing the statistics (e.g., [161]). Therefore, the double requirement of discreteness and Lorentz invariance seems to be inextricably related to a random structure at the fundamental level.

The calculation of geodesic distances [152] and of other geometric indicators such as the spacetime dimension [151, 178, 181], the Riemann curvature [173, 174, 177] and topology [163, 170] clarifies the relations between discrete ordered structures and macroscopic faithful embeddings. In general, these quantities depend on the number N of points in a given volume $\mathcal{V}$ and on the number R of relations within the interval $[x, y]$ between any two points in the set.

11.6.1.2 Dynamics

The dynamics of causets is under construction through different approaches. In one of them (*classical sequential growth*), classical dynamics is regarded as a process of stochastic growth of elements of the causet from existing elements [154, 162]. There is no actual direction of growth apart from that determined by the order relation $\prec$; however, one can take N as a monotonically growing "time" variable along which the birth process unfolds. The details of the process determine the probability of each transition from a set of N elements to one with $N+1$ elements and, in particular, the probability of forming specific causal sets for a given N. Each new element can either be in the future of an existing point or be space-like, that is, not related by $\prec$; no elements can be added to the past. Matter modes can emerge in the effective action without introducing them by hand.

The choice of action S is related to the recovery of the gravity+matter continuum limit $S[\mathcal{C}] \to S_{\rm eff}[g_{\mu\nu}, \phi]$. Call *link* the relation $y \prec x$ without other elements in between. For a Minkowski embedding, the probability to find n points y in a region A with volume $\mathcal{V}$ sharing a link with a given point x near A is finite due to discreteness. By Lorentz invariance, the same statement holds for any other region A' (disjoint from A) obtained by a sufficiently large boost on A, so that x actually has an infinite number of links, many of which at large spatial distances. The theory is therefore non-local at the kinematical level [150, 151]. If a continuum limit exists, then higher-order derivative terms in the dynamics are suppressed by powers of the discreteness fundamental length scale, $\mathcal{L} = a_0 l_*^{-4} + a_1 l_*^{-2} R + a_2 R^2 + O(l_*^2)$ (the a_i's are dimensionless constants), and the effective action $S_{\rm eff}$ is approximately local. General covariance then guarantees that the Einstein–Hilbert action is recovered [173, 181]. Evidence has been collected that kinematical Laplace–Beltrami operators can be constructed so that the effective dynamics is approximately local at sufficiently large scales, where the transition scale between the non-local and local regimes can be much larger than l_* [173, 174, 177]. These operators constitute the local limit of a family of highly non-local operators, which can capture salient features of causet dynamics in the continuum approximation but in regimes well below the non-locality length scale [179, 180].

At the quantum level, transition probabilities are replaced by transition amplitudes. The models of classical sequential growth are a somewhat intermediate step between classical and quantum dynamics, since their intrinsic stochastic character is akin to a quantum process; their relation to quantum causal sets is similar to the one between Brownian motion and quantum mechanics [186]. Since time (and emergent Lorentz invariance) is discrete and inevitably included in the picture via the ordered causal structure, the canonical formalism is perhaps less convenient than the study of the local properties of amplitudes in a sum-over-histories approach. In this case, the dynamics is governed by the path integral

$$Z[N] = \sum_{\mathcal{C}\in\Omega} e^{iS[\mathcal{C}]}\,, \tag{11.82}$$

where Ω is a sample of causets suitably chosen [154, 159, 175] and $S[\mathcal{C}]$ is an action dependent on the details of the causet $\mathcal{C}$, in particular N and R. In the classical growth model of [154], N acts as a stochastic time parameter, so that one does not expect to extend the sum in (11.82) to causets with different N. This is the reason why we made the N dependence explicit in the left-hand side.

11.6.2 Cosmology

Without entering into the details of the dynamics, classical sequential growth shows evidence of some cosmology, where a large homogeneous and isotropic universe emerges after a period of cycles of expansion and contraction with increasing maximal spatial volumes [155–157, 172]. The initial configuration of each cycle is a single causet element (called "post" in combinatorial language). It corresponds to what we would call, in a continuum geometry, a big bang or big crunch, since the whole universe is constituted by just one spacetime point. Each post is followed by an exponential volume expansion resembling a de Sitter phase. After each cycle, the dynamics repeats itself but with different values of the fundamental constants. Since there is no infinity associated with a post, the big-bang problem somehow disappears altogether. Such results have not been developed to the stage of a full, realistic cosmological scenario. Here we concentrate on two much more generic predictions of causets about, respectively, the production of B-modes in the CMB polarization [161, 169, 228] and the smallness of Λ [153, 158, 176, 181].

11.6.2.1 Generation of B-Modes

Since Lorentz invariance is preserved in the Minkowski embedding, the effects of microscopic discreteness and stochastic fluctuations can be encoded in Lorentz-invariant dynamical equations. In particular, one can study the diffusion of point particles, for instance the Brownian motion in a medium or the propagation of photons across cosmological scales. Consider the latter. The small random fluctuations in the momentum of the particle are reflected, in the continuum approximation, into a drift term in the diffusion equation of the particle in momentum space [161, 169]. For a polarized photon on a FLRW background, this drift or friction term depends on a coefficient d and on the first derivative of the azimuthal angle parametrizing the polarization space. It violates parity explicitly and causes a rotation of the polarization vector during the propagation of the photon by an angle $\chi = 2td/\nu$, where ν is the photon frequency. Therefore, B-modes are generated in the CMB from gradient E-modes and non-vanishing cross-spectra, even in the absence of primordial gravitational waves [228]:

$$C_\ell^{BB} \propto C_\ell^{EE} \sin^2(2\chi)\,, \qquad C_\ell^{TB} \propto C_\ell^{TE} \sin(2\chi)\,, \qquad C_\ell^{EB} \propto C_\ell^{EE} \sin(4\chi)\,. \tag{11.83}$$

When $d = 0$, one has $C_\ell^{TB} = C_\ell^{EB} = C_\ell^{BB} = 0$. The model of diffusion discussed here has a wider application than causal sets and holds in any Lorentz-invariant theory with an uncertainty in the spacetime structure at small scales. Observations on the CMB polarization spectra can in principle constrain the parameter space of causets in the continuum approximation. In Sect. 5.9.2, we saw other examples of Lagrangian models where non-vanishing TB and EB spectra can be originated [229].

11.6.2.2 Cosmological Constant

To address the Λ problem in causet theory, one first argues that an $O(1)$ cosmological constant in the effective dynamics is compensated by some mechanism, so that $\Lambda = 0$ in average. Since the observable patch of the universe is made of a random distribution of points n with mean N, this compensation does not take place exactly but within some statistical accuracy. The observed cosmological constant is given by this effect.

To begin with, one notes that the kinematical predetermination of N in the causet approach is closely akin to the volume fixing of unimodular gravity discussed in Sect. 7.6.3. In fact, in the quantum theory (11.82) the sum is taken over a family of causets with fixed N which, in the continuum or macroscopic approximation, corresponds to a partition function over D-dimensional geometries with a given volume $\mathcal{V}$. Classically and in the continuum approximation, it is clear that $\mathcal{V}$ and Λ are conjugate quantities: they appear in the effective action via the usual combination $\mathcal{V}\Lambda/\kappa^2 = \int d^D x\sqrt{-g}(\Lambda/\kappa^2)$, where $\kappa^2 \propto m_{\rm Pl}^{2-D}$. Since $\mathcal{V}$ is fixed, the parameter in front of it is a Lagrange multiplier which can be chosen arbitrarily to compensate a bare cosmological constant. However, we also have to take into account the statistical fluctuations of these variables. Just as time and energy are conjugate quantities in ordinary quantum mechanics, so are the volume and the cosmological constant in a statistical sense and their fluctuations obey an uncertainty principle:

$$\Delta\mathcal{V}\,\Delta\Lambda \gtrsim \kappa^2\,. \tag{11.84}$$

If these were just ordinary quantum fluctuations, we would be unable to determine either their magnitude or the mean $\langle\Lambda\rangle$. However, they are governed by the Poisson distribution (11.80), from which we know both the mean and the variance for $\mathcal{V}$. From the combination $\mathcal{V}\Lambda$ in the action, one can interpret Λ as the action of α^2 fundamental elements; supposing each contribution is independent and fluctuates in sign by $\pm\alpha^2$, then $\langle\Lambda\rangle$ is expected to be zero or very close to zero [158]:

$$\langle\Lambda\rangle = 0\,. \tag{11.85}$$

A further suggestion in favour of (11.85) is that only $\Lambda = 0$ configurations are stable against fluctuations of the causal set not corresponding to manifolds [153].

To summarize: (i) discreteness and Lorentz invariance require a Poisson sprinkling; (ii) this is assumed also in curved scenarios; (iii) the statistical uncertainty and the volume fixing imply the uncertainty relation (11.84), so that large Poisson fluctuations of the volume are compensated by small fluctuations of the cosmological constant and vice versa; (iv) from (11.81) and (11.84) and assuming (11.85), when the uncertainty principle is minimized one has the estimate

$$\rho_\Lambda = \frac{\Delta\Lambda}{\kappa^2} \sim \frac{1}{\Delta\mathcal{V}} = \frac{\alpha m_{\rm Pl}^{D/2}}{\sqrt{\mathcal{V}}} = \frac{\alpha^2 m_{\rm Pl}^D}{\sqrt{N}}\,. \tag{11.86}$$

In general, one can choose $\mathcal{V}$ as the volume of the past light cone of any representative point on the hypersurface for which one wants to calculate the value of $\mathcal{V}$. In a cosmological context, the total D-volume of the observable universe is, at any given time t since the big bang, $\mathcal{V} \sim tH^{-(D-1)} \sim H^{-D}$, so that $\rho_\Lambda \sim \alpha(m_{\rm Pl}H)^{D/2}$. Only in $D = 4$ dimensions ρ_Λ is, *at any given time*, of the same order of magnitude of the critical density,

$$\rho_\Lambda \sim \alpha m_{\rm Pl}^2 H^2 . \tag{11.87}$$

To explain dark energy, α should be in the range $\alpha = O(10^{-2}) - O(1)$. In particular, according to (2.11) and (2.13), if $\alpha = O(1)$ we get the very large number today ($t = t_0$)

$$N_0 \approx m_{\rm Pl}^4 \mathcal{V}_0 \sim 10^{244} . \tag{11.88}$$

From (11.86), one thus obtains the same order of magnitude as in (2.118), $\rho_\Lambda \sim 10^{-122} m_{\rm Pl}^4$.

Like asymptotic safety (Sect. 11.2.2), this scenario resembles phenomenological models where $\Lambda(t)$ is time dependent but without kinetic term. Equation (11.87) is the basis for the solution of the cosmological constant problems in causet theory. The old problem is solved because Λ acquires the observed value today, while the coincidence problem can be potentially addressed because (11.87) embodies an ever-present Λ component with tracking behaviour, as numerical solutions indicate [158, 176]. Note also that (11.87) is valid only in four dimensions; the resolution of the cosmological constant problem is therefore tightly related to the number of topological dimensions of the embedding. This is one of the many instances in theoretical physics where one question (Why is the cosmological constant so small?) is answered by replacing it by yet another question of different type (Why four dimensions?).

However, some fine tuning lurks behind the causet solution of the Λ problem. Fluctuations of the ever-present cosmological constant cannot be as large as $\alpha = O(10^{-2}) - O(1)$, lest they shift the main milestone epochs of cosmic evolution (matter-radiation equality, decoupling, and so on) to unacceptable values. Such shifts would displace the peak positions of the CMB and galaxy-clustering power spectra away from their observed values. In particular, the large-scale CMB spectrum constrains the parameter α to be very small, $\alpha < 10^{-7} - 10^{-5}$ depending on the assumptions [166, 167]. One can get a rough idea of such a bound by noting that the inhomogeneous fluctuation of Newton's potential Φ is proportional to the one of ρ_Λ at some cosmological scale k via the Poisson equation $\Delta\Phi \sim k^2\delta\rho_\Lambda$. Taking into account that these fluctuations are of order α times the critical density, the temperature fluctuation at last scattering is of order

$$\frac{\Delta T}{T} \sim \frac{\Delta\Phi}{\Phi} \sim \frac{\delta\rho_\Lambda}{\rho_{\rm tot}} \sim \alpha , \tag{11.89}$$

which implies that $\alpha \lesssim 10^{-5}$ [166]. This range is insufficient to explain late-time acceleration.

In order to make the considerations in favour of a viable solution of the Λ problem more robust and to better understand those against it, it should be made possible to follow the cosmic evolution through all its main stages. The argument leading to (11.87) should be embedded in a more complete framework where dynamics were under control. The mechanism by which $\Lambda = 0$ has to be worked out in detail; the quantum partition function (7.104) of unimodular gravity gives some support to a zero average cosmological constant, but results obtained in Euclidean signature are notoriously delicate. The role of matter should also be taken into account. It is not yet known whether emergent matter degrees of freedom [154] from pure "order and number" are viable from the point of view of both cosmological and particle physics. Furthermore, the stochastic process modeled by a constant α could be modified by dynamics. An effectively time-varying $\alpha(t)$ could accommodate both the need of small fluctuations at early times and of larger ones at late times. This resolution of the fine-tuning problem would not be dissimilar from the lattice-refinement picture in loop quantum cosmology, where the number of elementary nodes per unit volume changes dynamically. Its phenomenological origin might as well be the same, namely, the role of inhomogeneities in the full dynamics. All these elements may influence the results (11.86), (11.87) and (11.88) in ways which are presently difficult to assess.

11.7 Non-commutative Spacetimes

11.7.1 Framework

Divergences in ordinary quantum field theory may be cured if the structure of spacetime undergoes suitable modifications below some physical scale $l_\star$ often assumed to be the Planck length $l_{\rm Pl}$. On a smooth continuous manifold, one can identify classical events by points, which means that the uncertainty in the determination of time and space coordinates is exactly zero, $\Delta x^\mu \Delta x^\nu = 0$. In quantum mechanics this uncertainty does not vanish and acquires a lower bound, given by Heisenberg's principle. The idea of non-commutativity is to promote this feature of quantum mechanics to an intrinsic characteristic of spacetime, where the coordinates X^μ are the elements of a Lie algebra $\mathfrak{g}$ and quantum fields can be defined to live thereon [230–235]. The uncertainty $\Delta X^\mu \Delta X^\nu > 0$ is expressed by the commutation relation

$$[X^\mu, X^\nu] = \mathrm{i}\theta^{\mu\nu}(X)\,, \tag{11.90}$$

where $\theta^{\mu\nu}$ is an anti-symmetric tensor of dimension (length)^2. A special case is that of a linear algebra $[X^\mu, X^\nu] = C_\lambda^{\mu\nu} X^\lambda$, for some structure constants $C_\lambda^{\mu\nu}$. In particular, κ-Minkowski spacetime is defined by the algebra $[X^0, X^\alpha] = \mathrm{i}l_{\mathrm{Pl}}^2 X^\alpha$, $[X^\alpha, X^\beta] = 0$, for $\alpha, \beta = 1, 2, 3$.

Momentum space is made of the elements of the Lie group $\mathbb{G}$ dual to $\mathfrak{g}$. Since the measure on $\mathbb{G}$ will be in general non-trivial, non-commutative spacetimes are tightly related to a momentum space endowed with curvature and a non-standard dispersion relation.

The most conservative scenario where this structure makes its appearance is *classical* $(2+1)$-dimensional Einstein gravity coupled to point particles [236, 237]. In three dimensions, gravity is topological and particles are introduced as topological defects on the spacetime manifold [238]. The particle phase space is deeply modified by the curved background and it turns out that momenta are group valued and the particle spacetime coordinates have non-zero Poisson bracket.

The momentum space of GFT is also of this sort (see (11.51)) and, in fact, a non-commutative theory of matter fields emerges from the GFT action [117–120]. This model is obtained by expanding $\varphi(g) = \varphi^{(0)}(g) + \psi(g)$ in terms of a classical solution $\varphi^{(0)}$ describing flat building blocks and a perturbation ψ. Plugging this decomposition into the GFT action, the resulting effective action for ψ is a non-commutative field theory invariant under a deformation of the Poincaré group. It is no surprise, then, that a class of spin foams reproduces the Feynman-graph amplitudes of the same non-commutative field theory [222, 239, 240].

Non-commutativity shows up also in a variety of other contexts, from low-energy phenomenology of quantum gravity [241, 242] and heuristic approaches to quantum spacetime [243, 244] to deformations of relativistic symmetries [245, 246] and multi-scale spacetimes [247, 248]. In string theory, in the presence of a background Neveu–Schwarz field the end-points of open strings obey the non-commutative algebra (11.90) [249–252], where $\theta^{\mu\nu}$ is assumed to be constant in flat spacetime and covariantly constant in the presence of gravity. Since the closed-string sector and hence gravity is unaffected, the low-energy effective action is non-commutative only in matter fields. Independent field-theory models of a non-commutative gravitational action based on (11.90) can be constructed [253–256]. The same type of algebra (11.90) arises in compactifications of M-theory [257].

The treatment of field theories on non-commutative spacetimes is greatly simplified by the introduction of a Weyl map [258–260] putting in correspondence functions of the non-commuting variables X^μ with functions of the commuting "classical" coordinates x^μ. The map is invertible (i.e., the correspondence is one-to-one) provided the space of classical functions is equipped with a non-commutative $\star$-product. It is customary to define the Weyl map on plane waves, the basic building blocks of a field theory, and then to construct the functions via Fourier transform. The $\star$-product can also be defined explicitly: on a curved background [261],

$$(f \star g)(x) := \sum_{n=0}^{+\infty} \frac{(\mathrm{i}/2)^n}{n!} \theta^{\mu_1\nu_1}(x) \dots \theta^{\mu_n\nu_n}(x)(\nabla_{\mu_1} \dots \nabla_{\mu_n} f)(\nabla_{\nu_1} \dots \nabla_{\nu_n} g)\,, \tag{11.91}$$

for any f and g in the extended space of classical functions. When θ is constant and spacetime is flat, (11.91) is the associative Moyal product [262, 263]

$$(f * g)(x) := \exp\left(\frac{\mathrm{i}}{2}\,\theta^{\mu\nu}\frac{\partial}{\partial x^\mu}\frac{\partial}{\partial y^\nu}\right)[f(x)g(y)]\big|_{y=x}. \tag{11.92}$$

In general, however, the product (11.91) is non-associative, $(f \star g) \star h \neq f \star (g \star h)$. It is also an example of operator with an infinite number of derivatives; non-commutativity and non-locality are closely connected. When $\theta^{\mu\nu}$ is constant, the algebra (11.90) of non-commutative coordinates translates into

$$[x^\mu, x^\nu]_\star := x^\mu \star x^\nu - x^\nu \star x^\mu = \mathrm{i}\theta^{\mu\nu}\,, \qquad \theta^{\mu\nu} = \text{const}\,, \tag{11.93}$$

corresponding to the uncertainty relation

$$\Delta x^\mu \Delta x^\nu \geqslant \tfrac{1}{2}|\theta^{\mu\nu}|\,. \tag{11.94}$$

The action of a scalar field theory on a curved background and with a monomial potential is

$$S = -\int \mathrm{d}^4x\,\sqrt{-g}\left[\frac{1}{2}(\partial_\mu\phi) \star (\partial^\mu\phi) + \lambda\phi \star \cdots \star \phi\right], \tag{11.95}$$

where an operator ordering has been chosen such that there is no $\star$-product between the measure weight $\sqrt{-g}$ and the Lagrangian density.

11.7.2 Cosmology

In cosmological applications of the theory, new effects show only in inhomogeneous configurations, since the homogeneous background dynamics is unchanged. Therefore, non-commutativity can affect inflationary perturbations [264–266] and astrophysics [267].[10]

[10]For WDW quantum cosmology with non-commutative mini-superspace coordinates or non-commutative phase-space variables, see [268] and [269], respectively. In the second case, where not only mini-superspace coordinates but also their conjugate momenta obey a non-commutative algebra, also black-hole backgrounds have been studied and the wave-function found to vanish at the central singularity [270–272].

11.7.2.1 Non-commutative Inflation Without Inflaton

In one particular non-commutative model, early-time acceleration without a slow-rolling scalar field is obtained from semi-classical Einstein equations where the expectation value of the matter energy-momentum tensor is calculated on coherent states [273, 274].

Another, more developed scenario is based on a modified dispersion relation $-(k^0)^2 + f(k^0)|\mathbf{k}|^2 = 0$ inspired by non-commutative geometry [275–278], (5.199) with

$$f(k^0) = [1 + (\alpha k^0)^\gamma]^2 , \tag{11.96}$$

where α and γ are real parameters. Here, the content of the universe is pure radiation and cosmological perturbations are generated by thermal rather than quantum fluctuations. This mechanism of seeding of cosmic structures has been considered also in standard cosmology [279–281], but in that case it is difficult to produce inflation in the absence of a slow-rolling field or of quantum-gravity modifications to the dynamics [282]. On the other hand, radiation in non-commutative spacetime can sustain an accelerated expansion thanks to the deformed dispersion relation. The spectrum thus produced is almost scale invariant with a slight red tilt and can be compatible with observations.

11.7.2.2 Non-commutative Inflation with Inflaton

Another class of models has a more direct contact with non-commutative field theories. Here the inflationary mechanism is fairly standard, as it is driven by an ad hoc scalar in slow roll (scalars may appear in quite specific non-commutative geometries [283]). Expanding (11.91) and (11.95) to leading order in θ, one obtains effective models with small non-commutativity where standard slow-roll inflation takes place [284–286]. When non-commutative effects become appreciable, (11.93) may lead to large anisotropies (the anti-symmetric spatial tensor $\theta^{\alpha\beta}$ selects preferred directions) and non-Gaussianities [261, 287].

To make the problem tractable, the components of $\theta^{\mu\nu}$ are usually assumed to be constant in a particular coordinate frame. Two common choices are the comoving frame ($[x^\mu, x^\mu]_\star = \mathrm{i}\theta^{\mu\nu} = \text{const}$, with $x^0 = t$) [287–290] and the physical or proper frame ($[x_\mathrm{p}^\mu, x_\mathrm{p}^\mu]_\star = \mathrm{i}\theta_\mathrm{p}^{\mu\nu} = \text{const}$, with $x_\mathrm{p}^0 = t$ and $x_\mathrm{p}^\alpha = a(t)\,x^\alpha$) [261, 287]. The relation between the matrix elements of the two frames is dictated by the relative scaling of proper and comoving spatial coordinates: $\theta_\mathrm{p}^{0\alpha} = a\,\theta^{0\alpha}$, $\theta_\mathrm{p}^{\alpha\beta} = a^2\,\theta^{\alpha\beta}$.

Taking (11.93) to hold for comoving coordinates, the scalar spectrum reads [288–290]

$$\mathcal{P}_\mathrm{s}^{(\theta)}(k) = \mathcal{P}_\mathrm{s}^{(0)}[\beta \cosh(H\theta^{0\alpha}k_\alpha) + \mathrm{i}(1-\beta)\sinh(H\theta^{0\alpha}k_\alpha)] , \tag{11.97}$$

where $\mathcal{P}_{\rm s}^{(0)}$ is the standard commutative spectrum and $0 < \beta \leqslant 1$ is a free parameter. The non-linear parameter $f_{\rm NL}$ is modified by similar effects as well as by the contribution of the space-space part $\theta^{\alpha\beta}$. CMB data constrain the non-commutative length scale to be $\sqrt{\theta} < 10^{-19}$ m, corresponding to a lower bound on the energy of about 10 TeV [291].

Another possibility is to regard the coordinates in (11.94) to be the proper ones and consider a simplified model where $\theta^{\alpha\beta} = 0$ and $\sqrt{\theta^{0\alpha}} = l_\star$. Then, the spacetime uncertainty relation reduces to $\Delta t \Delta x_{\rm p} = \Delta\tilde{\tau}\Delta x \geqslant l_\star^2/2$ for all spatial directions, where $\mathrm{d}\tilde{\tau} := a(t)\mathrm{d}t$ and $\mathrm{d}x_{\rm p} = a(t)\mathrm{d}x$. The algebra of classical coordinates (11.93) is then [292]

$$[\tilde{\tau}, x]_\star = \mathrm{i}l_\star^2 \,. \tag{11.98}$$

In this setting (Brandenberger–Ho model), inflation driven by a scalar field has been considered [292–306]. The observed spectra can be produced in two different regimes. If the cosmological energy scale at which a perturbation with physical wave-number $p = k/a$ crosses the horizon is much smaller than the non-commutativity scale $E_\star = 1/l_\star$, $l_\star p = l_\star H \ll 1$, then non-commutative effects are soft. During inflation, $\tilde{\tau} \simeq a/H$, so that at horizon crossing $l_\star k_* = l_\star a_* H_* \simeq (H_* l_\star)^2 \tilde{\tau}_*/l_\star \ll \tilde{\tau}_*/l_\star$. These modes are contained in a UV region characterized by perturbations generated inside the Hubble horizon.

If the observed power spectra lie in this regime, one gets corrections to the standard observables suppressed as $(l_\star H)^4$. Omitting scalar and tensor subscripts, the power spectrum, spectral index and running read

$$\mathcal{P}^{(\theta)} = \mathcal{P}_{\rm s}^{(0)}\Sigma^2 \,, \qquad n^{(\theta)} = n^{(0)} + \sigma\epsilon \,, \qquad \alpha^{(\theta)} \simeq \alpha^{(0)} + \sigma O(\epsilon^2) \,, \tag{11.99}$$

with

$$\Sigma^2 \simeq 1 - b\delta_{\rm UV} \,, \qquad \sigma \simeq 4b\,\delta_{\rm UV} \,, \qquad \delta_{\rm UV} := (l_\star H)^4 \,, \tag{11.100}$$

where the numerical value of the coefficient $b = O(1) > 0$ depends on the operator ordering in the action [292, 301]. These corrections are similar to those of WDW quantum cosmology (Sect. 10.2.5), but more strongly suppressed. Moreover, their sign is fixed: power spectra are suppressed at large scales and the spectral index is blue tilted.

On the other hand, in the IR regime $l_\star H \ll 1$ the wave modes are generated outside the horizon. Since they are frozen until they enter the horizon for the first time, their magnitude depends on the conformal time τ_0 when they were generated. This corresponds to the time when the spacetime uncertainty relation is saturated and quantum fluctuations start out with their vacuum amplitude. Then, one can show that the IR region is characterized by $l_\star k_* \simeq \tilde{\tau}_0/l_\star$ at horizon crossing. Expanding the exact spectrum in the small parameter $(l_\star H)^{-2}$, the observables (11.99) now have

$$\Sigma^2 \simeq \delta_{\rm IR}^3 \quad \text{or} \quad \simeq \delta_{\rm IR}\,, \qquad \sigma = O(1)\,, \qquad \delta_{\rm IR} := \delta_{\rm UV}^{-1/2} = (l_\star H)^{-2}. \tag{11.101}$$

This regime is somewhat speculative since the Hubble radius is smaller than $l_\star$. Yet, it brings forward some non-trivial imprint of non-commutativity. While the UV corrections are akin to those in WDW quantum cosmology, where the corrections are of the "naive" kind $\sim (H/E_\star)^n$ and suppressed by the ratio of the Hubble parameter during inflation with respect to a fundamental energy scale $E_\star = 1/l_\star$, the IR corrections are just the opposite. This situation is loosely remindful of the relative status of inverse-volume and holonomy corrections, expressed by the heuristic relation (10.152).

At intermediate scales, the two types of spectra can be interpolated and they are of the form (11.99) with $\Sigma^2 \simeq 1 - c\sqrt{\delta_{\rm UV}}$ and $\sigma \simeq 2c\sqrt{\delta_{\rm UV}}$, for some model-dependent $c = O(1) > 0$ [293, 301].

The characteristic scale signaling the transition between the UV and the IR regime is not fixed by the model, which is rather phenomenological. Therefore, for the purpose of placing observational constraints one has the freedom to tune the free parameters so that to describe the whole power spectrum by only one of the three regimes (UV, intermediate or IR). Both in the UV [259, 294, 297, 299, 300, 304, 305] and intermediate [293] regime, the effect of new physics is typically small and can be rendered compatible with observations for realistic values of the non-commutativity scale $l_\star$. Assuming instead the strongly non-commutative IR regime, the blue tilt in the scalar index is large and such models are much more severely constrained, to the point of being almost ruled out for the simplest choices of scalar-field potential [303, 306].

11.8 Non-local Gravity

11.8.1 Non-locality

We saw in Sect. 7.5.1 that the presence of derivative operators of order higher than two in a Lagrangian often leads to ghosts at the perturbative level. Consider the instance of a massless scalar field in Minkowski spacetime with action

$$S_{\rm loc} = \int d^D x \left[\frac{1}{2}\phi\Box(1-\Box_M)\phi - V(\phi)\right], \qquad \Box_M := \frac{\Box}{M^2}\,, \tag{11.102}$$

where $M^2 > 0$ is a mass scale. The classical equation of motion is $\Box(1-\Box_M)\phi - V_{,\phi} = 0$. The propagator of the free theory ($V = 0$) obeys the Green equation $(\Box - \Box^2/M^2)G(x-x') = \delta(x-x')$ which, transformed to momentum space, gives

$(-p^2 - p^4/M^2)\tilde{G}(p^2) = 1$, i.e.,

$$\tilde{G}(p^2) = -\frac{1}{p^2} + \frac{1}{p^2 + M^2}\,. \tag{11.103}$$

Thus, the action (11.102) actually encodes two degrees of freedom, a massless ordinary scalar (residue -1) and a massive ghost (residue $+1$) with mass M. Higher-order derivatives do not inevitably give rise to ghosts: their presence can be easily avoided in Horndeski theory (Sect. 7.5.1) as well as in higher-spin models [307]. However, outside these special classes of dynamics one can reasonably expect to encounter instabilities whenever one has derivatives higher than second order.

In quantum gravity, the problem depends on the background one chooses to expand the metric in graviton modes. In a cosmological setting, there exist non-trivial regions in the parameter space where ghosts disappear. However, since ghosts persist on the Minkowski background where particle field theory is usually developed, it is important and perhaps more elegant to find a general mechanism preserving unitarity. The situation can radically change when non-perturbative effects are taken into account. For example, the unitarity problem in higher-order gravity (in particular, Stelle's theory) is resolved in a suitable strong-coupling limit in the presence of fermionic matter [308–311].

Another possibility is the following. Introducing operators $\Box^n$ in the example (11.102), one soon realizes that the higher the order of derivatives, the larger the number of degrees of freedom [312–314]. At least one ghost mode, sometimes called Ostrogradski instability [312, 315], is ever persistent. But what happens when the number of derivatives is infinite? These theories are called *non-local* or *non-polynomial*.

In systems defined in continuous spacetimes, non-locality is the appearance of an infinite number of derivatives in the kinetic terms $\phi f(\partial)\phi$ of fields or, which is equivalent, by interactions $\int \mathrm{d}x'\phi(x)F(x-x')\phi(x')$ dependent on non-coincident points. In fact,

$$\begin{aligned}\int \mathrm{d}x\,\mathrm{d}x'\,\phi(x)F(x-x')\phi(x') &= \int \mathrm{d}x\,\mathrm{d}z\,\phi(x)F(-z)\phi(x+z)\\ &= \int \mathrm{d}x\,\mathrm{d}z\,\phi(x)F(-z)\int \mathrm{d}p\,\mathrm{e}^{\mathrm{i}p(x+z)}\tilde{\phi}(p)\\ &= \int \mathrm{d}x\,\mathrm{d}z\,\phi(x)F(-z)\,\mathrm{e}^{z\partial_x}\int \mathrm{d}p\,\mathrm{e}^{\mathrm{i}px}\tilde{\phi}(p)\\ &= \int \mathrm{d}x\,\phi(x)\left[\int \mathrm{d}z\,F(-z)\,\mathrm{e}^{z\partial_x}\right]\phi(x)\\ &=: \int \mathrm{d}x\,\phi(x)f(\partial_x)\,\phi(x)\,. \end{aligned} \tag{11.104}$$

Historically, non-locality was first found in the interactions of quantum field theory [316–335].

Consider now the action

$$S_{\text{non-loc}} = \frac{1}{2}\int d^D x\, [\phi \Box W_e(-\Box_M)\phi - V(\phi)] \,, \tag{11.105}$$

where

$$W_e(\Box_M) := e^{-\Box_M} \tag{11.106}$$

is called *form factor* and V is a local interaction. A field redefinition

$$\tilde{\phi} := e^{-\Box_M/2}\phi \tag{11.107}$$

cannot reduce the action to a local one unless $V = 0$. The free propagator of ϕ is now

$$\tilde{G}(p^2) = -\frac{e^{-p^2/M^2}}{p^2} \,. \tag{11.108}$$

Adopting the series definition of the operator $e^{-\Box_M}$,

$$e^{-\Box_M} := \sum_{n=0}^{+\infty} \frac{(-\Box_M)^n}{n!} \,, \tag{11.109}$$

one can truncate up to order $n = 1$ and find out that, in this limit, (11.102) is an approximation of (11.105), $S_{\text{non-loc}} = S_{\text{loc}} + O(\Box^3)$. However, higher-order and non-local theories are physically quite distinct. The main difference with respect to the higher-order propagator (11.103) is that the exponential in (11.108) is an entire function which does not introduce extra poles. The spectrum of the model (11.105) is the same as in the local field theory with $M^{-2} = 0$ (this is obvious from the field redefinition (11.107), which trivializes the model in the free limit $V \to 0$) but, if M is real, the propagator (11.108) vanishes much more rapidly in the UV in the Euclidean limit $p^2 \to p_E^2 \to +\infty$. The problem, in general, is that solutions to non-local equations of motion cannot be expressed via series representations such as (11.109), unless certain restrictive conditions (for instance, slow variation of the fields or convergence of perturbative series on the functional space of solutions) are satisfied. It is therefore clear that, to solve this type of non-local models in the presence of non-trivial interactions, a perturbative truncation of the operators is inadequate and one should employ non-perturbative techniques.

Generally speaking, non-local operators are troublesome also for other reasons. On one hand, a particle interpretation of a quantum field theory may even be absent, due to the replacement of poles in the propagators with branch cuts for certain operators such as $\Box^\alpha$, where α is non-integer [336, 337]. Moreover, causality may be violated at microscopic scales [319]. On the other hand, the Cauchy problem can be ill defined or highly non-standard [319, 321]. It entails an infinite number of initial conditions $\phi(0)$, $\dot{\phi}(0)$, $\ddot{\phi}(0)$, ..., representing an infinite number of degrees of freedom. Since the Taylor expansion of $\phi(t)$ around $t = 0$ is given by the full set of initial conditions, specifying the Cauchy problem would be tantamount to knowing the solution itself, if analytic [338]. This makes it very difficult to find analytic solutions to the equations of motion, even on Minkowski spacetime.

Fortunately, the exponential operator (11.106) and its higher-derivative generalization $W_e[-(-\Box_M)^l]$ with $l \in \mathbb{N}_+$ are under much greater control than other non-local operators. In these cases, the diffusion-equation method is available to find analytic solutions [339–346] which are well defined when the perturbative expansion (11.109) is not [340]. The Cauchy problem can be rendered meaningful, both in the free theory [347] and in the presence of interactions, by transferring the infinite number of initial conditions $\phi(0)$, $\dot{\phi}(0)$, $\ddot{\phi}(0)$, ... into a finite number of initial conditions for an auxiliary scalar field living in an extended spacetime with a fictitious extra coordinate [341]. Causality, too, is respected in theories with exponential operators, both at the micro- and the macroscopic level [333, 348]. Exponential non-locality arises in string field theory, as we shall discuss in Sect. 13.7.4.

The non-locality (11.106) is also responsible for a feature of interest for gravity: *asymptotic freedom*, namely the fact that the theory becomes weakly coupled at short distances. If the toy model (11.105) schematically represented the action for the linearized graviton modes, the Fourier transform of the spatial counterpart of (11.108) would give the Newtonian potential $\Phi(r)$. While at large distances $\Phi(r) \sim 1/r$ in $D - 1 = 3$ spatial dimensions, in the limit $p \to \infty$ the potential does not diverge but reaches a constant value determined by the UV cutoff $r_0 \sim 1/M$, $\Phi(r) \sim 1/r_0$ [349]. Notice that $\Phi(r) \sim 0$ in asymptotic safety when $r \to 0$, as we saw in (8.17): asymptotic safety is therefore a stronger condition than asymptotic freedom.

11.8.2 Framework

The strong damping in the UV of the propagator (11.108) after Wick rotation suggests, together with other cumulative evidence, that theories with exponential non-locality have interesting renormalization properties. After early studies of quantum scalar field theories [327, 331–333, 350] (see also [351]) and gauge and gravitational theories [352–359], in recent years there has been a resurgence of interest in non-local classical and quantum gravity [348, 349, 360–369]. A non-local theory of gravity aims at fulfilling a synthesis of minimal requirements: (i) spacetime is a continuum where Lorentz invariance is preserved at all scales; (ii) classical local (super)gravity should be a good approximation at low energy; (iii) the theory has to be perturbatively super-renormalizable or finite at the quantum level; (iv) the model has to be unitary and ghost free, without extra degrees of freedom in addition to those present in the classical theory; (v) typical classical solutions must be singularity-free.

To meet these demands, one considers a general action combining curvature tensors, covariant derivatives of the curvature tensors and non-polynomial terms [363, 370]:

$$\begin{aligned} S_g &= \sum_{n=0}^{N+2} \alpha_{2n} M^{D-2n} \int d^D x \sqrt{-g}\, \mathcal{O}_{2n}(\partial_\rho g_{\mu\nu}) + S_{\rm NP} \\ &= \frac{1}{2\kappa^2} \int d^D x \sqrt{-g} \left\{ R - 2\Lambda - \sum_{n=0}^{N} \left[a_n R (-\Box_M)^n R + b_n R_{\mu\nu} (-\Box_M)^n R^{\mu\nu} \right] \right. \\ &\quad \left. - R_{\mu\nu}\, \mathcal{F}_2(-\Box_M)\, R^{\mu\nu} - R \mathcal{F}_0(-\Box_M) R + O(R^3) \right\} . \end{aligned} \tag{11.110}$$

Here, M is some characteristic mass-energy scale and $\mathcal{O}_{2n}(\partial_\rho g_{\mu\nu})$ denotes general-covariant operators containing $2n$ derivatives of the metric $g_{\mu\nu}$, while $S_{\rm NP}$ is a non-polynomial action defined in terms of two entire functions $\mathcal{F}_{2,0}$ which we will fix shortly [348]. The couplings and the non-local functions of the theory have the following dimensions in mass units: $[a_n] = [b_n] = -2$, $[\kappa^2] = 2 - D$, $[\mathcal{F}_2] = [\mathcal{F}_0] = -2$.

The maximal number of derivatives in the local part of the action is $2N + 4$. From the discussion of the local theory (7.79) in Sect. 8.2, renormalizability sets $2N+4 \geqslant D$. To avoid fractional powers of the Laplace–Beltrami operator $\Box$ (which do not even admit a series representation), we take $2N + 4 = D$ in even dimensions and $2N + 4 = D + 1$ in odd dimensions. For $N \geqslant 0$ and $n \geqslant 2$, only the operators $R_{\mu\nu}\Box^{n-2}R^{\mu\nu}$, $R\Box^{n-2}R$ and $R_{\mu\nu\sigma\tau}\Box^{n-2}R^{\mu\nu\sigma\tau}$ contribute to the graviton propagator (with $2n$ derivatives in total) [370], but using the Bianchi and Ricci identities the third operator can be eliminated. The last line of (11.110) includes only these leading-order terms, with the redefinitions $\Lambda = -\alpha_0\kappa^2 M^D \mathcal{O}_0$, $1/(2\kappa^2) = \alpha_2 M^{D-2}$, and so on.

The entire functions $\mathcal{F}_{0,2}(z)$ in (11.110) are (here $z := -\Box_M$)

$$\mathcal{F}_2(z) = \frac{W(z)^{-1} - 1 - M^2 z \sum_{n=0}^{N} \tilde{b}_n z^n}{M^2 z} , \tag{11.111}$$

$$\mathcal{F}_0(z) = -\frac{W(z)^{-1} - 1 + 2M^2 z \sum_{n=0}^{N} \tilde{a}_n z^n}{2M^2 z} , \tag{11.112}$$

for general parameters $\tilde{a}_n$ and $\tilde{b}_n$, where $W(z)$ is a non-polynomial entire function without zeros in the whole complex plane.

The Lagrangian can be expanded at second order in the graviton fluctuation (7.101) around the Minkowski background. Adding a gauge-fixing term $\mathcal{L}_{\rm GF}$ due to a local gauge symmetry under infinitesimal coordinate transformations [371], the linearized gauge-fixed Lagrangian reads $\mathcal{L}_{\rm lin} + \mathcal{L}_{\rm GF} = h^{\mu\nu}\mathcal{O}_{\mu\nu\rho\sigma}h^{\rho\sigma}/2$. Inverting the operator $\mathcal{O}$ [372], one finds the following two-point function in the harmonic gauge (3.18) and in momentum space:

$$\mathcal{O}^{-1}(p) = -\frac{P^{(2)}}{p^2\bar{\mathcal{F}}_2(p^2)} + \frac{P^{(0)}}{(D-2)p^2\bar{\mathcal{F}}_0(p^2)}, \tag{11.113a}$$

where

$$\bar{\mathcal{F}}_2(p^2) = 1 + p^2\beta(p^2), \tag{11.113b}$$

$$\beta(p^2) = \sum_{n=0}^{N} b_n\left(\frac{p^2}{M^2}\right)^n + \mathcal{F}_2\left(\frac{p^2}{M^2}\right), \tag{11.113c}$$

$$\bar{\mathcal{F}}_0(p^2) = 1 - p^2\frac{D\beta(p^2) + 4(D-1)\alpha(p^2)}{D-2}, \tag{11.113d}$$

$$\alpha(p^2) = \sum_{n=0}^{N} a_n\left(\frac{p^2}{M^2}\right)^n + \mathcal{F}_0\left(\frac{p^2}{M^2}\right). \tag{11.113e}$$

In (11.113a), we omitted the tensorial indices of the operator $\mathcal{O}^{-1}$ and of the projectors $P^{(0)}$ and $P^{(2)}$, defined as [372, 373] $P^{(2)}_{\mu\nu\rho\sigma}(p) = (\theta_{\mu\rho}\theta_{\nu\sigma} + \theta_{\mu\sigma}\theta_{\nu\rho})/2 - \theta_{\mu\nu}\theta_{\rho\sigma}/(D-1)$, $P^{(0)}_{\mu\nu\rho\sigma}(p) = \theta_{\mu\nu}\theta_{\rho\sigma}/(D-1)$, $\theta_{\mu\nu} = \eta_{\mu\nu} - p_\mu p_\nu/p^2$.

We now assume that the theory is renormalized at some scale μ_0. Setting $\tilde{a}_n = a_n(\mu_0)$ and $\tilde{b}_n = b_n(\mu_0)$, the bare propagator only possesses the gauge-invariant, physical massless spin-2 graviton pole, and $\bar{\mathcal{F}}_2 = \bar{\mathcal{F}}_0 = W^{-1}$. Choosing another renormalization scale, the bare propagator acquires poles which cancel with a shift in the self-energy in the dressed propagator. Thus, (11.113a) reads

$$\mathcal{O}^{-1}(p) = -\frac{W(p^2/M^2)}{p^2}\left(P^{(2)} - \frac{P^{(0)}}{D-2}\right). \tag{11.114}$$

A non-local field theory is unitary and micro-causal provided the following properties are satisfied by $W(z)$ [333]: (i) $W(z)$ is an entire analytic function in the complex z-plane and it has a finite order of growth $1/2 \leqslant \rho < +\infty$, i.e., $\exists\, b > 0, c > 0$ such that $|W(z)| \leqslant c\, e^{b\,|z|^\rho}$; (ii) when $\mathrm{Re}(z) \to +\infty$ ($p^2 \to +\infty$ or $p_{\rm E}^2 \to +\infty$), $W(z)$ decreases quite rapidly: $\lim_{\mathrm{Re}(z)\to+\infty} |z|^N |W(z)| = 0$, $\forall\, N > 0$; (iii) $[W(z)]^* = W(z^*)$ and $W(0) = 1$; (iv) the function $W^{-1}(z)$ is real and positive on the real axis and it has no zeros on the whole complex plane $|z| < +\infty$. This requirement implies that there are no gauge-invariant poles other than the transverse massless physical graviton pole.

11.8.2.1 Super-Renormalizable and Finite Quantum Gravities?

We now specialize to a form factor satisfying the above properties [348]:

$$W(z) = \mathrm{e}^{-\mathrm{H}(z)}\,, \tag{11.115}$$

where H is a polynomial. In particular, we consider $W(z) = \exp(-z^l)$, where $l \in \mathbb{N}_+$; the case $l = 1$ corresponds to the exponential form factor (11.106). The high-energy propagator takes the form $\mathcal{O}^{-1}(p) = -\exp[-(p^2/M^2)^l]/p^2$. The n-graviton interaction has the same scaling, since it can be written schematically as $\mathcal{L}^{(n)} \sim h^n \Box_\eta h \exp[(-\Box_\eta)^l/M^{2l}]h + \ldots$, where $\Box_\eta = \eta^{\mu\nu}\partial_\mu\partial_\nu$ and "…" are other sub-leading interaction terms. Placing an upper bound to the amplitude with L loops, one finds heuristically

$$\begin{aligned}\mathcal{A}^{(L)} &\sim \int (\mathrm{d}^D p)^L \left[\mathrm{e}^{-(p^2/M^2)^l} p^{-2}\right]^I \left[\mathrm{e}^{(p^2/M^2)^l} p^2\right]^V \\ &= \int (\mathrm{d}p)^{DL} \left[\mathrm{e}^{-(p^2/M^2)^l} p^{-2}\right]^{L-1}.\end{aligned} \tag{11.116}$$

In the last step, we used the topological identity (8.10). The L-loops amplitude is ultraviolet finite for $L > 1$ and it diverges at most as p^D for $L = 1$. Only one-loop divergences survive and, therefore, this theory is power-counting super-renormalizable and unitary, as well as micro-causal [331–333, 348, 350, 352, 363, 369, 374]. A more rigorous power counting than the one of (11.116) can be found in [369].

All of this applies to spacetimes of any dimension $D \geqslant 3$. In odd dimensions, there are no counter-terms for pure gravity at the one-loop level in dimensional regularization and the theory is finite; it stays so also when matter is added to fill up the supergravity multiplet [375, 376]. One may then infer that the amplitudes with an arbitrary number of loops are finite and all the beta functions vanish. In particular, we can fix all the coefficients of the higher-curvature terms in (11.110) to zero, while the couplings $a_n(\mu)$ and $b_n(\mu)$ do not run with the energy: $a_n(\mu) = \tilde{a}_n = \text{const}$, $b_n(\mu) = \tilde{b}_n = \text{const}$. Using (11.111) and (11.112) for the exponential form factor (11.106), the gravitational action (11.110) simplifies to

$$S_g = \int \mathrm{d}^D x\sqrt{-g}\left[\frac{R-2\Lambda}{2\kappa^2} - R_{\mu\nu}\,\mathcal{F}_2(-\Box_M)\,R^{\mu\nu} - R\mathcal{F}_0(-\Box_M)R\right] \tag{11.117}$$

$$= \frac{1}{2\kappa^2}\int \mathrm{d}^D x\sqrt{-g}\left[R - 2\Lambda + G_{\mu\nu}\,\gamma(\Box)\,R^{\mu\nu}\right], \tag{11.118}$$

where

$$\gamma(\Box) := \frac{\mathrm{e}^{-\Box_M} - 1}{\Box}\,. \tag{11.119}$$

The equations of motion are $e^{-\Box_M} G_{\mu\nu} + O(R^2) = \kappa^2 T_{\mu\nu}$. For generic form factors, they can be found in [349] when $\mathcal{F}_2 = 0$ and in [377, 378] for $\mathcal{F}_0 \neq 0 \neq \mathcal{F}_2$ (following the calculation for higher-order actions of [379]).

A non-perturbative analysis of the renormalization group flow of the theory has not been performed yet but the damping of the form factor already points to the fact that this class of non-local gravitational theories is asymptotically free and the only fixed point in the UV is Gaussian (relevant couplings go to zero in the ultraviolet). This should be contrasted with asymptotically-safe gravity, where the couplings do not vanish in the UV. The effective action of the two theories is quite different (the latter, in particular, is local) and, therefore, it does not come as a surprise that their UV behaviour and the way they resolve the infinities of perturbative Einstein gravity, do not match.

11.8.3 Cosmology

In the simplest classical cosmological applications of the operator (11.106), gravity is local and the only non-local content is a scalar field [340, 342, 345, 380–401]. When also the gravitational dynamics is dominated by exponential operators, the big-bang [349, 361, 362, 377, 402–405] and black-hole [406] singularities are apparently resolved, thanks to asymptotic freedom.[11]

The simplest means to find cosmological solutions is to require that curvature terms are eigenstates of the form factor. If, for instance, the Ricci tensor $R_{\mu\nu}$ is an eigenfunction of the Laplace–Beltrami operator, so will the Ricci scalar R in any theory where the metric is covariantly constant. One is therefore interested in finding profiles $a(t)$ such that

$$\Box R = \lambda_1 R + \lambda_2 , \tag{11.120}$$

where $\lambda_{1,2}$ are constants. Two such typical profiles in $D = 4$ are

$$a(t) = a_* \cosh\left(\sqrt{\frac{\omega}{2}}\, t\right), \tag{11.121a}$$

$$a(t) = a_* \exp\left(\frac{H_1}{2}\, t^2\right), \tag{11.121b}$$

where ω and H_1 are constants. From (2.72) and for zero intrinsic curvature $\textsc{k}$, $R = 6(2H^2 + \dot{H}) = 3\omega\{1 + [\tanh(\sqrt{\omega/2}t)]^2\}$, $\lambda_1 = \omega$ and $\lambda_2 = -6\omega^2$ in the first case, while $R = 6H_1(1 + 2H_1 t^2)$, $\lambda_1 = -6H_1$ and $\lambda_2 = 12H_1^2$ in the second. The scale factors (11.121) are non-singular at $t = 0$, hence the big bang is replaced by

[11] For other types of non-locality which include infrared modifications to gravity, see [407–429].

a classical bounce where $a(0) = a_* \neq 0$. The relation between ω or H_1 and the other parameters of the model (matter energy density and cosmological constant) is determined by the equations of motion. The solutions (11.121) are, in general, exact in vacuum and for certain classes of form factors $\mathcal{F}_{0,2}$ in (11.117),[12] while they can be regarded as a good approximation near the bounce in the presence of matter fields.

In the specific example (11.118), one can find a profile $a_{\rm af}(t)$ by non-perturbative methods exploiting asymptotic freedom [405]. Near the bounce, this scale factor takes the asymptotic form (11.121b), while at late times it matches the usual power law $a_{\rm af}(t) \simeq t^p$ (2.93). The shape (11.121b) is rather insensitive to the type of matter dominating at late times, which only determines the numerical value of $H_1 \propto p$. One can fit the solution $a_{\rm af}(t)$ with the effective Friedmann equation

$$H^2 = \frac{\kappa^2}{3}\rho_{\rm eff} := \frac{\kappa^2}{3}\rho\left[1-\left(\frac{\rho}{\rho_*}\right)^\beta\right], \tag{11.122}$$

where $\rho_* = \rho(a_*)$ is the critical energy density at which the bounce occurs and $\beta = O(1)$ is a real parameter. The exponent β is determined by plugging the profile $a_{\rm af}(t)$ into (11.122) for a given energy density profile $\rho(a_{\rm af})$. The effective equation (11.122) is similar to (10.121) for homogeneous LQC in the absence of inverse-volume corrections.

To solve the big-bang problem convincingly, one should show by a stability analysis that bouncing solutions such as (11.121) do not rely on special initial conditions; infinitely many solutions with a big bang (for instance, a power-law scale factor) are, after all, known for certain non-local actions [361, 377]. This task is still in progress but points towards a reassuring answer.

11.9 Comparison of Quantum-Gravity Models

The variety of approaches to quantum gravity and cosmology may confuse the reader accustomed to the clear-cut answers of traditionally accepted physics. This state of affairs is due partly to the lack of any recognized imprint of quantum gravity in the observable world and partly to the difficulty in developing all these lines of research to an adequate level of rigorousness *and* contact with phenomenology. It is therefore helpful to ask whether some of these theories are physically equivalent and, if not, in what they differ.

Just like LQG, the frameworks of asymptotic safety and causal dynamical triangulations are quantizations of standard general relativity, with no exotic modification to the classical dynamics. A difference among these approaches is

[12] Notice that $\mathcal{F}_2 = 0$ in [349, 362, 403] and $\mathcal{F}_0 \neq 0 \neq \mathcal{F}_2$ in [377, 405].

in the way quantization is carried out: via the Hamiltonian formalism in LQG, through the covariant functional renormalization approach in asymptotic safety, and via a discretization of the covariant path integral in causal dynamical triangulations. The choice of the gravitational degrees of freedom deeply affects the quantization procedure. In LQG, Ashtekar–Barbero variables and spin networks are used; in asymptotic safety, the covariant metric is quantized; in causal dynamical triangulations, discrete triangulations regularize the divergences of the continuum.

On the other hand, all the other approaches introduce some fundamental ingredients foreign to both classical general relativity and standard quantization frameworks. In spin foams, one reformulates gravity as a constrained topological theory and builds a combinatorial and group structure encoded in labelled simplicial complexes. Group field theory shares the same type of ingredients as in spin foams but with the addition of a group-manifold structure and, in principle, of non-gravitational degrees of freedom. In causal sets, spacetimes descend from a discrete ordered structure. In non-commutative geometry, spacetime has a non-trivial algebraic structure. In non-local gravity, operators with infinitely many derivatives are included in the Lagrangian.

All these methods except those based on non-commutativity are background-independent, although a $3+1$ splitting is necessarily assumed in LQG as well as in the best studied versions of group field theory and causal dynamical triangulations. Except non-commutative field theories and non-local gravity, they are also non-perturbative in the sense that no graviton expansion around a fixed background is used, although spin foams and group field theories can be treated via several perturbative techniques according to the needs.

Of these approaches, asymptotic safety, causal dynamical triangulations and non-local gravity essentially describe continuous spacetimes. In contrast, LQG is characterized by a discrete quantum geometry, which is a consequence of the choice of canonical quantum variables rather than an assumption; spin foams and group field theories inherit the same property. In non-commutative spacetimes, below the fundamental scale appearing in the coordinates algebra it is not possible to maintain the continuum picture.

In general, these theories are not physically equivalent despite some similarities. Non-commutativity and non-locality arise, in one form of another, as effective or intrinsic properties of the other theories, but they can be conceived as stand-alone, independent proposals as defined in Sects. 11.7 and 11.8. Asymptotic safety, CDT and LQG are all based on the same classical theory (general relativity), but the different quantization procedures lead to inequivalent quantum models. Both LQG and spin foams are apparently recovered in GFT under certain approximations, but GFT has, by construction, more degrees of freedom than any of these.

Nevertheless, almost all these theories share a rather typical characteristic worth mentioning, since it can have cosmological applications. The dimension of the effective spacetimes emerging as a suitable limit of quantum geometry (to be defined case by case) changes with the probed scale and its value in the far ultraviolet is often smaller than 4 [430–433]. This phenomenon, called *dimensional flow*,[13] has been detected in all the theories considered in this chapter: asymptotically-safe gravity (Hausdorff dimension $d_{\rm H} = D$ always, while the spectral dimension is $d_{\rm S} \stackrel{\rm UV}{\simeq} D/2$ in D topological dimensions at the non-Gaussian fixed point; analytic results) [21, 41]; causal dynamical triangulations (for phase-C geometries, $d_{\rm S} \stackrel{\rm UV}{\simeq} D/2$ [58, 59, 434, 435] or, more recently, $d_{\rm S} \stackrel{\rm UV}{\simeq} 3/2$ [72]; numerical results) and the related models of random combs [213, 436] and random multi-graphs [437, 438]; spin foams [439, 440]; causal sets [178]; non-commutative geometry [237, 441, 442] and κ-Minkowski spacetime [247, 443–446]; non-local super-renormalizable quantum gravity ($d_{\rm S} < 1$ in the UV in $D = 4$) [363]. In LQG, while there is no conclusive evidence of variations of the spectral dimension for individual quantum-geometry states based on given graphs or complexes [447], genuine dimensional flow has been encountered in non-trivial superpositions of spin-network states [448], as an effect of quantum discreteness of geometry. Since these states appear also in GFT, the latter inherits the same feature.

Other examples, all based on analytic results, are Hořava–Lifshitz gravity ($d_{\rm S} \stackrel{\rm UV}{\simeq} 2$ for any D) [41, 435, 449], spacetimes with black holes [450–452], fuzzy spacetimes [453], multi-fractal and multi-fractional spacetimes (variable model-dependent $d_{\rm H}$ and $d_{\rm S}$) [187, 188, 432, 454–456] and string theory [457]. When the spectral dimension $d_{\rm S}$ decreases to values lower than D at small scales, the UV properties of the theory usually improve; counter-examples, however, exist [458].

Some properties of dimensional flow are universal simply because one always reaches the IR in a smooth asymptotic way, a fact that has repercussions in the profile of the flow also at mesoscopic and small scales [456]. However, geometries with the same dimensionality profiles are not necessarily isomorphic. The same value of the spectral dimension may appear in widely different theories without implying a physical duality among them. This is due to the fact that the spectral dimension is obtained from a diffusion equation but, as is well known in transport theory, different diffusion equations can accidentally give rise to the same correlation properties at small scales [41, 187]. Therefore, there is no precise link between renormalizability or UV finiteness on one hand, and dimensional flow on the other hand.

In models with dimensional flow, however, the change of spacetime dimensionality can give rise to almost scale-invariant primordial spectra without invoking a slow-rolling scalar field. In one such case, one considers a phenomenological dispersion relation (5.199) as it could stem from quantum gravity. The correction function f is higher order in spatial momenta, as in Hořava–Lifshitz gravity. In particular, for

$$f(\mathbf{p}) = 1 + \beta|\mathbf{p}|^4 \tag{11.123}$$

[13]The alternative name of "dimensional reduction" is often employed in the quantum-gravity literature, despite the fact that it is already in use in Kaluza–Klein and string scenarios, where spacetime has $D > 4$ topological dimensions and compactification to four observable dimensions is performed. For this reason, and to include also all scenarios where the dimension in the UV is not smaller than in the IR, we prefer the naming "flow."

the leading UV term in the kinetic operator $\mathcal{K}$ in (11.2) is $(\nabla^2)^3$ and the spectral dimension in the UV is $d_S \simeq 2$. Cosmological perturbations generated by the same dispersion relation obey the Mukhanov–Sasaki equation (3.32) with $k^2 \to k^2 f(k)$, and they are scale invariant both outside and *inside* the Hubble horizon [459–461]. Deviations from scale invariance can be obtained easily by a modification of the dispersion relation; the resulting spacetime has a spectral dimension close but not exactly equal to 2 in the UV.

Another instance is provided by multi-fractional spacetimes [462]. There, the ordinary integration measures of position and momentum space are replaced by some measures with non-trivial weights, $\mathrm{d}^D x \to \mathrm{d}^D x\, v(x)$, $\mathrm{d}^D p \to \mathrm{d}^D p\, w(p)$. These weights contain a hierarchy of fundamental time and length scales such that both the Hausdorff and spectral dimension change continuously. The form of v is dictated by multi-fractal geometry and by the universal IR behaviour of dimensional flow [456], while that of w follows from the requirement of an invertible momentum transform. Kinetic operators are also modified according to the symmetries imposed on the Lagrangian. If the spectral dimension is sufficiently small in the UV, power spectra are almost scale invariant [463]. Other features of the multi-fractal measures can also help to reformulate the big-bang and cosmological constant problem and trigger a phase of cyclic evolution in the early universe.

References

1. S. Weinberg, Ultraviolet divergences in quantum gravity, in *General Relativity: An Einstein Centenary Survey*, ed. by S.W. Hawking, W. Israel (Cambridge University Press, Cambridge, 1979)
2. M. Reuter, Nonperturbative evolution equation for quantum gravity. Phys. Rev. D **57**, 971 (1998). [arXiv:hep-th/9605030]
3. D. Dou, R. Percacci, The running gravitational couplings. Class. Quantum Grav. **15**, 3449 (1998). [arXiv:hep-th/9707239]
4. W. Souma, Non-trivial ultraviolet fixed point in quantum gravity. Prog. Theor. Phys. **102**, 181 (1999). [arXiv:hep-th/9907027]
5. A. Bonanno, M. Reuter, Renormalization group improved black hole spacetimes. Phys. Rev. D **62**, 043008 (2000). [arXiv:hep-th/0002196]
6. A. Bonanno, M. Reuter, Cosmology of the Planck era from a renormalization group for quantum gravity. Phys. Rev. D **65**, 043508 (2002). [arXiv:hep-th/0106133]
7. O. Lauscher, M. Reuter, Ultraviolet fixed point and generalized flow equation of quantum gravity. Phys. Rev. D **65**, 025013 (2002). [arXiv:hep-th/0108040]
8. O. Lauscher, M. Reuter, Is quantum Einstein gravity nonperturbatively renormalizable? Class. Quantum Grav. **19**, 483 (2002). [arXiv:hep-th/0110021]
9. M. Reuter, F. Saueressig, Renormalization group flow of quantum gravity in the Einstein–Hilbert truncation. Phys. Rev. D **65**, 065016 (2002). [arXiv:hep-th/0110054]
10. O. Lauscher, M. Reuter, Towards nonperturbative renormalizability of quantum Einstein gravity. Int. J. Mod. Phys. A **17**, 993 (2002). [arXiv:hep-th/0112089]
11. O. Lauscher, M. Reuter, Flow equation of quantum Einstein gravity in a higher-derivative truncation. Phys. Rev. D **66**, 025026 (2002). [arXiv:hep-th/0205062]
12. R. Percacci, D. Perini, Constraints on matter from asymptotic safety. Phys. Rev. D **67**, 081503 (2003). [arXiv:hep-th/0207033]

13. R. Percacci, D. Perini, Asymptotic safety of gravity coupled to matter. Phys. Rev. D **68**, 044018 (2003). [arXiv:hep-th/0304222]
14. D. Perini, Gravity and matter with asymptotic safety. Nucl. Phys. Proc. Suppl. **127**, 185 (2004). [arXiv:hep-th/0305053]
15. M. Reuter, F. Saueressig, Nonlocal quantum gravity and the size of the universe. Fortsch. Phys. **52**, 650 (2004). [arXiv:hep-th/0311056]
16. D.F. Litim, Fixed points of quantum gravity. Phys. Rev. Lett. **92**, 201301 (2004). [arXiv:hep-th/0312114]
17. R. Percacci, D. Perini, Should we expect a fixed point for Newton's constant? Class. Quantum Grav. **21**, 5035 (2004). [arXiv:hep-th/0401071]
18. M. Reuter, H. Weyer, Quantum gravity at astrophysical distances? JCAP **0412**, 001 (2004). [arXiv:hep-th/0410119]
19. A. Bonanno, M. Reuter, Proper time flow equation for gravity. JHEP **0502**, 035 (2005). [arXiv:hep-th/0410191]
20. M. Reuter, F. Saueressig, From big bang to asymptotic de Sitter: complete cosmologies in a quantum gravity framework. JCAP **0509**, 012 (2005). [arXiv:hep-th/0507167]
21. O. Lauscher, M. Reuter, Fractal spacetime structure in asymptotically safe gravity. JHEP **0510**, 050 (2005). [arXiv:hep-th/0508202]
22. M. Reuter, J.-M. Schwindt, A minimal length from the cutoff modes in asymptotically safe quantum gravity. JHEP **0601**, 070 (2006). [arXiv:hep-th/0511021]
23. P. Fischer, D.F. Litim, Fixed points of quantum gravity in extra dimensions. Phys. Lett. B **638**, 497 (2006). [arXiv:hep-th/0602203]
24. M. Reuter, J.-M. Schwindt, Scale-dependent metric and causal structures in Quantum Einstein Gravity. JHEP **0701**, 049 (2007). [arXiv:hep-th/0611294]
25. B.F.L. Ward, Massive elementary particles and black holes. JCAP **0402**, 011 (2004). [arXiv:hep-ph/0312188]
26. A. Bonanno, M. Reuter, Entropy signature of the running cosmological constant. JCAP **0708**, 024 (2007). [arXiv:0706.0174]
27. M. Reuter, H. Weyer, Background independence and asymptotic safety in conformally reduced gravity. Phys. Rev. D **79**, 105005 (2009). [arXiv:0801.3287]
28. D. Benedetti, P.F. Machado, F. Saueressig, Taming perturbative divergences in asymptotically safe gravity. Nucl. Phys. B **824**, 168 (2010). [arXiv:0902.4630]
29. E. Manrique, M. Reuter, Bimetric truncations for quantum Einstein gravity and asymptotic safety. Ann. Phys. (N.Y.) **325**, 785 (2010). [arXiv:0907.2617]
30. J.E. Daum, U. Harst, M. Reuter, Running gauge coupling in asymptotically safe quantum gravity. JHEP **1001**, 084 (2010). [arXiv:0910.4938]
31. S. Weinberg, Asymptotically safe inflation. Phys. Rev. D **81**, 083535 (2010). [arXiv:0911.3165]
32. A. Bonanno, A. Contillo, R. Percacci, Inflationary solutions in asymptotically safe $f(R)$ theories. Class. Quantum Grav. **28**, 145026 (2011). [arXiv:1006.0192]
33. A. Contillo, M. Hindmarsh, C. Rahmede, Renormalisation group improvement of scalar field inflation. Phys. Rev. D **85**, 043501 (2012). [arXiv:1108.0422]
34. M. Reuter, F. Saueressig, Fractal space-times under the microscope: a renormalization group view on Monte Carlo data. JHEP **1112**, 012 (2011). [arXiv:1110.5224]
35. A. Bonanno, An effective action for asymptotically safe gravity. Phys. Rev. D **85**, 081503 (2012). [arXiv:1203.1962]
36. M. Hindmarsh, I.D. Saltas, $f(R)$ gravity from the renormalisation group. Phys. Rev. D **86**, 064029 (2012). [arXiv:1203.3957]
37. S. Rechenberger, F. Saueressig, R^2 phase-diagram of QEG and its spectral dimension. Phys. Rev. D **86**, 024018 (2012). [arXiv:1206.0657]
38. A. Kaya, Exact renormalization group flow in an expanding Universe and screening of the cosmological constant. Phys. Rev. D **87**, 123501 (2013). [arXiv:1303.5459]
39. A. Codello, G. D'Odorico, C. Pagani, Consistent closure of renormalization group flow equations in quantum gravity. Phys. Rev. D **89**, 081701 (2014). [arXiv:1304.4777]

40. Y.-F. Cai, Y.-C. Chang, P. Chen, D.A. Easson, T. Qiu, Planck constraints on Higgs modulated reheating of renormalization group improved inflation. Phys. Rev. D **88**, 083508 (2013). [arXiv:1304.6938]
41. G. Calcagni, A. Eichhorn, F. Saueressig, Probing the quantum nature of spacetime by diffusion. Phys. Rev. D **87**, 124028 (2013). [arXiv:1304.7247]
42. E.J. Copeland, C. Rahmede, I.D. Saltas, Asymptotically safe Starobinsky inflation. Phys. Rev. D **91**, 103530 (2015). [arXiv:1311.0881]
43. P. Donà, A. Eichhorn, R. Percacci, Matter matters in asymptotically safe quantum gravity. Phys. Rev. D **89**, 084035 (2014). [arXiv:1311.2898]
44. K. Falls, Asymptotic safety and the cosmological constant. JHEP **1601**, 069 (2016). [arXiv:1408.0276]
45. P. Donà, A. Eichhorn, R. Percacci, Consistency of matter models with asymptotically safe quantum gravity. Can. J. Phys. **93**, 988 (2015). [arXiv:1410.4411]
46. M. Niedermaier, The asymptotic safety scenario in quantum gravity: an introduction. Class. Quantum Grav. **24**, R171 (2007). [arXiv:gr-qc/0610018]
47. M. Niedermaier, M. Reuter, The asymptotic safety scenario in quantum gravity. Living Rev. Relat. **9**, 5 (2006)
48. M. Reuter, F. Saueressig, Functional renormalization group equations, asymptotic safety and quantum Einstein gravity. arXiv:0708.1317
49. A. Codello, R. Percacci, C. Rahmede, Investigating the ultraviolet properties of gravity with a Wilsonian renormalization group equation. Ann. Phys. (N.Y.) **324**, 414 (2009). [arXiv:0805.2909]
50. D.F. Litim, Renormalisation group and the Planck scale. Philos. Trans. R. Soc. Lond. A **369**, 2759 (2011). [arXiv:1102.4624]
51. M. Reuter, F. Saueressig, Asymptotic safety, fractals, and cosmology. Lect. Notes Phys. **863**, 185 (2013). [arXiv:1205.5431]
52. B.F.L. Ward, An estimate of Λ in resummed quantum gravity in the context of asymptotic safety. Phys. Dark Univ. **2**, 97 (2013)
53. J. Ambjørn, R. Loll, Non-perturbative Lorentzian quantum gravity, causality and topology change. Nucl. Phys. B **536**, 407 (1998). [arXiv:hep-th/9805108]
54. J. Ambjørn, J. Jurkiewicz, R. Loll, A non-perturbative Lorentzian path integral for gravity. Phys. Rev. Lett. **85**, 924 (2000). [arXiv:hep-th/0002050]
55. J. Ambjørn, J. Jurkiewicz, R. Loll, Dynamically triangulating Lorentzian quantum gravity. Nucl. Phys. B **610**, 347 (2001). [arXiv:hep-th/0105267]
56. J. Ambjørn, J. Jurkiewicz, R. Loll, Emergence of a 4D world from causal quantum gravity. Phys. Rev. Lett. **93**, 131301 (2004). [arXiv:hep-th/0404156]
57. J. Ambjørn, J. Jurkiewicz, R. Loll, Semiclassical universe from first principles. Phys. Lett. B **607**, 205 (2005). [arXiv:hep-th/0411152]
58. J. Ambjørn, J. Jurkiewicz, R. Loll, Spectral dimension of the universe. Phys. Rev. Lett. **95**, 171301 (2005). [arXiv:hep-th/0505113]
59. J. Ambjørn, J. Jurkiewicz, R. Loll, Reconstructing the universe. Phys. Rev. D **72**, 064014 (2005). [arXiv:hep-th/0505154]
60. J. Ambjørn, J. Jurkiewicz, R. Loll, The universe from scratch. Contemp. Phys. **47**, 103 (2006). [arXiv:hep-th/0509010]
61. J. Ambjørn, A. Görlich, J. Jurkiewicz, R. Loll, Planckian birth of the quantum de Sitter universe. Phys. Rev. Lett. **100**, 091304 (2008). [arXiv:0712.2485]
62. J. Ambjørn, J. Jurkiewicz, R. Loll, The self-organized de Sitter universe. Int. J. Mod. Phys. D **17**, 2515 (2009). [arXiv:0806.0397]
63. J. Ambjørn, A. Görlich, J. Jurkiewicz, R. Loll, The nonperturbative quantum de Sitter universe. Phys. Rev. D **78**, 063544 (2008). [arXiv:0807.4481]
64. J. Ambjørn, A. Görlich, J. Jurkiewicz, R. Loll, Geometry of the quantum universe. Phys. Lett. B **690**, 420 (2010). [arXiv:1001.4581]
65. J. Ambjørn, A. Görlich, S. Jordan, J. Jurkiewicz, R. Loll, CDT meets Hořava–Lifshitz gravity. Phys. Lett. B **690**, 413 (2010). [arXiv:1002.3298]

66. J. Ambjørn, A. Görlich, J. Jurkiewicz, R. Loll, J. Gizbert-Studnicki, T. Trześniewski, The semiclassical limit of causal dynamical triangulations. Nucl. Phys. B **849**, 144 (2011). [arXiv:1102.3929]
67. J. Ambjørn, S. Jordan, J. Jurkiewicz, R. Loll, A second-order phase transition in causal dynamical triangulations. Phys. Rev. Lett. **107**, 211303 (2011). [arXiv:1108.3932]
68. J. Ambjørn, S. Jordan, J. Jurkiewicz, R. Loll, Second- and first-order phase transitions in causal dynamical triangulations. Phys. Rev. D **85**, 124044 (2012). [arXiv:1205.1229]
69. S. Jordan, R. Loll, Causal dynamical triangulations without preferred foliation. Phys. Lett. B **724**, 155 (2013). [arXiv:1305.4582]
70. S. Jordan, R. Loll, De Sitter universe from causal dynamical triangulations without preferred foliation. Phys. Rev. D **88**, 044055 (2013). [arXiv:1307.5469]
71. J. Ambjørn, J. Gizbert-Studnicki, A. Görlich, J. Jurkiewicz, The effective action in 4-dim CDT. The transfer matrix approach. JHEP **1406**, 034 (2014). [arXiv:1403.5940]
72. D.N. Coumbe, J. Jurkiewicz, Evidence for asymptotic safety from dimensional reduction in causal dynamical triangulations. JHEP **1503**, 151 (2015). [arXiv:1411.7712]
73. J. Ambjørn, D.N. Coumbe, J. Gizbert-Studnicki, J. Jurkiewicz, Signature change of the metric in CDT quantum gravity? JHEP **1508**, 033 (2015). [arXiv:1503.08580]
74. D.N. Coumbe, J. Gizbert-Studnicki, J. Jurkiewicz, Exploring the new phase transition of CDT. JHEP **1602**, 144 (2016). [arXiv:1510.08672]
75. J. Ambjørn, J. Gizbert-Studnicki, A. Görlich, J. Jurkiewicz, N. Klitgaard, R. Loll, Characteristics of the new phase in CDT. arXiv:1610.05245
76. R. Loll, The emergence of spacetime, or, quantum gravity on your desktop. Class. Quantum Grav. **25**, 114006 (2008). [arXiv:0711.0273]
77. J. Ambjørn, J. Jurkiewicz, R. Loll, Causal dynamical triangulations and the quest for quantum gravity, in [78]. [arXiv:1004.0352]
78. G.F.R. Ellis, J. Murugan, A. Weltman (eds.), *Foundations of Space and Time* (Cambridge University Press, Cambridge, 2012)
79. J. Ambjørn, A. Görlich, J. Jurkiewicz, R. Loll, Nonperturbative quantum gravity. Phys. Rep. **519**, 127 (2012). [arXiv:1203.3591]
80. M.P. Reisenberger, C. Rovelli, "Sum over surfaces" form of loop quantum gravity. Phys. Rev. D **56**, 3490 (1997). [arXiv:gr-qc/9612035]
81. J.W. Barrett, L. Crane, Relativistic spin networks and quantum gravity. J. Math. Phys. **39**, 3296 (1998). [arXiv:gr-qc/9709028]
82. J.W. Barrett, L. Crane, A Lorentzian signature model for quantum general relativity. Class. Quantum Grav. **17**, 3101 (2000). [arXiv:gr-qc/9904025]
83. A. Perez, C. Rovelli, Spin foam model for Lorentzian general relativity. Phys. Rev. D **63**, 041501 (2001). [arXiv:gr-qc/0009021]
84. E.R. Livine, S. Speziale, New spinfoam vertex for quantum gravity. Phys. Rev. D **76**, 084028 (2007). [arXiv:0705.0674]
85. J. Engle, R. Pereira, C. Rovelli, Loop-quantum-gravity vertex amplitude. Phys. Rev. Lett. **99**, 161301 (2007). [arXiv:0705.2388]
86. J. Engle, R. Pereira, C. Rovelli, Flipped spinfoam vertex and loop gravity. Nucl. Phys. B **798**, 251 (2008). [arXiv:0708.1236]
87. L. Freidel, K. Krasnov, A new spin foam model for 4D gravity. Class. Quantum Grav. **25**, 125018 (2008). [arXiv:0708.1595]
88. E.R. Livine, S. Speziale, Consistently solving the simplicity constraints for spinfoam quantum gravity. Europhys. Lett. **81**, 50004 (2008). [arXiv:0708.1915]
89. R. Pereira, Lorentzian LQG vertex amplitude. Class. Quantum Grav. **25**, 085013 (2008). [arXiv:0710.5043]
90. J. Engle, E. Livine, R. Pereira, C. Rovelli, LQG vertex with finite Immirzi parameter. Nucl. Phys. B **799**, 136 (2008). [arXiv:0711.0146]
91. F. Conrady, L. Freidel, Path integral representation of spin foam models of 4D gravity. Class. Quantum Grav. **25**, 245010 (2008). [arXiv:0806.4640]

92. F. Conrady, L. Freidel, Semiclassical limit of 4-dimensional spin foam models. Phys. Rev. D **78**, 104023 (2008). [arXiv:0809.2280]
93. J.W. Barrett, R.J. Dowdall, W.J. Fairbairn, F. Hellmann, R. Pereira, Lorentzian spin foam amplitudes: graphical calculus and asymptotics. Class. Quantum Grav. **27**, 165009 (2010). [arXiv:0907.2440]
94. C. Rovelli, Discretizing parametrized systems: the magic of Ditt-invariance. arXiv:1107.2310
95. C. Rovelli, On the structure of a background independent quantum theory: Hamilton function, transition amplitudes, classical limit and continuous limit. arXiv:1108.0832
96. D. Oriti, Spacetime geometry from algebra: spin foam models for non-perturbative quantum gravity. Rep. Prog. Phys. **64**, 1489 (2001). [arXiv:gr-qc/0106091]
97. A. Perez, Spin foam models for quantum gravity. Class. Quantum Grav. **20**, R43 (2003). [arXiv:gr-qc/0301113]
98. C. Rovelli, A new look at loop quantum gravity. Class. Quantum Grav. **28**, 114005 (2011). [arXiv:1004.1780]
99. A. Perez, The spin-foam approach to quantum gravity. Living Rev. Relat. **16**, 3 (2013).
100. A. Ashtekar, M. Campiglia, A. Henderson, Loop quantum cosmology and spin foams. Phys. Lett. B **681**, 347 (2009). [arXiv:0909.4221]
101. C. Rovelli, F. Vidotto, On the spinfoam expansion in cosmology. Class. Quantum Grav. **27**, 145005 (2010). [arXiv:0911.3097]
102. A. Ashtekar, M. Campiglia, A. Henderson, Casting loop quantum cosmology in the spin foam paradigm. Class. Quantum Grav. **27**, 135020 (2010). [arXiv:1001.5147]
103. E. Bianchi, C. Rovelli, F. Vidotto, Towards spinfoam cosmology. Phys. Rev. D **82**, 084035 (2010). [arXiv:1003.3483]
104. A. Henderson, C. Rovelli, F. Vidotto, E. Wilson-Ewing, Local spinfoam expansion in loop quantum cosmology. Class. Quantum Grav. **28**, 025003 (2011). [arXiv:1010.0502]
105. G. Calcagni, S. Gielen, D. Oriti, Two-point functions in (loop) quantum cosmology. Class. Quantum Grav. **28**, 125014 (2011). [arXiv:1011.4290]
106. E. Bianchi, T. Krajewski, C. Rovelli, F. Vidotto, Cosmological constant in spinfoam cosmology. Phys. Rev. D **83**, 104015 (2011). [arXiv:1101.4049]
107. H. Huang, Y. Ma, L. Qin, Path integral and effective Hamiltonian in loop quantum cosmology. Gen. Relat. Grav. **45**, 1191 (2013). [arXiv:1102.4755]
108. F. Hellmann, Expansions in spin foam cosmology. Phys. Rev. D **84**, 103516 (2011). [arXiv:1105.1334]
109. L. Qin, G. Deng, Y.-G. Ma, Path integrals and alternative effective dynamics in loop quantum cosmology. Commun. Theor. Phys. **57**, 326 (2012). [arXiv:1206.1131]
110. J. Rennert, D. Sloan, A homogeneous model of spinfoam cosmology. Class. Quantum Grav. **30**, 235019 (2013). [arXiv:1304.6688]
111. D.V. Boulatov, A model of three-dimensional lattice gravity. Mod. Phys. Lett. A **07**, 1629 (1992). [arXiv:hep-th/9202074]
112. H. Ooguri, Topological lattice models in four dimensions. Mod. Phys. Lett. A **07**, 2799 (1992). [arXiv:hep-th/9205090]
113. R. De Pietri, L. Freidel, K. Krasnov, C. Rovelli, Barrett–Crane model from a Boulatov–Ooguri field theory over a homogeneous space. Nucl. Phys. B **574**, 785 (2000). [arXiv:hep-th/9907154]
114. M.P. Reisenberger, C. Rovelli, Space-time as a Feynman diagram: the connection formulation. Class. Quantum Grav. **18**, 121 (2001). [arXiv:gr-qc/0002095]
115. A.R. Miković, Quantum field theory of spin networks. Class. Quantum Grav. **18**, 2827 (2001). [arXiv:gr-qc/0102110]
116. D. Oriti, J. Ryan, Group field theory formulation of 3D quantum gravity coupled to matter fields. Class. Quantum Grav. **23**, 6543 (2006). [arXiv:gr-qc/0602010]
117. W.J. Fairbairn, E.R. Livine, 3D spinfoam quantum gravity: matter as a phase of the group field theory. Class. Quantum Grav. **24**, 5277 (2007). [arXiv:gr-qc/0702125]
118. E.R. Livine, Matrix models as non-commutative field theories on $\mathbb{R}^3$. Class. Quantum Grav. **26**, 195014 (2009). [arXiv:0811.1462]

119. F. Girelli, E.R. Livine, D. Oriti, Four-dimensional deformed special relativity from group field theories. Phys. Rev. D **81**, 024015 (2010). [arXiv:0903.3475]
120. D. Oriti, Emergent non-commutative matter fields from group field theory models of quantum spacetime. J. Phys. Conf. Ser. **174**, 012047 (2009). [arXiv:0903.3970]
121. R.J. Dowdall, Wilson loops, geometric operators and fermions in 3d group field theory. Centr. Eur. J. Phys. **9**, 1043 (2011). [arXiv:0911.2391]
122. A. Baratin, D. Oriti, Group field theory with noncommutative metric variables. Phys. Rev. Lett. **105**, 221302 (2010). [arXiv:1002.4723]
123. A. Baratin, B. Dittrich, D. Oriti, J. Tambornino, Non-commutative flux representation for loop quantum gravity. Class. Quantum Grav. **28**, 175011 (2011). [arXiv:1004.3450]
124. J. Ben Geloun, V. Bonzom, Radiative corrections in the Boulatov–Ooguri tensor model: the 2-point function. Int. J. Theor. Phys. **50**, 2819 (2011). [arXiv:1101.4294]
125. A. Baratin, D. Oriti, Quantum simplicial geometry in the group field theory formalism: reconsidering the Barrett–Crane model. New J. Phys. **13**, 125011 (2011). [arXiv:1108.1178]
126. A. Baratin, D. Oriti, Group field theory and simplicial gravity path integrals: a model for Holst–Plebanski gravity. Phys. Rev. D **85**, 044003 (2012). [arXiv:1111.5842]
127. J. Ben Geloun, Two- and four-loop β-functions of rank-4 renormalizable tensor field theories. Class. Quantum Grav. **29**, 235011 (2012). [arXiv:1205.5513]
128. S. Carrozza, D. Oriti, V. Rivasseau, Renormalization of tensorial group field theories: Abelian $U(1)$ models in four dimensions. Commun. Math. Phys. **327**, 603 (2014). [arXiv:1207.6734]
129. C. Guedes, D. Oriti, M. Raasakka, Quantization maps, algebra representation and non-commutative Fourier transform for Lie groups. J. Math. Phys. **54**, 083508 (2013). [arXiv:1301.7750]
130. S. Gielen, D. Oriti, L. Sindoni, Cosmology from group field theory formalism for quantum gravity. Phys. Rev. Lett. **111**, 031301 (2013). [arXiv:1303.3576]
131. S. Carrozza, D. Oriti, V. Rivasseau, Renormalization of an $SU(2)$ tensorial group field theory in three dimensions. Commun. Math. Phys. **330**, 581 (2014). [arXiv:1303.6772]
132. S. Gielen, D. Oriti, L. Sindoni, Homogeneous cosmologies as group field theory condensates. JHEP **1406**, 013 (2014). [arXiv:1311.1238]
133. S. Gielen, Quantum cosmology of (loop) quantum gravity condensates: an example. Class. Quantum Grav. **31**, 155009 (2014). [arXiv:1404.2944]
134. G. Calcagni, Loop quantum cosmology from group field theory. Phys. Rev. D **90**, 064047 (2014). [arXiv:1407.8166]
135. S. Gielen, D. Oriti, Quantum cosmology from quantum gravity condensates: cosmological variables and lattice-refined dynamics. New J. Phys. **16**, 123004 (2014). [arXiv:1407.8167]
136. D. Oriti, D. Pranzetti, J.P. Ryan, L. Sindoni, Generalized quantum gravity condensates for homogeneous geometries and cosmology. Class. Quantum Grav. **32**, 235016 (2015). [arXiv:1501.00936]
137. S. Gielen, Identifying cosmological perturbations in group field theory condensates. JHEP **1508**, 010 (2015). [arXiv:1505.07479]
138. D. Oriti, D. Pranzetti, L. Sindoni, Horizon entropy from quantum gravity condensates. Phys. Rev. Lett. **116**, 211301 (2016). [arXiv:1510.06991]
139. D. Oriti, L. Sindoni, E. Wilson-Ewing, Bouncing cosmologies from quantum gravity condensates. arXiv:1602.08271
140. D. Oriti, L. Sindoni, E. Wilson-Ewing, Emergent Friedmann dynamics with a quantum bounce from quantum gravity condensates. Class. Quantum Grav. **33**, 224001 (2016). [arXiv:1602.05881]
141. A.G.A. Pithis, M. Sakellariadou, P. Tomov, Impact of nonlinear effective interactions on group field theory quantum gravity condensates. Phys. Rev. D **94**, 064056 (2016). [arXiv:1607.06662]
142. L. Freidel, Group field theory: an overview. Int. J. Theor. Phys. **44**, 1769 (2005). [arXiv:hep-th/0505016]
143. D. Oriti, The group field theory approach to quantum gravity, in [144]. [arXiv:gr-qc/0607032]
144. D. Oriti (ed.), *Approaches to Quantum Gravity* (Cambridge University Press, Cambridge, 2009)

145. D. Oriti, The microscopic dynamics of quantum space as a group field theory, in [78]. [arXiv:1110.5606]
146. D. Oriti, Group field theory as the second quantization of loop quantum gravity. Class. Quantum Grav. **33**, 085005 (2016). [arXiv:1310.7786]
147. A. Baratin, D. Oriti, Ten questions on group field theory (and their tentative answers). J. Phys. Conf. Ser. **360**, 012002 (2012). [arXiv:1112.3270]
148. S. Gielen, L. Sindoni, Quantum cosmology from group field theory condensates: a review. SIGMA **12**, 082 (2016). [arXiv:1602.08104]
149. L. Bombelli, J. Lee, D. Meyer, R. Sorkin, Space-time as a causal set. Phys. Rev. Lett. **59**, 521 (1987)
150. C. Moore, Comment on "Space-time as a causal set". Phys. Rev. Lett. **60**, 655 (1988)
151. L. Bombelli, J. Lee, D. Meyer, R.D. Sorkin, Bombelli et al. reply. Phys. Rev. Lett. **60**, 656 (1988)
152. G. Brightwell, R. Gregory, Structure of random discrete spacetime. Phys. Rev. Lett. **66**, 260 (1991)
153. R.D. Sorkin, Forks in the road, on the way to quantum gravity. Int. J. Theor. Phys. **36**, 2759 (1997). [arXiv:gr-qc/9706002]
154. D.P. Rideout, R.D. Sorkin, Classical sequential growth dynamics for causal sets. Phys. Rev. D **61**, 024002 (2000). [arXiv:gr-qc/9904062]
155. R.D. Sorkin, Indications of causal set cosmology. Int. J. Theor. Phys. **39**, 1731 (2000). [arXiv:gr-qc/0003043]
156. X. Martín, D. O'Connor, D.P. Rideout, R.D. Sorkin, "Renormalization" transformations induced by cycles of expansion and contraction in causal set cosmology. Phys. Rev. D **63**, 084026 (2001). [arXiv:gr-qc/0009063]
157. A. Ash, P. McDonald, Moment problems and the causal set approach to quantum gravity. J. Math. Phys. **44**, 1666 (2003). [arXiv:gr-qc/0209020]
158. M. Ahmed, S. Dodelson, P.B. Greene, R. Sorkin, Everpresent Λ. Phys. Rev. D **69**, 103523 (2004). [arXiv:astro-ph/0209274]
159. G. Brightwell, H.F. Dowker, R.S. García, J. Henson, R.D. Sorkin, "Observables" in causal set cosmology. Phys. Rev. D **67**, 084031 (2003). [arXiv:gr-qc/0210061]
160. D. Rideout, Dynamics of Causal Sets. Ph.D. thesis, Syracuse University, Syracuse (2001). [arXiv:gr-qc/0212064]
161. F. Dowker, J. Henson, R.D. Sorkin, Quantum gravity phenomenology, Lorentz invariance and discreteness. Mod. Phys. Lett. A **19**, 1829 (2004). [arXiv:gr-qc/0311055]
162. S. Major, D. Rideout, S. Surya, Spatial hypersurfaces in causal set cosmology. Class. Quantum Grav. **23**, 4743 (2006). [arXiv:gr-qc/0506133]
163. S. Major, D. Rideout, S. Surya, On recovering continuum topology from a causal set. J. Math. Phys. **48**, 032501 (2007). [arXiv:gr-qc/0604124]
164. L. Bombelli, J. Henson, R.D. Sorkin, Discreteness without symmetry breaking: a theorem. Mod. Phys. Lett. A **24**, 2579 (2009). [arXiv:gr-qc/0605006]
165. D. Rideout, S. Zohren, Evidence for an entropy bound from fundamentally discrete gravity. Class. Quantum Grav. **23**, 6195 (2006). [arXiv:gr-qc/0606065]
166. J.D. Barrow, Strong constraint on ever-present Λ. Phys. Rev. D **75**, 067301 (2007). [arXiv:gr-qc/0612128]
167. J.A. Zuntz, The cosmic microwave background in a causal set universe. Phys. Rev. D **77**, 043002 (2008). [arXiv:0711.2904]
168. S. Johnston, Particle propagators on discrete spacetime. Class. Quantum Grav. **25**, 202001 (2008). [arXiv:0806.3083]
169. L. Philpott, F. Dowker, R.D. Sorkin, Energy-momentum diffusion from spacetime discreteness. Phys. Rev. D **79**, 124047 (2009). [arXiv:0810.5591]
170. S. Major, D. Rideout, S. Surya, Stable homology as an indicator of manifoldlikeness in causal set theory. Class. Quantum Grav. **26**, 175008 (2009). [arXiv:0902.0434]
171. S. Johnston, Feynman propagator for a free scalar field on a causal set. Phys. Rev. Lett. **103**, 180401 (2009). [arXiv:0909.0944]

172. M. Ahmed, D. Rideout, Indications of de Sitter spacetime from classical sequential growth dynamics of causal sets. Phys. Rev. D **81**, 083528 (2010). [arXiv:0909.4771]
173. D.M.T. Benincasa, F. Dowker, Scalar curvature of a causal set. Phys. Rev. Lett. **104**, 181301 (2010). [arXiv:1001.2725]
174. D.M.T. Benincasa, F. Dowker, B. Schmitzer, The random discrete action for two-dimensional spacetime. Class. Quantum Grav. **28**, 105018 (2011). [arXiv:1011.5191]
175. S. Surya, Evidence for the continuum in 2D causal set quantum gravity. Class. Quantum Grav. **29**, 132001 (2012). [arXiv:1110.6244]
176. M. Ahmed, R. Sorkin, Everpresent Λ. II. Structural stability. Phys. Rev. D **87**, 063515 (2013). [arXiv:1210.2589]
177. F. Dowker, L. Glaser, Causal set d'Alembertians for various dimensions. Class. Quantum Grav. **30**, 195016 (2013). [arXiv:1305.2588]
178. A. Eichhorn, S. Mizera, Spectral dimension in causal set quantum gravity. Class. Quantum Grav. **31**, 125007 (2014). [arXiv:1311.2530]
179. S. Aslanbeigi, M. Saravani, R.D. Sorkin, Generalized causal set d'Alembertians. JHEP **1406**, 024 (2014). [arXiv:1403.1622]
180. S. Johnston, Correction terms for propagators and d'Alembertians due to spacetime discreteness. Class. Quantum Grav. **32**, 195020 (2015). [arXiv:1411.2614]
181. R.D. Sorkin, Spacetime and causal sets, in *Relativity and Gravitation: Classical and Quantum*, ed. by J.C. D'Olivo, E. Nahmad-Achar, M. Rosenbaum, M.P. Ryan, L.F. Urrutia, F. Zertuche (World Scientific, Singapore, 1991)
182. D.D. Reid, Introduction to causal sets: an alternate view of spacetime structure. Can. J. Phys. **79**, 1 (2001). [arXiv:gr-qc/9909075]
183. J. Henson, The causal set approach to quantum gravity, in [144]. [arXiv:gr-qc/0601121]
184. J. Henson, Discovering the discrete universe. arXiv:1003.5890
185. S. Surya, Directions in causal set quantum gravity, in *Recent Research in Quantum Gravity*, ed. by A. Dasgupta (Nova Science, Hauppauge, 2011). [arXiv:1103.6272]
186. F. Dowker, Introduction to causal sets and their phenomenology. Gen. Relat. Grav. **45**, 1651 (2013)
187. G. Calcagni, Diffusion in multiscale spacetimes. Phys. Rev. E **87**, 012123 (2013). [arXiv:1205.5046]
188. G. Calcagni, G. Nardelli, Spectral dimension and diffusion in multiscale spacetimes. Phys. Rev. D **88**, 124025 (2013). [arXiv:1304.2709]
189. G. Calcagni, L. Modesto, G. Nardelli, Quantum spectral dimension in quantum field theory. Int. J. Mod. Phys. D **25**, 1650058 (2016). [arXiv:1408.0199]
190. J. Berges, N. Tetradis, C. Wetterich, Non-perturbative renormalization flow in quantum field theory and statistical physics. Phys. Rep. **363**, 223 (2002). [arXiv:hep-ph/0005122]
191. G. Calcagni, Multifractional spacetimes, asymptotic safety and Hořava–Lifshitz gravity. Int. J. Mod. Phys. A **28**, 1350092 (2013). [arXiv:1209.4376]
192. B.F.L. Ward, Planck scale cosmology in resummed quantum gravity. Mod. Phys. Lett. A **23**, 3299 (2008). [arXiv:0808.3124]
193. D. Kalligas, P.S. Wesson, C.W.F. Everitt, Bianchi type I cosmological models with variable G and Λ: a comment. Gen. Relat. Grav. **27**, 645 (1995)
194. B.F.L. Ward, Einstein–Heisenberg consistency condition interplay with cosmological constant prediction in resummed quantum gravity. Mod. Phys. Lett. A **30**, 1550206 (2015). [arXiv:1507.00661]
195. J. Ambjørn, S. Varsted, Three-dimensional simplicial quantum gravity. Nucl. Phys. B **373**, 557 (1992)
196. J. Ambjørn, D.V. Boulatov, A. Krzywicki, S. Varsted, The vacuum in three-dimensional simplicial quantum gravity. Phys. Lett. B **276**, 432 (1992)
197. J. Ambjørn, J. Jurkiewicz, Four-dimensional simplicial quantum gravity. Phys. Lett. B **278**, 42 (1992)
198. M.E. Agishtein, A.A. Migdal, Critical behavior of dynamically triangulated quantum gravity in four dimensions. Nucl. Phys. B **385**, 395 (1992). [arXiv:hep-lat/9204004]

199. J. Ambjørn, S. Jain, J. Jurkiewicz, C.F. Kristjansen, Observing 4d baby universes in quantum gravity. Phys. Lett. B **305**, 208 (1993). [arXiv:hep-th/9303041]
200. S. Catterall, J.B. Kogut, R. Renken, Phase structure of four-dimensional simplicial quantum gravity. Phys. Lett. B **328**, 277 (1994). [arXiv:hep-lat/9401026]
201. P. Bialas, Z. Burda, A. Krzywicki, B. Petersson, Focusing on the fixed point of 4D simplicial gravity. Nucl. Phys. B **472**, 293 (1996). [arXiv:hep-lat/9601024]
202. B.V. de Bakker, Further evidence that the transition of 4D dynamical triangulation is first order. Phys. Lett. B **389**, 238 (1996). [arXiv:hep-lat/9603024]
203. S. Catterall, R. Renken, J.B. Kogut, Singular structure in 4D simplicial gravity. Phys. Lett. B **416**, 274 (1998). [arXiv:hep-lat/9709007]
204. J.A. Wheeler, Geometrodynamics and the issue of the final state, in *Relativity, Groups and Topology*, ed. by C. DeWitt, B.S. DeWitt (Gordon and Breach, New York, 1964)
205. T. Regge, General relativity without coordinates. Nuovo Cim. **19**, 558 (1961)
206. R.M. Williams, P.A. Tuckey, Regge calculus: a bibliography and brief review. Class. Quantum Grav. **9**, 1409 (1992)
207. R.M. Williams, Discrete quantum gravity: the Regge calculus approach. Int. J. Mod. Phys. B **6**, 2097 (1992)
208. R.M. Williams, Recent progress in Regge calculus. Nucl. Phys. Proc. Suppl. **57**, 73 (1997). [arXiv:gr-qc/9702006]
209. F. David, What is the intrinsic geometry of two-dimensional quantum gravity? Nucl. Phys. B **368**, 671 (1992)
210. T. Jonsson, J.F. Wheater, The spectral dimension of the branched polymer phase of two-dimensional quantum gravity. Nucl. Phys. B **515**, 549 (1998). [arXiv:hep-lat/9710024]
211. J.D. Correia, J.F. Wheater, The spectral dimension of non-generic branched polymer ensembles. Phys. Lett. B **422**, 76 (1998). [arXiv:hep-th/9712058]
212. C. Destri, L. Donetti, The spectral dimension of random trees. J. Phys. A **35**, 9499 (2002). [arXiv:cond-mat/0206233]
213. B. Durhuus, T. Jonsson, J.F. Wheater, Random walks on combs. J. Phys. A **39**, 1009 (2006). [arXiv:hep-th/0509191]
214. B. Durhuus, T. Jonsson, J.F. Wheater, The spectral dimension of generic trees. J. Stat. Phys. **128**, 1237 (2007). [arXiv:math-ph/0607020]
215. J.J. Halliwell, J.B. Hartle, Wave functions constructed from an invariant sum over histories satisfy constraints. Phys. Rev. D **43**, 1170 (1991)
216. C. Teitelboim, Quantum mechanics of the gravitational field. Phys. Rev. D **25**, 3159 (1982)
217. J.J. Halliwell, Derivation of the Wheeler–DeWitt equation from a path integral for minisuperspace models. Phys. Rev. D **38**, 2468 (1988)
218. J.B. Hartle, K.V. Kuchař, Path integrals in parametrized theories: the free relativistic particle. Phys. Rev. D **34**, 2323 (1986)
219. J.J. Halliwell, M.E. Ortiz, Sum-over-histories origin of the composition laws of relativistic quantum mechanics and quantum cosmology. Phys. Rev. D **48**, 748 (1993). [arXiv:gr-qc/9211004]
220. R. Gurau, J.P. Ryan, Colored tensor models – a review. SIGMA **8**, 020 (2012). [arXiv:1109.4812]
221. V. Rivasseau, Quantum gravity and renormalization: the tensor track. AIP Conf. Proc. **1444**, 18 (2011). [arXiv:1112.5104]
222. L. Freidel, S. Majid, Noncommutative harmonic analysis, sampling theory and the Duflo map in 2+1 quantum gravity. Class. Quantum Grav. **25**, 045006 (2008). [arXiv:hep-th/0601004]
223. E. Joung, J. Mourad, K. Noui, Three dimensional quantum geometry and deformed symmetry. J. Math. Phys. **50**, 052503 (2009). [arXiv:0806.4121]
224. C.J. Pethick, H. Smith, *Bose–Einstein Condensation in Dilute Gases* (Cambridge University Press, Cambridge, 2002)
225. L.P. Pitaevskii, S. Stringari, *Bose–Einstein Condensation* (Clarendon Press, Oxford, 2003)
226. S. Giorgini, L.P. Pitaevskii, S. Stringari, Theory of ultracold atomic Fermi gases. Rev. Mod. Phys. **80**, 1215 (2008). [arXiv:0706.3360]

227. G. Calcagni, S. Gielen, D. Oriti, Group field cosmology: a cosmological field theory of quantum geometry. Class. Quantum Grav. **29**, 105005 (2012). [arXiv:1201.4151]
228. C.R. Contaldi, F. Dowker, L. Philpott, Polarization diffusion from spacetime uncertainty. Class. Quantum Grav. **27**, 172001 (2010). [arXiv:1001.4545]
229. A. Lue, L. Wang, M. Kamionkowski, Cosmological signature of new parity-violating interactions. Phys. Rev. Lett. **83**, 1506 (1999). [arXiv:astro-ph/9812088]
230. J. Madore, *An Introduction to Noncommutative Geometry and Its Physical Applications* (Cambridge University Press, Cambridge, 1999)
231. M.R. Douglas, N.A. Nekrasov, Noncommutative field theory. Rev. Mod. Phys. **73**, 977 (2001). [arXiv:hep-th/0106048]
232. R.J. Szabo, Quantum field theory on noncommutative spaces. Phys. Rep. **378**, 207 (2003). [arXiv:hep-th/0109162]
233. A. Connes, *Noncommutative Geometry* (Academic Press, San Diego, 2004)
234. P. Aschieri, M. Dimitrijevic, P. Kulish, F. Lizzi, J. Wess, *Noncommutative Spacetimes* (Springer, Berlin, 2009)
235. A.P. Balachandran, A. Ibort, G. Marmo, M. Martone, Quantum fields on noncommutative spacetimes: theory and phenomenology. SIGMA **6**, 052 (2010). [arXiv:1003.4356]
236. H.-J. Matschull, M. Welling, Quantum mechanics of a point particle in $(2+1)$-dimensional gravity. Class. Quantum Grav. **15**, 2981 (1998). [arXiv:gr-qc/9708054]
237. M. Arzano, E. Alesci, Anomalous dimension in three-dimensional semiclassical gravity. Phys. Lett. B **707**, 272 (2012). [arXiv:1108.1507]
238. S. Deser, R. Jackiw, G. 't Hooft, Three-dimensional Einstein gravity: dynamics of flat space. Ann. Phys. (N.Y.) **152**, 220 (1984)
239. L. Freidel, E. Livine, Ponzano–Regge model revisited III: Feynman diagrams and effective field theory. Class. Quantum Grav. **23**, 2021 (2006). [arXiv:hep-th/0502106]
240. L. Freidel, E.R. Livine, 3D quantum gravity and noncommutative quantum field theory. Phys. Rev. Lett. **96**, 221301 (2006). [arXiv:hep-th/0512113]
241. G. Amelino-Camelia, L. Smolin, A. Starodubtsev, Quantum symmetry, the cosmological constant and Planck-scale phenomenology. Class. Quantum Grav. **21**, 3095 (2004). [arXiv:hep-th/0306134]
242. L. Freidel, J. Kowalski-Glikman, L. Smolin, 2+1 gravity and doubly special relativity. Phys. Rev. D **69**, 044001 (2004). [arXiv:hep-th/0307085]
243. S. Doplicher, K. Fredenhagen, J.E. Roberts, Space-time quantization induced by classical gravity. Phys. Lett. B **331**, 39 (1994)
244. S. Doplicher, K. Fredenhagen, J.E. Roberts, The quantum structure of space-time at the Planck scale and quantum fields. Commun. Math. Phys. **172**, 187 (1995). [arXiv:hep-th/0303037]
245. S. Majid, H. Ruegg, Bicrossproduct structure of κ-Poincaré group and non-commutative geometry. Phys. Lett. B **334**, 348 (1994). [arXiv:hep-th/9405107]
246. A. Agostini, G. Amelino-Camelia, F. D'Andrea, Hopf algebra description of noncommutative space-time symmetries. Int. J. Mod. Phys. A **19**, 5187 (2004). [arXiv:hep-th/0306013]
247. M. Arzano, G. Calcagni, D. Oriti, M. Scalisi, Fractional and noncommutative spacetimes. Phys. Rev. D **84**, 125002 (2011). [arXiv:1107.5308]
248. G. Calcagni, M. Ronco, Deformed symmetries in noncommutative and multifractional spacetimes. arXiv:1608.01667
249. C.-S. Chu, P.-M. Ho, Non-commutative open string and D-brane. Nucl. Phys. B **550**, 151 (1999). [arXiv:hep-th/9812219]
250. V. Schomerus, D-branes and deformation quantization. JHEP **9906**, 030 (1999). [arXiv:hep-th/9903205]
251. N. Seiberg, E. Witten, String theory and noncommutative geometry. JHEP **0909**, 032 (1999). [arXiv:hep-th/9908142]
252. A. Matusis, L. Susskind, N. Toumbas, The IR/UV connection in the non-commutative gauge theories. JHEP **0012**, 002 (2000). [arXiv:hep-th/0002075]
253. A.H. Chamseddine, G. Felder, J. Fröhlich, Gravity in noncommutative geometry. Commun. Math. Phys. **155**, 205 (1993). [arXiv:hep-th/9209044]

254. P. Aschieri, C. Blohmann, M. Dimitrijevic, F. Meyer, P. Schupp, J. Wess, A gravity theory on noncommutative spaces. Class. Quantum Grav. **22**, 3511 (2005). [arXiv:hep-th/0504183]
255. P. Aschieri, M. Dimitrijevic, F. Meyer, J. Wess, Noncommutative geometry and gravity. Class. Quantum Grav. **23**, 1883 (2006). [arXiv:hep-th/0510059]
256. E. Harikumar, V.O. Rivelles, Noncommutative gravity. Class. Quantum Grav. **23**, 7551 (2006). [arXiv:hep-th/0607115]
257. A. Connes, M.R. Douglas, A.S. Schwarz, Noncommutative geometry and matrix theory: compactification on tori. JHEP **9802**, 003 (1998). [arXiv:hep-th/9711162]
258. J. Madore, S. Schraml, P. Schupp, J. Wess, Gauge theory on noncommutative spaces. Eur. Phys. J. C **16**, 161 (2000). [arXiv:hep-th/0001203]
259. P. Kosiński, J. Lukierski, P. Maślanka, Local field theory on κ-Minkowski space, star products and noncommutative translations. Czech. J. Phys. **50**, 1283 (2000). [arXiv:hep-th/0009120]
260. A. Agostini, F. Lizzi, A. Zampini, Generalized Weyl systems and κ-Minkowski space. Mod. Phys. Lett. A **17**, 2105 (2002). [arXiv:hep-th/0209174]
261. F. Lizzi, G. Mangano, G. Miele, M. Peloso, Cosmological perturbations and short distance physics from noncommutative geometry. JHEP **0206**, 049 (2002). [arXiv:hep-th/0203099]
262. H.J. Groenewold, On the principles of elementary quantum mechanics. Physica **12**, 405 (1946)
263. J.E. Moyal, Quantum mechanics as a statistical theory. Proc. Camb. Philos. Soc. **45**, 99 (1949)
264. A. Kempf, Mode generating mechanism in inflation with a cutoff. Phys. Rev. D **63**, 083514 (2001). [arXiv:astro-ph/0009209]
265. A. Kempf, J.C. Niemeyer, Perturbation spectrum in inflation with a cutoff. Phys. Rev. D **64**, 103501 (2001). [arXiv:astro-ph/0103225]
266. R. Easther, B.R. Greene, W.H. Kinney, G. Shiu, Inflation as a probe of short distance physics. Phys. Rev. D **64**, 103502 (2001). [arXiv:hep-th/0104102]
267. O. Bertolami, C.A.D. Zarro, Towards a noncommutative astrophysics. Phys. Rev. D **81**, 025005 (2010). [arXiv:0908.4196]
268. H. García-Compeán, O. Obregón, C. Ramírez, Noncommutative quantum cosmology. Phys. Rev. Lett. **88**, 161301 (2002). [arXiv:hep-th/0107250]
269. C. Bastos, O. Bertolami, N. Costa Dias, J.N. Prata, Phase-space noncommutative quantum cosmology. Phys. Rev. D **78**, 023516 (2008). [arXiv:0712.4122]
270. C. Bastos, O. Bertolami, N. Costa Dias, J.N. Prata, Black holes and phase space noncommutativity. Phys. Rev. D **80**, 124038 (2009). [arXiv:0907.1818]
271. C. Bastos, O. Bertolami, N. Costa Dias, J.N. Prata, The singularity problem and phase-space noncanonical noncommutativity. Phys. Rev. D **82**, 041502(R) (2010). [arXiv:0912.4027]
272. C. Bastos, O. Bertolami, N. Costa Dias, J.N. Prata, Noncanonical phase-space noncommutativity and the Kantowski–Sachs singularity for black holes. Phys. Rev. D **84**, 024005 (2011). [arXiv:1012.5523]
273. M. Rinaldi, A new approach to non-commutative inflation. Class. Quantum Grav. **28**, 105022 (2011). [arXiv:0908.1949]
274. H. Perrier, R. Durrer, M. Rinaldi, Explosive particle production in non-commutative inflation. JHEP **1301**, 067 (2013). [arXiv:1210.5373]
275. S. Alexander, J. Magueijo, Noncommutative geometry as a realization of varying speed of light cosmology. arXiv:hep-th/0104093
276. S. Alexander, R. Brandenberger, J. Magueijo, Noncommutative inflation. Phys. Rev. D **67**, 081301 (2003). [arXiv:hep-th/0108190]
277. S. Koh, R.H. Brandenberger, Cosmological perturbations in non-commutative inflation. JCAP **0706**, 021 (2007). [arXiv:hep-th/0702217]
278. U.D. Machado, R. Opher, Conceptual problem for noncommutative inflation and the new approach for nonrelativistic inflationary equation of state. Phys. Rev. D **87**, 123517 (2013). [arXiv:1211.6478]
279. P.J.E. Peebles, *Principles of Physical Cosmology* (Princeton University Press, Princeton, 1993)

280. J. Magueijo, L. Pogosian, Could thermal fluctuations seed cosmic structure? Phys. Rev. D **67**, 043518 (2003). [arXiv:astro-ph/0211337]
281. T. Biswas, R. Brandenberger, T. Koivisto, A. Mazumdar, Cosmological perturbations from statistical thermal fluctuations. Phys. Rev. D **88**, 023517 (2013). [arXiv:1302.6463]
282. J. Magueijo, P. Singh, Thermal fluctuations in loop cosmology. Phys. Rev. D **76**, 023510 (2007). [arXiv:astro-ph/0703566]
283. F. Lizzi, G. Mangano, G. Miele, G. Sparano, Inflationary cosmology from noncommutative geometry. Int. J. Mod. Phys. A **11**, 2907 (1996). [arXiv:gr-qc/9503040]
284. O. Bertolami, L. Guisado, Noncommutative scalar field coupled to gravity. Phys. Rev. D **67**, 025001 (2003). [arXiv:gr-qc/0207124]
285. E. Di Grezia, G. Esposito, A. Funel, G. Mangano, G. Miele, Spacetime noncommutativity and antisymmetric tensor dynamics in the early Universe. Phys. Rev. D **68**, 105012 (2003). [arXiv:gr-qc/0305050]
286. S.A. Alavi, F. Nasseri, Running of the spectral index in noncommutative inflation. Int. J. Mod. Phys. A **20**, 4941 (2005). [arXiv:astro-ph/0406477]
287. C.-S. Chu, B.R. Greene, G. Shiu, Remarks on inflation and noncommutative geometry. Mod. Phys. Lett. A **16**, 2231 (2001). [arXiv:hep-th/0011241]
288. E. Akofor, A.P. Balachandran, S.G. Jo, A. Joseph, B.A. Qureshi, Direction-dependent CMB power spectrum and statistical anisotropy from noncommutative geometry. JHEP **0805**, 092 (2008). [arXiv:0710.5897]
289. T.S. Koivisto, D.F. Mota, CMB statistics in noncommutative inflation. JHEP **1102**, 061 (2011). [arXiv:1011.2126]
290. A. Nautiyal, Anisotropic non-gaussianity with noncommutative spacetime. Phys. Lett. B **728**, 472 (2014). [arXiv:1303.4159]
291. E. Akofor, A.P. Balachandran, A. Joseph, L. Pekowsky, B.A. Qureshi, Constraints from CMB on spacetime noncommutativity and causality violation. Phys. Rev. D **79**, 063004 (2009). [arXiv:0806.2458]
292. R. Brandenberger, P.-M. Ho, Noncommutative spacetime, stringy spacetime uncertainty principle, and density fluctuations. Phys. Rev. D **66**, 023517 (2002). [arXiv:hep-th/0203119]
293. S. Tsujikawa, R. Maartens, R. Brandenberger, Non-commutative inflation and the CMB. Phys. Lett. B **574**, 141 (2003). [arXiv:astro-ph/0308169]
294. Q.-G. Huang, M. Li, CMB power spectrum from noncommutative spacetime. JHEP **0306**, 014 (2003). [arXiv:hep-th/0304203]
295. M. Fukuma, Y. Kono, A. Miwa, Effects of space-time noncommutativity on the angular power spectrum of the CMB. Nucl. Phys. B **682**, 377 (2004). [arXiv:hep-th/0307029]
296. Q.-G. Huang, M. Li, Noncommutative inflation and the CMB multipoles. JCAP **11**, 001 (2003). [arXiv:astro-ph/0308458]
297. Q.-G. Huang, M. Li, Power spectra in spacetime noncommutative inflation. Nucl. Phys. B **713**, 219 (2005). [arXiv:astro-ph/0311378]
298. H. Kim, G.S. Lee, Y.S. Myung, Noncommutative spacetime effect on the slow-roll period of inflation. Mod. Phys. Lett. A **20**, 271 (2005). [arXiv:hep-th/0402018]
299. H. Kim, G.S. Lee, H.W. Lee, Y.S. Myung, Second-order corrections to noncommutative spacetime inflation. Phys. Rev. D **70**, 043521 (2004). [arXiv:hep-th/0402198]
300. R.-G. Cai, A note on curvature fluctuation of noncommutative inflation. Phys. Lett. B **593**, 1 (2004). [arXiv:hep-th/0403134]
301. G. Calcagni, Noncommutative models in patch cosmology. Phys. Rev. D **70**, 103525 (2004). [arXiv:hep-th/0406006]
302. G. Calcagni, Consistency relations and degeneracies in (non)commutative patch inflation. Phys. Lett. B **606**, 177 (2005). [arXiv:hep-ph/0406057]
303. G. Calcagni, S. Tsujikawa, Observational constraints on patch inflation in noncommutative spacetime. Phys. Rev. D **70**, 103514 (2004). [arXiv:astro-ph/0407543]
304. Q.-G. Huang, M. Li, Running spectral index in noncommutative inflation and WMAP three year results. Nucl. Phys. B **755**, 286 (2006). [arXiv:astro-ph/0603782]

305. X. Zhang, F.-Q. Wu, Noncommutative chaotic inflation and WMAP three year results. Phys. Lett. B **638**, 396 (2006). [arXiv:astro-ph/0604195]
306. G. Calcagni, S. Kuroyanagi, J. Ohashi, S. Tsujikawa, Strong Planck constraints on braneworld and non-commutative inflation. JCAP **1403**, 052 (2014). [arXiv:1310.5186]
307. X. Bekaert, N. Boulanger, D. Francia, Mixed-symmetry multiplets and higher-spin curvatures. J. Phys. A **48**, 225401 (2015). [arXiv:1501.02462]
308. E. Tomboulis, $1/N$ expansion and renormalization in quantum gravity. Phys. Lett. B **70**, 361 (1977)
309. E. Tomboulis, Renormalizability and asymptotic freedom in quantum gravity. Phys. Lett. B **97**, 77 (1980)
310. M. Kaku, Strong-coupling approach to the quantization of conformal gravity. Phys. Rev. D **27**, 2819 (1983)
311. E.T. Tomboulis, Unitarity in higher-derivative quantum gravity. Phys. Rev. Lett. **52**, 1173 (1984)
312. M. Ostrogradski, Mémoire sur les équations différentielles relatives au problème des isopérimètres. Mem. Act. St. Petersbourg **VI 4**, 385 (1850)
313. D.A. Eliezer, R.P. Woodard, The problem of nonlocality in string theory. Nucl. Phys. B **325**, 389 (1989)
314. J.Z. Simon, Higher-derivative Lagrangians, nonlocality, problems, and solutions. Phys. Rev. D **41**, 3720 (1990)
315. R.P. Woodard, Ostrogradsky's theorem on Hamiltonian instability. Scholarpedia **10**, 32243 (2015). [arXiv:1506.02210]
316. G. Wataghin, Bemerkung über die Selbstenergie der Elektronen. Z. Phys. **88**, 92 (1934)
317. F. Bopp, Lineare Theorie des Elektrons. II. Ann. Phys. (Berlin) **434**, 573 (1943)
318. R.P. Feynman, A relativistic cut-off for classical electrodynamics. Phys. Rev. **74**, 939 (1948)
319. A. Pais, G.E. Uhlenbeck, On field theories with non-localized action. Phys. Rev. **79**, 145 (1950)
320. N. Shôno, N. Oda, Note on the non-local interaction. Prog. Theor. Phys. **8**, 28 (1952)
321. W. Pauli, On the hamiltonian structure of non-local field theories. Nuovo Cim. **10**, 648 (1953)
322. M. Chrétien, R.E. Peierls, Properties of form factors in non-local theories. Nuovo Cim. **10**, 668 (1953)
323. C. Hayashi, Hamiltonian formalism in non-local field theories. Prog. Theor. Phys. **10**, 533 (1953)
324. C. Hayashi, On field equations with non-local interaction. Prog. Theor. Phys. **11**, 226 (1954).
325. M. Chrétien, R.E. Peierls, A study of gauge-invariant non-local interactions. Proc. R. Soc. Lond. A **223**, 468 (1954) [World Sci. Ser. 20th Cent. Phys. **19**, 397 (1997)]
326. M. Meyman, The causality principle and the asymptotic behavior of the scattering amplitude. Zh. Eksp. Teor. Fiz. **47**, 1966 (1965) [Sov. Phys. JETP **20**, 1320 (1965)]
327. G.V. Efimov, Non-local quantum theory of the scalar field. Commun. Math. Phys. **5**, 42 (1967)
328. G.V. Efimov, On a class of relativistic invariant distributions. Commun. Math. Phys. **7**, 138 (1968)
329. M.Z. Iofa, V.Ya. Fainberg, Wightman formulation for a nonlocalizable field theory. I. Zh. Eksp. Teor. Fiz. **56**, 1644 (1969) [Sov. Phys. JETP **29**, 880 (1969)]
330. M.Z. Iofa, V.Ya. Fainberg, Wightman formulation for nonlocalizable field theories II. Theory of asymptotic fields and particles. Teor. Mat. Fiz. **1**, 187 (1969) [Theor. Math. Phys. **1**, 143 (1969)]
331. V.A. Alebastrov, G.V. Efimov, A proof of the unitarity of S matrix in a nonlocal quantum field theory. Commun. Math. Phys. **31**, 1 (1973)
332. V.A. Alebastrov, G.V. Efimov, Causality in quantum field theory with nonlocal interaction. Commun. Math. Phys. **38**, 11 (1974)
333. G.V. Efimov, Нелокальные взаимодействия квантованных полей [*Nonlocal Interactions of Quantized Fields* (in Russian)] (Nauka, Moscow, 1977)
334. V.Ya. Fainberg, M.A. Soloviev, How can local properties be described in field theories without strict locality? Ann. Phys. (N.Y.) **113**, 421 (1978)

335. V.Ya. Fainberg, M.A. Soloviev, Nonlocalizability and asymptotical commutativity. Teor. Mat. Fiz. **93**, 514 (1992) [Theor. Math. Phys. **93**, 1438 (1992)]
336. R.L.P. do Amaral, E.C. Marino, Canonical quantization of theories containing fractional powers of the d'Alembertian operator. J. Phys. A **25**, 5183 (1992)
337. D.G. Barci, L.E. Oxman, M. Rocca, Canonical quantization of non-local field equations. Int. J. Mod. Phys. A **11**, 2111 (1996). [arXiv:hep-th/9503101]
338. N. Moeller, B. Zwiebach, Dynamics with infinitely many time derivatives and rolling tachyons. JHEP **0210**, 034 (2002). [arXiv:hep-th/0207107]
339. G. Calcagni, G. Nardelli, Tachyon solutions in boundary and cubic string field theory. Phys. Rev. D **78**, 126010 (2008). [arXiv:0708.0366]
340. G. Calcagni, M. Montobbio, G. Nardelli, Route to nonlocal cosmology. Phys. Rev. D **76**, 126001 (2007). [arXiv:0705.3043]
341. G. Calcagni, M. Montobbio, G. Nardelli, Localization of nonlocal theories. Phys. Lett. B **662**, 285 (2008). [arXiv:0712.2237]
342. G. Calcagni, G. Nardelli, Nonlocal instantons and solitons in string models. Phys. Lett. B **669**, 102 (2008). [arXiv:0802.4395]
343. D.J. Mulryne, N.J. Nunes, Diffusing nonlocal inflation: solving the field equations as an initial value problem. Phys. Rev. D **78**, 063519 (2008). [arXiv:0805.0449]
344. G. Calcagni, G. Nardelli, Kinks of open superstring field theory. Nucl. Phys. B **823**, 234 (2009). [arXiv:0904.3744]
345. G. Calcagni, G. Nardelli, Cosmological rolling solutions of nonlocal theories. Int. J. Mod. Phys. D **19**, 329 (2010). [arXiv:0904.4245]
346. G. Calcagni, G. Nardelli, String theory as a diffusing system. JHEP **1002**, 093 (2010). [arXiv:0910.2160]
347. N. Barnaby, N. Kamran, Dynamics with infinitely many derivatives: the initial value problem. JHEP **0802**, 008 (2008). [arXiv:0709.3968]
348. E.T. Tomboulis, Super-renormalizable gauge and gravitational theories. arXiv:hep-th/9702146.
349. T. Biswas, A. Mazumdar, W. Siegel, Bouncing universes in string-inspired gravity. JCAP **0603**, 009 (2006). [arXiv:hep-th/0508194]
350. G.V. Efimov, Amplitudes in nonlocal theories at high energies. Teor. Mat. Fiz. **128**, 395 (2001) [Theor. Math. Phys. **128**, 1169 (2001)]
351. E.T. Tomboulis, Nonlocal and quasilocal field theories. Phys. Rev. D **92**, 125037 (2015). [arXiv:1507.00981]
352. N.V. Krasnikov, Nonlocal gauge theories. Teor. Mat. Fiz. **73**, 235 (1987) [Theor. Math. Phys. **73**, 1184 (1987)]
353. J.W. Moffat, Finite nonlocal Gauge field theory. Phys. Rev. D **41**, 1177 (1990)
354. B.J. Hand, J.W. Moffat, Nonlocal regularization and the one-loop topological mass in three-dimensional QED. Phys. Rev. D **43**, 1896 (1991)
355. D. Evens, J.W. Moffat, G. Kleppe, R.P. Woodard, Nonlocal regularizations of gauge theories. Phys. Rev. D **43**, 499 (1991)
356. N.J. Cornish, New methods in quantum nonlocal field theory. Mod. Phys. Lett. A **07**, 1895 (1992)
357. N.J. Cornish, Quantum nonlocal field theory: physics without infinities. Int. J. Mod. Phys. A **07**, 6121 (1992)
358. N.J. Cornish, Quantum non-local gravity. Mod. Phys. Lett. A **07**, 631 (1992)
359. J.W. Moffat, Ultraviolet complete quantum gravity. Eur. Phys. J. Plus **126**, 43 (2011). [arXiv:1008.2482]
360. J. Khoury, Fading gravity and self-inflation. Phys. Rev. D **76**, 123513 (2007). [arXiv:hep-th/0612052]
361. G. Calcagni, G. Nardelli, Nonlocal gravity and the diffusion equation. Phys. Rev. D **82**, 123518 (2010). [arXiv:1004.5144]
362. T. Biswas, T. Koivisto, A. Mazumdar, Towards a resolution of the cosmological singularity in non-local higher derivative theories of gravity. JCAP **1011**, 008 (2010). [arXiv:1005.0590]

363. L. Modesto, Super-renormalizable quantum gravity. Phys. Rev. D **86**, 044005 (2012). [arXiv:1107.2403]
364. T. Biswas, E. Gerwick, T. Koivisto, A. Mazumdar, Towards singularity- and ghost-free theories of gravity. Phys. Rev. Lett. **108**, 031101 (2012). [arXiv:1110.5249]
365. S. Alexander, A. Marcianó, L. Modesto, The hidden quantum groups symmetry of super-renormalizable gravity. Phys. Rev. D **85**, 124030 (2012). [arXiv:1202.1824]
366. L. Modesto, Super-renormalizable multidimensional quantum gravity: theory and applications. Astron. Rev. **8**, 4 (2013). [arXiv:1202.3151]
367. L. Modesto, Towards a finite quantum supergravity. arXiv:1206.2648
368. G. Calcagni, L. Modesto, Nonlocal quantum gravity and M-theory. Phys. Rev. D **91**, 124059 (2015). [arXiv:1404.2137]
369. L. Modesto, L. Rachwał, Super-renormalizable and finite gravitational theories. Nucl. Phys. B **889**, 228 (2014). [arXiv:1407.8036]
370. M. Asorey, J.L. López, I.L. Shapiro, Some remarks on high derivative quantum gravity. Int. J. Mod. Phys. A **12**, 5711 (1997). [arXiv:hep-th/9610006]
371. K.S. Stelle, Renormalization of higher-derivative quantum gravity. Phys. Rev. D **16**, 953 (1977)
372. A. Accioly, A. Azeredo, H. Mukai, Propagator, tree-level unitarity and effective nonrelativistic potential for higher-derivative gravity theories in D dimensions. J. Math. Phys. **43**, 473 (2002)
373. P. Van Nieuwenhuizen, On ghost-free tensor Lagrangians and linearized gravitation. Nucl. Phys. B **60**, 478 (1973)
374. S. Talaganis, T. Biswas, A. Mazumdar, Towards understanding the ultraviolet behavior of quantum loops in infinite-derivative theories of gravity. Class. Quantum Grav. **32**, 215017 (2015). [arXiv:1412.3467]
375. M.J. Duff, D.J. Toms, Kaluza–Klein–Kounterterms, in *Unification of Fundamental Particle Interactions II*, ed. by J. Ellis, S. Ferrara (Springer, Amsterdam, 1983)
376. S. Deser, D. Seminara, Tree amplitudes and two loop counterterms in $D = 11$ supergravity. Phys. Rev. D **62**, 084010 (2000). [arXiv:hep-th/0002241]
377. A.S. Koshelev, Stable analytic bounce in non-local Einstein–Gauss–Bonnet cosmology. Class. Quantum Grav. **30**, 155001 (2013). [arXiv:1302.2140]
378. T. Biswas, A. Conroy, A.S. Koshelev, A. Mazumdar, Generalized ghost-free quadratic curvature gravity. Class. Quantum Grav. **31**, 015022 (2014); Erratum-ibid. **31**, 159501 (2014). [arXiv:1308.2319]
379. H.-J. Schmidt, Variational derivatives of arbitrarily high order and multi-inflation cosmological models. Class. Quantum Grav. **7**, 1023 (1990)
380. I.Ya. Aref'eva, Nonlocal string tachyon as a model for cosmological dark energy. AIP Conf. Proc. **826**, 301 (2006). [arXiv:astro-ph/0410443]
381. I.Ya. Aref'eva, L.V. Joukovskaya, Time lumps in nonlocal stringy models and cosmological applications. JHEP **0510**, 087 (2005). [arXiv:hep-th/0504200]
382. I.Ya. Aref'eva, A.S. Koshelev, S.Yu. Vernov, Stringy dark energy model with cold dark matter. Phys. Lett. B **628**, 1 (2005). [arXiv:astro-ph/0505605]
383. G. Calcagni, Cosmological tachyon from cubic string field theory. JHEP **0605**, 012 (2006). [arXiv:hep-th/0512259]
384. I.Ya. Aref'eva, A.S. Koshelev, Cosmic acceleration and crossing of $w = -1$ barrier from cubic superstring field theory. JHEP **0702**, 041 (2007). [arXiv:hep-th/0605085]
385. I.Ya. Aref'eva, I.V. Volovich, On the null energy condition and cosmology. Theor. Math. Phys. **155**, 503 (2008). [arXiv:hep-th/0612098]
386. N. Barnaby, T. Biswas, J.M. Cline, p-adic inflation. JHEP **0704**, 056 (2007). [arXiv:hep-th/0612230]
387. A.S. Koshelev, Non-local SFT tachyon and cosmology. JHEP **0704**, 029 (2007). [arXiv:hep-th/0701103]
388. I.Ya. Aref'eva, L.V. Joukovskaya, S.Yu. Vernov, Bouncing and accelerating solutions in nonlocal stringy models. JHEP **0707**, 087 (2007). [arXiv:hep-th/0701184]

389. I.Ya. Aref'eva, I.V. Volovich, Quantization of the Riemann zeta-function and cosmology. Int. J. Geom. Methods Mod. Phys. **4**, 881 (2007). [arXiv:hep-th/0701284]
390. J.E. Lidsey, Stretching the inflaton potential with kinetic energy. Phys. Rev. D **76**, 043511 (2007). [arXiv:hep-th/0703007]
391. N. Barnaby, J.M. Cline, Large non-Gaussianity from non-local inflation. JCAP **0707**, 017 (2007). [arXiv:0704.3426]
392. L.V. Joukovskaya, Dynamics in nonlocal cosmological models derived from string field theory. Phys. Rev. D **76**, 105007 (2007). [arXiv:0707.1545]
393. L. Joukovskaya, Rolling solution for tachyon condensation in open string field theory. arXiv:0803.3484
394. I.Ya. Aref'eva, A.S. Koshelev, Cosmological signature of tachyon condensation. JHEP **0809**, 068 (2008). [arXiv:0804.3570]
395. L. Joukovskaya, Dynamics with infinitely many time derivatives in Friedmann–Robertson–Walker background and rolling tachyons. JHEP **0902**, 045 (2009). [arXiv:0807.2065]
396. N. Barnaby, N. Kamran, Dynamics with infinitely many derivatives: variable coefficient equations. JHEP **0812**, 022 (2008). [arXiv:0809.4513]
397. N.J. Nunes, D.J. Mulryne, Non-linear non-local cosmology. AIP Conf. Proc. **1115**, 329 (2009). [arXiv:0810.5471]
398. A.S. Koshelev, S.Yu. Vernov, Cosmological perturbations in SFT inspired non-local scalar field models. Eur. Phys. J. C **72**, 2198 (2012). [arXiv:0903.5176]
399. S.Yu. Vernov, Localization of non-local cosmological models with quadratic potentials in the case of double roots. Class. Quantum Grav. **27**, 035006 (2010). [arXiv:0907.0468]
400. S.Yu. Vernov, Localization of the SFT inspired nonlocal linear models and exact solutions. Phys. Part. Nucl. Lett. **8**, 310 (2011). [arXiv:1005.0372]
401. A.S. Koshelev, S.Yu. Vernov, Analysis of scalar perturbations in cosmological models with a non-local scalar field. Class. Quantum Grav. **28**, 085019 (2011). [arXiv:1009.0746]
402. A.S. Koshelev, S.Yu. Vernov, On bouncing solutions in non-local gravity. Phys. Part. Nucl. **43**, 666 (2012). [arXiv:1202.1289]
403. T. Biswas, A.S. Koshelev, A. Mazumdar, S.Yu. Vernov, Stable bounce and inflation in non-local higher derivative cosmology. JCAP **1208**, 024 (2012). [arXiv:1206.6374]
404. F. Briscese, A. Marcianò, L. Modesto, E.N. Saridakis, Inflation in (super-)renormalizable gravity. Phys. Rev. D **87**, 083507 (2013). [arXiv:1212.3611]
405. G. Calcagni, L. Modesto, P. Nicolini, Super-accelerating bouncing cosmology in asymptotically-free non-local gravity. Eur. Phys. J. C **74**, 2999 (2014). [arXiv:1306.5332]
406. C. Bambi, D. Malafarina, L. Modesto, Terminating black holes in quantum gravity. Eur. Phys. J. C **74**, 2767 (2014). [arXiv:1306.1668]
407. N. Arkani-Hamed, S. Dimopoulos, G. Dvali, G. Gabadadze, Nonlocal modification of gravity and the cosmological constant problem. arXiv:hep-th/0209227
408. M.E. Soussa, R.P. Woodard, A nonlocal metric formulation of MOND. Class. Quantum Grav. **20**, 2737 (2003). [arXiv:astro-ph/0302030]
409. A.O. Barvinsky, Nonlocal action for long-distance modifications of gravity theory. Phys. Lett. B **572**, 109 (2003). [arXiv:hep-th/0304229]
410. A.O. Barvinsky, On covariant long-distance modifications of Einstein theory and strong coupling problem. Phys. Rev. D **71**, 084007 (2005). [arXiv:hep-th/0501093]
411. H.W. Hamber, R.M. Williams, Nonlocal effective gravitational field equations and the running of Newton's constant G. Phys. Rev. D **72**, 044026 (2005). [arXiv:hep-th/0507017]
412. S. Deser, R.P. Woodard, Nonlocal cosmology. Phys. Rev. Lett. **99**, 111301 (2007). [arXiv:0706.2151]
413. S. Nojiri, S.D. Odintsov, Modified non-local-$F(R)$ gravity as the key for the inflation and dark energy. Phys. Lett. B **659**, 821 (2008). [arXiv:0708.0924]
414. S. Jhingan, S. Nojiri, S.D. Odintsov, M. Sami, I. Thongkool, S. Zerbini, Phantom and non-phantom dark energy: the cosmological relevance of non-locally corrected gravity. Phys. Lett. B **663**, 424 (2008). [arXiv:0803.2613]

415. T.S. Koivisto, Dynamics of nonlocal cosmology. Phys. Rev. D **77**, 123513 (2008). [arXiv:0803.3399]
416. T.S. Koivisto, Newtonian limit of nonlocal cosmology. Phys. Rev. D **78**, 123505 (2008). [arXiv:0807.3778]
417. S. Capozziello, E. Elizalde, S. Nojiri, S.D. Odintsov, Accelerating cosmologies from non-local higher-derivative gravity. Phys. Lett. B **671**, 193 (2009). [arXiv:0809.1535]
418. C. Deffayet, R.P. Woodard, Reconstructing the distortion function for nonlocal cosmology. JCAP **0908**, 023 (2009). [arXiv:0904.0961]
419. G. Cognola, E. Elizalde, S. Nojiri, S.D. Odintsov, S. Zerbini, One-loop effective action for non-local modified Gauss–Bonnet gravity in de Sitter space. Eur. Phys. J. C **64**, 483 (2009). [arXiv:0905.0543]
420. S. Nojiri, S.D. Odintsov, M. Sasaki, Y.-l. Zhang, Screening of cosmological constant in non-local gravity. Phys. Lett. B **696**, 278 (2011). [arXiv:1010.5375]
421. S. Nojiri, S.D. Odintsov, Unified cosmic history in modified gravity: from $F(R)$ theory to Lorentz non-invariant models. Phys. Rep. **505**, 59 (2011). [arXiv:1011.0544]
422. A.O. Barvinsky, Dark energy and dark matter from nonlocal ghost-free gravity theory. Phys. Lett. B **710**, 12 (2012). [arXiv:1107.1463]
423. Y.-l. Zhang, M. Sasaki, Screening of cosmological constant in non-local cosmology. Int. J. Mod. Phys. D **21**, 1250006 (2012). [arXiv:1108.2112]
424. E. Elizalde, E.O. Pozdeeva, S.Yu. Vernov, Y.-l. Zhang, Cosmological solutions of a nonlocal model with a perfect fluid. JCAP **1307**, 034 (2013). [arXiv:1302.4330]
425. M. Jaccard, M. Maggiore, E. Mitsou, Nonlocal theory of massive gravity. Phys. Rev. D **88**, 044033 (2013). [arXiv:1305.3034]
426. S. Deser, R.P. Woodard, Observational viability and stability of nonlocal cosmology. JCAP **1311**, 036 (2013). [arXiv:1307.6639]
427. L. Modesto, S. Tsujikawa, Non-local massive gravity. Phys. Lett. B **727**, 48 (2013). [arXiv:1307.6968]
428. S. Foffa, M. Maggiore, E. Mitsou, Apparent ghosts and spurious degrees of freedom in non-local theories. Phys. Lett. B **733**, 76 (2014). [arXiv:1311.3421]
429. A. Conroy, T. Koivisto, A. Mazumdar, A. Teimouri, Generalised quadratic curvature, non-local infrared modifications of gravity and Newtonian potentials. Class. Quantum Grav. **32**, 015024 (2015). [arXiv:1406.4998]
430. G. 't Hooft, Dimensional reduction in quantum gravity, in *Salamfestschrift*, ed. by A. Ali, J. Ellis, S. Randjbar-Daemi (World Scientific, Singapore, 1993). [arXiv:gr-qc/9310026]
431. S. Carlip, Spontaneous dimensional reduction in short-distance quantum gravity? AIP Conf. Proc. **1196**, 72 (2009). [arXiv:0909.3329]
432. G. Calcagni, Fractal universe and quantum gravity. Phys. Rev. Lett. **104**, 251301 (2010). [arXiv:0912.3142]
433. S. Carlip, The small scale structure of spacetime, in [78]. [arXiv:1009.1136]
434. D. Benedetti, J. Henson, Spectral geometry as a probe of quantum spacetime. Phys. Rev. D **80**, 124036 (2009). [arXiv:0911.0401]
435. T.P. Sotiriou, M. Visser, S. Weinfurtner, Spectral dimension as a probe of the ultraviolet continuum regime of causal dynamical triangulations. Phys. Rev. Lett. **107**, 131303 (2011). [arXiv:1105.5646]
436. M.R. Atkin, G. Giasemidis, J.F. Wheater, Continuum random combs and scale dependent spectral dimension. J. Phys. A **44**, 265001 (2011). [arXiv:1101.4174]
437. G. Giasemidis, J.F. Wheater, S. Zohren, Dynamical dimensional reduction in toy models of 4D causal quantum gravity. Phys. Rev. D **86**, 081503(R) (2012). [arXiv:1202.2710]
438. G. Giasemidis, J.F. Wheater, S. Zohren, Multigraph models for causal quantum gravity and scale dependent spectral dimension. J. Phys. A **45**, 355001 (2012). [arXiv:1202.6322]
439. F. Caravelli, L. Modesto, Fractal dimension in 3d spin-foams. arXiv:0905.2170
440. E. Magliaro, C. Perini, L. Modesto, Fractal space-time from spin-foams. arXiv:0911.0437
441. A. Connes, Noncommutative geometry and the standard model with neutrino mixing. JHEP **0611**, 081 (2006). [arXiv:hep-th/0608226]

442. A.H. Chamseddine, A. Connes, M. Marcolli, Gravity and the standard model with neutrino mixing. Adv. Theor. Math. Phys. **11**, 991 (2007). [arXiv:hep-th/0610241]
443. D. Benedetti, Fractal properties of quantum spacetime. Phys. Rev. Lett. **102**, 111303 (2009). [arXiv:0811.1396]
444. M. Arzano, T. Trześniewski, Diffusion on κ-Minkowski space. Phys. Rev. D **89**, 124024 (2014). [arXiv:1404.4762]
445. Anjana V., E. Harikumar, Spectral dimension of kappa-deformed spacetime. Phys. Rev. D **91**, 065026 (2015). [arXiv:1501.00254]
446. Anjana V., E. Harikumar, Dimensional flow in the kappa-deformed spacetime. Phys. Rev. D **92**, 045014 (2015). [arXiv:1504.07773]
447. G. Calcagni, D. Oriti, J. Thürigen, Spectral dimension of quantum geometries. Class. Quantum Grav. **31**, 135014 (2014). [arXiv:1311.3340]
448. G. Calcagni, D. Oriti, J. Thürigen, Dimensional flow in discrete quantum geometries. Phys. Rev. D **91**, 084047 (2015). [arXiv:1412.8390]
449. P. Hořava, Spectral dimension of the universe in quantum gravity at a Lifshitz point. Phys. Rev. Lett. **102**, 161301 (2009). [arXiv:0902.3657]
450. S. Carlip, D. Grumiller, Lower bound on the spectral dimension near a black hole. Phys. Rev. D **84**, 084029 (2011). [arXiv:1108.4686]
451. J.R. Mureika, Primordial black hole evaporation and spontaneous dimensional reduction. Phys. Lett. B **716**, 171 (2012). [arXiv:1204.3619]
452. M. Arzano, G. Calcagni, Black-hole entropy and minimal diffusion. Phys. Rev. D **88**, 084017 (2013). [arXiv:1307.6122]
453. L. Modesto, P. Nicolini, Spectral dimension of a quantum universe. Phys. Rev. D **81**, 104040 (2010). [arXiv:0912.0220]
454. G Calcagni, Geometry of fractional spaces. Adv. Theor. Math. Phys. **16**, 549 (2012). [arXiv:1106.5787]
455. G. Calcagni, Geometry and field theory in multi-fractional spacetime. JHEP **1201**, 065 (2012). [arXiv:1107.5041]
456. G. Calcagni, Multiscale spacetimes from first principles. arXiv:1609.02776
457. G. Calcagni, L. Modesto, Nonlocality in string theory. J. Phys. A **47**, 355402 (2014). [arXiv:1310.4957]
458. G. Calcagni, G. Nardelli, Quantum field theory with varying couplings. Int. J. Mod. Phys. A **29**, 1450012 (2014). [arXiv:1306.0629]
459. G. Amelino-Camelia, M. Arzano, G. Gubitosi, J. Magueijo, Dimensional reduction in the sky. Phys. Rev. D **87**, 123532 (2013). [arXiv:1305.3153]
460. G. Amelino-Camelia, M. Arzano, G. Gubitosi, J. Magueijo, Rainbow gravity and scale-invariant fluctuations. Phys. Rev. D **88**, 041303 (2013). [arXiv:1307.0745]
461. G. Amelino-Camelia, M. Arzano, G. Gubitosi, J. Magueijo, Dimensional reduction in momentum space and scale-invariant cosmological fluctuations. Phys. Rev. D **88**, 103524 (2013). [arXiv:1309.3999]
462. G. Calcagni, Multi-scale gravity and cosmology. JCAP **1312**, 041 (2013). [arXiv:1307.6382]
463. G. Calcagni, S. Kuroyanagi, S. Tsujikawa, Cosmic microwave background and inflation in multi-fractional spacetimes. JCAP **1608**, 039 (2016). [arXiv:1606.08449]

Chapter 12
String Theory

Quaedam processu priorem exuunt formam et in novam transeunt. Ubi aliquid animus diu protulit et magnitudinem eius sequendo lassatus est, infinitum coepit vocari; quod longe aliud factum est quam fuit cum magnum videretur sed finitum. Eodem modo aliquid difficulter secari cogitavimus: novissime crescente hac difficultate insecabile inventum est.

— Seneca, *Ad Lucilium Epistularum Moralium*, XX, 118, 17

Certain things undergo a transformation from their prior form into a new one. When the mind extends an object and gets tired of following its magnitude, one begins to call it infinity; which is by far a different thing than what it was when it was seen large but finite. In the same way, we reflect upon something which is difficult to dissect: as the difficulty of this sub-division increases, we invent the indivisible.

Contents

(continued)

G. Calcagni, *Classical and Quantum Cosmology*, Graduate Texts in Physics,
DOI 10.1007/978-3-319-41127-9_12

Between the late 1960s and the early 1970s, an attempt was made to describe the quantum interactions of hadrons via the so-called dual (or dual-resonance) models [1–16]. These were essentially S-matrix models, whose dynamics was encoded in effective scattering amplitudes characterized by certain symmetries and spectra. It was soon recognized that such dynamics, symmetries and spectra could be reproduced by a Hamiltonian-Lagrangian model of quantum mechanics where the fundamental objects were not point-wise particles but oscillating *strings* [17–29]. It did not take long for this framework to acquire an autonomous life as a fundamental theory including not only quark interactions but also the electroweak and the gravitational force [30–33].

In the widest acceptance of the term, string theory is not a model of Nature but a theoretical framework. Just as quantum field theory is a set of tools much wider than the Standard Model of electroweak and strong interactions, the immense apparatus of strings, branes and extra dimensions is much more than what we would need in order to describe the observed natural phenomena at high energies. At the same time, however, so far we have been unable to extract enough information from the theory to complete this task in a fully satisfactory way. String theory has more than four dimensions, much more than gravitational and matter degrees of freedom, more symmetries than $SU(3) \otimes SU(2) \otimes U(1)$ and diffeomorphisms, more than one graviton and the 17 particles of the Standard Model (12 fermions, 4 gauge bosons and one scalar). From all the bounty the theory can offer, it is difficult to reproduce these degrees of freedom in the low-energy limit while maintaining an acceptably low number of assumptions, unobserved side effects and undesired fine tunings.

Nevertheless, string theory continues to fascinate for its economy of thought, its mathematical beauty and the tantalizing possibility of solving many fundamental problems in a unified frame.

Contrary to the majority of the other theories so far examined in the book, the broad goal of string theory is not to quantize gravity but, rather, to have all interactions emerging from a more fundamental set-up that can be studied with a language closer to that of particle physics than to that of general relativity. Diffeomorphism invariance and the very concept of spacetime do not play a central role because the gravitational action arises only in the low-energy limit. Spacetime and the graviton stem from the degrees of freedom of a two-dimensional conformal field theory, while diffeomorphisms are a low-energy byproduct of a very large symmetry group. In this loose sense, string theory is an extremely sophisticated scenario of emergent gravity.

This chapter is meant to give the cosmologist a compact and intuitive bird's eye view of the main features of string theory. Sections 12.1 and 12.2 introduce the classical bosonic and supersymmetric strings and their quantizations, D-branes, fluxes and string field theory. Section 12.3 is dedicated to the compactification of string theory on torii, Calabi–Yau spaces and orbifolds and discusses the important problem of the stabilization of string moduli in flux compactifications. Section 12.4 deals with string dualities and M-theory.

We will omit several technical and conceptual details, which can be found in dedicated textbooks [34–37] and reviews [38]. Such details are important to take control of the theory. We therefore encourage the reader seriously interested in strings to consult first of all the just-mentioned books and, as far as the historical development of the theory is concerned, the references in the present text.

12.1 Bosonic String

12.1.1 Classical Free Strings and Branes

When a string propagates, its trajectory is not a one-dimensional world-line as for a point particle but a two-dimensional *world-sheet*, which is a strip for open strings and a tubular surface for closed strings (Fig. 12.1). The free propagation of the bosonic string in D-dimensional Minkowski spacetime is governed by the Polyakov action [39–41]

$$\begin{aligned} \mathcal{S}[\gamma, X] &= -\frac{1}{4\pi\alpha'}\int_{-\infty}^{+\infty} \mathrm{d}\tau\, \mathrm{d}\sigma\, \sqrt{-\gamma}\, \gamma^{ab}\eta_{\mu\nu}\partial_a X^\mu \partial_b X^\nu \\ &= \frac{1}{4\pi\alpha'}\int_{-\infty}^{+\infty} \mathrm{d}\tau \int_0^l \mathrm{d}\sigma\, \left(\partial_\tau X_\mu \partial_\tau X^\mu - \partial_\sigma X_\mu \partial_\sigma X^\mu\right), \end{aligned} \tag{12.1}$$

where α' is called Regge slope (the tension of the string is equal to $(2\pi\alpha')^{-1}$ [24]), γ_{ab} is the world-sheet metric ($a, b = 1, 2$) with signature $(-,+)$, $\sigma^a = \tau, \sigma$ are coordinates parametrizing, respectively, the (infinite) proper-time length and the

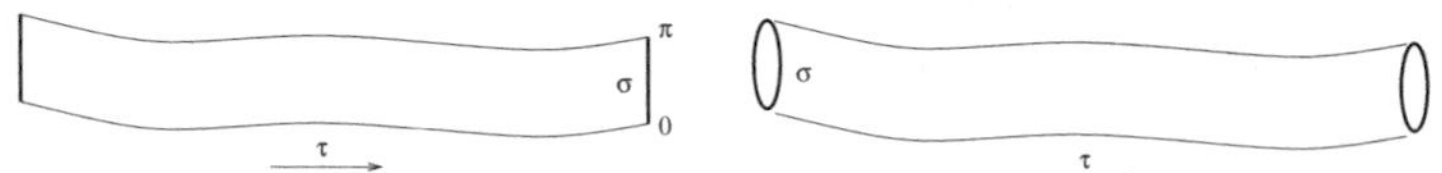

Fig. 12.1 Open- and closed-string world-sheet

(finite) spatial width of the world-sheet, $X^\mu(\tau,\sigma)$ are D scalar fields and Greek indices are contracted with the Minkowski metric. By convention, for the closed string $l=2\pi$ while $l=\pi$ for the open string. Since $[X^\mu]=-1$, the Regge slope has dimension $[\alpha']=-2$; it is the only free parameter of the theory and $l_s:=\sqrt{\alpha'}$ fixes the length scale of the string.

In the second line of (12.1), we gauge fixed the world-sheet metric to the flat one η_{ab} by combining world-sheet diffeomorphisms, $X'^\mu(\sigma'^a)=X^\mu(\sigma^a)$ and $\partial_a\sigma'^c\partial_b\sigma'^d\gamma'_{cd}=\gamma_{ab}$, and two-dimensional Weyl (or conformal) invariance, $\gamma'_{ab}=\Omega^2(\sigma^c)\gamma_{ab}$ [42]. In infinitesimal form,

$$\delta X^\mu=\xi^a\partial_a X^\mu\,,\qquad \delta\gamma_{ab}=\nabla_a\xi_b+\nabla_b\xi_a+\omega\gamma_{ab}\,,\tag{12.2}$$

where ξ^a and ω are, respectively, a vector and a scalar. Different gauge choices correspond to different ways to embed the world-sheet in spacetime, while conformally equivalent metrics represent the same embedding. In general, the metric $\gamma_{ab}(\tau,\sigma)$ is dynamical but it has no kinetic term because in two dimensions the Ricci tensor is proportional to the metric itself: $\mathcal{R}^{(2)}_{ab}=C\gamma_{ab}$. From the trace, it stems that $C=\mathcal{R}^{(2)}/2$ and the world-sheet Ricci scalar is a constant. The energy-momentum tensor

$$\mathcal{T}_{ab}:=-\frac{2}{\sqrt{-\gamma}}\frac{\delta\mathcal{S}}{\delta\gamma^{ab}}=\frac{1}{2\pi\alpha'}\left(\partial_a X_\mu\partial_b X^\mu-\frac{1}{2}\gamma_{ab}\partial_c X_\mu\partial^c X^\mu\right)\tag{12.3}$$

vanishes identically: $\mathcal{T}_{ab}=\mathcal{R}^{(2)}_{ab}-(1/2)\gamma_{ab}\mathcal{R}^{(2)}=0$. As a consequence of conformal invariance,

$$\mathcal{T}_a{}^a=0\,,\tag{12.4}$$

while diffeomorphism invariance guarantees that the energy-momentum tensor is covariantly conserved. These conditions will also be enforced at the quantum level.

From (12.1), the equation of motion for each X^μ is

$$(-\partial_\tau^2+\partial_\sigma^2)X^\mu=0\,.\tag{12.5}$$

We Fourier transform X in the world-sheet coordinate τ:

$$X^\mu(\tau,\sigma)=\int_{-\infty}^{+\infty}\frac{\mathrm{d}k}{2\pi}\,\mathrm{e}^{-ik\tau}\,g(k,\sigma)\,X^\mu_k\,,\tag{12.6}$$

where $g(k,\sigma)$ is some kernel function determined by the boundary conditions necessary to solve (12.5). The latter is $(\partial_\sigma^2+k^2)g(k,\sigma)=0$, the harmonic oscillator with solution $g(k,\sigma)=b_k\sin(k\sigma)+c_k\cos(k\sigma)$.

12.1.1.1 Open String

The free open string is described by the Neumann boundary conditions

$$\partial_\sigma X^\mu(\tau,\sigma)\Big|_{\sigma=0} = 0 = \partial_\sigma X^\mu(\tau,\sigma)\Big|_{\sigma=\pi}, \tag{12.7}$$

which require, respectively, that $b_k = 0$ and that $k = n$ be integer. Setting $c_k = 1$, the free open string solution is

$$X^\mu_{\text{Neu}}(\tau,\sigma) = x^\mu + 2\alpha' p^\mu \tau + \mathrm{i}\sqrt{2\alpha'} \sum_{0\neq n\in\mathbb{Z}} \mathrm{e}^{-\mathrm{i}n\tau} \cos(n\sigma)\,\frac{\alpha^\mu_n}{n}, \tag{12.8}$$

where we split the zero mode X^μ_0 into a constant x^μ and a linear term $p^\mu\tau$ and we rewrote the other modes as $X^\mu_n =: \mathrm{i}\sqrt{2\alpha'}\alpha^\mu_n/n$. The vectors x^μ and p^μ are interpreted as the position and momentum of the center of mass of the string. Other boundary conditions will be discussed in Sect. 12.1.2.

For open strings with Neumann boundary conditions, the Hamiltonian is

$$\begin{aligned} H &= \int_0^\pi \mathrm{d}\sigma\, \mathcal{T}_{\tau\tau} = \frac{1}{4\pi\alpha'}\int_0^\pi \mathrm{d}\sigma\, \left(\partial_\tau X_\mu \partial_\tau X^\mu + \partial_\sigma X_\mu \partial_\sigma X^\mu\right) \\ &= \frac{1}{2}\eta_{\mu\nu} \sum_{m\in\mathbb{Z}} \alpha^\mu_{-m}\alpha^\nu_m =: L_0, \end{aligned} \tag{12.9}$$

where in the last line we reserved a special symbol for the quantity

$$L_0 = \alpha' p^2 + \eta_{\mu\nu}\sum_{m=1}^{+\infty} \alpha^\mu_{-m}\alpha^\nu_m =: \alpha' p^2 + N. \tag{12.10}$$

The Hamiltonian H and the momentum $P = \int_0^\pi \mathrm{d}\sigma\, \mathcal{T}_{\tau\sigma}$ generate world-sheet reparametrizations of, respectively, the coordinate τ and σ. Due to the trace condition (12.4), there are no other independent components of the energy-momentum tensor. The conserved charges associated with reparametrizations of the string world-sheet are given by the Fourier (or, by recasting the system as a *conformal field theory* on the complex plane [43–48], the Laurent) modes of such components. We are at liberty to choose the linear combinations $\mathcal{T}_{++} := (\mathcal{T}_{\tau\tau} + \mathcal{T}_{\tau\sigma})/2$ and $\mathcal{T}_{--} := (\mathcal{T}_{\tau\tau} - \mathcal{T}_{\tau\sigma})/2$, so that instead of Fourier transforming P we can consider

$$L_n := \int_0^\pi \mathrm{d}\sigma\, (\mathrm{e}^{-\mathrm{i}n\sigma}\mathcal{T}_{--} + \mathrm{e}^{\mathrm{i}n\sigma}\mathcal{T}_{++}) = \frac{1}{2}\eta_{\mu\nu}\sum_{m\in\mathbb{Z}} \alpha^\mu_{n-m}\alpha^\nu_m. \tag{12.11}$$

Here we used the mode decomposition (12.8) and the definition of the Kronecker delta $\delta_{n,0} := (2\pi)^{-1}\int_0^{2\pi} \mathrm{d}\sigma\, \mathrm{e}^{-\mathrm{i}n\sigma}$. By definition, $\alpha^\mu_0 = \sqrt{2\alpha'}p^\mu$. The charges L_n

obey the Witt algebra

$$\{L_m, L_n\} = (m-n)L_{m+n}\,. \tag{12.12}$$

The classical equations of motion (12.5) then correspond to $H = 0 = P$, i.e., $L_n = 0$ for all $n \in \mathbb{Z}$. The condition $P = 0$ reflects the fact that we can choose a gauge (the so-called static gauge $X^0 = t = \tau$) where lines of constant σ and constant τ are orthogonal, $\partial_\tau X_\mu \partial_\sigma X^\mu = 0$. The latter condition allows one to write the equations of motion (12.5) as [37]

$$0 = (\partial_\tau X_\mu \pm \partial_\sigma X_\mu)(\partial_\tau X^\mu \pm \partial_\sigma X^\mu) = 4\alpha' \sum_{n\in\mathbb{Z}} L_n\, \mathrm{e}^{-\mathrm{i}n\sigma^\pm}\,, \qquad \sigma^\pm := \tau \pm \sigma\,. \tag{12.13}$$

This is another way to state that $L_n = 0$ classically.

12.1.1.2 Closed String

The closed string does not have end-points at which to specify boundary conditions. On the other hand, it has two sets of wave modes, one moving clockwise ("right", R) and the other moving counter-clockwise ("left", L). The general solution of (12.5) reads

$$X^\mu(\tau,\sigma) = X^\mu_{\mathrm{R}}(\sigma^-) + X^\mu_{\mathrm{L}}(\sigma^+)\,, \tag{12.14a}$$

$$X^\mu_{\mathrm{R,L}} = \frac{x^\mu}{2} + \frac{\alpha'}{2} p^\mu \sigma^\mp + \mathrm{i}\sqrt{\frac{\alpha'}{2}} \sum_{0\neq n\in\mathbb{Z}} \mathrm{e}^{-\mathrm{i}n\sigma^\mp} \frac{\alpha^\mu_{\mathrm{R,L}\,n}}{n}\,. \tag{12.14b}$$

In the following and as in the standard literature, we will call α and $\tilde{\alpha}$ the oscillators α_{R} and α_{L}, respectively. Using the coordinates $\sigma^\pm$, one finds that $\mathcal{T}_{++} = \partial_+ X^\mu \partial_+ X_\mu/(2\pi\alpha')$ and $\mathcal{T}_{--} = \partial_- X^\mu \partial_- X_\mu/(2\pi\alpha')$, which are the only non-vanishing components of the world-sheet energy-momentum tensor. All the open-string expressions hold for the closed string, with some differences. One is the definition of the zero modes, $\alpha^\mu_0 = \sqrt{\alpha'/2}\,p^\mu = \tilde{\alpha}^\mu_0$. The integration range of the Hamiltonian and of the total momentum is now $[0, 2\pi]$, so that $H = \int_0^{2\pi} \mathrm{d}\sigma\,(\mathcal{T}_{++} + \mathcal{T}_{--})$ and $P = \int_0^{2\pi} \mathrm{d}\sigma\,(\mathcal{T}_{++} - \mathcal{T}_{--})$, which also reflects in (12.11):

$$L_n := \int_0^{2\pi} \mathrm{d}\sigma\, \mathrm{e}^{-\mathrm{i}n\sigma} \mathcal{T}_{--} = \frac{1}{2}\eta_{\mu\nu} \sum_{m\in\mathbb{Z}} \alpha^\mu_{n-m}\alpha^\nu_m\,, \tag{12.15}$$

together with its left-moving counterpart $\tilde{L}_n$ defined with $\mathcal{T}_{++}$. The closed-string Hamiltonian and total momentum are then given by $H = L_0 + \tilde{L}_0$ and $P = L_0 - \tilde{L}_0$,

where

$$L_0 := \frac{\alpha'}{4}p^2 + \eta_{\mu\nu}\sum_{m=1}^{+\infty}\alpha^\mu_{-m}\alpha^\nu_m = \frac{\alpha'}{4}p^2 + N \tag{12.16}$$

and $\tilde{L}_0 = \alpha' p^2/4 + \tilde{N}$. Although we use the same symbols L_0 and L_n as in the open-string case, there will be no danger of confusion.

12.1.2 D-Branes

If one imposes the end-points of the open string to be fixed, then (12.7) is replaced by the Dirichlet boundary conditions

$$X^\mu(\tau,\sigma)\Big|_{\sigma=0} = x_1^\mu\,, \qquad X^\mu(\tau,\sigma)\Big|_{\sigma=\pi} = x_2^\mu\,. \tag{12.17}$$

At the end-points, the variation δX^μ is zero. Then, the general solution is

$$X^\mu_{\rm Dir} = x_1^\mu + (x_2^\mu - x_1^\mu)\frac{\sigma}{\pi} + \sqrt{2\alpha'}\sum_{0\neq n\in\mathbb{Z}} \mathrm{e}^{-in\tau}\sin(n\sigma)\,\frac{\alpha_n^\mu}{n}\,. \tag{12.18}$$

Each of the string end-points spans a spacetime object called *D-brane* [49] (see Fig. 12.2; "D" stands for Dirichlet). Branes can have different spatial dimensionality p, in which case they are called Dp-branes.

For string systems to be dynamical, the time direction must satisfy a Neumann boundary condition, otherwise $X^0 = t = \text{const}$. Therefore, $\mu \neq 0$ in (12.17) and

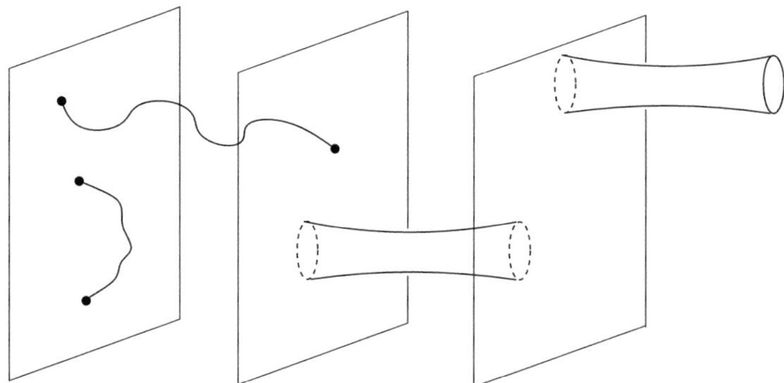

Fig. 12.2 Open strings with Dirichlet boundary conditions. The end-points are fixed on two D-branes or on the same brane, while the string can oscillate in the bulk. Closed strings are absorbed by branes which then acquire open-string excitations

(12.18). Branes do not preserve the global Lorentz invariance and the Lorentz group is broken into $SO(1,D-1) \to SO(1,p) \otimes SO(D-1-p)$.

An effective action for branes can be written down as an extension of (12.1) which includes a $U(1)$ gauge field A_a ($a = 0,1,\dots,p$). The *Dirac–Born–Infeld* (DBI) *action* for a Dp-brane with induced metric $\eta_{ab} = \eta_{\mu\nu}\partial_a X^\mu \partial_b X^\nu$ is

$$\mathcal{S}_p^{\rm DBI} = -\mathcal{T}_p \int d^{p+1}\sigma \sqrt{-\det(\eta_{ab} + 2\pi l_s^2 F_{ab})}\,, \qquad \mathcal{T}_p = \frac{1}{(2\pi)^p g_s l_s^{p+1}}\,, \tag{12.19}$$

where $\mathcal{T}_p$ is the *brane tension* and $F_{ab} = \partial_a A_b - \partial_b A_a$ is the field strength of the gauge field. The dependence of $\mathcal{T}_p$ on the inverse string coupling g_s, to be introduced later, indicates that branes are intrinsically non-perturbative. Dp-branes move through spacetime and interact with strings; therefore, they are fully dynamical objects [35].

12.1.3 Quantum Strings and Critical Dimension

There are different ways to quantize the string and they are all mutually consistent: old covariant quantization, light-cone quantization, path-integral covariant quantization and Becchi–Rouet–Stora–Tyutin (BRST) quantization.

In the so-called *old covariant quantization* of the string (e.g., [10, 11, 22, 24]), general covariance is preserved. After defining the classical momentum $\Pi^\mu := \delta\mathcal{S}/(\delta\partial_\tau X_\mu) = \partial_\tau X^\mu/(2\pi\alpha')$ and promoting X and Π to operators, one imposes the commutation relation

$$[\hat{X}^\mu(\tau,\sigma), \hat{\Pi}^\nu(\tau,\sigma')] = i\eta^{\mu\nu}\delta(\sigma-\sigma')\,. \tag{12.20}$$

Thus, the center of mass and the momentum are conjugate, $[\hat{x}^\mu, \hat{p}^\nu] = i\eta^{\mu\nu}$, and so are the oscillators. For the open string, $[\alpha_n^\mu, \alpha_m^\mu] = m\delta_{n,-m}\eta^{\mu\nu}$ and, for X^μ to be self-adjoint, $(\alpha_n^\mu)^\dagger = \alpha_{-n}^\mu$. Operators with negative and positive n are therefore interpreted as, respectively, creating and annihilating string modes. The vacuum $|N=0;p=0\rangle$ of the Fock space represents a configuration with no excitations and no momentum. The eigenvalues of $\hat{p}^\mu$ are indicated as p^μ.

To guarantee that all annihilation operators be on the right, one imposes a normal ordering on the *Virasoro operators*:

$$\hat{L}_n := \frac{1}{2}\sum_{m\in\mathbb{Z}} \eta_{\mu\nu} {:}\alpha_{n-m}^\mu \alpha_m^\nu{:} = \frac{1}{2}\eta_{\mu\nu}\left(\sum_{m=0}^{+\infty} \alpha_{n-m}^\mu \alpha_m^\nu + \sum_{m=-1}^{-\infty} \alpha_m^\mu \alpha_{n-m}^\nu\right). \tag{12.21}$$

For $n \neq 0$ the normal ordering is immaterial, since operators α^μ with different index commute. On the other hand, for $n = 0$ the ordering is important and distinguishes

$\hat{L}_0 = :L_0:$, which is the quantization of (12.10), from the operator

$$\hat{L}_0' = \alpha' \hat{p}^2 + \hat{N} - A\,, \qquad A = -\frac{D}{2}\sum_{m=1}^{+\infty} m\,, \tag{12.22}$$

which is the quantization of (12.9). The difference is in a formally divergent constant term, $\hat{L}_0 = \hat{L}_0' + A$. In general, there are infinitely many choices for the quantum Hamiltonian, which is parametrized by an *arbitrary* constant:

$$\hat{H} := \hat{L}_0 - a\,. \tag{12.23}$$

The operators $\hat{L}_n$ obey the *Virasoro algebra* (duplicated for the closed string) [4]

$$[\hat{L}_m, \hat{L}_n] = (m-n)\hat{L}_{m+n} + \frac{D}{12}(m^3 - m)\delta_{m,-n}\,, \tag{12.24}$$

which is the central extension of the classical Witt algebra (12.12). The central charge (i.e., the delta term in the right-hand side of (12.24)) is determined by using the Jacobi identity for nested commutators and calculating the vacuum expectation value of $[\hat{L}_2, \hat{L}_{-2}]$. A particularly important sub-algebra of (12.24) is generated by $\hat{L}_0$ and $\hat{L}_{\pm 1}$ and has vanishing central charge:

$$[\hat{L}_1, \hat{L}_{-1}] = 2\hat{L}_0\,, \qquad [\hat{L}_0, \hat{L}_{\pm 1}] = \mp \hat{L}_{\pm 1}\,. \tag{12.25}$$

This is the special linear (or conformal-anomaly) algebra $sl(2, \mathbb{R})$.

To get the spectrum of string excitations, one imposes conditions on physical (or admissible [37]) states $|\Phi\rangle$ corresponding to the vanishing of the energy-momentum tensor, $\hat{\mathcal{T}}_{++}|\Phi\rangle = 0 = \hat{\mathcal{T}}_{--}|\Phi\rangle$. In terms of the Virasoro operators, these conditions are compatible with the commutation relations (12.24) provided we do not impose them for all integer n:

$$\hat{L}_n|\Phi\rangle = 0 \quad \forall n > 0\,, \qquad (\hat{L}_0 - a)|\Phi\rangle = 0\,. \tag{12.26}$$

The further condition $\hat{L}_{n<0}|\Phi\rangle = 0$ would be in contradiction with the Virasoro algebra and it is not even necessary: the expectation values $\langle \Phi_1|\hat{L}_{n<0}|\Phi_2\rangle = \langle \Phi_1|\hat{L}_{n>0}^\dagger|\Phi_2\rangle$ on two physical states is zero by virtue of the first constraint in (12.26). States annihilated by $\hat{L}_{n>0}$ are called primary. There are also spurious states of the form $|\Phi'\rangle = \hat{L}_{n>0}|\Phi\rangle$ which are orthogonal to all the others. Two physical states are isomorphic if their difference is a null state, i.e., a spurious admissible state.

Physical states are eigenvectors of both operators $\hat{p}$ and $\hat{N}$ in (12.22), with eigenvalue p (the momentum of the mode) and N (the number of excitations in the state). The mass spectrum of physical states is given by the dispersion relation

$p^2 + M^2_{\rm open} = 0$, which yields

$$M^2_{\rm open} = \frac{N-a}{\alpha'}\,, \tag{12.27}$$

while for the closed string another copy of (12.26) holds for the $\tilde{L}_n$ and

$$M^2_{\rm closed} = \frac{4(N-a)}{\alpha'} = \frac{4(\tilde{N}-a)}{\alpha'}\,. \tag{12.28}$$

This implies that the number of right- and left-moving modes is the same for physical states, $N = \tilde{N}$ (level-matching condition).

Consider now the open-string spectrum.

- The ground state

$$N = 0: \qquad |0;p\rangle = |0\rangle_{\rm osc} \otimes |p\rangle \tag{12.29}$$

contains no oscillations, it is an eigenfunction of the string momentum, $\hat{p}^\mu\,|0;p\rangle = p^\mu\,|0;p\rangle$, and has a mass $M^2_{\rm open} = -a/\alpha'$. It has no Lorentzian indices and corresponds to a scalar particle. If $a > 0$, as it will soon turn out, then this is a negative-mass mode, i.e., a tachyon.

- The next state in the tower is

$$N = 1: \qquad |1;p\rangle = \epsilon_\mu(p)\,\alpha^\mu_{-1}\,|0;p\rangle\,, \tag{12.30}$$

where ϵ_μ is a polarization vector. This state is admissible if $\hat{L}_1\,|1;p\rangle = (\sqrt{2\alpha'}p_\mu\alpha^\mu_1 + \dots)\,|1;p\rangle = 0$, requiring $\epsilon_\mu p^\mu = 0$. Since $\hat{L}_{-1}\,|0;p\rangle = \sqrt{2\alpha'}p_\mu \times\alpha^\mu_{-1}\,|0;p\rangle$, $|1;p\rangle$ is a spurious state when $\epsilon^\mu \propto p^\mu$. Such state is null (i.e., spurious and admissible) when $\hat{L}_1\hat{L}_{-1}\,|0;p\rangle = 0$, which requires $p^2 = 0$. In this case, which is the only one compatible with the light-cone quantization discussed below, there is an isomorphism class of polarization vectors, $\epsilon_\mu \cong \epsilon_\mu + \text{const}\times p_\mu$, the mass of the state is zero and the state (12.30) corresponds to a massless Lorentz vector particle identified with the photon. The arbitrary constant in (12.23) is thus fixed to

$$a = 1\,. \tag{12.31}$$

The two conditions $\epsilon_\mu p^\mu = 0$ and $p^2 = 0$ remove pure gauge modes and one is left with $D-2$ degrees of freedom. These correspond to the "little group" of spatial rotations $SO(D-2)$ [8], the sub-group leaving this state invariant at rest.

- The $N = 2$ state $\alpha^\mu_{-1}\alpha^\nu_{-1}\,|0;p\rangle$ is a rank-2 tensor with mass $M^2_{\rm open} = 1/\alpha'$, invariant under $SO(D-1)$. The number of degrees of freedom is the dimension of this group, $(D-1)(D-2)/2$.

The quantum algebra of the system is not canonical, since commutators with $\mu, \nu = 0$ are negative definite. These are associated with an infinite tower of negative-norm states $\alpha^0_{-n}\,|0,p\rangle$, $n > 0$. Fortunately, such ghosts are non-physical, as a gauge fixing shows. Even if we have fixed γ_{ab} as the flat metric, there is still a residual gauge invariance: a conformal transformation of the coordinates X^μ yields a conformal factor that can be compensated by a Weyl transformation. This residual freedom can be spent by selecting the so-called light-cone gauge [9], which breaks manifest covariance. The advantage in doing so is that negative-norm states disappear from the game. In fact, the $\mu = 0$ and $\mu = D-1$ coordinates are combined together as $X^\pm := (X^0 \pm X^{D-1})/\sqrt{2}$, which is nothing but a frame whose axes coincide with the local light cone. The transverse coordinates X^i are indexed by $i = 1, \dots, D-2$. The light-cone gauge consists in setting to zero all the oscillators of X^+, which becomes $X^+ = x^+ + p^+\tau$. Therefore, only $D-2$ oscillators survive (per sector, in the case of the closed string) [8]. All that has been said above can be easily recast in light-cone gauge, resulting in the so-called *light-cone quantization* of the string [24]. The transverse operators

$$L_n^\perp := \frac{1}{2}\sum_{m\in\mathbb{Z}} \delta_{ij}{:}\alpha^i_{n-m}\alpha^\nu_j{:}\,, \qquad i,j = 1,\dots,D-2\,, \tag{12.32}$$

and in particular

$$L_0^\perp = \alpha'\hat{p}_i\hat{p}^i + \hat{N}^\perp\,, \qquad N^\perp := \delta_{ij}\sum_{m=1}^{+\infty}\alpha^i_{-m}\alpha^j_m\,, \tag{12.33}$$

obey the Virasoro algebra (12.24) with the replacement $D \to D-2$. The same replacement happens in the normal-ordering constant in (12.22), $A = -(D-2)\sum_{m=1}^{+\infty} m/2$, as well as in the condition (12.26) for admissible states: $L_n^\perp\,|\psi\rangle = 0$ for all $n > 0$, $(L_0^\perp - a^\perp)\,|\psi\rangle = 0$. A calculation of the commutator of the generators of the Lorentz algebra fixes the value of both $a^\perp$ and D [12, 24, 25]:

$$a^\perp = 1\,, \qquad D = 26\,. \tag{12.34}$$

To explain this result about the *critical dimension* of spacetime, for the sake of simplicity we sketch a heuristic argument. First of all, we identify the arbitrary constant $a^\perp$ in the Hamiltonian constraint with the zero-point energy A of the oscillators. The latter is the normal-ordering constant present in the alternative definition $L_0'^\perp = L_0^\perp - A$, so that $a^\perp = A = -(D-2)\sum_{m=1}^{+\infty} m/2$. Next, introducing a regularization parameter ϵ, one has $\sum_{m=1}^{+\infty} m\,\mathrm{e}^{-\epsilon m} = -\partial_\epsilon\sum_{m=1}^{+\infty}\mathrm{e}^{-\epsilon m} = -\partial_\epsilon[\mathrm{e}^{-\epsilon}/(1-\mathrm{e}^{-\epsilon})] = \partial_\epsilon[1/(1-\mathrm{e}^{\epsilon})] = \mathrm{e}^{\epsilon}/(1-\mathrm{e}^{\epsilon})^2 = 1/\epsilon^2 - 1/12 + O(\epsilon^2)$. The first term can be reabsorbed by adding a counter-term in the action, an operation which preserves Weyl invariance. Sending ϵ to zero, the final result is $a^\perp = (D-2)/24$. Then, one notices that the $N^\perp = 1$ state $\alpha^i_{-1}\,|0;p\rangle$ contains $D-2$ degrees of freedom.

Vectors in D dimensions have such counting only if they are massless, which implies $M^2_{\rm open} = (1 - a^\perp)/\alpha' = 0$, hence $a^\perp = 1$, hence $D = 26$.

The critical dimension of string theory is determined by the simultaneous requirement that the theory maintain its classical symmetries when quantized and that ghosts decouple from the physical spectrum. The latter is known as the *no-ghost theorem*: the bosonic-string spectrum is ghost free only if $D = 26$ [10, 11]. This can be proven also in the *BRST quantization* [50–53] of the bosonic string [48, 54–56].[1] Conformal invariance is of paramount importance for ensuring that the theory is unitary and, in fact, the techniques of conformal field theory (CFT) make these results particularly transparent.

Another way to quantize the string is via a *covariant path integral* (or, more precisely, partition function) of the form

$$Z = \int [\mathcal{D}h_{ab}][\mathcal{D}X^\mu]\,\mathrm{e}^{-\mathcal{S}[h,X]}\,. \tag{12.35}$$

Here, the world-sheet metric h_{ab} in the functional measure and in $\mathcal{S}$ has Euclidean signature and replaces γ_{ab}. This modification is convenient to make the path integral convergent and for the study of interactions [27, 28], but it does not change the physics. From now on, we will employ a Riemannian world-sheet.

Due to the redundancy of world-sheet diffeomorphisms and Weyl transformations (12.2), there are unwanted gauge degrees of freedom in (12.35). These can be canceled out by promoting ξ^a and ω to auxiliary fermionic fields with integer spin, i.e., Faddeev–Popov ghosts. The Jacobian resulting from integrating out the ghosts is something of the form $\int[\mathcal{D}\xi^a][\mathcal{D}\omega] \propto \exp[(D-26)C\int \mathrm{d}\tau\,\mathrm{d}\sigma\,(\partial\omega)^2]$, where C is some constant. If we want Weyl invariance to be preserved at the quantum level, this quantity must be independent of the Weyl factor and, hence, of ghosts. Therefore, it must be $D = 26$ [41]. For $D \neq 26$, there is a conformal anomaly and the commutator of the energy-momentum tensor $\mathcal{T}$ with itself does not vanish.

For the closed string, the arguments on the spacetime dimensionality are mainly unchanged, only doubly copied for the two sectors of the theory. In the old covariant quantization, physical states are determined by the constraints $\hat{L}_{n>0}\,|\Phi\rangle = 0 = \hat{\tilde{L}}_{n>0}\,|\Phi\rangle$, $(\hat{L}_0 - 1)\,|\Phi\rangle = 0 = (\hat{\tilde{L}}_0 - 1)\,|\Phi\rangle$. The Hamiltonian is $\hat{H} = \hat{L}_0 + \hat{\tilde{L}}_0 - 2$.

- The $N = \tilde{N} = 0$ ground state $|0, \tilde{0}, p\rangle := |0\rangle_{\rm osc} \otimes \widetilde{|0\rangle}_{\rm osc} \otimes |p\rangle$ is a tachyon with mass $M^2_{\rm closed} = -4/\alpha'$.
- The $N = \tilde{N} = 1$ state $|1, \tilde{1}, p\rangle = \alpha^i_{-1}\,|0\rangle_{\rm osc} \otimes \tilde{\alpha}^j_{-1}\widetilde{|0\rangle}_{\rm osc} \otimes |p\rangle$ (gauge modes have been removed already) is a rank-2 tensor given by the product of two massless vectors. It is invariant under the group $SO(D-2) \otimes SO(D-2)$,

[1] Strictly speaking, the no-ghost theorem forbids ghosts for the Veneziano dual model [1, 4] when $D \leqslant 26$. However, for $D < 26$ the physical spectrum is not the one of bosonic string theory since it contains also the longitudinal modes removed in the light-cone quantization [57, 58].

which can be decomposed in three irreducible representations [32]: a scalar called *dilaton*, a symmetric traceless tensor h_{ij} (the *graviton* [30, 31]) and an anti-symmetric tensor B_{ij} (the *Kalb–Ramond field* [26]). This is the massless sector of bosonic string theory. The total number of degrees of freedom is $1 + [(D-1)(D-2)/2 - 1] + (D-2)(D-3)/2 = (D-2)^2$. The tensor h_{ij} is identified with the graviton because gauge modes are modded out by a transformation $h_{ij} \to h_{ij} + p_i\xi_j + p_j\xi_i$, which is nothing but the spatial part of an infinitesimal diffeomorphism.

The above spectra are characteristic of *oriented* string theory, where the parameter σ fixes a direction along the string. There also exist *unoriented* strings which are invariant under a *twist* transformation

$$\sigma \to l - \sigma \,. \tag{12.36}$$

In this case, both the photon (12.30) of the open string and the anti-symmetric tensor $B_{\mu\nu}$ of the closed string disappear in the unoriented version of the theory. Even after removing these fields and the gauge group they carry, there is another non-trivial gauge group associated with the open string. Symmetry considerations require to introduce, at each end-point of the string, fermionic fields in the irreducible spinor representation of $SO(2^{N/2})$, where N is even. This representation has dimension $2^{N/2}$. Thus, each end-point is labelled by an index with $2^{N/2}$ values, called Chan–Paton factor [59, 60], which determines the type of brane attached to the string. One-loop tadpole divergences cancel for $N = D$ and the gauge group associated with the theory is therefore $SO(8192)$ [61, 62].

12.1.4 Interactions

In the old covariant quantization, the rigorous (and, actually, first) derivation of the critical dimension of spacetime entails interactions. When strings meet, their interactions are not point-wise but extended to a finite region of space; this feature is at the core of the ultraviolet finiteness that the theory is believed to have. Feynman diagrams consist of strips (for open strings) and tubular Riemann surfaces (for closed strings), endowed with handles (the number of handles is called genus). The Feynman expansion is governed by the dimensionless string coupling $g_s \propto l_{\rm Pl}/l_s$.

To string vertices, there correspond operators with which one can construct transition amplitudes between in- and out-states. It turns out that, while closed string theory is self-consistent, open string theory is not complete, since some diagrams correspond to closed-string features (an example is given in Fig. 12.3) and an open string can join its end-points to form a closed one. In other words, closed strings can be produced in open-string scattering. Therefore, closed strings must be included in open string theory. Interactions are consistent only when $D = 26$ [63].

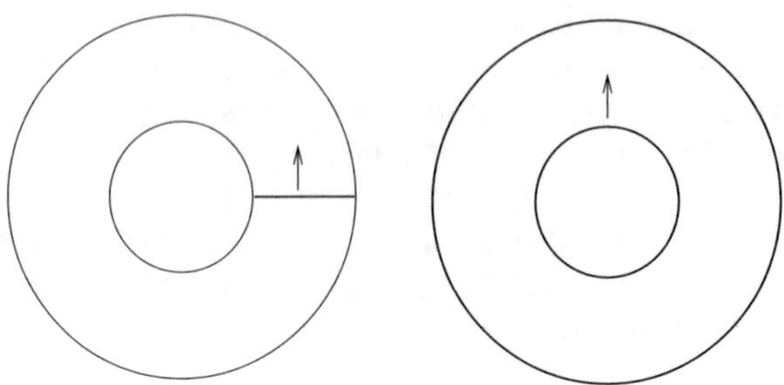

Fig. 12.3 The annulus, a one-loop diagram that can be seen either as an open string propagating around (*left*) or as a closed string expanding or contracting (*right*)

Like most of the features of string theory, the study of the Riemannian surfaces describing string interactions is conveniently approached [48] in the covariant formalism of conformal field theory [43–48]. In CFTs, massless interacting fields live in two dimensions (parametrized as a complex plane) and their dynamics is conformally invariant. The conformal structure of Riemann surfaces is fully encoded in a set of parameters called *moduli* [64–67], so that the set of Riemannian surfaces is also known as moduli space. Compared to the old covariant or light-cone formalisms we are adopting here, CFT is arguably a more economic and elegant way to introduce string theory. However, it is technically more involved with respect to the average presentation of this book and, for this reason, we will omit it from the chapter.

12.1.5 Low-Energy Limit

If we allow closed strings to move in a non-flat background, we have to replace the Minkowski metric $\eta_{\mu\nu}$ in (12.1) with an X-dependent rank-2 tensor. This tensor decomposes into a symmetric part $\bar{g}_{\mu\nu}(X)$, encoding the graviton mode, and an anti-symmetric tensor $B_{\mu\nu}(X)$ (Kalb–Ramond field) accounting for the other closed-string excitations:

$$\mathcal{S} = -\frac{1}{4\pi\alpha'}\int d\tau\, d\sigma\, \sqrt{h}\left[h^{ab}(\bar{g}_{\mu\nu} + iB_{\mu\nu})\partial_a X^\mu \partial_b X^\nu + \alpha' \mathcal{R}^{(2)}\Phi\right], \tag{12.37}$$

where the imaginary unit comes from the analytic continuation to the Euclidean world-sheet [35] and we included the world-sheet Ricci scalar term, now coupled to a non-constant field $\Phi(X)$. This field Φ gives the dilaton when combined with the trace of $\bar{g}_{\mu\nu}$, but often it is simply called dilaton. The kinetic term for X in the action is no longer quadratic; while dynamical systems with such property are called k-essence in cosmology, in field and string theory they are dubbed *non-linear sigma models* [68–70].

At the quantum level, the trace of the energy-momentum tensor is

$$2\alpha'\mathcal{T}_a{}^a = -(\beta^{\bar{g}}_{\mu\nu}s^{ab} + \mathrm{i}\beta^B_{\mu\nu}\epsilon^{ab})\partial_a X^\mu \partial_b X^\nu - \alpha\beta^\Phi \mathcal{R}^{(2)}, \tag{12.38}$$

where we have split the world-sheet metric into a symmetric and an anti-symmetric part, $h^{ab} = s^{ab} + \epsilon^{ab}$. At the one-loop level, the beta functions are [71]

$$\beta^{\bar{g}}_{\mu\nu} = \alpha'\bar{R}_{\mu\nu} + 2\alpha'\bar{\nabla}_\mu\bar{\nabla}_\nu\Phi - \frac{\alpha'}{4}H_{\mu\rho\sigma}H_\nu{}^{\rho\sigma} + O(\alpha'^2), \tag{12.39a}$$

$$\beta^B_{\mu\nu} = -\frac{\alpha'}{2}\bar{\nabla}^\sigma H_{\sigma\mu\nu} + \alpha' H_{\sigma\mu\nu}\bar{\nabla}^\sigma\Phi + O(\alpha'^2), \tag{12.39b}$$

$$\beta^\Phi = -\frac{\alpha'}{2}\bar{\Box}\Phi + \alpha'\bar{\nabla}_\sigma\Phi\bar{\nabla}^\sigma\Phi - \frac{\alpha'}{24}H_{\rho\sigma\tau}H^{\rho\sigma\tau} + \frac{D-26}{6} + O(\alpha'^2), \tag{12.39c}$$

where the 3-form $H_{\sigma\mu\nu} = \partial_\sigma B_{\mu\nu} + \partial_\mu B_{\nu\sigma} + \partial_\nu B_{\sigma\mu}$ (H_3 in short) is the Kalb–Ramond field strength. This is the string analogue of the electromagnetic field strength for point particles. Notice that $[H_3] = 1$ and $[\Phi] = 0$. In order to have conformal invariance, all beta functions must vanish simultaneously, thus giving the equations of motion for gravity, for the anti-symmetric tensor and for the dilaton at first order in α': $\beta^{\bar{g}}_{\mu\nu} = 0$, $\beta^B_{\mu\nu} = 0$, $\beta^\Phi = 0$. These equations are obtained also from the effective spacetime action [71–73]

$$S = \frac{1}{2\kappa_D^2}\int \mathrm{d}^D x\,\sqrt{-\bar{g}}\,\mathrm{e}^{-2\Phi}\left[\frac{2(26-D)}{3\alpha'} + \bar{R} - \frac{1}{2}H_3^2 + 4(\bar{\nabla}\Phi)^2 + O(\alpha')\right], \tag{12.40}$$

where κ_D^2 is the D-dimensional Newton's constant and $H_3^2 = H_{\mu\nu\rho}H^{\mu\nu\rho}/3!$. In $D = 26$ dimensions, a solution to the dynamical equations is the Minkowski metric $\bar{g}_{\mu\nu} = \eta_{\mu\nu}$, $B_{\mu\nu} = 0$ and a constant dilaton. The action (12.40) describes a scalar field non-minimally coupled to gravity. The metric $\bar{g}_{\mu\nu}$ plays the role of the Jordan metric of Sect. 7.4, hence the use of a bar. For this reason, the Jordan frame is often called string frame. After a conformal transformation $g_{\mu\nu} = \exp[-4\Phi/(D-2)]\bar{g}_{\mu\nu}$, one can recast (12.40) in the Einstein frame:

$$S = \frac{1}{2\kappa_D^2}\int \mathrm{d}^D x\,\sqrt{-g}\left[\frac{2(26-D)}{3\alpha'}\mathrm{e}^{4\Phi/(D-2)} + R - \mathrm{e}^{-8\Phi/(D-2)}\frac{1}{2}H_3^2 - \frac{4}{D-2}(\nabla\Phi)^2 + O(\alpha')\right], \tag{12.41}$$

where the indices in H_3^2 are now contracted with the Einstein-frame metric $g_{\mu\nu}$.

On a background with non-zero curvature and fields, the brane action (12.19) is modified as (fermionic part ignored)

$$S_p^{\mathrm{DBI},\bar{g}} = -\tilde{T}_p \int \mathrm{d}^{p+1}\sigma\, \mathrm{e}^{-\Phi} \sqrt{-\det(\bar{g}_{ab} + B_{ab} + 2\pi l_\mathrm{s}^2 F_{ab})}\,, \tag{12.42a}$$

$$\tilde{T}_p = g_\mathrm{s} T_p = \frac{1}{(2\pi)^p l_\mathrm{s}^{p+1}}\,, \tag{12.42b}$$

where $\bar{g}_{ab} = \bar{g}_{\mu\nu}\partial_a X^\mu \partial_b X^\nu$ and a similar expression for B_{ab} are the projections of, respectively, the metric and the Kalb–Ramond field on the brane. By definition, the vacuum expectation value of the dilaton measures the strength of the string coupling:

$$\mathrm{e}^{\Phi} = g_\mathrm{s}\,. \tag{12.43}$$

12.1.6 *String Field Theory*

The Polyakov action (12.1) is the string extension of the classical and quantum mechanical action for a free point particle. The main dynamical object is, in these cases, the *target* coordinate $X^\mu(\tau,\sigma)$ parametrized by the world-sheet and the particle position $q^\mu(\tau)$ parametrized by the world-line. Multi-particle and multi-string interactions are described by n-point correlation functions. For the particle, interactions among infinitely many fundamental objects are included by moving from classical or quantum mechanics to classical or quantum field theory. A similar generalization is possible also for strings: *string field theory* (SFT).

A string field Ψ is a mathematical object given by the superposition (called level expansion) of string oscillators applied to the Fock vacuum (or vacua, in the case of the closed string). The coefficients of the level expansion are particle-field modes of progressively higher mass. The interaction and precise field content depend on the type of string and on the composition rules of the theory. Open SFT is the simplest example of this construction [74, 75]. The bosonic open SFT action is [76]

$$S = -\frac{1}{g_\mathrm{o}^2}\int \left(\frac{1}{2\alpha'}\Psi * Q\Psi + \frac{1}{3}\Psi * \Psi * \Psi\right), \tag{12.44}$$

where g_o is the open string coupling,[2] Q is the BRST operator (which annihilates the field in the free theory, $Q\Psi = 0$) and $*$ is a non-commutative product describing how strings interact. A visual picture of the cubic interaction is obtained by gluing the halves of two different open strings, while the other two halves merge into a third string.

[2]In several circumstances like here, g_s will be further differentiated into the open-string and closed-string coupling g_o and g_c, respectively.

The equation of motion from (12.44) is $Q\Psi + \alpha'\Psi * \Psi = 0$. The action is invariant under the infinitesimal gauge transformation $\delta\Psi = Q\Psi + \Psi * \Gamma - \Gamma * \Psi$, where Γ is a zero-ghost-number state. The gauge group of string field theory is very large and not completely known. It includes at least world-sheet reparametrizations, spacetime diffeomorphisms and (in the supersymmetric extensions) supersymmetry transformations [77, 78].

Choosing a CFT representation of (12.44), fixing the gauge and keeping only some of the particle fields in Ψ (a technical assumption called level truncation), one can obtain from (12.44) an effective spacetime action for the fields which exhibits a non-local interaction. We only show the final result of this procedure when truncating up to the tachyonic mode ϕ [79, 80]:

$$\mathcal{S} = \frac{1}{g_o^2}\int d^D x \left[\frac{1}{2}\phi\left(\Box + \frac{1}{\alpha'}\right)\phi - \frac{e^{3/(\alpha' M_o^2)}}{3}\tilde{\phi}^3\right], \tag{12.45}$$

where

$$\tilde{\phi} := e^{\Box/M_o^2}\phi\,, \qquad \frac{1}{M_o^2} = \alpha' \ln\left(\frac{3^{3/2}}{4}\right) \approx 0.2616\,\alpha'\,. \tag{12.46}$$

Notice the appearance of the correct open-string tachyonic mass $-1/\alpha'$ in the free part. The value of M_o is dictated by conformal invariance, which is partly broken by level truncation. The equation of motion from (12.45) is

$$\left(\Box + \frac{1}{\alpha'}\right)e^{-2\Box/M_o^2}\tilde{\phi} - e^{3/(\alpha' M_o^2)}\tilde{\phi}^2 = 0\,. \tag{12.47}$$

The other fields of the level expansion, as well as a toy model called p-adic or non-Archimedean string [81], display the same type of non-locality.

The degrees of freedom corresponding to the graviton are included in the spectrum of closed SFT [82–96]. The action is

$$S = \int \frac{1}{\alpha'}\Phi \star Q b_0^- \Phi + g_c \sum_{N=3}^{+\infty} \frac{(g_c\,\alpha')^{N-3}}{2^{N-3}N!}\Phi \star [\Phi^{N-1}]\,, \tag{12.48}$$

where Φ is the closed-string field, $\star$ is a non-commutative product, Q is the BRST operator, $b_0^- = (b_0 - \bar{b}_0)/2$ is a combination of anti-ghost zero modes, g_c is the closed-string coupling and the symbol $[\Phi^{N-1}] = [\Phi_1 \star \cdots \star \Phi_{N-1}]$ represents the string field obtained combining $N-1$ terms $\Phi_1, \dots, \Phi_{N-1}$ using the N-string vertex function. Again, for physical processes dominated by light states one can truncate the string field in terms of oscillators and particle fields, $\Phi = \Phi_{\rm phys} + \Phi_{\rm aux}$, where $\Phi_{\rm aux}$ is made of various auxiliary fields,

$$\Phi_{\rm phys} = c_0^-\left(\phi + A_{\mu\nu}\,\alpha_{-1}^{\mu}\bar{\alpha}_{-1}^{\nu} + \dots\right)|0\rangle\,, \tag{12.49}$$

$c_0^- = c_0 - \bar{c}_0$ is a combination of the left and right ghost zero mode, ϕ is the bosonic tachyon field and the symmetric and anti-symmetric parts of the tensor $A_{\mu\nu}$ are, respectively, the graviton $\bar{g}_{\mu\nu}$ and the two-form $B_{\mu\nu}$. After gauge fixing and a rearrangement of the auxiliary fields, the resulting effective Lagrangian for the physical modes contains a kinetic and an interaction part, $\mathcal{L} = \mathcal{L}_{\rm free} + \mathcal{L}_{\rm int}$. The kinetic and mass terms read

$$\mathcal{L}_{\rm free} = -\frac{1}{2}\partial_\lambda A_{\mu\nu}\partial^\lambda A^{\mu\nu} - \frac{1}{2}\partial_\lambda\phi\partial^\lambda\phi - \frac{2}{\alpha'}\phi^2\,, \tag{12.50}$$

while there are about 50 cubic interaction terms:

$$\begin{aligned}\mathcal{L}_{\rm int} &= \frac{g_{\rm c}}{3!\epsilon^3}\tilde{\phi}^3 + \frac{g_{\rm c}\alpha'}{2^3\epsilon^2}(\partial\tilde{\phi})^2\tilde{A} + \frac{g_{\rm c}}{2\epsilon}\tilde{\phi}\tilde{A}^2 + \frac{g_{\rm c}\,\alpha'}{2^4\epsilon}\,\tilde{\phi}\,(\partial\tilde{A})^2\\ &\quad+\frac{g_{\rm c}\alpha'^2}{2^7\epsilon}\,\tilde{\phi}\,(\partial^2\tilde{A})^2 - \frac{g_{\rm c}\,\alpha'}{2^3}\,\tilde{A}^2(\partial^2\tilde{A}) + \dots\,,\end{aligned} \tag{12.51}$$

where $\epsilon = 2^4/3^3$ and fields with a tilde are "dressed" by the same exponential non-local operator (12.46) except that the constant $M_{\rm o}^2 \to M_{\rm c}^2 = 2M_{\rm o}^2$ has an extra factor of 2, which can be removed by a coordinate rescaling.

The same non-local structure in effective spacetime actions is expected also when open and closed strings are considered simultaneously, as proposals for the p-adic tachyonic action indicate [97–101]. One can conclude that, after a field redefinition [74, 79, 80, 85], the effective field actions on target spacetime stemming from string field theory have the following schematic structure for bosonic fields $\tilde{f}_i$:

$$\boxed{S_{\rm eff} = \sum_i \int {\rm d}^D x \left[\frac{1}{2}\tilde{f}_i(\Box - m_i^2){\rm e}^{-\Box/(c_i m_i^2)}\tilde{f}_i - U(\tilde{f}_i, \partial)\right],} \tag{12.52}$$

where the c_i are constants and U is a cubic potential of the dressed fields $\tilde{f}_i$ and their derivatives.

12.2 Superstring

A pressing problem of the bosonic case is that the ground state $|0;p\rangle$ is unstable since it corresponds to a scalar tachyon with mass $\propto -1/\alpha'$. Therefore, the theory has been quantized on a false vacuum. Superstring theory can address this issue. Tachyons will appear only for a class of unstable D-branes, which will decay into lower-dimensional stable branes.

12.2.1 Action

In the Ramond–Neveu–Schwarz formulation of the theory [5–7, 102, 103], the gauge-fixed action (12.1) is augmented by a fermionic sector [39, 40, 104, 105]:

$$\mathcal{S}[X,\psi] = -\frac{1}{4\pi\alpha'}\int \mathrm{d}\tau\,\mathrm{d}\sigma\,\eta_{\mu\nu}\left(\partial_a X^\mu \partial^a X^\nu - \mathrm{i}\bar{\psi}^\mu \rho^a \partial_a \psi^\nu\right), \tag{12.53}$$

where ψ^μ are D Majorana fermions made of two components and ρ^a are 2×2 matrices obeying the algebra

$$\{\rho^a, \rho^b\} = -2\eta^{ab}. \tag{12.54}$$

They are related to the Pauli matrices (5.208) as $\rho^0 = \sigma^2$ and $\rho^1 = \mathrm{i}\sigma^1$. Since $\mathrm{i}\rho^a$ is real, so are the two-component spinors ψ^μ.

The system described by (12.53) in invariant under: (a) the local world-sheet diffeomorphism and Weyl transformations (12.2), augmented by $\delta\psi^\mu = \xi^a\partial_a\psi^\mu$; (b) the global Poincaré invariance of the bosonic string; (c) the global supersymmetry transformations $\delta X^\mu = \bar{\epsilon}\psi^\mu$, $\delta\psi^\mu = -\mathrm{i}\rho^a\partial_a X^\mu\epsilon$ and $\delta\bar{\psi}^\mu = \mathrm{i}\bar{\epsilon}\rho^a\partial_a X^\mu$, where ϵ is a constant Majorana spinor. An action more general than (12.53), invariant under both world-sheet reparametrizations and local spacetime supersymmetry, is also possible [106, 107]. It can be quantized in the light-cone gauge and gives the same physics.

Using the equations of motion

$$\partial_a\partial^a X^\mu = 0\,, \qquad \mathrm{i}\rho^a\partial_a\psi^\mu = 0\,, \tag{12.55}$$

one can easily show that the supercurrent $\mathcal{J}^a = \rho^b\rho^a\partial_b X\cdot\psi/2$ is conserved on shell (Problem 12.1). Here and in what follows, $\cdot$ represents contraction of spacetime indices via $\eta_{\mu\nu}$. Supersymmetry transformations are then expressed by the commutator $[\mathcal{Q}\bar{\epsilon}, X^\mu] = \bar{\epsilon}\psi^\mu$, where $\mathcal{Q} := \int_0^l \mathrm{d}\sigma\,\mathcal{J}^0$ is the supercharge. The energy-momentum tensor is also conserved, $\partial_a\mathcal{T}^{ab} = 0$, and its trace vanishes, $\mathcal{T}_a{}^a = 0$.

In terms of the left- and right-moving coordinates defined in (12.13), the dynamics (12.55) is recast as $\partial_+\partial_- X^\mu = 0$ and $\partial_-\psi^\mu_+ = 0 = \partial_+\psi^\mu_-$, where we split the 2-spinors $\psi^\mu = (\psi^\mu_-, \psi^\mu_+)^{\mathrm{t}}$ into two Grassmann fields $\psi^\mu_\pm = \psi^\mu_\pm(\sigma^\pm)$. The constrains on the energy-momentum tensor components and the supercurrent are then $\mathcal{T}_{\pm\pm} = [\partial_\pm X\cdot\partial_\pm X + (\mathrm{i}/2)\psi_\pm\cdot\partial_\pm\psi_\pm]/(2\pi\alpha') = 0$ and $\mathcal{J}_\pm = \psi_\pm\cdot\partial_\pm X/(2\pi\alpha') = 0$.

These equations of motion are obtained by varying the action (12.53) and imposing suitable conditions to make boundary terms vanish. The conditions on the scalars X^μ are the same of Sect. 12.1 and define open or closed strings. The boundary condition $\psi_+\cdot\delta\psi_+ - \psi_-\cdot\delta\psi_- = 0$ on the spinorial fields defines distinct fermionic sectors of the theory, so that $\psi^\mu_+ = \pm\psi^\mu_-$ at the end-points (coincident, in the closed case) of the string. At $\sigma = 0$, by convention we set $\psi_+^\mu(\tau,0) = \psi_-^\mu(\tau,0)$, while for all μ

$$\text{Neveu–Schwarz (NS) sector:} \qquad \psi_+^\mu(\tau,l) = -\psi_-^\mu(\tau,l)\,, \tag{12.56a}$$

$$\text{Ramond (R) sector:} \qquad \psi_+^\mu(\tau,l) = +\psi_-^\mu(\tau,l)\,. \tag{12.56b}$$

For the open string there are only two sectors (NS or R), while for the closed string there are four (NS-NS, NS-R, R-NS and R-R) since we can impose the boundary conditions on the left- and right-moving parts of (12.14) independently.

The most general solutions of ψ are expanded in terms of harmonic oscillators. For the open string, they are

$$\text{NS sector:} \qquad \psi_\pm^\mu = \frac{\sqrt{\alpha'}}{2} \sum_{n-\frac{1}{2}\in\mathbb{Z}} b_n^\mu\, \mathrm{e}^{-\mathrm{i}n\sigma^\pm}\,, \tag{12.57a}$$

$$\text{R sector:} \qquad \psi_\pm^\mu = \frac{\sqrt{\alpha'}}{2} \sum_{r\in\mathbb{Z}} d_r^\mu\, \mathrm{e}^{-\mathrm{i}r\sigma^\pm}\,. \tag{12.57b}$$

The Virasoro generators are defined as in (12.11), augmented by a fermionic part:

$$\text{NS sector:} \qquad L_n^{(\mathrm{NS})} = \frac{1}{2}\eta_{\mu\nu}\left[\sum_{m\in\mathbb{Z}} \alpha_{n-m}^\mu \alpha_m^\nu + \sum_{k-\frac{1}{2}\in\mathbb{Z}} \left(k+\frac{n}{2}\right) b_{-k}^\mu b_{n+k}^\nu\right], \tag{12.58a}$$

$$\text{R sector:} \qquad L_n^{(\mathrm{R})} = \frac{1}{2}\eta_{\mu\nu}\left[\sum_{m\in\mathbb{Z}} \alpha_{n-m}^\mu \alpha_m^\nu + \sum_{r\in\mathbb{Z}} \left(r+\frac{n}{2}\right) d_{-r}^\mu d_{n+r}^\nu\right], \tag{12.58b}$$

where $n \in \mathbb{N}$. Fermionic generators are obtained by Fourier expanding the supercurrent in the periodic and anti-periodic sectors. In the Neveu–Schwarz sector, $\mathcal{G}_k = \sqrt{2}\int_0^\pi \mathrm{d}\sigma\,(\mathrm{e}^{-\mathrm{i}k\sigma}\mathcal{J}_- + \mathrm{e}^{\mathrm{i}k\sigma}\mathcal{J}_+) = \sum_m \alpha_{-m}\cdot b_{k+m}$ for $k \in \mathbb{Z}+1/2$. In the Ramond sector, $\mathcal{F}_r = \sqrt{2}\int_0^\pi \mathrm{d}\sigma\,(\mathrm{e}^{-\mathrm{i}r\sigma}\mathcal{J}_- + \mathrm{e}^{\mathrm{i}r\sigma}\mathcal{J}_+) = \sum_m \alpha_{-m}\cdot d_{r+m}$ for $r \in \mathbb{Z}$.

For the closed string,

$$\text{NS sector:} \qquad \psi_-^\mu = \sqrt{2\alpha'} \sum_{k+\frac{1}{2}\in\mathbb{Z}} b_k^\mu\, \mathrm{e}^{-2\mathrm{i}k\sigma^-}, \qquad \psi_+^\mu = \sum_{k+\frac{1}{2}\in\mathbb{Z}} \tilde{b}_k^\mu\, \mathrm{e}^{-2\mathrm{i}k\sigma^+}, \tag{12.59a}$$

$$\text{R sector:} \qquad \psi_-^\mu = \sqrt{2\alpha'} \sum_{r\in\mathbb{Z}} d_r^\mu\, \mathrm{e}^{-2\mathrm{i}r\sigma^-}, \qquad \psi_+^\mu = \sum_{r\in\mathbb{Z}} \tilde{d}_r^\mu\, \mathrm{e}^{-2\mathrm{i}r\sigma^+}, \tag{12.59b}$$

and the Virasoro generators follow from the definition (12.15) (the right-hand side having also fermionic oscillators) and its right-moving counterpart.

12.2.2 Quantization

The old covariant quantization proceeds by imposing (12.20) and equal-time anti-commutation relations for ψ: $\{\psi^\mu_A(\tau,\sigma),\,\psi^\nu_B(\tau,\sigma')\} = \pi\delta(\sigma-\sigma')\delta_{AB}$, where $A, B = \pm$. This results in the algebras $\{d^\mu_r,\,d^\nu_s\} = \eta^{\mu\nu}\delta_{r,-s}$, $\{b^\mu_k,\,b^\nu_l\} = \eta^{\mu\nu}\delta_{k,-l}$ and their counterparts in the right sector for the closed string. The Fock vacuum is annihilated by

$$\alpha^\mu_{n\geqslant 0}\,|0\rangle_{\rm osc} = 0\,, \qquad b^\mu_{k>0}\,|0\rangle_{\rm osc,\,NS} = 0\,, \qquad d^\mu_{r>0}\,|0\rangle_{\rm osc,\,R} = 0\,. \tag{12.60}$$

In the case of the NS oscillators b, there is no zero mode as k is semi-integer. There is a unique ground state $|0;p\rangle_{\rm NS}$ corresponding to a Lorentz scalar; the NS sector of the theory is therefore called also bosonic sector. In the case of d, the $r=0$ oscillator is excluded from the definition of the vacuum due to the non-vanishing anti-commutator $\{d^\mu_0,\,d^\nu_0\} = \eta^{\mu\nu}$. The irreducible representations of this Clifford algebra are $SO(1,9)$ spinors and, therefore, the vacuum $|0;p\rangle_{\rm R}$ of the Ramond sector is a fermionic state.

The Virasoro super-algebras of the NS and R sector generated, respectively, by the normal-ordered operators $\hat L^{\rm (NS)}_n$, $\hat{\mathcal{G}}_k$ and $\hat L^{\rm (R)}_n$, $\mathcal{F}_r$ can be found in [34]. As in the bosonic theory, these operators are applied to Fock-space states to define a set of physicality conditions. The counterpart of (12.26) in the NS and R sector are, respectively,

$$\text{NS sector:}\quad \hat L^{\rm (NS)}_n\,|\Phi\rangle = 0 = \mathcal{G}_k\,|\Phi\rangle \quad \forall n,k>0\,, \quad \left(\hat L^{\rm (NS)}_0 - \tfrac{1}{2}\right)|\Phi\rangle = 0\,, \tag{12.61a}$$

$$\text{R sector:}\quad \hat L^{\rm (R)}_n\,|\Phi\rangle = 0 = \mathcal{F}_r\,|\Phi\rangle \quad \forall n,r\geqslant 0\,, \tag{12.61b}$$

where, in the case of the open string,

$$\hat L^{\rm (NS)}_0 = \alpha'\hat p^2 + \eta_{\mu\nu}\left(\sum_{m>0}\alpha^\mu_{-m}\alpha^\nu_m + \sum_{k>0} k\, b^\mu_{-k}b^\nu_k\right) =: \alpha'\hat p^2 + \hat N_{\rm NS}\,, \tag{12.62a}$$

$$\hat L^{\rm (R)}_0 = \alpha'\hat p^2 + \eta_{\mu\nu}\left(\sum_{m>0}\alpha^\mu_{-m}\alpha^\nu_m + \sum_{r>0} r\, d^\mu_{-r}d^\nu_r\right) =: \alpha'\hat p^2 + \hat N_{\rm R}\,. \tag{12.62b}$$

On physical states, the number operators take eigenvalues $N_{\rm NS} = 0, 1/2, 1, 3/2, \dots$ and $N_{\rm R} = 0, 1, 2, \dots$. For the closed string, the coefficient in front of $\hat p^2$ changes as $\alpha' \to \alpha'/4$ and there is another copy of operators $\hat{\tilde L}^{\rm (NS,R)}_n$, $\hat{\tilde{\mathcal{G}}}_k$ and $\hat{\tilde{\mathcal{F}}}_r$ for right-moving modes. Therefore, the open and closed superstring spectra are

$$M^2_{\text{open, NS}} = \frac{2N_{\text{NS}} - 1}{2\alpha'}, \qquad M^2_{\text{open, R}} = \frac{N_{\text{R}}}{\alpha'}, \tag{12.63}$$

$$M^2_{\text{closed, NS}} = \frac{2(2N_{\text{NS}} - 1)}{\alpha'}, \qquad M^2_{\text{closed, R}} = \frac{4N_{\text{R}}}{\alpha'}, \tag{12.64}$$

plus independent copies of the second line in the right-moving sector.

The critical dimension of spacetimes where the superstring can live in is

$$\boxed{D = 10\,.} \tag{12.65}$$

This result can be obtained from the extension to superstrings of the proofs given in the bosonic case: among these, there are the no-ghost theorem for the string spectrum [10, 13] also in BRST quantization [47, 48], the requirement for the Lorentz generators of the theory to obey the Lorentz algebra [12, 24] and the cancellation of the conformal anomaly in the covariant path-integral quantization [105].

12.2.3 Type-I Superstring

The theory of open strings contains both fermions and bosons but also a tachyon. Furthermore, there are states that do not combine to supersymmetric multiplets. Both the tachyon and this class of states can be eliminated by imposing that all physical states are eigenvectors with eigenvalue 1 of the Gliozzi–Scherk–Olive (GSO) operators [108]

$$\mathcal{P}_{\text{NS}} := \frac{1 - (-1)^{\sum_{k>0} b_{-k} b_k}}{2}, \qquad \mathcal{P}_{\text{R}} := \frac{1 - (-1)^{\sum_{r>0} d_{-r} d_r}}{2}. \tag{12.66}$$

The lowest levels in the open-string spectrum (12.63) are:

- NS sector:
  - $N_{\text{NS}} = 0$, $M^2_{\text{open, NS}} = -1/(2\alpha')$: the ground state $|0; p\rangle_{\text{NS}}$ of the NS sector is a tachyon, which is projected out since $\mathcal{P}_{\text{NS}}\,|0; p\rangle_{\text{NS}} = 0$.
  - $N_{\text{NS}} = 1/2$, $M^2_{\text{open, NS}} = 0$: a massless vector $b^{\mu}_{-1/2}\,|0; p\rangle_{\text{NS}}$, corresponding to a non-Abelian gauge field A^a_{μ}. It survives the GSO projection, since $(\mathcal{P}_{\text{NS}} - 1)b^{\mu}_{-1/2}\,|0; p\rangle_{\text{NS}} = 0$.
  - $N_{\text{NS}} = 1$, $M^2_{\text{open, NS}} = 1/(2\alpha')$: a massive vector $\alpha^{\mu}_{-1}\,|0; p\rangle_{\text{NS}}$ and a massive rank-2 tensor $b^{\mu}_{-1/2} b^{\nu}_{-1/2}\,|0; p\rangle_{\text{NS}}$.

- R sector:
  - $N_\mathrm{R} = 0$, $M^2_{\text{open, R}} = 0$: a massless spinor $|0;p\rangle_\mathrm{R}$, the ground state of the Ramond sector, associated with a fermionic field we will denote as ψ. This state survives the GSO projection, since $(\mathcal{P}_\mathrm{R} - 1)\,|0;p\rangle_\mathrm{R} = 0$. In $D = 10$, the spinor representation of $SO(D)$ has dimension $2^{10/2} = 32$, corresponding to a complex 32-component spinor. The reality condition reduces it to a 32-component Majorana spinor, while the chirality condition (Weyl spinor) cuts it to 16 components. These components are not independent since they are constrained by the Dirac equation. Therefore, $|0;p\rangle_\mathrm{R}$ encodes 8 degrees of freedom. Also, defining the chirality matrix $\Gamma_{11} = \Gamma_0\Gamma_1\cdots\Gamma_9$ in terms of the $SO(1,9)$ gamma matrices, the ground state is defined to have positive chirality, $\Gamma_{11}\,|0;p\rangle_\mathrm{R} = +\,|0;p\rangle_\mathrm{R}$.
  - $N_\mathrm{R} = 1, M^2_{\text{open, R}} = 1/\alpha'$: D massive spinors $\alpha^\mu_{-1}\,|0;p\rangle_\mathrm{R}$. D extra spinors of the form $d^\mu_{-1}\,|0;p\rangle_\mathrm{R}$ are projected out: $\mathcal{P}_\mathrm{R} d^\mu_{-1}\,|0;p\rangle_\mathrm{R} = 0$. The number of physical spinors is reduced to $D-2$ in light-cone quantization [102, 109, 110].

One can show that the states that survive the projection form multiplets. For instance, we have seen that $|0;p\rangle_\mathrm{R}$ has 8 real independent components (symbolically indicated with $\mathbf{8}_\mathrm{s}$, where "s" stands for spinor), while it is easy to see that $b^\mu_{-1/2}\,|0;p\rangle_\mathrm{NS}$ represents 8 degrees of freedom corresponding to the transverse directions of a vector ($\mathbf{8}_\mathrm{v}$, where "v" stands for vector). This hints to the fact that the massless spectrum is a supersymmetric multiplet $\mathbf{8}_\mathrm{v} \otimes \mathbf{8}_\mathrm{s}$ [108, 111], actually describing the super-Yang–Mills theory with gauge group $SO(32)$ [108, 112]. A proof of supersymmetry at all mass levels exists [109].

When combined with the type-IIB closed-string modes described below, a deeper analysis of the spectrum leads to the formulation of *type-I* string theory (see also Sect. 12.3.3). This is a theory of unoriented open and closed strings with $\mathcal{N} = 1$ supersymmetry. Its low-energy limit is $SO(32)$ super-Yang–Mills theory [113] coupled to a supergravity sector [110, 114].

12.2.4 Type-II Superstrings

The GSO projection works also for the closed string and must be applied on the left- and right-moving sectors separately [108]. This results in *type-II* theories for oriented closed strings with 32 supercharges [108, 110]. The massless spectrum is:

- NS-NS sector:
  - $N_\mathrm{NS} = 0 = \tilde{N}_\mathrm{NS}$, $M^2_{\text{closed, NS–NS}} = -2/\alpha'$: the tachyon $|0,\tilde{0};p\rangle_\mathrm{NS} := |0\rangle_\mathrm{NS} \otimes \widetilde{|0\rangle}_\mathrm{NS} \otimes |p\rangle$ is projected out. From now on, we omit the momentum label in kets.
  - $N_\mathrm{NS} = \frac{1}{2} = \tilde{N}_\mathrm{NS}$, $M^2_{\text{closed, NS–NS}} = 0$: the rank-2 tensor $|\frac{1}{2},\tilde{\frac{1}{2}}\rangle_\mathrm{NS} = b^\mu_{-1/2}\,|0\rangle_\mathrm{NS} \otimes \tilde{b}^\nu_{-1/2}\widetilde{|0\rangle}_\mathrm{NS}$ is decomposed into a scalar Φ, a symmetric tensor

$\tilde{g}^{\mu\nu}$ and an anti-symmetric tensor $B^{\mu\nu}$. In terms of transverse components in 10 dimensions, $\mathbf{8}_\mathrm{v} \otimes \mathbf{8}_\mathrm{v} = \mathbf{1} \oplus \mathbf{35}_\mathrm{v} \oplus \mathbf{28}_\mathrm{a}$, where "a" stands for anti-symmetric. This supergravity sector is called universal because it is present in all closed superstring theories.

- R-R sector: $N_\mathrm{R} = 0 = \tilde{N}_\mathrm{R}$, $M^2_{\text{closed, R}-\text{R}} = 0$: the ground state is $|0, \tilde{0}\rangle_\mathrm{R} := |0\rangle_\mathrm{R} \otimes \widetilde{|0\rangle}_\mathrm{R}$. If one defines the right sector to have positive chirality ($\Gamma_{11}\,|0\rangle_\mathrm{R} = +\,|0\rangle_\mathrm{R}$), one can choose whether the left sector has same or opposite chirality, $\Gamma_{11}\widetilde{|0\rangle}_\mathrm{R} = \pm\widetilde{|0\rangle}_\mathrm{R}$. If one takes opposite chirality, the theory is called *type-IIA*, otherwise *type-IIB*.
  - Type-IIA: In terms of representations of $SO(8)$, the non-chiral ground state $|0, \tilde{0}\rangle_\mathrm{R}$ is decomposed as $\mathbf{8}_\mathrm{s} \otimes \mathbf{8}_\mathrm{c} = \mathbf{8}_\mathrm{v} \oplus \mathbf{56}_\mathrm{v}$ ("c" stands for conjugate spinor), a 1-form C_μ and an anti-symmetric 3-form $C_{\mu\nu\rho}$.[3] These fields are called *Ramond–Ramond potentials*.
  - Type-IIB: The decomposition of the chiral ground state is $\mathbf{8}_\mathrm{s} \otimes \mathbf{8}_\mathrm{s} = \mathbf{1} \oplus \mathbf{28}_\mathrm{a} \oplus \mathbf{35}_\mathrm{s}$ and the R-R potentials are a pseudo-scalar C_0 (axion), an anti-symmetric 2-form $C_{\mu\nu}$ (also dubbed $B'_{\mu\nu}$ in the literature) and the self-dual 4-form $C^{(+)}_{\mu\nu\rho\sigma}$.
- NS-R and R-NS sectors:
  - $N_\mathrm{NS} = \frac{1}{2}$, $\tilde{N}_\mathrm{R} = 0$, $M^2_{\text{closed, NS}-\text{R}} = 0$: the fermionic state $|\frac{1}{2}, \tilde{0}\rangle_{\mathrm{NS}-\mathrm{R}} = b^\mu_{-1/2}\,|0\rangle_\mathrm{NS} \otimes \widetilde{|0\rangle}_\mathrm{R}$ is decomposed as $\mathbf{8}_\mathrm{v} \otimes \mathbf{8}_\mathrm{s} = \mathbf{8}_\mathrm{c} \oplus \mathbf{56}_\mathrm{c}$. These degrees of freedom correspond to a *dilatino* and a *gravitino*, super-partners of, respectively, the dilaton and the graviton.
  - $N_\mathrm{R} = 0$, $\tilde{N}_\mathrm{NS} = \frac{1}{2}$, $M^2_{\text{closed, R}-\text{NS}} = 0$: The same analysis holds for the R-NS state $|0, \tilde{\frac{1}{2}}\rangle_{\mathrm{R}-\mathrm{NS}} = |0\rangle_\mathrm{NS} \otimes \tilde{b}^\mu_{-1/2}\widetilde{|0\rangle}_\mathrm{R}$. In type-IIA theory, the decomposition is $\mathbf{8}_\mathrm{c} \otimes \mathbf{8}_\mathrm{v} = \mathbf{8}_\mathrm{s} \oplus \mathbf{56}_\mathrm{s}$ and one has a dilatino and a gravitino of opposite chirality with respect to those in the NS-R sector. In type-IIB theory, the spectrum is chiral since $\mathbf{8}_\mathrm{s} \otimes \mathbf{8}_\mathrm{v} = \mathbf{8}_\mathrm{c} \oplus \mathbf{56}_\mathrm{c}$ as in the NS-R sector.

12.2.5 Interactions and Anomaly Cancellation

String interactions [102, 114–116] (also formulated with superfields [117, 118]) play an essential role in determining whether superstring theory is perturbatively finite. This was checked at one loop [102, 103, 119, 120] and up to two loops and four external states in string scattering amplitudes [121–127]. At arbitrary loop order, there are non-renormalization theorems stating that loop corrections to the vacuum amplitude and to the massless n-point functions for $n = 0, 1, 2, 3$ vanish to all orders in $D = 10$ flat backgrounds [48, 128, 129]. Although no formal proof exists, many independent arguments encourage the belief that superstring theory is indeed finite.

[3] The number of degrees of freedom of a p-form for the $SO(8)$ group is $\binom{8}{p} = 8!/[p!(8-p)!]$.

The recent super-moduli space formalism [130] and a new formalism using picture-changing operators [131, 132] permit to compute multi-loop off-shell amplitudes and, therefore, to have renormalization properties under a much better control. Ultimately, the problem of renormalizability is tightly related to the formulation of a supersymmetric field theory of strings, to which Sect. 12.2.9 is dedicated. Sen [133] offers a valuable pedagogical overview on the subject of divergences in superstring perturbation theory.

Two other applications of interactions are the establishing of the low-energy limits of superstrings, summarized in Sect. 12.2.7, and the calculation of anomalies.

A theory free from quantum anomalies is not only mathematically viable but also a useful framework wherein phenomenological predictions are made possible. If superstrings aim to describe sensible low-energy particle and gravitational physics, then unitarity, Lorentz invariance and general covariance must be preserved by quantum interactions. This may not be the case if anomalous Feynman diagrams involving chiral fields give a non-zero net contribution to scattering amplitudes. It is therefore necessary (and non-trivial) to check that the low-energy limit of type-I and type-II string theories be free from these anomalies, which are called chiral or gauge anomalies. The case of type-IIA is the simplest: the theory conserves parity and there are no anomalies. Less obvious is the case of type-IIB theory. Indeed, cancellations between fields of different spin guarantee that $D = 10, \mathcal{N} = 2$ chiral SUGRA is the only case free from one-loop gravitational anomalies coming from the coupling of Weyl fermions and of the anti-symmetric tensor with gravity [134].

Regarding type-I theory, we recall that the vector multiplet $\mathbf{8}_\text{v} \otimes \mathbf{8}_\text{s}$ of super-Yang–Mills theories (a gauge vector field plus its super-partner) lives in the adjoint representation of some group $\mathbb{G}$. When super-Yang–Mills is coupled to $D = 10$ chiral SUGRA, the so-called Green–Schwarz mechanism is enforced: the only two groups which allow the cancellation of gauge, gravitational and mixed anomalies are $\mathbb{G} = SO(32)$ and $\mathbb{G} = E_8 \otimes E_8$ [135–138]. The low-energy limit of the type-I open-string sector has precisely $\mathbb{G} = SO(32)$.

12.2.6 Heterotic Superstrings

Remarkably, both anomaly-free groups $E_8 \otimes E_8$ (henceforth denoted as $E_8 \times E_8$) and $SO(32)$ have realizations in terms of closed string theories known as *heterotic* [139–141]. In type-II theory, there are 16 supercharges from the left sector and 16 from the right one, for a total of 32 supercharges and $\mathcal{N} = 1 + 1 = 2$ supersymmetries. In the heterotic string, one picks a right sector with both bosons and fermions and a purely bosonic left sector with 16 directions compactified on a torus (more on compact directions later), for a total of $32 - (26 - 10) = 16$ supercharges (GSO projection) and $\mathcal{N} = 1$ supersymmetry. Due to this mismatch, which can always be prescribed since the left- and right-moving sectors are decoupled, this theory is a hybrid between bosonic and superstring theory, hence the name "heterotic."

The use of the critical dimensionality in each separate sector guarantees anomaly cancellation.

Depending on whether the right sector is NS or R, the spectrum of $SO(32)$ and $E_8 \times E_8$ heterotic string theory is determined by the mass formulæ

$$\frac{\alpha'}{2} M^2_{\text{closed NS}} = N_{\text{NS}} + \tilde{N} - \frac{3}{2} + \alpha' p_{\text{L}}^2 \,, \tag{12.67}$$

$$\frac{\alpha'}{2} M^2_{\text{closed R}} = N_{\text{R}} + \tilde{N} - 1 + \alpha' p_{\text{L}}^2 \,, \tag{12.68}$$

together with a condition guaranteeing that there is no distinguished σ point in the closed string:

$$\text{NS right sector:} \qquad \tilde{N} - 1 + \alpha' p_{\text{L}}^2 = N_{\text{NS}} - \tfrac{1}{2} \,, \tag{12.69}$$

$$\text{R right sector:} \qquad \tilde{N} - 1 + \alpha' p_{\text{L}}^2 = N_{\text{R}} \,. \tag{12.70}$$

Here $p_{\text{L}}^2 = \sum_{I=10}^{25} p_I^2$ is the momentum of the compact directions in the left sector, which takes discrete values, $\alpha' p_{\text{L}}^2 = 0, 1, \ldots$.

Let $|N, \tilde{N}; 0\rangle_{\text{NS,R}} := |N\rangle_{\text{NS,R}} \otimes |\tilde{N}; p_{\text{L}}^2\rangle_{\text{L}}$. The $N_{\text{NS}} = 0$ tachyon states $|0, \tilde{N}; 0\rangle_{\text{NS}} = |0\rangle_{\text{NS}} \otimes |\tilde{N}; 0\rangle_{\text{L}}$ (with zero winding number) do not respect the condition (12.69) for any $\tilde{N} \in \mathbb{N}$ and are removed from the picture.

The massless ($M^2_{\text{closed}} = 0$) spectrum levels are the following.

- NS sector, $N_{\text{NS}} = \frac{1}{2}$:
  - $\tilde{N} = 1$, $p_{\text{L}}^2 = 0$: the 10×10 rank-2 tensor $|\frac{1}{2}, \tilde{1}; 0\rangle_{\text{NS}} = b^\mu_{-1/2} |0\rangle_{\text{NS}} \otimes \tilde{\alpha}^\nu_{-1} |\tilde{0}; 0\rangle_{\text{L}}$ is split into the components of the universal sector of the closed superstring.
  - $\tilde{N} = 1$, $p_{\text{L}}^2 = 0$ and $\tilde{N} = 0$, $p_{\text{L}}^2 = 1/\alpha'$: the ($\tilde{N} = 1$) 16 neutral vectors $b^\mu_{-1/2} |0\rangle_{\text{NS}} \otimes \tilde{\alpha}^I_{-1} |\tilde{0}; 0\rangle_{\text{L}}$ and the ($\tilde{N} = 0$) 480 charged vectors $b^\mu_{-1/2} |0\rangle_{\text{NS}} \otimes |\tilde{0}; 1/\alpha'\rangle_{\text{L}}$ are 496 gauge bosons in the adjoint representation of the group $E_8 \times E_8$ or $SO(32)$.

- R sector, $N_{\text{R}} = 0$:
  - $\tilde{N} = 1, p_{\text{L}}^2 = 0$: the fermionic state $|0, \tilde{1}; 0\rangle_{\text{R}} = |0\rangle_{\text{R}} \otimes \tilde{\alpha}^\mu_{-1} |\tilde{1}; 0\rangle_{\text{L}}$ decomposes into the gravitino and the dilatino, super-partners of the universal sector.
  - $\tilde{N} = 1, p_{\text{L}}^2 = 0$ and $\tilde{N} = 0$, $p_{\text{L}}^2 = 1/\alpha'$: the fermionic states $|0\rangle_{\text{R}} \otimes \tilde{\alpha}^I_{-1} |\tilde{0}; 0\rangle_{\text{L}}$ and $|0\rangle_{\text{R}} \otimes |\tilde{0}; 1/\alpha'\rangle_{\text{L}}$ are gauginos, super-partners of the gauge bosons in the same irreducible adjoint representation.

12.2.7 Massless Spectra and Low-Energy Limits

Let us succintly denote with a number the indices of p-forms: $C_p = C_{\mu_1\dots\mu_p}$. Then, to summarize the massless content of the theory:

- **Type-IIA superstring theory** is non-chiral (spacetime parity is conserved), it has $\mathcal{N} = 2$ ten-dimensional supersymmetries, its R-R fields are forms of odd order (a 1-form C_1 and a 3-form C_3) and its gauge group is $U(1)$ as for the bosonic closed string (one Abelian vector field). Therefore, the type-IIA low-energy limit [110] is $D = 10$, $\mathcal{N} = 2$ non-chiral SUGRA [142, 143], with universal bosonic sector Φ, $\bar{g}_2$ and B_2. It is important to note that this model of supergravity arises via dimensional reduction [142–144] of $D = 11$, $\mathcal{N} = 1$ SUGRA [144], the only theory of supergravity in eleven dimensions. We will sketch this relation in Sect. 12.4.
- **Type-IIB superstring theory** is chiral (spacetime parity is violated), it has $\mathcal{N} = 2$ ten-dimensional supersymmetries of the same chirality, its R-R fields are forms of even order (a 0-form C_0, a 2-form C_2 and the self-dual 4-form $C_4^{(+)}$) and there is no gauge field. Its low-energy limit [110] is $D = 10$, $\mathcal{N} = 2$ chiral SUGRA [145–148], with universal bosonic sector Φ, $\bar{g}_2$ and B_2. Contrary to the non-chiral model, this one does not descend from the dimensional reduction of $D = 11$ supergravity.
- **Type-I superstring theory** has a gauge vector A_1 in the bosonic open spectrum. The closed-string sector is unoriented type-IIB theory and is therefore chiral. After the identification $\sigma \to \pi - \sigma$, only Φ and $\bar{g}_2$ (from the NS-NS sector) and C_2 (from the R-R sector) survive in the bosonic closed-string spectrum, plus their super-partners. The low-energy limit [110, 113, 114] of, respectively, the open- and closed-string sectors are $D = 10$ super-Yang–Mills [108, 112] with gauge group $SO(32)$ [113] and $D = 10$, $\mathcal{N} = 2$ chiral SUGRA.
- **Heterotic superstring theories** have $\mathcal{N} = 1$ ten-dimensional supersymmetry and their low-energy limit [139, 140] is $D = 10$, $\mathcal{N} = 1$ SUGRA [149] (with universal bosonic sector Φ, $\bar{g}_2$ and B_2) coupled with super-Yang–Mills theory with gauge group $E_8 \times E_8$ or $SO(32)$.

The supersymmetric generalization of the string action (12.37) on curved backgrounds, of the beta functions (12.39) and of the low-energy spacetime action (12.40) for these string theories was found in [71, 72, 150–152]. The massless spectrum of the five superstring theories is summarized in Table 12.1.

The SUGRA actions in $D = 10$ with this p-form population can be found in any textbook, e.g. [35]. Here we report only the type-II cases because they have applications in Sect. 12.4 and Chap. 13. The NS-NS universal sector is

$$S_{\text{NS-NS}}[\bar{g}, \Phi, B_2] = \frac{1}{2\kappa_{10}^2} \int \mathrm{d}^{10}x \sqrt{-\bar{g}}\, \mathrm{e}^{-2\Phi} \left[\bar{R} + 4(\nabla\Phi)^2 - \frac{1}{2} H_3^2 \right], \qquad (12.71)$$

Table 12.1 Massless spectrum (p-forms) of the five superstring theories. Gauge groups are indicated in boldface

	I	IIA	IIB	Heterotic
SUGRA sector (NS-NS type-II or NS heterotic)	Φ (dilaton), $\bar{g}_2$ (graviton)	Φ (dilaton), $\bar{g}_2$ (graviton), B_2 (Kalb–Ramond)		
NS gauge sector	A_1 [**SO(32)**]			496 vectors [$\mathbf{E_8} \times \mathbf{E_8}$ or **SO(32)**]
R-R sector (R-R potentials)	C_2	C_1 [**U(1)**], C_3	C_0 (axion), C_2, $C_4^{(+)}$	
R sector	ψ (gaugino)			dilatino, gravitino, 496 gauginos
NS-R and R-NS sectors		dilatino, gravitino (non-chiral)	dilatino, gravitino (chiral)	

while the R-R sectors are

$$S_{\text{R-R}}^{\text{IIA}}[B_2, C_1, C_3] = -\frac{1}{4\kappa_{10}^2} \int d^{10}x \sqrt{-\bar{g}} \left(F_2^2 + \tilde{F}_4^2\right), \tag{12.72}$$

$$S_{\text{R-R}}^{\text{IIB}}[B_2, C_0, C_2, C_4^{(+)}] = -\frac{1}{4\kappa_{10}^2} \int d^{10}x \sqrt{-\bar{g}} \left(F_1^2 + \tilde{F}_3^2 + \frac{1}{2}\tilde{F}_5^2\right). \tag{12.73}$$

Here $F_p^2 := F_{\mu_1 \dots \mu_p} F^{\mu_1 \dots \mu_p}/p!$, $\tilde{F}_3 := F_3 - C_0 \wedge H_3$, $\tilde{F}_4 := F_4 + C_1 \wedge H_3$ and the field strength $\tilde{F}_5 := F_5 - (1/2)C_2 \wedge H_3 + (1/2)B_2 \wedge F_3$ is constrained to be self-dual ($\tilde{F}_5 = *\tilde{F}_5$) in the equations of motion. Including also the Chern–Simons actions $S_{\text{CS}}^{\text{IIA}} = -(4\kappa_{10}^2)^{-1} \int B_2 \wedge F_4 \wedge F_4$ and $S_{\text{CS}}^{\text{IIB}} = -(4\kappa_{10}^2)^{-1} \int C_4 \wedge H_3 \wedge F_3$, the bosonic part of the low-energy type-II string actions reads

$$\boxed{S_{\text{IIA}} = S_{\text{NS-NS}}[\bar{g}, \Phi, B_2] + S_{\text{R-R}}^{\text{IIA}}[B_2, C_1, C_3] + S_{\text{CS}}^{\text{IIA}},} \tag{12.74}$$

$$\boxed{S_{\text{IIB}} = S_{\text{NS-NS}}[\bar{g}, \Phi, B_2] + S_{\text{R-R}}^{\text{IIB}}[B_2, C_0, C_2, C_4^{(+)}] + S_{\text{CS}}^{\text{IIB}}.} \tag{12.75}$$

Notice that these expressions are in the string (Jordan) frame, hence the use of bars as in (12.40).

12.2.8 Branes

Strings are not the only content of the theory. In general, type-IIA and type-IIB theory display, respectively, forms of odd and even order, corresponding to branes of even and odd order. In the above table, we appreciate that type-IIA theory has two odd-order forms, C_1 and C_3, while type-IIB theory has three even-order forms, C_0, C_2 and $C_4^{(+)}$. This pattern in the p-form content is extended to all admissible orders. For a $(p+1)$-form, let $F_{p+2} = dC_{p+1}$ be its strength (we use the compact notation for differential forms introduced in Sect. 9.1.1). Higher R-R fields are Hodge dual to lower R-R fields, since $*F_{p+2} = \tilde{F}_{D-2-p} = dC_{D-3-p}^{\text{M}}$. To distinguish fundamental forms from their duals, we call the first "electric" C_{p+1}^{E} forms and the second "magnetic" C_{n+1}^{M} forms.

A $(p+1)$-form naturally relates to a $(p+1)$-dimensional world-volume $\mathcal{V}_{p+1}$, in the sense that the integral $\int_{\mathcal{V}_{p+1}} C_{p+1}$ is covariant and well defined. Therefore, electric and magnetic forms are associated with extended objects [153–155]. In string theory, these are, respectively, Dp-branes and Dn-branes with $n = D-4-p$. The strength of an electric C_{p+1}^{E} form defines the Dp-brane charge

$Q_p := \int_{S^{D-2-p}_\infty} *F_{p+2}$ and its magnetic dual $g_{D-4-p} := \int_{S^{p+2}_\infty} F_{p+2}$, where these integrals are defined on a hypersphere at infinity (the ideal boundary of the brane). The DBI brane action (12.19) is augmented by the coupling between the brane and its $(p+1)$-form, represented by the Chern–Simons action $\mathrm{i}\tilde{\mathcal{T}}_p \int_{\mathcal{V}_{p+1}} C_{p+1}$. On a curved background with the forms Φ, $B_{\mu\nu}$ and $F_{\mu\nu}$, one has

$$\mathcal{S}_p = \mathcal{S}_p^{\mathrm{DBI},\bar{g}} + \mathrm{i}\tilde{\mathcal{T}}_p \int_{\mathcal{V}_{p+1}} C_{p+1} \wedge \mathrm{e}^{B_2^{(p)}+2\pi\alpha' F_2^{(p)}}, \tag{12.76}$$

where $\mathcal{S}_p^{\mathrm{DBI},\bar{g}}$ is given by (12.42) and $B_2^{(p)}$ and $F_2^{(p)}$ are the projections B_{ab} and F_{ab} of $B_{\mu\nu}$ and $F_{\mu\nu}$ on the brane.

For each $D=10$ type-II theory, the relations

$$\boxed{\mathrm{D}p\text{-brane} \leftrightarrow F_{p+2} = \mathrm{d}C^{\mathrm{E}}_{p+1} \quad \longleftrightarrow \quad \tilde{F}_{8-p} = \mathrm{d}C^{\mathrm{M}}_{7-p} \leftrightarrow \mathrm{D}(6-p)\text{-brane}} \tag{12.77}$$

eventually translate into the presence of all possible stable Dp-branes with world-volume $\mathcal{V}_{p+1}$. For instance, for $p=2$ one has $*(\mathrm{d}C^{\mathrm{E}}_3) = *F_4 = \tilde{F}_6 = \mathrm{d}C^{\mathrm{M}}_5$, so that C_5 is dual to C_3 and both D2- and D4-branes are automatically included in type-IIA theory. The dynamics of branes is governed by the action (12.76) appropriately decorated with fermionic fields.

To summarize, string theories are populated by the following branes [156]:

- Type-I string: D1-, D5- and D9-branes. The D9-brane is the whole target spacetime and is associated with freely propagating open strings.
- Type-IIA string: D0-, D2-, D4-, D6- and D8-branes (p even). The D8-brane is associated with a 9-form with constant field strength.
- Type-IIB string: D(−1)-, D1-, D3-, D5-, D7- and D9-branes (p odd). The D(−1)-brane is an instanton localized in space and time. D1-branes are $(1+1)$-dimensional objects called D-strings to differentiate them from fundamental (or F-) strings. D3-branes are called dyons, objects coupling electrically and magnetically at the same time.
- Heterotic string: no D-branes.

The 2-form associated with F-strings is the NS-NS Kalb–Ramond field B_2. The magnetic dual of an F-string is a five-dimensional soliton called Neveu–Schwarz 5-brane, NS5-brane in short. Its action is similar to (12.42) but with $\sqrt{-\det(\bar{g}+B_2)} \to \sqrt{-\det(\bar{g}+g_{\mathrm{s}}C_2)}$.

The dynamical degrees of freedom of all these branes are encoded in fundamental open strings attached to them. Branes of odd (respectively, even) order in type-IIA (type-IIB) string theory are characterized by a tachyonic scalar mode in the open-string spectrum which marks an instability. For instance, D5-branes are stable in type-IIB theory but unstable in type-IIA theory. Unstable branes naturally decay into branes of lower dimensionality according to some specific rules [157, 158].[4] This phenomenon is called *tachyon condensation* [161–180] and is described by the rolling of the tachyon from a local maximum of its potential (corresponding to the unstable brane configuration) down to the local minimum representing the lower-dimensional stable brane, where the tachyon decays into particles. A possible role of the tachyon in cosmology will be discussed in Sect. 13.7.2.

12.2.9 Superstring Field Theory

While there is only one proposal for the bosonic open string field, there are many for open superstring field theory. The first by Witten [181] was later modified in [182–184], depending on whether a certain picture-changing operator in the action is chiral and local [183, 184] or non-chiral and bilocal [182, 185, 186]. A non-polynomial version of open super-SFT is due to Berkovits [187, 188]. The Ramond sector has been implemented only very recently in a new theory with connections both with Witten's and Berkovits' proposals [189–193].

The non-local effective action for the tachyon, representing an unstable brane, has been constructed only for the non-chiral version [182, 185, 186]. In that case, at lowest order the interaction is slightly more complicated than that of (12.47):

$$\left(\Box + \frac{1}{2\alpha'}\right) e^{-2\Box/M_\mathrm{o}^2}\tilde{\phi} - \frac{e^{4/(\alpha' M_\mathrm{o}^2)}}{9}\tilde{\phi} e^{2\Box/M_\mathrm{o}^2}\tilde{\phi}^2 = 0\,. \tag{12.78}$$

The general form of the action (12.52) is the same as for the bosonic case, except that now the potential U depends on the fields $\tilde{f}_i$ further dressed by non-local operators.

For Berkovits' SFT, the low-energy spacetime limit has never been derived while retaining non-locality, so that it is not obvious whether the exponential operator would arise also in that case. However, several dualities between different open SFTs are known, including a mapping between supersymmetric and bosonic classical solutions [194, 195], cubic and Berkovits' supersymmetric SFTs (where classical solutions are mapped onto one another) [196, 197] and between different polynomial supersymmetric SFTs [198, 199]. In this sense, we can regard the non-locality (12.78) as a generic feature of open superstring field theory at the effective level.

Supersymmetric versions of both type-II and heterotic closed SFT have been studied comparatively less [200–203] but very recently there have been exciting developments. The computation of multi-loop off-shell amplitudes in covariant superstring theory is at a closer hand than before [131, 132] and so is a complete

[4]These rules, called Sen descent relations, were born as conjectures but they have been proven both numerically and analytically in string field theory [159, 160].

formulation of type-II and heterotic closed super-SFT [204–208]. One expects to find the same non-local structure in the low-energy effective spacetime actions.

The type of non-locality in (12.46) and (12.52) is particularly benign. Not only is the Cauchy problem of the initial conditions well defined in general [209, 210], but it also allows one to study tachyon condensation in all its phases [211–213]. Exact solutions of the full theory are available only for marginal deformations, i.e., configurations where the brane is starting to decay and the tachyon has just begun to roll down the potential [214]. At the level of target spacetime, one can go beyond marginal deformations and there exist approximate but very accurate non-perturbative solutions which describe the whole brane decay [212]. These properties are a direct reflection of the gauge symmetries of SFT [213].[5]

12.3 Compactification

The multiple derivation of the value of the spacetime critical dimension (Sects. 12.1.3 and 12.2.2) is of utmost importance since it is a prediction of a well-defined quantum theory of strings. The bosonic string can be consistently quantized in a Lorentz-invariant spacetime only when such spacetime has $D = 1 + 25$ directions, while the quantum superstring is consistent in $D = 1 + 9$. Such a prediction, unavailable in most of the quantum-gravity frameworks of the preceding chapters, is accessible thanks to the rigidity of CFTs and the virtual absence of fundamental free parameters. However, one has to explain the discrepancy with respect to the observed $D = 1 + 3$.

12.3.1 T-Duality

The most immediate possibility is to compactify 22 (in the bosonic case) or 6 (in the supersymmetric one) spatial directions. Consider for instance the compactification of one such direction on a circle S^1 with radius r, so that the direction $D - 1$ is periodic with period $2\pi r$:

$$X^{D-1} \cong X^{D-1} + 2\pi rn\,, \qquad n \in \mathbb{Z}\,. \tag{12.79}$$

[5] All known exact solutions in open SFT are superpositions of special states in the Fock space, called surface states, which obey a "diffusion" equation involving only Virasoro and ghost operators. This diffusion equation performs a change of gauge which reparametrizes a trivial non-normalizable solution of the equation of motion into a non-trivial normalized solution. At the level of spacetime fields, the same structure survives in (12.52) and, in fact, a spacetime diffusion equation holds for non-perturbative approximate solutions [213]. States similar to the open-string surface states exist also in the closed case [215], which is responsible of the fact that the closed-string non-locality in (12.51) is the same as in (12.52).

In this sub-section, we focus on the bosonic sector. The momentum of the center of mass of the string takes discrete values on a compact manifold, so that $p^{D-1} = n/r$. Closed strings also have an extra feature: they can wrap around the circle. This gives rise to another discrete quantity $w \in \mathbb{Z}$, the *winding number*, which counts how many times a closed string winds around the compact direction: $X^{D-1}(\tau, \sigma + 2\pi) = X^{D-1}(\tau, \sigma) + 2\pi r w$. Overall,

$$X^{D-1}(\tau, \sigma) = x^{D-1} + \alpha' \frac{n}{r}\tau + wr\sigma + (\text{oscillations}). \tag{12.80}$$

Again, one can split the closed string into left-moving and right-moving modes as in (12.14). In particular, the quantum momentum operator is

$$\hat{P} = \hat{L}_0 - \hat{\tilde{L}}_0 , \qquad \hat{L}_0 = \frac{\alpha'}{4} p_{\rm L}^2 + \hat{N} , \qquad \hat{\tilde{L}}_0 = \frac{\alpha'}{4} p_{\rm R}^2 + \hat{\tilde{N}} , \tag{12.81}$$

where

$$p_{\rm L,R} = \frac{n}{r} \pm \frac{wr}{\sqrt{\alpha'}} . \tag{12.82}$$

On a physical state and for a non-zero winding number, $\hat{P}\,|\psi\rangle = 0$ no longer implies the level-matching condition: $N - \tilde{N} = nw \neq 0$. These relations reveal a surprising symmetry of closed string theory which goes under the name of *T-duality* [216]: the system is invariant under the simultaneous interchange of the compactification radius with its inverse and of the winding number with momentum:

$$\text{T-duality:} \qquad r \leftrightarrow r' = \frac{\alpha'}{r} , \qquad w \leftrightarrow n . \tag{12.83}$$

When compactifying (12.40) or (12.71) on a circle, $\int d^D x\, e^{-2\Phi} R^{(D)} \propto \int d^{D-1} x\, r\, e^{-2\Phi} R^{(D-1)} + \dots$. T-duality must preserve the physical couplings, implying that $r e^{-2\Phi} = r' e^{-2\Phi'}$ and that the dilaton must transform as $\Phi' = \Phi - \ln(r/\sqrt{\alpha'})$.

The limit of decompactification $r \to +\infty$ is physically equivalent to sending the radius to zero. Therefore, the self-dual radius $r = \sqrt{\alpha'} = l_s$ sets the characteristic scale of perturbative string theory, as can also be seen via other arguments [217]. Below this scale, there is structure accessible with non-perturbative techniques [35]. The same arguments can be replicated for the other directions and T-duality is extended to the compact manifold $\mathbb{T}^{D-4} = S^1 \times \cdots \times S^1$, the $(D-4)$-dimensional torus.

Open strings do not have a conserved winding number because they can always be unwrapped from the compact space. Since w is T-dual to momentum, the center-of-mass momentum is not conserved either, which corresponds to a breaking of translation invariance. This breaking is due to the confinement of string end-points to branes. The limit $r \to 0$ is dual to the decompactification limit, so that open strings compactified on a torus of infinitely small radius are dual to strings in D dimensions attached to Dp-branes. This picture can be made rigorous in several ways.

12.3.2 Spontaneous Compactification

Although compactification can reduce the number of large spacetime dimensions down to four, one should also find a mechanism which somehow forced the extra dimensions to become compact. In other words, a viable compactification scheme must occur in the theory in a "natural" manner. This requirement can be characterized, first of all, by classical solutions to the equations of motion such that fields acquire position-dependent vacuum expectation values along certain directions where spacetime is strongly curved. In this case, the line element can be decomposed into a four-dimensional line element (times an overall positive function of the coordinates) plus a contribution from the extra dimensions:

$$\mathrm{d}^2 s_D^2 = f^2(x)\,\mathrm{d}s_4^2 + \mathrm{e}^{2u(x)}\mathrm{d}s_{D-4}^2\,, \tag{12.84}$$

for some functions f and u. If such solutions exist and are stable, the compactification is said to be *spontaneous* [218, 219].

However, this is not sufficient to obtain a realistic model of Nature. At the classical level, the compactified theory must have the correct global and local symmetries: it must recover both general-covariant Einstein gravity and the $SU(3)\otimes SU(2)\otimes U(1)$ Lorentz-invariant Standard Model, eventually embedded in a supersymmetric extension, with the correct number of generations of fermions and bosons. Undesired features such as a large effective cosmological constant should also be avoided. At the quantum level, the dimensionally reduced theory should be free from tachyons (meaning that quantization has been performed on a stable vacuum) as well as gauge and gravitational anomalies, which would result in a spoiling of symmetries and the appearance of negative-norm states. We will come back to spontaneous compactification and its relation with the observed value $D = 4$ in Sect. 12.4.

12.3.3 Calabi–Yau Spaces and Orbifolds

The stringent requirements listed in Sect. 12.3.2 are not met in the bosonic string and they provide further motivation for the superstring. There, not only is the dimensionality of target spacetime reduced from 26 to 10, but the remaining 6 extra directions can be compactified with a wealth of tools that allow for a comprehensive classification of those spaces which comply with our wish-list. The structure of the compact spatial manifold depends on a set of non-fundamental parameters for which there is a freedom of choice. This freedom will make string theory lose some predictive power but one will be able to get realistic phenomenological models of Nature.

The simplest compactification of the low-energy limit of superstring theories to four dimensions is on the flat hypertorus [102, 110, 220],

$$\mathcal{M}_{10} \cong \mathcal{M}_4 \times \mathbb{T}^6 . \tag{12.85}$$

When dimensionally reduced to $D = 4$, $SO(32)$ super-Yang–Mills theory (i.e., type-I superstring when $\alpha' \to 0$) generates $\mathcal{N} = 4$ super-Yang–Mills, while both non-chiral and chiral $D = 10$, $\mathcal{N} = 2$ SUGRA (type-IIA and IIB superstrings when $\alpha' \to 0$) both collapse to the $\mathcal{N} = 8$ SUGRA limit [142, 143, 221].[6] These relations are valid also beyond tree level [102]. If one does not require the dynamical effective field equations to be solved exactly, these four-dimensional limits can contain any number of fermion families.

Other types of compactifications, more involved than that on a 6-torus, are necessary to obtain realistic effective quantum field theories complying with the requirements of Sect. 12.3.2, valid at energies above those probed so far in accelerators ($\lesssim$10 TeV) but much lower than the string mass scale $m_s = l_s^{-1}$. In particular, it is important to find compactification schemes where the classical string equations are exactly soluble and the four-dimensional theory is anomaly-free. An early instance where anomaly cancellation is preserved in the dimensional reduction [136] consisted in compactifying four of the ten dimensions as $\mathcal{M}_{10} \cong \mathcal{M}_6 \times \mathrm{K3}$, where the compact space K3 is Kummer's quartic surface [222, 223]. To achieve this, it was not necessary to solve the classical string equations exactly. Two dimensions of the manifold $\mathcal{M}_6$ can be further wrapped but simple compact spaces such as $\mathbb{T}^2 \times \mathrm{K3}$ do not give rise to viable four-dimensional phenomenology.

A powerful way to restrict the allowed compact spaces is to sharpen the phenomenological demands imposed on the four-dimensional manifold $\mathcal{M}_4$. In particular, we may want to obtain the minimal extension of the Standard Model on a maximally symmetric spacetime (Minkowski, de Sitter or anti-de Sitter). These conditions are equivalent to have unbroken $\mathcal{N} = 1$ supersymmetry on $\mathcal{M}_4$ and $R_{acbd} \propto g_{ab}g_{cd} - g_{ad}g_{bc}$, where $a, b, \dots$ are the coordinates on $\mathcal{M}_4$. Then, starting from heterotic SUGRA in ten dimensions, the compactification scheme is uniquely selected as [224, 225]

$$\boxed{\mathcal{M}_{10} \cong \mathcal{M}_4 \times \mathcal{C}_3} \tag{12.86}$$

(up to overall conformal factors [226]), where $\mathcal{M}_4 = M_4$ is Minkowski spacetime (thus, the cosmological constant is zero) and $\mathcal{C}_3$ is a compact six-dimensional

[6] A different compactification reduces $D = 11$, $\mathcal{N} = 1$ SUGRA to $D = 10$, $\mathcal{N} = 1$ SUGRA, in turn reduced to $D = 4$, $\mathcal{N} = 4$ SUGRA [149].

Calabi–Yau space. Calabi–Yau spaces $\mathcal{C}_n$ are Kähler manifolds (Sect. 5.12) with n complex dimensions ($2n$ real dimensions) endowed with an $su(n)$ connection ($SU(n)$ holonomy).[7] In the present case, $n = 3$. The Riemannian and complex structures are mutually compatible and the metric can be defined via a Kähler potential, as described in Sect. 5.12. A consequence of having an $su(3)$ connection is that the Ricci tensor R_{mn} of $\mathcal{C}_3$ vanishes identically. Applying the dimensional reduction (12.86) to type-II theories, one ends up with $D = 4$, $\mathcal{N} = 2$ SUGRA; the four-manifold $\mathcal{M}_4$ in (12.86) can be more general than Minkowski spacetime and have a non-vanishing cosmological constant. Calabi–Yau spaces can be compact or non-compact.

K3 is a Calabi–Yau 2-fold and, in fact, it is the only compact simply-connected 2-fold. If the definition of Calabi–Yau spaces is extended also to manifolds with proper sub-groups of $SU(n)$, then $\mathbb{T}^6$ and $\mathbb{T}^2 \times$ K3 are compact 3-folds (which we have already discarded on phenomenological grounds). A less trivial example is the following. Let CP^{n+1} be the complex projective hyperplane, the space of $n+2$ complex variables $z_i \neq 0$ identified under dilations, $z_i \cong \lambda z_i$ for any $\lambda \in \mathbb{C} \setminus \{0\}$. This is a Kähler manifold and so is any sub-space defined by analytic equations on the z_i. In particular, K3 $\cong \mathrm{CP}^3$. The space of all the zeros of a homogeneous polynomial $P_{n+2}(z_i)$ of order $n+2$ in CP^{n+1} is a compact Calabi–Yau n-fold.

This and other Calabi–Yau spaces constructed from complex projective hyperplanes are discussed in the literature, together with the following example not stemming from CP^{n+1} [224]. Let us parametrize the torus $\mathbb{T}^2$ with a complex coordinate z under the identifications $z \cong z + 1 \cong z + \mathrm{e}^{\mathrm{i}\pi/3}$. This set is invariant under the $\mathbb{Z}_3$ symmetry generated by the transformation $\alpha(z) = \mathrm{e}^{2\mathrm{i}\pi/3} z$. On $\mathbb{T}^2$, this transformation has three fixed points, $z = 0, \mathrm{e}^{\mathrm{i}\pi/6}/\sqrt{3}, 2\mathrm{e}^{\mathrm{i}\pi/6}/\sqrt{3}$. On the 6-torus $\mathbb{T}^6 \cong \mathbb{T}^2 \times \mathbb{T}^2 \times \mathbb{T}^2$, the transformation $\alpha(z_1)\alpha(z_2)\alpha(z_3)$ generates a $\mathbb{Z}_3$ symmetry and has a total of $3^3 = 27$ fixed points. Let $\mathcal{F}$ be the set of these fixed points and $\tilde{T} = \mathbb{T}^6 \setminus \mathcal{F}$. Then, the quotient $\tilde{T}/\mathbb{Z}_3$ is a non-compact complex manifold which can be rendered compact by gluing, at the position of each singularity (former fixed point), copies of a non-compact manifold with certain properties and the correct asymptotic behaviour. The result, called Z, is a simply-connected Calabi–Yau 3-fold.

12.3.3.1 Orbifolds and Orientifolds

The explicit construction of Calabi–Yau spaces may pass through an intermediate step known as *orbifold*, a space $\mathcal{M}/\mathbb{G}$ (not necessarily a manifold) obtained by the quotient of a smooth manifold $\mathcal{M}$ with a discrete group $\mathbb{G}$ [231, 232]. The name orbifold comes from the fact that, by definition, for any group element $g \in \mathbb{G}$ and point $y \in \mathcal{M}$ there is a congruence class $y \cong gy = \alpha(y)$, such that each point in

[7] Kähler n-folds have a $U(n)$ holonomy. The group $U(n) \cong SU(n) \otimes U(1)$ can be reduced to $SU(n)$ under special provisions, as conjectured by Calabi [227, 228] and proved by Yau [229, 230].

$\mathcal{M}/\mathbb{G}$ is identified with its orbit $\{gy \,|\, g \in \mathbb{G}\}$ under the action of the group. Here and in the following, y_l are real coordinates on an orbifold or Calabi–Yau space. In physical applications, $l, m, n = 4, 5, \ldots, 9$ and the four-fold $\mathcal{M}_4$ is spanned by the coordinates x^a.

A trivial example of a smooth orbifold is the hypertorus $\mathbb{T}^N \cong \mathbb{R}^N/Z^N$, where Z^N is the discrete group of translations. Often, a less broad definition of orbifold is employed where the action of $\mathbb{G}$ is not free. In this case, the transformation $\alpha(y)$ has fixed points which constitute conical singularities in $\mathcal{M}/\mathbb{G}$. For instance, consider the discrete group $\mathbb{Z}_2$ defined by reflections, $\alpha(y) = -y$. On a unit circle S^1 parametrized by $y \in [0, 2\pi]$, the only two fixed points are $y = 0$ and $y = \pi$, so that $S^1/\mathbb{Z}_2$ is an orbifold with two singularities.[8]

Amusingly, the flat tetrahedron so much used in quantum gravity is an orbifold [233]: it is $\mathbb{T}^2/\mathbb{Z}_2$, where the origin of the reflection is the centre of the fundamental lattice cell defining the torus. $\tilde{T}/\mathbb{Z}_3$ is another example of orbifold. Asymmetric orbifolds where the left- and right-moving sectors live on different spaces can also be conceived [234].

A generalization of orbifold of importance for type-I theory is the *orientifold*, an orbifold where the twist symmetry (12.36) is implemented [235]. Thus, strings on orientifolds are unoriented. Type-I superstring theory is obtained by a $\mathbb{Z}_2$ twist on the left and right modes of the closed string [235–238]. The twist mixes the two sectors of the theory, which have therefore to be symmetric. In particular, left and right sectors must have the same chirality and the closed string from which the open one descends must be of type IIB. Colloquially, the procedure to get the type-I superstring may be described as placing type-IIB theory on an orientifold. The bosonic open string descends in a similar way from the bosonic closed string.

The regularization (or "blowing up") of the singularities of orbifolds can lead to Calabi–Yau spaces, as in the case of Z. Similarly, K3 can be obtained by regularizing the 16 singularities of the orbifold $(\mathbb{T}^4 \setminus \mathcal{F})/\mathbb{Z}_2$, where the group (isomorphic to) $\mathbb{Z}_2$ is generated by the transformation $\alpha(y_l) = -y_l$ on the periodic coordinates $y_l \cong y_l + 1$ on the four-torus [231]. Strings can wrap around singularities, which gives rise to twisting modes apart from the usual winding modes in non-trivial topologies.

The regularization of singularities is a procedure more general than its application to orbifolds and can produce compact Calabi–Yau spaces from non-compact ones. An important example is the *Klebanov–Strassler throat* [239], a conical singularity [240] on a non-compact Calabi–Yau space. Deforming the throat to a cone with a rounded tip, one obtains a compact Calabi–Yau n-fold with smooth geometry [241]. The deformed Klebanov–Strassler throat is an important ingredient in flux compactification and cosmological scenarios, as we will describe in Sects. 12.3.7 and 13.5.

An advantage of dealing with orbifolds is that, on one hand, there are not many six-dimensional orbifolds compatible with the string and spacetime symmetries.

[8]The discrete group $\mathbb{Z}_N$ is defined by rotations of a vector by an angle of $2\pi/N$ around the origin, so that it has N fixed points on a circle.

On the other hand, the study of compactification schemes becomes fairly tractable and singularities are relatively innocuous insofar as the string spectrum, dynamics [232, 233] and interactions [233, 242] are concerned. On Z, on its generalizations and on other Calabi–Yau spaces, the string classical equations are soluble with arbitrarily good approximation (i.e., exactly soluble on the corresponding orbifolds [231–233, 243]), which is one of the requisites for a spontaneous compactification. When compactifying the $E_8 \times E_8$ heterotic string on some of these spaces, one can obtain effective four-dimensional models with an almost realistically low number of fermion generations (2, 4 or 5, but not 3) [224, 225, 232].

In the light of the web of dualities described in Sect. 12.4 between string theories on one hand and an 11-dimensional theory of supergravity and membranes on the other hand, compactification schemes from 11 to 4 dimensions are also of interest. In that case, the compact space can be constructed from a Calabi–Yau four-fold C_4 in the limit where one of the extra dimensions is unfolded [244]. For certain choices of the seven-dimensional compact manifold, one can obtain $\mathcal{N} = 1$ or $\mathcal{N} = 2$ supersymmetry in $D = 4$ [245].

12.3.4 Cycles and Fluxes

Calabi–Yau spaces usually have compact n-dimensional sub-spaces Γ_n called *n-cycles*. The integrals

$$\frac{\tilde{\mathcal{T}}_p}{2\pi} \int_{\Gamma_{p+2}} F_{p+2} = k_F \in \mathbb{Z} \tag{12.87}$$

of field strengths of the NS-NS sector and of the R-R potentials, performed on non-trivial cycles Γ_{p+2} of the compact manifold, are called $(p+2)$-form *fluxes* and play a crucial role in compactification schemes.[9] An analogue of the Dirac quantization condition on the electric and magnetic charge of Maxwell's theory holds for branes, so that fluxes are quantized by an integer k_F (the "quantum number" of the brane flux). This explains the appearance of the Dp-brane tension $\tilde{\mathcal{T}}_p = g_s \mathcal{T}_p = [(2\pi)^p \ell_s^{p+1}]^{-1}$ in (12.87) (the left-hand side of (12.87) is dimensionless, since $[F_{p+2}] = 1$).

In the universal supergravity sector, one has a 1-form flux associated with the dilaton and the 3-form flux of the Kalb–Ramond field strength $H_3 = \mathrm{d}B_2$; this is the quantum number associated with fundamental strings. The R-R fluxes are sourced by the p-branes of the theory. For instance, in type-IIB string theory, the R-R fluxes are the surface integrals of the 1-, 3- and 5-forms associated with, respectively, the 0-, 2- and 4-form R-R potentials. D3-branes are a source for the R-R electric and

[9] An informal use of the term "flux" often applies to the field strengths themselves.

magnetic 5-form flux, D5-branes (more precisely, their dual D1-branes) source the R-R magnetic 3-form flux, and so on.

12.3.5 Moduli

The shape and size of Calabi–Yau spaces are governed by a set of parameters called, as in the case of Riemannian surfaces, *moduli* [249]. These parameters are actually fields, since they depend on the position on $\mathcal{C}_3$. In the most general stringy use of the word, moduli are spacetime-dependent fields in 10 dimensions with a well-defined Lorentz structure. Upon compactification, they usually reduce to a number of scalars on the four-dimensional manifold $\mathcal{M}_4$. In multi-brane configurations, moduli are also the separations of branes. These will be extensively used by some important cosmological models of Chap. 13.

We left the discussion in Sect. 5.12 at the point where one had a supergravity no-scale model of the form (5.230) but apparently no justification for its origin from first principles. Supersymmetry is spontaneously broken at the tree level at a scale which cannot be determined due to an invariance of the potential under field rescalings, hence the name "no-scale." String theory completes the picture in a remarkable way: compactifying the $D = 10$ supergravity action on a Calabi–Yau space $\mathcal{C}_3$, the resulting low-energy four-dimensional model is precisely (5.230). This was first shown for $E_8 \times E_8$ SUGRA [246–248]. The classical scale invariance of the SUGRA action is an inheritance of conformal invariance of the superstring vertex diagrams.

The fields which appear in the super- and Kähler potential are the string moduli. In this sub-section, we overview the main types of moduli without entering into precise technical details, which the reader can find in [246, 247, 249, 250] (a more pedagogical introduction is in [251]). To classify the spacetime moduli on $\mathcal{M}_4$, we consider the metric decomposition

$$\mathrm{d}s_{10}^2 = \bar{g}_{\mu\nu}\mathrm{d}x^\mu\mathrm{d}x^\nu = \mathrm{e}^{-6u(x)}\,\mathrm{d}s_4^2 + \mathrm{e}^{2u(x)}g_{mn}(y)\mathrm{d}y^m\mathrm{d}y^n\,, \tag{12.88a}$$

$$\mathrm{d}s_4^2 = g_{ab}(x)\mathrm{d}x^a\mathrm{d}x^b\,, \tag{12.88b}$$

where we have assumed for simplicity that all compactification radii are equal and u is a scalar related to the radius by $r = l_\mathrm{s}\exp u$. The metric g_{ab} is the one in the Einstein frame (hence the Weyl prefactor e^{-6u}). Energy units are absorbed in the definitions so that all moduli are dimensionless.

12.3.5.1 Axions

String axions are real pseudo-scalar fields which arise from the p-forms (the NS-NS and R-R potentials) of the theory [252]. In type-IIB string compactifications, where most of the cosmological model we will see in Chap. 13 live, we have:

- The R-R 0-form C_0.
- The axion $\tilde{\phi}_B$ of the universal SUGRA sector, coming from the $D = 4$ Hodge dual of the four-dimensional NS-NS 2-form $B_2^{(4)} = B_{ab}\mathrm{d}x^a \wedge \mathrm{d}x^b$: $\mathrm{d}\tilde{\phi}_B = \mathrm{e}^{-8u} * \mathrm{d}B_2^{(4)}$.
- The axion $\tilde{\phi}_C$ coming form the $D = 4$ Hodge dual of the four-dimensional R-R 2-form $C_2^{(4)} = C_{ab}\mathrm{d}x^a \wedge \mathrm{d}x^b$: $\mathrm{d}\tilde{\phi}_C = \mathrm{e}^{-8u} * \mathrm{d}C_2^{(4)}$.
- The dimensionally reduced R-R 4-form $C_4^{(+)}$ produces a 2-form $a_{ab} \propto C^{(+)}_{abmn}$ and one defines an axion θ as the Hodge dual $\mathrm{e}^{-8u} * \mathrm{d}\theta = \mathrm{d}a_2$. The factor $\exp(-8u)$ pops up after transforming from the string to the Einstein frame (see below).
- The axions arising from the fluxes of the $D = 10$ forms B_2 and C_2 over 2-cycles of $\mathcal{M}_6$ and of $C_4^{(+)}$ over 4-cycles. For N_2 and N_4 such cycles,

$$\theta_{B,i_2} = \frac{1}{2\pi\alpha'}\int_{\Gamma_2^{i_2}} B_2\,, \qquad \theta_{C,i_2} = \frac{1}{2\pi\alpha'}\int_{\Gamma_2^{i_2}} C_2\,, \qquad \theta_{i_4} = \frac{1}{2\pi\alpha'^2}\int_{\Gamma_4^{i_4}} C_4^{(+)}\,, \tag{12.89}$$

where $i_2 = 1, \dots, N_2$ and $i_4 = 1, \dots, N_4$.

Some of these axions (e.g., $\tilde{\phi}_B$ and $\tilde{\phi}_C$) are absent in type-IIB orientifold compactifications. Type-IIA and the heterotic strings have other axionic spectra. In general, we will collect all axions of a model under the same symbol θ_i, $i = 1, \dots, N_{\text{axions}}$ where the total number of axions N_{axions} can be rigorously determined for a given compact manifold and flux population. Typically, N_{axions} is very large.

The actions (12.74) and (12.75) depend only on the field strength of the p-forms, so that axions enjoy the classical shift symmetry

$$\theta_i \to \theta_i + \text{const}\,, \tag{12.90}$$

where the constant is arbitrary. Actually, the shift (12.90) is the definition of classical axion. However, at the quantum level non-perturbative instantonic effects, such as Euclidean branes and the wrapping of string world-sheets on non-trivial cycles, break this continuous symmetry to a discrete one, $\theta_i \to \theta_i + 2\pi n$ [253, 254]. The period 2π can be intuitively understood by noting that the world-sheet wrapping on a 2-cycle is governed by a Euclidean instanton action $\mathcal{S}$ given by the exponentiation of the world-sheet action of the 2-form B_2, the imaginary term in (12.37). In practice, the exponent is a 2-form flux on the world-sheet Σ_2 where the integrand is a combination of B_2 and another (the Kähler) 2-form. With the normalization convention as in (12.89), $\mathcal{S} \propto \exp[-\mathrm{i}(2\pi\alpha')^{-1}\int_{\Sigma_2} B_2] = \mathrm{e}^{-\mathrm{i}\theta_B}$. Therefore, θ_B has a periodic potential with period 2π (restoring length units, $2\pi l_s$). We have seen an example of such potential already in Chap. 5: it is the cosine profile (5.90) of natural inflation. We will come back to string axions and their potentials in Sect. 13.4.

Axions appear not only as independent moduli but also as components of the complex-structure moduli, the axio-dilaton and the Kähler moduli.

12.3.5.2 Complex-Structure Moduli

For a Calabi–Yau space, *shape moduli* or *complex-structure moduli* are coordinates $z_i(x)$ locally parametrizing the complex structure. If $\mathcal{C}_3$ has handles, cycles or other non-trivial topology features, the number of complex-structure moduli can be very large.

The Kähler potential associated with the shape moduli is [241, 249]

$$K_{\rm cs}(z_i) = -\ln\left(-i\int_{\mathcal{C}_3}\Omega\wedge\Omega^\dagger\right), \tag{12.91}$$

where Ω_{lmn} is the holomorphic $(3,0)$-form of the Calabi–Yau space. In special geometry, $K_{\rm cs}$ depends only on the imaginary part of shape moduli and the real parts ${\rm Re}z_i$ are axions (see, e.g., [255]).

12.3.5.3 Axio-Dilaton

String axions can be combined with other real scalars into complex moduli, so that the low-energy effective action becomes a simple expression. Here we will see a first example, the type-IIB *axio-dilaton* (or simply dilaton)

$$\tau(x) = e^{-\Phi(x)} + iC_0(x)\,. \tag{12.92}$$

Note that, thanks to the choice of warping factors in (12.88), the ten-dimensional dilaton Φ is the same as the four-dimensional dilaton $\Phi_{\rm 4D}$ (Problem 12.2). Making the conformal transformation $g_{\mu\nu} = \exp(-\Phi/2)\bar{g}_{\mu\nu}$ and defining

$$G_3 := F_3 - i\tau H_3\,, \tag{12.93}$$

the type-IIB low-energy action (12.75) in the Einstein frame becomes

$$S_{\rm IIB} = \frac{1}{2\kappa_{10}^2}\int d^{10}x\sqrt{-g}\left[R - \frac{\partial_\mu\tau\partial^\mu\tau^*}{2({\rm Re}\tau)^2} - \frac{|G_3|^2}{2{\rm Re}\tau} + \dots\right], \tag{12.94}$$

where we wrote only the part of interest for the cosmological models of Chap. 13. Other moduli which we will not meet during our journey are the 2-form scalars $G_{i_2} = \theta_{C,i_2} + \mathrm{i}\tau\theta_{B,i_2}$.

The axio-dilaton contributes to the $D = 10$ Kähler potential with the term [241, 246–249]

$$\boxed{K(\tau) = -\ln(\tau + \tau^*)\,.} \tag{12.95}$$

Often in the literature, the definition $\tau \to -\mathrm{i}\tau$ is employed as well as the symbol S. We will reserve S (as the $D = 4$ chiral-supergravity scalar which appears in (5.230)) to the axio-dilaton of the heterotic string. In this case, there is no axion C_0 and the imaginary part of S is supplemented by the axion $\tilde{\phi}_B$ of the universal SUGRA sector. The heterotic axio-dilaton is

$$S(x) = \mathrm{e}^{-\Phi(x)} + \mathrm{i}\tilde{\phi}_B(x)\,. \tag{12.96}$$

The Kähler potential is as in (12.95), $K(S) = -\ln(S + S^*)$.

12.3.5.4 Kähler Moduli

The *Kähler moduli* determine the scales and the total volume of $\mathcal{C}_3$. A Kähler modulus ϱ is made of a *size modulus* σ and an axion θ.[10] The square of the radius in (12.88), u itself or functions of u such as σ are indistinctly called *radion*. In the case of the isotropic metric (12.88), there is only one radion u and one size modulus $\sigma = \exp(4u)$, related to the volume $\mathcal{V}_6 = \int_{\mathcal{C}_3} \mathrm{d}^6 y\, \sqrt{g^{(6)}}$ of the Calabi–Yau manifold by

$$\mathcal{V} = (\mathrm{Re}\varrho)^{3/2} = \sigma^{3/2}\,, \qquad \mathcal{V} := l_\mathrm{s}^{-6}\mathcal{V}_6\,. \tag{12.97}$$

Then, the Kähler modulus is

$$\varrho(x) = \mathrm{e}^{4u(x)} + \mathrm{i}\theta(x) = \sigma(x) + \mathrm{i}\theta(x)\,, \tag{12.98}$$

where θ was defined among the axions in the list above as the dualized 4-form. In the heterotic case (chiral supergravity), the field ϱ is called T. The Kähler potential

[10] About nomenclature in the literature. The term "size modulus" is often used as a synonym for the composite modulus $\sigma + \mathrm{i}\theta$, while "Kähler modulus" is often (and more correctly) used as a synonym for the size modulus σ, since a general definition of Kähler modulus is the flux of the Kähler form over a 2-cycle, $t_i := \alpha'^{-1} \int_{\Gamma_2} J$ (the σ_i below are functions of t_i). In this book, we distinguish these two denominations as per our declaration above (Kähler modulus $= \sigma + \mathrm{i}\theta$), which appears in the literature as often as the others.

is augmented by the term [241, 246–248]

$$\boxed{\tilde{K}(\varrho) = -3\ln(\varrho + \varrho^*) = -2\ln \mathcal{V}\,,} \tag{12.99}$$

and the total Kähler potential is

$$\boxed{K = K_{\rm cs}(z_i) + K(\tau) + \tilde{K}(\varrho)\,.} \tag{12.100}$$

With these ingredients, it is not difficult to see that the four-dimensional action is

$$S_{\rm IIB}^{4D} = \frac{1}{2\kappa_4^2}\int \mathrm{d}^4x\sqrt{-g^{(4)}}\left[R^{(4)} - 2\frac{\partial_\mu\tau\partial^\mu\tau^*}{(\mathrm{Re}\tau)^2} - 6\frac{\partial_\mu\varrho\partial^\mu\varrho^*}{(\mathrm{Re}\varrho)^2} + \dots\right], \tag{12.101}$$

plus the complex-structure term. Consistently, this action can also be obtained by compactifying the $D = 10$ expression (12.94) on (12.88).

Calabi–Yau spaces can be anisotropic and have six different radions. Moreover, in type-IIB theory they have 2-cycles and 4-cycles and the total volume $\mathcal{V}$ is made of all these contributions. Ignoring 2-cycle terms dependent on the 2-form scalars G_{i_2}, Kähler moduli take the general form

$$\varrho_{i_4}(x) = \sigma_{i_4}(x) + \mathrm{i}\theta_{i_4}(x)\,, \qquad i_4 = 1,\dots,N_4\,, \tag{12.102}$$

where σ_{i_4} is the volume of the i_4-th 4-cycle and θ_{i_4} is defined in (12.89). The Kähler potential remains the same as in (12.99), $\tilde{K}(\mathcal{V}) = -2\ln\mathcal{V}$, where $\mathcal{V} = \mathcal{V}(\varrho_{i_4})$.

12.3.5.5 Stabilization of the Moduli

One of the major goals of modern string theory and string cosmology is to stabilize the moduli. Moduli that do not acquire a fixed value can lead to trouble. The compact manifold can decompactify and a higher-dimensional spacetime unfold. The evolution of the universe can be disrupted at very early stages as well as at late times. A running dilaton, for instance, would not respect the tight observational constraints on the variation of Newton's coupling G.

It turns out that there are two non-perturbative mechanisms that can perform the task of freezing these fields: flux compactification and gaugino condensation. The complex-structure moduli and the axio-dilaton are stabilized when the topology of the Calabi–Yau space allows several flux fields to acquire non-zero expectation values (Sect. 12.3.7). Kähler moduli are not stabilized by fluxes and the problem of radion stabilization requires a different approach: the condensation of the supersymmetric partners of the gauge fields (Sect. 12.3.9).

12.3.6 Stacking Branes

A technique to obtain a four-dimensional field theory with little or no supersymmetry is to stack D-branes and other localized extended objects on compact spaces and invoke the extremely powerful *AdS/CFT correspondence*. The AdS/CFT correspondence [256–258] and its extensions (see [259–261] for reviews) state that a given superstring theory on a suitable background induces a given supersymmetric conformal field theory on the boundary of spacetime. To pass from one theory to the other, one relates the string background to some brane configuration.

Let us first make a short digression about gauge groups and branes. In the case of a single D-brane, one immediately notices that there is a $U(1)$ gauge theory associated with it. Let $g_{ab} = \eta_{ab} + (2\pi l_s^2)^2 \partial_a \phi^m \partial_b \phi_m$ be the induced metric on the brane in flat spacetime, where ϕ^m are $D-1-p$ scalars representing the transverse coordinates of the embedding. Expanding (12.19) in the fields, the quadratic terms combine into the Lagrangian $\mathcal{L}^{(2)} = -g_{\rm YM}^{-2}(F_{ab}F^{ab}/4 + \partial_a\phi^m\partial^a\phi_m/2)$, where the effective Yang–Mills coupling is $g_{\rm YM}^2 = 2(2\pi)^{p-2} l_s^{p-3} g_s$. For D3-branes,

$$g_{\rm YM}^2 = 4\pi g_s\,. \tag{12.103}$$

In superstring theory, additional spinor fields in the DBI action provide the fermionic degrees of freedom of the gauge theory. For N branes, the fields A_μ and ϕ^m become matrices which transform in the adjoint representation of $U(N)$. The $U(1)$ excitations on an individual brane decouple from inter-brane modes (the non-diagonal elements of the matrix fields) and the gauge group is therefore $SU(N)$. In the large-N limit, the product $g_{\rm YM}^2 N$ is kept fixed, which leads to the identification $g_s \sim 1/N$. Thus, on a stack of branes one has a gauge theory. Embedding this stack in a $D = 10$ spacetime is equivalent to consider a string theory in a spacetime (usually the product of anti-de Sitter and a five-manifold) with boundaries. The AdS/CFT insight consists in recognizing that the degrees of freedom on the boundary are described by the gauge theory.

The first established AdS/CFT correspondence is between type-IIB strings on $\mathrm{AdS}_5 \times S^5$ (five-dimensional anti-de Sitter (AdS) spacetime times the 5-sphere) and the large-N limit of $SU(N)$ super-Yang–Mills gauge theory in $D = 4$ with $\mathcal{N} = 4$ supersymmetries (hence 8 supercharges) [256–258]. It is easy to see that $\mathrm{AdS}_5 \times S^5$ spacetime is conveniently described by N parallel D3-branes on a smooth ten-dimensional manifold. An $SU(N)$ super-CFT lives on the brane stack and, in the low-energy limit, it decouples from bulk degrees of freedom. Since the Minkowski spacetime M_4 on which the CFT lives is identified (up to some points at infinity) with the boundary of AdS_5, the AdS/CFT correspondence is *holographic*: in the large-N limit, the physics in the bulk is described by that of its boundary. The dimension of operators in the gauge theory are then related to the modes of the string spectrum, which is of type IIB in this case.

More typically, branes are placed at the singularity of the transverse space [262–266]. In one explicit example, AdS_5 times a certain five-dimensional manifold $\mathcal{M}_5$

is equivalent to a stack of parallel D3-branes at the conical singularity of a non-compact Calabi–Yau 3-fold, which eventually helps to show that type-IIB string theory on $\mathrm{AdS}_5 \times \mathcal{M}_5$ is dual to an $\mathcal{N} = 1$ super-CFT in four dimensions [265]. These correspondences are carried out by using classical SUGRA actions, since string loop corrections go as $g_s \sim 1/N$ and are sub-dominant in the large-N limit.

12.3.7 Flux Compactification

Since D-branes carry fluxes, the stacking of branes on a compact manifold can be regarded as a case of *flux compactification* [226, 241, 267–270]. If the expectation value of gauge forms is non-zero in the compact space, background fluxes are turned on and supersymmetry is partially broken.[11] This helps in reducing the number of supercharges in the final configuration. Moreover, fluxes generate effective potentials for the axio-dilaton and the shape moduli of the compact space. The *Gukov–Vafa–Witten superpotential* associated with G_3 is [244, 272, 273]

$$W = W_G := \int_{\mathcal{C}_3} \Omega \wedge G_3 \,, \tag{12.104}$$

where Ω is the holomorphic form in (12.91) and G_3 is the complex 3-flux (12.93) stemming from the combination of the NS-NS and R-R 3-fluxes together with the $D = 10$ axio-dilaton in type-IIB string theory. In the flux compactification of [241], W_G combines with the Kähler potential (12.100) (given by (12.91), (12.95) and (12.99)) into the analogue of the potentials (5.225) and (5.231):

$$\begin{aligned}\kappa_4^4 V &= \mathrm{e}^K \left[\mathcal{G}^{ij} D_i W_G (D_j W_G)^* - 3|W_G|^2\right]\\ &= \mathrm{e}^K \sum_{i,j\neq\varrho} \mathcal{G}^{ij} D_i W_G (D_j W_G)^* .\end{aligned} \tag{12.105}$$

In the second equality, the sum is over all moduli except ϱ, since W_G is independent of ϱ and the Kähler potential (12.99) is such that the F-term $\mathcal{G}^{\varrho\varrho^*} D_\varrho W_G (D_\varrho W_G)^* = (|\varrho|^2/3)|W_G \partial_\varrho \tilde{K}|^2 = (|\varrho|^2/3)| - 3W_G/\varrho|^2 = 3|W_G|^2$ exactly cancels the term $-3|W_G|^2$. We follow the convention (5.224), so that $[K] = 0 = [W]$ and all moduli fields are dimensionless.

Supersymmetry is spontaneously broken by W_G: for a *given* flux background,

$$\boxed{W_G = W_0 = \mathrm{const} \neq 0 \,.} \tag{12.106}$$

[11] In $D = 11$, vacuum solutions that respect the full supersymmetry on M_4 times a compact space do not admit non-trivial fluxes [271].

Once the flux is fixed, the complex structure z_i and the axio-dilaton τ adjust to minimize the F-terms (12.105) associated with (12.104) [241, 268]. The moduli space is therefore partially stabilized.

The presence of fluxes is important but not sufficient to guarantee a finite scale hierarchy which could explain why the Planck mass $m_{\rm Pl}$ is much larger than the electroweak energy scale. In addition, one must also ensure that spacetime be *warped*. A warped spacetime [274], either non-compact [239, 275] or compact [226, 269, 276–280], is a special case of (12.88) such that the normalization of the four-dimensional metric $g_{ab}(x^a, y^m) = \exp[2\omega(y^m)]\tilde{g}_{ab}(x^a)$ varies in the transverse directions y^m:

$$\mathrm{d}s_D^2 = \mathrm{e}^{2\omega(y)-(D-4)u(x)}\,\mathrm{d}s_4^2 + \mathrm{e}^{-2\omega(y)+2u(x)} g_{mn}(y)\mathrm{d}y^m\mathrm{d}y^n\,, \qquad (12.107\mathrm{a})$$

$$\mathrm{d}s_4^2 = \tilde{g}_{ab}(x)\,\mathrm{d}x^a\mathrm{d}x^b\,. \qquad (12.107\mathrm{b})$$

We have already mentioned an example of warped space: the deformed Klebanov–Strassler throat [241], which we will discuss in Sect. 13.5.2. Typically, Calabi–Yau 3-folds possess several such warped regions and it is easy to place fluxes on a warped background without fine tuning.

The electroweak scale can be generated by the suppression of the D-dimensional Planck scale by the position-dependent warping factor ω. This mechanism can be seen in action, for instance, by plugging the *Ansatz* (12.107) with $\tilde{g}_{ab} = \eta_{ab}$ in the equations of motion of classical $D = 10$ SUGRA with the R-R and NS-NS fluxes ($F_3 = \mathrm{d}C_2$ and $H_3 = \mathrm{d}B_2$) of type-IIB supergravity, and then finding solutions $\omega(y)$ and $g_{mn}(y)$ on a compact manifold [241]. For a smooth energy-momentum tensor, all solutions on compact warped spacetimes are such that all fluxes vanish and ω is constant [276, 277]. To obtain non-trivial solutions, it is necessary to include localized matter sources such as D-branes and objects with negative tension called Op-planes (the "O" stands for orientifold); their $(p+1)$-dimensional world-volume actions are hence added to the low-energy SUGRA action. The solutions of [241] have vanishing four-dimensional cosmological constant and are invariant under the rescaling $g_{mn} \to \lambda^2 g_{mn}$. Non-supersymmetric solutions (more generic than supersymmetric ones, which have $\mathcal{N} = 1$ in this case) with $\Lambda = 0$ and a dynamical radial modulus u (in practice, the Kähler modulus ϱ) are nothing but no-scale models, as it can be checked by noting that the superpotential generated by the 3-fluxes of the solutions obeys the condition (5.223).

No-scale models are, in general, scenarios where size (Kähler) moduli are unfixed. In all solutions, a large hierarchy of scales can be generated by reasonably small flux quantum numbers, since the warp factors in (12.107) depend exponentially on these. Both the flux content and the presence of localized sources is determined by the string and brane spectra of superstring theory, which thus provides a concrete realization of the models of Sect. 5.12. Notice, however, that α'-corrections to the SUGRA action, quantum corrections and non-perturbative effects can spoil the no-scale properties of solutions and hence stabilize the radion (Sect. 12.3.9).

There exist a wealth of flux compactifications involving different compact spaces and superstring theories (often related by the dualities described in Sect. 12.4) [226, 241, 267, 268, 281–296]. The goal, in general, is to promptly stabilize all the moduli and to obtain as much a realistic low-energy four-dimensional field theory as possible.

12.3.8 String Theory and the Standard Model

In the quest for the Standard Model, much of the attention has been drawn to models in type-I and heterotic string theory [297–303] and, especially, to type-II orientifold compactifications with branes intersecting at angles. Orientifolds are a somewhat special class of Calabi–Yau spaces but they permit to obtain controllable realizations of both the non-supersymmetric Standard Model with three generations [304–318] and its minimal $\mathcal{N} = 1$ supersymmetric extension [304, 319–335]. At the intersection of D-branes [336], chiral spinorial degrees of freedom emerge which mimic the fermionic families of the $SU(3) \otimes SU(2) \otimes U(1)$ Standard Model.

Usually, the result is not completely satisfactory due to the presence of a "hidden sector" of extra particles that cannot be removed from the spectrum. When the hidden sector has masses above the present observational limits, it becomes either innocuous or a valuable asset, inasmuch as it can give characteristic predictions that could be tested in accelerators. However, if these particles have light masses, the model is regarded only as semi-realistic. In their open-string sector, non-supersymmetric models closely resemble the Standard Model but they have the additional problem that their vacuum is not protected from instabilities by exact or softly-broken supersymmetry. The consequence is that some moduli in the closed-string sector (including the dilaton) are not stabilized [308]. Moreover, it is difficult to reconcile particle-physics models with cosmology, mostly because the parameter ranges required by moduli stabilization and by a small positive Λ do not overlap.

All these reasons stimulate the further study of the embedding of the Standard Model in string theory in compactification schemes with background fluxes and brane stacks, intersecting branes or branes at singularities [289, 295, 337–352]. A very important element of the discussion, related to the way moduli are stabilized, is how to determine the vacuum of the theory in such schemes. In the next sub-section, we will tackle this problem in general terms, without reference to specific proposals for a Standard Model.

12.3.9 Anti-de Sitter Vacua

In type-IIA superstring theory, all moduli can be stabilized by fluxes [291–296]. The resulting $D = 4$ stable vacua are either non-supersymmetric or $\mathcal{N} = 1$ supersymmetric, with a negative cosmological constant. In type-IIB and heterotic theories, classical fluxes alone cannot give a mass to Kähler moduli and one has to include quantum correction to the super- and Kähler potentials. String loop corrections to the Kähler potential [353] can give a mass to all the moduli in type-IIB theory to yield an AdS vacuum [354], but at the price of some fine tuning. Barring this perturbative effect, it is necessary to introduce a non-perturbative mechanism which we describe presently. As in the type-IIA case, the resulting stable vacua are anti-de Sitter.

12.3.9.1 Gaugino Condensation and String Instantons

We have seen examples of condensates in Sects. 7.6.4, 9.4.2 and 11.5.2. Bilinears (fermionic, in the first two cases) acquired a non-trivial expectation value $\langle\psi\bar{\psi}\rangle \neq 0$, leading to a more favourable vacuum configuration and the breaking of certain symmetries of the system. The second of these properties can play a role in the stabilization of Kähler moduli and it is based on the observation that local supersymmetry of SUGRA models can be broken by the condensation of gauginos, the super-partners of the gauge bosons [355–359]. When this happens, a mass $m_{3/2}$ for the gravitino and a superpotential for the Kähler moduli are induced [355, 358].

Condensation is also a rather generic feature of string theory. For instance, among the type-IIB brane configurations appearing in the AdS/CFT correspondence of Sect. 12.3.6, consider the $\mathcal{N}=1$ $SU(N)$ super-Yang–Mills theory living on a stack of N D7-branes. The only modulus not fixed by the flux background is the radion $\sigma = \mathrm{Re}\varrho$ and it determines the gauge coupling via

$$\frac{8\pi^2}{g_{\mathrm{YM}}^2} = 2\pi\sigma\,. \tag{12.108}$$

If gluino condensation takes place, then the superpotential term $W_{\mathrm{cond}} = \mathcal{A}\exp(-2\pi\varrho/N)$ is generated, where $\mathcal{A}$ is a positive constant determined by the UV cut-off scale of the gauge theory. The non-perturbative nature of this correction is clear from (12.103), since $\exp(-\mathrm{Re}\varrho) = \exp(-1/g_{\mathrm{s}})$. A similar superpotential W arises also from the contribution of string instantons [254, 360–363], in particular type-IIB Euclidean D3-branes [361, 362], and gaugino condensation in four dimensions actually descends from an M5-brane instanton in M-theory [364]. In general, the normalization $\mathcal{A} = \mathcal{A}(z_i, \tau)$ depends on the other moduli and it becomes a constant when these are stabilized.

In $E_8 \times E_8$ heterotic string theory, the breaking of one of the E_8 groups down to a product of smaller non-Abelian groups is sufficient to trigger gaugino condensation, which, in turn, breaks supersymmetry via an effective superpotential that features the axio-dilaton S rather than the modulus T [365–373],

$$W_{\mathrm{cond}}(S) = \mathcal{A}\mathrm{e}^{-\beta S}\,, \tag{12.109}$$

where β is a constant. In general, one sees the emergence of multi-exponential superpotentials of the form [368, 369]

$$W_{\rm cond}(S) = \sum_m c_m \mathrm{e}^{-\beta_m S}, \qquad W_{\rm cond}(\varrho_i) = \sum_n c_n \mathrm{e}^{-\sum_l \alpha_{l,n}\varrho_l}, \tag{12.110}$$

which are called *racetrack*.

12.3.9.2 KKLT Stabilization Scenarios

The stabilization of Kähler moduli by these non-perturbative mechanisms is well illustrated by the *Kachru–Kallosh–Linde–Trivedi model* (𝕂KLT in short) [374]. For simplicity, one considers only one Kähler modulus ϱ in type-IIB string theory and combines the tree-level contribution (12.106) with $W_{\rm cond}$ into the generic superpotential

$$W(\varrho) = W_0 + W_{\rm cond}(\varrho) = W_0 + \mathcal{A}\,\mathrm{e}^{-\alpha\varrho}\,, \tag{12.111}$$

where $W_0 < 0$, $\mathcal{A} > 0$ and $\alpha > 0$ are real-valued constants. Remarkably, both (12.111) and a two-field racetrack potential arise, independently from flux compactification, as a phenomenological requirement to address the moduli problem (Sect. 13.2) in the supergravity scenario with a heavy gravitino [375].

One assumes that the dilaton and complex-structure moduli have already been stabilized by fluxes, so that the potential only depends on ϱ:

$$\boxed{\kappa_4^4 V = \mathrm{e}^{\tilde K(\varrho)}\left\{\mathcal{G}^{\varrho\varrho^*} D_\varrho W(\varrho)[D_\varrho W(\varrho)]^* - 3|W(\varrho)|^2\right\},} \tag{12.112}$$

where $\tilde K$ is given by (12.99) and we have absorbed an overall numerical normalization factor in the definition of W. From $\partial_\varrho W = -\alpha\mathcal{A}\exp(-\alpha\varrho)$ and $\partial_\varrho \tilde K = -3/(2{\rm Re}\varrho)$, the definition (5.221) is

$$D_\varrho W = \partial_\varrho W + W\partial_\varrho \tilde K = -\frac{3W_0}{2{\rm Re}\varrho} - \mathcal{A}\,\mathrm{e}^{-\alpha\varrho}\left(\alpha + \frac{3}{2{\rm Re}\varrho}\right). \tag{12.113}$$

After decomposing ϱ as

$$\varrho = \sigma + \mathrm{i}\theta\,, \tag{12.114}$$

where $\sigma = \mathrm{Re}\varrho$ and $\theta = \mathrm{Im}\varrho$, and noting that $\mathcal{G}^{\varrho\varrho^*} = (\varrho + \varrho^*)^2/3 = 4\sigma^2/3$, the potential (12.112) becomes

$$\kappa_4^4 V = \frac{\alpha \mathcal{A}}{2\sigma^2}\,\mathrm{e}^{-2\alpha\sigma}\left[\left(1+\frac{\alpha\sigma}{3}\right)\mathcal{A} + W_0 \mathrm{e}^{\alpha\sigma}\cos(\alpha\theta)\right]. \tag{12.115}$$

For the sole purpose of keeping the presentation pedagogical, take now the axionic part of the Kähler modulus to be zero, $\varrho = \sigma$, and repeat the discussion around (7.24). The condition $D_\varrho W = 0$ for a supersymmetric vacuum is met at some critical value $\sigma_{\min} > 0$ such that

$$W_0 = -\left(1+\frac{2\alpha\sigma_{\min}}{3}\right)\mathcal{A}\,\mathrm{e}^{-\alpha\sigma_{\min}}. \tag{12.116}$$

Correspondingly, the minimum of (12.115) with $\theta = 0$ is negative:

$$\kappa_4^4 V_{\min} = -3|W(\sigma_{\min})|^2 \mathrm{e}^{\tilde{K}(\sigma_{\min})} = -\frac{(\alpha\mathcal{A})^2}{6\sigma_{\min}}\,\mathrm{e}^{-2\alpha\sigma_{\min}} < 0. \tag{12.117}$$

Notice that the gravitino mass $m_{3/2} := \mathrm{e}^{\tilde{K}/2}|W||_{\sigma=\sigma_{\min}} M_{\mathrm{Pl}}$ is determined by the scale of the minimum

$$V_{\min} = -3m_{3/2}^2 M_{\mathrm{Pl}}^2, \tag{12.118}$$

where $M_{\mathrm{Pl}} = \kappa_4^{-1}$ is the reduced Planck mass in four dimensions.

For the SUGRA approximation to be valid and to maintain α'-corrections to the Kähler potential suppressed, the volume of compact space must be sufficiently large, $\sigma \gg 1$ (small Yang–Mills coupling g_{YM}, perturbative field-theory regime). Also, non-perturbative corrections to W are under control provided $\alpha\sigma > 1$. Both conditions are met in (12.116) if $|W_0| \ll 1$, which can be obtained by tuning the fluxes to preserve supersymmetry. The potential (12.115) for typical values of the parameters is depicted in Fig. 12.4. For $|W_0| \approx 10^{-13}$, the gravitino mass is $O(\mathrm{TeV})$ and the string scale is close to the grand-unification scale.

To summarize, instantonic effects or gaugino condensation in the $\mathbb{K}$LT model preserve supersymmetry but lower the minimum of the potential to negative values, thus violating the property $V_{\min} = 0$ of no-scale solutions. This results in a four-manifold $\mathcal{M}_4$ with a negative cosmological constant: the supersymmetric vacuum is anti-de Sitter spacetime.

Realistic type-IIB models with many light moduli are more complicated than the original $\mathbb{K}$LT construction [376–400] (see [377, 378, 381, 389] for especially hands-on constructions). In the $\mathbb{K}$LT model, we have integrated out the axio-dilaton τ and the complex-structure moduli z_i before including non-perturbative quantum effects. This procedure, which resulted in the effective superpotential (12.111), is not justified in general. For instance, when one tries to stabilize the dilaton τ and the Kähler modulus ϱ simultaneously, the resulting AdS minimum is unstable;

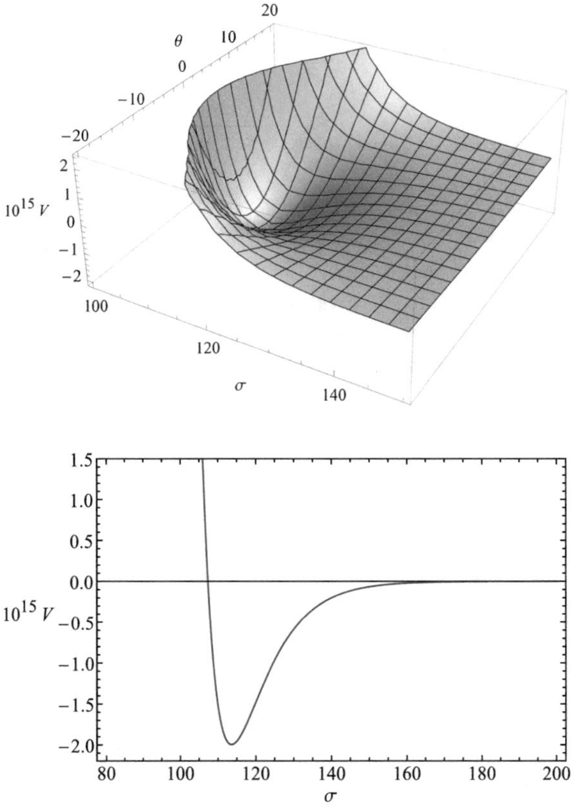

Fig. 12.4 The $\mathbb{K}$LT potential (12.115) multiplied by 10^{15} and in Planck-mass units $M_{\rm Pl} = \kappa_4^{-1} = 1$, with $W_0 = -10^{-4}$, $\mathcal{A} = 1$ and $\alpha = 0.1$, in the (σ, θ) plane (*top panel*) and for $\theta = 0$ (*bottom panel*). The local minimum is at $\sigma_{\rm min} \approx 113.6$

then, one has to turn on the complex-structure moduli as well [378, 388, 392]. Other moduli arise as follows. Chiral fields mimicking the Standard Model can be generated in three steps, by placing a singularity near the tip of the throat, modding out the discrete symmetry associated with such a singularity and placing an anti-D3-brane (indicated as $\overline{\rm D3}$-brane) on one of the fixed points [387]. For consistency, the resulting Calabi–Yau space must be populated by D7-branes [401, 402] and other D3- and $\overline{\rm D3}$-branes. The presence of gauge fields on the branes at the fixed points introduces new moduli to be stabilized.

In many of these generalizations, the contribution $W_{\rm cond}$ from gaugino condensation is of the racetrack form (12.110) and the potential V may acquire several minima.

Gaugino condensation and instantonic branes have also been considered in heterotic string theory [376, 386, 392, 403, 404]. Also in these cases the stable vacua are anti-de Sitter or, under some tuning, Minkowski. All models where moduli are fixed to an AdS or Minkowski minimum will be called $\mathbb{K}$LT *stabilization scenarios*.

12.3.9.3 Large-Volume Stabilization Scenarios

An important offspring of the $\mathbb{K}$LT construction are type-IIB *large-volume stabilization scenarios* [382, 383, 390, 393, 395, 405–411]. The combined presence of α'^3-corrections [412] and many Kähler moduli (as well as g_s loop corrections to the Kähler potential, in certain cases [408, 410]) generically leads to a non-supersymmetric AdS vacuum corresponding to a Calabi–Yau space with an exponentially large volume $\mathcal{V}_6$. The leading α'-corrections modify the total Kähler potential (12.100) as [412]

$$K = K_{\rm cs}(z_i) + K(\tau) + \tilde{K}^{(\alpha')}(\varrho) = K_{\rm cs}(z_i) + K(\tau) - 2\ln\left(\mathcal{V} + g_s^{-\frac{3}{2}}\xi\right), \qquad (12.119)$$

where $0 < \xi = O(1)$ is a constant dependent on the characteristics of $\mathcal{C}_3$. These corrections are perturbative in the string coupling and hence in a large-volume expansion, since $g_s = 1/\mathrm{Re}\varrho \propto \mathcal{V}^{-2/3}$ for one size modulus. The next-to-leading term in the logarithm in (12.119) is $O(\mathcal{V}^{-1/3})$ [395]. Viable models have more than one size modulus. The case of the orientifold $\mathbb{P}^4_{[1,1,1,6,9]}$ (a Calabi–Yau manifold constructed in [389, 413]) features just two fields $\varrho_\gg$ and $\varrho_\ll$, where the labels stand for large and small, respectively. The "small" modulus is assumed to give a negative contribution to the total dimensionless volume

$$\mathcal{V} \propto (\mathrm{Re}\varrho_\gg)^{\frac{3}{2}} - (\mathrm{Re}\varrho_\ll)^{\frac{3}{2}} = \sigma_\gg^{\frac{3}{2}} - \sigma_\ll^{\frac{3}{2}}. \qquad (12.120)$$

If

$$\sigma_\gg \gg \sigma_\ll > 1, \qquad (12.121)$$

then $\mathcal{V}_6$ is positive and large compared to the string scale. More small moduli would represent holes in a "Swiss-cheese" compact space. The superpotential is of racetrack type. Assuming, as in the $\mathbb{K}$LT scenario, that the dilaton and the complex-structure moduli have been fixed previously, one has

$$W = W_0 + \mathcal{A}_\gg e^{-\alpha_\gg \varrho_\gg} + \mathcal{A}_\ll e^{-\alpha_\ll \varrho_\ll} \simeq W_0 + \mathcal{A}_\ll e^{-\alpha_\ll \varrho_\ll}. \qquad (12.122)$$

Using (12.112), (12.119), (12.121) and (12.122), one finds the effective potential

$$\boxed{\kappa_4^4 V = \beta_1 \frac{\sqrt{\sigma_\ll}}{\mathcal{V}}\, e^{-2\alpha_\ll\sigma_\ll} + \beta_2 \frac{|W_0|\sigma_\ll}{\mathcal{V}^2}\, e^{-\alpha_\ll\sigma_\ll}\cos(\alpha_\ll\theta_\ll + \psi_\ll) + \beta_3 \frac{\xi |W_0|^2}{g_s^{3/2}\mathcal{V}^3}\,,} \qquad (12.123)$$

where $\beta_{1,2,3}$ are model-dependent numerical coefficients that also depend on $\alpha_\ll \mathcal{A}_\ll$, $\psi_\ll$ is defined as $\exp(\mathrm{i}\psi_\ll) := \mathcal{A}_\ll W_0^*/|\mathcal{A}_\ll W_0|$ and $\theta_\ll = \mathrm{Im}\varrho_\ll$ is the axion of the "small" mode. Minimizing V with respect to the axion at the global minimum $\theta_\ll^{\min} = (\pi - \psi_\ll)/\alpha_\ll$, we get

$$\kappa_4^4 V = \beta_1 \frac{\sqrt{\sigma_\ll}}{\mathcal{V}}\, \mathrm{e}^{-2\alpha_\ll \sigma_\ll} - \beta_2 \frac{|W_0|\sigma_\ll}{\mathcal{V}^2}\, \mathrm{e}^{-\alpha_\ll \sigma_\ll} + \beta_3 \frac{\xi |W_0|^2}{g_\mathrm{s}^{3/2}\mathcal{V}^3} \,. \tag{12.124}$$

The range of the constants is such that the potential has a local negative minimum, as in the KKLT case. A qualitative estimate runs as follows. The minimum of V in the $\sigma_\ll$ direction is found by solving the transcendental equation $\partial V/\partial \sigma_\ll = 0$. If $\alpha_\ll \sigma_\ll$ is moderately large but $(\beta_2/\beta_1)^2 \sigma_\ll = O(1)$, one has $\alpha_\ll \sigma_\ll^{\min} \propto \ln \mathcal{V}$ and $\mathrm{e}^{-\alpha_\ll \sigma_\ll} \propto \sqrt{\sigma_\ll}|W_0|/\mathcal{V}$ approximately. Plugging this back into (12.124) and minimizing in $\mathcal{V}$ (i.e., $\sigma_\gg$), one gets

$$\sigma_\gg^{\min} \propto \frac{\xi^{2/3}}{g_\mathrm{s}} \,, \qquad \mathcal{V}^{\min} \propto |W_0| \exp\left(\frac{\alpha_\ll \sigma_\ll^{\min}}{g_\mathrm{s}}\right) , \tag{12.125}$$

while the potential mininum is

$$V_{\min} = V_{\mathrm{np}} + V_{\alpha'} = -\frac{|W_0|^2}{\kappa_4^4 \mathcal{V}^3} \left[\gamma_1 (\ln \mathcal{V})^{3/2} - \gamma_2 \xi\right] , \qquad \mathcal{V} = \mathcal{V}^{\min} \,, \tag{12.126}$$

where $\gamma_{1,2}$ are constants. Thus, with a relatively small string coupling g_s we can obtain a moderately large modulus $\sigma_\ll$, an exponentially large volume of the compact space and, hence, an exponentially small negative cosmological constant $V_{\min}$.

The string and gravitino mass are related to the four-dimensional reduced Planck mass by

$$m_\mathrm{s} = \frac{g_\mathrm{s}}{\sqrt{4\pi}} \frac{M_\mathrm{Pl}}{\sqrt{\mathcal{V}}} \,, \qquad m_{3/2} = \frac{g_\mathrm{s}^2 W_0}{\sqrt{4\pi}} \frac{M_\mathrm{Pl}}{\mathcal{V}} \,, \tag{12.127}$$

and a large hierarchy of scales is generated. Therefore, string scenarios with large extra dimensions [414–416] can be embedded concretely in this large-volume stabilization scheme. In specific type-IIB examples, $\mathcal{V}_6 = 10^4$–$10^{15}\, l_\mathrm{s}^6$ and, for $g_\mathrm{s} = 0.1$ and $W_0 = 10^2$, the string scale and the gravitino mass are $m_\mathrm{s} \approx 10^{15}$–$10^9$ GeV and $m_{3/2} \approx 10^{14}$–10^3 GeV. Very large volumes $\mathcal{V}_6 = 10^{15} l_\mathrm{s}^6$ explain the Planck-to-electroweak hierarchy with a supersymmetry breaking scale $M_\mathrm{SUSY} \sim m_{3/2} = O(\mathrm{TeV})$.

A difference with respect to $\mathbb{K}$LT scenarios is that there is no strong theoretical constraint on W_0, which can take non-large values $W_0 < 10$–100. The fine tuning $|W_0| \ll 1$ in the $\mathbb{K}$LT construction is mainly due to the limited range of applicability of the approximations, where one must balance the size between tree-level and non-perturbative contributions relative to ignored α'-corrections. Supersymmetry breaking is also managed differently. In $\mathbb{K}$LT stabilization scenarios, the AdS minimum is supersymmetric; supersymmetry will be broken only later, by the uplifting mechanisms described in Sect. 13.1.1. On the other hand, the AdS minimum of large-volume scenarios is already non-supersymmetric, due to the F-terms of Kähler moduli.

12.4 Dualities and M-Theory

The five $D = 10$ superstring theories are equivalent to one another by dualities, relations that hold in certain compactification limits or manipulations of the parameters. These equivalences are physical, meaning that the mass spectrum, the brane population, the low-energy effective action, the partition function and other features of two dual theories coincide.

A hint of a yet deeper structure is the existence of dualities also between $D = 10$ and $D = 11$ supergravity. The bosonic sector of $D = 11$ SUGRA is sketched as

$$\boxed{S_{11} = \frac{1}{2\kappa_{11}^2}\int d^{11}x\,\sqrt{-g}\left[R - 2\Lambda - \frac{1}{2}F_4^2 + \mathcal{L}(A_3)\right],} \tag{12.128}$$

where Λ is a bare cosmological constant, F_4 is the field strength of the three-form A_3 and $\mathcal{L}(A_3)$ is the Chern–Simons Lagrangian. Notice the absence of a dilaton field. Both $D = 10$ and $D = 11$ SUGRA are special for different reasons. Nahm showed that $D = 9 + 1$ is the highest spacetime dimensionality where super-Yang–Mills theories can be constructed, while supergravity cannot exist in more than $10 + 1$ dimensions [145]. Moreover, an elegant theorem by Freund and Rubin shows that D-dimensional Einstein gravity with a rank-$(s-1)$ anti-symmetric tensor field A_{s-1} compactifies spontaneously (in the sense of Sect. 12.3.1) as $\mathcal{M}_s \times \mathcal{M}_{D-s}$, where one of the manifolds is non-compact and contains the time direction and the other is a Riemannian compact manifold [417]. In the absence of supersymmetry, nothing more can characterize the number of compact dimensions, since one can introduce tensors of arbitrary rank $1 \leqslant s \leqslant D$. However, in $D = 11$ supergravity there is only one s-form, with $s = 4$. Therefore, there is a preferential compactification $\mathcal{M}_4 \times \mathcal{M}_7$ and the compact space is either four- or seven-dimensional. Notably, in the second case one has just a four-dimensional spacetime. Eleven dimensions are special also in a cosmological context not related to supersymmetry, for two independent reasons. First, $D = 11$ marks a transition in the way to approach the big-bang singularity in general relativity without matter, oscillatory chaotic á la BKL for $D \leqslant 10$ and monotonic otherwise [418–420] (Sect. 6.3.5). Second, in general relativity with p-forms the approach to the singularity is chaotic for $D \geqslant 3$ and the structure of chaos depends on D: it cannot be a Kac–Moody billiard for $D > 11$ and, in the presence of just one 3-form field, it is a Coxeter polyhedron only in $D = 11$ [421].

Below, we will see that different compactifications of $D = 11$ SUGRA coincide with the compactifications (in one dimension less) of $D = 10$ SUGRA. Given that $\mathcal{N} = 2$ and $\mathcal{N} = 1$ supergravity in $D = 10$ are the low-energy limit of superstring theories, one may suspect a connection between the latter and an 11-dimensional structure, conventionally called *M-theory*. The "M" [422] stands for "mother" (it is supposed to be the fundamental theory unifying all superstrings), "magic" (for its marvels), "mysterious" (because it is known only through certain features, for instance its low-energy limit and its brane content), "membrane" (as its building blocks are identified with branes [423–425]) and "matrix" (as there is evidence that its degrees of freedom are the same of a matrix model [426]).

In particular, the low-energy limit of M-theory is postulated to be $D = 11$ SUGRA. Recall that, to a C^{E}_{p+1} form, there corresponds a dual form C^{M}_{D-3-p} and that these are interpreted, in string theory, as Dp- and D$(D-4-p)$-branes. Then, one can conjecture that, just like $D = 10$ SUGRA is the low-energy limit of superstring theory with a certain brane population, the structure whose low-energy limit is $D = 11$ SUGRA is a quantum theory of super-membranes [423, 424]. Since in (12.128) there is a 3-form and its magnetic dual is a 6-form, M-theory is characterized by M2- and M5-branes.[12] The classical and quantum dynamics of these branes can be studied via the construction of a class of $(p+1)$-dimensional actions [423–425] and via the dualities between $D = 10$ superstrings and the eleven-dimensional theory.

To help the reading, we mark a duality between $D = 10$ superstrings with a bullet • and a duality between a $D = 10$ superstring theory and the $D = 11$ theory with a triangle ►.

The first type of relation between string theories is T-duality, introduced in Sect. 12.3.1. • (a) The $E_8 \times E_8$ and $SO(32)$ heterotic strings are related by a twist in the boundary conditions [232]. Upon toroidal compactification (in particular, on $M^9 \times S^1$), they collapse into each other [427–429]. Since taking large and small radii corresponds to perform gauge transformations, this equivalence can be interpreted as the fact that the two heterotic superstrings are different ground states of the same theory. • (b) Also type-IIA and type-IIB theories are mutually equivalent when dimensionally reduced on a hypertorus (in particular, on S^1) [49, 430]. As in the previous case, type-II theories are interpreted as different ground states of a larger structure.

[12] We reserve the "D" for the Dirichlet branes of string theories, while the p-super-membranes of the $D = 11$ theory are called Mp-branes.

Another relation between superstrings is *S-duality*, which maps one theory with coupling g_s (the parameter governing the loop expansion) to another (or the same) theory with coupling g_s^{-1} [431–435]:

$$\text{S-duality:} \qquad g_s \longleftrightarrow \frac{1}{g_s} . \tag{12.129}$$

The letter "S" indicates that this duality is at the level of *s*tring *s*tates rather than of *t*arget spacetime. T-duality is a target duality. While T-duality is perturbative in g_s and non-perturbative in α', S-duality is non-perturbative in g_s and is valid order by order in α' (so that it can be checked at the level of low-energy effective actions).[13]

Finally, the generalization of S-duality to type-II theories is called *U-duality* [436]. "U" stands for *u*nified, since it involves both S and T dualities. However, to stress the non-perturbative relation (12.129) between the strong- and weak-coupling sector of mathematically different theories, U-dualities are often called S-dualities.

• (c) Type-I string theory and $SO(32)$ heterotic string theory are S-dual [437], while • (d) type-IIB string theory is S-self-dual [436]. ▶ (e) A U-duality also holds between type-IIA theory and 11-dimensional SUGRA on $M^{10} \times S^1$ [436, 438, 439], and ▶ (f) between $E_8 \times E_8$ heterotic theory and 11-dimensional SUGRA on $M^{10} \times S^1/\mathbb{Z}_2$ [422, 440–442].[14]

To give a flavour of (e) and of how the low-energy limit of type-IIA string theory connects with 11-dimensional gravity, we compactify the action (12.128) on a circle S^1 with radius r. The eleven-dimensional metric g_{MN} decomposes into a scalar $\Phi(x^\mu)$, a vector $A_\mu(x^\mu)$ and the ten-dimensional metric $\bar{g}_{\mu\nu}(x^\sigma)$, with $\mu, \nu, \sigma = 0, \dots, 9$:

$$\mathrm{d}s^2 = g_{MN}\mathrm{d}x^M\mathrm{d}x^N = \mathrm{e}^{-\frac{2}{3}\Phi}\bar{g}_{\mu\nu}\mathrm{d}x^\mu\mathrm{d}x^\nu + \mathrm{e}^{\frac{4}{3}\Phi}(\mathrm{d}x^{10} + A_\nu\mathrm{d}x^\nu)^2, \tag{12.130}$$

where $M, N = 0, \dots, 10$. Replacing this decomposition into (12.128) and ignoring Λ, we obtain exactly (12.74). All the low-energy ingredients of type-IIA string theory are in place: the universal SUGRA sector (graviton $\bar{g}_2$, dilaton Φ and Kalb–Ramond field B_2) and the R-R fields C_1 and C_3, corresponding to D0-, D6-, D2- and D4-branes. Intuitively, one direction of the M2- and M5-brane of the $D = 11$ theory wraps around the circle to give, respectively, the D1- and D4-brane in $D = 10$. If the M2-brane has toroidal spatial sections, the resulting D1-branes are tubular surfaces interpreted as type-IIA strings, as the dimensional reduction of the three-dimensional brane action to the string action strongly indicates [425]. More involved

[13]Originally, T and S referred to the T and S fields in the chiral multiplet composing the superpotential and Kähler potential (5.230) in $D = 4$ SUGRA (regarded as the dimensional reduction of $D = 10$ heterotic string on a six-torus). The decompactification limit corresponds to Re$T \to \infty$, while the real part of the complex dilaton field S is $1/g_c^2$ [431].

[14]An early study of solitons in the low-energy limit of the heterotic string suggested that, at strong coupling, the latter admits a dual description as a weakly interacting theory of D5-branes [443].

arguments at the quantum level confirm the conclusion that the strong-coupling behaviour of type-IIA string theory is weakly-coupled $D = 11$ supergravity.

In ten dimensions, the gravitational coupling constant for all closed string theories is

$$2\kappa_{10}^2 = (2\pi)^7\, l_s^8\, g_c^2\,, \tag{12.131}$$

while the gravitational coupling in eleven and ten dimensions are related by

$$2\kappa_{11}^2 = (2\pi r)2\kappa_{10}^2 = (2\pi)^8\, l_s^9\, g_c^3\,. \tag{12.132}$$

Defining the 11-dimensional Planck mass and length as $2\kappa_{11}^2 = (2\pi)^8 M_{11}^{-9} = (2\pi)^8 l_{11}^9$, we get

$$l_{11} = g_c^{\frac{1}{3}}\, l_s\,, \qquad g_c = (M_{11} r)^{\frac{3}{2}}\,, \qquad l_s^2 = \frac{l_{11}^3}{r}\,. \tag{12.133}$$

The $D = 11$ theory is the non-perturbative decompactification limit $g_c \to \infty$ ($r \to \infty$) of the $D = 10$ theory.

Different compactifications give rise to a wealth of U- and S-dualities between the heterotic theories (which coincide when dimensionally reduced, according to (a)) and type-II strings [436, 438, 444–449]. ► (g) On $M^7 \times \mathbb{T}^3$, heterotic string theory is dual to $D = 11$ SUGRA on $M^7 \times$ K3 (here, a closed string is obtained by wrapping an M5-brane with topology K3 $\times S^1$ around K3); • (h) on $M^6 \times \mathbb{T}^4$, it is dual to type-IIA string theory on $M^6 \times$ K3; • (i) on $M^5 \times \mathbb{T}^5$, it is dual to type-IIB string theory on $M^6 \times$ K3; • (j) on $M^4 \times \mathbb{T}^6$ and consistently with (a), it is dual to itself. The $E_8 \times E_8$ heterotic theory on $M^6 \times$ K3 is both • (k) dual to itself and ► (l) dual to the $D = 11$ theory compactified on $M^6 \times$ K3 $\times S^1/\mathbb{Z}_2$, consistently with (f).

Some derived dualities follow from those above. By virtue of (e), (g) and (h), ► (m) both heterotic and type-IIA string theory in six dimensions are dual to a model of super-membranes in $D = 11$ on $M^6 \times$ K3 $\times S^1$ [450]. Also, by virtue of (b) and (e), ► (n) type-II string theory on $M^9 \times S^1$ is dual to the $D = 11$ theory on $M^9 \times \mathbb{T}^2$ [451–453].

Finally, Calabi–Yau compactifications give rise to more threads of the web of dualities constituting M-theory. Compatibly with (e), ► (o) the low-energy limit of type-IIA theory compactified on $M^4 \times \mathcal{C}_3$ is equivalent to $D = 11$ SUGRA on $M^4 \times \mathcal{C}_3 \times S^1$, where $\mathcal{C}_3$ is the same Calabi–Yau space [454, 455]. Moreover, ► (p) $D = 11$ SUGRA on $M^5 \times \mathcal{C}_3$ [454, 455] is dual to heterotic theory on $M^5 \times$K3$\times S^1$ [456–458]. The fundamental heterotic string is identified with the M5-brane wrapped around a four-cycle of the Calabi–Yau space. ► (q) The $D = 11$ theory on $M^5 \times \mathcal{C}_3$ is also equivalent to type-II SUGRA on $M^4 \times \mathcal{C}_3$ when one of the compact directions is appropriately decompactified [457]. Last but not least, we recall from Sect. 12.3.3 that (r) type-I theory is type-IIB theory on an orientifold [235–238].

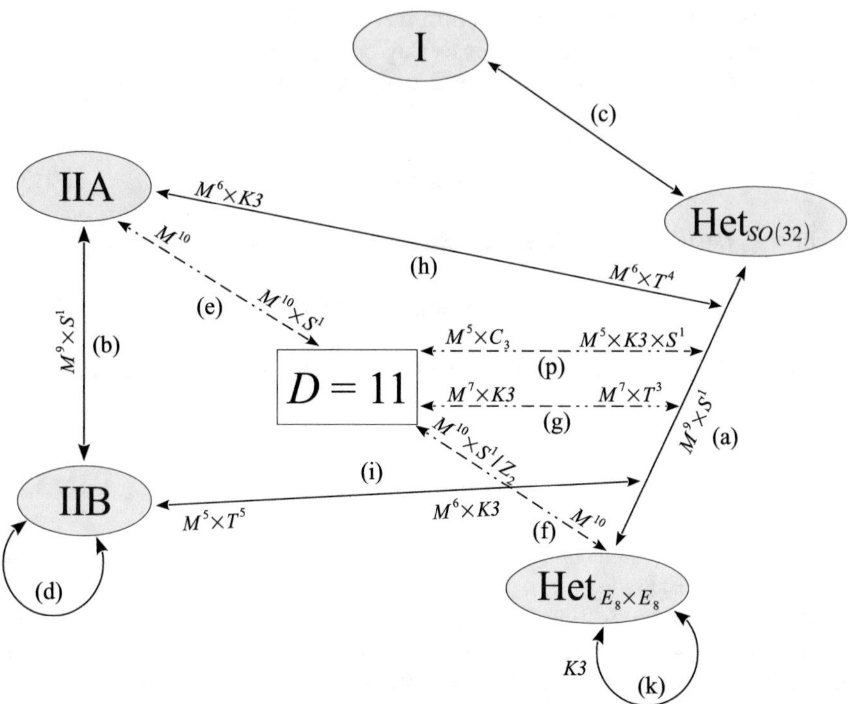

Fig. 12.5 M theory: the web of dualities among string theories (*solid lines*) and between string theories and the $D = 11$ super-membrane model (*dot-dashed lines*). Along each connection, the topology of the compactification scheme at each end-point is indicated. The dualities (a) and (b) are also valid on $M^{9-d} \times \mathbb{T}^d$ for $d \geqslant 2$

These 18 dualities, and many more we have not discussed here, marked one of the deepest discoveries in string theory and a breakthrough in our understanding of its non-perturbative structure. The partial list (a)-(i), (k) and (p) is summarized in Fig. 12.5.

The flux compactification of M-theory [269, 270, 459–470] follows a scheme similar to that in string theory. Non-perturbative effects stabilize the moduli on Minkowski or anti-de Sitter vacua.

12.5 Problems and Solutions

12.1 Conserved supercurrent. Show that $\partial_a \mathcal{J}^a = 0$ on shell, where $\mathcal{J}^a = \rho^b \rho^a (\partial_b X^\mu) \psi_\mu / 2$ and the ρ^a obey the Clifford algebra (12.54).

Solution From (12.54), one notes that $\rho_a\rho^b\rho^a = -\rho^b\rho_a\rho^a - 2\eta_a{}^b\rho^a = -2\rho^b + 2\rho^b = 0$, so that

$$2\partial_a\mathcal{J}^a = \rho^b\rho^a(\partial_a\partial_b X^\mu)\psi_\mu + \rho^b\rho^a(\partial_b X^\mu)(\partial_a\psi_\mu) \overset{(12.55)}{=} 0\,.$$

12.2 Dilaton in 10 and 4 dimensions. Modify (12.88) with a generic warp factor $e^{2u}g_{mn} \to e^{2\beta u}g_{mn}$ in the six-dimensional metric, with β a constant. Show that for $\beta \neq 1$ the ten-dimensional dilaton Φ is not the same as the four-dimensional dilaton Φ_{4D}.

Solution The $D = 10$ Ricci scalar in the low-energy action (12.75) decomposes as $\bar{R}[\bar{g}_{\mu\nu}] = \bar{R}^{(4)}[\bar{g}_{ab}] + \bar{R}^{(6)}[\bar{g}_{mn}] + \dots = e^{6u}R^{(4)}[g_{ab}] + e^{-2\beta u}R^{(6)}[g_{mn}] + \dots$ (from (7.150), with $\Omega = e^{3u}$ in the four-dimensional part and $\Omega = e^{-\beta u}$ in the six-dimensional part), while the ten-dimensional volume factor is split as $\sqrt{-\bar{g}} = (e^{-12u}\sqrt{-g^{(4)}})(e^{6\beta u}\sqrt{g^{(6)}}) = e^{6(\beta-2)u}\sqrt{-g^{(4)}}\sqrt{g^{(6)}}$, so that

$$\sqrt{-\bar{g}}e^{-2\Phi}\bar{R} = \sqrt{g^{(6)}}\sqrt{-g^{(4)}}e^{-2\Phi}\left[e^{6(\beta-1)u}R^{(4)} + e^{-4(3-\beta)u}R^{(6)} + \dots\right].$$

This expressions defines the four-dimensional dilaton $\Phi_{4D} := \Phi - 3(\beta - 1)u$. Therefore, a choice of warping factors different than (12.88) leads, in general, to $\Phi \neq \Phi_{4D}$.

References

1. G. Veneziano, Construction of a crossing-simmetric, Regge-behaved amplitude for linearly rising trajectories. Nuovo Cim. A **57**, 190 (1968)
2. S. Fubini, G. Veneziano, Level structure of dual-resonance models. Nuovo Cim. A **64**, 811 (1969)
3. M.A. Virasoro, Alternative constructions of crossing-symmetric amplitudes with Regge behavior. Phys. Rev. **177**, 2309 (1969).
4. M.A. Virasoro, Subsidiary conditions and ghosts in dual-resonance models. Phys. Rev. D **1**, 2933 (1970)
5. P. Ramond, Dual theory for free fermions. Phys. Rev. D **3**, 2415 (1971)
6. A. Neveu, J.H. Schwarz, Factorizable dual model of pions. Nucl. Phys. B **31**, 86 (1971)
7. A. Neveu, J.H. Schwarz, Quark model of dual pions. Phys. Rev. D **4**, 1109 (1971)
8. E. Del Giudice, P. Di Vecchia, S. Fubini, General properties of the dual resonance model. Ann. Phys. (N.Y.) **70**, 378 (1972)
9. E. Del Giudice, P. Di Vecchia, S. Fubini, R. Musto, Light-cone physics and duality. Nuovo Cim. A **12**, 813 (1972)
10. P. Goddard, C.B. Thorn, Compatibility of the dual Pomeron with unitarity and the absence of ghosts in the dual resonance model. Phys. Lett. B **40**, 235 (1972)

11. R.C. Brower, Spectrum-generating algebra and no-ghost theorem for the dual model. Phys. Rev. D **6**, 1655 (1972)
12. P. Goddard, C. Rebbi, C.B. Thorn, Lorentz covariance and the physical states in dual-resonance models. Nuovo Cim. A **12**, 425 (1972)
13. R.C. Brower, K.A. Friedman, Spectrum-generating algebra and no-ghost theorem for the Neveu–Schwarz model. Phys. Rev. D **7**, 535 (1973)
14. J.H. Schwarz, Dual resonance theory. Phys. Rep. **8**, 269 (1973)
15. G. Veneziano, An introduction to dual models of strong interactions and their physical motivations. Phys. Rep. **9**, 199 (1974)
16. J. Scherk, J.H. Schwarz, Dual field theory of quarks and gluons. Phys. Lett. B **57**, 463 (1975)
17. Y. Nambu, Quark model and the factorization of the Veneziano amplitude, in *Symmetries and Quark Models*, ed. by R. Chand (Gordon and Breach, New York, 1970) [Reprinted in *Broken Symmetry. Selected Papers of Y. Nambu*, ed. by T. Eguchi, K. Nishijima (World Scientific, Singapore, 1995)]
18. L. Susskind, Structure of hadrons implied by duality. Phys. Rev. D **1**, 1182 (1970)
19. T. Takabayasi, Relativistic quantum mechanics of a mechanical continuum underlying the dual amplitude. Prog. Theor. Phys. **44**, 1429 (1970)
20. O. Hara, On origin and physical meaning of Ward-like identity in dual-resonance model. Prog. Theor. Phys. **46**, 1549 (1971)
21. T. Gotō, Relativistic quantum mechanics of one-dimensional mechanical continuum and subsidiary condition of dual resonance model. Prog. Theor. Phys. **46**, 1560 (1971)
22. L.N. Chang, F. Mansouri, Dynamics underlying duality and gauge invariance in the dual-resonance models. Phys. Rev. D **5**, 2535 (1972)
23. F. Mansouri, Y. Nambu, Gauge conditions in dual resonance models. Phys. Lett. B **39**, 375 (1972)
24. P. Goddard, J. Goldstone, C. Rebbi, C.B. Thorn, Quantum dynamics of a massless relativistic string. Nucl. Phys. B **56**, 109 (1973)
25. S. Mandelstam, Interacting-string picture of dual-resonance models. Nucl. Phys. B **64**, 205 (1973)
26. M. Kalb, P. Ramond, Classical direct interstring action. Phys. Rev. D **9**, 2273 (1974)
27. M. Kaku, K. Kikkawa, Field theory of relativistic strings. I. Trees. Phys. Rev. D **10**, 1110 (1974)
28. M. Kaku, K. Kikkawa, Field theory of relativistic strings. II. Loops and Pomerons. Phys. Rev. D **10**, 1823 (1974)
29. C. Rebbi, Dual models and relativistic quantum strings. Phys. Rep. **12**, 1 (1974)
30. T. Yoneya, Quantum gravity and the zero-slope limit of the generalized Virasoro model. Lett. Nuovo Cim. **8**, 951 (1973)
31. T. Yoneya, Connection of dual models to electrodynamics and gravidynamics. Prog. Theor. Phys. **51**, 1907 (1974)
32. J. Scherk, J.H. Schwarz, Dual models for non-hadrons. Nucl. Phys. B **81**, 118 (1974)
33. J. Scherk, J.H. Schwarz, Dual models and the geometry of space-time. Phys. Lett. B **52**, 347 (1974)
34. M.B. Green, J.H. Schwarz, E. Witten, *Superstring Theory* (Cambridge University Press, Cambridge, 1987)
35. J. Polchinski, *String Theory* (Cambridge University Press, Cambridge, 1998)
36. K. Becker, M. Becker, J.H. Schwarz, *String Theory and M-Theory* (Cambridge University Press, Cambrdige, 2007)
37. B. Zwiebach, *A First Course in String Theory* (Cambridge University Press, Cambridge, 2009)
38. D. Tong, String theory. arXiv:0908.0333
39. L. Brink, P. Di Vecchia, P.S. Howe, A locally supersymmetric and reparametrization invariant action for the spinning string. Phys. Lett. B **65**, 471 (1976)
40. S. Deser, B. Zumino, A complete action for the spinning string. Phys. Lett. B **65**, 369 (1976)
41. A.M. Polyakov, Quantum geometry of bosonic strings. Phys. Lett. B **103**, 207 (1981)

42. A.M. Polyakov, Conformal symmetry of critical fluctuations. Pisma Zh. Eksp. Teor. Fiz. **12**, 538 (1970) [JETP Lett. **12**, 381 (1970)]
43. A.M. Polyakov, Non-Hamiltonian approach to conformal quantum field theory. Zh. Eksp. Teor. Fiz. **66**, 23 (1974) [Sov. Phys. JETP **39**, 10 (1974)]
44. A.A. Belavin, A.M. Polyakov, A.B. Zamolodchikov, Infinite conformal symmetry in two-dimensional quantum field theory. Nucl. Phys. B **241**, 333 (1984)
45. D. Friedan, Z. Qiu, S. Shenker, Conformal invariance, unitarity, and critical exponents in two dimensions. Phys. Rev. Lett. **52**, 1575 (1984)
46. D. Friedan, Z. Qiu, S. Shenker, Superconformal invariance in two dimensions and the tricritical Ising model. Phys. Lett. B **151**, 37 (1985)
47. D. Friedan, S. Shenker, E. Martinec, Covariant quantization of superstrings. Phys. Lett. B **160**, 55 (1985)
48. D. Friedan, E. Martinec, S. Shenker, Conformal invariance, supersymmetry and string theory. Nucl. Phys. B **271**, 93 (1986)
49. J. Dai, R.G. Leigh, J. Polchinski, New connections between string theories. Mod. Phys. Lett. A **04**, 2073 (1989)
50. C. Becchi, A. Rouet, R. Stora, The abelian Higgs Kibble model, unitarity of the S-operator. Phys. Lett. B **52**, 344 (1974)
51. C. Becchi, A. Rouet, R. Stora, Renormalization of the abelian Higgs–Kibble model. Commun. Math. Phys. **42**, 127 (1975)
52. C. Becchi, A. Rouet, R. Stora, Renormalization of gauge theories. Ann. Phys. (N.Y.) **98**, 287 (1976)
53. I.V. Tyutin, Gauge invariance in field theory and statistical physics in operator formalism, unpublished (1975). arXiv:0812.0580
54. K. Fujikawa, Path integral of relativistic strings. Phys. Rev. D **25**, 2584 (1982)
55. M. Kato, K. Ogawa, Covariant quantization of string based on BRS invariance. Nucl. Phys. B **212**, 443 (1983)
56. S. Hwang, Covariant quantization of the string in dimensions $D \leqslant 26$ using a Becchi–Rouet–Stora formulation. Phys. Rev. D **28**, 2614 (1983)
57. J.L. Gervais, A. Neveu, The dual string spectrum in Polyakov's quantization (I). Nucl. Phys. B **199**, 59 (1982)
58. J.L. Gervais, A. Neveu, Dual string spectrum in Polyakov's quantization (II). Mode separation. Nucl. Phys. B **209**, 125 (1982)
59. J.E. Paton, H.M. Chan, Generalized Veneziano model with isospin. Nucl. Phys. B **10**, 516 (1969)
60. N. Marcus, A. Sagnotti, Group theory from "quarks" at the ends of strings. Phys. Lett. B **188**, 58 (1987)
61. M.R. Douglas, B. Grinstein, Dilaton tadpole for the open bosonic string. Phys. Lett. B **183**, 52 (1987); Erratum-ibid. B **187**, 442 (1987)
62. S. Weinberg, Cancellation of one-loop divergences in $SO(8192)$ string theory. Phys. Lett. B **187**, 278 (1987)
63. C. Lovelace, Pomeron form factors and dual Regge cuts. Phys. Lett. B **34**, 500 (1971)
64. L. Bers, Uniformization, moduli, and Kleinian groups. Bull. Lond. Math. Soc. **4**, 257 (1972)
65. S. Wolpert, On the homology of the moduli space of stable curves. Ann. Math. **118**, 491 (1983)
66. D. Friedan, S. Shenker, The integrable analytic geometry of quantum string. Phys. Lett. B **175**, 287 (1986)
67. D. Friedan, S. Shenker, The analytic geometry of two-dimensional conformal field theory. Nucl. Phys. B **281**, 509 (1987)
68. D. Friedan, Nonlinear models in $2 + \epsilon$ dimensions. Phys. Rev. Lett. **45**, 1057 (1980)
69. C. Lovelace, Strings in curved space. Phys. Lett. B **135**, 75 (1984)
70. D.H. Friedan, Nonlinear models in $2 + \varepsilon$ dimensions. Ann. Phys. (N.Y.) **163**, 318 (1985)
71. C.G. Callan, D. Friedan, E.J. Martinec, M.J. Perry, Strings in background fields. Nucl. Phys. B **262**, 593 (1985)

72. E.S. Fradkin, A.A. Tseytlin, Effective field theory from quantized strings. Phys. Lett. B **158**, 316 (1985)
73. E.S. Fradkin, A.A. Tseytlin, Quantum string theory effective action. Nucl. Phys. B **261**, 1 (1985)
74. K. Ohmori, A review on tachyon condensation in open string field theories. arXiv:hep-th/0102085
75. E. Fuchs, M. Kroyter, Analytical solutions of open string field theory. Phys. Rep. **502**, 89 (2011). [arXiv:0807.4722]
76. E. Witten, Non-commutative geometry and string field theory. Nucl. Phys. B **268**, 253 (1986)
77. Z.-a. Qiu, A. Strominger, Gauge symmetries in (super)string field theory. Phys. Rev. D **36**, 1794 (1987)
78. Y. Okawa, L. Rastelli, B. Zwiebach, Analytic solutions for tachyon condensation with general projectors. arXiv:hep-th/0611110
79. V.A. Kostelecký, S. Samuel, The static tachyon potential in the open bosonic string theory. Phys. Lett. B **207**, 169 (1988)
80. V.A. Kostelecký, S. Samuel, On a nonperturbative vacuum for the open bosonic string. Nucl. Phys. B **336**, 263 (1990)
81. L. Brekke, P.G.O. Freund, M. Olson, E. Witten, Nonarchimedean string dynamics. Nucl. Phys. B **302**, 365 (1988)
82. M. Saadi, B. Zwiebach, Closed string field theory from polyhedra. Ann. Phys. (N.Y.) **192**, 213 (1989)
83. T. Kugo, H. Kunitomo, K. Suehiro, Nonpolynomial closed string field theory. Phys. Lett. B **226**, 48 (1989)
84. T. Kugo, K. Suehiro, Nonpolynomial closed string field theory: action and its gauge invariance. Nucl. Phys. B **337**, 434 (1990)
85. V.A. Kostelecký, S. Samuel, Collective physics in the closed bosonic string. Phys. Rev. D **42**, 1289 (1990)
86. B. Zwiebach, Closed string field theory: quantum action and the Batalin–Vilkovisky master equation. Nucl. Phys. B **390**, 33 (1993). arXiv:hep-th/9206084
87. A. Sen, B. Zwiebach, A proof of local background independence of classical closed string field theory. Nucl. Phys. B **414**, 649 (1994). [arXiv:hep-th/9307088]
88. A. Sen, B. Zwiebach, Quantum background independence of closed string field theory. Nucl. Phys. B **423**, 580 (1994). [arXiv:hep-th/9311009]
89. A. Sen, B. Zwiebach, A note on gauge transformations in Batalin–Vilkovisky theory. Phys. Lett. B **320**, 29 (1994). [arXiv:hep-th/9309027]
90. Y. Okawa, B. Zwiebach, Twisted Tachyon condensation in closed string field theory. JHEP **0403**, 056 (2004). [arXiv:hep-th/0403051]
91. H. Yang, B. Zwiebach, Dilaton deformations in closed string field theory. JHEP **0505**, 032 (2005). [arXiv:hep-th/0502161]
92. H. Yang, B. Zwiebach, A closed string tachyon vacuum? JHEP **0509**, 054 (2005). [arXiv:hep-th/0506077]
93. Y. Michishita, Field redefinitions, T-duality and solutions in closed string field theories. JHEP **0609**, 001 (2006). [arXiv:hep-th/0602251]
94. N. Moeller, Closed bosonic string field theory at quintic order: five-tachyon contact term and dilaton theorem. JHEP **0703**, 043 (2007). [arXiv:hep-th/0609209]
95. N. Moeller, Closed bosonic string field theory at quintic order. II: marginal deformations and effective potential. JHEP **0709**, 118 (2007). [arXiv:0705.2102]
96. N. Moeller, A tachyon lump in closed string field theory. JHEP **0809**, 056 (2008). [arXiv:0804.0697]
97. L. Brekke, P.G.O. Freund, p-adic numbers in physics. Phys. Rep. **233**, 1 (1993)
98. N. Moeller, M. Schnabl, Tachyon condensation in open-closed p-adic string theory. JHEP **0401**, 011 (2004). [arXiv:hep-th/0304213]
99. V. Vladimirov, Nonlinear equations for p-adic open, closed, and open-closed strings. Theor. Math. Phys. **149**, 1604 (2006). [arXiv:0705.4600]

100. K. Ohmori, Toward open-closed string theoretical description of rolling tachyon. Phys. Rev. D **69**, 026008 (2004). [arXiv:hep-th/0306096]
101. T. Biswas, J.A.R. Cembranos, J.I. Kapusta, Thermal duality and Hagedorn transition from *p*-adic strings. Phys. Rev. Lett. **104**, 021601 (2010). [arXiv:0910.2274]
102. J.H. Schwarz, Superstring theory. Phys. Rep. **89**, 223 (1982)
103. M.B. Green, Supersymmetrical dual string theories and their field theory limits —a review. Surveys High Energy Phys. **3**, 127 (1983)
104. L. Brink, J.H. Schwarz, Local complex supersymmetry in two dimensions. Nucl. Phys. B **121**, 285 (1977)
105. A.M. Polyakov, Quantum geometry of fermionic strings. Phys. Lett. B **103**, 211 (1981)
106. M.B. Green, J.H. Schwarz, Covariant description of superstrings. Phys. Lett. B **136**, 367 (1984)
107. M.B. Green, J.H. Schwarz, Properties of the covariant formulation of superstring theories. Nucl. Phys. B **243**, 285 (1984)
108. F. Gliozzi, J. Scherk, D.I. Olive, Supersymmetry, supergravity theories and the dual spinor model. Nucl. Phys. B **122**, 253 (1977)
109. M.B. Green, J.H. Schwarz, Supersymmetrical dual string theory. Nucl. Phys. B **181**, 502 (1981)
110. M.B. Green, J.H. Schwarz, Supersymmetrical string theories. Phys. Lett. B **109**, 444 (1982)
111. F. Gliozzi, J. Scherk, D.I. Olive, Supergravity and the spinor dual model. Phys. Lett. B **65**, 282 (1976)
112. L. Brink, J.H. Schwarz, J. Scherk, Supersymmetric Yang–Mills theories. Nucl. Phys. B **121**, 77 (1977)
113. N. Marcus, A. Sagnotti, Tree-level constraints on gauge groups for type I superstrings. Phys. Lett. B **119**, 97 (1982)
114. M.B. Green, J.H. Schwarz, Supersymmetric dual string theory: (II). Vertices and trees. Nucl. Phys. B **198**, 252 (1982)
115. M.B. Green, J.H. Schwarz, Supersymmetric dual string theory: (III). Loops and renormalization. Nucl. Phys. B **198**, 441 (1982)
116. M.B. Green, J.H. Schwarz, Superstring interactions. Nucl. Phys. B **218**, 43 (1983)
117. M.B. Green, J.H. Schwarz, L. Brink, Superfield theory of type (II) superstrings. Nucl. Phys. B **219**, 437 (1983)
118. M.B. Green, J.H. Schwarz, Superstring field theory. Nucl. Phys. B **243**, 475 (1984)
119. M.B. Green, J.H. Schwarz, Infinity cancellations in $SO(32)$ superstring theory. Phys. Lett. B **151**, 21 (1985)
120. E. D'Hoker, D.H. Phong, Momentum analyticity and finiteness of the 1-loop superstring amplitude. Phys. Rev. Lett. **70**, 3692 (1993). [arXiv:hep-th/9302003]
121. E. D'Hoker, D.H. Phong, Two-loop superstrings. I. Main formulas. Phys. Lett. B **529**, 241 (2002). [arXiv:hep-th/0110247]
122. E. D'Hoker, D.H. Phong, Two-loop superstrings. II. The chiral measure on moduli space. Nucl. Phys. B **636**, 3 (2002). [arXiv:hep-th/0110283]
123. E. D'Hoker, D.H. Phong, Two-loop superstrings. III. Slice independence and absence of ambiguities. Nucl. Phys. B **636**, 61 (2002). [arXiv:hep-th/0111016]
124. E. D'Hoker, D.H. Phong, Two-loop superstrings. IV. The cosmological constant and modular forms. Nucl. Phys. B **639**, 129 (2002). [arXiv:hep-th/0111040]
125. E. D'Hoker, D.H. Phong, Two-loop superstrings. V. Gauge slice independence of the *N*-point function. Nucl. Phys. B **715**, 91 (2005). [arXiv:hep-th/0501196]
126. E. D'Hoker, D.H. Phong, Two-loop superstrings VI: Non-renormalization theorems and the 4-point function. Nucl. Phys. B **715**, 3 (2005). [arXiv:hep-th/0501197]
127. E. D'Hoker, D.H. Phong, Two-loop vacuum energy for Calabi–Yau orbifold models. Nucl. Phys. B **877**, 343 (2013). [arXiv:1307.1749]
128. E.J. Martinec, Nonrenormalization theorems and fermionic string finiteness. Phys. Lett. B **171**, 189 (1986)

129. E.P. Verlinde, H.L. Verlinde, Multiloop calculations in covariant superstring theory. Phys. Lett. B **192**, 95 (1987)
130. E. Witten, Superstring perturbation theory revisited. arXiv:1209.5461
131. A. Sen, Off-shell amplitudes in superstring theory. Fortsch. Phys. **63**, 149 (2015). [arXiv:1408.0571]
132. A. Sen, E. Witten, Filling the gaps with PCO's. JHEP **1509**, 004 (2015). [arXiv:1504.00609]
133. A. Sen, Ultraviolet and infrared divergences in superstring theory. arXiv:1512.00026
134. L. Alvarez-Gaumé, E. Witten, Gravitational anomalies. Nucl. Phys. B **234**, 269 (1984)
135. M.B. Green, J.H. Schwarz, Anomaly cancellation in supersymmetric $D = 10$ gauge theory and superstring theory. Phys. Lett. B **149**, 117 (1984)
136. M.B. Green, J.H. Schwarz, P.C. West, Anomaly-free chiral theories in six dimensions. Nucl. Phys. B **254**, 327 (1985)
137. M.B. Green, J.H. Schwarz, The hexagon gauge anomaly in type 1 superstring theory. Nucl. Phys. B **255**, 93 (1985)
138. A. Sagnotti, A note on the Green–Schwarz mechanism in open string theories. Phys. Lett. B **294**, 196 (1992). [arXiv:hep-th/9210127]
139. D.J. Gross, J.A. Harvey, E. Martinec, R. Rohm, Heterotic string. Phys. Rev. Lett. **54**, 502 (1985)
140. D.J. Gross, J.A. Harvey, E. Martinec, R. Rohm, Heterotic string theory: (I). The free heterotic string. Nucl. Phys. B **256**, 253 (1985)
141. D.J. Gross, J.A. Harvey, E. Martinec, R. Rohm, Heterotic string theory: (II). The interacting heterotic string. Nucl. Phys. B **267**, 75 (1986)
142. E. Cremmer, B. Julia, The $N = 8$ supergravity theory. I. The Lagrangian. Phys. Lett. B **80**, 48 (1978)
143. E. Cremmer, B. Julia, The $SO(8)$ supergravity. Nucl. Phys. B **159**, 141 (1979)
144. E. Cremmer, B. Julia, J. Scherk, Supergravity in theory in 11 dimensions. Phys. Lett. B **76**, 409 (1978)
145. W. Nahm, Supersymmetries and their representations. Nucl. Phys. B **135**, 149 (1978)
146. M.B. Green, J.H. Schwarz, Extended supergravity in ten dimensions. Phys. Lett. B **122**, 143 (1983)
147. J.H. Schwarz, P.C. West, Symmetries and transformations of chiral $N = 2, D = 10$ supergravity. Phys. Lett. B **126**, 301 (1983)
148. P.S. Howe, P.C. West, The complete $N = 2, d = 10$ supergravity. Nucl. Phys. B **238**, 181 (1984)
149. A.H. Chamseddine, $N = 4$ supergravity coupled to $N = 4$ matter and hidden symmetries. Nucl. Phys. B **185**, 403 (1981)
150. E.S. Fradkin, A.A. Tseytlin, Effective action approach to superstring theory. Phys. Lett. B **160**, 69 (1985)
151. A. Sen, Equations of motion for the heterotic string theory from the conformal invariance of the sigma model. Phys. Rev. Lett. **55**, 1846 (1985)
152. A. Sen, Heterotic string in an arbitrary background field. Phys. Rev. D **32**, 2102 (1985)
153. R.I. Nepomechie, Magnetic monopoles from antisymmetric tensor gauge fields. Phys. Rev. D **31**, 1921 (1985)
154. C. Teitelboim, Gauge invariance for extended objects. Phys. Lett. B **167**, 63 (1986)
155. C. Teitelboim, Monopoles of higher rank. Phys. Lett. B **167**, 69 (1986)
156. J. Polchinski, Dirichlet branes and Ramond–Ramond charges. Phys. Rev. Lett. **75**, 4724 (1995). [arXiv:hep-th/9510017]
157. A. Sen, Descent relations among bosonic D-branes. Int. J. Mod. Phys. A **14**, 4061 (1999). [arXiv:hep-th/9902105]
158. A. Sen, Non-BPS states and branes in string theory. Class. Quantum Grav. **17**, 1251 (2000). [arXiv:hep-th/9904207]
159. M. Schnabl, Analytic solution for tachyon condensation in open string field theory. Adv. Theor. Math. Phys. **10**, 433 (2006). [arXiv:hep-th/0511286]

160. T. Erler, C. Maccaferri, String field theory solution for any open string background. JHEP **1410**, 029 (2014). [arXiv:1406.3021]
161. A. Sen, SO(32) spinors of type I and other solitons on brane-antibrane pair. JHEP **9809**, 023 (1998). [arXiv:hep-th/9808141]
162. A. Sen, BPS D-branes on non-supersymmetric cycles. JHEP **9812**, 021 (1998). [arXiv:hep-th/9812031]
163. A. Sen, Supersymmetric world-volume action for non-BPS D-branes. JHEP **9910**, 008 (1999). [arXiv:hep-th/9909062]
164. A. Sen, Universality of the tachyon potential. JHEP **9912**, 027 (1999). [arXiv:hep-th/9911116]
165. M.R. Garousi, Tachyon couplings on non-BPS D-branes and Dirac–Born–Infeld action. Nucl. Phys. B **584**, 284 (2000). [arXiv:hep-th/0003122]
166. E.A. Bergshoeff, M. de Roo, T.C. de Wit, E. Eyras, S. Panda, T-duality and actions for non-BPS D-branes. JHEP **0005**, 009 (2000). [arXiv:hep-th/0003221]
167. J. Klusoň, Proposal for non-Bogomol'nyi–Prasad–Sommerfield D-brane action. Phys. Rev. D **62**, 126003 (2000). [arXiv:hep-th/0004106]
168. G.W. Gibbons, K. Hori, P. Yi, String fluid from unstable D-branes. Nucl. Phys. B **596**, 136 (2001). [arXiv:hep-th/0009061]
169. D. Kutasov, M. Mariño, G.W. Moore, Some exact results on tachyon condensation in string field theory. JHEP **0010**, 045 (2000). [arXiv:hep-th/0009148]
170. D. Kutasov, M. Mariño, G.W. Moore, Remarks on tachyon condensation in superstring field theory. arXiv:hep-th/0010108
171. P. Kraus, F. Larsen, Boundary string field theory of the $D\bar{D}$ system. Phys. Rev. D **63**, 106004 (2001). [arXiv:hep-th/0012198]
172. T. Takayanagi, S. Terashima, T. Uesugi, Brane-antibrane action from boundary string field theory. JHEP **0103**, 019 (2001). [arXiv:hep-th/0012210]
173. A. Sen, Rolling tachyon. JHEP **0204**, 048 (2002). [arXiv:hep-th/0203211]
174. A. Sen, Tachyon matter. JHEP **0207**, 065 (2002). [arXiv:hep-th/0203265]
175. A. Sen, Field theory of tachyon matter. Mod. Phys. Lett. A **17**, 1797 (2002). [arXiv:hep-th/0204143]
176. F. Leblond, A.W. Peet, SD-brane gravity fields and rolling tachyons. JHEP **0304**, 048 (2003). [arXiv:hep-th/0303035]
177. N.D. Lambert, H. Liu, J.M. Maldacena, Closed strings from decaying D-branes. JHEP **0703**, 014 (2007). [arXiv:hep-th/0303139]
178. M.R. Garousi, Off-shell extension of S-matrix elements and tachyonic effective actions. JHEP **0304**, 027 (2003). [arXiv:hep-th/0303239]
179. M.R. Garousi, Slowly varying tachyon and tachyon potential. JHEP **0305**, 058 (2003). [arXiv:hep-th/0304145]
180. A. Sen, Tachyon dynamics in open string theory. Int. J. Mod. Phys. A **20**, 5513 (2005). [arXiv:hep-th/0410103]
181. E. Witten, Interacting field theory of open superstrings. Nucl. Phys. B **276**, 291 (1986)
182. C.R. Preitschopf, C.B. Thorn, S.A. Yost, Superstring field theory. Nucl. Phys. B **337**, 363 (1990)
183. I.Ya. Aref'eva, P.B. Medvedev, A.P. Zubarev, Background formalism for superstring field theory. Phys. Lett. B **240**, 356 (1990)
184. I.Ya. Aref'eva, P.B. Medvedev, A.P. Zubarev, New representation for string field solves the consistency problem for open superstring field theory. Nucl. Phys. B **341**, 464 (1990)
185. I.Ya. Aref'eva, A.S. Koshelev, D.M. Belov, P.B. Medvedev, Tachyon condensation in cubic superstring field theory. Nucl. Phys. B **638**, 3 (2002). [arXiv:hep-th/0011117]
186. I.Ya. Aref'eva, L.V. Joukovskaya, A.S. Koshelev, Time evolution in superstring field theory on nonBPS brane. 1. Rolling tachyon and energy momentum conservation. JHEP **0309**, 012 (2003). [arXiv:hep-th/0301137]
187. N. Berkovits, Super-Poincaré invariant superstring field theory. Nucl. Phys. B **459**, 439 (1996). [arXiv:hep-th/9503099]

188. N. Berkovits, A. Sen, B. Zwiebach, Tachyon condensation in superstring field theory. Nucl. Phys. B **587**, 147 (2000). [arXiv:hep-th/0002211]
189. T. Erler, S. Konopka, I. Sachs, Resolving Witten's superstring field theory. JHEP **1404**, 150 (2014). [arXiv:1312.2948]
190. H. Kunitomo, Y. Okawa, Complete action of open superstring field theory. Prog. Theor. Exp. Phys. **2016**, 023B01 (2016). [arXiv:1508.00366]
191. T. Erler, Y. Okawa, T. Takezaki, Complete action for open superstring field theory with cyclic A_∞ structure. JHEP **1608**, 012 (2016). [arXiv:1602.02582]
192. H. Matsunaga, Comments on complete actions for open superstring field theory. arXiv:1510.06023
193. S. Konopka, I. Sachs, Open superstring field theory on the restricted Hilbert space. JHEP **1604**, 164 (2016). [arXiv:1602.02583]
194. T. Erler, Marginal solutions for the superstring. JHEP **0707**, 050 (2007). [arXiv:0704.0930]
195. E. Fuchs, M. Kroyter, Marginal deformation for the photon in superstring field theory. JHEP **0711**, 005 (2007). [arXiv:0706.0717]
196. E. Fuchs, M. Kroyter, On the classical equivalence of superstring field theories. JHEP **0810**, 054 (2008). [arXiv:0805.4386]
197. I.Ya. Aref'eva, R.V. Gorbachev and P.B. Medvedev, Pure gauge configurations and solutions to fermionic superstring field theories equations of motion. J. Phys. A **42**, 304001 (2009). [arXiv:0903.1273]
198. M. Kroyter, Superstring field theory equivalence: Ramond sector. JHEP **0910**, 044 (2009). [arXiv:0905.1168]
199. M. Kroyter, Comments on superstring field theory and its vacuum solution. JHEP **0908**, 048 (2009). [arXiv:0905.3501]
200. Y. Okawa, B. Zwiebach, Heterotic string field theory. JHEP **0407**, 042 (2004). [arXiv:hep-th/0406212]
201. N. Berkovits, Y. Okawa, B. Zwiebach, WZW-like action for heterotic string field theory. JHEP **0411**, 038 (2004). [arXiv:hep-th/0409018]
202. B. Jurčo, K. Münster, Type II superstring field theory: geometric approach and operadic description. JHEP **1304**, 126 (2013). [arXiv:1303.2323]
203. H. Matsunaga, Construction of a gauge-invariant action for type II superstring field theory. arXiv:1305.3893
204. T. Erler, S. Konopka, I. Sachs, NS-NS sector of closed superstring field theory. JHEP **1408**, 158 (2014). [arXiv:1403.0940]
205. T. Erler, S. Konopka, I. Sachs, Ramond equations of motion in superstring field theory. JHEP **1511**, 199 (2015). [arXiv:1506.05774]
206. A. Sen, BV master action for heterotic and type II string field theories. JHEP **1602**, 087 (2016). [arXiv:1508.05387]
207. A. Sen, Covariant action for type IIB supergravity. JHEP **1607**, 017 (2016). [arXiv:1511.08220]
208. A. Sen, Wilsonian effective action of superstring theory. arXiv:1609.00459
209. N. Barnaby, N. Kamran, Dynamics with infinitely many derivatives: the initial value problem. JHEP **0802**, 008 (2008). [arXiv:0709.3968]
210. G. Calcagni, M. Montobbio, G. Nardelli, Localization of nonlocal theories. Phys. Lett. B **662**, 285 (2008). [arXiv:0712.2237]
211. G. Calcagni, G. Nardelli, Tachyon solutions in boundary and cubic string field theory. Phys. Rev. D **78**, 126010 (2008). [arXiv:0708.0366]
212. G. Calcagni, G. Nardelli, Kinks of open superstring field theory. Nucl. Phys. B **823**, 234 (2009). [arXiv:0904.3744]
213. G. Calcagni, G. Nardelli, String theory as a diffusing system. JHEP **1002**, 093 (2010). [arXiv:0910.2160]
214. N. Moeller, B. Zwiebach, Dynamics with infinitely many time derivatives and rolling tachyons. JHEP **0210**, 034 (2002). [arXiv:hep-th/0207107]

215. I. Kishimoto, Y. Matsuo, E. Watanabe, A universal nonlinear relation among boundary states in closed string field theory. Prog. Theor. Phys. **111**, 433 (2004). [arXiv:hep-th/0312122]
216. A. Giveon, M. Porrati, E. Rabinovici, Target space duality in string theory. Phys. Rep. **244**, 77 (1994). [arXiv:hep-th/9401139]
217. G. Calcagni, L. Modesto, Nonlocality in string theory. J. Phys. A **47**, 355402 (2014). [arXiv:1310.4957]
218. E. Cremmer, J. Scherk, Spontaneous compactification of space in an Einstein–Yang–Mills–Higgs model. Nucl. Phys. B **108**, 409 (1976)
219. E. Cremmer, J. Scherk, Spontaneous compactification of extra space dimensions. Nucl. Phys. B **118**, 61 (1977)
220. M.B. Green, J.H. Schwarz, L. Brink, $N = 4$ Yang–Mills and $N = 8$ supergravity as limits of string theories. Nucl. Phys. B **198**, 474 (1982)
221. B. de Wit, D.Z. Freedman, On $SO(8)$ extended supergravity. Nucl. Phys. B **130**, 105 (1977)
222. G.W. Gibbons, S.W. Hawking, Classification of gravitational instanton symmetries. Commun. Math. Phys. **66**, 291 (1979)
223. D.N. Page, A physical picture of the K3 gravitational instanton. Phys. Lett. B **80**, 55 (1978)
224. P. Candelas, G.T. Horowitz, A. Strominger, E. Witten, Vacuum configurations for superstrings, Nucl. Phys. B **258**, 46 (1985)
225. A. Strominger, E. Witten, New manifolds for superstring compactification. Commun. Math. Phys. **101**, 341 (1985)
226. A. Strominger, Superstrings with torsion. Nucl. Phys. B **274**, 253 (1986)
227. E. Calabi, The space of Kähler metrics. Proc. Int. Congr. Math. **2**, 206 (1954)
228. E. Calabi, On Kähler manifolds with vanishing canonical class, in *Algebraic Geometric and Topology: A Symposium in Honor of S. Lefschetz*, ed. by R.H. Fox et al. (Princeton University Press, Princeton, 1957)
229. S.-T. Yau, Calabi's conjecture and some new results in algebraic geometry. Proc. Natl. Acad. Sci. **74**, 1798 (1977)
230. S.-T. Yau, On the Ricci curvature of a compact Kähler manifold and the complex Monge–Ampère equation, I. Commun. Pure Appl. Math. **31**, 339 (1978)
231. L. Dixon, J.A. Harvey, C. Vafa, E. Witten, Strings on orbifolds. Nucl. Phys. B **261**, 678 (1985)
232. L. Dixon, J.A. Harvey, C. Vafa, E. Witten, Strings on orbifolds (II). Nucl. Phys. B **274**, 285 (1986)
233. L. Dixon, D. Friedan, E. Martinec, S. Shenker, The conformal field theory of orbifolds. Nucl. Phys. B **282**, 13 (1987)
234. K.S. Narain, M.H. Sarmadi, C. Vafa, Asymmetric orbifolds. Nucl. Phys. B **288**, 551 (1987)
235. A. Sagnotti, Open strings and their symmetry groups, in *Nonperturbative Quantum Field Theory*, ed. by G. 't Hooft, A. Jaffe, G. Mack, P.K. Mitter, R. Stora (Plenum, New York, 1988). [arXiv:hep-th/0208020]
236. M. Bianchi, A. Sagnotti, On the systematics of open-string theories. Phys. Lett. B **247**, 517 (1990)
237. M. Bianchi, A. Sagnotti, Twist symmetry and open string Wilson lines. Nucl. Phys. B **361**, 519 (1991)
238. M. Bianchi, G. Pradisi, A. Sagnotti, Toroidal compactification and symmetry breaking in open string theories. Nucl. Phys. B **376**, 365 (1992)
239. I.R. Klebanov, M.J. Strassler, Supergravity and a confining gauge theory: duality cascades and χSB-resolution of naked singularities. JHEP **0008**, 052 (2000). [arXiv:hep-th/0007191]
240. P. Candelas, X.C. de la Ossa, Comments on conifolds. Nucl. Phys. B **342**, 246 (1990)
241. S.B. Giddings, S. Kachru, J. Polchinski, Hierarchies from fluxes in string compactifications. Phys. Rev. D **66**, 106006 (2002). [arXiv:hep-th/0105097]
242. S. Hamidi, C. Vafa, Interactions on orbifolds. Nucl. Phys. B **279**, 465 (1987)
243. E. Witten, New issues in manifolds of $SU(3)$ holonomy. Nucl. Phys. B **268**, 79 (1986)
244. S. Gukov, C. Vafa, E. Witten, CFT's from Calabi–Yau four-folds. Nucl. Phys. B **584**, 69 (2000); Erratum-ibid. B **608**, 477 (2001). [arXiv:hep-th/9906070]

245. P. Kaste, R. Minasian, A. Tomasiello, Supersymmetric M theory compactifications with fluxes on seven-manifolds and G structures. JHEP **0307**, 004 (2003). [arXiv:hep-th/0303127]
246. E. Witten, Dimensional reduction of superstring models. Phys. Lett. B **155**, 151 (1985)
247. S. Ferrara, C. Kounnas, M. Porrati, General dimensional reduction of ten-dimensional supergravity and superstring. Phys. Lett. B **181**, 263 (1986)
248. M. Cvetič, J. Louis, B.A. Ovrut, A string calculation of the Kähler potentials for moduli of Z_N orbifolds. Phys. Lett. B **206**, 227 (1988)
249. P. Candelas, X.C. de la Ossa, Moduli space of Calabi–Yau manifolds. Nucl. Phys. B **355**, 455 (1991)
250. T.W. Grimm, J. Louis, The effective action of $N = 1$ Calabi–Yau orientifolds. Nucl. Phys. B **699**, 387 (2004). [arXiv:hep-th/0403067]
251. D. Baumann, L. McAllister, *Inflation and String Theory* (Cambridge University Press, Cambridge, 2015). [arXiv:1404.2601]
252. P. Svrček, E. Witten, Axions in string theory. JHEP **0606**, 051 (2006). [arXiv:hep-th/0605206]
253. X.-G. Wen, E. Witten, World-sheet instantons and the Peccei–Quinn symmetry. Phys. Lett. B **166**, 397 (1986)
254. M. Dine, N. Seiberg, X.-G. Wen, E. Witten, Nonperturbative effects on the string world sheet. Nucl. Phys. B **278**, 769 (1986)
255. R. Blumenhagen, D. Herschmann, E. Plauschinn, The challenge of realizing F-term axion monodromy inflation in string theory. JHEP **1501**, 007 (2015). [arXiv:1409.7075]
256. J.M. Maldacena, The large-N limit of superconformal field theories and supergravity. Adv. Theor. Math. Phys. **2**, 231 (1998) [Int. J. Theor. Phys. **38**, 1113 (1999)]. [arXiv:hep-th/9711200]
257. S.S. Gubser, I.R. Klebanov, A.M. Polyakov, Gauge theory correlators from non-critical string theory. Phys. Lett. B **428**, 105 (1998). [arXiv:hep-th/9802109]
258. E. Witten, Anti de Sitter space and holography. Adv. Theor. Math. Phys. **2**, 253 (1998). [arXiv:hep-th/9802150]
259. O. Aharony, S.S. Gubser, J.M. Maldacena, H. Ooguri, Y. Oz, Large N field theories, string theory and gravity. Phys. Rep. **323**, 183 (2000). [arXiv:hep-th/9905111]
260. A.V. Ramallo, Introduction to the AdS/CFT correspondence. Springer Proc. Phys. **161**, 411 (2015). [arXiv:1310.4319]
261. V.E. Hubeny, The AdS/CFT correspondence. Class. Quantum Grav. **32**, 124010 (2015). [arXiv:1501.00007]
262. S. Kachru, E. Silverstein, 4D conformal field theories and strings on orbifolds. Phys. Rev. Lett. **80**, 4855 (1998). [arXiv:hep-th/9802183]
263. A.E. Lawrence, N. Nekrasov, C. Vafa, On conformal field theories in four dimensions. Nucl. Phys. B **533**, 199 (998). [arXiv:hep-th/9803015]
264. A. Kehagias, New type IIB vacua and their F-theory interpretation. Phys. Lett. B **435**, 337 (1998). [arXiv:hep-th/9805131]
265. I.R. Klebanov, E. Witten, Superconformal field theory on threebranes at a Calabi–Yau singularity. Nucl. Phys. B **536**, 199 (1998). [arXiv:hep-th/9807080]
266. B.S. Acharya, J.M. Figueroa-O'Farrill, C.M. Hull, B.J. Spence, Branes at conical singularities and holography. Adv. Theor. Math. Phys. **2**, 1249 (1998). [arXiv:hep-th/9808014]
267. J. Polchinski, A. Strominger, New vacua for type II string theory. Phys. Lett. B **388**, 736 (1996). [arXiv:hep-th/9510227]
268. O. DeWolfe, S.B. Giddings, Scales and hierarchies in warped compactifications and brane worlds. Phys. Rev. D **67**, 066008 (2003). [arXiv:hep-th/0208123]
269. K. Becker, M. Becker, M theory on eight manifolds. Nucl. Phys. B **477**, 155 (1996). [arXiv:hep-th/9605053]
270. K. Dasgupta, G. Rajesh, S. Sethi, M-theory, orientifolds and G-flux. JHEP **9908**, 023 (1999). [arXiv:hep-th/9908088]
271. P. Candelas, D.J. Raine, Compactification and supersymmetry in $d = 11$ supergravity. Nucl. Phys. B **248**, 415 (1984)

272. T.R. Taylor, C. Vafa, RR flux on Calabi–Yau and partial supersymmetry breaking. Phys. Lett. B **474**, 130 (2000). [arXiv:hep-th/9912152]
273. M. Haack, J. Louis, M theory compactified on Calabi–Yau fourfolds with background flux. Phys. Lett. B **507**, 296 (2001). [arXiv:hep-th/0103068]
274. B. de Wit, H. Nicolai, A new SO(7) invariant solution of $d = 11$ supergravity. Phys. Lett. B **148**, 60 (1984)
275. L. Randall, R. Sundrum, An alternative to compactification. Phys. Rev. Lett. **83**, 4690 (1999). [arXiv:hep-th/9906064]
276. B. De Wit, D.J. Smit, Residual supersymmetry of compactified $d = 10$ supergravity. Nucl. Phys. B **283**, 165 (1987)
277. J.M. Maldacena, C. Nuñez, Supergravity description of field theories on curved manifolds and a no go theorem. Int. J. Mod. Phys. A **16**, 822 (2001). [arXiv:hep-th/0007018]
278. L. Randall, R. Sundrum, Large mass hierarchy from a small extra dimension. Phys. Rev. Lett. **83**, 3370 (1999). [arXiv:hep-ph/9905221]
279. H.L. Verlinde, Holography and compactification. Nucl. Phys. B **580**, 264 (2000). [arXiv:hep-th/9906182]
280. C.S. Chan, P.L. Paul, H.L. Verlinde, A note on warped string compactification. Nucl. Phys. B **581**, 156 (2000) [arXiv:hep-th/0003236]
281. M. Graña, J. Polchinski, Gauge-gravity duals with a holomorphic dilaton. Phys. Rev. D **65**, 126005 (2002). [arXiv:hep-th/0106014]
282. S. Kachru, M.B. Schulz, S. Trivedi, Moduli stabilization from fluxes in a simple IIB orientifold. JHEP **0310**, 007 (2003). [arXiv:hep-th/0201028]
283. A.R. Frey, J. Polchinski, $N = 3$ warped compactifications. Phys. Rev. D **65**, 126009 (2002). [arXiv:hep-th/0201029]
284. S. Gurrieri, J. Louis, A. Micu, D. Waldram, Mirror symmetry in generalized Calabi–Yau compactifications. Nucl. Phys. B **654**, 61 (2003). [arXiv:hep-th/0211102]
285. S. Kachru, M.B. Schulz, P.K. Tripathy, S.P. Trivedi, New supersymmetric string compactifications. JHEP **0303**, 061 (2003). [arXiv:hep-th/0211182]
286. P.K. Tripathy, S.P. Trivedi, Compactification with flux on K3 and tori. JHEP **0303**, 028 (2003). [arXiv:hep-th/0301139]
287. K. Becker, M. Becker, K. Dasgupta, P.S. Green, Compactifications of heterotic theory on non-Kähler complex manifolds, I. JHEP **0304**, 007 (2003). [arXiv:hep-th/0301161]
288. R. Blumenhagen, D. Lüst, T.R. Taylor, Moduli stabilization in chiral type IIB orientifold models with fluxes. Nucl. Phys. B **663**, 319 (2003). [arXiv:hep-th/0303016]
289. J.F.G. Cascales, A.M. Uranga, Chiral 4d string vacua with D-branes and NSNS and RR fluxes. JHEP **0305**, 011 (2003). [arXiv:hep-th/0303024]
290. K. Becker, M. Becker, P.S. Green, K. Dasgupta, E. Sharpe, Compactifications of heterotic strings on nonKahler complex manifolds II. Nucl. Phys. B **678**, 19 (2004). [arXiv:hep-th/0310058]
291. J.-P. Derendinger, C. Kounnas, P.M. Petropoulos, F. Zwirner, Superpotentials in IIA compactifications with general fluxes. Nucl. Phys. B **715**, 211 (2005). [arXiv:hep-th/0411276]
292. S. Kachru, A.-K. Kashani-Poor, Moduli potentials in type-IIA compactifications with RR and NS flux. JHEP **0503**, 066 (2005). [arXiv:hep-th/0411279]
293. G. Villadoro, F. Zwirner, $N = 1$ effective potential from dual type-IIA D6/O6 orientifolds with general fluxes. JHEP **0506**, 047 (2005). [arXiv:hep-th/0503169]
294. O. DeWolfe, A. Giryavets, S. Kachru, W. Taylor, Type IIA moduli stabilization. JHEP **0507**, 066 (2005). [arXiv:hep-th/0505160]
295. P.G. Cámara, A. Font, L.E. Ibáñez, Fluxes, moduli fixing and MSSM-like vacua in a simple IIA orientifold. JHEP **0509**, 013 (2005). [arXiv:hep-th/0506066]
296. B.S. Acharya, F. Benini, R. Valandro, Fixing moduli in exact type IIA flux vacua. JHEP **0702**, 018 (2007). [arXiv:hep-th/0607223]
297. R. Donagi, B.A. Ovrut, T. Pantev, D. Waldram, Standard models from heterotic M theory. Adv. Theor. Math. Phys. **5**, 93 (2001). [arXiv:hep-th/9912208]

298. R. Blumenhagen, L. Görlich, B. Körs, D. Lüst, Noncommutative compactifications of type I strings on tori with magnetic background flux. JHEP **0010**, 006 (2000). [arXiv:hep-th/0007024]
299. R. Blumenhagen, B. Körs, D. Lüst, Type I strings with F- and B-flux. JHEP **0102**, 030 (2001). [arXiv:hep-th/0012156]
300. V. Braun, Y.-H. He, B.A. Ovrut, T. Pantev, A heterotic standard model. Phys. Lett. B **618**, 252 (2005). [arXiv:hep-th/0501070]
301. V. Braun, Y.-H. He, B.A. Ovrut, T. Pantev, A standard model from the $E_8 \times E_8$ heterotic superstring. JHEP **0506**, 039 (2005). [arXiv:hep-th/0502155]
302. V. Braun, Y.-H. He, B.A. Ovrut, T. Pantev, Vector bundle extensions, sheaf cohomology, and the heterotic standard model. Adv. Theor. Math. Phys. **10**, 525 (2006). [arXiv:hep-th/0505041]
303. H.P. Nilles, S. Ramos-Sánchez, M. Ratz, P.K.S. Vaudrevange, From strings to the MSSM. Eur. Phys. J. C **59**, 249 (2009). [arXiv:0806.3905]
304. G. Aldazabal, L.E. Ibáñez, F. Quevedo, A.M. Uranga, D-branes at singularities: a bottom-up approach to the string embedding of the standard model. JHEP **0008**, 002 (2000). [arXiv:hep-th/0005067]
305. G. Aldazabal, S. Franco, L.E. Ibáñez, R. Rabadán, A.M. Uranga, $D = 4$ chiral string compactifications from intersecting branes. J. Math. Phys. **42**, 3103 (2001). [arXiv:hep-th/0011073]
306. G. Aldazabal, S. Franco, L.E. Ibáñez, R. Rabadán, A.M. Uranga, Intersecting brane worlds. JHEP **0102**, 047 (2001). [arXiv:hep-ph/0011132]
307. L.E. Ibáñez, F. Marchesano, R. Rabadán, Getting just the standard model at intersecting branes. JHEP **0111**, 002 (2001). [arXiv:hep-th/0105155]
308. R. Blumenhagen, B. Körs, D. Lüst, T. Ott, The Standard Model from stable intersecting brane world orbifolds. Nucl. Phys. B **616**, 3 (2001). [arXiv:hep-th/0107138]
309. D. Bailin, G.V. Kraniotis, A. Love, Standard-like models from intersecting D4-branes. Phys. Lett. B **530**, 202 (2002). [arXiv:hep-th/0108131]
310. D. Cremades, L.E. Ibáñez, F. Marchesano, Intersecting brane models of particle physics and the Higgs mechanism. JHEP **0207**, 022 (2002). [arXiv:hep-th/0203160]
311. D. Cremades, L.E. Ibáñez, F. Marchesano, Standard model at intersecting D5-branes: lowering the string scale. Nucl. Phys. B **643**, 93 (2002). [arXiv:hep-th/0205074]
312. C. Kokorelis, New Standard Model vacua from intersecting branes. JHEP **0209**, 029 (2002). [arXiv:hep-th/0205147]
313. C. Kokorelis, Exact Standard Model compactifications from intersecting branes. JHEP **0208**, 036 (2002). [arXiv:hep-th/0206108]
314. C. Kokorelis, Exact Standard Model structures from intersecting D5-branes. Nucl. Phys. B **677**, 115 (2004). [arXiv:hep-th/0207234]
315. D. Bailin, G.V. Kraniotis, A. Love, New standard-like models from intersecting D4-branes. Phys. Lett. B **547**, 43 (2002). [arXiv:hep-th/0208103]
316. D. Bailin, G.V. Kraniotis, A. Love, Standard-like models from intersecting D5-branes. Phys. Lett. B **553**, 79 (2003). [arXiv:hep-th/0210219]
317. D. Bailin, G.V. Kraniotis, A. Love, Intersecting D5-brane models with massive vector-like leptons. JHEP **0302**, 052 (2003). [arXiv:hep-th/0212112]
318. C. Kokorelis, Standard model compactifications of IIA $Z_3 \times Z_3$ orientifolds from intersecting D6-branes. Nucl. Phys. B **732**, 341 (2006). [arXiv:hep-th/0412035]
319. D. Berenstein, V. Jejjala, R.G. Leigh, Standard model on a D-brane. Phys. Rev. Lett. **88**, 071602 (2002). [arXiv:hep-ph/0105042]
320. M. Cvetič, G. Shiu, A.M. Uranga, Three-family supersymmetric standardlike models from intersecting brane worlds. Phys. Rev. Lett. **87**, 201801 (2001). [arXiv:hep-th/0107143]
321. M. Cvetič, G. Shiu, A.M. Uranga, Chiral four-dimensional $N = 1$ supersymmetric type IIA orientifolds from intersecting D6-branes. Nucl. Phys. B **615**, 3 (2001). [arXiv:hep-th/0107166]

322. M. Cvetič, P. Langacker, G. Shiu, Phenomenology of a three-family standardlike string model. Phys. Rev. D **66**, 066004 (2002). [arXiv:hep-ph/0205252]
323. M. Cvetič, P. Langacker, G. Shiu, A three-family standard-like orientifold model: Yukawa couplings and hierarchy. Nucl. Phys. B **642**, 139 (2002). [arXiv:hep-th/0206115]
324. R. Blumenhagen, L. Görlich, T. Ott, Supersymmetric intersecting branes on the type 2A $T^6/\mathbb{Z}_4$ orientifold. JHEP **0301**, 021 (2003). [arXiv:hep-th/0211059]
325. M. Cvetič, I. Papadimitriou, G. Shiu, Supersymmetric three family $SU(5)$ grand unified models from type IIA orientifolds with intersecting D6-branes. Nucl. Phys. B **659**, 193 (2003); Erratum-ibid. B **696**, 298 (2004). [arXiv:hep-th/0212177]
326. D. Cremades, L.E. Ibáñez, F. Marchesano, Yukawa couplings in intersecting D-brane models. JHEP **0307**, 038 (2003). [arXiv:hep-th/0302105]
327. G. Honecker, Chiral supersymmetric models on an orientifold of $\mathbb{Z}_4 \times \mathbb{Z}_2$ with intersecting D6-branes. Nucl. Phys. B **666**, 175 (2003). [arXiv:hep-th/0303015]
328. M. Cvetič, I. Papadimitriou, Conformal field theory couplings for intersecting D-branes on orientifolds. Phys. Rev. D **68**, 046001 (2003); Erratum-ibid. D **70**, 029903(E) (2004). [arXiv:hep-th/0303083]
329. M. Cvetič, I. Papadimitriou, More supersymmetric standardlike models from intersecting D6-branes on type IIA orientifolds. Phys. Rev. D **67**, 126006 (2003). [arXiv:hep-th/0303197]
330. R. Blumenhagen, D. Lüst, S. Stieberger, Gauge unification in supersymmetric intersecting brane worlds. JHEP **0307**, 036 (2003). [arXiv:hep-th/0305146]
331. M. Cvetič, T. Li, T. Liu, Supersymmetric Pati–Salam models from intersecting D6-branes: a road to the Standard Model. Nucl. Phys. B **698**, 163 (2004). [arXiv:hep-th/0403061]
332. T.P.T. Dijkstra, L.R. Huiszoon, A.N. Schellekens, Chiral supersymmetric standard model spectra from orientifolds of Gepner models. Phys. Lett. B **609**, 408 (2005). [arXiv:hep-th/0403196]
333. G. Honecker, T. Ott, Getting just the supersymmetric standard model at intersecting branes on the $\mathbb{Z}_6$ orientifold. Phys. Rev. D **70**, 126010 (2004); Erratum-ibid. D **71**, 069902 (2005). [arXiv:hep-th/0404055]
334. D. Lüst, P. Mayr, R. Richter, S. Stieberger, Scattering of gauge, matter, and moduli fields from intersecting branes. Nucl. Phys. B **696**, 205 (2004). [arXiv:hep-th/0404134]
335. M. Cvetič, P. Langacker, T. Li, T. Liu, D6-brane splitting on type IIA orientifolds. Nucl. Phys. B **709**, 241 (2005). [arXiv:hep-th/0407178]
336. M. Berkooz, M.R. Douglas, R.G. Leigh, Branes intersecting at angles. Nucl. Phys. B **480**, 265 (1996). [arXiv:hep-th/9606139]
337. R. Blumenhagen, B. Körs, D. Lüst, S. Stieberger, Four-dimensional string compactifications with D-branes, orientifolds and fluxes. Phys. Rep. **445**, 1 (2007). [arXiv:hep-th/0610327]
338. P.G. Cámara, L.E. Ibáñez, A.M. Uranga, Flux-induced SUSY-breaking soft terms on D7-D3 brane systems. Nucl. Phys. B **708**, 268 (2005). [arXiv:hep-th/0408036]
339. F. Marchesano, G. Shiu, Minimal supersymmetric standard model string vacua from flux compactifications. Phys. Rev. D **71**, 011701(R) (2005). [arXiv:hep-th/0408059]
340. L.E. Ibáñez, Fluxed minimal supersymmetric standard model. Phys. Rev. D **71**, 055005 (2005). [arXiv:hep-ph/0408064]
341. M. Cvetič, T. Liu, Supersymmetric standard models, flux compactification and moduli stabilization. Phys. Lett. B **610**, 122 (2005). [arXiv:hep-th/0409032]
342. F. Marchesano, G. Shiu, Building MSSM flux vacua. JHEP **0411**, 041 (2004). [arXiv:hep-th/0409132]
343. M. Cvetič, T. Li, T. Liu, Standard-like models as type IIB flux vacua. Phys. Rev. D **71**, 106008 (2005). [arXiv:hep-th/0501041]
344. R. Blumenhagen, M. Cvetič, P. Langacker, G. Shiu, Toward realistic intersecting D-brane models. Ann. Rev. Nucl. Part. Sci. **55**, 71 (2005). [arXiv:hep-th/0502005]
345. H. Verlinde, M. Wijnholt, Building the standard model on a D3-brane. JHEP **0701**, 106 (2007). [arXiv:hep-th/0508089]
346. F. Marchesano, Progress in D-brane model building. Fortsch. Phys. **55**, 491 (2007). [arXiv:hep-th/0702094]

347. J.P. Conlon, A. Maharana, F. Quevedo, Towards realistic string vacua. JHEP **0905**, 109 (2009). [arXiv:0810.5660]
348. M.J. Dolan, S. Krippendorf, F. Quevedo, Towards a systematic construction of realistic D-brane models on a del Pezzo singularity. JHEP **1110**, 024 (2011). [arXiv:1106.6039]
349. M. Cicoli, C. Mayrhofer, R. Valandro, Moduli stabilisation for chiral global models. JHEP **1202**, 062 (2012). [arXiv:1110.3333]
350. M. Cicoli, S. Krippendorf, C. Mayrhofer, F. Quevedo, R. Valandro, D-branes at del Pezzo singularities: global embedding and moduli stabilisation. JHEP **1209**, 019 (2012). [arXiv:1206.5237]
351. M. Cicoli, D. Klevers, S. Krippendorf, C. Mayrhofer, F. Quevedo, R. Valandro, Explicit de Sitter flux vacua for global string models with chiral matter, JHEP **1405**, 001 (2014). [arXiv:1312.0014]
352. R. Blumenhagen, A. Font, M. Fuchs, D. Herschmann, E. Plauschinn, Y. Sekiguchi, F. Wolf, A flux-scaling scenario for high-scale moduli stabilization in string theory. Nucl. Phys. B **897**, 500 (2015). [arXiv:1503.07634]
353. M. Berg, M. Haack, B. Körs, String loop corrections to Kähler potentials in orientifolds. JHEP **0511**, 030 (2005). [arXiv:hep-th/0508043]
354. M. Berg, M. Haack, B. Körs, Stabilization of the compactification volume by quantum corrections. Phys. Rev. Lett. **96**, 021601 (2006). [arXiv:hep-th/0508171]
355. S. Ferrara, L. Girardello, H.P. Nilles, Breakdown of local supersymmetry through gauge fermion condensates. Phys. Lett. B **125**, 457 (1983)
356. I. Affleck, M. Dine, N. Seiberg, Supersymmetry breaking by instantons. Phys. Rev. Lett. **51**, 1026 (1983)
357. I. Affleck, M. Dine, N. Seiberg, Dynamical supersymmetry breaking in supersymmetric QCD. Nucl. Phys. B **241**, 493 (1984)
358. I. Affleck, M. Dine, N. Seiberg, Dynamical supersymmetry breaking in four dimensions and its phenomenological implications. Nucl. Phys. B **256**, 557 (1985)
359. M.A. Shifman, A.I. Vainshtein, On gluino condensation in supersymmetric gauge theories with SU(N) and O(N) groups. Sov. Phys. JETP **66**, 1100 (1987) [Nucl. Phys. B **296**, 445 (1988)]
360. M. Dine, N. Seiberg, X.-G. Wen, E. Witten, Nonperturbative effects on the string world sheet (II). Nucl. Phys. B **289**, 319 (1987)
361. K. Becker, M. Becker, A. Strominger, Fivebranes, membranes and non-perturbative string theory. Nucl. Phys. B **456**, 130 (1995). [arXiv:hep-th/9507158]
362. E. Witten, Non-perturbative superpotentials in string theory. Nucl. Phys. B **474**, 343 (1996). [arXiv:hep-th/9604030]
363. J.A. Harvey, G.W. Moore, Superpotentials and membrane instantons. arXiv:hep-th/9907026
364. S.H. Katz, C. Vafa, Geometric engineering of $N = 1$ quantum field theories. Nucl. Phys. B **497**, 196 (1997). [arXiv:hep-th/9611090]
365. J.-P. Derendinger, L.E. Ibáñez, H.P. Nilles, On the low energy $d = 4$, $N = 1$ supergravity theory extracted from the $d = 10$, $N = 1$ superstring. Phys. Lett. B **155**, 65 (1985)
366. M. Dine, R. Rohm, N. Seiberg, E. Witten, Gluino condensation in superstring models. Phys. Lett. B **156**, 55 (1985)
367. J.-P. Derendinger, L.E. Ibáñez, H.P. Nilles, On the low-energy limit of superstring theories. Nucl. Phys. B **267**, 365 (1986)
368. N.V. Krasnikov, On supersymmetry breaking in superstring theories. Phys. Lett. B **193**, 37 (1987)
369. J.A. Casas, Z. Lalak, C. Muñoz, G.G. Ross, Hierarchical supersymmetry breaking and dynamical determination of compactification parameters by non-perturbative effects. Nucl. Phys. B **347**, 243 (1990)
370. A. Font, L.E. Ibáñez, D. Lüst, F. Quevedo, Supersymmetry breaking from duality invariant gaugino condensation. Phys. Lett. B **245**, 401 (1990)

371. S. Ferrara, N. Magnoli, T.R. Taylor, G. Veneziano, Duality and supersymmetry breaking in string theory. Phys. Lett. B **245**, 409 (1990)
372. T.R. Taylor, Dilaton, gaugino condensation and supersymmetry breaking. Phys. Lett. B **252**, 59 (1990)
373. B. de Carlos, J.A. Casas, C. Muñoz, Supersymmetry breaking and determination of the unification gauge coupling constant in string theories. Nucl. Phys. B **399**, 623 (1993). [arXiv:hep-th/9204012]
374. S. Kachru, R. Kallosh, A.D. Linde, S.P. Trivedi, de Sitter vacua in string theory. Phys. Rev. D **68**, 046005 (2003). [arXiv:hep-th/0301240]
375. M. Endo, M. Yamaguchi, K. Yoshioka, Bottom-up approach to moduli dynamics in heavy gravitino scenario: superpotential, soft terms, and sparticle mass spectrum. Phys. Rev. D **72**, 015004 (2005). [arXiv:hep-ph/0504036]
376. G. Curio, A. Krause, D. Lüst, Moduli stabilization in the heterotic/IIB discretuum. Fortsch. Phys. **54**, 225 (2006). [arXiv:hep-th/0502168]
377. F. Denef, M.R. Douglas, B. Florea, A. Grassi, S. Kachru, Fixing all moduli in a simple *F*-theory compactification. Adv. Theor. Math. Phys. **9**, 861 (2005). [arXiv:hep-th/0503124]
378. D. Lüst, S. Reffert, W. Schulgin, S. Stieberger, Moduli stabilization in type IIB orientifolds (I). Nucl. Phys. B **766**, 68 (2007). [arXiv:hep-th/0506090]
379. E. Dudas, S.K. Vempati, Large *D*-terms, hierarchical soft spectra and moduli stabilisation. Nucl. Phys. B **727**, 139 (2005). [arXiv:hep-th/0506172]
380. G. Aldazabal, P.G. Cámara, A. Font, L.E. Ibáñez, More dual fluxes and moduli fixing. JHEP **0605**, 070 (2006). [arXiv:hep-th/0602089]
381. D. Lüst, S. Reffert, E. Scheidegger, W. Schulgin, S. Stieberger, Moduli stabilization in type IIB orientifolds (II). Nucl. Phys. B **766**, 178 (2007). [arXiv:hep-th/0609013]
382. J.P. Conlon, S.S. Abdussalam, F. Quevedo, K. Suruliz, Soft SUSY breaking terms for chiral matter in IIB string compactifications. JHEP **0701**, 032 (2007). [arXiv:hep-th/0610129]
383. R. Blumenhagen, S. Moster, E. Plauschinn, Moduli stabilisation versus chirality for MSSM like type IIB orientifolds. JHEP **0801**, 058 (2008). [arXiv:arXiv:0711.3389]
384. A.R. Frey, M. Lippert, B. Williams, Fall of stringy de Sitter spacetime. Phys. Rev. D **68**, 046008 (2003). [arXiv:hep-th/0305018]
385. C. Escoda, M. Gómez-Reino, F. Quevedo, Saltatory de Sitter string vacua. JHEP **0311**, 065 (2003). [arXiv:hep-th/0307160]
386. C.P. Burgess, R. Kallosh, F. Quevedo, de Sitter string vacua from supersymmetric D-terms. JHEP **0310**, 056 (2003). [arXiv:hep-th/0309187]
387. J.F.G. Cascales, M.P. García del Moral, F. Quevedo, A.M. Uranga, Realistic D-brane models on warped throats: fluxes, hierarchies and moduli stabilization. JHEP **0402**, 031 (2004). [arXiv:hep-th/0312051]
388. R. Brustein, S.P. de Alwis, Moduli potentials in string compactifications with fluxes: mapping the discretuum. Phys. Rev. D **69**, 126006 (2004). [arXiv:hep-th/0402088]
389. F. Denef, M.R. Douglas, B. Florea, Building a better racetrack. JHEP **0406**, 034 (2004). [arXiv:hep-th/0404257]
390. V. Balasubramanian, P. Berglund, Stringy corrections to Kähler potentials, SUSY breaking, and the cosmological constant problem. JHEP **0411**, 085 (2004). [arXiv:hep-th/0408054]
391. R. Kallosh, A.D. Linde, Landscape, the scale of SUSY breaking, and inflation. JHEP **0412**, 004 (2004). [arXiv:hep-th/0411011]
392. K. Choi, A. Falkowski, H.P. Nilles, M. Olechowski, S. Pokorski, Stability of flux compactifications and the pattern of supersymmetry breaking. JHEP **0411**, 076 (2004). [arXiv:hep-th/0411066]
393. V. Balasubramanian, P. Berglund, J.P. Conlon, F. Quevedo, Systematics of moduli stabilisation in Calabi–Yau flux compactifications. JHEP **0503**, 007 (2005). [arXiv:hep-th/0502058]
394. K. Choi, A. Falkowski, H.P. Nilles, M. Olechowski, Soft supersymmetry breaking in KKLT flux compactification. Nucl. Phys. B **718**, 113 (2005). [arXiv:hep-th/0503216]
395. J.P. Conlon, F. Quevedo, K. Suruliz, Large-volume flux compactifications: moduli spectrum and D3/D7 soft supersymmetry breaking. JHEP **0508**, 007 (2005). [arXiv:hep-th/0505076]

396. S.B. Giddings, A. Maharana, Dynamics of warped compactifications and the shape of the warped landscape. Phys. Rev. D **73**, 126003 (2006). [arXiv:hep-th/0507158]
397. G. Villadoro, F. Zwirner, de Sitter vacua via consistent D terms. Phys. Rev. Lett. **95**, 231602 (2005). [arXiv:hep-th/0508167]
398. J.J. Blanco-Pillado, R. Kallosh, A.D. Linde, Supersymmetry and stability of flux vacua. JHEP **0605**, 053 (2006). [arXiv:hep-th/0511042]
399. A. Achúcarro, B. de Carlos, J.A. Casas, L. Doplicher, de Sitter vacua from uplifting D-terms in effective supergravities from realistic strings. JHEP **0606**, 014 (2006). [arXiv:hep-th/0601190]
400. O. Lebedev, H.P. Nilles, M. Ratz, de Sitter vacua from matter superpotentials. Phys. Lett. B **636**, 126 (2006). [arXiv:hep-th/0603047]
401. P. Ouyang, Holomorphic D7 branes and flavored $N = 1$ gauge theories. Nucl. Phys. B **699**, 207 (2004). [arXiv:hep-th/0311084]
402. S. Kuperstein, Meson spectroscopy from holomorphic probes on the warped deformed conifold. JHEP **0503**, 014 (2005). [arXiv:hep-th/0411097]
403. G.L. Cardoso, G. Curio, G. Dall'Agata, D. Lüst, Heterotic string theory on non-Kähler manifolds with H-flux and gaugino condensate. Fortsch. Phys. **52**, 483 (2004). [arXiv:hep-th/0310021]
404. B. de Carlos, S. Gurrieri, A. Lukas, A. Micu, Moduli stabilisation in heterotic string compactifications. JHEP **0603**, 005 (2006). [arXiv:hep-th/0507173]
405. J.P. Conlon, F. Quevedo, Gaugino and scalar masses in the landscape. JHEP **0606**, 029 (2006). [arXiv:hep-th/0605141]
406. J.P. Conlon, Moduli stabilisation and applications in IIB string theory. Fortsch. Phys. **55**, 287 (2007). [arXiv:hep-th/0611039]
407. A. Westphal, de Sitter string vacua from Kähler uplifting. JHEP **0703**, 102 (2007). [arXiv:hep-th/0611332]
408. M. Berg, M. Haack, E. Pajer, Jumping through loops: on soft terms from large volume compactifications. JHEP **0709**, 031 (2007). [arXiv:0704.0737]
409. J.P. Conlon, C.H. Kom, K. Suruliz, B.C. Allanach, F. Quevedo, Sparticle spectra and LHC signatures for large volume string compactifications. JHEP **0708**, 061 (2007). [arXiv:0704.3403]
410. M. Cicoli, J.P. Conlon, F. Quevedo, General analysis of LARGE volume scenarios with string loop moduli stabilisation. JHEP **0810**, 105 (2008). [arXiv:0805.1029]
411. D. Ciupke, J. Louis, A. Westphal, Higher-derivative supergravity and moduli stabilization. JHEP **1510**, 094 (2015). [arXiv:1505.03092]
412. K. Becker, M. Becker, M. Haack, J. Louis, Supersymmetry breaking and α' corrections to flux induced potentials. JHEP **0206**, 060 (2002). [arXiv:hep-th/0204254]
413. P. Candelas, A. Font, S.H. Katz, D.R. Morrison, Mirror symmetry for two-parameter models—II. Nucl. Phys. B **429**, 626 (1994). [arXiv:hep-th/9403187]
414. N. Arkani-Hamed, S. Dimopoulos, G.R. Dvali, The hierarchy problem and new dimensions at a millimeter. Phys. Lett. B **429**, 263 (1998). [arXiv:hep-ph/9803315]
415. N. Arkani-Hamed, S. Dimopoulos, G.R. Dvali, Phenomenology, astrophysics and cosmology of theories with submillimeter dimensions and TeV scale quantum gravity. Phys. Rev. D **59**, 086004 (1999). [arXiv:hep-ph/9807344]
416. I. Antoniadis, N. Arkani-Hamed, S. Dimopoulos, G.R. Dvali, New dimensions at a millimeter to a Fermi and superstrings at a TeV. Phys. Lett. B **436**, 257 (1998). [arXiv:hep-ph/9804398]
417. P.G.O. Freund, M.A. Rubin, Dynamics of dimensional reduction. Phys. Lett. B **97**, 233 (1980)
418. J. Demaret, M. Henneaux, P. Spindel, Non-oscillatory behavior in vacuum Kaluza–Klein cosmologies. Phys. Lett. B **164**, 27 (1985)
419. J. Demaret, Y. De Rop, M. Henneaux, Are Kaluza–Klein models of the universe chaotic? Int. J. Theor. Phys. **28**, 1067 (1989)
420. T. Damour, M. Henneaux, B. Julia, H. Nicolai, Hyperbolic Kac–Moody algebras and chaos in Kaluza–Klein models. Phys. Lett. B **509**, 323 (2001). [arXiv:hep-th/0103094]

421. T. Damour, S. de Buyl, M. Henneaux, C. Schomblond, Einstein billiards and overextensions of finite-dimensional simple Lie algebras. JHEP **0208**, 030 (2002). [arXiv:hep-th/0206125]
422. P. Hořava, E. Witten, Heterotic and type I string dynamics from eleven dimensions. Nucl. Phys. B **460**, 506 (1996). [arXiv:hep-th/9510209]
423. E. Bergshoeff, E. Sezgin, P.K. Townsend, Supermembranes and eleven-dimensional supergravity. Phys. Lett. B **189**, 75 (1987)
424. E. Bergshoeff, E. Sezgin, P.K. Townsend, Properties of the eleven-dimensional supermembrane theory. Ann. Phys. (N.Y.) **185**, 330 (1988)
425. M.J. Duff, P.S. Howe, T. Inami, K.S. Stelle, Superstrings in $D = 10$ from supermembranes in $D = 11$. Phys. Lett. B **191**, 70 (1987)
426. T. Banks, W. Fischler, S.H. Shenker, L. Susskind, M theory as a matrix model: a conjecture. Phys. Rev. D **55**, 5112 (1997). [arXiv:hep-th/9610043]
427. K.S. Narain, New heterotic string theories in uncompactified dimensions < 10. Phys. Lett. B **169**, 41 (1986)
428. K.S. Narain, M.H. Sarmadi, E. Witten, A note on toroidal compactification of heterotic string theory. Nucl. Phys. B **279**, 369 (1987)
429. P.H. Ginsparg, On toroidal compactification of heterotic superstrings. Phys. Rev. D **35**, 648 (1987)
430. M. Dine, P. Huet, N. Seiberg, Large and small radius in string theory. Nucl. Phys. B **322**, 301 (1989)
431. A. Font, L.E. Ibáñez, D. Lüst, F. Quevedo, Strong-weak coupling duality and non-perturbative effects in string theory. Phys. Lett. B **249**, 35 (1990)
432. J.H. Schwarz, A. Sen, Duality symmetric actions. Nucl. Phys. B **411**, 35 (1994). [arXiv:hep-th/9304154]
433. J.H. Schwarz, A. Sen, Duality symmetries of 4D heterotic strings. Phys. Lett. B **312**, 105 (1993). [arXiv:hep-th/9305185]
434. A. Sen, Strong-weak coupling duality in four-dimensional string theory. Int. J. Mod. Phys. A **9**, 3707 (1994). [arXiv:hep-th/9402002]
435. M.J. Duff, Strong/weak coupling duality from the dual string. Nucl. Phys. B **442**, 47 (1995). [arXiv:hep-th/9501030]
436. C.M. Hull, P.K. Townsend, Unity of superstring dualities. Nucl. Phys. B **438**, 109 (1995). [arXiv:hep-th/9410167]
437. J. Polchinski, E. Witten, Evidence for heterotic – type I string duality. Nucl. Phys. B **460**, 525 (1996). [arXiv:hep-th/9510169]
438. E. Witten, String theory dynamics in various dimensions. Nucl. Phys. B **443**, 85 (1995). [arXiv:hep-th/9503124]
439. P.K. Townsend, The eleven-dimensional supermembrane revisited. Phys. Lett. B **350**, 184 (1995). [arXiv:hep-th/9501068]
440. P. Hořava, E. Witten, Eleven-dimensional supergravity on a manifold with boundary. Nucl. Phys. B **475**, 94 (1996). [arXiv:hep-th/9603142]
441. E. Witten, Strong coupling expansion of Calabi–Yau compactification. Nucl. Phys. B **471**, 135 (1996). [arXiv:hep-th/9602070]
442. T. Banks, M. Dine, Couplings and scales in strongly coupled heterotic string theory. Nucl. Phys. B **479**, 173 (1996). [arXiv:hep-th/9605136]
443. A. Strominger, Heterotic solitons. Nucl. Phys. B **343**, 167 (1990); Erratum-ibid. B **353**, 565 (1991)
444. N. Seiberg, Observations on the moduli space of superconformal field theories. Nucl. Phys. B **303**, 286 (1988)
445. J.A. Harvey, A. Strominger, The heterotic string is a soliton. Nucl. Phys. B **449**, 535 (1995); Erratum-ibid. B **458**, 456 (1996). [arXiv:hep-th/9504047]
446. C. Vafa, E. Witten, A one-loop test of string duality. Nucl. Phys. B **447**, 261 (1995). [arXiv:hep-th/9505053]
447. M.J. Duff, J.T. Liu, R. Minasian, Eleven-dimensional origin of string-string duality: a one-loop test. Nucl. Phys. B **452**, 261 (1995). [arXiv:hep-th/9506126]

448. E. Witten, Some comments on string dynamics, in *Future Perspectives in String Theory: Strings'95*, ed. by I. Bars, P. Bouwknegt, J. Minahan, D. Nemeschansky, K. Pilch, H. Saleur, N.P. Warner (World Scientific, Singapore, 1996). [arXiv:hep-th/9507121]
449. M.J. Duff, R. Minasian, E. Witten, Evidence for heterotic/heterotic duality. Nucl. Phys. B **465**, 413 (1996). [arXiv:hep-th/9601036]
450. P.K. Townsend, String-membrane duality in seven dimensions. Phys. Lett. B **354**, 247 (1995). [arXiv:hep-th/9504095]
451. J.H. Schwarz, An $SL(2, Z)$ multiplet of type IIB superstrings. Phys. Lett. B **360**, 13 (1995); Erratum-ibid. B **364**, 252 (1995). [arXiv:hep-th/9508143]
452. P.S. Aspinwall, Some relationships between dualities in string theory. Nucl. Phys. Proc. Suppl. **46**, 30 (1996). [arXiv:hep-th/9508154]
453. J.H. Schwarz, The power of M theory. Phys. Lett. B **367**, 97 (1996). [arXiv:hep-th/9510086]
454. M. Bodner, A.C. Cadavid, S. Ferrara, $(2, 2)$ vacuum configurations for type IIA superstrings: $N = 2$ supergravity Lagrangians and algebraic geometry. Class. Quantum Grav. **8**, 789 (1991)
455. A.C. Cadavid, A. Ceresole, R. D'Auria, S. Ferrara, Eleven-dimensional supergravity compactified on Calabi–Yau threefolds. Phys. Lett. B **357**, 76 (1995). [arXiv:hep-th/9506144]
456. G. Papadopoulos, P.K. Townsend, Compactification of $D = 11$ supergravity on spaces of exceptional holonomy. Phys. Lett. B **357**, 300 (1995). [arXiv:hep-th/9506150]
457. I. Antoniadis, S. Ferrara, T.R. Taylor, $N = 2$ heterotic superstring and its dual theory in five dimensions. Nucl. Phys. B **460**, 489 (1996). [arXiv:hep-th/9511108]
458. S. Ferrara, R.R. Khuri, R. Minasian, M-theory on a Calabi–Yau manifold. Phys. Lett. B **375**, 81 (1996). [arXiv:hep-th/9602102]
459. G. Curio, A. Krause, G-fluxes and nonperturbative stabilization of heterotic M-theory. Nucl. Phys. B **643**, 131 (2002). [arXiv:hep-th/0108220]
460. B.S. Acharya, A moduli fixing mechanism in M theory. arXiv:hep-th/0212294
461. E.I. Buchbinder, B.A. Ovrut, Vacuum stability in heterotic M theory. Phys. Rev. D **69**, 086010 (2004). [arXiv:hep-th/0310112]
462. M. Becker, G. Curio, A. Krause, de Sitter vacua from heterotic M-theory. Nucl. Phys. B **693**, 223 (2004). [arXiv:hep-th/0403027]
463. E.I. Buchbinder, Raising anti-de Sitter vacua to de Sitter vacua in heterotic M theory. Phys. Rev. D **70**, 066008 (2004). [arXiv:hep-th/0406101]
464. B.S. Acharya, F. Denef, R. Valandro, Statistics of M theory vacua. JHEP **0506**, 056 (2005). [arXiv:hep-th/0502060]
465. P.S. Aspinwall, R. Kallosh, Fixing all moduli for M-theory on K3×K3. JHEP **0510**, 001 (2005). [arXiv:hep-th/0506014]
466. L. Anguelova, K. Zoubos, Flux superpotential in heterotic M-theory. Phys. Rev. D **74**, 026005 (2006). [arXiv:hep-th/0602039]
467. V. Braun, B.A. Ovrut, Stabilizing moduli with a positive cosmological constant in heterotic M-theory. JHEP **0607**, 035 (2006). [arXiv:hep-th/0603088]
468. G. Curio, A. Krause, S-Track stabilization of heterotic de Sitter vacua. Phys. Rev. D **75**, 126003 (2007). [arXiv:hep-th/0606243]
469. A. Krause, Supersymmetry breaking with zero vacuum energy in M-theory flux compactifications. Phys. Rev. Lett. **98**, 241601 (2007). [arXiv:hep-th/0701009]
470. F. Paccetti Correia, M.G. Schmidt, Moduli stabilization in heterotic M-theory. Nucl. Phys. B **797**, 243 (2008). [arXiv:0708.3805]

Chapter 13
String Cosmology

Así que, casi me es forzoso seguir por su camino, y por él tengo de ir a pesar de todo el mundo, y será en balde cansaros en persuadirme a que no quiera yo lo que los cielos quieren, la fortuna ordena y la razón pide, y, sobre todo, mi voluntad desea.
— Miguel de Cervantes, *Segunda Parte del Ingenioso Cavallero Don Quixote de la Mancha*, II – 6

Therefore, it is almost forced upon me to follow its path, and because of it I must go despite everybody, and in vain will you tire yourselves out to persuade me not to want what heavens want, fate orders and reason demands and, most of all, what my will desires.

Contents

(continued)

G. Calcagni, *Classical and Quantum Cosmology*, Graduate Texts in Physics,
DOI 10.1007/978-3-319-41127-9_13

After introducing some basic aspects of strings and branes in Chap. 12, we move to cosmological models arising in or motivated by the theory, concentrating on those based upon 𝕂LT and *large-volume uplifting scenarios* [1–28] embedded in the *string landscape* [29–59] (reviews are [60–64]). Uplifting scenarios realize a de Sitter spacetime in mechanisms of moduli stabilization in M-theory and string theory, with two main consequences:

- The old cosmological constant problem is reinterpreted in a manner quite unique to string theory and a possible resolution is proposed [65, 66] (Sect. 13.1). A dynamical approach to the Λ problem where string axions play the role of dark energy is also possible [67–69] (Sect. 13.4.5).
- Inflation can be realized by several models: (i) *size moduli inflation* (Sect. 13.3), divided into large-volume models [70–80] (Sect. 13.3.1), volume-modulus inflation [81–83] (Sect. 13.3.2) and fluxless inflation [84, 85] (Sect. 13.3.3); (ii) *axion inflation* (Sect. 13.4), divided into racetrack axion inflation [86, 87] (Sect. 13.4.1), the axion valley [88, 89] (Sect. 13.4.2), N-flation [89–105] (Sect. 13.4.3), aligned and hierarchical axion inflation [106–119] (Sect. 13.4.4; see also [104, 105, 120]) and monodromy inflation [121–143] (Sect. 13.4.5; see also [116]); (iii) *warped D-brane inflation*, divided into *slow-roll D-brane inflation* [144–167] (Sect. 13.5) and *DBI inflation* [168–191] (Sect. 13.6); some references discuss both classes [192–194]. The scenarios (i)–(iii) are reviewed in [88, 195–202].

The above list is not exhaustive and neglects many other proposals, some of which are reported in Sect. 13.7:

- Models with standard acceleration mechanisms (inflation or dark energy):
  - *Braneworld* (Sect. 13.7.1). Inflation is realized in a brane and the cosmological constant may relax to small values naturally. The big-bang problem is not addressed.
  - *Cosmological tachyon* (Sect. 13.7.2). The inflaton is identified with the string tachyon. Neither the cosmological-constant nor the big-bang problem are solved.
  - *Higher-order gravity* models (Sect. 13.7.3), stringy or string-inspired realizations of some of the scenarios of Sect. 7.5.
  - *Non-local models* (Sect. 13.7.4). These are essentially a sub-set of the models discussed in Sects. 11.8.2 and 11.8.3. The big-bang problem is addressed already at the classical level.
  - *Pre-big-bang cosmology* (Sect. 13.7.5), an early attempt to resolve the big-bang singularity with the string low-energy effective action of gravity. Inflation is driven by the dilaton.
- Models with alternatives to inflation:
  - *String-gas cosmology* (Sect. 13.7.6). The big-bang problem is solved by T-duality while the early-universe spectra acquire a characteristic prediction on the sign and magnitude of the tensor spectral index.
  - *Cyclic ekpyrotic universe* (Sect. 13.7.7). Inflation is replaced by a mechanism of cyclic contractions and expansions. The cosmological-constant and big-bang problems may be addressed.

All these scenarios are summarized in Sect. 13.8. The big-bang problem in string theory is discussed in Sect. 13.9 under the perspective of chaotic billiards.

13.1 String Landscape

The topology, geometry, shape and size of the compact space on which the theory is dimensionally reduced are parametrized by moduli. The symmetry groups and the number of moduli are enlarged when space is compactified and non-perturbative gauge symmetries (for instance, from branes) are included. The main effect of all these hidden (i.e., not directly observable) sectors is to increase the multiplicity of vacua in the four-dimensional effective, low-energy, particle-field-theory limit of string or M-theory. By "vacuum," we mean a local minimum in the effective potential of such effective limit. The increase in the number of vacua happens independently of whether the hidden sectors induce a change in the number of particles (matter or gauge carriers) in the low-energy limit. If there are $n_{\rm hid}$ hidden sectors each giving rise to $n_{\rm vac}$ distinct vacua not in causal contact, the number of vacua goes as

$$N_{\text{vac}} \sim (n_{\text{vac}})^{n_{\text{hid}}} . \tag{13.1}$$

The set of vacua with cardinality N_{vac} is called string *landscape* [29].

The landscape of string and M-theory can be mapped by cataloguing Calabi–Yau spaces, fluxes and moduli fields systematically [30, 31, 34, 37, 40, 41, 43]. Under several assumptions and approximations, one can even come to estimate the number of vacua that can give rise to the Standard Model of particles [42, 49, 52–54]. Another requirement one can impose on the moduli space is that it reproduces a cosmological background with a small but non-vanishing cosmological constant. In practice, it is much easier to ask whether, how and how often string theory realizes a de Sitter background in $3+1$ non-compact dimensions. Even if a de Sitter universe does not describe what we observe, it is a first step towards understanding the cosmological constant problem in string theory.

In Sects. 12.3.7 and 12.3.9, we have illustrated how flux compactification and gaugino condensation fix all the moduli to an AdS vacuum. The number of such vacua can be estimated in each model. For instance, in type-IIA string theory the number of vacua with a cosmological constant larger than a certain value $\bar{\Lambda}$ is $N_{\text{vac}}(|\Lambda| \geqslant |\bar{\Lambda}|) \sim |\bar{\Lambda}|^{-2/9}$: minima with a small negative cosmological constant are favoured [203]. In M-theory, vacua are not uniformly distributed in Λ and statistical results depend on the type of compactification [46].

Vacua with $\Lambda \geqslant 0$ do exist, as we will see in Sect. 13.1.1, and N_{vac} is possibly very large. Within a given set of N_{vac} minima with a roughly uniform distribution of Λ, one expects to find some vacua with a cosmological constant as small as $\Lambda/\kappa_4^2 \sim m_{\text{Pl}}^4/N_{\text{vac}}$. The observed vacuum would then be realized if $N_{\text{vac}} \gtrsim 10^{124}$. This cardinality can be easily reached by flux compactification (Sect. 12.3.7), where there can be as many as $n_{\text{hid}} \sim 10-500$ fluxes wrapping around Calabi–Yau spaces, each allowing for $n_{\text{vac}} \sim 10$ phases:

$$N_{\text{vac}} \sim 10^{10}-10^{500} , \tag{13.2}$$

with yet larger numbers for specific compactification schemes. Most of these vacua do not realize the observed universe but, still, the number of minima that can describe physics with the observed value of the constants of Nature can be impressively large. In this context, the concept of naturalness of couplings in quantum field theory gives way to another perspective: a field-theory limit A is more natural in string theory than another limit B if the number of phenomenologically acceptable vacua leading to A is larger than the number of vacua leading to B [30].

In the absence of a guiding principle of super-selection of all these vacua, a deterministic resolution of the old cosmological constant problem is out of the question. Nevertheless, a probabilistic resolution would be at hand if one had a sufficiently large number of vacua, so that vacua with $\Lambda \approx \Lambda_{\text{obs}} \sim 10^{-123} m_{\text{Pl}}^2$ would occur with high frequency. The smallness of Λ would then amount to a statistical-selection effect rather than to some unwanted fine tuning [29, 30].

13.1.1 de Sitter Vacua

We have seen in Sects. 12.3.3 and 12.3.7 that the cosmological constant on the visible four-manifold $\mathcal{M}_4$ vanishes identically in Calabi–Yau compactifications (equation (12.86)). Complete moduli stabilization is achieved by gaugino condensation or instantonic branes, which lower the minimum to an AdS configuration (Sect. 12.3.9). The lifting of the minimum to a vanishing or positive value is achieved by combining these ingredients with a third one, described later, which typically relies on placing branes at some key points of the Calabi–Yau manifold. The resulting $\mathbb{K}$LT *uplifting scenarios* [1–4, 6, 7, 9, 11, 12, 14, 17–20] and their cousins the *large-volume uplifting scenarios* [10, 13, 15, 21, 22] can be summarized as the realization of a four-dimensional de Sitter background in string theory, with applications to inflation and to late-time acceleration. They are the next step beyond, respectively, $\mathbb{K}$LT and large-volume stabilization scenarios.

The original $\mathbb{K}$LT construction [1] is based on the warped compactification of type-IIB superstring theory with 3-form fluxes (Sect. 12.3.7). In this model, the dilaton and complex-structure moduli are stabilized by fluxes (Sect. 12.3.5) [204], while the only Kähler modulus, the axio-radion ϱ, is stabilized by gaugino condensation (Sect. 12.3.9). At this point, all non-supersymmetric and $\mathcal{N} = 1$ supersymmetric moduli are stabilized at the classical level to an anti-de Sitter vacuum configuration.

To break supersymmetry and lift the minimum to positive values, it is sufficient to add a small number of anti-D3-branes to the deformed Klebanov–Strassler throat in the Calabi–Yau space $\mathcal{C}_3$ [1] (Fig. 13.1). While D-branes preserve supersymmetries of the same chirality as the supersymmetries of the compactification, by definition $\overline{\text{D}}$-branes preserve supersymmetries with opposite chirality, since they carry charges opposite to those of the background fluxes. Therefore, anti-branes break the supersymmetries of the system. $\overline{\text{D3}}$-branes, which are transverse to $\mathcal{C}_3$ and are completely embedded in the four-manifold, must be added to cancel tadpoles of the NS-NS B_2 field [205] when too many fluxes are turned on.

Just like in the non-compact version of this construction [206], the volumes of the $\overline{\text{D3}}$-branes are stabilized by the background fluxes and no additional moduli are introduced. Let us split the Kähler modulus ϱ as in (12.114). The positive tension of the anti-branes over-compensates the AdS minimum [1] by a contribution [144, 206]

$$\Delta V_1 = \frac{\beta_1}{\sigma^2}\,, \tag{13.3}$$

where the coefficient $\beta_1 \propto \mathcal{T}_3 > 0$ depends on the number of $\overline{\text{D3}}$-branes, on their tension $\mathcal{T}_3$ and on the warp factor in the throat [206, 207]. Supersymmetry is explicitly broken by ΔV_1.

A similar inverse-power-law potential arises from wrapping a D7-brane (or more) around a 4-cycle Γ_4 of the Calabi–Yau manifold and turning on the fluxes of the

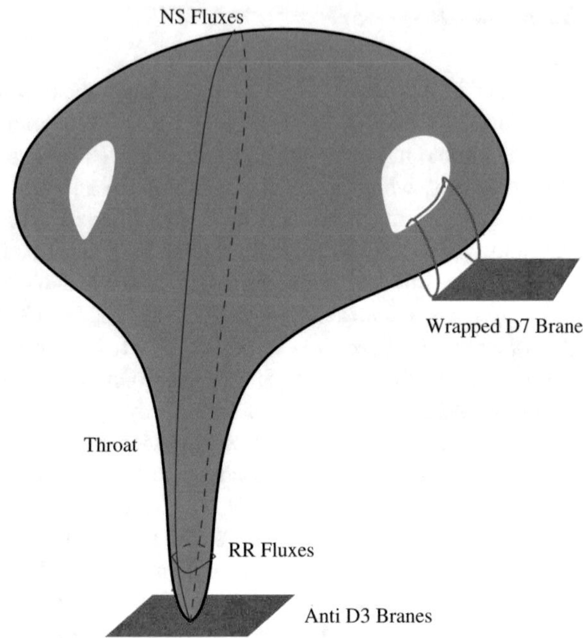

Fig. 13.1 A deformed Klebanov–Strassler throat embedded in a Calabi–Yau space with fluxes around non-trivial cycles. In the figure, a stack of $\overline{\text{D3}}$-brane are trapped near the tip of the throat, while a D7-brane is wrapped around a 4-cycle (Credit: [6])

gauge fields living inside the brane [4]. These induce a D-term potential

$$\Delta V_2 = \frac{g_{\rm YM}^2}{2} D^2 = \mathcal{T}_7 \int_{\Gamma_4} \mathrm{d}^4 y\, \sqrt{g_8}\, F_{mn} F^{mn} = \frac{\beta_2}{\sigma^3}\,, \tag{13.4}$$

where $\mathcal{T}_7$ is the brane tension, g_8 is the determinant of the brane metric, F_{mn} is the field strength of the gauge field and the magnitude of $\beta_2 > 0$ depends on the intensity of the flux. Also matter fields charged under $U(1)$ may contribute to (13.4) but we shall ignore them. Contrary to (13.3), the D-term (13.4) is supersymmetric and one can employ all the tools of $D = 4$ supergravity. These tools show that the D-term (13.4) can improve the stability of the AdS minimum but cannot uplift it [14], unless α'-corrections are also included.

Since $\overline{\text{D3}}$-branes and wrapped D7-branes with fluxes coexist in the same scenario, in general one will have both contributions, which we parametrize with a generic inverse power-law

$$\Delta V_{n-1} = \frac{\beta}{\sigma^n} \propto \frac{1}{\mathcal{V}^{2n/3}}\,, \qquad n \geqslant 2\,, \tag{13.5}$$

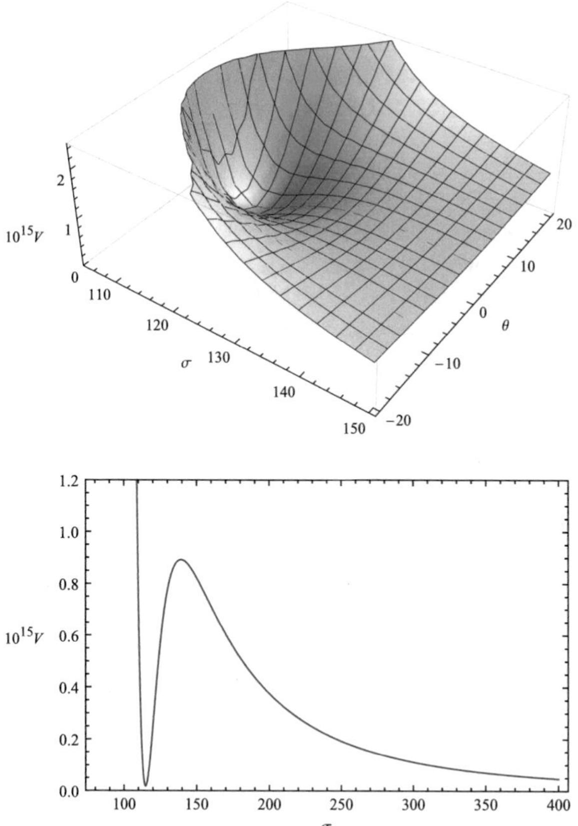

Fig. 13.2 The KKLT potential (13.6) multiplied by 10^{15} and in Planck-mass units $M_{\rm Pl} = \kappa_4^{-1} = 1$, with $n = 3$, $W_0 = -10^{-4}$, $\mathcal{A} = 1$, $\alpha = 0.1$ (as in Fig. 12.4) and $\beta = 3 \times 10^{-9}$, in the (σ, θ) plane (*top panel*) and for $\theta = 0$ (*bottom panel*). The global minimum is at $\sigma_{\rm min} \approx 114.9$, not far away from the AdS minimum $\sigma_{\rm min} \approx 113.6$ of (12.115)

where we used (12.97) and (12.99).The total potential is the sum of (12.115) and ΔV_{n-1}:[1]

$$\kappa_4^4 V_{\rm KKLT} = \frac{\alpha \mathcal{A}}{2\sigma^2}\, {\rm e}^{-2\alpha\sigma} \left[\left(1 + \frac{\alpha\sigma}{3}\right)\mathcal{A} + W_0 {\rm e}^{\alpha\sigma} \cos(\alpha\theta)\right] + \frac{\beta}{\sigma^n}\,, \qquad (13.6)$$

where $W_0 < 0$. The last contribution is positive and the AdS minimum can be lifted to a positive value for a certain critical range of β (Fig. 13.2). Below this range, the local minimum stays negative, while above it it disappears and only

[1] Yet another way to generate this class of potentials is to find local minima in the complex-structure and dilaton directions in moduli space. Around these points, V develops a de Sitter minimum in the σ direction [208].

the runaway minimum at $\sigma \to +\infty$ remains. Inside the critical range, $\Lambda > 0$ and the non-compact four-manifold where the observable universe lives in is de Sitter spacetime. Note that the position of the minimum does not change much from the value (12.117), provided β is sufficiently small; therefore, the large-volume approximation $\sigma \gg 1$ maintains its validity.

An important feature of de Sitter vacua, which will bear consequences for the cosmological constant problem, is that there is an asymptotic minimum at $V = 0$ in the limit $\sigma \to +\infty$, separated from the de Sitter minimum by a potential barrier. Due to quantum fluctuations of moduli fields, one expects de Sitter vacua to decay to more favourable ground states (not necessarily Minkowski spacetime) via quantum tunneling. The tunneling process is described by a Coleman–de Luccia instanton [209, 210]. However, the lifetime of de Sitter vacua is much greater than the age of the universe and a full cosmic evolution can take place in each vacuum [1].[2] A vacuum state which is unstable but on a very large time scale is called *metastable*. This result of string theory is in agreement with general thermodynamic arguments that suggest that de Sitter spacetime is metastable in any theory of quantum gravity where the entropy inside the dS horizon is finite [211].

As we have seen in Sect. 12.3.9, the $\mathbb{K}$LT model of moduli stabilization can be heavily affected by the details of the concrete Calabi–Yau compactification and of the dynamical stabilization mechanism. In general, models with only one light modulus ϱ do not work [7, 12, 212] and one has to stabilize up to $O(10)$ Kähler moduli and $O(10^2)$ complex-structure moduli. In particular, the complex-structure fields should be stabilized simultaneously with the other light moduli, since otherwise both the AdS minimum and the corresponding uplifted de Sitter minimum are unstable [12]. The range of admissible values for W_0 can only be determined by a detailed study of a multi-moduli stabilization scheme; in many situations, such range does not include values for which a de Sitter minimum is produced.

The uplifting mechanism can also be adapted to heterotic string theory [4, 7, 12] (where gaugino condensation generates the potential (12.109) or its racetrack extension (12.110) for the axio-dilaton S), type-IIA string theory [213] (where the AdS minimum, obtained only via fluxes, is raised to positive values by perturbative corrections to the Kähler potential) and M-theory [5, 8, 214, 215].

Moreover, alternative or concomitant sources breaking supersymmetry can appear, such as corrections to the F-term (5.222) [7, 13, 15, 216], the already mentioned D-terms such as (13.4) [4, 17, 19, 216, 217], or perturbative α'^3-corrections to the Kähler potential [13]. Depending on the model, supersymmetry breaking can take place before or after lifting the minimum.[3] The first case corresponds to large-volume stabilization and uplifting scenarios. We have already mentioned that $W_0 \propto$

[2]The decay rate has the same expression of the tunneling probability (10.48), which is of order of $\exp(-10^{123})$.

[3]Thanks to the interplay of non-perturbative and α'^3-corrections to the Kähler potential, it is also possible to hit a de Sitter minimum without passing through an AdS vacuum (thus avoiding the introduction of anti-branes) [23].

$\mathcal{V}_6/l_s^6$ can be much larger than in the original $\mathbb{K}$LT scenario. This datum is important when considered in the string landscape, where (assuming the complex-structure moduli have already been integrated out) $e^{K(\varrho_I,\tau)}W_0^2$ is uniformly distributed in the choices of flux [34, 43]. The number of type-IIB vacua with a supersymmetry breaking scale $M_{\rm SUSY}$ smaller than some given scale $\bar{M}$ is $N_{\rm vac}(M_{\rm SUSY} < \bar{M}) \sim \bar{M}^{12}$ [43] and low-$M_{\rm SUSY}$ models are, so to speak, disfavoured. On the other hand, in large-volume scenarios $M_{\rm SUSY} \sim m_{3/2}$ can be high, as we saw below (12.127). Therefore, large values of W_0 are more frequent in string theory and large-volume scenarios are, in this sense, favoured in the landscape. However, the number of vacua with small W_0 can still be large enough to support also models with a low supersymmetry scale. Backed by the typicality of $\mathbb{K}$LT and large-volume uplifting scenarios, the string landscape is pocked with enough de Sitter vacua to sustain an interesting hypothesis on the origin of the old cosmological constant problem.

13.1.2 Cosmological Constant

In supergravity, the strategy towards a solution of the cosmological constant problem is to overshoot the value (7.11) to zero or to some negative value and then to lift the potential by some non-perturbative mechanism. String theory starts with the same procedure but does not stop there.

The value of $\Lambda = \kappa^2 V_{\rm KKLT}$ obtained in the previous sub-section depends on the free parameters of the model. The $\mathbb{K}$LT scenario has two types of free parameters: the number of $\overline{\rm D3}$-branes at the deformed Klebanov–Strassler throat and the number of flux quanta at the 3-cycles. Fluxes push $\overline{\rm D3}$-branes towards the end of the throat, where the warp factor (and, consequently, β in (13.6)) is exponentially small. By tuning the number of fluxes not too severely, one can even obtain a very fine-tuned value for ΔV, thus leaving a very small positive cosmological constant. This tuning cannot be arbitrarily high, since the number of branes and fluxes can only take discrete values. Then, one could obtain the observed value of Λ for certain choices of the parameters in $\mathbb{K}$LT and large-volume string models.

Such a mildening of the fine tuning in the cosmological constant problem is not dissimilar to analogous situations in some of the models of Chap. 7. The many ways one can compactify the theory give rise to free parameters. Perhaps, for this reason the above conclusion would not be regarded as particularly elegant, even if it is the result of a top-down construction from a theory of everything. However, so far we have neglected the fact that the de Sitter minimum is metastable. As it happens, the decay of the vacuum has such a dramatic consequence on the physical picture as to change the approach to the Λ problem completely.

Assume that the Universe has reached a vacuum state with some $\Lambda = \Lambda_1 \neq 0$. As we have seen, moduli fields do not sit indefinitely at this vacuum and eventually tunnel to another minimum with $0 < \Lambda = \Lambda_2 < \Lambda_1$. Here they stay until they tunnel again to a neighboring valley with a yet smaller minimum $\Lambda_3 < \Lambda_2$; and so on. This scenario bears resemblance with eternal inflation (Sect. 5.6.5), to the point where one can describe both with the same picture; vice versa, eternal inflation is naturally realized by the string landscape, at least qualitatively. A local observer would see the cosmological constant in their causal patch decrease in a series of events [29]. Adopting a global view, causal patches (or metastable "bubbles") with progressively smaller Λ form within a larger patch with larger cosmological constant. The bubbles expand at a slower rate than the space between them, so that they evolve independently from one another and never enter into causal contact. Although the probability that one patch goes through a phase with $\Lambda \sim 10^{-123} m_{\rm Pl}^2$ is very small, among the infinitely many transitions occurring in the Universe there will most likely be some bubbles in an acceptable vacuum.

A concrete model of quantum fluctuations driving the decay of the vacuum can be understood as follows. We have seen in Chap. 7 and Sect. 10.2.4 that the cosmological constant is effectively determined by the contribution of different sources, from symmetry-breaking effective potentials to dynamical fields, including anti-symmetric tensor fields [218–220]. As we have remarked in the course of Chap. 12, p-forms are associated with branes, so that one can expect that quantum fluctuations of a p-form field gives rise to the non-perturbative creation of virtual pairs of branes–anti-branes. Once these pairs are created, they do not annihilate because the space separating them expands fast enough. The net consequence is that the energy density of the field decreases, thus reducing the value of the effective cosmological constant [221, 222]. The process can be described in terms of compact spatial 2-branes arising from a 3-form field (i.e., a 4-flux), which form the walls of the de Sitter bubbles continuously nucleating via quantum tunneling.

This mechanism can be adapted to and improved in M-theory, where the presence of many 4-fluxes in compactified configurations gives rise to a discrete spectrum with thinly spaced energy levels [65, 66]. Consider the compactification of the action (12.128) on a 7-manifold with volume $\mathcal{V}_7$:

$$S_{11} = \frac{1}{2\kappa_4^2} \int d^4x \sqrt{-g} \left[R - 2\Lambda_0 - \frac{1}{2} F_4^2 + \dots \right], \tag{13.7}$$

where we keep the symbols Λ_0 and F_4 to denote the bare cosmological constant and the 4-flux after dimensional reduction. The four-dimensional Newton's constant is $\kappa_4^{-2} = \kappa_{11}^{-2} \mathcal{V}_7$. A solution for the gauge-field equations of motion $\nabla \cdot F_4 = 0$ is $F^{\rm sol}_{\mu\nu\sigma\tau} = c\epsilon_{\mu\nu\sigma\tau}$, where c is a mass coefficient. This produces an effective cosmological constant $\Lambda_{\rm eff} = \Lambda_0 - (F_4^{\rm sol})^2/4 = \Lambda_0 + c^2/4$, which can be positive if $\Lambda_0 \lessgtr 0$. At the quantum level, $(F_4^{\rm sol})^2 \to \langle (F_4^{\rm sol})^2 \rangle$ becomes the expectation value of the gauge field strength. In string theory, this expectation value is quantized due to the simultaneous presence of electric and magnetic sources and for a consistent quantization of branes [65, 223]. Then, $c = 2\kappa_4^2 en$, where e is the charge of the M2-brane (with dimension $[e] = 3$) and $n \in \mathbb{Z}$. Creation of brane pairs changes the value of c, which jumps to lower and lower values due to the repeated tunneling towards more favourable vacua. Notice that these branes are not the fundamental ones of the theory; they correspond to instantonic solutions of the Euclidean action [224].

A first requisite one demands to obtain a solution of the old cosmological constant problem is that the spacing between the allowed values of $\Lambda_{\rm eff}$ be of order of $10^{-123}m_{\rm Pl}^2$ or smaller. Such a requisite translates into asking that the minimal spacing $\Delta\Lambda$ in the cosmological-constant spectrum be no larger than the observed value,

$$\Delta\Lambda \lesssim \Lambda_{\rm obs} . \tag{13.8}$$

This cannot be achieved by a single 4-flux, but many fluxes arise if the compact manifold contains non-trivial 3-cycles. If there are $N-1$ 3-cycles, one produces N fluxes associated with N different types of branes with charges $c_i \propto \kappa_4^2 e_i$: these are either M2-branes completely embedded in the non-compact manifold $\mathcal{M}_4$ or M5-branes with three directions wrapped around a 3-cycle. The effective cosmological constant reads

$$\Lambda_{\rm eff} = \Lambda_0 + \sum_{i=1}^{N} n_i^2 q_i^2 , \tag{13.9}$$

where $q_i := c_i/(2n_i)$ is a constant with dimension $[q_i] = 1$.

We can represent the flux spectrum as an N-dimensional regular lattice with spacing q_i; the volume of an elementary cell is $\mathcal{V}_q \sim \prod_i q_i$. On this grid, the observational range of the cosmological constant is represented by a thin shell centered at $n_i = 0$ of average radius $r = \sqrt{|\Lambda_0|}$ and thickness $\Delta r = \Lambda_{\rm obs}/r$ (Fig. 13.3). The volume of the shell is $\Delta\mathcal{V} = \Omega_{N-1} r^{N-1}\,\Delta r$, where $\Omega_{N-1} = 2\pi^{N/2}/\Gamma(N/2)$ is the area of a unit N-sphere. The inequality (13.8) is satisfied if at least one point of the lattice falls in the shell, which occurs if $d\mathcal{V}_q \lesssim \Delta\mathcal{V}$, where d is the typical degeneracy of states per charge level. This relation yields the minimal spacing [65]

$$\Delta\Lambda = \frac{d\prod_{i=1}^{N} q_i}{\Omega_{N-1}|\Lambda_0|^{\frac{N}{2}-1}} . \tag{13.10}$$

For $\Lambda_0 = O(m_{\rm Pl}^2)$ and small degeneracy $d = O(1)$, only about $N = 10^2$ fluxes are sufficient to produce the desired spacing if the charges $q_i \sim 10^{-1}$–$10^0 m_{\rm Pl}$ are not much smaller than the Planck mass.

In the simple case of a 7-torus with radii r_l ($l = 1, \ldots, 7$), there are 35 3-cycles with volume $\mathcal{V}_{3,i} = (2\pi)^3 r_{m_i} r_{m_i'} r_{m_i''}$ (one for each of the unordered triplets (m, m', m'')). If one writes the gravitational coupling in 11 dimensions as $\kappa_{11}^{-2} = 4\pi M_{11}^9$, where M_{11} is the fundamental mass scale of M-theory, then the four-dimensional reduced Planck mass is $M_{\rm Pl}^2 = \kappa_4^{-2} = 4\pi M_{11}^9 \mathcal{V}_7$, the tension of an M5-brane wrapped around the i-th cycle and the tension of the M2-brane are, respectively, $\mathcal{T}_i = 2\pi M_{11}^6 \mathcal{V}_{3,i}$ for $i < N$ and $\mathcal{T}_N = 2\pi M_{11}^3$, and the brane charges are $e = 2\pi M_{11}^3$. By taking into proper account the volume factors of the 3-cycles and

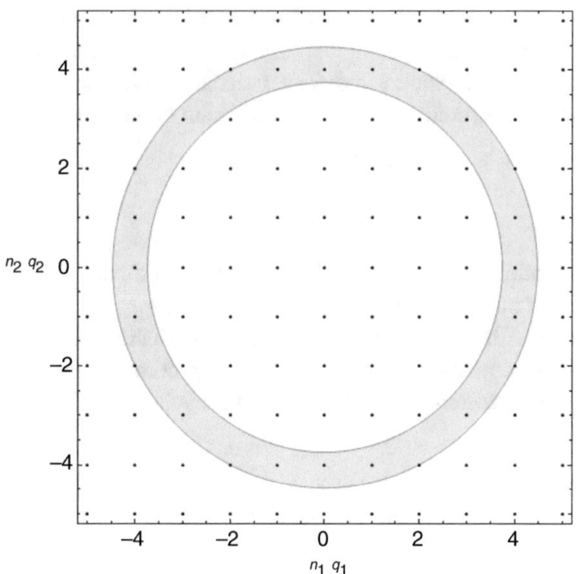

Fig. 13.3 The allowed values of the energy density of $N = 2$ 4-forms as described in the text. A negative bare cosmological constant Λ_0 is compensated up to the observed value $\Lambda_{\rm obs}$ if at least one point lies in the shell (*shaded region*) of average radius $r = \sqrt{|\Lambda_0|}$

of the compact 7-manifold, one finds that $q_i = \kappa_4^2 \mathcal{T}_i$. Assuming $\mathcal{V}_{3,i} \approx l_{\rm Pl}^3$, we have

$$q_{i\neq N} \simeq (4\pi)^2 \left(\frac{M_{11}}{m_{\rm Pl}}\right)^6 m_{\rm Pl}\,, \qquad q_N \simeq (4\pi)^2 \left(\frac{M_{11}}{m_{\rm Pl}}\right)^3 m_{\rm Pl} \tag{13.11}$$

and, to obtain (13.8), the 11-dimensional mass must be close to the Planck mass.

For large degeneracies d or small N, the value of the charges is reduced, but still to a level acceptable in string phenomenology. Clearly, one flux is not enough, as the brane charge would be fine-tuned in that case, $q \lesssim \pi\Lambda_{\rm obs}$. This *gap* (or *small-step*) *problem* [221, 222] is not present in string and M-theory [65], where the compact space usually has many 3-cycles. Furthermore, the nucleation rate of bubbles with small cosmological constant is tiny compared to the age of the universe. Therefore, the landscape mechanism of quantum dynamical suppression of the cosmological constant is capable of producing an observationally viable phase with $\Lambda \approx \Lambda_{\rm obs}$ which is stable on cosmological time scales.

Even if there were enough realizations of the $\Lambda_{\rm obs} \sim 10^{-123} m_{\rm Pl}^4$ vacuum, to address the cosmological constant problem one should ask whether and why such vacuum should be a preferred outcome for an observer. The *anthropic principle* [225–227] has been proposed as a selection rule coming to the fore at the time of answering these questions [29, 65]. In its weakest form, it is based on the following observation. If the cosmological constant were too large, the accelerated expansion at large redshift would hinder gravitational collapse and the formation of galaxies. On the other hand, if Λ were too negative, the universe would have recollapsed before galaxies and stars could form. These two conditions give, respectively, an upper and a lower bound for the observationally admissible values of Λ [226]:

$$-10^{-123}m_{\rm Pl}^4 < \rho_\Lambda < 10^{-121}m_{\rm Pl}^4\,. \tag{13.12}$$

The weak anthropic principle states that the observed Λ is so small because galaxy and star formation, both prerequisites for the emergence of life (and, eventually, of an observer), would not have taken place otherwise. This is not a mere tautology: the interval (13.12), which includes the point $\rho_\Lambda = \Lambda_{\rm obs}/\kappa^2$, was computed before any evidence of dark energy was known and may be regarded as an actual prediction on the existence and range of values of a cosmological constant. This argument can be made more compelling, and the range (13.12) further restricted, by more sophisticated analyses [228, 229] or by stronger versions of the anthropic principle.

Coming back to the string landscape, the Universe is made of patches (cosmologically large, causally disconnected regions) with different cosmological constants, but only in those regions where the cosmological constant is sufficiently small does galaxy and star formation take place. Therefore, we observe a small cosmological constant simply because we inhabit one of the statistically allowed bubbles where organic life is made possible.

Similar reasonings can also be applied to the parameters of the Standard Model, in particular the Higgs mass. In this case, the requirement that atomic nuclei be stable (*atomic principle*) yields a prediction for the range of the Higgs mass [230]. Applying the weak anthropic principle and the atomic principle to string theory, one can scan the cosmological constant and the Higgs mass across the landscape and address the cosmological constant problem and the hierarchy problem simultaneously [45]. This gives rise to stimulating consequences on the supersymmetry breaking scale $M_{\rm SUSY}$ (typically very high) and the observability of new O(TeV) particle phenomenology at the LHC, on which we will comment in Chap. 14.

An alternative which does not rely on the anthropic principle is *statistical selection*, which aims at extracting predictions from the analysis of joint probabilities in the landscape [30, 40, 63]. An intuitive example is the following. Imagine there were two sets of vacua A and B, one with $N_{\rm vac}^{\rm A} \sim 10^{160}$ elements and a low-scale supersymmetry (say, $M_{\rm SUSY} = 10\,{\rm TeV}$) and the other with $N_{\rm vac}^{\rm B} \sim 10^{120}$ elements and a high-scale supersymmetry (say, $M_{\rm SUSY} = 10^9$–$10^{17}\,{\rm GeV}$). Suppose that both sets reproduce all known couplings except for the observed cosmological constant $\Lambda_{\rm obs}$. By comparing $N_{\rm vac}^{\rm A}$ and $N_{\rm vac}^{\rm B}$, one cannot conclude convincingly that "string theory predicts low-scale supersymmetry," since both numbers are very large. However, the requirement that these vacua also reproduce $\Lambda \sim \Lambda_{\rm obs}$ would make the argument more stringent. If both sets realized a uniform distribution of cosmological constants, then we would expect about $N_{\rm vac}^{{\rm A},\Lambda} \sim 10^{40}$ low-supersymmetry viable vacua but only $N_{\rm vac}^{{\rm B},\Lambda} \sim 1$–$10^3$ viable vacua with high-energy supersymmetry. Both alternatives would describe the observed universe but only the one with low-scale supersymmetry would be realized with significant frequency in the string landscape.

In actual procedures to survey the landscape, in reality one devises a set of vacua with given supersymmetry breaking scale and determines the probability distribution of these models as a function of Λ. Conversely, the distribution of models with $\Lambda_{\rm obs}$ and the observed couplings of the Standard Model can be analyzed as a function of the supersymmetry scale.

Actual determinations of the vacua distribution in scenarios with flux compactification are much more refined that the crude estimates based on (13.1) [40]. Notwithstanding the problems related to an incomplete control over the moduli space and to a lack of experimental evidence for supersymmetry (which considerably increases the number of possibilities), these studies fuel the hope that statistical selection in string theory may be a viable explanation of the smallness of Λ.

13.1.3 Open Problems

The impact of the string landscape on epistemology and physics has been compared, in its importance, to the one caused by the discovery of a large number of solar systems apart from ours [45]. Observers believing in the uniqueness of our Solar System would try to understand why the Earth-Sun distance takes exactly the value we measure. As soon as the observational range is expanded to galactic scales and a landscape of 10^{11} stars is found, it becomes clear that just historic accidents, rather than any fundamental principle, have determined our planetary distances. The deep mystery of the value of the Astronomical Unit dissolves into a reassuring statistical explanation, gladly helped by astrobiology considerations on the conditions for carbon-based life in habitable orbital belts.

In the opinion of some, however, the existence of as many vacua as (13.2) places a high stake on the possibility that string theory may not be falsifiable. Of course, unique predictivity is not required for a theory to be falsifiable experimentally; the Standard Model and general relativity are strikingly predictive despite the presence of a number of free parameters. In the case of string theory, born as a fundamental framework with no free parameters, demands and expectations are perhaps higher, especially now that we still lack a distinctive signature of stringy effects in particle physics or cosmology. We do not wish to enter a debate about the feasibility of string theory, since it would involve philosophical issues and technical caveats on the assumptions leading to (13.2); these are discussed, for instance, in [30, 32]. Suffice it to say that, as we have seen, a large statistics such as (13.2) can actually help in reinterpreting the cosmological constant problem in an exceptional way.

Two basic requirements for a genuine solution of the Λ problem are a robust counting of the vacua and a coherent description of the evolution of an observer in the string landscape. Much speculation is involved in the present understanding of these features. On one hand, the absence of a rigorous global description of bubble nucleation (due to the presence of de Sitter horizons, which define causal patches) makes the models discussed above somewhat too heuristic to extract solid information on the probabilities involved in the nucleation process. This is nothing but the measure problem in eternal inflation [231] (Sect. 5.6.5) or in multiverse scenarios [232, 233]. Estimates of such probability distribution have been made nonetheless and, in the absence of better theoretical grounds, they represent the state of the art. When they do not involve anthropic arguments, these estimates work towards a solution of the Λ problem by statistical selection; however, they are not conclusive yet. On the other hand, there is no compelling and universally accepted principle or mechanism preferring one vacuum over another. Originally invoked to supply a guidance in the vast string landscape, the anthropic principle has been the subject, in virtually all of its forms, of diversified and complex criticism (a non-exhaustive sample can be tasted in [32, 44, 234]), its actual effectiveness has been questioned [235] and alternative criteria for vacuum selection have been proposed [236].

Rather than an intrinsic flaw in string theory, these issues demonstrate the need to extend our knowledge farther than the present point. In this respect, and until such progress is made, any comparison between the string framework and the theories of quantum gravity of Chap. 11 will be forcefully limited by the degree of development of the single proponents.

13.2 Inflation in the Landscape

Within the landscape, there are several models of string cosmology that include an early phase of acceleration. Although the step from a perfect de Sitter vacuum to inflation seems small, it entails a number of subtleties concerning moduli stabilization. In fact, the latter interferes with a viable period of inflation and Kähler and superpotentials must be hand-picked with care. For any given vacuum in the landscape, we will describe two classes of inflationary scenarios: moduli inflation (Sects. 13.3 and 13.4) and D-brane inflation (Sects. 13.5 and 13.6). Different regions of the landscape may offer a favourable habitat for any of these scenarios. In some cases, the phenomenology of string inflation was developed some years before its actual embedding in the theory.

Taken as a whole, these models are the string representative of quintessential inflation, where one or more scalar fields are responsible for both the early-universe and late-time acceleration (Sect. 7.3.6). As we will see, a modulus or several moduli drive the inflationary era and then relax, at late times, at a metastable de Sitter minimum of their potential, without relying on severely fine-tuned initial conditions.

Before going into the details of how inflation takes place in flux compactifications, let us discuss the general phenomenology of inflation in the big picture of the string landscape, where the dynamical field or fields pass through many metastable vacua. Concrete realizations of inflation are the subject of Sects. 13.3, 13.4, 13.5 and 13.6. In the rest of the chapter, the inflaton will be denoted as ϕ when identified with any of the moduli except for the position of branes, in which case we will employ the symbol φ.

13.2.1 Single-Field Inflation

The string landscape can relate the string or M-theory couplings to the inflationary energy scale, thus connecting fundamental micro-physics with observations of the early universe. To argue this, we introduce a novel ingredient with respect to the picture described in Sect. 13.1: we assume that inflation takes place between a given vacuum $V_{\rm dS}^{(1)}$ and the next one $V_{\rm dS}^{(0)} = \rho_\Lambda < V_{\rm dS}^{(1)}$ with the value ρ_Λ of the cosmological constant observed today. The sequence of tunnelings through the preceding metastable vacua is not important in what follows and one can consider the simplified context of a false-vacuum decay followed by a period of inflation. A potential barrier separates the inflationary trough in the landscape from the higher vacuum $V_{\rm dS}^{(1)}$. The final inflaton vacuum is $V_{\rm dS}^{(0)}$.

The single-field case is one of the simplest representatives of landscape inflation. One assumes (or shows, in specific models) that all moduli but the inflaton scalar have been stabilized at an early stage. Multi-field constructions support such a dynamical situation to some extent. Many viable single-field models have three main properties: a very small negative curvature, just enough e-foldings and suppression of the power spectrum at low multipoles. The scalar spectrum is suppressed at large scales due to a steep feature at the beginning of inflation, the potential barrier separating $V_{\rm dS}^{(0)}$ from $V_{\rm dS}^{(1)}$.

Let the bare cosmological constant be constituted by the inflaton potential, $\Lambda_0 \simeq \kappa_4^2 V(\phi)$. During inflation, the cosmological dynamics is governed by the effective Friedmann equation

$$H^2 \simeq \frac{\Lambda_{\rm eff}}{3}, \tag{13.13}$$

where $\Lambda_{\rm eff}$ is given by (13.9). Consider a large-field model with $V_{,\phi} > 0$. As we know from (5.131) in Sect. 5.6.2, in quasi-de Sitter spacetime the quantum fluctuation of the scalar field during a Hubble time $\Delta t \sim H^{-1}$ is $|\delta\phi| \simeq H/(2\pi)$. On the other hand, during the same interval the field displacement due to classical slow rolling is $|\Delta\phi| \simeq V_{,\phi}\Delta t/(3H) \simeq V_{,\phi}/(3H^2)$. Quantum fluctuations dominate the evolution of the inflaton if $|\Delta\phi| < |\delta\phi|$, which gives an upper bound for $V_{,\phi}$:

$$V_{,\phi} < \frac{\Lambda_{\rm eff}^{3/2}}{2\sqrt{3}\pi}. \tag{13.14}$$

In this case, bubble (i.e., brane) nucleation takes place and the effective cosmological constant undergoes a sequence of suppressions. When the inflaton approaches the minimum of its potential, inequality (13.14) is violated and reheating occurs. At this point, the nucleation stops and the cosmological constant (the actual constant in $\Lambda_{\rm eff}$) takes its final value $\Lambda_{\rm final}$. Let $i = N$ be the bubble representing our causal patch. Inside it, the value of n_N in (13.9) is lowered by 1, so that the value of the cosmological constant is lowered by

$$[n_N^2 - (n_N^2 - 1)^2]q_N^2 = (2n_N - 1)q_N^2 =: \Lambda_* \,. \tag{13.15}$$

This is the cosmological constant at the penultimate nucleation. If $\Lambda_{\rm final} \approx \Lambda_{\rm obs}$, we have seen that $q^2 \sim 10^{-2} - 10^{-1} m_{\rm Pl}^2$, so that $\Lambda_* \sim m_{\rm Pl}^2$ and it dominates over any other contribution in $\Lambda_{\rm eff}$ at the end of inflation. By this approximation and using (13.11), inequality (13.14) reads

$$V_{,\phi}(\phi_{\rm e}) < \frac{(2n_N - 1)^{3/2}}{2\sqrt{3}\pi} q_N^3 \simeq \frac{(4\pi)^5}{\sqrt{3}} \left(\frac{M_{11}}{m_{\rm Pl}}\right)^9 m_{\rm Pl}^3 \,,$$

where we have considered that, for large N, the flux numbers are $n_i = O(1)$. The lower bound for M_{11} is

$$\frac{M_{11}}{m_{\rm Pl}} > 0.3 \frac{V_{,\phi}^{1/9}(\phi_{\rm e})}{m_{\rm Pl}^{1/3}} \,. \tag{13.16}$$

The quadratic potential $V = m^2\phi^2/2$ is under strong observational pressure and toroidal compactification is not the best choice one can make in string theory, but the combination of these ingredients can give the reader an idea of the typical energy scale of this scenario. According to (5.87c), a minimum number of 60 e-foldings is guaranteed if $m \approx 0.5 m_{\rm Pl}$. Then, $V_{,\phi}(\phi_{\rm e}) = m^2\phi_{\rm e} \approx 0.5 m_{\rm Pl}^3/\sqrt{4\pi}$ and $M_{11} > 0.2 m_{\rm Pl}$ is around or somewhat above the grand-unification scale. In models with large compact dimensions [237–239], this scale can be lowered as much as $O(\text{TeV})$ [65].

Scenarios more detailed than (13.13) take into account the consequences of a tunneling origin of the universe in the landscape [240–244]. The Coleman–de Luccia instanton has the following peculiarity: an observer inside the newly nucleated bubble would locally see an infinite open universe [209, 210]. Anthropic bounds coupled with statistical analyses show that an inflationary period with just the minimum number of e-foldings $\mathcal{N}_{\rm e} \approx 60$ may be theoretically favoured [240].[4] In fact, on one hand a long stage of inflation may be improbable in this scenario and curvature would not be washed away. On the other hand, an $\Omega_0 \ll 1$ would impede regular structure formation. The weak anthropic principle then places an upper bound on the curvature $\Omega_{\rm K}$ close to the current experimental bound (2.128). The combination of these arguments gives just enough inflation.

These are special features of landscape-based single-field inflation that lie at the border of observability and can place strong constraints on the parameter space of the models. For instance, despite the tight bound (2.128), there may be a non-negligible but strongly measure-dependent theoretical chance to detect a negative small curvature [240, 242]. Moreover, the CMB temperature and polarization

[4]Throughout this chapter, we use the definition (2.48) for the number of e-folds and omit the subscript a. This quantity coincides with the improved definition (5.12) during inflation.

spectra C_ℓ^{TT}, C_ℓ^{BB} and C_ℓ^{EE} can be suppressed or enhanced at large scales depending on whether the phase just after nucleation is, respectively, of fast or slow roll [241, 243, 244]. The second case is already ruled out by observations, while the first [240, 243] is compatible with the loss of power that has actually been found since WMAP.[5] The theoretical chance to observe such suppression is larger than the one of seeing curvature [243]. Although cosmic-variance effects are most prominent at these scales and the observed suppression is not in contradiction with the standard ΛCDM model, new polarization measurements can improve our knowledge of the low-ℓ region.

13.2.2 Large-Field Models and the Weak Gravity Conjecture

Any string model of inflation where the effective inflaton ϕ takes values comparable with or larger than the Planck energy falls into the class of large-field models introduced in Sect. 5.5.1. However, large-field inflation in the landscape can potentially suffer from a severe constraint coming from the *weak gravity conjecture* [246–250]. This conjecture, stemming from the assumption that the number of stable particles not protected by a symmetry is finite, elevates the observation that gravity is the weakest force in Nature to the status of principle. Then, in any theory of quantum gravity bending to such principle, elementary charged objects must feel a gauge force stronger than their mutual gravitational attraction; in other words, the mass-to-charge ratio $|m/q|$ is bounded from above. For a dimensionless q,

$$\left|\frac{m}{q}\right| \lesssim M_{\rm Pl} \,. \tag{13.17}$$

(The right-hand side should not be taken literally; it can be $m_{\rm Pl}$ or $O(1)$ variations of the same.) There are a mild and a strong version of the principle. In the mild version, the above inequality is enforced only on the charged object which minimizes the mass-to-charge ratio, so that $|m/q|_{\rm min} \lesssim M_{\rm Pl}$. In the strong version, which implies the mild one, (13.17) holds for the lightest particle, so that $|m_{\rm min}/q| \lesssim M_{\rm Pl}$. The bound (13.17) is motivated by several facts in string theory and, to date, there are no string counter-examples to the conjecture. Evidence for the mild version is more abundant than for the strong one.

The weak gravity conjecture has consequences that go against the intuition of traditional low-energy effective field theory. For instance, in four dimensions and in the presence of a $U(1)$ gauge field with coupling q_0, the conjecture implies that there exists a UV cut-off scale $\Lambda_{\rm UV} \sim q_0 M_{\rm Pl}$, smaller than the Planck energy, beyond which the effective field theory breaks down. If q_0 is one of the Standard-

[5] A fast-rolling inflaton is a possible source for large-scale power suppression independently of its realization in the string landscape [245].

Model gauge couplings near the grand-unification scale, then Λ_{UV} is close to the heterotic string scale 10^{17} GeV. Moreover, the sub-millimeter observation of any tiny gauge coupling $q_0 \ll 1$ would imply per force the existence of a low-energy cut-off scale $\Lambda_{UV} \ll M_{Pl}$ well below the Planck or even the GUT scale. A cut-off $\Lambda_{UV} < M_{Pl}$ has an impact on the class of large-field models of inflation, where the effective theory is expected to receive corrections at the Planck scale, not below it. Axion monodromy models coming from consistent string-theory compactifications (Sect. 13.4.5) provide a UV completion of large-field inflation but they are heavily penalized for exactly the same reason (Sect. 13.4.6).

Recalling the great effort spent in Chaps. 9, 10 and 11, it is amusing to note that the weak gravity conjecture, if true, excludes all theories of quantum gravity where matter fields are introduced by hand after formulating the gravity sector. In those cases (that include loop quantum gravity, spin foams, asymptotic safety and CDT in their current formulation, but perhaps not group field theory), one has the freedom to consider gauge fields with arbitrarily small coupling.

13.2.3 Multi-field Inflation

If moduli are not stabilized before inflation, it is possible to have several dynamical scalar fields and a multi-dimensional potential. Multi-field inflation (Sect. 5.5.3) can occur in an ample variety of situations within string theory and models can range from landscape-related [161–165, 251] and generic moduli-related [252] (including N-flation [90] and chain inflation [253–257]) to generic string-inspired ones [258–260].

From a statistical point of view, small-field potentials with slow-roll near a saddle point are more likely to be realized in the landscape [161–163, 251, 259, 260]. Moduli inflation (Sect. 13.3) and the infrared DBI model (Sect. 13.6) are examples of string inflation at inflection points. As the inflaton evolves, however, it becomes unlikely to get trapped in a metastable vacuum with positive energy for a sufficient time [163]. To avoid this problem, which would partly invalidate the single-field scenario outlined above, anthropic arguments can select the characteristics of the saddle point so that the latter is in the vicinity of a hole in the landscape with a positive-valued bottom. Inflation terminates when the scalar field rolls down the hole and one ends up with a small $\Lambda > 0$. The same anthropic arguments also help to realize enough e-foldings and a sufficiently long inflationary period.

These models do not generate a high level of non-Gaussianity during slow rolling [164, 260] but the hole must be shallow in order to avoid production of large non-Gaussianities at the end of inflation [165]. To avoid fine tuning, one must conclude that the inflaton potential has mild slopes and that ragged landscapes with deep minima and steep slopes do not lead, in general, to viable scenarios without a certain amount of fine tuning.

13.2.4 Moduli Problem and η-Problem

The models we will describe below exemplify the subtle points involved in moduli stabilization and multi-field inflation in string theory. Excepting a few cases, the underlying Calabi–Yau space is not known and there is no way to determine the total number of complex-structure and Kähler moduli: one must make a phenomenological choice on their number and type. Then, to make the problem tractable one stabilizes all but a few fields, either by mere assumption or by $\mathbb{K}$LT-like mechanisms. Here are the first two caveats to bear in mind in string cosmology. One, which we already had occasion to appreciate, is that stabilizing a modulus for real usually leads to very different results from assuming that it has been stabilized by some unknown mechanism. The second is that stabilizing n moduli simultaneously can lead to very different outcomes from doing so at stages (i.e., stabilizing first k moduli and then the other $n-k$). Both points will be illustrated by the various shapes the moduli + inflaton potential V will take in each model.

Even retaining as many dynamical fields as possible with an act of good will, one must make sure to avoid the *moduli problem*, which is in fact a set of problems [261–265]. Light moduli can disrupt the inflaton dynamics and bar the way to flat directions; or, if inflation happens, they may freeze too late and spoil the standard and well-constrained nucleosynthesis scenario; or, when they decay they can produce too much entropy and, again, jeopardize nucleosynthesis. A task of any string model is to produce a potential which avoids such problems. In general, models of string inflation are successful in this respect but in a non-unique way: each individual potential comes from a different cascade of stabilizations.

Finally, whether and how moduli are stabilized before or during inflation may determine the onset of the notorious η-problem plaguing generic supergravity models of inflation (Sect. 5.12.3).

13.3 Size Moduli Inflation

Historically, the presence of light degrees of freedom in supergravity and string compactifications suggested quite early to identify the inflaton with one of the moduli, typically the dilaton or the Kähler modulus, while all the other fields were assumed to be stabilized by some unspecified mechanism [266–270]. Despite its attractive simplicity, the idea did not work due to the difficulty to obtain slow rolling in the typical non-flat potentials of the moduli. The main culprit of such steepness was the exponential form (12.109) and (12.110) of the non-perturbative superpotential. Thanks to the recent progress in understanding moduli stabilization in $\mathbb{K}$LT and related constructions, it has been possible to revisit *moduli inflation* as a candidate model of the early universe.

In this section, we mainly explore scenarios, each based on its own set of assumptions, where the inflaton is a size modulus σ or a combination of size moduli in type-IIB and heterotic string theory. In general, these belong to the class of small-field models introduced in Sect. 5.5.2, where the inflaton rolls down a flat local maximum.

13.3.1 Large-Volume Inflation

13.3.1.1 Blow-Up Inflation

When many Kähler moduli ϱ_i are stabilized explicitly, one can identify the last one to freeze with the inflaton. In type-IIB flux compactifications, this can be achieved [70–72, 74] in the large-volume stabilization and uplifting scenarios [10, 13, 15] described in Sects. 12.3.9.3 and 13.1.1. One takes a simplified racetrack superpotential (12.110),

$$W = \sum_i \mathcal{A}_i e^{-\alpha_i \varrho_i}, \tag{13.18}$$

and generalizes the potential (12.124) accordingly. While in the models of Sect. 13.4.1 the inflaton will be the axionic part $\theta_\ll$ of a Kähler modulus $\varrho_{\rm infl} = \sigma_\ll + i\theta_\ll$, here it is the last of the "small" or "blow-up" moduli $\sigma^i_\ll$ (describing the size of the 4-cycles, the "holes" of $\mathcal{C}_6$) to roll down to the minimum. All the Kähler moduli except the inflaton are stabilized dynamically. These models are of small-field type (Sect. 5.5.2) and nowadays they are called of blow-up inflation.

After freezing the dilaton, the potential for many Kähler moduli is the generalization of (12.123):

$$\frac{V(\mathcal{V},\sigma^i_\ll)}{M_{\rm Pl}^4} = \sum_{i=1}^{I}\left[\beta_1^i \frac{\sqrt{\sigma^i_\ll}}{\mathcal{V}}\, e^{-2\alpha^i_\ll \sigma^i_\ll} + \beta_2^i \frac{|W_0|\sigma^i_\ll}{\mathcal{V}^2}\, e^{-\alpha^i_\ll\sigma^i_\ll}\cos(\alpha^i_\ll\theta^i_\ll + \psi^i_\ll)\right]$$
$$+\beta_3 \frac{\xi |W_0|^2}{g_s^{3/2}\mathcal{V}^3}, \tag{13.19}$$

where $\exp(i\psi^i_\ll) := \mathcal{A}^i_\ll W_0^*/|\mathcal{A}^i_\ll W_0|$ and $\mathcal{A}^i_\ll$ is the amplitude in (13.18) associated with the $\ll$ modes. Here we follow [70] for simplicity and stabilize also the axionic part $\theta_\ll$. (In [71], θ is switched on and a scenario intermediate between the axionic and the large-volume model arises.) Minimizing with respect to all the axions at $\theta^i_\ll = (\pi - \psi^i_\ll)/\alpha^i_\ll$ for all i and all $\sigma^i_\ll$ for $i \neq I$, one obtains a constant (and partly non-perturbative) contribution (12.126) for each i. Adding also the uplifting term (13.5) with $n = 9/2$, we get a constant $V_0 = \sum_{i\neq I} V^i_{\rm min} + \Delta V \propto (\mathcal{V}^{\rm min})^{-3}$, where $\mathcal{V}^{\rm min}$ is the minimized Calabi–Yau volume. Calling $\phi = \sigma^I_\ll$ the surviving modulus (not normalized canonically) and ignoring the heavily suppressed $\exp(-2\alpha_I\phi)$ term in (13.19), the total potential reads schematically

$$\boxed{V(\phi) = V_0 - V_1\,\phi\,\mathrm{e}^{-\alpha_I\phi}\,,} \tag{13.20}$$

where $\alpha_I > 0$ and $0 < V_1 \propto (\mathcal{V}^{\min})^{-2}$. The shape of (13.20) is shown in Fig. (13.4).

Viable inflation does not require to fine tune the parameters of the potential [70]. The slow-roll parameters at horizon-crossing are $\epsilon_* < 10^{-12}$ and $\eta_* \simeq -2/\mathcal{N}_k$. The observed normalization of the scalar spectrum restricts the size of the Calabi–Yau volume,

$$10^5 \lesssim \mathcal{V}^{\min} \lesssim 10^7\,. \tag{13.21}$$

For $50 < \mathcal{N}_k < 70$, the scalar spectral index $n_\mathrm{s} \simeq 1 + 2\eta_*$ is

$$0.96 \lesssim n_\mathrm{s} \lesssim 0.97\,. \tag{13.22}$$

The index running is negative and small, $\alpha_\mathrm{s} = O(10^{-4})$. Since the inflationary scale is rather low, $V^{1/4} \sim 10^{13}\,\mathrm{GeV}$, the tensor-to-scalar ratio (5.156) is unobservable, $r \lesssim 10^{-10}$. The tensor index $n_\mathrm{t} \simeq -2\epsilon_*$ is nearly zero, too. These numbers persist in a complete model with axions [71]. In general, two Kähler moduli may be not enough to stabilize the volume.

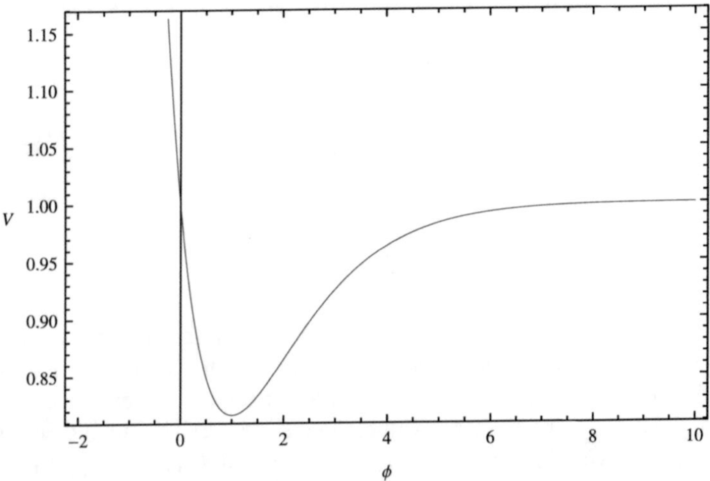

Fig. 13.4 The potential (13.20) of large-volume moduli inflation in $M_\mathrm{Pl} = 1$ units and for casual choices of the parameters $V_{0,1}$ and α_I

13.3.1.2 Fibre Inflation

Blow-up inflation can suffer from the η-problem because loop corrections can dominate over non-perturbative effects, as we will see presently. A scenario capable of producing a larger tensor spectrum and to avoid this problem is fibre inflation [73, 75, 76, 79, 80]. Fibre moduli are Kähler moduli whose potential is generated at one loop in string quantum corrections. There are two such moduli $\sigma_{1,2}$ in the model representative of this class [73], living on a Calabi–Yau space with a K3 fibration in a type-IIB large-volume compactification. An ordinary Kähler modulus $\sigma_\ll$ is also included. The internal volume is $\mathcal{V} = \alpha(\sqrt{\sigma_1}\sigma_2 - \gamma\sigma_\ll^{3/2})$, where $\alpha > 0$ and $\gamma > 0$ are model-dependent constants. Stabilization is sought in the large-volume limit $\mathcal{V} \gg 1$, so that $\sqrt{\sigma_1}\sigma_2 \gg \gamma\sigma_\ll^{3/2}$. In the absence of quantum corrections, the potential $V(\mathcal{V}, \sigma_\ll)$ depends only on two of the three moduli. The flat direction can be parametrized by $\sigma_\ll$ and σ_1 at fixed $\mathcal{V}$. Therefore, while in the blow-up case the inflaton is the size of a small cycle, in fibre inflation it is the size of a large one. Let us call $V_* = V(\mathcal{V}^{\rm min}, \sigma_\ll^{\rm min})$ the value of the potential at the minimum.

One-loop corrections modify $V(\mathcal{V}, \sigma_\ll)$ with a term of the form

$$\Delta V_{g_s} = \frac{W_0^2}{\mathcal{V}^2}\left(\frac{A}{\sigma_1^2} - \frac{B}{\mathcal{V}\sqrt{\sigma_1}} + \frac{C\sigma_1}{\mathcal{V}^2}\right), \tag{13.23}$$

where $A, C \propto g_s^2 > 0$ and B can be of either sign. This contribution deforms the flat direction and stabilizes σ_1. In the single-field approximation, one freezes $\mathcal{V}$ and $\sigma_\ll$ and studies the dynamics of σ_1 in the effective potential $V_* + \Delta V_{g_s}$. If $0 < A, C \ll B$, the potential $V(\phi) = V_* + \Delta V_{g_s}(\phi)$ for the canonically normalized field $\kappa_4\phi = \sqrt{3/4}\ln\sigma_1$ is

$$\boxed{V(\phi) \simeq V_{g_s}\left(3 - 4\,e^{-\kappa_4\phi/\sqrt{3}} + e^{-4\kappa_4\phi/\sqrt{3}} + \beta\,e^{2\kappa_4\phi/\sqrt{3}}\right),} \tag{13.24}$$

where $\beta = O(g_s^2) \ll 1$ and the constant

$$V_{g_s} \propto \frac{1}{\mathcal{V}^{10/3}}, \qquad \mathcal{V} = \mathcal{V}^{\rm min}, \tag{13.25}$$

depends on $\mathcal{V}^{\rm min}$ and $\sigma_\ll^{\rm min}$. All the other coefficients of the potential are independent of the compactification parameters. This potential has a global minimum reached from a flat plateau, where inflation occurs (Fig. 13.5).

In the slow-roll phase, one can neglect the second and third exponentials in (13.24) and compute the primordial spectra. One finds $\epsilon_V \simeq 3\eta_V^2/2$ and $r \simeq 6(n_s - 1)^2$. For $50 \leqslant \mathcal{N}_k \leqslant 60$ e-foldings, the theoretical points in the (n_s, r) plane are well within the 1σ-level likelihood contour,

$$0.965 \leqslant n_s \leqslant 0.970, \qquad 0.005 \leqslant r \leqslant 0.007. \tag{13.26}$$

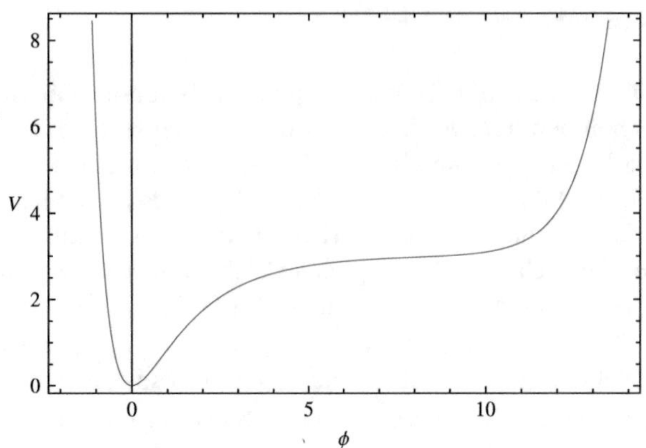

Fig. 13.5 The potential (13.24) in $k_4^{-1} = M_{\rm Pl} = 1$ units with $\beta = 10^{-6}$

The tensor-to-scalar ratio is not as small as in the other string models of inflation and, in fact, the inflationary scale is relatively high, $V^{1/4} \sim 10^{15}$–10^{16} GeV. Relaxing the single-field approximation and including the dynamics of the moduli $\mathcal{V}$ and $\sigma_{\ll}$, one obtains similar results.

Post-inflationary conversion of isocurvature perturbations into adiabatic modes can generate observable non-Gaussianities of the local form (Sect. 4.6.3.1), if the minimum internal volume is not too large. The non-linear parameter is

$$f_{\rm NL}^{\rm local} \simeq 10^5 \frac{(32\pi\beta_3\xi W_0^2/3)^{1/3}}{g_{\rm s}^{1/6}\mathcal{V}^{\rm min}}, \tag{13.27}$$

where the constants β_3 and ξ are those of equation (13.19). To realize an efficient curvaton scenario, $10^3 \lesssim \mathcal{V}^{\rm min} \lesssim 10^8$. Taking $\mathcal{V}^{\rm min} = 10^3$, $g_{\rm s} = 10^{-2}$, $W_0 = 10^{-1}$ and $\xi = 10^{-1}$ (β_3 depends on ξ, on the parameters of the superpotential and on the expectation value of one of the fields), one obtains $f_{\rm NL}^{\rm local} = O(10)$. Choosing a larger volume $\mathcal{V}^{\rm min} = 10^6$ and $g_{\rm s} = 10^{-2}$, $W_0 = 10$ and $\xi = 1$, non-Gaussianities in the squeezed limit are within the experimental bound, $f_{\rm NL}^{\rm local} = O(1)$.

Blow-up and fibre inflation yield attractive predictions but, like most string cosmological models, they meet with some issues at the time of reading the fine print. The inflaton tends to reheat into the hidden sector of the theory unless the latter is severely constrained. This problem restricts the viable parameter space of both classes of models [76] without ruling them out.

A third large-volume model (poly-instanton inflation) produces an index $n_{\rm s} \approx 0.96$ and a tensor-to-scalar ratio $r \sim 10^{-5}$ [77]; it may also serve as a quintessence scenario [271]. In a fourth model, the α'^3-corrections considered so far (Sect. 12.3.9.3) can be combined with other α'^3-contributions coming from higher-order curvature terms in $\mathcal{N} = 1$ SUGRA superspace, to give Starobinsky-like potentials for the fibre modulus [78]. String loop $g_{\rm s}$-corrections do not drive inflation but generate post-inflationary minima for the potential.

13.3.2 Volume-Modulus Inflation

In this sub-section, we present a scenario complementary to large-volume inflation: while in the latter case the inflaton was a blow-up or a fibre modulus, in the present one the inflaton is the volume modulus $\mathcal{V} = \mathcal{V}_6/l_s^6$ of the Calabi–Yau internal space. We already had this modulus play the role of the inflaton in several of the SUGRA models of Sect. 5.12.4, where we called it T. The similarities are many: K is a no-scale Kähler potential or a related deformation and the inflaton potential V is the sum of exponential terms. However, both the details of the deformations and the coefficients in V differ. While the models of SUGRA inflation proceed from hand-picked super- and Kähler potentials, volume-modulus inflation finds its own microscopic origin in string scenarios of flux compactification, which give rise to very specific potentials and phenomenology.[6]

The concrete embedding of this mechanism in the SUGRA limit of string theory can proceed in different ways [81–83]. We will see in Sect. 13.4.1 that, in type-IIB flux compactification with a racetrack superpotential, the flattest directions in the total potential are typically along $\theta = \mathrm{Im}\varrho$. However, for certain values of the parameters [11, 18, 51], inflation can be driven by $\sigma = \mathrm{Re}\varrho = \mathcal{V}^{2/3}$ rather than the axionic part of the Kähler modulus ϱ, at an inflection point of the potential [81]. Predictions range in the interval $0.93 \lesssim n_s < 1$, while the tensor-to-scalar ratio is negligible, $r < 10^{-6}$. Viable inflation is characterized by some strong fine tuning of the initial conditions which, however, might not be a problem since the model supports eternal inflation. Regions with exponentially growing volume are continuously produced and quantum fluctuations make the inflaton start always at the saddle point of the potential. For long or eternal inflation, however, the scalar spectral index is either $n_s \approx 0.93$ (if V is not $\mathbb{Z}_2$-symmetric) or $n_s \approx 0.95$ (if V is $\mathbb{Z}_2$-symmetric), somewhat redder than the preferred PLANCK 2015 value 0.97 (see (4.70) and (4.72)).

Another possibility is to perform, as in the large-volume scenarios, the $\mathcal{V} \gg 1$ expansion of the SUGRA limit of string theory [82, 83]. In this case, the volume becomes large after, not before, inflation. Consider first a single-field scenario where all moduli but $\mathcal{V}$ have been stabilized and let $\kappa_4\phi = \sqrt{2/3}\ln\mathcal{V}$ be the canonically normalized volume field. The total potential is given by six terms: the $V_{\alpha'} \propto \mathcal{V}^{-3} = \exp(-\sqrt{27/2}\kappa_4\phi)$ perturbative $O(\alpha'^3)$ contribution in (12.126), the $V_{\rm np} \propto -\mathcal{V}^{-3}(\ln\mathcal{V})^{3/2} \propto -\phi^{3/2}\exp(-\sqrt{27/2}\kappa_4\phi)$ non-perturbative contribution in (12.126) (but now with $\mathcal{V}$ not minimized), the $V_{g_s} \propto \mathcal{V}^{-10/3} = \exp[-(10/\sqrt{6})\kappa_4\phi]$

[6] A somewhat hybrid between theory and phenomenology is the no-scale SUGRA model of [272–275], which deploys the Kähler potential $K = -3\ln(K+K^*) - \ln(S+S^*) + |\Phi|^2(T+T^*)^{-n}$ and the superpotential $W \propto \Phi(T-\mathrm{const})$. Here, S is the dilaton and Φ is a chiral field with modular weight $n = 1, 2, 3, 4, 5$. The two no-scale logarithmic contributions [276–278] and the last term [279–281] stem from the orbifold compactification of the heterotic string to four dimensions. The inflaton can be either the imaginary or the real part of T. However, to the best of our knowledge the superpotential W does not arise naturally in string theory.

perturbative string-loop correction (13.25) (again, with variable $\mathcal{V}$), the $\Delta V_2 \propto \mathcal{V}^{-2} = \exp(-\sqrt{6}\kappa_4\phi)$ D-term correction (13.4) ((13.5) with $n = 3$) and two more terms we have ignored so far, a correction $V_{F^4} \propto \mathcal{V}^{-11/3} = \exp[-(11/\sqrt{6})\kappa_4\phi]$ from the $\alpha'^3 R^4$ higher-derivative term in the $D = 10$ type-IIB SUGRA action [28] and the contribution $V_{\rm hid} \propto \mathcal{V}^{-8/3} = \exp[-(8/\sqrt{6})\kappa_4\phi]$ of a possible hidden sector of charged matter fields [27, 282]. Overall [83],

$$\begin{aligned} V(\phi) &= V_{\alpha'} + V_{\rm np} + V_{g_s} + \Delta V_2 + V_{F^4} + V_{\rm hid} \\ &= V_0\left[(1 - c_{\rm np}\phi^{3/2})e^{-\sqrt{\frac{27}{2}}\kappa_4\phi} + c_{g_s}e^{-\frac{10}{\sqrt{6}}\kappa_4\phi} + c_{\Delta V}e^{-\sqrt{6}\kappa_4\phi}\right. \\ &\quad \left. + c_{F^4}e^{-\frac{11}{\sqrt{6}}\kappa_4\phi} + c_{\rm hid}e^{-\frac{8}{\sqrt{6}}\kappa_4\phi}\right], \end{aligned} \tag{13.28}$$

where $V_0 \propto |W_0|^2$, $c_{\rm np}$, c_{g_s}, $c_{\Delta V}$, c_{F^4} and $c_{\rm hid}$ are tunable constant coefficients. In the absence of blow-up modes $\sigma_\ll$, the $\phi^{3/2}$ term in the first contribution disappears, while in the presence of one frozen blow-up mode $\sigma_\ll$, one can neglect the last term in (13.28) (or the D-term ΔV_2, with qualitatively similar results). The typical shape of (13.28) can be appreciated in Fig. 13.6.

Requiring V to have a de Sitter minimum and an inflection point where inflation lasts about $\mathcal{N}_e = 60$ e-foldings, one can work out the cosmological primordial spectra. For single-field inflation, the first slow-roll parameter ϵ is always much smaller than η, so that $n_s \simeq 1 + 2\eta_*$ and tensor modes are negligible. The normalization of the scalar spectrum fixes V_0 in (13.28) and, hence, W_0. The magnitude

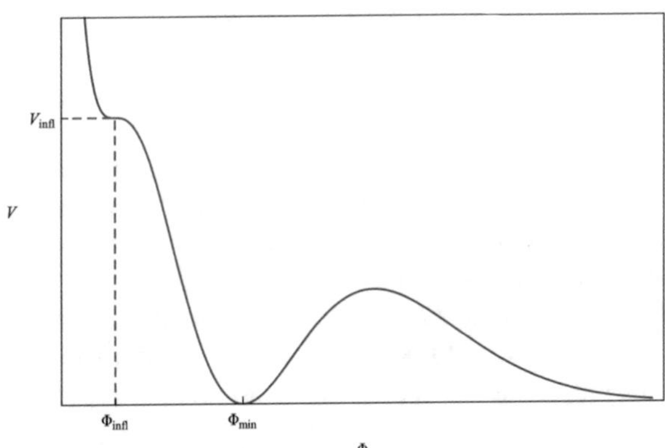

Fig. 13.6 Log-linear plot of the volume-modulus potential (13.28) [82, 83] in $k_4^{-1} = M_{\rm Pl} = 1$ units, with $V_0 = 10^{10}$ (for convenience), $c_{\rm np} = 0$ (this is the case without frozen blow-up moduli), $c_{g_s} \approx 3.485$, $c_{\Delta V} \approx 4.7 \times 10^{-4}$, $c_{F^4} \approx 4.282$ and $c_{\rm hid} \approx 0.107$. The parameters have been determined by fixing the position of the inflection point and of the de Sitter minimum at, respectively, $\phi_{\rm infl} = 3.5$ and $\phi_{\rm min} = 4.2$. Inflation takes place at the inflection point at an energy density $V_{\rm infl} \approx 3.7 \times 10^{-10}$, while the dark-energy era is around $\phi_{\rm min}$. These values are not realistic but allow us to see the main qualitative features of the potential. In general, the barrier between $\phi_{\rm min}$ and the runaway Minkowski minimum at infinity is much smaller compared with $V_{\rm infl}$

of W_0 depends on whether the visible matter sector is in communication with the hidden sources of supersymmetry breaking. If it is (non-sequestered models), the soft terms are of order of the gravitino mass, while if it is not (sequestered models) these terms are much smaller than $m_{3/2}$ and $M_{\rm SUSY} = O(\text{TeV})$. In particular, there are two cases [83]:

- *High-scale supersymmetry.* This is a non-sequestered model where the volume is of order 100 both during inflation (i.e., at the inflection point) and at late times, $\mathcal{V}_{\rm infl} \sim 10^2 \sim \mathcal{V}^{\rm min}$. Since $\mathcal{V}$ is not very large, the validity of the effective-field-theory approximation is not guaranteed and this case is not under good control. Still, one can explore its main features in the absence of frozen blow-up moduli. The flux contribution to the superpotential is $W_0 = O(10^{-5})$, the inflationary scale is $V_{\rm infl}^{1/4} \sim 10^{14}$ GeV, the tensor-to-scalar ratio is $r \sim 10^{-10}$ and the spectral index is $n_{\rm s} \approx 0.960$.
- *Low-scale supersymmetry.* In the non-sequestered case, the value of the de Sitter minimum is $\mathcal{V}^{\rm min} \sim 10^{12} - 10^{15}$, while $\mathcal{V}^{\rm min} \sim 10^6 - 10^7$ in the sequestered case. All the other details are the same, with or without the frozen blow-up modulus: the coefficients $c_{g_{\rm s}}$ and $c_{\rm hid}$ must be negative, the flux term is $W_0 \sim 10^{-2} - 10^2$, during inflation $\mathcal{V}_{\rm infl} \sim 10^3 - 10^5$, the tensor-to-scalar ratio is $r \sim 10^{-9}$ and the scalar index is $n_{\rm s} \approx 0.967$.

To check the actual level of fine tuning on the parameter range required to obtain the right amount of e-foldings, one can turn on the blow-up mode $\sigma_{\ll}$ and consider two-field inflation. It turns out that the fine tuning is about 10^{-3}.

13.3.3 Fluxless Inflation

Another model of moduli inflation [84, 85] is in the context of the heterotic string at weak coupling without fluxes ($W_0 = 0$ in the superpotential). The inflaton Φ is a generic complex modulus field and its dynamics is described by $\mathcal{N} = 1$ SUGRA in four dimensions. The Kähler potential is

$$K = -3\ln(T + T^*) - \ln(S + S^*) + |\Phi|^2 , \tag{13.29a}$$

while the superpotential is

$$W = A_1\, \mathrm{e}^{-S/N_1} - A_2\, \mathrm{e}^{-S/N_2}(\alpha + \beta\Phi)^{-\delta} . \tag{13.29b}$$

The Kähler modulus T is stabilized by the kinetic terms of matter fields, which induce a mass breaking the flatness of the T direction. The inflationary mechanism and gaugino condensates can trap also T. We do not write these non-perturbative contributions here and assume that T has been frozen already, or can be frozen in a more complicated scenario. The axio-dilaton S has a double exponential superpotential coming from the gaugino condensation of two hidden SUGRA sectors with gauge groups $SU(N_1)$ and $SU(N_2)$. The constants $A_{1,2} = O(1)$ and $0 < \delta \lesssim O(1)$ depend on $N_{1,2}$ and on the string cut-off scale. The term $\alpha + \beta\Phi$, with $\alpha = O(1) = \beta$, is the mass of a non-singlet gauge field, which depends on the vacuum expectation value of moduli coupled with such field. In this way, W acquires a dependence from the inflaton.

The field Φ starts near a saddle point ($V_{,\Phi\Phi} = 0$) where the slow-roll conditions hold, then it rolls down a local minimum. This configuration guarantees that the model is of eternal inflation. This detail is important inasmuch as it solves a problem typical of models where inflation occurs before stabilizing the dilaton [84, 269]: the barrier between the $\mathbb{K}$LT vacuum and the Minkowski vacuum at Re$S \to \infty$ (Fig. 13.2) is several orders of magnitude smaller than the Planck-scale energy of generic initial conditions set at some time t_i after the big bang. An inflaton starting from $\Phi(t_i) = O(m_{Pl})$ corresponds to a dilaton ReS starting at about the same energy, which overshoots the $\mathbb{K}$LT minimum and rolls over the barrier. This observation is known as the *runaway dilaton problem* or, more generally, *overshooting problem*.[7]

Eternal inflation sets the initial conditions near a saddle point of the total effective potential $V(S, \Phi)$. Consequently, S does not acquire enough kinetic energy to climb the barrier and it gets trapped in a minimum away from the asymptotic vacuum at $|S| \to \infty$. Thus is the runaway dilaton problem avoided. Without fine tuning, the scalar spectrum is compatible with observations. For $30 < \mathcal{N}_e < 60$,

$$0.97 < n_s \lesssim 0.98 . \tag{13.30}$$

The running of the scalar index is negligible, $\alpha_s < 10^{-5}$. The tensor spectrum is too small to be observed.

13.4 Axion Inflation

In this section, we present some ways to employ SUGRA and string axions in cosmology. After checking on models of racetrack potentials in Sect. 13.4.1, we will see cases of axion inflation capable of reproducing, under suitable approximations, the single-field cosine potential (5.90) of natural inflation. To generate viable perturbation spectra, the axion decay constant f must be greater than the Planck mass (Sect. 5.9.1) but $f < M_{Pl}$ in superstring theory [284]. How do strings face this interesting challenge?

[7] We already had occasion to comment on the necessity of stabilizing the dilaton, or any other non-minimally coupled scalar, at early times to respect stringent "fifth-force" constraints on the variation of Newton's coupling [283] (Sects. 7.4.5 and 12.3.5.5).

13.4.1 Racetrack Axion Inflation

The first model of axionic inflation we examine [86, 87] is based on a minimal modification of the $\mathbb{K}$LT flux-compactification scenario of Sect. 12.3.9.2 [3]. The first difference is that the superpotential is not (12.111) but of racetrack type (12.110) in the Kähler modulus,

$$W = W_0 + \mathcal{A}e^{-\alpha\varrho} + \mathcal{B}e^{-\beta\varrho}. \tag{13.31}$$

The constant term W_0 represents the contribution of previously-stabilized moduli, including the dilaton. Second, the axion $\theta = \mathrm{Im}\varrho$ is not assumed to be stabilized and, in fact, it is identified with the inflaton. The reason for this choice is that the Kähler potential (12.99) does not depend on θ, so that the total potential $V(\sigma, \theta)$ may have flat regions in the θ direction. Eventually, this helps to address the η-problem. The details of the stabilization mechanism, absent in old models of moduli inflation, mitigate the steepness of the racetrack superpotential to acceptable levels.

The potential $V(\sigma, \theta)$ can be readily written down; we leave this task to the reader, plugging (13.31) in (12.112). It has several positive or negative minima (Fig. 13.7). The kinetic term for the fields can be obtained from the kinetic matrix in (5.220a).

As in the fluxless model of Sect. 13.3.3, eternal inflation avoids the overshooting problem arising when fields with large kinetic energy reach the asymptotic limit $\mathrm{Re}\varrho \to +\infty$, where space rapidly decompactifies. The number of regions with initial conditions $\theta(t_\mathrm{i}) \ll m_\mathrm{Pl}$, are replenished near the saddle point $V_{,\theta\theta} = 0$ by the mechanism of eternal reproduction and slow roll is then guaranteed. Inflation takes place near the saddle point, away from the $\mathbb{K}$LT minimum. Since the effective dynamics of the system is well described by a field theory on a curved background, all the inflationary observables are calculated as in Chap. 5. For $\mathcal{N}_\mathrm{e} > 40$ e-foldings, the scalar spectral index is

$$n_\mathrm{s} \lesssim 0.97\,. \tag{13.32}$$

The inflationary scale $V^{1/4} \sim 10^{14}\,\mathrm{GeV}$ is too low to produce an observable tensor spectrum. Cosmic strings are not produced, either; see, in contrast, the case (13.63) of warped D-brane inflation.

Other criteria that successful inflation should satisfy are a low amplitude for density perturbations, $\delta\rho/\rho \sim 10^{-5}$, and the observed value of the vacuum, $\rho_\Lambda \sim 10^{-123} m_\mathrm{Pl}^4$. Both criteria are met by a fine tuning of the parameters of the potential of one part over 10^3, provided the ratio α/β is irrational [86].

An explicit model on the orientifold $\mathbb{P}^4_{[1,1,1,6,9]}$ has a similar fine tuning, of order of percent [87]. On this space, there are only two Kähler moduli, whose superpotential is

$$W = W_0 + \mathcal{A}e^{-\alpha\varrho_1} + \mathcal{B}e^{-\beta\varrho_2}. \tag{13.33}$$

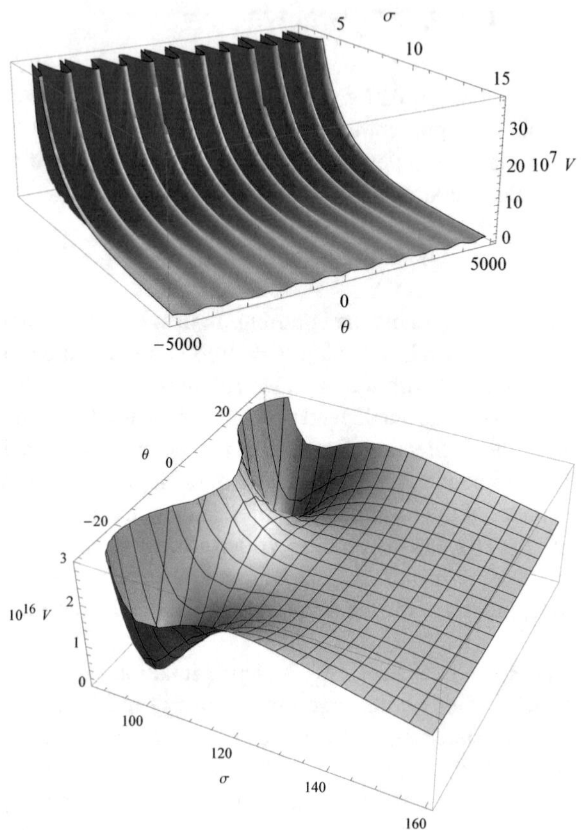

Fig. 13.7 *Top*: the total potential $V(\sigma, \theta)$ for the axionic inflaton model of [86], in $M_{\rm Pl} = 1$ units and for the parameter choice of [86]. The potential is periodic and nearly flat in the $\theta = {\rm Im}\varrho$ (inflaton) direction, while it is steep along the $\sigma = {\rm Re}\varrho$ direction. *Bottom*: a zooming in of the potential at a saddle point where inflation occurs; the inflaton θ rolls down one of the two minima

The dilaton and the complex-structure moduli have been frozen by fluxes into the constant contribution W_0. A region of the potential $V(\sigma_1, \theta_1, \sigma_2, \theta_2)$ is shown in Fig. 13.8.

13.4.2 Axion Valley

As a further example of how the inflaton potential changes when a different cascade of stabilizations is enforced, we mention the *axion valley* model of [88, 89], an interesting and very simple way to obtain natural inflation (Sect. 5.5.2) in $\mathcal{N} = 1$ supergravity. Whether this model can be derived from the low-energy limit of string theory is still unclear [89].

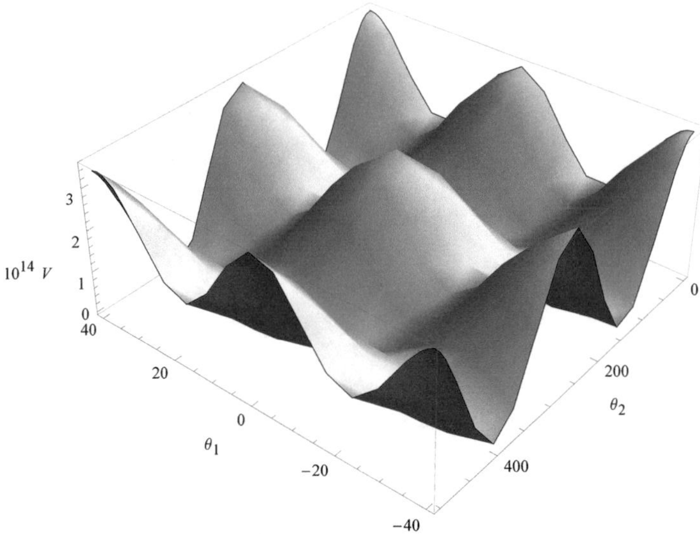

Fig. 13.8 The total potential $V(\sigma_{1,\min}, \theta_1, \sigma_{2,\min}, \theta_2)$ for the axionic-inflaton model of [87], in $M_{\rm Pl} = 1$ units and for the parameter choice of [87]. The size moduli $\sigma_{1,2} = \mathrm{Re}\varrho_{1,2}$ have been fixed to their minimum. The role of the inflaton is played by the flattest direction, in this case θ_1

Here, the dilaton and all complex-structure and Kähler moduli are assumed to be stabilized by the 𝕂LT mechanism. As in the fluxless model (13.29), we have a complex modulus Φ, this time with a Kähler potential symmetric under shifts and a generic non-perturbative superpotential:

$$K = \tfrac{1}{4}(\Phi + \Phi^*)^2\,, \qquad W = W_0 + \mathcal{A}\,\mathrm{e}^{-\alpha\Phi}\,, \tag{13.34}$$

where W_0 is real. The superpotential can be generalized to a racetrack type in order to keep multi-instanton corrections under control [89] but for illustrative purposes (13.34) will suffice. Splitting the scalar as $\Phi = \chi + \mathrm{i}\theta$ and noting that $D_\Phi W = \partial_\Phi W + W\partial_\Phi K = W_0\chi + \mathcal{A}(\chi - \alpha)\,\mathrm{e}^{-\alpha(\chi+\mathrm{i}\theta)}$, $\mathcal{G}_{\Phi\Phi^*} = \partial^2 K/(\partial\Phi\partial\Phi^*) = 1/2 = (\mathcal{G}^{\Phi\Phi^*})^{-1}$ and $|W|^2 = W_0^2 + \mathcal{A}^2\mathrm{e}^{-2\alpha\chi} + 2W_0\mathcal{A}\mathrm{e}^{-\alpha\chi}\cos(\alpha\theta)$, the F-term (12.105) reads

$$\boxed{V(\chi,\theta) = \mathrm{e}^K(2|D_\Phi W|^2 - 3|W|^2) = V_1(\chi) + V_2(\chi)\,\cos(\alpha\theta)\,,} \tag{13.35a}$$

where

$$V_1(\chi) = W_0^2(2\chi^2 - 3)\,\mathrm{e}^{\chi^2} + \mathcal{A}^2[2(\chi-\alpha)^2 - 3]\,\mathrm{e}^{\chi(\chi-2\alpha)}\,, \tag{13.35b}$$

$$V_2(\chi) = -2W_0\mathcal{A}[3 - 2\chi(\chi-\alpha)]\,\mathrm{e}^{\chi(\chi-\alpha)}\,. \tag{13.35c}$$

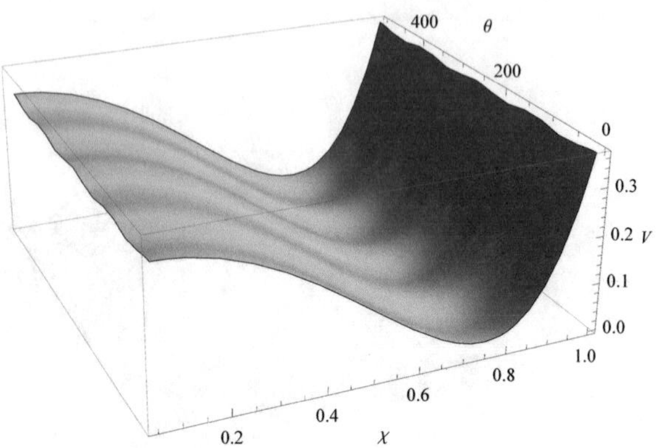

Fig. 13.9 The axion valley (13.35) [88] in $M_{\rm Pl} = 1$ units, with $W_0 = -10^{-3}$, $\mathcal{A} = 1$ and $\alpha = 0.05$. The potential is periodic along the θ direction

The $\mathbb{K}$LT uplifting mechanism adds a positive constant contribution V_0 to (13.35), so that the minima $V_{\rm min} = V(\chi_{\rm min}, \theta_{\rm min}) + V_0$ of the total potential are positive and close to zero. Since it is possible to obtain a potential steep in χ and nearly flat in θ, the latter plays the role of the inflaton (Fig. 13.9).

Both the axion-valley potential in Fig. 13.9 and the axionic model in Fig. 13.7 are steep in the χ direction and nearly flat and periodic in the axion (the inflaton) direction, while the $\mathbb{K}$LT model in Fig. 13.2a is relatively steep both in the radion (the inflaton) and the axion direction. Despite the similarity, the axion-valley potential (13.35) is very different from the one in Fig. 13.7 inasmuch as it has a sharp minimum in the χ direction, while there is no valley in the axionic-moduli models.

The cosine potential (5.90) of natural inflation is obtained by fixing the parameters V_0, W_0, $\mathcal{A}$ and α so that $V_1(\chi_{\rm min}) + V_0 = V_2(\chi_{\rm min}) = \Lambda^4/2$ and $\alpha = 1/f$. A normalization Λ of order of the GUT scale and values above the PLANCK 2015 constraint $f \gtrsim 6.9\,M_{\rm Pl}$ [285] can be easily obtained in the context of supergravity.

13.4.3 *N-flation*

Another model populated by axions is *N-flation* [89–105]. We have already seen, both in the axion-valley example and in the other scenarios of moduli inflation, how a cosine dependence on axion moduli arises in supergravity. One can take N axions θ_n each with a periodic potential similar to the potential (5.90) of the pseudo-Nambu–Goldstone boson in natural inflaton [90]:

$$V(\theta_1, \ldots, \theta_N) = \sum_{n=1}^{N} V_n(\theta_n)\,, \qquad V_n(\theta_n) = \Lambda_n^4 \cos\frac{\theta_n}{f_n}\,. \tag{13.36}$$

Instanton corrections proportional to $\cos(2\theta_n/f_n)$ and $\cos(\theta_n/f_n)\cos(\theta_{n'}/f_{n'})$ are dropped from (13.36) (we will see such corrections more in detail in Sect. 13.4.6). For $\theta_n/f_n \sim \pi$, we have $V_n(\theta_n) \simeq m_n^2\theta_n^2/2$, where $m_n = \Lambda_n^2/f_n$. In the typical string landscape, no one of the individual scalars θ_n rolls slowly for enough e-foldings but, taken collectively, the radial degree of freedom $\phi^2 := \sum_n \theta_n^2$ behaves as an inflaton. If all masses are equal, $m_n = m$, then $V(\theta_1,\dots,\theta_N) \simeq m^2\phi^2/2 = V(\phi)$ and the predictions of N-flation are very close to standard chaotic inflation with a quadratic potential [90]. The number of e-foldings is proportional to the number of axions. Take an initial condition $\theta_{n,\mathrm{i}} = \alpha M_{\mathrm{Pl}}$ and $f_n = f$ for all n, where the constant $\alpha = \alpha'^{3/2}\sigma/\sqrt{\mathcal{V}_6} = \sigma/\sqrt{\mathcal{V}} \leqslant f/M_{\mathrm{Pl}} \lesssim O(1)$ is determined by the string compactification and it is bounded from above by the decay constant f. The initial condition for the inflaton is $\phi_{\mathrm{i}}^2 = N\alpha^2 M_{\mathrm{Pl}}^2$. From (5.85e), one has $\mathcal{N}_{\mathrm{e}} \simeq \phi_{\mathrm{i}}^2/(4M_{\mathrm{Pl}}^2) = N\alpha^2/4$.

Contrary to its phenomenological cousin (5.87), N-flation is not fine tuned because radiative corrections are under control. Cross-couplings between the axions, coming from quantum string corrections, are suppressed. An interesting feature is that the maximum number of e-folds is limited by N and the number of complex-structure moduli in the Calabi–Yau compactification. In general, both this upper bound and N are large enough to fuel a long inflationary era.

For different masses m_n, the analysis of the dynamics is more complicated and shows that N-flation usually produces a redder scalar spectrum [91, 93]. Nearly scale-invariant cosmological scalar perturbations are compatible with observations only if $N \gtrsim O(10^3)$ (easily realized in string theory) and the axion mass spectrum is densely packed [92]. Parametrizing this spectrum as $m_n^2 = m^2\exp[(n-1)/\Delta]$, one has $\Delta \gtrsim 300$; this restriction is confirmed [97] in a more sophisticated parametrization of m_n [91].

Inflationary observables do not deviate from the single-field case and are, therefore, compatible with the data. The first and second slow-roll parameters are of the same size, $\epsilon_* \sim \eta_* = O(10^{-2})$, while the spectral index and tensor-to-scalar ratio are

$$0.93 < n_{\mathrm{s}} \leqslant 0.95\,, \qquad r \lesssim 10^{-3}\,. \tag{13.37}$$

r is smaller than the current bounds (4.71), (4.73) and (4.74) but larger than in other moduli-inflation scenarios and in D-brane inflation. Non-Gaussianity is too small to be observed if inflation takes place at small field values (quadratic-potential approximation) [94–96], while it can be much larger near the hilltop [100, 101].

The main assumption underlying N-flation is that radions are stabilized before inflation. However, we have seen that in most examples of moduli inflation radions and axions are stabilized more or less simultaneously, since they have masses of the same order. Near a local minimum, the steepness of the multi-field potential in the radion and axion directions is similar. This will be true also for the $\mathbb{KLMT}$ scenario (Sect. 13.5) and it implies a fine tuning.

One exception is the axion valley but this model has never been fully derived from string theory [89]. Reheating in N-flation may also be problematic [98]. The N-flaton ϕ couples with all matter sectors, including the hidden ones originated by supersymmetry breaking. Unless the couplings are fine tuned, ϕ can decay into hidden-sector particles and give rise to cosmological relics incompatible with constraints on dark matter; we have mentioned a similar reheating problem for blow-up and fibre inflation. Other issues are discussed in [104, 196].

Another possibility is to consider the large-volume potential (13.19). In this case, the axion masses can be hierarchically smaller than the other moduli and, with some tuning, one can stabilize the volume and the blow-up modes in advance, leaving an effective potential which differs from (13.36) by a constant additive term and phases:

$$V(\theta_{\ll}) = V_0 + \sum_{i=1}^{I} c_i \cos(\alpha^i_{\ll}\theta^i_{\ll} + \psi^i_{\ll}) . \tag{13.38}$$

The parameters in (13.38) are tunable to give viable acceleration with essentially the same properties as N-flation [102]. The volume is stabilized at $\mathcal{V} = O(10^2)$ with a large number of axions $N \sim 10^5$. Then, during inflation $H \sim 10^{15}$ GeV. Unfortunately, too many axions correspond to a large density of cycles per unit volume, which may give rise to a loss of control of the string compactification. Moving away from the large-volume expansion entails other problems [99] which are, in fact, only the tip of a grim iceberg. An analysis on simulated landscapes show that N-flation is under strong theoretical pressure in type-II and heterotic string compactifications, unless extreme fine tuning is invoked [104].

13.4.4 Aligned and Hierarchical Axion Inflation

From what discussed above, one can evince that N-flation can be safely regarded as a supergravity model of inflation but its embedding in string theory is subject to many theoretical constraints [89, 99, 102–105, 120]. However, this is not the end of the story for string axions. The N-flationary potential (13.36) (sum of single-field cosines) is a special case of the more general multi-axion profile [106–117]

$$\boxed{V(\theta_1, \dots, \theta_N) = \sum_{j=1}^{J} \Lambda_j^4 \left[1 - \cos\left(\sum_{n=1}^{N} \frac{p_{jn}\theta_n}{f_n}\right)\right].} \tag{13.39}$$

Specific relations among the decay constants $p_{jn}/f_n < M_{\rm Pl}$ (all smaller than the Planck mass) give rise to flat directions in the axion space. This can be achieved in two ways, which we illustrate for two axions and $J = N = 2$: either by an alignment mechanism such that $|p_{11}/p_{12} - p_{21}/p_{22}| \ll 1$ [106] (for perfect alignment, the inflaton is $p_{11}\theta_1 + p_{12}\theta_2$) or by establishing a hierarchy of the form $p_{12} = 0$, $p_{11} \ll p_{22}$ [107, 111]. Integrating out the most massive modes, one obtains the effective single-field inflationary potential (5.90) with an effective super-Planckian decay constant $f_{\rm eff} \gtrsim M_{\rm Pl}$ and the usual properties of natural inflation (almost scale invariance, small but non-negligible tensor-to-scalar ratio r, small non-Gaussianity, and so on).

The potential (13.39) can be embedded in four-dimensional supergravity and in type-IIB string theory compactified on a Calabi–Yau space with fluxes [107, 111–115], as well as in type-IIA theory [119]. The axions θ_n in (13.39) are given by linear combinations of the Kähler-moduli axions θ_i, while the coefficients Λ_j depend on the expectation values of the size moduli σ_i. Just like for N-flation, the way in which size moduli are stabilized (with the $\mathbb{K}$LT mechanism or with the large-volume scheme) and the region where inflation occurs determine the level of tuning of the parameters (of order of $10^{-4}-10^{-2}$) required to suppress unwanted features such as the large non-Gaussianities typical of multi-field and axionic models [286]. However, the actual moduli space in the landscape giving rise to viable inflation is likely to be quite restricted, at least in type-II theory [104, 105, 116, 119, 120]. This result does not rule out the alignment and hierarchical mechanisms but it limits their scope to certain types of compactifications.

13.4.5 *Monodromy Inflation*

13.4.5.1 *D*-Term Monodromy

A careful inspection of flux compactifications suggests a way to modify the axion potential to obtain large-field inflation. We saw from (12.89) that the axions θ_{B,i_2}, θ_{C,i_2} and θ_{i_4} arise in type-IIB Calabi–Yau compactifications from the integration of, respectively, the forms B_2, C_2 and $C_4^{(+)}$ over the 2- and 4-cycles of the Calabi–Yau space (recall that the indices i_2 and i_4, omitted from now on, run over the cycles). If we also wrap branes around these cycles, one discovers that, when moving around a cycle, the brane energy increases. This phenomenon is called a *monodromy*.

We omit the details of the model, which involve many extended sources and stabilization ingredients. We shall just say that the effect of the wrapped brane is to break the periodicity (i.e., the shift symmetry) of the axionic potential. For instance, a D5-brane wrapped around a 2-cycle Γ_2 of length $L(x) = \sqrt{\alpha'\theta_{\rm crit}(x)}$ generates a potential for θ_B proportional to $V(\theta_B) \propto (g_s\alpha'^2)^{-1}\sqrt{\theta_B^2+\theta_{\rm crit}^2} \sim \theta_B$ for $\theta_B \gg \theta_{\rm crit}$. This happens because the DBI action (12.42) is evaluated over the wrapping configuration and the flux of the Kalb–Ramond field B_{ab} through Γ_2 is nothing but θ_B. The linear potential [122] dominates over the periodic contributions. Similarly, a wrapped NS5-brane gives a dominant linear potential to θ_C, for θ_C larger than the critical value $\theta_{\rm crit}$ (assuming the length of the cycle to be the same). If L is of the same order of the compactification scale $R = \sqrt{\alpha' l(x)}$, the relations between the axions and the canonically normalized inflaton ϕ are $\phi \sim M_{\rm Pl}\theta_B/\theta_{\rm crit}$ and $\phi \sim M_{\rm Pl} g_s \theta_C/\theta_{\rm crit}$, so that

$$V(\phi) \sim \phi \,. \tag{13.40}$$

The potential can be further flattened by the back-reaction of the inflaton on the geometry. If the cycle length L is much smaller than R (for instance, when the 2-forms are localized in a throat), then $\phi \sim M_{\rm Pl}\theta_B l^{1/2}/\theta_{\rm crit}^{3/2}$ and $\phi \sim M_{\rm Pl} g_s \theta_C l^{1/2}/\theta_{\rm crit}^{3/2}$. By back-reaction of the flux, the size R depends on the axion and increases dynamically. In particular, one can argue that $l^2 \propto \theta_B$, so that $\phi \sim \theta_B^{5/4}$ and the inflaton potential becomes [126]

$$V(\phi) \sim \theta_B(\phi) \sim \phi^{\frac{4}{5}} \,. \tag{13.41}$$

This is only one example of flattening mechanism.

So far in this chapter, we have discussed models in type-IIB string theory and one heterotic case. Type-IIA moduli inflation [287], based on stabilization scenarios in the corresponding theory [203, 288, 289], is difficult to achieve and it will not be discussed here. However, instead of compactifying on a warped Calabi–Yau space one can decide to consider other internal manifolds with special properties that give rise to a rather different phenomenology. On a fairly wide class of orbifolds, branes can wrap around cycles as usual, but the length of the cycle increases while the brane moves and so is the brane energy. Again we have monodromies, although not necessarily of axionic type.

For instance, de Sitter vacua in type-IIA flux compactification can be obtained in one such special orbifold $\mathcal{M}_6 = \mathcal{N}_3 \times \tilde{\mathcal{N}}_3$, the product of two identical Nil 3-manifolds (also known as *twisted tori*) [290]. Each twisted torus is parametrized by, respectively, a triplet of coordinates (y_1, y_2, y_3) and $(\tilde{y}_1, \tilde{y}_2, \tilde{y}_3)$ [121]. The cross section of $\mathcal{N}_3$ along a certain direction y_1 is isomorphic to a 2-torus $\mathbb{T}^2$. The direction y_1 is rendered compact by periodically identifying tori at different points by a discrete group. The direction y_2 is a 1-cycle of $\mathcal{M}_6$ whose length depends on y_1.

Having this background, we can wrap a D4-brane on the above 1-cycle and note that, while the brane moves along y_1, the cycle becomes longer and the brane tension increases [121]. After compactifying the DBI 4-brane action on $\mathcal{M}_6$ and expanding it to lowest order in the derivatives, one identifies the canonically normalized inflaton $\phi(x) = \phi[y_1(x)]$ with a function of y_1. Once all the moduli but ϕ are stabilized, the resulting total Lagrangian is $\mathcal{L} \simeq R/(2\kappa_4^2) - (\nabla\phi)^2/2 - V(\phi)$, where the effective potential $V(\phi)$ has a global minimum at $\phi = 0$ and grows as $\sim \phi^2$ near the origin and as $\sim \phi^{2/3}$ at large ϕ. The transition scale $\phi_{\rm crit}$ depends on the parameters of the string compactification and it is sub-Planckian. However, the region $\phi \ll \phi_{\rm crit}$ near the origin cannot sustain chaotic inflation because the field can attain trans-Planckian values only by violating the consistency requirement that $V^{1/4}$ does not exceed the energy used to stabilize the moduli.[8]

In the other asymptotic region $\phi \gg \phi_{\rm crit}$, one realizes the large-field model of inflation (5.81) with $n = 2/3$, while wrapping the D4-brane around a different direction results in an effective potential with $n = 2/5$ [121]. Yet different brane wrappings induce a linear potential for the C_3 axion [122], just like in type-IIB theory.

An $n = 2/3$ or $n = 4/3$ potential arises also by flattening a quadratic potential in the case where inflation (driven by the Kalb–Ramond axion θ_B) happens during, not after, moduli stabilization [133]. The vacuum expectation value of the moduli changes adiabatically while deforming the inflationary potential. The cases $n = 2$ and $n = 3$ are also obtained, from the flattening of a quartic potential [133] or when the candidate inflaton is a complex-structure modulus [131, 133–135].

Therefore, in general, *monodromy inflation* [121–143] is a class of large-field models giving rise to a monomial potential (5.81):

$$\boxed{V(\phi) = V_0\phi^n + \frac{\Lambda^4}{2}\cos\frac{\phi}{f} + \cdots \simeq V_0\phi^n\,, \qquad n = \frac{2}{5}, \frac{2}{3}, \frac{4}{5}, 1, \frac{4}{3}\,.} \tag{13.42}$$

All the constants V_0, n, Λ and f are determined by the parameters of the string compactification (including the vacuum expectation values of the stabilized moduli) and are such that the oscillatory term does not spoil the slow-roll approximation. That is because the instantonic effects giving rise to the periodic potential are exponentially suppressed with respect to the leading monomial term. In practice, aligned and hierarchical axion inflation (Sect. 13.4.4) can be regarded as the multi-field generalization of monodromy inflation [107, 108], since the periodicity of the inflaton potential is broken by the mixing with heavier axions.

In the region $\phi \gg \phi_{\rm crit}$, all these models are self-consistent inasmuch as moduli, the inflationary trajectory and the effective background are not destabilized by the evolution of the inflaton field, by the monodromy correction $V(\phi)$ to the total moduli potential or by the back-reaction of the wrapped brane.[9] The inflationary potential is also robust against α'-corrections in the brane curvature and against string loop contributions, which are highly damped.

---

[8]In other words, V must be smaller than the potential barrier separating the system from the runaway vacuum of weak coupling and of the decompactification limit.

[9]By *back-reaction*, one means the effect of the relative size of the brane core with respect to the curvature radius of the internal manifold. For instance, in the $n = 2/3$ model on twisted tori discussed here, the size $r_{\rm brane}$ of the D4-brane core can be determined by evaluating the gravitational potential $\propto r_{\rm brane}/|\vec{r}|$, where $\vec{r} \in \{(y_1, \tilde{y}_1, \tilde{y}_2)\}$ is a vector in the sub-space spanned by three of the six internal directions (the other three are dimensionally reduced). $|\vec{r}|$ is the distance from the intersection point with the D4-brane. The constant $r_{\rm brane}$ is proportional to the 7-dimensional Newton constant; if it is smaller than the other sizes of $\mathcal{M}_6$, then the brane is a good probe of the geometry [121].

Brane (or D-term) monodromy faces a number of issues. In the $n = 2/3$ type-IIA model, large corrections to the slow-roll parameters can come from an arbitrary orientation of the D4-brane from other extended sources. A symmetric orientation solves the problem but it seems difficult, if not impossible, to construct a global embedding with this feature. The model does not admit anti-de Sitter or Minkowski vacua and the spectrum of allowed de Sitter vacua has a lower bound higher than the observed value $\Lambda_{\rm obs}$ [140].

In the $n = 1$ type-IIB model, stabilization of the Kähler moduli steepens the potential of θ_B (roughly speaking, non-perturbative superpotentials such as (13.18) create an η-problem for θ_B) but not that of θ_C. Thus, θ_C can play the role of the inflaton more likely than θ_B. However, the NS5-brane must be paired (to ensure tadpole cancellation) with a distant $\overline{\rm NS5}$-brane and their interaction is logarithmic with the distance. This interaction term is not negligible with respect with $V(\phi)$ and it can destabilize the inflationary dynamics.

13.4.5.2 F-Term Monodromy

An inflationary potential generated by an F-term can bypass these and other issues [129–132, 134, 135, 137, 138]. In this case, supersymmetry is broken spontaneously and there is an effective SUGRA description, not available in previous models where supersymmetry is broken explicitly at the string scale. Using D7-branes instead of 5-branes in type-IIB theory yields a model with an $n = 2$ quadratic inflationary potential, where the inflaton is an axion among the complex-structure moduli [131, 132]. The $n = 2$ case also emerges from the F-term scenarios of [129, 130]. One identifies the axio-inflaton with specific integrals of a p-form over a p-cycle, called massive Wilson lines. Then, the effective potential $V(\phi)$ behaves as a power law ϕ^n asymptotically and interpolates between $n = 2/3$ and $n = 2$.

Other mechanisms involve background fluxes or torsion and produce polynomial potentials with $n \geqslant 2$. In a type-IIB toy model, one is able to move away from the polynomial trend of F-term axion monodromy and to obtain, in the large-field regime, the Starobinsky potential (5.234), where M depends on the parameters of the flux compactification [137]. In type-IIA theory, it is possible to obtain an inflaton potential interpolating between $n = 1$ and $n = 2$ [139, 141].

13.4.5.3 Inflationary Predictions

The parameter space of monodromy models is the set of vacuum expectation values of the moduli, in turn dependent on brane charges and quantized fluxes. Within the theoretically allowed region in this parameter space, there is a window for generating observationally acceptable observables. For instance, in the type-IIA $n = 2/3$ case the fine tuning on the parameters is $O(10^{-2})$ and corresponds to an $O(100)$ anisotropy in the radii of $\mathcal{M}_6$ and to a reasonable $k_F \propto \int_{\mathcal{M}_6} F_6 = O(100)$

number of flux units of the R-R 6-form F_6. Type-IIB brane models have the same amount of tuning and so do F-term monodromy scenarios; the issue of tuning can be explored also by counting viable vacua in the landscape [135].

Within these windows of viability, one has almost scale-invariant spectra and a non-negligible tensor-to-scalar ratio. As we saw in Sect. 5.5.1.2, the slow-roll parameters (5.85a) and (5.85b) are small as long as $\phi \gg m_{\rm Pl}$, a condition respected by monodromy scenarios. Equation (5.197) with $\mathcal{N}_k \gg n = O(1)$ are

$$n_{\rm s} \simeq 1 - \frac{n+2}{2\mathcal{N}_k}\,, \qquad r \simeq \frac{4n}{\mathcal{N}_k}\,. \tag{13.43}$$

For the values of n given in (13.42) and for $\mathcal{N}_k = 60$,

$$0.97 < n_{\rm s} \lesssim 0.98\,, \qquad 0.03 < r < 0.09\,, \tag{13.44}$$

compatible with PLANCK 2015 data [285]. The linear case $n = 1$ corresponds to $(n_{\rm s}, r) \approx (0.975, 0.07)$. The sub-dominant periodic part of the potential (13.42) can create an oscillatory modulation of the perturbative correlation functions (spectrum, bispectrum, trispectrum, and so on) [123–125, 136, 286]. These oscillations are strongly constrained by observations and are virtually undetectable in the CMB temperature spectrum, but they are also responsible for non-Gaussianities of "resonant" type, encoded in a non-linear parameter $f_{\rm NL}^{\rm res} \sim 10$. The present sensitivity on $f_{\rm NL}^{\rm res}$ does not allow to check this prediction yet. With an eye to the future, we also mention that a coupling between the axion and gauge matter breaks the shift symmetry spontaneously and can generate a *TB* polarization pattern in the CMB, as explained in Sect. 5.9.2. In turn, observations of non-Gaussian signals constrain the magnitude of such coupling and of the axion decay constant $f \sim g_{\rm s} M_{\rm Pl}/\mathcal{V}^{1/3}$ in (13.42) [285] (the expression for f is more complicated in the presence of many moduli).

13.4.5.4 Quintessence

Finally, axion monodromy has been also proposed as a source of quintessence [68, 69]. Placing 5-branes in a Klebanov–Strassler throat not containing the Standard Model breaks the shift symmetry, while the warp factor at the bottom of the throat provides the $\sim 10^{-120}$ damping of the potential necessary for dark energy. There is no tracking behaviour for the linear potential and, hence, the coincidence problem is not solved. To get viable dark energy (but indistinguishable from the ΛCDM model [69]), one must fine tune the initial conditions, a plague of the quintessence models of Sect. 7.3 that might find an accommodation in some corners of the string landscape. More general axion-driven realizations of quintessence in string theory have a similar issue [67].

13.4.6 Problems with Axion Inflation and Ways Out

Let us now come back to the question, raised at the beginning of this section, of how string theory can manage to get large axion decay constants. The bottom line is that it cannot.

For each of the axion-inflation models listed here, we have mentioned the presence of a variety of issues and their proposed cures. Unfortunately, these problems might be not point-wise occurrences due to the technical limitations of individual models, but a reflection of an obstacle endemic to the string landscape: large axion decay constants are in contrast with the weak gravity conjecture (Sect. 13.2.2).

We know from Sect. 12.2.8 that, given a R-R $(p+1)$-form C_{p+1} in ten dimensions, we have a p-brane with charge q_p (this is the dimensionless version of Q_p). For $p=0$ in type-IIB theory, the D(-1)-brane is an instanton with mass $m>0$ and action $S_{\rm inst}\sim m/M_{\rm Pl}$. This instanton couples with the 0-form axion $\theta=\phi/f$ by a positive $U(1)$ coupling $q_0\sim M_{\rm Pl}/f$ which breaks the shift symmetry of the four-dimensional potential:

$$V(\phi)\simeq\sum_l \Lambda_l^4 e^{-lS_{\rm inst}}\cos\left(\frac{l\phi}{f}\right). \tag{13.45}$$

The weak gravity conjecture states that $m/q_0\lesssim M_{\rm Pl}$, so that $S_{\rm inst}\sim m/M_{\rm Pl}\lesssim q_0\sim M_{\rm Pl}/f$. Therefore, $M_{\rm Pl}/f$ cannot be too small (a condition for successful inflation) lest the higher-order l-terms in (13.45) become more important and the effective theory fail, at energies much lower than desired by theoretical consistency [104, 246]. The extension of the conjecture to many copies of $U(1)$ [248, 249] leads to a similar hindrance for multi-axion models such as N-flation, aligned and hierarchical axion inflation [104, 105, 116, 119, 120, 291]. In these cases, there is an instanton for each axion and the axion decay constants are individually bounded by the Planck mass, $f_n<M_{\rm Pl}$ for all n. If the weak gravity conjecture holds, the constraint $f_{\rm eff}<M_{\rm Pl}$ obtained in these models for a collective, effective decay constant $f_{\rm eff}$ must eventually receive yet-ignored corrections that bring down the collective bound to the individual one. Performing such a check would contribute to verify the validity of the conjecture.

A possible loophole in the above arguments would open up if only the mild version of the conjecture held. Then, the constraints would apply only to the particle minimizing $|m/q|$, while another axion would be free to drive inflation. Consider the case of two axions $\phi_{1,2}$ such that ϕ_2 couples with an instanton with mass $m_2>m_1$ and charge $q_2=kq_1>q_1$, where k is a positive integer. If the lighter particle does not obey the weak-gravity bound but the heavier one does, this is a configuration with k stable states and we are in the presence of the mild version of the weak gravity conjecture. For a potential given by the sum of N copies of (13.45) and taking the lowest-order contribution $l = 1$, we have $f_2 = f_1/k < f_1$ and

$$V(\phi_1, \phi_2) \simeq \Lambda_1^4 e^{-m_1/M_{\rm Pl}} \cos\left(\frac{\phi_1}{f_1}\right) + \Lambda_2^4 e^{-m_2/M_{\rm Pl}} \cos\left(\frac{k\phi_2}{f_1}\right). \tag{13.46}$$

The heavier particle has a smaller decay constant and cannot enjoy the super-Planckian enhancement in axion inflation. Its contribution to (13.46) is suppressed by an exponential factor $\exp(-m_2/M_{\rm Pl}) \ll \exp(-m_1/M_{\rm Pl})$ and the dominant contribution is by ϕ_1, which can play the role of the inflation [105].

However, there are no counter-examples to the strong weak gravity conjecture in string theory and it is non-trivial to find realistic embeddings of the suppression mechanism leading to (13.46). Attempts to realize it in aligned/hierarchical models [117, 118] face a series of obstacles [120] which, nevertheless, might be circumvented; see the type-IIB construction of [118], where a potential different from (13.39) is obtained.

To summarize, in Sect. 13.4 we have examined cosmological string scenarios aiming to recover the periodic potential (5.90) in some limit and to explain inflation with such potential. Starting from a population of axions, the continuous shift symmetry (12.90) is explicitly broken by non-perturbative effects and many string embeddings of the potential can be found in the literature. Examples of this class of models are N-flation, aligned and hierarchical inflation. The majority of these cases are constrained by the weak gravity conjecture and either fine tuning or special hand-picked Calabi–Yau spaces are the price to pay in order to attain viable inflation. Some models [112] avoid this theoretical constraint but not other issues [116]. Other multi-axion models under construction might survive the screening of the mild version of the weak gravity conjecture [105, 117, 118, 292] but not easily [120]. We also mention that one can give up the alignment mechanism and consider other types of mixing in type-II compactifications with intersecting D-branes, giving rise to the natural-inflation potential (5.90) for a certain linear combination ξ of axions [293, 294]. The resulting effective decay constant f_ξ is different from the $f_{\rm eff}$ of previous models and it may avoid the weak-gravity bound. A particular case of natural inflation [295] can evade the weak gravity conjecture but not all its versions [291].

Finally, a model-independent study of configurations with N axions shows that, at large N, the effective enhanced decay constant $f_{\rm eff}$ converges to a finite value (hence the scenario is consistent with the weak gravity conjecture) only if the number of instantons in the inflationary potential grows fast with N, more than quadratically and perhaps exponentially [296]. Since multi-axion large-field models such as N-flation and aligned/hierarchical inflation predict, by their assumptions, a different (much slower) scaling of $f_{\rm eff}(N)$, one can reach two mutually exclusive conclusions: either the weak gravity principle holds and these models are inconsistent at large N, or we are witnessing a violation of the weak gravity principle. The latter interpretation is suggested by the lack of such a huge number of unsuppressed instantons in controlled string compactifications.

All in all, the debate on axion inflation with periodic potentials and the weak gravity principle is fairly recent and wide open.

In axion monodromy inflation, the discrete shift symmetry is further broken, explicitly (with branes) or spontaneously (by F-terms), by perturbative mechanisms and inflation is driven by the symmetry breaking term, which dominates over the sinusoidal term. The instantonic corrections to the periodic part of the potential are therefore innocuous. However, the weak gravity principle may still pose a problem, since it implies the existence of a UV cut-off smaller than the Planck scale. The problem of large-field scenarios is to control the perturbative corrections $\sum_n (\phi/m_{\rm Pl})^n$ to the inflationary potential. The shift symmetry of axions protects their potential from such corrections but this symmetry is broken by stringy non-perturbative effects. In turn, these effects strongly depend on the details of the UV model, so that the challenge is to balance realistic UV physics on one hand and good inflation on the other hand. The weak gravity principle can constrain the UV cut-off in axion monodromy models and, from that, the inflationary parameter space. Large-field displacements are allowed but there is a lower bound for the axion decay constant [142].

So far, axion monodromy inflation seems safe from weak-gravity assaults. However, we can appreciate other difficulties in obtaining a viable UV model of monodromy inflation by recalling that even F-term monodromy, proposed to bypass the problems of brane monodromy, has its own issues. Single-field inflation from F-term axion monodromy in type-IIB orientifold compactifications is possible but subject to several theoretical constraints [134, 135, 138]. In F-term models, the inflaton does not appear in the Kähler potential and there is no danger to have an η-problem, provided the inflaton be a linear combination of axions only. Also, identifying the inflaton with the universal axion C_0 [130] may be problematic due to the guaranteed presence of other light axions, which cannot be stabilized beforehand. Moreover, the fine tuning of the parameters may be severe or even impossible in certain cases, such as in the weak-coupling regime $g_{\rm s} \ll 1$ of type-IIB theory on an orientifold. The conclusion from various negative and positive examples is that not all regions in the string landscape can support F-term monodromy inflation and care must be exercised in model building and in the choice of the Calabi–Yau space.

13.5 Slow-Roll D-Brane Inflation

13.5.1 Early Brane-Inflation Models

In old Kaluza–Klein scenarios, compactification of a higher-dimensional space down to $3 + 1$ directions is the most direct way to obtain a world with the observed number of dimensions. Essentially the same could be told about string theory but with a notable difference: the extra dimensions can be much larger than the Planck scale [237–239] and the observer may live in a D3-brane.

An interesting situation is that of a stack of D-branes at a non-minimal energy configuration, that is to say, the branes are displaced from one another with respect to the transverse extra dimensions [297]. The displacement is governed by the scalar mode φ of an open string connecting two separate branes. For an observer in a brane, this field plays the role of the inflaton. As the branes approach one another but while still being at a distance, the weakly-coupled scalar rolls slowly down its potential and causes the non-compact spatial directions in the branes to expand exponentially. At some critical inter-brane distance, the potential $V(\varphi)$ becomes too steep and inflation ends. At small distances compared to the string scale, the tachyonic mode switches on and triggers the brane decay. Reheating consists in the collision of the branes and the consequent release of energy as radiation.

Consider one extra direction and assume, for the moment, that all moduli but the brane separation have been stabilized. For Hubble patches $H^{-1} \gg R$ much larger than the size R of the extra dimension, the dynamics on one brane is described by the first Friedmann equation $H^2 \propto \rho_{\rm eff}$, where the effective four-dimensional energy density $\rho_{\rm eff} \simeq \mathcal{T}_3\dot{\varphi}^2 + V(\varphi)$ (from the DBI action; $\mathcal{T}_3$ is the 3-brane tension) encodes the kinetic energy of the relative brane motion and the inter-brane interaction potential $V(\varphi)$. At zero separation, the potential vanishes, $V(0) = 0$. At distances $\varphi \neq 0$, the coupling between modes on different branes is suppressed at least exponentially, so that there is a short-range attractive contribution $\sim\mathcal{T}_3[1 - \exp(-|\varphi/\varphi_0|)]$ (or a more strongly damped profile) to $V(\varphi)$. This adds to other terms including a potential $\sim\exp(-m|\varphi|)|\varphi|^{2-N_{{\rm D}p}}$ which describes the exchange of massive (or massless, when $m = 0$) bulk modes, where $N_{{\rm D}p}$ is the number of branes. Due to the strong fall-off of brane interactions, no mass term for φ is generated in the potential and a slow-roll regime is easily achieved.

From the total $V(\varphi)$, one can work out the details of the inflationary era. Brane–anti-brane interactions are under better control than those of branes of the same orientation and one can realize *brane inflation* [297] in string theory [298–303]. In this case, the effective potential is of the form [299–301]

$$V_{{\rm D}p\text{-}\overline{{\rm D}p}}(\varphi) = V_0\left(1 - \frac{\beta}{\varphi^{d_\perp - 2}}\right). \tag{13.47}$$

where $V_0 \propto m_{\rm s}^{p+1} = l_{\rm s}^{-(p+1)}$ and $\beta \propto m_{\rm s}^{d_\perp - 2}$ are dimensionful constants and $d_\perp = 9-p$ is the number of large dimensions transverse to the Dp-branes. One has $d_\perp = 6$ if all extra dimensions are large; if $d_\perp = 2$, the potential becomes logarithmic. For $d_\perp = 6$ and assuming the branes are at the maximal possible distance $R \simeq \mathcal{V}_6^{1/6}$ (the size of the compact space), inflation lasts about $\mathcal{N}_{\rm e} \approx 80$ e-foldings and generates a scalar spectrum with index $n_{\rm s} - 1 \approx 0.97$. The typical compactification scale is about $R^{-1} \approx 10^{12}$ GeV, while the string scale is at $m_{\rm s} \sim 10^{15} - 10^{16}$ GeV. However, non-extremal but more natural cases with $R \ll \mathcal{V}_6^{1/6}$ do not give enough e-folds $\mathcal{N}_{\rm e} \propto R^6/\mathcal{V}_6 \propto |\eta|^{-1}$ and the perturbation spectrum strongly deviates from scale invariance. This is nothing but the η-problem.

Other configurations with branes intersecting at an angle θ can solve this problem and appear in many constructions of the Standard Model (Sect. 12.3.9). The coefficients V_0 and β in (13.47) depend on θ and the brane–anti-brane case corresponds to $\theta = \pi$. More general potentials $V = V_0(1-\beta\varphi^{-n})$ are also possible.

Apart from the specific form of the potential and a limited range of realistic couplings, brane inflation has other characteristic features that distinguish it from a standard inflationary mechanism. For instance, cosmic strings are produced at the end of acceleration; their fate depends on the details of the model [304–306].

These possibilities were recognized before the advent of $\mathbb{K}$LT scenarios [297–299, 304, 305] and promptly embedded in them soon afterwards. In concrete compactification schemes of moduli stabilization, pairs of D3-branes and $\overline{\text{D3}}$-branes are located at the tip of a deformed Klebanov–Strassler throat, where the warped geometry naturally leads to acceleration. The Randall–Sundrum braneworld scenario we will see in Sect. 13.7.1 [307] can nest in the full theory via the $\mathbb{K}$LT mechanism of moduli stabilization: the braneworld then coincides with a stack of $\overline{\text{D3}}$-branes (for the Standard Model) or D3-branes (for supersymmetric extensions of the Standard Model) at the tip of the throat [6]. Then, the scalar φ corresponds to the radial direction along the throat.

13.5.2 *Warped D-Brane Inflation and* $\mathbb{K}$L$\mathbb{M}$T *Model*

A crucial assumption of brane inflation is that all moduli have been stabilized, so that V_0 and β in (13.47) are constant. If the internal space were not static, the potential $V(\varphi, \mathcal{V}_6) \propto \mathcal{V}_6^{-2}$ would be too steep along the direction of the Kähler modulus and brane inflation would end quickly. The challenge of obtaining viable inflation well illustrates the huge gap between cosmological models where moduli stabilization is assumed and top-down scenarios where such stabilization is carried out explicitly. To differentiate modern cosmological models with respect to early proposals, we call the former *warped D-brane inflation*. We begin with models where the slow-roll conditions are satisfied, *slow-roll D-brane inflation* [144–167, 193, 194] and their prototype $\mathbb{K}$L$\mathbb{M}$T *inflation* (pronounced "KKLMMT" or "KLMT," from the acronym of the six authors who first studied this scenario [144]). The $\mathbb{K}$L$\mathbb{M}$T construction is fairly simple: in the $\mathbb{K}$LT setting, one adds a mobile D3-brane to the Klebanov–Strassler throat which makes the de Sitter vacuum dynamical. The model must be engineered carefully in order to make the inter-brane potential flat and the inflaton sufficiently light. First, warping of the background space flattens the potential much more than brane-inflation models on flat spaces. Second, gaugino condensation gives a heavy mass to the inflaton and one should consider stabilization mechanisms alternative to that of Sect. 12.3.9. α'-corrections to the Kähler potential (although not of the form computed in [308]) can stabilize the Kähler modulus while avoiding this problem. Third, while the anti-brane is automatically fixed at the tip of the throat by the dynamics, the position of the mobile brane must be subject to a suitable non-perturbative superpotential. Let us see these ingredients in detail in type-IIB flux compactification, starting from the background geometry.

13.5.2.1 Step 1: Klebanov–Strassler Geometry

The Klebanov–Strassler throat in six dimensions can be parametrized by a conical metric

$$ds^2 = e^{2\omega(r)}\, ds_4^2 + e^{-2\omega(r)} ds_6^2\,, \tag{13.48}$$

where the internal line element ds_6^2 depends on the radial coordinate r (spanning the length of the throat) and on angular coordinates. The conifold singularity corresponds to $r = 0$. When deforming the throat into a smooth compact space, the singularity is replaced by a smooth tip isomorphic to S^3 and located at some $r = \bar{r} \ll 1$. $\omega(r)$ increases with r monotonically and tends to a constant at small r.

For $r > \bar{r}$ but smaller than the gluing point $r_{\rm max}$ of the conifold with the rest of the Calabi–Yau space, the cross section of the throat is the coset space $T^{1,1} = [SU(2) \otimes SU(2)]/U(1)$, with topology $S^2 \times S^3$, while away from the tip the throat is isomorphic to $\mathrm{AdS}_5 \times S^2 \times S^3$ spacetime with characteristic scale $R_{\rm AdS}$. More generically, the cross section of a warped space can be an Einstein manifold X_5. Along the throat, there is a large gravitational redshift $e^{\omega(r)} \gg 1$ in the warped metric (13.48). The tip and the gluing point act as, respectively, an IR and a UV cut-off for AdS_5. For this reason, the tip is sometimes called the infrared end of the throat, while the gluing point is the ultraviolet one.

The scale $R_{\rm AdS}$ is related to the location of the tip at $\bar{r}$. As a matter of fact, the deformation of the singular Klebanov–Strassler conifold into a compact space is achieved thanks to the presence of the background 3-forms F_3 and H_3, which carry the quantized fluxes

$$\frac{\tilde{\mathcal{T}}_3}{2\pi}\int_{S^3} F_3 = k_F\,, \qquad \frac{\tilde{\mathcal{T}}_3}{2\pi}\int_{\Gamma_3} H_3 = -k_H\,. \tag{13.49}$$

As in (12.87), $\tilde{\mathcal{T}}_3 = [(2\pi)^3\alpha'^2]^{-1}$, while S^3 and Γ_3 are, respectively, the sphere S^3 at the tip and its dual 3-cycle; $k_F, k_H \gg 1$ are positive integers. Then, the minimum warp factor in (13.48) (i.e., the ratio between the tip position and the AdS radius) is finite [204]:

$$e^{2\omega(\bar{r})} = \left(\frac{\bar{r}}{R_{\rm AdS}}\right)^2 = \frac{\bar{\varphi}^2}{\lambda\sqrt{\tilde{\mathcal{T}}_3}} = \exp\left(-\frac{4\pi k_H}{3 g_s k_F}\right), \tag{13.50}$$

where, for later convenience, we have defined $\bar{\varphi} := \sqrt{\mathcal{T}_3}\bar{r}$ and a dimensionless coupling $\lambda := \sqrt{\mathcal{T}_3}R^2_{\rm AdS}$. The radius $R_{\rm AdS} = (4\pi g_s N\alpha'^2)^{1/4}$ depends on the number

$$N = k_F k_H \gg 1 \tag{13.51}$$

of D-brane charges of the background. This numerology gets a fresh insight in the AdS/CFT correspondence and $\lambda \gg 1$ is interpreted as the strong-coupling limit of the dual CFT (see (12.103) and [196]).

An exact metric describing the whole deformed throat exists but for our purposes the above rough decomposition into tip, throat and gluing point will be enough.

13.5.2.2 Step 2: $\overline{\text{D3}}$-Brane

Consider now a D3-brane with tension $\mathcal{T}_3$ (see (12.42)) placed in the throat at a point $r = r_{\rm D3}$, with $\bar{r} < r_{\rm D3} < r_{\rm max}$. Its action on $\text{AdS}_5 \times S^5$ [309, 310] well approximates the brane dynamics away from tip and gluing point:

$$\boxed{\begin{aligned} \mathcal{S}_{\rm DBI} &= \int d^4x\,\sqrt{-g}\mathcal{L}_{\rm DBI}, \qquad f(\varphi) = \frac{\lambda^2}{\varphi^4}, \\ \mathcal{L}_{\rm DBI} &= -f^{-1}(\varphi)\left[\sqrt{1+f(\varphi)\,g^{\mu\nu}\partial_\mu\varphi\partial_\nu\varphi} - 1\right], \end{aligned}} \tag{13.52}$$

where $\varphi := \sqrt{\mathcal{T}_3}r_{\rm D3}$ and we have neglected fermionic and gauge fields. Notice the non-perturbative dependence $1/g_s \propto 1/\lambda^2$ of the action from the string coupling, when written in terms of $r_{\rm D3}$. The strong-coupling regime corresponds to $\lambda \gg 1$, but (13.52) is valid also at weak coupling [311].

The action $\bar{\mathcal{S}}_{\rm DBI} = \int d^4x\,\sqrt{-g}\bar{\mathcal{L}}_{\rm DBI}$ for a dynamical $\overline{\text{D3}}$-brane is the same as (13.52) except for the last term, which is $+1$:

$$\bar{\mathcal{L}}_{\rm DBI} = -f^{-1}(\tilde{\varphi})\left[\sqrt{1+f(\tilde{\varphi})\,g^{\mu\nu}\partial_\mu\tilde{\varphi}\partial_\nu\tilde{\varphi}} + 1\right], \tag{13.53}$$

where $\tilde{\varphi} := \sqrt{\mathcal{T}_3}r_{\overline{\rm D3}}$. Expanding this action for a small derivative term, one finds a quartic potential with a global minimum at the origin, i.e., near the tip of the Klebanov–Strassler throat:

$$\bar{\mathcal{L}}_{\rm DBI} = -\frac{2}{f(\tilde{\varphi})} + O[(\partial\tilde{\varphi})^2] = -\frac{2\tilde{\varphi}^4}{\lambda^2} + O[(\partial\tilde{\varphi})^2]\,. \tag{13.54}$$

This means that any $\overline{\text{D3}}$-brane placed in the deformed throat, and actually onto any spot on the Calabi–Yau space [150], will be eventually drawn towards the region $\tilde{\varphi} \to \bar{\varphi}$ with smaller warp factor, i.e., down the throat. This is an intuitive justification of an ingredient of the $\mathbb{KLT}$ stabilization of Sect. 13.1.1, where a stack of anti-branes was put in the throat to cancel tadpole anomalies and to break supersymmetry. Similarly, in the $\mathbb{KLMT}$ construction one adds a $\overline{\text{D3}}$-brane at the tip $r = \bar{r}$ of the Klebanov–Strassler throat, assuming that it was fixed there dynamically. Then, $r_{\overline{\text{D3}}} = \bar{r}$ and $\tilde{\varphi} = \bar{\varphi}$.

13.5.2.3 Step 3: D3-Brane

In type-IIB theory, D3-branes are stable and there is no potential term in the DBI action (13.52). The presence of a $\overline{\text{D3}}$-brane trapped at the tip, parallel to the D3-brane, induces an attractive potential $V(\varphi)$ which pulls the D3-brane towards the tip. This situation is depicted in Fig. 13.10a.

If we truncate the action (13.52) to second order in the derivatives, we obtain a canonical kinetic term for the inflaton φ:

$$\mathcal{L}_\varphi \simeq -\frac{1}{2}\partial_\mu\varphi\partial^\mu\varphi - V_{\text{D3-}\overline{\text{D3}}}(\varphi)\,, \tag{13.55}$$

where the potential is (13.47) with $d_\perp = 6$:

$$\boxed{V_{\text{D3-}\overline{\text{D3}}}(\varphi) = 2\frac{\bar{\varphi}^4}{\lambda^2}\left(1 - \frac{1}{N}\frac{\bar{\varphi}^4}{\varphi^4}\right).} \tag{13.56}$$

Note that the constant term is nothing but $-\bar{\mathcal{L}}_{\text{DBI}}$ for a fixed anti-brane position.

To the brane–anti-brane potential (13.56), one should also add the contribution $V_{\text{CY}}(\varphi)$ from all corrections coming from the specific embedding in a Calabi–Yau

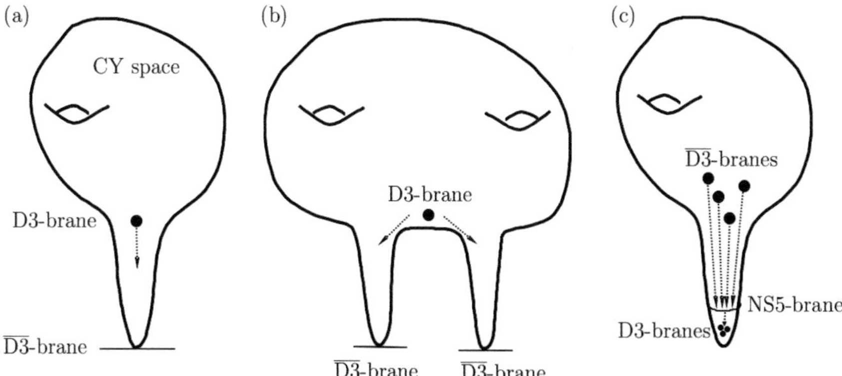

Fig. 13.10 Warped D-brane inflation (**a**) in the $\mathbb{KLMT}$ model [144], (**b**) with a $\mathbb{Z}_2$ symmetry [151] and (**c**) with only $\overline{\text{D3}}$-branes [150]

space, populated by fluxes, branes of different kind (in particular, D7-branes) at different positions and more moduli that are stabilized eventually. The total potential then reads

$$\boxed{V(\varphi) = V_{\text{D3-}\overline{\text{D3}}}(\varphi) + V_{\text{CY}}(\varphi)\,.} \tag{13.57}$$

13.5.3 *Cosmological* 𝕂L𝕄T *Dynamics*

Thanks to (13.50), for certain choices of fluxes V_0 can be very small and the potential extremely flat. The negative sign in (13.56) indicates an attractive force between the two branes. When the latter are too close and $\bar{r} \sim r_{\text{D3}}$, the potential (13.56) breaks down; this stage corresponds to reheating.

Assuming that the compactification is stable, viable inflation follows through. The first slow-roll parameters of Sect. 5.4.2.2 are $\epsilon_V \simeq (8\kappa^2\bar{\varphi}^8/N^2)\varphi^{-10}$ and $\eta_V \simeq -(20\kappa^2\bar{\varphi}^4/N)\varphi^{-6} \simeq -5/(6\mathcal{N}_k)$, which are both small if φ is sufficiently large. Since $N \gg 1$, $\epsilon_V \ll \eta_V$ and the tensor spectrum is unobservable. In fact, it is easy to see that for $\mathcal{N}_k = 60$ e-foldings both the tensor-to-scalar ratio (4.66) and the tensor index $n_{\text{t}} = -2\epsilon_*$ are extremely small,

$$r = \frac{\mathcal{P}_{\text{t}}}{\mathcal{P}_{\text{s}}} = O(\epsilon_*) \approx 10^{-11}\,, \tag{13.58}$$

while the inflaton potential at horizon crossing is lower than the GUT scale by a couple of orders of magnitude, $V^{1/4} \sim 10^{14}$ GeV. The scalar spectral index $n_{\text{s}} = 1 + 2\eta_{V*} - 6\epsilon_{V*} \simeq 1 + 2\eta_{V*}$ is well within the observed range,

$$n_{\text{s}} \simeq 1 - \frac{5}{3\mathcal{N}_k} \approx 0.97 \tag{13.59}$$

for 60 e-folds.

The potentials (13.56) (which appears in old brane inflation [299] and in the 𝕂L𝕄T model [144]) and (13.47) with $d_\perp = 4$ (typical of old brane inflation with brane at angles [303]) are both compatible with PLANCK 2015 data [285]. However pleasing this result may be, the real challenge of string cosmology is not to fit data per se but, rather, to do so in the natural parameter space of the theory. The remainder of this section is dedicated to this goal.

To ensure that inflation actually takes place, the moduli must be stabilized. However, the 𝕂LT scheme of Sects. 12.3.9 and 13.1.1 receives corrections that risk to spoil the cosmological scenario just detailed. The Kähler potential (12.99) is augmented non-trivially by the contribution of the brane moduli,

$$\tilde{K}(\varrho, \phi) = -3\ln[2\text{Re}\varrho - c(\phi, \phi^*)] = -2\ln\mathcal{V}\,, \tag{13.60}$$

where ϕ is a multiplet of fields consisting in the position of the D-branes on the Calabi–Yau space. The superpotential $W = W(\varrho)$ can be taken for the moment to be dependent only on the Kähler modulus, as in (12.111) (but it will be generic in what follows). Then, one can compute the F-term (12.112) with $\tilde{K}$ given by (13.60). To this F-term, one must add the D-term (13.3) with the replacement $\mathrm{Re}\varrho \to \mathrm{Re}\varrho - c(\phi, \phi^*)/2$. In a neighborhood in the moduli space where $c(\phi, \phi^*) \simeq |\phi|^2$ and assuming that $\mathrm{Im}\varrho = 0$ and that $W(\varrho) = W(\sigma)$ is real, the total potential reads

$$V(\sigma, \phi) = \frac{\sigma^2 V_0(\sigma)}{(\sigma - |\phi|^2/2)^2}, \qquad V_0(\sigma) = \frac{1}{\kappa_4^4 \sigma^2}\left(\sigma W_{,\sigma}^2 - 3W W_{,\sigma} + \beta_1\right). \tag{13.61}$$

In the putative de Sitter minimum $\sigma = \sigma_{\min}$, the potential can be expanded to obtain the effective mass for the inflaton $\varphi := \phi\sqrt{3/(2\mathrm{Re}\varrho)}/\kappa_4$:

$$\boxed{V_{\mathrm{CY}}(\varphi) = V(\sigma_{\min}, \varphi) \simeq V_0(\sigma_{\min})\left(1 + \frac{2\kappa_4^2}{3}|\varphi|^2\right),} \tag{13.62}$$

which yields $m_\varphi^2 = 2\kappa_4^2 V_0(\sigma_{\min})/3 \simeq 2H^2$ and a large slow-roll parameter $\eta_V \simeq m_\varphi^2/(3H^2) \simeq 2/3$. This relation does not agree with the slow-roll condition (5.59), $m_\varphi^2 \ll H^2$: the η-problem of old brane inflation and of F-term SUGRA inflation is brought to the surface also in slow-roll D-brane inflation. Ultimately, the problem is due to the fact that the inflaton couples non-minimally with gravity in the total effective action in four dimension, via a $|\varphi|^2 R$ term [144, 312] (the relation $m_\varphi^2 = 2H^2$ is an indication of this conformal coupling).

13.5.3.1 Gravitino Mass

In the 𝕂𝕃𝕄𝕋 model, the scale of inflation is related to the gravitino mass by (12.118). General considerations show that $V \lesssim (m_{3/2} M_{\mathrm{Pl}})^2$, hence $H \lesssim m_{3/2}$ [11]. The same relation holds in monodromy inflation [121]. Thus, to have a large inflationary scale (and observable tensor modes), one should develop phenomenological models where supersymmetry is broken at high energies. This is not an issue by itself but it does not produce interesting phenomenology at accelerators, apart in specific scenarios (for instance, the split supersymmetry mentioned later in Chap. 14). Conversely, for $m_{3/2} = O(\mathrm{TeV})$ there is the problem of obtaining long slow-roll inflation with a very low scale. Although one can engineer low-scale inflation [313–315], its typical parameter space cannot be reproduced in a string-theory embedding. The gravitino-mass problem can be circumvented in racetrack inflation (13.31) [11, 86, 87], where one can have a low gravitino mass and a high inflationary scale at the same time; a discussion of these issues in large-volume inflation can be found in [82].

13.5.3.2 Cosmic Strings

At the end of inflation, lower-dimensional wrapped branes and metastable fundamental and Dirichlet strings are produced during brane collision [199]. These long strings interact in a complex network of cosmological size and they eventually decay into shorter strings and gravitational radiation [305, 306, 316–321]. The evolution and characteristic tension μ of cosmic strings produced inside the throat are constrained by CMB and large-scale-structure observations [322, 323]. No signature of a string network has been detected so far. The prediction

$$G\mu \sim 10^{-10} - 10^{-9} \tag{13.63}$$

of the $\mathbb{KLMT}$ model [317] is compatible with the experimental bound

$$G\mu < 10^{-7} - 10^{-6} \tag{13.64}$$

and within the detection threshold of future experiments dedicated to gravitational waves and B-mode polarization [154, 199, 324, 325] (see Chap. 14).

13.5.4 *Refinements and Related Models*

So far, the form of the total potential (13.57) has been determined only for the SUGRA terms (13.62). Taking into account more curvature, flux and brane effects, in the conifold approximation of the warped throat, the potential for the D3-brane radial position is [155, 157, 160, 166]

$$V_{\rm CY}(\varphi) = V_0 + c_1\varphi + c_{\frac{3}{2}}\varphi^{\frac{3}{2}} + c_2\varphi^2 + c_{\frac{5}{2}}\varphi^{\frac{5}{2}} + b\varphi^{2\sqrt{7}-\frac{5}{2}} + \dots\,, \tag{13.65}$$

where the coefficients c_i and b depend on the angular position on the conifold and the ellipsis denotes higher-order terms negligible in a small-φ expansion. The η-problem arises from the supergravity coupling of the inflaton to the four-dimensional curvature. In fact, the leading contribution to c_2 is the $O(H^2)$ mass of the $\mathbb{KLMT}$ model from curvature effects, which produce terms $\propto \varphi^n$ with $n = 2, 3, 7/2, 4, \dots$. The last term in (13.65), with irrational exponent $2\sqrt{7}-5/2 \approx 2.79$, is a flux-related contribution stemming from the compactification of the non-compact throat. Terms of order less than 2 help to alleviate the η-problem and, in general, the coefficients can be adjusted to give viable slow-roll inflation, possibly without much fine tuning [158, 159, 167].

The $\mathbb{KLMT}$ example is quite instructive of the importance of constructing, rather then assuming, a robust moduli stabilization scheme. To have a frozen internal volume, a dynamical field φ and viable inflation at the same time, one can relax the assumption that the superpotential depends only on the Kähler moduli and consider the more general case

$$W = W(\varrho, \varphi) \,. \tag{13.66}$$

A suitable dependence on the brane moduli can compensate the large inflaton mass coming from the Kähler modulus and lower it to an acceptable level. Since viable models have $\eta_V = O(0.01)$, the tuning required on fluxes and brane positions to solve the η-problem may be modest, of order of percent. Constructions with specific choices of (13.66) confirm this expectation [144]. Furthermore, symmetry arguments [145, 147, 148] and one-loop corrections to the superpotential from the open-string channel [152] and the closed-string channel [155, 157, 326] can lower the inflaton mass to a viable value. However, at least for models with closed-string corrections, a relaxation of the η-problem is possible only in a limited portion of the landscape, for certain embeddings of $\overline{\text{D3}}$/D3-branes and wrapped D7-branes [155–157, 327]. In certain cases, one can skip Step 2 (Sect. 13.5.2.2) and sustain inflation only with D3-branes: a running dilaton can suffice to uplift the minimum of the potential to de Sitter or Minkowski spacetime without $\overline{\text{D3}}$-branes in the throat [327].

Extensions of the $\mathbb{KLMT}$ model (Fig. 13.10a) to a multi-moduli dynamics usually entail a solid Calabi–Yau construction motivated from the bottom, for instance requiring to have Standard-Model-like chiral families on the anti-brane. The type-IIB flux compactification of [6], mentioned in Sect. 12.3.9.2, is of this sort. The fixed points on the throat are associated with additional moduli such as chiral scalars, chiral matter fields and gauge multiplets. These fields ϕ^i contribute to the total potential $V(\varphi, \phi^i)$ by an F-term (with a racetrack superpotential W), a D-term and a supersymmetry-breaking term coming from anti-branes. For suitable initial conditions, the trajectories of the fields collectively drive the inflaton φ along a sufficiently flat trough. To obtain enough e-foldings, all parameters must be tuned by one part in 10^3 [149], about the same order of magnitude of the tuning in volume-modulus inflation (Sect. 13.3.2). The observed amplitude of the scalar spectrum is recovered if the string scale is close to the GUT scale, $m_s \sim 10^{15}$ GeV, while the scalar spectral index n_s turns out to be slightly blue-tilted. The modest fine tuning and the wrong sign of the typical $n_s - 1$ can be overcome by a modification of the physical picture. In particular, the presence of many fields suggest the possibility to realize inflation not by the slow rolling of the inter-brane separation φ but through a sequence of short stages of acceleration driven by different fields, as in the model of chain inflation [253–257] (Sect. 13.2).

A similar alleviation of the η-problem via fixed points has been explored in [151]. In that case, one considers a Calabi–Yau space with a $\mathbb{Z}_2$ symmetry and a D3-brane located at the fixed point. Symmetric with respect to this fixed point, there are two Klebanov–Strassler throats and a $\overline{\text{D3}}$-brane at the bottom (i.e., tip) of each (Fig. 13.10b). The D3-brane feels an attractive force towards the anti-branes, which results in a total hilltop potential

$$V(\varphi) = V_0 - \tfrac{1}{2} m^2 \varphi^2 \,, \tag{13.67}$$

where φ is the D$\overline{\text{D}}$-brane separation (identical for both throats) and $m^2 \ll H^2$ is a small effective mass. The inflaton slowly rolls down the local maximum of the potential and $\eta_* \sim -10^{-2}$. The scalar index $n_s \simeq 1 + 2\eta_*$ is red-tilted, the inflationary scale is $V^{1/4} \sim 10^{14}\,\text{GeV}$ and the gravitational-wave spectrum is suppressed below any future detection level (remember the discussion about the Lyth bound (5.161)). Also in this case, the string scale is $m_s \sim 10^{15}\,\text{GeV}$.

A third possibility, based on the brane dynamics developed in [206], is to start with $\overline{\text{D3}}$-branes only and to identify the inflaton with their collective motion near the tip [150]. When the anti-branes approach the tip of the throat and stack together, due to the presence of the background R-R 3-form flux they coalesce into an NS5-brane (this is known as Myers effect [328]). This NS5-brane has topology $\mathbb{R}^4 \times S^2$: it is wrapped on an S^2 sphere inside S^3, leaving four directions non-compact. Parametrizing the tip S^3 with the metric $\text{d}\Omega_3^2 = \text{d}\psi^2 + (\sin\psi)^2 \text{d}\Omega_2^2$, the NS5-brane action is [206]

$$S_{\text{NS5}} = -\frac{T_3 k_F \alpha_0^4}{g_s} \int \text{d}^4x \sqrt{-g} \left[V_2(\psi) \sqrt{1 + \frac{g_s k_F}{\alpha_0^2} \partial_\mu \psi \partial^\mu \psi} + \frac{1}{\pi} U(\psi) \right], \tag{13.68}$$

$$V_2(\psi) = \frac{1}{\pi} \sqrt{(\sin\psi)^4 + U^2(\psi)}, \qquad U(\psi) = \frac{\pi N_{\bar{3}}}{k_F} - \psi + \frac{1}{2}\sin(2\psi), \tag{13.69}$$

where k_F are the R-R flux units in (13.49) associated with the F_3 form, $\alpha_0 = \exp\omega(\bar{r})$ is the warp factor of (13.48) at the tip and $N_{\bar{3}}$ is the number of $\overline{\text{D3}}$-branes in the condensate. The effective potential

$$V_{\text{eff}}(\psi) = \frac{T_3 k_F \alpha_0^4}{g_s} \left[V_2(\psi) + \frac{1}{\pi} U(\psi) \right] \tag{13.70}$$

has a metastable vacuum at some $\psi_{\min}$ for $N_{\bar{3}}/k_F \lesssim 0.08$, while for $N_{\bar{3}}/k_F > 0.08$ it is monotonic and the potential develops a plateau around $\psi_{\min}$. In both cases, the local vacuum or the plateau correspond to the NS5-brane state. When enough $\overline{\text{D3}}$-branes have condensed into the metastable NS5-brane, the latter acquires enough energy to unwrap itself and pass through S^3, eventually decaying (via quantum tunneling or classically) into a cluster of $N_3 = k_F - N_{\bar{3}}$ D3-branes (Fig. 13.10c). This final configuration is supersymmetric.

The flatness of the potential near $\psi_{\min}$ is a promising region where the inflation ψ can roll slowly enough to sustain a prolonged era of acceleration. In the case of a local minimum, from the point of view of a four-dimensional observer, a metastable bubble is created by the brane condensation. If the potential is flat enough, the bubble inflates. The typical inflationary scale is rather high, of order of the string energy m_s. This renders moduli stabilization more difficult, as one should fix the dilaton and the Kähler modulus already at these high scales. Some fine tuning may be in action here: viable inflation occurs for choices of the parameters which lie just beyond the range allowed by the approximations of the model.

Brane inflation has also been constructed in the low-energy limit of M-theory compactified on an $S^1/\mathbb{Z}_2$ orbifold (a configuration described more in detail in Sect. 13.7.1), using the non-perturbative dynamics of M5-branes in the presence of fluxes [329–334]. If a single M5-brane approaches the brane at the boundary of the orbifold, the inflaton is the distance $\phi \propto \Delta y$ between the boundary and the M5-brane. More generally, one can distribute a stack of M5-branes along the orbifold; equally-separated branes are a dynamical attractor solution and the inflaton is the distance $\phi \propto \Delta y = \Delta y_1 = \Delta y_2 = \dots$. Therefore, this is a case of assisted inflation [331] (Sect. 5.5.3). Corrections to the superpotential can destabilize the inflationary trajectory unless one fine tunes the parameters of the model [332], in particular the number of M5-branes [334].

13.5.5 Why the Tensor Spectrum Is Small

All warped D-brane inflationary models in string theory predict a negligible amplitude for the tensor spectrum. This general result can be explained with disarming simplicity by the Lyth bound (5.161) [193]. After calculating the scalar and tensor spectra, one realizes that, as a function of the number of e-foldings, the tensor-to-scalar ratio $r(\mathcal{N}) = 8[\dot{\varphi}/(HM_{\rm Pl})]^2 = 8(\varphi_{,\mathcal{N}}/M_{\rm Pl})^2$ is the same as in standard inflation. Thus, the inequality (5.161) holds. On the other hand, the field excursion $\Delta\varphi$ during inflation has an upper bound determined by the throat geometry. It is not difficult to show that $\Delta\varphi$ is limited by the number of background charges (13.51) in a generic flux compactification, $(\Delta\varphi/M_{\rm Pl})^2 < 4/N$ [193]. From (5.161),

$$r < \frac{32}{30^2 N} \approx \frac{0.036}{N} \,. \tag{13.71}$$

Since N is large in configurations where the AdS/CFT correspondence holds and the SUGRA approximation is under control, the field variation during inflation is usually much smaller than the reduced Planck mass and $r \ll 10^{-3}$. On the other hand, a positive detection of gravitational waves at the level $r \gtrsim 0.01$ would paint warped D-brane inflation into a very tight corner because it would imply $N \gtrsim 4$, an almost unwarped Calabi–Yau space which cannot be described consistently with the tools employed so far.

In contrast, multi-brane inflation in M-theory can give rise to a detectable r because the bound on the tensor-to-scalar ratio applies not to the inflation φ but to the inter-brane separation Δy. φ and Δy are related to each other by a proportionality factor that weakens the Lyth bound [333].

13.6 DBI Inflation

13.6.1 Setting

In the previous section, we introduced a model of inflation with the canonical Lagrangian (13.55) and discussed the challenges it has to face in order to sustain a sufficiently long period of acceleration. An assumption we have kept so far, and that we will abandon now, is to neglect higher-order derivatives in the DBI action. The dynamics governed by (13.52) with an effective potential,

$$\boxed{S = S_g + S_\varphi \,, \qquad S_\varphi = \int \mathrm{d}^4x \sqrt{-g}\,[\mathcal{L}_{\mathrm{DBI}} - V(\varphi)] \,,} \tag{13.72}$$

is capable of addressing the technical issues of the $\mathbb{KLMT}$ scenario related to moduli stabilization.

The warp factor $f(\varphi)$ in (13.52), valid for an exact $\mathrm{AdS}_5 \times S^5$ background, receives corrections from the actual shape of the Klebanov–Strassler geometry. In particular, the effect of the IR cut-off of AdS_5 can be modeled by

$$f(\varphi) = \frac{\lambda^2}{(\varphi^2 + \varphi_0^2)^2} \,, \qquad \varphi_0^2 > 0 \,. \tag{13.73}$$

The effective potential $V(\varphi)$ is generated by the interaction of the D3-brane with the matter and Kaluza–Klein modes of the theory. Since conformal invariance of $\mathrm{AdS}_5 \times X_5$ is broken by the cut-offs, a mass term for φ can be generated.

We will call the cosmological model of the early universe based on the action (13.72) *DBI inflation* [168–194]. (The NS5-brane action (13.68) is a special model of DBI inflation.) It has several points of contact with other scenarios mentioned throughout the book, mainly k-essence (Sect. 7.5.1, characterized by Lagrangians $\mathcal{L}(\phi, \nabla\phi)$ which are higher order in $(\nabla\phi)^2$), k-inflation [335, 336] (the same type of model as k-essence but applied to the early universe) and early models based on the DBI action for an unstable brane (which will be discussed in Sect. 13.7.2).[10]

On a FLRW background, $\mathcal{L}_\varphi = (1 - \gamma^{-1})f^{-1} - V$, where the factor

$$\gamma := \frac{1}{\sqrt{1 - f\dot{\varphi}^2}} \tag{13.74}$$

[10]DBI inflation is not different, conceptually, from slow-roll D-brane inflation: the setting is about the same (branes and anti-branes moving on a warped Calabi–Yau space) and, just like in the comparison between k- and standard inflation, all the changes stem from the kinetic term.

places an upper limit on the kinetic energy of the field. Slow-roll models correspond to $\gamma = O(1)$; models with $\gamma \gtrsim 1$ and $\gamma \gg 1$ are called, respectively, relativistic and ultra-relativistic.

Since the field momentum is $\Pi_\varphi = \partial\mathcal{L}_\varphi/\partial\dot{\varphi} = \gamma\dot{\varphi}$, the energy density and pressure of the scalar are

$$\rho = \Pi_\varphi\dot{\varphi} - \mathcal{L}_\varphi = \frac{\gamma-1}{f} + V\,, \qquad P = \mathcal{L}_\varphi = \frac{\gamma-1}{f\gamma} - V\,. \tag{13.75}$$

13.6.2 UV Model

There are mainly two versions of DBI inflation. In the first (UV model) [168, 169, 173–175, 177, 178, 191–194], the D3-brane travels from the UV region of AdS_5 towards the tip of the throat. For a quadratic potential

$$V(\varphi) = \tfrac{1}{2}m^2\varphi^2\,, \tag{13.76}$$

one obtains a sustainable era of inflation along the AdS part of the throat. The mass in (13.76) is not very small, in general, due to the relative position of the anti-branes with respect to the moving brane. The warp factor f^{-1} at the IR end of the throat takes its minimum value, so that, in the absence of a potential, the energy of the $\overline{\mathrm{D3}}$-branes at the tip is lowered to levels that cannot sustain inflation. Therefore, the extra potential term $V(\varphi)$ in (13.72) must be sufficiently steep. This will limit the freedom in the choice of parameters in V, a fact that we will better appreciate when considering the IR model of Sect. 13.6.3.

Fortunately, a steep potential does not spoil inflation. While in the $\mathbb{KLMT}$ model the D3-brane must travel slowly down the throat, in DBI inflation the D3-brane moves fast. The inflaton potential can be steep without spoiling the slow-roll condition, due to the upper bound $\dot{\varphi} < f^{-1/2}$ on the speed of φ. When the brane reaches the tip ($\varphi \simeq \varphi_0$), it is slowed down by the production of virtual light particles. Subsequently, the scalar field decays into dust matter and radiation and inflation ends.

These features can help to alleviate the technical difficulties met by the models of Sect. 13.5, where we saw that, in general, moduli stabilization induces a large mass for the inflaton (η-problem). This is no longer an issue if the kinetic term takes the DBI form. Therefore, both the suppression of the conformal coupling $\varphi^2 R$ and the fast-roll phase conspire to give viable inflation without conflicting with the stabilization of the moduli fields.

Let us look at the model in detail. In the AdS part of the Klebanov–Strassler geometry, away from the cut-offs,

$$\varphi_0 \ll \varphi \ll M_{\mathrm{Pl}}\,. \tag{13.77}$$

The upper bound (sub-Planckian vacuum expectation values) is necessary to avoid a proliferation of quantum corrections to the gravitational Lagrangian $\mathcal{L}_g$ in (13.72) [168]. The latter is of the form $2\mathcal{L}_g = (M_{\rm Pl}^2+\xi\varphi^2)R+O(R/\varphi^2)+\dots$ (bars omitted). In the regime (13.77), the conformal coupling is sub-dominant with respect to the Einstein–Hilbert term, provided $\xi = O(1)$, while terms of the form R/φ^2 are of order of $H^2/\varphi^2 \propto (m/M_{\rm Pl})^2$, which are negligible if the inflaton mass is sufficiently small. This is indeed the case, as we shall see shortly.

With an Einstein–Hilbert gravitational sector, the cosmological dynamics is given by the standard Friedmann equations with the energy density and pressure (13.75). The scalar field approaches zero as $\varphi \simeq \lambda/t$, while the scale factor obeys an approximate power law,

$$a(t) \propto t^p\,, \qquad p = \frac{1}{\epsilon} \simeq \frac{\lambda}{\sqrt{6}}\frac{m}{M_{\rm Pl}}\,. \tag{13.78}$$

In the strong-coupling regime $\lambda \gg 1$, one can obtain enough acceleration ($p \gg 1$ and $\mathcal{N}_{\rm e} = O(p) = O(100)$) for a mass m lower than the reduced Planck mass. However, and contrary to standard inflation, m does not have to be *much* lower than $M_{\rm Pl}$.

The inflationary observables are not difficult to compute [169, 336]. Linear scalar perturbations $u_k = z\mathcal{R}_k$ obey the Mukhanov–Sasaki equation

$$u_k'' + \left(c_{\rm s}^2k^2 - \frac{z''}{z}\right)u_k = 0\,, \qquad c_{\rm s} = \frac{1}{\gamma}\,, \qquad z = \gamma^{3/2}\frac{a\dot\varphi}{H}\,. \tag{13.79}$$

Note that the propagation speed of the fluctuations is much smaller than the speed of light, $c_{\rm s} \ll 1$. In the limit $\gamma \to 1$, one recovers the usual result (5.118). On the adiabatic vacuum and in the slow-roll approximation, the power spectrum is $\mathcal{P}_{\rm s} \propto H^2/(c_{\rm s}\epsilon)$. To obtain a small-enough amplitude, for $0.05 < \epsilon < 0.2$ the parameter $\lambda \sim (R_{\rm AdS}/l_{\rm s})^4$ must be quite large, $\lambda = 10^{10} - 10^{12}g_{\rm s}$, implying for $g_{\rm s} = O(1)$ that the curvature radius of the throat is about 100–1000 times larger than the string length scale $l_{\rm s}$.

The deviation from a pure de Sitter expansion is compensated by a shrinking in the sound horizon, which entails a freezing of perturbation modes at progressively smaller scales. The scalar tilt is

$$n_{\rm s} - 1 = 2\eta_* - 4\epsilon_* - 2s_*\,, \tag{13.80}$$

where $s := \dot c_{\rm s}/(Hc_{\rm s})$.

As in $\mathbb{KLMT}$ models, DBI inflation produces a tensor spectrum with a small index $n_{\rm t} = -2\epsilon_*$ and a small amplitude,

$$r = 16\epsilon_* c_{\rm s}\,, \tag{13.81}$$

which is suppressed by a factor $c_s = \gamma^{-1}$. However, there is a lower bound on r which makes the tensor signal potentially observable. In fact, another characteristic effect of the model is a large non-Gaussianity [169, 337]. When expanding the action (13.72) in the fluctuations $\delta\varphi$, higher-order terms acquire higher-order coefficients in $\gamma \gg 1$. This enhances non-linear effects and deviations from Gaussian statistics. The equilateral non-linear parameter is

$$f_{\rm NL}^{\rm equil} \simeq -\frac{35}{108}\left(\frac{1}{c_s^2}-1\right). \tag{13.82}$$

The 1σ-level PLANCK 2015 constraint $f_{\rm NL}^{\rm equil} = 11 \pm 69$ on this model gives a lower bound

$$c_s \gtrsim 0.07 \quad (95\,\%\ \text{CL}) \tag{13.83}$$

on the sound speed [338] which, for $\epsilon = O(10^{-2})$, implies $r > 0.1$. For a constant non-linear parameter, there is a more precise lower bound on the tensor-to-scalar ratio, written in terms of $f_{\rm NL}^{\rm equil}$ and the scalar index [177]:

$$r > \frac{4(1-n_s)}{\sqrt{1+3|f_{\rm NL}^{\rm equil}|}}. \tag{13.84}$$

For $1 - n_s = O(10^{-2})$ and $|f_{\rm NL}^{\rm equil}| \lesssim O(10)$, one has

$$r > 10^{-2}, \tag{13.85}$$

not far from the upper limits (4.73) and (4.74).

However, there also exists an upper bound on the observed r which depends on the volume $\text{Vol}(X_5)$ (in string length units) of the Einstein-manifold cross-section of the Calabi–Yau space $\mathcal{C}_6$ [177]:

$$r < \frac{10^{-6}}{\text{Vol}(X_5)}. \tag{13.86}$$

The volume of the cross section of the Klebanov–Strassler throat is $\text{Vol}(T^{1,1}) = (16\pi^3/27)l_s^5$, which fixes a typical order of magnitude for $\text{Vol}(X_5) = O(10)$ and the upper bound (13.86): $r < 10^{-7}$, incompatible with (13.85) by several orders of magnitude. This means that the UV DBI model can accommodate observations only if the compactification volume of the X_5 cross section of $\mathcal{C}_6$ is unnaturally small, $\text{Vol}(X_5) \sim 10^{-5}$.

In this sense, the UV DBI model is under very strong observational pressure. Together, non-Gaussianity and the Lyth bound (valid not only for standard and D-brane inflation but also for k-inflation and, in particular, DBI inflation [193]) limit the number of background fluxes. The quadratic potential (13.76) is already excluded, since $N < r\gamma^2/8 \simeq 27r|f_{\rm NL}^{\rm equil}|/70 < 3$ [193]. Similarly, for large but not too large values of γ one obtains a relativistic scenario situated between the $\mathbb{KLMT}$ slow-roll scenario and the UV DBI model [192]. This case is also unfeasible because a scalar index compatible with observations is usually associated with a large tensor-to-scalar ratio. Also inflation too close to the tip of the Klebanov–Strassler throat is ruled out, as it produces too large non-Gaussianities [174].

Generalizations and modifications of basic DBI inflation do not alleviate these problems in a substantial way. Relaxing the assumption that the D3-brane moves towards the throat tip along the radial direction and allowing for a more generic spiral motion, one obtains a multi-field model where the dynamics of the fields is governed by the DBI Lagrangian [181, 182, 184–190]

$$\mathcal{L}_{\rm DBI}[\varphi_i] = -f^{-1}\left(\sqrt{1+f\sum_{i=1}^{6}\partial_\mu\varphi_i\partial^\mu\varphi_i}-1\right) - V(\{\varphi_i\})\,.$$

The non-zero angular momentum of the brane turns out to provide a few extra e-folds of inflation but the level of non-Gaussianity is still large, since the leading contribution is the single-field one. A large $f_{\rm NL}$ is also obtained when one considers many non-interacting D-branes, which give rise to several copies $\sum_n \mathcal{L}_{\rm DBI}[\varphi_n]$ of the single-field DBI Lagrangian (13.52) [339, 340].

On the other hand, replacing the D3-brane with a D5-brane wrapped on a 2-cycle or a D7-brane wrapped on a 3-cycle increases the upper bound (13.86) but there are other issues to solve, first of all how to obtain a sufficient number of background charges (limited by a maximum value in Calabi–Yau compactifications) [179, 180].

13.6.3 IR Model

Generically, Calabi–Yau spaces are multi-throat and many $\overline{\rm D3}$-branes live in the low-curvature regions. Depending on the number and position of these branes and of the background fluxes, one can imagine a qualitative scenario combining bits of the information we already gathered. Due to the attraction (13.54) towards the warped regions, some of these anti-branes settle down some throats (Fig. 13.11a), while others decay into D3-branes passing through the NS5-brane condensate (13.68). The lifetime of the anti-branes and their condensates depend on the ratio N/k_F. After some time, the typical snapshot will show stacks of $\overline{\rm D3}$-branes in some throats and stacks of D3-branes in other throats with different shallowness (warping factors). These D3-branes may exit their throat attracted by the anti-branes in the region with the largest warp factor, which we call A-throat (A stems from anti-brane; see Fig. 13.11b).

The surviving $\overline{\rm D3}$-branes lift the AdS vacuum to de Sitter and pave the ground for inflation. In the so-called IR model of DBI inflation [170–174, 176, 183, 191,

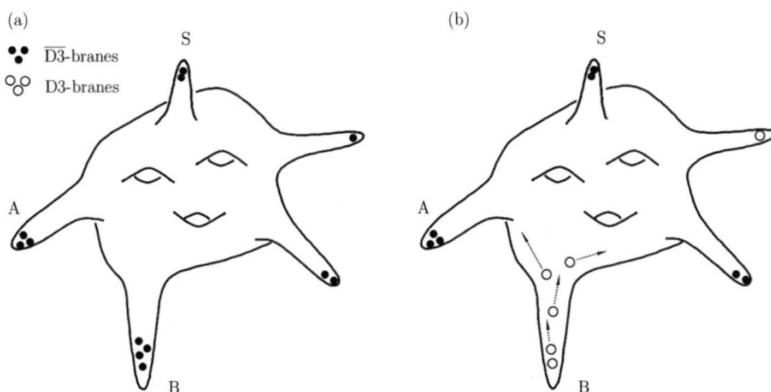

Fig. 13.11 IR DBI inflation as described in the text

194], D3-branes are placed in a throat in a stabilized Calabi–Yau space (dubbed brane-throat or B-throat) and the motion is from the IR to the UV end. The inflaton is the AdS radial coordinate of the D3-stack along the B-throat and inflation occurs when the branes exit the IR region. Once got away from the B-throat, the branes spread towards different throats (not necessarily the A-throat) and inflation ends when they annihilate or collide with the $\overline{\text{D}}$-branes therein. The Standard Model lives in one of these throats (S-throat).

From the $\mathbb{K}$LT moduli stabilization, we saw that, in the absence of special precautions the inflaton tends to acquire a mass $m^2 = O(H^2)$ [144]. The mass of the UV DBI model may be not very small, either, due to the suppression of the $\overline{\text{D3}}$-branes energy by the warp factor f^{-1}. On the other hand, by placing the anti-branes in or past the UV part of the B-throat one avoids such suppression and their energy is enough to pull the D3-branes at the IR end efficiently. As a consequence, there is more freedom in the shape of the potential $V(\varphi)$, which we will encode in a phenomenological parameter β. The sign of the mass term changes if we switch the direction of motion from UV$\to$IR to IR$\to$UV, so that we can describe the force sucking N_3 D3-branes from the B-throat by the potential

$$\boxed{V(\varphi) = V_0 - \tfrac{1}{2}\beta H^2\varphi^2\,,} \tag{13.87}$$

where $\beta > 0$ and the Hubble parameter H is approximately constant. This potential covers several scenarios with quadratic potentials: for $-1 \ll \beta < 0$ and $\gamma \approx 1$, the slow-roll (small mass) $\mathbb{KLMT}$ model; for $\beta \approx -1$ and $\gamma \gtrsim 1$, the intermediate model of [192]; for $\beta \gg 1$ and $\gamma \gg 1$, ultra-relativistic UV DBI inflation; for $\beta \ll 1$ and $\gamma \approx 1$, the slow-roll model (13.67) of [151].

For the potential (13.87) with $\beta > 0$ and the dynamics (13.72), the attractor solution is $\varphi \simeq -\lambda/t$ (here the initial conditions are set at $t = -\infty$) and $H \approx$ const, followed by a non-relativistic phase with $\phi \propto \exp[(\sqrt{9+4\beta}-3)Ht/2]$ (which is of slow rolling only if $\beta \ll 1$).

Inflation lasts about $\Delta t \sim N_B/m = N_B/(\beta H)$, where N_B (estimated to be $\gtrsim g_s N_3^2$ to keep the warp factor (13.50) at the tip small) is the number of background charges in the B-throat. The DBI action is valid as long as one can ignore the back-reaction of the background with respect to the brane energy, which translates into the inequality $\gamma \ll N_B$. Since $N_B = O(10^6)$ or higher is easily achieved in flux compactifications, it is not difficult to have enough e-foldings,

$$\mathcal{N}_e \sim H\Delta t \sim \frac{N_B}{\beta}\,, \tag{13.88}$$

even if β is large. The propagation speed of cosmological perturbations is

$$c_s \simeq \frac{3}{\beta \mathcal{N}_e} \sim \frac{3}{N_B} \ll 1\,, \tag{13.89}$$

while the scalar spectral index is

$$n_s - 1 \simeq -\frac{4}{\mathcal{N}_e} + \frac{2}{\mathcal{N}_e + \sqrt{27N_3N_B/8}}\,. \tag{13.90}$$

The first term in (13.90) always dominates and the spectrum is red-tilted. For $50 < \mathcal{N}_e < 70$, $N_3N_B \sim N_B^{3/2} \sim (\beta\mathcal{N}_e)^{3/2}$ and $\beta < 10^6$, we obtain

$$0.92 < n_s \lesssim 0.97\,, \tag{13.91}$$

consistent with PLANCK 2015 data at the 3σ-level. The scalar running α_s is negative.

Since the cosmic expansion is exponential during inflation ($H \approx$ const), the first slow-roll parameter ϵ is negligibly small and so are the tensor-to-scalar ratio r and the tensor index n_t. Again, the Lyth bound provides an enlightening explanation of the suppression of primordial gravitational waves [193].

Finally, non-Gaussianity constrains the range of β through the non-linear parameter (13.82). Using (13.89),

$$f_{NL}^{equil} \simeq -\frac{35}{108}\left(\frac{\beta^2\mathcal{N}_e^2}{9} - 1\right). \tag{13.92}$$

When combined with PLANCK data, this formula enforces a strong restriction on the parameter space [338]:

$$\beta \lesssim 0.7 \qquad (95\,\%\ \text{CL})\,. \tag{13.93}$$

This is the same constraint as (13.83) for $\mathcal{N}_e = 60$. As in the UV case, non-Gaussianities are excessive if inflation takes place close to the tip of the throat [174], since their magnitude increases with the decrease of the warp factor f^{-1}. At the UV end of the throat, the warp factor increases and non-Gaussianities are lowered down to acceptable levels.

The details of the reheating phase depend on how the D3-branes end their life, by collision or annihilation with anti-branes. In general, cosmic strings will be produced after inflation in the A-throat with $N_{\bar{3}}$ anti-branes, with a string tension in the range

$$3 \times 10^{-12} \sqrt{g_s} < G\mu < 9 \times 10^{-6} \sqrt{\frac{g_s}{N_{\bar{3}}}}, \tag{13.94}$$

compatible with the experimental bound (13.64).

13.7 Other Models

We continue with a brief mention of models which appeared during the years and tackled inflation and the cosmological constant problem independently of the landscape picture.[11] The interested reader should look into the literature for more details. In this section, we do not attempt to provide an exhaustive list of references.

13.7.1 Braneworld

Before modern models of moduli stabilization came to prominence, an alternative to the brane scenarios of Sect. 13.5 attracted the attention of the community for several years. Regarded as the $D = 11$ strong-coupling limit of the $E_8 \times E_8$ heterotic string [341, 342], M-theory admits a consistent compactification on a Calabi–Yau space, resulting in a five-dimensional non-compact spacetime [343, 344]. This model, called *heterotic M-theory* [345], consists in two 9-branes at the boundary of an eleven-dimensional bulk, each carrying an $\mathcal{N} = 1$ super-Yang–Mills theory with group E_8. The inter-brane direction has the structure of an orbifold $S^1/\mathbb{Z}_2$. Upon compactification, the dynamics of this spacetime admits a solution with two parallel D3-branes at the end-points of the fifth dimension [346, 347], on which

[11] The research on some of these proposals (such as braneworld or tachyon cosmology) has been discontinued for several reasons, mainly for observational or theoretical difficulties. Experimental problems arose when WMAP, PLANCK and supernovæ data begun to rule out cosmological models of inflation or dark energy, making the parameter space of rigidly string-motivated approaches unviable. Theoretical difficulties include the presence of undesired collateral effects or an incomplete control on moduli stabilization, the latter being improved in the $\mathbb{K}$LT construction. Concerning this point, one might differentiate between string approaches to cosmology and models which are only *string-inspired*, i.e., that contain certain ingredients borrowed from or similar to some features of string theory but that do not meet the highest standard of rigor in their assumptions, their justification of their starting point or their controllability within the fundamental theory. Such a discrimination [198] would be unavoidably subjective and we will not attempt it.

one can construct a cosmology [348, 349]. The distance between the boundary branes is larger, or even much larger, than the size of the Calabi–Yau manifold [343, 344, 350, 351].

This set-up inspired a plethora of *braneworld* scenarios where the observed universe with the Standard Model is assumed to live in a brane embedded in a non-compact AdS_5 five-dimensional bulk (see [352, 353] for reviews). One of the first phenomenological models proposed is the Randall–Sundrum braneworld [307]. The invisible brane with negative tension is sent away to infinity, while the observed universe is on the brane with positive tension $\lambda > 0$.[12] The strong curvature of AdS_5 localizes gravity mainly on the visible brane but there is some leakage of information between brane and bulk. The extra non-compact dimension can communicate with the matter confined in the brane and the cosmological evolution thereon can change substantially from the standard four-dimensional case [355–362]. If the brane energy density is comparable with its tension, $\rho/\lambda \gtrsim 1$, then the bulk back-reacts for the presence of the brane matter and, for an FLRW metric with a perfect fluid, quadratic corrections to the Friedmann equation arise [349, 356, 357]:

$$H^2 = \frac{\kappa_4^2}{6\lambda}\rho(2\lambda + \rho) + \frac{\mathcal{E}}{a^4}\,, \tag{13.95}$$

where $\mathcal{E} = \text{const}$ is a "dark radiation" term arising from the brane-bulk gravitational interaction. Gravity experiments impose the bulk curvature scale to be $\lesssim 1$ mm, that is, $\kappa_5^{-1} \sim m_5 \gtrsim 10^8$ GeV and $\lambda^{1/4} \gtrsim 10^3$ GeV.

In another braneworld scenario [363], one includes also the α'-leading-order quantum corrections to the heterotic low-energy effective action. These corrections take the form of the Gauss–Bonnet term (7.78) [364]. Up to a boundary term, the five-dimensional bulk Lagrangian with a negative cosmological constant $\Lambda_5 < 0$ reads $[R - 2\Lambda_5 + (8g_s^2)^{-1}\mathcal{L}_{\text{GB}}]/(2\kappa_5^2)$. Gauss–Bonnet braneworld cosmology differs from (13.95) in the modifications at high energy density [365–373]. A cosmological solution of the theory can be found via a five-dimensional warped metric such that its projection on the 3-brane is FLRW-like. The effective Friedmann equation on the brane is [368, 370, 371]

$$H^2 = g_s^2(c_+ + c_- - 2)\,, \qquad c_\pm = \left[\sqrt{\left(1 + \frac{\Lambda_5}{6g_s^2}\right)^{3/2} + \left(\frac{\delta}{\delta_0}\right)^2} \pm \frac{\delta}{\delta_0}\right]^{\frac{2}{3}}\,, \tag{13.96}$$

[12] In an earlier version where the Standard Model is on the brane with negative tension, the extra direction is compact [354]; however, this model is not realistic since the visible brane turns out to be an anti-gravity world [355].

where $\delta_0^{-1} := \kappa_5^2/(4g_s)$ and $\delta = \rho + \lambda$. Expanding (13.96) to quadratic order in δ, one recovers (13.95) provided some relations between κ_5, κ_4, g_s and λ are satisfied. When $\delta/\delta_0 \gg 1$, one has a high-energy non-standard cosmology $H^2 \simeq (g_s^2\kappa_5^2/2)^{2/3}\rho^{2/3}$. When the energy density is far below the string scale ($\lambda/\delta_0 \ll \delta/\delta_0 \ll 1$) but $\rho \gg \lambda$, we have $H^2 \simeq [\kappa_4^2/(6\lambda)]\rho^2$, the high-energy regime of (13.95). Finally, when $\rho \ll \lambda$ the brane grows stiff with respect to its matter content and the standard evolution is recovered, $H^2 \simeq (\kappa_4^2/3)\rho$.

13.7.1.1 Inflation

These results from Randall–Sundrum and Gauss–Bonnet braneworlds indicate that, in general, the physics in the bulk can influence the effective dynamics induced on the brane, giving rise to modifications of standard cosmology. The dynamics can be studied with a deformed Friedmann equation [374–376]

$$H^2 \propto \rho^q\,, \qquad q > 0\,, \tag{13.97}$$

where different values of $q = 1, 2, 2/3, \ldots$ describe high-energy patches of the early universe valid in certain epochs. Many models of braneworld inflation have been studied in the background dynamics (13.95), (13.96) or (13.97) and including inhomogeneous cosmological perturbations [377–382]. The problems of the hot-big-bang models are solved via the traditional inflationary mechanism, the only difference being in the effective dynamics the scalar field obeys.

13.7.1.2 Big-Bang Problem

A resident on a brane may or may not experience a singular cosmological evolution, depending on the immense wealth of details that characterize brane cosmological models. The cosmological solutions of braneworld scenarios with dynamics (13.95), (13.96) and (13.97) have, in general, a big bang, which is not approached with the mixmaster behaviour discussed in Sects. 6.3.4 and 13.9.2 (at least in the Randall–Sundrum case [383]). Nevertheless, when considering the effects of the bulk, there are hints about a singularity resolution. In the simplified model of (13.48), a stable D3-brane travels towards the tip of the Klebanov–Strassler throat of a type-IIB Calabi–Yau compactification and bounces back. An observer on the brane would start with a contraction period, see a bounce instead of a bang and then go through an epoch of expansion. The presence of a bounce can be understood by looking at the metric (13.48), where ds_4^2 is chosen to be the four-dimensional Minkowski metric. Parametrizing the brane trajectory with the proper time t of the in-falling brane, the brane observer lives in a flat FLRW universe with scale factor $a(t) = \exp \omega[r(t)]$. Since ω is finite at the tip of the throat, $a(t)$ acquires a finite minimum there [384]. However, issues appear when trying to connect non-singular solutions with a realistic cosmological evolution [385–387].

13.7.1.3 Self-Tuning Cosmological Constant

In certain solutions of five-dimensional Randall–Sundrum brane models [307, 354], the zero mode of the warp factor ω in (12.107) is moderately adjusted to suppress the four-dimensional cosmological constant to zero. Poincaré invariance is therefore preserved in the brane.

This result, which is the stringy version of an earlier proposal in generic Kaluza–Klein compactifications [388], can be obtained in two ways. One is to consider the renormalization-group flow of the couplings of the theory and to recognize the existence of solutions such that $\Lambda_k = 0$ throughout the trajectory [389, 390].[13] The other is to check the stability of $\Lambda = 0$ against radiative corrections in perturbative quantum field theory on the brane [391, 392]. In both cases, however, these solutions are in general not preferred, since their consistency requires some fine tuning on other parameters of the model [393, 394] and they possess singularities if the brane is flat [395]. Also, the cosmological evolution dictated by the effective Friedmann equations on the brane may require some extra tuning, at least in some toy models [396]. A generalization of this setting to six-dimensional supergravity, where supersymmetry is broken by the branes, can circumvent the fine tuning; its chance of success depends on the details of the construction [397–399].

13.7.1.4 Open Problems

Nowadays, braneworld scenarios have partly lost their appeal mainly because they are not as phenomenologically robust as the warped D-brane models of Sect. 13.5. To begin with, the majority of braneworld models are not based on concrete schemes of moduli stabilization, as they predate the $\mathbb{K}$LT construction. Also, contrary to its compact but yet more unrealistic version [307], the non-compact Randall–Sundrum scenario [354] does not solve the hierarchy problem. Moreover, for the values $q = 2, 2/3$, the inflationary dynamics associated with (13.97) is limited by the most recent data even more severely than standard general-relativity models [400]. Despite its drastic change with respect to (2.81), a modified Friedmann equation such as (13.97) does not lead, once all parameter constraints are met, to characteristic features in the cosmic microwave background that could strongly differentiate them from Einstein gravity.

[13]We met a similar type of trajectory in the RG flow of asymptotic safety (Sect. 11.2.2).

13.7.2 Cosmological Tachyon

In Sect. 12.2.8, we mentioned that unstable Dp-branes (p even and odd in, respectively, type-IIB and type-IIA theory) can decay into lower-dimensional branes or into the closed-string vacuum via tachyon condensation. Tachyon condensation on an unstable Dp-brane with metric $g_{\mu\nu}$ is described, up to fermionic fields, by the DBI action [401–403]

$$\mathcal{S}_{\rm DBI,T} = \int d^{p+1}x\, \mathcal{L}_{\rm DBI,T}\,, \tag{13.98a}$$

$$\mathcal{L}_{\rm DBI,T} = -V(T)\sqrt{-\det(g_{\mu\nu} + 2\pi l_s^2 F_{ab} - \partial_\mu T \partial_\nu T)}\,. \tag{13.98b}$$

Here, the scalar field T has dimension $[T] = -1$ and V is its potential, which is calculated exactly to all orders in the Regge slope α' but at the tree level in g_s. On the other hand, the SFT tachyonic action (12.52) is non-perturbative also in g_s. To recapitulate, in this book we have seen four different DBI actions: the actions (12.19) and (12.42) for a stable Dp-brane in flat or curved spacetime, the action (13.52) for a stable D3-brane on $\mathrm{AdS}_5 \times S^5$ and the action (13.98) for a unstable Dp-brane on a generic curved background.

Expression (13.98) can be easily compared with the action (12.42) for a stable brane: the tension $\mathcal{T}_p$ has been replaced by the tachyon potential and a kinetic term has been introduced. On the other hand, the difference between (13.98) and (13.52) is the regime of the string coupling: strong in the model of DBI inflation of Sect. 13.6 (where the D3-$\overline{\rm D3}$ pair is unstable but has no tachyonic mode), weak in the present case. Note that (13.98) can be mapped to the square-root part of (13.52) with the field redefinition $T = \lambda/\varphi$ and the identification $V(T) = \lambda^2/T^4$.

The tachyon potential has some universal features. There is a maximum at $T_{\rm max} = 0$, near which $V(T) = 1 - M^2T^2/2 + O(T^4)$, where $M^2 < 0$ is the negative tachyon mass. A local minimum $V(T_{\rm min}) = 0$ is located at infinity, $T_{\rm min} = \pm\infty$. Potentials interpolating between the maximum and the minimum are $V(T) = \exp(-M^2T^2/2)$ [404–409] and $V(T) = 1/\cosh(MT)$ [410, 411].[14]

13.7.2.1 Inflation

In the brane inflation models of Sect. 13.5, the inflationary era takes place before the onset of the tachyon instability, i.e., when the colliding branes are still at a distance larger than the string length. In warped D-brane inflation, the inflaton is

[14] Another but rather different formalism where one can analyze tachyon condensation is string field theory (Sects. 12.1.6 and 12.2.9), in which case the kinetic term for the tachyon is canonical and the potential is dressed with exponential non-local operators. The local minimum is at some finite $T_{\rm min}$ in this case.

the distance between the $\overline{\text{D3}}$-brane of the 𝕂LT construction and a mobile D3-brane is added on the warped throat. However, one can try to postpone inflation to tachyon condensation in a setting where the individual branes are unstable. The first example of this scenario is a type-IIA brane scenario where one considers the annihilation of a D5- and a $\overline{\text{D5}}$-brane into an unstable 3-brane [298]. On the 3-brane, the tachyon scalar field vanishes, $T = 0$. This is the location of the local maximum of its potential. At this point, the tachyon rolls down and drives or helps inflation on the 3-brane [298, 412].

The tachyon field as an inflationary agent has been studied in a more cosmological fashion, often quite independently from string theory, both in four dimensions and on a braneworld [413–436] (see also [374–376]). The idea is simply to take (13.98) with $p = 3$ as the matter source for the Friedmann equations.

13.7.2.2 Dark Energy

Tachyonic models with inverse-power-law and exponential potentials have also been proposed as a source of dark energy [417, 418, 432, 437–443]. They can be viewed as a special case of k-essence (Sect. 7.5.1), they have quite similar phenomenology to canonical quintessence and suffer from the same issues of fine tuning [443]. Therefore, there is no apparent gain in choosing a DBI scalar field instead of a Klein–Gordon one. Also, future singularities can appear in the late-time evolution [431].

13.7.2.3 Open Problems

Cosmology based upon a rolling tachyon suffers from a number of problems, including a small number of e-foldings (insufficient slow rolling), a large amplitude $\delta\rho/\rho \gg 1$ for density perturbations and the formation of caustics [418, 419, 421, 424, 426, 428, 433]. A longer inflationary period can be obtained but only in a configuration with 10^6–10^{13} D-branes.[15] Moreover, if the local minimum of $V(T)$ is at infinity, there are no oscillations at the end of inflation and a reheating mechanism appears difficult [421, 427].

These issues arise if one does not embed the model in 𝕂LT scenarios, where the compactification scheme and moduli stabilization follow a certain procedure. In this case, the parameters of the tachyon potential and the number of branes must be fine tuned. Attempts to realize tachyon inflation in 𝕂LT scenarios is attractive because, instead of adding a mobile D3-brane and facing the stabilization issues of standard

[15] Although a tachyonic inflationary period with not many branes is too short, it can be a viable means to provide natural initial conditions for a standard scalar inflationary period, similarly to what happens in fast-roll inflation [444]. Then, this standard inflation lasts a sufficient number of e-foldings and dilutes the non-linear perturbation structure generated by the tachyonic phase.

$\mathbb{K}$LT inflation, acceleration is driven by the open-string tachyon of the $\overline{\text{D3}}$-brane and no new moduli are introduced. In practice, in the $\mathbb{K}$LT construction of Sect. 13.1.1, the constant brane tension in (13.3) is replaced by (13.98) [432],

$$\beta_1 \propto \mathcal{T}_3 \to \mathcal{L}_{\text{DBI},T}\,. \tag{13.99}$$

Then, the problem of large density perturbations is solved by considering a small warp factor in a warped metric, while the problem of reheating is overcome by accounting for the negative cosmological constant arising from gaugino condensation. However, the modification (13.99) can spoil the local minimum of the total SUGRA potential and jeopardize the viability of the model as a competitive string-theory candidate.[16] A more effective use of the DBI action, where the scalar field is the inter-brane separation rather than a tachyonic mode, has been introduced in Sect. 13.6.

If, on the other hand, one gives complete freedom on the choice of the potential $V(T)$ and ignores any other stringy detail such as the compactification scheme and the moduli problem, the tachyon is not distinguishable phenomenologically from a canonical inflaton due to a field redefinition [417, 422]. Conversely, given the same potential, one can distinguish the tachyon from a standard canonical scalar but it does not give any appreciable advantage over the latter, even on a braneworld [435]. Overall, the connection with string theory becomes progressively weaker as cosmological applications of the tachyon advance.

13.7.3 Modified Gravity

String theory can give a motivation to some (but very few) of the modified gravity models considered in Sect. 7.5.

13.7.3.1 Gauss–Bonnet Gravity

In order to be of use for phenomenology, the low-energy limit of string theory should be free from instabilities. This is not guaranteed in principle, since the $O[(\alpha' R)^n]$ corrections to the beta functions in the universal SUGRA sector introduce higher-order derivatives and, hence, possible Ostrogradski instabilities (Sect. 11.8.1). These corrections are known up to $O(R^4)$ [446–451]. In particular, the correction at quadratic order is the Gauss–Bonnet term (7.78), which is non-topological in $D > 4$ and ghost-free on a Minkowski target spacetime [447]. When compactifying the low-energy action $S = \alpha'^{-1} \int d^{10}x\, \sqrt{-g}[O(\alpha')\mathcal{L}_{\text{EH}} + O(\alpha'^2)\mathcal{L}_{\text{GB}}]$ to four dimensions, the couplings of the Einstein–Hilbert term $\mathcal{L}_{\text{EH}} \propto R$ and of the Gauss–

[16]On top of that, it is not clear whether the promotion of the classical field T to a quantum object correctly describes quantum string theory [445]. This makes the quantization of the tachyon Lagrangian (13.98) and the interpretation of cosmological perturbations a delicate subject.

Bonnet invariant $\mathcal{L}_{\rm GB}$ acquire a dependence on the moduli fields. For only one such field ϕ, the gravitational action is of the form [450]

$$S = \int \mathrm{d}^4x\sqrt{-g}\left\{f_1(\phi)\frac{R}{2\kappa_4^2} + f_2(\phi)[\mathcal{L}_{\rm GB} + a_4(\nabla\phi)^4] + \mathcal{L}_\phi + \mathcal{L}_{\rm mat}\right\}, \tag{13.100}$$

where $\mathcal{L}_\phi = -\omega(\phi)(\nabla\phi)^2/2 - V(\phi)$ includes a quadratic kinetic term and a potential for ϕ and $\mathcal{L}_{\rm mat}$ is the ϕ-dependent matter Lagrangian. If the field $\kappa_4\phi$ is the dilaton $\Phi_{\rm 4D}$ arising in the loop expansion of the low-energy effective action, then

$$\kappa_4\phi = \Phi_{\rm 4D}, \qquad f_1 = -\omega = \mathrm{e}^{-\kappa_4\phi}, \qquad f_2 = \frac{\lambda}{2}\mathrm{e}^{-\kappa_4\phi} \tag{13.101}$$

in the string (Jordan) frame and at the tree level, where $\lambda = 1/4$, $1/8$ or 0 for the bosonic, heterotic and type-II string, respectively.[17] The coefficient a_4 is fixed to recover the three-point scattering amplitude for the graviton [364, 450]. For the heterotic string with stabilized dilaton ReS, stabilized axions and only one radion modulus $\mathrm{Re}T = \exp(2\kappa_4\phi)$, one has $f_1 = 1$, $\omega = 3/2$, $f_2 \propto -\ln[2\mathrm{e}^{\kappa_4\phi}\eta^4(\mathrm{i}\mathrm{e}^{\kappa_4\phi})]$ and $a_4 = 0$, where η is the Dedekind function [452]. At large $|\phi|$, $f_2 \sim \cosh(\kappa_4\phi)$ [453]. The potential $V(\phi)$ in $\mathcal{L}_\phi$ is non-zero by the instantonic and condensation effects described in Sect. 12.3.9.

The main problem of (13.100) is its loose connection with detailed flux-compactification scenarios of moduli stabilization, so crucial for phenomenology. The use of this action in modern string cosmology suffers from various limitations. In Sect. 7.5.1, we have discussed the inadequacy of (13.100) to explain the fine tuning of dark energy in the parameter space allowed by the no-ghost, classical-stability and sub-luminal conditions which arise on a perturbed FLRW background [454–456]. As a model of the early universe, (13.100) has been employed to study the dynamics of perturbations across the bounce (Sect. 13.7.5). A class of bouncing cosmological solutions with a modulus give circumstantial evidence that the big-bang singularity may be resolved in string theory [452]. However, as repeated time and again, the existence of bouncing solutions in any system which also has singular attractor solutions is insufficient to address the big-bang problem. Moreover, the curvature expansion in low-energy string actions is expected to break down at such high densities and other stringy effects, of which we will see a sample in Sects. 13.7.4, 13.7.5, 13.7.6 and 13.7.7, may become predominant. Issues related to inflation will be discussed in Sect. 13.7.5.

[17] In type-II theories, the Gauss–Bonnet term is removed after a field redefinition. Therefore, the first higher-curvature correction to the Einstein–Hilbert action is $O(\alpha'^3R^4)$.

13.7.3.2 Inverse-Power-Law Gravity

The compactifications used in the string cosmological models of this chapter rely on Calabi–Yau spaces. Two alternatives are: (i) hyperbolic compactifications, where a $(4+d)$-dimensional spacetime is reduced to $M^4 \times H_d$ and H_d is a d-dimensional compact hyperbolic manifold with $d \geqslant 2$ and time-dependent volume $\mathcal{V}_d$; (ii) product-space compactifications, where the internal space is the product of flat, spherical and hyperbolic spaces. Hyperbolic and product-space compactifications produce exponential potentials of the type (7.42), which have been used in models of cosmic acceleration (inflation or dark energy) [457–463]. Presently, it is not clear whether a viable parameter space is generated in a natural way in these scenarios, which suffer from some yet unresolved issues [461, 464].

A curious application of hyperbolic compactifications concerns the model

$$f(R) = R - \frac{\beta^4}{R}, \tag{13.102}$$

equation (7.87) with $c_n = -\beta^4 < 0$ and $n = -1$ [465]. Compactifying the $D = 10$ Einstein–Hilbert action on a six-dimensional hyperbolic manifold, after a few field redefinitions one obtains an effective action $\propto -1/R$ [466]. Presumably, the Einstein–Hilbert term can be recovered in the complete SUGRA action.

As we remarked in Sect. 7.5.4, Minkowski spacetime is not a solution of systems with inverse powers of the Ricci scalar, an aspect that would worry most particle physicists. However, a conformal transformation recasts (13.102) into $2\kappa_4^2\bar{\mathcal{L}} = \bar{R} - (\nabla\phi)^2/2 \mp 2\beta^2 e^{-2\phi/3}\sqrt{1-e^{\phi/3}}$, where $\phi = 3\ln(1+\beta^4/R^2)$ and Minkowski is a solution in the Einstein frame. Playing around with the compactification scheme can yield more general inverse powers of R. This may be a tenuous indication that special compactifications in string theory can give rise to specific classes of $f(R)$ modified gravity models in four dimensions, similar to (7.91). However, the bound (7.92) is typically violated in these cases.

13.7.4 Non-local Models

The theories of non-polynomial gravity described in Sect. 11.8 have, by construction, the same type of non-locality as the one realized in the effective spacetime limit (12.52) of string field theory. To see if they can be derived as effective models stemming from SFT, one should face the yet-unsolved difficult problem of constructing a well-defined string field theory from a conformal field theory representation. From that, one should derive an effective non-linear action for the graviton while retaining all non-local effects. As we mentioned in Sect. 12.1.6, the gauge symmetry group of string field theory is wide. It includes spacetime diffeomorphisms and supersymmetry transformations [467] but the way in which such symmetries arise in the low-energy field-theory limit is not straightforward. For instance, in order to identify general coordinate transformations in closed string theory one has to make field redefinitions order by order in the string coupling g_s [468]; the same holds for Abelian transformations in the open case [469]. Therefore, it is not even obvious whether one can obtain, directly from closed SFT, an effective action of gravity which is simultaneously non-local (with the correct type of form factors), covariant and non-linear.[18]

The action (11.118) has these three properties, with the form factor (11.119) chosen to match with the graviton propagator of (12.51) in the linear limit. In $D = 11$ and with the inclusion of the 3-form A_3 and of fermionic super-partners, it also has other features compatible with string theory, including S-duality and the correct local SUGRA limit when compactified to $D = 10$ [472]. Since (11.118) and the other spacetime actions discussed in Sect. 11.8 with exponential non-locality are not derived from SFT, they are commonly referred to as string-inspired models [473]. When properly motivated, they can help us to understand the effect of exponential operators on the evolution of the early universe. In particular, we have seen that the form factors (11.106) and (11.119) can remove the classical singularities of local gravity. It may therefore be possible to address the big-bang problem in string field theory. The cosmological dynamics of non-local gravity is under study.

We conclude this sub-section by quoting some bounds on the scale $M \sim 1/\sqrt{\alpha'}$ appearing in the exponential operator (11.106) using simplified particle-physics models and the non-local dispersion relation characteristic of these scenarios. LHC data bound the scale of non-locality as [474]

$$M > 2.5 \times 10^3\,\mathrm{GeV}\,, \tag{13.103}$$

roughly corresponding to $\sqrt{\alpha'} = l_s < 10^{-19}\,\mathrm{m}$. Observations on opto-mechanical heavy quantum oscillators are less constraining, $l_s < 10^{-15}\,\mathrm{m}$ [475]. The range (13.103) is not surprising since it is close to the lowest energy scale for supersymmetry breaking. However, table-top high-precision experiments already under construction will be able to place much stronger bounds $l_s < 10^{-22} - 10^{-29}\,\mathrm{m}$ [475], thus providing independent information on the string scale.

13.7.5 Pre-Big-Bang and Dilaton Cosmology

The T-duality (12.83) is strongly suggestive of the existence of an effective minimal length scale, the self-dual radius $l_s = \sqrt{\alpha'}$. This critical scale hints that singularities may be resolved in string theory: a big-bang initial configuration would be as much unlikely as an infinitely large newborn Universe.

If the effective dynamics obeys T-duality, the evolution of the universe is symmetric under the mapping [476, 477]

[18] The novel formulation of closed super-SFT advanced in [470, 471] may open up new possibilities.

$$\ln a(t) \quad \longleftrightarrow \quad -\ln a(-t) \tag{13.104}$$

and the big bang of classical general relativity can be replaced by a bounce at $a(0)$. Imposing the scale-factor duality (13.104) to the solutions of effective equations of motion leads to generic *pre-big-bang* scenarios where the dynamics is well defined before and after a cosmic bounce [478–501]. The bounce can take place only in the presence of inhomogeneous perturbations, since the junction of the solutions at $t = 0$ is singular in purely homogeneous backgrounds [481, 485–487]. The trans-Planckian problem is also solved, since the perturbation modes we observe today started with an initial wave-length much larger than the Planck length.

The low-energy dilaton-gravity action (12.40) respects the scale-factor duality (13.104) and is often used in these scenarios, including in many early studies on string-gas cosmology (Sect. 13.7.6). CMB anisotropies are generated during the deflationary contracting phase and, although this might be considered an alternative to inflation, the ingredients producing inhomogeneities are the same: a fluctuating scalar field (in this case, the dilaton) and the back-reaction of the metric.

The regime of applicability of perturbation theory is reduced by the extreme conditions near the bounce and, in fact, dilaton deflation generates scale-dependent spectra. In models where the gravitational action is linear in the Ricci scalar R, the typical prediction is $n_{\rm s} - 1 \simeq n_{\rm t} \approx 3$ [484, 491]. Inclusion of the Gauss–Bonnet curvature correction and of the quartic term for the dilaton [488, 490, 492–496, 498] (equations (13.100) and (13.101)) flattens the spectra slightly but not enough [496]:

$$2 \leqslant n_{\rm s} - 1 \simeq n_{\rm t} \leqslant 3 \,. \tag{13.105}$$

These results will be discussed more in detail in the ekpyrotic scenario of Sect. 13.7.7.

We take this opportunity to stress the most formidable obstacle dilaton-based scenarios must face to explain the cosmic acceleration, at either early or late times: the overshooting problem cursorily mentioned in Sect. 13.3 [84, 269, 283]. For definiteness, consider the low-energy limit of the $E_8 \times E_8$ superstring and recall, from (12.92), that $\Phi_{\rm 4D} = -\ln({\rm Re}S)$ and that (12.43) holds.[19] The non-perturbative potential $V(\Phi_{\rm 4D})$ of the dilaton has an asymptotic absolute minimum at $\Phi_{\rm 4D} \to -\infty$, corresponding to the weak-coupling limit of string theory at low energy [502]. Unless $V(\Phi_{\rm 4D})$ has other minima, the field will run away to infinity where $V(\Phi_{\rm 4D} \to -\infty) = 0$, either by classical motion (for a high kinetic energy) or by thermal effects [503]. Usually the runaway is classical, since the typical potential is very steep due to gaugino condensation; see the racetrack superpotential (12.110). If the dilaton is identified with the inflaton, it is difficult to find a regime of slow rolling and the perturbation spectrum strongly deviates from scale invariance. If the dilaton is not the inflaton ϕ, it couples to it via a non-minimal interaction. The

[19] Note that a relation analogous to (12.108) holds in the $SU(3)$ super-Yang–Mills sector, so that (12.108) and (12.103) give (12.43).

kinetic energy of Φ_{4D} dominates the evolution of ϕ and, again, any slow-roll regime is quickly disrupted.

In both cases, a dynamical dilaton during or after inflation may give rise to variations of the fundamental couplings [504] and violations of the weak equivalence principle larger than experimental constraints [505],[20] unless the dilaton couples universally to all matter sectors [506, 507] (i.e., for all matter species i the couplings b_i are the same, $b_i(\Phi_{4D}) = b(\Phi_{4D})$) or the chameleon mechanism is enforced (Sect. 7.4.5). In the scenarios of moduli and D-brane inflation, we have seen that the above problems are avoided: the dilaton can be trapped in a local minimum while inflation is successfully driven by other fields.

The dilaton has also been considered as a quintessence candidate [508, 509]. The full form of the couplings b_i is obtained at any given perturbative order by the string loop expansion in Riemann surfaces of higher genus (Sect. 12.1.4). In the weak-coupling limit $g_s \ll 1$, this expansion is $b_i(\Phi_{4D}) = e^{-\Phi_{4D}} + c_0^i + c_1^i e^{\Phi_{4D}} + c_2^i e^{2\Phi_{4D}} + \dots$, for some constants c_n^i. Under certain assumptions on the quantum corrections to the string effective action [510], one may expect the opposite $g_s \gg 1$ limit to take the form

$$b_i(\Phi_{4D}) \simeq C_i + O(e^{-\Phi_{4D}}),$$

so that these couplings tend to a constant C_i when $\Phi_{4D} \to +\infty$. In particular, the dilaton can evolve to strong coupling at late times and all its field couplings except the one with dark matter become trivial in this limit [508]. The equivalence principle is still violated but by effects below or near present constraints [511–513]. The resulting model has all the properties of tracking solutions, including their shortcomings.

13.7.6 String-Gas Cosmology

The resolution of the big bang in string theory could make use of the thermodynamical properties of superstring winding modes in the weak-coupling regime $g_s \ll 1$ [514–516]. In a compact space, the excitation modes of a thermal ensemble of strings are momentum modes and winding modes. According to (12.82), the energy of winding modes decreases with the size of the space, so that, for an adiabatic process, they will dominate the thermal bath in small boxes. The temperature of this bath of light modes cannot rise indefinitely and, in fact, the total energy and the partition function diverge as the system approaches a critical temperature T_H called *Hagedorn temperature* [517]. This is the maximum attainable temperature physically.

[20]The bound (7.73) from big-bang nucleosynthesis and post-Newtonian solar-system tests is $\omega > 4 \times 10^4$, while for the dilaton $\omega = -1$ (apply the field redefinition $\varphi = e^{-\kappa_4 \phi/2}$ to (13.101)). The actual check of violations is much more involved than this back-of-the-envelope observation.

13.7.6.1 Big-Bang Problem

One can exploit this phenomenon to study the very early Universe at the string scale, a scenario dubbed *string-gas cosmology* [514–516, 518–542] (reviews are [543, 544]).[21] In a non-compact space, the specific heat is not positive definite; a realistic scenario thus requires space to be compact. A nine-dimensional torus is the simplest choice but orbifold and Calabi–Yau compactifications are also possible which do not alter the main features of the proposal [521, 522].

As in pre-big-bang cosmology, T-duality is invoked to replace the big bang with a bounce where the temperature is very close to $T_{\rm H}$ [514]. The radius at which the bounce phase is triggered increases with the entropy of the Universe. Three of the spatial directions then inflate enough to push the radius of the compact space beyond the particle horizon. Intuitively, this may be achieved by the tendency of the system to attain thermal equilibrium, which is reached only if the world-sheets of winding strings interact. However, world-sheets can intersect effectively only in a spacetime with four or less dimensions. Therefore, it may be possible for a compact Universe to have three large spatial directions, while the remaining six remain compact [514, 516, 518]. The same mechanism can be realized also in the more general case of a gas of D-branes [519] and in the low-energy limit of M-theory [524, 526].

It has been verified that, for reasonable initial conditions, three spatial directions can grow large (and become isotropic), while the six-dimensional compact space is stabilized around the string scale $l_{\rm s}$ by a mechanism involving the massless modes of the string spectrum [523, 525, 527, 528, 530, 531]. This mechanism takes care of the Kähler moduli, while fluxes stabilize the shape moduli [530]. Due to the absence of R-R fluxes in the simplified treatment of these models, the dilaton must be stabilized separately. In toroidal compactification, the dilaton sits at the minimum of a non-perturbative potential generated by gaugino condensation [540]. The same mechanism breaks supersymmetry in a way compatible both with late-time cosmology and with particle physics [541].

13.7.6.2 Alternative to Inflation

String-gas cosmology offers an alternative to inflation based on string thermodynamics. By T-duality, both the big-bang and the trans-Planckian problem can be solved in principle, since scales smaller than the string scale are not physically within reach. The horizon problem is solved by an early stage of contraction when the Hubble horizon is larger than the physical Universe [520].

For three large spatial dimensions, one can also obtain an almost scale-invariant primordial scalar spectrum from thermal fluctuations. The mean square energy-

[21] String-gas cosmology has been one of the very first applications of string theory to the early universe [545–548].

density fluctuations in a 3-torus of radius ϱ and volume $\mathcal{V}$ are $\langle\delta\rho^2\rangle \propto T^2 C_{\mathcal{V}}/\varrho^6$, where $C_{\mathcal{V}} = \delta E/\delta T|_{\mathcal{V}=\text{const}}$ is the specific heat capacity of the string gas in the volume $\mathcal{V}$. From the string partition function in the Hagedorn phase, one finds that the heat capacity is $C_{\mathcal{V}} \propto \varrho^2/[T(1-T/T_{\text{H}})]$ [515]. Intuitively, it scales as a surface rather than a volume because winding modes (which dominate the Hagedorn phase) appear with one dimension less to a bulk observer. On the other hand, the Poisson equation $\nabla^2\Phi \propto \delta\rho$ (the 00 component of Einstein's equations, which are assumed) reads $\Phi_k \propto k^{-2}\delta\rho_k$ in momentum space. Therefore, $\mathcal{P}_{\text{s}} \propto k^3\langle\zeta_k^2\rangle \propto k^3\langle\Phi_k^2\rangle \propto k^{-1}\langle\delta\rho_k^2\rangle \propto k^{-4}\langle\delta\rho^2\rangle$ and, replacing $\varrho \sim k^{-1}$, one has [533, 535]

$$\mathcal{P}_{\text{s}} \propto \frac{T(k)}{T_{\text{H}}}\frac{1}{1-T(k)/T_{\text{H}}}, \tag{13.106}$$

where the temperature T is evaluated at the time when the mode with comoving wave-number k exits the horizon. Since T is almost constant during the Hagedorn phase, the spectrum is almost constant. The spectral index is

$$n_{\text{s}} - 1 \simeq \frac{1}{1-T(k)/T_{\text{H}}}\frac{\text{d}(T/T_{\text{H}})}{\text{d}\ln k} < 0, \tag{13.107}$$

where the approximation considers that $T(k)$ is very close to the Hagedorn temperature. A small red tilt is produced because scalar modes are generated by the energy density, which increases with T; as $\text{d}T/\text{d}k < 0$, an increase of power is observed for small k.

Contrary to the standard inflationary paradigm, the tensor spectrum is blue tilted, since [534, 535, 542]

$$\mathcal{P}_{\text{t}} \propto \frac{T(k)}{T_{\text{H}}}\left[1-\frac{T(k)}{T_{\text{H}}}\right]\ln^2\left\{\frac{1}{l_{\text{s}}^2k^2}\left[1-\frac{T(k)}{T_{\text{H}}}\right]\right\} \tag{13.108}$$

and

$$n_{\text{t}} \simeq 1 - n_{\text{s}} > 0. \tag{13.109}$$

Intuitively, tensor modes are generated by anisotropic pressure terms in the energy-momentum tensor, but near the Hagedorn temperature the thermal bath is dominated by winding modes and the pressure decreases. There is thus a decrease of power at low k. A blue-tilted tensor spectrum is one of the characteristic predictions of string-gas cosmology that could be tested if a primordial gravitational signal was discovered. The tensor-to-scalar ratio can be computed from (13.108) and (13.106) and its magnitude depends on the string scale l_{s}. The latter can be tuned (to about $l_{\text{s}} \sim 10^{-3}l_{\text{Pl}}$) to obtain an observable $r = O(0.1)$.

The level of non-Gaussianity of the model depends on the string scale. The local non-linear parameter is [539]

$$f_{\rm NL}^{\rm local} \simeq \left(\frac{l_{\rm s}}{l_{\rm Pl}}\right)^2 \frac{H}{T} \simeq 10^{-30} \left(\frac{l_{\rm s}}{l_{\rm Pl}}\right)^2 k\tau_0 \,. \tag{13.110}$$

At scales comparable with the horizon today, $f_{\rm NL}$ is negligible if the string length is close to the Planck scale, while $f_{\rm NL} = O(1)$ if the string energy is O(TeV).

13.7.6.3 Open Problems

The lack of an analytic treatment of the Hagedorn phase forbids a direct determination of the effective dynamical equations in the very early universe. Dilaton gravity, given by the low-energy effective actions (12.40) and (12.41), is unviable. A background with a non-stabilized dilaton puts the whole string-gas scenario in jeopardy due to several issues, including cosmic spectra with a strong scale dependence and a difficulty in smoothly connecting the Hagedorn phase with a late-time sensible evolution of the universe [536, 538]. Einstein gravity is excluded per se because it does not exhibit T-duality, but it can be regarded as a dynamics with frozen dilaton. However, higher-order curvature terms are expected to dominate at such high-density regimes. In other words, the assumption of weak coupling $g_{\rm s} \ll 1$ may break down near the singularity or the bounce. Therefore, expressions such as (13.106) and (13.108), obtained in Einstein gravity (ideally, in dilaton gravity in a quasi-static configuration where the dilaton is constant), should be taken *cum grano salis* until a more fundamental setting, with a robust mechanism for dilaton stabilization, becomes available. The discovery of a blue-tilted tensor spectrum would stimulate further research in this direction.

The non-local models of Sect. 13.7.4 can provide bouncing scenarios wherein to embed the Hagedorn phase [537]. However, in this case the scalar spectrum is very nearly Harrison–Zel'dovich, unless the string scale were O(TeV). Also, a sufficiently long Hagedorn phase (necessary to maintain thermal equilibrium over the observed scales) requires some fine tuning.

The understanding of moduli stabilization is limited by an incomplete knowledge of string and brane dynamics at strong coupling. Some results in $D = 11$ SUGRA (regarded as the low-energy limit of M-theory) confirm that three spatial directions decompactify but the compact directions grow slowly in time instead of being stabilized [524]. The number of decompactified directions and their evolution is also sensitive to the initial conditions, both in eleven and ten dimensions, which points towards some fine tuning [526, 529]. Typical initial conditions in $D = 10$ are such that either there are too few strings to wrap around space and all dimensions grow (strong coupling), or string interactions switch off too rapidly to allow the system to thermalize and all directions remain wrapped by a large number of strings (weak coupling). Massless string modes can indeed help to stabilize the moduli but, again, the fate of the dilaton is a separate and important question, as we just remarked. Although several ingredients have already been proposed to get viable stabilization mechanisms, they are still to be harmonized rigorously in a realistic effective evolution of the early universe.

13.7.7 Cyclic Ekpyrotic Universe

A stringy version of cyclic scenarios (Sect. 6.2.4.1) and an alternative to inflation is the *cyclic ekpyrotic universe* [549–582], strongly based on an older non-cyclic scenario [583–587] (see [588, 589] for reviews).[22] For the greater richness of the details of its embedding in string theory, this setting may be considered as an evolution of the pre-big-bang scenario of Sect. 13.7.5 [498] and, in fact, the analyses of that bounce apply also to the ekpyrotic case [497, 499–501]. We will first describe the original proposal and then introduce some modifications necessary to make it compatible with observations.

Imagine two flat 3-branes constituting the boundary of a five-dimensional spacetime and interacting with an attractive potential $V(\varphi)$ along a compact fifth dimension parametrized by the radion φ. These branes can be the orbifold planes of heterotic M-theory (M-theory compactified on $\mathcal{C}_3\times S^1/\mathbb{Z}_2$) [346–349] or the $(3+1)$-dimensional manifolds of a Randall–Sundrum setting. The radion potential takes the general form

$$V(\varphi) = V_0\left(\mathrm{e}^{2\varphi/\varphi_1} - \mathrm{e}^{-2\varphi/\varphi_0}\right)F(\varphi)\,, \tag{13.111}$$

where $-\infty < \varphi \leqslant \varphi_{\max}$ is dimensionless, $\varphi/\varphi_1 \ll 1$ and $\varphi/\varphi_0 \gg 1$. The function $F(\varphi)$ represents non-perturbative effects and scales as $\exp(-1/g_s)$ or $\exp(-1/g_s^2)$. In M-theory, φ is the 11th dimension and $g_s \propto \exp(\gamma\varphi)$, with $\gamma > 0$. The potential (13.111) has a local negative minimum and is very flat at the sides, tending to $V_0 > 0$ for $\varphi \to \varphi_{\max} \gg 1$ and to 0 for $\varphi \to -\infty$ (Fig. 13.12).

As the branes get closer, the gravitational energy in the bulk is converted into brane kinetic energy. Since the branes are boundary ones, instead of collapsing via tachyon condensation they collide and oscillate back and forth their center of mass along the extra direction. During the collision at coincident branes ($\varphi = -\infty$), part of the brane kinetic energy is converted into matter and radiation. An observer on one of the branes experiences the brane collision as a big bang (vanishing scale factor a_{E} in the Einstein frame) after a period of contraction. Even if the fifth dimension does experience a big crunch and it collapses to a point, and even if $a_{\mathrm{E}}(t) = 0$ for the brane observer, the brane metric in the Jordan frame, the local temperature and the energy density on the brane remain finite at the event. This

[22]The term “ekpyrotic” is inspired by a cosmogonic model, attributed to Greek Stoicism, of cyclic destruction (*ekpyrosis*) and creation (*palingenesis*) of the world in and from a great fire.

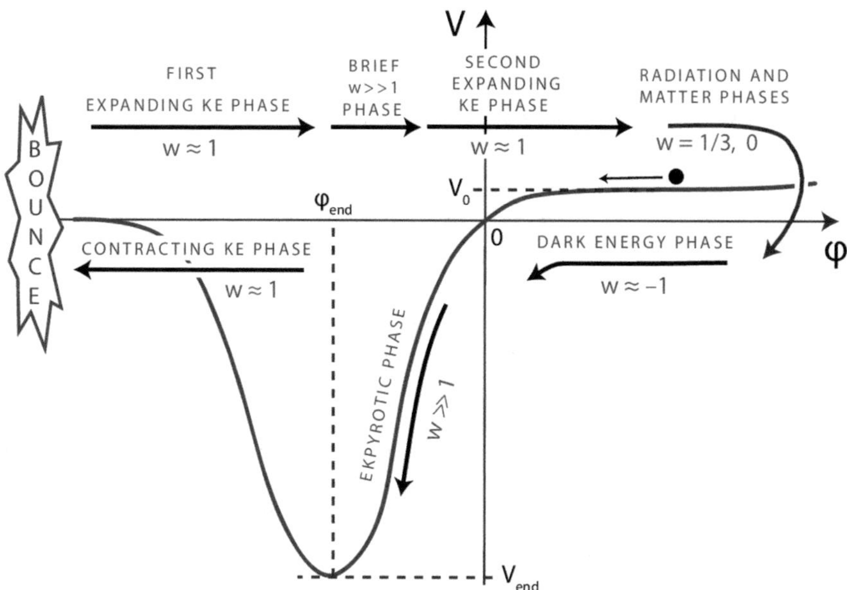

Fig. 13.12 Typical single-field potential in ekpyrotic scenarios and dynamical stages in a cycle (Reprinted figure with permission from [562]. ©2013 by the American Physical Society)

bounce is not symmetric, an essential feature for realizing the difference in scales between primordial and late-time acceleration.

13.7.7.1 Alternative to Inflation

The evolution of the universe can be described purely in terms of four-dimensional variables (Fig. 13.12). Just after the brane collision and the onset of a new separation, the observable universe expands in a sequence of eras dominated by radiation. The radion φ turns around and rolls back along its potential during radiation domination, at which point $w_\varphi = -1$ (zero kinetic energy) but ρ_φ is still sub-dominant. There follows matter domination and, when the matter energy density is red-shifted away, ρ_φ dominates and a quintessence era begins, with a barotropic index $w_\varphi \approx -1$ increasing to less negative values. After several trillion years of acceleration, the kinetic energy of φ starts to dominate over the potential, the observed universe decelerates and then contracts slowly, entering a fast-roll ekpyrotic phase with $V(\varphi) \simeq -V_0 \mathrm{e}^{-2\varphi/\varphi_0}$ and $\epsilon = 3(w_\varphi + 1)/2 \gg 1$. The horizon and flatness problems are solved at this stage. Eventually, at the end of the ekpyrotic phase the contraction becomes faster ($w_\varphi \approx 1$) until the system collapses into what the observer would perceive as a big crunch (but not with BKL oscillations, which are suppressed by the $w_\varphi \gg 1$ ekpyrotic phase [558]). A big bang and a new era of expansion follow.

During the contraction phase, a pattern of inhomogeneities is developed. In fact, due to quantum fluctuations the branes are not parallel at all points and they collide at slightly different times in different places. These patches begin their evolution and cooling down from the bounce out of sync, which causes the anisotropies observed in the sky. The spectral properties of this pattern depend on the specific model of ekpyrosis. We discuss three models: (i) perturbations generated during the ekpyrotic phase by a single field, (ii) perturbations generated during the ekpyrotic phase by two or more scalar fields and (iii) perturbations generated before ekpyrosis by a single field.

(i) The original single-field model does not work. For large $\varphi > 0$, the potential (13.111) can be approximated by the exponential potential (5.78) with negative-definite normalization, which is a solution of the Friedmann equations for a power-law expansion $a \sim (-t)^p$ with $p = \varphi_0^2/2$ (equation (5.79)) and spectral indices (5.143) and (5.154). Since in this phase gravitational effects are weak, the tensor spectrum is suppressed. However, the universe is quasi static during contraction and $0 \lesssim p < 1$, so that both spectra are strongly blue-tilted, $n_s - 1 = n_t \gtrsim 2$ (in the Einstein frame). This blue-tilted deviation from scale invariance is a typical issue of steep exponential potentials and it also excludes dilaton inflation in pre-big-bang models, for which (13.105) holds [552]. The problem of all these single-field models is that linear perturbation theory breaks down near the bounce in most gauge choices and the power spectra, which are produced in the contracting phase, acquire a strong scale dependence [551–553, 560, 584, 586] unless specific potentials and matching conditions at the bounce are chosen [499, 500, 554, 556].

(ii) Apart from extending the choice of matching conditions at the bounce, one can relax the assumption that all moduli are frozen except the radion. Including also a dynamical axion or the Kähler modulus of the Calabi–Yau space, one is capable to generate red-tilted, almost scale-invariant scalar perturbations through the bounce [554, 555, 563, 566, 567]. The effective potential $V(\varphi_1, \varphi_2)$ during this ekpyrotic phase can take a variety of forms. Typically, it can be parametrized by a double exponential

$$V(\varphi_1, \varphi_2) = -V_1\, e^{-\int d\varphi_1\, c_1(\varphi_1)} - V_2\, e^{-\int d\varphi_2\, c_2(\varphi_1)}, \tag{13.112}$$

where $V_{1,2} > 0$ and $c_{1,2}$ are generic functions which become almost constant during ekpyrosis. Defining $\sigma := (\dot\varphi_1\varphi_1 + \dot\varphi_2\varphi_2)/\dot\sigma$ and $s := (\dot\varphi_1\varphi_2 - \varphi_1\dot\varphi_2)/\dot\sigma$, up to third order in s, the potential during ekpyrosis reads [567]

$$V(\sigma, s) \simeq -V_0\, e^{\sqrt{2\epsilon}\,\sigma}\left(1 + \epsilon s^2 + \frac{\kappa_3}{3!}\,\epsilon^{3/2} s^3\right),$$

where κ_3 is a constant. Entropy perturbations produced at a first stage are subsequently converted into curvature perturbations by the interaction of the two fields. The typical range of the scalar spectral index $n_s - 1 = 2/\epsilon - d\ln\epsilon/d\mathcal{N}$ (where $\mathcal{N}$ is the number of e-folds before the end of the ekpyrotic phase) is

$$0.97 < n_s < 1.02 \,; \tag{13.113}$$

the observed value lies in this interval within the experimental uncertainty. While scale invariance in standard inflation is a consequence of the slow-roll approximation, the cyclic ekpyrotic model achieves this goal when $\epsilon \gg 1$.

The typical level of non-Gaussianity of this scenario is at least one order of magnitude larger than the one of inflation, $|f_{\rm NL}^{\rm local}| \sim 5-100$ [567–569, 571, 572]. The actual range of values depends on when entropy perturbations were converted into curvature perturbations. A conversion during the ekpyrotic phase is ruled out, since $f_{\rm NL}^{\rm local} < -40$. If the conversion took place during kinetic-energy domination after the ekpyrotic phase, then [570]

$$f_{\rm NL}^{\rm local} \simeq 5 + \tfrac{3}{2}\kappa_3\sqrt{\epsilon} \,. \tag{13.114}$$

Since $\epsilon = O(10^2)$, the parameter space giving rise to acceptable local non-Gaussianity is severely constrained, $-0.8 < \kappa_3 < 0.5$ at the 95 % CL [338]. The model is not ruled out but specific potentials must be chosen [582].

In an alternative model where φ_1 drives ekpyrosis and φ_2 generates entropy perturbations, $V_1 > 0$ and $V_2 = 0$ in (13.112) and φ_2 has a non-canonical kinetic term $\propto f(\varphi_1)(\nabla\varphi_2)^2$ [577, 578]. This configuration suppresses any non-Gaussian signal down to acceptable levels (in particular, $f_{\rm NL}^{\rm local} \approx 5$) and reduces the level of fine tuning [580, 581].

(iii) An alternative which employs only one scalar field is to assume that the observed perturbations were created before the ekpyrotic phase, during the transition phase when the kinetic energy $\dot\varphi^2/2$ of the field almost cancels the second exponential term in (13.111) [573, 575]. The potential is simplified to $V(\varphi) = V_0\,(1-{\rm e}^{-c_0\varphi})$, where $c_0 := 2/\varphi_0$, and the nearly constant term $V \sim V_0$ fuels the cosmic evolution. At this stage (which is a dynamical attractor), both the scale factor and the Hubble parameter are almost constant. The resulting scalar spectrum is adiabatic and scale invariant. In order to obtain a reheating phase at a scale between electroweak and GUT, $c_0 = 10^{28}-10^{40}$, but for a constant c_0 scale invariance is almost exact. To get an observable red tilt, it is sufficient to generalize the potential in the transition phase as

$$V(\varphi) \simeq V_0\left[1-{\rm e}^{\int {\rm d}\varphi\, c(\varphi)}\right], \tag{13.115}$$

where $c(\varphi)$ is slowly varying around $c \approx c_0$. Then,

$$n_s - 1 \simeq -\frac{2}{\ln(c_0^2/2)} = O(10^{-2}) \,. \tag{13.116}$$

Unfortunately, a simple exponential potential strongly suffers from a trans-Planckian problem [574]. It also gives rise to momentum-dependent large non-Gaussianities, with an equilateral non-linear parameter $f_{\rm NL}^{\rm equil} \simeq 5k^2/(6c_0^2 V_0)$ [573]. To overcome these issues, one has to ensure that the transition phase does not last too long, so that to avoid the late-time creation of modes with strong non-Gaussian statistics. For example, one can consider a $c(\varphi)$ which decreases to some value $\ll c_0$ just after nearly-Gaussian scale-invariant fluctuations have been generated [575].

To summarize, all models (i)–(iii) are characterized by a tensor spectrum with negligible amplitude and a strongly blue tilt:

$$n_{\rm t} \gtrsim 2\,, \qquad r \approx 0\,. \tag{13.117}$$

On the other hand, in those models which achieve near scale invariance in the scalar sector a conspicuous non-Gaussian signal is typically produced, unless the potential is carefully tailored or the kinetic term of the entropy-generating field modified.

After the ekpyrotic phase, the qualitative features of the evolution of the universe are the same for all models. The transition between a big crunch and the next big bang is governed by the smooth brane dynamics. At the bounce, matter and radiation are created on the brane and the universe begins with high density again. One should take into account the effect of non-linearities at this moment [554, 556, 576, 579, 585, 587].

When the branes are sufficiently far apart, the inter-brane potential $V(\varphi)$ becomes positive and fuels a dark-energy dominated era. The late-time accelerated expansion washes away any relic produced since the preceding big bang and drives the universe to a nearly empty state, thus restoring the local conditions in existence exactly one cycle before. The total entropy increases at each new cycle but the entropy density is cyclically diluted from a bang to the following crunch. The horizon problem is automatically solved in this scenario and inflation is not needed. The flatness problem is also nullified, since the brane universe remains flat all the time either because it began on an almost stable (hence almost flat) brane or, more generally, due to the curvature dilution by the preceding cycles of expansion.

13.7.7.2 Big-Bang Problem

The cycles of the ekpyrotic universe are asymmetric and they spend most of the time expanding. Therefore, the averaged expansion condition (6.14) is satisfied both in a cycle and when averaging over an arbitrary number of identical cycles, $\mathcal{H}_{\rm av} > 0$. Thus, the BGV theorem of Sect. 6.2.3 applies: the ekpyrotic universe is geodesically past-incomplete and there must have been a primordial Big Bang where and when everything begun.

However, the cyclic solution is a dynamical attractor and the system becomes insensitive to the initial conditions. Our cycle is only the last of a very long sequence which will continue in the future indefinitely. The probability that the observed particles have been created at the beginning of a past cycle is argued to become exponentially smaller the longer the time interval between that cycle and ours. Therefore, the issue of the past-incompleteness of geodesics becomes physically irrelevant in this scenario. In this sense, but modulo a measure problem yet to be addressed, this may be regarded as a solution to the big-bang problem [550].

13.7.7.3 Cosmological Constant Problem

The ekpyrotic scenario embeds a relaxation mechanism for the cosmological constant problem [561] via an adaptation of an older proposal by Abbott [590]. Consider an axion field ϕ coupled with a non-Abelian gauge sector (e.g., the strong force) through an interaction $\propto (\phi/f)\tilde{F}F$, where f is a mass scale. Integrating out the gauge fields, one obtains a cosine potential similar to (5.90). It is also assumed that the discrete symmetry $\phi \to \phi + 2\pi f n$ is softly broken by a term with no minimum in the range of interest in ϕ. For simplicity, one can take a linear term:

$$U(\phi) = A\cos\left(\frac{\phi}{f}\right) + B\frac{\phi}{f}\,, \tag{13.118}$$

where $B \ll A$. Since radiative corrections to B are proportional to B, a very small value of B would be protected against quantum effects. This potential is depicted in Fig. 13.13 and has a typical "washboard" form with gradually decreasing periodic minima $V_n^{\rm min}$. As the universe evolves from some vacuum with large $\Lambda = V_{n\gg 1}^{\rm min}$, the axion undergoes a sequence of quantum tunneling events $V_n^{\rm min} \to V_{n-1}^{\rm min} \to V_{n-2}^{\rm min} \to \dots$ describing de Sitter bubbles, eventually relaxing the effective cosmological

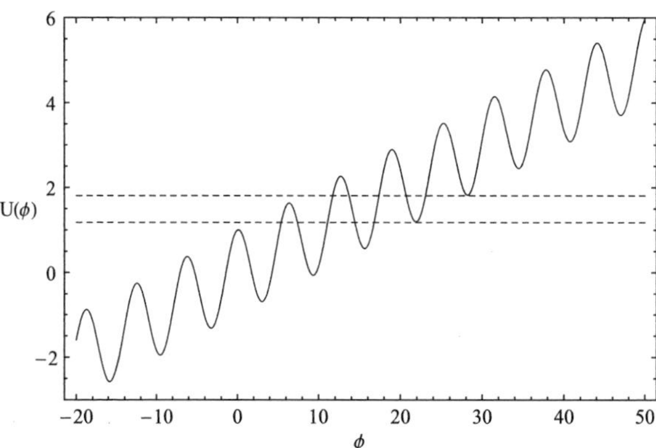

Fig. 13.13 Relaxation of the cosmological constant in the ekpyrotic cyclic universe. The difference $V_n^{\rm min} - V_{n-1}^{\rm min}$ between two adjacent vacua (*dashed lines*, n decreasing from right to left) must be small enough to hit the observed value of Λ within the experimental uncertainty

constant to a very small value $0 < V_0^{\rm min} < B$. A transition to negative values $V_{-1}^{\rm min}$, $V_{-2}^{\rm min}$, ... would make a bubble collapse in about one Hubble time, while bubbles in other regions would continue to expand with $\Lambda > 0$. The time spent in each local vacuum becomes large according to an exponential-of-exponential progression, and configurations with a small Λ are most favored.

Abbott's model was proposed within a non-cyclic, standard big-bang cosmology. The problem is that the tunneling sequence occurs too slowly compared with the Hubble expansion and the universe becomes "empty" by over-acceleration well before the formation of structures and the reaching of the desired vacuum $V_0^{\rm min}$. However, by including the non-Abelian sector in the ekpyrotic scenario the empty-universe problem is easily solved, since matter and radiation are created at every bounce event. Also, plausibly the Universe is much older than the age of the present cycle and the relaxation mechanism can take place on a time scale far greater than H^{-1}. Moreover, since the brane does not become singular at the bounce, the kinetic energy of the axion never diverges and the tunneling events down the washboard take place from one minimum to the adjacent one, without erratic jumps. Cycle after cycle, the universe spends an exponentially longer time in a state with smaller and smaller cosmological constant.

13.7.7.4 Open Problems

Like most braneworld scenarios, the ekpyrotic proposal does not include the details of the compactification scheme and of the stabilization of non-dynamical moduli. A posteriori, this turns out to be a disadvantage because it introduces free parameters (types of potentials, number of dynamical moduli, type of matching conditions, and so on) and a loss of contact with string theory, as we discussed in Sect. 13.7.1. The choice of potential is especially crucial for the avoidance of trans-Planckian issues and unacceptably large non-Gaussianities. By itself, $V(\varphi)$ does not entail any fine tuning of the initial conditions [582] but, without any input from the full theory, it remains purely phenomenological. Also, moduli stabilization would become even more challenging in more realistic scenarios where the bulk is populated by branes, the latter being necessary to get viable particle physics.

Moreover, the fact that one of the compact directions becomes singular cyclically may not be attractive in a theory supposed to be finite, even if the type of singularity involved is much milder than the outright disappearance of the whole spacetime as in general relativity.

Regarding the cosmological constant problem, the relaxation mechanism described above does not provide a complete solution because the parameter B must be fine tuned to a very small value to guarantee that the interval $0 < V_0^{\rm min} < B$ is compatible with the experimental uncertainty on Λ. At present, a theoretical justification for this tuning within string theory is still missing. This gap problem and further criticism are discussed in [591].

13.8 Inflation and Alternatives: Compact Summary

Table 13.1 is a grand summary of the strongest contenders for a scenario of the early universe motivated by string theory. We have left out models which are in contrast with experiments for string-motivated values of their parameters ($\mathbb{K}\mathrm{L}\mathbb{M}\mathrm{T}$ models with large η, tachyon inflation and several versions of the ekpyrotic scenario), phenomenological models which do not have full control on moduli stabilization (the braneworld, among others) and models which have not been developed to the point of giving predictions for the cosmological observables (such as non-local gravity). The level of fine tuning of all these scenarios is similar and modest.

- Qualitatively, models of moduli inflation all share the same predictions of single-field standard inflation: a nearly scale-invariant scalar spectrum, a negligible tensor spectrum and a low level of non-Gaussianity. Contrary to the brane-based models of Sects. 13.5 and 13.6, these scenarios do not involve complicated brane dynamics. Since they are completely described by low-energy $D = 4$ supergravity, after writing down the total potential they can be treated with the field-theory formalism of Chap. 5, at least during inflation and perhaps also in the reheating phase. The main difference with respect to the models of Chap. 5 is that both the form and the parameters of the inflaton potential are motivated by string theory, since the details of the super- and Kähler potentials depend on the compactification scheme and on the geometry of the underlying Calabi–Yau space.
- N-flation is not among the most favoured models due to the low scalar index (13.37), a possible $f_{\rm NL} = O(1)$ non-Gaussianity and its uncertain embedding in string theory. It is not excluded either, since it is not in clear tension with observations.
- D-brane inflation has three characteristics which immediately tell it apart from moduli inflation: (a) it is more involved due to the fundamental role played by branes and anti-branes, (b) it takes more efforts to obtain a flat potential and to solve the η-problem and (c) the typical inter-brane potential cannot sustain eternal inflation (this property has some consequences on the choice of initial conditions but it does not worsen the level of fine tuning of the model). The richness of this proposal makes it a tantalizing example of how string theory can get in touch with observations.
- String-gas cosmology exploits string thermodynamics to manufacture a cosmological model of density fluctuations, while the ekpyrotic universe exploits the dynamics of brane collisions to sustain cyclic fast-rolling eras of acceleration and contraction. With respect to moduli, D-brane and DBI inflation, both proposals rely much less heavily on the details of flux compactification and moduli stabilization, they focus more on a resolution of the big-bang problem and they offer an alternative to inflation.

All scenarios entail a few simplifications and many open problems. The details of moduli stabilization are taken care of but only partially. In fact, there are not many

Table 13.1 Typical values of the cosmological observables and level of fine tuning of some early-universe string models compatible with observations. The interval $0.95 < n_s < 0.98$ roughly corresponds to the 3σ-level range allowed by observations. "Moduli inflation" includes racetrack axion inflation, fluxless inflation, large-volume blow-up inflation (fibre inflation yields $r \sim 10^{-3}$) and volume-modulus inflation. "Axion inflation" includes N-flation and aligned/hierarchical multi-axion models. "Warped D-brane inflation" includes the $\mathbb{KLMT}$ model and its most promising modifications. "Ekpyrotic universe (ii)" refers to models of type (ii) with entropy-to-curvature conversion of perturbations after the ekpyrotic phase. Very low values of ϵ_* (the first slow-roll parameter at horizon crossing) or of the tensor-to-scalar ratio, undetectable by present or near-future experiments, have been approximated to zero. "Small" and "large" indicate the typical level of non-Gaussianity. Models with potentially dangerous non-Gaussianities are tuned to be compatible with observations. UV DBI inflation is almost ruled out due to some problems in its theoretical consistency

Model	ϵ_*	$\lvert\eta_*\rvert$	$0.95 < n_s < 0.98$	n_t	r	f_{NL}	Fine tuning
Moduli inflation	~ 0	$\sim 10^{-2}$	✓	$\lesssim 0$	~ 0	Small	$10^0 - 10^{-3}$
Axion inflation	$\sim 10^{-2}$	$\gtrsim 10^{-2}$	✓	$\sim -10^{-2}$	$\lesssim 0.001$	Small?	$10^{-2} - 10^{-4}$
Monodromy inflation	$\sim 10^{-2}$	$\gtrsim 10^{-2}$	✓	$\sim -10^{-2}$	$0.01 - 0.1$	Small	10^{-2}
Warped D-brane inflation	~ 0	$\sim 10^{-2}$	✓	$\lesssim 0$	~ 0	Small	$10^{-2} - 10^{-3}$
UV DBI inflation	$\sim 10^{-1}$	$\sim 10^{-1}$	✓	$\sim -10^{-1}$	$\gtrsim 0.01$	Large	10^{-2}
IR DBI inflation	~ 0	$\sim 10^{-2}$	✓	$\lesssim 0$	~ 0	Large	$10^{-1} - 10^{-2}$
String gas	–	–	✓	$\sim +10^{-2}$	$\lesssim 0.1$	Small	10^{-2}
Ekpyrotic universe (ii)	$\sim 10^2$	$\sim 10^2$	✓	$\gtrsim 2$	~ 0	Large	$10^{-1} - 10^{-2}$

examples based on concrete Calabi–Yau spaces and, in the absence of an explicit construction, an unspecified number of moduli is usually assumed to be frozen. For the same reason, there is some margin of freedom in the choice of the super- and Kähler potentials. The making of Calabi–Yau spaces populated by Klebanov–Strassler throats with a suitable brane population must pass through a series of technical steps and consistency checks, including the stability of supergravity solutions.[23] The multi-throat, multi-brane scenario of IR DBI inflation is only qualitative and still under study. One of the consequences is that the potential attracting the D-branes out of their throats is phenomenological.

Despite all their limitations, models of string cosmology are rigid enough to be falsifiable by current experiments. As a matter of fact, a number of them have already been ruled out in the last decade, not long after their birth.

While all the viable models pass the test of near scale invariance of curvature perturbations, they greatly differ in the tensor sector. Moduli inflation, warped D-brane inflation, IR DBI inflation and the ekpyrotic universe predict a negligible tensor-to-scalar ratio, either because the amplitude of primordial gravitational waves is too low or due to a strong blue tilt. A detection of a non-zero $r = O(10^{-2})$ would rule out these models, while it would restrict the parameter space of string-gas cosmology. The level of non-Gaussianity of DBI inflation and of the ekpyrotic universe is typically large and current estimates of the non-linear parameter $f_{\rm NL}$ place tight constraints on the parameters.

These predictions can be assessed within the bigger picture of the string landscape. Hoping to confront string theory with experiments soon, one may wonder whether large-r scenarios are "more common" than the others. We do not know the answer yet but a preliminary counting of the relative frequency of large-field and small-field models indicates that string theory has no preference towards a large or small tensor-to-scalar ratio r [56]. However, this estimate has been made locally in a mini-landscape of controllable compactification models which does not encompass all the cases in the table.

The issue of fine tuning is also open. There are indications that the initial conditions chosen for inflation may be not so special, at least in the mini-landscape of one model of volume-modulus inflation [57]. Counting the number of viable

[23]The inclusion of $\overline{\rm D}$-branes in the throat is a delicate issue, since the stability of the de Sitter vacuum may be compromised by the brane back-reaction [592–600]. Supergravity solutions describing branes in a throat have, in general, a divergent flux density. If the singularity is not physical, it indicates that the configuration is not static: the anti-branes annihilate with the fluxes. In this case, one cannot uplift the potential with the $\mathbb{KLT}$ mechanism and the viability of $\mathbb{KLMT}$ inflation, based exquisitely on the supergravity approximation of string theory, is put to question. This problem is not settled in stone for several reasons. On one hand, it is possible (but very difficult to prove, technically) that stable branes-in-a-throat configurations could be realized beyond the SUGRA low-energy limit. On the other hand, we have seen that one can realize D-brane inflation with widely different brane configurations (even without $\overline{\rm D}$-branes), some of which may be free from singularities even if the original $\mathbb{KLMT}$ set-up were in danger. Third, there are arguments which regard these flux singularities as physical and, hence, resolvable [601–603]. In this case, the de Sitter vacuum would be metastable and D-brane inflation could safely take place.

configurations for type-IIB F-term monodromy inflation shows that these models should be realized with enough frequency in the landscape [135]. The mapping of the landscape of multi-axion potentials similar to (13.39) is in progress [604]. The numbers in the last column of Table 13.1 should then be interpreted with care.

To conclude, cosmological string models have enough predictive power to go beyond simple compatibility checks with present bounds. The challenge they have to face, also in the light of upcoming or near-future data on the tensor spectrum and on non-Gaussianity, is to give robust theoretical motivations to the ever-shrinking viable parameter space.

13.9 Big-Bang Problem

Having discussed how inflation is realized in string theory and string-inspired scenarios, and how string theory can address the cosmological constant problem, we turn to the big-bang problem. The subject is surrounded by a perhaps inevitable halo of incompleteness due to the very fragmentary information collected from the theory during the years. Still, it is important to assess this information and place it side by side, either as a follow-up or as an alternative path, with the results on classical singularities in Chap. 6 and with the big-bang resolution proposed by loop quantum cosmology in Sect. 10.3. After summarizing the study of singularities in string theory, we divide the rest of the section in two parts, which cover scales larger or smaller than the string scale $l_s = \sqrt{\alpha'}$: a classical chaotic approach to the putative singularity in the low-energy limit, valid for scales $> l_s$ (Sect. 13.9.2), and a quantum approach at scales $\lesssim l_s$ (Sect. 13.9.3).

13.9.1 Big Bang in String Theory

In this chapter, we have seen three mechanisms that could remove the cosmic singularity:

- *Non-locality.* Non-locality is implemented as exponential differential operators $\exp(\alpha'\Box)$ in an effective low-energy dynamics (Sect. 13.7.4, non-perturbative in α' and g_s).
- *T-duality.* T-duality is implemented in the FLRW solutions of pre-big-bang cosmology (Sect. 13.7.5, perturbative in α' and g_s) and in string-gas cosmology (Sect. 13.7.6);
- *Ekpyrotic mechanism.* The colliding branes of the ekpyrotic paradigm (Sect. 13.7.7, perturbative in α' and non-perturbative in g_s).

All these frameworks have some drawbacks. Realistic non-local cosmological models have not yet been derived from full string field theory. Both the pre-big-bang and the ekpyrotic scenarios have been developed much more extensively than non-local models but they are not quite embedded in a flux compactification scheme of moduli stabilization.

Other approaches probe different corners of the string parameter space and open widely different views on the big-bang problem (see [605–608] for reviews).

- *Orbifold singularities.* For instance, in contrast with non-local scenarios and the ekpyrotic solution of [559], non-perturbative in α', time-dependent orbifolds with a space-like or light-like singularity have been used as models of cosmological singularities[24] in perturbative string theory (in α' and g_s) [609–619]. In a class of cases, divergences in scattering amplitudes and various instabilities can arise and signal a fundamental singularity in the geometry. At the intuitive level of general relativity, near the big-bang singularity the kinetic terms in the non-linear sigma model (12.37) are suppressed (some coefficients in $\bar{g}_{\mu\nu}$ and $B_{\mu\nu}$ go to zero), leading to non-suppressed terms in the world-sheet path integral (i.e., violent fluctuations of the fields) and to divergent amplitudes in the genus expansion. However, ingredients such as orientifold planes [616] or the contribution of twisted states [617, 619] can resolve the singularity of specific orbifold models into a smooth bounce.
- *Tachyon condensation.* A positive outcome is also achieved in another type-II or heterotic setting perturbative in α' and g_s [620]. Closed-string tachyon condensation introduces action terms in (12.37) that grow towards the space-like singularity, thus balancing the suppression of the non-linear sigma model and controlling field fluctuations. As a consequence, the big bang is replaced by a phase characterized by a thermal distribution of closed-string modes and where spacetime simply ends through a topology change [621, 622]. This "Nothing state" [620, 623, 624] is a string realization of the Hartle–Hawking no boundary proposal (Sect. 9.2.3). The same mechanism has been utilized also to resolve the singularity inside black holes [625].
- *AdS/CFT correspondence.* Going beyond perturbation theory and using non-perturbative (in g_s) techniques such as the AdS/CFT correspondence, it was found that black-hole space-like singularities are resolved in $D = 3$ [626], while the $D > 3$ case is more complicated [627, 628]. Contrary to the cosmological singularity, this one is hidden beyond an event horizon. To obviate this problem, solutions in $\mathcal{N} = 8$ SUGRA compactified to $D = 4$ and $D = 5$ have been constructed such that smooth, asymptotically AdS initial data evolve, without fine tuning, to a space-like big-crunch singularity [629]. This can open up the study of the naked big-bang singularity in these regimes [630, 631]. Examples of light-like singularities in a type-IIB AdS bulk show hints of a resolution in the super-Yang–Mills sector of the theory, which is particularly well behaved [632, 633].

[24]In particular, the two-dimensional Misner space $\mathbb{R}^{1,1}$/boost has been a popular subject of study, where $\mathbb{R}^{1,1} = M_2$ is $D = 2$ Minkowski spacetime and "boost" is a finite boost transformation of the light-cone coordinates.

- *Matrix theory.* In the matrix incarnation of M-theory, as g_s increases one promotes a light-like linear-dilaton background of type-IIA string theory to a solution of M-theory [634, 635]. Both solutions have a light-like singularity but, in the $D = 11$ extension, near the singularity the notion of spacetime is replaced by that of non-commuting matrices. The physics at the big bang turns out to be under control because the Yang–Mills gauge sector is weakly coupled there. Also in this case, the analysis of the singularity is facilitated.

There is another context, perturbative in α' and non-perturbative in g_s, indicating that cosmological singularities may be resolved. It is the study of early-time homogeneous solutions (including the Kasner metric (6.20)) to the Einstein equations in the low-energy limit of M-theory [636, 637]. Compactifying $D = 11$ gravity on hypertori of various dimensions, one notices that the ensuing cosmology has a big-bang singularity which, however, disappears by U-duality in vacuum. (Recall from Sect. 12.4 that U-duality combines S- and T-duality to map $D = 11$ supergravity into weakly-coupled type-II string theories.) While vacuum solutions are U-dual invariant, the matter sector is not. The physical interpretation is that, near the singularities found in the $D = 11$ low-energy bosonic sector, matter decays into new states that cannot be captured by the SUGRA field theory and that require a dual, more efficient description in terms of type-II string modes. This is the subject of the rest of the chapter.

13.9.2 Classical Billiards: Living with the Singularity

In Sect. 6.3.5, we had occasion to remark that the BKL oscillatory behaviour in general relativity is typical for 9 or fewer spatial dimensions, while for $D = 10 + 1$ the approach to the singularity is monotonic. In Sect. 12.4, we met again the magic value $D = 11$, this time in the context of supergravity and M-theory. Before looking for a singularity resolution in string theory, it is interesting to explore this coincidence of numbers and check whether the BKL behaviour detailed in Sect. 6.3 persists in supergravity, i.e., in the low-energy limit of string theories and M-theory. Such is the subject of this section.

A central part of the discussion will be based on a characterization of the BKL singularity we have not explicitly used in Sects. 6.3.4–6.3.5, but that is hidden in the formulæ therein. Instead as "oscillations," the chaotic sequence of Kasner epochs can be described as collisions on the walls of a *billiard* [638]. The term billiard in cosmological applications was introduced in the case of $D = 4$ general relativity in vacuum [639] and, more extensively, in D dimensions either without matter or with several perfect fluids [640–642] (other early works are cited in Sect. 6.3.4.1). The most interesting case for string theory is when matter is constituted by several p-forms [643–650] and fermions [651–656]. Reference [657] is a comprehensive introduction and summary to general billiards and their applications to string and M-theory.

To check whether space-like singularities are approached chaotically, consider first the D-dimensional toy model [644]

$$S = \frac{1}{2}\int d^D x\,\sqrt{-g}\left[\frac{R}{\kappa_D^2} - \partial_\mu\phi\partial^\mu\phi - \sum_n e^{\lambda_n\kappa_D\phi}F_{n+1}^2\right], \tag{13.119}$$

where λ_n is dimensionless and F_{n+1} is the field strength of an n-form A_n (we use n instead of p to avoid confusion in what follows). The action (13.119) is a schematic representation of several cases. For $D = 11$, one has the low-energy M-theory (SUGRA) action (12.128), with $\phi = 0$ and $n = 3$. For $D = 10$ and $\phi \propto \Phi$ the dilaton, one gets the massless bosonic sector of the low-energy string-theory actions in the Einstein frame, including type-I theory ($n = 2$), the heterotic theories ($n = 2$) and the transformed versions of both the type-IIA action (12.74) ($n = 1, 2, 3$) and the type-IIB action (12.75) ($n = 0, 2, 2, 4$). Other terms such as the Chern–Simons action and interactions among the n-forms do not change the results below [649]. The values and signs of the couplings λ_n are given in [644] (or can be derived by the reader for type-II theories).

In vacuum, the Kasner metric (6.20)–(6.21) and the profile $\kappa_D\phi = \varphi(\mathbf{x}) + p_\phi \ln(t/t_0)$ are solutions to the Einstein equations stemming from (13.119), provided the generalization of conditions (6.25) holds:

$$\sum_{i=1}^{D-1} p_i = 1\,, \qquad p_\phi^2 + \sum_{i=1}^{D-1} p_i^2 = 1\,. \tag{13.120}$$

The singularity at $t = 0$ is approached monotonically if the Kasner solution is stable against the contributions $\propto \sum_{n,I} t^{2w_n^I}$ of the n-forms to the Einstein equations and those $\propto \sum_J t^{2w_\nabla^J}$ of the spatial derivatives of the metric, where I and J are generic labels. This happens if such contributions drop faster than the $\ddot{g}_{\mu\nu}/g_{\mu\nu}+(\partial\phi)^2 \propto t^{-2}$ leading terms in the limit $t \to 0$, which is guaranteed if all the exponents w_n^I and w_∇^J are positive. Adopting a generalized Kasner metric (6.19) and the same *Ansätze* above for a_i and ϕ (where now, approximately, the exponents $p_i(\mathbf{x})$ and $p_\phi(\mathbf{x})$ depend on spatial coordinates), the exponents w_∇^J have been computed in [658–660] and actually have a triple index:

$$w_\nabla^{ijk} = 1 + p_i - p_j - p_k > 0\,. \tag{13.121}$$

For every n in the spectrum, the n-form contribution is split into two parts (one "electric" and the other "magnetic") which are sub-dominant if [644]

$$w_n^{i_1\dots i_n} = p_{i_1} + \dots + p_{i_n} - \tfrac{1}{2}\lambda_n p_\phi > 0\,, \tag{13.122a}$$

$$w_n^{j_1\dots j_{D-2-n}} = p_{j_1} + \dots + p_{j_{D-2-n}} + \tfrac{1}{2}\lambda_n p_\phi > 0\,, \tag{13.122b}$$

where the indices in each expression are all different. Overall, one can define the set of exponents

$$w_K = \{w_\nabla^{ijk},\, w_n^{i_1\ldots i_n},\, w_n^{j_1\ldots j_{D-2-n}}\}\,, \qquad K = 1,\ldots,K_{\rm tot}\,. \tag{13.123}$$

Their number is typically large. In M-theory, $K_{\rm tot} = 690$.

In the absence of n-forms (Einstein gravity plus a scalar field), the Kasner solution is asymptotically stable for any D [658], while in the absence also of the dilaton the solution is asymptotically stable for $D \geqslant 11$ [659–661]. Turning on just a vector field is sufficient to make the BKL dynamics appear [658]. Adding several n-forms generates a behaviour more complex than the usual BKL approach.

In M-theory, there is only a 3-form and no dilaton and it is possible to show that the electric exponent $\alpha_n^{i_1 i_2 i_3} = p_{i_1} + p_{i_2} + p_{i_3}$ is always negative semi-definite. With S-duality ($D = 11$ SUGRA compactified on a circle), one can map this condition to the type-IIA case, which is in turn mapped to type IIB by T-duality. A direct calculation in heterotic theory and its S-dualization to type-I theory close the web of relations among the theories and eventually lead to the conclusion that none of them admits an asymptotically stable Kasner solution [644]. Monotonic solutions do exist but they are non-generic [645].

Having excluded a monotonic solution in the presence of inhomogeneities and n-forms, one may apply the analysis of the mixmaster dynamics of Sect. 6.3.4 to the string case. Kasner epochs with different exponents $\tilde p^\mu := (\tilde p_\phi, \tilde p_i)$ are joined together at the points where the conditions (13.121)–(13.122) are violated, to produce a pattern of oscillations towards the singularity. Each Kasner epoch is analogous to the free motion of a billiard ball interrupted by collisions against the $K_{\rm tot}$ walls $w_K = 0$ of the pool table, hence the nickname billiards to the model of approach to the singularity.

The law $p^\mu \to \tilde p^\mu(p^\nu)$ can be written down in a covariant form and governs iteratively a chaotic motion on the sphere (13.120). To describe relativistic billiards, we define the ten Einstein-frame variables

$$\beta^i := -\ln a_i \tag{13.124}$$

in M-theory ($i = 1,\ldots,10$) or $\beta^i = -\ln a_i$ and $\beta^{10} := -\kappa_{10}\phi$ in string theories ($i = 1,\ldots,9$). In each Kasner epoch, $\beta^i \simeq -p_i$. In the string frame, $\beta^{10} \to \beta^0 := -\ln(\sqrt{g_s}\mathrm{e}^{-2\Phi})$. Then, in the limit $t \to 0^+$ the action (13.119) evaluated on the generalized Kasner metric (6.19) with scale factors (13.124) can be approximated by [646, 649]

$$S[\beta] = \int \mathrm{d}\tau \left[\mathcal{G}_{\mu\nu}\frac{\mathrm{d}\beta^\mu}{\mathrm{d}\tau}\frac{\mathrm{d}\beta^\nu}{\mathrm{d}\tau} - V(\beta)\right], \tag{13.125}$$

where $\mu = 1,\ldots,10$ or $\mu = 0,\ldots,9$ depending on the frame, $\mathrm{d}\tau = -\mathrm{d}t/\sqrt{g_{D-1}}$ and $\mathcal{G}_{\mu\nu}$ is a metric with Lorentzian signature $(-,+,\cdots,+)$ determined by the model. In $D = 11$ SUGRA, $\mathcal{G}_{\mu\nu}\mathrm{d}\beta^\mu\mathrm{d}\beta^\nu = \sum_\mu(\beta^\mu)^2 - (\sum_\mu \beta^\mu)^2$, while in string theories $\mathcal{G}_{\mu\nu} = \eta_{\mu\nu}$ in the string frame. The exponents (13.123) can be rewritten as the scalar product $w_K = w_K(\beta) := \mathcal{G}_{\mu\nu} w_K^\mu \beta^\nu$ between the model-dependent vectors β^μ and w_K^μ. All the walls have space-like gradients, $\mathcal{G}_{\mu\nu} w_K^\mu w_K^\nu > 0$. The effective potential V reads[25]

$$V(\beta) \simeq \sum_{K=1}^{K_{\rm tot}} c_K \, e^{-2w_K(\beta)}, \tag{13.126}$$

where $c_K > 0$ are approximately constant coefficients all positive in the models of interest; therefore, all the walls are repulsive.

Fortunately, most of the walls are not involved in the billiard motion when approaching the singularity and they can be neglected to a first approximation. It turns out that only 10 walls out of $O(700)$ play a relevant role in this otherwise tremendously complicated game. We will denote these walls by $\alpha_I(\beta) = \mathcal{G}_{\mu\nu} \alpha_I^\mu \beta^\nu$, $I = 1, \dots, 10$. For the block of $\mathcal{N} = 2$ supergravities (M-theory, type-IIA and type-IIB superstrings), they are

$$\alpha_i(\beta) = \beta^{i+1} - \beta^i\,, \qquad i = 1, \dots, 9\,, \tag{13.127a}$$

$$\alpha_{10}(\beta) = \beta^1 + \beta^2 + \beta^3\,, \tag{13.127b}$$

while for the block of $\mathcal{N} = 1$ supergravities (type-I, heterotic $SO(32)$ and heterotic $E_8 \times E_8$ superstrings) the relevant walls in the string frame are

$$\alpha_1(\beta) = \beta^1\,, \qquad \alpha_i(\beta) = \beta^i - \beta^{i-1}\,, \qquad i = 2, \dots, 9\,, \tag{13.128a}$$

$$\alpha_{10}(\beta) = \beta^0 - (\beta^7 + \beta^8 + \beta^9)\,. \tag{13.128b}$$

When normalized to 1, the relevant walls (13.127) and (13.128) define unit vectors normal to the faces of a simplex in the 9-dimensional hyperbolic space H^9 (loosely speaking, a maximally symmetric Riemannian manifold with constant negative curvature). The angles between faces are given by the scalar products $\alpha_I \cdot \alpha_{I'} = \mathcal{G}_{\mu\nu} \alpha_I^\mu \alpha_{I'}^\nu$. The set $\{\alpha_I\}$ forms a basis of simple roots for a rank-10 hyperbolic Kac–Moody algebra, while the vectors β^μ parametrize the associated Cartan sub-algebra.[26] For the $\mathcal{N} = 2$ and $\mathcal{N} = 1$ SUGRA blocks, this algebra is, respectively, E_{10} and BE_{10} [646]. The reflections in the walls of the cosmic billiard form a group corresponding to the Weyl group of E_{10} or BE_{10}, i.e., the sub-group of the isometry group of the root system $\{\alpha_I\}$ generated by reflections with respect to the

[25]The dynamics (13.125) with an exponential potential of the form (13.126) is said to be Toda-like and has been considered in classical and quantum cosmology since the first descriptions of billiard systems [638]; see also [662–664].

[26]This sudden escalation in the terminology would deserve a digression for which we lack space here. Instead, we refer to [649, 657] for a brief introduction to Kac–Moody algebras and to [665] for an in-depth study, in particular Sects. 6.2, 6.4–6.7 and 7.8 therein.

hyperplanes orthogonal to the root vectors. From these properties, it is possible to show that the evolution towards the singularity is chaotic in a precise mathematical sense [646]. The emergence of chaos relies on the facts that the billiard H^9 is hyperbolic and has a finite volume.

More generic billiards and their underlying algebras can be classified according to the content of the action and to the dimensionality of spacetime [647, 649, 657]. Dimensional reduction of any given model preserves its chaotic properties. The bosonic sector of $D = 4$, $\mathcal{N} = 1, 2, 3, 4, 5, 6, 8$ SUGRA has been studied in [650]. Fermions have been included in [651–656].

The physical interpretation of the billiard in M-theory is that of an ensemble of interacting branes [666–671]. This result relies on an extension of the dynamics of the above simple billiard model with infinite-potential dominant walls. Including also the walls neglected at first and, going beyond the Cartan sub-algebra, all the roots of E_{10}, one finds a correspondence between the roots of E_{10} and the content of M-theory. Arguments in $D = 11$ SUGRA compactified on hypertori and various orbifolds suggest the following picture. The walls of the billiard correspond to background fluxes, with the orientation and position of each wall determined by the type and magnitude of the associated flux. Fluxes are changed by discrete units via instantons, among which we can find Kaluza–Klein particles and Euclidean M2- and M5-branes. Real roots of E_{10} (i.e., those such that $\alpha\cdot\alpha > 0$; they are all mapped into simple roots by a transformation $\alpha \to \alpha_I$ in the Weyl group) represent the instantons changing the fluxes of the model or, in orbifold compactifications, extra D-branes contained in the twisted sector. Imaginary roots ($\alpha \cdot \alpha \leqslant 0$; they are not mappable to simple roots through the Weyl group) correspond to physical branes with Lorentzian world-volume. Inner products $\alpha \cdot \alpha'$ encode brane interactions.[27]

13.9.3 Quantum Billiards: Avoiding the Singularity?

The analysis of Sect. 13.9.2 relies on low-energy actions and is, therefore, classical and perturbative in α'. It is thus expected to be modified, or to break down altogether, at time scales $t \lesssim t_s = \sqrt{\alpha'}$ of order of the string scale. Assuming $t_s = t_{\rm GUT} \approx 10^{-40}$ s, the number of collisions (the analogue of the BKL oscillations in Sect. 6.3.4) from today $t_0 \approx 10^{18}$ s until t_s is about

$$\mathcal{N}_{\rm coll} \simeq \ln\tau \simeq \ln\left(\ln\frac{t_0}{t_s}\right) \approx 5\,. \tag{13.129}$$

[27] In a complementary approach, real and imaginary roots have been identified with, respectively, the spatial gradients of SUGRA fields and the higher-order corrections $O(R^m)$ to the leading-α'-order low-energy action [648, 657, 672, 673].

Although $\mathcal{N}_{\rm coll}$ is quite small, the mixing properties of the system are, to quote the suggestive words of Damour and Henneaux [646], "enough for churning up the fabric of spacetime and transforming any [...] patch of space into a turbulent foam at $t_{\rm s}$." Below the string scale, non-perturbative and quantum corrections to the low-energy dynamics could spoil the chaotic model of billiards and even remove the singularity.

A first and still preliminary inspection of the problem has been conducted on a model of canonical quantum cosmology on a mini-superspace [674–679]. Consider pure bosonic gravity in D dimensions, no dilatons and the generalized Kasner metric (6.19) with scale factors (13.124) (the role of fermionic partners is discussed in [676]). Take also the decomposition $\beta^i = \rho\omega^i$, where ρ is the radial direction in the future light cone and the $\omega^i(z)$ depend on the coordinates z on the hyperbolic space H^{D-2} [649]. Since the unit hyperboloid is $\omega_i\omega^i = -1$, one has $\rho^2 = -\beta_i\beta^i > 0$. The classical Hamiltonian $\mathcal{H}_{\rm SUGRA}$ in the canonical variables ρ and ω^i and their momenta can be found readily [649]. The Wheeler–DeWitt equation stems from a canonical quantization of $\mathcal{H}_{\rm SUGRA}$:

$$\hat{\mathcal{H}}\Psi[\rho, z] = 0\,, \qquad \hat{\mathcal{H}} \simeq G^{ij}\partial_i\partial_j = -\rho^{2-D}\partial_\rho\left(\rho^{D-2}\partial_\rho\right) + \rho^{-2}\nabla^2_{H^{D-2}}\,, \tag{13.130}$$

where $\nabla^2_{H^{D-2}}$ is the Laplacian operator in H^{D-2}. The effect of matter fields is approximated by sharp potential walls, which we do not see in (13.130) because the exponential potential terms are negligible in the BKL limit $\beta^i \to +\infty$. As in early studies of cosmological billiards, the walls are encoded in boundary conditions on the wave-function [680]. Separating variables as $\Psi[\rho, z] = r(\rho)f(z)$ and imposing the wave-function to vanish at the boundary, one finds that

$$r(\rho) = \rho^{-\frac{D-3}{2}}\,\mathrm{e}^{\pm\mathrm{i}\sqrt{E-\left(\frac{D-3}{2}\right)^2}\ln\rho}\,, \tag{13.131}$$

where E is the eigenvalue in $\nabla^2_{H^{D-2}}f(z) = -Ef(z)$ and is bounded from below by $E \geqslant [(D-3)/2]^2$.

The wave-function Ψ is complex and oscillating. In general, wave-packets travelling towards the singularity at $\rho \to +\infty$ spread across the whole billiard domain and, noting that $\lim_{\rho\to+\infty}\mathrm{Re}[r(\rho)] = 0^+$ for $D > 3$, the probability density $|\Psi|^2$ vanishes at the singularity. This result looks similar to the familiar case of WDW quantum cosmology with three spatial directions, presented in Sect. 10.2. In particular, in Sect. 10.2.3 we remarked that having $\Psi = 0$ at the singularity is not enough to solve the big-bang problem, mainly because such wave-functions are not typical and the lowest eigenvalue of the volume operator is usually zero. Here the situation is somewhat different because the boundary conditions are fixed by the billiard problem and having $\Psi = 0$ at the singularity seems to be a robust feature across several modifications of (13.130) [676–679].

However, this is not the end of the story. In $D = 11$ SUGRA and its compactifications, one further requires that the scalar product in H^{D-2} is invariant under the action of the Weyl group of E_{10}, which translates into a symmetry condition on the wave-function,

$$\Psi[\rho, z] = \pm\Psi[\rho, \alpha_I \cdot z]\,. \tag{13.132}$$

With this condition imposed on the eigenfunction $f(z)$, Ψ becomes a *Maass waveform*, whose actual precise behaviour near the singularity is still shrouded in mystery. What emerges from this datum and the results below is a purely algebraic structure near the singularity, which suggests a novel scenario for the birth of the Universe: the big bang is neither smoothed out nor skipped, it simply cannot be reached in an ordinary sense [681].

Let us explain. One of the most characteristic features of the BKL conjecture is that spatial points are completely decoupled near the singularity. When approaching the big bang, spacetime becomes nearly homogeneous and it experiences a sort of simplification in the type and number of degrees of freedom. This phase transition is brought to an extreme in higher-dimensional supergravity models. In [648, 657, 672, 673] and the fermionic extension [651–656], evidence has been gathered that, near the singularity ($t \lesssim t_s$) and in the strong-coupling limit $G_D \to \infty$, all the $D = 11$ SUGRA dynamical fields (including the n-forms and the spatial metric g_{ij}) are replaced by just one degree of freedom in one dimension, a homogeneous dynamical variable $\upsilon(t) \in E_{10}$. Denoting with $\mathcal{E}_{10}$ the group related to the algebra E_{10}, the group element associated with υ is $\nu(t) = \exp[\int \mathrm{d}t\, \upsilon(t)] \in \mathcal{E}_{10}$. The one-parameter dynamics at scales $t \lesssim t_s$ is governed by the action

$$\mathcal{S}[\nu] = \int \frac{\mathrm{d}t}{n(t)} \langle \mathcal{P}(t)|\mathcal{P}(t)\rangle\,, \qquad \mathcal{P}(t) := \frac{1}{2}\left[\frac{\mathrm{d}\ln\nu(t)}{\mathrm{d}t} + \left(\frac{\mathrm{d}\ln\nu(t)}{\mathrm{d}t}\right)^{\mathrm{t}}\right], \tag{13.133}$$

where $n(t)$ is the lapse function, $\langle\cdot|\cdot\rangle$ is the standard bilinear form on $E_{10}/K(E_{10})$ and the suffix t denotes the transpose. The solution $\upsilon(t)$ to the dynamics parametrizes a null-geodesic motion in the infinite-dimensional quotient space $\mathcal{E}_{10}/K(\mathcal{E}_{10})$, where $K(\mathcal{E}_{10})$ is the maximal compact sub-group of $\mathcal{E}_{10}$. Elements ek of $\mathcal{E}_{10}/K(\mathcal{E}_{10})$ are group elements of $\mathcal{E}_{10} \ni e$ left-invariant under $K(\mathcal{E}_{10}) \ni k$. Equation (13.133) is known as the $E_{10}/K(E_{10})$ *coset model*. The classical Hamiltonian $\mathcal{H}_{\rm coset}$ associated with (13.133) is formally identical to that of $D = 11$ SUGRA with a 3-form, although written with different variables. By studying the fundamental billiard model with simple roots and its extension to dozens of non-fundamental extra walls, and including quantum corrections to the SUGRA action, it is possible to establish a dynamics-preserving map between the SUGRA variables and the coset-space canonical variables. In particular, one can write $\nu(t) = \beta^i h_i(t)$ for some specific $h_i(t)$. A notable feature of the dynamics is that $\mathcal{H}_{\rm coset}$ vanishes only if the geodesics are future-oriented, which implies that trajectories cannot go back in β-space.

From the properties of the $E_{10}/K(E_{10})$ coset model, a new picture of the cosmic singularity emerges [681]. As one gets close to the big bang, space dissolves into a time-dependent description in terms of a Lie algebra. The notion of intrinsic time disappears upon quantization, when the system (13.130) (and its generalizations) deparametrizes and the role of evolution variable is played by one of the canonical fields.

In WDW quantum cosmology, the classical trajectory in the space of 3-geometries passing through the singularity is smeared at $\rho = +\infty$ and the distribution Ψ vanishes and is continuous therein (Sect. 10.2.3). In LQC, the point $\rho = +\infty$ is altogether removed from this trajectory, since Ψ evolves discretely and jumps over the singular point $\rho = +\infty$ through a cosmic bounce (Sect. 10.3.7). Here, in contrast, the wave-function Ψ neither hits nor jumps through the singularity: spacetime melts away to be replaced by a purely algebraic harmony. This would be an original way to realize background independence in quantum gravity: at the string scale, the Universe would become just symmetry.

In particular, it is possible that E_{10} is a key symmetry not only of supergravity (as first conjectured in [682]) but also of full M-theory, both being characterized by a very rich structure in addition to supersymmetry [683–691]. Once again, the study of the very early Universe in extreme regimes is inextricably entangled with the quest for a fundamental theory.

References

1. S. Kachru, R. Kallosh, A.D. Linde, S.P. Trivedi, de Sitter vacua in string theory. Phys. Rev. D **68**, 046005 (2003). [arXiv:hep-th/0301240]
2. A.R. Frey, M. Lippert, B. Williams, Fall of stringy de Sitter spacetime. Phys. Rev. D **68**, 046008 (2003). [arXiv:hep-th/0305018]
3. C. Escoda, M. Gómez-Reino, F. Quevedo, Saltatory de Sitter string vacua. JHEP **0311**, 065 (2003). [arXiv:hep-th/0307160]
4. C.P. Burgess, R. Kallosh, F. Quevedo, de Sitter string vacua from supersymmetric D-terms. JHEP **0310**, 056 (2003). [arXiv:hep-th/0309187]
5. E.I. Buchbinder, B.A. Ovrut, Vacuum stability in heterotic M theory. Phys. Rev. D **69**, 086010 (2004). [arXiv:hep-th/0310112]
6. J.F.G. Cascales, M.P. García del Moral, F. Quevedo, A.M. Uranga, Realistic D-brane models on warped throats: fluxes, hierarchies and moduli stabilization. JHEP **0402**, 031 (2004). [arXiv:hep-th/0312051]
7. R. Brustein, S.P. de Alwis, Moduli potentials in string compactifications with fluxes: mapping the discretuum. Phys. Rev. D **69**, 126006 (2004). [arXiv:hep-th/0402088]
8. M. Becker, G. Curio, A. Krause, de Sitter vacua from heterotic M-theory. Nucl. Phys. B **693**, 223 (2004). [arXiv:hep-th/0403027]
9. F. Denef, M.R. Douglas, B. Florea, Building a better racetrack. JHEP **0406**, 034 (2004). [arXiv:hep-th/0404257]
10. V. Balasubramanian, P. Berglund, Stringy corrections to Kähler potentials, SUSY breaking, and the cosmological constant problem. JHEP **0411**, 085 (2004). [arXiv:hep-th/0408054]
11. R. Kallosh, A.D. Linde, Landscape, the scale of SUSY breaking, and inflation. JHEP **0412**, 004 (2004). [arXiv:hep-th/0411011]
12. K. Choi, A. Falkowski, H.P. Nilles, M. Olechowski, S. Pokorski, Stability of flux compactifications and the pattern of supersymmetry breaking. JHEP **0411**, 076 (2004). [arXiv:hep-th/0411066]
13. V. Balasubramanian, P. Berglund, J.P. Conlon, F. Quevedo, Systematics of moduli stabilisation in Calabi–Yau flux compactifications. JHEP **0503**, 007 (2005). [arXiv:hep-th/0502058]

14. K. Choi, A. Falkowski, H.P. Nilles, M. Olechowski, Soft supersymmetry breaking in KKLT flux compactification. Nucl. Phys. B **718**, 113 (2005). [arXiv:hep-th/0503216]
15. J.P. Conlon, F. Quevedo, K. Suruliz, Large-volume flux compactifications: moduli spectrum and D3/D7 soft supersymmetry breaking. JHEP **0508**, 007 (2005). [arXiv:hep-th/0505076]
16. S.B. Giddings, A. Maharana, Dynamics of warped compactifications and the shape of the warped landscape. Phys. Rev. D **73**, 126003 (2006). [arXiv:hep-th/0507158]
17. G. Villadoro, F. Zwirner, de Sitter vacua via consistent *D* terms. Phys. Rev. Lett. **95**, 231602 (2005). [arXiv:hep-th/0508167]
18. J.J. Blanco-Pillado, R. Kallosh, A.D. Linde, Supersymmetry and stability of flux vacua. JHEP **0605**, 053 (2006). [arXiv:hep-th/0511042]
19. A. Achúcarro, B. de Carlos, J.A. Casas, L. Doplicher, de Sitter vacua from uplifting D-terms in effective supergravities from realistic strings. JHEP **0606**, 014 (2006). [arXiv:hep-th/0601190]
20. O. Lebedev, H.P. Nilles, M. Ratz, de Sitter vacua from matter superpotentials. Phys. Lett. B **636**, 126 (2006). [arXiv:hep-th/0603047]
21. J.P. Conlon, F. Quevedo, Gaugino and scalar masses in the landscape. JHEP **0606**, 029 (2006). [arXiv:hep-th/0605141]
22. J.P. Conlon, Moduli stabilisation and applications in IIB string theory. Fortsch. Phys. **55**, 287 (2007). [arXiv:hep-th/0611039]
23. A. Westphal, de Sitter string vacua from Kähler uplifting. JHEP **0703**, 102 (2007). [arXiv:hep-th/0611332]
24. M. Berg, M. Haack, E. Pajer, Jumping through loops: on soft terms from large volume compactifications. JHEP **0709**, 031 (2007). [arXiv:0704.0737]
25. J.P. Conlon, C.H. Kom, K. Suruliz, B.C. Allanach, F. Quevedo, Sparticle spectra and LHC signatures for large volume string compactifications. JHEP **0708**, 061 (2007). [arXiv:0704.3403]
26. M. Cicoli, J.P. Conlon, F. Quevedo, General analysis of LARGE volume scenarios with string loop moduli stabilisation. JHEP **0810**, 105 (2008). [arXiv:0805.1029]
27. M. Cicoli, D. Klevers, S. Krippendorf, C. Mayrhofer, F. Quevedo, R. Valandro, Explicit de Sitter flux vacua for global string models with chiral matter. JHEP **1405**, 001 (2014). [arXiv:1312.0014]
28. D. Ciupke, J. Louis, A. Westphal, Higher-derivative supergravity and moduli stabilization. JHEP **1510**, 094 (2015). [arXiv:1505.03092]
29. L. Susskind, The anthropic landscape of string theory, in *Universe or Multiverse?* ed. by B. Carr (Cambridge University Press, Cambridge, 2007). [arXiv:hep-th/0302219]
30. M.R. Douglas, The statistics of string/M theory vacua. JHEP **0305**, 046 (2003). [arXiv:hep-th/0303194]
31. S. Ashok, M.R. Douglas, Counting flux vacua. JHEP **0401**, 060 (2004). [arXiv:hep-th/0307049]
32. T. Banks, M. Dine, E. Gorbatov, Is there a string theory landscape? JHEP **0408**, 058 (2004). [arXiv:hep-th/0309170]
33. M.R. Douglas, B. Shiffman, S. Zelditch, Critical points and supersymmetric vacua. Commun. Math. Phys. **252**, 325 (2004). [arXiv:math/0402326]
34. F. Denef, M.R. Douglas, Distributions of flux vacua. JHEP **0405**, 072 (2004). [arXiv:hep-th/0404116]
35. A. Giryavets, S. Kachru, P.K. Tripathy, On the taxonomy of flux vacua. JHEP **0408**, 002 (2004). [arXiv:hep-th/0404243]
36. L. Susskind, Supersymmetry breaking in the anthropic landscape, in *From Fields to Strings*, ed. by M. Shifman (World Scientific, Singapore, 2005). [arXiv:hep-th/0405189]
37. M.R. Douglas, Statistical analysis of the supersymmetry breaking scale. arXiv:hep-th/0405279.
38. M. Dine, E. Gorbatov, S.D. Thomas, Low energy supersymmetry from the landscape. JHEP **0808**, 098 (2008). [arXiv:hep-th/0407043]
39. B. Freivogel, L. Susskind, Framework for the string theory landscape. Phys. Rev. D **70**, 126007 (2004). [arXiv:hep-th/0408133]

40. M.R. Douglas, Basic results in vacuum statistics. C. R. Phys. **5**, 965 (2004). [arXiv:hep-th/0409207]
41. J.P. Conlon, F. Quevedo, On the explicit construction and statistics of Calabi–Yau flux vacua. JHEP **0410**, 039 (2004). [arXiv:hep-th/0409215]
42. R. Blumenhagen, F. Gmeiner, G. Honecker, D. Lüst, T. Weigand, The statistics of supersymmetric D-brane models. Nucl. Phys. B **713**, 83 (2005). [arXiv:hep-th/0411173]
43. F. Denef, M.R. Douglas, Distributions of nonsupersymmetric flux vacua. JHEP **0503**, 061 (2005). [arXiv:hep-th/0411183]
44. T. Banks, Landskepticism or why effective potentials don't count string models. arXiv:hep-th/0412129
45. N. Arkani-Hamed, S. Dimopoulos, S. Kachru, Predictive landscapes and new physics at a TeV. arXiv:hep-th/0501082
46. B.S. Acharya, F. Denef, R. Valandro, Statistics of M theory vacua. JHEP **0506**, 056 (2005). [arXiv:hep-th/0502060]
47. M.R. Douglas, B. Shiffman, S. Zelditch, Critical points and supersymmetric vacua, III: string/M models. Commun. Math. Phys. **265**, 617 (2006). [arXiv:math-ph/0506015]
48. J. Gomis, F. Marchesano, D. Mateos, An open string landscape. JHEP **0511**, 021 (2005). [arXiv:hep-th/0506179]
49. F. Gmeiner, R. Blumenhagen, G. Honecker, D. Lüst, T. Weigand, One in a billion: MSSM-like D-brane statistics. JHEP **0601**, 004 (2006). [arXiv:hep-th/0510170]
50. K.R. Dienes, Statistics on the heterotic landscape: gauge groups and cosmological constants of four-dimensional heterotic strings. Phys. Rev. D **73**, 106010 (2006). [arXiv:hep-th/0602286]
51. A. Ceresole, G. Dall'Agata, A. Giryavets, R. Kallosh, A.D. Linde, Domain walls, near-BPS bubbles, and probabilities in the landscape. Phys. Rev. D **74**, 086010 (2006). [arXiv:hep-th/0605266]
52. M.R. Douglas, W. Taylor, The landscape of intersecting brane models. JHEP **0701**, 031 (2007). [arXiv:hep-th/0606109]
53. F. Gmeiner, G. Honecker, Millions of standard models on $\mathbb{Z}_6'$? JHEP **0807**, 052 (2008). [arXiv:0806.3039]
54. H.P. Nilles, S. Ramos-Sánchez, M. Ratz, P.K.S. Vaudrevange, From strings to the MSSM. Eur. Phys. J. C **59**, 249 (2009). [arXiv:0806.3905]
55. C. Asensio, A. Seguí, Applications of an exact counting formula in the Bousso–Polchinski landscape. Phys. Rev. D **82**, 123532 (2010). [arXiv:1003.6011]
56. A. Westphal, Tensor modes on the string theory landscape. JHEP **1304**, 054 (2013). [arXiv:1206.4034]
57. J.J. Blanco-Pillado, M. Gómez-Reino, K. Metallinos, Accidental inflation in the landscape. JCAP **1302**, 034 (2013). [arXiv:1209.0796]
58. A.P. Braun, T. Watari, Distribution of the number of generations in flux compactifications. Phys. Rev. D **90**, 121901 (2014). [arXiv:1408.6156]
59. Y.-H. He, V. Jejjala, L. Pontiggia, Patterns in Calabi–Yau distributions. arXiv:1512.01579
60. M. Graña, Flux compactifications in string theory: a comprehensive review. Phys. Rep. **423**, 91 (2006). [arXiv:hep-th/0509003]
61. M.R. Douglas, S. Kachru, Flux compactification. Rev. Mod. Phys. **79**, 733 (2007). [arXiv:hep-th/0610102]
62. R. Blumenhagen, B. Körs, D. Lüst, S. Stieberger, Four-dimensional string compactifications with D-branes, orientifolds and fluxes. Phys. Rep. **445**, 1 (2007). [arXiv:hep-th/0610327]
63. F. Denef, M.R. Douglas, S. Kachru, Physics of string flux compactifications. Ann. Rev. Nucl. Part. Sci. **57**, 119 (2007). [arXiv:hep-th/0701050]
64. F. Denef, Course 12 – Lectures on constructing string vacua. Les Houches **87**, 483 (2008). [arXiv:0803.1194]
65. R. Bousso, J. Polchinski, Quantization of four-form fluxes and dynamical neutralization of the cosmological constant. JHEP **0006**, 006 (2000). [arXiv:hep-th/0004134]

66. J.L. Feng, J. March-Russell, S. Sethi, F. Wilczek, Saltatory relaxation of the cosmological constant. Nucl. Phys. B **602**, 307 (2001). [arXiv:hep-th/0005276]
67. N. Kaloper, L. Sorbo, Where in the string landscape is quintessence? Phys. Rev. D **79**, 043528 (2009). [arXiv:0810.5346]
68. S. Panda, Y. Sumitomo, S.P. Trivedi, Axions as quintessence in string theory. Phys. Rev. D **83**, 083506 (2011). [arXiv:1011.5877]
69. G. Gupta, S. Panda, A.A. Sen, Observational constraints on axions as quintessence in string theory. Phys. Rev. D **85**, 023501 (2012). [arXiv:1108.1322]
70. J.P. Conlon, F. Quevedo, Kähler moduli inflation. JHEP **0601**, 146 (2006). [arXiv:hep-th/0509012]
71. J.R. Bond, L. Kofman, S. Prokushkin, P.M. Vaudrevange, Roulette inflation with Kähler moduli and their axions. Phys. Rev. D **75**, 123511 (2007). [arXiv:hep-th/0612197]
72. Z. Lalak, D. Langlois, S. Pokorski, K. Turzynski, Curvature and isocurvature perturbations in two-field inflation. JCAP **0707**, 014 (2007). [arXiv:0704.0212]
73. M. Cicoli, C.P. Burgess, F. Quevedo, Fibre inflation: observable gravity waves from IIB string compactifications. JCAP **0903**, 013 (2009). [arXiv:0808.0691]
74. J.J. Blanco-Pillado, D. Buck, E.J. Copeland, M. Gómez-Reino, N.J. Nunes, Kähler moduli inflation revisited. JHEP **1001**, 081 (2010). [arXiv:0906.3711]
75. C.P. Burgess, M. Cicoli, M. Gómez-Reino, F. Quevedo, G. Tasinato, I. Zavala, Non-standard primordial fluctuations and nongaussianity in string inflation. JHEP **1008**, 045 (2010). [arXiv:1005.4840]
76. M. Cicoli, A. Mazumdar, Reheating for closed string inflation. JCAP **1009**, 025 (2010). [arXiv:1005.5076]
77. M. Cicoli, F.G. Pedro, G. Tasinato, Poly-instanton inflation. JCAP **1112**, 022 (2011). [arXiv:1110.6182]
78. B.J. Broy, D. Ciupke, F.G. Pedro, A. Westphal, Starobinsky-type inflation from α'-corrections. JCAP **1601**, 001 (2016). [arXiv:1509.00024]
79. C.P. Burgess, M. Cicoli, S. de Alwis, F. Quevedo, Robust inflation from fibrous strings. JCAP **1605**, 032 (2016). [arXiv:1603.06789]
80. M. Cicoli, F. Muia, P. Shukla, Global embedding of fibre inflation models. arXiv:1611.04612
81. A.D. Linde, A. Westphal, Accidental inflation in string theory. JCAP **0803**, 005 (2008). [arXiv:0712.1610]
82. J.P. Conlon, R. Kallosh, A.D. Linde, F. Quevedo, Volume modulus inflation and the gravitino mass problem. JCAP **0809**, 011 (2008). [arXiv:0806.0809]
83. M. Cicoli, F. Muia, F.G. Pedro, Microscopic origin of volume modulus inflation. JCAP **1512**, 040 (2015). [arXiv:1509.07748]
84. Z. Lalak, G.G. Ross, S. Sarkar, Racetrack inflation and assisted moduli stabilisation. Nucl. Phys. B **766**, 1 (2007). [arXiv:hep-th/0503178]
85. L. Alabidi, D.H. Lyth, Inflation models and observation. JCAP **0605**, 016 (2006). [arXiv:astro-ph/0510441]
86. J.J. Blanco-Pillado, C.P. Burgess, J.M. Cline, C. Escoda, M. Gómez-Reino, R. Kallosh, A.D. Linde, F. Quevedo, Racetrack inflation. JHEP **0411**, 063 (2004). [arXiv:hep-th/0406230]
87. J.J. Blanco-Pillado, C.P. Burgess, J.M. Cline, C. Escoda, M. Gómez-Reino, R. Kallosh, A.D. Linde, F. Quevedo, Inflating in a better racetrack. JHEP **0609**, 002 (2006). [arXiv:hep-th/0603129]
88. R. Kallosh, On inflation in string theory. Lect. Notes Phys. **738**, 119 (2008). [arXiv:hep-th/0702059]
89. R. Kallosh, N. Sivanandam, M. Soroush, Axion inflation and gravity waves in string theory. Phys. Rev. D **77**, 043501 (2008). [arXiv:0710.3429]
90. S. Dimopoulos, S. Kachru, J. McGreevy, J.G. Wacker, N-flation. JCAP **0808**, 003 (2008). [arXiv:hep-th/0507205]
91. R. Easther, L. McAllister, Random matrices and the spectrum of N-flation. JCAP **0605**, 018 (2006). [arXiv:hep-th/0512102]
92. S.A Kim, A.R. Liddle, N-flation: multi-field inflationary dynamics and perturbations. Phys. Rev. D **74**, 023513 (2006). [arXiv:astro-ph/0605604]

93. Y.-S. Piao, Perturbation spectra of "N-flation". Phys. Rev. D **74**, 047302 (2006). [arXiv:gr-qc/0606034]
94. S.A Kim, A.R. Liddle, N-flation: non-Gaussianity in the horizon-crossing approximation. Phys. Rev. D **74**, 063522 (2006). [arXiv:astro-ph/0608186]
95. T. Battefeld, R. Easther, Non-Gaussianities in multi-field inflation. JCAP **0703**, 020 (2007). [arXiv:astro-ph/0610296]
96. D. Battefeld, T. Battefeld, Non-Gaussianities in N-flation. JCAP **0705**, 012 (2007). [arXiv:hep-th/0703012]
97. S.A Kim, A.R. Liddle, N-flation: observable predictions from the random matrix mass spectrum. Phys. Rev. D **76**, 063515 (2007). [arXiv:0707.1982]
98. D.R. Green, Reheating closed string inflation. Phys. Rev. D **76**, 103504 (2007). [arXiv:0707.3832]
99. T.W. Grimm, Axion inflation in type II string theory. Phys. Rev. D **77**, 126007 (2008). [arXiv:0710.3883]
100. S.A Kim, A.R. Liddle, D. Seery, Non-Gaussianity in axion N-flation models. Phys. Rev. Lett. **105**, 181302 (2010). [arXiv:1005.4410]
101. S.A Kim, A.R. Liddle, D. Seery, Non-Gaussianity in axion N-flation models: detailed predictions and mass spectra. Phys. Rev. D **85**, 023532 (2012). [arXiv:1108.2944]
102. M. Cicoli, K. Dutta, A. Maharana, N-flation with hierarchically light axions in string compactifications. JCAP **1408**, 012 (2014). [arXiv:1401.2579]
103. T.W. Grimm, Axion inflation in F-theory. Phys. Lett. B **739**, 201 (2014). [arXiv:1404.4268]
104. T. Rudelius, On the possibility of large axion moduli spaces. JCAP **1504**, 049 (2015). [arXiv:1409.5793]
105. J. Brown, W. Cottrell, G. Shiu, P. Soler, Fencing in the swampland: quantum gravity constraints on large field inflation. JHEP **1510**, 023 (2015). [arXiv:1503.04783]
106. J.E. Kim, H.P. Nilles, M. Peloso, Completing natural inflation. JCAP **0501**, 005 (2005). [arXiv:hep-ph/0409138]
107. M. Berg, E. Pajer, S. Sjors, Dante's Inferno. Phys. Rev. D **81**, 103535 (2010). [arXiv:0912.1341]
108. K. Choi, H. Kim, S. Yun, Natural inflation with multiple sub-Planckian axions. Phys. Rev. D **90**, 023545 (2014). [arXiv:1404.6209]
109. T. Higaki, F. Takahashi, Natural and multi-natural inflation in axion landscape. JHEP **1407**, 074 (2014). [arXiv:1404.6923]
110. R. Kappl, S. Krippendorf, H.P. Nilles, Aligned natural inflation: monodromies of two axions. Phys. Lett. B **737**, 124 (2014). [arXiv:1404.7127]
111. I. Ben-Dayan, F.G. Pedro, A. Westphal, Hierarchical axion inflation. Phys. Rev. Lett. **113**, 261301 (2014). [arXiv:1404.7773]
112. C. Long, L. McAllister, P. McGuirk, Aligned natural inflation in string theory. Phys. Rev. D **90**, 023501 (2014). [arXiv:1404.7852]
113. X. Gao, T. Li, P. Shukla, Combining universal and odd RR axions for aligned natural inflation. JCAP **1410**, 048 (2014). [arXiv:1406.0341]
114. I. Ben-Dayan, F.G. Pedro, A. Westphal, Towards natural inflation in string theory. Phys. Rev. D **92**, 023515 (2015). [arXiv:1407.2562]
115. T.C. Bachlechner, C. Long, L. McAllister, Planckian axions in string theory. JHEP **1512**, 042 (2015). [arXiv:1412.1093]
116. T. Rudelius, Constraints on axion inflation from the weak gravity conjecture. JCAP **1509**, 020 (2015). [arXiv:1503.00795]
117. M. Montero, A.M. Uranga, I. Valenzuela, Transplanckian axions!? JHEP **1508**, 032 (2015). [arXiv:1503.03886]
118. A. Hebecker, P. Mangat, F. Rompineve, L.T. Witkowski, Winding out of the swamp: evading the weak gravity conjecture with F-term winding inflation? Phys. Lett. B **748**, 455 (2015). [arXiv:1503.07912]
119. E. Palti, On natural inflation and moduli stabilisation in string theory. JHEP **1510**, 188 (2015). [arXiv:1508.00009]

120. J. Brown, W. Cottrell, G. Shiu, P. Soler, On axionic field ranges, loopholes and the weak gravity conjecture. JHEP **1604**, 017 (2016). [arXiv:1504.00659]
121. E. Silverstein, A. Westphal, Monodromy in the CMB: gravity waves and string inflation. Phys. Rev. D **78**, 106003 (2008). [arXiv:0803.3085]
122. L. McAllister, E. Silverstein, A. Westphal, Gravity waves and linear inflation from axion monodromy. Phys. Rev. D **82**, 046003 (2010). [arXiv:0808.0706]
123. R. Flauger, L. McAllister, E. Pajer, A. Westphal, G. Xu, Oscillations in the CMB from axion monodromy inflation. JCAP **1006**, 009 (2010). [arXiv:0907.2916]
124. S. Hannestad, T. Haugbølle, P.R. Jarnhus, M.S. Sloth, Non-Gaussianity from axion monodromy inflation. JCAP **1006**, 001 (2010). [arXiv:0912.3527]
125. R. Flauger, E. Pajer, Resonant non-Gaussianity. JCAP **1101**, 017 (2011). [arXiv:1002.0833]
126. X. Dong, B. Horn, E. Silverstein, A. Westphal, Simple exercises to flatten your potential. Phys. Rev. D **84**, 026011 (2011). [arXiv:1011.4521]
127. J.P. Conlon, Brane-antibrane backreaction in axion monodromy inflation. JCAP **1201**, 033 (2012). [arXiv:1110.6454]
128. H. Peiris, R. Easther, R. Flauger, Constraining monodromy inflation. JCAP **1309**, 018 (2013). [arXiv:1303.2616]
129. F. Marchesano, G. Shiu, A.M. Uranga, F-term axion monodromy inflation. JHEP **1409**, 184 (2014). [arXiv:1404.3040]
130. R. Blumenhagen, E. Plauschinn, Towards universal axion inflation and reheating in string theory. Phys. Lett. B **736**, 482 (2014). [arXiv:1404.3542]
131. A. Hebecker, S.C. Kraus, L.T. Witkowski, D7-brane chaotic inflation. Phys. Lett. B **737**, 16 (2014). [arXiv:1404.3711]
132. M. Arends, A. Hebecker, K. Heimpel, S.C. Kraus, D. Lüst, C. Mayrhofer, C. Schick, T. Weigand, D7-brane moduli space in axion monodromy and fluxbrane inflation. Fortsch. Phys. **62**, 647 (2014). [arXiv:1405.0283]
133. L. McAllister, E. Silverstein, A. Westphal, T. Wrase, The powers of monodromy. JHEP **1409**, 123 (2014). [arXiv:1405.3652]
134. R. Blumenhagen, D. Herschmann, E. Plauschinn, The challenge of realizing F-term axion monodromy inflation in string theory. JHEP **1501**, 007 (2015). [arXiv:1409.7075]
135. A. Hebecker, P. Mangat, F. Rompineve, L.T. Witkowski, Tuning and backreaction in F-term axion monodromy inflation. Nucl. Phys. B **894**, 456 (2015). [arXiv:1411.2032]
136. R. Flauger, L. McAllister, E. Silverstein, A. Westphal, Drifting oscillations in axion monodromy. arXiv:1412.1814
137. R. Blumenhagen, A. Font, M. Fuchs, D. Herschmann, E. Plauschinn, Towards axionic Starobinsky-like inflation in string theory. Phys. Lett. B **746**, 217 (2015). [arXiv:1503.01607]
138. R. Blumenhagen, A. Font, M. Fuchs, D. Herschmann, E. Plauschinn, Y. Sekiguchi, F. Wolf, A flux-scaling scenario for high-scale moduli stabilization in string theory. Nucl. Phys. B **897**, 500 (2015). [arXiv:1503.07634]
139. D. Escobar, A. Landete, F. Marchesano, D. Regalado, Large field inflation from D-branes. Phys. Rev. D **93**, 081301 (2016). [arXiv:1505.07871]
140. D. Andriot, A no-go theorem for monodromy inflation. JCAP **1603**, 025 (2016). [arXiv:1510.02005]
141. D. Escobar, A. Landete, F. Marchesano, D. Regalado, D6-branes and axion monodromy inflation. JHEP **1603**, 113 (2016). [arXiv:1511.08820]
142. A. Hebecker, F. Rompineve, A. Westphal, Axion monodromy and the weak gravity conjecture. JHEP **1604**, 157 (2016). [arXiv:1512.03768]
143. A. Hebecker, J. Moritz, A. Westphal, L.T. Witkowski, Axion monodromy inflation with warped KK-modes. Phys. Lett. B **754**, 328 (2016). [arXiv:1512.04463]
144. S. Kachru, R. Kallosh, A.D. Linde, J.M. Maldacena, L.P. McAllister, S.P. Trivedi, Towards inflation in string theory. JCAP **0310**, 013 (2003). [arXiv:hep-th/0308055]
145. J.P. Hsu, R. Kallosh, S. Prokushkin, On brane inflation with volume stabilization. JCAP **0312**, 009 (2003). [arXiv:hep-th/0311077]

146. A. Buchel, R. Roiban, Inflation in warped geometries. Phys. Lett. B **590**, 284 (2004). [arXiv:hep-th/0311154]
147. H. Firouzjahi, S.-H.H. Tye, Closer towards inflation in string theory. Phys. Lett. B **584**, 147 (2004). [arXiv:hep-th/0312020]
148. J.P. Hsu, R. Kallosh, Volume stabilization and the origin of the inflaton shift symmetry in string theory. JHEP **0404**, 042 (2004). [arXiv:hep-th/0402047]
149. C.P. Burgess, J.M. Cline, H. Stoica, F. Quevedo, Inflation in realistic D-brane models. JHEP **0409**, 033 (2004). [arXiv:hep-th/0403119]
150. O. DeWolfe, S. Kachru, H.L. Verlinde, The giant inflaton. JHEP **0405**, 017 (2004). [arXiv:hep-th/0403123]
151. N. Iizuka, S.P. Trivedi, An inflationary model in string theory. Phys. Rev. D **70**, 043519 (2004). [arXiv:hep-th/0403203]
152. M. Berg, M. Haack, B. Körs, Loop corrections to volume moduli and inflation in string theory. Phys. Rev. D **71**, 026005 (2005). [arXiv:hep-th/0404087]
153. K. Dasgupta, J.P. Hsu, R. Kallosh, A.D. Linde, M. Zagermann, D3/D7 brane inflation and semilocal strings. JHEP **0408**, 030 (2004). [arXiv:hep-th/0405247]
154. U. Seljak, A. Slosar, *B* polarization of cosmic microwave background as a tracer of strings. Phys. Rev. D **74**, 063523 (2006). [arXiv:astro-ph/0604143]
155. D. Baumann, A. Dymarsky, I.R. Klebanov, L. McAllister, P.J. Steinhardt, A delicate universe: compactification obstacles to D-brane inflation. Phys. Rev. Lett. **99**, 141601 (2007). [arXiv:0705.3837]
156. A. Krause, E. Pajer, Chasing brane inflation in string theory. JCAP **0807**, 023 (2008). [arXiv:0705.4682]
157. D. Baumann, A. Dymarsky, I.R. Klebanov, L. McAllister, Towards an explicit model of D-brane inflation. JCAP **0801**, 024 (2008). [arXiv:0706.0360]
158. S. Panda, M. Sami, S. Tsujikawa, Prospects of inflation in delicate D-brane cosmology. Phys. Rev. D **76**, 103512 (2007). [arXiv:0707.2848]
159. L. Hoi, J.M. Cline, How delicate is brane-antibrane inflation? Phys. Rev. D **79**, 083537 (2009). [arXiv:0810.1303]
160. D. Baumann, A. Dymarsky, S. Kachru, I.R. Klebanov, L. McAllister, Holographic systematics of D-brane inflation. JHEP **0903**, 093 (2009). [arXiv:0808.2811]
161. N. Agarwal, R. Bean, L. McAllister, G. Xu, Universality in D-brane inflation. JCAP **1109**, 002 (2011). [arXiv:1103.2775]
162. M. Dias, J. Frazer, A.R. Liddle, Multifield consequences for D-brane inflation. JCAP **1206**, 020 (2012); Erratum-ibid. **1303**, E01 (2013). [arXiv:1203.3792]
163. D. Battefeld, T. Battefeld, S. Schulz, On the unlikeliness of multi-field inflation: bounded random potentials and our vacuum. JCAP **1206**, 034 (2012). [arXiv:1203.3941]
164. L. McAllister, S. Renaux-Petel, G. Xu, A statistical approach to multifield inflation: many-field perturbations beyond slow roll. JCAP **1210**, 046 (2012). [arXiv:1207.0317]
165. D. Battefeld, T. Battefeld, A smooth landscape: ending saddle point inflation requires features to be shallow. JCAP **1307**, 038 (2013). [arXiv:1304.0461]
166. D. Baumann, A. Dymarsky, S. Kachru, I.R. Klebanov, L. McAllister, D3-brane potentials from fluxes in AdS/CFT. JHEP **1006**, 072 (2010). [arXiv:1001.5028]
167. A. Ali, A. Deshamukhya, S. Panda, M. Sami, Inflation with improved D3-brane potential and the fine tunings associated with the model. Eur. Phys. J. C **71**, 1672 (2011). [arXiv:1010.1407]
168. E. Silverstein, D. Tong, Scalar speed limits and cosmology: acceleration from D-cceleration. Phys. Rev. D **70**, 103505 (2004). [arXiv:hep-th/0310221]
169. M. Alishahiha, E. Silverstein, D. Tong, DBI in the sky: non-Gaussianity from inflation with a speed limit. Phys. Rev. D **70**, 123505 (2004). [arXiv:hep-th/0404084]
170. X. Chen, Multithroat brane inflation. Phys. Rev. D **71**, 063506 (2005). [arXiv:hep-th/0408084]
171. X. Chen, Inflation from warped space. JHEP **0508**, 045 (2005). [arXiv:hep-th/0501184]
172. X. Chen, Running non-Gaussianities in Dirac–Born–Infeld inflation. Phys. Rev. D **72**, 123518 (2005). [arXiv:astro-ph/0507053]

173. X. Chen, M.-x. Huang, S. Kachru, G. Shiu, Observational signatures and non-Gaussianities of general single field inflation. JCAP **0701**, 002 (2007). [arXiv:hep-th/0605045]
174. S. Kecskemeti, J. Maiden, G. Shiu, B. Underwood, DBI inflation in the tip region of a warped throat. JHEP **0609**, 076 (2006). [arXiv:hep-th/0605189]
175. G. Shiu, B. Underwood, Observing the geometry of warped compactification via cosmic inflation. Phys. Rev. Lett. **98**, 051301 (2007). [arXiv:hep-th/0610151]
176. S. Thomas, J. Ward, IR inflation from multiple branes. Phys. Rev. D **76**, 023509 (2007). [arXiv:hep-th/0702229]
177. J.E. Lidsey, I. Huston, Gravitational wave constraints on Dirac–Born–Infeld inflation. JCAP **0707**, 002 (2007). [arXiv:0705.0240]
178. H.V. Peiris, D. Baumann, B. Friedman, A. Cooray, Phenomenology of D-brane inflation with general speed of sound. Phys. Rev. D **76**, 103517 (2007). [arXiv:0706.1240]
179. T. Kobayashi, S. Mukohyama, S. Kinoshita, Constraints on wrapped DBI inflation in a warped throat. JCAP **0801**, 028 (2008). [arXiv:0708.4285]
180. M. Becker, L. Leblond, S.E. Shandera, Inflation from wrapped branes. Phys. Rev. D **76**, 123516 (2007). [arXiv:0709.1170]
181. D.A. Easson, R. Gregory, D.F. Mota, G. Tasinato, I. Zavala, Spinflation. JCAP **0802**, 010 (2008). [arXiv:0709.2666]
182. M.-x. Huang, G. Shiu, B. Underwood, Multifield Dirac–Born–Infeld inflation and non-Gaussianities. Phys. Rev. D **77**, 023511 (2008). [arXiv:0709.3299]
183. R. Bean, X. Chen, H. Peiris, J. Xu, Comparing infrared Dirac–Born–Infeld brane inflation to observations. Phys. Rev. D **77**, 023527 (2008). [arXiv:0710.1812]
184. D. Langlois, S. Renaux-Petel, D.A. Steer, T. Tanaka, Primordial fluctuations and non-Gaussianities in multifield Dirac–Born–Infeld inflation. Phys. Rev. Lett. **101**, 061301 (2008). [arXiv:0804.3139]
185. D. Langlois, S. Renaux-Petel, D.A. Steer, T. Tanaka, Primordial perturbations and non-Gaussianities in DBI and general multifield inflation. Phys. Rev. D **78**, 063523 (2008). [arXiv:0806.0336]
186. F. Arroja, S. Mizuno, K. Koyama, Non-Gaussianity from the bispectrum in general multiple field inflation. JCAP **0808**, 015 (2008). [arXiv:0806.0619]
187. D. Langlois, S. Renaux-Petel, D.A. Steer, Multi-field DBI inflation: introducing bulk forms and revisiting the gravitational wave constraints. JCAP **0904**, 021 (2009). [arXiv:0902.2941]
188. S. Mizuno, F. Arroja, K. Koyama, T. Tanaka, Lorentz boost and non-Gaussianity in multifield DBI inflation. Phys. Rev. D **80**, 023530 (2009). [arXiv:0905.4557]
189. S. Mizuno, F. Arroja, K. Koyama, Full quantum trispectrum in multifield DBI inflation. Phys. Rev. D **80**, 083517 (2009). [arXiv:0907.2439]
190. S. Renaux-Petel, Combined local and equilateral non-Gaussianities from multifield DBI inflation. JCAP **0910**, 012 (2009). [arXiv:0907.2476]
191. L. Lorenz, J. Martin, J. Yokoyama, Geometrically consistent approach to stochastic DBI inflation. Phys. Rev. D **82**, 023515 (2010). [arXiv:1004.3734]
192. S.E. Shandera, S.-H.H. Tye, Observing brane inflation. JCAP **0605**, 007 (2006). [arXiv:hep-th/0601099]
193. D. Baumann, L. McAllister, A microscopic limit on gravitational waves from D-brane inflation. Phys. Rev. D **75**, 123508 (2007). [arXiv:hep-th/0610285]
194. R. Bean, S.E. Shandera, S.-H.H. Tye, J. Xu, Comparing brane inflation to WMAP. JCAP **0705**, 004 (2007). [arXiv:hep-th/0702107]
195. S.-H.H. Tye, Brane inflation: string theory viewed from the cosmos. Lect. Notes Phys. **737**, 949 (2008). [arXiv:hep-th/0610221]
196. L. McAllister, E. Silverstein, String cosmology: a review. Gen. Relat. Grav. **40**, 565 (2008). [arXiv:0710.2951]
197. M. Cicoli, F. Quevedo, String moduli inflation: an overview. Class. Quantum Grav. **28**, 204001 (2011). [arXiv:1108.2659]
198. C.P. Burgess, L. McAllister, Challenges for string cosmology. Class. Quantum Grav. **28**, 204002 (2011). [arXiv:1108.2660]

199. E.J. Copeland, L. Pogosian, T. Vachaspati, Seeking string theory in the cosmos. Class. Quantum Grav. **28**, 204009 (2011). [arXiv:1105.0207]
200. E. Pajer, M. Peloso, A review of axion inflation in the era of Planck. Class. Quantum Grav. **30**, 214002 (2013). [arXiv:1305.3557]
201. C.P. Burgess, M. Cicoli, F. Quevedo, String inflation after Planck 2013. JCAP **1311**, 003 (2013). [arXiv:1306.3512]
202. D. Baumann, L. McAllister, *Inflation and String Theory* (Cambridge University Press, Cambridge, 2015). [arXiv:1404.2601]
203. O. DeWolfe, A. Giryavets, S. Kachru, W. Taylor, Type IIA moduli stabilization. JHEP **0507**, 066 (2005). [arXiv:hep-th/0505160]
204. S.B. Giddings, S. Kachru, J. Polchinski, Hierarchies from fluxes in string compactifications. Phys. Rev. D **66**, 106006 (2002). [arXiv:hep-th/0105097]
205. S. Sethi, C. Vafa, E. Witten, Constraints on low-dimensional string compactifications. Nucl. Phys. B **480**, 213 (1996). [arXiv:hep-th/9606122]
206. S. Kachru, J. Pearson, H.L. Verlinde, Brane/flux annihilation and the string dual of a non-supersymmetric field theory. JHEP **0206**, 021 (2002). [arXiv:hep-th/0112197]
207. J.M. Maldacena, H.S. Nastase, The supergravity dual of a theory with dynamical supersymmetry breaking. JHEP **0109**, 024 (2001). [arXiv:hep-th/0105049]
208. A. Saltman, E. Silverstein, The scaling of the no-scale potential and de-Sitter model building. JHEP **0411**, 066 (2004). [arXiv:hep-th/0402135]
209. S.R. Coleman, Fate of the false vacuum: semiclassical theory. Phys. Rev. D **15**, 2929 (1977); Erratum-ibid. D **16**, 1248 (1977)
210. S.R. Coleman, F. De Luccia, Gravitational effects on and of vacuum decay. Phys. Rev. D **21**, 3305 (1980)
211. N. Goheer, M. Kleban, L. Susskind, The trouble with de Sitter space. JHEP **0307**, 056 (2003). [arXiv:hep-th/0212209]
212. D. Robbins, S. Sethi, A barren landscape?—Metastable de Sitter vacua are nongeneric in string theory. Phys. Rev. D **71**, 046008 (2005). [arXiv:hep-th/0405011]
213. F. Saueressig, U. Theis, S. Vandoren, On de Sitter vacua in type IIA orientifold compactifications. Phys. Lett. B **633**, 125 (2006). [arXiv:hep-th/0506181]
214. V. Braun, B.A. Ovrut, Stabilizing moduli with a positive cosmological constant in heterotic M-theory. JHEP **0607**, 035 (2006). [arXiv:hep-th/0603088]
215. E.I. Buchbinder, Raising anti-de Sitter vacua to de Sitter vacua in heterotic M theory. Phys. Rev. D **70**, 066008 (2004). [arXiv:hep-th/0406101]
216. R. Blumenhagen, S. Moster, E. Plauschinn, Moduli stabilisation versus chirality for MSSM like type IIB orientifolds. JHEP **0801**, 058 (2008). [arXiv:arXiv:0711.3389]
217. E. Dudas, S.K. Vempati, Large D-terms, hierarchical soft spectra and moduli stabilisation. Nucl. Phys. B **727**, 139 (2005). [arXiv:hep-th/0506172]
218. M.J. Duff, P. van Nieuwenhuizen, Quantum inequivalence of different field representations. Phys. Lett. B **94**, 179 (1980)
219. A. Aurilia, H. Nicolai, P.K. Townsend, Hidden constants: the θ parameter of QCD and the cosmological constant of $N = 8$ supergravity. Nucl. Phys. B **176**, 509 (1980)
220. M. Henneaux, C. Teitelboim, The cosmological constant as a canonical variable. Phys. Lett. B **143**, 415 (1984)
221. J.D. Brown, C. Teitelboim, Dynamical neutralization of the cosmological constant. Phys. Lett. B **195**, 177 (1987)
222. J.D. Brown, C. Teitelboim, Neutralization of the cosmological constant by membrane creation. Nucl. Phys. B **297**, 787 (1988)
223. J. Polchinski, A. Strominger, New vacua for type II string theory. Phys. Lett. B **388**, 736 (1996). [arXiv:hep-th/9510227]
224. M.J. Duff, R.R. Khuri, J.X. Lu, String solitons. Phys. Rep. **259**, 213 (1995). [arXiv:hep-th/9412184]
225. B. Carter, Large number coincidences and the anthropic principle in cosmology, in *IAU Symposium 63: Confrontation of Cosmological Theories with Observational Data*, ed. by M.S. Longair (Reidel, Dordrecht, 1974)

226. S. Weinberg, Anthropic bound on the cosmological constant. Phys. Rev. Lett. **59**, 2607 (1987)
227. S. Weinberg, The cosmological constant problem. Rev. Mod. Phys. **61**, 1 (1989)
228. H. Martel, P.R. Shapiro, S. Weinberg, Likely values of the cosmological constant. Astrophys. J. **492**, 29 (1998). [arXiv:astro-ph/9701099]
229. J. Garriga, A. Vilenkin, On likely values of the cosmological constant. Phys. Rev. D **61**, 083502 (2000). [arXiv:astro-ph/9908115]
230. V. Agrawal, S.M. Barr, J.F. Donoghue, D. Seckel, Viable range of the mass scale of the standard model. Phys. Rev. D **57**, 5480 (1998). [arXiv:hep-ph/9707380]
231. S. Winitzki, *Eternal Inflation* (World Scientific, Singapore, 2009)
232. W.R. Stoeger, G.F.R. Ellis, U. Kirchner, Multiverses and cosmology: philosophical issues. arXiv:astro-ph/0407329
233. B. Freivogel, Making predictions in the multiverse. Class. Quantum Grav. **28**, 204007 (2011). [arXiv:1105.0244]
234. T. Banks, M. Dine, L. Motl, On anthropic solutions of the cosmological constant problem. JHEP **0101**, 031 (2001). [arXiv:hep-th/0007206]
235. M.L. Graesser, S.D.H. Hsu, A. Jenkins, M.B. Wise, Anthropic distribution for cosmological constant and primordial density perturbations. Phys. Lett. B **600**, 15 (2004). [arXiv:hep-th/0407174]
236. H. Firouzjahi, S. Sarangi, S.-H.H. Tye, Spontaneous creation of inflationary universes and the cosmic landscape. JHEP **0409**, 060 (2004). [arXiv:hep-th/0406107]
237. N. Arkani-Hamed, S. Dimopoulos, G.R. Dvali, The hierarchy problem and new dimensions at a millimeter. Phys. Lett. B **429**, 263 (1998). [arXiv:hep-ph/9803315]
238. N. Arkani-Hamed, S. Dimopoulos, G.R. Dvali, Phenomenology, astrophysics and cosmology of theories with submillimeter dimensions and TeV scale quantum gravity. Phys. Rev. D **59**, 086004 (1999). [arXiv:hep-ph/9807344]
239. I. Antoniadis, N. Arkani-Hamed, S. Dimopoulos, G.R. Dvali, New dimensions at a millimeter to a Fermi and superstrings at a TeV. Phys. Lett. B **436**, 257 (1998). [arXiv:hep-ph/9804398]
240. B. Freivogel, M. Kleban, M. Rodríguez Martínez, L. Susskind, Observational consequences of a landscape. JHEP **0603**, 039 (2006). [arXiv:hep-th/0505232]
241. D. Yamauchi, A. Linde, A. Naruko, M. Sasaki, T. Tanaka, Open inflation in the landscape. Phys. Rev. D **84**, 043513 (2011). [arXiv:1105.2674]
242. A. De Simone, M.P. Salem, Distribution of Ω_k from the scale-factor cutoff measure. Phys. Rev. D **81**, 083527 (2010). [arXiv:0912.3783]
243. R. Bousso, D. Harlow, L. Senatore, Inflation after false vacuum decay: observational prospects after Planck. Phys. Rev. D **91**, 083527 (2015). [arXiv:1309.4060]
244. M. Cicoli, S. Downes, B. Dutta, F.G. Pedro, A. Westphal, Just enough inflation: power spectrum modifications at large scales. JCAP **1412**, 030 (2014). [arXiv:1407.1048]
245. C.R. Contaldi, M. Peloso, L. Kofman, A.D. Linde, Suppressing the lower multipoles in the CMB anisotropies. JCAP **0307**, 002 (2003). [arXiv:astro-ph/0303636]
246. N. Arkani-Hamed, L. Motl, A. Nicolis, C. Vafa, The string landscape, black holes and gravity as the weakest force. JHEP **0706**, 060 (2007). [arXiv:hep-th/0601001]
247. T. Banks, M. Johnson, A. Shomer, A note on gauge theories coupled to gravity. JHEP **0609**, 049 (2006). [arXiv:hep-th/0606277]
248. G. Dvali, Black holes and large N species solution to the hierarchy problem. Fortsch. Phys. **58**, 528 (2010). [arXiv:0706.2050]
249. C. Cheung, G.N. Remmen, Naturalness and the weak gravity conjecture. Phys. Rev. Lett. **113**, 051601 (2014). [arXiv:1402.2287]
250. C. Cheung, G.N. Remmen, Infrared consistency and the weak gravity conjecture. JHEP **1412**, 087 (2014). [arXiv:1407.7865]
251. F.G. Pedro, A. Westphal, The scale of inflation in the landscape. Phys. Lett. B **739**, 439 (2014). [arXiv:1303.3224]
252. C.P. Burgess, R. Easther, A. Mazumdar, D.F. Mota, T. Multamäki, Multiple inflation, cosmic string networks and the string landscape. JHEP **0505**, 067 (2005). [arXiv:hep-th/0501125]

253. K. Freese, D. Spolyar, Chain inflation in the landscape: 'Bubble bubble toil and trouble'. JCAP **0507**, 007 (2005). [arXiv:hep-ph/0412145]
254. K. Freese, J.T. Liu, D. Spolyar, Chain inflation via rapid tunneling in the landscape. arXiv:hep-th/0612056
255. Q.-G. Huang, Simplified chain inflation. JCAP **0705**, 009 (2007). [arXiv:0704.2835]
256. D. Chialva, U.H. Danielsson, Chain inflation revisited. JCAP **0810**, 012 (2008). [arXiv:0804.2846]
257. J.M. Cline, G.D. Moore, Y. Wang, Chain inflation reconsidered. JCAP **1108**, 032 (2011). [arXiv:1106.2188]
258. R. Easther, Folded inflation, primordial tensors, and the running of the scalar spectral index. arXiv:hep-th/0407042.
259. J. Frazer, A.R. Liddle, Exploring a string-like landscape. JCAP **1102**, 026 (2011). [arXiv:1101.1619]
260. J. Frazer, A.R. Liddle, Multi-field inflation with random potentials: field dimension, feature scale and non-Gaussianity. JCAP **1202**, 039 (2012). [arXiv:1111.6646]
261. G.D. Coughlan, W. Fischler, E.W. Kolb, S. Raby, G.G. Ross, Cosmological problems for the Polonyi potential. Phys. Lett. B **131**, 59 (1983)
262. J.R. Ellis, D.V. Nanopoulos, M. Quirós, On the axion, dilaton, Polonyi, gravitino and shadow matter problems in supergravity and superstring models. Phys. Lett. B **174**, 176 (1986)
263. T. Banks, D.B. Kaplan, A.E. Nelson, Cosmological implications of dynamical supersymmetry breaking. Phys. Rev. D **49**, 779 (1994). [arXiv:hep-ph/9308292]
264. B. de Carlos, J.A. Casas, F. Quevedo, E. Roulet, Model-independent properties and cosmological implications of the dilaton and moduli sectors of 4D strings. Phys. Lett. B **318**, 447 (1993). [arXiv:hep-ph/9308325]
265. M. Endo, M. Yamaguchi, K. Yoshioka, Bottom-up approach to moduli dynamics in heavy gravitino scenario: superpotential, soft terms, and sparticle mass spectrum. Phys. Rev. D **72**, 015004 (2005). [arXiv:hep-ph/0504036]
266. K.-i. Maeda, M.D. Pollock, On inflation in the heterotic superstring model. Phys. Lett. B **173**, 251 (1986)
267. J.R. Ellis, K. Enqvist, D.V. Nanopoulos, M. Quiros, Evolution with temperature and the possibility of inflation from the superstring in four dimensions. Nucl. Phys. B **277**, 231 (1986)
268. P. Binétruy, M.K. Gaillard, Candidates for the inflaton field in superstring models. Phys. Rev. D **34**, 3069 (1986)
269. R. Brustein, P.J. Steinhardt, Challenges for superstring cosmology. Phys. Lett. B **302**, 196 (1993). [arXiv:hep-th/9212049]
270. T. Banks, M. Berkooz, S.H. Shenker, G.W. Moore, P.J. Steinhardt, Modular cosmology. Phys. Rev. D **52**, 3548 (1995). [arXiv:hep-th/9503114]
271. M. Cicoli, F.G. Pedro, G. Tasinato, Natural quintessence in string theory. JCAP **1207**, 044 (2012). [arXiv:1203.6655]
272. J.A. Casas, Baryogenesis, inflation and superstrings, in *International Europhysics Conference on High Energy Physics*, ed. by D. Lellouch, G. Mikenberg, E. Rabinovici (Springer, Berlin, 1999). [arXiv:hep-ph/9802210]
273. J. Ellis, M.A.G. García, D.V. Nanopoulos, K.A. Olive, A no-scale inflationary model to fit them all. JCAP **1408**, 044 (2014). [arXiv:1405.0271]
274. J. Ellis, M.A.G. García, D.V. Nanopoulos, K.A. Olive, Two-field analysis of no-scale supergravity inflation. JCAP **1501**, 010 (2015). [arXiv:1409.8197]
275. J. Ellis, M.A.G. García, D.V. Nanopoulos, K.A. Olive, Phenomenological aspects of no-scale inflation models. JCAP **1510**, 003 (2015). [arXiv:1503.08867]
276. E. Witten, Dimensional reduction of superstring models. Phys. Lett. B **155**, 151 (1985)
277. S. Ferrara, C. Kounnas, M. Porrati, General dimensional reduction of ten-dimensional supergravity and superstring. Phys. Lett. B **181**, 263 (1986)
278. M. Cvetič, J. Louis, B.A. Ovrut, A string calculation of the Kähler potentials for moduli of Z_N orbifolds. Phys. Lett. B **206**, 227 (1988)

279. L.J. Dixon, V. Kaplunovsky, J. Louis, On effective field theories describing $(2, 2)$ vacua of the heterotic string. Nucl. Phys. B **329**, 27 (1990)
280. J.P. Derendinger, S. Ferrara, C. Kounnas, F. Zwirner, On loop corrections to string effective field theories: field-dependent gauge couplings and σ-model anomalies. Nucl. Phys. B **372**, 145 (1992)
281. A. Brignole, L.E. Ibáñez, C. Muñoz, Towards a theory of soft terms for the supersymmetric Standard Model. Nucl. Phys. B **422**, 125 (1994); Erratum-ibid. B **436**, 747 (1995)
282. M. Cicoli, S. Krippendorf, C. Mayrhofer, F. Quevedo, R. Valandro, D-branes at del Pezzo singularities: global embedding and moduli stabilisation. JHEP **1209**, 019 (2012). [arXiv:1206.5237]
283. E. Witten, The cosmological constant from the viewpoint of string theory. arXiv:hep-ph/0002297
284. T. Banks, M. Dine, P.J. Fox, E. Gorbatov, On the possibility of large axion decay constants. JCAP **0306**, 001 (2003). [arXiv:hep-th/0303252]
285. P.A.R. Ade et al. [Planck Collaboration], Planck 2015. XX. Constraints on inflation. Astron. Astrophys. **594**, A20 (2016). [arXiv:1502.02114]
286. N. Barnaby, M. Peloso, Large non-Gaussianity in axion inflation. Phys. Rev. Lett. **106**, 181301 (2011). [arXiv:1011.1500]
287. M.P. Hertzberg, M. Tegmark, S. Kachru, J. Shelton, O. Özcan, Searching for inflation in simple string theory models: an astrophysical perspective. Phys. Rev. D **76**, 103521 (2007). [arXiv:0709.0002]
288. G. Villadoro, F. Zwirner, $N = 1$ effective potential from dual type-IIA D6/O6 orientifolds with general fluxes. JHEP **0506**, 047 (2005). [arXiv:hep-th/0503169]
289. M. Ihl, T. Wrase, Towards a realistic type IIA $T^6/\mathbb{Z}_4$ orientifold model with background fluxes, part 1. Moduli stabilization. JHEP **0607**, 027 (2006). [arXiv:hep-th/0604087]
290. E. Silverstein, Simple de Sitter solutions. Phys. Rev. D **77**, 106006 (2008). [arXiv:0712.1196]
291. B. Heidenreich, M. Reece, T. Rudelius, Weak gravity strongly constrains large-field axion inflation. JHEP **1512**, 108 (2015). [arXiv:1506.03447]
292. T.C. Bachlechner, C. Long, L. McAllister, Planckian axions and the weak gravity conjecture. JHEP **1601**, 091 (2016). [arXiv:1503.07853]
293. G. Shiu, W. Staessens, F. Ye, Widening the axion window via kinetic and Stückelberg mixings. Phys. Rev. Lett. **115**, 181601 (2015). [arXiv:1503.01015]
294. G. Shiu, W. Staessens, F. Ye, Large field inflation from axion mixing. JHEP **1506**, 026 (2015). [arXiv:1503.02965]
295. A. de la Fuente, P. Saraswat, R. Sundrum, Natural inflation and quantum gravity. Phys. Rev. Lett. **114**, 151303 (2015). [arXiv:1412.3457]
296. D. Junghans, Large-field inflation with multiple axions and the weak gravity conjecture. JHEP **1602**, 128 (2016). [arXiv:1504.03566]
297. G.R. Dvali, S.-H.H. Tye, Brane inflation. Phys. Lett. B **450**, 72 (1999). [arXiv:hep-ph/9812483]
298. S.H.S. Alexander, Inflation from $\mathrm{D}-\mathrm{D}^{-}$ brane annihilation. Phys. Rev. D **65**, 023507 (2002). [arXiv:hep-th/0105032]
299. G.R. Dvali, Q. Shafi, S. Solganik, D-brane inflation. arXiv:hep-th/0105203
300. C.P. Burgess, M. Majumdar, D. Nolte, F. Quevedo, G. Rajesh, R.J. Zhang, The inflationary brane-antibrane universe. JHEP **0107**, 047 (2001). [arXiv:hep-th/0105204]
301. G. Shiu, S.-H.H. Tye, Some aspects of brane inflation. Phys. Lett. B **516**, 421 (2001). [arXiv:hep-th/0106274]
302. B. Kyae, Q. Shafi, Branes and inflationary cosmology. Phys. Lett. B **526**, 379 (2002). [arXiv:hep-ph/0111101]
303. J. García-Bellido, R. Rabadán, F. Zamora, Inflationary scenarios from branes at angles. JHEP **0201**, 036 (2002). [arXiv:hep-th/0112147]
304. N.T. Jones, H. Stoica, S.-H.H. Tye, Brane interaction as the origin of inflation. JHEP **0207**, 051 (2002). [arXiv:hep-th/0203163]

305. S. Sarangi, S.-H.H. Tye, Cosmic string production towards the end of brane inflation. Phys. Lett. B **536**, 185 (2002). [arXiv:hep-th/0204074]
306. N.T. Jones, H. Stoica, S.-H.H. Tye, The production, spectrum and evolution of cosmic strings in brane inflation. Phys. Lett. B **563**, 6 (2003). [arXiv:hep-th/0303269]
307. L. Randall, R. Sundrum, An alternative to compactification. Phys. Rev. Lett. **83**, 4690 (1999). [arXiv:hep-th/9906064]
308. K. Becker, M. Becker, M. Haack, J. Louis, Supersymmetry breaking and α' corrections to flux induced potentials. JHEP **0206**, 060 (2002). [arXiv:hep-th/0204254]
309. R.G. Leigh, Dirac–Born–Infeld action from Dirichlet σ-model. Mod. Phys. Lett. A **4**, 2767 (1989)
310. O. Aharony, S.S. Gubser, J.M. Maldacena, H. Ooguri, Y. Oz, Large N field theories, string theory and gravity. Phys. Rep. **323**, 183 (2000). [arXiv:hep-th/9905111]
311. J.M. Maldacena, The large-N limit of superconformal field theories and supergravity. Adv. Theor. Math. Phys. **2**, 231 (1998) [Int. J. Theor. Phys. **38**, 1113 (1999)]. [arXiv:hep-th/9711200]
312. N. Seiberg, E. Witten, The D1/D5 system and singular CFT. JHEP **9904**, 017 (1999). [arXiv:hep-th/9903224]
313. G.G. Ross, S. Sarkar, Successful supersymmetric inflation. Nucl. Phys. B **461**, 597 (1996). [arXiv:hep-ph/9506283]
314. G. German, G.G. Ross, S. Sarkar, Low-scale inflation. Nucl. Phys. B **608**, 423 (2001). [arXiv:hep-ph/0103243]
315. R. Allahverdi, K. Enqvist, J. García-Bellido, A. Jokinen, A. Mazumdar, MSSM flat direction inflation: slow roll, stability, fine tunning and reheating. JCAP **0706**, 019 (2007). [arXiv:hep-ph/0610134]
316. G. Dvali, A. Vilenkin, Formation and evolution of cosmic D strings. JCAP **0403**, 010 (2004). [arXiv:hep-th/0312007]
317. E.J. Copeland, R.C. Myers, J. Polchinski, Cosmic F- and D-strings. JHEP **0406**, 013 (2004). [arXiv:hep-th/0312067]
318. M.G. Jackson, N.T. Jones, J. Polchinski, Collisions of cosmic F- and D-strings. JHEP **0510**, 013 (2005). [arXiv:hep-th/0405229]
319. T. Damour, A. Vilenkin, Gravitational radiation from cosmic (super)strings: bursts, stochastic background, and observational windows. Phys. Rev. D **71**, 063510 (2005). [arXiv:hep-th/0410222]
320. T. Damour, A. Vilenkin, Gravitational wave bursts from cosmic strings. Phys. Rev. Lett. **85**, 3761 (2000). [arXiv:gr-qc/0004075]
321. E.J. Copeland, T.W.B. Kibble, Cosmic strings and superstrings. Proc. R. Soc. Lond. A **466**, 623 (2010). [arXiv:0911.1345]
322. P.A.R. Ade et al. [Planck Collaboration], Planck 2013 results. XXV. Searches for cosmic strings and other topological defects. Astron. Astrophys. **571**, A25 (2014). [arXiv:1303.5085]
323. P.A.R. Ade et al. [Planck Collaboration], Planck 2015 results. XIII. Cosmological parameters. Astron. Astrophys. **594**, A13 (2016). [arXiv:1502.01589]
324. S. Kuroyanagi, K. Miyamoto, T. Sekiguchi, K. Takahashi, J. Silk, Forecast constraints on cosmic string parameters from gravitational wave direct detection experiments. Phys. Rev. D **86**, 023503 (2012). [arXiv:1202.3032]
325. S. Kuroyanagi, K. Miyamoto, T. Sekiguchi, K. Takahashi, J. Silk, Forecast constraints on cosmic strings from future CMB, pulsar timing and gravitational wave direct detection experiments. Phys. Rev. D **87**, 023522 (2013); Erratum-ibid. D **87**, 069903(E) (2013). [arXiv:1210.2829]
326. D. Baumann, A. Dymarsky, I.R. Klebanov, J.M. Maldacena, L.P. McAllister, A. Murugan, On D3-brane potentials in compactifications with fluxes and wrapped D-branes. JHEP **0611**, 031 (2006). [arXiv:hep-th/0607050]
327. C.P. Burgess, J.M. Cline, K. Dasgupta, H. Firouzjahi, Uplifting and inflation with D3 branes. JHEP **0703**, 027 (2007). [arXiv:hep-th/0610320]
328. R.C. Myers, Dielectric branes. JHEP **9912**, 022 (1999). [arXiv:hep-th/9910053]

329. B. de Carlos, J. Roberts, Y. Schmohe, Moving five-branes and membrane instantons in low energy heterotic M-theory. Phys. Rev. D **71**, 026004 (2005). [arXiv:hep-th/0406171]
330. E.I. Buchbinder, Five-brane dynamics and inflation in heterotic M-theory. Nucl. Phys. B **711**, 314 (2005). [arXiv:hep-th/0411062]
331. K. Becker, M. Becker, A. Krause, M-theory inflation from multi M5-brane dynamics. Nucl. Phys. B **715**, 349 (2005). [arXiv:hep-th/0501130]
332. J. Ward, Instantons, assisted inflation and heterotic M-theory. Phys. Rev. D **73**, 026004 (2006). [arXiv:hep-th/0511079]
333. A. Krause, Large gravitational waves and the Lyth bound in multi-brane inflation. JCAP **0807**, 001 (2008). [arXiv:0708.4414]
334. P. Vargas Moniz, S. Panda, J. Ward, Higher order corrections to heterotic M-theory inflation. Class. Quantum Grav. **26**, 245003 (2009). [arXiv:0907.0711]
335. C. Armendáriz-Picón, T. Damour, V.F. Mukhanov, *k*-inflation. Phys. Lett. B **458**, 209 (1999). [arXiv:hep-th/9904075]
336. J. Garriga, V.F. Mukhanov, Perturbations in *k*-inflation. Phys. Lett. B **458**, 219 (1999). [arXiv:hep-th/9904176]
337. D. Babich, P. Creminelli, M. Zaldarriaga, The shape of non-Gaussianities. JCAP **0408**, 009 (2004). [arXiv:astro-ph/0405356]
338. P.A.R. Ade et al. [Planck Collaboration], Planck 2013 Results. XXIV. Constraints on primordial non-Gaussianity. Astron. Astrophys. **571**, A24 (2014). [arXiv:1303.5084]
339. Y.-F. Cai, W. Xue, *N*-flation from multiple DBI type actions. Phys. Lett. B **680**, 395 (2009). [arXiv:0809.4134]
340. Y.-F. Cai, H.-Y. Xia, Inflation with multiple sound speeds: a model of multiple DBI type actions and non-Gaussianities. Phys. Lett. B **677**, 226 (2009). [arXiv:0904.0062]
341. P. Hořava, E. Witten, Heterotic and type I string dynamics from eleven dimensions. Nucl. Phys. B **460**, 506 (1996). [arXiv:hep-th/9510209]
342. P. Hořava, E. Witten, Eleven-dimensional supergravity on a manifold with boundary. Nucl. Phys. B **475**, 94 (1996). [arXiv:hep-th/9603142]
343. E. Witten, Strong coupling expansion of Calabi–Yau compactification. Nucl. Phys. B **471**, 135 (1996). [arXiv:hep-th/9602070]
344. T. Banks, M. Dine, Couplings and scales in strongly coupled heterotic string theory. Nucl. Phys. B **479**, 173 (1996). [arXiv:hep-th/9605136]
345. B.A. Ovrut, Lectures on heterotic M-theory, in *Strings, Branes and Extra Dimensions*, ed. by S.S. Gubser, J.D. Lykken (World Scientific, Singapore, 2004). [arXiv:hep-th/0201032]
346. A. Lukas, B.A. Ovrut, K.S. Stelle, D. Waldram, Universe as a domain wall. Phys. Rev. D **59**, 086001 (1999). [arXiv:hep-th/9803235]
347. A. Lukas, B.A. Ovrut, K.S. Stelle, D. Waldram, Heterotic M-theory in five dimensions. Nucl. Phys. B **552**, 246 (1999). [arXiv:hep-th/9806051]
348. A. Lukas, B.A. Ovrut, D. Waldram, Cosmological solutions of Hořava–Witten theory. Phys. Rev. D **60**, 086001 (1999). [arXiv:hep-th/9806022]
349. A. Lukas, B.A. Ovrut, D. Waldram, Boundary inflation. Phys. Rev. D **61**, 023506 (2000). [arXiv:hep-th/9902071]
350. I. Antoniadis, M. Quirós, Large radii and string unification. Phys. Lett. B **392**, 61 (1997). [arXiv:hep-th/9609209]
351. K. Benakli, Phenomenology of low quantum gravity scale models. Phys. Rev. D **60**, 104002 (1999). [arXiv:hep-ph/9809582]
352. P. Brax, C. van de Bruck, A.C. Davis, Brane world cosmology. Rep. Prog. Phys. **67**, 2183 (2004). [arXiv:hep-th/0404011]
353. R. Maartens, K. Koyama, Brane-world gravity. Living Rev. Relat. **13**, 5 (2010)
354. L. Randall, R. Sundrum, Large mass hierarchy from a small extra dimension. Phys. Rev. Lett. **83**, 3370 (1999). [arXiv:hep-ph/9905221]
355. T. Shiromizu, K.-i. Maeda, M. Sasaki, The Einstein equation on the 3-brane world. Phys. Rev. D **62**, 024012 (2000). [arXiv:gr-qc/9910076]

356. P. Binétruy, C. Deffayet, D. Langlois, Non-conventional cosmology from a brane universe. Nucl. Phys. B **565**, 269 (2000). [arXiv:hep-th/9905012]
357. P. Binétruy, C. Deffayet, U. Ellwanger, D. Langlois, Brane cosmological evolution in a bulk with cosmological constant. Phys. Lett. B **477**, 285 (2000). [arXiv:hep-th/9910219]
358. C. Csáki, M. Graesser, C. Kolda, J. Terning, Cosmology of one extra dimension with localized gravity. Phys. Lett. B **462**, 34 (1999). [arXiv:hep-ph/9906513]
359. J.M. Cline, C. Grojean, G. Servant, Cosmological expansion in the presence of extra dimensions. Phys. Rev. Lett. **83**, 4245 (1999). [arXiv:hep-ph/9906523]
360. É.É. Flanagan, S.-H.H. Tye, I. Wasserman, Cosmological expansion in the Randall–Sundrum brane world scenario. Phys. Rev. D **62**, 044039 (2000). [arXiv:hep-ph/9910498]
361. R. Maartens, D. Wands, B.A. Bassett, I.P.C. Heard, Chaotic inflation on the brane. Phys. Rev. D **62**, 041301 (2000). [arXiv:hep-ph/9912464]
362. E.J. Copeland, A.R. Liddle, J.E. Lidsey, Steep inflation: ending brane world inflation by gravitational particle production. Phys. Rev. D **64**, 023509 (2001). [arXiv:astro-ph/0006421]
363. J.E. Kim, B. Kyae, H.M. Lee, Effective Gauss–Bonnet interaction in Randall–Sundrum compactification. Phys. Rev. D **62**, 045013 (2000). [arXiv:hep-ph/9912344]
364. D.J. Gross, J.H. Sloan, The quartic effective action for the heterotic string. Nucl. Phys. B **291**, 41 (1987)
365. J.E. Kim, B. Kyae, H.M. Lee, Various modified solutions of the Randall–Sundrum model with the Gauss–Bonnet interaction. Nucl. Phys. B **582**, 296 (2000); Erratum-ibid. B **591**, 587 (2000). [arXiv:hep-th/0004005]
366. S. Nojiri, S.D. Odintsov, Brane world cosmology in higher derivative gravity or warped compactification in the next-to-leading order of AdS/CFT correspondence. JHEP **0007**, 049 (2000). [arXiv:hep-th/0006232]
367. I.P. Neupane, Consistency of higher derivative gravity in the brane background. JHEP **0009**, 040 (2000). [arXiv:hep-th/0008190]
368. C. Charmousis, J.-F. Dufaux, General Gauss–Bonnet brane cosmology. Class. Quantum Grav. **19**, 4671 (2002). [arXiv:hep-th/0202107]
369. S. Nojiri, S.D. Odintsov, S. Ogushi, Friedmann–Robertson–Walker brane cosmological equations from the five-dimensional bulk (A)dS black hole. Int. J. Mod. Phys. A **17**, 4809 (2002). [arXiv:hep-th/0205187]
370. S.C. Davis, Generalized Israel junction conditions for a Gauss–Bonnet brane world. Phys. Rev. D **67**, 024030 (2003). [arXiv:hep-th/0208205]
371. E. Gravanis, S. Willison, Israel conditions for the Gauss–Bonnet theory and the Friedmann equation on the brane universe. Phys. Lett. B **562**, 118 (2003). [arXiv:hep-th/0209076]
372. J.E. Lidsey, N.J. Nunes, Inflation in Gauss–Bonnet brane cosmology. Phys. Rev. D **67**, 103510 (2003). [arXiv:astro-ph/0303168]
373. J.-F. Dufaux, J.E. Lidsey, R. Maartens, M. Sami, Cosmological perturbations from brane inflation with a Gauss–Bonnet term. Phys. Rev. D **70**, 083525 (2004). [arXiv:hep-th/0404161]
374. G. Calcagni, Slow-roll parameters in braneworld cosmologies. Phys. Rev. D **69**, 103508 (2004). [arXiv:hep-ph/0402126]
375. G. Calcagni, Consistency relations and degeneracies in (non)commutative patch inflation. Phys. Lett. B **606**, 177 (2005). [arXiv:hep-ph/0406057]
376. G. Calcagni, Braneworld Cosmology and Noncommutative Inflation. Ph.D. thesis, Parma University, Parma (2005). [arXiv:hep-ph/0503044]
377. R. Maartens, Cosmological dynamics on the brane. Phys. Rev. D **62**, 084023 (2000). [arXiv:hep-th/0004166]
378. C. van de Bruck, M. Dorca, R.H. Brandenberger, A. Lukas, Cosmological perturbations in brane-world theories: formalism. Phys. Rev. D **62**, 123515 (2000). [arXiv:hep-th/0005032]
379. K. Koyama, J. Soda, Evolution of cosmological perturbations in the brane world. Phys. Rev. D **62**, 123502 (2000). [arXiv:hep-th/0005239]
380. D. Langlois, Evolution of cosmological perturbations in a brane universe. Phys. Rev. Lett. **86**, 2212 (2001). [arXiv:hep-th/0010063]

381. D. Langlois, R. Maartens, M. Sasaki, D. Wands, Large-scale cosmological perturbations on the brane. Phys. Rev. D **63**, 084009 (2001). [arXiv:hep-th/0012044]
382. D. Langlois, R. Maartens, D. Wands, Gravitational waves from inflation on the brane. Phys. Lett. B **489**, 259 (2000). [arXiv:hep-th/0006007]
383. A. Coley, No chaos in brane-world cosmology. Class. Quantum Grav. **19**, L45 (2002). [arXiv:hep-th/0110117]
384. S. Kachru, L. McAllister, Bouncing brane cosmologies from warped string compactifications. JHEP **0303**, 018 (2003). [arXiv:hep-th/0205209]
385. S. Mukherji, M. Peloso, Bouncing and cyclic universes from brane models. Phys. Lett. B **547**, 297 (2002). [arXiv:hep-th/0205180]
386. P. Kanti, K. Tamvakis, Challenges and obstacles for a bouncing universe in brane models. Phys. Rev. D **68**, 024014 (2003). [arXiv:hep-th/0303073]
387. J.L. Hovdebo, R.C. Myers, Bouncing brane worlds go crunch! JCAP **0311**, 012 (2003). [arXiv:hep-th/0308088]
388. V.A. Rubakov, M.E. Shaposhnikov, Extra space-time dimensions: towards a solution to the cosmological constant problem. Phys. Lett. B **125**, 139 (1983)
389. E.P. Verlinde, H.L. Verlinde, RG flow, gravity and the cosmological constant. JHEP **0005**, 034 (2000). [arXiv:hep-th/9912018]
390. S.P. de Alwis, Brane world scenarios and the cosmological constant. Nucl. Phys. B **597**, 263 (2001). [arXiv:hep-th/0002174]
391. N. Arkani-Hamed, S. Dimopoulos, N. Kaloper, R. Sundrum, A small cosmological constant from a large extra dimension. Phys. Lett. B **480**, 193 (2000). [arXiv:hep-th/0001197]
392. S. Kachru, M.B. Schulz, E. Silverstein, Self-tuning flat domain walls in 5D gravity and string theory. Phys. Rev. D **62**, 045021 (2000). [arXiv:hep-th/0001206]
393. S. Förste, Z. Lalak, S. Lavignac, H.P. Nilles, A comment on self-tuning and vanishing cosmological constant in the brane world. Phys. Lett. B **481**, 360 (2000). [arXiv:hep-th/0002164]
394. S. Förste, Z. Lalak, S. Lavignac, H.P. Nilles, The cosmological constant problem from a brane-world perspective. JHEP **0009**, 034 (2000). [arXiv:hep-th/0006139]
395. I. Antoniadis, S. Cotsakis, I. Klaoudatou, Brane singularities and their avoidance. Class. Quantum Grav. **27**, 235018 (2010). [arXiv:1010.6175]
396. S.M. Carroll, L. Mersini-Houghton, Can we live in a self-tuning universe? Phys. Rev. D **64**, 124008 (2001). [arXiv:hep-th/0105007]
397. Y. Aghababaie, C.P. Burgess, S.L. Parameswaran, F. Quevedo, Towards a naturally small cosmological constant from branes in 6D supergravity. Nucl. Phys. B **680**, 389 (2004). [arXiv:hep-th/030425]
398. Y. Aghababaie, C.P. Burgess, J.M. Cline, H. Firouzjahi, S.L. Parameswaran, F. Quevedo, G. Tasinato, I. Zavala, Warped brane worlds in six dimensional supergravity. JHEP **0309**, 037 (2003). [arXiv:hep-th/0308064]
399. C.P. Burgess, Supersymmetric large extra dimensions and the cosmological constant: an update. Ann. Phys. (N.Y.) **313**, 283 (2004). [arXiv:hep-th/0402200]
400. G. Calcagni, S. Kuroyanagi, J. Ohashi, S. Tsujikawa, Strong Planck constraints on braneworld and non-commutative inflation. JCAP **1403**, 052 (2014). [arXiv:1310.5186]
401. A. Sen, Universality of the tachyon potential. JHEP **9912**, 027 (1999). [arXiv:hep-th/9911116]
402. M.R. Garousi, Tachyon couplings on non-BPS D-branes and Dirac–Born–Infeld action. Nucl. Phys. B **584**, 284 (2000). [arXiv:hep-th/0003122]
403. E.A. Bergshoeff, M. de Roo, T.C. de Wit, E. Eyras, S. Panda, T-duality and actions for non-BPS D-branes. JHEP **0005**, 009 (2000). [arXiv:hep-th/0003221]
404. D. Kutasov, M. Mariño, G.W. Moore, Some exact results on tachyon condensation in string field theory. JHEP **0010**, 045 (2000). [arXiv:hep-th/0009148]
405. D. Kutasov, M. Mariño, G.W. Moore, Remarks on tachyon condensation in superstring field theory. arXiv:hep-th/0010108
406. P. Kraus, F. Larsen, Boundary string field theory of the $D\bar{D}$ system. Phys. Rev. D **63**, 106004 (2001). [arXiv:hep-th/0012198]

407. T. Takayanagi, S. Terashima, T. Uesugi, Brane-antibrane action from boundary string field theory. JHEP **0103**, 019 (2001). [arXiv:hep-th/0012210]
408. M.R. Garousi, Off-shell extension of S-matrix elements and tachyonic effective actions. JHEP **0304**, 027 (2003). [arXiv:hep-th/0303239]
409. M.R. Garousi, Slowly varying tachyon and tachyon potential. JHEP **0305**, 058 (2003). [arXiv:hep-th/0304145]
410. F. Leblond, A.W. Peet, SD-brane gravity fields and rolling tachyons. JHEP **0304**, 048 (2003). [arXiv:hep-th/0303035]
411. N.D. Lambert, H. Liu, J.M. Maldacena, Closed strings from decaying D-branes. JHEP **0703**, 014 (2007). [arXiv:hep-th/0303139]
412. A. Mazumdar, S. Panda, A. Pérez-Lorenzana, Assisted inflation via tachyon condensation. Nucl. Phys. B **614**, 101 (2001). [arXiv:hep-ph/0107058]
413. G.W. Gibbons, Cosmological evolution of the rolling tachyon. Phys. Lett. B **537**, 1 (2002). [arXiv:hep-th/0204008]
414. M. Fairbairn, M.H.G. Tytgat, Inflation from a tachyon fluid? Phys. Lett. B **546**, 1 (2002). [arXiv:hep-th/0204070]
415. S. Mukohyama, Brane cosmology driven by the rolling tachyon. Phys. Rev. D **66**, 024009 (2002). [arXiv:hep-th/0204084]
416. A. Feinstein, Power-law inflation from the rolling tachyon. Phys. Rev. D **66**, 063511 (2002). [arXiv:hep-th/0204140]
417. T. Padmanabhan, Accelerated expansion of the universe driven by tachyonic matter. Phys. Rev. D **66**, 021301 (2002). [arXiv:hep-th/0204150]
418. D. Choudhury, D. Ghoshal, D.P. Jatkar, S. Panda, On the cosmological relevance of the tachyon. Phys. Lett. B **544**, 231 (2002). [arXiv:hep-th/0204204]
419. A. Frolov, L. Kofman, A. Starobinsky, Prospects and problems of tachyon matter cosmology. Phys. Lett. B **545**, 8 (2002). [arXiv:hep-th/0204187]
420. G. Shiu, I. Wasserman, Cosmological constraints on tachyon matter. Phys. Lett. B **541**, 6 (2002). [arXiv:hep-th/0205003]
421. L. Kofman, A. Linde, Problems with tachyon inflation. JHEP **0207**, 004 (2002). [arXiv:hep-th/0205121]
422. H.B. Benaoum, Accelerated universe from modified Chaplygin gas and tachyonic fluid. arXiv:hep-th/0205140
423. M. Sami, Implementing power law inflation with rolling tachyon on the brane. Mod. Phys. Lett. A **18**, 691 (2003). [arXiv:hep-th/0205146]
424. M. Sami, P. Chingangbam, T. Qureshi, Aspects of tachyonic inflation with exponential potential. Phys. Rev. D **66**, 043530 (2002). [arXiv:hep-th/0205179]
425. G. Shiu, S.-H.H. Tye, I. Wasserman, Rolling tachyon in brane world cosmology from superstring field theory. Phys. Rev. D **67**, 083517 (2003). [arXiv:hep-th/0207119]
426. Y.-S. Piao, R.-G. Cai, X. Zhang, Y.-Z. Zhang, Assisted tachyonic inflation. Phys. Rev. D **66**, 121301(R) (2002). [arXiv:hep-ph/0207143]
427. J.M. Cline, H. Firouzjahi, P. Martineau, Reheating from tachyon condensation. JHEP **0211**, 041 (2002). [arXiv:hep-th/0207156]
428. M.C. Bento, O. Bertolami, A.A. Sen, Tachyonic inflation in the braneworld scenario. Phys. Rev. D **67**, 063511 (2003). [arXiv:hep-th/0208124]
429. G.W. Gibbons, Thoughts on tachyon cosmology. Class. Quantum Grav. **20**, S321 (2003). [arXiv:hep-th/0301117]
430. D.A. Steer, F. Vernizzi, Tachyon inflation: tests and comparison with single scalar field inflation. Phys. Rev. D **70**, 043527 (2004). [arXiv:hep-th/0310139]
431. V. Gorini, A. Kamenshchik, U. Moschella, V. Pasquier, Tachyons, scalar fields and cosmology. Phys. Rev. D **69**, 123512 (2004). [arXiv:hep-th/0311111]
432. M.R. Garousi, M. Sami, S. Tsujikawa, Cosmology from a rolling massive scalar field on the anti-D3 brane of de Sitter vacua. Phys. Rev. D **70**, 043536 (2004). [arXiv:hep-th/0402075]
433. J. Raeymaekers, Tachyonic inflation in a warped string background. JHEP **0410**, 057 (2004). [arXiv:hep-th/0406195]

434. H. Yavartanoo, Cosmological solution from D-brane motion in NS5-branes background. Int. J. Mod. Phys. A **20**, 7633 (2005). [arXiv:hep-th/0407079]
435. G. Calcagni, S. Tsujikawa, Observational constraints on patch inflation in noncommutative spacetime. Phys. Rev. D **70**, 103514 (2004). [arXiv:astro-ph/0407543]
436. A. Ghodsi, A.E. Mosaffa, D-brane dynamics in RR deformation of NS5-branes background and tachyon cosmology. Nucl. Phys. B **714**, 30 (2005). [arXiv:hep-th/0408015]
437. J.-g. Hao, X.-z. Li, Reconstructing the equation of state of the tachyon. Phys. Rev. D **66**, 087301 (2002). [arXiv:hep-th/0209041]
438. J.S. Bagla, H.K. Jassal, T. Padmanabhan, Cosmology with tachyon field as dark energy. Phys. Rev. D **67**, 063504 (2003). [arXiv:astro-ph/0212198]
439. L.R.W. Abramo, F. Finelli, Cosmological dynamics of the tachyon with an inverse power-law potential. Phys. Lett. B **575**, 165 (2003). [arXiv:astro-ph/0307208]
440. L.P. Chimento, Extended tachyon field, Chaplygin gas and solvable k-essence cosmologies. Phys. Rev. D **69**, 123517 (2004). [arXiv:astro-ph/0311613]
441. J.M. Aguirregabiria, R. Lazkoz, Tracking solutions in tachyon cosmology. Phys. Rev. D **69**, 123502 (2004). [arXiv:hep-th/0402190]
442. E.J. Copeland, M.R. Garousi, M. Sami, S. Tsujikawa, What is needed of a tachyon if it is to be the dark energy? Phys. Rev. D **71**, 043003 (2005). [arXiv:hep-th/0411192]
443. G. Calcagni, A.R. Liddle, Tachyon dark energy models: dynamics and constraints. Phys. Rev. D **74**, 043528 (2006). [arXiv:astro-ph/0606003]
444. A.D. Linde, Fast-roll inflation. JHEP **0111**, 052 (2001). [arXiv:hep-th/0110195]
445. A. Sen, Time and tachyon. Int. J. Mod. Phys. A **18**, 4869 (2003). [arXiv:hep-th/0209122]
446. E.S. Fradkin, A.A. Tseytlin, Effective field theory from quantized strings. Phys. Lett. B **158**, 316 (1985)
447. B. Zwiebach, Curvature squared terms and string theories. Phys. Lett. B **156**, 315 (1985)
448. C.G. Callan, D. Friedan, E.J. Martinec, M.J. Perry, Strings in background fields. Nucl. Phys. B **262**, 593 (1985)
449. D.J. Gross, J.H. Sloan, The quartic effective action for the heterotic string. Nucl. Phys. B **291**, 41 (1987)
450. R.R. Metsaev, A.A. Tseytlin, Order α' (two-loop) equivalence of the string equations of motion and the σ-model Weyl invariance conditions: dependence on the dilaton and the antisymmetric tensor. Nucl. Phys. B **293**, 385 (1987)
451. I. Jack, D.R.T. Jones, D.A. Ross, On the relationship between string low-energy effective actions and $O(\alpha'^3)$ σ-model β-functions. Nucl. Phys. B **307**, 130 (1988)
452. I. Antoniadis, J. Rizos, K. Tamvakis, Singularity-free cosmological solutions of the superstring effective action. Nucl. Phys. B **415**, 497 (1994). [arXiv:hep-th/9305025]
453. R. Easther, K.-i. Maeda, One loop superstring cosmology and the nonsingular universe. Phys. Rev. D **54**, 7252 (1996). [arXiv:hep-th/9605173]
454. S. Kawai, M.-a. Sakagami, J. Soda, Instability of 1-loop superstring cosmology. Phys. Lett. B **437**, 284 (1998). [arXiv:gr-qc/9802033]
455. S. Kawai, J. Soda, Evolution of fluctuations during graceful exit in string cosmology. Phys. Lett. B **460**, 41 (1999). [arXiv:gr-qc/9903017]
456. G. Calcagni, B. de Carlos, A. De Felice, Ghost conditions for Gauss–Bonnet cosmologies. Nucl. Phys. B **752**, 404 (2006). [arXiv:hep-th/0604201]
457. N. Kaloper, J. March-Russell, G.D. Starkman, M. Trodden, Compact hyperbolic extra dimensions: branes, Kaluza–Klein modes, and cosmology. Phys. Rev. Lett. **85**, 928 (2000). [arXiv:hep-ph/0002001]
458. G.D. Starkman, D. Stojkovic, M. Trodden, Large extra dimensions and cosmological problems. Phys. Rev. D **63**, 103511 (2001). [arXiv:hep-th/0012226]
459. G.D. Starkman, D. Stojkovic, M. Trodden, Homogeneity, flatness, and "large" extra dimensions. Phys. Rev. Lett. **87**, 231303 (2001). [arXiv:hep-th/0106143]
460. P.K. Townsend, M.N.R. Wohlfarth, Accelerating cosmologies from compactification. Phys. Rev. Lett. **91**, 061302 (2003). [arXiv:hep-th/0303097]

461. R. Emparan, J. Garriga, A note on accelerating cosmologies from compactifications and S-branes. JHEP **0305**, 028 (2003). [arXiv:hep-th/0304124]
462. C.-M. Chen, P.-M. Ho, I.P. Neupane, J.E. Wang, A note on acceleration from product space compactification. JHEP **0307**, 017 (2003). [arXiv:hep-th/0304177]
463. C.-M. Chen, P.-M. Ho, I.P. Neupane, N. Ohta, J.E. Wang, Hyperbolic space cosmologies. JHEP **0310**, 058 (2003). [arXiv:hep-th/0306291]
464. M. Gutperle, R. Kallosh, A.D. Linde, M/string theory, S-branes and accelerating universe. JCAP **0307**, 001 (2003). [arXiv:hep-th/0304225]
465. S.M. Carroll, V. Duvvuri, M. Trodden, M.S. Turner, Is cosmic speed-up due to new gravitational physics? Phys. Rev. D **70**, 043528 (2004). [arXiv:astro-ph/0306438]
466. S. Nojiri, S.D. Odintsov, Where new gravitational physics comes from: M-theory? Phys. Lett. B **576**, 5 (2003). [arXiv:hep-th/0307071]
467. Z.-a. Qiu, A. Strominger, Gauge symmetries in (super)string field theory. Phys. Rev. D **36**, 1794 (1987)
468. D. Ghoshal, A. Sen, Gauge and general coordinate invariance in nonpolynomial closed string theory. Nucl. Phys. B **380**, 103 (1992). [arXiv:hep-th/9110038]
469. J.R. David, $U(1)$ gauge invariance from open string field theory. JHEP **0010**, 017 (2000). [arXiv:hep-th/0005085]
470. A. Sen, BV master action for heterotic and type II string field theories. JHEP **1602**, 087 (2016). [arXiv:1508.05387]
471. A. Sen, Covariant action for type IIB supergravity. JHEP **1607**, 017 (2016). [arXiv:1511.08220]
472. G. Calcagni, L. Modesto, Nonlocal quantum gravity and M-theory. Phys. Rev. D **91**, 124059 (2015). [arXiv:1404.2137]
473. T. Biswas, A. Mazumdar, W. Siegel, Bouncing universes in string-inspired gravity. JCAP **0603**, 009 (2006). [arXiv:hep-th/0508194]
474. T. Biswas, N. Okada, Towards LHC physics with nonlocal Standard Model. Nucl. Phys. B **898**, 113 (2015). [arXiv:1407.3331]
475. A. Belenchia, D.M.T. Benincasa, S. Liberati, F. Marin, F. Marino, A. Ortolan, Tests of quantum gravity induced nonlocality via optomechanical quantum oscillators. Phys. Rev. Lett. **116**, 161303 (2016). [arXiv:1512.02083]
476. G. Veneziano, Scale factor duality for classical and quantum strings. Phys. Lett. B **265**, 287 (1991)
477. K.A. Meissner, G. Veneziano, Symmetries of cosmological superstring vacua. Phys. Lett. B **267**, 33 (1991)
478. M. Gasperini, G. Veneziano, Pre-big-bang in string cosmology. Astropart. Phys. **1**, 317 (1993). [arXiv:hep-th/9211021]
479. M. Gasperini, G. Veneziano, $O(d,d)$-covariant string cosmology. Phys. Lett. B **277**, 256 (1992). [arXiv:hep-th/9112044]
480. M. Gasperini, G. Veneziano, Inflation, deflation, and frame independence in string cosmology. Mod. Phys. Lett. A **8**, 3701 (1993). [arXiv:hep-th/9309023]
481. R. Brustein, G. Veneziano, The graceful exit problem in string cosmology. Phys. Lett. B **329**, 429 (1994). [arXiv:hep-th/9403060]
482. E.J. Copeland, A. Lahiri, D. Wands, Low-energy effective string cosmology. Phys. Rev. D **50**, 4868 (1994). [arXiv:hep-th/9406216]
483. E.J. Copeland, A. Lahiri, D. Wands, String cosmology with a time-dependent antisymmetric tensor potential. Phys. Rev. D **51**, 1569 (1995). [arXiv:hep-th/9410136]
484. R. Brustein, M. Gasperini, M. Giovannini, V.F. Mukhanov, G. Veneziano, Metric perturbations in dilaton driven inflation. Phys. Rev. D **51**, 6744 (1995). [arXiv:hep-th/9501066]
485. N. Kaloper, R. Madden, K.A. Olive, Towards a singularity-free inflationary universe? Nucl. Phys. B **452**, 677 (1995). [arXiv:hep-th/9506027]
486. R. Easther, K.-i. Maeda, D. Wands, Tree level string cosmology. Phys. Rev. D **53**, 4247 (1996). [arXiv:hep-th/9509074]

487. N. Kaloper, R. Madden, K.A. Olive, Axions and the graceful exit problem in string cosmology. Phys. Lett. B **371**, 34 (1996). [arXiv:hep-th/9510117]
488. M. Gasperini, M. Maggiore, G. Veneziano, Towards a non-singular pre-big-bang cosmology. Nucl. Phys. B **494**, 315 (1997). [arXiv:hep-th/9611039]
489. E.J. Copeland, R. Easther, D. Wands, Vacuum fluctuations in axion-dilaton cosmologies. Phys. Rev. D **56**, 874 (1997). [arXiv:hep-th/9701082]
490. R. Brustein, R. Madden, Model of graceful exit in string cosmology. Phys. Rev. D **57**, 712 (1998). [arXiv:hep-th/9708046]
491. J.-c. Hwang, Gravitational wave spectra from pole-like inflations based on generalized gravity theories. Class. Quantum Grav. **15**, 1401 (1998). [arXiv:gr-qc/9710061]
492. S. Foffa, M. Maggiore, R. Sturani, Loop corrections and graceful exit in string cosmology. Nucl. Phys. B **552**, 395 (1999). [arXiv:hep-th/9903008]
493. M. Gasperini, Tensor perturbations in high-curvature string backgrounds. Phys. Rev. D **56**, 4815 (1997). [arXiv:gr-qc/9704045]
494. J.-c. Hwang, H. Noh, Conserved cosmological structures in the one-loop superstring effective action. Phys. Rev. D **61**, 043511 (2000). [arXiv:astro-ph/9909480]
495. C. Cartier, E.J. Copeland, R. Madden, The graceful exit in string cosmology. JHEP **0001**, 035 (2000). [arXiv:hep-th/9910169]
496. C. Cartier, J.-c. Hwang, E.J. Copeland, Evolution of cosmological perturbations in nonsingular string cosmologies. Phys. Rev. D **64**, 103504 (2001). [arXiv:astro-ph/0106197]
497. F. Finelli, Assisted contraction. Phys. Lett. B **545**, 1 (2002). [arXiv:hep-th/0206112]
498. M. Gasperini, G. Veneziano, The pre-big bang scenario in string cosmology. Phys. Rep. **373**, 1 (2003). [arXiv:hep-th/0207130]
499. M. Gasperini, M. Giovannini, G. Veneziano, Perturbations in a non-singular bouncing Universe. Phys. Lett. B **569**, 113 (2003). [arXiv:hep-th/0306113]
500. M. Gasperini, M. Giovannini, G. Veneziano, Cosmological perturbations across a curvature bounce. Nucl. Phys. B **694**, 206 (2004). [arXiv:hep-th/0401112]
501. L.E. Allen, D. Wands, Cosmological perturbations through a simple bounce. Phys. Rev. D **70**, 063515 (2004). [arXiv:astro-ph/0404441]
502. M. Dine, N. Seiberg, Is the superstring weakly coupled? Phys. Lett. B **162**, 299 (1985)
503. W. Buchmüller, K. Hamaguchi, O. Lebedev, M. Ratz, Dilaton destabilization at high temperature. Nucl. Phys. B **699**, 292 (2004). [arXiv:hep-th/0404168]
504. E. Witten, Some properties of $O(32)$ superstrings. Phys. Lett. B **149**, 351 (1984)
505. T.R. Taylor, G. Veneziano, Dilaton couplings at large distances. Phys. Lett. B **213**, 450 (1988)
506. T. Damour, A.M. Polyakov, The string dilaton and a least coupling principle. Nucl. Phys. B **423**, 532 (1994). [arXiv:hep-th/9401069]
507. T. Damour, A.M. Polyakov, String theory and gravity. Gen. Relat. Grav. **26**, 1171 (1994). [arXiv:gr-qc/9411069]
508. M. Gasperini, F. Piazza, G. Veneziano, Quintessence as a runaway dilaton. Phys. Rev. D **65**, 023508 (2002). [arXiv:gr-qc/0108016]
509. F. Piazza, S. Tsujikawa, Dilatonic ghost condensate as dark energy. JCAP **0407**, 004 (2004). [arXiv:hep-th/0405054]
510. G. Veneziano, Large-N bounds on, and compositeness limit of, gauge and gravitational interactions. JHEP **0206**, 051 (2002). [arXiv:hep-th/0110129]
511. T. Damour, F. Piazza, G. Veneziano, Runaway dilaton and equivalence principle violations. Phys. Rev. Lett. **89**, 081601 (2002). [arXiv:gr-qc/0204094]
512. T. Damour, F. Piazza, G. Veneziano, Violations of the equivalence principle in a dilaton runaway scenario. Phys. Rev. D **66**, 046007 (2002). [arXiv:hep-th/0205111]
513. C.J.A.P. Martins, P.E. Vielzeuf, M. Martinelli, E. Calabrese, S. Pandolfi, Evolution of the fine-structure constant in runaway dilaton models. Phys. Lett. B **743**, 377 (2015). [arXiv:1503.05068]
514. R. Brandenberger, C. Vafa, Superstrings in the early universe. Nucl. Phys. B **316**, 391 (1989)
515. N. Deo, S. Jain, O. Narayan, C.-I Tan, Effect of topology on the thermodynamic limit for a string gas. Phys. Rev. D **45**, 3641 (1992)

516. A.A. Tseytlin, C. Vafa, Elements of string cosmology. Nucl. Phys. B **372**, 443 (1992). [arXiv:hep-th/9109048]
517. R. Hagedorn, Statistical thermodynamics of strong interactions at high energies. Nuovo Cim. Suppl. **3**, 147 (1965)
518. M. Sakellariadou, Numerical experiments on string cosmology. Nucl. Phys. B **468**, 319 (1996). [arXiv:hep-th/9511075]
519. S. Alexander, R.H. Brandenberger, D.A. Easson, Brane gases in the early universe. Phys. Rev. D **62**, 103509 (2000). [arXiv:hep-th/0005212]
520. R. Brandenberger, D.A. Easson, D. Kimberly, Loitering phase in brane gas cosmology. Nucl. Phys. B **623**, 421 (2002). [arXiv:hep-th/0109165]
521. D.A. Easson, Brane gases on K3 and Calabi–Yau manifolds. Int. J. Mod. Phys. A **18**, 4295 (2003). [arXiv:hep-th/0110225]
522. R. Easther, B.R. Greene, M.G. Jackson, Cosmological string gas on orbifolds. Phys. Rev. D **66**, 023502 (2002). [arXiv:hep-th/0204099]
523. S. Watson, R.H. Brandenberger, Isotropization in brane gas cosmology. Phys. Rev. D **67**, 043510 (2003). [arXiv:hep-th/0207168]
524. R. Easther, B.R. Greene, M.G. Jackson, D.N. Kabat, Brane gas cosmology in M theory: late time behavior. Phys. Rev. D **67**, 123501 (2003). [arXiv:hep-th/0211124]
525. S. Watson, R.H. Brandenberger, Stabilization of extra dimensions at tree level. JCAP **0311**, 008 (2003). [arXiv:hep-th/0307044]
526. R. Easther, B.R. Greene, M.G. Jackson, D.N. Kabat, Brane gases in the early universe: thermodynamics and cosmology. JCAP **0401**, 006 (2004). [arXiv:hep-th/0307233]
527. S.P. Patil, R. Brandenberger, Radion stabilization by stringy effects in general relativity. Phys. Rev. D **71**, 103522 (2005). [arXiv:hep-th/0401037]
528. S. Watson, Moduli stabilization with the string Higgs effect. Phys. Rev. D **70**, 066005 (2004). [arXiv:hep-th/0404177]
529. R. Easther, B.R. Greene, M.G. Jackson, D.N. Kabat, String windings in the early universe. JCAP **0502**, 009 (2005). [arXiv:hep-th/0409121]
530. Y.-K.E. Cheung, S. Watson, R. Brandenberger, Moduli stabilization with string gas and fluxes. JHEP **0605**, 025 (2006). [arXiv:hep-th/0501032]
531. S.P. Patil, R. Brandenberger, The cosmology of massless string modes. JCAP **0601**, 005 (2006). [arXiv:hep-th/0502069]
532. T. Battefeld, S. Watson, String gas cosmology. Rev. Mod. Phys. **78**, 435 (2006). [arXiv:hep-th/0510022]
533. A. Nayeri, R.H. Brandenberger, C. Vafa, Producing a scale-invariant spectrum of perturbations in a Hagedorn phase of string cosmology. Phys. Rev. Lett. **97**, 021302 (2006). [arXiv:hep-th/0511140]
534. R.H. Brandenberger, A. Nayeri, S.P. Patil, C. Vafa, Tensor modes from a primordial Hagedorn phase of string cosmology. Phys. Rev. Lett. **98**, 231302 (2007). [arXiv:hep-th/0604126]
535. R.H. Brandenberger, A. Nayeri, S.P. Patil, C. Vafa, String gas cosmology and structure formation. Int. J. Mod. Phys. A **22**, 3621 (2007). [arXiv:hep-th/0608121]
536. N. Kaloper, L. Kofman, A.D. Linde, V. Mukhanov, On the new string theory inspired mechanism of generation of cosmological perturbations. JCAP **0610**, 006 (2006). [arXiv:hep-th/0608200]
537. T. Biswas, R. Brandenberger, A. Mazumdar, W. Siegel, Non-perturbative gravity, the Hagedorn bounce and the cosmic microwave background. JCAP **0712**, 011 (2007). [arXiv:hep-th/0610274]
538. N. Kaloper, S. Watson, Geometric precipices in string cosmology. Phys. Rev. D **77**, 066002 (2008). [arXiv:0712.1820]
539. B. Chen, Y. Wang, W. Xue, R. Brandenberger, String gas cosmology and non-Gaussianities. arXiv:0712.2477
540. R.J. Danos, A.R. Frey, R.H. Brandenberger, Stabilizing moduli with thermal matter and nonperturbative effects. Phys. Rev. D **77**, 126009 (2008). [arXiv:0802.1557]

541. S. Mishra, W. Xue, R. Brandenberger, U. Yajnik, Supersymmetry breaking and dilaton stabilization in string gas cosmology. JCAP **1209**, 015 (2012). [arXiv:1103.1389]
542. R.H. Brandenberger, A. Nayeri, S.P. Patil, Closed string thermodynamics and a blue tensor spectrum. Phys. Rev. D **90**, 067301 (2014). [arXiv:1403.4927]
543. R.H. Brandenberger, String gas cosmology: progress and problems. Class. Quantum Grav. **28**, 204005 (2011). [arXiv:1105.3247]
544. R.H. Brandenberger, Unconventional cosmology. Lect. Notes Phys. **863**, 333 (2013). [arXiv:1203.6698]
545. E. Alvarez, Superstring cosmology. Phys. Rev. D **31**, 418 (1985); Erratum-ibid. D **33**, 1206 (1986)
546. M.J. Bowick, L.C.R. Wijewardhana, Superstrings at high temperature. Phys. Rev. Lett. **54**, 2485 (1985)
547. E. Alvarez, Strings at finite temperature. Nucl. Phys. B **269**, 596 (1986)
548. D. Mitchell, N. Turok, Statistical properties of cosmic strings. Nucl. Phys. B **294**, 1138 (1987)
549. P.J. Steinhardt, N. Turok, A cyclic model of the universe. Science **296**, 1436 (2002). [arXiv:hep-th/0111030]
550. P.J. Steinhardt, N. Turok, Cosmic evolution in a cyclic universe. Phys. Rev. D **65**, 126003 (2002). [arXiv:hep-th/0111098]
551. D.H. Lyth, The failure of cosmological perturbation theory in the new ekpyrotic and cyclic ekpyrotic scenarios. Phys. Lett. B **526**, 173 (2002). [arXiv:hep-ph/0110007]
552. S. Tsujikawa, Density perturbations in the ekpyrotic universe and string inspired generalizations. Phys. Lett. B **526**, 179 (2002). [arXiv:gr-qc/0110124]
553. J. Martin, P. Peter, N. Pinto Neto, D.J. Schwarz, Passing through the bounce in the ekpyrotic models. Phys. Rev. D **65**, 123513 (2002). [arXiv:hep-th/0112128]
554. F. Finelli, R. Brandenberger, Generation of a scale-invariant spectrum of adiabatic fluctuations in cosmological models with a contracting phase. Phys. Rev. D **65**, 103522 (2002). [arXiv:hep-th/0112249]
555. S. Gratton, J. Khoury, P.J. Steinhardt, N. Turok, Conditions for generating scale-invariant density perturbations. Phys. Rev. D **69**, 103505 (2004). [arXiv:astro-ph/0301395]
556. A.J. Tolley, N. Turok, P.J. Steinhardt, Cosmological perturbations in a big-crunch–big-bang space-time. Phys. Rev. D **69**, 106005 (2004). [arXiv:hep-th/0306109]
557. J. Khoury, P.J. Steinhardt, N. Turok, Designing cyclic universe models. Phys. Rev. Lett. **92**, 031302 (2004). [arXiv:hep-th/0307132]
558. J.K. Erickson, D.H. Wesley, P.J. Steinhardt, N. Turok, Kasner and mixmaster behavior in universes with equation of state $w \gtrsim 1$. Phys. Rev. D **69**, 063514 (2004). [arXiv:hep-th/0312009]
559. N. Turok, M. Perry, P.J. Steinhardt, M theory model of a big crunch/big bang transition. Phys. Rev. D **70**, 106004 (2004); Erratum-Ibid. D **71**, 029901 (2005). [arXiv:hep-th/0408083]
560. P. Creminelli, A. Nicolis, M. Zaldarriaga, Perturbations in bouncing cosmologies: dynamical attractor versus scale invariance. Phys. Rev. D **71**, 063505 (2005). [arXiv:hep-th/0411270]
561. P.J. Steinhardt, N. Turok, Why the cosmological constant is small and positive. Science **312**, 1180 (2006). [arXiv:astro-ph/0605173]
562. J.K. Erickson, S. Gratton, P.J. Steinhardt, N. Turok, Cosmic perturbations through the cyclic ages. Phys. Rev. D **75**, 123507 (2007). [arXiv:hep-th/0607164]
563. J.-L. Lehners, P. McFadden, N. Turok, P.J. Steinhardt, Generating ekpyrotic curvature perturbations before the big bang. Phys. Rev. D **76**, 103501 (2007). [arXiv:hep-th/0702153]
564. E.I. Buchbinder, J. Khoury, B.A. Ovrut, New ekpyrotic cosmology. Phys. Rev. D **76**, 123503 (2007). [arXiv:hep-th/0702154]
565. K. Koyama, D. Wands, Ekpyrotic collapse with multiple fields. JCAP **0704**, 008 (2007). [arXiv:hep-th/0703040]
566. A.J. Tolley, D.H. Wesley, Scale-invariance in expanding and contracting universes from two-field models. JCAP **0705**, 006 (2007). [arXiv:hep-th/0703101]
567. E.I. Buchbinder, J. Khoury, B.A. Ovrut, On the initial conditions in new ekpyrotic cosmology. JHEP **0711**, 076 (2007). [arXiv:0706.3903]
568. K. Koyama, S. Mizuno, F. Vernizzi, D. Wands, Non-Gaussianities from ekpyrotic collapse with multiple fields. JCAP **0711**, 024 (2007). [arXiv:0708.4321]

569. E.I. Buchbinder, J. Khoury, B.A. Ovrut, Non-Gaussianities in new ekpyrotic cosmology. Phys. Rev. Lett. **100**, 171302 (2008). [arXiv:0710.5172]
570. J.-L. Lehners, P.J. Steinhardt, Non-Gaussian density fluctuations from entropically generated curvature perturbations in ekpyrotic models. Phys. Rev. D **77**, 063533 (2008); Erratum-ibid. D **79**, 129903(E) (2009). [arXiv:0712.3779]
571. J.-L. Lehners, P.J. Steinhardt, Intuitive understanding of non-Gaussianity in ekpyrotic and cyclic models. Phys. Rev. D **78**, 023506 (2008); Erratum-ibid. D **79**, 129902(E) (2009). [arXiv:0804.1293]
572. J. Khoury, F. Piazza, Rapidly-varying speed of sound, scale invariance and non-Gaussian signatures. JCAP **0907**, 026 (2009). [arXiv:0811.3633]
573. J. Khoury, P.J. Steinhardt, Adiabatic ekpyrosis: scale-invariant curvature perturbations from a single scalar field in a contracting universe. Phys. Rev. Lett. **104**, 091301 (2010). [arXiv:0910.2230]
574. A. Linde, V. Mukhanov, A. Vikman, On adiabatic perturbations in the ekpyrotic scenario. JCAP **1002**, 006 (2010). [arXiv:0912.0944]
575. J. Khoury, P.J. Steinhardt, Generating scale-invariant perturbations from rapidly-evolving equation of state. Phys. Rev. D **83**, 123502 (2011). [arXiv:1101.3548]
576. Y.-F. Cai, D.A. Easson, R. Brandenberger, Towards a nonsingular bouncing cosmology. JCAP **1208**, 020 (2012). [arXiv:1206.2382]
577. T. Qiu, X. Gao, E.N. Saridakis, Towards anisotropy-free and nonsingular bounce cosmology with scale-invariant perturbations. Phys. Rev. D **88**, 043525 (2013). [arXiv:1303.2372]
578. M. Li, Note on the production of scale-invariant entropy perturbation in the ekpyrotic universe. Phys. Lett. B **724**, 192 (2013). [arXiv:1306.0191]
579. B. Xue, D. Garfinkle, F. Pretorius, P.J. Steinhardt, Nonperturbative analysis of the evolution of cosmological perturbations through a nonsingular bounce. Phys. Rev. D **88**, 083509 (2013). [arXiv:1308.3044]
580. A. Fertig, J.-L. Lehners, E. Mallwitz, Ekpyrotic perturbations with small non-Gaussian corrections. Phys. Rev. D **89**, 103537 (2014). [arXiv:1310.8133]
581. A. Ijjas, J.-L. Lehners, P.J. Steinhardt, General mechanism for producing scale-invariant perturbations and small non-Gaussianity in ekpyrotic models. Phys. Rev. D **89**, 123520 (2014). [arXiv:1404.1265]
582. A.M. Levy, A. Ijjas, P.J. Steinhardt, Scale-invariant perturbations in ekpyrotic cosmologies without fine-tuning of initial conditions. Phys. Rev. D **92**, 063524 (2015). [arXiv:1506.01011]
583. J. Khoury, B.A. Ovrut, P.J. Steinhardt, N. Turok, Ekpyrotic universe: colliding branes and the origin of the hot big bang. Phys. Rev. D **64**, 123522 (2001). [arXiv:hep-th/0103239]
584. D.H. Lyth, The primordial curvature perturbation in the ekpyrotic universe. Phys. Lett. B **524**, 1 (2002). [arXiv:hep-ph/0106153]
585. J. Khoury, B.A. Ovrut, N. Seiberg, P.J. Steinhardt, N. Turok, From big crunch to big bang. Phys. Rev. D **65**, 086007 (2002). [arXiv:hep-th/0108187]
586. R. Brandenberger, F. Finelli, On the spectrum of fluctuations in an effective field theory of the ekpyrotic universe. JHEP **0111**, 056 (2001). [arXiv:hep-th/0109004]
587. J. Khoury, B.A. Ovrut, P.J. Steinhardt, N. Turok, Density perturbations in the ekpyrotic scenario. Phys. Rev. D **66**, 046005 (2002). [arXiv:hep-th/0109050]
588. J.-L. Lehners, Ekpyrotic and cyclic cosmology. Phys. Rep. **465**, 223 (2008). [arXiv:0806.1245]
589. J.-L. Lehners, Ekpyrotic nongaussianity: a review. Adv. Astron. **2010**, 903907 (2010). [arXiv:1001.3125]
590. L.F. Abbott, A mechanism for reducing the value of the cosmological constant. Phys. Lett. B **150**, 427 (1985)
591. R. Bousso, The cosmological constant. Gen. Relat. Grav. **40**, 607 (2008). [arXiv:0708.4231]
592. I. Bena, M. Graña, N. Halmagyi, On the existence of meta-stable vacua in Klebanov–Strassler. JHEP **1009**, 087 (2010). [arXiv:0912.3519]

593. A. Dymarsky, On gravity dual of a metastable vacuum in Klebanov–Strassler theory. JHEP **1105**, 053 (2011). [arXiv:1102.1734]
594. I. Bena, G. Giecold, M. Graña, N. Halmagyi, S. Massai, On metastable vacua and the warped deformed conifold: analytic results. Class. Quantum Grav. **30**, 015003 (2013). [arXiv:1102.2403]
595. J. Bläbäck, U.H. Danielsson, T. Van Riet, Resolving anti-brane singularities through time-dependence. JHEP **1302**, 061 (2013). [arXiv:1202.1132]
596. I. Bena, M. Graña, S. Kuperstein, S. Massai, Anti-D3 branes: singular to the bitter end. Phys. Rev. D **87**, 106010 (2013). [arXiv:1206.6369]
597. D. Junghans, Dynamics of warped flux compactifications with backreacting antibranes. Phys. Rev. D **89**, 126007 (2014). [arXiv:1402.4571]
598. D. Junghans, D. Schmidt, M. Zagermann, Curvature-induced resolution of anti-brane singularities. JHEP **1410**, 34 (2014). [arXiv:1402.6040]
599. J. Bläbäck, U.H. Danielsson, D. Junghans, T. Van Riet, S.C. Vargas, Localised anti-branes in non-compact throats at zero and finite T. JHEP **1502**, 018 (2015). [arXiv:1409.0534]
600. U.H. Danielsson, T. Van Riet, Fatal attraction: more on decaying anti-branes. JHEP **1503**, 087 (2015). [arXiv:1410.8476]
601. B. Michel, E. Mintun, J. Polchinski, A. Puhm, P. Saad, Remarks on brane and antibrane dynamics. JHEP **1509**, 021 (2015). [arXiv:1412.5702]
602. G.S. Hartnett, Localised anti-branes in flux backgrounds. JHEP **1506**, 007 (2015). [arXiv:1501.06568]
603. J. Polchinski, Brane/antibrane dynamics and KKLT stability. arXiv:1509.05710
604. G. Wang, T. Battefeld, Vacuum selection on axionic landscapes. JCAP **1604**, 025 (2016). [arXiv:1512.04224]
605. H. Liu, G.W. Moore, N. Seiberg, The challenging cosmic singularity. arXiv:gr-qc/0301001
606. L. Cornalba, M.S. Costa, Time-dependent orbifolds and string cosmology. Fortsch. Phys. **52**, 145 (2004). [arXiv:hep-th/0310099]
607. B. Craps, Big bang models in string theory. Class. Quantum Grav. **23**, S849 (2006). [arXiv:hep-th/0605199]
608. M. Berkooz, D. Reichmann, A short review of time dependent solutions and space-like singularities in string theory. Nucl. Phys. Proc. Suppl. **171**, 69 (2007). [arXiv:0705.2146]
609. H. Liu, G.W. Moore, N. Seiberg, Strings in a time-dependent orbifold. JHEP **0206**, 045 (2002). [arXiv:hep-th/0204168]
610. B. Craps, D. Kutasov, G. Rajesh, String propagation in the presence of cosmological singularities. JHEP **0206**, 053 (2002). [arXiv:hep-th/0205101]
611. A. Lawrence, On the instability of 3d null singularities. JHEP **0211**, 019 (2002). [arXiv:hep-th/0205288]
612. H. Liu, G.W. Moore, N. Seiberg, Strings in time-dependent orbifolds. JHEP **0210**, 031 (2002). [arXiv:hep-th/0206182]
613. M. Fabinger, J. McGreevy, On smooth time-dependent orbifolds and null singularities. JHEP **0306**, 042 (2003). [arXiv:hep-th/0206196]
614. G.T. Horowitz, J. Polchinski, Instability of spacelike and null orbifold singularities. Phys. Rev.D **66**, 103512 (2002). [arXiv:hep-th/0206228]
615. M. Berkooz, B. Craps, D. Kutasov, G. Rajesh, Comments on cosmological singularities in string theory. JHEP **0303**, 031 (2003). [arXiv:hep-th/0212215]
616. L. Cornalba, M.S. Costa, On the classical stability of orientifold cosmologies. Class. Quantum Grav. **20**, 3969 (2003) .[arXiv:hep-th/0302137]
617. M. Berkooz, B. Pioline, Strings in an electric field, and the Milne universe. JCAP **0311**, 007 (2003). [arXiv:hep-th/0307280]
618. M. Berkooz, B. Pioline, M. Rozali, Closed strings in Misner space: cosmological production of winding strings. JCAP **0408**, 004 (2004). [arXiv:hep-th/0405126]

619. B. Durin, B. Pioline, Closed strings in Misner space: a toy model for a big bounce? in *String Theory: From Gauge Interactions to Cosmology*, ed. by L. Baulieu, J. de Boer, B. Pioline, E. Rabinovici (Springer, Berlin/Germany, 2006); NATO Sci. Ser. II **208**, 177 (2006). [arxiv:hep-th/0501145]
620. J. McGreevy, E. Silverstein, The tachyon at the end of the universe. JHEP **0508**, 090 (2005). [arXiv:hep-th/0506130]
621. A. Adams, X. Liu, J. McGreevy, A. Saltman, E. Silverstein, Things fall apart: topology change from winding tachyons. JHEP **0510**, 033 (2005). [arXiv:hep-th/0502021]
622. G.T. Horowitz, Tachyon condensation and black strings. JHEP **0508**, 091 (2005). [arXiv:hep-th/0506166]
623. A. Strominger, T. Takayanagi, Correlators in timelike bulk Liouville theory. Adv. Theor. Math. Phys. **7**, 369 (2003). [arXiv:hep-th/0303221]
624. Y. Nakayama, S.J. Rey, Y. Sugawara, The Nothing at the beginning of the universe made precise. arXiv:hep-th/0606127
625. G.T. Horowitz, E. Silverstein, The inside story: quasilocal tachyons and black holes. Phys. Rev. D **73**, 064016 (2006). [arXiv:hep-th/0601032]
626. P. Kraus, H. Ooguri, S. Shenker, Inside the horizon with AdS/CFT. Phys. Rev. D **67**, 124022 (2003). [arXiv:hep-th/0212277]
627. L. Fidkowski, V. Hubeny, M. Kleban, S. Shenker, The black hole singularity in AdS/CFT. JHEP **0402**, 014 (2004). [arXiv:hep-th/0306170]
628. G. Festuccia, H. Liu, Excursions beyond the horizon: black hole singularities in Yang–Mills theories (I). JHEP **0604**, 044 (2006). [arXiv:hep-th/0506202]
629. T. Hertog, G.T. Horowitz, Towards a big crunch dual. JHEP **0407**, 073 (2004). [arXiv:hep-th/0406134]
630. T. Hertog, G.T. Horowitz, Holographic description of AdS cosmologies. JHEP **0504**, 005 (2005). [arXiv:hep-th/0503071]
631. B. Craps, T. Hertog, N. Turok, Quantum resolution of cosmological singularities using AdS/CFT correspondence. Phys. Rev. D **86**, 043513 (2012). [arxiv:hep-th/0712.4180]
632. C.-S. Chu, P.-M. Ho, Time-dependent AdS/CFT duality and null singularity. JHEP **0604**, 013 (2006). [arXiv:hep-th/0602054]
633. S.R. Das, J. Michelson, K. Narayan, S.P. Trivedi, Time-dependent cosmologies and their duals. Phys. Rev. D **74**, 026002 (2006). [arXiv:hep-th/0602107]
634. B. Craps, S. Sethi, E.P. Verlinde, A matrix big bang. JHEP **0510**, 005 (2005). [arXiv:hep-th/0506180]
635. B. Craps, A. Rajaraman, S. Sethi, Effective dynamics of the matrix big bang. Phys. Rev. D **73**, 106005 (2006). [arXiv:hep-th/0601062]
636. T. Banks, W. Fischler, L. Motl, Dualities versus singularities. JHEP **9901**, 019 (1999). [arXiv:hep-th/9811194]
637. A. Feinstein, M.A. Vázquez-Mozo, M-theory resolution of four-dimensional cosmological singularities via U-duality. Nucl. Phys. B **568**, 405 (2000). [arXiv:hep-th/9906006]
638. C.W. Misner, Mixmaster universe. Phys. Rev. Lett. **22**, 1071 (1969)
639. A.A. Kirillov, On the nature of the spatial distribution of metric inhomogeneities in the general solution of the Einstein equations near a cosmological singularity. Zh. Eksp. Teor. Fiz. **103**, 721 (1993) [Sov. Phys. JETP **76**, 355 (1993)]
640. V.D. Ivashchuk, V.N. Melnikov, A.A. Kirillov, Stochastic properties of multidimensional cosmological models near a singular point. Pis'ma Zh. Eksp. Teor. Fiz. **60**, 225 (1994) [JETP Lett. **60**, 235 (1994)]
641. V.D. Ivashchuk, V.N. Melnikov, Billiard representation for multidimensional cosmology with multicomponent perfect fluid near the singularity. Class. Quantum Grav. **12**, 809 (1995)
642. A.A. Kirillov, V.N. Melnikov, Dynamics of inhomogeneities of metric in the vicinity of a singularity in multidimensional cosmology. Phys. Rev. D **52**, 723 (1995). [arXiv:gr-qc/9408004]
643. V.D. Ivashchuk, V.N. Melnikov, Billiard representation for multidimensional cosmology with intersecting p-branes near the singularity. J. Math. Phys. **41**, 6341 (2000). [arXiv:hep-th/9904077]

644. T. Damour, M. Henneaux, Chaos in superstring cosmology. Phys. Rev. Lett. **85**, 920 (2000). [arXiv:hep-th/0003139]
645. T. Damour, M. Henneaux, Oscillatory behaviour in homogeneous string cosmology models. Phys. Lett. B **488**, 108 (2000); Erratum-ibid. B **491**, 377 (2000). [arXiv:hep-th/0006171]
646. T. Damour, M. Henneaux, E_{10}, BE_{10} and arithmetical chaos in superstring cosmology. Phys. Rev. Lett. **86**, 4749 (2001). [arXiv:hep-th/0012172]
647. T. Damour, S. de Buyl, M. Henneaux, C. Schomblond, Einstein billiards and overextensions of finite-dimensional simple Lie algebras. JHEP **0208**, 030 (2002). [arXiv:hep-th/0206125]
648. T. Damour, M. Henneaux, H. Nicolai, E_{10} and a small tension expansion of M theory. Phys. Rev. Lett. **89**, 221601 (2002). [arXiv:hep-th/0207267]
649. T. Damour, M. Henneaux, H. Nicolai, Cosmological billiards. Class. Quantum Grav. **20**, R145 (2003). [arXiv:hep-th/0212256]
650. M. Henneaux, B. Julia, Hyperbolic billiards of pure $D = 4$ supergravities. JHEP **0305**, 047 (2003). [arXiv:hep-th/0304233]
651. T. Damour, A. Kleinschmidt, H. Nicolai, Hidden symmetries and the fermionic sector of eleven-dimensional supergravity. Phys. Lett. B **634**, 319 (2006). [arXiv:hep-th/0512163]
652. S. de Buyl, M. Henneaux, L. Paulot, Extended E_8 invariance of 11-dimensional supergravity. JHEP **0602**, 056 (2006). [arXiv:hep-th/0512292]
653. T. Damour, A. Kleinschmidt, H. Nicolai, $K(E_{10})$, supergravity and fermions. JHEP **0608**, 046 (2006). [arXiv:hep-th/0606105]
654. T. Damour, A. Kleinschmidt, H. Nicolai, Constraints and the E_{10} coset model. Class. Quantum Grav. **24**, 6097 (2007). [arXiv:0709.2691]
655. T. Damour, C. Hillmann, Fermionic Kac–Moody billiards and supergravity. JHEP **0908**, 100 (2009). [arXiv:0906.3116]
656. T. Damour, A. Kleinschmidt, H. Nicolai, Sugawara-type constraints in hyperbolic coset models. Commun. Math. Phys. **302**, 755 (2011). [arXiv:0912.3491]
657. M. Henneaux, D. Persson, P. Spindel, Spacelike singularities and hidden symmetries of gravity. Living Rev. Relat. **11**, 1 (2008)
658. V.A. Belinskiĭ, I.M. Khalatnikov, Effect of scalar and vector fields on the nature of the cosmological singularity. Zh. Ehsp. Teor. Fiz. **63**, 1121 (1972) [Sov. Phys. JETP **36**, 591 (1973)]
659. J. Demaret, M. Henneaux, P. Spindel, Non-oscillatory behavior in vacuum Kaluza–Klein cosmologies. Phys. Lett. B **164**, 27 (1985)
660. J. Demaret, Y. De Rop, M. Henneaux, Are Kaluza–Klein models of the universe chaotic? Int. J. Theor. Phys. **28**, 1067 (1989)
661. T. Damour, M. Henneaux, B. Julia, H. Nicolai, Hyperbolic Kac–Moody algebras and chaos in Kaluza–Klein models. Phys. Lett. B **509**, 323 (2001). [arXiv:hep-th/0103094]
662. V.D. Ivashchuk, V.N. Melnikov, Perfect-fluid type solution in multidimensional cosmology. Phys. Lett. A **136**, 465 (1989)
663. V.D. Ivashchuk, V.N. Melnikov, A.I. Zhuk, On Wheeler–DeWitt equation in multidimensional cosmology. Nuovo Cim. B **104**, 575 (1989)
664. V.D. Ivashchuk, V.N. Melnikov, Multidimensional cosmology with m-component perfect fluid. Int. J. Mod. Phys. D **3**, 795 (1994). [arXiv:gr-qc/9403064]
665. J. Fuchs, C. Schweigert, *Symmetries, Lie Algebras and Representations* (Cambridge University Press, Cambridge, 2003)
666. N.A. Obers, B. Pioline, E. Rabinovici, M-theory and U-duality on T^d with gauge backgrounds. Nucl. Phys. B **525**, 163 (1998). [arXiv:hep-th/9712084]
667. N.A. Obers, B. Pioline, U-duality and M-theory. Phys. Rep. **318**, 113 (1999). [arXiv:hep-th/9809039]
668. J. Brown, O.J. Ganor, C. Helfgott, M theory and E_{10}: billiards, branes, and imaginary roots. JHEP **0408**, 063 (2004). [arXiv:hep-th/0401053]
669. J. Brown, S. Ganguli, O.J. Ganor, C. Helfgott, E_{10} orbifolds. JHEP **0506**, 057 (2005). [arXiv:hep-th/0409037]

670. M. Henneaux, M. Leston, D. Persson, P. Spindel, Geometric configurations, regular subalgebras of E_{10} and M-theory cosmology. JHEP **0610**, 021 (2006). [arXiv:hep-th/0606123]
671. M. Bagnoud, L. Carlevaro, Hidden Borcherds symmetries in $\mathbb{Z}_n$ orbifolds of M-theory and magnetized D-branes in type 0′ orientifolds. JHEP **0611**, 003 (2006). [arXiv:hep-th/0607136]
672. T. Damour, H. Nicolai, Eleven dimensional supergravity and the $E_{10}/K(E_{10})$ σ-model at low A_9 levels. arXiv:hep-th/0410245
673. T. Damour, H. Nicolai, Higher-order M-theory corrections and the Kac–Moody algebra E_{10}. Class. Quantum Grav. **22**, 2849 (2005). [arXiv:hep-th/0504153]
674. V.D. Ivashchuk, V.N. Melnikov, Multidimensional classical and quantum cosmology with intersecting p-branes. J. Math. Phys. **39**, 2866 (1998). [arXiv:hep-th/9708157]
675. L.A. Forte, Arithmetical chaos and quantum cosmology. Class. Quantum Grav. **26**, 045001 (2009). [arXiv:0812.4382]
676. A. Kleinschmidt, M. Koehn, H. Nicolai, Supersymmetric quantum cosmological billiards. Phys. Rev. D **80**, 061701 (2009). [arXiv:0907.3048]
677. M. Koehn, Relativistic wavepackets in classically chaotic quantum cosmological billiards. Phys. Rev. D **85**, 063501 (2012). [arXiv:1107.6023]
678. V.D. Ivashchuk, V.N. Melnikov, Quantum billiards in multidimensional models with fields of forms. Grav. Cosmol. **19**, 171 (2013). [arXiv:1306.6521]
679. V.D. Ivashchuk, V.N. Melnikov, Quantum billiards in multidimensional models with branes. Eur. Phys. J. C **74**, 2805 (2014). [arXiv:1310.4451]
680. C.W. Misner, Minisuperspace, in *Magic Without Magic*, ed. by J.R. Klauder (Freeman, San Francisco, 1972)
681. T. Damour, H. Nicolai, Symmetries, singularities and the de-emergence of space. Int. J. Mod. Phys. D **17**, 525 (2008). [arXiv:0705.2643]
682. B. Julia, Kac–Moody symmetry of gravitation and supergravity theories, in *Applications of Group Theory in Physics and Mathematical Physics*, ed. by M. Flato, P. Sally, G. Zuckerman. Lectures in Applied Mathematics, vol. 21 (AMS, Providence, 1985)
683. G.W. Moore, Finite in all directions. arXiv:hep-th/9305139
684. S. Mizoguchi, E_{10} symmetry in one-dimensional supergravity. Nucl. Phys. B **528**, 238 (1998). [arXiv:hep-th/9703160]
685. S. Elitzur, A. Giveon, D. Kutasov, E. Rabinovici, Algebraic aspects of matrix theory on T^d. Nucl. Phys. B **509**, 122 (1998). [arXiv:hep-th/9707217]
686. O.J. Ganor, Two conjectures on gauge theories, gravity, and infinite dimensional Kac–Moody groups. arXiv:hep-th/9903110
687. P.C. West, E_{11} and M theory. Class. Quantum Grav. **18**, 4443 (2001). [arXiv:hep-th/0104081]
688. I. Schnakenburg, P.C. West, Kac–Moody symmetries of 2B supergravity. Phys. Lett. B **517**, 421 (2001). [arXiv:hep-th/0107181]
689. F. Englert, L. Houart, A. Taormina, P.C. West, The symmetry of M-theories. JHEP **0309**, 020 (2003). [arXiv:hep-th/0304206]
690. P.C. West, E_{11}, $SL(32)$ and central charges. Phys. Lett. B **575**, 333 (2003). [arXiv:hep-th/0307098]
691. F. Englert, L. Houart, $\mathcal{G}$+++ invariant formulation of gravity and M-theories: exact BPS solutions. JHEP **0401**, 002 (2004). [arXiv:hep-th/0311255]

Chapter 14
Perspective

And I said to my spirit, When we become the enfolders of those orbs and the pleasure and knowledge of every thing in them, shall we be filled and satisfied then?
And my spirit said No, we level that lift to pass and continue beyond.

— Walt Whitman, *Leaves of Grass* (1855 edition)

Observations have verified the cosmic concordance model to a high degree of accuracy. We have detailed information about the accelerating phase of the early universe and the formation of cosmic structures since then, from super-clusters of galaxies down to our solar system. Still, the theoretical framework is largely incomplete and many question marks remain. We do not know whether the universe begun with a bang or passed through a state of minimal size. We do not know what triggered primordial inflation, whether a scalar field or gravity itself or some other mechanism. We do not know what constitutes two thirds of the energy density of the universe and is responsible for late-time acceleration. The impressive number of quantum-gravity and string scenarios proposed in the last years is a manifestation of this remarkable juncture of questions. Any such model should be able to address a large set of issues: Does it solve the big-bang problem? Does it explain the cosmological constant? Does it solve the coincidence problem? What is the nature of the agent responsible for primordial inflation? Do the trans-Planckian and η problems naturally disappear? Does the model leave characteristic signatures in the sky, something different from the standard cosmological theory?

Most scenarios answer only a minimal part of these and other questions. For instance, while the problem of the initial singularity has been attacked in many different ways and found possible resolutions, the cosmological constant problem is still alive and in good health. We do not observe a big bang; we do observe a cosmological constant. Getting the measured value of Λ in a robust theory without fine tuning is proving to be a formidable task.

Eventually, gathering more data will help to disentangle those We Don't Knows. The increased precision of our knowledge of the primordial CMB correlation functions thanks to WMAP and PLANCK have already ruled out whole classes of inflationary models while putting others under tight pressure, most of which had been thriving for years. Further progress in the same direction via a better

G. Calcagni, *Classical and Quantum Cosmology*, Graduate Texts in Physics,
DOI 10.1007/978-3-319-41127-9_14

determination by PLANCK of the polarization spectra and of the level of non-Gaussianity, as well as via other experiments such as BICEP3 [1, 2] (in operation) and LiteBIRD [3] (to be launched in 2022) will stimulate the study of the most favoured scenarios of inflation [4, 5] and their origin from a fundamental theory. The detection of a B-mode polarization signal in the CMB is particularly important because it would make contact with the physics of gravitational waves and vorticity (vector) perturbations and, perhaps, also with quantum gravity. In fact, a large tensor-to-scalar ratio r can be produced not only by specific models of slow-roll inflation (such as those with quadratic and cosine potential), but also in scenarios beyond the standard big-bang model, for instance in the presence of a cosmological bounce [6] or provided the dispersion relation of perturbation modes be modified as in the large class of quantum-gravity models of Sect. 11.9. Also, non-zero TB and EB spectra would also announce parity violation and call for new, possibly gravity-related phenomenology (Sects. 5.9.2 and 11.6.2). The forecast sensitivity of near-future experiments, both ground-based and air-borne, can saturate the lower bound (5.162) and permit a detection of a tensor-to-scalar ratio $r = O(10^{-3})$ [7, 8], up to two orders of magnitude smaller than the current bounds (4.71) and (4.73).

Together with polarization, the search for gravitational waves will receive the lion's share of the community attention in the next few years, especially after the observation of the black-hole mergers GW150914 and GW151226 by Advanced LIGO [9, 10]. A battery of experiments will scan the sky for the direct detection of signals originated from various sources, including primordial inflation and cosmic strings [11, 12]. Observations will be carried on by ground-based interferometers (Advanced LIGO [13, 14], in operation; Advanced Virgo [15], to be in operation by 2017; KAGRA [16], to be in operation by 2018; LIGO-India [17, 18], under construction), space-borne laser interferometer antennæ (eLISA [19, 20], to be in operation by 2034; DECIGO [21, 22], proposed) and pulsar timing array projects (PPTA [23], in operation; EPTA [24], in operation; NANOGrav [25], in operation; SKA [26], to be in operation by 2020). Some of these experiments will be sensitive to a range of frequencies overlapping with the inflationary spectrum (e.g., DECIGO [27]) and can give valuable information on the post-inflationary reheating epoch [28]. As GW150914 showed, it is possible to use gravitational waves to constrain different kinds of deviation from general relativity, including violations of Lorentz symmetry from quantum-gravity or exotic geometry effects [29–32]. The independent discovery of gravitational waves, or the determination of strong upper bounds on the signal, will allow us to test several models beyond standard cosmology, *in primis* string theory. In particular, the cosmic strings produced in D-brane inflationary models (equation (13.63)) will be within the range of detectability.

In parallel, the composition of the universe and the behaviour of the gravitational interaction at different curvature and energy scales will be clarified by different experiments. Multi-purpose missions such as the Fermi Gamma-ray Space Telescope [33] will study dark matter and highly energetic astrophysical processes. The nature of dark energy is expected to be pinned down by detailed surveys of cosmic structures, baryon acoustic oscillations and galaxies at a wide range of redshift spanning 10 billion years, such as the one to be performed by the EUCLID satellite to be launched in 2020 [34, 35]. And, although we celebrated the ΛCDM model from page 1, it must also be said that not all observations are in that much concordance, since the estimates of some PLANCK 2015 parameters have not been reproduced by other data sets. This discrepancy can disappear if a dynamical dark-energy component is assumed instead of a cosmological constant [36]. Near-future data will settle this interesting issue.

Probably, cosmology alone cannot provide the smoking gun of quantum gravity or string theory. As a matter of fact, as the concept of falsifiability evolves with the challenges offered by modern theories, a smoking gun of such theories may assume rather subtle shapes. Consider for instance the issue of whether and how one can test the string landscape. The Standard Model of electroweak and strong interactions works extremely well also at scales above the electroweak scale $\sim m_W$, but it cannot be extended arbitrarily: if E was the energy scale below which the Standard Model is valid, then the Higgs mass should be fine tuned to an accuracy of order $(m_W/E)^2$. To avoid such a fine tuning, a low-energy scale $E \sim M_{\rm SUSY}$ for the onset of supersymmetry has been traditionally invoked. However, the string landscape has revolutionized the concept of naturalness (Sect. 5.10.4) and our expectations about how small the coupling constants of Nature can be. Supersymmetry with a large energy scale $M_{\rm SUSY}$ is no longer excluded on account of an unwanted fine tuning and there may be sufficiently many vacua with a large supersymmetry scale where the observed Higgs mass and cosmological constant are realized. A supersymmetric extension of the Standard Model with large $M_{\rm SUSY}$, such as the scenario of split supersymmetry [37–42], can have measurable consequences which include a prediction range for the Higgs mass and the observability of the gluino at the LHC. The observed Higgs mass falls within the predicted range; the appearance of long-lived gluinos in LHC particle collisions can be the next signature of a high supersymmetry scale at $10-10^9$ TeV. Indirectly, the string landscape would gain support from this observation: the large number of vacua of the landscape, coupled with weak anthropic arguments of super-selection [43], can accommodate the overt "unnaturalness" of split supersymmetry in a natural albeit peculiar way. At the time of completion of this book, no significant evidence for new particles has been observed and previous hints about a resonance at 750 GeV [44–46] have been recognized as statistical fluctuations.

Information on quantum gravity is leaking also from experiments with analogue-gravity systems (Sect. 7.6.4). In one of the most famous semi-classical approximations to quantum gravity, Hawking predicted the evaporation of a black hole via the production of virtual quantum particles at the horizon and the subsequent splitting into two real particles: one falling back into the black hole and the other escaping away [47, 48]. Soon after, it was realized that the propagation of particles travelling at the speed of light near a black-hole horizon is described by the same equations governing the motion of sound waves in a convergent quantum-fluid flow (in other words, phonons play the role of photons) [49], in particular a Bose–Einstein condensate [50, 51]. The possibility to create an analogue black hole in the laboratory has been stimulating experimental research and, recently, the analogue of the thermal spectrum of Hawking radiation has been finally observed in a low-temperature Bose–Einstein condensate [52, 53]. This step forward in the empirical determination of simulated quantum-gravity effects in controlled conditions might just as well be the first of a long series, hopefully leading us to uncharted territory away from semi-classicality and giving us a valuable insight into new gravitational physics.

String and quantum-gravity cosmologies have started to produce characteristic and sophisticated predictions. A non-ambiguous physical evidence for exotic states of matter or geometry in the universe would open up a new season for our view of the high-energy and gravitational structure of spacetime and would dramatically boost the research for a viable, completely consistent theory of fundamental interactions. Whatever the final answer (if any) turns out to be, our journey through modern cosmology has just begun.

References

1. Z. Ahmed et al. [BICEP3 Collaboration], BICEP3: a 95GHz refracting telescope for degree-scale CMB polarization. Proc. SPIE Int. Soc. Opt. Eng. **9153**, 91531N (2014). [arXiv:1407.5928]
2. W.L.K. Wu et al., Initial performance of BICEP3: a degree angular scale 95 GHz band polarimeter. J. Low Temp. Phys. **184**, 765 (2016). [arXiv:1601.00125]
3. T. Matsumura et al., Mission design of LiteBIRD. J. Low. Temp. Phys. **176**, 733 (2014). [arXiv:1311.2847]
4. J. Martin, C. Ringeval, V. Vennin, Encyclopædia inflationaris. Phys. Dark Univ. **5–6**, 75 (2014). [arXiv:1303.3787]
5. J. Martin, C. Ringeval, R. Trotta, V. Vennin, The best inflationary models after Planck. JCAP **1403**, 039 (2014). [arXiv:1312.3529]
6. J.-Q. Xia, Y.-F. Cai, H. Li, X. Zhang, Evidence for bouncing evolution before inflation after BICEP2. Phys. Rev. Lett. **112**, 251301 (2014). [arXiv:1403.7623]
7. P. Creminelli, D. López Nacir, M. Simonović, G. Trevisan, M. Zaldarriaga, Detecting primordial B-modes after Planck. JCAP **1511**, 031 (2015). [arXiv:1502.01983]
8. M. Remazeilles, C. Dickinson, H.K.K. Eriksen, I.K. Wehus, Sensitivity and foreground modelling for large-scale cosmic microwave background B-mode polarization satellite missions. Mon. Not. R. Astron. Soc.**458**, 2032 (2016). [arXiv:1509.04714]
9. B.P. Abbott et al. [LIGO Scientific and Virgo Collaborations], Observation of gravitational waves from a binary black hole merger. Phys. Rev. Lett. **116**, 061102 (2016). [arXiv:1602.03837]
10. B.P. Abbott et al. [LIGO Scientific and Virgo Collaborations], GW151226: observation of gravitational waves from a 22-solar-mass binary black hole coalescence. Phys. Rev. Lett. **116**, 241103 (2016). [arXiv:1606.04855]
11. S. Kuroyanagi, K. Miyamoto, T. Sekiguchi, K. Takahashi, J. Silk, Forecast constraints on cosmic string parameters from gravitational wave direct detection experiments. Phys. Rev. D **86**, 023503 (2012). [arXiv:1202.3032]
12. S. Kuroyanagi, K. Miyamoto, T. Sekiguchi, K. Takahashi, J. Silk, Forecast constraints on cosmic strings from future CMB, pulsar timing and gravitational wave direct detection experiments. Phys. Rev. D **87**, 023522 (2013); Erratum-ibid. D **87**, 069903(E) (2013). [arXiv:1210.2829]
13. J. Aasi et al. [The LIGO Scientific Collaboration], Advanced LIGO. Class. Quantum Grav. **32**, 074001 (2015). [arXiv:1411.4547]
14. https://www.advancedligo.mit.edu
15. T. Accadia et al., Status of the Virgo project. Class. Quantum Grav. **28**, 114002 (2011)
16. K. Kuroda [LCGT Collaboration], Status of LCGT. Class. Quantum Grav. **27**, 084004 (2010)

17. C.S. Unnikrishnan, IndIGO and LIGO-India: scope and plans for gravitational wave research and precision metrology in India. Int. J. Mod. Phys. D **22**, 1341010 (2013). [arXiv:1510.06059]
18. https://www.ligo.caltech.edu/news/ligo20160217
19. https://www.elisascience.org
20. P. Amaro-Seoane et al., eLISA/NGO: astrophysics and cosmology in the gravitational-wave millihertz regime. GW Notes **6**, 4 (2013). [arXiv:1201.3621]
21. N. Seto, S. Kawamura, T. Nakamura, Possibility of direct measurement of the acceleration of the universe using 0.1-Hz band laser interferometer gravitational wave antenna in space. Phys. Rev. Lett. **87**, 221103 (2001). [arXiv:astro-ph/0108011]
22. S. Kawamura et al., The Japanese space gravitational wave antenna: DECIGO. Class. Quantum Grav. **28**, 094011 (2011)
23. http://www.atnf.csiro.au/research/pulsar/ppta
24. R.D. Ferdman et al., The European Pulsar Timing Array: current efforts and a LEAP toward the future. Class. Quantum Grav. **27**, 084014 (2010). [arXiv:1003.3405]
25. P.B. Demorest et al., Limits on the stochastic gravitational wave background from the North American Nanohertz Observatory for Gravitational Waves. Astrophys. J. **762**, 94 (2013). [arXiv:1201.6641]
26. http://www.skatelescope.org
27. S. Kuroyanagi, S. Tsujikawa, T. Chiba, N. Sugiyama, Implications of the *B*-mode polarization measurement for direct detection of inflationary gravitational waves. Phys. Rev. D **90**, 063513 (2014). [arXiv:1406.1369]
28. S. Kuroyanagi, K. Nakayama, S. Saito, Prospects for determination of thermal history after inflation with future gravitational wave detectors. Phys. Rev. D **84**, 123513 (2011). [arXiv:1110.4169]
29. S.B. Giddings, Gravitational wave tests of quantum modifications to black hole structure. Class. Quantum Grav. **33**, 235010 (2016). [arXiv:1602.03622]
30. J. Ellis, N.E. Mavromatos, D.V. Nanopoulos, Comments on graviton propagation in light of GW150914. Mod. Phys. Lett. A **31**, 1675001 (2016). [arXiv:1602.04764]
31. M. Arzano, G. Calcagni, What gravity waves are telling about quantum spacetime. Phys. Rev. D **93**, 124065 (2016). [arXiv:1604.00541]
32. G. Calcagni, Lorentz violations in multifractal spacetimes. arXiv:1603.03046
33. http://www.nasa.gov/mission_pages/GLAST/main/index.html
34. http://www.euclid-ec.org
35. L. Amendola et al. [Euclid Theory Working Group Collaboration], Cosmology and fundamental physics with the Euclid satellite. Living Rev. Relat. **16**, 6 (2013). [arXiv:1206.1225]
36. S. Joudaki et al., KiDS-450: testing extensions to the standard cosmological model. arXiv:1610.04606
37. N. Arkani-Hamed, S. Dimopoulos, Supersymmetric unification without low energy supersymmetry and signatures for fine-tuning at the LHC. JHEP **0506**, 073 (2005). [arXiv:hep-th/0405159]
38. G.F. Giudice, A. Romanino, Split supersymmetry. Nucl. Phys. B **699**, 65 (2004); Erratum-ibid. B **706**, 65 (2005). [arXiv:hep-ph/0406088]
39. N. Arkani-Hamed, S. Dimopoulos, G.F. Giudice, A. Romanino, Aspects of split supersymmetry. Nucl. Phys. B **709**, 3 (2005). [arXiv:hep-ph/0409232]
40. I. Antoniadis, S. Dimopoulos, Splitting supersymmetry in string theory. Nucl. Phys. B **715**, 120 (2005). [arXiv:hep-th/0411032]
41. N. Arkani-Hamed, S. Dimopoulos, S. Kachru, Predictive landscapes and new physics at a TeV. arXiv:hep-th/0501082
42. N. Arkani-Hamed, A. Gupta, D.E. Kaplan, N. Weiner, T. Zorawski, Simply unnatural supersymmetry. arXiv:1212.6971
43. V. Agrawal, S.M. Barr, J.F. Donoghue, D. Seckel, Viable range of the mass scale of the standard model. Phys. Rev. D **57**, 5480 (1998). [arXiv:hep-ph/9707380]
44. G. Aad et al. [ATLAS Collaboration], Search for high-mass diphoton resonances in pp collisions at $\sqrt{s} = 8$ TeV with the ATLAS detector. Phys. Rev. D **92**, 032004 (2015). [arXiv:1504.05511]

45. V. Khachatryan et al. [CMS Collaboration], Search for diphoton resonances in the mass range from 150 to 850 GeV in pp collisions at $\sqrt{s} = 8$ TeV. Phys. Lett. B **750**, 494 (2015). [arXiv:1506.02301]
46. A. Strumia, Interpreting the 750 GeV digamma excess: a review. arXiv:1605.09401
47. S.W. Hawking, Black hole explosions. Nature **248**, 30 (1974)
48. S.W. Hawking, Particle creation by black holes. Commun. Math. Phys. **43**, 199 (1975); Erratum-Ibid. **46**, 206 (1976)
49. W.G. Unruh, Experimental black hole evaporation. Phys. Rev. Lett. **46**, 1351 (1981)
50. L.J. Garay, J.R. Anglin, J.I. Cirac, P. Zoller, Black holes in Bose–Einstein condensates. Phys. Rev. Lett. **85**, 4643 (2000). [arXiv:gr-qc/0002015]
51. C. Barceló, S. Liberati, M. Visser, Analog gravity from Bose–Einstein condensates. Class. Quantum Grav. **18**, 1137 (2001). [arXiv:gr-qc/0011026]
52. J. Steinhauer, Observation of self-amplifying Hawking radiation in an analog black hole laser. Nat. Phys. **10**, 864 (2014). [arXiv:1409.6550]
53. J. Steinhauer, Observation of thermal Hawking radiation and its entanglement in an analogue black hole. Nature Phys. **12**, 959 (2016). [arXiv:1510.00621]

Index

G. Calcagni, *Classical and Quantum Cosmology*, Graduate Texts in Physics,
DOI 10.1007/978-3-319-41127-9

S

inted in the United States
/ Bookmasters